Gorilla Pathology and Health

John E. Cooper DTVM, FRCPath, FRSB, CBiol, FRCVS
RCVS Specialist in Veterinary Pathology
Diplomate, European College of Veterinary Pathologists
Diplomate, European College of Zoological Medicine
Honorary Research Fellow, Durrell Institute of Conservation and Ecology,
The University of Kent, Canterbury, United Kingdom
Wildlife Health, Forensic and Comparative Pathology Services, United Kingdom

With a Catalogue of Preserved Materials

Gordon Hull
Gorilla Pathology Study Group,
London, United Kingdom

ELSEVIER

AMSTERDAM • BOSTON • HEIDELBERG • LONDON
NEW YORK • OXFORD • PARIS • SAN DIEGO
SAN FRANCISCO • SINGAPORE • SYDNEY • TOKYO

Academic Press is an imprint of Elsevier

Academic Press is an imprint of Elsevier
125 London Wall, London EC2Y 5AS, United Kingdom
525 B Street, Suite 1800, San Diego, CA 92101-4495, United States
50 Hampshire Street, 5th Floor, Cambridge, MA 02139, United States
The Boulevard, Langford Lane, Kidlington, Oxford OX5 1GB, United Kingdom

Notices
Knowledge and best practice in this field are constantly changing. As new research and experience broaden our understanding, changes in research methods, professional practices, or medical treatment may become necessary.

Practitioners and researchers must always rely on their own experience and knowledge in evaluating and using any information, methods, compounds, or experiments described herein. In using such information or methods they should be mindful of their own safety and the safety of others, including parties for whom they have a professional responsibility.

To the fullest extent of the law, neither the Publisher nor the authors, contributors, or editors, assume any liability for any injury and/or damage to persons or property as a matter of products liability, negligence or otherwise, or from any use or operation of any methods, products, instructions, or ideas contained in the material herein.

British Library Cataloguing-in-Publication Data
A catalogue record for this book is available from the British Library

Library of Congress Cataloging-in-Publication Data
A catalog record for this book is available from the Library of Congress

ISBN: 978-0-12-802039-5

For Information on all Academic Press publications
visit our website at https://www.elsevier.com

  Working together
to grow libraries in
developing countries

www.elsevier.com • www.bookaid.org

Publisher: Sara Tenney
Acquisition Editor: Kristi Gomez
Editorial Project Manager: Pat Gonzalez
Production Project Manager: Melissa Read
Designer: Matthew Limbert

Typeset by MPS Limited, Chennai, India

This book is dedicated to those of all backgrounds, beliefs and races who have contributed to our knowledge and understanding of gorillas, and thereby enhanced the health, welfare and conservation of these endangered species. We remember particularly those who have died during the course of their work.

Thank you very much

Merci beaucoup

Asante sana

Murakoze cyane

Read not to contradict and confute; nor to believe and take for granted; nor to find talk and discourse; but to weigh and consider.

Sir Francis Bacon (1561–1626)

Almost all the wise world is little else in nature but parasites and sub-parasites.

Ben Jonson (1606)

Contents

Foreword   xv
Preface   xvii
Acknowledgements   xxiii
About the Authors   xxix
About the Contributors   xxxi
List of Abbreviations   xxxiii

Part I
Gorilla Pathology and Health

1. The Genus *Gorilla* – Morphology, Anatomy and the Path to Pathology

John E. Cooper

Introduction   3
Primate and Comparative Anatomy   5
The Current Status of Anatomy   8
Terminology   9
Conclusions   10

2. The Growth of Studies on Primate Pathology

John E. Cooper

Introduction – The Growth of Pathology   11
Studies on the Pathology of Primates   12
African Studies   12
The Importance of Comparative Pathology   14
Primates in Biomedical Research   15
Studies on Microscopic Anatomy and Cellular Pathology   15
The Continuing Relevance of Studies on Pathology   16
The Advantages and Dangers of Extrapolation   16
Access to Data   18
The Relevance of Pathology to Health   18

3. Infectious Disease and Host Responses

John E. Cooper

Introduction   19
Disease and Death in Gorillas   19
The Aetiology of Disease   19
What Is 'Pathology'?   23
Understanding Host Responses – A Comparative Approach   23
Endogenous Factors Influencing Host Responses   24
Discussion of Infectious and Noninfectious Diseases   24
Infectious Conditions   24
Bacteria   25
Viruses   27
Retroviruses   27
Filoviruses   28
Poxviruses   29
Parasites (By Ian Redmond)   29
Microbiota, Microbiome and Normal Flora   30
Emerging Diseases   31

4. Noninfectious Disease and Host Responses

John E. Cooper

Introduction – Noninfectious Conditions   33
Trauma – Causes and Pathogenesis   33
Specific Responses – Trauma   33
Fighting   37
Infected Wounds   38
Attacks on Humans   38
Wounds – General   40
Ageing of Wounds   40
Healing and Regeneration   41
Sepsis   41
Snares   42

Firearms, Exposure to Blasts, Explosions
and Unexploded Ordnance   44
Predation   45
Exposure to Temperature Changes   45
Stress and Stressors   45

5. Methods of Investigation – Observation, Clinical Examination and Health Monitoring

John E. Cooper

Introduction   47
Surveillance of Syndromes   47
Clinical Work With Gorillas   47
The Importance of Observation   48
Indicators of Ill Health   50
Restraint and Handling   50
Clinical Examination   50
Anaesthesia   51
Medical and Surgical Treatment   51
The Importance of Multidisciplinary
Collaboration   52
Health Monitoring of Gorillas   52
Why Is Health Monitoring Important?   53
Health Monitoring Techniques   53
Some Conclusions   55

6. Methods of Investigation – Postmortem Examination

John E. Cooper

Introduction   57
The Nature and Role of the Postmortem
Examination   57
Before Embarking on a Necropsy   57
Personnel and Training   58
Interdisciplinary Studies   59
Access to Material for Postmortem
Examination   60
Postmortem Procedures   61
Equipment   65
Recording of Findings   65
Performance   65
External Examination and Sampling   65
Internal Examination and Sampling   66
Description of Lesions   68
Forensic Postmortem Examinations   68
Postmortem Changes and Determination
of Time of Death/PMI   69
Causes of Death Diagnosis   70

7. Methods of Investigation – Sampling and Laboratory Tests

John E. Cooper

Introduction   73
Sampling   73
Methods   75
Swabs   75
Saliva   75
Faeces   75
Hair   78
Urine   78
DNA and Other Molecular Techniques   78
Blood   79
Biochemistry   80
Cytology   80
Biopsies   81
Histology   81
Reading of Slides   81
Microbiology   82
Serology   82
Additional Tests   82
Environmental Testing   83
Quality Control   84
Recording and Reporting of Findings   84
Personnel and Equipment   85
Safety in Laboratory Work   85
Storage of Samples and Reference Collections 85

8. Nonspecific Pathology

John E. Cooper

Introduction   87
General Considerations in Pathology,
Not Related to Organ Systems   87
Age-Related Changes   87
Developmental Abnormalities   90
Genetic Anomalies   90
Behavioural Pathology   91
Aggression   92
Effects of Capture, Transportation
and Captivity   93
Welfare   93
Pain   95
Neoplasia   95
Poisoning   96

9. Skin and Integument

John E. Cooper

Introduction   97
Biology of the Skin   97

Investigative Methods 98
   Sampling 98
   Laboratory Tests 100
Diseases and Pathology of the
   Skin – General 100
Noninfectious Causes 101
Infectious Causes 102
Ectoparasites (By Ian Redmond) 102
   Bacteria 105
   Fungi 106
   Viruses 106
Diseases of Uncertain Aetiology 106
Conclusions 106

10. Respiratory and Cardiovascular Systems

John E. Cooper

Introduction 107
Respiratory System 107
   Respiratory Pathology 107
   Infectious Diseases 110
   Lung Parasites (By Ian Redmond) 110
   Noninfectious Diseases 111
Investigation of the Respiratory Tract 111
Cardiovascular System 112
Investigation of the Cardiovascular System 112
Cardiovascular Disease 113
Research on Heart Disease in Great Apes,
   Including Gorillas 115
Pathogenesis of Cardiovascular Disease 116
Other Pathology of the Vascular System 117
   Shock 117

11. Alimentary Tract and Associated Organs

John E. Cooper and Ian Redmond

Introduction 119
Oral Cavity 119
   Teeth 120
Dental Disease and Pathology 121
Some Specific Conditions and Their
   Pathogenesis 122
   Tartar and Calculus 122
   Periodontal Disease 123
   Dental Caries 123
   Enamel Hypoplasia 123
   Other Pathology 123
   Effects of Ageing (Age-Related Changes) 123
   Halitosis 124
   The Ongoing Importance of Studies on Teeth 124

The Gastrointestinal Tract 124
   Investigation of the GIT 124
   Gut Flora and Fauna 125
   Parasites 125
   Pathology and Pathogenesis 128
   Colitis 129
   Assessment of Faeces 130
   Diet, Feeding and Disease 131
   Inanition, Starvation and Cachexia 132
   Obesity 133
   Condition 133
   Regurgitation and Reingestion 133
   Coprophagy 134
   Malabsorption 134
   Cystic Fibrosis 134
Investigation of the Liver 134
Pathology and Pathogenesis of Hepatic
   Disorders 135
Investigation of the Biliary System 136
Investigation of the Exocrine Pancreas 136

12. Lymphoreticular and Haemopoietic Systems and Allergic Conditions

John E. Cooper

Introduction 139
The Lymphoreticular System 139
Immunity 139
Immunological Disorders 140
Hypersensitivities and Allergies 141
Immunodeficiency (By Geoffrey
   Pearson and John E. Cooper) 142
Blood and Bone Marrow 146
Blood Parasites (By Ian Redmond
   and John E. Cooper) 147
Neoplasia 148

13. Urinary and Reproductive Systems

John E. Cooper

Introduction 149
Urinary System 149
   Urinary Disease and Pathology 149
   Sampling Techniques 150
   Methods of Investigation 151
   Specific Considerations 151
Reproductive System 151
   Mammary Gland 152
   Reproductive Physiology 152
   Copulation 153
   Pregnancy 153
   Parturition 153

Fetal and Placental Pathology   154
    Placenta   155
    Neonatal Pathology and Health   155
    Orphans and Hand-Rearing   157
    Infanticide   157
    Developmental Abnormalities   158
    Female Reproductive Pathology   158
    Male Reproductive Pathology   159
    Birth Control and Contraception   160

14. Musculoskeletal System

John E. Cooper

Introduction   161
Movement Disorders   161
The Pathology of Bone   162
The Importance of the Skeleton   163
Skull   164
Vertebral Column   164
Joints   165
Investigation of Bones   165
    Basic Rules   165
Skeletal Disease and Pathology   169
    Infectious Diseases of Bones   171
    Developmental Diseases of Bones   172
Fractures and Their Repair   172
Specific Diagnosis of Diseases of the Skeleton 174
    Infectious Disease   174
    Noninfectious Disease   175
    Metabolic Bone Disease   175
Muscle   175
Connective Tissue   175
Adipose Tissue   175

15. Nervous System and Special Senses

John E. Cooper

Introduction   177
Psychological Studies on Gorillas   177
Studies on the Brain of Gorillas   178
Investigation of the Nervous System   178
Pathology and Diseases of the Nervous
    System   180
    Mental Disorders   184
    Peripheral Nerves   184
The Special Senses   185
    The Eye   185
    The Nose   187
    The Ear   187

16. Endocrinological and Associated Conditions

John E. Cooper

Introduction   189

The Endocrine System   189
    Pituitary (Hypophysis)   189
    Thyroid   190
    Parathyroid   190
    Adrenal   192
    Pancreas   193
    Pineal   194
Sleep   194
Stress   194

17. Field Studies in Pathology and Health Monitoring

Jane Hopper, Gladys Kalema-Zikusoka, Celsus Sente and Jenny Jaffe

Introduction   197
Pre-Release Health Considerations
    for Gorillas   197
    Hazard Identification   198
    Risk Assessment   198
    Risk Management   199
    Testing for Pathogens and Disease   201
    Preventive Healthcare   202
Quarantine   202
    Budgets   202
    Risk Communication   203
Setting-Up Long-Term Health Monitoring
    Systems for Mountain Gorillas Using
    a One Health Approach   203
    Assessment of Disease Risks to Gorillas 203
    Building a Gorilla Research Clinic for
        Analysis of Pathogens From Wildlife
        and Domestic Animals   203
    Linking With Local Human Health Centres
        Through Comparative Disease
        Investigations   204
    Linking With Local Human Health Centres
        Through Community-Based Health
        Promotion   204
    Conclusion: Impact on Conservation 204
    Future Recommendations for
        Strengthening the Programme   205
Cryptosporidium, Giardia and Helminths
    of Habituated Nyakagezi Mountain
    Gorilla Group in Mgahinga Gorilla
    National Park   205
    Identification of Cryptosporidium,
        Giardia and Helminths   205
    Prevalence and Means   206
    Discussion   206
Working With Great Apes During an Ebola
    Outbreak   206
    Great Apes and Ebola   207
    Realities of Working at a Great Ape
        Sanctuary During a Major EVD Outbreak 207

Precautions at Work   207
Precautions Outside the Sanctuary   208
Challenges During the Ebolavirus Disease
    Outbreak   209
Precautions Taken to Prevent Ebolavirus
    Disease Affecting the Sanctuary   209

18. Legal Considerations

Margaret E. Cooper

Introduction   211
General   211
Definition of 'Wild'   211
IUCN Red List of Threatened Species Status   212
Nomenclature   212
Legislation   212
Free-Living Gorillas   215
General   215
International Legislation   218
National Legislation   221
Gorillas in Captivity   223
General   223
Authorisation to Keep Gorillas   223
Gorillas in Research   223
Animal Welfare   224
Responsibility for Damage Caused
    by Captive Gorillas   225
Other Relevant Law   225
Occupational Health and Safety   225
Movement of Gorillas and Samples   226
Law Regulating the Veterinary Profession   226
Veterinary Medicines   227
Permission, Reference Material, Data,
    Intellectual Property and Copyright   228

19. Pathology, Health and Conservation – The Way Forward

John E. Cooper and Gordon Hull

Introduction   229
Knowledge of the Genus and Its Biology   229
Threats to Survival   229

The Implications and Importance
    of Studies on Pathology and Health   229
Advances in Health Care   230
Collaboration and Sharing of Information   231
A Greater Role for Africa   231
Involvement of African Academic
    Institutions   232
Scientific Advances   233
Human Health   233
Attention to Welfare   233
Continued Search for Gorilla Material   233
Greater Dissemination of Information   234
Broadening the Holistic Approach   234
Concluding Remarks   235

Appendix 1: Glossary of Terms   237
Appendix 2: Protocols and Reports   243
Appendix 3: Field Pathology   287
Appendix 4: Hazards, Including Zoonoses   293
Appendix 5: Case Studies – Museums
                and Zoological Collections   301
Appendix 6: Scientific Names of Species
                and Taxa Mentioned in Text   315
References and Further Reading   317

Part II
A Catalogue of Preserved Materials

Gordon Hull

Introduction to the Catalogue   359

Catalogue of Preserved Gorilla
Materials   379

Appendix CA1: Use of Collections and Handling
                of Biological Material   603
Appendix CA2: Retrieval, Preparation and Storage
                of Skeletal and Other Material   609
Index   617

Foreword

This is a book about gorillas, those close relatives of human beings which are confined to the continent of Africa, that have become emblematic of the remarkable conservation and sustainable development effort, and yet whose future survival remains largely uncertain.

Gorillas have only been known to science for a little over 150 years, whereas other great apes, such as the chimpanzee and orangutan, were recognised several centuries ago.

During the first 100 years, knowledge of the biology of gorillas progressed from meticulous studies on their anatomy (often comparing their morphology with that of other extant and extinct hominins), followed by research on behaviour — both in the wild and in captivity. Only over the past few decades has serious attention been paid to the factors that influence the survival of gorillas in the wild.

It was Dr Louis (LSB) Leakey who first advocated that a greater understanding of gorillas (and other great apes) and their relevance to the evolution of *Homo sapiens* required detailed fieldwork. Encouraged by the early studies of Carl Akeley and the seminal research of George Schaller, Louis Leakey recruited the late Dian Fossey, who subsequently played an instrumental role in bringing the plight of the species to the world's attention. Alongside the achievements of these scientists, the work of Central African conservation practitioners has played the critical role in the survival of the species, albeit less recognised and celebrated. The outcome has seen some remarkable successes, such as the dramatic increase in mountain gorilla populations since their catastrophic low point in 1985. The circumstances of this achievement, which include the Rwandan genocide and the horrors of four civil wars in the eastern Democratic Republic of Congo, make it one of the great miracles of modern conservation. But this success has come at a tragic price, such as the 140 of Virunga National Park's rangers who have lost their lives while protecting these species and their ecosystems.

The conservation effort has had multiple outcomes, of which the most important has been the contribution of gorilla populations across their range, in supporting the livelihoods of the people living in and around their habitats. The combined efforts of African and international conservationists has seen the growth of a remarkable industry generating tens of millions in revenue for post-conflict economies and offering quality employment for a new generation of young professionals from rural communities. Yet this opportunity carries with it enormous risks. While gorilla viewing plays a determining role in their conservation, tourism also implies the close proximity of human visitors to gorilla groups, exposing them to the greatest of all threats, disease. It is comforting to see the depth of understanding contained within this work that complements the substantial efforts by veterinarians and rangers from the various park services and highly committed partners, such as the Mountain Gorilla Veterinary Project, who provide the data, knowledge and experience that are the foundation of good science.

In this book, John Cooper and Gordon Hull bring together much of the available information about the pathology of gorillas — that is, how infectious and non-infectious diseases (including injuries) manifest themselves as well as how gorillas respond — and the influence of such complex interactions on the health, welfare and conservation of these threatened creatures. The chapters of the book are written mainly by John Cooper, a specialist wildlife veterinary pathologist, supported by contributions from Margaret Cooper, Ian Redmond, Jane Hopper, Gladys Kalema-Zikusoka, Jenny Jaffe, Celsus Sente and Ogeto Mwebi. The text is supplemented with a unique catalogue of information as to the whereabouts internationally of skulls, skeletons and skins, fluid-preserved specimens, casts, histological sections and samples, together with images and clinical, postmortem and laboratory records. Such data will be of great value to primatologists, veterinarians, evolutionary biologists and osteologists, as well as to field conservationists.

We welcome this book, not only because it is an important contribution to gorilla and great ape biology but also because it should help in much-needed attempts to save these magnificent animals from extinction. It also will assist those who are involved in human medicine, not least of all because large numbers of free-living gorillas have died in West Africa from *Ebolavirus* infection in recent years; a proper understanding of the epidemiology of this distressing, highly fatal, disease requires knowledge of wildlife that may either harbour or transmit the causal organism.

Living and working in Africa, as we both do, we appreciate the emphasis throughout the text on the importance, in studies on gorillas and other wildlife, of involving local people, ranging from veterinarians and scientists to trackers and village leaders. Only with such an environmentally sensitive approach, coupled with interdisciplinary scientific research, will the vital work of saving and conserving the two species of gorilla prove a success.

Louise Leakey
Research Professor, Stony Brook University,
Turkana Basin Institute

Emmanuel de Merode
Director, Virunga National Park/
Parc National des Virunga
January 2016

Preface

The more you know, the more you know you don't know.

Aristotle 384−322 BC

Why a book about the pathology of gorillas? And why is it accompanied by a catalogue of materials and resources? The rationale is straightforward. We know so little about the pathology of gorillas. Yet the answers to some of the most pressing questions posed by conservationists may lie hidden in existing collections of material, which can provide valuable data on pathology and diseases. What can be learned? Where can collections be found? This publication is intended to answer these questions and play a part in rectifying the current dearth of information.

The two species of gorilla that exist in Africa today are, like all the world's great apes, threatened by persecution, habitat destruction, illegal trade and diseases.

It is conventional wisdom that attempts to halt the decline of any species require an understanding of the biology of that taxon and, based on this knowledge, the application of appropriate conservation measures. An aspect of the biology that has tended to be overlooked in the past is the susceptibility of a species to infectious and noninfectious diseases, how it responds to such 'insults' and, insofar as pathogens are concerned, the intricacies of host-parasite relationships. In the case of the genus *Gorilla*, these matters have only recently attracted serious scientific attention, and much remains to be learned.

The genesis of the book goes back to 1994 when, following the evacuation of the senior author and his wife (the Coopers) from the fighting and genocide in Rwanda, they embarked upon a study of skeletal and dental pathology of the mountain gorilla (*Gorilla beringei*) in the collections of the Royal College of Surgeons of England and the National Museums of Kenya (Cooper and Cooper, 2007, 2008). This led to a project involving the gross examination of material from both species of gorilla in museums and collections in Belgium, France, Germany, the Netherlands, Kenya, Rwanda, South Africa and Uganda; this was accompanied by appeals for information to colleagues who may have had access to uncatalogued skeletal material. Those studies revealed that the gorilla

skeletal material from collections in Europe and Africa showed a similar range of pathological changes to those reported by Lovell (1990a) in her review of great ape skeletons in the United States.

It soon became apparent that not only skeletons but also other material could provide valuable data on the pathology and diseases of gorillas. As Part II: A Catalogue of Preserved Materials, amply illustrates, various specimens are to be found in the world's scientific collections and many are available for investigation, provided that appropriate arrangements are made. Most of the material is from gorillas that died, or were killed, decades ago and they therefore provide valuable reference (baseline) data. Gross and microscopic studies can, increasingly, be supplemented with imaging and DNA and other molecular studies.

While some of these collections have been extensively investigated, many have not, and studies on them by medical scientists are few and far between.

On their return to Rwanda after the cessation of hostilities, the Coopers resumed their research on gorilla material there. Some specimens had been destroyed or damaged, and gorilla bodies and tissues that had been left for "safe keeping" at the Laboratoire Nationale de Rubilizi in Kigali prior to the outbreak of the war were not to be found.

The compilation of this book and its accompanying catalogue are essentially the result of the belief by the two authors that there is a need a) to collate, as far as possible, published and unpublished data about the pathology of *Gorilla* spp. and to attempt to relate this information to the health, welfare and conservation of gorillas in the wild and in captivity and b) to produce a global list of skeletal and other material resources. Our aim to produce a global list has certainly been achieved. The catalogue that comprises Part II of this book, is, we believe, the first such database of its kind and owes its existence to years of dedicated work by Gordon Hull.

Our awareness of how gorillas respond to infectious and noninfectious conditions, including injuries, has advanced over the past three decades, largely driven by proponents of high quality zoo animal medicine and by far-sighted field projects. It remains a fact, though, that in contrast to what we know about humans, domesticated species and laboratory animals, there is still a paucity of properly collated, scientifically assessed, information about the causes and pathogenesis of diseases in gorillas. The need to have access to such data has, however, never been greater. Gorillas are a threatened species and both infectious and noninfectious diseases contribute to their ill health and death. In the case of some infectious conditions, such as *Ebolavirus* disease (EVD), whole populations are at risk and there are profound health implications for humans.

In his book *Listellany* (2014), John Rentoul claimed that 'Lists are the future of journalism, the internet and the world. Also they are the past'. Veterinary and medical pathologists love lists, however – of differential diagnoses, for example – but Part II: A Catalogue of Preserved Materials, is already a list and an additional one, detailing all the reports of pathological changes in gorillas, would not make interesting, or necessarily useful, reading. Instead, therefore, the approach in Part I, Gorilla Pathology and Health, is an applied and practically orientated one, with an emphasis on how pathology affects the individual animal, its group and its species.

At this stage our definition of the term 'pathology' needs to be explained. The etymology is clear from its Greek roots – *pathos*, suffering/disease and *logos*, reason/plan. In this book we use the term 'pathology' to mean the science of the study of disease, with particular reference to its causes, its development and its appearance. In common (lay) parlance 'pathology' tends to be linked with death, carcases and postmortem examination and not, as it should, to the investigation also of clinical disease and the application of laboratory tests. This book intentionally incorporates the latter as well as the former.

The authors are two British enthusiasts with a shared interest in gorillas but rather different backgrounds.

John Cooper (JEC) is a member of a husband and wife team, from the United Kingdom. He and his wife Margaret (a lawyer, who is a contributor in her own right to this book) have travelled widely and lectured together in many countries. They have spent nearly 20 years living overseas, mainly in Africa, the Middle East and the Caribbean. The Coopers worked with mountain gorillas in Rwanda from 1993 to 1995 (see below) and they continue to study these and other primates. The Coopers are now based in Britain but are also engaged in voluntary work with wildlife, domesticated animals and communities in East Africa. JEC trained as a veterinary surgeon (veterinarian), has had a lifelong interest in natural history and is now a specialist veterinary (comparative) pathologist with particular involvement in studies on wildlife and so-called 'exotic' species, tropical diseases and

comparative medicine. His attraction to what is now termed 'One Health' is in part because, as a veterinary student, he was taught by medical lecturers: his books on physiology, histology and pathology were primarily concerned with *Homo sapiens*, not conventional domesticated animals. It was, therefore, perhaps providential that he should spend 13 years of his professional life at the Royal College of Surgeons of England (RCS) as Senior Lecturer in Comparative Pathology. The RCS had a 'Hunterian' approach to comparative medicine at least 250 years before the current mantra of 'One Health' became so popular.

In the early 1970s, at the request of the renowned palaeontologist Dr Louis S.B. Leakey, John served as honorary veterinary surgeon for the National Primate Research Centre (now the Institute of Primate Research) in Kenya. Later, from 1993 to 1995, he was the Director of the Centre Vétérinaire des Volcans/Volcano Veterinary Center (CVV/VVC) in Rwanda, responsible for the health of the mountain gorillas.

Gordon Hull (GH) is an amateur naturalist with a deep interest in primates and certain other mammals. He has specialised in the study of gorillas over many years, during which time he has amassed a great deal of technical and historical information about specimens in zoos, museums, and other institutions throughout the world. Although unaffiliated, he has been, and remains, an assiduous and effective researcher, able to elicit excellent responses and cooperation from professional scientists and lay persons alike. Gordon is a member of the Gorilla Pathology Study Group (GPSG) and was a coauthor with John Cooper and Gladys Kalema-Zikusoka of the paper 'Diseases and Pathology of the Genus *Gorilla*: The Need for a Database of Material and Resources', that was presented at an International Scientific Conference in Nairobi, Kenya, in September 2013, and which arguably set the scene for the production of this book and its accompanying catalogue.

The VVC, where the Coopers were based in the 1990s, was established in 1986 at the request of Dian Fossey, with moral and financial support from Mrs Ruth Morris Keesling, and administered by the Morris Animal Foundation (MAF). Dr James Foster established the VVC and served as its Director from 1986 to 1988 (see picture below). Several of the subsequent Directors are referred to in this book but the Coopers would like to pay particular tribute to their friend the late Barkley Hastings. Barkley enthusiastically contributed to RCS/ZSL clinicopathological seminars in his London days, then played a pivotal role in initiating research at the VVC that ranged from extensive and enterprising studies on protozoal and helminth parasites to morphometrics on dead and immobilised animals and serology, microbiology, haematology and blood biochemistry.

A picture by a local artist, Sinanenye J.M.V., on paper and cardboard. It depicts the first CVV/VVC veterinarian, James Foster, treating a mountain gorilla on Sabinyo in the Parc National des Volcans/Volcanoes National Park, Rwanda.

Following the metamorphosis of the CVV/VVC into an independent body, the Mountain Gorilla Veterinary Project (MGVP), now often termed 'Gorilla Doctors', it was not infrequently suggested that whereas the VVC was originally a 'mostly clinical programme' (Cranfield and Minnis, 2007), the MGVP had changed it into 'a holistic conservation medicine programme'. One feature published in 2006 in an eminent American veterinary journal even stated that 'During its early years, the project's (VVC's) efforts were small in scope, limited to up to four medical interventions a year and the occasional necropsy to determine how a gorilla died'. This is all very misleading. It is not only an unjustifiable criticism of MAF's innovative and humane management of a unique project but totally overlooks the sterling work of Barkley Hastings and others who, under the most difficult of circumstances (social unrest, sometimes war, no mobile telephones or emails, no running water, often working in three languages) managed between 1986 and 1996 to carry out much significant work in such fields as parasitology and cytology. In the 6 months prior to the outbreak of the 1994 genocide, the VVC was embarking on a whole range of progressive initiatives with the aim of providing a veterinary service, including laboratory investigations, for the whole of the Parc National des Volcans and intended to assist various kinds of wildlife as well as local people.

Gorillas are only found in Africa and their long-term survival in the wild depends on those who live and work there. Throughout this book, therefore, there is an emphasis on the need to involve African scientists (veterinarians, pathologists, primatologists and biologists), as well as African institutions, in studies on the health and pathology of gorillas and in locating and conserving specimens within that continent. This approach is coupled with championing the need for tissues and other samples from gorillas to be retained within the species' range states, where access can be readily provided to local students and researchers, rather than overseas. This is in keeping with the spirit of the 1992 Convention on Biological Diversity (CBD), which deals with the fair and equitable sharing of genetic resources (see Chapter 18, Legal considerations). The essence of that Convention is that scientists from western (richer) countries should be taking their expertise to poorer nations, not removing material from range states so that it can be 'safely' processed and archived in North America or Europe. In respect of specimens from mountain gorillas, the importance of this approach to African countries was emphasised at the Population and Habitat Viability Assessment held in Uganda in December 1997 (Werikhe et al., 1998) and discussed further at a C.L. Davis Foundation Symposium in 1999, also in Uganda. At the latter the attendees had been told of plans to establish a database/reference collection of mountain gorilla material in Colorado, USA (described in a brochure a year later as 'the Biological Resource Center. Here, mountain gorilla samples can be stored in one central facility. These samples are available to scientists worldwide whose studies focus on mountain gorilla health issues'). The C.L. Davis Foundation Symposium in Uganda resolved that records and specimens concerning mountain gorillas should be duplicated and one set kept in the host country (Rwanda, Uganda and DRC), but whether this was done routinely in the succeeding years is not clear.

For some time prior to the completion of this book, reference had been made in various quarters to 'a biobank inventory' of gorilla material held by MGVP. Some limited information about this, made available to us courtesy of Gorilla Doctors, is to be found in Part II, A Catalogue of Preserved Materials. However, it should be said that earlier enquiries about its whereabouts and contents largely proved unfruitful. This was not only disappointing but raised the question of how modestly funded African scientists, from whose homelands all this gorilla material originated, could hope to gain access to such resources.

The vital role of local Africans in studies on gorillas was, of course, recognised long ago. Early hunters, explorers and biologists would not have succeeded in their quest to find gorillas without the help of village people who had knowledge of the animals and the terrain. This was exemplified by the life of Reuben (sometimes spelt Rueben or Ruben) Rwanzagire in Uganda: the finding of his last resting place by a group from Makerere University was recounted by Cooper and Cooper (2001a).

In more recent years Africans have played an increasingly major role in the science of conservation and healthcare of gorillas. Their task is not always easy. A letter published in *Veterinary Times* (Cooper and Cooper, 2004) paid tribute to African veterinarians who had worked with mountain gorillas and, mindful of history, concluded 'They are a constant source of support to the expatriates with whom they work, not least because of their local standing and their facility with vernacular languages: however, unlike the expatriate vets, they cannot depart when the working environment becomes dangerous'.

Over the succeeding 20 years the situation has certainly improved insofar as involvement of local people is concerned; but for as long as most of the funding for gorilla projects comes from the United States and Europe, there will be a tendency for expatriates to take the lead over policy and for Africans to have to follow. Gorilla organisations stress that they are 'international' but in reality 'internationalism' means 'existing or carried on between nations; agreed on by many nations' (Oxford English Dictionary), not just having members of staff who come from other countries.

One initiative to involve African colleagues more fully in their own right was the formation of the Gorilla Pathology Study Group (GPSG) in 2008. The GPSG receives no specific funding but small grants are attracted for training purposes. Thus, for example, in November 2011 the GPSG organised workshops on primate pathology in both Kenya and Uganda.

Apart from their intrinsic importance, as endangered species in precarious habitats, gorillas are of considerable relevance to the health and wellbeing of humans — extending from being natural 'models' of cardiovascular and osteoarthritic conditions to their (regrettable) susceptibility to *Ebolavirus* infection. Working with live gorillas in the wild or in captivity provides opportunities to learn about the role of disease in these species and the concurrent collection, collation and publication of data helps to fill gaps in our knowledge of relevant epidemiology and pathology in both *Gorilla* and *Homo*.

The background to Part II: A Catalogue of Preserved Materials, is detailed in its Introduction. As far as Part I of the book is concerned, when the project started it was hoped that the chapters would include reference to the vast majority of published papers about pathogenesis of known diseases of gorillas, together with a comprehensive (meta) analysis of causes of death and pathological findings in gorillas from the majority of the world's zoos that keep this species. Neither has proved possible. The number of references relating directly to gorillas, or relevant to them, is enormous and continuously growing. In addition, though, it has been disappointing to note how many scientists representing zoo collections have declined to share basic information, or even failed to respond to our polite request for collaboration. This is puzzling. Such a refusal to cooperate not only denies others, including Africans from range countries, the chance to see, perhaps to use, samples, but also dents the claim of modern zoos to be contributing to conservation, research and education — the three tenets of the European Zoos Directive.

How different this is from the kindness we have received from the majority of our correspondents, and indeed, the experiences of the British (medical) pathologist GBD Scott, who wrote in his 1992 book: '... Dr Kurt Benirschke, the then Director of Research at the San Diego Zoo, placed the Zoo's records of autopsies on primates at my disposal ... he gave me every encouragement ...'.

Alas, attempts at collaboration over gorillas and their conservation appear often to engender competition and possessiveness rather than openness. Two veterinary colleagues who have generously contributed material for this book are worth quoting:

Problems only arise when people don't share information.
Sharon Redrobe, UK

I personally feel no one should have ownership over gorilla-related information and you are fighting the good fight getting the information out there...
Yvonne Schulmann, USA

Another unexpected difficulty that we have encountered is that the accuracy of diagnosis is not always taken seriously by those who hold specimens. A senior curator at a prestigious London museum told one of us (JEC) that providing accurate catalogue descriptions of lesions of gorilla skulls was not essential 'because someone else will come along later and make corrections'! Such an attitude does not augur well for the scientific documentation of material that, as the species face increasing pressure in the wild, becomes more and more important in terms of providing valuable background data. Perhaps the answer to the question of unreliable diagnoses is for museums to restrict their catalogue entries to descriptions, supplemented whenever possible with photographs, and to leave attempts at interpretation of any morphological abnormalities or distinct lesions to veterinary, medical or dental pathologists.

This book is written in British English. This not only expresses the background of the authors, but also reflects the fact that many people working with gorillas in their range states use European terminology and syntax.

The word 'native' (or its equivalent in French or German) was employed routinely in earlier reports when Europeans and other expatriates were referring to local African people but the term is never used in this book (including the catalogue) except when it forms part of a quotation or citation.

Abbreviations that cause confusion are excluded from this work, in particular 'cbc' which is often used, without explanation, in North American texts. It means 'complete blood count' but in many other parts of the world different terms are employed, especially 'total blood count'.

The word 'quite' is avoided throughout the text, intentionally, except where it is part of a quotation. In British English 'quite' generally means 'fairly; to a certain extent or degree' whereas in American English it usually equates with 'very; to the utmost or most absolute extent or degree'. In a scientific text such a difference in connotation could be dangerous!

A few terms need specific explanation. In this book the term 'free-living' is used to describe gorillas that are living in the wild. 'Free-ranging' is favoured by some authors, especially American, but that term is used rather differently in Britain and other parts of the world.

The international nature of studies on endangered species means that the metric system should be used. This is the case in all gorilla range states. Standard international units (Système International d'Unités, SI) were introduced in 1960 as the result of an initiative that started in 1948. Animals should be weighed in grams (or kilograms) and measured in millimetres, centimetres or metres. However, the reader should be aware that in the United States and a few other countries, 'conventional' units are still routinely employed, and be prepared to request conversion or to do so oneself. An excellent explanation of the SI system was given by Douglas (1977), whose short paper concluded with the following words 'Data from one laboratory can only be compared with those from another laboratory if identical methods are used'. That statement, and the clear description of the system by Douglas, are as true today as they were nearly 40 years ago.

The need also to use international dates was well illustrated in Rwanda where a British veterinarian's field notes (using the European system) were put on to a computer by an American colleague using North American (USA) dates; subsequent analysis of them by a third person resulted in considerable confusion because one of the days cited (3/6/94, translated into 3rd June 1994) was when the Rwandan civil war was in full force and there were no veterinarians in the field!

In this book reference is made to both recent and previous (older) literature. There is an increasing tendency for journals and editors to insist on the inclusion in manuscripts only of the former, suggesting that this is more relevant and reliable. Such is a mistake. In a book such as this it is important that the references quoted reflect the history and development of the subject, not just 'new', 'cutting edge' publications. An appreciation of how our understanding of gorilla pathology and health has evolved helps put current work in context and elucidates where there are still significant gaps in our knowledge. Our friend Andy Richford, who guided us in the formulation of this book, advised us against including too much 'grey literature': in other words, material that is anecdotal or based upon an individual's observations and deduction, rather than properly executed scientific study (see elsewhere regarding 'hypothesis-based' research). This was sound advice but we have not been able to follow it *in toto*. Many significant reports on gorillas and their health, especially in the earlier days, were the result of meticulous observation and record keeping by trackers of free-living gorillas or keepers of those in captivity. More recently too, there has been an abundance of relevant published case reports and observations in this book in various journals and magazines. These should not be overlooked.

Part I of this book is intended to be a monograph on the pathology (in its broadest sense) of gorillas and relevant implications for the health of these animals, both in the wild and in captivity. It is therefore particularly, but not exclusively, aimed at primatologists, veterinarians, biologists, osteologists and conservationists − especially African colleagues with very limited resources.

In order to save space, the full English names of the different species and subspecies of gorilla are not used in the following text. The two species are referred to broadly as 'lowland' (or '*gorilla*') and 'mountain' (or '*beringei*') except where it is necessary to distinguish them more specifically. The scientific names of all species mentioned in the text are listed in Appendix 6: Scientific Names of Species.

In photomicrographs, the stain used was haematoxylin and eosin (H&E), generally Cole's, unless otherwise indicated. Except where specifically stated, the magnification of photomicrographs is either ×100 (low power) or 400 (high power).

There is an ancient proverb that states: *I hear ~ ~ I forget, I see ~ ~ I remember, I do ~ ~ I understand.* In this book, therefore, the reader will find not just information about the pathogenesis of disease in the genus *Gorilla* but also practical guidance, coupled with appropriate references, as to how to investigate animals, how to select samples and how to examine them. In other words, how to **do** it.

We have also, intentionally, included in the book some vignettes — brief, hopefully evocative, descriptions of incidents, mainly based on the Coopers' experiences in Rwanda, that help illustrate the pleasures and the challenges of work with wildlife in Africa.

To conclude, the gorilla has attracted a great deal of scientific interest since its discovery and recognition as a separate genus, *Gorilla*, in the mid-19th century — less than 200 years ago. Nevertheless, it took nearly a century for an authoritative textbook to be published detailing the anatomy of gorillas. Notwithstanding the appearance in recent years of a substantial number of papers on infectious diseases of the genus, remarkably little is yet known about the gross and microscopic pathology of gorillas and how they respond to infectious and noninfectious insults. Our joint publication is intended to play a part in rectifying the situation. The centenary volume of The Pathological Society of Great Britain and Ireland in 2006 had as its title 'Understanding Disease' — that is the main aim of this work.

John E. Cooper
Gordon Hull
London, England
April 2016

Acknowledgements

Nia zikiwa moja, kilicho mbali huja
(Swahili: When minds are one, what is far comes near)

The successful compilation of this book would not have been possible without the help, advice and support of many people. Panegyrics, lofty writings in praise of a person, are not popular nowadays but it is important that those who have helped us in the immediate or distant past are acknowledged.

First, we both (JEC and GH) thank Louise Leakey and Emmanuel de Merode for writing the Foreword. They are a husband and wife team who have not only contributed much to Africa, which is their home, but also have many links with gorillas and their conservation. Louise's grandfather, Dr LSB (Louis) Leakey is extolled in Chapter 1: The genus *Gorilla* – Morphology, Anatomy and the Path to Pathology. In addition to his contributions to palaeontology (and those of his wife, Mary, and other members of the family), Louis Leakey recruited Dian Fossey to study mountain gorillas in Central Africa.

When the Coopers first lived in Kenya, from 1969 to 1973, JEC served as (honorary) veterinary surgeon to Louis's Primate Centre at Tigoni. Following Louis's death in 1972 JEC and Peter Holt (thanked elsewhere in these pages) continued to tend the monkeys and to support the Centre until its future was assured. Ultimately it metamorphosed into the National Museums of Kenya's Institute of Primate Research (IPR) (see Chapter 2: The Growth of Studies on Primate Pathology).

It is an honour to use Figs. 1.4 and 13.5, together with the front cover picture, that were taken in Rwanda by the late Robert (Bob) Campbell and were kindly provided by Heather Campbell, his widow and a friend and former colleague from the Coopers' days in Kenya. Other images are supplied by Margaret E. Cooper or John E. Cooper except for those specifically credited beneath the figure or attributed below:

- Cambridge Veterinary School/Kate Hughes, Fernando Constantino-Casas, Michael Day (Figs. 2.5, 9.1, 9.2, 9.4, 9.6, 10.1, 10.2, 10.3, 10.4, 11.5, 13.1, 14.5, 14.6, 15.3, 16.3)
- Paddy Mannion (Figs. 6.2, 11.2 and 14.1)
- David Perpiñán (Figs. 10.7–10.9)
- Sam Young (Fig. 15.4)
- Carina Phillips (Figs. A.5.14–5.16)
- Michelle Burrows (Fig. A.5.17).

These images have greatly enhanced the book and complemented our own.

We express our sincere appreciation to Melissa Read (freelance Project Manager assisting the Elsevier Production team in Oxford) who, during the final stages of completing this book, has dealt with two idiosyncratic authors and their queries with much calmness, kindness and humour. There are many who, throughout the gestation of this publication, have given us wise counsel. We particularly thank Colin Groves for his interest in – and support for – our project and for sending us interesting images. Despite his many commitments he has always responded promptly to our email messages and furnished us with sound advice, especially but not exclusively about taxonomy. In his own inimitable way, early in this venture, he suggested 'For the benefit of those who have been asleep over the past 30 years, you might explain that the family Hominidae used to be restricted to humans only, but now includes great apes'. In the same vein we are grateful to David Chivers and Roger Short for their advice and support.

We thank our contributors: Margaret Cooper, Ian Redmond, Jane Hopper, Gladys Kalema-Zikusoka, Jenny Jaffe, Ogeto Mwebi, and Celsus Sente; the 'teams' who worked with us on (1) 'Guy' at the NHM, Brian Livingstone, Paolo Viscardi, Keith Maybury, Allen Goodship, Richard Sabin and Roberto Portela Miguez, (2) skeletons in the Osman Hill Collection at the RCS, Carina Phillips, Martyn Cooke and Paul Budgen and (3) the student projects at Bristol Zoo Gardens, Sophia Keen, Dermot McInerney, Michelle Barrows, Christoph Schwitzer, Bryan Carroll, Kate Robson-Brown, Jonathan Musgrave and Allen Goodship. Mick Carman and Don Cousins also provided helpful information relating to our study of 'Guy'.

Sally Dowsett, family friend and long-term supporter of our projects in Africa, typed drafts and for some years served as honorary secretary to the Gorilla Pathology Study Group. Pauline Muirhead kindly gave similar assistance while Gemma Saunders and Jenny Cocking cheerfully prepared laminates.

Many people kindly agreed to read (review) parts of this book before publication and/or provided helpful comments on content and orthography. We thank (arranged alphabetically) David Alderton, Ekane Humphrey Anoah (Wildlife Law Expert of the Last Great Ape Organisation, Cameroon), Helen Chatterjee, Vanessa Cooper, Max Cooper, Sarah Cooper, Chris Daborn, Jeremy Dearling, Ellie Devenish Nelson, Scott Dillon, Ofir Drori, Keith Hardy, Mamun Jeneby, Peter Kertesz, Arthur Kemoli, Brian Livingstone, Keith Maybury, Jaimie Morris, Jonathan Musgrave, Emily Neep, Howard Nelson, David Ojigo, Caroline Pond, Ian Redmond, Sharon Redrobe, Jenny Rees Davies, Rob Shave, Nick Short, Victoria Strong, Simon King, and David Williams.

Many others who kindly helped are cited as a 'personal communication' in the text or acknowledged in the catalogue. In addition, literature, contacts and/or helpful suggestions were received at various times from Wilbur Amand, Frances Barr, Donald M. Broom, Debra C. Bourne, Tom Butynski, Héctor Sanz Cabañes, Sarah Chapman, Christopher P. Conlon, Chris Furley, Hayley Murphy, Rick Murphy, Navarro Serra, Norma Chapman, Andrew Dixson, James Hassell, the late Barkley Hastings, Richard Jakob-Hoff, Jesús Maria Pérez Jiménez, Ian Keymer, Sascha Knauf, Felix Lankester, Michael Marks, Shelly Masi, Nelly Ménard, Dick Montali, Torsten Morner, Stuart Nixon, John Bosco Nizeyi (JBN), Richard J. de Norman, Marc Nussbaumer, Mark Rose, Bruce Rothschild, Peter W Scott, Angelique Todd, Anna and Steve Tolan, Lydia Tong, Tonino Van Wonterghem, Roberta Wallace, David Warrell, Michael Woodford and Sam Young.

Copies of postmortem reports were made available to us by many kind colleagues, including Rachael Liebmann and other medical pathologists.

Colleagues in the Gorilla Pathology Study Group (GPSG) have not only contributed to this book but also collaborated enthusiastically in relevant primate workshops in Africa. We are indebted to the late Professor Phillip V. Tobias, the eminent South African palaeontologist, GPSG Patron, for his encouragement, and we wish to record his role, for nearly half a century, in promoting the involvement of Africans in studies on the mountain gorilla and in helping to facilitate access to gorilla skulls at 'Wits'.

Dr Sam Thompson, the lifeblood of the C. L. Davis Foundation in the United States, died in 2014. JEC remembers well his kindness and support, especially in respect of training programmes for veterinary pathologists in East Africa and the opportunity for JEC to record lectures for teaching purposes.

JEC is grateful to former Volcano Veterinary Center (Centre Vétérinaire des Volcans) (VVC)/Mountain Gorilla Veterinary Project (MGVP) veterinarians who responded personally to his letters, in particular Jonathan Sleeman, Antoine Mudakikwa and Ken Cameron. Officially sanctioned, summarised, information about MGVP findings was kindly furnished by Linda Lowenstine and she and Karen A. Terio provided the most recent versions of their general ape necropsy form, the GAHP recommended cardiac necropsy (prosection) guide and the guide for pathologists.

More than 1000 people have contributed to Part II: A Catalogue of Preserved Materials. We would like to thank them collectively here, since it has not been possible to mention everyone by name. Museum curators and collection managers, zoo vets, university academics, secretaries and others have selflessly dealt with many time-consuming enquiries (particularly from GH), despite already being under ample pressure from their everyday duties. Without their cheerful collaboration, this book would never have come about. In addition, GH would like to thank Spartaco Gippoliti for his help in contacting Italian collections, and Colin Groves, Rebecca Jabbour, Andrew Kitchener, Peter Holt, James L. Newman and Esteban Sarmiento for kindly reviewing the text for the Introduction to Part II. Paolo Viscardi provided some useful text (Tyranny of the test tube) for the Introduction. GH would also like to thank the experts in linguistics and the classics who have contributed in no small measure to the Introduction: their enthusiastic willingness to help with a project lying outside their normal sphere of activity is both exemplary and laudable.

We are grateful to Rob Shave for information about the International Primate Heart Project. The Coopers are indebted to Terry B. Kensler, Collection Manager, Laboratory of Primate Morphology, Caribbean Primate Research Center, Puerto Rico, for her welcome and for sharing information about skeletal methods so freely.

Both John and Margaret Cooper thank Professor Richard Griffiths for permission to use the DICE (Durrell Institute of Conservation and Ecology, University of Kent) address for this publication. Since DICE's inception, it has trained hundreds of postgraduate students, from over 80 countries, and the Coopers are proud to have been associated with it, as Honorary Research Fellows, for many years.

JEC owes a great deal to his many mentors, both as a schoolboy and in later years. He particularly appreciates the training and guidance he received in respect of:

- natural history from Major Maxwell Knight (former M15 agent, the prototype for 'M' in the James Bond books), Mr Gerald Durrell (animal collector extraordinaire and Founder of the Jersey Zoo/Durrell Wildlife Conservation Trust/'Durrell' in the (British) Channel Islands), Mr Henry Berman (teacher extraordinaire). These and others with a love of the countryside enabled him to apply lessons about the natural world learned in the heaths and woodlands of England to wild places in Africa and elsewhere.

- osteology and its links with primatology from Dr Louis S.B. Leakey and Professor Phillip Tobias.
- odontology methods from Professor Loma Miles, Professor David Poswillo and Dr Caroline Grigson at the RCS and Professor Arthur Kemoli in Nairobi.
- comparative pathology (creatures great and small) from Dr Edward Elkan, Professor Peer Zwart, Professor Norman Ashton, Professor Gerry Slavin, Drs Ariela Pomerance, Mike Bennett, Ashley Price, Rodney Finlayson and R.N. T-W. Fiennes.

JEC also acknowledges the help of former colleagues in respect of primates and other matters at the University of the West Indies (UWI) — Paluri Murti, Rahul Naidu, Haytham Al-Bayaty, Lee Koma, Ravi Seebaransingh, Rod Suepaul and Richard Spence.

JEC owes a particular debt of gratitude to his teachers over the years, especially those who taught him medical and veterinary pathology at Bristol University and tropical veterinary pathology at Edinburgh University.

From 1978 to 1991, when JEC was Senior Lecturer in Comparative Pathology at the RCS, working with, and teaching, both medical and dental graduates, he was able to practise 'One Health' on a daily basis. He also received much support from RCS staff members Nina Wedderburn, Martyn Cooke, Liz Allen, John Turk, Ian Lyle, Derek Manning, Ashley Miles, Bari Logan, Ken Applebee, Alan Graham and Brian Eaton and Professor AJE Cave (who was publishing material on gorillas before JEC was born). He received encouragement from former Presidents such as Sir Alan Parks and Sir Reginald Murley. The many veterinary students who 'saw practice' at the RCS helped JEC with numerous relevant projects.

During his time as Guest Professor in Berlin, at the Institut für Zoo und Wildtierforschung/Institute for Zoo Biology and Wildlife (IZW), JEC and his wife received much support from the Director, Professor Reinhold R. Hofmann, an old friend from Kenya, and a warm welcome from colleagues there. Amongst those who contributed notably to studies on gorillas was Dr Roland Frey who performed a skilled dissection of a gorilla's hand that was transported by car, with all necessary permits, across the length of Western Europe!

Over the years very many people have helped nurture the Coopers' studies on skeletal lesions in gorillas.

In Rwanda, before the outbreak of genocide, they included staff of the VVC, notably Innocent Kiragi. Afterwards, following the Coopers' evacuation to England in 1994 prior to returning to the VVC, Elizabeth Allen, John Turk and Martyn Cooke at the RCS provided facilities and support. The Coopers' strong links with Jersey Zoo/Durrell Wildlife Conservation Trust/'Durrell', ranging from helping run their Summer Schools to JEC's period as consultant veterinary pathologist to the Trust, greatly helped in this respect too. The support at Jersey Zoo of Gerald and Lee Durrell, Jeremy Mallinson, John Hartley, Nick Le Q. Blampied, Tony Allchurch, Richard Johnstone-Scott and other Trust members was much appreciated.

Following their departure from the VVC in Rwanda in 1995, the Coopers extended their skeletal research by visiting numerous museums and collections in different parts of the world. Those institutions where they worked — and the people who offered them a warm welcome — are too many to mention individually but are amongst the establishments listed in Part II: A Catalogue of Preserved Materials. The Coopers are grateful to staff of the Medical School of Witwatersrand, South Africa, for their support and assistance with mountain gorilla specimens, especially Professor B Kramer who gave permission for the examination of the material and Peter Dawson, Mary-Ann Costello, Llewellyn Sinclair and Peter Montja for the preparation and radiographing.

Most recently we have received help and been given access to material (see also Part II: A Catalogue of Preserved Materials): at the Muséum National d'Histoire Naturelle, Paris (Jacques Cuisin); Booth Museum (John A. Cooper); Royal Zoological Society of Scotland (RZSS) (Simon Girling, Lynda Burrill, Roslin Talbot and Jo Elliott); the Zoological Society of London (Edmund Flach and Tom Kearns); Bristol Zoo Gardens (Michelle Barrows, Allen Goodship, Rowena Killick, Lynsey Bugg, Sarah Gedman, Ryan Walker and Sam Matthews). Moses Cooper, probably the youngest contemporaneous gorilla researcher (he was born in 2006), has for 2 years ably assisted his grandfather in skeletal studies.

Over the years colleagues at Cambridge University Department of Veterinary Medicine have been generous with their time and facilities. These include Michael Herrtage, Fernando Constantino-Casas, Kate Hughes, Paddy Mannion, Chrissie Willers, Anna Dussek, Lucy Webb, Vicky O'Mahony, Mathew Rhodes, Madeline Fordham, Rayna Skoyles, Scott Dillon, Louise Grimson and Paul Tonks. Three colleagues at Cambridge have helped with translation of papers — Emilie Cloup (*merci*!), Heidi Radke and Angelika Rupp (*danke*!) and Carolina Arenas (*gracias*!).

The cordial welcome and generosity provided by Michael Day at the School of Veterinary Sciences, Langford, University of Bristol, has been particularly appreciated.

The authors are grateful to the librarians at the RCVS (Clare Boulton), Linnean Society (Lynda Brooks and Gina Douglas), National Museums of Kenya (Asha Owano) and Institute of Primate Research (Grace Mathani) for their help and use of their reading and literature-searching facilities.

By contrast, the keepers of some collections and databases either failed to reply to our polite requests or declined to help. Those attitudes only serve to accentuate the kind, generous, responses of many others. Two examples illustrate this:

Always unrestricted access to the bones
Ogeto Mwebi, National Museums of Kenya

And, last but not least, please remember that the doors of the collections stay open for further investigations!
Jacques Cuisin, Muséum National d'Histoire Naturelle.

Such is the spirit of true scientific collaboration.

Having read widely in order to compile this book, we realise more than ever the debt owed to earlier pioneers, most of whom we never met but who contributed to a greater understanding of gorillas. For instance, in giving the Coopers a copy of her 1991 book, Penelope Bodrey-Sanders opened our eyes even wider in respect of the achievements of Carl Akeley. As she put it: *Nearly three quarters of a century ago Akeley became a pioneer in African conservation through his art and love of gorillas. Stimulated directly or indirectly by him, many others have over the years accepted the challenge and moral obligation to fight on behalf of Africa's wildlife. The mountain gorilla is Akeley's living monument, and those that follow his lead remain his most enduring legacy.* There are other, often forgotten, pioneers — for example, Kenneth Carr, hunter and explorer, whose papers and photographs were passed to us by Edward Le Conte, and Jean-Marie Eugène Derscheid the Belgian zoologist who became the initial director of Africa's first national park. During the Second World War Derscheid served in the Belgian Resistance and helped Allied soldiers and airmen escape from occupied Europe. He was shot as a spy by the Nazis on 13 March 1944, 5 days after JEC was born.

While the origins of the catalogue, go back to GH's absorbing interest in gorillas that began in the 1960s, the book has a much more recent origin — the Coopers' time (1993–95) with the Morris Animal Foundation's (MAF's) Volcano Veterinary Center (Centre Vétérinaire des Volcans) in Rwanda. More is said about this in the Preface, but this is an opportunity to pay tribute to many affected by the outbreak of violence (the 'Rwandan genocide') in April 1994 and its aftermath. Some Rwandans who helped the VVC or collected samples for the project were killed. Others survived but with terrible legacies. Etched in the Coopers' memories forever are the words of one tracker who, in faltering English, wrote: *Do not forget we.* They will, indeed, not forget him nor many, many others.

The Coopers particularly remember those who did so much to keep them safe and to help them escape from the fighting and from Rwanda. The United Nations troops, so maligned by many and so small in numbers, must never be forgotten. Those from Ghana and Bangladesh who served in Kinigi were the Coopers' friends, not just their peacekeepers. Some were never to return to their home countries. Three other blue-hatted UN soldiers, amidst the carnage going on around them, with only a pistol apiece, helped 169 evacuees, including the Coopers, to cross the border to relative safety in Zaire (now DRC). Those who criticise the United Nations usually do so from their comfortable office and forget Dag Hammarskjöld's words: *The UN was not created to take mankind to heaven, but to save humanity from hell.*

In the context of those days, John and Margaret Cooper particularly wish to acknowledge the support of the staff of the Centre Vétérinaire des Volcans before, during and following the genocide in 1994 and to recognise in these pages the friendship and courage of JEC's Rwandan counterpart. Other thanks and poignant memories, can only be summarised here: the safe haven provided (when they had crossed into Zaire) by Alyette de Munck; support and friendship from Nicole Merlo; the Belgian government plane to Burundi; the flight with the US Marines to Kenya (Coopers' *nyumba ya pili* — second home); the Institute of Primate Research who provided a secure temporary home for them and JEC's counterpart; the good-humoured comradeship of Garry Richardson with whom JEC returned to Rwanda very soon after the cessation of hostilities; the Coopers' exciting but unpredictable road trips back into Rwanda over the succeeding months, with Gillian Njeru of the British High Commission tracking their whereabouts (no mobile phones or emails); the NGOs who provided overnight accommodation; the support of Canadian, British and Australian UN forces; return to Rwanda to restart gorilla health operations, with support from the US Embassy, the British Consul, Georges Gérin and the reassuring visit to make a film of their work by Christian Hermann (the accompanying Mozart clarinet concerto continues to serve as a moving reminder of 'camping' in a house in war-torn Kigali).

No mention of the Coopers' time in Rwanda would be complete without expressing their heartfelt thanks to Rob Hilsenroth ('Dr Lobb') then Executive Director of the Morris Animal Foundation. Rereading faxes and JEC's diaries from 1994 brings memories, good and bad, flooding back and makes them realise how much practical and emotional support Rob gave during those tumultuous times. *Asante sana!*

Recollections of those difficult, but exhilarating, days were rekindled on 23 October 2016 when John and Margaret Cooper were guests of honour at a memorable dinner held by Rwanda's Gorilla Doctors in the garden of the Hotel Muhabura in Ruhengeri, Rwanda. This open-hearted welcome, led by Dr Mike Cranfield and Dr Jean Bosco Noehli, seemed to symbolise a passing of the torch from the wazee (the old ones) to the new generation of African gorilla veterinarians.

The Coopers also had a moving reunion, after 22 years, with some of the surviving members of their staff at the VVC. *Asanteni sana.*

The authors conclude with a tribute to someone who is symbolic of the sacrifices made by those living in Africa to the interlinked causes of conservation and development. The name of Rosamond Carr (1912−2006) is familiar to all those with an interest in Rwanda and the mountain gorillas. Madame Carr, as she was known to local people, was confidante and friend to Dian Fossey, a role immortalised in the book and film *Gorillas in the Mist*. In her later years Madame Carr became internationally recognised for the remarkable work of her orphanage, Imbabazi, that cared for victims of the genocide. To the Coopers, Roz Carr was a personal friend with whom they shared many happy times but also periods of grief. It was they who broke the news to her, on 6 April 1994, that the President of Rwanda had been assassinated and bloodshed had started. They were amongst those who urged her, when the killings intensified, to leave Rwanda and they were part of the welcoming party on her return, via Nairobi, when the fighting finished. The Coopers shared her sadness at the loss of friends and innocent people and rejoiced with her when she discovered those, including some of her devoted staff, who had, after all, survived the bloodshed. The Coopers last saw Madame Carr at her 90th birthday party in Rwanda in 2002. There, in a speech in English, French and Swahili, on behalf of all who had served with the VVC or MGVP, they thanked her and her staff (especially her devoted headman, Sembagare Munyamboneza) for their friendship and support and praised her own personal faith and courage. All who have worked with the mountain gorillas owe her an enduring debt of gratitude.

John E. Cooper
Gordon Hull
November 2016

About the Authors

John E. Cooper, a specialist veterinary pathologist, is a member of a husband and wife team, from the United Kingdom. He and his wife Margaret (a lawyer, who is a contributor to this book) travel widely and lecture together in many countries. They have spent nearly 20 years living overseas, mainly in Africa, the Middle East and the Caribbean. Professor Cooper was the veterinarian for the Virunga gorillas from 1993 to 1995 and continues to study the health of great apes and other species. The Coopers are now based in Britain, where they have visiting academic commitments, but are also engaged in voluntary work with wildlife, domesticated animals and communities in East Africa.

Gordon Hull is an amateur naturalist with a keen interest in mammals, and primates in particular. Over a period of many years, he has compiled a great deal of information about gorillas in zoos, museums and other scientific institutions throughout the world, which he is now making available in this book. Gordon is a member of the Gorilla Pathology Study Group and enjoys a continuing, wide-ranging and productive collaboration with gorilla experts and enthusiasts alike.

About the Contributors

Paul Budgen, FLS, Gorilla Pathology Study Group, London, United Kingdom

Martyn Cooke, Head of Conservation, The Royal College of Surgeons of England, United Kingdom

John E. Cooper, DTVM, FRCPath, FRSB, CBiol, FRCVS, The University of Kent, Canterbury, United Kingdom

Margaret E. Cooper, LLB, FLS, The University of Kent, Canterbury, United Kingdom

Allen Goodship, BVSc, PhD, MRCVS, Professorial Research Associate, Department of Med Phys & Biomedical Eng UCL, Faculty of Engineering Science, Emeritus Professor of Orthopaedic Science, Institute of Orthopaedics & Musculoskeletal Science UCL, Division of Surgery & Interventional Science, Royal National Orthopaedic Hospital, Stanmore Campus University College, United Kingdom

Jane Hopper, MA, VetMB, CertZooMed, MRCVS, Head of Veterinary Services, Port Lympne Wild Animal Park, Lympne, United Kingdom

Gordon Hull, Gorilla Pathology Study Group, London, United Kingdom

Jenny Jaffe, DVM, MSc (WAH), MRCVS, Tacugama Chimpanzee Sanctuary, Freetown, Sierra Leone, Institute of Zoology, ZSL, London, United Kingdom

Gladys Kalema-Zikusoka, BVet Med, MRCVS, MsPVM, Conservation Through Public Health, Entebbe, Uganda

Sophia Keen, Human Sciences BSc, BVSc Veterinary Science, University of Bristol, Bristol, United Kingdom

Brian N. Livingstone, FRCS, FLS (retired), Formerly Consultant Orthopaedic Surgeon, Wrightington, Wigan & Leigh, NHS Trust, Organising Secretary North West Regional Bone Tumour Register 1995–2002, United Kingdom

N. Keith Maybury, DM, FRCS, FLS, Consultant General and Vascular Surgeon (retired)

Dermot McInerney, University of Bristol, Bristol, United Kingdom

Roberto P. Miguez, BA, Senior Curator, Mammal Section Life Sciences – Vertebrate Division, The Natural History Museum, Cromwell Road, London, United Kingdom

Jaimie Morris, Doctoral candidate, BSc (Hons), Ecology Research Group, School of Human and Life Sciences, Canterbury Christ Church University, Canterbury, Kent, United Kingdom

Ogeto Mwebi, PhD, MSc, Post-graduate Diploma, Ordinary Diploma, Osteology Section, National Museums of Kenya, Nairobi, Kenya

Geoffrey R. Pearson, BVMS, PhD, FRCPath, FRCVS, Professor Emeritus, University of Bristol, School of Veterinary Science, Langford, Bristol, United Kingdom

Carina Phillips, BA (Hons), MA, AMA, Curator, Royal College of Surgeons of England, United Kingdom

Ian Redmond, OBE, CBiol, HonDUni, DSc (h.c.), FLS, Ambassador UNEP Convention on Migratory Species, Co-founder, UN Great Apes Survival Partnership, Chairman, Ape Alliance, Chairman, The Gorilla Organization, Senior Wildlife Consultant, Born Free Foundation

Celsus Sente, BVM, MSc, Assistant Lecturer, Department of Wildlife and Aquatic Animal Resources, School of Veterinary Medicine and Animal Resources, College of Veterinary Medicine, Animal Resources and Bio-security, Makerere University, Kampala, Uganda

Paolo Viscardi, MPhil, BSc, FLS, Chair of the Natural Sciences Collections Association. Curator of Zoology, Natural History Division, National Museum of Ireland – Natural History, Dublin, Ireland

List of Abbreviations

See also: http://en.wikipedia.org/wiki/List_of_medical_abbreviations.

AAZV	American Association of Zoo Veterinarians
ACVP	American College of Veterinary Pathologists
AD	anno domini
AMR	antimicrobial resistance
BC	before Christ
BCE	before the Common Era
BINP	Bwindi Impenetrable National Park
CCCU	Canterbury Christ Church University
CE	Common Era or Current Era
CR	Congo Republic (Republic of Congo, West Congo, Congo-Brazzaville)
CTPH	Conservation Through Public Health
DFGF-Europe	Dian Fossey Gorilla Fund Europe
DFGFI	Dian Fossey Gorilla Fund International
DRA	disease risk analysis
DRC	Democratic Republic of the Congo
EAZWV	European Association of Zoo and Wildlife Veterinarians
ECVP	European College of Veterinary Pathologists
ECZM	European College of Zoological Medicine
EEP	European Endangered Species Programme
ELISA	enzyme-linked immunosorbent assay
EU	European Union
FAO	Food and Agriculture Organization of the United Nations
FFI	Fauna and Flora International
GRASP	Great Apes Survival Partnership
HIC	high-income country
ICCN	Institut Congolais pour la Conservation de la Nature
IGCP	International Gorilla Conservation Programme
IPR	Institute of Primate Research
ISIS	International Species Information System
LMIC	low- and middle-income country
MAF	Morris Animal Foundation
mDNA	mitochondrial DNA
MGVP	Mountain Gorilla Veterinary Project

MRCVS/FRCVS	Member/Fellow of the Royal College of Veterinary Surgeons
MRI	Magnetic resonance imaging
MSF	Médecins Sans Frontières
nDNA	nuclear DNA
NGO	nongovernmental organisation
NHP	nonhuman primate
NMK	National Museums of Kenya
NSAID	nonsteroidal anti-inflammatory drug
ORTPN	Office Rwandaise du Tourisme et des Parcs Nationaux
PASA	Pan African Sanctuary Alliance
PCM	The Powell-Cotton Museum
PCR	polymerase chain reaction
PCV	packed cell volume
PNV	Parc National des Volcans
POC	point of care
PPP	personal protective equipment
RCPath	Royal College of Pathologists
RCVS	Royal College of Veterinary Surgeons
SI Units	Standard international units (Système International d'Unités)
SSP	Species Survival Plan
UK	United Kingdom
UN	The United Nations
UNEP	United Nations Environment Programme
UNESCO	United Nations Educational, Scientific and Cultural Organization
USA	United States of America
USSR	The Union of Soviet Socialist Republics
UWA	Uganda Wildlife Authority
VVC/CVV	Volcano Veterinary Center/Centre Vétérinaire des Volcans
WHO	World Health Organization of the United Nations
WWF	World Wide Fund for Nature
ZIMS	Zoological Information Management System

Part I

Gorilla Pathology and Health

Chapter 1

The Genus *Gorilla* — Morphology, Anatomy and the Path to Pathology

John E. Cooper

I have found the existence of an animal of an extraordinary character in this locality and which I have reason to believe is unknown to the naturalist.

Thomas Savage (1847), letter to Richard Owen

INTRODUCTION

The animal that we now know as a lowland or mountain gorilla, *Gorilla gorilla* or *Gorilla beringei*, was not recognised by scientists until halfway through the 19th century. In 1847 Thomas S. Savage, an American medical missionary, then based in Boston, Massachusetts, United States, was sent some skeletal remains from Gabon. They were thought to be from a new species of chimpanzee. As a result of careful examination of these bones, Savage and Wyman (1847) published an account of 'the external characters and habits of *Troglodytes gorilla*, a new species of orang from the Gaboon River'. As explained in the Introduction to Part II: A Catalogue of Preserved Materials, Savage and Wyman gave the animal the generic name *Troglodytes* because they mistakenly believed that it was a new species of chimpanzee, then known as *Troglodytes niger*, and *gorilla* because they surmised that the creature seen on the coast of West Africa in the 6th century BC by Hanno, the Carthaginian navigator and described in his *Periplus of Hanno* was 'probably one of the species of the Orang'.

The next stage in unravelling the taxonomic status and anatomical features of the gorilla followed dissections by Richard Owen (Fig. 1.1) of specimens he received at the British Museum, which enabled him to publish his 'Memoir on the Gorilla (*Troglodytes gorilla*, Savage)' (Owen, 1865). Owen went about obtaining specimens of his own via Samuel Stutchbury, Curator of the museum of the Bristol Institution for the Advancement of Science, Literature and the Arts, who encouraged the captains of ships to procure specimens on their travels. Work in

France by Isidore Geoffroy Saint-Hilaire at the National Museum of Natural History in Paris led to the removal of the gorilla from the genus *Troglodytes* and the introduction of the genus *Gorilla* (Geoffroy St Hilaire, 1852, 1853).

Specimens obtained by brave hunters — who quickly discovered that gorillas are not invulnerable to bullets — that reached Europe as skins, skeletons or cadavers, provided much information about the gorilla's structure.

This sequence of events illustrates well how the discovery of a new taxon starts with observations and individual reports, usually of varying accuracy, that describe the basic morphology of the animal — in the case of Hanno's 'savage people, the greater part of whom were women, whose bodies were hairy'. Morphology concerns the form, the visible shape or configuration, of an organism without necessarily any handling or physical investigation. The next step in the case of the gorilla — and, indeed, other new species also, was anatomical study. This usually comprises dissection, attempting to link structure with function and classifying the animal(s) on the basis of anatomy and, nowadays, molecular tests.

In 1965, a century after the 'discovery' of the gorilla, in his Inaugural Lecture as Professor of Zoology at Imperial College, London, H. R. Hewer tracked the history and development of zoological science from Aristotle (384—322 BC) to the mid-20th century. He described taxonomy, morphology and anatomy as the 'older branches of zoology', that laid the foundations for the development of new disciplines such as embryology and ecology — and on to molecular biology and genetics.

So what is the relevance of Hewer's vision of zoology in the new millennium to the pathology and health of gorillas? His 'Tree of Zoology' is an important reminder of the importance of the inter-relatedness of the different disciplines and of how advances in one can provide a ready and accessible bridge to the others. In the context of this book, the message is that morphological and anatomical

J. E. Cooper & G. Hull: Gorilla Pathology and Health. DOI: http://dx.doi.org/10.1016/B978-0-12-802039-5.00001-9

FIGURE 1.1 The bust of Richard Owen at the Royal College of Surgeons of England survived the London blitz, and remains as a testimony to his work. *Image courtesy of the Royal College of Surgeons.*

FIGURE 1.2 Early naturalists learned much about the habits of gorillas as a result of sound observation. These mountain gorillas have moved into an area of bamboo to feed. *Image courtesy of Maxwell Cooper.*

studies on the gorilla have provided essential information as to what is 'normal' in gorillas and thereby paved the way to understanding what is 'abnormal' – the basis of pathology. An appreciation of pathology and the pathogenesis of disease furnishes in turn a framework for scientifically ('evidence') based medical and surgical interventions and for comprehensive health surveillance of these animals, both in the wild and in captivity.

It is important also to remember that a knowledge of the basic biology – alas, so often neglected in teaching nowadays in favour of modern molecular science and 'problem-based learning' – remains a vital tenet to working with live or dead whole animals. In Thomas Huxley's book *The Crayfish* (1879), he famously stated 'For, whoever will follow its pages, crayfish in hand ... will find himself brought face to face with all the great zoological questions ... by completing the history of one group of animals, secure the foundation of the whole of biological science'. In contrast, this book is concerned with only one genus, *Gorilla*, which comprises two extant species, *G. gorilla* and *G. beringei*, but our approach to promoting a better understanding of the pathology and health of the taxon will, we hope, prove of assistance also to those studying other groups of animals. Our text will also emphasise the importance in all endeavours of a sound understanding of an animal's anatomy, physiology, behaviour and natural history. As Diderot stated 'Il faut être

profond dans l'art ou dans la science pour en bien posséder les éléments' [We must be deeply familiar with art and science if we are properly to understand their components].

How do wildlife veterinarians, pathologists, primatologists and others concerned with the health of gorillas gain the necessary knowledge of the biology of these animals? Books, scientific publications and 'popular' articles are a good start. There are many texts that can be advocated (see References and Further Reading) of which particular attention is drawn to the seminal works of Akeley (1920, 1929), Schaller (1963, 1964), Fossey (1983), Meder (1993), Napier and Napier (1967), Dixson (1981) and Taylor and Goldsmith (2003).

Exposure to live gorillas is the best way to learn about their natural history. So many of the earlier observers, even if not trained biologists, had a naturalist's approach to the world around them and to the places where gorillas lived. Thus, Fred Merfield also had an interest in entomology (Loxdale, 2013) and Charles (CRS) Pitman, an all-round naturalist and a senior Game Warden in Uganda during colonial times (Pitman, 1931, 1942) remains best known for his work on Ugandan snakes. It is often such an observer, alluded to by Owen (1859) as a 'competent and candid naturalist', who first describes accurately the characteristics and habits of wildlife. In contrast, many modern gorilla researchers are sound, formally trained, scientists, with impressive lists of publications and large grants, but do not always have a wider grounding in, nor an empathy with, other fauna and flora. That is despite the clear attraction of gorilla habitat in Central Africa, even without the sighting of a gorilla, as emphasised by Cooper and Cooper (1994), in their 'A trip to the mountain gorillas: a naturalist's perspective' (Fig. 1.2).

Live gorillas in zoos also provide opportunities for study (Fig. 1.3) – and began to do so in a limited way more than a century ago (in France, e.g., Milne-Edwards,

FIGURE 1.3 A silverback lowland gorilla in a modern European zoo. Such animals offer opportunities for study and public education.

1884; Patit, 1926). Gippoliti (2006) discussed the historical and present role of zoos in relation to education and suggested that such institutions can and should also contribute to promoting public awareness of the welfare needs of primates (see also Chapter 8: Nonspecific Pathology). Writing more generally, Doolittle and Grand (1995) argued that 'captive exotic species enrich and clarify the major principles of anatomy (form and function, adaptive behaviour and evolutionary process) and encourage students to integrate gross anatomy with behaviour, including that associated with exhibit design'.

The question of gorillas in zoos has been – and, in some quarters, remains – a contentious issue (see also Chapter 8: Nonspecific Pathology). Cousins (2015) pointed out that 'Early menageries knew little about the requirements of our nearest living kin: bad diets, bleak housing, lack of proper husbandry, woefully inadequate healthcare and high mortality....' and, like Gippoliti (2006) and others, believes that there is still much room for improvement. The adverse effects of poor management include both physical and psychological ill health, which are referred to elsewhere in this book.

Notwithstanding concerns about zoological collections, past and present, this book deals with gorillas both in the wild and in captivity. One of the main aims of our publication is to promote the welfare of gorillas and this is as important, if not more so, in respect of animals in zoos.

The apparently successful and humane care of gorillas under captive conditions is a relatively new development. It was only 35 years ago that Dixson (1981) wrote 'even nowadays it is no easy matter to keep gorillas alive and healthy in captivity; only during the last twenty-five years have they reproduced successfully in zoos'. Most of the experience gained in this area over the past four decades

relates to lowland gorillas. Mountain gorillas do not appear to tolerate captivity well.

As far as lowland gorillas are concerned, though, knowledge and expertise have most certainly advanced, as explained elsewhere in this book (see, e.g., under Stress and stressors in Chapter 16: Endocrinological and Associated Conditions), but much remains to be learned.

PRIMATE AND COMPARATIVE ANATOMY

Comparative anatomy was a subject of great interest over 2000 years ago. It led to a proper recognition and understanding of the importance of the organs of animals, including primates. Comparative anatomy is not just of historical interest but also affords a basis for studies today, not least of all because some of the earlier publications provide excellent descriptions of structures and, with the aid of illustrations (often line drawings), assistance in how to locate them.

It is interesting to note that some of the earliest naturalists, such as Gaius Plinius Secundus (Pliny the Elder) (AD 23–79), explored the comparative aspects of structures centuries prior to this term being recognised and before there was any clearly defined philosophy as to why morphological studies might be of importance to humans as well as to other taxa. Early investigators saw merit in exploring anatomy for its own sake, not necessarily because it might throw light on human health and wellbeing. Nevertheless, the emergence of anatomical knowledge about animals, especially monkeys and apes, began to go hand in hand with developments in human anatomy with the studies of Galen of Pergamum (AD 129–c.216), arguably the single most influential figure in Western medicine.

Galen performed many 'medical demonstrations', including dissections on living monkeys (probably Barbary macaques). His work provided information on the anatomy of these poor beasts – Galen drew on many centuries of tradition before him, including that of Hippocrates (BC 460–377), who also dissected monkeys – but it has been argued by some that the demonstrations were largely aimed at impressing those observing ('Elders of the physicians') with his surgical skills (Mattern, 2013). Nevertheless, as will be illustrated below, Galen's work drove anatomical and physiological wisdom for over 1000 years.

Fifteen hundred years later, Andreas Vesalius (Andre van Wesele) (1514–64) was born in Brussels, then part of the Holy Roman Empire. Vesalius became dissatisfied with learning from animals and criticised Galen for basing his anatomical theories on such. Vesalius' subsequent dissections of the human body helped to correct misconceptions dating from ancient times. When only 29 years old, he published what was to be the first edition of his book on human anatomy, *De humani corporis fabrica libri VII (The Fabric of the Human Body)* (now accessible in an annotated translation by Garrison and Hast, 2014). This proved to be the most comprehensive and accurate account of human anatomy of its day, correcting the errors in the traditional teachings of Galen and advocating that the dissection of cadavers should be carried out by physicians.

The origins of comparative primate anatomy go back to the Italian anatomist Marco Aurelio Severino, whose life spanned the 16–17th centuries. Severino drew specific attention to the anatomical similarities between monkeys/apes and humans and advocated that, because of this, 'apes' should be exploited for medical purposes (Severino, 1645). The Dutch physician Nicolaas Tulp had already published a short description of an anthropoid ape (presumably a chimpanzee) from Angola and later in the century the English medical graduate Edward Tyson carried out what was probably the first detailed dissection of an ape, a juvenile chimpanzee, in the Western World (Tyson, 1699). A comprehensive and critical history of these and other events concerning primates is provided by Gibbs et al. (2002) in their paper on the soft-tissue anatomy of the extant hominoids.

These studies on primates were an important part of a long established interest by scholars in the comparative aspects of structure in animals. A useful starting point to understanding the evolution of the subject – and, indeed, how the term 'comparative' came into vogue – is the book *A History of Comparative Anatomy from Aristotle to the Eighteenth Century* (Cole, 1949).

Interestingly, the first detailed comprehensive study of comparative skeletal structure by Belon (1555) focused on the anatomy of humans and birds. In 1623 Lord Francis Bacon used the term 'comparative anatomy', apparently for the first time in a European language but he employed it in the sense of comparison of members of the same species (individual variations), not different species. Fifty years later, Grew (1681) wrote about comparative anatomy, but his treatise dealt with plants not with animals. Gradually, recognition of the practical importance of comparative anatomical studies grew and in 1674 Bartholin wrote 'Zootomia, on the neglected' arguing that animal diseases may resemble those of humans.

As stressed earlier, knowledge of not only the superficial morphology but also the detailed anatomy of an animal remains the cornerstone of investigation of its whole biology. Thus, within a few years of the recognition of the genus *Gorilla*, Bennett (1884) published on its skull and Chapman (1892) on its brain. Likewise, only a decade after the discovery of the mountain gorilla, *beringei*, in Deutsch-Ostafrika (German East Africa) (Figs. 1.4 and 1.5), Swedish collector Arrhenius was recording anatomical data about it in the field (Lönnberg, 1917; see also Part II: A Catalogue of Preserved Materials). Distinguishing characteristics of *Homo* species were further explored by Schultz

FIGURE 1.4 The mountain gorilla. *Picture taken in the Virungas in 1970 by wildlife photographer Bob Campbell.*

FIGURE 1.5 Examination of the skull and mandible of the holotype of *Gorilla beringei* with the Curator at the Berlin Museum.

(1934) and, 20 years later, by Steiner (1954) – both, again, on the basis of examination of preserved material.

Gorillas quickly began to feature in comparative anatomical studies. These included research on, for example, structure and evolution by Sonntag (1924); the endocranial form of skulls by Harris (1926), with special reference to dolichocephaly; the laryngeal sacs by Miller (1941); the mastoid process by Ashton and Zuckerman (1952); variations in the venous systems of mammals (Barnett et al., 1958); the visceral organs of the mountain gorilla by Hosokawa and Kamiya (1961); the brain of the mountain gorilla by Hosokawa et al. (1965) and the frontal sinus in *Gorilla, Pan* and *Pongo* by Blaney (1986).

FIGURE 1.6 With Phillip Tobias at the University of Witwatersrand.

Traditionally many anatomical studies on gorillas have been orientated towards attempts to elucidate evolutionary similarities and differences between the extant 'great apes', modern humans and fossil hominids. Far-sighted scientists who have addressed this subject over the past one-and-a-half centuries are many but the names (each with a relevant publication) include Owen (1861), Huxley (1863), Keith (1896, 1926, 1899), Dart (1925), Jones (1938), Leakey (1970), Tobias (1991) (Fig. 1.6) and Walker and Shipman (1998).

According to Spikins (2015), it was the 'father of palaeoanthropology', Louis (LSB) Leakey (Fig. 1.7) – see also Chapter 2: The Growth of Studies on Primate Pathology – who first thought, in addition to studying skeletal remains of primates, of looking at the lives of extant great apes to see what they might tell us about our ancestry.

Such an approach is not without its challenges, however. Some indication of the passionate feelings engendered in the 19th century, even amongst scientists, is illustrated by what came to be known as the 'Great Hippocampus Question'. This revolved around the dispute between Huxley and Owen concerning the anatomy of apes and 'human uniqueness'. Owen had asserted that only the brains of humans possessed a hippocampus minor (now termed the calcar avis) and claimed that it was this that explained the unique traits of *Homo*. In fact, nonhuman primates (apes and monkeys) also have a calcar avis.

It was not only such scientific disagreement that lent controversy to comparisons between humans and apes. Racial prejudices (based largely on ignorance at the time) led to, for example, Owen giving an exposition in 1861 at the Royal Institution, London, on 'the distinctive characters between the Negro (or lowest variety of Human Race) and the Gorilla' (Owen, 1861). Interestingly from the perspective of this book – especially its emphasis on the importance of sharing of resources – the skeletal and brain material from that gorilla had been brought over to

FIGURE 1.7 Louis Leakey who pioneered gorilla research and who, with his family, revolutionised our understanding of the origin of humans.

Britain from France by Paul Belloni Du Chaillu (see Part II: A Catalogue of Preserved Materials), the man who was chiefly responsible for drawing the attention of the Western world to this new species (Reel, 2013).

The intense interest in the 19th century in comparing humans with apes was probably largely prompted by Huxley's 1863 book *Evidence as to Man's place in Nature* and, of course, by Darwin's *The Descent of Man* (Darwin, 1871). In the latter Darwin made clear his belief was that apes and humans had evolved from a common ancestor. The same volume contains Darwin's moving words that 'Man with all his noble qualities, with sympathy that feels for the most debased, with benevolence which extends not only to other men but to the humblest living creature, with his god-like intellect which has penetrated into the movements and constitution of the solar system – with all these exalted powers – still bears in his bodily frame the indelible stamp of his lowly origin.'

It is often assumed that it was the joint seminal presentation by Charles Darwin and Alfred Russel Wallace ('Theory of Evolution by Natural Selection') to a meeting of the Linnean Society of London on 1 July 1858 that set in motion open debate about man as an animal and his/

her relationship to other species. In fact, exactly a century before, Linnaeus (Carl von Linné) had included man (humans) in his classification of the Animal Kingdom. Notwithstanding his strong religious convictions, Linnaeus had argued that the anatomical characters of *Homo sapiens* compelled him to do so. He cited the words *Nosce te ipsum* (Know thyself) (originally in Greek γνῶθι σεαυτόν) that were engraved above the Temple of Apollo at Delphi and argued that, as humans are capable of studying their own anatomy, physiology, morals and politics, the species is on the road to ever greater wisdom.

So, how similar was the recently discovered species of gorilla to *Homo* in terms of its anatomy? In his 'Memoir on the Gorilla', Owen (1865) listed the changes he considered necessary to 'transmute a gorilla into a man'. Owen's list is short and largely comprises anatomical changes to the brain, teeth and intestines. In fact, as many authors over the years have pointed out (see, e.g., Dixson, 1981), humans possess few if any physical characters that were not anticipated in, or a feature of, their primate ancestors, a point made clear when Mivart published his 'Man and Apes: An Exposition of Structural Resemblances and Differences Bearing upon Questions of Affinity and Origin' in 1873. Mivart described primates, including humans, as being 'unguiculate, claviculate, placental mammals, with orbits encircled by bone; three kinds of teeth, at least at one time of life; brain always with a posterior lobe and calcarine fissure; the innermost digit of at least one pair of extremities opposable; hallux with a flat nail or none; a well-developed caecum; penis pendulous; testes scrotal; always two pectoral mammae' (Mivart, 1873). These profound observations are in stark contrast to the wit of the librettist W. S. Gilbert who said 'Man however well-behaved, at best is only a monkey shaved'.

Interest in such a comparative, often osteological, approach, comparing and contrasting the different hominids, continues. Just one example in the past decade that included studies on *Gorilla* was the work by Williams and Orban (2007) on the ontogeny and phylogeny of the pelvis in *Gorilla, Pongo, Pan, Australopithecus* and *Homo*. Palaeontologists use a comparative approach in their studies of hominin remains and their findings excite the general public. However, as Johanson and Edgar (1996) pointed out in their book, with its exquisite photographs of skulls (including *G. beringei*), discussion by palaeoanthropologists of 'our closest living relatives' can still be very contentious. While all this was happening, public interest in 'apes and monkeys' continued undiminished as these animals were portrayed in different roles in art, culture and religion (Morris and Morris, 1966). Indeed, from the Nile Valley 5000 years ago to the present, nonhuman primates have fascinated and inspired *Homo sapiens* (Zuckerman, 1998).

THE CURRENT STATUS OF ANATOMY

The formal teaching of anatomical skills has, regrettably, declined in recent years but the significance of knowing which structures and appearances are normal (and, conversely, which are not) is no less important than it was in the past for medical, dental and veterinary practitioners. Interestingly, but perhaps not surprisingly, a number of those who have published over the years on primate anatomy have had a background of training, sometimes practice, in human medicine and this has influenced their thinking and work — Napier and Hill being two examples. The relevance of a sound knowledge of comparative anatomy to clinical work on *Gorilla* was well illustrated by Jandial et al. (2004) who, in the context of lumbar disc surgery in a gorilla, stated 'For best ... results, one needs to consider the similarities and differences between the gorilla and human vertebral anatomy'. This is an echo of the prayer of the English surgeon Kay (1744) 'Lord, give me skill in Nature, and a good and useful Knowledge in Anatomy'.

While the teaching of anatomy may have waned in some quarters, the introduction of new preparative and investigative techniques, such as plastination (see Appendix 1: Glossary of Terms) and geometric morphometrics, has helped encourage anatomical studies on nonhuman primates, including gorillas. The standard textbook on primate anatomy, the eight volumes by Osman Hill (Hill, 1953—1970) remain the gold standard although more recent texts have appeared — for instance, the books by Ankel-Simons (2007) and Gebo (2014).

Insofar as the genus *Gorilla* is concerned, attention was drawn earlier in this chapter to the many fairly specialised anatomical studies, on various organs, that followed the discovery of the lowland and mountain gorillas. The first comprehensive work was 'The Anatomy of the Gorilla', edited by Gregory (1950) and referred to in many of the chapters that follow in this book. Specific studies on anatomical features continue to this day and it is particularly encouraging to note the publication of a series of monographs describing musculoskeletal structures of gorillas, chimpanzees, hylobatids and orangutans. One is a photographic and descriptive musculoskeletal atlas of *Gorilla* (Diogo et al., 2011) while the latest in the series details the anatomy in a baby gorilla and includes CT scans (Diogo et al., 2014). Other recent publications attest to the need for veterinarians who work with apes to have a sound knowledge of gross morphology in comparative studies — for example, the report of Kawashima and Sato (2012) on the extrinsic cardiac nerve plexus (ECNP) of orangutans, gorillas and chimpanzees. Quantitative, rather than qualitative, research has also been undertaken — for instance, the studies on body mass in lowland gorillas by Zihlman and McFarland (2000).

The foregoing has concentrated on gross anatomy. It took nearly 100 years for there to be significant studies on the microscopic anatomy (histology) of gorillas – for example, by Straus (1950) on the skin – and this, together with developments in our understanding of the gross and microscopic pathology of this genus, will be discussed in Chapter 2: The Growth of Studies on Primate Pathology.

In considering the evolution of medical knowledge over the past two millennia, from anatomy (the theme of this chapter), through pathology (the focus of the next), to clinical care, one must never forget the debt owed to Arab study and writings. For over 700 years, Arabic was the language of science. In AD 813 the Caliph Abu Ja'far Abdullah al Ma'mun created in Baghdad a flourishing centre of learning called Bayt-al-Hikma (House of Wisdom) – this at a time when the peoples of the British Isles were preoccupied with tribal divisions, war and plundering. Caliph al Ma'mun's pupils included Nestorian Christian Hunain ibn Ishaq, an illustrious physician and a leading translator of Galen. One hundred and seventy years later the Persian philosopher Avicenna was born and his 'Canon of Medicine' was to become the standard medical text, not only in the Islamic world but throughout Europe, until their 16th century.

None of the works of the great Islamic scholars can be said to have had a direct influence on our knowledge of gorilla pathology but their studies ensured the growth of medical knowledge, leading eventually to the 'discovery' by Europeans of many essential fundamentals. As Withington (1894) said in his book: 'It was this people who took from the hands of the unworthy successors of Galen and Hippocrates the flickering torch of Greek medicine. They failed to restore its ancient splendour, but they at least prevented its extinction and they handed it back after five centuries burning more brightly than before'.

It is worth mentioning that in the 9th century Iraqi zoologist al-Jahiz propounded a rudimentary theory of selection, explaining how environmental factors can affect species, forcing them to adapt and then pass on their characteristics to their offspring. This was 1000 years before Darwin and Wallace.

It is of considerable interest and significance that for nearly 1500 years Galen's writings were held to be the ultimate authority on matters of anatomy and physiology. According to Morris and Morris (1966), no one after Galen seems to have dissected monkeys until the 16th century.

TERMINOLOGY

Before moving on to Chapter 2: The Growth of Studies on Primate Pathology, a word needs to be said about wording and phraseology in this book.

The use of precise, properly defined, terminology is important in all scientific endeavour. Such use is crucial when working with primates, including gorillas, and their derivatives but unfortunately has not always been widely practised and promoted. The appeal of these animals to people from all walks of life, from different countries and from disparate disciplines, has meant that words and terms relating to description and observations on the genus are often not standardised.

First and foremost, it is important to state that, in this book, we attempt to adhere to modern taxonomical thinking insofar as gorillas are concerned. The family Hominidae used to be restricted to humans only, but it now includes great apes. Within that family, the subfamily Homininae is now considered to include the tribes Hominini (humans and their close, now extinct, ancestors), Panini (chimpanzees) and Gorillini (gorillas) while orangutans are in the other subfamily, Ponginae. An early, but very helpful, profile of the genus *Gorilla*, outlining much of its anatomical and other characteristics, is to be found in Napier and Napier (1967) while an excellent recent study of the taxon is presented by Butynski et al. (2013).

Standard medical terminology has been adhered to throughout this book. The primatological literature not infrequently refers to medical matters using incorrect, ambiguous or inadequately defined terms: 'disease infection' is an oft used example in one journal. Even medical and veterinary personnel can be guilty, however. de Lahunta (2014) urged all authors, including pathologists, to adhere to acceptable nomenclature and drew attention to the guidance given in the *Nomina Anatomica Veterinaria* (2005): 'There is no excuse for continuing to refer to anatomical structures using eponyms. These are now archaic'. Despite widespread use of the term 'pathologic(al) lesion', especially in lay literature, it is avoided in this book because it is tautologous; by definition, a lesion is abnormal (Jortner, 2012). As a substitute, 'pathologic(al) change' is sometimes used. Throughout the book it is assumed that a disease cannot be transmitted (it is the result of an interaction between a pathogen or physical insult and an animal) but an organism can.

For those without a grounding in Latin and Greek, the book *Veterinary Medical Terminology* by Taibo (2014) is recommended. Each medical word is broken down into syllables to enable the reader to understand the meaning of the word. These definitions are discussed in the text and then summarised in tables.

Differences in usage of words in variants of English and between scientists from distinct disciplines can cause confusion. This is especially the case when working in the range states where gorillas are found (or, in some cases, kept in captivity) but where English is not the first language. For instance, injuries or signs suggestive of disease may be noted in habituated gorillas by persons from diverse backgrounds – visitors (tourists), guides, guards, research workers or veterinarians – but maximum value can only be obtained from such reports if standard, recognised, wording and descriptions are used.

It was with this in mind that the author (JEC), when working in Rwanda, compiled a list of definitions to help those without a formal veterinary medical background (Cooper, 1997b). Some were also translated into French and Swahili. These included, for example, insistence that the term 'clinical signs' should be used for the gorillas rather than 'symptoms'. The former are those features of a disease that can be seen by an observer while the latter are experienced and can only be recounted by the (human) patient.

In the same document it was suggested that lay terms should be avoided whenever possible because they are so often ambivalent or inaccurate. An example given was the ambiguity of describing a skin lesion on a gorilla as a 'cut', unless the wound is known to have been caused by a cutting injury. Another was the word 'belching', often used by gorilla biologists working in the field – see, for example, Lanjouw (2002). A 'belch' is actually an 'eructation', the passing of gas from the stomach through the mouth (Schaller, 1963). The low grunting sound made by a gorilla is not eructation.

And what about such commonplace, but important, terms as 'wild' and 'free-living'? They are often used in different ways, by disparate authors. In this book all gorillas, whether in their natural habitat in Africa or in captivity elsewhere in the world are considered to be 'wild' animals on the grounds that they are not domesticated (see Appendix 1: Glossary of Terms, for definition of 'domesticated' and Chapter 18: Legal Considerations). Those gorillas that are not in captivity are referred to as 'free-living.' A complication is that the term 'free-ranging' is often used for such gorillas in the United States, but in Europe 'free-ranging' has a different meaning, viz., animals such as deer or ducks that are free to roam but return to the paddock or their housing at the end of the day. For that reason, 'free-ranging' is not used at all in this book except when quoting the writings of others.

Even the description 'free-living' might sometimes present problems. Jesús Pérez (personal communication) points out that the term is also used in studies on certain organisms such as amoebae, to differentiate those that are found in the environment from others that are parasitic on animals or plants!

In a further comment on wording, Jesús Pérez suggests that scientists should avoid terms that are 'anthropic or have human connotations': for example, when he speaks in Spanish about a wild animal, he prefers to use the word 'silvestre' rather than 'salvage'. The same point can be made about descriptions of animals, including gorillas, which are judgemental in terms of behaviour. Therefore, in this book, the only mention of such terms as 'vicious' is when earlier authors are being cited.

There is no easy solution to the problems that different uses of language can cause.

In this book the authors have tried to be consistent in terminology and, despite their own shared deep respect for gorillas and their wellbeing, have endeavoured to avoid phraseology that is anthropomorphic. The inclusion of an extensive Glossary (see Appendix 1) intended to encourage consistency and to assist those readers who are relatively unfamiliar with medical or biological terms or not primarily anglophone.

CONCLUSIONS

To return to the main theme of this chapter, in studying and advancing the pathology and health of *Gorilla*, knowledge not only of the history of its discovery but also of its relationships and probable evolution is important. Current thinking is that the earliest primitive ape may have been *Kamoyapithecus hamiltoni*, the jaws, teeth and long bones of which were found in a 27 million-year-old site to the west of Lake Turkhana in present-day Kenya. At least a dozen genera and many species of extinct primitive apes have been identified from other sites in Kenya. The most famous is probably *Proconsul africanus*, a generalised tailless quadruped, possibly the ancestor of modern hominins, discovered by Louis and Mary Leakey in 1942. The diversity of apes in the Early Miocene was not, alas, followed by any discovery of their successors in Africa but fossil apes became increasingly common in Europe and Asia, probably because high global temperatures encouraged the growth of forests. Apes had disappeared from Eurasia by the end of the Miocene other than *Gigantopithecus* which survived until 400,000 years ago. Some believe that the origin of modern gorillas was *Samburupithecus*, known only from a 9 million-year-old fragment of upper jaw from Samburu – again in modern day Kenya.

The authors make no apologies for this brief incursion into the evolution of the great apes. Some knowledge of this subject is relevant to understanding the biology of extant primate species. Even more important, though, is that the skills and perseverance that many palaeontologists bring to the description of fossil specimens and to formulating theories about speciation and behaviour are an exemplar to other scientists as to how primate specimens, large or small, however diminutive, should be investigated. Notwithstanding the caveats expressed elsewhere in this book about the need for caution when it comes to making diagnoses, much can be learned from palaeontologists by veterinarians, comparative pathologists and others who are attempting to glean information from present-day skulls, skeletons and soft tissue remains of gorillas. It emphasises the importance of interdisciplinary research, a theme that is pursued further in Chapter 2: The Growth of Studies on Primate Pathology.

Chapter 2

The Growth of Studies on Primate Pathology

John E. Cooper

...... he is more like a giant in stature than a man; for he is very tall, and hath a man's face, hollow-eyed, with longe haire upon his browes.

Andrew Battell (1625)

INTRODUCTION – THE GROWTH OF PATHOLOGY

Until about 1990, much of what we knew about the diseases and pathology of gorillas was based on postmortem and clinical work in zoological collections. For example, earlier issues of the *Transactions of the Zoological Society of London* confirm that most animals that died in that collection (the 'London Zoo') were examined postmortem and that detailed records were maintained (see, e.g., Hamerton, 1935, 1938, 1939a,b, 1943). Comparable data were amassed by other zoos in Europe and in North America – for example, the Philadelphia Zoo in the United States (Fox, 1923). Such reports from zoological collections remain very useful sources of information on causes of morbidity and mortality of captive animals but, as examples in this book will illustrate, the quality and reliability of such records vary greatly.

In 1989 Gerald Durrell, founder of the Jersey Wildlife Preservation Trust (now the Durrell Conservation Trust, 'Durrell') in the (British) Channel Islands stated 'In any well conducted zoological collection the importance of studying, recording and disseminating the knowledge obtained from the animals of which you are custodians cannot be overemphasised. Since the inception of the Trust in 1963, we have endeavoured to keep detailed records of the animals in our care, which we publish'.

And, indeed, in its relatively short life, JWPT's house journal *Dodo*, which is cited several times in this book, included numerous publications about threatened and endangered species, including lowland gorillas, many (appropriately) compiled by the Trust's keepers and field staff. Alas, *Dodo* is not only no longer produced but, apparently, unavailable electronically. However, some indication as to what those earlier issues contained can be gleaned from (Anonymous, 1994) a *Cumulative Index of Jersey Wildlife Preservation Trust's Annual Reports and Dodo Journal 1964–1994, Volumes 1–30*.

Reports of zoological collections continue to provide some pickings but they rarely recount detailed clinical and postmortem findings nowadays. Other publications relating to zoos, for example, the *International Zoo Yearbook, International Zoo News* and *Der Zoologische Garten*, have regularly included contributions about gorillas but the papers are not necessarily rigorously peer-reviewed. More in-depth, usually peer-reviewed, accounts of postmortem and clinical findings in gorillas are to be found in past and present proceedings of conferences (including the long-running series in German edited by Ippen). *Index Veterinarius*, which has for many years listed publications on veterinary topics, yields some relevant data. Sources of information, including the Internet, are discussed in more detail in the introduction to the References and Further Reading, at the end of this book.

Specific textbooks on the pathology or diseases of gorillas appear not to have been published – at least not in English nor in other languages that have formed part of the authors' literature searches. Data are, however, to be found in more general publications about captive and free-living wildlife – for example, in Schmidt and Hubbard (1987), *Atlas of Zoo Animal Pathology, Volume 1*, which includes descriptions of lesions seen in various organs of captive lowland gorillas.

More information about literature is to be found in Chapter 3: Infectious Disease and Host Responses.

J. E. Cooper & G. Hull: Gorilla Pathology and Health. DOI: http://dx.doi.org/10.1016/B978-0-12-802039-5.00002-0

STUDIES ON THE PATHOLOGY OF PRIMATES

As was pointed out in Chapter 1: The Genus *Gorilla* — Morphology, Anatomy and the Path to Pathology, increasingly more detailed and varied anatomical studies, based on a mixture of wild-caught and captive gorillas, became possible as the 20th century progressed. Studies on anatomy led inevitably to awareness of, and interest in, the gross pathology of the species — for example, the natural history of upper respiratory disease and how it might relate to the structure of sinuses and the effect of trauma (for instance, following trapping or snaring) on osseous structures. Similar studies on the gross anatomy of soft tissues of gorillas encouraged accurate description and weighing/measuring of internal organs.

These studies in turn encouraged pathological investigations, which in recent years have become increasingly sophisticated as molecular and imaging techniques (Fig. 2.1) have been added to more traditional methods of gross, histological and microbiological investigation (see below and Chapter 7: Methods of Investigation — Sampling and Laboratory Tests). This book integrates data from both 'traditional' and 'modern' schools and stresses the importance of combining the two.

Interest in the pathology of all nonhuman primates (NHPs), not just apes, very much followed in the wake of

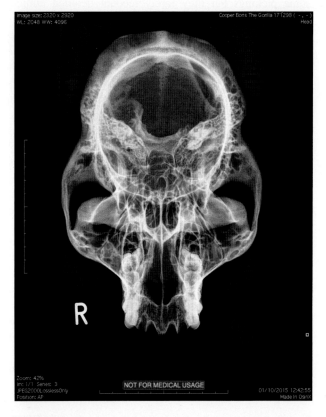

FIGURE 2.1 Imaging has always played an important part in the study of the pathology and health of gorillas. This is a dorsoventral radiograph of a gorilla skull.

the growth of knowledge of the anatomy of these animals. In the 18th century the Scottish surgeon, anatomist and natural historian John Hunter (1728—93) pioneered studies on comparative pathology (see below) and NHPs were amongst the many species he studied — see later. Despite the ravages of the Second World War the remains of Hunter's museum is still housed in London, at the Royal College of Surgeons of England (RCS). The Hunterian Collection includes 25 NHPs (see Turk et al., 2000, for some pathology data), but not from gorillas. There are, however, gorilla specimens elsewhere in the College: 25 gorilla specimens in the Hunterian Museum collection (these were collected by the RCS to supplement Hunter's specimens) and 58 gorilla specimens in the Odontological Collection, of which 9 are from Osman Hill's original collection (see Appendix 5: Case Studies — Museums and Zoological Collections). All the foregoing figures refer to a mixture of wet and dry tissues. The RCS long played a central role in the examination of gorilla material (see Chapter 1: The Genus *Gorilla* — Morphology, Anatomy and the Path to Pathology, for example, for references to Owen and Keith, and entries in Part II: A Catalogue of Preserved Materials) (Fig. 2.2).

Museums have continued to provide specimens and inspiration for research on the pathology of primates, including gorillas. There are many such cases referred to in this book; one notable example (Lovell, 1990a,b) was based on extensive use of great ape material in the Smithsonian Institution, United States, and culminated in the production of a book *Patterns of Injury and Illness in Great Apes. A Skeletal Analysis*.

A useful starting point in tracing earlier studies on the general and specific pathology of NHPs is Part I (General Pathology) of *Pathology of Simian Primates* edited by Fiennes more than 40 years ago (Fiennes, 1972). Medical researchers, such as Scott (1992) made substantial contributions to the subject over the succeeding decades.

It must, however, be appreciated that such advances were often going on hand-in-hand with developments in the broader field of zoological pathology (Montali and Migaki, 1978; Munson, 1990). As Munson pointed out, while early zoo veterinarians appreciated the value of pathology in the development of animal health programmes in their collections, they often could not find pathologists with experience or interest in disease problems of captive wild animals. Also, convincing directors of zoos of the importance of pathology was sometimes difficult because animal deaths were perceived to be a failure on the part of a zoo rather than a source of information.

AFRICAN STUDIES

The part played by scientists in Africa in the development of primate pathology as a discipline must not be overlooked. For much of the 20th century, East and South Africa were

FIGURE 2.2   Statue of John Hunter, the father of comparative pathology, at the Royal College of Surgeons of England.

the focus of palaeontological studies relating, inter alia, to the origins of humans — see Chapter 1: The Genus *Gorilla* — Morphology, Anatomy and the Path to Pathology. Many of those involved in such work had an interest in non-human, as well as human, primates. Dr Louis Leakey, Kenyan palaeoanthropologist and archaeologist, put this into practice in two ways. First, he recruited three researchers, Dian Fossey, Jane Goodall and Birutė Galdikas (sometimes referred to as 'The Trimates') to study gorillas, chimpanzees and orangutans, respectively. Secondly, in 1958, with Cynthia Booth, he founded the National Primate Research Centre at Tigoni, Kenya. Subsequently, under the direction of Richard Leakey, James Else and others, this evolved (appropriately for a Leakey enterprise!) into the existing Institute of Primate Research (IPR) in Karen, Kenya.

Both the Centre and Institute offered a much-needed focus for work on the biology of NHPs (monkeys, not apes) and for local African scientists to train and to carry out research. Even in the early days, there were opportunities for studies on the diseases and pathology of NHPs (see, e.g., Cooper and Holt, 1975) and in recent years IPR has generously provided facilities and support for workshops organised by the GPSG (see Preface) and for studies on gorilla material curated by its parent body, the National Museums of Kenya (Cooper and Cooper, 2008) (Fig. 2.3).

Insofar as gorillas are concerned, credit must be given to those in Central Africa who, even before the establishment of the VVC (later the MGVP) collated data on the diseases and pathology of mountain gorillas. This began

with the establishment of a Gorilla Research Unit in Uganda, charged with (amongst other things) the 'recovery of most valuable anatomical material...to supplement the growing corpus of knowledge about...the "last British Gorillas"' (Tobias, 1961). Seminal work followed by Schaller (1963, 1964), Dian Fossey (earlier years) and various medical personnel, such Galloway and Allbrook (1959), who undertook postmortem examinations of gorillas, recorded their findings, saved specimens and in some cases published their results (see reference to this elsewhere in the book).

It was the establishment in 1986 of the Centre Vétérinaire des Volcans/Volcano Veterinary Center (CVV/VVC) in Rwanda that really set the scene for focused studies on health and disease of (mountain) gorillas. The history of this is outlined in the Preface. The Morris Animal Foundation (MAF) in the United States administered and nurtured this new venture. Under the initial leadership and subsequent watchful eye of James Foster, DVM (Fig. 2.4), during its first 10 years the VVC not only provided a veterinary service for the gorillas but also pioneered work on wildlife health and field diagnostic techniques — see reference to Barkley Hastings in the Preface. In 1993 MAF, for the first time, recruited as Director of the VVC a Specialist in Veterinary Pathology who introduced a whole new raft of investigative procedures for the health monitoring of the gorillas and other species, incorporating techniques developed in neighbouring Tanzania and equipment from Kenya. Alas, after an encouraging start these ventures foundered because most

FIGURE 2.3   The National Museum of Kenya in Nairobi founded over a century ago.

FIGURE 2.4   The VVC in 1993. James Foster hands over responsibility to the author.

of the equipment and samples were destroyed during the Rwandan conflict in 1994.

In 1996 the VVC left the umbrella of the MAF and became a separate body, the Mountain Gorilla Veterinary Project (MGVP). Over the succeeding years, the MGVP grew and, in addition to its numerous contributions to mountain gorilla veterinary care and research (see many references to its staff and publications elsewhere in this book) broadened the original VVC's remit into more formal studies on the environment, people and other animals.

Over these same years progress was being made in respect of pathology and health of lowland gorillas but this, understandably, was less focused because of the countries and numbers of gorillas involved. Credit for the advances is largely due to zoological gardens and organisations in Europe and North America who have funded projects in West Africa — the Aspinall Foundation (Howletts and Port Lympne Wild Animal Parks) in the United Kingdom, for instance (see Chapter 17: Field Studies in Pathology and Health Monitoring).

THE IMPORTANCE OF COMPARATIVE PATHOLOGY

As implied earlier, studies on the pathology of gorillas are very much intertwined with our understanding of the pathogenesis of disease in other species, including

Homo sapiens. John Hunter, referred to earlier, is generally regarded as the 'father of comparative pathology'. He trained as a (human) surgeon, but started life as a schoolboy naturalist and maintained, throughout his life, an interest in all types of animals. He studied species as diverse as honey bees and hominids and his interests even extended into the Plant Kingdom. Such a broad approach is the real mark of a comparative pathologist.

Comparative pathology can probably be best defined as 'the discipline that is concerned with the study of the responses of any taxon of animal, including humans, to infectious or non-infectious insults and comparing and contrasting these with those responses observed in other species' (Cooper, 2012). Gresham and Jennings (1962), a medical and veterinary pathologist respectively, explained its importance in studies on diseases of both humans and (vertebrate) animals but went to pains to emphasise that it was not only the similarities that were important: 'The reactions of human tissues to various forms of injury are by no means always similar to those seen in other mammals so that a great deal can be learned about the pathogenesis of disease from a study of disorders produced by the same agent in a variety of other animal species'.

PRIMATES IN BIOMEDICAL RESEARCH

In any consideration of the evolution and refinement of studies on the pathology of NHPs, it is important not to underestimate the part played by those working with animals maintained for research. Monkeys — and to a much lesser extent apes — have been used for studies on basic biology, physiology, the pathogenesis of disease and the development and testing of medicines for decades. Their use has often been controversial but that will not be discussed here. What cannot be disputed, however, is that a great deal about the gross and microscopical pathology of NHPs has been learned from such work and much of this is applicable to the same, and some other, species in zoos, rehabilitation centres and in the wild. Not only are the data derived from biomedical studies extensive; they are also, in most cases, based on the detailed and meticulous examination of many animals and proper, scientific, collation and analysis of the findings. This is well exemplified by the books edited by Abee et al. (2012). Disappointingly these volumes do not include the word 'gorilla' in the index but do provide extensive information on a wide range of naturally occurring infectious and noninfectious diseases of NHPs, including some excellent sections on aetiology, pathogenesis and pathology. Meetings between those who work with primates in laboratories and those involved with free-living and captive apes also provide an opportunity for valuable debate

about primate pathology (Lowenstine, 2003; Sasseville et al., 2012, 2013).

STUDIES ON MICROSCOPIC ANATOMY AND CELLULAR PATHOLOGY

As pointed out in Chapter 1, The Genus *Gorilla* — Morphology, Anatomy and the Path to Pathology, it took nearly a century before serious attention began to be paid to the microscopic anatomy (histology) of gorillas, although there had been one or two enterprising incursions into the topic a few decades earlier, including study of the brain by Campbell (1916).

Straus (1950) arguably set the scene for a modern approach to microscopical studies with descriptions of the skin and a few years later Inoue and Hayama (1961) reported histopathological findings in on two mountain gorillas. A year before, Galloway and Allbrook (1959), writing from Uganda, had bemoaned that knowledge of microscopic anatomy, neuroanatomy and systematic morphology of *beringei* was greatly hampered by the paucity of skeletons available for study and of accurate dissections of soft parts and urged the collection of gorilla tissues for study. They did, however, pay tribute to earlier African explorers and their staff who had been '... assiduous both in the collection of osteological material and the presentation of their trophies to their national or private museums'.

Since then, a whole range of laboratory investigative procedures has come into routine use in diagnostic work with gorillas — see below. However, histopathology, often using time-honoured techniques (Fig. 2.5) remains an essential component, as numerous citations in this book attest.

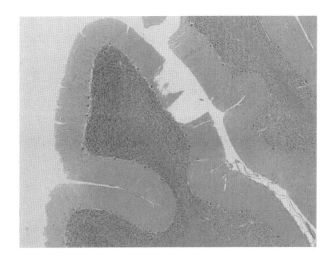

FIGURE 2.5 An early histological section of the brain of a gorilla, examined postmortem, stained with haematoxylin and eosin.

Data on the physiology and related features (e.g., blood biochemistry) of gorillas have taken longer to collect and reference values are still largely based on small numbers of animals, mainly lowland gorillas, examined routinely during health checks or veterinary investigation, in zoological collections. This situation is slowly changing as portable equipment becomes available that permits accurate testing of blood and other samples in the field (Cooper, 2013).

As explained elsewhere, this book is primarily concerned with gross pathology and with changes in cells and tissues. Brief mention is made, when appropriate, to ultrastructural and biochemical changes and to molecular techniques (see Chapter 7: Methods of Investigation — Sampling and Laboratory Tests). This is important because an early deterioration in function of the cells of a tissue may initially not be detectable using morbid anatomical or histological techniques: what was originally, several decades ago, termed 'the biochemical lesion', initial degenerative changes in cells, will precede anything visible with the microscope. An emphasis in these pages on gross pathology and histopathology not only fits the author's (JEC's) training and expertise but also is likely to be of most value to those working with gorillas, especially in the field. Any detailed coverage of other 'pathology' disciplines, such as biochemistry, would warrant a far longer book.

THE CONTINUING RELEVANCE OF STUDIES ON PATHOLOGY

Pathology, sometimes described as 'the science of tissue injury and of the host adaptive responses that characterise disease' (Kelly, 1994), remains at the core of veterinary investigation. It is the bridge between the basic sciences and clinical medicine. For example, it links an understanding of the normal anatomy of a gorilla with an appreciation of what is abnormal in (for example) a damaged limb — and thereby assists in surgical correction. Likewise, detection by a pathologist of abnormal or excessive numbers of certain cells in the blood of a live gorilla will facilitate diagnosis and treatment of an infection or neoplasm.

Postmortem work (necropsy) is, regrettably, often regarded as the main (or only) activity of the pathologist — as Kelly (above) pointed out, the professional who has been caricatured as one 'who knows everything, but always too late'. It is clear, however, that the pathologist is an integral part of the team that is attempting to diagnose or to treat gorillas — or any other species. Necropsy does, indeed, play an indispensable role and has a vital function in clinical audit (a

continuous process of quality control) but it is only one part of the pathologist's armoury.

The recognition and interpretation of both gross and microscopic changes provide a basis for the prevention and treatment of clinical conditions in gorillas, as in humans and other animals (Kelly, 1994; Callanan, 2012). The pertinence of this was made very clear a century ago, when the Canadian physician Sir William Osler stated, in the context of human medicine: 'As is your pathology, so is your practice'.

The same argument applies to research on the health and diseases of gorillas. Without critical pathology, no such research can be considered adequate: pathology is of fundamental importance to recognition of tissue changes and evaluation of their significance. An exciting range of new techniques has evolved in recent years, such as immunocytochemistry, specific cell markers, polymerase chain reaction, in situ hybridisation and morphometry. These techniques do not replace gross and microscopical pathology; they enhance their value (Table 2.1).

THE ADVANTAGES AND DANGERS OF EXTRAPOLATION

The paucity of detailed information and almost total absence of experimental data on the pathophysiology of the genus *Gorilla* means that those working with these species frequently have need to extrapolate from other species when interpreting clinical signs and lesions and laboratory results — or formulating treatment protocols. Comparative pathology as a discipline often facilitates a better understanding of disease in one species by extrapolation from another. However, caution has to be exercised as not only manifestations of a disease but even its pathogenesis may differ markedly between species.

One can either approach the investigation of diseases of gorillas in terms of traditional veterinary pathology or by reference to human pathology. Both approaches can yield dividends. Veterinary knowledge that encompasses an understanding of host:parasite relations in wild animals, especially NHPs, can assist in understanding how free-living gorillas are exposed to and respond to different organisms. On the other hand, the anatomical similarities of *Gorilla* to *H. sapiens*, plus increasing evidence of comparable disease pathogenesis in the two genera, means that extrapolation from humans can sometimes be fruitful. It should not, however, be assumed that the evolutionary and morphological similarities between gorillas and humans inevitably implies that their diseases are the same. In his paper about cellular glycan coating and the role of sialic acids, Varki (2010) pointed out that, in respect of health, there are some uniquely human features. As a result, differences exist between humans and (for

TABLE 2.1 The Role of Pathology in Diagnosis, Health Monitoring and Research

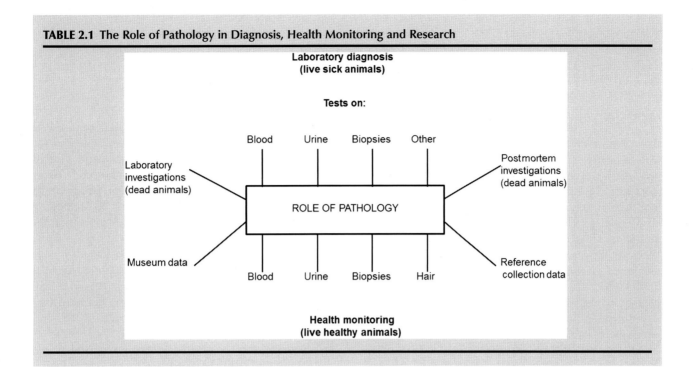

example) gorillas in respect of susceptibility, incidence and severity of various diseases 'beyond those explained by anatomical reasons'.

Gorillas are, of course, but one taxon of primates. As Ian Redmond pointed out in a review in *Primate Eye* of the *Handbook of the Mammals of the World. Vol. 3. Primates* (Mittermeier et al., 2013) 'The diversity of primates continues to astound. From the diminutive Madame Berthe's mouse lemur (weighing just 31 g) to the Grauer's gorilla (males of which exceed 200 kg), they have evolved an astonishing variety of size and form in almost every habitat on earth'.

In theory, of course, as with any other taxon, the best route to an understanding of gorilla pathology is an in-depth study of the subject. This, however, is easier said than done. As explained earlier, there is a paucity of scientific information about gorillas and experimental work on the species is not acceptable. The situation is compounded by reluctance or failure to publish records or to share research on captive or free-living gorillas and a growing trend to consider as intellectual property what could, in the interests of the species, become open-source scientific information (see Preface). The greater good of gorillas should provide incentive enough for scientists to pool their information and collaborate. Full use must, therefore, be made of all available material – an important reason for publishing Part II: A Catalogue of Preserved Materials, in conjunction with this book – and at the same time *cautiously* extrapolating from other species – see later.

'Thinking human' was very much the rule in the early days (19th century) of study of the newly discovered gorilla because of its apparent close similarity to the human. Little wonder therefore that for many decades work on live and dead gorillas remained largely the preserve of medically trained men (literally) and of comparative anatomists and physiologists. Such people inevitably related the gorilla to the one species that they knew well, *H. sapiens*. The veterinary profession were slow to become involved in this new (to them) area of medicine. For veterinarians the starting point was often what they knew about the anatomy, physiology, pathology and clinical medicine of domesticated livestock rather than 'thinking human'. Part of the dilemma for veterinarians was – and to a certain extent remains – whether attention to gorillas should be based on treating individual patients, as is standard when treating a companion animal (dog, cat, horse) or should be orientated towards 'herd/flock medicine' that focuses more on the group than on one sick creature. Reports in the early days were not always complimentary about the veterinary profession or their skills; for example, when 'Princess Topsy Caralinda', a female gorilla which was part of the Noells' gorilla show in the United States died 'despite the efforts of six veterinarians' (Morris and Morris, 1966).

The need for studies on gorilla health and pathology to remain interdisciplinary, not just confined to veterinarians, is emphasised elsewhere in this book. Medical and dental pathologists have contributed significantly to our understanding of diseases in gorillas and other NHPs and the value of such collaboration was emphasised by Allchurch (2001). The author (JEC) has worked with such colleagues for much of his professional career — both in East Africa and Europe — and in the production of this book.

ACCESS TO DATA

In a review of infectious diseases of the great apes of Africa nearly 40 years ago, Kalter (1980) discussed the chimpanzee at great length, drawing attention to its susceptibility to human diseases, but stated that 'very few infectious disease data are available on the gorilla'.

Obtaining reliable information to guide clinical and postmortem work on gorillas is still not easy. For instance, there are relatively few reliable reference values published or even readily available on request. In some institutions enquiries for such are met with the demand that a formal written application must be made to the appropriate office or committee! As this book shows, although apparently numerous, published data are scattered and disparate, most relating to only the one species, *Gorilla gorilla*, and then largely based on studies in zoos.

Despite this relative paucity of reliable information on the pathology of gorillas and the reluctance of some individuals and organisations to share what they have, there are many opportunities to learn more. Collecting and collating the available data (see above) is one route. The other is to make better use of the tissues from gorillas available in museums and elsewhere in many parts of the world — see Part II: A Catalogue of Preserved Materials. The most prevalent material in the world's museums is skeletal but zoos and, in Africa, primate centres and sanctuaries are potentially a rich source of histological and other samples and archived necropsy reports.

THE RELEVANCE OF PATHOLOGY TO HEALTH

The aim of pathology as far as this book is concerned is to understand better the pathogenesis of diseases that affect gorillas and to use this information to promote their health, both in captivity and in the wild. Health and health monitoring were outlined by Foster (1993) and are discussed in detail in Chapter 5, Methods of Investigation — Observation, Clinical Examination and Health Monitoring.

The health of gorillas cannot be considered in isolation. It is relevant to other species, including humans, and to their environment. More widely, it is a component of global health and 'One Health'. The need for a more international and holistic approach to health is recognised by the medical profession as well as by veterinarians and ecologists. 'Pathology is Global' was the title of the (British) Royal College of Pathologists' 5-year strategy launched 3 years ago (Fleming, 2013). Relevant to this book is that it reflected concerns by the medical profession that 'global health' cannot be achieved without alleviating disease and improving healthcare services and that these in turn depend upon adequate pathology services. This is true too in work on other species. If gorillas are to be subject to the highest levels of healthcare, those working with them in zoos, in rescue centres and in the wild need access to first-class pathology facilities.

Chapter 3

Infectious Disease and Host Responses

John E. Cooper

A mole with fungus disease of its lungs is as interesting as a cow with tuberculosis and a man with stomach cancer. Each is a living thing responding to injury

G Austin Gresham (1971)

INTRODUCTION

In 1920 Cark Akeley wrote: 'One of the most interesting facts in this account … is the fact that the baby gorilla caught cold from him (Mr Foster). Animals usually do not catch man's diseases'.

In fact, gorillas are not 'free of disease' but susceptible to a wide range of infectious and noninfectious conditions, both in captivity and in the wild. The causes of infectious conditions and how gorillas may respond to them are discussed below. Noninfectious diseases and host responses are covered in Chapter 4: Noninfectious Disease and Host Responses.

DISEASE AND DEATH IN GORILLAS

First it is important to define terms used in this chapter and later throughout the book (see also Appendix 1: Glossary of Terms).

A 'disease' is defined as 'an unhealthy condition of the body or a part thereof, disordered state of an organism or organ' or, more succinctly, as suggested by Alley (2000) 'anything that causes impairment of normal function'. 'Health' is defined as 'a state of physical and mental well-being of the body, not merely the absence of disease' (see Chapter 5: Methods of Investigation — Observation, Clinical Examination and Health Monitoring). A pathogen is 'any living organism or other agent that is the cause of disease'.

Causes of ill-health ('morbidity') and/or death ('mortality') in *Gorilla* spp. have in the past been broadly divided into those that are (1) **infectious** (caused by organisms such as viruses or helminth worms), and (2) **noninfectious** (caused by factors such as trauma, drowning or genetic abnormalities) (see Table 3.1). In practice, however, it is

often not so clear-cut. Thus, for instance, an infant gorilla may become chilled (hypothermia — noninfectious) but then develop pneumonia, a secondary disease, caused by bacteria or viruses (infectious). Geriatric disorders are commonly a combination too (see Chapter 8: Nonspecific Pathology). An alternative approach therefore is to say that diseases of gorillas can be caused by any combination of (1) infectious agents, (2) noninfectious factors and (3) environmental/managemental factors.

THE AETIOLOGY OF DISEASE

So what are the causes of disease and of death in gorillas? How do they differ between free-living and captive animals?

Nearly 80 years ago, Kennard and Willner (1941) reported on necropsy findings in 70 anthropoid apes and there were occasional publications on individual gorillas (Fitzgibbons et al., 1971; Fitzgibbons and Simmons, 1975) but it was not really until the early 1980s that serious interest in causes of disease and of death in gorillas started to grow. Dixson (1981) provided a useful but very general summary about what was known concerning gorilla diseases and health, although the reader should be forewarned that neither of those two words, nor 'veterinary', appears in the Index! At about the same time, Benirschke and Adams (1980) started their landmark paper on diseases and causes of death with the words 'A comprehensive review of gorilla mortality has not been undertaken'.

Most available information about causes of death of gorillas is qualitative, not quantitative. Published information was initially based on zoo records; see, for example, Gucwinski et al. (1975), chapters by Janssen and Bush, Paul-Murphy, and Swenson in Fowler (1993) and the review of medical literature of great apes in the 1980s by Janssen and Bush (1990). Benirschke and Adams (1980), quoted earlier, reported on the causes of 48 gorilla deaths.

Zoos continue to produce information that can be added to the growing database of causes of morbidity and

J. E. Cooper & G. Hull: Gorilla Pathology and Health. DOI: http://dx.doi.org/10.1016/B978-0-12-802039-5.00003-2

TABLE 3.1 Diseases of Gorillas

Infectious	Noninfectious
Viruses	'Environmental'
Bacteria	Behavioural
Fungi	Trauma
Protozoa	Electrocution
Metazoan parasites	Hyperthermia
.... *plus*	Hypothermia
Others	Malnutrition
	Metabolic
	Toxicity
	Genetic
	Developmental

mortality. In preparing this book we have corresponded with, or visited, many zoological collections in different parts of the world. A small number have been unwilling to share any information (see Preface and Introduction to the Part II: A Catalogue of Preserved Materials) but the vast majority have been helpful and transparent. Thus, for example, Roberta Wallace (personal communication) provided us with a list of gorillas for which Milwaukee County Zoo, United States, has necropsy reports and diagnoses. These are referred to elsewhere in this book.

An important need for zoos which have kept gorillas is to put resources into finding, collating, analysing and disseminating information from their archives. The latter often includes handwritten notes. An excellent example of the value of such retrospective surveys is the recent study by a veterinary student (Keen, 2015), reviewing data relating to management and welfare of gorillas at Bristol Zoo Gardens (BZG), United Kingdom from 1930 to 2015 (see Appendix 5: Case Studies — Postmortem Investigations and Studies on Health).

More specific studies in zoos have provided valuable information relating to particular aspects of gorilla pathology. Thus Dixson (1981), quoted earlier, was able to publish histological details of testes removed postmortem from six adult male great apes at the London Zoo. Debyser (1995) reported on an epidemiological and pathological study on captive primates in Dutch and Belgian zoological gardens and a German primate centre. She described the findings in terms of pathology and vitamins, trace elements and heavy metals. The report did not detail the species investigated in the study other than in terms of their taxonomic families.

Reports on individual captive gorillas, often well-known animals, have appeared over the years, in different languages, and often give an insight into factors that influenced health during captivity. For example, Kruger (1979) provided *ein Epilog* on 'Coco' and 'Pucker', two of Dian Fossey's gorillas, that lived 10 years in captivity and then died within 1 month of each other at the Cologne Zoo.

Data on free-living gorillas are not easy to collect, nor to analyse. Often in the past they have usually been qualitative observations, not based on proper surveys. Thus Groves (1970) stated that 'arthritis, bony tumours, fractures, yaws and leprosy have all been identified among wild gorillas. About one in five lose a few teeth, probably through infection; a few get caries and about 60% suffer from dental abscesses in old age, causing infection of the maxillary sinuses in many cases'. Early postmortem reports on mountain gorilla were published by Gaffikin (1949), Galloway and Allbrook (1959) and Hall-Craggs (1961a).

Both Schaller (1963, 1964) and Fossey (1983) included useful observations and some data on diseases, parasites and pathology in their books. Fifteen (*sic*) postmortem reports are listed by Fossey, many performed by investigators with varying medical knowledge and before protocols were developed.

A brief review of pathology data from 1998 to 2000 was made available in a document provided to NGOs by the Mountain Gorilla Veterinary Project (Cranfield, 2001). This identified respiratory disease as the single most significant health problem in mountain gorillas. Sometimes deaths occurred during respiratory disease outbreaks in which morbidity (illness) was high and morbidity (death) was relatively low. Infant mortality also contributed to losses. The exact cause of infant deaths was often difficult to determine as the carcases were decomposed when dropped by the mother.

The late James Foster (see Preface) kept copious notes about his observations on *beringei* in the Virungas and was generous in his willingness to exchange information with colleagues regardless of their affiliations or possible motives. He cared for gorillas, not for gorilla politics. Reference is made below to some of those personal communications and to unpublished data about his observations in *Karisoke Archives*.

Thus, for example, in 1993 (personal communication) James Foster noted that the 'condition' in gorillas that was most often reported by staff at Karisoke and researchers was trauma, but that the extent of injury was not usually indicated in those reports. Respiratory diseases and diarrhoea were second and third in the list. He referred also to individual cases such as the silverback 'Peanuts', who had intermittent swelling and fluid seepage from below the right eye (*Karisoke Archives*, 1983); a tentative diagnosis was made of a malar abscess.

Notwithstanding those early contributions, an indication of the imprecise nature of the records available until relatively recently was illustrated by the discussion at the PHVA (Population and Habitat Viability Assessment) for the mountain gorilla held in Uganda nearly 20 years ago (Werikhe et al., 1997). Those attending the PHVA were asked to collate personal experiences, based on their work with mountain gorillas, and to present it to the group in

BOX 3.1 Potential Sources of Disease

- Humans: tourists, local population, transient populations, field staff (vets, researchers, trackers, etc.), other primates: chimps, baboons, monkeys.
- Other animals: dogs, bushpigs/duikers/buffalo/elephants, cows (sic), goats, birds.
- Noninfectious diseases.
- Snares, projectile weapons, mines, fires, intra- and interspecific aggression, human attacks.

BOX 3.2 Predisposing Factors of Disease (Conditions)

1. Habituation – increased likelihood of human contact.
2. Habituation – increased likelihood of stress, in particular during the habituation process.
3. Tourist/gorilla ratio and the number of visits/day.
4. Climate – cold and wet weather.
5. Disturbance (e.g., refugee movement, gunfire).
6. Gorilla home ranges overlapping with areas of human activity, including that outside the protected areas.
7. Genetic predisposition to disease (including inbreeding).
8. Group composition; smaller groups or those with larger numbers of infants more susceptible?
9. Population density.
10. Habitat destruction.
11. Snares; direct physical injury and risk of serious secondary infection/possible death.
12. Human/human waste contact.

BOX 3.3 List Conditions Diagnosed

1. Scabies – well-documented.
2. Respiratory disease – upper respiratory disease, including air sacculitis, lower respiratory disease.
3. Diarrhoea with abdominal pain.
4. Eye infection.
5. Snares.
6. Vesicular dermatitis (limited to head).
7. Tapeworm, nodular worm, louse, mite.
8. Serologic evidence only of alpha herpesvirus, measles, mycoplasma, possibly others.
9. Osteoarthritis.
10. Facial abscess/tooth infection.
11. Spondylitis/spondylosis.
12. Healed fractures.
13. Ankylosis.
14. Strabismus.

terms of 'potential sources of disease' and 'predisposing factors of disease (conditions)'. The results are summarised in Boxes 3.1 and 3.2.

Box 3.3 lists conditions that were diagnosed by the group.

The 'Actual disease reports from field veterinarians' were perhaps the nearest to evidence-based science that the group could offer. All of the data above relate to habituated animals, but the point was made at the PHVA that in nonhabituated gorillas 'the same issues are relevant, but there are additional limitations due to lack of accessibility to these animals'.

As far as 'actual diseases' were concerned, the final PHVA Report stated that 'Very little actual prevalence data exists (sic)' and listed what was thought, or known, at the time.

So what has been learned since the PHVA (on *Gorilla beringei*) in 1997? Have achievements since then lived up to the PHVA Recommendations? Despite the work of the MGVP and others, reliable quantitative data on morbidity and mortality rates in free-living gorillas are still not openly available but hopefully this will be rectified soon

when records, mainly about *beringei*, collected over several decades, are finally published.

Detailed analysis of the various postmortem examinations of free-living gorillas carried out over those several decades would contribute greatly to our understanding of morbidity and mortality, especially if coupled with the results of supporting tests such as bacteriology and histopathology. This need was put forward by the author (JEC) originally in 1996, at the Annual Meeting of the European College of Veterinary Pathologists (ECVP) and repeated at the PHVA on the mountain gorilla in Uganda in 1997. A decade later Cranfield and Minnis (2007) were able to announce 'several new insights' following a review (in a paper presented at a conference in Australia) of 100 mountain gorilla deaths by Nutter et al. (2005). Unfortunately that 2005 report was no more than an expanded abstract, and (in marked contrast to the comparable paper on immobilisations by Sleeman et al. (2000)) had not been compiled in collaboration with those VVC/MGVP veterinarians who had carried out the work in the preceding years. Nutter et al. (2005) cautioned that some necropsies had been performed by people with limited medical training and that, although postmortem examination protocols for their project were standardised in 1988 (and revised in 1994 by J.E. Cooper, then Director of the VVC, Diplomate of the ECVP), they had been implemented with varying rigour since then. The majority of cases (66/100) had at least some tissues examined histologically, but 12 diagnoses were made on gross examination only.

The causes of death by age and sex in the survey by Nutter et al. are presented in a modified form in Table 3.2.

The authors reported that trauma was the leading cause of death in all age groups of gorillas – in infants, the result

TABLE 3.2 Causes of Death in 100 Mountain Gorillas

Cause	% of Total
Trauma	40
Respiratory	24
Undetermined	17
Multifactorial	5
Gastrointestinal	4
Metabolic	3
Cardiac	3
Infectious – other	1
Developmental	1
Neurological	1
Parasitic	1
Total	100

After Nutter, F.B., Whittier, C.A., Cranfield, M.R., Lowenstine, L.J., 2005. Causes of death for mountain gorillas (*Gorilla beringei beringei* and *Gorilla beringei* undecided) from 1968–2004: an aid to conservation programs. In: Proceedings of the Wildlife Disease Association International Conference, Cairns, Queensland, Australia, 200–201.

FIGURE 3.1 Young gorillas appear to be particularly susceptible to pneumonia. In the case of habituated mountain gorillas in the Virungas, such infections may be associated with movement to higher altitude and/or close proximity to humans.

of infanticide. Within the juvenile and adult age groups, trauma sometimes followed aggression between gorillas but more often was the result of injury caused by snares, occasional poaching for infants and, very rarely, a demand for gorillas as bushmeat. The second most prevalent cause of death in all age groups was pneumonia (Fig. 3.1). While outbreaks of respiratory disease in gorillas can be considered a normal phenomenon (see Chapter 10: Respiratory and Cardiovascular Systems), there were two factors that suggested human-induced components: (1) the number of medically unscreened tourists that visited the groups, each with the potential to spread upper respiratory viruses and (2) the fact that the gorillas had had much of the lower-altitude areas of their habitat taken away for pyrethrum farming, and as a result were now inhabiting colder and moister regions.

Other causes of mortality reported by Nutter et al. (2005) were considered by Cranfield and Minnis (2007) to be significantly less prevalent or difficult to treat or manage under field conditions – for example, 'old-age arthritic changes that may cause problems for an individual when it is trying to keep-up with the group'.

At the time of writing (January 2016), 11 years after the review by Nutter et al. (2005), the MGVP still considers the major cause of death in free-living *beringei* to be trauma, including infanticide, other conspecific trauma (male–male interactions), human-related trauma (snares, intentional killing) and a few accidents (mainly falls)

(L.J. Lowenstine, Gorilla Doctors, personal communication). Lowenstine goes on to say 'Respiratory diseases continue to be an important cause of morbidity and mortality (Spelman et al., 2013), much more so than GI (*gastrointestinal*) issues, in spite of the large numbers of parasites and bacterial pathogens encountered on published coprological examinations'.

The direct and indirect effect of humans on gorillas has long been emphasised. Morris and Morris (1966) wrote 'Man is without doubt the gorilla's major predator.' In their paper four decades later Cranfield and Minnis (2007) stated that 'the largest threat to gorilla species in general is probably logging, which in turn provides greater access for the bushmeat industry but mountain gorillas living in four national parks, are more at risk from human and livestock diseases (Werikhe et al., 1998)' (Fig. 3.2).

To a certain extent one can predict morbidity and mortality rates in gorillas. In their paper 'Living fast and dying young: a comparative analysis of life history variation among mammals', Promislow and Harvey (1990) examined the way in which life history traits relate to juvenile mortality as opposed to adult mortality, and found that juvenile mortality is more highly correlated with life-history than is adult mortality. Mammals with high natural mortality tend to mature early and give birth to small offspring in large litters after a short gestation. In contrast, gorillas are long-lived, mature late and invest substantially in each offspring (see also Chapter 8: Nonspecific Pathology).

FIGURE 3.2   Monitoring of livestock health on the edge of Bwindi Impenetrable National Park in Uganda.

There is clearly a need for a published survey of diseases of free-living *Gorilla* spp. of the quality of that carried out on baboons in the Kruger National Park 40 years ago (McConnell et al., 1974). The South African study resulted in a scholarly paper comprising 71 pages, 214 (black and white) photographs of gross and histological findings, a breakdown of organ systems affected and a list of parasites identified.

WHAT IS 'PATHOLOGY'?

As indicated in Appendix 1: Glossary of Terms, pathology is 'the science of the study of disease'. The definition can be expanded to include that it 'involves the essential nature of disease … the study of the functional and morphological changes in the tissues and fluids of the body during disease'.

Thomson (1978) emphasised the pivotal role of the pathologist in diagnosing disease in both live and dead animals: he stated 'Both diagnosis and prognosis require comprehension and recognition of lesions and their pathogenesis.'

UNDERSTANDING HOST RESPONSES – A COMPARATIVE APPROACH

Scott (1992), in his book *Comparative Primate Pathology*, discussed the various pathological changes seen in the organs and tissues of NHPs and emphasised the value of comparative studies. His approach was very much that of a 'human pathologist' – taking disease in *Homo* as his starting point and then extending the study into other species of primate. Thus, understandably, initial chapters in Scott's book cover such topics as 'Ageing in humans and primates', 'Thrombosis, disturbances of circulation, and vascular disease' and 'The comparative oncology of humans and primates'.

The approach is different in this book. Here the focus is on the genus *Gorilla* and the pathological changes that have been reported in *Gorilla gorilla* and *G. beringei*. Nevertheless, there is a strong emphasis on similarities in such responses between the genera *Gorilla* and *Homo*.

Gorillas and other apes exhibit the same spectrum of organ, tissue and cellular changes as do other mammals. The main categories of responses seen in gorillas in response to infectious or noninfectious insults are degeneration, regeneration, inflammation and neoplasia.

Inflammatory changes are a feature of many responses in gorillas. The relationship of inflammation to phylogeny was explained in a scholarly paper by Montali (1988). Experimental studies on inflammation have not been possible in the genus *Gorilla* but observation and evaluation of the cellular responses seen in both live and dead gorillas suggest that these animals respond to infectious, traumatic and toxic insults as do humans and other great apes. It should be noted that fibrosis, an essential part of healing (especially of chronic lesions), is also an age-related change (see Chapter 8: Nonspecific Pathology). As gorillas get older, fibrosis is seen increasingly in histological sections of diverse organs and — judged from studies on human fibrotic disorders in the United States and Europe — is likely to contribute to overt or covert disability.

Many cellular factors play a part in defence against pathogens and other insults in humans and other primates. One example is autophagy, a process that helps regulate inflammation and the degradation of microorganisms. So too do molecules such as proteins.

The acute phase response (APR) is part of the innate immune system and acute phase proteins (APPs) are the basis of the early response to stimuli such as trauma, infection, neoplasia and autoimmune disease. APPs are mainly produced in the liver and they circulate in the blood. They modulate the innate immune system response to tissue injury. Studies on APPs in nonhuman primates (NHPs) have to date focused upon macaque monkeys (Krogh et al., 2014).

ENDOGENOUS FACTORS INFLUENCING HOST RESPONSES

The temperature-dependence of tissue responses is all important when comparing responses in endothermic ('warm-blooded') and ectothermic ('cold-blooded') animals (see Chapter 4: Noninfectious Disease and Host Responses). The size of the animal is also relevant. As an animal's size increases, its metabolic rate falls; this forms the basis of biological scaling in clinical veterinary medicine (Kirkwood, 1983) as well as being an important factor in healing times in animals following trauma, both accidental and non-accidental (see Chapter 4: Noninfectious Disease and Host Responses). In the case of gorillas, with a body mass broadly similar to that of humans, the speed and progression of metabolic processes appear to parallel those seen in *Homo*.

There are various other factors that can influence how rapidly and effectively a gorilla responds to a traumatic insult, many comparable to those that play a part in *Homo* — the health, age and sex of the patient, its immune status, adverse stressors, the presence of infection, the quality of the vascular supply to the affected area and the effects of clinical investigation and treatment. The microbiome may also be of importance — see later.

DISCUSSION OF INFECTIOUS AND NONINFECTIOUS DISEASES

Infectious Conditions

As emphasised earlier, until fairly recently, information about infectious diseases of gorillas was confined almost entirely to individual case reports, mainly captive animals that became ill or died in zoos or were examined in the wild. Sometimes full microbiological investigation was performed but often it was not.

Benirschke and Adams (1980), quoted earlier, reporting on the causes of 48 gorilla deaths, included 'Primary cause of death' amongst animals up to the age of 8 years as hepatitis, shigellosis, salmonellosis, hyperthermia (environmental), pneumonia, amoebiasis, strongyloidosis, ileus, balantidiasis, verminosis, cytomegalovirus infection, viral hepatitis, bronchopneumonia and pyelonephritis; and amongst animals of 8–30 years of age enterocolitis, appendicitis, salmonellosis, strongyloidiasis, ileus, balantidiasis, coccidioidomycosis, subdural haemorrhage, fatty liver and haemosiderosis, dissecting aneurysm of aorta, thrombophlebitis with embolism and pyelonephritis. Benirschke and Adams specifically stated that 'A comprehensive review of gorilla mortality has not been undertaken' and lamented that 'It was surprising to find relatively little of this vital information published, particularly since a large number of animals have died in the past in zoological gardens'.

In recent years the situation has changed in respect of both captive and free-living gorillas. Data are collected by most zoos and some are shared at meetings and in databases (although not always with open access). For example, in a Pre-meeting Workshop (Sasseville et al., 2012), Lowenstine referred to geographically restricted infections, such as coccidioidomycosis in gorillas, which is a limiting factor in exhibiting this species in zoos in the southwestern United States. Some parasites are common or cause serious infections in zoo apes, such as *Balamuthia mandrillaris*, *Strongyloides stercoralis*, *Balantidium* sp. and *Enterobius vermicularis*. Diseases of free-living apes that are not seen in (Western) zoos include *Ebolavirus* disease (EVD). She explained that in North American zoos necropsy reports are compiled by veterinary and pathology advisors to Species Survival Plans. In contrast, for free-living apes a long-term pathology database exists only for Virunga and Bwindi mountain gorillas; she did not explain how to gain access to this database.

Access to more sophisticated and sensitive tests, some of them applicable to fieldwork (see Cooper, 2013a,b), has resulted in the recognition of a range of pathogens in gorillas. This does not mean that earlier reports of, say, bronchopneumonia or enteritis in an animal are not valid (possibly valuable) records, but care must be taken in interpretation (see Chapter 6: Methods of Investigation — Postmortem Examination), especially if only limited microbiological investigation was possible at the time. Material from a few of these earlier cases survives — fixed or frozen — in zoo hospitals or research institutes (see Part II: A Catalogue of Preserved Materials) and this is worthy of retrospective study using modern techniques such as PCR (see Chapter 7: Methods of Investigation — Sampling and Laboratory Tests).

The isolation of pathogenic organisms from a sick or dead gorilla always raises questions as to which (if any) was responsible for an infectious condition (see Chapter 7: Methods of Investigation — Sampling and Laboratory Tests). In respiratory disease a virus may initiate tissue damage and this can predispose to a bacterial infection (see Chapter 10: Respiratory and Cardiovascular Systems). In the past, however, such conditions were often attributed to bacteria because there was no evidence, such as inclusion bodies or syncytia in histological sections, of a viral infection.

In captivity gorillas may be particularly susceptible to viral infections because of the relatively intensive conditions in which they are kept and their regular close proximity to humans. Equally, however, such animals can be a source of infection to humans (see Appendix 4: Hazards, Including Zoonoses) and therefore prophylactic protection may be desirable (see Chapter 12: Lymphoreticular and Haemopoietic Systems and Allergic Conditions).

In recent years the part played by infectious agents in causing morbidity and/or mortality in Africa's great apes has been recognised. The fact that some of the conditions are 'emerging diseases' and are zoonotic has added extra impetus to field and laboratory studies. Thus, over a decade ago Kilbourn et al. (2002), at a large veterinary meeting, outlined 'disease trends' in Central Africa and discussed the implications on great ape health and conservation. A year later Walsh et al. (2003) warned of a catastrophic ape decline in western equatorial Africa. These alerts were repeated by, for example, Hoffmann et al. (2008) who outlined action needed to prevent extinctions caused by disease. More recently Keita et al. (2014) discussed apes as a source of human pathogens and McCabe et al. (2014) the links between infectious disease, behavioural flexibility and the evolution of culture in primates. Interactions between primates and their pathogens have influenced the molecular evolution of both host and parasite (Brinkworth and Pechenkina, 2013).

There follows a review of infectious diseases diagnosed and/or reported in captive or free-living gorillas. This is intentionally brief and not exhaustive because the thrust of this book is on the *pathology* of diseases of gorillas, how they and infectious agents/noninfectious factors interact and how to investigate live animals, dead animals and samples, not an inventory of diseases.

Diseases and/or their associated causal organisms that are important in gorillas in terms of either prevalence or potential seriousness are listed in Box 3.4. In many cases additional information is given either in the succeeding text or elsewhere in the book (see Index).

Diagnostic investigations depend upon the organism involved and the material available; see Chapter 7: Methods of Investigation Sampling and Laboratory Tests. Important considerations include not only the type of sample but also the amount needed (e.g., volume of blood), the method of preservation (freezing, transport medium, etc.), and where the investigations are to be carried out (laboratories have their own requirements: samples to be sent to another country will need permits — see Chapter 18: Legal Considerations).

More information on some of the organisms in Table 7.1 in Chapter 7 may be sought in various books and peer-reviewed papers (see References and Further Reading) and on the Internet. The former includes Bourne, 2009 and publications by IUCN, PASA, GRASP and others — see References and Further Reading.

BACTERIA

Bacteria have long been implicated in diseases of NHPs, mainly based on data from captive animals (Abee et al., 2012). The text below discusses a few genera only.

Some species of bacteria were considered particularly dangerous in captive apes; Benirschke and Adams (1980) wrote that 'Shigellosis and salmonellosis are probably the most feared infections in gorillas because of their acute nature and severity'. These same organisms continue to be a cause of concern, even insofar as free-living animals are concerned; Nizeyi et al. (2001), working with *beringei*, reported that the prevalence of campylobacteriosis and salmonellosis had doubled in the previous 4 years, and that *Shigella* spp. were found for the first time in habituated mountain gorillas; although contact with humans was the probable source, the researchers could not confirm this. Other enteric bacteria have also been investigated, including *Helicobacter* species (Flahou et al., 2014).

Another disease that has plagued captive NHPs in the past — and sometimes now — is tuberculosis (mycobacteriosis). The disease was of great concern in zoos and affected a range of primate species, including gorillas. In his autobiography, Graham-Jones (2001) recorded 'respiratory tuberculosis' in three young gorillas at the London Zoo. These animals, he said, came from 'collecting centres in Africa'. Other popular accounts also refer to tuberculosis; for instance Prince-Hughes (2001), writing about the circus gorilla 'Gargantua' who died in 1949, stated that the cause of death was 'double pneumonia compounded by tuberculosis, and cancer of the lip'.

Early definitive diagnosis of tuberculosis in a mountain gorilla was by Fiennes (1966). This juvenile female had only been at the London Zoo for 8 months. Liver, lungs, spleen and kidneys all contained caseous tubercles of varying size from 1 to 8 mm. The mesenteric glands

BOX 3.4 Diseases and/or Their Associated Causal Organisms in Gorillas

Category	Organism or Disease	Comments
	General	Sasseville et al. (2013)
Bacteria	*Campylobacter* spp.	Nizeyi et al. (2001), Whittier et al. (2010)
	Enteropathogenic *E. coli*	See text
	Helicobacter spp.	Flahou et al. (2014)
	Mycoplasmas	Cole et al. (1970)
	Salmonella spp.	See text
	Shigella spp.	Nizeyi et al. (2001), Banish et al. (1993)
	Staphylococcus aureus	Schaumburg et al. (2012a,b, 2013), Unwin et al. (2012)
	Clostridium spp. *Clostridium septicum* *Clostridium tetani*	Fontenot et al. (2005)
	Tuberculosis (mycobacteriosis)	See text
	Other bacteria	See text
Viruses	Adenovirus	See Wevers et al. (2010, 2011)
	Ebola/Marburg (Filoviridae)	See text
	Encephalomyocarditis (EMC) virus	See Bourne (2009): http://wildpro.twycrosszoo.org/S/virus/picornaviridae/picornaviridae_emcv.htm
	Hepatitis (various)	See text
	Herpesvirus	Seimon et al. (2015), Marennikova et al. (1973/1974)
	Measles (morbillivirus)	Hastings et al. (1991), Cranfield and Minnis (2007), Jaffe (2012)
	Parainfluenza III (paramyxovirus)	Köndgen et al. (2008)
	Poliomyelitis (enterovirus)	Allmond et al. (1965)
	Rabies	See Chapter 15: Nervous System and Special Senses
	Respiratory syncytial virus "(pneumovirus)"	See Palacios et al. (2011) and Chapter 8: Nonspecific Pathology
	SIV/HIV	See text
	STLV	Masters et al. (2010)
	Varicella	White et al. (1972), Kornecva et al. (1975), Myers et al. (1987), Bourne (2009)
	Cytomegalovirus	Reynolds et al. (2013)
Fungi	Candidiasis	Jaffery et al. (1976), Abee et al. (2012)
	Coccidioidomycosis	See text and Sasseville et al. (2012)
	Dermatophytosis	See Chapter 9: Skin and Integument
	Geotrichosis	Dolensek et al. (1977)
	Pneumocystis	Demanche et al. (2003)

were enlarged and caseated. Acid-fast organisms were demonstrated in smears, and cultures in modified Dubos medium became positive in 10 days. The growth characteristics of the organism indicated it to be of bovine type.

Mycobacteriosis remains a threat to simian primates (Thoen et al., 2014; Wolf et al., 2013). Gorillas and other great apes are susceptible to it and its prevention and detection are an important part of health programmes for rescue centres and sanctuaries; PASA has paid

considerable attention to this. Insofar as humans are concerned, tuberculosis is responsible for significant morbidity and mortality, especially in the tropics. Active tuberculosis (TB) develops when the host's defences are impaired. In their paper 'The double burden; a new-age pandemic meets an ancient infection', Bridson et al. (2014) demonstrated a clear association between diabetes mellitus and susceptibility to TB, treatment failure and complications. Similar interactions may be important in gorillas but little is known.

Other bacterial diseases reported in gorillas range from streptococcal pneumonias and enteritides (see relevant chapters), the latter associated with various members of the Enterobacteriaceae, to probably less frequent, but significant, infections that present health hazards both to great apes and to humans – for example, anthrax (Leendertz et al., 2004, 2006). For fungi, see Box 3.4.

VIRUSES

Although cases of varicella and generalized cytomegalovirus infection were reported in captive gorillas in the 1970s, serious investigation of the possible role of viruses in morbidity and mortality of *Gorilla* and other great apes dates back less than three decades.

Even in the early 1990s the emphasis was on captive (zoo) animals. In 1993 Brack surveyed virus infections in NHPs and 2 years later (Brack et al., 1995) included relevant data in a text in German about diseases of zoo and free-living primates.

Over the subsequent 20 years the subject has expanded exponentially. The book by Voevodin and Marx (2009) was probably the first comprehensive text that covered all (then) known simian viruses. Soon afterwards, over a hundred pages of text and references concerning viral diseases on NHPs, with particular (but not exclusive) reference to those animals used in research, comprised the first chapter of the 852-page tome by Abee et al. (2012). During this same period, awareness of the significance of viral infections in free-living apes, including gorillas, surfaced and grew. For example, the paper by Köndgen et al. (2008) described how pandemic human viruses might cause decline of endangered great apes. Close contact between humans and habituated animals during 'ape tourism' and observational research studies had raised concerns that the risks of transmission of pathogens might outweigh the benefits of such activities. Up until that time only bacterial and parasitic infections of typically low virulence had been shown to move from humans to free-living great apes but tissue samples taken from habituated chimpanzees that died during three respiratory disease outbreaks at the authors' research site in Côte d'Ivoire were found to contain two common human paramyxoviruses.

Interspecies transmission of various pathogens between humans and NHPs is enhanced by their phylogenetic similarity. For example, the worldwide pandemic of human immunodeficiency virus (HIV) was the result of transmission of the simian version of the virus, simian immunodeficiency virus (SIV) – see later. There are implications for all, human and nonhuman (Ryan and Walsh, 2011).

In the early 2000s interest in the possible role of viral, as opposed to bacterial, pathogens in free-living mountain

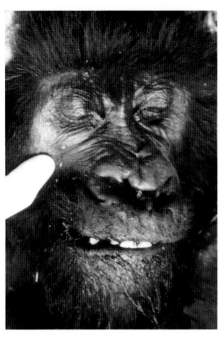

FIGURE 3.3 The face of gorilla 'Inyuma' at the time of her death. Careful examination of the buccal cavity did not reveal any lesions but inclusion bodies, suspicious of a herpesvirus, were detected on histology in the oesophagus and tongue.

gorillas prompted the MGVP to start looking for evidence of such infections. A summary of some preliminary seroprevalence figures for infectious agents in free-living mountain gorillas was presented at the Wildlife Disease Association International Conference in Australia (Whittier et al., 2005).

Retrospective examination of stored tissues (see Part 2: A Catalogue of Preserved Materials) is proving an excellent way of detecting viruses (frozen material) or signs of viral infection (histology) in gorillas that died some years ago. For example, intranuclear inclusion bodies, suspicious of a herpesvirus, in the oesophagus and tongue of mountain gorilla 'Inyuma', examined postmortem by the author (JEC) in 1995 (L.J. Lowenstine, Gorilla Doctors, personal communication) (Fig. 3.3).

Some specific viruses are now discussed in more detail.

Retroviruses

See also Chapter 12: Lymphoreticular and Haemopoietic Systems and Allergic Conditions.

The genus of retroviruses that is of most relevance to this book is *Lentivirus*. As the generic name suggests, one of its characteristics is a long incubation period.

HIV is a lentivirus. It and its associated acquired immunodeficiency syndrome (AIDS) were first recognised by scientists in the 1980s. HIV, the HIV-1 strain, is thought to have first entered the human population early in the 20th century in what is now the Democratic Republic of the Congo (DRC), following contact with bushmeat infected with an SIV (Michael et al., 1999). Then Peeters et al. (2002) examined 788 monkeys that were hunted in the rainforests of Cameroon for bushmeat or were kept as pets. Serological evidence of SIV infection was found in 13 of 16 primate species. Molecular analysis identified five new phylogenetic SIV lineages. These data not only documented for the first time that a substantial proportion of wild monkeys in Cameroon were SIV-infected but also confirmed that people who hunt and handle bushmeat are exposed to a range of genetically highly divergent viruses. Research carried out several years later by Kazanji et al. (2015) added weight to this by implicating the hunting of NHPs as being the origin of human T-cell lymphotropic virus type 1.

SIVs have been reported in other primates, including gorillas (Neel et al., 2009; Peeters and Delaporte, 2012). Gorillas are also a source of HIV (D'arc et al., 2015). SIVs of some species, such as African green (vervet) and sooty mangabey monkeys, do not cause immunodeficiency in contrast to experimental SIV infection of some other susceptible primate species, such as macaques, which can cause chronic wasting syndromes and a disease similar to human AIDS (Lowenstine and Lerche, 1993).

The pathogenesis of SIV infection in gorillas is still not fully understood but an interesting recent finding by Moeller et al. (2015) was that such an infection has relatively little effect on the gorilla microbiome — see later. Considerable research has been carried out on the virus–host relationship in SIV-infected chimpanzees and some other simians — see, for example, Greenwood (2014) and Greenwood et al. (2014, 2015).

The similarities between the human and nonhuman lentiviruses, SIVs, have led to the use of the latter for the study of the pathogenesis of AIDS. Gorillas are not involved in this type of research but the health and safety implications are relevant to all those who work with NHPs. Khabbaz et al. (1994) were the first to describe a laboratory worker in the United States in whom antibodies cross-reactive to HIV-2 and SIV developed after percutaneous exposure to blood from a macaque experimentally infected with SIV14. They tested 60 serum samples from workers in two SIV research facilities and found a scientist in another laboratory who had antibodies to HIV-2 and SIV.

The implications of exposure to SIVs and other retroviruses from free-living NHPs in Africa are of great importance. Wolfe et al. (2004) investigated this in rural villages in Cameroon and detected antibodies to simian foamy virus (SFV), a retrovirus endemic in most Old World primates, in people living in central African forests who reported direct contact with blood and body fluids of free-living NHPs. Sequence analysis revealed three geographically independent human SFV infections, each of which was acquired from a distinct NHP lineage — De Brazza's guenon, mandrill and lowland gorilla. Their findings showed that retroviruses are actively crossing into human populations, and demonstrate that people in Central Africa are currently infected with SFV.

Filoviruses

Marburgvirus and *Ebolavirus* (family Filoviridae) cause haemorrhagic fevers in humans and NHPs in Africa. At the time of writing (2015), there has been an epidemic of *Ebolavirus* infection in humans in West Africa — see below and the account by Jenny Jaffe in Chapter 17: Field Studies in Pathology and Health Monitoring.

International concern about the effects of *Ebolavirus* infection on free-living lowland gorillas started to surface in 2003 when a statement from the Zoological Society of London, widely reported in the world's media, suggested that between 600 and 800 gorillas, almost two-thirds of the population, had died from Ebolavirus disease (EVD) in the Lossi Sanctuary in the Congo Republic (RC). Leroy et al. (2004) then announced that some recent human outbreaks in Gabon and the Congo came soon after marked declines in the numbers of animals, chimpanzees, gorillas and duikers, which had been killed by different strains of the virus. According to the scientists, there had been five human Ebola outbreaks in western central Africa in the previous 2 years; 313 people had been affected and 264 died. The outbreaks began after people had handled dead animals and recounted that there had been many animal carcases in the forests just before and during the 2001 human EVD outbreaks in Gabon. In one sanctuary, signs of gorillas and duikers fell by 50% from 2002 to 2003. Eight gorilla groups, a total of 143 animals, had apparently disappeared. The researchers wrote: 'Our data ... confirm that Ebola outbreaks occur abruptly, exterminating exposed animal populations very rapidly and very locally'. Bermejo et al. (2006) described the death of 5000 gorillas in an EVD.

At the time of the report by Leroy et al. (2004), the belief was that outbreaks of EVD in animals were the result of multiple infections (see also Walsh et al., 2005) from 'still unknown natural hosts' but a year later Leroy et al. (2005) discussed African fruit bats as reservoirs of *Ebolavirus* followed by a paper by Biek et al. (2006) that reported the finding of Ebola Zaire virus in a bat reservoir.

At an early stage, concern was expressed in both medical and conservation circles regarding the possible adverse effects of EVD on gorillas and other great apes.

Leroy et al. (2004), cited earlier, stated that 'the slow reproductive cycle of the great apes, together with hunting and poaching, may lead to their extinction in western central Africa' and Sleeman (2004) drew attention to the probable role of *Ebolavirus* in the decline of great ape populations. Subsequent papers have echoed this fear — see, for instance, Walsh et al. (2007) — and ecologists have contributed to gaining a better understanding of the impact of *Ebolavirus* on gorillas and other apes (Rouquet et al., 2005; Pourrut et al., 2005; Devos et al., 2008; Genton et al., 2015).

The history outlined very briefly above quickly led to an awareness that wild animals, alive and dead, could serve as sentinels of EVD. In particular, those killed by the *Ebolavirus* could help provide warnings of a possible outbreak. Studies on the harvesting and distribution of bushmeat could, likewise, be of value to those striving to control the human disease. As Leroy et al. (2004) pointed out over a decade ago in their studies 'Almost all human Ebola outbreaks in Gabon and RC have been linked to the handling of dead animals by villagers or hunters, and increased animal mortality always preceded the first human cases'. Nevertheless, despite the established significance of animals, including gorillas, in EVD this appears still not to be fully recognised by those involved with human patients. Thus, Carson et al. (2014), in a paper entitled 'Ebola: controlling the nightmare' published following several thousand deaths in Liberia, Guinea and Sierra Leone, made no mention at all of animals, apes or bushmeat. As EVD is inextricably linked with animals, the veterinary profession should be an integral part of its surveillance, control and research. Cooper (2015) pointed out that the disease provides an excellent opportunity for the much-vaunted concept of 'One Health' to be applied to a situation and lamented that the (British) veterinary press had, so far, paid scant attention to the role that its members should be playing.

This book is concerned primarily with the pathology and, to a certain extent, the diagnosis of infectious diseases; relatively little will be said about the latter here because of the highly dangerous nature of Ebolaviruses and the need, therefore, to follow rigorous precautions. Basic considerations in sample taking are considered in Chapter 7: Methods of Investigation — Sampling and Laboratory Tests. Field diagnosis of EVD presents challenges since, by definition, access to appropriate protection for staff and onlookers is unlikely to be immediately available. It is worth mentioning that when this is the case but there is a pressing need to test for *Ebolavirus*, a skin snip may suffice for diagnosis. Careful hygienic precautions are still necessary, however (Cooper and Cooper, 2013) — a mock-up of the procedure should be carried out beforehand to familiarise those involved with the necessary safeguards — and, of course, the skin snip sample is usually inadequate to make any other diagnosis. According to Leroy et al. (2004), animal carcasses in the forest are not infectious after 3−4 days.

A breakthrough in monitoring for the presence of *Ebolavirus* in great apes was reported by Reed et al. (2014) who described noninvasive detection of antibodies against *Ebolavirus* in the faeces of free-living apes. They stressed that the method would contribute to early detection, and identification of immunologically naïve at-risk populations that might be subjects for vaccination, provision of a means to monitor the efficacy of such immunisation and — relevant to points made above — provide information of relevance to public health in local African communities.

Poxviruses

Monkeypox is caused by an orthopoxvirus with a wide host range, including lowland gorillas and chimpanzees (Williams and Barker, 2008) as well as humans (Reynolds et al., 2013). There are useful data on pathogenesis, clinical signs and diagnosis on the Wildpro website page (http://wildpro.twycrosszoo.org/S/00dis/viral/Monkeypox.htm) (Bourne, 2009). Records on the website include an outbreak at Rotterdam Zoo, where there were fatal infections in six orangutans, clinical disease in two lowland gorillas but only mild disease in several chimpanzees. The source of infection was apparently two giant anteaters that developed vesicular lesions affecting the snout, tongue and soles of the feet, from which monkeypox virus was isolated. Support for the theory that gorillas may not be particularly susceptible to monkeypox comes from correspondence (Peer Zwart, personal communication, 20 July 2015) who recalls that, during the outbreak of monkeypox around 1962 in Blijdorp Zoo, Rotterdam, The Netherlands, orangs, chimpanzees and squirrel monkeys were affected but the two gorillas present in the same house remained healthy. Generalised cytomegalovirus infection has been reported in *Gorilla* (Tsuchiya et al., 1970).

PARASITES (BY IAN REDMOND)

Specific types of parasite are covered in respective chapters; see Chapter 9: Skin and Integument; Chapter 10: Respiratory and Cardiovascular Systems; Chapter 11: Alimentary Tract and Associated Organs and Chapter 12: Lymphoreticular and Haemopoietic Systems and Allergic Conditions.

The dictionary definition of a parasite is, 'An organism which lives in or on another organism (its host) and benefits by deriving nutrients at the other's expense' (Oxford

Dictionary). The study of parasites has traditionally been dominated by the desire to understand their life cycle in order to kill them or to prevent them from gaining access to potential hosts – especially when those hosts are humans and/or their domestic animals and crops. The fact that parasites and their hosts have often coevolved over very long periods of time, however, means that host–parasite relations may be more complex than the simple definition implies. The third factor involved is the environment, which adds its own evolutionary pressures affecting the strength and specificity of interactions between hosts and parasites.

Hosts are under selective pressure to resist parasites, and parasites are selected to overcome host defences. This creates a special case of coevolution, defined as the reciprocal adaptive genetic change of two antagonists (e.g., different species or genes, in this case the host and its parasite) through reciprocal selective pressures. As Cooper (2009a) put it, 'The close association between a parasite and its host can produce a myriad of effects at the physiological, biochemical and pathological levels. At one end of the spectrum, an infected animal will develop clinical signs of disease (morbidity) and may die as a result (mortality). At the other end, the parasite may become resident in the host, permanently or transiently, but produce no apparent clinical signs of disease'.

Indeed, further along that same spectrum, a resident parasite may actually bring benefits to the host in some way, moving past mere commensalism (where one party benefits without causing damage to the other) into a mutually beneficial relationship more akin to symbiosis. This spectrum of relations has been explored by authors from Ainsworth (1955, looking mainly at fungal examples) and Thomson (1978, in the context of veterinary pathology) to use of game-theory in seeking to understand the evolutionary forces on both parasite and host (Renaud and De Meeus, 1991) and arguments for their conservation (Pérez, 2009). One of the most interesting aspects of this relationship is where the parasite triggers changes in the host's behaviour that may be advantageous to the parasite but injurious or even deadly to the host. This can be something as simple as female pinworms, *Enterobius vermicularis*, emerging to lay eggs on the perianal skin, causing itching, which when the host scratches increases the chance of reinfection by the faecal-oral route, or the more complex behavioural changes induced by the rabies virus causing rabid animals to bite others, thereby enabling infected saliva to enter a new host's tissues. Such changes have not yet been documented in gorillas, but it is reasonable to assume that parasites shared by gorillas and humans would have a similar effect.

The fact that every wild animal is also a habitat for parasites, many of them species-specific, has led to a reappraisal of estimates of biodiversity loss; clearly, every endangered species is host to numerous other taxa whose survival depends upon the host species remaining extant – leading Stork and Lyal (1993) to coin the term 'coextinction' rates and suggest 'There may be conflicts in conservation needs, forcing us to bid farewell to the gorilla louse or the lice of the Californian condor while retaining their hosts'. As Pizzi (2009) asks, if biodiversity preservation is one of the fundamental objectives of conservation... the loss of affiliated species such as parasites, needs consideration if conservation is not to fall victim to taxonomic chauvinism.

MICROBIOTA, MICROBIOME AND NORMAL FLORA

Twenty years ago the author expressed surprise at the paucity, almost total absence, of reliable data on the 'normal' bacterial flora of *Gorilla* species (Cooper, 1997c). He stressed that information on this subject was needed because (1) interpretation of the significance of isolates from animals with clinical disease (dermatitis, enteritis, etc.) depends upon awareness of which organisms are part of the resident flora, (2) rehabilitation/release programmes need to be aware that, in returning gorillas to the wild, they may inadvertently also be introducing bacteria (and other organisms) into the species environment (IUCN/SSC, 2013) and (3) close proximity between captive gorillas and *Homo sapiens*, especially when orphaned babies are hand-reared, probably means that the bacterial flora of such gorillas are radically altered.

In recent years the situation is being rectified. Whittier et al. (2010), for instance, were able to use a portable, real-time, polymerase chain reaction (PCR) instrument to detect *Campylobacter* spp. in the faeces of free-living mountain gorillas and speculated that this reflected previously uncharacterised intestinal flora. Tsuchida et al. (2014) described a new species of *Lactobacillus* in the faeces of captive and free-living western lowland gorillas.

The 'microbiota' are the microorganisms that live on the skin, in the buccal cavity, on the conjunctiva and in the gastrointestinal tract. The term 'microbiome' refers to the genomes of these organisms. 'Normal flora' is an older term, but still often used, for bacteria (especially) that are commonly present on or in the body and which usually do not cause disease/may assist in maintaining health. However, 'flora' refers to plants, prompting some scientists to want to replace the term.

Hoffmann et al. (2016) wrote of the microbiome as 'the trillions of microorganisms that maintain health and cause disease in humans and companion animals' and explained that the term refers to the complex collection of microorganisms, their genes, and their metabolites that colonise the mucosal surfaces, digestive tract, and skin. The microbiome interacts with its host, assisting in digestion and detoxification, supporting immunity, protecting against pathogens and thereby maintaining health. Diseased and healthy humans and other animals are often colonised with different microbiomes.

Much of the research relating to the microbiome involves studies on mice and on human volunteers and is indicating that gut flora can affect health by (1) reducing the numbers of potential pathogens, such as *Salmonella* spp. and (2) changing the balance between bacterial taxa in the GIT, including those that in humans are associated with diabetes, obesity and chronic inflammatory bowel disease and in animals with obesity. The possible role of gut-residing bacteria in autoimmune arthritis is discussed in Chapter 12: Musculoskeletal System.

There is also growing evidence that helminths may be important. Some seem to have a beneficial effect on inflammatory disorders of the intestinal tract of humans, including inflammatory bowel disease and coeliac disease. However, the mechanisms by which helminths regulate immune responses, leading to the amelioration of symptoms and signs of such chronic inflammation are not clear. It has been suggested that helminth-induced modifications of the gut commensal flora (microbiota) may be responsible.

There are already some published studies on the microbiomes of NHPs. A special topic issue of *Folia Primatologica* included a paper comparing the gut microbiome of two species of free-living lemurs (Fogel, 2015). McCord et al. (2014) reported that faecal microbiomes of NHPs in Western Uganda showed that species-specific communities are largely resistant to habitat perturbation. Yildirim et al. (2010) pointed out that host-associated microbes are an integral part of an animal's digestive system and that these interactions have a long evolutionary history. They presented a comparative analysis of gastrointestinal microbial communities from three different species of Old World monkeys and showed that the microbial organisms within each species were more similar to each other than to those from other primate species. Comparison of faecal microbiota from NHPs with microbiota of humans revealed that the gut microbiota of these species are distinct and reflect host phylogeny.

Gomez et al. (2015) specifically looked at the gut microbiome composition of free-living western lowland gorillas and were able to show that this reflected the host's ecology.

Notwithstanding the harmless, sometimes beneficial, nature of intestinal (and other) organisms, the enteric carriage of potentially pathogenic bacteria remains an important consideration in zoological collections. Banish et al. (1993) described an eradication programme for *Shigella flexneri* in primates that involved the use of enrofloxacin and rigorous cleaning and disinfection of animal housing.

EMERGING DISEASES

In his book *De L'expectation en Médecine* Charcot (1857) declared: 'Disease is very old, and nothing about it has changed. It is we who change, as we learn to recognise what was formerly imperceptible'. So what is an 'emerging' disease? The term is occasionally applied to a noninfectious condition, such as the exposure of gorillas to new types of trapping or other physical hazards, but is usually used for infectious diseases. The term 'emerging infectious disease' (EID) came into vogue during the 1970s–80s and was initially used for conditions that were considered to be a new threat to public health. EIDs were then described as those infectious diseases that had (1) newly appeared, (2) rapidly increased in incidence, (3) expanded in geographical range and/or (4) developed increasing or novel mechanisms of antimicrobial resistance (IOM, 1992). Many in the first three categories were of animal origin and zoonotic. More traditional zoonoses, ranging from shigellae to rabies (Schwabe, 1969, see Appendix 4: Hazards, Including Zoonoses) remain a challenge but they are now being supplemented by a whole spectrum of novel diseases of relevance to human health (Fig. 3.4).

Many factors have been implicated in the emergence of 'new' diseases and in the unprecedented spread of others. Global climate change is one such factor (see Chapter 4: Noninfectious Disease and Host Responses). Another is increased frequency of contact between animals of different species because of habitat destruction and fragmentation; this may, for example, help explain the appearance and spread of EVD and other diseases in humans in Africa (see earlier). That zoonotic infections can appear and increase in prevalence following changes to the environment is not a new concept: the seminal work by Pavlovsky (1966), in the former USSR, culminating in his thesis on the 'natural nidality of diseases', was a direct result of observations on the zoonoses encountered when the Russians extended their agricultural operations into previously uninhabited regions.

FIGURE 3.4 The presence of a zoonotic 'emerging disease' in gorilla habitat presents a particular threat to local people, such as these children who live in the vicinity of Bwindi Impenetrable National Park in Uganda.

Chapter 4

Noninfectious Disease and Host Responses

John E. Cooper

Up to the present time it has been customary to believe that wild animals possess a high standard of health, which is rigidly maintained by the action of natural selection, and which serves as the general, though unattainable, ideal of bodily health for a highly diseased human civilization. This belief is partly true and partly false.

Charles Elton (1931)

INTRODUCTION – NONINFECTIOUS CONDITIONS

Important and common examples of noninfectious factors that may cause pathological changes in gorillas are various forms of trauma, electrocution, burning (heat, friction or chemicals), cold, irradiation and exposure to water. These are discussed in the text and in Table 4.1. Their incidence, prevalence and significance are closely related to environmental factors in free-living gorillas and to management practices in captive gorillas.

Studies on mountain gorilla skeletons 25 years ago (Lovell, 1990a,b) revealed that 22% showed either trauma, inflammation, arthritis, dental abnormalities or periodontal diseases. Developmental anomalies (6%) and bone neoplasm (6%) were also detected. Trauma mainly occurred in the 'postcranium' and longbones and some of the latter showed signs of natural healing. A higher percentage of males showed broken canine teeth and skull fractures, which Lovell attributed to physical aggression between individuals.

TRAUMA – CAUSES AND PATHOGENESIS

The pathogenesis of traumatic injury is twofold: (1) damage locally of essential epithelial, connective, muscular and nervous tissues, affecting the body surface, head,

CNS, ear, eye, thorax or abdomen, and (2) secondary systemic effects such as loss of blood, fluid and electrolyte disturbances, thrombosis, fibrinolysis, fat embolism and metabolic changes. There may also be endocrine responses, toxaemia and infection.

Both physical and chemical injuries usually cause recognisable wounds, defined here as 'an interruption or break in the continuity of the surface of the body or of an internal organ caused by trauma or other forms of injury'.

Trauma is now covered in more detail. This is because (1) traumatic injuries are amongst the most prevalent causes of morbidity in gorillas, especially in the wild, and (2) much new information about the pathogenesis, diagnosis and management of trauma has been amassed in recent years (Mason and Purdue, 2000). According to the WHO Global Burden of Disease, trauma is now responsible for 5 million human deaths each year.

Wounds are discussed in more detail later.

SPECIFIC RESPONSES – TRAUMA

Trauma is now used to illustrate host responses in *Homo* that are, it appears, very similar in *Gorilla*.

Leadbetter (2011) summarised the response of humans to trauma. He stressed that, while these are often primarily considered at the local level (the injured organ), there is also a systemic reaction which itself can influence the local response. This systemic response involves a complex interaction of the neural, endocrine and immune systems. The clinical presentation depends upon the interaction of primary inflammatory and compensatory anti-inflammatory responses and such genetic factors as the sex of the patient. Systemic inflammation can occur after trauma (Lenz et al., 2007).

Similar organ and tissue responses to injuries occur in animals and have been described and discussed by

J. E. Cooper & G. Hull: Gorilla Pathology and Health. DOI: http://dx.doi.org/10.1016/B978-0-12-802039-5.00004-4

TABLE 4.1 Pathogenesis of Noninfectious Diseases of Gorillas

Factor	Significance or Example	Direct Effects	Secondary Effects	Other Comments
Trauma (physical)	Fighting	See text	See text	See text
	Snares	See text	See text	See text
	Weapons	See text	See text	See text
	Firearms, exposure to blasts, explosions and unexploded ordnance	See text	See text	See text
Electrocution	Most likely to occur in captivity although some NHPs are regularly electrocuted by powerlines in the wild, for example, colobus monkeys in Kenya	Direct damage to tissues, including burns. Apnoea and ventricular fibrillation (usually the cause of death in other species)	Lymph nodes may be haemorrhagic. Free blood in the respiratory tract. May be discoloration and rigor of muscles and black unclotted blood	See Cooper (1996b)
Burning	Burns can be caused by ultraviolet light (see below) heat, chemicals or friction. Most likely to occur in captivity. Bush and grass fires are a cause of death from burning and suffocation in some free-living primates. Burning by friction is possible when a gorilla is held in a snare (see text) or as a sequel to stereotyped behaviour, for example, persistent rubbing on a rough surface	Local tissue damage. Burns varying in severity from first- to third-degree. First-degree burns involve only the outer epidermis and are associated with pain, heat and oedema. Second-degree burns affect the whole epidermis and vesicles are characteristic. In third-degree burns both epidermis and dermis are destroyed because nerve endings are lost, pain is not a major feature. Dehydration, pyrexia due to pyrogens	Secondary infection by organisms such as *Pseudomonas* spp. Dehydration, secondary (hypovolemic) shock and if the animal survives, toxic changes, detectable histopathologically, in the liver and kidneys	See Cooper (1996b) Gardner-Roberts et al. (2007) referred to lesions In gorillas that resembled 'an electric burn' or the effects of a 'chemical detergent'
Exposure to water	Gorillas in captivity and in the wild exhibit varying propensities to enter and use water. In zoos gorillas regularly play with and drink water (a form of environmental enrichment?) but in the wild are often assumed to obtain fluid from their food. Schaller (1963) reported never seeing free-living mountain gorillas drink water but James Foster (personal communication) observed young animals drinking from puddles and believed that older gorillas did too if unwell	Fluid loss may result in dehydration (see Carter, 1974) In drowning the hair is usually wet or dry and matted. May be water in the buccal or nasal cavities	In fluid loss signs secondary to dehydration may be seen If water has been inhaled, respiratory signs predominate. Aquatic organisms may be detected microscopically in inhaled fluids	See Cooper (1996b) and Cooper and Cooper (2007) Brown et al. (1982) reported on 'water-contact by a captive male lowland gorilla'. The use of a stick to check the depth of water in the Nouabalé-Ndoki National Park (CR) was described by Breuer et al. (2005)

(*Continued*)

TABLE 4.1 (Continued)

Factor	Significance or Example	Direct Effects	Secondary Effects	Other Comments
	Imprecise early reports state that 'captive apes' drowned in deep water Free-living (mountain) gorillas appear to dislike rain and zoo staff working with *Gorilla gorilla* confirm that they will retreat into their quarters if it rains			
Ionising radiation	Overexposure of the skin to ultraviolet rays from the sun can cause skin lesions in most species. 'Sunburn' should therefore be a differential diagnosis — for example, as it was at Chester Zoo, United Kingdom, in the case of 'Mukisi' who had a 'sore' inside the lower lip — actually a herpesvirus lesion (Liz Ball and Stephanie Sanderson, personal communication). Animals in captivity are also exposed to X-rays during radiography	Dermatitis due to damage of tissues by either particulate or electromagnetic radiation Lesions depend on the dose of radiation and the area affected. The more rapid the rate of mitosis of a cell, the greater the sensitivity to radiation damage. Thus, the haemopoietic tissues usually are affected such that within 1–3 weeks of exposure the animal develops leukopoenia, thrombocytopoenia, and anaemia. Lymphoid tissues show aplasia and destruction of bone marrow	Common sequelae include bacterial infections, diarrhoea with loss of fluids and electrolytes, damage to germinal cells of testes or ovaries and developmental effects in the fetus. Skin neoplasia and photosensitisation may occur	
Rise in temperature (hyperthermia)	The most common cause of hyperthermia is pyrexia, a rise of body temperature in ill-health, especially infection. Overheating can also occur in captive gorillas if environmental control is inadequate. Free-living animals can become overheated by the sun; see text regarding 'siesta'. Hyperthermia can develop as a result of a chase, including immobilisation — especially when the ambient temperature is high As a general rule, above a certain point ('critical high temperature') the animal will lose its ability to thermoregulate; irreversible changes take place that can	Acute heat stress can produce similar effects to burns (above) but may, in addition, cause damage to kidneys, liver, heart, and brain, and changes in serum enzymes and electrolytes	Reduced spermatogenesis. fetal abnormalities Heat stress can be acute or chronic. Acute signs include panting or other signs of overheating, a raised body temperature and tachycardia. In severe sunburn, cutaneous erythema is present. The animal may be prostrate, in a stupor, dehydrated. Chronic heat stress is more subtle in its effects, and animals may show only poor growth and high embryonic mortality. A postmortem diagnosis depends largely on clinical history. Visceral congestion and petechiae are usually a feature	See Cooper (1996b)

(Continued)

TABLE 4.1 (Continued)

Factor	Significance or Example	Direct Effects	Secondary Effects	Other Comments
	result in death. Spontaneous hyperthermia in the gorilla was reported by Guilloud and FitzGerald (1967)			
Drop in temperature (hypothermia)	A drop in body temperature can follow immobilisation or clinical anaesthesia of a gorilla. Chilling can occur in captive animals if environmental controls fail in cold weather Fiennes (1966) reported on two young captive gorillas that 'suffered during the blizzard and died from different causes' Low environmental temperatures affect morbidity and mortality of free-living gorillas, particularly infants (see, for example, Byers and Hastings, 1991). More severe exposure to cold can result in frostbite, with dehydration, disruption of cells by ice crystals, thrombosis and ischaemia that can progress to gangrene and sloughing (see also snares)	Few clinical signs other than a drop in body temperature, especially in young animals. An infant may become weak and unable to feed. Even a slight drop in temperature can be an important stressor	Cold stress in captive great apes, including gorillas, can be detected by incremental lines of dental cementum (Cipriano, 2002) In postmortem cases there may be only mild, nonspecific pathological changes in cases of chilling Freezing, however, may be associated with areas of hair loss and oedema. Tissues may be avascular or necrotic In mild chilling, the animal may appear cold to the touch, depressed in appearance, and shiver; often no other clinical signs are present. In the case of freezing, the body surface initially feels cold and appears pale; later it may become erythematous and oedematous with the animal exhibiting pain. As with burns, pruritus may accompany healing. In severe cases the affected area becomes ischaemic and sloughs	See Cooper (1996b)
Slipping, tripping, falling	Captive and free-living gorillas may sustain injuries as a result of slipping, tripping, or falling while climbing or playing Young gorillas, like children, sustain injuries during 'rough and tumble play'. In captivity attention to the design of enclosures minimises such injuries			Although the gorilla did not specifically feature, earlier work on the pathogenesis of slipping, tripping and falling accidents involved studies on zoo animals (Manning et al., 1985)

many authors over the years. Cooper (2011) emphasised that the effects of tissue damage on most vertebrates are not markedly different from those in humans. Primary responses to trauma in mammals, birds, reptiles, amphibians and fish can include inflammation, degeneration, proliferation (including regeneration), hypo- and hyperpigmentation, metaplasia and, on occasion, neoplasia. Secondary effects in these taxa include infection and its sequelae, circulatory shock and organ damage or failure.

However, various factors influence the speed or type of the host response. The first of these is whether

FIGURE 4.1   A captive lowland gorilla investigates its hand.

the animal is cold-blooded or warm-blooded (see Chapter 3: Infectious Disease and Host Responses). The majority of vertebrate animals are ectothermic and even the two main groups of endotherms, the mammals and the birds, can show degrees of poikilothermy (fluctuation in body temperature) associated with immaturity, hibernation and aestivation. Since healing and other tissue responses are, in most cases, temperature-related, the higher the temperature of an animal's body (regardless of whether this is a result of endogenous or exogenous, usually behavioural, thermoregulation), the more rapid are such processes as inflammation, epithelialisation and the production of granulation tissue. Some specific examples of trauma in gorillas will now be discussed in more detail.

As in other species, some injuries of gorillas are iatrogenic. Injections can cause tissue damage, as can surgery (see Chapter 5: Methods of Investigation − Observation, Clinical Examination and Health Monitoring).

Gorillas in the wild and in captivity will investigate even minor wounds on themselves, sometimes on other animals, in great detail (Fig. 4.1). This can lead to exacerbation of the damage. Some specific examples of trauma in gorillas will now be discussed in more detail.

FIGHTING

In a discussion about 'Injuries and fatalities caused by conspecifics', Meder (1994) stated that, while humans are probably responsible for most killings of gorillas in the wild, many individuals die during interactions with conspecifics. She went on to say that at least 37% of all mountain gorillas in Rwanda that die within the first 3 years of life succumb to injuries incurred by (mainly male) conspecifics − infanticide (see Chapter 13: Urinary and Reproductive Systems).

Adult gorillas attack other adults and may both receive and inflict wounds during intergroup encounters. Meder stated that it is estimated that 62% of all injuries to mountain gorillas in the Virunga Volcanoes are the result of such intergroup confrontations. Her comments on the injuries inflicted or sustained in this way are interesting: 'The wounds usually heal quickly, but they may also lead to death. During fights between silverback males, their canines may break off and sometimes even get stuck in the skull of the opponent. Injuries incurred during such fights have been fatal in several cases'.

It is well-recognised that adult male gorillas may both sustain and dispense wounds during such encounters. As a prelude to their postmortem report, Galloway and Allbrook (1959) explained that the animal that they examined was one of a pair who were said to have been fighting during the preceding 10 days. Their findings helped confirm this: 'there were full-depth skin injuries, looking like bites, on the backs of both hands. The right eye appeared bruised. The left upper canine was broken'. Forty years later an interesting first-hand description of a battle between two silverbacks was penned by Christopher Whittier (Digit News (DFGF) issue 21 December 2001); written in popular terms, it provides useful information from a veterinary observer and includes a photograph of extensive wounds on the back of the animal known as 'Ryango'.

Fossey (1983) described lesions in the gorilla 'Whinny' as: 'Skeletal Findings. Conducted by Jay Matternes. Extensive pathology on right side of skull indicative of a localised infection resulting from a wound, probably a bite. Infection appeared to be spreading to the occipital through the pneumatised area around the mastoid process, and if the lesions had entered the endocranium, meningitis would have resulted'.

The anatomy of the male gorilla reflects such a propensity to fight, as well as the need to be able to support its head, jaws and masticatory apparatus. Napier and Napier (1967) pointed out that 'The nuchal area of the skull is enormously expanded to provide an additional area of attachment for the bulky posterior neck muscles' and referred to the studies in *Gorilla* and *Pongo* on the nuchal crest by Ashton and Zuckerman (1956b); 'The crests are most powerfully developed in the gorilla, especially in males, where a nuchal crest may eventually become a uniform shelf of bone as much as 4 cm wide. A sagittal crest ... is present in all males, and may be as much as 5 cm high. It is found in only about two in every five females'.

The effect of fight wounds depends on their location and severity. Facial lesions may affect sight or feeding. Clinical signs may include trailing behind or staying on the periphery of the group.

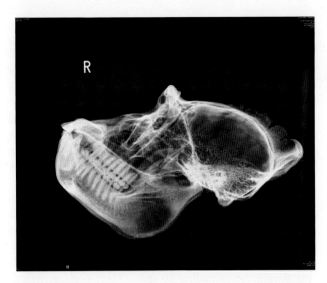

FIGURE 4.2 A lateral radiograph of the skull of a silverback, showing deep fissures in the sagittal crest (right of picture) — probably due to fighting — and demonstrating excellent dentition!

FIGURE 4.3 Observation, in good sunlight, is necessary if even minor blemishes are not to be missed during health monitoring of habituated animals.

Groves (1970) described the possible sequelae of fighting between gorillas as including 'cuts, swellings ... toothmarks on the sagittal crest'. In fact, damage to the sagittal crest can be more severe (Fig. 4.2).

The investigation of injuries in gorillas, whatever their cause, should follow standard procedures in terms of collecting population data, defining environmental variables, collecting samples, and keeping records (Wobeser, 2010).

Notwithstanding some of the observations above, proper analysis of fight and other injuries in gorillas has been surprisingly limited. This is in marked contrast to the situation in humans, especially children, and in companion and working animals where patterns of injury are recognised as being associated with particular types of abuse. Patterning can also help in the recognition of individual traits on the part of the perpetrator and assist in forensic cases.

In order to develop a workable system for assessment of injuries in gorillas, it is necessary to record routinely and systematically the size, nature and position of the lesions. Pre-prepared charts with dorsal and ventral outlines of the body are used in human pathology. They are of value in gorilla work but it is easy to miss a wound in these species because of the hair coverage. An ultraviolet light can help, as can clipping of the hair. When photographs of wounds are taken, a measuring scale should be included in the images.

Detection of wounds in free-living gorillas is not always easy. An animal may hide its injury. Careful observation, in good sunlight, is important (Fig. 4.3).

If the gorilla is dead, it is important to skin the carcase to look for subcutaneous haemorrhages. Careful dissection of the cranium and spinal cord are necessary to detect localised damage to the central nervous system. Radiography, if available, should be used for dead gorillas before the necropsy starts.

INFECTED WOUNDS

Any fight wound can introduce infection but, as in other species, those due to bites are a particular source (Cooper, 2005; Cooper and Cooper, 2007). Culture and cytology will help in assessment and, in captive animals which can be swabbed regularly, bacterial counts can facilitate in the monitoring of healing. The author's (JEC) approach to bite-wounds is given in Box 4.1.

ATTACKS ON HUMANS

Gorillas will also bite humans, including poachers (especially if they are trying to capture a baby). According to Morris and Morris (1966), the man largely responsible for spreading popular knowledge of the gorilla in Europe and American was Paul Du Chaillu, an American citizen. In his *Explorations and Adventures in Equatorial Africa*, Du Chaillu (1861) conveyed a distorted picture of the gorilla, emphasising its horrific appearance, ferocity and malignity. But to his credit, he said that 'many of the actions reported of it are the false and vain imaginings of ignorant Negroes and credulous travellers ... the gorilla does not lurk in trees by the roadside and drag up unsuspicious passers-by in its claws, and choke them to death in its vice-like paws; it does not attack the elephant, and beat

BOX 4.1 Approach to Bite-Wounds

1. Obtain as full a history of the alleged incident as possible.
2. If a live or dead victim is available, investigate the presumed bite wound as follows:
   a. Observe without touching. If available, use ultraviolet light.
   b. Photograph in black and white as well as colour/draw the wound(s) from different angles.
   c. Record any 'pattern'.
   d. Measure the external dimensions and depth of the visible wound using sterile techniques.
   e. Describe the lesion in medical (clinical/pathological) terms – see text.
   f. Take samples:
      i. multiple swabs, in appropriate transport medium, for microbiological examination and DNA studies, including the area surrounding the wound. Also search for other material that may harbour DNA from the predator(s) or from subsequent scavengers, for example, saliva or scratch marks elsewhere on the victim's body,
      ii. touch preparations for cytological examination and, if the victim is dead, a wedge of tissue for histological examination.

FIGURE 4.4 A relatively minor wound on the leg of a captive silverback lowland gorilla.

him to death with sticks; it does not carry off women from the native villages . . .'

Carl Akeley was perhaps the first Westerner to counter claims that gorillas were a danger to humans. In a letter to an official in the Belgian Congo in 1922, he stated that the gorilla is 'a wholly acceptable citizen and not the wicked villain of popular belief . . . a splendid animal in every sense, in no sense aggressive or inclined to look for trouble'.

Earlier Wood (1893) had described a man who had been wounded by a gorilla 'His left hand was completely crippled, and the marks of teeth were visible on the wrist'. However, Wood pointed out the essentially peaceful nature of gorillas with a quotation from local West African hunters. 'Leave Ngina alone and Ngina leave you alone'. Local knowledge of gorilla behaviour was described by Jenks (1911). Schaller (1964) described incidents when humans, usually Africans, were attacked by male (occasional female) mountain gorillas. He concluded that usually the relationships between gorillas and humans 'consist of respect and avoidance'. Sabater (1966) discussed lowland gorilla attacks against humans in Rio Muni, West Africa.

Veterinarians and others may be bitten during the course of their work, including investigative procedures (interventions) but, according to Cranfield and Minnis (2007), 'the incidence is low and the wounds, to date, have been minor'.

Even in captivity, where animals are carefully monitored for signs of aggression, injuries, sometimes necessitating veterinary intervention, often occur (Fig. 4.4). Fatalities may ensue. One such case was vividly described by Fowler (1999) in respect of aggression shown by a male gorilla at the Sacramento Zoo towards the female with which he shared a cage. During one anoestrous period, he 'picked up the female and flung her across the enclosure. He also bit her and inflicted skin wounds 25 cm in length. These became infected and she died'.

Introducing a new gorilla into a group in a zoo can be especially difficult. Fights often ensue, sometimes warranting veterinary attention. These squabbles need not involve the newcomer as such: sometimes they are between the dominant silverback and a subordinate blackback, prompted by the arrival of a strange female.

The effects of such incidents can be stressful to zoo staff as well as to the gorillas themselves. There were media reports in May 2015 that 'zookeepers need counselling' after a silverback, 'Otana', sent to Australia from Howletts Wild Animal Park in the United Kingdom had killed a gorilla at Melbourne Zoo in front of primary school pupils.

A better understanding of the causes of fights and subsequent injuries may result from ethological studies such as the recently published observations on the relationships between adult male and maturing mountain gorillas by Rosenbaum et al. (2016). There is also interest in comparing and contrasting sexual aggression in the great apes with comparable activity in humans (Nadler, 1988).

WOUNDS – GENERAL

As stressed in the context of fighting, above, it is important that wounds in gorillas are both described and classified (Box 4.2.) properly, not only for scientific reasons but because (in live animals) treatment options are improved by understanding the nature of the wound. The type of wound (Box 4.2) provides information on its aetiology and pathogenesis (Box 4.3).

Although gorillas appear hardy, their injuries sometimes take a long time to resolve. Factors that contribute to this include those in Box 4.4.

BOX 4.2 Wound Classification

- **Clean** – a surgical wound, with no contamination.
- **Clean contaminated** – a wound that occurred less than 2 hours ago with minimal dirt.
- **Contaminated** – chronic open wound, a traumatic wound 2–4 hours old or a wound involving body fluids such as urine or faeces.
- **Dirty** – a traumatic wound greater than 4 hours old and/or with contamination (mud, plant material).

BOX 4.3 Wound Types

- **Incision** – slicing trauma from a sharp object, including a surgical wound.
- **Laceration** – tearing of the skin with associated damage, including bruising, around the skin edges.
- **Abrasion** – caused by friction.
- **Degloving** –removal of the skin like a glove. Usually the result of severe trauma such as that inflicted by a mechanical implement.
- **Puncture** –a wound that is deeper than it is wide; a fight wound, for example.
- **Burns** – chemical, thermal or electrical.

BOX 4.4 Factors That Influence Wound Healing

Local	Systemic
- Trauma - Movement - Infection - Desiccation - Oxygen - Presence of foreign material	- Haemorrhage - Malnutrition - Concurrent disease - Administration of medication – NSAIDs may hasten healing while corticosteroids are likely to delay it

---

Infection of wounds occurs easily, especially in free-living gorillas. Wound infections can be either (1) endogenous (organisms originate from within the patient), and (2) exogenous (organisms originate from the environment).

Infection readily delays healing, especially if bacteria are present that produce collagenases. The first indication of infection may be local signs, such as a purulent exudate. This was the case, for instance, when the author reported on a gorilla in Zaire (now the DRC) in 1994. 'I am worried about gorilla Rafiki ... he has already a bad wound caused by a buffalo and the Zairois Guards reported *usaha* (pus) today'. Alternatively, or in addition, there may be systemic signs suggestive of infection such as lethargy due to a bacteraemia or toxaemia.

Factors that commonly predispose to wound infections in gorillas in the wild include foreign bodies, devitalised tissue, dead space and haematoma formation. If the animal can be handled the removal of foreign material (extraneous and dead tissues) is essential in order to reduce the bacterial count – but doing this and keeping the wound clean is far from easy in free-living animals with limb injuries. Although treatment is not, strictly, part of this book, it is worth mentioning the possible value of medicinal plants (Cousins and Huffman, 2002). When modern antimicrobials are not available in the field, inexpensive topical agents may prove helpful. Specific reference to the use of (for example) honey or silver-impregnated dressings to treat wounds in gorillas has not to date been found but such products are used in some parts of Africa for humans and domestic livestock and may find favour in the future if lesions fail to respond to conventional antibiotics. Maggots are found on wounds in live gorillas – see, for example, the rectal prolapse case described by Kalema-Zikusoka and Lowenstine (2001) but have not yet, apparently, found a place in wound management!

Mudakikwa et al. (2001) referred to the taking of wound swabs from gorillas but did not specify which tests should be performed. Cytology can be important – as can bacterial counts (Box 4.1). Any injuries detected during the course of a necropsy should be investigated thoroughly to ascertain whether they occurred in life, at the time of death, or postmortem (Fig. 4.5).

AGEING OF WOUNDS

The ageing (dating) of skin and soft tissue wounds, contusions and fractures in both live and dead gorillas is important, especially in forensic cases. However, as medical forensic texts testify, the topic is fraught with difficulties.

Approximate ageing of certain traumatic and inflammatory lesions is possible in humans, domesticated livestock and laboratory mammals (DiMaio and DiMaio,

FIGURE 4.5 A damaged digit on the hand of an infant mountain gorilla. This occurred postmortem but must, nevertheless, be carefully examined, described and photographed as it may be relevant.

BOX 4.5 Some Criteria for Ageing of Wounds in Gorillas

- Organisation of blood clots.
- Inflammatory infiltrates – acute versus chronic, numbers of pyknotic/karyorrhectic cells.
- Presence or absence of phagocytosed organisms or other material.
- Fibroplasia.
- Evidence of healing, including, in fractures, presence of callus and its degree of ossification.
    The above can vary significantly; the result of such factors as:
- The age, possibly the sex, of the gorilla – some of this a result of differences in metabolic rate.
- Whether a lesion is infected or subjected to movement/trauma, which can delay 'normal' healing.

BOX 4.6 Factors Affecting Healing

- Organ/tissue affected
- Age
- Health
- Immune status
- Genetics
- Movement of affected parts
- Administration of medication, especially antiinflammatory and nonsteroidal
- Infection

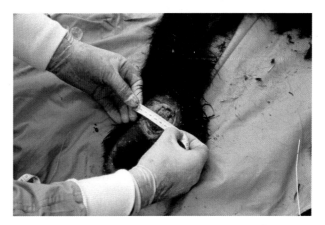

FIGURE 4.6 A granulating wound of at least 6 weeks duration is measured as part of investigation during anaesthesia/immobilisation.

2001; Cooper and Cooper, 2007). Criteria that can be used include those shown in Box 4.5.

HEALING AND REGENERATION

'Regeneration' implies the production of the same tissue, whereas 'repair' can include other tissues that were not necessarily a significant part of the original. Thus, for example, a skin wound in a gorilla may heal by 'first intention', in which case the original elements are replaced in toto, or it may 'granulate' and be characterised by fibrosis and scarring. Some of the factors affecting such healing in primates are shown in Box 4.6.

Pronounced granulation will impede the migration of epithelial cells across the wound surface and thereby retard full wound healing. Such was the case with

'Inyuma' (Fig. 4.6), a female gorilla in the Sabinyo Group in Rwanda who had a chronic lesion on her foot (see also elsewhere). No one cause of death was identified and on the postmortem report it was given as 'multifactorial'.

SEPSIS

Sepsis and septicaemia are considered to be synonymous in this book. Little appears to have been published specifically about this syndrome in gorillas, but it is alluded to, directly or indirectly, in case reports relating to infectious diseases in these species. Fossey (1983) was one of the first to record a probable case, 'Kweli', who 'died of subacute pulmonary congestion, edema, and microfocal hemorrhages compatible with heart failure associated with bacterial septicemia'.

Elsewhere in this book (Parts I and II) references can be found to 'Pucker' and 'Coco' who both died of a Gram-negative septicaemia and shock, as well as severe cystic dysplasia of the thymus (Krueger et al., 1980). Lowenstine (1990) recorded two animals dying of sepsis secondary to trauma.

Sepsis in humans and some other species is usually secondary to infection and characterised by dysregulation of inflammation, the immune system and haemostasis. The outcome can be multiorgan dysfunction and disseminated intravascular coagulation.

SNARES

In her notes from Rwanda on 16 March 1994 Amy Bevis, then working at the Karisoke Research Center, wrote (personal communication): *We find anywhere from 10 to 90 traps a week. The poachers are after buffalo, hyrax and antelopes. Every year four or five gorillas get caught in poachers' traps and two or three die as a result.*

Snares continue to be a major cause of injury and death in gorillas, both in Central and West Africa. These are set by local people to catch ungulates, such as antelope (Fig. 4.7), but from time to time they trap a gorilla (Fossey, 1983, Cooper and Cooper, 2007, 2013). They were the main reason for the establishment of the VVC in Rwanda in 1986 (Fig. 4.8). Such is the frequency of snare-associated injuries that sometimes habituated gorillas are named after their disability. One of the best known – and greatly loved by rangers and visitors alike – was 'Karema' (Congolese Swahili – 'handicapped'), who had not only lost his left hand, probably as a result of a snare, but (following the death of his mother) was cared for when young by his father. He was killed and butchered, probably by rebel soldiers, at the age of 18.

Snares are made of a variety of materials including sisal, nylon rope, wire and local fibrous vegetation (Fig. 4.9). Nylon rope appears to be particularly abrasive to the skin while multistrand wire is more likely at an early stage to cut into underlying tissues.

Snares are usually easily detached from the ground or undergrowth by an entrapped gorilla. The trapped animal is often able to break the wire or rope from the sapling or bamboo pole to which it is attached, but in doing so pulls the noose tighter around its limb. Surprisingly, the ability of gorillas to extricate themselves (or their limbs) from snares is very limited; attempts to do so by pulling usually result in a tightening of the constriction and exacerbation of tissue damage. The late James Foster suggested that

FIGURE 4.8 Auscultation following removal of a snare from an immobilised young mountain gorilla. The whole procedure took only 4 minutes. Vigilance was essential as the group's silverback was threatening.

FIGURE 4.7 The author and a tracker detect a snare, set on a track to catch duiker, in Rwanda.

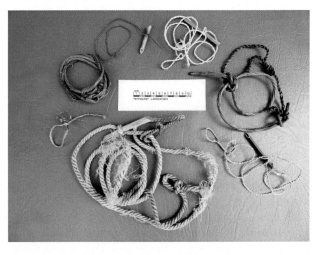

FIGURE 4.9 Types of snare used in the Virungas.

training habituated gorillas to remove snares might significantly reduce the toll of injury, suffering and death that these devices cause. The apparent inability to remove a snare is interesting in view of the comment of Hockings (2015) that chimpanzees, bonobos and gorillas show 'snare awareness'. Some animals deactivate snares and even remove them from the limbs of other animals. For example, a female bonobo at Wamba, DRC was observed examining and trying to remove a snare from the finger of a conspecific, and succeeded in removing the piece of wood from the metallic wire (Tokuyama et al., 2012).

Sometimes snares can be removed from gorillas by humans by hand, not requiring immobilisation of the animal. Such was the case, for example, with a 4-year-old animal in the DRC (then Zaire) September 1994. The youngster had been trapped by a snare attached to a tree; it had circled the tree several times and become firmly held. It was a relatively easy task to restrain the gorilla by hand and remove the offending *mtego*. The wound was minor and needed no specific treatment. Snares cause direct physical injury and risk of serious secondary infection. There is also the question of stress and the animal's welfare. Nevertheless, there has apparently been little or no proper research on snare injuries in gorillas. Areas that need study include the types of tissue damage that different snares cause; which if any are most likely to become infected; how snare damage progresses in terms of both local and systemic pathology; and the extent to which lesions differ depending upon the species, subspecies, age and sex of the gorilla.

Studies on snares in some other countries (see, e.g., in England and Wales: http://www.antisnaring.org.uk/snares/snare-laws/defra-snares-code-of-practice) have concentrated on determining the humaneness of snares but often contain information that might be applied to research on snare injuries in animals in Africa.

A detailed, graphic description of snare injuries detected postmortem in an unnamed male gorilla was included in the necropsy reports in Fossey (1983): 'There was a gangrenous bone infection of lower left leg that had spread from the ankle, where the trap wire was still embedded, into the flesh. The wounds were at least 4 months old and had caused the loss of all toes, resulting in a contracted, hairless stump and extreme atrophy of the entire extremity'. A second case ('Lee') was accompanied by a sketch showing abnormalities that were possibly related to the animal's left leg having been damaged by a snare injury. Ankylosis was thought to have developed in the left knee joint, which the gorilla had used as a substitute foot because he kept his left lower leg in a flexed position to avoid contact between the injured foot and the ground.

The effect of a snare and the severity of any injuries it causes appear to depend upon (1) the *type of material* involved, (2) *where* the snare is attached to the gorilla,

(3) *how tightly* it is attached and (4) *how long* it has been in place. However, as explained above, these personal observations (JEC, Ian Redmond) need proper verification.

Initially a snare creates tension on the tissues, which increases as the snare tightens, acting as a tourniquet, reducing or completely cutting off the blood supply to the distal end of the limb. Partial or total ischaemia is the result. The apparent sequence of events in nonfatal cases is: Ischaemia … necrosis … sometimes gangrene … infection … sloughing … impaired locomotion/prehension.

The wounds can be deep and secondary infection and dehydration often supervene. If a snare is attached to the gorilla's wrist or ankle, the hand or foot will eventually become necrotic and, if untreated, slough as the wire or rope passes between the carpal or tarsal bones; skin then grows over the stump and many such amputee gorillas survive and live near normal lives with such a disability. Dislocations sometimes also occur. Occasionally a snare becomes enclosed in granulation tissue. James Foster (personal communication) observed this in a habituated silverback in Rwanda in 1995 but gave no further details; it would be interesting to know whether the snare in question was made of wire, rope, synthetic twine or vegetation.

If the snare is around the long bones of the arm or leg, the snare cannot cut through the bone so the source of trauma (and usually associated infection) remains and the gorilla dies a slow, undoubtedly painful, death. Snare injuries are clearly very painful from the outset: an afflicted youngster being carried by its mother may whimper as it changes position.

Snare injuries from which a gorilla has survived will have a permanent effect. For example, mountain gorilla 'Gashunguri' had a snare successfully removed from a hand but, following this, started to feed with the hand clenched while using her knuckles for walking. She also showed a nervous disposition when the veterinarian appeared: this was attributed to the fact that she was followed for a long time when carrying the snare prior to successful immobilisation. Byrne and Stokes (2002) described how chimpanzees and mountain gorillas in Uganda have adapted their feeding techniques to their disabilities. Some animals not only survive the loss of part of a limb but manage competently (see Fig. 4.10).

It is not just the snare that causes tissue damage. In 1995 Jonathan Sleeman examined a 6-week-old infant mountain gorilla that had (apparently) a snare wound affecting two digits; there were deep wounds. The mother was immobilised so that the infant could be treated. No snare was present but the animal's hair had become twisted, cut deep into the tissues and produced a strangulating lesion (Cooper and Cooper, 1995).

Following successful sloughing of an extremity, granulation tissue usually covers the bone. Norton (1990)

FIGURE 4.11 Thoracic vertebra of a gorilla. There is a piece of steel embedded in the body. *Image courtesy of Marc Nussbaumer.*

FIGURE 4.10 Mukono, a sub-adult Eastern lowland gorilla, lost his hand in a snare as an infant, but copes well. Kahuzi-Biega NP, DRC. *Image courtesy of Ian Redmond.*

described the healed stump that replaced the right hand of 'Ndumi', who had earlier been 'tangled with a poacher's wire snare'. Two other animals in that group also had injuries associated with snares.

Quite apart from local and systemic infection, it is assumed that tetanus may be a sequel to snare (and other) injuries in gorillas because of the anaerobic nature of ischaemic lesions. The administration of tetanus toxoid may be wise — and should also be made available for staff working in gorilla habitat (see, e.g., Stoinski et al., 2008).

FIREARMS, EXPOSURE TO BLASTS, EXPLOSIONS AND UNEXPLODED ORDNANCE

Most firearm injuries to free-living gorillas are caused by rifles. Some animals are killed by a single shot through the heart, probably by professional killers (poachers or military) — 'Luwawa' in the Virungas in 1993, for example (Cooper and Cooper, 1995). Even in captivity a gorilla may suffer firearm injuries. In 1997, in Johannesburg Zoo, South Africa, gorilla 'Max' was shot twice (but not fatally) by a criminal who had jumped into his enclosure (Phillip Tobias, personal communication).

Unfortunately, findings of postmortem investigation of gorillas that have been killed by shooting are rarely detailed. For example, accessible information about

'Rugabo' and 'Luwawa', both found dead in Zaire (DRC) in 1995 stated only that they had died of 'gunshot'. As with other traumatic deaths (see earlier), detailed descriptions of the injuries in such cases are most important, especially if legal action is possible.

In forensic cases an important part of investigation of gunshot wounds is to ascertain the type of weapon that was used. The pathologist must be able to describe and investigate correctly, including differentiating between entry and exit gunshot wounds. Radiography is important in investigation of the wound and in recovering fragments from tissues. The pathologist must carefully remove, handle and store all relevant material. Plastic forceps should be used to remove bullets before gently rinsing and drying them, to avoid oxidation, and wrapping them individually in soft material in a firm container.

Gorillas in war-torn areas are also vulnerable to hand grenades or blasts from ordnance.

In Rwanda, following the 1994 genocide, humans and cattle were injured by unexploded mines and there were indications that gorillas might also have been injured in this way — for example, the lone silverback 'Mikono'. The death of 'Mikono' was reported at the time, by the International Gorilla Conservation Programme (IGCP), to be the result of his stepping on a land mine. A planned postmortem examination by the author (JEC) was not permitted for political reasons and so this scenario could neither be confirmed nor refuted.

An example of fatal injury to a gorilla following a blast is illustrated by the thoracic vertebra depicted in Fig. 4.11. This is one of a number of images of the case kindly sent by Marc Nussbaumer from the Institute of Anatomy Universität Bern, Switzerland. While working on a gorilla skeleton, the taxidermist found some metal (steel) shrapnel in this vertebra and, implanted in his skull, several pieces of lead. The assumption is that the gorilla was first wounded by an explosive device (steel is not used in guns as it damages the barrel) and then dispatched with a shotgun.

Exposure to blasts may cause a gorilla's body to be thrown about and there can be both external and internal damage and probably concussion. In humans, there can be pressure wave damage, especially to CNS, sometimes burns and psychological effects (Cooper, 1996; Cooper and Cooper, 2007).

Injuries in gorillas can be the result of other weapons. Schaller (1964) put 'general causes of mortality' into three categories and, under 'Injuries', listed predation by humans (snares, pitfalls, nets, guns, spears) — all used in hunting.

In Africa, machetes (cutlasses, *pangas*), spears and arrows are still often used for hunting, especially poaching. Machetes usually produced a characteristic 'chop wound' (DiMaio and DiMaio, 2001). The features of an arrow or spear attack are characteristic skin wounds (Fig. 4.12). These were clear features in four mountain gorillas examined postmortem by the author (JEC) in Bwindi (Cooper, 1995c). Bite wounds characteristic of a canid were also present; the involvement of dogs was later confirmed.

Fossey (1983) described injuries in the world-famous 'Digit', who was killed by poachers in 1977: 'External Examination. Five spear wounds, any one of which could have been fatal, into ventral and dorsal body surfaces; decapitated and hands hacked off'. Part II: A Catalogue of Preserved Materials lists a gorilla's skeleton in the Institute for Anatomy and Cell Biology, Greifswald University, Germany, with an arrowhead embedded in its left nuchal crest.

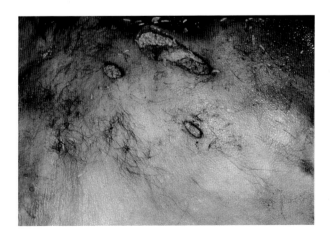

FIGURE 4.12 The skull of this badly decomposed mountain gorilla (note maggots), one of four killed by poachers, shows two types of skin wound, characteristic of those caused by spears and arrows.

PREDATION

There is a long history of claims that leopards might prey on gorillas. Dixson (1981) includes in his book a plate featuring a gorilla battling with a leopard, reproduced from Brehm's *Life of Animals* (1895). More recently Tobias (1961), in describing the work of the South African Gorilla Research Unit in Uganda, was explicit that a major threat to mountain gorillas in Uganda was predation by leopards. Schaller (1963, 1964) described his examinations of faeces for evidence of 'bones, black hair, or other signs to indicate that the cat had preyed upon gorillas'. He found only the remains of duiker and hyrax but later discovered evidence that leopards do on occasion prey on and kill gorillas.

EXPOSURE TO TEMPERATURE CHANGES

The effects of hyperthermia and hypothermia on individual animals are discussed in Table 4.1. There are broader considerations also. Concern is increasingly being expressed about environmental changes, such as depletion of the ozone layer and emission of 'greenhouse gases' with resultant global warming and the adverse effects that this might have on the health of humans, other animals and plants (Cooper, 2010; Jonsson and Reid, 2000; Summers, 2009).

Temperature changes correlated with climate and weather have long been considered to be important in population dynamics. Charles Darwin (1859) wrote: 'Climate plays an important part in determining the average number of species, and periodical seasons of extreme drought or cold, I believe to be the most effective of all checks'. More specific interest in climate change goes back over a century, when many papers on the topic were published in Europe, especially Germany. The subject even formed the basis of a lecture to the Linnean Society of London during the first year of the Second World War (Simpson, 1940). At the present time, the diverse effects of a rise in global temperature are the subject of much concern and some concerted international action.

STRESS AND STRESSORS

Although many of the adverse factors covered in this chapter may serve as stressors to gorillas, stress is primarily covered in Chapter 16: Endocrinological and Associated Conditions.

The adverse effects of capture, transportation and captivity are referred to under Behavioural pathology in Chapter 8: Nonspecific Pathology.

Chapter 5

Methods of Investigation – Observation, Clinical Examination and Health Monitoring

John E. Cooper

Observe, record, tabulate, communicate. Use your five senses. Learn to see, learn to hear, learn to feel, learn to smell, and know that by practice alone you can become expert.

Sir William Osler

INTRODUCTION

In this chapter clinical observation, examination and health monitoring of gorillas are discussed, with particular emphasis on the need to integrate them with other areas of investigation, ranging from postmortem examinations to environmental studies and attention to human health. The role of pathology – and, indeed, of pathologists – in clinical work is sometimes overlooked.

Mudakikwa et al. (2001) provided a useful background to the programme of clinical medicine, preventive health care and research on mountain gorillas carried out by him and his MGVP colleagues in the Virunga Volcanoes region. Six years later Cranfield and Minnis (2007) described an integrated health approach to the conservation of mountain gorillas and explained that the MGVP health monitoring field programme for mountain gorillas was a syndromic surveillance system (see later) with input by (1) trackers and guides, in English or French, recording abnormalities under seven headings and (2) a veterinarian recording clinical signs. They stressed that their (MGVP) health monitoring system 'uses stringent definitions, thereby ensuring consistent data from the collectors'. If animals are observed with clinical signs, the clinical decision tree (Decision Tree Writing Group, 2006) is used either to plan an intervention or to continue monitoring.

More recently, Kalema-Zikusoka (2013) outlined the integrated programme of CTPH insofar as the health of mountain gorillas is concerned and the implications of these activities for conservation.

SURVEILLANCE OF SYNDROMES

Syndromic surveillance can be defined as 'the tracking of syndromes, groups of symptoms or signs of disease, rather than the tracking of specific diseases'. It can help in the detection of outbreaks of disease and thereby provide an early warning system. An early form of syndromic surveillance was described by Hastings (1988). It is already being widely used in work with mountain gorillas by MGVP (Innocent Rwego, personal communication) as part of the clinical decision tree (Decision Tree Writing Group, 2006). Data sheets are used to record the gorillas seen and not seen together, with any abnormalities in respect of activity, breathing, eating, defaecation, urination, discharges or skin/hair appearance. A global positioning system (GPS) device is used to record the location of the gorillas.

CLINICAL WORK WITH GORILLAS

A brief discussion of clinical techniques follows, with particular reference to the implications insofar as pathology is concerned.

In clinical work with gorillas the author recommends the steps depicted in Box 5.1. A 'triage' system – the assignment of degrees of urgency to decide the order of treatment of wounds or attention to clinical signs – is applicable in all clinical work.

J. E. Cooper & G. Hull: Gorilla Pathology and Health. DOI: http://dx.doi.org/10.1016/B978-0-12-802039-5.00005-6

BOX 5.1 Steps in Clinical Investigation

Observe
↓
Examine
↓
Describe
↓
Investigate
↓
Explore
↓
Interpret and/or diagnose
↓
Treat and/or manage

FIGURE 5.1 A mountain gorilla, unaware of a human observer, lies on its back and thereby allows an unimpaired view of the animal's abdomen — important in health monitoring.

THE IMPORTANCE OF OBSERVATION

Claudius Galen (see Chapter 1: The Genus *Gorilla* — Morphology, Anatomy and the Path to Pathology) always emphasised the importance of clinical observation in medicine. Likewise, the astronomer Galileo Galilei famously prefaced pronouncements about his scientific discoveries with the word 'Observavi ...'. He did not however, at that stage, expound upon his interpretation or theorise about what he had observed. As Richard Owen was to say in the context of gorillas 400 years later, 'The accurate record of facts in natural history is the first duty of the observer; the true deduction of their consequences is his next aim'.

Observational studies are invaluable in clinical diagnosis. They require in-depth knowledge of what constitutes normal activity and behaviour — part of the 'natural history' referred to in Chapter 1: The Genus *Gorilla* — Morphology, Anatomy and the Path to Pathology (see also Some Conclusions at the end of this chapter). Early reports on the behaviour of free-living gorillas were made by people such as Carl Akeley, Walter Baumgärtel, Fred Merfield and Charles Pitman who had a deep interest in natural history, or hunting, or both, but who lacked any formal training in ethology or scientific method. Nevertheless, their meticulous observations were seminal, not least of all because they countered previous claims that gorillas were ferocious beasts. Subsequent field studies by experienced ecologists also encompassed observational studies; as an example, Schaller (1963), who was also an excellent field naturalist, recorded such basic, but important, facts as 'Gorillas sneeze, cough, yawn, hiccup, burp, and scratch in a manner similar to man'.

Before any gorilla, captive or free-living, is examined clinically it should be observed. Clinical signs may be best detected before an animal is even aware of the observer's presence (Fig. 5.1). Dian Fossey drew attention to this in her work with free-living mountain gorillas when she described contact as either 'obscured', when gorillas did not know she was watching them, or 'open' when they were aware of her presence. In respect of the former, she stated 'Obscured contacts were especially valuable in revealing behavior that otherwise would have been inhibited by my presence' (Fossey, 1983). This, as the author (JEC) knows from his own work in Central Africa, includes their hiding and disguising any signs of injury or ill health. Indeed though free-living gorillas may exhibit specific clinical signs, such as nasal discharge, they often only show *nonspecific* evidence of ill health, such as lagging behind their group.

The observations of trackers and rangers can be exemplary. Often (as the author knows to his cost!) such people detect early signs of ill health long before the college-trained veterinarian. Their skills are not just a question of formal instruction. Much depends upon the individual. In a fax (no access to emails or mobile phones in Rwanda in those days!), at the height of the genocide in 1994, the author commented favourably on trackers who had accompanied him to see an injured gorilla in Zaire (now the DRC): 'guides are excellent ...... they know the names of animals and plants and take time to point out things en route'.

The same skills can apply in captivity where observant zoo staff may report abnormal behaviour in a gorilla before it attracts the attention of a veterinarian. However, Don Cousins (personal communication, 2013), who knew many of the earlier, post Second World War, keepers well believes that in those days some gorillas and other apes failed to receive veterinary attention at an early stage because they (the keepers) regarded the animals in their care as 'theirs' and preferred to tend them themselves.

Detection of clinical signs is one thing; recording them properly is another. In a modern 'western' zoo setting, keepers are well trained and able to record relevant observations in a systematic fashion. Such was not always the case, however. Records from British zoos dating back only to the 1960s and 1970s are littered with imprecise terms. For instance, the expression 'off colour' was used regularly to indicate that a gorilla might need veterinary examination. Other descriptions were more helpful (e.g., 'runny nose') but usually lacked additional data such as whether the discharge was serous, mucous or purulent and how many nostrils were affected.

Records of observations on free-living gorillas are, of course, not all written in English. Many are in French and some in African languages, such as (in Central Africa) Swahili, Lingala or Kinyarwanda. Prior to the outbreak of fighting in Rwanda in 1994, laboratory records at the VVC/CVV were in French, Swahili, or English depending upon who was dealing with the material and his/her proficiency in language. So, too, was instructional literature about health monitoring (Fig. 5.2).

Imprecision in reporting can perpetuate poor terminology. An otherwise excellent Fauna and Flora International (FFI) report in 2015 depicted a young gorilla that 'bears the scars of an encounter with a snare' but the open lesions on the animal's knuckles are clearly still healing wounds, not scars.

Checklists (check sheets) of observations have much to commend them. They enable trackers in the field and

keepers in zoos to record findings easily and consistently. They were introduced into zoos in the United Kingdom in the 1970s. Thus, for example, those looking after the famous gorilla 'Guy' (see Appendix 5: Case Studies – Postmortem Investigations and Studies on Health) were able to tick boxes to indicate that the animal had shown 'difficulty in eating' before death but was not (for instance) lame or apparently injured. Some but not all, in situ gorilla projects use checklists. It is important to remember that the largest single cause of mistakes made by medical, dental and veterinary practitioners is human error, commonly caused by memory and attention lapses. The latter are a particular feature when the operator is distracted or under stress – scenarios that are all too common when working with gorillas, live or dead.

It must be clear from the foregoing why an understanding of behaviour of gorillas is relevant to the study of their pathology. Not only can changes in behaviour be important indicators of physical disease but also abnormal behaviour can itself be considered 'pathological' (see Chapter 8: Nonspecific Pathology). Correlation of behaviour and pathology in captive animals, including primates, was largely pioneered by the University of Oklahoma Medical School and Oklahoma City Zoo (Stout et al., 1969).

Observation with the naked eye can be supplemented with the use of field glasses (Fig. 5.3) but Schaller (1963) suggested that some gorillas may consider 'the unblinking stare of binoculars' to be a threat. This may also apply sometimes to a camera lens; there was a report in the British Press in February 2015 of a habituated silverback attacking a tourist's camera. Despite this, photographic images can be invaluable; the book by Norton (1990) provides useful advice on technique.

Camera trap photography can provide information about gorillas, especially those that are nonhabituated. The main aim is usually to detect and identify them – for example, current work on Cross River gorillas in the forested mountains on the Nigeria-Cameroon border – but

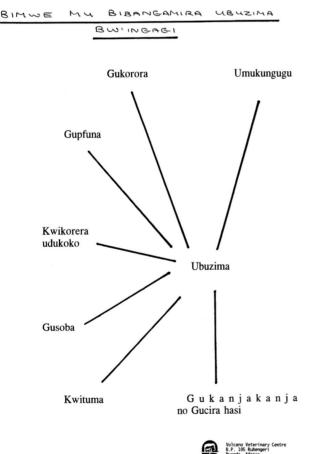

FIGURE 5.2 One of the original instructional leaflets about health monitoring of gorillas, in Kinyarwanda, that was used by the VVC in the months leading up to the genocide in 1994.

FIGURE 5.3 Use of binoculars permits detailed observation of the soles of the feet of a mountain gorilla.

careful analysis of the photographs can sometimes permit detection of skin lesions or show that an animal is not keeping up with members of the group.

Thermal imaging may also have potential, as reportedly used by a BBC team to obtain images of gorillas sleeping in Kahuzi-Biega National Park in the DRC.

INDICATORS OF ILL HEALTH

It is important to realise that gorillas show 'clinical signs' of disease, not 'symptoms' (see Appendix 1: Glossary of Terms). There is a third term that is rarely heard in gorilla circles but ought to be better known. This is 'semeions'. A semeion is what the clinician writes down in his/her notes (Worthington, 2006). As Wilbush (1984) put it 'Symptoms advertise illness ... semeions are the evidentiary data discovered by the doctor when the patient is questioned', adding that 'few clinicians openly dispute the value of a medical history ... many nonetheless appear to regard it as time-consuming and, in a world where time is money, obviously not cost effective'.

RESTRAINT AND HANDLING

Physical restraint (keeping the animal under control) and handling (feeling, manipulating and palpating the animal) are together part of examining a gorilla but both can also cause injuries and be stressful.

Much useful information can, however, be obtained during restraint and handling — and often even more so if the animal is anaesthetised (see later). The importance of weighing gorillas (and at the same time recording measurements, morphometrics), cannot be overemphasised. Understandably, this often has to be opportunistic when the animal is handled for some other reasons but the increasing trend in zoos to include weighbridges or similar devices within cages or enclosures means that regular recordings are possible. Some collections, such as the Zoological Society of London, have an annual 'weigh-in day' when all animals have their weights recorded.

Full value can only be obtained from observation and clinical examination of gorillas if (1) findings are recorded and reported promptly and (2) standard descriptions are given, using accepted medical terminology. Insofar as the latter are concerned, when working in Rwanda, the author compiled a list of basic definitions to help those without a formal veterinary medical background. These were subsequently published (Cooper, 1997b) and some are given in Appendix 1: Glossary of Terms.

It was also stressed that, in addition to using a standard description, as above, it was helpful to quantify the problem — for example, by estimating the size of a lesion in centimetres or by grading a clinical sign, for example,

diarrhoea or coughing, using a scale from 1 (mild) to 3 (severe). An alternative to the above in respect of visible lesions was to draw the affected part (e.g., a leg) and to include on it an outline of the lesion. When a gorilla is being clinically examined, naked eye assessment of lesions should be supplemented by magnification, using at least a simple lens, if not a surgical loupe.

Assessment of a gorilla's 'condition' is not easy on observation alone and needs to be properly calculated when an animal is being handled (see Chapter 11: Alimentary Tract and Associated Organs).

CLINICAL EXAMINATION

Captive gorillas provide unequalled opportunities for study but caution must be exercised in interpreting findings because these may not truly reflect the situation in the wild. For example, the close contact between captive gorillas and their keepers means that bacterial flora may be acquired from the latter (see Chapter 9: Skin and Integument). Diets in captivity may also play a part. If a zoo feeds pellets, vegetables and its own 'gorilla formula' (see Chapter 11: Alimentary Tract and Associated Organs), this may (1) alter the intestinal flora and fauna and (2) have an effect on gut morphology. The subject is ripe for study and very relevant to the release of gorillas to the wild. As a general rule, clinical examination will necessitate immobilisation/anaesthesia of the animal. The exceptions are (1) young gorillas, especially those that are being hand-reared, (2) captive animals of any age that will permit (sometimes following operant conditioning) the performance of palpation, auscultation and other clinical investigative techniques and (3) captive or free-living gorillas that are weak or moribund. Interventions without anaesthesia may also be carried out because the gorilla has respiratory problems (Cranfield and Minnis, 2007). It is in the interests of animal welfare that noninvasive

FIGURE 5.4 A mountain gorilla rests with its mouth open, exposing for observation its teeth and part of its buccal cavity.

diagnostic tests should be used whenever possible. Work with live gorillas, especially in captivity, provides opportunities to evaluate these. Some are hardly 'new' in that they have long been used in human medicine — for example, breath analysis (Wyse et al., 2004). Others have been employed for NHPs — for instance, ultrasonography to evaluate echinococcosis (O'Grady et al., 1982). In the field, simple but well-established ways of assessing the animal remain important. Mucous membrane appearance is one important example (Fig. 5.4).

Support staff can be taught how to use mucous membrane colour in diagnosis — jaundice, of course, and also degrees of anaemia. It may be possible to devise for gorilla workers a system similar to 'FAMACHA', which uses a colour code to enable African farmers of any level of education to evaluate the degree of anaemia in their sheep and goats.

Colour is always important in clinical work. Changes in skin colour of gorillas are not easy to detect but buccal, rectal and other mucous membranes are usually accessible in live animals. Photographs may help but do not replace visual assessment.

ANAESTHESIA

Methods of restraining and anaesthetising NHPs have evolved rapidly over the past 40 years (Cerveny and Sleeman, 2014). Many of the advances can be traced to work in zoos (Vercruysse and Mortelmans, 1978; Ludders et al., 1982) and laboratories (Sainsbury et al., 1989).

Immobilisation and anaesthesia are not, per se, part of pathological examination. However, most anaesthetic agents affect the cardiovascular system, respiration and thermoregulation and therefore have the potential to damage organs and tissues. Emergencies can occur during anaesthesia or surgery and if death results a pathologist is likely to be involved (Cooper and Cooper, 2007). In addition, injection sites (including those involving darts) may be associated with tissue damage and sometimes infection.

Anaesthetic procedures for captive gorillas in a zoological setting have been described by a number of authors — see, for example, the series on zoo and wildlife medicine initiated and largely edited by Murray Fowler (see Miller and Fowler, 2012 and earlier volumes). Inhalation techniques were introduced many years ago (Cook and Clarke, 1985).

Field immobilisation presents its own challenges. Sleeman et al. (2000) discussed this in respect of mountain gorillas and described 26 anaesthetic procedures involving 24 free-living animals. The report by Sleeman et al. was, at the time, the most extensive and comprehensive account of immobilisation of free-living mountain gorillas. It had the added value that it was written in

consultation with VVC/MGVP veterinarians who had carried out the procedures over the preceding years. The paper also recognised the contribution of two veterinarians who had since died. Subsequently, in an as yet unpublished thesis, Jaffe et al. (2012) reported on 37 anaesthetic procedures involving 30 free-living and four orphaned captive mountain gorillas. Twenty-nine of the interventions used medetomidine-ketamine, one medetomidine only and five dexmedetomidine-ketamine.

Ketamine hydrochloride has been the mainstay of gorilla work, both in the wild and in captivity. Recent concerns and controls make it less easy to obtain and to use — and to carry from one country to another (see Chapter 18: Legal Considerations).

MEDICAL AND SURGICAL TREATMENT

Clinical care is not of immediate significance to gorilla pathology other than in respect of the effects such care may have (beneficial or deleterious) on lesions. The extent to which clinical intervention should or should not be undertaken has been the subject of much debate and clinical response decision trees (see earlier) have been developed for some situations (Mayigane, 2013). The role of the diagnostic pathologist in clinical cases is to be part of the team, not just providing backup support. The prompt, correct production and interpretation of laboratory results is a particularly vital constituent of reaching a diagnosis and taking appropriate action.

As examples in these pages illustrate, individual captive and free-living gorillas receive medical attention (Fig. 5.5). Groups of mountain gorillas have been treated for sarcoptic mange and vaccinated against measles; see, for example, Kalema et al. (1998) and Sholley (1989). Immunisation is discussed in Chapter 12: Lymphoreticular and Haemopoietic Systems and Allergic Conditions.

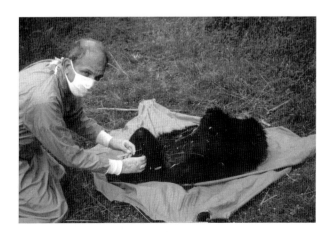

FIGURE 5.5 An immobilised mountain gorilla receives treatment.

The practice of 'evidence-based medicine' (evidence-based veterinary medicine), that is, the use of current best evidence in making decisions about the care of a patient, is increasingly promoted for the treatment of humans and animals. This has led in the (human) medicine field to guidelines for the management of many conditions that also reduce the risk of legal or disciplinary action. The concept is a good one but in work on gorillas there is little statistically significant 'evidence' to support most methods of treatment. The limited number of cases in gorilla work means that there is a tendency to ascribe success to a particular form of therapy on the basis of only a few experiences. The clinician and the pathologist should be ever mindful of the danger of following the maxim 'post hoc, ergo propter hoc' ('after this, therefore, because of this') — in other words, because two events occurred in succession, the former event (in this case, treatment) did not necessarily cause the latter.

Surgery will not be discussed in any detail here, other than to emphasise the importance of (1) knowing the normal anatomy of the species and its relevant organs — a reason for having access to preserved 'reference' material (see Part II: A Catalogue of Preserved Materials) and (2) understanding that pathological changes, both traumatic and infectious, can result from surgical procedures, especially if all does not proceed satisfactorily.

Surgical site infection (SSI) is an important example of an unsatisfactory surgical outcome that has been reported in both captive and free-living gorillas. Common signs are local inflammation (heat, pain, swelling) and a discharge (exudate) from the wound. Culture of any discharge will usually yield a selection of bacteria, many of them secondary contaminants (see Chapter 7: Methods of Investigation — Sampling and Laboratory Tests). It is better to take a sample from freshly expressed discharge, following aseptic cleaning of the site, or from deeper tissues, if necessary, using fine-needle aspiration or biopsying. In the author's experience the bacteria involved in SSI are usually the gorilla's own endogenous organisms, especially skin staphylococci. Biofilms (Clutterbuck et al., 2007) are important in SSI. Standard methods of minimising SSI in human and veterinary medicine are (1) careful preoperative preparation, (2) provision of a 'sterile' theatre environment and (3) appropriate antibiotic prophylaxis. Neither (1) nor (2) is easy under field conditions and it is commendable, probably a reflection of the resistance of the gorilla rather than the surgeon's skills, that so relatively few postoperative infections occur under such circumstances.

Techniques for taking biopsies from 'exotic' and wildlife species were described by Cooper (1994). By definition, a 'biopsy' can only be taken from a live animal, but some of the techniques outlined by Cooper are applicable to sampling dead gorillas — for instance, bone and hair samples.

THE IMPORTANCE OF MULTIDISCIPLINARY COLLABORATION

The close anatomical similarity of *Gorilla* to *Homo* (see Chapter 1: The Genus *Gorilla* — Morphology, Anatomy and the Path to Pathology), coupled sometimes with veterinarians' lack of experience or confidence in dealing with such a species, led at an early stage to involvement of the medical and dental professions in clinical and other work on captive gorillas.

Notwithstanding the comment of the late Gerald Durrell that 'The two most dangerous animals to let loose unsupervised in a zoo are an architect and a veterinary surgeon', veterinarians gradually began to take a more leading role in the care of apes. Involvement of the medical and dental professions continues to prove valuable, sometimes synergistic, in clinical and postmortem work with gorillas and other great apes. This relationship was discussed by Allchurch (2001).

As the concept of 'One Health' (see Chapter 19: Pathology, Health and Conservation — The Way Forward) gains momentum, the value of such collaboration in primate work, in terms of both animal and human health, becomes increasingly apparent (Cooper, 2010c).

HEALTH MONITORING OF GORILLAS

This book is entitled *Pathology and Health* and the inclusion of the second noun is to recognise the role of pathology in promoting the health of these endangered animals.

So what is 'health'? It is derived from old English — 'hal', which meant 'whole' or 'sound', as in the expression 'hale and hearty'. In the Preamble to the Constitution of the World Health Organization in 1946 the following definition was introduced: 'Health is a state of complete physical, mental and social well-being and not merely the absence of disease or infirmity'. More recently, Lebel (2003) suggested that it is better defined as 'a harmonious participation in the resources of the environment, which allows individuals the full play of their functions and aptitudes'.

The word 'monitoring' comes from the Latin *monere* (to warn) and in modern parlance can be defined as 'to warn, check, control or to keep a continuous record'. Health monitoring is therefore essentially a form of preventive medicine, with its most important function being to detect ill health before distinct clinical signs of disease, or death, occur. It implies ongoing evaluation of health status using various investigations and tests.

The current scientific approach to the health monitoring of animals owes its origins to laboratory animal work and it was this background that prompted the author to develop 'health monitoring' programmes for his work with the Durrell Wildlife Conservation Trust on threatened species in the Mascarene Islands in the 1970s and 1980s. His time in Rwanda working with mountain gorillas led to an extension of this concept, with a particular emphasis on the use of noninvasive techniques (Cooper, 1998).

The aim of such monitoring is to ascertain as accurately as possible the health status of animals or groups of animals and to specify (describe and quantify) this. Monitoring also permits the collection and recording of biological data on the individual, group or species. Although there is sometimes overlap, monitoring is different from diagnosis:

- Monitoring — supposedly healthy animals
- Diagnosis — probably diseased animals

The term 'screening' is also sometimes used (see Appendix 1: Glossary of Terms). Following the author's system, this can be considered a form of health monitoring but screening usually implies the testing of an animal or group of animals for a specific organism (e.g., *Shigella*), lesion (e.g., vesicles) or evidence of disease (e.g., raised antibody titres to a known pathogen) rather than the 'broad brush' approach that usually typifies health monitoring of a population of wild animals. The term 'screening' is also sometimes used when an individual animal is monitored but only a limited number of tests or samples is possible. The latter is often the case when individual primates are given a 'wellness examination' in captivity (Weiss, 2015).

That said, interpretations of what health monitoring is — and how it should be practised — differ, even amongst those who work with gorillas. The term is sometimes used rather loosely in documents. Thus, Mudakikwa et al. (2001) defined 'health monitoring' in a more focused way, relating it to their own work on mountain gorillas, stating that 'health monitoring means regular visits to the group to assess health status by observation, as well as collecting and screening biological samples, collected noninvasively'. They further defined 'disease surveillance' as 'the proactive testing for high-risk diseases to which a population is susceptible'. In a document directed at wildlife managers rather than 'health professionals', Gilardi et al. (2015) presented 'Best Practice Guidelines for Health Monitoring and Disease Control in Great Ape Populations'; the content is generally excellent but definitions and terminology are sometimes ambiguous. Other approaches to health monitoring include those presented by Jane Hopper, Gladys Kalema-Zikusoka and Celsus Sente in this book (see Chapter 17: Field Studies in Pathology and Health Monitoring).

WHY IS HEALTH MONITORING IMPORTANT?

Safeguarding human health is an important *raison d'être* of health monitoring (see also Appendix 5: Hazards,

---

BOX 5.2 The Importance of Health Monitoring

1. To provide background data on the individual, group or species.
2. To reduce the risk of introducing potentially pathogenic organisms into other populations of the species or the environment in which it lives.
3. To minimise risks to the individual or species itself if introduced, reintroduced or translocated.
4. To protect the health of:
   a. Other wild (free-living) animals
   b. Domesticated animals
   c. Humans.
5. To satisfy legal and ethical requirements relating to animal health, welfare and conservation (see Chapter 18: Legal Considerations).
6. For reasons of animal welfare.

Including Zoonoses and Chapter 19: Pathology, Health and Conservation: The Way Forward) (see also Box 5.2).

This has been long recognised in the context of captive gorillas and other primates and was certainly an important everyday consideration, especially in respect of protecting local African staff, when the author worked in Rwanda. Cranfield (2004) discussed links between human and mountain gorilla health and reported that in 2000 the MGVP's human health working group had identified strategies that might both reduce the risk of disease (*sic* = pathogen) transmission between humans and gorillas and enhance the health of humans living in close proximity to the gorillas. Their 'employee health programme' focused on park conservation employees as they were the people who had the most frequent contact with the gorillas. Subsequently, Sleeman et al. (2002) and Guerrera et al. (2003) reported on a medical survey of the local human population to determine possible health risks to the mountain gorillas of Bwindi Impenetrable Forest National Park, Uganda. Other work on gorilla—human health is referred to elsewhere in this book.

The animal welfare implications of using minimally invasive health monitoring are also most important. The techniques outlined in this chapter — and regularly used, with modifications, in work on gorillas in Africa — do not require an animal to be captured or even touched.

HEALTH MONITORING TECHNIQUES

This section describes techniques that can prove useful in health monitoring of gorillas, but which can be considered 'minimally invasive' insofar as their impact on the animals is concerned. The term 'minimally invasive' is used in preference to 'noninvasive' since even observation or collection of naturally voided samples can have an adverse effect on the behaviour of an animal. 'Invasive' is described in the Oxford Dictionary as 'any intrusion or invasion of privacy' and is a reminder that in biological terms also it need not involve physical contact.

Substantial scientific information can be obtained from a live free-living gorilla if it is examined 'clinically' and samples are taken. This, however, involves capture and restraint, with the attendant disadvantages and dangers listed earlier. Postmortem examination also yields valuable data – sometimes more than 'clinical' investigation – but, in the case of gorillas, where euthanasia is rarely acceptable, means that a specimen has to be found dead. Despite these caveats, 'minimally invasive' techniques have an essential role to play in health assessment of free-living gorillas.

So, how do we monitor the health of free-living gorillas? See Box 5.3.

Techniques for health monitoring always benefit from reference to texts (books and papers) relating to studies in the wild. Some of these are concerned principally with one taxon (*Gorilla*, *Pan*, etc.) but others are more general in scope. A recent volume by Sterling et al. (2013) provides information based on methods used to study primates over several years, including sampling and zoonoses.

Veterinarians who are working with free-living gorillas in the wild can learn much from primatologists, ecologists and behaviourists. Rangers, trackers and other 'support' staff are equally important as it is often they, not the scientists, who know and understand the animals. To be of optimal value, health monitoring needs to be interdisciplinary.

By definition, health monitoring of free-living gorillas meets the criteria of 'fieldwork' (see Appendix 3: Field Pathology) and therefore reference should be made to published work on the collection and examination of samples under such circumstances (see, e.g., Cooper, 2009a, 2013b; Kalema-Zikusoka and Rubanga, 2013).

The components of health monitoring will only be mentioned briefly here as they are covered in papers referred to above. Analysis of history, records and other data is important and is likely to include study of the population dynamics of the species. For example, a reduction in numbers of animals may indicate an increase in death rate, prompting a search for carcases for postmortem examination. Other indicators may include the failure of young animals to survive or a change in the age structure of a group. Data can be of value retrospectively. For example, following an outbreak of infectious disease, records of movement of animals from one group or locality to another may provide useful clues on how and when pathogens may have spread. Valuable information may also be obtained from other records. In the context of habituated gorillas, if an infectious disease of possible human origin is detected, careful analysis of the records of visits made by tourists, research workers and visitors may permit the source of the outbreak to be traced (Foster, 1993).

Routine observation was described earlier in this chapter. Mudakikwa et al. (2001) listed important clinical signs in gorillas as 'including discharge from body orifices, abnormal respiration rates, injuries, poor hair condition, swollen navel (in newborns), scratching and abnormal manipulation behavior (transport of infants, plant manipulation, grooming)'.

Postmortem and pathological examinations are discussed in Chapter 6: Methods of Investigation – Postmortem Examination and Chapter 7: Methods of Investigation – Sampling and Laboratory Tests.

It is in the detection, collection and examination of samples that health monitoring really comes into its own, with the opportunity to link clinical acumen (based on observation) with sound investigative science (Cooper, 1998).

The reliability of health monitoring depends upon the quality of samples, procedures, facilities and staff, the number/proportion of animals sampled and the prevalence of the organism, antibody or lesion (see Chapter 7: Methods of Investigation – Sampling and Laboratory Tests).

When working with gorillas, the rules in Box 5.4 apply to the collection, handling and processing of samples.

And what types of noninvasive sample or location are likely to provide useful information? See Box 5.5.

Faeces, hair, saliva and night nests are discussed in Chapter 7: Methods of Investigation – Sampling and Laboratory Tests, also urine in Chapter 13: Urinary and Reproductive Systems. A hand lens is an essential companion when examining material in night nests (Fig. 5.6).

Heymann (1999) drew attention, with references to the potential value of shed hair and saliva (adhering to feeding residues), in the discovery of pathogens or other possible indicators of disease. Smiley Evans et al. (2015) recently opened up a new method of monitoring – the detection of viruses on discarded plants from free-living mountain gorillas (Fig. 5.7).

BOX 5.3 Monitoring the Health of Free-Living Gorillas

1. By analysing history, especially clinical, postmortem records, published and unpublished data.
2. By routine observation and clinical examination of live animals.
3. By routine postmortem and pathological examination of dead animals.
4. By routine examination of samples.
5. By 'environmental monitoring'.

BOX 5.4 The Collection, Handling and Processing of Samples From Gorillas

Sampling

- Take any sample that is available, even if at the time it appears inadequate or of poor quality. It may not be easy or possible to obtain another later.
- Collect samples hygienically and use the appropriate container, swab, transport medium or fixative. Seal containers, especially bags, securely to prevent escape of parasites. Follow a standard technique that minimises error at each stage of collection.
- If there is likely to be a substantial delay before processing the specimen, consider carrying out some investigations (e.g., examination of wet preparations of faeces for fragile parasites) in the field. Special equipment, for example, a battery-operated microscope, may be needed for this (see Appendix 3: Field Pathology).
- If there is insufficient material to carry out a range of tests, choose those that are most relevant. Draw up protocols to facilitate this choice.
- Where appropriate, seek help from colleagues with specialist knowledge or facilities.
  Remember that submission to another country of 'recognisable derivatives' from certain protected species may necessitate CITES permits (see Chapter 18: Legal Considerations). Keep careful records, with backup, including photographs of important findings.

BOX 5.5 Noninvasive Health Monitoring – Types of Sample

Faeces	Can provide much useful information on gross and microscopical examination and laboratory analysis
Urine	
Hair	As above
Saliva	As above
Night nests	Has potential in detecting parasites and viruses (see text) and possibly in providing other information
Food remains	
Tracks (footprints)	May yield hairs (and/or ectoparasites), faeces, urine and blood
	If fresh, may provide information on feeding behaviour and dentition and permit collection of saliva. Bitemarks may permit identification (Cooper and Cooper, 2007)
	May reveal pedal injury, lameness, haemorrhage. Like faeces, recent tracks may have a characteristic odour; often pungent if a silverback, 'similar … to those of horses' in other gorillas (Fossey, 1983)

FIGURE 5.6 Close investigation of mountain gorilla faeces in a night nest. The proximity of the observer also permits assessment of odour.

FIGURE 5.7 A mountain gorilla eats plant material. If this is subsequently discarded, it can be retrieved and used for laboratory tests.

SOME CONCLUSIONS

Health monitoring is an integral and essential part of the management of both captive and free-living gorillas. It should be performed routinely and prior to translocation or reintroduction of animals. The rules for health monitoring are summarised in Box 5.6.

BOX 5.6 Rules for Health Monitoring

1. Some monitoring is better than none.
2. Plan carefully — prepare a protocol and follow it.
3. Be prepared to be selective — a cost:benefit analysis is usually necessary.
4. Standardise techniques, following recognised techniques, and have a system of quality control.
5. Collect and collate data systematically.
6. Analyse and interpret results carefully — especially those that are negative.
7. Produce a written 'specification' (profile) for the animal or population in the form of an electronic report and possibly (especially captive specimens) a printed 'passport'.
8. Save material where possible — for retesting and subsequent research (establish a Reference Collection).
9. Supplement preplanned health monitoring programmes with the 'opportunistic' collection of samples whenever this is possible — for example, clinically when an animal has to be handled or moved, postmortem if an animal is found dead.
10. Disseminate results. Publish. Encourage others to use similar health monitoring programmes and promote the development of improved protocols.

Veterinarians and others who are involved in noninvasive health monitoring of free-living gorillas need to able to observe. They must also have a sound knowledge of the natural history of the species. The importance of these complementary traits was elucidated very well in the book *Be a Nature Detective*, by Knight (1968b), published half a century ago. In it the author teaches readers how to learn about the lives of wild animals by observation and, in particular, by examination of their signs, for example, footprints, partly gnawed food items, dropped feathers and regurgitated pellets. In this context, the words of the great scientist Paracelsus are apposite: 'The physician must know the invisible as well as the visible'. This should be the aim too of those engaged in the health assessment of gorillas.

Chapter 6

Methods of Investigation – Postmortem Examination

John E. Cooper

INTRODUCTION

Postmortem examinations are also referred to in this book as 'necropsies' (singular, 'necropsy'). In the past, the word 'autopsy' (Greek *autopsia*, meaning 'I see with my eyes') has generally been reserved in the English-speaking world for the postmortem examination of a human being but in French 'l'autopsie' can be used for both humans and animals and some veterinary pathologists are beginning to use the word 'autopsy' for animals too (Law et al., 2012).

THE NATURE AND ROLE OF THE POSTMORTEM EXAMINATION

In the context of this book necropsy is taken to include not only full postmortem examination of a recently deceased animal but also the gross examination of gorilla remains or derivatives, such as parts of carcasses, organs and skeletal structures. Some of these, particularly bones, may be from animals that died some time previously – material in museum collections, for instance (see later) – when special techniques and precautions are often necessary.

Postmortem examination of gorillas can be performed for various purposes (see Box 6.1).

A necropsy can afford information not only in respect of the general purposes above but, moreover, provide detailed answers to specific questions, such as whether the gorilla had any underlying developmental abnormalities and its reproductive status, together with the provision of data on organ weights and organ:bodyweight ratios.

In gorilla work the usual overriding reason for performing a necropsy, both in zoos and in the wild, is to throw light on why a gorilla has been unwell, or died, and how such knowledge might be used to advantage in respect of the care or management of other animals in the collection or free-living group. The rationale was put well by Sir Thomas Beddoes in 1808 in his comments on the uncertainty of medicine: 'Nor do I see how it should cease, till careful observation of the sick be combined, beyond any example, past or present, with early and exact dissection of the dead' – a sentiment wryly expressed much earlier in London mortuaries in the late Middle Ages by the display of a notice stating that *Hic locus est ubi mors gaudet succurrere vitae* (this is the place where death delights to help the living).

BEFORE EMBARKING ON A NECROPSY

History is important. Guidance regarding the collection of background information is given in Box 6.2.

The finding of a carcase in the field must prompt careful consideration of the action that should be taken. Consider, for example, a scenario in which the dead animal is noted to have fresh blood around one or more body orifices. There are two initial considerations:

1. Might this animal be infected with an organism that is transmissible and dangerous to humans, such as *Bacillus anthracis* or *Ebolavirus*?
2. Could it have been killed using physical means, such as a firearm, by a human (poacher, guerrilla or soldier)?

The first of these necessitates the instigation of strict biosecurity and, whether or not a necropsy is to be carried out, appropriate protection of staff and local people. The second requires a full forensic investigation.

The initial action to be taken is similar in both cases – cordoning off of the area around the carcase followed by adherence to the appropriate protocol.

Other decisions may have to be made at an early stage. For example, if a live baby is still hanging on to the carcase (its mother), should it be removed?

J. E. Cooper & G. Hull: Gorilla Pathology and Health. DOI: http://dx.doi.org/10.1016/B978-0-12-802039-5.00006-8

From the outset, whatever the circumstances, strict biosecurity is vital and the number of people involved must be kept to a minimum.

PERSONNEL AND TRAINING

Who should perform a postmortem examination? Ideally, whether captive or free-living, a gorilla should be necropsied by a specialist veterinary pathologist – for example, a Diplomate of the ECVP or ACVP, preferably one with experience and training in primate disease. The value of involving a specialist pathologist in both clinical and postmortem investigations is emphasised by reference to the training programme for, and expectations of, a Fellow of the Royal College of Pathologists in the United Kingdom. The specialist is expected, amongst other things, to have factual knowledge of general pathology, interpretive skills at both macroscopic and microscopic levels, enough technical knowledge of the processing, sectioning and staining of histological sections and of cytological preparations to be able to interact appropriately with colleagues responsible for such work, understanding of information technology sufficient to produce databases, critical skills for the assessment of published literature and familiarity with health and safety. Mistakes are often made, especially errors in interpretation (and therefore in diagnosis) if the necropsy is performed by a nonveterinarian or by a veterinarian with little experience in pathology.

Often, however, especially in the field, a specialist pathologist is not available and the necropsy has to be carried out by a veterinary clinician, a biologist, a technician or a member of field staff. If properly prepared and aware of their limitations any of these people can perform a satisfactory dissection of a gorilla, *describe* the findings and *collect* samples for laboratory investigation. They should NOT be expected to make a diagnosis.

Postmortem examinations performed by knowledgeable biologists who are not veterinarians may not adhere to 'standard' necropsy protocols. This point has been recognised by some such workers themselves, for example, Sikes (1969) in her studies on African elephants: 'This procedure is described in some detail here, as it differs in certain important parts from that generally used in standard veterinary practice ...'.

Dian Fossey too was aware of her limitations (she was neither a veterinarian nor a pathologist) and understood the need, in the absence of specialist assistance, for her to carry out necropsies of gorillas to as high a standard as possible. She also insisted on such discipline in respect of her staff and students. In a letter to Joan and Alan Root in 1975 (shown to the author by the late Joan Root in 2004 and reproduced here with both Roots' permission), Fossey stated:

'Not even a brain capacity was done. I can't believe such ignorance, such a waste when the facilities were available'

'These damn kids made no attempt to preserve any tissues or even get the body on to the University of Butare to (sic) the Veterinary School'

'What makes me so bloody mad is that slobby students seem to feel that they can simply take over and waste the value of a gorilla's carcase.'

Although a basic necropsy can, with tuition, be performed by a person who is not medically or veterinary trained, lesions can easily be missed. An inexperienced person may, for example, not know how to locate certain organs, such as the pancreas and lymph nodes, or not be able to dissect a heart so that all four chambers, valves and great vessels are properly inspected. Certain postmortem techniques are complex and require special knowledge. Some relate to specific organ systems (e.g., the correct dissection and examination of joints, the central nervous system, or the middle ear), others relate to scenarios that need a different, sometimes unorthodox, approach, such as assessment of multiple traumatic injuries, including those caused by ballistics, correct sampling for toxicology, systematic investigation of sudden death or appropriate investigation of a badly decomposed carcase (Cooper and Cooper, 2013).

A word should be said about necropsies on gorillas performed by medical (as opposed to veterinary) pathologists. These have generally been of high standard and, on a number of occasions, introduced procedures that, while standard in human autopsies, are often overlooked or inadequately followed by veterinary pathologists – extensive and systematic measuring and weighing of organs, for instance. An early example was the dissection of a mountain gorilla by Galloway and Allbrook (1959); they not only took many measurements but were able to comment (Galloway had never necropsied a gorilla before) 'Liver ... differed from man's in having well-marked notches about 2 inches deep in the anterior edges of the right lobe (lateral to the gall bladder) and in the left lobe ... the lungs looked more normal than most human lungs'. The value and sometimes disadvantages of involving the medical profession are alluded to below and elsewhere in the book.

Schaller's (1963) description and measurements of the deformed arm bones of a mountain gorilla provided a pertinent reminder of the importance of simple morphometrics – easily performed by a competent biologist such as Schaller, even when working in the field.

It is most important that morphometric data on gorillas are collected and collated. Unfortunately, for obvious reasons, the extensive databases on organ weights that were established in the 1970s and 1980s for some NHPs such as laboratory marmosets still do not exist for *Gorilla*. It is all the more disappointing, therefore, that weights and measurements are not comprehensively recorded in all gorilla necropsies, including some in zoos. Such data are of great value in research. As the physicist William Thomson, 1st Baron Kelvin, said in the 19th century: 'When you can measure what you are speaking about, and express it in numbers, you know something about it'.

Although opinions amongst biologists differ, the author advocates sticking to 'weight' (a *force)*, not 'mass' (an *absolute quantity of matter* that cannot be measured directly with scales or spring balance). Both solid organs and viscera should be measured, the latter in terms of length. Such an approach has long proved useful in humans and in domesticated animals (see, e.g., Saxena et al., 1998) but has not generally been applied in detail to wildlife. Measuring takes time but, in the author's experience, a task that can easily be delegated to junior staff, especially in the field.

Both weights and measurements should be metric and the equipment used should be accurate and regularly calibrated; as the Old Testament (Proverbs) reminds us: 'Dishonest scales are an abomination to the Lord, But a just weight is His delight'.

Postmortem work in the field needs special skills and traits – see Appendix 3: Field Pathology. In particular, it is not for the squeamish. As Cooper and Cooper (2007) put it, the investigator requires 'a spirit of enthusiasm and curiosity, the hallmarks of a detective, which rapidly numbs the senses to foetid smells, maggot-infested viscera or an uninviting pile of arachnid-ridden crumbling vertebrae'. In such work *inquisitiveness* is essential. In a recent paper in the *Journal of the Royal Society of Medicine*, Schattner (2015) lamented that a sense of curiosity is not evaluated amongst (British) medical school candidates – nor, one assumes,

amongst their veterinary counterparts. Curiosity is associated with a desire to learn more and implies attention to detail and heightened powers of observation – both very valuable in a pathologist dealing with iconic species (gorillas) about which so little is still known.

An additional asset in anyone working with gorillas especially performing postmortem or clinical work, is 'aequanimitas'. This term is emblazoned on the shield of The Johns Hopkins University Department of Medicine, United States. It means imperturbability, calmness amid storm and was regarded by the great Canadian physician Sir William Osler as the premier quality of a good clinician. In Africa it helps when dealing with people and even more so if facing a charging gorilla (where the time-honoured advice is 'calmly sink to a crouching position')!

And what about other staff who are involved in work on a dead gorilla? Rangers, trackers, and other field staff remain the cornerstone of studies on free-living animals. At times, in addition to their prime task of locating and observing, they may have to perform at least a basic necropsy on a dead gorilla and/or take samples from it. Even when this contingency does not arise, they are important members of the team whenever a necropsy is being performed by a veterinarian. It is essential, therefore, that Dian Fossey's example is followed and such 'support staff' receive some training in (1) dissection techniques – see later, (2) recognition of organs, both normal and abnormal, (3) sampling techniques and (4) health and safety. Insofar as the first two are concerned, local Africans are often already competent in identifying organs and pathology because they keep (and possibly kill) animals at home and are *au fait* with butchering and apportioning appropriate organs for consumption to different groups of the community.

When circumstances permit, it can be good public relations to let interested parties observe a necropsy on a gorilla. This can be achieved in modern postmortem rooms in Europe where a 'viewing room', with observers protected by glass, is usually standard. In Africa it may be less easy to allow such viewing without exposing observers to pathogens but is important to attempt it so that local people feel some 'ownership' of gorillas that have died in their locality and understand the value of a necropsy.

INTERDISCIPLINARY STUDIES

As was pointed out in Chapter 2: The Growth of Studies on Primate Pathology, and mentioned above, medical and dental pathologists have often contributed significantly to our understanding of diseases in gorillas. Such collaboration should be fostered (see Appendix 5: Case Studies – Museums and Zoological Collections, regarding work on gorilla 'Guy').

The anatomical similarity of gorillas to humans, coupled sometimes with a lack of experience or confidence in dealing with such a species on the part of veterinarians, led at an early stage to involvement of the medical and dental professions in clinical and other work on captive gorillas. Indeed, at some zoos there was a tendency to consult a 'medic' before calling a vet (Kirk, 1940).

Veterinarians gradually began to take a more leading role but involvement of these and other disciplines continues to prove valuable, sometimes synergistic, in clinical and postmortem work with gorillas and other great apes. This relationship was discussed by Allchurch (2001) and there have been many examples insofar as gorilla studies are concerned. The legal implications are briefly mentioned in Chapter 18: Legal Considerations.

Necropsies were in the past sometimes carried out in a hospital mortuary — less so nowadays because of concerns about health and safety. One drawback of the practice, relevant to entries in Part II: A Catalogue of Preserved Materials, was that hospital pathologists did not always want to store gorilla tissues in a (human) medical environment. They often took samples for histology and then discarded the rest.

King et al. (2007) made the point that students (or assistants) should not, as a general rule, be left to perform a necropsy on their own. In the absence of an experienced pathologist, they may miss a lesion, or fail to examine an organ properly. It is far better, whenever possible, for the pathologist to remain with the group and supervise. This counsel of perfection cannot usually be applied in the field, however; there the postmortem examination may have to be performed by junior staff and, at best, samples taken by them for submission to the pathologist.

Osteologists, archaeologists and palaeontologists can also contribute to the examination of gorilla remains, and their praises in terms of investigative precision were sung at the end of Chapter 1: The Genus *Gorilla* — Morphology, Anatomy and the Path to Pathology. However, it must be borne in mind that methods, terminology and interpretation of findings by persons in those disciplines may differ markedly from those followed by those with a human, veterinary or dental background (Cooper and Cooper, 2007). An indication of this is seen in how often the word 'pathologies' is used by osteologists, archaeologists and palaeontologists; it is not traditional medical terminology — and in this author's opinion an abuse of the Greek roots of the word!

Having said this, it is time for veterinary pathologists, including those working with gorillas, to avail themselves of the new techniques that other disciplines are using in studies on long-dead humans that might be applied to skeletal and other remains from gorillas. As Cooper and Cooper (2007) demonstrated, there is extensive literature on the pathological investigation of Egyptian mummies and other material, most of which dates back thousands of years. Amongst many new disciplines that may be relevant to *Gorilla* is archaeogenetics which is primarily concerned with studies of past and present genetic variations, with particular reference to the extraction and analysis of DNA.

ACCESS TO MATERIAL FOR POSTMORTEM EXAMINATION

Gorillas or their derivatives are likely to come from zoological collections (anywhere in the world), from rescue/rehabilitation centres (in Africa), from confiscations (e.g., of organs destined for bushmeat — anywhere in the world) and from the wild (habituated or nonhabituated, found dead). They may (should) be delivered to the pathologist properly wrapped following standard health and safety precautions but this is not always the case in the field (Fig. 6.1).

Sometimes a gorilla or its parts have been buried and need to be exhumed. Exhumation for autopsy of a human body is rare, even in forensic practice, and entails considerable bureaucracy in order to obtain authorisation (Gresham and Turner, 1979). The situation is rather different in work with gorillas. Permission may be required (see Chapter 18: Legal Considerations) but growing interest in examining the carcases, or the remains, of animals

FIGURE 6.1 An infant mountain gorilla brought in from the forest as it was found and, here, presented for postmortem examination.

that have been buried in the past has led to the disinterring of substantial numbers of gorillas in recent years. This work presents some particular challenges, as discussed by Ogeto Mwebi (see Appendix CA2: Retrieval, Preparation and Storage of Skeletal and Other Material).

POSTMORTEM PROCEDURES

A discussion of postmortem techniques follows but the reader should also refer to Appendix 2: Protocols and Reports.

Postmortem examination must be carried out systematically and thoroughly (see Box 6.3), with appropriate attention to the health and safety of those involved. It is important at the outset to ascertain the background history including any clinical data and in the field to request appropriate protection if Ebola (EVD) or Marburg disease is suspected.

Precautions that are necessary during the necropsy are shown in Box 6.4.

Health hazards associated with the clinical or postmortem examination of primates have been recognised for some decades. They can be noninfectious or infectious (see Chapter 18: Legal Considerations and Appendix 4: Hazards, Including Zoonoses) which summarise these and provide an introduction to risk assessment. The susceptibility of free-living gorillas to human pathogens and vice versa has prompted the concept of 'integrated health' (Cranfield and Minnis, 2007; Kalema-Zikusoka, 2013) and of carefully evaluating the benefit and dangers of gorilla research (Cranfield, 2008).

The main categories of hazard when performing a necropsy on a gorilla are chemical (e.g., formalin), physical (e.g., slippery floors, sharp instruments) and infectious (e.g., salmonellae). Appropriate hygienic precautions

should be taken and the following rules adhered to at all times:

1. No smoking, eating or drinking.
2. Follow a strict 'clean and dirty' system to avoid contamination of papers or writing materials.
3. Always wear protective clothing
4. Avoid generating splashes by using gently running water and handling all tissues gently.
5. Place all discarded tissues and discarded 'sharps' in appropriately marked containers (in the EU, yellow 'clinical waste' bags or containers).

Similar hazards are present during 'in the field' necropsy procedures (see Appendix 3: Field Pathology) and are best avoided in much the same way.

BOX 6.3 Stages in a Necropsy

- Obtain adequate assistance (see personnel).
- Ensure health and safety and introduce/regularly review risk assessments.
- Perform the postmortem examination, including record-keeping.
- Take laboratory samples.
- Retain material and store carcase.
- Clear and clean the facilities or (free-living animals) the site, including further attention to health and safety.
- Collate records.
- Receive and analyse laboratory findings.
- Produce report.
- Retain all forms, other records (e.g., tapes), specimens and samples.

BOX 6.4 Precautions That are Necessary During the Necropsy

- The table, tray or postmortem area should be as clean as possible (including freedom from insecticides or disinfectants), before the examination starts.
- Each carcase or separate derivative should have a distinct reference number and be marked as such.
- If not constructed of material which is easily cleaned, the necropsy table, tray or postmortem area should be covered with plastic or similar sheeting that can be readily changed or discarded.
- The surface should be washed and cleaned after each examination.
- All instruments and protective clothing should be changed between examinations. If necessary, several sets of instruments should be available.
- The necropsy technique should minimise contamination, especially spillage of fluids from one organ to another. Tying of loops of intestines or other organs with coloured string will help to reduce this risk and can help subsequently in locating the organs.
- All samples for toxicological assay (blood, urine, etc.) should be placed in separate containers (this is different from histological or electron microscopical examination where tissues can, if really necessary, be placed in the same container of fixative).
- The remains of each animal should be placed in separate storage bags at the end of the postmortem examination.
- The side (not the lid) of each sample container should be marked with the relevant information and a useful extra precaution, especially when using alcoholic fixatives, is to place a paper label, with data inscribed in pencil, inside the specimen jar.
- Those involved in the necropsy work should not handle mobile phones. A 'non-touch' speaker can be used or a member of the team can be designated 'clean' for this purpose.

There are numerous ways of carrying out a postmortem examination and specific methods have been described in the literature for domesticated animals. In addition, in recent years recommended approaches for wild species have been published, in some cases promulgated as part of the Action Plan of the appropriate Taxon Advisory Group (in Australasia, e.g., www.arazpa.org.au).

Access to appropriate literature is an essential prerequisite to good necropsy technique. A much loved publication in veterinary circles, *The Necropsy Book* (King et al., 2007) is recommended. It is primarily intended for the practising veterinarian and veterinary pathologist working with domesticated species but its no-nonsense, down-to-earth, approach (very much in the tradition of Galen and Vesalius − see Chapter 1: The Genus *Gorilla*. Morphology, Anatomy and the Path to Pathology) has long ensured it a place on the shelves of wildlife pathologists. Features that appeal to the present author (JEC) are its specimen necropsy forms, the line drawings of specific locations of lesions and its 'Greek and Latin combining form word list'.

There is, surprisingly, relatively little written guidance to the dissection of gorillas or great apes. Terio and Kinsel (2008) provided a useful, step-by-step, pictorial guide but the material lacks detailed text and applies generally to NHPs. Methods for the postmortem examination of NHPs, including apes, have however been described and discussed by pathologists, including the present author, at conferences and workshops.

When necropsying gorillas and other great apes the author has found it useful to refer to (and carry to the necropsy site) a relevant human medical book. A particularly helpful one is *Post-Mortem Procedures* by Gresham and Turner (1979). Although published over 30 years ago and containing only black and white photos, this book is a mine of valuable information, not least because the authorship combined the knowledge and training of a highly respected pathologist with the practical experience of a senior postmortem technician. Some other human texts, especially those concerned with forensics, also include useful practical hints on dissection or handling of organs in order to detect lesions or to answer questions about the circumstances of death (see, e.g., Rutty, 2004).

Interestingly, not surprisingly to the present author (JEC), some of the best, practical guidance is to be found in very early works on *Homo*. Thus, the first (1858) edition of *Gray's Anatomy*, still available in facsimile, included in bold font, before anatomical features were discussed, advice on how best to get access to the organ in question. The following is an example:

The Spinal Cord and Its Membranes

Dissection. *To dissect the cord and its membranes, it will be necessary to lay open the whole length of the spinal canal. For this purpose, the muscles must be separated from the vertebral grooves, so as to expose the spinous processes and laminae to the vertebrae; and the latter must be sawn through on each side, close to the roots of the transverse processes, from the third or fourth cervical vertebrae, above, to the sacrum below. The vertebral arches having been displaced, by means of a chisel, and the separate fragments removed, the dura matter will be exposed, covered by a plexus of veins and a quality of loose areolar tissue, often infiltrated with a serous fluid. The arches of the upper vertebrae are best divided by means of a strong pair of cutting bone-forceps.*

This same procedure, with some modification, can be applied to the examination of a gorilla.

When needing help with specific features of gorilla anatomy, one should not forget that some of the earlier publications about gorillas, referred to in Chapter 1, The Genus *Gorilla* − Morphology, Anatomy and the Path to Pathology, provide excellent descriptions of structures and, with the aid of illustrations (often line-drawings), provide some assistance in how to locate them and explore them. The book by Swindler and Wood (1982), essentially an atlas of the gross anatomy of the baboon, chimpanzee and human, is also relevant.

A full gross postmortem examination of a gorilla can be divided into three stages: (1) external examination and sampling, (2) internal examination (opening the carcase) and sampling, and (3) examination of organs and tissues (dissecting organs in situ or removing them or their contents for laboratory tests).

As an alternative to a full postmortem examination it is possible to carry out a 'partial necropsy' on a gorilla. Galloway and Allbrook (1959), retrieving a partly autolysed dead gorilla from the field (see later), were perhaps the first to explain the reasons for, and the limitations of, what they called 'a limited postmortem'.

Even a gorilla from a zoo may need such a 'limited' necropsy if, for instance, the remains are required for display in a museum or other similar purposes; in this case a 'cosmetic' postmortem examination must be carried out in order to obtain the necessary information and material while inflicting minimum external damage. This necessitates skill and experience. In some cases, an endoscope may prove useful (Cooper and Cooper, 2007; 2013).

In human medicine the performance of autopsies is no longer necessarily considered the 'gold standard' − or even necessarily essential − in postmortem diagnosis. This is strongly contested by some pathologists and clinicians (Ayoub and Chow, 2008). As reference to the literature, especially the *Journal of Forensic Radiology and Imaging* shows, postmortem computed tomography (PMCT) and magnetic resonance imaging (MRI) are increasingly being used − not only to determine the cause of death (COD) but also to help identify remains at accidents (see also Appendix CA2: Retrieval, Preparation and Storage of Skeletal and Other Material). It may involve both PMCT imaging (Figs. 6.2 and 6.3) and

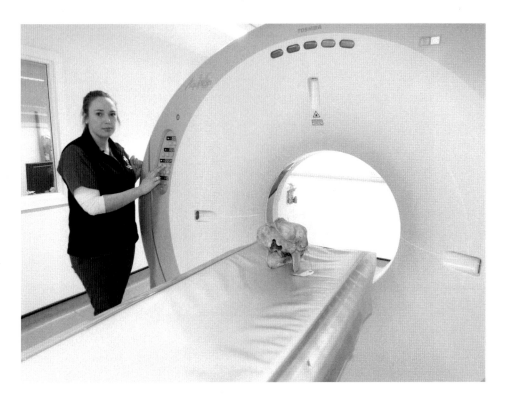

FIGURE 6.2   A gorilla skull being scanned (CT) as part of a retrospective study of osseous lesions.

postmortem magnetic resonance imaging. In adults CT is more accurate than MRI and the accuracy of CT is improved if angiography is employed.

The 'digital autopsy' has value and appeal in human medicine for many reasons, including cultural and religious considerations, in addition to practicability. Colleagues with an aversion to odours also point out that in digital examinations there is no smell! It probably has few applications in gorilla work except where, for reasons of health (certain zoonoses), a necropsy is either impossible or unwise.

Space does not permit detailed analysis of the many postmortem examinations that have been carried out on gorillas in diverse parts of the world and by people from different disciplines. As mentioned earlier, the reports listed by Fossey (1983) include excellent examples of how necropsies can be performed and data recorded. The 14 postmortem reports in Fossey's (1983) book are remarkably comprehensive and informative, especially bearing in mind that those performed in Rwanda were several years before the establishment there of the Centre Vétérinaire des Volcans/Volcano Veterinary Centre, later the MGVP. Features of the Fossey reports that warrant particular mention are:

1. In a number of the cases a whole set of body measurements was taken and these are depicted in simple, but very clear, accompanying sketches.

2. Dimensions of organs and lesions were also often provided by medical practitioners, not necessarily pathologists, who were involved in several of the examinations and, probably as a result, those reports also provide detailed descriptions of wounds and other lesions.

3. In the majority of the cases, supporting laboratory tests were performed – parasitological, microbiological and histopathological examinations. In some instances the name of the person reporting on those tests is given – for example ('Kweli') D.L. Graham, D.V.M.

4. In three cases ('Rafiki', 'Whinny' and 'Thor'), skeletal examination was carried out by Jay Matternes; this included detailed investigation of dentition and of degree of ossification of long bones.

5. Some extra investigations were performed in individual cases, for example, myoglobin examination of 'Thor' by the University of Cambridge. An example of opportunistic sampling and testing – always prudent when dealing with scarce material.

6. Some, not all, of the reports give 'Conclusions', rather than a 'Diagnosis' or 'COD'. Such conclusions are often a wise approach when the COD may be multifunctional and/or when subsequent investigations on samples may yield further, unexpected, information.

Some indication of techniques used and how deductions were made can also often be gleaned from more recent publications. The survey of causes of death of

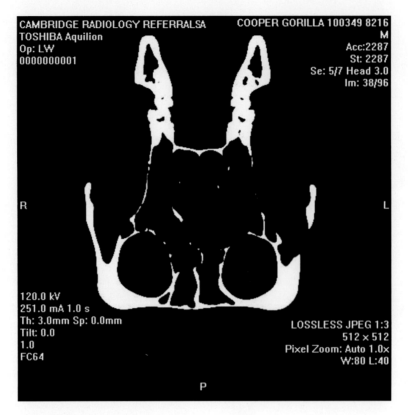

FIGURE 6.3 One slice of the scan of the above gorilla skull.

mountain gorillas by Nutter et al. (2005) is, unfortunately, available only as an expanded abstract and the histological findings to which the authors alluded in their presentation have not, it seems, been subsequently published and disseminated, neither to those VVC/MGVP veterinarians who performed most of the original postmortem examinations nor in the public domain. Nutter et al. stated that 'Ideally, thorough gross and histopathology examinations would be conducted on all dead gorillas. In reality, work is limited by the ability to recover dead gorillas, the field conditions under which gross postmortem exams are often conducted, and the quality of tissues available for histopathologic exam and ancillary testing'. In addition they said: 'Recurrent episodes of war and instability have also caused the loss of valuable records and samples'. This is a recurring theme when endeavouring to collate material in some parts of Africa and has to be borne in mind when assessing the achievements of a project or its veterinarians.

The ability to dissect is as much an art as a science. As pointed out earlier, in Africa local staff are usually far more confident prosectors than are their expatriate colleagues because they use such techniques in food preparation at home. Colleagues from Europe and North America though are sometimes woefully inadequate. In the United Kingdom, for example, dissection of animals is often not encouraged or permitted at school and learning in private

at home on road-kills (as did the author, at an early age) is all too often discouraged on grounds of 'health and safety'. Roscoe (2007) lamented this, calling dissection 'a dying art' and argued that it complements educational teaching with realistic experience.

Skills such as dissection are not easily or quickly perfected. In the earlier days of the British veterinary profession, the certificate issued by the Royal College of Veterinary Surgeons (RCVS) read that the candidate had 'satisfied the examiners in the Art and Science of Veterinary Medicine and Surgery'. The 'art' includes observation, 'hands-on' examination and investigation of the live or dead animal. Being both an artist and a scientist marks out a really good pathologist. Two simple illustrations, sometimes overlooked, are the importance when performing a necropsy of (1) observing the carcase carefully before handling – or even touching – it, and (2) manipulating tissues with sensitivity and using the correct instruments. Those with a sense of history are mindful of the battles that the 'giants' of the past, such as Harvey and Willis, faced because of State and Church condemnation of autopsy and the care they had to exercise when dealing with a dead body. A more contemporaneous consideration is that, in this Digital Age, which includes an almost universal possession of mobile phones with cameras, photos quickly circulate and there can be serious repercussions if a much loved gorilla from a zoo is seen to be necropsied in a clumsy and insensitive fashion. As the author advises his students: *Be a veterinary surgeon, not a backwoods butcher.*

Actually seeing and doing postmortem examinations is so important. And practice makes perfect as Galen (Comp. Med.Gen. 3.2) stressed: '... of those who have seen the parts of the body shown by a teacher, none is able to remember accurately having seen them only once or twice, but it is necessary to see them many times'.

EQUIPMENT

A postmortem examination on a valuable species such as a gorilla demands high standards and use of appropriate equipment. Forceps are an example of where a choice is necessary:

- Rat-toothed — used for grasping and restraining tissues that are not likely to be used for histology and similar investigations.
- Nontoothed — for grasping and restraining tissues that are either easily torn or which are destined for histology or electron microscopy where forcep damage can cause processing artefacts.
- Plastic — for removing lead shot or bullets (metal forceps can damage the surface of shot or bullets, making identification of the weapon difficult) or for taking samples for analysis of metallic compounds.

Good illumination is always important in postmortem work. Specialist magnification glasses help provide an enhanced image. While an ordinary magnifying lens (glass) will help, a proper loupe is even better, especially as it can be worn rather than held. Loupes are available as either 'Flip-up' or 'Through-the-lens' (TTL) and each has its own advantages. An LED light unit is a useful addition; these are now available in a small lightweight form. Loupes and light units can, of course, be used for a variety of purposes other than postmortem examination, ophthalmological investigations, for example.

RECORDING OF FINDINGS

Throughout the examination it is important to record findings. This can either be done by two people, one of whom writes while the other dissects, or singlehandedly using a tape-recorder, preferably with a voice-activated microphone. Properly labelled photographs should be taken at regular intervals. Clear line-drawings (diagrams) are valuable for recording details, especially the position of lesions.

PERFORMANCE

A gorilla is best necropsied on a table. This must be of adequate size so that the animal's limbs do not drop or dangle from it. A metal (easily disinfected) tray can be used for neonates (Fig. 6.4).

In the absence of a suitable table, so often the case in the field, an appropriate 'postmortem area' can be established at floor (ground) level, but this is less satisfactory and very tiring for those involved in the necropsy as they have to stoop or kneel. A solution, if the carcase is to be buried after examination, is to dig the hole first and then for the pathologist to carry out the examination while standing in the hole (Fig. 6.5). The animal should, if possible, be downwind from where the pathologist is working, to reduce both odour and exposure to pathogens.

During and following the necropsy it is important that unwanted animal material and 'clinical waste' are disposed of hygienically and legally (see earlier and Chapter 18: Legal Considerations). In the case of captive gorillas in zoos, this usually means incineration or maceration. In the field burying or burning are likely.

It is wise for the pathologist to adhere to a familiar postmortem procedure. A sudden change to another method can result in mistakes or carry the risk that a lesion is missed. It must never be forgotten that a full postmortem examination of a gorilla, if done correctly, can be a lengthy process. A thorough necropsy of an infant in the field can take several hours. As Akeley (1920) reported:

I had skinned the old gorilla roughly in the field the day before. If I wanted properly to preserve the specimen, there was no time to be lost. The brains and internal organs I had to preserve in formalin. The whole business was a full hard day's work. One of the chief difficulties with scientific collecting is the necessity for doing all the skinning, cleaning, measuring, and preserving at once.

EXTERNAL EXAMINATION AND SAMPLING

The carcase should always be carefully examined externally. Skinning is usually an important part of a forensic necropsy, in order to detect injury to the integument and superficial tissues. Most pathologists tend to leave removal of the pelt until after the internal examination is completed but some prefer skinning the entire animal at the outset.

If properly done, skinning is a lengthy process, especially if the specimen is required for taxidermy. Ionides (1965) gave a graphic description of such a dissection in the field:

For the skinning Williams made his cuts at the back of the body so that the bare parts were never touched. He also made plaster casts of the face, hands and feet as a guide for the taxidermist. As it was raining a good deal of the time the initial curing was done by steeping the skin for long periods during the first three days in a saturated alum bath. After that it dried out quite easily. The entire skeleton was

FIGURE 6.4 A simple system for necropsying a neonate, applicable to work in the field. A metal tray and basic equipment.

preserved — the Museum wanted that; the organs were kept by Sebley as they were required for medical research; and the meat went to the Batwa for a feast.

Scoring of condition is wise rather than merely recording subjectively such comments as: 'The carcase of a well-nourished female gorilla'. Gorillas that have become bloated following death can be particularly difficult to assess. As in clinical work, all senses except taste should be used when examining a carcase — eyes, ears (crepitation, etc.), nose (odours associated with certain infections, parasites and poisons). Fossey (1983) wrote about 'Quince': '... strong odor (resembled silverback odor) noted around right axillary region, though total lack of residue or odor from left axillary'. The sense of smell is superior in women. A female colleague may detect odours that a man misses.

Gentle palpation should be carried out.

Photographs and drawings (outlines) of animals are valuable to show lesions. Medical forensic pathologists take fingerprints from cadavers as part of identification (DiMaio and DiMaio, 2001); comparable records can be assembled from nose prints of gorillas.

If available, radiography and other imaging techniques should be used before the carcase is opened, especially in suspected cases of trauma.

Determining the sex of dead animals in the field is sometimes difficult. This was the situation when four dead gorillas killed by poachers were examined by the author in Bwindi, Uganda, in 1995. The history and circumstances strongly suggested that they were females but marked decomposition made it impossible to locate genital organs and, as the animals had to be buried immediately after the necropsy, investigation of the skeletons was, regrettably, not possible.

The gorilla's skin should be carefully assessed and not just described in such terms as 'healthy' or 'normal'. The same applies to all mucous membranes.

INTERNAL EXAMINATION AND SAMPLING

Once internal organs are visible they should be observed in situ before being touched (Fig. 6.6). When they are manipulated it should be for a purpose; avoid what the author (JEC) calls the 'dab syndrome'! Treat tissues with respect. Initially the organs should be carefully moved aside to reveal deeper structures (Fig. 6.7).

Odours may be detected immediately after the carcase (or a hollow organ, such as the stomach) is opened. The smells should be noted and recorded.

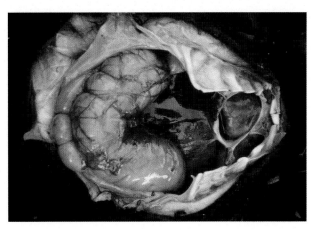

FIGURE 6.6 The initial internal appearance of a neonate gorilla. This animal probably died of inanition. As a result, there is very little intraabdominal fat and the viscera are readily observed.

FIGURE 6.5 A (very autolysed) gorilla is placed on the edge of a specially dug hole so that the pathologist, subsequently standing in the hole, will be able to necropsy it without the need to stoop.

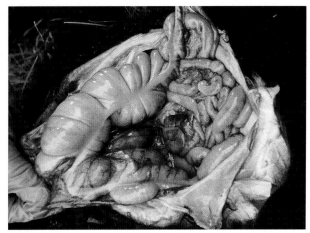

FIGURE 6.7 The same animal as above, following gentle moving of GIT. Deeper organs are now easily identified.

When examining body organs in situ one should employ eyes, ears, nose, fingers, plus a hand lens (magnifying glass) or surgical/dissecting loupe, and appropriate use of lighting — sun, artificial sources — and making use of both reflected and transmitted light ('transillumination').

Incisions must always be carried out with care, especially in cases where there is already traumatic injury or, perhaps, a ruptured aneurysm is suspected and therefore dissection may damage important evidence.

In human autopsies all viscera may be removed en masse and this technique can have merit in gorilla work. Whatever the procedure followed, the use of coloured string to signify what is (for example) pylorus, and what is terminal rectum, is recommended (Cooper and Cooper, 2007).

Very gentle washing of the outside or opened/cut surfaces of organs will facilitate examination.

When recording the appearance of internal organs, knowledge of what is 'normal' (within a standard range) is essential. Photographs may help but do not replace visual assessment.

Advice on sampling and laboratory investigation is given in the next chapter and is very well described in the *Standardized Necropsy Report Form for Gorillas* — see Appendix 2: Protocols and Reports. Increasingly, new methods of investigation are being introduced. For example, even acute phase proteins may have their uses in dead gorilla work. They are beginning to be studied in abattoirs as a possible means of detecting disease in carcases, using meat juice, and a similar approach might be useful in cases where a dead gorilla shows postmortem decomposition, making standard microbiological tests of little value.

In theory, of course, all material from such an important species as a gorilla warrants 'rapid action' insofar as getting laboratory tests done but the list in Box 6.5 may assist those working in the field.

DESCRIPTION OF LESIONS

Perhaps the best introduction to the terms used in pathology, in the English language at any rate, remains Thomson's *General Veterinary Pathology* (1978). It not only provides clear, usually succinct, definitions but also explains, in 'reader-friendly' prose, the mechanisms underlying clinical disease and pathological changes.

All lesions should be photographed in colour, with a nonreflecting scale and other relevant information, particularly the gorilla's name or reference number, date and (where appropriate) 'left' and 'right'.

Describing lesions should follow a system (see Box 6.6).

Colour is important but often overlooked. It has long been used as an indicator of pathological change. Changes in skin colour of gorillas are not easy to detect but buccal, rectal and other mucous membranes are usually readily accessible in live and dead animals. It is wise to standardise descriptions using a key — either by compiling a system based on, say, a chart of paint colours (available through most 'Do it yourself' shops) or one designed for checking photographic colours.

Bruises also produce colour changes. They are important clues in any postmortem examination but their detection in gorillas can be difficult without skinning because of pigmentation.

Pathologists are generally excellent observers who enjoy descriptions. Often they are interested in the natural world. As a result, over the years, both animals (Allen and Cooper, 1983) and plants (Cooper, 2009) have provided the medical and veterinary professions with evocative terminology, ranging from skin lesions described as 'ichthyosis' to 'apple-green' staining reactions. The use of such similes should be avoided, however, especially when working overseas, as they are not only subjective but assume that everyone is familiar with the skin of fish or the appearance of one of many varieties of fruit.

Touch can be very important: a slight change in the surface morphology of an organ can be best detected by gently running an index finger over it with the eyes closed — as if one is blind. Proper description of the 'feel' of organs or lesions is not easy, but important.

FORENSIC POSTMORTEM EXAMINATIONS

The author (JEC) advocates that any free-living gorilla that is found dead should be treated as a forensic case (Cooper and Cooper, 2007; 2013). So too should any captive gorilla that dies under suspicious circumstances.

In Africa, intentional trapping or killing of gorillas is still common and, in theory at any rate, the arrest and prosecution of the person(s) responsible is possible. Although this is often not enforced (see Chapter 18: Legal Considerations), the assumption should always be that 'this is a legal case' and appropriate steps taken to ensure that the examination of the animal, record keeping and the collection, labelling and submission of samples/evidence (especially the chain of custody) are consistent with those expected in a court of law.

The examination of a dead gorilla for signs of human interaction may be particularly important in the wild, where trapping or poaching may be involved, but it can also from time to time be applicable to captive animals in cases of intentional assault (Cooper and Cooper, 2007). In either case, a forensic approach is needed (Cooper and Cooper, 2013). The 'crime scene' must be secured and access strictly limited. Special attention should be paid to anything that may indicate human involvement, such as the removal of limbs, head or skin.

The area where the animal is found needs to be methodically searched, not only for evidence of human proximity (e.g., footprints, dropped cigarette ends) but also for discarded knives, ropes or bags. The protocols devised for use when examining other wildlife for signs of human interaction can fairly readily be adapted for the investigation of gorillas.

Forensic entomology is briefly mentioned in Chapter 7: Methods of Investigation — Sampling and Laboratory Tests. It is vital to collect invertebrates found associated with, or near, a dead gorilla. They need to be processed correctly (Barnes, 2013).

A youngster examined by the author in March 1994 was a case where this was done but, alas, the specimens disappeared during the Rwanda conflict: 'PM of baby gorilla — very decomposed, measurements taken, plus attempts at sampling. Stomach empty … many fly eggs and larvae'.

Examples of incidents where specific forensic investigations were very much justified include the author's (JEC's) investigation of four dead gorillas, killed by spears, arrows and dogs, in Bwindi in 1995 (see elsewhere in text) and a number of cases about which he was consulted by colleagues. The bare facts of two of the latter are given here, to illustrate the scenarios, without details of findings. The first (Jane Hopper, personal communication) concerned gorillas in a project in the Congo where rescued orphans are rehabilitated and re-released. It had gone very well but there were too many males and so a very large island was created for them. Suddenly and unexpectedly two silverbacks were found dead. Both were found face down, facing in the same direction and about 20 m apart from each other. They were very close to the feeding station where they received a small amount of food daily to allow checking and observations. A veterinarian performed necropsies but, 'due to the risk of Ebola and the fact that the bodies were 5 days old and very bloated he did not turn them over to see the ventral side or take any samples apart from one piece of muscle. The only testing done was for Ebola and this was negative. The backs and heads showed no sign of injury or fight marks'.

The second example relates to four mountain gorillas found dead and examined by Chris Whittier (personal communication). They showed different degrees of postmortem change. Two 'were mostly skin, bones and maggots' while the others 'were fairly intact except for some maggots and ants in the feces and bullet wounds'. In such a case a full wildlife forensic investigation is required, as described by Cooper and Cooper (2013). In particular, an estimation of time of death would need (among other things) very careful collection of information about environmental factors, namely ambient temperature, recent fluctuations in temperature, relative humidity (RH) and recent fluctuations in RH. Collection of maggots and other invertebrates would probably also help. Any histological evaluation of the gorillas' tissues would need to be coupled with the analyses of temperature and RH.

POSTMORTEM CHANGES AND DETERMINATION OF TIME OF DEATH/PMI

It is preferable that a carcase be fresh or only chilled prior to examination as methods of 'preserving' carcases and tissues inevitably have some adverse effects (see Table 6.1).

Accurate differentiation of antemortem and postmortem changes is important in all necropsies but especially so in a forensic case. It presents challenges, even in a zoo setting. An example of how antemortem lesions can be

TABLE 6.1 Storage and Preservation of Carcases and Tissues From Gorillas

Method of Treatment	Advantages	Disadvantages	Comments
Chilled at +4°C	Good preservation for a few days	Autolysis continues to take place	The autolysis is slow and generally predictable
Frozen	Can be stored indefinitely. An excellent way of preserving carcases and tissue for subsequent toxicological or (if the temperature is sufficiently low — 20°C or more) virological investigation	Freezing and thawing cause tissue damage and therefore artefacts in terms of both gross and histological changes, making interpretation of histological sections difficult. There may be a haemorrhagic or congested appearance to the viscera. Pathogens generally not killed; there may be rapid spread and multiplication of bacteria during thawing	Some allowance can be made for gross artefacts (e.g., corneal opacity) but confusion can still arise
Fully fixed in formalin or similar fixative	As above. Pathogens are usually killed (but prions are not)	Affects appearance of organs — colour changes. Permits histopathology but makes DNA studies difficult	Careful interpretation needed, as in embalmed cadavers
Fixed in ethanol or methanol	As above, but DNA extraction is little affected	As above — colour changes	As above

confused with postmortem changes was given by Rutty (2004) who provided an excellent review of shock (in its sense of 'inadequate perfusion of the tissues') and drew attention to how pathological changes, in the lung for example, can be mistaken for autolysis.

There are various specialised methods for attempting to assess the time of death and thereby 'postmortem interval' (PMI) most of them developed for human work (DiMaio and DiMaio, 2001). In gorilla work one has to include analysis of the history and the circumstances following death, coupled with assessment of *rigor mortis*, *livor mortis* (pooling of blood) and *algor mortis* (cooling), degree of desiccation, discoloration of skin and internal organs, infestation by maggots, carrion beetles, arachnids and scavenging patterns (jackals, rodents).

Remarkably few necropsy reports on gorillas provided adequate information about the presence or absence of *rigor mortis*. Fossey (1983) certainly recorded this — as in the case of 'Lee' who had in life had an injured foot: 'Following death, and before *rigor mortis*, the right leg could be fully extended; the left leg had a 45 percent degree of contraction'.

Early field records often provided interesting data on the speed of postmortem change under different circumstances. Galloway and Allbrook (1959), for example, wrote: 'That evening I was told that the body had been brought 50 miles over the mountains to Kabale, that it was reputed to have been dead 3 days, and to be rather "blown up", and that the skin was coming off its chest'.

The time between death and necropsy is usually recorded by zoos but how the carcase had been stored (e.g., at 4°C in a fridge) is not always recorded.

Determining the time of death (PMI) is linked with decomposition changes. Progression of postmortem changes can depend upon various factors (see Box 6.7).

Organs and tissues of gorillas, like other species, undergo postmortem autolysis at different rates. Box 6.8 provides some indication, based on the author's (JEC's) work with other primates, of the sequence of such changes, starting with those organs that rapidly autolyse and finishing with those that show relatively little decomposition.

CAUSES OF DEATH DIAGNOSIS

Ideally, both the necropsy and follow-up laboratory investigations of a dead gorilla should be overseen by the same person, who should be a specialised veterinary (sometimes medical) pathologist. Such an arrangement is often the case in a modern zoo but often not possible in the field.

Ascertaining the cause of death/making a diagnosis (the two are not always the same) requires an understanding of physiology, pathology and clinical medicine. Anyone involved in a necropsy who is not so experienced should heed the words of Richard Owen (1859): 'The accurate record of facts in natural history is the first duty of the observer; the true deduction of their consequences is his next aim'.

Munro and Munro (2012) warned of the many pitfalls in the interpretation of pathological changes. They were writing in the context of forensic work but many of their comments and cautionary remarks are relevant also to diagnosis. Common challenges in postmortem work include estimation of the age of skin wounds and bruises and the estimation of the time since death. They argued that the 'multispecies' nature of veterinary pathology, combined with the preponderance of published observations originating from animal experimentation, rather than casework, poses challenges. Extrapolation of results from one species to another may jeopardise the reliability of the pathologist's opinion and while experimental studies may not truly reflect the spectrum of changes seen in actual cases (e.g., extent of injuries, infection, age and health of victim). Munro and Munro who suggested that methods for estimation of PMI — see earlier — might in future involve novel procedures such as diagnostic imaging.

BOX 6.7 Factors Affecting Postmortem Changes

- Size of the gorilla
- Health status of the animal before death
- Presence or absence of ingesta in gastrointestinal tract (GIT)
- Manner of death
- Environment — temperature and humidity, including any method of storage
- Location and position of the body, subsequent movement or interventions.

BOX 6.8 Stages of Autolysis

- Brain
- Testis
- Gastrointestinal tract
- Pancreas
- Liver
- Kidney
- Skin
- Skeletal and heart muscle

Earlier reports on gorillas that died in zoos often gave 'cardiac arrest' as a cause of death, especially when the animal died under anaesthesia or suddenly and unexpectedly. The author (JEC) prefers to follow the approach used in modern (human) forensic medicine, whereby three aspects have to be considered (DiMaio and DiMaio, 2001):

- The cause of death (COD) – for example, a gunshot wound or pneumonia.
- The mechanism of death – the physiological changes produced by the gunshot wound or pneumonia that made it fatal. Examples could include haemorrhage or a blood-borne infection.
- The manner of death – how the COD came about – 'accident', etc.

A more easily understood approach is to ask:

- How did this gorilla die?
- Why did this gorilla die?
- How long did it take this gorilla to die?
- When did this gorilla die?
- Where did this gorilla die?
- Who was involved, intentionally or inadvertently, in its death?

Not all of these questions will be easy to answer and very often the veterinary pathologist will only be able to offer limited evidence, coupled with opinion.

Specifying the COD has long taxed the medical profession, especially because of its legal implications (Cooper and Cooper, 2007). Age-related changes, in old animals, present particular challenges. Thirty years ago an eminent British forensic pathologist (Knight, 1986) wrote: 'In old age, the frequent plethora of potential lesions makes the choice difficult, especially where the relative contribution of each is hard to assess. Notwithstanding the views of some practitioners, "senility" is a perfectly acceptable COD, though it is best to qualify it by reference to a vital organ affected such as the myocardium, using a term such as "senile myocardial degeneration"'. A recent issue of *Veterinary Pathology* (Ward et al., 2016) includes a thought-provoking Editorial that, among other things, debates the role of pathology in determining the COD (Ward et al., 2016).

Sometimes a provisional diagnosis has to be based on a combination of professional and non-professional contributions. Thus, for instance, a basic field postmortem examination may have been carried out by a nonveterinary fieldworker with no formal training in the recognition of lesions, who may or may not have selected the correct organs or tissues for laboratory investigation by a specialist pathologist. On such occasions the latter may be able to provide a reason as to why the gorilla died (or, at

BOX 6.9 Types of Diagnosis

Morphological	(the lesion)
Aetiological	(the cause)
Definitive	(specifics)

least, contributory factors) based on histological, microbiological or other tests. Failure to do so may, however, indicate that the samples submitted were not appropriate or adequate. The advent of molecular techniques that can detect organisms does not obviate the need to examine tissues and to study host responses. The detection of a pathogenic organism is not the same as a diagnosis.

In any report, correct use of descriptive pathology is important. As emphasised by Thomson (1978), diagnoses in reports can be morphological, such as abscessation of the liver, aetiological, which often consists of two words, for example, mycobacterial enteritis, or final/definitive (see Box 6.9).

Safe storage of reports, including original notes, is vital. Even 'paper' records should be retained. Often these are still kept in boxes or cupboards (Fig. 6.8), without backup copies. There is an urgent need to ensure that such records are scanned, or at the very least photocopied. Old typewritten reports often have unnumbered pages and lost staples, meaning that sheets can be lost or transposed.

Not all postmortem examinations provide a diagnosis (Cooper, 2002). Examples of diseases in which no specific macroscopical lesions are likely to be visible include rabies, listeriosis, tetanus and various types of poisoning. The pathologist must be resigned to the fact that some cases will be inconclusive, even if a whole battery of supporting laboratory tests is carried out.

Failure to make a diagnosis may be on account of human factors. In the past, sometimes still, free-living gorillas are not examined fully postmortem, either because the body is not found, or because the remains are considered 'too decomposed for a necropsy' (never an excuse for not carrying out at least some basic investigations) or because of social/political/military factors. The last of these was often the case during the tumultuous years in Rwanda in the 1990s. Thus, neither 'Effie' (the oldest known female mountain gorilla – see Chapter 8: Nonspecific Pathology), nor her mate, the enormous silverback 'Ziz', could be examined in 1994/95 because of the constraints of the war that was in progress at this time. As a result, speculative diagnosis ('natural causes', Effie, 'respiratory failure', Ziz) could neither be confirmed nor refuted.

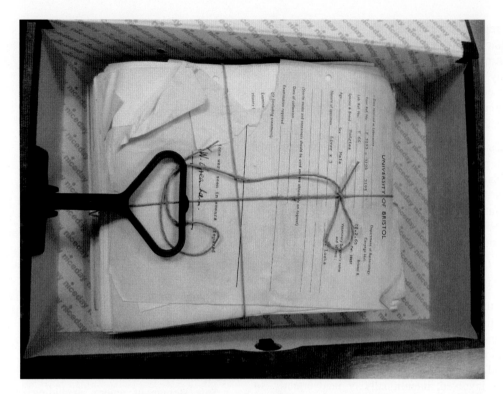

FIGURE 6.8 Type-written postmortem records, several of them concerning gorillas, that have been stored in this box file for 50 years.

Chapter 7

Methods of Investigation — Sampling and Laboratory Tests

John E. Cooper

This day relenting God Hath placed within my hand, a wondrous thing: and God Be praised. I know this little thing a myriad men will save. O Death, where is thy sting? Thy Victory, O Grave?

Sir Ronald Ross, August 20th, 1897
(on discovering the malaria parasite)

INTRODUCTION

As was pointed out in Chapter 6, Methods of Investigation — Postmortem Examination, data relevant not only to diagnosis but also to a better understanding of gorilla pathology can be obtained from three main sources — live animals, dead animals and samples (which may in turn come from either live or dead gorillas), or from the environment in which they live. The material that provides this information is in the form of samples.

The taking of samples from captive gorillas has been a standard procedure for decades but specimens from free-living gorillas raise all kinds of practical, legal and ethical dilemmas (see Chapter 16: Legal Considerations). This was, for example, a stumbling block in work with mountain gorillas when the author was based in Rwanda from 1993 to 1995 (Cooper, 1995b). Policies began to change and so in 2008, Eckhart and Lanjouw were able to write 'When gorilla immobilizations occur, the veterinarians are now allowed to collect biological samples, which could include blood, urine, feces, hair, biopsies, and culture swabs'.

The value of having samples increases as improved microtechnology, aimed at obtaining more data from even smaller samples, evolves (see Appendix 4: Field Pathology).

Investigative methods for examining samples can be traditional, in the sense of gross, microscopical and microbiological/parasitological, techniques, or more modern tests involving molecular biology (PCR, for instance — see later). Laboratory testing is the single largest activity in

human medical work — in the British National Health Service, for example — and is essential in gorilla studies for fast, accurate, diagnosis and health monitoring.

SAMPLING

In wildlife disease investigations, the term 'sample' is used in two ways. The first is as a synonym for a 'specimen', as in 'a blood sample'; the second in the statistical sense of a subcollection or subset of units from the population. Wobeser (2010) emphasised that the collection and analysis of samples is the basis of investigation and that the validity of the results and conclusions of a study, including diagnosis or health monitoring, is wholly dependent on the quality of those samples.

It is important to ensure that the samples collected are representative of the situation or population from which they are taken. Errors can occur during sampling (Wobeser, 2010).

The choice always needs careful consideration and consultation. A checklist is useful. It is always most important to consult, in advance, the person(s) who will do the investigations on the specimen(s). The circumstances of ill health or death will also influence this. When the body of 'Nkuringo', the silverback aged at least 50 years, was found in Bwindi in 2008, the Chief Warden said 'Mzee Nkuringo had lost most of his teeth and used to move slowly ... the essential parts of his body are being studied in the Government laboratory in Kampala...'.

Table 7.1 gives examples of samples that may be obtainable from live or dead gorillas.

A useful account of the value of samples ('biomaterials') in work with gorillas was the chapter by Lehn (2008) in the book edited by Stoinski et al. (2008). Lehn argued that, while biomaterials can be defined as 'any organic piece or derivative of a plant or animal', a sample is 'the same except

J. E. Cooper & G. Hull: Gorilla Pathology and Health. DOI: http://dx.doi.org/10.1016/B978-0-12-802039-5.00007-X

that this implies that it is destined for examination and probable analysis'. In this book the term 'sample' (occasionally 'specimen') is used throughout and any relevant differences in respect of usage are explained at the time.

Mudakikwa et al. (2001) outlined the basic sample collection protocol used by the MGVP when an animal is sedated, including methods used for blood, urine, faeces, hair, rectal and wound swabs and 'live' genetic material

such as lymphocytes and epithelial cells. The present author's approach to samples was described in Cooper and Cooper (2007, 2013). Minimally or noninvasive methods of collecting samples are discussed in Chapter 4: Methods of Investigation – Observation, Clinical Examination and Health Monitoring.

Methods have been devised for calculating the minimum sample size required to detect a disease in a population in which the prevalence may be extremely low.

TABLE 7.1 Samples From Gorillas

Sample	Live Animal	Dead Animal
Hair	Yes	Yes
Faeces	Yes	Yes
Urine	Yes	Yes
Blood	Yes	Sometimes, if soon after death
Pus or other discharges	Yes	Yes
Swabs from external lesions	Yes	Yes
Touch preparations for cytology	Yes	Yes
Swabs from internal lesions	Sometimes, via an endoscope or a surgical incision	Yes
Saliva	Yes	Yes
Tissue samples for histology, electron microscopy, etc., of external lesions	Yes (biopsy)	Yes
Tissue samples for histology, etc., of internal lesions	Sometimes (biopsy)	Yes
Larger tissue samples for (for example) toxicology	Usually not feasible	Yes
Stomach contents for toxicology	Sometimes by lavage (gastric washing) or following regurgitation or vomiting	Yes
Ectoparasites	Yes	Sometimes
Invertebrates (see text)	Yes	Yes

Source: Adapted from Cooper, J.E., Cooper, M.E., 2013. Wildlife Forensic Investigation: Principles and Practice. Taylor & Francis/CRC Press, Boca Raton, FL.

Minimal Sampling Size (95% Confidence Limits, to Find One Infected Animal)

Population Size	2% Disease Incidence	10% Disease Incidence
50	46	20
250	110	25
1000	136	27
5000	142	27
10,000	147	27

A modified summary of the points made by Wobeser (2010) is shown in Box 7.1.

BOX 7.1 Sampling

- Ensure that the samples are representative of the population from which they are drawn.
- Samples may be nonrepresentative because of random error or because of bias.
- Of the three basic types of bias (selection, measurement, confounding), selection bias is most common in wild animal work.
- It may be impossible to prevent bias but try to determine its direction and to use this information in interpreting results.
- Consult in advance the person who will do the analysis regarding methods for collection and preservation of specimens.
- Specimens that are unlabelled or lose their labels are of no value.
- No analytical test should be used without a full understanding of its suitability for the species being studied.
- Sampling design should always be developed in consultation with a knowledgeable biometrician. Sample size must be determined prior to starting data collection.
- Extraneous variability may be reduced by restriction in choosing specimens for sampling.
- Specialised methods are often more appropriate than simple random sampling.
- Replication is required with groups if comparisons are to be made between those groups.

(Source: After Wobeser, 2010).

METHODS

A résumé of investigative procedures is given in Table 7.2.

More detailed information on some methods, many of them the 'standard laboratory techniques' referred to above, is given in Table 7.3.

Swabs

Swabs are used for many investigations but the correct choice of swab is important, sometimes vital. There are many sorts; some include transport medium.

In gorilla work the important considerations when purchasing and using swabs are: (1) type; (2) applicability to the work in progress; (3) portability and use under field conditions; and (4) cost. Consistency of product and method are essential if findings are to be compared and contrasted.

Colour-coded swabs have much to recommend them in the field, especially when working with the assistance of local villagers who are not familiar with the technology and may not even be able to read what is written on the outside of the tube. The type of swab to be used in gorilla work will depend on the indications; for example, the author (JEC) recommends calcium alginate-impregnated swabs on a (flexible) aluminium stick for sampling the nasal cavity and turbinates of nonhuman primates (NHPs).

Transport media also vary. Some contain antibiotics, to suppress bacteria, and are intended for isolation of viruses, chlamydiae or protozoa. Glycerol should not be used when taking samples for virology, as it is virucidal. Whittier et al. (2004) compared storage methods for reverse-transcriptase PCR amplification of rotavirus RNA from gorillas and showed that only samples stored in guanidine thiocyanate (GT) buffer gave 100% positive results after 180 days. They recommended that faecal samples were collected in GT for viral RNA analysis.

Saliva

Saliva has many uses. Buccal swabs, for example, are regularly taken from gorillas for diagnostic purposes. Noninvasive saliva collection techniques for free-living mountain gorillas and captive eastern gorillas were described by Smiley et al. (2010), and more recently methods that could be used in laboratory monkeys, using rope and oral swabs, by Smiley Evans et al. (2015).

Faeces

Early investigators provided helpful descriptions of normal defaecation in gorillas – see Schaller (1963) and Chapter 11: Alimentary Tract and Associated Organs. Faeces can usually be easily collected from known individual animals. They are used for not only veterinary investigation but also other research such as steroid and DNA studies.

Much can be learned from proper investigation of faeces. A proper description is important but often poorly done. Methods of recording colour and consistency were formulated for mountain gorillas in Rwanda (Cooper, 1996; Cooper and Nizeyi, unpublished) and later, in collaboration with Chris Dutton, for captive primates at Jersey Zoo (DWCT Veterinary Research Projects, 1999, number 10). The paper about bovine faeces by Stober and Serrano (1974) has largely been forgotten or ignored but it deserves to be read (or re-read) by all those working with animals, domesticated or wild. It describes how faeces should be observed, described and used in diagnosis, with particular reference to assessment of the quantity, colour, odour, consistency, degree of comminution (broken into smaller fragments – a rough indicator of extent of digestion) and admixtures (combinations) of other substances. Not all of the text, of course, is directly relevant to gorillas but the approach and the sound practical advice it offers makes it pertinent to veterinarians, field biologists, ranges and trackers.

Careful examination is crucial when trying to ascertain the age of gorilla faeces; this can provide information in forensic cases (Cooper and Cooper, 2013). Important variables in appearance and consistency include: (1) loose 'normal' faeces passed by a gorilla (often the group) when alarmed by humans or other animals; (2) loose 'abnormal' faeces containing excess mucus or undigested blood; (3) 'normal' faeces with a pale exterior, usually produced by lactating females and cautiously attributed by Fossey (1983) to 'the tendency gorilla mothers have to eat the feces of their offspring during the infant's first 4–6 months'. *Anoplocephala gorillae* proglottides are sometimes visible with the naked eye (see Chapter 11: Alimentary Tract and Associated Organs).

Careful selection of faecal samples is essential. There may be significant differences in findings between the surface and the interior and close inspection will provide information about the age of the sample (Cooper and Cooper, 2013). It is often wise to take the whole faecal deposit and store different portions under disparate conditions (chilled, frozen, fixed). As Part II: A Catalogue of Preserved Materials, indicates, Conservation Through Public Health (CTPH) in Uganda holds approximately 6000 dung (faecal) samples for future research and study.

When a faecal deposit is picked up it is important to check whether there are invertebrates under the material and to see if these are alive or dead – and, if appropriate, collect them (Cooper and Cooper, 2007).

Much can be learned from a comprehensive investigation of faeces, not just checking them for parasites. The appearance of voided faeces in the wild depends upon

TABLE 7.2 Investigation of Samples from Gorillas – General

Technique	Comments
Direct examination with hand lens, magnifying loupe or dissecting microscope	Recommended initially for any tissue or sample from live or dead gorillas. Direct examination must first be with hand lens or magnifying loupe. Then a stereomicroscope (dissecting microscope) should be used to examine specimens under both reflected (episcopic) and transmitted (diascopic) illumination. Then a compound (optical, 'light') microscope can be used, possibly followed by electron microscopy
Radiography (of certain samples, especially those containing bone – see below)	Portable machines can be used. Digital kits are increasingly available
Histology	Biopsies can be taken from live animals, either by excision (external lesions) or using endoscopic techniques. Tissue samples can be similarly removed from carcases. Buffered formol saline (BFS) is the standard fixative. In the field substitute fixatives may need to be sought (Cooper and Cooper, 2007, 2013). Methods of fixation must be appropriate, possibly involving inflation or perfusion (Cooper et al., 1978)
Bacteriology	If a bacterial infection is suspected. Standard techniques and PCR
Mycology	If a fungal infection is suspected. Standard techniques and PCR
Virology	If a viral infection is suspected. Standard techniques and PCR. Virological investigation is neither easy nor cheap. See text, including DNA tests
Parasitology	An important useful routine (ectoparasites and endoparasites) when examining live and dead gorillas. Combine faecal examination with cytology, below
Cytology of specific organs, intestinal contents, exudates, etc.	A whole range of simple and rapid tests can provide useful information, especially under field conditions (Cooper, 2013). See text
Haematology	Blood values and visual detection of some parasites. Standard laboratory techniques, adapted as necessary
Biochemistry	Biochemical values. Standard laboratory techniques, adapted as necessary. See text
Toxicology	Detection of poisons (see Chapter 6: Methods of Investigation – Postmortem Examination). Standard laboratory techniques, adapted as necessary. A valuable publication on the taking of samples from wild animals for toxicological analysis is by Rotstein (2008)
DNA studies	Identification, sexing. Valuable. Appropriate material for DNA studies should always be stored for subsequent reference. Standard laboratory techniques, adapted as necessary
Scanning electron microscopy (SEM)	Of value in specific cases. Expensive. Standard laboratory techniques, adapted as necessary
Transmission electron microscopy (TEM)	As above. Standard laboratory techniques, adapted as necessary
Bone preparation, examination and analysis	Bone density and other studies. See Radiography, above. Standard laboratory techniques, adapted as necessary
Digestion	Studies on soft tissues. Standard laboratory techniques, adapted as necessary
Ashing and mineral analysis	Studies on hard tissues. Standard laboratory techniques, adapted as necessary
Radioisotope studies	Age determination of remains. Standard laboratory techniques, adapted as necessary
Ballistic and allied investigation	Poaching cases. Specialised forensic techniques may be required (Cooper and Cooper, 2007, 2013).
Invertebrates in the vicinity	Can include flies (see Table 7.4)

many factors, including the weather and the amount of overhead vegetation. As Fossey (1983) pointed out, flies quickly swarm around fresh (mountain) gorilla faeces and lay 'countless hundreds of small white eggs … within minutes following defecation'. The faeces of normal free-living gorillas contain seeds, fruits and fibrous vegetation; a record should be made, with description and quantification of such material. Careful note should also be taken of

TABLE 7.3 Laboratory Techniques for Samples from Gorillas

Technique	Standard Recommended Methods
Aerobic bacteriology	Blood agar plates and MacConkey's – incubated at 37°C for 24 h (72 h if no growth). Samples may also be incubated at 24°C. Other media as appropriate
Anaerobic bacteriology	Same as above, but incubated anaerobically at 37°C for 48 h (and possibly 24°C)
Mycology	Sabouraud dextrose agar plates – one at 37°C for 7 days, the other at 22°C for up to 21 days. Direct microscopy of fresh or prepared material
Mycoplasmology	*Mycoplasma* growth medium
Virology	Tissue culture Egg inoculation Animal inoculation Fluorescent antibody technique (FAT) Molecular tests – see below
Serology	Serum neutralisation Gel diffusion precipitation Complement fixation ELISA FAT Molecular tests
Parasitology Skin scrapings and/or hair Faeces/intestinal contents Blood	 Direct microscopy plus treatment of sample with 10% KOH prior to examination Wet preparations in physiological saline plus salt or sugar flotation using McMaster counting chamber Lugol's iodine: identification of cysts ZnSO$_4$ flotation: amoebae, cysts Thin smears fixed in methanol and stained with Giemsa or air-dried and stained with a rapid stain. See text
Histology	Tissues dehydrated in alcohol. Embedded in paraffin wax, and sectioned at 6 μm or less. Stained routinely with haematoxylin and eosin. Frozen sections may be rapidly prepared using cryostat
Immunohistochemistry	Detecting and identifying endogenous and exogenous antigens, including microorganisms. See, for example, Fontenot et al. (2005) – immunohistochemical fluorescent antibody staining of muscle from a wound for *Clostridium septicum*
Cytology touch preparations, washings, brushings and aspirates	Air-dried for Romanowsky staining. Certain rapid stains include a fixative. See text
Haematology smears, blood in EDTA or lithium heparin	See text. Examination of smears for abnormalities, differential counts. PCV estimation using standard techniques
Clinical chemistry blood in lithium heparin or EDTA (depends on the species and test required)	Standard procedures for electrolytes and enzymes (colour reaction systems and atomic absorption or flame spectrophotometer methods) Automatic analysers
Electron microscopy tissue samples	Standard TEM and SEM procedures (see also below)
Chemical analysis/toxicology tissue samples	Standard gas-liquid chromatographic and other procedures. Animal studies
DNA and other molecular studies	Various – depend on samples. See text
Stable isotope studies, on hair	Detecting diet changes and thus patterns of movement
Hair	Gross examination (and with hand lens) Direct microscopy with or without KOH treatment Culture (bacteriological or mycological) Virology/PCR Hair shafts for mtDNA, follicles for genomic DNA Chemical analysis. See text DNA techniques

(Continued)

TABLE 7.3 (Continued)

Technique	Standard Recommended Methods
	Histology Electron microscopy
Semen	Direct microscopy — wet or fixed (stained) preparations. Bacteriology
Uroliths and other chemical deposits	Gross examination (and with hand lens) SEM. Optical crystallographic analysis. X-ray diffraction. Infrared spectroscopy. Chemical analysis
Invertebrates removed from a carcase or found in its vicinity	Standard forensic entomological techniques (see Barnes, 2013)
Bones and teeth	Histology, usually following decalcification (see Cooper and Cooper, 2007) Culture, usually following grinding DNA techniques

EDTA, ethylene diamine tetra-acetic acid; KOH, potassium hydroxide; SEM, scanning electron microscopy; TEM, transmission electron microscopy; PCR, polymerase chain reaction.

material that sinks in flotation test. This may include siliceous or other mineralised material suggestive of ingestion of soil.

Poor comminution of gorilla faeces — excess 'undigested' food — reflects either inadequate digestion (e.g., an enzymic deficiency) or accelerated passage of ingesta through the stomach or small intestine.

Obvious changes or abnormalities in faeces should be quantified: for example, faecal consistency of gorillas can be graded from 1 to 5 (Cooper, 1996; Cooper and Nizeyi, unpublished).

There are many unanswered questions about faeces of gorillas. For instance, it is axiomatic that consistency depends largely on their fluid content — but have estimates of water been systematically calculated in gorillas? How does the colour of gorilla faeces reflect the presence of bile or change as a result of ingesting certain plants? Gorilla faeces vary in odour but have such variations been linked with diet and health? Each of these questions would provide fieldworkers, including guides and trackers, with an opportunity to apply their powers of observation. It is another reason why 'routine' examination of faeces of gorillas by veterinarians should not just focus on parasites.

Hair

There are many uses of hair (Cooper and Cooper, 2013). The most basic one (important in ecological and forensic investigations) is to ascertain that the hair is, indeed, that of *Gorilla*. Furuichi et al. (1997) described the use of scanning electron microscopy for this purpose, to differentiate chimpanzee and gorilla samples but this is not always feasible, especially in Africa, and it is preferable to use simpler, cheaper techniques. This implies gross and microscopical examination — a skilled task but one that can be performed by a suitably trained person in a simple laboratory. Hair samples should be handled with care, especially if they are evidence in a suspected poaching or bushmeat case. Plastic forceps should be used and gloves worn. Hair should first be observed dry with the naked eye and a hand lens, dissecting microscope or magnifying loupe.

For examination hairs should be free of dirt and debris; cleaning with a fine brush may be needed; the dirt and debris should be retained in legal cases as it may be of evidential importance. Then microscopical examination (dry and wet preparations, reflected and transmitted light) can commence. Cut sections (prepared manually or using a microtome) may also be needed (Cooper and Cooper, 2013).

Urine

Urine collection and examination is discussed in Chapter 11: Urinary and Reproductive Systems.

DNA and Other Molecular Techniques

The use of molecular techniques is well established in work with captive gorillas and, increasingly, in the field. Whittier et al. (2010), for instance, described testing of faecal samples collected from free-living mountain gorillas for *Campylobacter* spp. using a portable, real-time polymerase chain reaction (PCR) instrument.

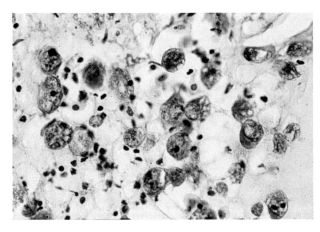

FIGURE 7.1 Immunocytochemistry — *Entamoeba histolytica* in liver.

Pathologists use molecular diagnostic techniques, but to varying degrees. Some haematologists use molecular tests as well as more traditional methods, much of genetic testing is based on multiple sequencing and most virological tests are molecular. Histopathological studies may, when appropriate, include molecular testing on formalin-fixed, paraffin-embedded, material but essentially still rely on conventional methods.

The use of molecular testing in pathology is increasing though. It now includes such procedures as immunocytochemistry (Fig. 7.1), hybridisation-based techniques and low-throughput PCR-based techniques (such as Q-PCR, Sanger sequencing and pyrosequencing). The elucidation of the human genome in 2003 (see later) has been followed by the introduction of techniques that can identify specific genes and analyse the signalling pathways involved. All of these have great potential in work on gorillas.

The most relevant breakthrough has been the generation of a reference genome assembly for *Gorilla* using DNA sampled from a single lowland animal. Sequence data for three other gorillas permitted a study of diversity within the genus *Gorilla*. See Scally et al. (2012) and the resources available at http://www.sanger.ac.uk/resources/downloads/gorilla/.

One of the most important molecular diagnostic tests, applicable to various types of sample, is the PCR. This offers many advantages over traditional methods of identifying bacteria and other organisms — see Box 7.2.

PCR has been used to advantage in studies on wildlife — for example, in the detection of *Mycobacterium* and *Leptospira* spp. and in the monitoring of spread of such organisms. PCR is not, however, a panacea. It often does not work as well as some other techniques when, for example, searching for intestinal organisms such as *Giardia* spp., probably because of the presence of PCR inhibitors in faeces. It also does not function adequately

on formalin-fixed samples; the deleterious effects of formalin are both time- and concentration-dependent.

The development of whole genome amplification and high-throughput sequencing techniques permits the sequencing of DNA samples to a far greater depth and at a lower cost. This facilitates the characterisation of the genomes of pathogens and the study of organisms in old material such as teeth and bones.

The taking and use of samples for DNA testing is covered in more detail elsewhere — see the section by Jaimie Morris in the preliminary report on skeletons in the Osman Hill Collection, Royal College of Surgeons — Appendix 6: Case Studies — Postmortem Investigations and Studies on Health.

Blood

Blood work is not covered in detail here. More information is in Chapter 10: Lymphoreticular and Haemopoietic Systems and Allergic Conditions.

Blood is a very important part of investigation — both for diagnosis and health monitoring — but obtaining a sample from a gorilla is usually an invasive technique.

Unfortunately, for obvious reasons, the extensive databases on haematology that were established for some NHPs, such as laboratory marmosets, in the 1970s and 1980s still do not exist for the genus *Gorilla*. Zoo and field records provide some information but still little on 'normal' healthy animals.

In monkeys changes in haematological values take place as a young animal ages (Sugimoto et al., 1986) but there appear to be no comparable studies on gorillas.

The late Christine Hawkey, haematologist at the London Zoo for many years, always emphasised the importance of the *art* of haematology — the correct preparation and staining of smears and, especially, meticulous microscopical examination, initially using low power, on all samples. She, for example, commented on a smear from 'Guy' (see Appendix 5: Case Studies − Postmortem Investigations and Studies on Health) '... hypochromic erythrocytes ... may provide a useful early indication of iron deficiency'.

Preparation of samples for examination will be mentioned briefly.

Blood smears are used for haematological studies and also to detect blood parasites. An old proverb states that 'Bon sang ne peut mentir' ('good blood cannot lie'). Although meant in a different context, those five French words (which will be understood by many gorilla workers in West and Central Africa) are a reminder that blood smears − and other samples − need to be of the highest quality if reliable results are to be obtained. As some biologists with limited training in haematology are beginning to realise, failure to detect parasites in smears may be because the preparation was poorly made, stained incorrectly or not properly examined (Cooper and Anwar, 2001). For that reason, a word will be said about making and staining blood smears.

There are three main ways of producing blood smears − see Box 7.3.

There are various ways of staining blood smears but no critical comparative studies appear to have been carried out in gorillas. 'Rapi-Diff II' is very quick and therefore of particular value in the field (see Appendix 4: Field Pathology) but the author advocates modifying the recommended (5-second) times to enhance staining and thereby helping to demonstrate organisms and changes. His particular preference, however, following several years of working in Africa, is Giemsa stain − 10% for 30 minutes. A thin film on a slide is fixed in pure methanol. The smear is then immersed in or flooded with a freshly prepared Giemsa stain solution (10% for 30 minutes) after which it is flushed with tap water and left to dry.

Techniques for the taking and examination of bone marrow from humans are well described by Islam (2013) and include aspiration and trephine biopsy, both applicable to gorillas. The need for correct preparation of a bone marrow aspirate film is carefully explained, with emphasis on the importance of producing a film that is narrower than the glass slide so that the edges can be readily viewed (also advisable with blood smears). Squash preparations are best made using two glass slides, not a slide and a coverslip.

BOX 7.3 Methods of Producing Blood Smears

1. Classic technique, using the edge of a glass slide.
2. Coverslip technique, using the surface of a coverslip.
3. Slide to slide technique, using the surface of a glass slide.

Biochemistry

Biochemical techniques are important diagnostic and prognostic aids and have been used for gorillas in captivity for many years. As a general rule the techniques employed and the reference values closely follow those in humans. There are many examples of their value. Elevated creatinine may indicate renal disease while raised bilirubin values may support a diagnosis of hepatic damage. Standard investigations for blood glucose can assist in diagnosing diabetes mellitus, liver or kidney dysfunction and tumours. Total serum protein values may assist in detecting dehydration and provide information about the animal's immune status.

Studies in companion, farm and laboratory animals have demonstrated many clinical applications for inflammatory biomarkers, including diagnosis, prognosis, detection of subclinical disease and chronic inflammation and monitoring stress. There is a need for similar research in gorillas.

Researchers used biochemistry in their work over half a century ago; for instance, Finlayson (1965) assayed serum cholesterol in primates at the London Zoo, between 1958 and 1963; his data included figures for two mountain gorillas.

Cytology

Cytology is a much-overlooked technique − cheap, rapid, and revealing information that may not be apparent in histological preparations. It should be used on all suitable occasions − live gorillas, dead gorillas and tissues from gorillas − even if more complex tests are envisaged later.

Cytological techniques can only be discussed briefly here. The essence in gorillas, as in other species, is to prepare, stain and examine touch preparations/impression smears of solid tissues or smears of fluids.

The key to successful cytological investigation is accurate sampling and tissue preparation. As in haematology (itself a form of cytology), the essential prerequisite is a monolayer. A useful analogy is an egg, where the yolk is the nucleus and the albumen is the cytoplasm. In cytological preparations, the cells should be like a fried egg − well spread and thin (Cooper and Sylvanose, 2016). Preparations that are too thick are not only difficult to read but also more likely to attract foreign bodies, especially in the field (Fig. 7.2).

At least two preparations, preferably more, should always be taken even if not all are stained and examined. It is far better to have an excess than to rely on only one smear and to have misgivings about sending it to a colleague for a second opinion because it may be lost or broken.

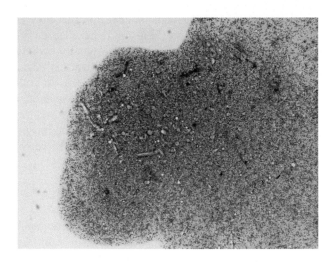

FIGURE 7.2 A cytological touch preparation of the spleen of a gorilla. It is too thick and has attracted vegetation and other debris.

Cytological preparations should be retained after initial examination, as they may be needed later. They should be stored in the dark, especially in Africa. It is often wise and a safety precaution to protect them by applying a cover slip (cover glass) to the surface but this makes it difficult to restain them at a later date.

Specific texts on human cytology can be used in work on gorilla material — for example, *Head and Neck Cytohistology* (Baloch et al., 2014). This is part of a series, sponsored by the Papanicolaou Society, that aims to provide an organ-based approach to the diagnosis of cytological and small biopsy samples.

Biopsies

A biopsy, by definition, is a tissue sample from a *live* animal. There are many entries in Part II: A Catalogue of Preserved Materials, that refer to 'biopsies' but some are actually tissue samples from dead animals.

Biopsy techniques suitable for exotic and wild animals were detailed by Cooper (1994).

Surgical pathology is discussed briefly in Chapter 5: Methods of Investigation — Observation, Clinical Examination and Health Monitoring.

Histology

The preparation of histological sections remains an important part of diagnostic pathology and plays a crucial role in veterinary studies on gorillas. Other material, such as smears of faeces or deposits in urine, can also be 'permanently' mounted — thereby presenting less of a health hazard in terms of residual pathogens and being available as reference material for review again later if a diagnosis is queried or a new disease is identified. Such specimens

can be adequately prepared, with or without staining, by applying a mounting fluid, and then a coverslip. It is particularly important in Africa to keep histological and other mounted preparations in the dark at an even temperature.

The preparation of histological sections is standard and will not be discussed here but specific points are made where appropriate elsewhere in the book. The central nervous system (CNS) presents particular challenges — see Chapter 15: Nervous System and Special Senses.

Histological techniques used are essentially the same for tissue specimens from live or dead gorillas. Both paraffin-embedded and frozen sections can be processed. Once a block has been made, serial (or step) sections can be prepared in order to search further for lesions and parasites.

Careful selection of samples and correct trimming are essential if stained histological sections of high quality are to be produced. At the time of sampling careful attention must be paid to (1) the site(s) of sampling (gorilla organs are large!) and (2) the number of samples to take from a particular organ or lesion. As a general rule, it is best to sample as many organs and sites as possible; the material can always be stored until needed or, as sometimes is the case in Africa, until funding or chemicals become available.

Tissues for histopathological examination must be fixed as soon as possible. The volume of fixative must exceed the volume of tissues by at least ten. Small pieces of tissue fix more rapidly and efficiently than do large. The standard histological fixative is 10% formalin, preferably neutral buffered, but other fluids can be used including, in an emergency in the field, alcoholic drinks such as spirits or beer (Spratt, 1993).

Trimming is the process whereby fixed tissues collected during postmortem examination or surgery are further dissected prior to their being embedded in paraffin wax. The person who does the trimming should ideally be the pathologist or clinician who took the tissues from the gorilla.

Reading of Slides

Adequate, competent, review of slides is time-consuming but essential. It needs to be properly planned. It is important to incorporate clinical and other information into interpretation of changes seen in histological sections. A certain amount of 'blind' reading of slides may sometimes be necessary — for example, in legal cases where a gorilla or its tissues are found and there is little or no history — but is not generally recommended.

When reading multiple blood smears or cytological preparations, one should look at them all even if this appears to be tedious and repetitious. The last slide may reveal something missed or not present in all the previous slides.

Awareness of artefacts is important when reading slides (see Fig. 7.2). Thompson and Luna (1978) first drew attention to the seriousness of such in the context of histological preparations but the message is equally true for other techniques including cytology. An excellent illustration of this was in the paper by Chang et al. (2014) who described how microorganisms and contaminants can be difficult to distinguish in various cytological preparations, including specimens from the gastrointestinal tract or the respiratory tract (where they may have been aspirated). Vegetable contaminants in particular can resemble squamous and columnar cells, viral inclusions, fungi and even nematodes and *Toxoplasma*.

A very helpful, well-illustrated guide to 'false' parasites in tissue sections was by Baird (1978).

Careful comparison of histopathological findings in different animals plays an important part in veterinary pathology. It enables lesions to be assessed more rigorously than does examination and description of changes in one animal only and makes it easier to apply a scoring system (see later).

In work on gorillas such direct comparisons are rarely feasible. The pathologist is likely to be examining tissues from one animal, not a series, and has either to describe lesions in the gorilla de novo or try to relate them to published descriptions by others. There are exceptions, of course — for example, in the investigation of a group of free-living animals that have been killed by poachers, where a pattern of injuries may be recognised. The situation is changing, however, as more and more samples begin to be collected in zoos and elsewhere (see Part II: A Catalogue of Preserved Materials) and because images of lesions can be shared electronically. There remains a need, nevertheless, for such material to be made more freely available — an outcome, it is hoped, of publishing this book.

The scoring of lesions is routine and well developed in diagnostic pathology. It owes its development to early work in the field of toxicological investigation (Scudamore, 2014). Linear and nonlinear grading schemes can be used. Such methods are applicable to work on gorillas and help to ensure consistency in reporting, both between animals and between observers. Scores may be represented in various ways — see Box 7.4.

BOX 7.4 Scoring of Lesions

+, ++, +++
(numerically) 1, 2, 3, 4, 5
(description) minimal, moderate, marked

Such methods are 'semiquantitative', in that they are based on relative severity of a lesion rather than on precise measurements. Image analysis is a rapidly evolving technique in diagnostic pathology. It automatically detects certain types of cell or tissue (e.g., goblet cells or collagen) or lesions and brings greater precision to quantification in diagnosis. A useful, recent, paper illustrates well the whole concept and advantages of whole slide imaging and digital pathology-based image analysis in the context of colitis in mice (Rogers et al., 2016).

Microbiology

Microbiology and microbiological techniques cannot be discussed in detail. A recommended text for work with gorillas is Topley and Wilsons' *Microbiology and Microbial Infections*. This was first published, as a single volume, in 1929 and now comprises eight volumes covering bacteriology (two volumes), mycology, virology (two volumes), parasitology and immunology. An excellent introduction to bacteria, their ecosystems, epidemiology, transmission and role in organ and system infections is provided in Volume I (Borriello et al., 2005).

Until relatively recently, diagnosis of viral infections was performed by virus isolation in cell lines, eggs or animals, electron microscopy, immunofluorescence and serological tests. Work on HIV in the 1980s led to the development of enzyme immunoassays, followed a decade later by PCR assays. Now molecular tests enable any known virus to be detected, including subtypes and variant strains. Such tests are of great value in rapid identification of viruses in outbreaks but, of course, the detection of an organism is not the same as determining its role as a pathogen — for which other investigations are needed.

Serology

Various serological tests (see Table 7.3) are a key part of diagnosis, health monitoring and surveillance. Mudakikwa et al. (2001) described serological studies in free-living gorillas as 'the investigation of disease by the measurement of variables present in serum'.

Additional Tests

Many simple techniques are applicable in the field — see Appendix 4: Field Pathology. Additionally, King et al. (2007) provided valuable advice on 'ancillary testing'. These ranged from the use of 'Pandy Reagent' for testing for excess protein in CSF to the collection of 'heavy, tiny particles' from a carcase, such as pieces of lead (essentially a flotation test); assessment of motility and appearance of intestinal villi by gentle application of warm saline; warming a (cytology) slide by breathing on it (or, in Africa, exposing it to the sun), before applying a touch preparation/impression smear: this accelerates the drying of cells and minimises artefactual change.

ENVIRONMENTAL TESTING

Environmental monitoring includes not only the examination and analysis of ambient materials such as water, soil and air but also examination of other species, animal or plant that may provide useful information on pathogens and toxic or radioactive substances in the locality.

The planting of 'sentinels' (susceptible animals of another species, such as rodents), to see if they develop clinical or subclinical disease or acquire parasites has found favour in epidemiological research on some human and animal species but does not appear yet to have constituted a part of health studies on gorillas.

Environmental samples include those listed in Table 7.4.

Novel techniques for environmental monitoring are constantly evolving. For example, Torondel et al. (2015) described the field testing of sentinel toys to assess environmental faecal exposure of children in rural India.

In a zoo setting in Europe, North America or Australasia proper water quality tests can and should be performed on a regular basis, as well as when there is concern over, for example, an enteric disorder.

The testing of water in gorilla habitat in Africa is less straightforward. Millions of humans there do not have access to clean, potable, water. Even a modern city such as Nairobi was in 2015 displaying notices stating that 'Safe drinking water cannot be taken for granted'. It is not surprising that responsible wildlife projects are, increasingly, concerned with the health of local people, especially trackers and rangers and their families, not just with gorillas and other animals.

Local zoos and rehabilitation centres need to try to ensure that they are using clean water for their animals and their staff. The monitoring of streams, rivers and pools in free-living gorilla habitat is important. When funding permits, use should be made of modern methods of water testing such as enzyme-specific technology but often only traditional ways are available – for example, membrane filtration and multitube fermentation methods.

When out in the field, where the gorillas live, things can be very different – see Appendix 4: Field Pathology. Even if no sophisticated quality testing is feasible, steps can be taken to check water from pools and streams using (1) basic test kits such as the dip sticks favoured by aquarists and (2) simple 'water purity' investigations. The latter can include the description of the water sample – for instance, the presence or absence of foam and froth and the listing and grading of macro- and microscopic live organisms present. There is also a simple test for

TABLE 7.4 Samples From the Environment

Sample	Free-Living Gorillas	Captive Gorillas
Hair, faeces, urine, blood, pus or other discharge – for standard laboratory tests	From night nests and locations where gorillas have been feeding or (trails) travelling	From nesting areas and elsewhere in the animals' cage or enclosure
Invertebrates, including flies. Some species of fly can carry eggs of intestinal helminths and may be involved in their spread (Adenusi and Adewoga, 2013)	In vicinity of night nests or trails	In sleeping quarters or enclosures See Barnes (2013) regarding forensic identification of invertebrates
Water – for standard tests, for example, bacteriology, and chemical/toxicological analysis	Ground water – or pools/streams with which gorillas have contact. See text	Water provided in containers (or piped) or remaining after disinfection of cages. See text
Food – for standard tests, plus special investigations, for example, toxicology, stable isotope analysis	Leaves of plants and other material seen to be ingested in the wild	Samples of items provided or foraged in the enclosure, plus any other nutritional supplements
Soil – for standard tests, plus special investigations, for example, toxicology, radioactivity measurements. Should include a search for invertebrates and helminth eggs	Collected in situ from areas frequented by gorillas. See Chapter 17 (Field Studies in Pathology and Health Monitoring) re value of knowledge of physicochemical and biological properties of soil – correlation with *Burkholderia pseudomallei*	Collected from enclosures
Air – for various tests	Not usually applicable but may become more relevant as pollution increases (see text)	Air sampling techniques can be applied in gorilla cages

clarity that relies upon looking down a tube at a cross mark on a piece of paper and, after slowly filling the tube with a sample of the test water, recording at which point the cross can no longer be seen. To the reader who has access to laboratory facilities such investigations will appear crude. However, in the author's experience, simple tests yield useful comparative data on different water sources and thereby provide training, relevant to human health as well as that of gorillas, for trackers, villagers and local school projects.

QUALITY CONTROL

Some laboratory tests on gorillas are likely to be performed in-house, by the zoo or the project; others will need the assistance of outside laboratories. There are pros and cons to each method (Cooper and Cooper, 2015; Duncan, 2015).

Whichever, the laboratory should have its own system of quality control (QC); that is, procedures that provide for a continual check on the accuracy of testing methods, equipment, reagents and working practices. These might range from the simple inclusion of standards in every run of biochemical analyses at one extreme to the adoption of a complete set of procedures covering all aspects of the laboratory's work at the other. It is up to the individual laboratory to decide on the extent of the QC procedures it will apply, but the overriding objective must be that they are sufficient to ensure that there is every confidence in the accuracy and reliability of the results produced. Not always easy in the field!

As stressed by Wobeser, unlabelled specimens have very little scientific value (other than as teaching material). Containers should be marked in such a way that the label will not be lost during handling or by contact with fluids (Cooper and Cooper, 2007).

RECORDING AND REPORTING OF FINDINGS

The proper collation of data from gorillas is important. Such data may relate to clinical health, postmortem findings or laboratory tests.

The International Species Information System (ISIS) is an international nonprofit organization serving more than 800 zoos and aquariums in approximately 80 countries. ISIS provides its members with zoological data collection and sharing software called ZIMS – the Zoological Information Management System. The ZIMS database holds information about thousands of taxa and individual animals and members of ISIS are able to use this to assist in the management of their collections.

For zoos in wealthier countries, ZIMS is a great asset. The fact that it is online means that the data can be accessed wherever there is an Internet connection; users can check their records even when they are away from the collection. For animal records such zoos may use ZIMS, ARKS, MEDARKS, SPARKS or REGASP.

Such is not always the case in Africa or in other less 'developed' parts of the world where gorillas are kept or rehabilitated. Their records may or may not be electronic and even if they are, access to them may be difficult if there is an electricity failure – in the rainy season, for instance. Under such circumstances paper records are always needed.

Some projects in Africa were fortunate in that they had overseas support and, through this, the opportunity to develop electronic databases. Thus, Cranfield and Minnis (2007) described the evolution of collections of mountain gorilla health data from handwritten notes (many lost when the VVC was looted during the 1994 genocide), their transfer to Microsoft Word and then their encoding for entry into a database. Initially MedARKS was used but this could not handle all the databases needed and so IMPACT, which can cope with multiple databases relating to a variety of species, was developed.

It is not just in areas of Africa that gorilla records are largely paper-based. Research for this book revealed that in the older zoological collections and institutions much of the available information about the health of gorillas is still in the form of hand or typewritten notes (see Chapter 6: Methods of Investigation – Postmortem Examination). See also the report by Sophia Keen – Postmortem Investigations and Studies on Health – in Appendix 5: Case Studies.

The records of zoological collections that have kept gorillas are a rich source of information about health, diseases, longevity and reproductive performance. In practice such data are not always easy to locate. Earlier (paper) records can sometimes no longer be found. Others are difficult to decipher because they are handwritten and not consistent in terms of the information they contain. More recent records, covering the past 30 years, are generally much more complete, reliable and readily accessible. This is in part because they are electronic and generally follow a standard system (e.g., ZIMS in the United Kingdom). In addition, in Europe at any rate, the requirements of the European Directive on Zoos and the (British) Zoo Licensing Act (ZLA) – see Chapter 18: Legal Considerations – mean that comprehensive records on health and behaviour are the order of the day. This does not mean that older, paper records are forgotten, however. Preinspection audit for an inspection of a British zoo under the ZLA includes questions about records and asks 'Are they [old records] readily accessible to bona fide enquirers?'

That said, gaining access to zoo records is not always easy. During the compilation of this book, polite enquiries were sent to large numbers of zoological collections throughout the world. Some otherwise highly respected institutions either did not reply or responded in unequivocal terms that they were not willing to divulge (= share) information.

Keeping handwritten notes in the form of a 'day book' is also a requirement under the British ZLA.

Such records are likely to remain an important part of monitoring the health and welfare of zoo animals, including gorillas, for some time to come. They should at the very least be photocopied and the copy kept in a safe location. Faxing the document to a third party helps ensure that the data are relatively secure. Even if the notes are put in an electronic form, original records should be retained. Apart from being an important historical archive, they can serve as vital evidence in the event of a legal action or claim of malpractice (Cooper and Cooper, 2007, 2013).

Reporting of results of necropsies was covered in some detail in Chapter 6, Methods of Investigation – Postmortem Examination.

As far as laboratory reports are concerned, the wording used should be precise, especially if there may be legal implications (e.g., a poached animal in the wild or a gorilla that dies under strange circumstances in captivity) (Cooper and Cooper, 2013). For instance, a report should say 'No parasites seen', not 'No parasites present'. Ambiguity must also be avoided, especially in non-Anglophone countries: one otherwise very proficient project in Central Africa has issued reports stating 'Faeces: normal for parasites'; what does this mean? Apparently negative results may only reflect the quality of the sample, poor laboratory technique or shortage of trained personnel (Cooper and Anwar, 2001).

Interpretation of findings always has to be carried out with caution. An example is the isolation of bacteria from a lesion on or in a gorilla. Questions that must be asked are as follows:

1. Was the sample that yielded the organism properly taken and transported? In other words, did the material received and processed by the laboratory represent, as far as possible, the sample taken from the gorilla?
2. Was the organism isolated in pure culture or was it a mixed growth?
3. How profuse was the growth?
4. What was the isolate? Is it an organism that is a recognised primary or facultative pathogen of gorillas?
5. Do clinical and other findings (e.g., haematology) support implication of the organism as the cause of the gorilla's ill health?

It is sometimes argued that the only person who can really interpret laboratory results with reference to the individual patient is the clinician. In fact, in the author's experience, the best outcome usually follows discussions between clinician and pathologist. This is reinforced by recent studies in human medicine, where different disciplines are involved in the diagnostic process. For example, Galloway (2010) highlighted differences in the interpretation of reports between clinicians and pathologists and suggested that it might be prudent to record whether a diagnosis was made solely on histological grounds or as a result of clinico–pathological correlation.

PERSONNEL AND EQUIPMENT

A veterinarian will have broad knowledge of diagnostic and investigative techniques and should be able to link laboratory findings with clinical signs of disease or pathological changes, and is therefore the ideal person to coordinate all investigative work relating to live or dead gorillas.

SAFETY IN LABORATORY WORK

This is discussed in more detail in Appendix 4: Hazards, Including Zoonoses and (in respect of fieldwork) in Appendix 3: Field Pathology.

Microbiological safety cabinets are a standard feature of modern laboratories and offer a safe environment in which to deal with potentially hazardous samples. Laminar flow cabinets provide an opportunity to protect samples (e.g., those for DNA extraction) and establish a clean environment in which to work.

When taking samples for both diagnostic purposes and for the extraction of DNA carefully planned SOPs (standard operating procedures) are needed. The requirements of the diagnostician (to collect material and to minimise health risks to those handling it) are rather different from those of the molecular biologist (to obtain samples that are as free as possible from extraneous DNA).

STORAGE OF SAMPLES AND REFERENCE COLLECTIONS

The importance of retaining and storing animal material in the form of 'reference collections' is emphasised throughout this book and, indeed, is one of the *raisons d'être* for combining the text with a catalogue of preserved specimens. Such collections provide opportunities for basic studies (e.g., anatomy), for research on genetics and for retrospective investigation of disease. The methods used for retention and processing of material are important and there are challenges in terms of health and safety requirements and CITES requirements (see Chapter 18: Legal Considerations).

Reference collections of biological material have existed for many centuries in the form of private and public museums. For over 200 years these provided an opportunity for zoological and botanical material to be both displayed and studied. The emphasis in early museums was usually on taxonomy and morphology. Collections concerned with pathology and disease were less numerous, but those that existed provided an important resource for the teaching of human, veterinary and comparative medicine. Some of the early medical museums were, and remain, of great value in promoting an understanding of concepts and in encouraging research on health and disease.

As pointed out in the Introduction to Part II: A Catalogue of Preserved Materials, the curation of gametes is a relatively new concept but can play an important part in wildlife conservation (Holt et al., 1996).

Cranfield (2002) stressed that it is important to collect and store biological specimens from gorillas for later use, reiterating the pleas by the author (JEC) in 1996, at the Annual Meeting of the European College of Veterinary Pathologists (ECVP) and repeated at the PHVA on the mountain gorilla in Uganda in 1997. Both Cooper and Cranfield emphasised that these samples need to be collected, processed and stored in a proper and uniform fashion to assure quality, and therefore accuracy and reliability, of the resulting tests and procedures.

As is obvious from Part II: A Catalogue of Preserved Materials, many collections of gorilla material are to be found in different parts of the world. They range from microscope slides and DNA samples to whole skeletons and skins. Similar assemblages have already proved their worth in conservation programmes – for example, those established for Mascarene specimens 30 years ago (Cooper et al., 1998).

The retention of material from gorillas following clinical intervention or necropsy is routinely practised in many zoological collections (see Part II: A Catalogue of Preserved Materials). For example, at Twycross Zoo in the United Kingdom (Sarah Chapman, personal communication), where all postmortem examinations are performed in-house but histology is carried out elsewhere, duplicate sets of fixed

> **BOX 7.5 Reference Collections – Important Criteria**
> - Legal acquisition of all material.
> - Meticulous provenancing of all specimens.
> - Proper cataloguing, including unique reference numbers and reliable labelling of specimens.
> - Well-publicised information on how specimens can be studied or used, allowing open access to bona fide investigators from anywhere in the world.
> - Strict adherence to national and international laws when material is moved.

tissues and some, such as muscle, liver and kidney (stored fresh at $-80°C$), are retained at the zoo.

Various criteria are important in the establishment and maintenance of reference collections and these are listed in Box 7.5.

As Part II: A Catalogue of Preserved Materials, shows, much contemporaneous material from gorillas is maintained outside Africa. This is unsatisfactory, particularly when it concerns specimens from free-living gorillas. In an unpublished presentation in Kenya in 1997, John and Margaret Cooper first drew attention to the need to establish reference collections in *Africa* where the material can be readily examined by local scientists. The suggestion was that the Institute of Primate Research, based near Nairobi and affiliated to the National Museums of Kenya (see Chapter 2: The Growth of Studies on Primate Pathology) might be an ideal location for this. Discussion continues and is likely to intensify as African countries push for more control over the use of their own sovereign natural resources.

Chapter 8

Nonspecific Pathology

John E. Cooper

The wounds that cannot be seen are more painful than those that can be treated by a doctor.

Nelson Mandela

INTRODUCTION

In the succeeding eight chapters the particular features of pathological change in various organ systems are discussed. In each of these the approach is broadly that used by the author in his lectures to veterinary, medical and dental students; where applicable, it looks at each organ system in terms of the headings in Box 8.1.

The headings used in subsequent chapters are similar to those employed by Scott (1992) and McGee et al. (1992). However, where it seems more helpful to the reader, or only limited information relating to gorillas is available, certain organ systems or types of tissue have been subsumed into chapters of a more general nature — e.g., adipose tissue is covered in Chapter 14, Musculoskeletal System.

In this chapter nonspecific pathology of the gorilla, not relating to particular organ systems, is discussed, together with some important concepts that have a pathological relevance such as pain and welfare. Environmental pathology (relating to such factors as dust, air or water pollution and dietary contaminants) is touched upon in other chapters.

GENERAL CONSIDERATIONS IN PATHOLOGY, NOT RELATED TO ORGAN SYSTEMS

It is simplistic to assume that in gorillas, or in any other animal species, pathological changes occur independently in different organs. Some infectious diseases are systemic and affect multiple organs or have profound physiological sequelae, for example, electrolyte imbalance (Soifer, 1971). Shock affects most parts of the body (see Chapter 10: Respiratory and Cardiovascular Systems) and toxaemia/

septicaemia can cause necrosis of different organs. Metabolic disorders can damage organs and tissues ranging from the parathyroid to the appendicular skeleton. Sometimes there is interplay between two systems — for example, the 'cardiorenal syndrome' (CRS) in humans where acute or chronic dysfunction in one organ may induce acute or chronic dysfunction of the other (Ronco et al., 2010). It is likely that there are other examples of where dysfunction in one organ causes dysfunction in another; it and other syndromes could well be detected in gorillas in due course: both the clinician and the pathologist should be aware of them. All scientific research starts with observation!

Awareness of the nonspecificity of certain diseases is growing. In human medicine, not only surgeons but, increasingly, surgical pathologists may concentrate on an area of the body of the patient rather than an organ system. Such 'regional pathology' is illustrated by the book edited by Baloch et al. (2014) which is concerned solely with cytology and histology of the (human) head and neck.

AGE-RELATED CHANGES

When the eminent (medical) primatologist W.C. Osman Hill was Prosector to the Zoological Society of London nearly 70 years ago, he listed 'causes of death' in the collection's animals under nearly 20 headings, of which the first was 'Senility'. Findings were then given that contributed to this opinion/diagnosis (no specific criteria were used).

The term 'age-related changes' refers here to *old* age but some pathology can be specific to other stages of life, such as the neonatal period — see Chapter 13: Urinary and Reproductive Systems. As such it focuses on geriatric ('elderly') gorillas and gerontology — the scientific study of old age and the process of ageing. Ageing (or aging — USA) is the process of becoming older.

First, one may need to decide what an 'elderly' gorilla is. Figures on longevity exist for zoo gorillas and for relatively small numbers of free-living animals — see below. An indication of the former can be gleaned from 'Mouila'

J. E. Cooper & G. Hull: Gorilla Pathology and Health. DOI: http://dx.doi.org/10.1016/B978-0-12-802039-5.00008-1

BOX 8.1 Approach to Each Organ System

- A brief review of anatomy and key microscopical features.
- Embryological origin (ectoderm, mesoderm, endoderm).
- Overview of the types of response to infectious and non-infectious insults shown by the relevant organ system.
- Examples of diseases and pathological changes seen in the organ system usually under the following headings:
  - congenital/developmental
  - inflammatory (infectious and noninfectious)
  - traumatic
  - neoplastic
  - other

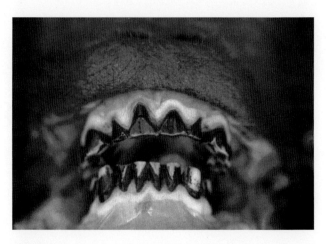

FIGURE 8.1 An elderly (dead) female mountain gorilla. Note the worn teeth (and ample tartar) and the swollen gingivae. *Image courtesy Ian Redmond.*

who died aged 54 in April 2014 at Howletts Wild Animal Park: at the time she was the oldest western lowland gorilla in the United Kingdom. It is less easy to obtain accurate statistics for free-living gorillas although habituation of some groups has helped provide some such data. Earlier figures quoted in books were usually only estimates; Verrengia (1997), for example, writing about Central Africa, stated 'They typically live about 35 years until they are weakened by parasites, food shortages and disease'.

As the Introduction to Part II: A Catalogue of Preserved Materials, points out, some authors describe gorillas as having faster 'life histories', possibly as a consequence of the lower energetic risk in their environment, than do chimpanzees and orangutans. Numerous studies have been made of the life history traits (rate of maturation and reproduction, age at primiparity, gestation length, age at weaning, interbirth interval, etc.) of gorillas but these studies are ongoing and the hypotheses put forward are often revised when new or additional data come to light.

Studies comparing western lowland and mountain gorillas (Robbins et al., 2004, 2007) suggest that, because the food resources of the former are seasonally less reliable and more widely dispersed, lowland gorillas tend to have greater day journey lengths and home ranges than do mountain gorillas. The starvation risk for infants means that mothers wean them about a year later and their interbirth interval is correspondingly increased by a year. When in captivity, however, lowland gorillas receive good-quality food on a regular basis and the lower level of energetic risk enables females to wean offspring at an earlier age. This may influence longevity and health. As in humans — and, indeed, other animals — features associated with ageing are physical and can be seen — change in hair colour or loss, reduced activity and altered behaviour are examples. If the gorilla is handled, or examined postmortem, other features will also be noted, such as wear of teeth (Fig. 8.1).

Most features of the ageing process as we understand it from *Homo* and domesticated species are invisible; they are

internal. Systemic hypertension is a good example in humans — a not uncommon finding in elderly people when their blood pressure is checked. Atsalis et al. (2004), Atsalis and Margulis (2006) and Margulis et al. (2007) studied reproductive cyclicity in geriatric gorillas; it continues into the fifth decade and then culminates in a menopause (see also Chapter 13: Urinary and Reproductive Systems).

Information on the clinical effects of old age on free-living gorillas is, at present, restricted to observations on individual animals. Such was the case with the silverback 'Ruhondeza' (Kalema-Zikusoka, 2012), arguably the most famous mountain gorilla in Bwindi Impenetrable National Park, Uganda, who died on 27 June 2012 when over 50 years old. Kalema-Zikusoka recounted how, 4 months before his death, following the take-over of his (small) group by another silverback, Ruhondeza had become a solitary male, living on community land in a local village. She checked the animal a few weeks before he died and reported 'Ruhondeza seemed fairly active, though very shy and wanting to be left alone'. Though Ruhondeza had destroyed a few banana crops, she and her colleagues were able to convince the community that they should tolerate and respect him on the grounds that he had in the past generated much revenue for them through tourism. He was followed and tracked until his death.

Other old animals in Bwindi also provide data on the appearances that might be expected in old age. When silverback 'Mugurus' (old man), believed to be about 50 years old, started to deteriorate he was described by Gladys Kalema-Zikusoka (personal communication) as 'looking very old with a lot of white hair in his face … haggard-looking … loose skin on his face and neck'. He had shown signs of ill health for 2 months — weak and finding it difficult to keep up with the rest of the group.

When the body of 'Nkuringo', a silverback aged at least 50 years, was found in Bwindi in 2008, the Chief Warden said 'Gorillas have a life expectancy of 60 years.

Mzee Nkuringo had lost most of his teeth and used to move slowly'.

As far as postmortem signs are concerned, Scott (1992) discussed some of the pathological changes associated with ageing in both 'humans and primates'. A general account by Meder (1994) reported that, while old gorillas require more time to forage, feed and move, healthy group members usually adjust their activities to meet such needs. According to Meder animals aged 35 years or more frequently suffer from arthritis, particularly affecting the joints of their hands and feet. There is some, but relatively little, attrition of teeth as gorillas age and 'whereas western lowland gorillas sometimes get caries, the mountain gorillas' diet is low in sugar, and caries is therefore rare'. She added that it is possible that older animals actually starve to death because they are unable to chew their food.

The report by Jones and Cave (1960) dealt with chimpanzees, not gorillas, but is worthy of study because it links diet and dental disease with increased longevity.

Specific reports on the pathological effects of ageing in gorillas were, until very recently, rare. One exception was the volume edited by Erwin and Hof (2002). Various contributors discussed gerontological studies on great apes and other primates and Nicholas and Zihlman proffered a 'preliminary report' on skeletal and dental evidence of ageing in captive lowland gorillas. Other papers cover 'skeletal ageing' in chimpanzees at Gombe, the comparative neuropathology of brain ageing in primates (a strong emphasis on gorillas – see Chapter 15: Nervous System and Special Senses) and comparative models of cognitive decline in ageing great apes.

Paixão et al. (2004), writing from Brazil, described their findings in a captive senile western lowland gorilla with chronic renal failure and septic polyarthritis. The (apparently) most extensive study to date (2016) is the paper by Lowenstine et al. (2016) reviewing the comparative pathology of ageing bonobos, chimpanzees, gorillas and orangutans. The three authors pointed out that there are significant differences in changes seen during ageing in these species – and differences from the 'human ape'. Common to all of them are dental attrition, periodontitis, tooth loss, osteopoenia and arthritis. Gout is (appears to be?) unique to humans while spondyloarthropathy is more prevalent in apes than humans. Humans are more prone to frailty, sarcopoenia, osteoporosis, postreproductive senescence, loss of brain volume and Alzheimer dementia. Cerebral vascular disease occurs in both humans and apes. Cardiovascular disease mortality increases in ageing humans and apes, but coronary atherosclerosis is the most significant type in humans. In captive great apes, idiopathic myocardial fibrosis and cardiomyopathy predominate, with arteriosclerosis of intramural coronary arteries. Similar cardiac lesions are occasionally seen in free-living apes. Lowenstine et al. (2016) suggested that

the finding of vascular changes in heart and kidneys and aortic dissections in gorillas and bonobos suggest that hypertension may be involved. Chronic renal disease is common in elderly humans and some ageing apes. Neoplasms common in both humans and apes include uterine leiomyomas in chimpanzees but other tumours of *Homo* in old age are uncommon in apes.

The paper above provides useful hard data on pathological changes in ageing gorillas, to which can be added data and observations from other sources. Some further, careful, extrapolation from studies on humans is probably also acceptable – for example, published evidence that geriatric diseases compromise host responses to stress and impair immunity. As in humans, ageing has biological, social and psychological implications for gorillas. Neurological and psychological changes associated with ageing are discussed in Chapter 15: Nervous System and Special Senses. Getting old is a complex process, neither specifically infectious nor noninfectious.

Individual observations on postmortem findings in old gorillas can also add to the sum of knowledge and, perhaps, suggest patterns. For example, Kalema-Zikusoka (personal communication) in her postmortem report on 'Mugurus' said 'I think Mugurus died of heart failure as there was a pericardial effusion and fat tissue in the heart and attached to the AV valves and chordae tendineae'.

Lowenstine et al. (2016) stated categorically that 'it is clear that great ape gerontology is in its infancy as compared with the understanding of ageing processes in the human ape'. There is clearly a need for clinicians and pathologists to focus more attention on elderly gorillas and to obtain as many data as possible from them. Some investigations can be easily performed as part of routine health checks, for example, cardiac assessment, including blood lipid values (Baitchman et al., 2003). Biomarkers related to the function of kidneys and liver are used to measure the ageing process in humans. Some such studies indicate that age-related changes start to occur in young people and can be used to predict the signs of advanced ageing that are seen in adults. It will be interesting to see if a similar picture emerges in gorillas.

Unfortunately, it is not always possible to perform a basic necropsy on a dead geriatric free-living gorilla, let alone take the various samples that might throw more light on the animal's demise. Such was the case with 'Effie', probably the oldest known female mountain gorilla, who was found dead in the Rwanda rainforest in 1994. She had been first identified by Dian Fossey in 1967, when 'Effie' was estimated to be 13 years old. The author (JEC) planned to make a visit to the forest to perform a postmortem examination but was prevented from doing so because of the military situation following the outbreak of genocide. The recovery and curation of skeletons of some such animals in Rwanda (see Appendix CA2: Retrieval, Preparation and Storage of Skeletal and

Other Material) means that skeletal information can be garnered but in gerontological studies soft tissues are also of great importance.

Research on ageing in mammals is now interdisciplinary and those interested in geriatric changes in gorillas need to collaborate with others working with different species. A recent issue of *Veterinary Pathology* (Ward et al., 2016) had as its special focus 'Pathology of ageing' and included a thought-provoking Editorial that discussed the pathogenesis of old age and the role of pathology in determining the cause of death (COD) (Ward et al., 2016). The organisation ShARM (http://www.sharmuk. org/), funded by Britain's Wellcome Trust, is open to all scientific investigators in the United Kingdom and overseas and aims to encourage research into ageing by facilitating the sharing of resources, including access to a biorepository of flash-frozen and formalin-fixed, paraffin-embedded, aged murine tissues.

DEVELOPMENTAL ABNORMALITIES

Developmental pathology is defined here as pathology of (1) fetal development (a form of fetopathy, disease or disorder of a fetus; see also Appendix 1: Glossary of Terms, and Chapter 13: Urinary and Reproductive Systems), (2) parturition (such as asphyxia or trauma), (3) adaptation to extra-uterine existence and (4) developmental tumours. The first of these is briefly considered here.

Developmental abnormalities have been recognised and described in various species of nonhuman primate (NHP). A useful starting point is the chapter by Schultz in Fiennes (1972). Schultz divided developmental abnormalities into those involving the head and skull, the dentition, the vertebral column and thorax, and the limbs (see also Chapter 14: Musculoskeletal System).

Detailed investigations of developmental abnormalities have usually involved NHPs in captivity rather than those in the wild. Hendrickx and Binkerd (1993) provided a useful analysis of findings in laboratory species but their monograph, not surprisingly, does not specifically mention gorillas.

Particular interest in malformations in laboratory primates was prompted by the suggested use of New World callitrichids as models for teratological research (Hetherington et al., 1975).

The author has been unable to locate specific information on how best to examine neonatal gorillas for developmental anomalies. However, use can be made of various texts covering human teratology and the excellent book on developmental abnormalities in domesticated livestock (Szabo, 1989). Despite its title, there is mention in Szabo's work of supernumerary incisors and premolars and the absence of a third molar in gorillas and his

Appendix on 'Examination of the Newborn' is of considerable value in work with NHPs.

Asymmetry of the cranium of mammals has attracted attention for many decades (see Part II: A Catalogue of Preserved Materials). In some species, including primates, it is considered to be indicative of stress and/or a high inbreeding coefficient. Groves and Humphrey (1973) demonstrated its existence in gorillas with lateralised brain function.

The few reports of developmental abnormalities in gorillas include syndactyly — see photograph of 'webbed toes of a mountain gorilla' in the *Gorilla Journal*, June 2015 (number 50). Syndactyly was first described in mountain gorillas by Schultz (1934) and then by Fossey (1983). In the early 1990s Foster (personal communication) said that syndactyly of the toes was especially prevalent in one research group of mountain gorillas (Group 5) but that he saw no evidence that it affected foraging. A survey by Routh and Sleeman (1997) showed a prevalence of 52.7% (39/74). Syndactyly of the toes was the standard finding (only one animal showed syndactyly of the fingers) and, of these, 36/39 involved digits III and IV. Kalema-Zikusoka and Lowenstine (2001) reported that syndactyly was present in three digits on the feet of a free-living mountain gorilla treated surgically for a rectal prolapse.

Cleft palate is another anomaly of interest. A baby gorilla with a palatine abnormality was necropsied in Rwanda by the author in 1994, just before the civil war started. The initial report has not been traced and the remains of the gorilla, which awaited further investigation and were probably stored frozen in the National Veterinary Laboratory, Kigali, were never found. Both were probably lost as a result of the war and widespread looting. Nutter et al. (2005) presented a brief account of a case of cleft palate in a neonatal mountain gorilla at the Wildlife Disease Association Conference.

Fowler (1999) described examination of a 2-year-old infant gorilla that had failed to mature, remaining at the developmental stage of a 2-month-old infant. Hydrocephalus was confirmed by CT scanning at the University of California Medical Center — 'one of the first ever CT scans on a gorilla'. A case of hydrocephaly was referred to in the context of the carriage of *Shigella flexneri* by primates in the National Zoological Park, Washington, DC, United States (Banish et al., 1993).

GENETIC ANOMALIES

The eminent British veterinary pathologist, the late Jack (JT) Done, frequently stated in his lectures that 'all disease has a genetic component'. Proven instances of genetic anomalies in gorillas are, however, few.

Errors of pigmentation are discussed in Chapter 9: Skin and Integument, with particular reference to albinism. Bradford et al. (2013) described an unusual Turner syndrome mosaic in a western lowland gorilla.

Inbreeding is often implicated as a cause of developmental and other disorders in gorillas but, as Mudakikwa et al. (2001) pointed out in the context of syndactyly (see above), no conclusion can be made concerning aetiologies of syndactyly until molecular genetic studies have been completed. Uncontrolled inbreeding in many other mammalian species is known to reduce fertility, maternal care, growth rate and general health.

The effects of inbreeding on juvenile mortality in small populations of ungulates were addressed by Ralls et al. (1979). This and subsequent publications have led to the belief that a high inbreeding coefficient is, per se, undesirable and deleterious. It is interesting, therefore, that a recent study (Xue et al., 2015) involving the sequencing of whole genomes from free-living mountain gorillas and comparison of all four *Gorilla* subspecies revealed that many harmful genetic variations have been removed from the mountain gorilla population through inbreeding. It appears that mountain gorillas are genetically adapted to living in small populations and have survived in such small numbers for millennia. Concerns remain, however, that the low level of genetic diversity may make the mountain gorillas more vulnerable to environmental change and to infectious disease, including human viruses.

BEHAVIOURAL PATHOLOGY

See also Chapter 15: Nervous System and Special Senses.

Interest in the 'normal' behaviour of gorillas and other great apes has a long history. Nearly 70 years ago Morris (1958) wrote of 'the *Inborn Activities* which develop in the same way in all members of a species, even in an unnatural environment'. Since then a remarkable amount has been published about the behaviour of gorillas, both in the wild and in captivity. Some of this has been prompted by a specific interest in the ethology of these species, both for their own sake (e.g., the welfare of captive gorillas) and in order to explore the possible origins of different traits in *Homo sapiens*. Insofar as this book is concerned, relatively little will be said about behaviour other than (1) the pathology associated with abnormal behaviour, particularly in captivity and (2) (briefly) those 'behaviours' that may be pathological in origin, either because they are secondary to organ dysfunction or because of some underlying psychotic condition. At present little is known about the latter but it is not unreasonable to suggest that, in due course, a number of psychological disorders will be recognised in the genus, some similar to those seen in humans. Behavioural ecology and health were the theme for a far-sighted paper by Heymann (1999) who,

in his 'glimpse into the future', raised several questions about diseases in free-living primates, including their role, if any, in mate selection and reproductive success.

The importance of self-mutilation or injury by other animals in captivity was emphasised over 70 years ago (Fox, 1923). In a recent review, Cousins (2015) discussed 'disturbed behaviours' in captive great apes. He was frank in his comments about early menageries and argued that those responsible for such collections knew little about the requirements of these animals — 'bad diets, bleak housing, lack of proper husbandry, woefully inadequate healthcare and high mortality'. He categorised abnormal behaviour under the headings of self-mutilation, stereotypies, regurgitation and reingestion, coprophagy and depression.

Cousins explained that self-mutilation ('autoaggression') expresses itself in self-abuse, ranging from hair plucking (or overgrooming) to deliberate self-wounding. Little is understood about the aetiology, although it is surmised to be pathological cleaning behaviour. The causes appear to include social deprivation, possibly a reflection of the absence of physical and social intimacy over an extended period. Self-wounding was observed in two female gorillas in Wroclaw Zoo, Poland. Both suffered from peripheral circulation disorders that arose from a combination of stress and flawed husbandry, and both responded by destroying the nails and phalanges of their hands and feet (Dorobisz et al., 1983).

Cousins cited also the case of 'Xara', a female gorilla in Rotterdam Zoo, The Netherlands, who gnawed off most of her right leg, which resulted in osteomyelitis. The remains of the leg were amputated and Xara was not only successfully reintegrated, but she went on to become a breeding animal (Meder, 1989).

Earlier writers listed such behavioural problems as apathy, overeating and overdrinking, coprophagy, stereotyped behaviour, self-directed aggression and infanticide as being linked with inactivity. Early authors of books wrote in general terms about the apparent gentle lifestyle of free-living animals; Verrengia (1997), for example, said 'An adult gorilla's life is spent in a quiet routine of eating, breeding, and sleeping'. In an early study Herbers (1981) analysed 'time resources and laziness in animals' and suggested that gorillas in the wild spent 25% of their time foraging, in contrast to 55% in chimpanzees). These figures have been greatly refined by recent ethological studies, both in the wild and in captivity (Less et al., 2012).

Stereotypies are well recognised in many species of mammal, including great apes. They can be defined as 'external forces fostering internal problems'. Stereotypical movements in NHPs include body-rocking, head-bobbing, pacing, weaving and head-banging, all equally unsettling to witness. Davenport (1979) reported a broad range of stereotypic patterns in nursery-reared chimpanzees and gorillas but found it absent in orangutans.

FIGURE 8.2 Environmental enrichment for gorillas in a European zoo.

The 'abnormal behaviour' that has attracted the greatest attention in captive gorillas is undoubtedly the practice of regurgitation and reingestion (sometimes referred to as 'R/R'. This is discussed in Chapter 11: Alimentary Tract and Associated Organs.

Space does not permit more than a brief mention of attempted prevention or treatment of behavioural disorders in gorillas. However, there is ample literature — see, for example, Nadler (1989) concerning the psychological well-being of captive gorillas and Brown and Gold (1997) and Rooney and Sleeman (1998) regarding behavioural enrichment devices (Fig. 8.2).

Positive reinforcement training (PRT) originated in zoos from which it has spread to other areas of care, including laboratory animals. It has its origins in observations many years ago. Dixson's (1981) book includes a plate showing 'Congo', a young female mountain gorilla, stacking boxes to reach a suspended food reward (after Yerkes, 1928).

There are various ways in which pathological changes can follow or be associated with abnormal behaviour. Stereotypies can cause injury to limbs; the insertion of foreign material into body orifices can result in infection (see vaginitis).

AGGRESSION

Aggressive behaviour commonly leads to injury in gorillas. It is usually sexual in origin and has been studied in the context of similar aggression in humans (Nadler, 1988). It may be directed at other gorillas or other species, including humans (see 'Fighting' section in Chapter 4: Noninfectious Disease and Host Responses).

Concern about belligerence in captive great apes in captivity has led to studies on management and methods of therapy (Burks and Maple, 1995; Redrobe, 2008).

In the wild, aggressive behaviour by a male gorilla is impressive. It has been described as 'psychological warfare'. The display was described at length by Schaller (1963, 1964). It comprises hooting, symbolic feeding, rearing up on to hind legs, tearing up vegetation, chestbeating and leg-kicking. This is followed by a rapid sideways run with more destruction of vegetation, and finally a thump on the ground with the palm of the hand.

In the field, attacks on veterinarians are not uncommon. Habituated animals seem able to spot someone who has tried to immobilise them, even if the veterinarian is differently dressed, blackens his face and (as the author tried in Rwanda) hides the blowpipe in his trouser leg! The exhortation 'don't run/*msikimbie*' is heard when a silverback charges. One (Rwandan) veterinarian suffered substantial damage to his thigh when the silverback 'Guhonda', recovering from anaesthesia, was startled by a tracker who started to run.

Local wildlife may also be attacked by great apes but, according to Ross et al. (2009) gorillas were the least likely to engage in such aggressive interactions. However unlike free-living gorillas, captive gorillas were reported to kill (and in one case, eat) local wildlife. This may have implications in terms of possible pathogen transmission.

EFFECTS OF CAPTURE, TRANSPORTATION AND CAPTIVITY

The history of bringing gorillas into captivity is a sorry one. Many died before they left Africa; others succumbed en route to Europe or North America. A mounted skin referred to in Part II, A Catalogue of Preserved Materials, is from a juvenile male purchased alive, with a young chimpanzee, in Gabon in 1851. Both died on their way to France later that year. Even if gorillas endured the (sea) journey, many only survived for a short time. For example, one juvenile male that arrived in the Mènagerie du Jardin des Plantes in Paris on 17 January died on 21 April 1884 (see Part II: A Catalogue of Preserved Materials). Those immigrés that survived a while in a zoo usually died of unknown causes and were rarely exposed to a professional necropsy. Morris and Morris (1966) related that 'Jenny', the first gorilla to reach Europe alive, was exhibited in England by the famous showman George Wombell in 1855 but was not destined to remain in his menageries for long. By the time the travelling troupe had reached Warrington, 'without previous symptoms of decay, Jenny fell sick and breathed her last'.

A few decades later the situation had not markedly improved. Burbridge (1928), for example, reported that of eight young gorillas captured, three survived, one being the mountain gorilla 'Congo' later studied by Yerkes.

Transportation of live gorillas presents challenges in terms of both the health and welfare of the gorilla and the safety of those persons coming into contact with it. The first animals to arrive at zoos in Europe and North America were transported by sea, as indicated above. They were exposed to various stressors over a long period of time, their diet was probably inadequate, and hygiene would have been basic.

In recent years gorillas have generally been moved by air, usually in specially designed crates. As recently as the mid-1990s it was possible for a 10-week-old lowland gorilla to be transported on the lap of a zookeeper in the first-class cabin of a commercial airline (Verrengia, 1997).

It is interesting that mountain gorillas have never been successfully kept in zoos although orphans have adapted to captivity in the Senkwekwe Centre, Virunga National Park, DRC, where their diet includes natural food-plants (Ian Redmond, personal communication).

Even reintroduction of gorillas brings with it concerns about the adverse ('stressful') effects of confinement, movement and release (see Jane Hopper's contribution in Chapter 17: Field Studies in Pathology and Health Monitoring). Other pertinent references include Pearson

et al. (2007) and Teixeira et al. (2007). Nor even is the rescue and care of 'orphan' gorillas without its challenges. Helen Attwater, who worked at the PPG Brazzaville orphanage from 1989 to 1995, during which time 64 gorilla orphans arrived, wrote: 'a pathologist from Kyoto University ... confirmed the conclusions we had drawn from our experiences to date: that the gorilla's two main areas of vulnerability were his complex, finely tuned digestive system, that could be easily and fatally disrupted, and his fragile emotional make-up.... Emotional stress, exposure to human disease and malnutrition that resulted in the breakdown of a fine balance of intestinal parasites held in check under natural conditions, made the survival of newly arrived orphans unlikely' (Attwater, 1999).

The various behavioural disorders and other factors discussed above are relevant to welfare − see later − and in most cases comprise stressors. Stress is discussed in detail in Chapter 12: Lymphoreticular and Haemopoietic Systems and Allergic Conditions.

Some other 'nonspecific' considerations relevant to the health and well-being of captive gorillas include environmental influences on reproductive and maternal behaviour (Miller-Schroeder and Paterson, 1989), interactions between gorillas and their keepers (Chelluri et al., 2012) and the effects of visitors in zoos (Carder and Semple, 2008; Wells, 2005; Stoinski et al., 2012). In addition to published papers, studies on visitors have formed part of a number of undergraduate and postgraduate student projects in the United Kingdom; summaries of some of the findings appear regularly in 'Primate Eye' (*Journal of the Primate Society of Great Britain, PSGB*) and the records of the Universities Federation for Animal Welfare (UFAW).

WELFARE

Regrettably, for reasons of space, welfare can only be covered per se in very general terms. Nevertheless, it is a very important consideration in all studies on gorillas, both in captivity and in the wild. Veterinarians in particular should regard welfare as an integral part of their responsibilities. In the United Kingdom, when veterinary graduates are admitted as Members of the Royal College of Veterinary Surgeons (RCVS), each person pledges that his/her 'constant endeavour will be to ensure the welfare of the animals committed to my care'.

The welfare of gorillas is linked to health and also to stress (see Broom and Johnson, 1993, and Chapter 16: Endocrinological and Associated Conditions).

Animal welfare is not easy to define. Broom (1996) suggested that it means how an individual animal manages 'to cope with its environment' and this is a useful general approach, applicable to all species. For a higher primate such as a gorilla, however, with presumed greater sentience (see elsewhere in this book), either Duncan's (1996) definition 'Animal welfare is to do with the feelings experienced by animals: the absence of strong negative feelings, usually called suffering, and (probably) the presence of positive feelings, usually called pleasure', or Webster's (2005) 'Living a natural life, being fit and healthy, and being happy' are probably more appropriate. The term 'well-being' is also to be found in this book and is used in the rather imprecise sense of 'a state of being comfortable, healthy, or happy' — see later.

Methods of assessing welfare have been proposed and reassessed over the years. Some criteria that can be applied to gorillas are:

- Behaviour
- Health and disease
- Physiological values
- Immunological status
- Reproductive performance
- Response to treatment

Not all of these can be applied to free-living gorillas but observation is of the greatest importance (Fig. 8.3). Welfare assessment depends upon being able to recognise individual animals: in the case of free-living gorillas this may rest on the knowledge of experienced field staff of animals' nose prints (Fig. 8.4).

The 'Five Freedoms' can be applied to the care of gorillas in captivity — see Chapter 18: Legal Considerations. The same Farm Animal Welfare Council (FAWC) that formulated those 'Freedoms' in 1993 produced in 2007 a report that emphasised that there are three 'essentials of stockmanship' that help ensure the welfare of captive animals. The 'essentials' are (1) knowledge of animal husbandry, (2) skills in animal husbandry and (3) personal qualities, such as affinity and empathy with animals, dedication and patience. These three traits are equally

germane to the care of gorillas and can usefully be applied to the recruitment and training of staff in zoos and rehabilitation centres.

Much has been published about the welfare of gorillas and other NHPs. Awareness of the benefits of high standards of welfare in terms of legal and ethical responsibilities as well as compassion has helped improve methods of housing gorillas. Gorillas remain popular amongst visitors to zoos. This has encouraged the construction of high-quality enclosures that cater both for the welfare of the animals and the expectations of the public. In 2007, for example, London Zoo's 'Gorilla Kingdom' was opened, at a cost of £5.3 million. Within 3 weeks it had attracted 120,000 people.

However, programmes aimed at promoting the welfare of zoo animals need to be based on sound science, not just sentiment and subjectivity. Social network analysis (SNA) has recently been advocated as an aid to improved management of zoo animals. SNA allows quantification of animal sociality and population structure and thereby guides decisions about husbandry. It appears to have potential in the care of gorillas and in promoting welfare.

One important contribution to animal welfare is euthanasia — humane killing. The euthanasia of great apes is always a sensitive and contentious matter but it is — and has long been — part of the decision-making process in respect of captive gorillas. It is unfortunate that this option is so often denied to veterinarians, even though their overwhelming remit is welfare (see earlier), when it comes to free-living animals, when other interventions have been approved.

Assessing how gorillas are 'coping with their environment' remains problematic. In the case of captive gorillas some insight into pleasurable and adversive stimuli can be gained from observational studies on captive animals, but this can be subjective and how it relates to the situation in the wild is not always clear. And what about emotions? Charles Darwin (1872), in his book *The Expression*

FIGURE 8.3 A free-living mountain gorilla is followed by her infant. Assessment of welfare will include observing both animals and how they behave and interact.

FIGURE 8.4 Identification of individual free-living gorillas often depends on such criteria as recognising their individual nose print.

of the Emotions in Man and Animals was one of the first to investigate this – looking at NHPs as well as other animals. Scott (2015) referred to 'laughter' by the famous gorilla 'Koko' and describes this as important 'social bonding behaviour'. Measuring emotions such as joy and anger in monkeys and apes may be possible using such modern noninvasive techniques as facial thermal imaging.

Apparent grief in animals has also attracted considerable interest amongst behavioural scientists. Cousins (2015) stated that it could trigger self-destructive behaviour, citing the adult female gorilla, 'Moina' in the London Zoo, United Kingdom, in 1938. After losing her long-term male companion she became depressed and began to self-mutilate, pulling strips of flesh from her fingers. Despite the introduction of a young chimpanzee for company, the behaviour continued until she developed septicaemia, to which she succumbed. Apparent concern, possibly grief, is often shown by gorillas in response to a death (Less et al., 2010) and the author (JEC) and his wife observed behaviour, perhaps best described as consternation, in a group of mountain gorillas following the immobilisation of one of their number, 'Bahati'.

The book about grief by Alderton (2011) is particularly well written and has the added attraction that it features an image of a gorilla on its front cover! Alderton contends that emotions – including grief – are linked with social awareness and can have survival value. From the viewpoint of this book, they may also provide valuable clues to health and welfare.

Emotions are often linked with compassion and the evolutionary origins of this all-important trait in *Homo* are increasingly being explored (Spikins, 2015).

PAIN

Pain is defined by the International Association for the Study of Pain (IASP) as 'an unpleasant sensory and emotional experience associated with actual or potential tissue damage, or described in terms of such damage'.

The definition is a reminder that, quite apart from its welfare implications, pain is associated with tissue damage and therefore can involve a pathologist as well as a clinician. The pathologist may, for example, be able to demonstrate postmortem lesions that are unequivocally associated with pain or discomfort, such as the stretching of meninges over nerve roots or chronic skin lesions. The pathology of pain has been considered by many authors, in both human and veterinary publications. Pain management has attracted great interest in recent years, especially in respect of domesticated animals and those kept for research. The emphasis has been on preventing, recognising and assessing pain and, according to the Global Pain Council of the World Small Animal Veterinary Association, 'working towards global freedom from pain'. Reducing the 'pain incidence – pain treatment gap' – that is, the time between pain occurring and adequate treatment being given – has been an important focus.

The initiative is relevant to all species of animal including, of course, gorillas. This book is not the place to dwell too much on the clinical aspects of pain but, since pain is associated with noxious stimuli and 'pathology' (physiological responses, such as changes in heart rate and blood pressure, elevated cortisol values and weight changes), it is mentioned briefly. It is also a pillar of clinical medicine, as the French maxim reminds us – 'guerir quelquefois, soulager souvent, consoler toujours' (to heal sometimes, to relieve often, to console always).

Pain is often assessed subjectively, perhaps particularly in great apes because of their similarity to humans. The author (JEC) was probably guilty of this when describing the field appearance of 'Rafiki', who had both buffalo wounds and probable bee stings: 'Rafiki is ... lying on his elbows, left hand clenched on his chin, looking for all the world like a pensive professor! I can see his (buffalo) wound, but not well. Now he gets up and walks a few paces ...'.

Objective measures of pain in animals and their interpretation are increasingly attracting attention. Criteria for the recognition of pain on the basis of facial expressions have been established for laboratory rodents and studies are in progress on domestic cats. Those working with gorillas, both in the wild and in captivity, could usefully develop a scoring system ('pain scale') for the genus *Gorilla*.

NEOPLASIA

Neoplasms occur in gorillas and examples are in other chapters. The comparative oncology of 'humans and primates' was discussed by Scott (1992) and a useful survey of neoplasia in NHPs was provided by Abee et al. (2012).

Early accounts of presumed neoplasms in captive gorillas rarely included a proper diagnosis. For example, Prince-Hughes (2001) stated that the COD of the circus gorilla 'Gargantua' in 1949 was 'double pneumonia compounded by tuberculosis, and cancer of the lip'. Part II: A Catalogue of Preserved Materials, includes reference (see Index) to cases of 'brain cancer', 'a large brain tumour', 'an inoperable tumour in ... stomach' and 'a nonoperable malignant tumour' but no further information appears to be available.

There are also anecdotal, unsubstantiated, reports of possible neoplasms in free-living gorillas: for example, Schaller (1964) described in an elderly mountain gorilla 'a cancerous-looking eye, the other a skin rash'.

Recent data from the MGVP (personal communication L.J. Lowenstine, Gorilla Doctors) refers to neoplasms but without any detail: 'via-a-vie (*sic*) the mountain gorillas is the information on neoplasms .... We are working with the African Gorilla Doctors to publish many of these cases and for that reason have not included descriptions of the pathology so as not to jeopardise our veterinarians' ability to publish the material'. Lowenstine also made mention of 'cancers — (female reproductive tract cancers)'.

In an as-yet unpublished thesis about the immobilisation of mountain gorillas, Jaffe et al. (2012) referred to an elderly female 'that died soon after induction due to underlying neoplastic disease'.

The most detailed analysis of neoplasms in gorillas to date is by Lowenstine et al. (2016) in their study of ageing in great apes. Reference is made to various types of neoplasia; some of these are alluded to elsewhere in this book.

POISONING

Although poisoning of NHPs was recognised many years ago — see the chapter on poisons and toxins in Fiennes (1972) — surprisingly little has been published about gorillas and toxicants.

It is assumed that gorillas in captivity are susceptible to a similar range of toxicoses as are humans — and, to a certain extent, domesticated animals.

Again there are some anecdotal or unsubstantiated reports of possible poisoning incidents in captive gorillas: for example, Prince-Hughes (2001) wrote 'Lintz fed the little gorilla (Buddy) a bottle of poison, which damaged his intestines' (see also Part II: A Catalogue of Preserved Materials). An entry in Part II: A Catalogue of Preserved Materials refers to a hand-reared infant who 'died of pesticide poisoning' but no further information appears to be available. A more authoritative contribution to the scant literature was by Furley (1997) who, in his 12-year review of clinical conditions in the Howletts Zoo (United Kingdom) gorilla colony reported that one baby 'gained access to leaf tobacco, ate some, and promptly developed an acute attack of dizziness, ataxia, loss of pupillary reflex and laboured breathing for a few minutes'.

There are at least three reports of lead poisoning (plumbism, le saturnisme) in captive gorillas. Fisher (1954) described one case, Bisschop (1956) another, Sill et al. (1996) the third.

As far as free-living gorillas are concerned, there are early references to possible toxicoses, some based on local information. Merfield and Miller (1956) wrote: 'The dead female ... I found, had a crippled right foot and a diseased and withered left fore-arm, the hand being bent inwards and the elbow rigid. Besalla said that the disease was called Chully-Chang and that it was caused by the poison of the liana or vine known as Nqua-Zock'. Part II: A Catalogue of Preserved Materials, refers to one animal 'killed by a poisoned dart from a bow'.

Free-living gorillas that encroach on human habitation may eat unfamiliar plants and materials and might be poisoned. Occasional reports in the press and verbal comments by local people in Rwanda talk of gorillas becoming 'intoxicated by a diet of fermenting bamboo stems'. Burbridge (2013) writing about blood samples from the remaining Mount Tshiaberimu gorillas, living on a forested 'island' in the western sector of the Virunga National Park, DRC, referred to 'changes to the liver suggestive of exposure to a toxin, perhaps from a plant not formerly utilised by the apes, and that may have contributed to the deaths of an adult male and female'.

Although so little appears to have been published on poisoning in gorillas, it is always wise to include toxicosis in the differential diagnosis when dealing with unexplained ill health and/or death in free-living or captive animals. It is also good practice to take appropriate samples from such cases in case they are needed; if necessary they can be stored for analysis at a later date. The difficulty, of course, is in the choice of samples. One cannot 'test for poison' as such as there are thousands of potential toxicants. An eminently practical publication about postmortem technique and the taking of samples from wild animals for toxicological analysis is Rotstein (2008).

Chapter 9

Skin and Integument

John E. Cooper

Immediately I was struck by the physical magnificence of the huge jet black bodies blended against the green palette wash of the thick forest foliage.

Dian Fossey (1983)

INTRODUCTION

This chapter covers the skin (epidermis and dermis) and keratinous structures such as the nails. They are all of ectodermal origin except the dermis, which is derived from mesoderm. They are collectively termed 'integument' here. The mammary gland is discussed in Chapter 13: Urinary and Reproductive Systems.

The skin is the largest and most readily visible organ of the body. In an adult male gorilla it weighs about 5 kg and postmortem before preparation, while still fresh, has a surface area of approximately 1.75 m² (Akeley, 1920).

BIOLOGY OF THE SKIN

Owen (1859) described the hair of the gorilla but not its microscopic appearance. A detailed description, with particular reference to histology, was by Straus (1950).

Both skin and hair were discussed in general terms by Napier and Napier (1967) and Dixson (1981).

The majority of studies on skin to date appear to have been restricted to specimens of the lowland gorilla. Mountain gorillas have generally provided less opportunity for dermatological investigation because they have never been successfully kept in zoos (but see Chapter 8: Nonspecific Pathology).

Mountain gorillas clearly exhibit differences from their lowland relatives in terms of the length and appearance of their hair and other features may also not be identical.

Macroscopically, the skin of gorillas appears black, but neonatal animals are paler in colour: pigmentation increases as the youngster matures. Some young gorillas show marked absence of melanin from digits or limbs, but usually these areas darken with age; a few remain with nonpigmented patches. Schaller (1964) reported that the growth of 'silver' hair on the backs of male free-living mountain gorillas started when the animals were about 9 years of age and continued for 2−3 years.

The epidermis and dermis of gorillas appear to resemble those of other primates, including humans (Scott, 1992). Gorillas have both apocrine and eccrine sweat glands. The latter are particularly prevalent on the distal ends of the limbs and on the palmar and the plantar surfaces of hands and feet. The sweat glands are associated with thickening and patterning of the epidermis − so-called dermatoglyphics (dermatoglyphs). These dermatoglyphics are important in enhancing sensitivity and grip and the eccrine glands that are associated with them provide lubrication. Interestingly, they were the subject of research nearly 80 years ago (Wolff, 1938). Wolff postulated that differences in the papillary ridges on the gorilla's thumb and fingers are of psychological significance. She suggested that, if there is a correlation of the gyri of the skin and those of the brain, the discovery may indicate a different mentality, perhaps temperament, of the gorilla and possibly some highly developed mental functions in gorillas that are not found in humans (see Chapter 15: Nervous System and Special Senses).

The characteristic mode of locomotion of gorillas − so-called 'knuckle-walking' − was described by Ellis and Montagna (1962), who pointed out that the interphalangeal joints that are so important in such movement are well-endowed with eccrine sweat glands.

The habit of the animal to apply those parts to the ground, in occasional progression, is manifested by these callosities. The back of the hand is hairy as far as the divisions of the fingers; the palm is naked and calloused (Owen, 1859). Beautiful pictures of the hands feature in Norton (1990).

Sweat glands are primarily present in the axilla. In the case of mature male gorillas the 'axillary organ',

J. E. Cooper & G. Hull: Gorilla Pathology and Health. DOI: http://dx.doi.org/10.1016/B978-0-12-802039-5.00009-3

composed of equal numbers of apocrine and eccrine glands, is large and usually very active; the characteristic smell of an adult male gorilla, especially when stressed, is attributed to these glands. The body odours produced by gorillas can be discriminated by humans (Hepper and Wells, 2010). The axillary organ is also present in female and immature gorillas, but less well-developed. Dixson (1981) included a figure depicting a section of skin from the armpit of an adult male lowland gorilla to show this organ.

The hair of the gorilla is essentially black, but variants of this occur that are associated with normal physiological changes or with ill health. Some animals, for example, naturally have a reddish brown pelage, particularly on the crown in lowland gorillas, sometimes extending over the neck and shoulders. Maturity in male gorillas is characterised by the development of silver-grey hairs in the saddle region, one of the suite of secondary sexual characteristics that puberty brings — hence the term 'silverback'. The silver appearance is particularly pronounced in the mountain gorilla where there is a marked contrast between the animal's saddle and the long black hairs of the body and arms. Adult male lowland gorillas also have a silver back, but the rest of their hair coat is relatively short. In both species, males lose the hair on their chest at puberty (around 10–12 years), after which the skin is bare from the neck to the bottom of the rib cage; at the same time, they grow longer hair on the arms which serves to exaggerate their apparent size when adopting the characteristic strut display. In adult females the breasts are hairless but the extent of bare skin on the chest is less than in adult males.

The long hair of mountain gorillas is assumed to reflect the need for this subspecies to tolerate low temperatures at the high altitudes where they often live (Schaller, 1963). Gorillas of both sexes will develop grey hair as they age, as do human beings, but this feature is again particularly pronounced in males.

Gorillas usually have no hair on the palmar surfaces of the limbs, the back of the knuckles, the nose, lips or ears. The hair is sparse in the axilla, in the groin and on the chest (Schaller, 1963).

Infant gorillas up to the age of about 3 years have a white anal tuft, about 3 cm in diameter. This, it is postulated, makes them more obvious to their mother when they venture off her body into dense vegetation. Rarely, additional white patches have been reported in lowland gorillas (Ian Redmond, personal observation).

INVESTIGATIVE METHODS

Techniques used for dermatological investigation of gorillas include observation, assessment and/or sampling of the animal's environment, clinical or postmortem examination and various laboratory tests (see Chapter 5: Methods of Investigation — Observation, Clinical Examination and Health Monitoring). As in standard veterinary work, trial therapy may assist in diagnosis.

Important points when initially examining gorillas with skin lesions include liberal use of photography and 'mapping', employing a 'skin lesion chart' similar to those used in routine human and veterinary dermatological work. Records should say what is seen, not bland statements such as 'Skin and coat healthy'. Lesions of the skin may be depressed or raised. They should be described in a systematic way (see Box 9.1 for raised lesions).

Skin lesions should be carefully measured before excision (from a live or dead gorilla), not least because of 'sample shrinkage'.

Sampling

Skin may be sampled using touch preparations, swabs, washings, brushings, scrapings and biopsies.

Hair samples from gorillas can be obtained in a number of ways (see Box 9.2).

Preserved skins from gorillas (see Part II: A Catalogue of Preserved Materials) offer an opportunity to obtain samples for retrospective diagnosis and research, even if the skin is in poor condition. A number of laboratory tests that are applicable to a fresh carcase may also be applied to a mounted specimen or a study skin. Portions of tissue can be rehydrated and examined histologically.

BOX 9.1 Describing Raised Skin Lesions

- Size
- Shape
- Associated alopecia, lesions, including colour of skin, at rest and when depressed
- Local lymph node involvement?

BOX 9.2 Sources of Hair

- Dropped naturally by live animals
- Brushed out or removed by keepers of live animals (Figs. 9.1 and 9.2)
- Plucked or cut from live animals (Fig. 9.3)
- Plucked or cut from dead animals or skins
- Collected from vacated night nests
- Collected from gorilla trails after violent interactions (silverbacks sometimes pull handfuls of hair out of opponents when fighting).

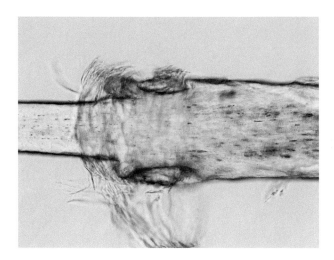

FIGURE 9.1 Normal arm hair removed by grooming at Bristol Zoo Gardens. Note the bulbous base, indicative that the hair was plucked not cut.

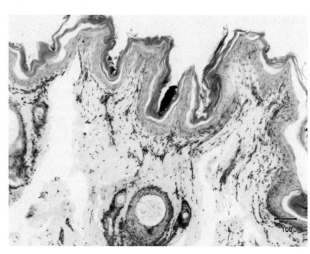

FIGURE 9.4 Portion of skin clipped from a lowland gorilla skin, rehydrated and processed for histology. Note the excellent preservation of the tissues.

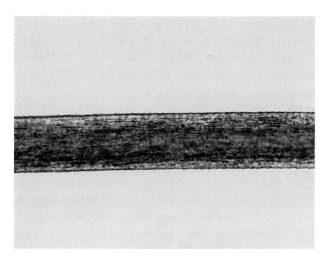

FIGURE 9.2 Another normal hair, showing the cortex and medulla which are important in identification (in forensic cases especially).

FIGURE 9.5 Examining gorilla skins in a Belgian museum.

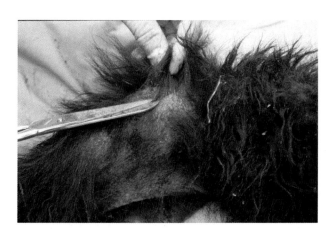

FIGURE 9.3 Hair being clipped for examination from an anaesthetised free-living gorilla.

The quality of the latter, taken in 2000 from specimens at the Powell-Cotton Museum (see Part II: A Catalogue of Preserved Materials), animals that had been killed in Africa 65 years earlier (1935), can be seen in Fig. 9.4.

Hairs from preserved skins can also be investigated (Fig. 9.5). Caution must be exercised in recording the presence of unusual ectoparasites on such material as contamination from humans or other species is possible between museum specimens (and, indeed, between live captive animals).

Nails can be clipped from live or dead animals and processed for histological and other investigations following the guidelines recommended by Cooper (1994).

Other aspects of taking and processing samples are discussed in Chapter 7: Methods of Investigation — Sampling and Laboratory Tests.

Laboratory Tests

Diagnostic procedures appropriate to samples from the integument of live or dead gorillas include use of ultraviolet light, cytology, potassium hydroxide preparations, bacterial and/or mycological culture, histopathological examination, PCR or other molecular techniques and blood tests (haematology, biochemistry, serology, endocrinological assays). In the case of large skin samples (usually from a dead gorilla), pinning-out will facilitate fixation of the skin and subsequent interpretation of histological changes.

Histological examination plays an important part in diagnosis. Skin responses in gorillas appear to resemble those in humans and other primates and the author uses standard methods, including pattern analysis of histological changes, to describe and analyse them. Pattern analysis helps in differentiating tissue reactions — for example, perivascular as opposed to perifollicular inflammatory cell 'cuffing', and the type of cell(s) involved is also relevant, for example, eosinophil, mast cell.

The particular considerations when examining the skin of a fetus or neonate are covered in Chapter 13: Urinary and Reproductive Systems.

DISEASES AND PATHOLOGY OF THE SKIN — GENERAL

The chapter on the skin and appendages of simian primates by Fiennes (1962) provided a useful introduction to dermatology in these taxa. More recently Abee et al. (2012) included a chapter on the integumentary system diseases of nonhuman primates (NHPs).

Much of the text that follows is an updated and expanded version of the paper presented to the British Veterinary Dermatological Study Group (BVDSG) in October 1997 and subsequently published (Cooper, 1997).

The data presented in 1997 were based primarily on (1) the author's experiences in Central Africa, working with free-living mountain gorillas from 1993 to 1995 and (2) published information in papers, books and reports supplemented with (a) a literature search carried out by the Library of the Royal College of Veterinary Surgeons, London, and (b) correspondence with approximately 50 individuals and organisations. The latter included a request for any records, published or unpublished, on skin diseases in the wild. No attempt was made to analyse the findings quantitatively. A letter, not a questionnaire, was used to solicit information and it was not always easy to interpret the results as the information was provided in diverse ways, sometimes in different languages, by different people. Respondents frequently used such terms as 'common', 'prevalent' or 'rarely seen' but such terminology can be very subjective. Therefore, both in the original paper

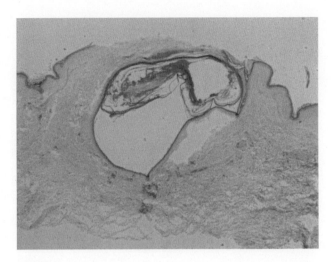

FIGURE 9.6 Keratinous cyst in the skin of an otherwise normal gorilla.

and here, only a qualitative account of skin diseases of gorillas was possible.

At the PHVA (Population and Habitat Viability Assessment) for the mountain gorilla held in Uganda nearly 20 years ago (Werikhe et al., 1997) the following conditions relevant to skin pathology were listed:

- Scabies — 'well-documented'
- Snares
- Vesicular dermatitis ('limited to head').

Recently the author has reviewed many unpublished histological reports from laboratories in different countries. These confirm a wide range of changes in the skin of gorillas, including cases of mild (moist) dermatitis, severe necrotising dermatitis and epidermal cysts, with hair follicles distended by whorled keratin (Fig. 9.6). Biopsies have also exhibited changes that are suggestive of a drug or food hypersensitivity reaction (see Chapter 12: Lymphoreticular and Haemopoietic Systems and Allergic Conditions), such as erosive and ulcerative dermatitis with acanthosis, parakeratosis and hyperkeratosis.

Diseases affecting the skin of the gorilla can be broadly divided into those that are:

1. noninfectious, and
2. infectious

However, there is often overlap between the two categories. For example, an injury due to a snare on the limb of a gorilla (a traumatic injury — noninfectious) — see Chapter 4: Noninfectious Disease and Host Responses — may initially cause abrasions and ulceration, but subsequent infection can result in the development of a pyoderma, cellulitis or gangrene.

Some infectious lesions are secondary to other conditions — for example, scabs on the ankles, wrists, near the left palpebra and ear, right ear, nape of the neck and

lower back in a case of pneumonia and meningoencephalitis associated with amoebae in a captive lowland gorilla
(Anderson et al., 1986).

NONINFECTIOUS CAUSES

According to Cousins (1972), the most widely reported noninfectious cause of skin disease in gorillas, both in captivity
and in the wild, is trauma, an opinion echoed by many subsequent authors and observers. Injuries are sometimes self-
inflicted, especially in captivity, but are more often caused
by other gorillas – including puncture wounds and gashes
inflicted by an opponent's canine teeth, scratches inflicted
by an another gorilla's nails and occasionally skin torn off
with a handful of hair – see Chapter 4: Noninfectious
Disease and Host Responses. 'Orphaned' gorillas in Africa
are frequently found with wounds caused by guns, snares,
spears, machetes (*pangas*), cigarettes or dogs (Chris Furley,
personal communication; Ian Redmond, personal
observation).

As emphasised elsewhere, injuries in free-living gorillas are often due to fighting between individuals (Fossey,
1983; Meder, 1994).

Hyperthermia in captive gorillas was reported by
Guilloud and FitzGerald (1967). Cousins (1974) remarked
that 'the colouring of the gorilla's skin and hair constitutes a heat trap in direct sunlight'.

Many anecdotal reports in captive gorillas implicate
bedding and other substrates as a cause of skin and other
lesions but it is rarely clear whether such cases have
an immunological ('allergic') pathogenesis or are merely
the result of physical irritation (see Chapter 12:
Lymphoreticular and Haemopoietic Systems and Allergic
Conditions).

Peripheral circulation disorders in captive lowland
gorillas were reported by Dorobisz et al. (1983): the
clinical features were of peripheral ischaemia with skin
necrosis. A lowland gorilla at Jersey Zoo/Durrell Wildlife
Conservation Trust/'Durrell' developed what appeared to
be Stevens-Johnson syndrome, with oedema and ulceration
of extremities – see Chapter 12: Lymphoreticular and
Haemopoietic Systems and Allergic Conditions.

Alopecia and scaliness of the skin are not uncommon
in free-living gorillas and there is some evidence that
these changes are associated with hormonal changes,
particularly in ageing females. In captivity similar alopecia and scaliness can occur, but here substrate, cage
design and exposure to chemical agents, such as disinfectants, may be involved – see above. Hair-plucking of
forearms and shoulders of captive gorillas is attributed
by Furley (personal communication) and Clark et al.
(2012) to stress. An example is when a gorilla is introduced into new groups; the animal stops plucking when
the stressors cease.

FIGURE 9.7 The hand of a young mountain gorilla caught in a snare.
There is severe skin damage associated with necrotic digits.

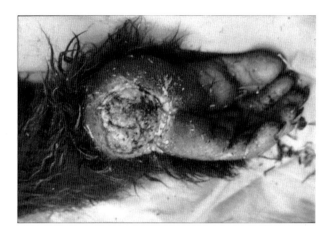

FIGURE 9.8 Chronic granulating lesion on the foot of an anaesthetised
adult, free-living, mountain gorilla.

Alopecia scoring, a quantitative assessment of hair
loss, has been used in captive macaques (Honess et al.,
2005) and something comparable might prove useful in
gorillas.

In free-living and captive gorillas a poor coat and an
associated build-up of keratinous debris will occur if
grooming (by the animal itself or by other gorillas) is inadequate (John Bosco Nizeyi, personal communication).

A frequent cause of traumatic injury in free-living
mountain gorillas is damage by snares (see Chapter 4:
Noninfectious Disease and Host Responses). Often skin
damage is severe – both initially, on account of tissue
damage, and, later, ischaemia, and sometimes in the long
term if only a bare stump remains (Fig. 9.7).

Chronic pedal lesions occur in captivity, usually as a
result of trauma. They take time to heal if the substrate is
abrasive. A case treated by the author was a free-living
mountain gorilla, 'Inyuma' (see also elsewhere in this
book). The cause of the lesion was unknown but gradually
it granulated (Fig. 9.8).

Ian Redmond (personal communication) points out that gorillas usually pay considerable attention to wounds, both their own and other group members' (see Fig 4.1 in Chapter 4: Noninfectious Disease and Host Responses), inspecting, grooming and licking the affected area if within reach; in captivity, self-mutilation may accentuate the condition (Jarofke and Klös, 1975). The extent to which this is 'psychological', as opposed to reflecting pain, pruritus or inflammation, is unclear. It is an important consideration whenever a captive gorilla has a skin lesion.

Johnstone-Scott (1998) and subsequently Allchurch and Dutton (2000) described possible toxic epidermal necrolysis or Stevens-Johnson syndrome in a captive lowland gorilla, probably as a reaction to the neuroleptic drug zuclopenthixol, and this case is discussed, with Figures, in Chapter 12: Lymphoreticular and Haemopoietic Systems and Allergic Conditions.

Colour changes may occur. Hair discoloration and alopecia were described by Mundy et al. (1998) in a colony of lowland gorillas with a protein deficiency. In 2009 there were reports of a free-living gorilla with possible vitiligo (a pathological reduction of melanin) in the DRC (Rick Murphy, personal communication).

Albino gorillas have been reported. Dixson (1981) claimed that the only individual captured and examined was the famous white gorilla 'Copito de Nieve' (or 'Snowflake') in Barcelona Zoo — technically a blonde, not albino, as he had blue eyes (see also Chapter 15: Nervous System and Special Senses) — but the account by Sabater (1967) of an albino lowland gorilla from Rio Muni, West Africa, with notes on its adaptation to captivity (1967), predates him. Chris Furley (personal communication) reported a young albino animal in the Congo that had to shade its eyes when venturing out into the sunlight. The youngster had a normal black mother.

Much has been written about 'Snowflake' ranging from popular accounts (Riopelle, 1967) through descriptive prose (Jonch, 1967) and postmortem findings (Márquez et al., 2008) to genome sequencing (Prado-Martinez et al., 2013). Such was the interest in the animal that the Barcelona Zoo enlisted a British zoo veterinarian to collect sperm from 'Snowflake' for artificial insemination (Taylor, 1991).

In his 12-year review of clinical conditions in the Howletts Zoo (United Kingdom) gorilla colony, Furley (1997) reported a number of skin conditions, including a possible chondroma or fibroma.

A neurodermatitis (atopic dermatitis) of captive gorillas — characterised by eczema and apparently a result of psychological disturbance (conflict with a partner) — was reported by Hog and Schindera (1989) — see Chapter 12: Lymphoreticular and Haemopoietic Systems and Allergic Conditions.

The nails of gorillas do not appear to have attracted much attention but the author (JEC) has observed colour variants in museum specimens. In appearance the nails of gorillas are similar to those of humans but thicker, with the protruding edge worn naturally to a slight curve. They are used in many ways such as scraping (e.g., bark, termitaria), excavating (e.g., soil, rotten wood) and scratching (both delicately when grooming and violently when fighting, as a weapon).

INFECTIOUS CAUSES

Studies on various taxa, including humans, indicate that, with the possible exception of the surface of the nails, all areas of the integument have both a resident microbial flora and a transient or contaminant flora of bacteria, fungi and possibly viruses.

The resident microbial flora of *Homo* was reviewed in some detail by Noble (1981). Forearm bacteria were subsequently studied by Zhan Gao et al. (2007) who showed that the bacterial biota in normal human skin is highly diverse, with evidence for 182 species of bacteria. Eight percent were unknown species that had never before been described. The first study to identify the composition of bacterial populations on the skin of domestic dogs was by Harvey and Lloyd (1995) who investigated the distribution of bacteria (other than staphylococci and *Propionibacterium acnes*) on the hair, at the skin surface and within the hair follicles of dogs. In defence of their study they pointed out that 'although the presence of organisms other than staphylococci has been reported, the details of their distribution ... has received little attention'.

Similar studies on gorillas to those outlined above would be of value, especially in view of concerns about the transfer of bacteria from humans to captive, especially hand-reared, gorillas. Some minimally invasive samples could be taken from live gorillas but most, such as punch samples, would need to be from dead animals. Processing could be as described by Harvey and Lloyd.

ECTOPARASITES (BY IAN REDMOND)

There is much interest in host—ectoparasite relationships in Africa (Cooper, 2003c). Published work in various parts of the world has included studies on the immunology of scabies and on other topics of relevance to the genus *Gorilla*.

A number of species of ectoparasites have been isolated from gorillas (Redmond, 1983). Those of particular significance are:

- Class Arachnida
- Order Acarina (mites and ticks):

    Mites are the smallest arthropods, but probably the most numerous and widely distributed. Parasitic forms often differ only slightly from free-living species, and parasitism is thought to have evolved many times

(Ewing, 1929). *Sarcoptes scabiei* is discussed in more detail below.

Pangorillalges gorillae

This tiny mite was first described as *Psoroptoides gorilla* (Gaud and Till, 1957). The genus *Pangorillalges* was erected by Fain (1962) to accommodate a new species of mite found on chimpanzees from Kivu (in the east of what is now the DRC), *Pangorillalges pani*.

Not yet reported but considered likely parasites of great apes are follicle *Demodex* sp. which are virtually ubiquitous in humans. If they have a deep evolutionary relationship with their host as their anatomy implies, it would be surprising not to find them in our closest relatives, derived from our common ancestor, but DNA analysis suggests an alternative hypothesis. The origin of the species found in humans might be from dogs (which have a closely related species of *Demodex*) during the period of domestication of the wolf (Thoemmes et al., 2014).

Large, pale mites were observed (and filmed in 2012) encrusted around the eyes of an otherwise apparently healthy blackback male lowland gorilla, habituated to human observers near the Mondika Research Centre, Congo, but it was not possible to collect specimens.

Rhipicephalus appendiculatus

Ixodid ticks might be expected to feed on gorillas because they have been seen occasionally on human observers in the bamboo forest of the Virunga Volcanoes (Redmond, personal observation) and so might also transmit tick-borne diseases, but these have not yet been recorded. *Rhipicephalus appendiculatus* has been reported to have been found occasionally on gorillas (Meder, 1994; Walker et al, 2000) though this tick species' distribution is such that only eastern gorillas could be affected. Schaller (1963), however, cites Dekeyser (1955) stating that 'In West Africa gorillas are infected with the tick *R. appendiculatus* which carries tickbite fever in man'. It seems there is some confusion over the species of tick involved.

- Insecta (lice and fleas):

Sucking lice, Order Anoplura

Pthirus gorillae

The gorilla louse is the larger of only two species known in the genus *Pthirus*, the other being the human pubic or crab louse *Pthirus pubis* (for a discussion of the spelling, sometimes given as *Phthirus* or *Phthirius*, see Kim and Emerson, 1968). It was first described in 1927 by Henry Ellsworth Ewing from nymphs extracted from eggs found on eastern gorillas collected in what was then the Belgian Congo on a hunting trip (though

curiously missing from Ewing's landmark *Manual of External Parasites*, published in 1929). It was not until 1968 that the adult female was described by Kim and Emerson, and formal descriptions of the adult male and instars have yet to be published (Lyal and Redmond, in preparation) though a drawing of the male illustrates Redmond (1978a). The gorilla louse is distinguished from the human crab louse by its larger size (adult females are 2.25 mm in length) and prominent eyes situated on protuberances on either side of the head; it is most often found around the groin, armpits and thick hair on the back of the legs and arms (where the eggs, or nits, are also found cemented to individual hairs). Close and careful grooming by other group members, especially by mothers of their infants, means that louse populations are kept in check. On the other hand, the fact that gorillas often rest together, play and embrace provides ample opportunity for lice to transfer from one host to another by clambering between hairs (Figs. 9.9 and 9.10),

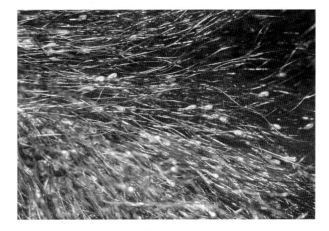

FIGURE 9.9 Eggs of *Pthirus gorilla*, the gorilla louse, on the skin of a free-living gorilla. *Image courtesy of Ian Redmond.*

FIGURE 9.10 Western lowland gorilla with mites around eyes. *Image courtesy of Ian Redmond.*

the long curved claws and crab-like shape being adaptations for grasping the thick hair shafts of gorillas (as with *P. pubis* in human pubic hair).

Lice appear to be no more than a minor irritation to a healthy, well-groomed, gorilla, but ill health or the loss of a close family member is correlated with an increase in louse numbers. For example, a postmortem examination of 'Mwelu', an 8-month-old infant mountain gorilla infanticide victim (Redmond, 1978b) revealed no sign of lice or eggs, but numerous mites *P. gorillae*, especially on the crown, groin and axillae. This contrasted with a necropsy of 'Kweli', a 39-month-old infant who died 3 months after sustaining a bullet wound to the clavicle during a poaching incident in which his mother and father were both killed (Redmond, 1978a). Wounded, sick and deprived of maternal comfort and grooming, when he died he was heavily infested with lice, with concentrations around the groin and axillary regions. The difference between these two infants was inferred to be due to 'Kweli's illness-induced loss of host resistance (a concept outlined by Richler, 1942) and a lack of maternal grooming. Adult lice were photographed feeding on the soft skin of the armpit, where even to a trained eye they were difficult to see, having the appearance of a flake of skin. Eggs (nits) were attached to hairs all over the body, with up to seven on one hair in some areas (one is the norm).

The genus *Pediculus*, which includes the human head and body lice, has not been recorded on gorillas, with the exception of a single dried specimen found in the fur of a western lowland gorilla in the drawers of the Natural History Museum in London (Redmond, unpublished); this is thought likely to have been acquired while the host was alive in captivity, or possibly by contamination of the skin during or after preparation (see earlier).

Fleas, Order Siphonaptera

The life cycle of fleas does not suit the peripatetic habits of gorillas. Fleas are laterally flattened ectoparasites, the larvae of which feed on organic detritus on the nest or bedding; because gorillas build a new nest every night, there is no species-specific gorilla flea, but it is likely that fleas of other species might occasionally feed on a passing gorilla, and in captivity human fleas *Pulex irritans*, cat fleas *Ctenocephalides felis* and dog fleas *Ctenocephalides canis* are likely to avail themselves of a meal. *Tunga penetrans*, the chigoe flea or jigger, is native to South and Central America, and so is not a natural parasite of gorillas, but has been inadvertently introduced to sub-Saharan

Africa where it parasitises various species including humans who walk barefoot — and therefore, in theory, gorillas being held in captivity or ranging naturally through infested habitat such as forest *shambas*. After mating, the female burrows into a fold or crack in the host's skin, most commonly under a toenail or a crack in the sole, and slowly expands to the size of a small pea, full of eggs. Pressure when walking then disperses the eggs into the soil, especially in dirt-floor dwellings. Meder (1994) mentioned that they are occasionally found on gorillas.

Some invertebrates are not parasites *sensu stricto* but can be a cause of injury to gorillas.

Stinging insects (Hymenoptera), such as bees (JEC) — see Chapter 12: Lymphoreticular and Haemopoietic Systems and Allergic Conditions. In March 1973 the senior author examined a weak captive monkey that had been killed by safari ants (*Dorylus* sp.) ('*siafu*' in Swahili) at the Primate Research Centre in Kenya (see Chapter 2: The Growth of Studies on Primate Pathology). It is possible that a similar fate could befall an incapacitated gorilla. Myiasis occurs from time to time in free-living and recently captured gorillas and the larvae of *Oesophagostomum* worms can cause subcutaneous lesions (Chris Furley, personal communication).

The ectoparasite of greatest clinical significance in gorillas is probably *S. scabiei*. This mite has been listed in the past as a cause of disease in both captive and free-living animals (Meder, 1994). In his book Cousins (1990) referred to *Sarcoptes* infection in the gorilla 'Guy' in the London Zoo in 1948 — see Appendix 5: Case Studies — Museums and Zoological Collections — and said of the disease that it caused 'such irritation that he scratched off almost all his hair, while the skin became thickened and rugose over the areas within most easy reach of his hands'. The origin of this observation has not been traced but the author (Don Cousins, personal communication) recalls receiving the information from the London Zoo and an old photograph of Guy, taken in November 1947, shows signs of hair loss — possible evidence to support the claim.

Sarcoptes in free-living gorillas attracted particular interest 20 years ago following the report from Uganda of an outbreak of mange in mountain gorillas in the Bwindi Impenetrable Forest National Park (Macfie, 1996; Kalema-Zikusoka et al., 1998). In the initial outbreak four habituated animals showed signs of marked pruritus and hair loss, and one of these, an infant, died. The origin of the infection remains unclear, but it has been suggested that the mites may have been acquired from domestic livestock — as has been reported between ruminants and humans in India (Mitra et al., 1995) — or from local people. Three months before the outbreak the affected group of gorillas had raided banana plants from an area where farmers were reported to have rough skin and where scabies was commonly diagnosed (Kalema-Zikusoka et al., 2002; Jaffe, 2012). The suggestion that immunosuppression might have contributed to the appearance of clinical disease is discussed in Chapter 12: Lymphoreticular and Haemopoietic Systems and Allergic Conditions.

The histology of *Sarcoptes* infection has been well studied in humans, dogs and some other species. It can range from small pruritic macules to severe skin lesions. Finding the mites in skin scrapings in severely affected animals is often difficult and this was the experience of the author (JEC) when he performed the first histological examination of skin, using facilities at Makerere University, from one of the Bwindi gorillas.

Diagnosis of sarcoptic mange in the gorilla is, as in other species, based initially upon detection of the mites in skin scrapings or biopsies, but the identification of the mites can be enhanced by DNA studies. The microscopical pathological changes seen in gorillas from Bwindi were similar to those described in other species (see, e.g., Morris and Dunstan, 1996; Arlian et al., 1990). Alternative diagnostic tests would be useful; use of the ELISA to detect antibodies to *S. scabiei* in pigs (Bornstein and Wallgran, 1997) suggests that this technique, as well as other serological studies (Bornstein, 1995), may be worthy of investigation in NHPs.

At the PHVA (Population and Habitat Viability Assessment) for the mountain gorilla held in Uganda nearly 20 years ago (Werikhe et al., 1997) the following conditions relevant to skin pathology were listed:

- Scabies —'well-documented'
- Snares
- Vesicular dermatitis ('limited to head').

Bacteria

Different types of organism can cause, or be associated with, skin disease in gorillas (see Table 3.1 in Chapter 3: Infectious Disease and Host Responses). Here only a few will be mentioned specifically as the focus of the book is on pathology and pathogenesis, not on microbiology or epidemiology/epizootiology.

Skin diseases associated with bacteria have been recorded in gorillas by a number of authors and unpublished data implicating various organisms were provided by several respondents to the survey in 1997. Dermatitis in free-living gorillas was reported by Blancou (1951) and Cousins (1972) and staphylococcal infections (which were localised, alopecic, erythematous and pruritic) in captive animals by Selbitz et al. (1980).

An early report of a pyoderma in a young gorilla was described, with details of its treatment, by Ostenrath (1978).

Treponema pallidum subsp. *pertenue* is the cause of 'yaws' (treponematosis) in humans. This affects skin, bones and joints. It was suggested long ago that it might also produce disease in gorillas. Photographs published by Schultz (1950) showed lowland gorillas with bone lesions similar to those of humans with treponematosis. Schaller (1964) referred to these pictures as being 'of lowland gorillas with what appears to be yaws'. The pioneer wildlife photographer Denis (1963) described a possible case in a free-living gorilla as follows:

The lips were gone. The nostrils were eaten almost away ….. Only the eyes were untouched and they glared at me with indescribable fury.

Chris Furley (personal communication) reported 'yaws' in the Congo in recently captured baby gorillas and in free-living animals. He described lesions consisting of crusty ulcerations on the face and hands and genital organs. The infection responded to treatment with penicillin.

More specific, evidence-based, information has appeared in recent years. Levréro et al. (2007) described skin lesions in a gorilla population in the Republic of Congo that, they argued, were typical of yaws. They based this on the 'symptoms' (*sic*) seen and the serological tests previously carried out in the park by Karesh (2000). Karesh had been able to detect treponemal antibodies (rapid plasma reagin and fluorescent treponemal antibodies tests) in gorillas in the Odzala-Kokoua National Park (PNOK), in the northwest of the Republic of Congo; two of the animals had skin lesions consistent with yaws. In the study by Levréro et al. 17% of 377 gorillas had skin lesions, mainly on their faces. Lesions appeared when individuals were young; in animals of more than 8 years they were more prevalent in males than in females. The authors argued that this was a result of behavioural traits of males — leading to more injuries. There were other interesting prevalences too relating to age, sex, and reproductive status and Levréro et al. used their findings to analyse 'the spread of a disease in a wild population'. Following this, Woodford et al. (2002) listed yaws as a possible disease risk when 'habituating' great apes. Other NHPs have been found to be infected with *T. pallidum* (see, e.g., Knauf et al., 2013) but whether they are a true reservoir is unclear (Marks et al., 2015).

Rousselot and Pellissier (quoted in Ruch, 1959) noted that gorillas suffer from a bacterial skin infection which resembles leprosy. Without a specific reference Cousins (1990) referred to the possibility that leprosy, due to

Mycobacterium leprae, might occur in gorillas, as claimed by local people in Cameroon. A brief note in the Annual Report for 1998 of the German Primate Centre (DPZ) referred to investigation of 'skin alterations in a gorilla: leprosy?' but it has not proved possible to locate further information.

Fungi

Dermatophytosis ('ringworm') was reported in captive gorillas 90 years ago (Weiman, 1927) and on occasion in various zoos since then (see, for instance, Cousins, 1984; Saez and Chauvier, 1977). Of the four species to which reference has been located, two, *Trichophyton mentagrophytes* and *Microsporum canis*, may cause clinical disease. *Chrysosporium keratinophilum* and *Scopulariopsis brevicaulis* appear to be harboured by gorillas.

Self-wounding can sometimes surface as a secondary complication to skin diseases. Cousins (1984) recounted that an adult male gorilla in the Jardin des Plantes, Paris, France suffered recurring attacks of ringworm (*Trichophyton mentagrophytes* infection) over many years. The infection was localised to the limbs and caused the animal so much aggravation that it resorted to severe self-mutilation.

Skin lesions were recorded in a case of coccidioidomycosis in a mountain gorilla reported by McKenney et al. (1944).

Viruses

One of the few viruses conclusively identified as a cause of skin disease in gorillas is *Varicella*, the cause of human 'chicken-pox' (Myers et al., 1987) – see Chapter 3: Infectious Disease and Host Responses. Chris Furley (personal communication) reported that captive gorillas under the age of 4 years nearly all seroconvert and some show lesions, mainly macules on the face.

There have been unsubstantiated reports of other viral skin lesions in gorillas, including claims that poxvirus infection occurs. Raised macular or papular lesions are occasionally seen on the faces of free-living mountain gorillas and usually resolve spontaneously. These may be due to a *Herpesvirus* – see Chapter 3: Infectious Disease and Host Responses.

Infectious diseases of the nails of gorillas do not appear to have been reported but are likely to exist, as in other primates.

DISEASES OF UNCERTAIN AETIOLOGY

As pointed out earlier, nonspecific signs such as alopecia and furfuraceous (scaly) lesions are seen in both captive and free-living gorillas and often there is no specific diagnosis, even when investigations are performed.

One example of such a picture was provided by Sarah Chapman (personal communication), veterinary surgeon at Twycross Zoo, England. The affected animal, 'Biddy', a 40-year-old lowland gorilla, had a long-standing (four year) history of chronic dermatitis which appeared to respond well to changes in diet, substrate and skin supplements. Earlier antibiotics and antifungals had been used 'with varying effects'. Investigations carried out on the gorilla had included biopsies, skin scrapes and microbiological culture. Another with undiagnosed skin condition was 'Rafiki', a male captive-bred lowland gorilla at Jersey Zoo/DWCT/'Durrell'. Between November 1985 and his movement to a North American Zoo in June 1991, 'Rafiki' had localised alopecia with occasional apparent pruritus. A whole battery of clinical and laboratory investigations, some involving (human) consultant dermatologists, together with trial systemic and local medication, failed to produce a definitive diagnosis nor an outright cure (Chris Dutton and Richard Johnstone-Scott, personal communication).

CONCLUSIONS

Both infectious agents and noninfectious factors cause skin lesions in captive and free-living gorillas and there is some published information about the pathology of such conditions. Unfortunately, however, very little specific research has been reported that might yield useful, practical, data on the skin of the two species of *Gorilla* and its response to different insults. Such a dearth of information on dermatology is not, however, confined to gorillas. In his book *Pathology of Simian Primates*, Fiennes (1972) stated: '. . . very few systematic studies have been made on the pathology of primate skin and unfortunately it has not been possible to interest any of the few primate dermatologists in contributing a specialised chapter on the subject'. Twenty years later Scott (1992) wrote: 'Whereas there are numerous detailed accounts on diseases of human skin, the primate skin has been, in comparison, largely ignored'.

These two statements remain broadly true today, especially in respect of gorillas.

Chapter 10

Respiratory and Cardiovascular Systems

John E. Cooper

Just as our soul (psyche)... holds us together, so do breath (pneuma) and air (aer) encompass the whole world.

Anaximenes of Miletus (c.550 BC)

The heart is the root of life and causes the versatility of the spiritual faculties. The heart influences the face and fills the pulse with blood.

Huang Ti (The Yellow Emperor) [2697−2597 BC]

INTRODUCTION

This chapter covers the respiratory and cardiovascular systems. For the purpose of this book, the former includes in part the nasal cavity ('nose'), sinuses and pharynx, trachea, larynx, air sacs, bronchial tree and lungs. There is also discussion of the nose and olfaction in Chapter 15, Nervous System and Special Senses.

RESPIRATORY SYSTEM

The larynx, trachea, lungs and associated structures are endodermal derivatives. The first scientific description of their anatomy in the gorilla was by Washburn (1950). The laryngeal sacs of *beringei* were the focus of a specific paper by Miller (1941). Cave (1961) described the frontal sinus of the gorilla, and Blaney (1986) reported comparative studies on the frontal sinus of *Gorilla*, *Pan* and *Pongo*.

Over the past 30 years it has become apparent that both free-living and captive gorillas are susceptible to not just a few but a whole range of respiratory conditions, ranging from 'colds' and 'coughs' to rhinitis, sinusitis, inflammation of the laryngeal sac and various types of pneumonia. The pleurae may be involved, either as such or by extension of lesions from the lungs.

Respiratory Pathology

The respiratory tract of *Homo* is subject to a wide range of pathological processes (McGee et al., 1992), a number of which are seen in *Gorilla*. Respiratory diseases have long been recognised in both captive and free-living non-human primates (NHPs) (Cooper et al., 1978). In captive animals they are responsible for considerable morbidity and mortality and it is increasingly recognised that they are of significance in the wild. Some are zoonoses and thus may pose a threat to those working with these species (Fiennes, 1967).

From the earliest days of keeping gorillas in captivity there were clinical and postmortem accounts of respiratory disease. For example, Fiennes (1964), in his Report as the Zoological Society of London's Pathologist for the year 1962, wrote of two mountain gorillas:

(A) Young male (DB 456/62). Duration of stay ... five months. The gorilla "Rundi" died as a result of pneumonia associated with infection by Pneumococcus *type 3. The posterior lobe of the right lung was in a condition of red hepatisation, the remaining lobes of both lungs showed a condition of patchy broncho-pneumonia.*

(B) Immature male (DB 510.62). Duration of stay ...... two years seven and a half months. The gorilla "Rueben" died as a result of lobar pneumonia associated with infection by Corynebacterium pyogenes. *The whole of both lungs showed pneumonic changes evenly distributed and associated with red hepatisation. The lungs were partially consolidated and portions floated submerged. There was great dilatation of the right heart and a small quantity of fluid in the pericardial sac. The infection evidently arose from two small organizing pyaemic foci in the right kidney, probably as a result of chilling during the blizzard of 26th−29th December.*

Such early investigations usually included bacteriological tests but rarely a specific search for viruses. Thus,

J. E. Cooper & G. Hull: Gorilla Pathology and Health. DOI: http://dx.doi.org/10.1016/B978-0-12-802039-5.00010-X

Fiennes reported that 'Throat swabs from the young female gorilla, which came at the same time as "Rundi", gave cultures of *Pneumococcus* type 2 ... that commonly associated with epidemics of pneumonia in humans ... *Corynebacterium pyogenes* commonly causes pneumonia associated with pyaemic abscesses in ruminants. Pneumonia due to this organism appears to be rather uncommon in other groups of animals.'

A review of pathology reports from (mainly British) zoos has revealed many findings relating to the respiratory tract, some 'incidental findings'. Examples of some of these are shown in Figs. 10.1–10.4. They include mucopurulent rhinitis, laryngitis, tracheitis, degrees of (not hypostatic) congested lung, acute bronchopneumonia (alveoli containing polymorphs, oedema fluid and red blood cells, airways filled with polymorphs), pleurisy, subpleural petechiae, pneumonic changes associated with inhaled vegetable matter and microabscesses in the lung parenchyma. Bacterial colonies are often seen in bronchiolar lumina and alveolar spaces and occasionally macrophages within alveoli are found to contain iron — sometimes reflecting chronic pulmonary congestion due to a degree of left-sided heart failure.

Epistaxis (nasal haemorrhage) occurs occasionally in gorillas, both in the wild and in captivity. In the former it may follow a fight or (presumed) damage from vegetation. Ebolavirus disease can be a differential diagnosis. A captive gorilla with epistaxis warrants careful observation, clinical examination and taking of swabs. Epistaxis of uncertain, but possibly bacterial, origin was first reported in a group of cynomolgus monkeys some years ago (Cooper and Baskerville, 1976), and cases continue to be described, some associated with irritant substrate or nasal foreign bodies.

Not infrequently pulmonary pathology in gorillas is secondary to other lesions. For example, 'Samson' at Bristol

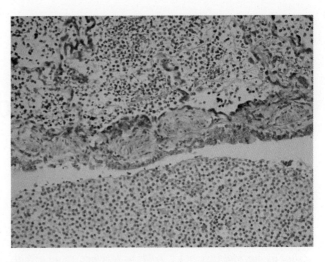

FIGURE 10.2 Inhalation pneumonia in a lowland gorilla with airways full of inflammatory cells but lining epithelium intact.

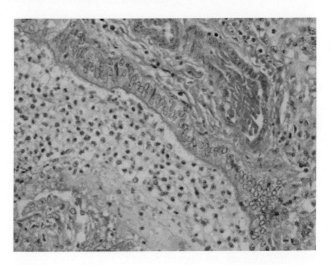

FIGURE 10.3 Higher power of the inhalation pneumonia above with airways full of inflammatory cells but lining epithelium intact.

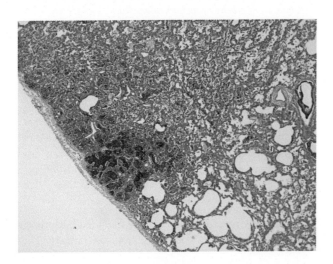

FIGURE 10.1 Intraalveolar haemorrhage with subpleural petechiae in a lowland gorilla.

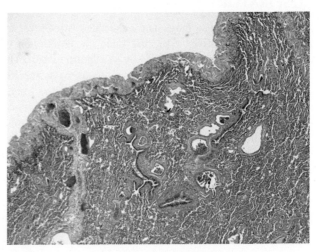

FIGURE 10.4 Chronic pleurisy and collapsed parenchyma in a lowland gorilla.

Zoo Gardens (see Chapter 11: Alimentary Tract and Associated Organs) had a large tumour of the lower oesophagus that not only interfered with cardiac function but also probably caused inhalational acute bronchopneumonia.

Various papers over the years have reported specific conditions affecting the respiratory system of captive gorillas. For example, Antonius et al. (1971) described pulmonary embolus and testicular atrophy, and Anderson et al. (1986) recounted pneumonia and meningoencephalitis due to an amoeba in a lowland gorilla (see Chapter 14: Musculoskeletal System).

Different types of organism can cause respiratory disease in captive gorillas. Some examples are to be found in Chapter 3, Infectious Disease and Host Responses.

Coccidioidomycosis is a primary respiratory infection caused by the fungus *Coccidioides immitis*. It infects a wide variety of wildlife species (McKenney et al., 1944) as well as humans. Gross lesions in a captive gorilla that had died of coccidioidomycosis included enlargement and caseation of the peribronchial lymph nodes and a soft and pulpy spleen that contained numerous tubercles scattered over the pleural surface. Histologically, there were irregular areas of necrosis resembling tubercles in the spleen, the liver contained more well-defined tubercles with giant cells and the lungs had similar tubercles which contained fewer neutrophils than the spleen (McKenney et al., 1944).

Hypertrophic pulmonary osteoarthropathy has been reported in an orangutan (Hime et al., 1972) but appears not to have been documented in *Gorilla*.

In so far as free-living gorillas are concerned, in her popular account Meder (1994) wrote of diseases of the respiratory tract with particular reference to *beringei* and postulated that a chilly and damp climate might be a predisposition to respiratory conditions, describing pneumonia as 'the most common cause of death in the mountain gorillas of the Virunga Volcanoes'. She pointed out that, if the weather is fine, mountain gorillas will lie out in the sun while lowland gorillas generally seek the coolness of the shade.

To this observation might be added the point that gorillas, both in the wild and in captivity, avoid rain (see Chapter 3: Infectious Disease and Host Responses). In this vein, Masi et al. (2012) discussed seasonal effects on the health of free-living western gorillas and chimpanzees.

Most information about free-living animals comes from studies on mountain gorillas. Fossey (1983) set the scene by recording relevant necropsy findings. 'Idano' was described as having an 'upper lobe that showed many fibrous adhesions suggestive of pleurisy of long duration'; 'Rafiki' (believed to be 55−60 years old) had the 'entirety of left lung adhered to chest wall, suggesting pleurisy of long duration. Both lungs heavily edematous, the left more so than the right, symptomatic of pneumonia'.

In a subsequent abstract, Lowenstine (1990) reported that long-term studies on the mountain gorilla had shown that most deaths occurred in the months of February−May. In 1988 and 1990 a severe respiratory disease outbreak was reported in the habituated groups and 'the most significant lesions were in the respiratory system in four animals'. She also stated that arteriosclerosis (see later), of the hyaline type, was a common finding in the older adult gorillas.

Cranfield (2001) included brief notes on some respiratory cases, including 'Kato'—'atypical interstitial pneumonia (measles and mycoplasma, presumptive)'; 'Zahabu'—'chronic resolving bronchopneumonia, emaciation...'; 'Umuyombyi'—'emaciation, laryngeal air sac abscess...'; 'Mukanza'—'atypical interstitial pneumonia (measles and mycoplasma, presumptive'; and 'Imbaraga'—'laryngeal airsacculitis and bronchiolitis associate with syncytial cells (paramyxoviral nucleocapsids on TEM; measles or RSV infection) with severe secondary bacterial bronchopneumonia (*Klebsiella oxytoca*) cultured.'

The prevalence of respiratory disease and its importance, especially in infants, was frequently emphasised by the author (JEC) when he was working with mountain gorillas in Rwanda. For example, after he had necropsied 5- to 6-month-old 'Insinzi' in March 1995, the Morris Animal Foundation issued a report: 'Dr Cooper said, "My preliminary diagnosis is pneumonia. It is not unusual for mountain gorilla infants to die from pneumonia or infanticide". According to Dr Cooper, about 40% die before the age of 3 years and about 65% of those that do die succumb to pneumonia.'

This view was reflected in the list of diseases recognised and listed by veterinarians at the Mountain Gorilla PHVA 2 years later (Werikhe et al., 1997): 'Respiratory disease, upper respiratory disease, including air sacculitis, lower respiratory disease. Serologic evidence only of alpha herpesvirus, measles, mycoplasma...'.

In a recent note from January 2016 (L.J. Lowenstine, Gorilla Doctors, personal communication), Lowenstine stated that respiratory diseases continue to be an important cause of morbidity and mortality in mountain gorillas, 'more so than gastro-intestinal conditions'. She cited Spelman et al. (2013) who had reported 18 outbreaks of respiratory disease in Rwanda's free-living human-habituated mountain gorillas between 1990 and 2010. Outbreaks lasted for between 2 weeks and 4 months and affected up to five different gorilla family groups, either concurrently or sequentially. Clinical signs in these animals ranged from nasal discharge, sneezing and mild intermittent coughing, to dyspnoea, tachypnoea, respiratory distress, weakness, complete anorexia and occasionally death. Although most animals recovered without treatment, 41 veterinary interventions were required for 35 severely ill gorillas. The observed clinical signs, the response to antimicrobial therapy amongst the sickest animals and the postmortem findings were considered most consistent with viral upper respiratory tract disease, complicated in some cases by secondary bacterial infections.

Clinical signs of respiratory disease in free-living mountain gorillas can range from minor nasal discharge to prolonged bouts of coughing and readily detectable hyperpnoea, sometimes dyspnoea, especially on exertion. Affected animals are often lethargic and may spend long periods in the night nest or have difficulty in keeping up with the group. Such was the case with 'Inyuma', an adult female immobilised, treated and finally necropsied by the author (JEC) in Rwanda in 1995 (although in her case the lethargy was exacerbated by a chronic foot lesion). She showed gross changes consistent with pneumonia, chole-cystitis, cystitis and endometritis.

Some respiratory conditions have prompted specific publications. Hastings (1991) described the veterinary management of a laryngeal air sac infection in a free-living mountain gorilla. Janssen (1993) reported the diagnosis of air sacculitis 'by needle aspiration of exudate'.

Infectious Diseases

The potential importance of viral infections in free-living gorillas was highlighted by the report of Palacios et al. (2011) who described human metapneumovirus in free-living habituated mountain gorillas that died during a respiratory disease outbreak in Rwanda in 2009. An 'outbreak' was defined as more than one-third of animals in a group exhibiting signs of respiratory disease (coughing, oculonasal discharge and/or lethargy). Palacios et al. stated that infectious diseases, primarily respiratory, accounted for 20% of mountain gorilla deaths and are second in terms of importance only to trauma. Mountain gorillas are, in many respects, immunologically naive and susceptible to infection with human pathogens.

Palacios et al. further suggested that human-to-gorilla transmission may explain the outbreaks that they had described in the Virunga National Park, where 75% of mountain gorillas are habituated to the presence of humans (local people and tourists), and where the frequency of human-to-gorilla closeness is particularly pronounced.

The account above is a reminder of the ease with which respiratory pathogens can be spread. As the eminent microbiologist Cedric A. Mims (Mims et al., 2000) put it: 'Respiratory spread is unique. Material from one person's respiratory tract can straight away be taken up fresh and unchanged into the respiratory tract of other individuals, in striking contrast to material expelled from the gastrointestinal tract'.

In 1998 Kalina and Butynski drew attention to the long-standing concerns among veterinarians and biologists concerning close human−gorilla contact and stressed

the importance, for visitors and research workers alike, to observe at least a minimum viewing distance of 5 m. Actual physical contact does sometimes occur because the gorilla approaches the human, not the other way round (the author was welcomed to his post in Rwanda in 1993 with a thump on his shoulder from the 'le responsable', the Sabinyo group's silverback!).

A subsequent questionnaire "Habituating" the great apes: the disease risks', produced by Michael Woodford and Tom Butynski, provided an opportunity for members of the Veterinary Specialist Group (VSG) of IUCN and readers of *African Primates* to submit their considered opinions on such matters as the diseases (*sic*) likely to be carried by humans or free-living apes and reasonable measures that might be taken to minimise the risks to both parties. This led to a publication on the topic (Woodford et al., 2002) and subsequent comment from other quarters (see, e.g., Sandbrook and Semple, 2006).

This is not the place to discuss in detail precautions to minimise contact between habituated gorillas and humans other than to emphasise in the context of respiratory diseases the importance of (1) restricting tourist numbers and proximity and (2) wearing face masks when visiting gorillas (but this probably depends upon the type of mask and how it is worn). Much remains to be learned about the transmission of pathogens in a forest environment, and there is a continuous need to educate those who work with or visit habituated gorillas.

Precautions to minimise contact between habituated gorillas and humans are often best incorporated into more general guidelines for visitors. The Uganda Wildlife Authority (UWA) produced a delightful little flyer with cartoons a few years ago entitled 'Gorilla Rules O.K. Be Responsible Ecotourists' and listing precautions under three headings (1) On the way to the gorillas, (2) When you are with the gorillas and (3) General health rules. The last of these provides opportunities for information to be included that helps protect local people and their communities.

LUNG PARASITES (BY IAN REDMOND)

Kondo et al. (1996) described alveolar hydatidosis in a gorilla and in a ring-tailed lemur in Japan.

There is a genus of mite, *Pneumonyssus*, in the Family Halarachnidae, that lives in the respiratory system of certain mammals, notably dogs and some primates. Several species have been reported from great apes, but only two from gorillas, and it is unclear whether either is a natural gorilla parasite or a result of accidental infestation as a result of contact between host species that only experience close proximity in captivity.

Innes and Hull (1972) recorded *Pneumonyssus oudemansi* in gorillas but, as this species is normally a parasite of rhesus macaques, it seems likely that its appearance in a gorilla was a result of cross-infestation in a zoological collection. Clinical signs of lung mites in apes include paroxysms of coughing and sneezing; postmortem, the lungs display yellow−white nodules 1−2 mm in diameter filled with yellow-brown pigment and from which adult mites can be recovered (Kaandorp, 2010).

Fain (1957) described a new species of lung mite, *Pneumonyssus pangorillae*, from specimens recovered from lungs of several chimpanzees and a single lung of a gorilla infant from Walikale, DRC. The chimpanzees examined were found to have two species of mites; the other one was *Pneumonyssus duttoni*, already known from red-tailed guenons, a monkey species sometimes preyed upon by chimpanzees. The dismembering and eating of monkeys by chimpanzees presents a clear means of transmission of mites, but not one open to nonpredatory primates such as gorillas. This raises the question: Do gorillas in the wild have lung mites? The 2-year-old gorilla examined by Fain had arrived at Antwerp Zoo only 5 weeks after being captured in the wild but died soon afterwards. It is not known whether it had lung mites when captured, but if housed with chimpanzees or monkeys, it too could be a result of cross-contamination. Lung mites have not yet been reported from free-living gorillas in their natural habitat.

One other record exists of a parasite found in the lung of a gorilla: during the postmortem examination of 'Digit', the famous young silverback mountain gorilla killed by poachers, a large trematode, a liver fluke, was observed (32 mm × 19 mm with a median ventral sucker 14 mm × 6 mm, Redmond, 1983). The necropsy took place in January 1978, 6 days after Digit's violent death, and it seems likely that the trematode had found its way to the lung from the liver after sensing the death of its host. Unfortunately the whereabouts of the specimen are not known and no record exists of its species identification.

Noninfectious Diseases

Respiratory disease need not be infectious. Inhalation or ingestion of an irritant or toxic material may cause damage to the tract. Inhaled particulate material can sometimes be seen in the lung, grossly, or on histological examination. For example, the lungs of gorillas kept in the London Zoo (Zoological Society of London, ZSL) in the early years showed dark particles of carbon, typical of anthracosis, while animals at the ZSL's park in rural Bedfordshire (Whipsnade) rarely did so. 'Guy' (see Appendix 5: Case Studies: Museums and Zoological Collections) was a case in point (Fig. 10.5).

Naturally occurring diatomaceous pneumoconiosis was described in the lungs of monkeys from Tchad and Nigeria (Dayan et al., 1978).

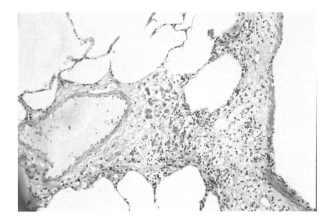

FIGURE 10.5 Anthracosis (dark black particulate material) in the lung of Guy, an elderly lowland gorilla from a city (London) zoo.

Asbestosis would seem an unlikely diagnosis in a gorilla, but it is worth bearing in mind that it has been reported in a free-living baboon in South Africa (Webster, 1963). Histological examination revealed interstitial fibrosis associated with asbestos bodies and fibres.

Possibly relevant to (mountain) gorillas is coltan, the source of tantalum, used in smartphones. Coltan is mainly to be found in the Democratic Republic of Congo (DRC), which is home to gorillas. If inhaled, coltan can irritate the mucous membranes and the upper respiratory tract. There appear to be no reports of this yet in gorillas in the region.

INVESTIGATION OF THE RESPIRATORY TRACT

Knowledge of special anatomy is important when the respiratory tract has to be examined (clinical work − auscultation) or examined postmortem. For example, a useful diagram of the laryngeal sac of an adult male gorilla is to be found in the book by Dixson (1981) (see also Postmortem Examination of the Air Sacs of Apes in Appendix 2: Protocols and Reports).

In both live and dead gorillas careful attention should be paid to the upper respiratory tract. The pharynx is often overlooked, especially in a dead animal that exhibits rigor mortis. A strong light source, preferably an auriscope with a long speculum, will assist. A captive gorilla that died at Bristol Zoo Gardens showed pharyngeal pathology that could easily have been missed if the pathologist had not carefully investigated it at the start of the necropsy. The histological findings are of interest. Folds of connective tissue supporting the pharyngeal epithelium were oedematous and contained increased numbers of small blood vessels consistent with chronic inflammation. There was diffuse infiltration by moderate

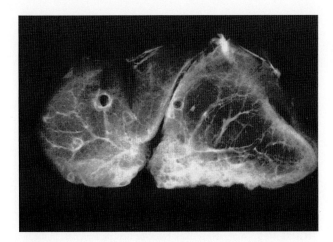

FIGURE 10.6 Radiograph of inflated primate lung showing areas of consolidation.

numbers of chronic inflammatory cells including plasma cells and some Russell body cells (large eosinophilic immunoglobulin-containing inclusions).

The correct, careful processing of respiratory tissue is essential for an accurate diagnosis. Fixation of pulmonary tissue is not a simple matter. The lung consists of 90% air and 10% solid tissue. If pieces of lung are placed directly into a fixative, deflation and collapse take place; a histological section may show parenchyma with apparent hypercellularity and alveolar wall thickening (van Kuppeveldt et al., 2000) that is not always easy to differentiate from early interstitial pneumonia. For this reason, methods of inflating the lungs have always been a particular interest of the author (JEC).

A method of fixing lungs by inflation with heated formalin vapour was first described in Europe by Wright et al. (1974) and subsequently applied to both human and nonhuman material by the author (JEC) and his colleagues. Following fixation, the lungs were radiographed before and after slicing at 1 cm and the images carefully examined for specific lesions (Fig. 10.6). Blocks of both visible lesions and apparently normal areas were then taken for histological investigation.

Cooper et al. (1978) presented the results of a survey of pulmonary lesions in captive primates based on a postmortem examination of formalin-inflated lungs. One case in the series that was examined using inflation was a 19-year-old lowland gorilla with a clinical history of bronchopneumonia, empyema, and pulmonary oedema. Histologically, the trachea was found to be hyperaemic with focal degeneration and sloughing of the surface epithelium. In some areas there was active regeneration. These features were considered to be those of a severe tracheitis. The lungs showed an acute bronchiolitis with sloughing of the epithelium and an associated pneumonia characterised by a haemorrhagic fibrinous

exudate and large areas of pyogenic inflammation. These features of a very florid pneumonia were considered consistent in many respects with changes in human influenza (the study was a collaborative one, with both medical and veterinary pathologists participating).

Inflation of lungs with heated formalin vapour had to be discontinued later on health and safety grounds but the procedure continued using 10% formal-saline fluid instead. The author (JEC) still strongly advocates fixing primate lungs by introducing fixative directly into the airways, especially if histological sections of high quality are sought. It is also helpful to radiograph the fixed lungs; this permits improved diagnosis (clinical and postmortem) because the radiological appearance can be correlated with histological findings. Such comparisons are also proving valuable in other fields of veterinary medicine (Scott, 2013).

CARDIOVASCULAR SYSTEM

This section covers the heart and blood vessels. The blood and bone marrow are discussed in Chapter 12: Lymphoreticular and Haemopoietic Systems and Allergic Conditions.

The heart and blood vessels are derived from embryonic mesoderm germ-layer cells that differentiate after gastrulation into mesothelium, endothelium and myocardium. Mesothelial pericardium forms the outer lining of the heart.

The first scientific description of the anatomy of the heart and aortic arch of the genus *Gorilla* was by Washburn (1950). Hall-Craggs (1961), lamenting the absence of anything in Washburn's chapter about the vessels of the heart, described these in two specimens of *beringei*.

The structure and function of the primate cardiovascular system were discussed by Scott (1992) who, amongst other things, commented that 'Details of the weight of the heart in gorillas are, not unnaturally, in short supply'. He did, however, include a table of heart weights of five gorillas at the Yerkes Regional Primate Research Center (YRPRC). In one male animal of unknown age, the heart weighed 1140 g and the case is marked 'obese + + + '.

More comprehensive anatomical studies have also been published. For example, Kawashima and Sato (2012) detailed the comparative anatomy of the extrinsic cardiac nerve plexus and postnatal reorganisation of the cardiac position and innervation in orangutans, gorillas, and chimpanzees.

INVESTIGATION OF THE CARDIOVASCULAR SYSTEM

Standard clinical diagnostic procedures, many of which have been used for (usually captive) gorillas, will not be discussed, as the emphasis here is on pathology. The pathophysiology of cardiovascular disease in great apes remains unclear, so it is prudent to perform a full cardiac assessment, including echocardiography, as a bare minimum, and ideally also electrocardiography and biomarkers. Cardiac troponins have been used for many years as biomarkers of heart disease in humans.

Increasingly sensitive imaging methods provide an opportunity to identify and track cardiac pathology in live gorillas, but as yet few studies seem to have been performed comparing the results of imaging with tissue morphology (see earlier: Cooper et al., 1978).

The detection of cardiovascular lesions often requires a detailed, skilled examination of blood vessels and the heart — and knowledge of the normal anatomy of these structures. Necropsies performed by inexperienced personnel (see Chapter 6: Methods of Investigation — Postmortem Examination) may result in changes being missed. At the very least, such people should be familiar with how to remove and preserve the heart for subsequent detailed investigation. Veterinary pathologists who are accustomed to working with domestic animals, may also be at a disadvantage as the latter species are more prone to such conditions as valvular diseases or cardiomyopathy, each with overt structural changes.

It seems that variation in how hearts of captive gorillas and other great apes have been examined in the past has contributed to making it difficult to analyse the true extent of cardiac pathology in these species. In order to help rectify this, the Great Ape Heart Project (GAHP) has produced and distributed within Europe a protocol with basic instructions. Essentially, zoos are asked to remove and fix the heart and then to send it to GAHP for a more detailed examination (Victoria Strong, personal communication). The definitive dissection is performed in accordance with a protocol based on methods used in humans (Sheppard, 2012). It is also important that the appropriate samples are collected for histopathology.

See Appendix 2, Protocols and Reports:

GAHP Recommended Cardiac Necropsy Prosection Guide
GAHP Recommended Cardiac Trimming Protocol for Pathologists

Field dissection of the mammalian heart and great vessels is outlined by a number of (nonveterinary/medical) authors; see, for instance, Sikes (1969).

It is important not to deal too hastily with the pericardial sac. It may be thickened, inflamed, or adherent to the left thoracic wall. It may contain an excess of fluid; this should be measured and a record made of whether it is whole blood/blood-tinged/colourless/straw-coloured. The PCV and specific gravity of this fluid should be recorded. The epicardium also warrants careful inspection; it should

be smooth and glistening and free of haemorrhage. The 'feel' of the heart can be a rough guide to its status; a 'flabby' heart warrants particular attention. It must be weighed and the chambers measured. The appearance, including the colour of (incised) myocardium should be recorded. The heart valves require meticulous examination and palpation. Like the epicardium, endocardial surfaces should be smooth and glistening. Both aorta and venae cavae must be thoroughly investigated. Any comments regarding the heart weight/size, chamber capacity, or wall thickness must be expressed and made with caution and in context; for example, a heart from an animal that died in systole can give the impression of being thickened in the left ventricle. Heart-weight/body-weight ratios and comparative left/right septal wall measurements are important.

CARDIOVASCULAR DISEASE

One of the earliest accounts of cardiovascular disease in a captive gorilla was by Steiner et al. (1955), who described cardiopathy and haemosiderosis coupled with neuropathological changes and testicular atrophy. Lesions were reported in lowland gorillas by Steiner et al. (1955) (histological evidence of intimal thickenings in the aorta and medial sclerosis/intimal degeneration of small coronary arteries) and Ratcliffe (1961) (infarct of the left ventricle, mural thrombus, thrombosis arteriosclerosis, and atheroma). Stehbens (1963) described atherosclerosis and both intimal and medial defects, 'similar to those observed in man', in a female lowland gorilla said to be only 14 months old.

In 1965 Rodney Finlayson, a medically qualified comparative pathologist working in London, published a monumental account of the pathology and incidence of spontaneous arterial disease in a series of about 2000 vertebrates, the majority of which died in captivity from natural causes. He stated that 'mild lipid infiltration of the walls of the large arteries is fairly common' in captive primates and expressed surprise that Fox had reported no evidence of arteriosclerosis in great apes examined at the Philadelphia Zoological Garden. Finlayson found 'arteriosclerotic changes in the gorilla, orangutan and chimpanzee . . .. The condition of the arteries in the relatively few gorillas has, however, occasioned little comment'. Finlayson described his postmortem findings in four mountain gorillas, all less than 4 years of age. Lesions seen included a very mild degree of aortic fatty streaking and, in two animals, intimal changes in the major coronary arteries. Postmortem serum cholesterol values of two of these mountain gorillas were higher than those usually observed in 'subhuman primates'.

Over the succeeding decades interest in cardiovascular disease of apes grew and saw the appearance of increasing numbers of reports. These included an account of congestive heart failure and nephritis (Robinson and Benirschke, 1980) and of chronic hypertension with subsequent congestive heart failure (Miller et al., 1999), both in western lowland gorillas. Lee et al. (1981) published data on the electrocardiogram of the lowland gorilla. 'A ventricular septal defect ... was diagnosed using in vivo oximetry' (Machado et al., 1989). Allchurch (1993) reported in detail on the sudden, unexpected, death of 'Jambo', the silverback male lowland gorilla at the Durrell Wildlife Conservation Trust ('Jersey Zoo') in the (British) Channel Isles. The animal was 31 years old and the cause of death was a ruptured aortic aneurysm, a condition that is well recognised as a factor (sudden death) in humans, a form of aortic 'dissection'. This finding prompted a review of the historical data available at the time, published in the *International Studbook of the Gorilla* and other literature. Out of a total of 54 deaths attributed to cardiovascular disorders, there were nine cases of ruptured aortic aneurysm.

In 1995 Schulman et al. reported on fibrosing cardiomyopathy (defined as myocardial replacement fibrosis with atrophy and hypertrophy of cardiac myocytes, absent to mild myocardial inflammation and no apparent aetiology or associated disease) in captive lowland gorillas in the United States. This was a retrospective study and the condition was identified in 11 animals, all male and with ages ranging from 11 to 37 years. In eight instances, gorillas of 16−37 years of age, the cardiac scarring was considered to be the cause of death; seven of these animals died suddenly. The pathogenesis of this condition is reported later.

By the 1990s there was growing interest in the possibility of using both spontaneous and induced cardiovascular disease in NHPs to study comparable conditions in humans. Amongst relevant papers was the review of spontaneous lesions in purpose-bred laboratory primates by Chamanza et al. (2006).

In his 12-year review of clinical conditions in the Howletts Zoo (United Kingdom) gorilla colony, Furley (1997) reported a case of cardiomyopathy. In their book Schmidt and Hubbard (1987) depicted a large area of arterial atherosclerosis in a gorilla. In the succeeding years interest in cardiovascular disease in great apes has burgeoned (McManamon and Lowenstine, 2012) leading to the establishment of a number of research projects on the subject (see later).

In a recent (2016) personal communication, L.J. Lowenstine, Gorilla Doctors, stated that 'In captive lowland gorillas in US zoos, heart disease still continues to be a major cause of mortality. Heart disease seems more prevalent in the Grauer's gorillas than in the mountain gorillas'. In 2014 Kambale et al. had presented an abstract, summarising findings in respect ofcardiovascular (and hepatic) disease in free-living *graueri*.

In a recent comprehensive review of the comparative pathology of ageing bonobos, chimpanzees, gorillas and orangutans, Lowenstine et al. (2016) confirmed that aortic dissections are the second-most common cardiovascular problem in gorillas and a major cause of sudden death. Of eight affected animals in an earlier report by Kenny et al. (1994), six were males and their ages ranged from 16 to 43 years. All dissections were in the ascending aorta (DeBakey type II, Stanford type A) with rupture of the aorta just distal to the aortic valves or within an aortic sinus. Plaques/grade I atheromas were demonstrated by transoesophageal echocardiography in the ascending and descending aorta in gorillas aged 10, 26, 27 and 36 years, and grade II atheroma in a 40-year-old male, suggesting that the severity increases with age (Baitchman et al., 2003).

Lowenstine et al. (2016) went on to state that, in their experience, coronary artery atherosclerosis is rare in gorillas, although in the past coronary artery disease was reported to have caused myocardial infarction in gorillas (Gray et al., 1981). Lowenstine et al. had seen atherosclerosis involving only small-calibre intrinsic coronary arterioles in two gorillas: a 34-year-old male with chronic heart failure, and a 39-year-old female with hypertension and stroke.

Valvular disease has not been frequently described in great apes. Isolated lesions were very rare in echocardiograms of 99 lowland gorillas, although mild mitral regurgitation was noted (Murphy et al., 2011). In the apes that Lowenstine et al. (2016) examined, infective valvular disease was seen in a few mountain gorillas and myxomatous degeneration of a mitral valve was also observed. A single case of aortic valve fibrosis and mineralisation was diagnosed clinically and confirmed at necropsy in a 47-year-old male zoo gorilla.

Lowenstine et al. went on to discuss the 'important and not yet completely answered question' of whether cardiovascular disease, particularly 'fibrosing cardiomyopathy' (see later), occurs in free-living apes or is only a feature of captivity. In a review of causes of mortality in mountain gorillas, cardiovascular disease was considered to be the cause of death in only 3% of animals (Nutter et al., 2005). More recently, a higher prevalence was noted in *graueri* from an isolated population on Mount Tshiaberimu in the DRC (Kambale et al., 2014; see earlier). Lesions seen in that study included myocardial fibrosis and arteriosclerosis in intrinsic coronary arteries, together with marked hypertrophy of myofibres similar to that described in zoo-housed lowland gorillas. Atherosclerosis (see later) has not been seen by Lowenstine et al. in coronary arteries of free-living mountain gorillas but fatty streaks and plaques have been detected in the aortae of two geriatric females aged 35 and 39 years, both of which were cachectic and had advanced neoplasia.

Reviews of pathology reports from diagnostic laboratories confirm the wide range of changes that are seen in the cardiovascular system of captive gorillas in the United Kingdom. For instance, amongst reports held by the School of Veterinary Sciences, Langford, University of Bristol, are descriptions of a thickened and opaque pericardial sac with an excess of clear, watery, straw-coloured fluid with fine fibrin strands on the epicardial surface in the pericardial cavity; serous atrophy of epicardial fat with areas of atheroma; an atheromatous lesion characterised by roughened, thickened, luminal surfaces to coronary arteries and with histological evidence of vacuolation of myo-intimal cells ('foam cell' formation) and a small area of calcification; and excess fibrous connective tissue and some fatty infiltration of the right ventricular myocardium (consistent with scar formation, possibly reflecting a previous myocardial infarct), together with an accumulation of haemosiderin-laden macrophages in the lung alveoli.

Important points arising from — and raised in — the paper by Lowenstine et al. (2016) are the need for (1) more necropsies with detailed examination of the heart, based on standard, agreed, protocols (see later: GAHP) in order to determine the nature and extent of cardiovascular disease in free-living apes and (2) research to define better the lesions and pathogenesis of the heart diseases that are so prevalent in captive (zoo) apes. The latter will now be discussed.

RESEARCH ON HEART DISEASE IN GREAT APES, INCLUDING GORILLAS

The GAHP is based at Zoo Atlanta in the United States (see Part II: A Catalogue of Preserved Materials). GAHP has developed standardised protocols for ante- and post-mortem evaluation of ape hearts and a database that permits analysis of clinical and necropsy data. It also maintains a database of information about cardiac disease in great apes (research requests for access to this information can be directed to gahpinfo@gmail.com) and, following the February 2011 GAHP workshop, established a reference list relating to the subject. The long-term aim of GAHP is to be better able to understand, diagnose, treat, and ideally prevent great ape heart disease (see http://greatapeheartproject.org/).

Twycross Zoo in the United Kingdom has a particular interest in cardiovascular disease in great apes and, in collaboration with The University of Nottingham, United Kingdom, is involved in research involving veterinarians, cardiologists and medical researchers, with a view to learning more about great ape heart disease. This project is part of an initiative led by the EAZA (European Association of Zoos and Aquaria) Great Ape TAG (Taxon Advisory Group), which is striving to develop a collaborative and cooperative approach to the investigation of cardiovascular disease among the European captive great ape population. Twycross Zoo's Ape Heart Project is run under the auspices of the EAZA Great Ape TAG and also collaborates with the USA-based GAHP at Zoo Atlanta (see above).

The International Primate Heart Project (initially established as the European Great Ape Heart Project) (IPHP) is based at Cardiff Metropolitan University, Wales, United Kingdom (http://primateheartproject.co.uk). Some information about the Project, particularly in respect of gorillas, can be found in Part II: A Catalogue of Preserved Material. The IPHP's work has primarily focused on chimpanzees, for which they have completed in excess of 350 assessments (Robert Shave, personal communication).

IPHP believes strongly that standardised approaches are needed in clinical work as well as postmortem studies if cardiovascular disease is to be properly understood. In this respect, Shave et al. (2014) have already proposed an approach for an echocardiographic assessment of cardiac structure and function in great apes. More work, with larger numbers of animals, is needed if reliable reference values are to be produced. Shave (personal communication) is concerned that the available literature regarding cardiopathology in great apes 'is sparse at best, and is littered with case studies, or case series that have become, in my opinion, dogma'. He urges a more guarded approach to declarations as to what exactly is causing heart-related deaths or at least to couple claims with a strong caveat that the existing literature may not fully represent what is actually occurring. In particular, data from elderly animals in zoo or research collections should be treated with caution because age and stress are likely to be confounding variables. There are many, varied, causes of sudden cardiac death in humans and it is probably unwise for veterinarians to try to pinpoint only one or two for comparable deaths in great apes. This echoes the thoughts of Schulman et al. (1995), who urged further research to identify the underlying cause(s) of fibrosing cardiomyopathy; this remains a need 20 years later.

Careful terminology is also important. As Part II: A Catalogue of Preserved Materials, shows, terms such as 'acute heart attack', 'heart failure' and 'an apparent heart attack' are used rather loosely in reports.

It is clear that much remains to be learned about the pathogenesis of cardiovascular disease in gorillas and other great apes. Once again, this is a truly interdisciplinary area of research that should involve ethologists, nutritionists and physiologists as well as veterinarians and comparative physiologists. A particularly exciting prospect is that of monitoring free-living animals, as has been done on a small scale in respect of total cholesterol triglyceride, high-density lipoprotein cholesterol and low-density lipoprotein cholesterol values in gorillas and orangutans (Schmidt et al., 2006).

As far as research on the microscopic pathology of cardiovascular disease is concerned, it is encouraging to note how many specimens of heart and aorta are stored in various collections throughout the world (see numerous entries in Part II: A Catalogue of Preserved Materials) and, in most cases, are available for study.

Assessment of live gorillas in captivity is, of course, also important, for example, studies on echocardiographic parameters (Murphy et al. 2011; Shave et al., 2014) — see later.

PATHOGENESIS OF CARDIOVASCULAR DISEASE

This section will discuss specific pathogenesis in some detail and then give some actual examples of the effects of cardiovascular disease.

Atherosclerosis is a specific form of arteriosclerosis, the occurrence of which in wild animals in zoological gardens attracted interest several decades ago (Fox, 1923). Comparable occurrences were reported by Ratcliffe and Cronin (1958) and Fiennes (1964).

During the same period, there was a rapidly increasing amount of research into human atherosclerosis resulting in much literature, often from diverse disciplines, on the subject (e.g., see the *Journal of Atherosclerosis Research*, 1961−67).

The word 'atherosclerosis' was not then used uniformly by all investigators, some even employing the terms 'atheroma' and 'arteriosclerosis' synonymously (Finlayson, 1965). The definition of atherosclerosis in humans that became accepted internationally was the one that was propounded by the World Health Organization in 1958: 'Atherosclerosis is a variable combination of changes of the intima of arteries...consisting of the focal accumulation of lipids, complex carbohydrates, blood and blood products, fibrous tissue, and calcium deposits, and associated with medial changes.'

Finlayson (1965) described vascular lesions, with particular attention to the comparative pathology of degenerative arterial disease and, more especially, atherosclerosis, in fish, amphibians, reptiles, birds and mammals. He showed that minor degrees of lipid infiltration of the arterial wall are seen commonly in many taxa of mammals, but more advanced atheromatous plaques occur frequently only in primates, including humans. He hypothesised on the histogenesis of atherosclerosis and concluded that it is a multifactorial disease. The possible role of dietary and other factors were discussed. Age was shown to be related to atherosclerosis, but the almost complete absence of lesions in some comparatively old animals suggested that

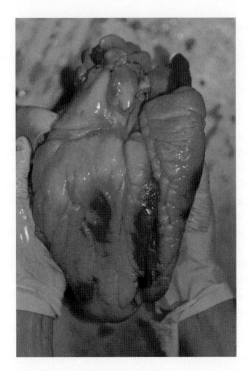

FIGURE 10.7 Heart of a male, obese, captive western lowland gorilla that had previous episodes of 'fainting' and then died suddenly following a fight. *Image Courtesy of David Perpiñán.*

atheroma is not an inevitable concomitant of senescence. The morphological pattern of arterial disease appeared to be similar in both free-living and captive animals. Atherosclerosis can affect blood vessels in various parts of the body, including the brain, mesentery, and choroid (see mention of these in entries in Part II: A Catalogue of Preserved Materials).

The pathogenesis of fibrosing cardiomyopathy remains unclear. The gross appearance is characteristic, usually marked pallor of heart muscle (Figs.10.7−10.9). Histologically, the features are usually those of multifocal, sometimes coalescing, moderate-to-marked myocardial fibrosis with atrophic and hypertrophied cardiac myocytes. There may be inflammatory changes.

The effects of cardiovascular pathology can be profound, affecting many organ systems. This is illustrated in data provided by Roberta Wallace, Milwaukee County Zoo (personal communication), which included: (1) a male gorilla in his 30s that died of 'massive' myocardial fibrosis, with acute myocardial necrosis and extensive systemic haemosiderosis; (2) another male in his 30s with congestive heart failure, chronic active nodular valvular endocarditis and, again, marked hepatic haemosiderin deposition; (3) a male, 14 years old, with granulomatous myocarditis, pancreatitis and pneumonia associated with granulomatous amoebic encephalitis due to *Balamuthia mandrillaris*; (4) a 44-year-old female, euthanased on

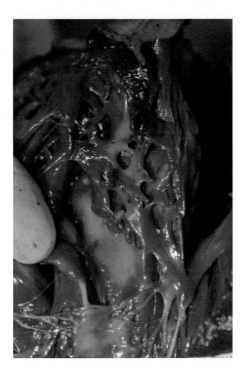

FIGURE 10.8 Right auricle opened. Note the pallor of underlying muscle. Histological diagnosis was myocardial fibrosis. *Image courtesy of David Perpiñán.*

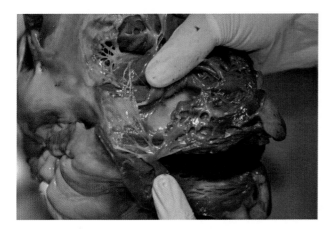

FIGURE 10.9 Another view of the opened heart. *Image courtesy of David Perpiñán.*

account of a cerebral vascular accident, underlying hypertension, and cardiovascular lesions (coronary atherosclerosis, biventricular myocardial hypertrophy with focal fibrosis of the left ventricle) and (5) a male aged 28 years with cardiomegaly, cardiomyopathy, thromboembolic disease, with severe thrombotic bronchopneumonia. In respect of the last of these, Roberta Wallace added: 'This animal developed a severe anemia of unknown origin for which we could never find a cause. We believe that the cardiac problems may have developed secondary to that.

I have seen and/or heard of this anemia of unknown origin in several gorillas. Sometimes it resolves on its own, and sometimes it ends in death of the animal.'

Not only can cardiovascular pathology have effects on other organs but it in turn may be affected by disease in other organs. An example, perhaps, was 'Guhuma' (Cranfield, 2001), who was reported to have vegetative endocarditis associated with Gram-positive bacteria and septic pneumonia. In addition, the cardiovascular system can be damaged by medication and routine procedures such as anaesthesia (see, e.g., Horne et al., 1997).

OTHER PATHOLOGY OF THE VASCULAR SYSTEM

A clear distinction must be made between arteriosclerotic diseases, of which atherosclerosis, as defined above, is one, and inflammation of arteries (arteritis) caused by infective and inflammatory agents or, possibly, an autoimmune response (see vasculitis in Chapter 12: Lymphoreticular and Haemopoietic Systems and Allergic Conditions). No parasitic arteritides appear to have been reported in gorillas.

The aorta has been described as the 'orphan organ' in medical circles in Britain on the grounds that it lacks a natural home within a defined medical or surgical specialty. In the context of gorillas, it is clearly a very significant site for pathological changes that warrants far more attention from clinicians and pathologists alike.

Shock

Although pathologists have focused largely on lesions of the heart and great vessels, it should not be forgotten that the functioning and failings of the cardiovascular system as a whole are of equal, if not greater, significance in terms of day-to-day health of gorillas. Chief amongst these is the maintenance of blood pressure and perfusion of tissues. Dysfunction can occur for many reasons but amongst the most important are those circulatory changes that can follow haemorrhage, infection, endotoxaemia, anaphylaxis, and certain other insults commonly termed 'shock', a syndrome recognised in all vertebrate species.

The word 'shock', like 'stress' (see Chapter 12: Lymphoreticular and Haemopoietic Systems and Allergic Conditions) is often misused. In the past, far too many veterinary reports referred to animals dying of 'shock' without detailed description of the pathological changes involved.

So what is the pathogenesis of shock? Essentially, it is a defect of the haemodynamics of vital organs. Circulation failure within the cardiovascular system results in hypovolaemic, distributive, or cardiogenic shock (see Boxes 10.1 and 10.2).

BOX 10.1 Causes of Hypovolaemic Shock

- Blood loss
- Plasma loss (burns, dysentery)
- Sequestration of body fluids (ascites)
- Fluid and electrolyte loss (vomiting, diarrhoea, dehydration)
- Diuresis (hypoadrenocorticism, diabetes insipidus, diabetes mellitus)

BOX 10.2 Causes of Distributive (Vasculogenic) Shock

Maldistribution of cardiac output away from vital organs/peripheral pooling of blood, as in:
- surgery
- anaphylaxis
- endotoxaemia
- sepsis
- anaesthetic overdose

Clinical reports often describe circulatory shock (of various degrees) in gorillas, but published reports are surprisingly rare. One account (Dorobisz et al., 1983) described circulatory disorders in lowland gorillas.

The postmortem appearance of shock is familiar to most pathologists. Rutty (2004), writing about *Homo*, provided an excellent review of the features of shock in its sense of 'inadequate perfusion of the tissues', and drew attention to how antemortem lesions associated with this can be mistaken for postmortem autolysis. The necropsy reports in Fossey (1983) contain references to septicaemia and shock, including 'shock lungs.'

Chapter 11

Alimentary Tract and Associated Organs

John E. Cooper and Ian Redmond

The Gorilla is a huge feeder, as its vast paunch, protruding when it stands upright, shows.

Sir Richard Owen (1859)

INTRODUCTION

This chapter covers the alimentary (gastrointestinal) tract, including the mouth, salivary glands, jaws and teeth, peritoneum, mesenteries, liver, biliary tract and exocrine pancreas.

The first detailed scientific description of the anatomy of the alimentary (gastrointestinal) tract in *Gorilla* was by Elftman and Atkinson in Gregory (1950). Their structure and function in NHPs were discussed by Scott (1980, 1992). The liver of African anthropoid apes was described by Stunkard and Goss (1950b). Gastrointestinal allometry has been extensively studied; see Martin et al. (1985) and MacLarnon et al. (1986a,b).

The origins of these structures relate to the embryonic stomodaeum, the production of a coelomic cavity and the development of an intestinal canal ('gut'), including embryological periods of canalisation, recanalisation, dilatation and exvagination. The gastrointestinal tract (GIT) and its derivatives — liver, gall bladder and pancreas (and lungs) — are endodermal but the serous membranes around organs are mesodermal derivatives.

The cardinal features of the GIT of the gorilla are similar to those in *Homo* (see Box 11.1).

ORAL CAVITY

Examination of the oral cavity should be a routine part of clinical examination of gorillas in captivity, especially in zoos but also when animals are inspected on arrival or for health checks in rehabilitation centres. The oral cavity can provide important medical information about the general health of the individual and is sometimes referred to as the 'mirror' of what is happening in the rest of the body.

The examination of the oral cavity should follow standard (human) odontological protocols, including the use of dental charts. Direct vision with the aid of mouth mirrors is usually ideal for examination. Occasionally an endoscope is helpful for accurate diagnosis of dental disorders. However, other techniques are often necessary, including imaging procedures such as radiography and computed tomography.

The importance of very meticulous and detailed examination of the buccal cavity at necropsy is well illustrated by the following extract from a report on a 2-year-old female lowland gorilla at Bristol Zoo Gardens (BZG), United Kingdom: 'External examination. The top incisors have been removed and the gum appears thickened and red. The palate is covered with cream white plaques. There is an abscess in the soft tissue medial to the left mandible'. Oral lesions can be missed if the pathologist or prosector is in too much of a rush to examine internal organs.

The examination of teeth and the oral cavity benefits greatly from an input by a human dentist or a specialist odontologist and interpretation in a young gorilla is helped if the dentist is experienced in paediatric work. A veterinarian with training in dentistry is also often an asset but usually such a person has specialist knowledge of domesticated species, such as dogs or horses, rather than primates.

An ideal situation is when the veterinarian and the dentist/odontologist work together, as depicted in Fig. 11.1.

In most cases the animal will need to be sedated or anaesthetised for a full dental examination, a potential risk if there are concurrent health problems such as cardiomyopathy (see Chapter 10: Respiratory and Cardiovascular Systems).

Free-living gorillas that have to be immobilised on account of ill health will usually have their oral cavity examined while under anaesthesia but unless dental disease is specifically suspected, the teeth are not generally subjected to a detailed and systematic assessment because of the exigencies of time. Experience in assessing dental

J. E. Cooper & G. Hull: Gorilla Pathology and Health. DOI: http://dx.doi.org/10.1016/B978-0-12-802039-5.00011-1

BOX 11.1 Features of the GIT of the Gorilla

- Oral cavity and oesophagus — stratified squamous lining
- Stomach — specialised mucosa, with glands for (inter alia) acid production
- Small intestine — thin walled, distinct mucosal lining with villi
- Caecum — well-developed
- Appendix — a blind sac
- Liver lobes variable in number: up to six reported

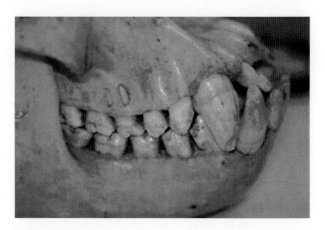

FIGURE 11.2 A museum specimen showing a full set of teeth and some alveolar bone resorption.

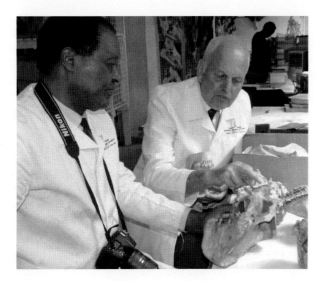

FIGURE 11.1 A veterinary pathologist consults a dental professor regarding lesions in the jaw of a mountain gorilla. Interdisciplinary collaboration in Nairobi, Kenya.

structures in gorillas can be gained by careful study of museum material (Fig. 11.2).

Gorillas in zoos are not infrequently reported with oral conditions. The latter include stomatitis, glossitis, gingivitis (often manifest by swollen gums) and damage to the buccal cavity. A few cases have been reported; for instance, gingival hyperplasia induced by diphenylhydantoin (Fagan and Oosterhuis, 1979), and mycotic gingivitis and mycotic infection ('thrush') of the oesophagus (Schmidt and Hubbard, 1987).

Gorillas in the wild also present with lesions of the oral cavity, ranging from erosions and ulcerations to localised proliferative changes. Infectious agents may be involved — see Gardner-Roberts et al. (2007) — 'examination of biopsies of the (stomatitis) lesions by light microscopy suggested a herpesvirus infection'. Oral disease in free-living gorillas may have far-reaching effects. For instance, 'Mwirima', the leader of a mountain gorilla group who became unwell with dysphagia in 2014; he

was found to have a lesion in his mouth. Within 2 weeks he was dead.

Cleft palate and other developmental conditions are discussed in Chapter 13: Urinary and Reproductive Systems.

Teeth

'The huge canines in the male give a most formidable aspect to the beast' wrote Sir Richard Owen when he first described a gorilla. Some decades earlier the famous French zoologist Georges Cuvier had said 'Show me your teeth and I will tell you what you are'. This remark was intended to relate to the identity of a species or individual, but is applicable also to health, including modern methods of dental investigation — as illustrated by the findings of Cipriano (2002) that 'cold stress' could be detected by recording incremental lines in dental cementum.

The dental formula of *Gorilla* is the same as that of *Homo* — 2123/2123 (Napier and Napier, 1967) but, as in humans, there are variations. In 1899 Selenka reported that fourth molar teeth are present in about 8% of adult gorillas, a frequency confirmed in similar comparative studies by Duckworth (1915). Supernumerary premolars are very rare (Schultz, 1964).

The relative accessibility and visibility of the teeth of gorillas, in both live and dead animals, and their durability (often, postmortem, long after soft tissue and some skeletal elements have disappeared — see Appendix CA2: Retrieval, Preparation and Storage of Skeletal and Other Material) means that odontological studies have attracted the attention of primatologists, veterinarians and odontologists. Dental development is of interest also to palaeontologists and evolutionary biologists, who still often report on individual animals; see, for instance, Joganic (2016). A series of papers based on substantial numbers of known-age mountain gorillas illustrate the value of retrospective research on free-living animals. For example, see Galbany et al. (2015) on how age and diet are related to tooth wear and Glowacka et al. (2016) on age-related changes in molar topography and shearing crest length.

Soon after the discovery of the mountain gorilla, too, there were published dental data on the species, based on examination of shot specimens, by Lönnberg (1917) in specimens collected by Captain Elias Arrhenius (see Part II: A Catalogue of Preserved Materials). The data on the teeth are of especial interest as they comprise a description in an old male of normal wear of incisors and canines and previous damage, attributed to a 'fight with a rival', to canines, including the filling of an 'alveole' [sic] with bone following the loss of a tooth. Five years later Barns (1922) described the skull of a 'very old male gorilla' and commented that, 'As the teeth are quite worn away, the animal was possibly of very great age'.

Gorilla teeth even played a small, but significant, part in the famous 'Piltdown fraud' when Charles Dawson sketched the skull of a female gorilla and its canines at the Royal College of Surgeons in London to illustrate that the wear visible on the Piltdown teeth was what would be expected on an ape-like human ancestor.

The continued interest in anthropological circles of comparative studies was reflected by Randall (1943), who reported on 'skeletal and dental development and variability of the gorilla'.

Rather more academically directed, often comparative, research on dentition was carried out in the 1980s and 1990s by various authors. For example, Ryan (1981) investigated anterior dental microwear and its relationships to diet and feeding behaviour in three African primates − *Pan, Gorilla* and *Papio*. Dean et al. (1992) explored the natural history of tooth wear, continuous eruption and periodontal disease in a number of wild-shot great apes. Lovell (1990a,b) included assessment of dental disease in her extensive survey of skeletal pathology (see later and Chapter 14: Musculoskeletal System). Gebo (2014) provided a very good overview of the morphology of teeth and the evolution of cusp patterns in primates and how this may correlate with diet adaptation.

DENTAL DISEASE AND PATHOLOGY

There are various ways of categorising types of dental disease in gorillas, largely because odontological studies on *Gorilla* have been carried out by persons from disparate disciplines, many without specific training in odontology. In fact, however, no method of classifying dental pathology can fully take into account the complexity of disease in an animal such as a gorilla where, particularly in the wild, there may be interactions between infectious and noninfectious factors, including self-inflicted trauma

or introduction of organisms on account of pain or dysphagia. The picture presented in clinical and postmortem cases often reflects a series of changes, including attempts at healing and regeneration of soft tissues and bone. Caution in interpretation is always wise. As Francis Bacon counselled, in a very different context, 'If a man will begin with certainties, he shall end in doubts; but if he will be content to begin with doubts he shall end in certainties'.

Interest in dental abnormalities of gorillas goes back almost to the dawn of discovery of this taxon. Thus, in 1885, at a meeting of the Odontological Society of Great Britain, Mr Storey Bennett (1885) exhibited a gorilla skull that showed antemortem injury to the facial bones, nares and right zygomatic and missing incisors with their 'sockets absorbed' and a supernumerary tooth 'on the inside of the ascending ramus of the right lower jaw'. Sir Frank Colyer (1931) described abnormal conditions of the teeth of a number of taxa of animals and related them to similar conditions in humans. Meanwhile, veterinarians and others working with captive, mainly lowland, gorillas in zoos began to report systematically on oral diseases in these animals. Dental case reports for gorillas between 1982 and 1988 were recounted by Scheels (1989). Individual case studies included accounts of localised periodontitis (McManamon et al., 1989) and of root canal therapy for a fractured canine tooth by Tomson and Schulte (1978). A generalised, but useful, review of dental disease recognised in gorillas at that time was produced by Cousins (1982).

Other than gross damage, such as fracture of teeth following trauma (see, e.g., Colyer, 1915; Cave and Jones, 1992; Meder, 1994), it is the outer layer of the tooth, the enamel, the layer that protects the underlying dentine and pulp, that is most important in terms of pathology. Like bone, enamel is composed of calcium phosphate; it is the most mineralised tissue in the body. Amelogenesis (enamel production) occurs only once. Defects of the enamel in gorillas are likely to be similar to those in humans. Sometimes enamel is not formed correctly and may thereby be damaged during mastication or a fight. Alternatively − and this seems to be more common in gorillas − the enamel is sound but becomes secondarily demineralised due to abrasion or bacterial invasion. This results in the formation of areas of weak enamel and dental caries may follow. In humans, genetic mutations may lead to enamel hypoplasia (amelogenesis imperfecta) but this appears not yet to have been definitively reported in gorillas.

In 1952 Schultz listed the number of cases of 'caries', 'abscess' and 'closed alveolus' that he found in a series of 352 gorillas. Interestingly his specimens included only four captive animals and these were all reported to have 'perfectly sound dentitions'. Schultz further commented that 'juveniles have incomplete second dentitions, adults show only slight to moderate degrees of attrition of teeth and in senile specimens the teeth are much to extremely worn...'.

Comments by those working in the field continued to augment our knowledge of normal and abnormal features of the dentition of gorillas. Pitman (1942) stated that 'many of the skulls of *Gorilla g. beringei*, from the Birunga regions, show considerable spaces between the teeth, due, possibly, to a preponderance of bamboo diet, whereas this spacing is absent in gorilla skulls from other localities in the eastern Congo, where there are no bamboos'. He referred to the work by Colyer 'which revealed that those from localities in which the bamboo is absent from the gorilla's habitat are readily recognisable, owing to the freedom from appreciable spacing between the teeth ... a result in infancy of food packing, tough fibrous pieces of the bamboo shoots becoming wedged between the teeth and gradually pushing them apart'.

Studies at the same time on Sierra Leone chimpanzees by Jones and Cave (1960) threw light on links between diet, longevity and dental disease, which is relevant to gorillas (see later). Silverback canine teeth, having evolved as weapons, are prone to damage. Schaller (1963) published data on a small series of mountain gorilla skulls, while Fossey (1983), also working with *beringei*, reported that 'From 64 skeletal specimens collected throughout six Virunga volcanoes, I found that ... 80 percent had either missing or broken canines' — including two silverback skulls with the canine cusp of another silverback broken off in the supra-orbital crest. The injuries had evidently been caused years before the animal's death judging from 'the extent of the bony tissue growth surrounding the region of penetration into their skulls'.

In addition though, as illustrated by gorilla specimens in collections in many parts of the world (see Part II: A Catalogue of Preserved Materials), a tooth infection can extend into surrounding bone, sinuses and retrobulbar spaces. It can also be assumed that, as in humans, infectious dental disorders, such as untreated chronic periodontal disease, can initiate a bacteraemia and thereby have systemic effects.

SOME SPECIFIC CONDITIONS AND THEIR PATHOGENESIS

The pathogenesis of infectious diseases of the oral cavity has been well explored in humans, over many years, and it is reasonable to assume that similar mechanisms operate in other apes, including gorillas. The teeth and their supporting structures provide an excellent environment for the growth and replication of transient and commensal microorganisms within the mouth. Food particles that collect between the teeth serve as nutrients for the development of a bacterial biofilm (plaque) on the surface of the tooth. Minerals in the saliva, such as calcium, combine with plaque to form calculus (tartar). Plaque initiates an immune response resulting in inflammation of the gingiva (gingivitis), which may progress to periodontitis (inflammation of the nongingival periodontal tissues — see later) and other local and systemic sequelae. Each of these will be discussed later.

Gorillas in zoos are not infrequently reported with oral conditions. The latter include gingivitis (often manifest by swollen gums) and various types of dental disease. Before the involvement of veterinarians with training in odontology, or of dental surgeons willing to look at great apes, damaged or infected teeth in these species were usually removed. When 'Guy', the famous gorilla at London Zoo, died in 1978 after 31 years of captivity, he had only his four canines remaining. Even those teeth show abnormal positioning and wear (see Appendix 5: Case Studies — Museums and Zoological Collections).

Forty years ago, Dolphin et al. (1976) explained dental prophylaxis for a lowland gorilla. In his 12-year review of clinical conditions in the Howletts Zoo (UK) gorilla colony, Furley (1997) listed various dental disorders.

Tartar and Calculus

Dental plaque appears on the surfaces of the teeth of humans and other species in the form of biofilms — complex mixed microbial communities in an extracellular polysaccharide matrix (Clutterbuck et al., 2007). The deposit becomes visible with time, usually as a yellowish discoloration, that later changes to a brown or pale yellow as it is converted into calculus or tartar. In his beautifully illustrated book about the mountain gorilla, Norton (1990) includes an impressive demonstration of tartar on the teeth of 'Imbaraga'.

The microbial flora of the gingival crevice of humans, presumably also of gorillas, tend to be proteolytic rather than saccharolytic. The result of this can be a rise in pH, which encourages mineralisation of plaque leading to the formation of calculus, the hard, bone-like tooth coating, which persists on teeth for prolonged periods, both antemortem and postmortem. However, it should be noted that Miles and Grigson (1990) commented that calculus on primate teeth tends to be detached after a time and 'may even have been cleared off when the cranium was prepared' — a point worth remembering when examining skeletal material.

In humans the rate of calculus formation is variable and is associated with individual differences in diet, salivary flow, local pH and genetic factors. One assumes that a similar situation pertains in *Gorilla*.

Periodontal Disease

Periodontal disease is the most prevalent oral infection in humans, where the aetiological importance of specific pathogens in the initiation and progression of periodontal disease has been well established. The microbial ecology of the oral cavity is diverse, with aerobic bacteria predominating in the early phase of gingivitis, followed by a predominantly anaerobic and Gram-negative bacterial profile when periodontitis becomes established.

Amongst specific pathogens are *Porphyromonas gingivalis* and *Prevotella intermedia* (Fukui et al., 1999). These bacteria are usually found in small numbers in the oral cavity of healthy humans and certain animals. But data on their role in *Gorilla* appear to be lacking.

Periodontal disease is characterised by inflammation of the tooth support structures leading to damage and loss of the periodontal membrane, the cementum, alveolar bone and adjacent soft tissues; tooth loss may be a sequel. In traditional veterinary medicine the severity of periodontal disease and its prognosis is related to such factors as the species of animal, age, breed, oral microbial profile and concurrent disease. The role of such factors in gorillas appears not to have been investigated but there is no doubt about the importance of periodontal disease, especially but not exclusively, in captivity. It is well recognised by veterinary clinicians. Frances Barr, for example (personal communication), recounts from her work in the 1980s at the BZG, United Kingdom, 'significant dental and periodontal disease in the …… adult male gorilla'. Huff (2010) described surgical extractions for periodontal disease in a gorilla.

Dental Caries

Caries is a progressive acidic demineralisation process of the inorganic matrix of dental tissues. It is secondary to bacterial breakdown of carbohydrate that in turn produces acids that damage hard dental tissues. Colyer (1931) found various cavities in five adults out of a total of 689 museum specimens. Meder (1994) made an interesting, but apparently unsubstantiated, comment that 'Whereas western lowland gorillas sometimes get caries, the mountain gorillas' diet is low in sugar, and caries is therefore rare in these animals'.

Jones and Cave's 1960 paper (see later) is concerned with chimpanzees, not gorillas, and was published more than 50 years ago, but their descriptions of caries, illustrated with excellent black and white photographs, still provides excellent groundwork for those embarking on studies on diseases of the teeth of gorillas.

Glick et al. (1979) reported dental caries, periodontal abscesses and extensive cranial osteitis in a captive lowland gorilla.

Enamel Hypoplasia

This is characterised by the presence of enamel that is hard but thin due to defective enamel matrix formation with a deficiency in the cementing substance. Linear enamel hypoplasia (LEH), a developmental defect, is a marker of physiological stress during tooth formation. LEH appears as lines or grooves in the enamel surface and represents the early cessation of enamel matrix production caused by systemic growth disruptions.

Nonmedical osteologists, archaeologists and palaeopathologists lay considerable weight on enamel hypoplasia and attribute its presence in human teeth to various factors including trauma, bacterial infection, 'slow enamel formation' and coeliac (celiac) disease. The apparent value of LEH in detecting stress in prehistoric, historic and modern populations has prompted some researchers to investigate it in other species of primate (Moggi−Cecchiand Crovella, 1991; McGrath et al., 2015).

Other Pathology

Many other lesions have been reported in captive and free-living gorillas and cannot all be discussed here. Alveolar abscesses are common in old gorillas (Schultz; in Gregory, 1950) and Gaffikin (1949) noted a periodontal abscess around the lower left lateral incisor of a male shot in Uganda. Similar abscesses are reported in live, captive, animals and are a feature of museum specimens (see Part II: A Catalogue of Preserved Materials).

Effects of Ageing (Age-Related Changes)

There is a well-established association between changes in the teeth of gorillas and old age (see Chapter 8: Nonspecific Pathology). Meder (1994) discussed this in a general review 20 years ago. Amongst other things, she referred to tartar (calculus) (see earlier) saying of mountain gorillas that: '… their teeth commonly have a blackish colour, and tartar often leads to periodontal disease and gingivitis, reabsorption of the jaw bone and eventually tooth loss. As a consequence of tooth decay, the sinuses of the upper jaw can become inflamed, especially in males. Frequently in adult gorillas the teeth of the upper jaw may be so affected that the dental roots are exposed. It is possible that older animals actually starve to death (see later) because they are unable to chew their food. Lowland gorillas, on the other hand, deposit considerably less tartar and rarely have periodontal disease'. No specific references are given.

Lowenstine et al. (2016) reported significant differences in the ageing process amongst apes, including *Homo*. What is common to all are dental attrition, periodontitis and tooth loss.

The complexity of dental disease in great apes — and how it can be related to anthropogenic factors — was highlighted in the paper about chimpanzees (Jones and Cave, 1960) — see earlier. The authors pointed out that the Sierra Leone chimpanzee had adapted its dietary and feeding habits to changed ecology resulting from replacement of high forest by secondary forest and extensive development of agriculture. Much of the animal's food was being procured from local farms and gardens. One consequence was that an abundant and different food supply appeared to be increased longevity. Another was the high incidence of what the authors called 'dental mischief from the mastication of infected food material obtained from the refuse heaps around native [*sic*] habitations'.

Those observations over 50 years ago may be relevant to gorillas, especially as habituated groups become more accustomed to entering *shambas* and coming into closer contact with humans.

Halitosis

Halitosis, caused by bacterial multiplication in the buccal cavity, can be a useful diagnostic feature in humans and domesticated animals, but it has been poorly documented in great apes. Gorillas in the wild and in captivity present with lesions of the oral cavity ranging from erosions and ulcerations to localised proliferative changes, and some of these lead to halitosis and other sequelae.

Breath analysis may be useful in detecting mild halitosis (see Chapter 5: Methods of Investigation — Observation, Clinical Examination and Health Monitoring).

The Ongoing Importance of Studies on Teeth

Most of our scientific understanding of dental pathology in gorillas comes from an investigation of jaws of dead specimens, especially those in museum collections in many parts of the world (see Part II: A Catalogue of Preserved Materials). In some cases the dental specimens have been properly examined and described by experienced, trained, odontologists - for example, at the Odontological Collection of the Royal College of Surgeons of England (Miles and Grigson, 1990). Often, however, especially in smaller museums, there is either no report or the scientific quality and veracity of what is recorded is very questionable because those compiling the catalogue did not have an odontological (or even medical/ veterinary) background (see comments in Preface). It cannot be overemphasised that correct interpretation of changes, physiological and pathological, in teeth requires a sound understanding of these unique structures.

Much of this book could have been devoted to teeth and dental pathology. The relative accessibility and visibility of the teeth of gorillas, in both live and dead animals, and their durability (often postmortem, long after soft tissue and some skeletal elements have

disappeared — see Appendix CA2: Retrieval, Preparation and Storage of Skeletal and Other Material) means that odontological studies have long attracted the attention of primatologists and veterinarians.

Quite apart from its interesting aetiologies and pathogeneses, dental disease in gorillas is a cause of concern per se, on welfare grounds. Shakespeare emphasised this in humans when (*Much Ado about Nothing*) he wrote: 'For there was never yet philosopher that could endure the toothache patiently'. Nowadays, veterinarians with training in odontology and/or dental surgeons willing to look at great apes perform root fillings on gorillas in order to preserve the dentition. However, if the prognosis of a tooth is poor, extractions will be carried out to avoid repeated anaesthesia. In the past, damaged or infected teeth were often removed; see 'Guy' (see Appendix 6: Case Studies — Postmortem Investigations).

THE GASTROINTESTINAL TRACT

Investigation of the GIT

In clinical cases (live gorillas) the GIT offers numerous opportunities for investigation. Both the oral cavity and the anus/terminal rectum can be thoroughly examined, using in the field at the very least an otoscope (auriscope) with a long speculum/telescope attachment. In a zoo environment, both rigid and flexible endoscopies are possible, coupled with plain and contrast radiography and other forms of imaging. Biopsies can be taken *per os*, per rectum and following laparoscopy or laparotomy.

When examining the alimentary tract postmortem, eyes, ears, nose and fingers all have a part to play. Investigation must always start with the orifices — buccal cavity and perianal region. The former may show erosion/ ulceration which may or may not be related to the teeth; the perineum may exhibit soiling that may or may not be significant. As far as internal organs are concerned, a good light source is essential - using sunshine, when present, in Africa, and artificial sources otherwise. Both reflected and transmitted light ('transillumination') are needed and, as in all types of examination, a hand lens (magnifying glass) or surgical/dissecting loupe is of great assistance, and sometimes essential.

As in all gorilla work, it is important to practise a proven 'clean-dirty' technique. In the case of the, GIT the material is likely to harbour both normal (resident) gut organisms (see below — 'gut microbial community') and potential pathogens (some zoonotic). In addition the odour of spilled ingesta tends to linger!

Careful examination of the GIT before fixation is essential, using the naked eye and a lens; the colour and appearance of the mucosa and lesions, such as small erosions and ulcers, will not look the same subsequently. Transillumination will assist.

Basic rules when performing gross examination of the GIT are 'wash thoroughly; handle with care avoiding touching the mucosa; compare with normal'. More detailed guidelines are given below:

- Observe externally and internally before touching, while the serosa and mucosa are still moist
- Initially make a small incision that can be enlarged later
- Handle with care, avoiding the mucosal surface
- Ligate portions of the tract (using coloured cord to facilitate recognition) in order to separate the stomach and other parts of the tract
- Gently wash; hold near tap to avoid damage to tissues and splashing
- Use transillumination and reflected light to examine
- Detect ulcers or parasitic nodules with iodine or stain, which can be poured off
- Take samples as necessary; handle with plastic, not rat-toothed, forceps.

Always record accurately whether the stomach contains food. Is it really 'full'?

Expressions such as 'stomach contained a recent meal' are helpful but give no indication of quantity or appearance.

Phosphate-buffered saline can be used to rinse the GIT before fixation. Pieces of the tract can then be placed on cardboard (mucosa side up) for fixation.

Whenever possible, tissue should be fixed separately, in labelled containers so that they can be identified when sections are prepared. This helps avoid uncertainty as to the identity of a tissue or a lesion when reading histological sections. While some structures are, of course, immediately identifiable down the microscope, others (e.g., GIT) can prove difficult, especially if a lesion masks recognisable architecture.

Histological techniques are standard for most of the GIT but teeth need special attention (Cipriano, 2002; McFarlin et al., 2014).

Gut Flora and Fauna

The vertebrate intestine is colonised by hundreds of species of microorganism. Faecal bacterial diversity was studied in a free-living gorilla by Frey et al. (2006).

Most gut organisms have a mutually beneficial relationship with the host. They are known to influence the development and balance of the host's immune system.

There is currently great interest in the 'gut microbial community' and, indeed, in the whole 'microbiome' of humans and other animals – see Hoffmann et al. (2016) and Chapter 3: Infectious Disease and Host Responses. Nonpathogenic segmented filamentous bacteria, for example, are found in the GIT of many species of mammal and bird and are probably ubiquitous in the Animal Kingdom (Wu et al., 2010) (see Chapter 14: Musculoskeletal System).

Parasites

The gut contents of any animal present a nutrient-rich habitat that numerous species have evolved to exploit. These include free-swimming helminths and protozoa that cause little apparent ill-effect and may even bring benefits to the host, as well as species that attach to the gut wall and cause damage.

Records of gorilla parasites were of a largely ad hoc nature for much of the 20th century, with a brief list of diseases and parasites appearing in Schaller (1963). Meder (1994) summarised the variety of parasites (e.g., *Balantidium coli*, *Entamoeba histolytica* and *Strongyloides fuelleborni*, the hookworm *Ancylostoma duodenale*, various *Necator* species, and *Onchocerca volvulus*), which both infect humans and gorillas. The pinworm *Enterobius lerouxi* has only been found in the lowland gorilla, while the tapeworm *Anoplocephala gorillae* has only been reported in the mountain gorilla (Fig. 11.3). Various African primates are also hosts to the threadworm *Oesophagostomum stephanostomum*, the most prevalent intestinal parasite of the lowland gorilla. Besides, many species are parasites in the stomach and small intestine of gorillas, e.g., *Probstmayria gorillae*, *Murshidia devians* and several species of trichostrongylides, worms of the intestinal tract. A heavy infection with those parasites may result in gastritis. Protozoa, e.g., of the genus *Troglodytella* (Fig. 11.4) are very common; however, it is not yet clear whether they are intestinal parasites or actually aid digestion. *Balantidium* is a significant parasite, however, and is being studied in free-living gorillas (Hassell et al., 2013).

Occasionally the filarial worm *Loa loa*, also found in humans, lives in the connective tissue of the abdominal cavity and in the musculature of the gorilla; the filaria *Mansonella perstans* occurs in the connective tissue and various species of *Dipetalonema* in the skin (Meder, 1994). The threadworm *Tetrapetalonema vanhoofi* may be found in the lymphatic and the blood system of gorillas (also in the gall bladder of chimpanzees).

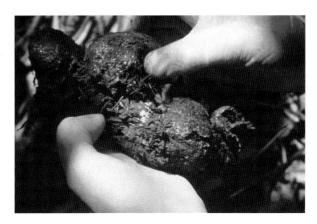

FIGURE 11.3 Proglottides of *Anoplocephala* are readily visible in faeces.

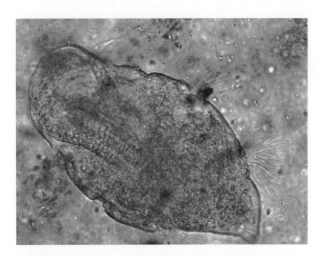

FIGURE 11.4 A ciliate, *Troglodytella abrassarti*, in the faeces of a captive lowland gorilla. *Image courtesy of Ann Thomasson/Durrell.*

Attitudes to parasites have traditionally been dominated by a conventional veterinary approach, which is to kill them by treating the host with anthelmintic drugs. Research into naturally occurring parasites in free-living gorillas has been limited, and the life cycles and intermediate hosts of most species have yet to be worked out. Faecal sampling provides the simplest means of establishing which species are present in a gorilla population, either by observing the presence of eggs passed in the faeces or adult parasites being lost through natural mortality.

In a study of 80 individual mountain gorillas in the Virunga Volcanoes (32 sampled regularly, 48 occasionally), Redmond (1983) reported high rates of infestation with three species of nematode, a cestode, and what appeared to be entodiniomorph protozoa. In the gorillas that were sampled regularly between 1976 and 1978, the rates were:

Probstmayria-like Type A nematodes:	90.7%
P. gorillae-like Type B nematodes:	100%
Strongyloid nematode eggs:	100%
Cestode *A. gorillae*, eggs and proglottides:	40.6%
Entodiniomorph ciliate-like protozoa:	93.3%

Samples were processed using a series of sieves, with 20 mg from each being examined under a microscope using a simple wet-mount method.

Improved protocols for faecal studies were introduced by Cooper and Nizeyi (1995, unpublished) and continue to be developed by successive generations of veterinarians at the MGVP.

Subsequent studies using different techniques to isolate specimens have revealed a more diverse parasite load in mountain gorillas. Sleeman et al. (2000) collected 98 faecal samples from 74 free-living mountain gorillas (*G. beringei*) from the Parc National des Volcans,

Rwanda, between July 1995 and January 1997. These were examined for parasites by Sheather's sugar and zinc sulphate flotation methods, trichrome staining and larval cultures. All samples contained at least one parasite. Seventeen endoparasites were identified, including eight protozoa, seven nematodes, one cestode and one trematode. Two species of arthropod mite were also recovered. Parasites observed on faecal examinations included strongyle/trichostrongyle-type eggs (72/74) (representing *Oesphagostomum* sp., *Trichostrongylus* sp., *Hyostrongylus* spp. and possibly *Murshidia* sp.), *Strongyloides* sp. (1/74), *Trichuris trichiura* (2/74), *Probstmayria* sp. (7/74), *Anoplocephala* sp. (63/74), *Entamoeba hartmanni* cysts and trophozoites (19/70), *Endolimax nana* cysts (31/70), *Iodamoeba büetschlii* cysts (11/70), *E. nana* or *I. buetschlii* trophozoites (63/70), *Entamoeba coli* cysts and trophozoites (14/70), *E. histolytica* trophozoite (1/70), *Chilomastix* sp. cysts and trophozoites (31/70) and *Giardia* sp. cysts (2/70). In addition, one ascarid and one trematode egg were seen. There were no significant differences in the prevalence of parasites between males and females and between age groups; however, infants and juveniles appeared to have a lower prevalence of *A. gorillae*, and the silverback males appeared to have a higher prevalence of *Probstmayria* sp. Parasite prevalence was consistent among the five social groups studied except that the Susa group had a significantly lower prevalence of *A. gorillae*. *T. trichiura*, *Strongyloides* sp., *Chilomastix* sp. and *E. nana* were identified for the first time in this population, and it is possible that these parasites were of human origin. Although there were no obvious clinical effects due to the presence of these parasites, six taxa identified (*T. trichiura*, *Strongyloides* sp., *Oesphagostomum* sp., *Trichostrongylus* sp., *E. histolytica* and *Giardia* sp.) could be pathogenic. Some of the parasite products and cultured larvae could not be identified to species.

It is interesting to note the low prevalence of *Probstmayria* using these methods as compared with the near universal presence reported by Redmond, who lacked equipment for flotation methods but closely examined the residue from a fine sieve in a black plastic tray against which the white nematodes (4-mm-long Type As and 2-mm Type Bs) were clearly visible. Ashford et al. (1990) also reported *Probstmayria* in almost all stools examined from 41 mountain gorillas in Bwindi Impenetrable National Park, Uganda, as well as high 'loads' (numbers) of *Oesophagostomum*-like ova (also reported by Hastings et al., 1992; Kalema, 1995).

Two new species of trichostrongylid nematodes were described from the stomach of an adult female mountain gorilla found dead, in a wasted condition, in Bwindi (Durette-Desset et al., 1992).

Mudakikwa et al. (1998) listed lesions associated with parasites in mountain gorillas, including small intestinal cestodiasis (*Anoplocephala*), peritoneal adhesions (abdominal nematodiasis), proliferative gastritis (trichostrongyles), periportal hepatic fibrosis (*Capillaria hepatica*), colonic mural nodules (*Oesophagastomum* sp.) and ectoparasites (lice).

The parasites of eastern lowland gorillas were studied in Kahuzi-Biega National Park by Eilenberger (1997a,b) and found to have similar parasites but in smaller numbers than either Virunga or Bwindi gorillas.

The most common nematode parasite in humans is *Ascaris lumbricoides*; given that humans and gorillas are physiologically similar sympatric species it is not surprising that it has been reported in gorillas in some locations, e.g., Lilly et al. (2002) in the Central African Republic.

Clearly the species detected will vary according to methods used; van Zijll et al. (2010) demonstrated the use of multiple diagnostic techniques to detect *Cryptosporidium* sp. and *Giardia* sp. in free-ranging western lowland gorillas and observations on the prevalence of these protozoan infections in two populations in Gabon. To identify suitable protozoa diagnostic techniques for wild gorillas, 95 faecal specimens were collected in Lopé National Park and east of Moukalaba-Doudou National Park in Gabon, areas with high and low levels of human activity, respectively. The samples were examined for *Cryptosporidium* sp. and *Giardia* sp. by using the following diagnostic techniques: a commercially available immunofluorescent antibody test kit, Merifluor, and a rapid immune-assay, ImmunoCard STAT!, to detect *Cryptosporidium* sp. and *Giardia* sp., and a modified Ziehl-Neelsen stain to detect *Cryptosporidium* sp. oocysts. The results obtained from the Merifluor test, considered the 'gold standard' in human studies, were used to estimate the prevalence of *Cryptosporidium* sp. and *Giardia* sp. infections in Lopé National Park (19.0% and 22.6%, respectively) and east of Moukalaba-Doudou National Park (0% and 9.1%, respectively). The difference in prevalence in these areas may be associated with differing levels of anthropogenic disturbance. The sensitivity and specificity of the latter two diagnostic techniques were calculated by using the Merifluor test as a control. The ImmunoCard STAT! was found suitable for *Giardia* sp. antigen detection (specific but not sensitive) and inappropriate for *Cryptosporidium* sp. antigen detection (not specific or sensitive). The modified Ziehl-Neelsen stain was found to be highly specific but not sensitive in the detection of *Cryptosporidium* sp. oocysts. These results underline the necessity of using ancillary tests and concentration methods to correctly identify positive samples. This is the first report of *Cryptosporidium* sp. and *Giardia* sp. infections in free-ranging western lowland gorillas and highlights the importance of verifying the accuracy of diagnostic techniques developed for human use before applying these to nonhuman primates.

Lankester et al. (2010) presented initial evidence of pathogenicity of *Dientamoeba fragilis* in the western lowland gorilla. Sente (2015) studied *Cryptosporidium, Giardia* and helminths of the habituated Nyakagezi mountain gorilla group in Mgahinga Gorilla National Park (see Chapter 17: Field Studies in Pathology and Health Monitoring).

Teare et al. (1982) described an epizootic of balantidiasis in lowland gorillas. A single case was documented in detail by Mainka (1990).

As well as new records of well-known species being reported from gorillas, original ones are still being discovered. Garin et al. (1982) described an apparently new entodiniomorph ciliate from faecal samples collected in Gabon. Although described as parasites, it has been speculated by Landsoud-Soukate et al. (1995) that such ciliates, which occurred frequently in both chimpanzees and gorillas in the Lopé Reserve (and in all previous coprological surveys of wild apes), may be symbionts involved in cellulose digestion.

Endoparasite control, however, has been one of the major challenges faced by PPG - Gabon, where ex-captive gorillas are prepared for a life in the wild − particularly during pre-release, with *O. stephanostomum* being found responsible for one mortality pre-release and suspected to be the underlying cause of several long-term health problems. Oesophagostomes have proved particularly hard to treat compared with other nematode worms due to their unique lifestyle − with the larva-forming nodules within the intestinal wall. These nodules appear to protect them to some extent from several standard anthelmintic treatments, with albendazole the most effective, and can develop into small abscess-like lesions resulting in intestinal inflammation and oedema. One way of equipping candidates for release would be to show them food-plant species in the release site that can be used as self-medication against oesophagostomes and other internal parasites, such as *Aspilia* and *Vernonia* spp. (Mahé, 2006). The success of such reintroduction programmes has been assessed by King et al. (2013) using Vortex population models, giving a 91% and 95% chance of their persisting over 200 years in Congo and Gabon, respectively.

Parasitic lesions reported by Lowenstine (1990) included ulcerative and proliferative gastritis, hepatic capillariasis, periportal fibrosis and colonic abscessation.

Hydatidosis is referred to elsewhere in this book. Amongst others, Bernstein (1972) an epizootic hydatid disease in captive apes, while Schmidt and Hubbard (1987) depicted in photographs echinococcosis in a gorilla − the peritoneal cavity, including the omentum, was full of *Echinococcus* cysts associated with a chronic inflammatory response.

In an attempt to standardise noninvasive parasitological research methods across primate species, Gillespie (2006) has produced practical guidelines which − if widely adopted − will greatly improve the potential for comparative studies between research sites and across different taxa. Such studies are important for a full understanding of biodiversity conservation because, as discussed in Chapter 9, Skin and Integument (Ectoparasites), each endangered species of host animal provides a habitat for species-specific parasites that are, self-evidently, equally endangered. This question of host—parasite coevolution and conservation was the subject of a special edition of Biodiversity and Conservation in 1996 (vol. 5, No. 8).

Pathology and Pathogenesis

Alimentary system disorders have been widely reported in both free-living and captive gorillas. Various 'enteric' conditions presenting as disturbances to gut function and faecal appearance have been attributed to changes in diet or to bacteria and parasites that have been detected in, or isolated from, the GIT. The pathology of many of them is still not fully understood. Abee et al. (2012) included a chapter on the digestive system diseases of NHPs.

Different organisms can cause, or be associated with, alimentary disease in captive gorillas. Examples are to be found in Box 3.5 in Chapter 3: Infectious Disease and Host Responses. Only a few are mentioned here as the focus of this book is on pathology and pathogenesis, not on microbiology or epidemiology (epizootiology).

Resident gut flora, faecal contamination and postmortem changes often make interpretation of isolates problematic. A mixed bag of *Escherichia coli*, *Streptococcus faecalis*, *Proteus* spp. and *Candida albicans* may or may not be significant.

It is always wise to do faecal cultures (or, if available, to use PCR) to check for salmonellae and shigellae. Even shortly after the genocide in Rwanda it proved possible to do cultures (thanks to the Australian Medical Support Force, UNAMIR II); *Salmonella* spp. were isolated — but under the prevailing, postgenocide conditions could not be further investigated.

Shigellosis and the carriage of *Shigella* spp. are a matter of concern in both captive and free-living gorillas. Forty years ago Lemen et al. (1974) described shigellosis in two infant gorillas. Stetter et al. (1993, 1995) discussed the epidemiology of *Shigella flexneri* infection in captive lowland gorillas and reported on a 10-year retrospective study assessing the prevalence and implications of shigellosis, including (by analysis of plasmid and chromosomal DNA) research on antimicrobial resistance.

Shigella spp. were amongst organisms that attracted particular interest in terms of possible human-gorilla-human transmission following the Rwandan conflict in 1994 when an estimated one million Rwandans fled to Goma in Zaire, taking shelter in makeshift camps where the water supply and sanitation facilities were almost nonexistent (Islam et al., 1995).

In free-living gorillas, alimentary disease recorded ranges from mild disturbances to fatal conditions such as the intestinal torsion that killed mountain gorilla 'Mishaya' in 2014. Fossey (1983) reported a pseudolymphoma of the appendix in mountain gorilla 'Pucker' and a gastric adenoma of the Brunner's glands in 'Coco'. She also referred to the histopathology report on a mass from the stomach of 'Quince' by pathologist D.L. Graham, DVM: 'Chronic granulomatous inflammatory tissue composed of scattered fibroblasts, capillaries and numerous macrophages, many of which contain hematin pigment'. Over the past few years field reports have frequently alluded to GIT disorders in both live and dead gorillas; for instance, in his brief review of the MGVC pathology data from 1998 to 2000, Cranfield (2001) listed under 'Peanuts' 'severe typhlocolitis and colonic and cecal tympany (combined Gram-negative bacteria and clostridia)' and under 'Arusha' 'gastric atony and peracute clostridial gastritis'.

Various parasites and bacterial pathogens are isolated from faeces. A useful reference insofar as *beringei* is concerned is the dissertation on digestive tract lesions by Muhangi (2009). Peden (1960) diagnosed the cause of death of a male at Kisoro as gastroenteritis; the stomach and intestine showed inflammation and petechial haemorrhages on the mucous membrane. Amongst a list of diseases of mountain gorillas recognised at the PHVA (Werikhe et al., 1997) (see Chapter 3: Infectious Disease and Host Responses) were '3. Diarrhoea with abdominal pain and 7. Tapeworm [*sic*]'.

Peritonitis has been reported in free-living gorillas — for example, in a silverback mountain gorilla found dead in July 2014. This animal had previously shown no signs of injury or ill health but, on postmortem examination, a perforation of the small intestine (cause uncertain) and acute peritonitis were diagnosed.

There were several early accounts of alimentary disease in captive gorillas (Schaller, 1963). A captive infant died of an inflammation of the appendix (Grzimek, 1957). Weinberg (1908) described appendicitis with adhesions in a gorilla.

Functional megacolon was reported in a young gorilla in Zurich Zoo by Baumgartner and Jörger (1988); the large intestine was grossly dilated and applying pressure on other organs and the body wall. On histology no abnormalities were detected (NAD) and no parasites were seen (NPS). In the same year, from Bristol Barr et al. (1988) described an oesophageal leiomyoma in a 23-year-old lowland gorilla 'Samson' that had a history of dysphagia, vomiting and lethargy. Endoscopic examination revealed an oesophageal mass proximal to the cardia. An oesophageal biopsy was taken. This showed necrotic tissue, inflammatory cells (both viable and degenerate), and granulation tissue. Postmortem, the presence of a very firm mass of whorled yellow and white tissue ($5 \times 3.5 \times 4$ cm), protruding into the lumen of the lower oesophagus, was confirmed. This had clearly interfered with cardiac function and caused vomiting, reflux oeso-phagitis and hyporexia. Histological examination identi-fied the mass as a leiomyoma composed of interlacing bundles of smooth muscle fibres, with nuclear pleomor-phism but no hypercellularity and few mitoses. Adjacent oesophagus showed full-thickness ulceration of the squa-mous epithelium, presumably the result of reflux.

Early reports on GIT disorders in young captive goril-las included three presentations by Simmons (1972a, 1972b and 1972/73) on, respectively, fatal gastric dilata-tion, gastrointestinal problems encountered in baby goril-las and fatal shigellosis.

Farnsworth et al. (1979) described the conservative treatment of an intestinal foreign body. Schmidt and Hubbard (1987) depicted a diffuse appendicitis with extensive mucosal necrosis.

In his 12-year review of clinical conditions in the Howletts Zoo (UK) gorilla colony, Furley (1997) reported severe diarrhoea and lethargy associated with *S. flexneri* and a similar acute case associated with *Shigella sonnei*, *Campylobacter jejuni* infection with severe diarrhoea and a hepatic abscess leading to a fatal purulent peritonitis.

Reviewing pathology reports from (mainly British) zoos has revealed many interesting findings relating to the alimen-tary tract, some being 'incidental findings'. These include (1) former localised peritonitis with extensive fibrous adhesions between the ascending and transverse colons and the ventral abdominal wall, (2) gastric mucosal hyperplasia − nodules of hyperplastic glands in the pylorus (Fig. 11.5) and (3) changes in the colon characterised by dilatation of submucosal lym-phatic vessels surrounded by fibrosis and calcification, and associated with portions of nematode parasites (probably *Strongyloides* spp.). Data provided by Roberta Wallace, Milwaukee County Zoo (personal communication), included a female animal in her 20s with severe fibrino-suppurative peri-tonitis secondary to hydatid disease and a 5-year-old female

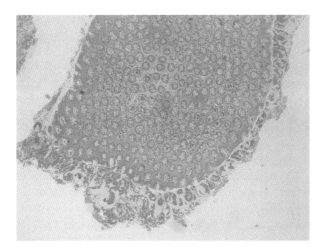

FIGURE 11.5 Gastric mucosal hyperplasia in the stomach of a captive lowland gorilla.

with severe chronic active peritonitis secondary to ulceration and perforation of the ileum caused by *B. coli*.

Kalema−Zikusoka and Lowenstine (2001) discussed rectal prolapse in mountain gorillas and listed possible causes as:

- Enteric parasitism leading to constipation/diarrhoea
- A 'weak rectum' due to a genetic defect
- Stress
- Trauma resulting from intergroup fighting
- Inappropriate diet

Furley (1997) also reported on a rectal prolapse that was successfully treated surgically.

The pathogenesis of GIT disease is essentially due to:

1. malabsorption
2. dehydration and electrolyte imbalance

Malabsorption can be:

1. acute, due to mucosal damage
2. chronic, due to inflammation of gut wall

Dehydration and electrolyte imbalance are follow:

1. vomiting − fluid and chloride lost; body compensates, leading to alkalosis
2. diarrhoea − fluid, N, Na and K lost; body tries to con-serve Na, complex biochemical changes.

During a postmortem examination, one should try to assess hydration status (often difficult) and to link nec-ropsy findings with (for example) urinalysis.

Clinical evidence of alimentary disease in gorillas may also include weight loss. In a study by Scott (1992), a female lost 40% of its weight following diarrhoea and another, male, 38% following 'intestinal infestation'.

Colitis

Scott and Keymer (1975) compared ulcerative colitis in apes, including gorillas, with the disease in humans. They pointed out that in apes both the acute changes (crypt abscesses and superficial mucosal ulceration) and the chronic changes (mucosal atrophy and sparse branched glands) are indistin-guishable from the lesions seen in humans. Scott (1979) discussed the comparative pathology of the primate colon, and Scott (1992) published a (black and white) photomicrograph showing acute changes of ulcerative colitis (crypt abscesses and superficial necrosis with acute inflammatory exudate) in a young female gorilla. The same volume included a photo-micrograph depicting the protozoon *B. coli* in a microherni-ation of the colon; such herniations can play a part in the passage of bacteria and parasites from the mucosa to the dee-per layers of the colon (Scott, 1982; Scott and Keymer, 1976).

Molteni et al. (1980) described chronic idiopathic diar-rhoea with enterocolitis and malabsorption in a captive lowland gorilla. The lamina propria of the colon was

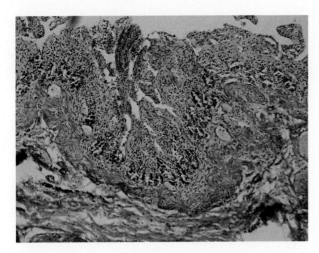

FIGURE 11.6 Colitis in a captive lowland gorilla associated with large numbers of *Strongyloides* worms.

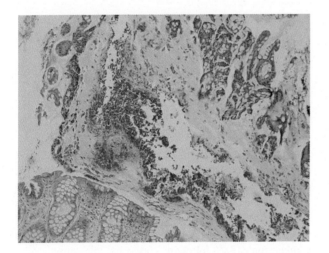

FIGURE 11.7 Ulcerative colitis, with extensive haemorrhage, in a captive lowland gorilla.

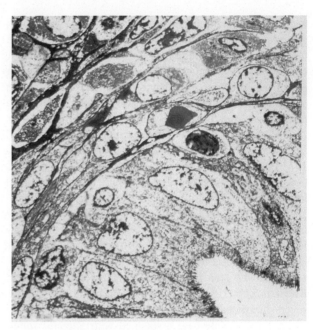

FIGURE 11.8 Transmission electron microscopy of normal colon. The lumen is at the bottom right.

early work by Scott and his colleagues, the relationship of colitis to inflammatory bowel diseases (IBD) of humans — ulcerative colitis and Crohn's disease — still require investigation. In a similar vein, it would be interesting to see if some of the apparently nonspecific GIT signs often seen in captive gorillas are in any way similar to human irritable bowel syndrome (IBS), a common, chronic, condition associated with gastric pain, flatulence and diarrhoea/constipation.

As discussed in Chapter 7: Methods of Investigation — Sampling and Laboratory Tests, image analysis-based approaches are used for scoring mouse models of colitis and could in due course permit objective assessment of histological changes of the GIT in gorillas.

Assessment of Faeces

The collection and examination of faecal material is discussed in Chapter 7: Methods of Investigation — Sampling and Laboratory Tests.

Faeces have long attracted the attention of primatologists and continue to do so (Kuehl et al., 2007). Half a century ago, Napier and Napier (1967) wrote 'Highland gorillas, which are bulk eaters, defaecate five or six times a day whereas the Western race, whose diet contains a higher proportion of fruit, seldom defaecate more than once daily ... nests of Western gorillas are usually free of excreta: the Highland race almost invariably foul their nests'. Redmond (1978, unpublished notes) did a 24-hour follow of a blackback male gorilla collecting all faeces

distended with inflammatory cells and consequential splaying of colonic glands. Macrophages in the mucosa contained large bacilli. The microscopic and ultrastructural changes in this case were those of a diffuse bacterial histiocytic colitis, similar to Whipple's disease.

Van Kruiningen et al. (1991) also described bacterial histiocytic colitis while a fatal ulcerative colitis, also in a lowland gorilla, was the subject of a report by Lankester et al. (2008).

Colitis is clearly a significant condition in gorillas and many laboratory reports testify to its presence in various degrees (Figs. 11.6 and 11.7) but more is needed to differentiate types and elucidate aetiologies and pathogeneses. Investigations by pathologists should include an electron-microscopical study of normal and abnormal colons (Fig. 11.8) but, for reliable interpretation, this requires biopsies taken from live gorillas or removed and fixed promptly from recently dead animals. Notwithstanding the produced. He recorded three major bowel movements, one during the night which was slept upon, one first thing in the morning upon rising from the night nest, and one on moving off after the day-rest siesta period around midday. Other smaller defaecations occurred while travel-feeding, and the total weight produced was 12 kg.

Diarrhoea — loose faeces — (see Glossary) may be due to enteritis but can be associated with other factors. Frightened or disturbed gorillas may pass profuse, watery, faeces ('fear dung'). Careful observation of free-living gorillas (Fig. 11.9) allows defaecation to be observed and samples to be taken from known individuals. Faeces can readily be collected from night nests (Fig. 11.10).

Diet, Feeding and Disease

The role of food in the pathogenesis of alimentary (GIT) disease is an interesting one. An unsuitable diet (in terms of consistency or composition) can produce changes in the GIT, ranging from excessive flatulence to inflammation of the lining mucosa. Longer-term effects may involve the skeletal system resulting, for example, in metabolic bone disease (MBD) if the calcium:phosphorus ratio is incorrect (see Chapter 14: Musculoskeletal System).

Our understanding of the nutritional requirements of gorillas remains limited. When the first gorilla was brought into captivity in 1887, it was fed only on fruit and bread. Despite recognition by Bartlett (1899) that the larger primates would benefit from receiving animal protein, this was not routinely given. According to Brambell and Mathews (1976) 'in the immediate postwar years the primates were relying for most of their protein and much of their minerals on the food being thrown into their cages by the visitors. Virtually nothing but fruit and cabbage was being provided by the Society and the diet lacked many essential elements'. The author (JEC) recalls from his childhood taking food for animals at the London Zoo in the late 1940s/early 1950s.

Modern zoos usually feed a diet of commercial pellets and vegetables, with fruit offered mainly as a reward (when training, for example, to present a limb). Individual animals may have their own, formulated meals (Fig. 11.11).

Developments in the feeding of captive gorillas have proceeded with — regrettably not always hand-in-hand with — a better understanding of diets consumed in the wild. Several years ago Bloomsmith (1989) described feeding enrichment for captive great apes. Cousins (2015) suggested that shortened feeding time in zoos

FIGURE 11.9 A free-living, habituated, mountain gorilla defaecates, permitting a sample to be collected from a known individual.

FIGURE 11.10 Faeces in a night nest. These have been flattened overnight but are sufficiently fresh to justify a number of laboratory investigations.

FIGURE 11.11 Individually formulated and prepared rations for gorillas in a European zoo.

could be deleterious to social apes and advocated that food should be concealed in hollow logs and artificial termite mounds.

As far as data on free-living animals are concerned, the book by Hohmann et al. (2006) discussed feeding ecology in apes. Rothman et al. (2007) investigated the nutritional composition of the diet of *beringei* while Rothman et al. (2008) explored the nutritional quality of gorilla diets and the implications of age, sex and season.

As pointed out in Part II, A Catalogue of Preserved Materials, mountain gorillas are specialised folivores; their primary diet of herbaceous vegetation remains seasonally abundant and of constant high quality (Watts, 1984). Western lowland gorillas, however, are frugivorous and have to rely on herbaceous vegetation when fruits are not available. Periodic dietary variability was studied in mountain gorillas using stable isotope analysis of faeces (Blumenthal et al., 2012). Eckhart and Lanjouw (2008) included in their book some useful information about diet, including tables of plants eaten by gorillas in both the Virungas and Bwindi.

Gorillas also ingest invertebrates, sometimes accidentally in vegetation, sometimes deliberately: Western gorillas eat termites *Cubitermes* on an almost daily basis at some sites and weaver ants *Oecophilla* at others (Cipolletta et al., 2007); mountain gorillas eat safari ants *Dorylus* less frequently (Watts, 1989), just when they come across them, which varies according to altitude.

It is well recognised by zoo personnel that changes of diet can precipitate bouts of diarrhoea, sometimes anorexia, in gorillas. In respect of animals destined for rehabilitation and release, Hopper (2009) wrote 'Milk powder was gradually changed to a brand readily available in Gabon. Some edible plants endemic to Gabon were grown by the zoo's horticulture department and fed to the gorillas before their departure'.

Relatively little is known of mineral deficiencies/imbalances in *Gorilla* although hypocalcaemia is recognised (see Chapter 14: Musculoskeletal System).

Schaller (1963) noted three locations where gorillas had eaten soil that was high in sodium and potassium, but this was an occasional not habitual occurrence; Fossey (1983) reported 'soil-eating binges' in the dry months and the behaviour has been filmed and photographed on numerous occasions since.

Cancelliere et al. (2014) studied mineral nutrients in the foods eaten by mountain gorillas. Rothman et al. (2006) reported on decaying wood as a sodium source. Geophagia (ingestion of soil) is recognised in many taxa and may indicate a craving for certain minerals. In East and Central Africa local women eat or suck *udongo* (specially selected stones), especially during pregnancy. Reynolds et al. (2015) reported that chimpanzees ate mineral-rich clay and drank clay-water.

Occasionally a captive gorilla will indulge in 'pica' — the ingestion of items of no significant nutritional value,

such as rubbish in its enclosure (see also Chapter 8: Nonspecific Pathology).

The importance of understanding the feeding habits of free-living gorillas cannot be overemphasised. This is not only relevant to their well-being in captivity; it may have ecological relevance, as apes are crucial to seed dispersal of many plants (Redmond, 2008; Beaune, 2015).

Inanition, Starvation and Cachexia

Hyporexia (decreased food intake) or anorexia (no food intake) can be caused by various factors — see below.

'Starvation' is a lack of essential nutrients, usually over a prolonged period, and is characterised clinically by profound physiological and metabolic disturbances. The term 'inanition' is also sometimes used but has a slightly different meaning; it is a condition resulting from not just lack of food but also water and/or a defect in assimilation of nutrients. It was a term used regularly by W.C. Osman Hill (ZSL). 'Cachexia' usually implies general ill health and malnutrition, marked by body weakness, emaciation and secondary diseases.

A protein deficiency in a colony of lowland gorillas was described, with details of clinical and biochemical features, by Mundy et al. (1998).

Gorillas that have starved or are cachectic, for whatever reason, captive or free-living, may be examined/seen alive, necessitating clinical investigation, or presented dead, necessitating detailed postmortem examination. There may be forensic implications (Cooper and Cooper, 2013), but sometimes the cause is chronic disease, such as neoplasia (Lowenstine et al., 2016).

Live gorillas that are anorexic will often show a flat abdomen instead of one that is distended with food (Fig. 11.12).

A pathologist may be asked to examine an animal postmortem and to give an opinion as to whether or not it has

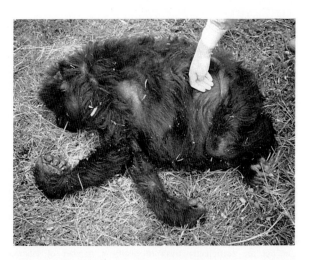

FIGURE 11.12 Gentle palpation of the abdomen of an anaesthetised (immobilised) mountain gorilla.

FIGURE 11.13 A field necropsy of a healthy gorilla killed by poachers. The large intestine, full of ingesta, spills out and obscures other organs.

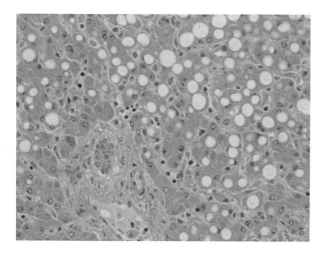

FIGURE 11.14 Photomicrograph of a lowland gorilla with a moderate degree of fatty change in the liver.

starved. An emaciated appearance, with a paucity of subcutaneous fat (see Chapter 14: Musculoskeletal System), reduced size, weight of organs and a relatively empty GIT, are strongly suggestive of this. Usually the large intestine dominates the opening and examination of the abdomen of a gorilla, obscuring other organs (Fig. 11.13).

A thorough necropsy then provides an opportunity to examine internal organs. Macroscopical examination will reveal the size of organs, their appearance (e.g., if a liver is pale and therefore possibly fatty) and is the time to inspect, weigh and assess any internal body fat (see Chapter 14: Musculoskeletal System). The GIT must be inspect in detail for ingesta (and its appearance), lesions or parasites.

Microscopical (histological) examination allows assessment of the appearance of organs — for example, whether the liver shows evidence of lipid (Fig. 11.14).

Starvation and inanition in captive gorillas can be due to a variety of factors (see Box 11.2).

BOX 11.2 Causes of Starvation and Inanition

- Failure to provide any food
- Failure to provide adequate food
- Failure of the animal to accept food
- Failure of the animal to masticate and swallow
- Food not remaining within the stomach (on account of regurgitation)
- Food not being adequately digested because of, for example, malabsorption, increased gut transit time or neoplasia
- Food that has been ingested and absorbed but is not being utilised because of a metabolic disorder or neoplasia

Obesity

See also Chapter 14: Musculoskeletal System. Obesity can be defined as 'abnormal or excessive fat accumulation that may impair health'. It has probably been seen far more frequently in gorillas than has inanition but usually only in captive animals. Schaller (1963) wrote, 'Gorilla males are often said to weigh 600 pounds or more, but these are the weight of obese zoo animals … of ten adult male mountain gorillas killed and weighed in the wild by hunters and collectors, the heaviest animal reached 480 pounds, and the average was about 375 pounds'.

Obese captive animals of both sexes may appear to have pronounced mammary gland development. Postmortem such animals display large quantities of subcutaneous and intraabdominal fat (this should be measured or weighed). Sometimes the liver is pale, a yellowish-brown colour. In severe cases in monkeys (and probably also in gorillas) the liver may sink with difficulty in formalin and exude fat droplets when incised. Fatty liver - kidney syndrome has been recognised in monkeys (Laber-Baird et al, 1987) and probably occurs in gorillas.

Condition

'Condition' is an important factor that affects an animal's chances of survival and its potential to breed successfully. There is no absolute measure of condition. It usually has to be assessed in relation to the demands facing that animal at a particular time, including pregnancy, lactation, movement ('migration'), exercise and cold weather.

Condition scores (body condition score, BCS) are increasingly used in both domesticated and wild animals (Cooper and Cooper, 2007) but appear not to have been widely adopted for gorillas.

Regurgitation and Reingestion

Regurgitation is different from vomiting (see Glossary). It is usually immediate and effortless.

The syndrome of regurgitation and reingestion (R&R) is seen in all captive great apes but especially so in gorillas (Loeffler, 1982; Akers and Schildkraut, 1985; Gould and Bres, 1986a,b). R&R appears never to have been recorded in wild apes. R&R in gorillas was discussed in the 1980s by a number of authors, including Gould and Bres (1986a) and Akers and Schildkrant (1985) (whose study included coprophagy). Gould and Bres (1986b) proposed that it might serve as a model for certain eating disorders of *Homo*, including human rumination syndrome, which in adults has been linked to a state of anxiety (Gould and Bres, 1986a).

Much has been written about the syndrome and how it might be prevented or treated. For instance, Ruempler (1992) advocated the addition of browse to ape diets to mitigate its frequency. More recently, Hill (2009) reported large amounts of stomach acid in the regurgitant of gorillas − but not in the saliva − and suggested this could be detrimental to health.

It has been postulated that R&R in captive gorillas is a strategy to prolong feeding time (Gould and Bes, 1986b; Cox et al., 2002). Foraging is an extended activity in free-living animals but space for this is limited in captivity. An interesting unpublished research proposal, worthy of more study, was by Elsner (2000).

R&R is probably multifactorial.

Coprophagy

Coprophagy, the ingestion of faeces, is often classed with R&R as abnormal, unwanted, behaviour (Cousins, 2015). It became such a problem in great apes at the San Diego Zoo that questionnaires were sent to collections and specialists in North America, Europe and Australia in order to gain a better understanding of the condition (Hill, 1966). Responses as to its cause provided a range of experiences and opinions, ranging from nutritional deficiency to psychological pressure of a confined environment. In a later survey of 56 North American collections, Akers and Schildkraut (1985) found that 56% of the animals were coprophagic to some extent.

Sometimes captive infant gorillas are seen to eat their mother's faeces. Such coprophagy may help infants establish an appropriate intestinal flora.

Occasional coprophagy is observed in mountain gorillas (Harcourt and Stewart, 1979; Fossey, 1983) and it was reported in free-living gorillas in Gabon (Krief et al., 2004). Redmond (1983) postulated that coprophagy in mountain gorillas might be an essential part of the life cycle of *Probstmayria*, which is ovoviviparous and breeds in the host's gut once established, apparently by infants ingesting warm dung containing live specimens. Coprophagy may also provide gorillas with vitamin B12, produced by bacteria in the hindgut, but only able to be absorbed in the foregut.

Malabsorption

Malabsorption, as recognised in humans and some other species, is characterised by a failure to digest or absorb nutrients; it may affect various organs. Scott (1992) discussed 'malabsorption syndrome' in humans and NHPs and described the clinical signs (such as chronic diarrhoea and steatorrhoea) and the pathogenesis − primary and secondary.

Wasting marmoset syndrome (WMS) is the most widely recognised form of NHPs. It affects New World marmosets and tamarins in captivity. Histopathological changes in both the human syndrome and WMS include decreased villous height to crypt depth, hyperplasia of crypts and increased lymphocytic infiltration of the mucosa.

Failure to thrive and chronic weight loss have long been recognised in captive monkeys and apes and protein-losing enteropathy in macaques was described more than 30 years ago (Rodger et al., 1980).

Although a specific 'wasting syndrome' has not, apparently, been described in gorillas, individual reports, some published, suggest that such may exist. For example, Stolpe (1986) reported a chronic enteropathy.

Cystic Fibrosis

Cystic fibrosis has been suspected in gorillas on a few occasions but apparently not confirmed. One example was a three-month-old female at BZG, United Kingdom, in 1985 with suspicious changes in the pancreas and thymus. However, a (human) paediatric pathologist felt that the changes seen on histology were not typical and commented that 'cystic fibrosis in humans is usually associated with blood IRT levels of >80 mg/ml' (Geoffrey Pearson, personal communication).

INVESTIGATION OF THE LIVER

Useful guidance to the anatomy of the liver and associated organs is to be found in the seminal work by Gregory (1950) − see earlier.

In a live gorilla a hepatic disorder is likely to be suspected following clinical examination followed by investigative tests (e.g., elevated liver enzymes). Clinical signs of hepatic dysfunction/failure will possibly range from enlargement of liver to vomiting and diarrhoea and/or jaundice. Laboratory investigations may reveal hypoalbuminaemia, hypoglycaemia, coagulopathy and elevated serum enzymes. Confirmation may be conducted by examining liver biopsies (Fig. 11.15).

Hepatic disease in a dead gorilla is usually diagnosed on the basis of observation of gross abnormalities (see Box 11.3).

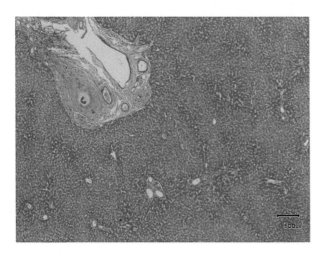

FIGURE 11.15   Histological appearance of normal liver of a lowland gorilla.

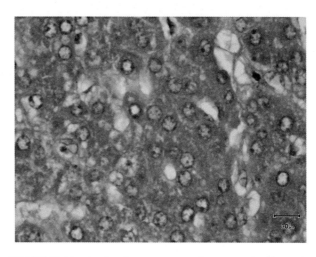

FIGURE 11.16   A refractile appearance to hepatocytes in this section of a liver of a lowland gorilla is suggestive of haemosiderosis.

BOX 11.3 Examination of the Liver

- Observe − changes in size, lobulation, colour, consistency (be wary of changes associated with postmortem autolysis)
- Examine (using lens if possible) − abscesses, foci, biliary hyperplasia
- Listen and feel when cutting − consistency, fibrosis, calcification
- Smell − pus? gas?
- Measure/weigh − important if a series of cases (compare/contrast)
- Take samples
    - impression smears (cytology)
    - tissues (histology, electron microscopy etc)
    - swabs (microbiology)
- Laboratory tests, especially histopathology, microbiology and electron microscopy
- Save material, frozen, for future reference and tests (e.g., toxicology)

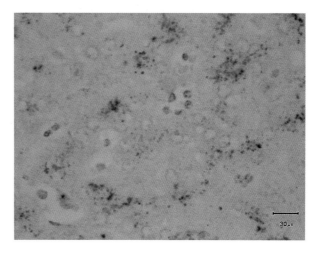

FIGURE 11.17   A Perls' stain confirms (a blue coloration) that the pigment is, indeed, haemosiderin.

BOX 11.4 Responses of the Liver

- Degeneration
- Necrosis
- Regeneration
- Fibrosis
- Bile-duct proliferation

It is important to examine and handle the liver *before* embarking on the GIT tract to avoid bacterial contamination.

Histological examination of the liver − if necessary together with electron microscopy − is always crucial as it may reveal a range of changes that are not visible grossly − haemosiderosis, for instance, is very common in captive gorillas (Figs. 11.16 and 11.17).

PATHOLOGY AND PATHOGENESIS OF HEPATIC DISORDERS

The pathogenesis of diseases of the liver in humans was clearly detailed by McGee et al. (2005), who stressed the importance of clinicopathological correlation in assessing common patterns of hepatic damage.

In *Gorilla*, as in *Homo*, the liver responds in different but generally consistent, ways to insults (see Box 11.4).

Necrosis may lead to toxaemia. In many primates, including humans, ascites is a feature of hepatic damage, both acute and chronic.

Hepatic disease in gorillas may be primary or secondary. The liver is a key organ that reflects general (systemic) disease as well as local pathology. Benirschke and Adams (1980), reporting on the causes of 48 gorilla deaths, included hepatitis as a 'primary cause of death' in animals up to the age of 8 years, and fatty liver and haemosiderosis in animals of 8−30 years of age.

Earlier, Nall et al. (1972) had reported death of a 3-year-old female gorilla with hepatic lesions that were suggestive of viral hepatitis in humans. The liver was not weighed but 'appeared small relative to the animal's body size ... . The margins were rounded'. There were reddened areas on the surface with loss of the normal lobular pattern. On histology the affected regions resembled the necrosis seen in viral hepatitis in humans, with loss of hepatocytes and collapse of the reticular framework. There was moderate steatosis, cholestasis and biliary changes.

The case above is covered in some detail because, in common with many of the early papers about gorillas, it contained sound descriptive information about gross and microscopical features of the case. Much has been published about hepatitis of gorillas since then − see, for example, Linnemann et al. (1984) and Makuwa et al. (2006) − but the emphasis is on molecular biology and serology, which are not the focus of this book.

Pathology reports from European zoos indicate that lesions in the liver of gorillas are common, even in animals with no history of hepatic disease, and include foci of inflammation, fibrosis, degeneration or necrosis, and degrees of vacuolation of hepatocytes. Areas of sinusoidal dilatation are reported as well as more extensive telangiectasis. Hepatocytes are often recorded as containing granular brown pigment − sometimes haemosiderin, sometimes lipofuscin.

Multinucleated hepatocytes (MNHs), a spontaneous incidental feature of laboratory macaques (Novilla et al, 2014) have been reported in gorillas (Lowenstine, 2003) (and recorded thirty years' ago by Dian Fossey in "Idano" − see later). Their significance is uncertain.

The liver of neonate and fetal gorillas normally shows haematopoiesis.

Three species of liver fluke, Trematoda, have been reported from gorillas (see list of gorilla endoparasites in Redmond, 1983).

INVESTIGATION OF THE BILIARY SYSTEM

Bile has attracted little attention from pathologists who work with gorillas but its history, being one of the four bodily humours in the Middle Ages, is a reminder that this was not always the case. The biliary tract − and bile itself − can still divulge important information.

In a live gorilla, ultrasonography will yield details on the volume of the gall bladder; other imaging techniques are used to investigate its status further.

Investigation postmortem should follow the following aphorism: 'Find, observe, handle, dissect'. Important steps include (1) measuring the gall bladder, (2) recording the colour and consistency of the bile, (3) assessing the patency of the duct and (4) noting the contents − signs of autolysis, debris 'stones'. There may be bile-staining of internal organs and (occasionally) overlying skin; this should be documented, especially in forensic cases, as it may help determine how long the gorilla has been dead.

An early report in Fossey (1983) ('Idano') provided a useful example of the range of pathological changes that may be seen in the liver of gorillas. 'Heavy cholestasis all over the lobules; sinuses had many polymorphonuclear and mononuclear white blood cells, Kupffer cells heavily reactive, several focal spots of necrosis.'

Jaundice (icterus) has been described in gorillas, including cholestatic jaundice of pregnancy (Cook and Clarke, 1985).

INVESTIGATION OF THE EXOCRINE PANCREAS

For the endocrine pancreas, see Chapter 16, Endocrinological and Associated Conditions.

The pancreas is sometimes given only a cursory examination during necropsy. Like the gall bladder, it is important (1) to locate it, (2) (before handling) to observe it with both the naked eye and with a lens or magnifying loupe and (3) to handle and palpate it *gently*. Confirmation that an unidentified piece of tissue is pancreas − in a field necropsy, for example − may be obtained histologically (Fig. 11.18).

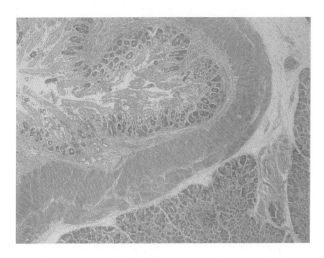

FIGURE 11.18 A histological section from a lowland gorilla shows the proximity (and normal appearance) of the pancreas and duodenum (above).

Diagnosis of exocrine pancreatic disease requires sound observation and techniques. Pancreatic insufficiency is one form of malabsorption, manifest clinically by diarrhoea, weight loss, and (often) pale, foetid, faeces (steatorrhoea). When conducting a postmortem analysis, an indication of any changes in shape, size or appearance should be recorded and the colour and consistency described accurately — for instance, 'firm, cream and lobulated'.

Chapter 12

Lymphoreticular and Haemopoietic Systems and Allergic Conditions

John E. Cooper

INTRODUCTION

This chapter is primarily concerned with the lymphoreticular tissues and disorders involving the immune system. Blood and haematozoa (blood parasites) are also covered here, briefly.

THE LYMPHORETICULAR SYSTEM

This is also sometimes referred to as the reticuloendothelial system. In *Gorilla*, as in *Homo*, it comprises a network of cells, tissues and organs distributed throughout the body that includes the lymph nodes, spleen, bone marrow and mucosa-associated lymphoid tissue/gut-associated lymphoid tissue (see Fig. 12.1). It is responsible for the development and maintenance of the animal's immune system.

The structure and function of the primate lymphoreticular system was discussed by Scott (1992). Abee et al. (2012) also included a useful chapter on the haemopoietic, lymphoid and mononuclear phagocyte systems of nonhuman primates (NHPs). The lymphoid tissue is an endodermal derivative. The thymus includes also an ectodermal component, from at least one pharyngeal pouch.

Changes in the lymphoid tissues of gorillas, as in other species, are commonly a response to antigenic stimulation. Sometimes the reaction is marked — for example, gorilla 'Jitu' at San Diego Zoo, California, in 1994, whose death was due to infection with an as yet unidentified free-living amoeba and whose tissues showed lymphoid hyperplasia, plasma cell infiltrates plus eosinophilic infiltration in lymph nodes (Philip Ensley, personal communication).

The spleen lies in the upper left quadrant of the abdomen, alongside the greater curvature of the stomach. It is composed of 'red' and 'white' pulp. The former consists essentially of a vascular network and sinuses. The white pulp comprises lymphoid tissue.

Changes in the size of the spleen can occur for many reasons, both physiological (e.g., congestion) and pathological (e.g., neoplasia). Data on spleen size in *Gorilla* are disappointingly few. Perhaps databases in North America might one day be made freely available to help provide such information. Unlimited access would be so helpful; a leaf could perhaps be taken out of the book of the medical pathologist Galloway over half a century ago (Galloway and Allbrook, 1959), who, in an account of his first ever postmortem examination of a (mountain) gorilla, openly included the following 'raw' but helpful field notes: 'Spleen — Crude weight ½ lb. Size of that normal in man (5 × 2½ × 1½ ins.). Grey with wrinkled capsule. No evidence of abnormality'.

The thymus and bone marrow, which produce T- and B-lymphocytes, respectively, are presumed to be the primary lymphoid organs in gorillas, as in humans. There are sometimes variations in the morphological appearance of the different lymphoid tissues in *Gorilla* but essentially they consist of (1) lymphatic tissues (lymphocytes, macrophages and antigen-presenting cells), and (2) reticular connective tissue (see Fig. 12.7 later).

IMMUNITY

For those with limited knowledge of immunology and its application to clinical disease, a useful and very readable introduction is provided by Day and Schultz (2010).

Immunisation by vaccination can only be mentioned briefly here. It is relevant to both captive and free-living gorillas because, in theory at any rate, it has the potential to protect these animals against a number of

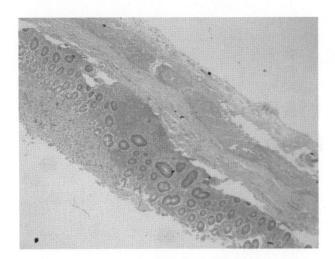

FIGURE 12.1 Portion of alimentary tract of a lowland gorilla showing well defined, normal, lymphoid tissue in the mucosa (bottom of picture).

infectious diseases. The caveat 'in theory at any rate' is included because:

1. Despite some interest in immunisation of captive 'exotic' species (Loomis, 1990; Bittle, 1993), very few studies have been performed on the efficacy of vaccines in free-living wildlife, either in terms of stimulating the production of protective antibodies or of actually protecting animals in the face of disease.
2. Recent work, primarily but not exclusively in the human field, has illustrated the complexity of 'vaccination'. While not doubting the great achievements over the decades, researchers have questioned the assumption that a potent vaccine, properly administered, is invariably immunogenic. Particular attention has focused on the role of nutrition and on gender differences (Flanagan, 2015; Jensen et al., 2015). There has also been a reappraisal of BCG and the role it appears to play in 'heterologous' protection (Kleinnijenhuis et al., 2015).

Immunisation of gorillas against infectious diseases has only been carried out on a limited scale. Hastings et al. (1991) described the 'ontogeny of a wildlife vaccination program' with respect to mountain gorillas and measles. There were subsequent reviews (Blasier, 2001) and evaluations (Blasier and Travis, 2005). Cranfield and Minnis (2007) expanded on the background, explaining that in 1989—90 there was an outbreak in the mountain gorillas of what was believed to be measles and 460 animals were vaccinated. The outbreak subsided; this may or may not have been the result of the vaccination but the incident 'confirmed that a sizeable portion of the ... population could be vaccinated in the face of a disease outbreak'.

The safety and efficacy in gorillas of vaccines (or, indeed, any other medicines) designed for use in *Homo*

sapiens or other species cannot be assumed — because proper testing, which would involve experimental studies, is neither feasible nor generally acceptable. Those wishing to use vaccines in gorillas need to consult widely (including discussions with the manufacturers) and to perform a proper cost:benefit analysis. Extrapolation from one species to another is often practised in some areas of veterinary medicine (Merck, 2007) but can be dangerous (Munro and Munro, 2012). It is often the only option available when dealing with a rare species such as *Gorilla gorilla* or *Gorilla beringei* and then *Homo* may be the most appropriate model.

There has been interest in the immune responses of gorillas and other anthropoid primates for some years. Semple et al. (2002) investigated immune system evolution and concluded that their analysis 'suggests that the risk of disease infection (*sic*) from the environment and the risk of injury have played a key role in immune system evolution among anthropoid primates'. Both macro- and microparasites could be relevant too. In respect of the former, it has long been believed that the behaviour of individual wild animals might play a part in the spread of pathogens. An example cited by Nunn and Haymann (2005) related to sleeping site preferences in New World monkeys; the choice may affect the risk of their being infected with malaria parasites. This contributes in part to the 'pathogen defence optimization hypothesis', which postulates that individual animals balance investment in costly behavioural and immunological defences against infection by pathogens.

The link between adipose tissue and the immune system (Pond, 2005) is discussed in Chapter 14: Musculoskeletal System.

As far as microparasites are concerned, there is also the 'bionic depletion theory'. This supposes that the immune system evolved in order to function in company with a microbiome (see elsewhere in text). There is increasing evidence for this — studies on ascarid and other worms and their relationship to malaria in Africa, for instance (see Shapiro et al., 2005), and experimental research (see, e.g., McKenney et al., 2015) showing that helminths in laboratory rats affect their microbiome and help 'temper immune reactions that might otherwise cause deleterious effects during development'. It will be fascinating to see if there are similar interactions in the genus *Gorilla*.

Pathologists who are relatively unfamiliar with reading slides and cytological preparations from lymphoid tissues will find helpful the textbook by Field and Geddie (2014).

IMMUNOLOGICAL DISORDERS

Relatively little has been published about immunological disorders in gorillas and current thinking is based largely on extrapolation from other species, mainly great apes and humans. Insofar as disorders involving the immune system of other NHPs are concerned, a useful starting point is Part I (General Pathology) of *Pathology of Simian Primates* (Fiennes, 1972), which contains two chapters about the immune system, one referring to autoimmune disorders.

In humans immune-mediated disease can affect the blood, skin, musculoskeletal, neurological, respiratory, endocrine, renal and reproductive systems and even the immune system itself. One assumes that the same may apply to gorillas — but so much remains to be learned.

Once again (see other chapters in this book), there may be links with the microbiome — that is, apparently commensal organisms on or in the gorilla's body. Studies by Wu et al. (2010) in mice showed that autoimmune arthritis was strongly attenuated in these animals when a single species of enteric segmented filamentous bacterium was introduced, suggesting (that at least in this murine model) a single commensal microbe, on account of its ability to promote a specific Th cell subset, can 'drive' an autoimmune disease. This and similar findings may or may not be applicable to gorillas.

Unpublished reports on gorillas in British zoos, reviewed by the author (JEC) have revealed a whole range of pathological changes in lymphoid tissue, some 'incidental' findings, others of specific significance. For example, one case in the archives at the School of Veterinary Science, Bristol, contained the following observations on an animal that died of tuberculosis: 'The spleen contains many rounded nodules of solid/firm white tissue (up to 1.5 cm diameter). The lesions ... are areas of fibrosis and granulomatous inflammation. There are scattered granulomas within areas of fibrosis with occasional foci of necrosis. ZN staining reveals occasional rod-shaped bacteria'.

Another was a description of changes in a gorilla with thymic abscesses. The only features noted on gross necropsy were mucopurulent exudate around both eyes and a fluctuating soft subcutaneous mass containing cloudy mucoid, red/brown fluid with a connective tissue wall up to 0.5 cm thick, within the right ventral neck. In the outer mediastinum there was an irregular, rounded, mass, a pus-filled structure, extending into the left pleural cavity. Histologically the mass appeared to be of thymic origin: 'There is extensive necrosis and inflammation within the tissue. At the periphery there is more normal thymic tissue with lymphoid and epithelial elements. Microbiological examination ... yielded *Salmonella typhimurium*'.

HYPERSENSITIVITIES AND ALLERGIES

So little in the way of authoritative text has been written about 'allergic' responses in gorillas that this section is largely only a small collection of case histories and reports. Wallach (1968) recounted a case of angioedema in a gorilla associated with strawberry ingestion. This might well have been a true allergy (see Glossary), a reaction to strawberries or something they contained, but there are other causes of angioedema — drug-induced, a genetic predisposition and idiopathic.

Atopic dermatitis in captive gorillas, characterised by eczema and apparently a result of psychological disturbance was reported by Hog and Schindera (1989) who suggested that the syndrome may be widespread. They discussed the possible role of psychological factors (a 'partner conflict') in the aetiology of the condition and the need for more immunological studies. Hayman et al. (2010) described the successful treatment with steroids and epinephrine of acute systemic anaphylaxis in a wild-born but captive infant lowland gorilla in the Republic of Congo (CR). The animal showed clinical signs of acute respiratory distress, lingual swelling and reaction to intradermal tuberculin that had been given 55 hours earlier. The authors discussed possible antigens that may have prompted the anaphylactic shock.

Reactions to insect bite and stings should be mentioned. Some elicit an allergic response; others presenting with tissue damage and clinical signs may only be toxic reactions to venom. Such cases can be severe and a cause of concern, especially in free-living animals. For example, the author (JEC) wrote from Goma (DRC, formerly Zaire) in 1994: 'I am worried about gorilla Rafiki .... Today he was badly stung by '*nyuki*' (bees): his face is swollen'.

Postmortem histology sometimes shows lesions suggestive of an autoimmune reaction. For example, Cranfield (2001), referring to mountain gorilla 'Kudatinya': '...vasculitis involving brain, lung and eye (viral/autoimmune?)'.

Gorillas, like other species, will react adversely to some substances (see Chapter 9: Skin and Integument). Many anecdotal reports about captive gorillas implicate bedding and other substrates as a cause of skin and other lesions but it is rarely clear whether such cases have a truly immunological ('allergic') pathogenesis or are merely the result of physical irritation. Whichever, bedding materials need to be chosen with care. For example, chopped straw often contains dust and peat moss is very dusty when dry. Hardwoods contain considerable quantities of tannins and alkaloids. Any wood product may be contaminated with wood preservative and some may constitute a health hazard to humans (and gorillas?): Wirth (1983) stated that some 60 species of tree had been implicated in the pathogenesis of human dermatoses, asthma, dizziness, diarrhoea, vomiting and loss of consciousness.

Reports of adverse reactions to medicinal products in gorillas are, perhaps not surprisingly, rare. An interesting incident, tentatively attributed to the use of the neuroleptic agent zuclopenthixol, was reported briefly by Allchurch and Dutton (2000). This was attributed to Stevens-Johnson Syndrome, a rare, life-threatening, drug-induced cutaneous reaction that has primarily been reported in humans and is associated with the use of various medications, including antibiotics, anticonvulsants and allopurinol. The affected animal was a female lowland gorilla, 'G-Ann', aged 16 years. Following the administration of zuclopenthixol she developed an acute reaction, resulting in multiple, severe, necrotic lesions of the skin and underlying tissues (see Figs. 12.2 and 12.3). Subsequent examination under general anaesthesia revealed classic signs of toxic epidermal necrolysis affecting the torso, a hand and a foot, and there were characteristic petechiae on the buccal mucosae. There was severe gangrenous change involving the left hand, with anticipated extensive loss of tissue associated with the thumb and the palmar surface of the hand, several discrete

FIGURE 12.2 The hand of gorilla 'G-Ann' when examined under anaesthesia. Much of the palmar surface is beginning to slough.

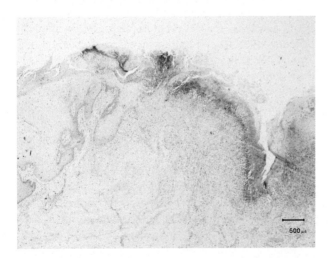

FIGURE 12.3 Histological section from gorilla 'G-Ann's' hand. Dermis and epidermis are necrotic.

necrotic lesions affecting arms and legs and extensive necrolysis of the left foot. There was also a deep induration of the right breast with necrosis of the nipple.

Despite therapy, viability of the left hand could not be restored and surgical amputation by wrist disarticulation was performed; G-Ann was eventually restored to good health (Johnstone-Scott, 1998). The amputated hand was subsequently transported across Europe by car to the Institut für Zoo-und Wildtierforschung (IZW) in Berlin, where it was dissected by Dr Roland Frey and remains as part of the Institute's collection (see Part II: A Catalogue of Preserved Materials).

Pathologists who examine tissues from gorillas sometimes make a tentative or definitive diagnosis of an allergic reaction but such reports are not always published. Thus, for example (see also Part II: A Catalogue of Preserved Materials), Drury R. Reavill and Robert E. Schmidt, California, USA (personal communication), described two cases that were examined using punch biopsy skin sections:

1. a 40-year-old male gorilla with chronic, dry, scaled and pruritic skin eruptions on its ventrum was diagnosed as having 'a severe acute erosive and ulcerative dermatitis with acanthosis, parakeratosis, and hyperkeratosis ... suspected to represent a drug or food hypersensitivity reaction',
2. another male gorilla, 17 years old, also with a chronic skin condition and pruritus attracted diagnoses of multifocal moderate ulcerative dermatitis with suppurative and septic folliculitis, and mild multifocal eosinophilic and acute perifolliculitis, also thought to be a hypersensitivity lesion but with a 'secondary bacterial dermatitis/folliculitis'.

One area of clinical medicine in which better understanding of immune reactions in great apes may be needed is that of blood transfusion. No relevant published data have been located in respect of gorillas but Debenham and Atencia (2014) described successful and apparently safe whole blood transfusion during treatment of severe anaemia in an orphaned, wild-born, chimpanzee in the Republic of Congo (CR).

IMMUNODEFICIENCY (BY GEOFFREY PEARSON AND JOHN E. COOPER)

In humans, causes of immunodeficiency can be physiological (neonatal and old age), congenital (disorders affecting granulocytes, T- and B-lymphocytes or complement) or acquired (infections, diabetes mellitus, neoplasia, medication, radiation, splenectomy, malnutrition). In all species opportunistic organisms can take advantage of an immunocompromised host and local or systemic infection may result. The respiratory tract is particularly often affected (see Chapter 10: Respiratory and Cardiovascular Systems) but other organ systems may also be involved. It can probably be assumed that a similar situation applies to immunodeficiency in gorillas but there are as yet few reliable data, as opposed to speculative observations, on this subject other than the work on retroviruses (see Chapter 3: Infectious Disease and Host Responses).

Assessing immune function in nondomesticated species goes back some years.

In 1997 Kennedy-Stoskopf discussed evaluation of immunodeficiency disorders in captive wild animals and drew attention to the difference between primary and secondary manifestations. She emphasised to the paucity of data at that time — a situation that has not changed markedly since (but see Chapter 3: Infectious Disease and Host Responses, regarding retroviruses).

Interest in the immune competence of mountain gorillas surfaced following the outbreak of sarcoptic mange (scabies) in gorillas in Uganda, reported by Kalema-Zikusoka et al. (2002). One of the authors (JEC) performed the first histological examination of skin from one of these cases, using facilities at Makerere University and a field microscope. He noted that the animal showed very little inflammatory response, even in the vicinity of *Sarcoptes* mites, and suggested that the gorillas in question might be immunocompromised and thus more susceptible to sarcoptic mange. This situation is wellrecognised in humans where *Sarcoptes* infections are frequently seen in AIDS patients. Any role for immunodeficiency in the outbreak could not be confirmed but the suggestion led to lively debate. That gorillas might be 'stressed' and thereby immunocompromised as a result of close contact with humans and other factors was an attractive, but then unproven, proposition.

The thymus may be important in the detection of immune disease in gorillas. It should always be examined carefully and a note made of its size and any evidence of atrophy. Neonates and fetuses occasionally show haemorrhage in the cortex and/or medulla in an otherwise normal thymus — a feature also recognised in human neonates and infants. Other changes are sometimes seen in the thymus during routine necropsy and histology — cysts, for example (Fig. 12.4).

Thymic lesions may be infectious in aetiology. As mentioned in Chapter 6: Methods of Investigation — Postmortem Examination, an 18-month-old female gorilla that died under general anaesthesia at Bristol Zoo Gardens had what was believed by the pathologist to be an enlarged, abscessed, thymus which yielded a heavy, pure, growth of *S. typhimurium*.

A paper presented to the October 1978 Symposium on the Comparative Pathology of Zoo Animals (National

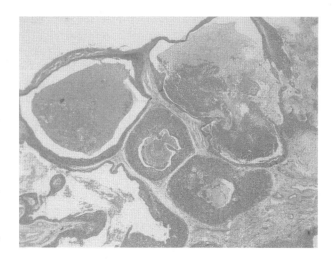

FIGURE 12.4 Thymus of a lowland gorilla showing multiple cysts. This animal also had pancreatic fibrosis.

Zoological Park, Washington, DC, USA) described the case of a 13-year-old female mountain gorilla, originally from Rwanda, which died at Cologne Zoological Garden after 9 years without any signs of disease. The Abstract at the time stated that necropsy of the animal had revealed a severe cystic dysplasia of the thymus with atrophy of thymus-dependent parts of peripheral lymphoid tissues. The kidneys showed an extensive haemorrhagic and necrotising nephritis caused primarily by Gram-negative organisms. The residual organs exhibited various signs of a shock syndrome with paralytic ileus and cardiac failure. The changes, especially in respect of the thymus, 'resembled an inherited T-cell immune deficiency syndrome of man'. The case was compared with similar ones in humans, and the type of immune deficiency syndrome was classified according to the WHO classification for such diseases. Two years later Krueger et al. (1980), in the published *Proceedings of the Symposium*, described cystic thymic dysplasia in two mountain gorillas, 'Coco' and 'Pucker','each with a fatal Gram-negative septicaemia (see Part II: A Catalogue of Preserved Materials). The two animals also featured in the reports reproduced by Fossey (1983): 'Autopsies done eight hours following deaths.... Both gorillas had severe thymic lesions. Pucker had a large multicystic thymic tumor weighing 385 mg with essentially no thymic tissue. The multicystic body and Hassall's corpuscles were absent. Coco had four small pea-sized thymic remnants weighing a total of 15 g, cystic dysplasia and occasional Hassall's corpuscles representing a degeneration product consequential of extensive involution'.

Postmortem examinations were carried out by one of the authors (GP) at Bristol University School of Veterinary Sciences, on two, infant, lowland gorillas. These were from the same sibling parents, from Bristol

Zoological Gardens; an 8-week-old male and a 12-week-old female, born in 1984 and 1985, respectively. The female died from bronchopneumonia, but the cause of death of the male was not established. In the female, patchy consolidation of the lungs was observed. Remaining postmortem findings were similar in both animals and were confined to the thymus. Grossly, the thymus was not definitely identified in either animal. In the female, a thin-walled cyst containing 20 mL of brown, mucoid fluid was present in the thorax anterior to the heart (Fig. 12.5). Small, cystic structures, up to 5 mm in diameter, containing similar material, were found in the cervical connective tissue adjacent to the trachea, extending from the ventral larynx to the thoracic inlet. In the male, small, yellow-brown cysts, in dark, red, connective tissue, were discovered in the cervical region, similar to the female. Detectable lymph nodes, spleen and remaining organs were considered to be within normal limits. Blocks of fixed tissues from a normal, 5-month-old, female, lowland gorilla (A11180), kindly supplied by Julia Tagg, Jersey Wildlife Preservation Trust (now Durrell Wildlife Conservation Trust/'Durrell'), were used for comparison. On histological examination of formalin-fixed tissues of the two infant gorillas, bronchopneumonia was detected in the lungs of the female (Fig. 12.6). In the thymus, changes were similar in both animals, and were compared with normal thymic architecture, which comprised lobules of lymphoid tissue with germinal centres and scattered Hassall's corpuscles in a delicate connective tissue stroma (Fig. 12.7). The thymus, in the cervical region of the male and female gorillas and the thoracic cyst in the female, was replaced by cystic spaces, containing amorphous eosinophilic material and occasionally haemorrhage, surrounded by a narrow rim of lymphocytes, in a loose, connective tissue stroma (Figs. 12.8 and 12.9). In the iliac

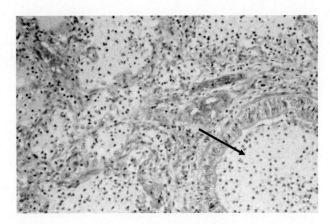

FIGURE 12.6 The same female gorilla. Lung, bronchopneumonia. Alveoli and a bronchiole (*arrow*) contain mixed inflammatory cells.

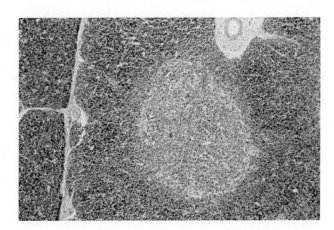

FIGURE 12.7 Five-month-old, normal, female, lowland gorilla. Thymus. Ample lymphoid tissue contains germinal centres.

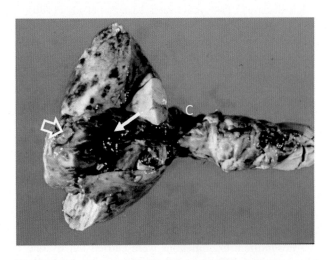

FIGURE 12.5 Twelve-week-old, female, lowland gorilla. Lung and cervical tissues. Patchy consolidation of the lung. Adjacent, red tissue (*arrow*) with small cysts (*open arrow*), represents vestigial thymus and extends into the cervical region (C).

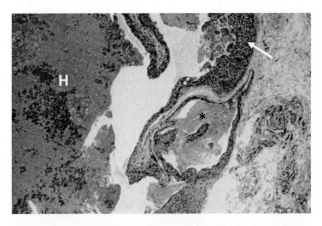

FIGURE 12.8 Eight-week-old, male, lowland gorilla. Cervical thymus. Cysts containing amorphous material (*) and haemorrhage (H), surrounded by a narrow rim of lymphocytes (*arrow*).

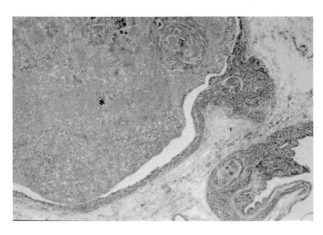

FIGURE 12.9 Twelve-week-old, female, lowland gorilla. Thoracic cyst wall. A large cyst within the wall contains amorphous, eosinophilic material (*) and has a narrow rim of lymphocytes.

FIGURE 12.11 Twelve-week-old, female, lowland gorilla. Iliac lymph node. The cortex contains sparse lymphocytes and germinal follicles are absent. (Capsule, *arrow*).

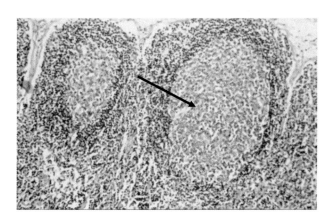

FIGURE 12.10 Five-month-old, normal, female, lowland gorilla. Lymph node. The cortex contains ample lymphocytes with germinal follicles (*arrow*).

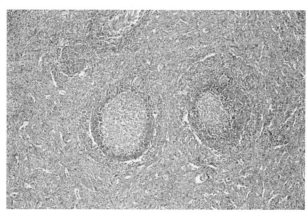

FIGURE 12.12 Five-month-old, normal, female, lowland gorilla. Spleen. White pulp contains germinal centres.

lymph node of the female gorilla, the normal architecture, as seen in the 5-month-old lowland gorilla with scattered, germinal follicles in the cortex (Fig. 12.10), was replaced by reduced numbers of lymphocytes, and germinal follicles were not observed (Fig. 12.11). In the spleen of the female gorilla, compared with the normal, 5-month-old gorilla, which had well-defined, germinal follicles in the white matter (Fig. 12.12), the white matter was poorly defined (Fig. 12.13). Based on the gross and histological findings in the lymphoid tissues of these two infant gorillas, a suggested diagnosis of immunodeficiency was made, possibly with a congenital cause, given the sibling relationship of the parents. It may also represent a thymic dysplasia, similar to that reported by Krueger et al. (1980). For a definitive diagnosis, more detailed, modern techniques may be required.

Other authors have implicated immunodeficiency, or at least some compromising of the gorilla's immune

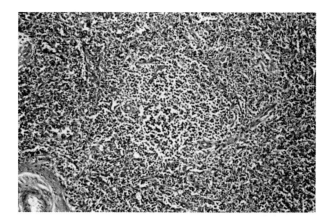

FIGURE 12.13 Twelve-week-old, female, lowland gorilla. White pulp (*centre*) is poorly defined.

system, in ill health and/or death. Anderson et al. (1986) recounted a case of pneumonia and meningoencephalitis associated with amoebae in a captive lowland gorilla and concluded 'Why the gorilla should have developed this amoebic disease is not known; the lymphoid tissues appeared morphologically adequate but may, perhaps, have been functionally deficient'.

The words 'stress' and 'immunosuppression' are mentioned from time to time in postmortem records, especially in zoos, but usually, in the authors' experience, without clear supporting evidence that either of these has been diagnosed scientifically (see Chapter 16: Endocrinological and Associated Conditions, regarding stress).

Awareness of possible immunosuppression in a gorilla may be prompted by the examination of a blood smear or lymph node biopsy or because an animal has an intractable infection. The bacterium *Pseudomonas aeruginosa* is a useful 'sentinel' in this respect. Repeated isolation of *P. aeruginosa* from a sick gorilla may sometimes only reflect environmental contamination by the organism, or poor sampling technique, but it can also serve as an early warning that the host is immunocompromised.

Generally, however, data on the lymphoreticular system of gorillas submitted for clinical or postmortem examination are sadly lacking — a situation that needs to be rectified. A factor contributing to the paucity of potentially useful data was for long the marked variability in clinical and postmortem protocols for gorillas. This meant that appropriate observations and collection of samples that might have thrown light on an animal's immune system were not collated. This situation has changed for the better in recent years with the introduction and adoption of more standardised procedures (see Chapters 6 and 7: Methods of Investigation — Postmortem Examination; Methods of Investigation — Sampling and Laboratory Tests).

In *Homo* the causes of immunodeficiency can be physiological (neonatal and old age), congenital (disorders affecting granulocytes, T- and B-lymphocytes or complement) or acquired (infections, diabetes mellitus, neoplasia, medication, radiation, splenectomy, malnutrition). In all species opportunistic organisms can take advantage of an immunocompromised host and local or systemic infection may result.

If a captive or free-living gorilla has to be immobilised and blood is to be taken, samples for lymphocyte and other function tests should be considered, in addition to standard haematology. Ultrasonography can provide information on the size of the spleen — enlargement may be an indicator of functional hypertrophy or splenic pathology (see earlier).

Likewise, postmortem examinations of gorillas should, as a routine, include the collection of samples from the lymphoreticular system of the animal. In particular, the spleen, lymph nodes, thymus (if present), palatine tonsil and other lymphoid organs should be carefully examined, described, weighed and/or measured and then sampled for histological and other investigation.

Many questions remain unanswered, especially insofar as immunocompetence is concerned. In particular, is immunodeficiency a truly significant factor in the health and wellbeing of gorillas? How prevalent is immunodeficiency — and what causes it? In this regard, Lowenstine et al. (2016), in their paper on aging in great apes stated that 'immunosenescence has been examined in baboons but not in aging apes'. How important might immunocompetence be in respect of macroparasites of gorillas? The organisms *Cryptosporidium*, *Giardia*, *Entamoeba histolytica* and *Strongyloides* occur in both humans and gorillas in Africa. In humans *Cryptosporidium* and *Strongyloides* are often associated with HIV infection (an important cause of immunodeficiency) but *Giardia* and *Entamoeba* are not. Is that true also in retrovirus-infected gorillas?

It seems probable that the genus *Gorilla*, like other species, has a finely tuned relationship with micro- and macroparasites. A better understanding of this host:parasite 'balance' is essential if we are to be better prepared to diagnose and to take appropriate action over infectious disease. The pathogen is only one part of the equation; the other is the ability of the host, the gorilla itself, to contend with the challenge.

BLOOD AND BONE MARROW

The haemopoietic system is responsible for the production of blood cells (haemopoiesis). The blood is covered in this chapter because, in addition to haemopoietic constituents, it contains reticuloendothelial cells and others that play a part in inflammation and immunity. Also, the spleen, sometimes haemal lymph nodes, participates in haemopoiesis.

Morphologically the blood of *Gorilla* is indistinguishable from that of *Homo*. An early study was by McClure et al. (1972) while a more general account of the primate haemopoietic system was described by Scott (1992). Interest in blood groups of NHPs, including apes, goes back to work in the 1920s; see, for example, Landsteiner and Miller (1925). A recent review was by Gamble et al. (2011).

Space does not permit detailed discussion of the blood. As stressed in Chapter 7: Methods of Investigation — Sampling and Laboratory Tests, laboratory examination often plays a key role in diagnosis and/or health monitoring and relates to a whole range of disciplines and organ systems. Indeed, it is increasingly being argued that use of the term 'pathology of the blood' might be more appropriate than subdividing techniques and skills into haematology, biochemistry and immunology (to name only three). The pathology of most diseases necessitating the examination of blood are multifactorial (Blann and Ahmed, 2014).

The importance of blood examination in diagnosis and health monitoring is alluded to, with some discussion of sampling, in Chapter 7: Methods of Investigation − Sampling and Laboratory Tests. Blood is an important indicator of systemic disease, infectious and noninfectious. An example of the latter was the normocytic, normochromic, anaemia that characterised protein deficiency in one colony of lowland gorillas (Mundy et al., 1998).

Histological sections not infrequently reveal changes in blood vessels, such as vasculitis and atheromatosis (see Chapter 10: Respiratory and Cardiovascular Systems). Part II, A Catalogue of Preserved Materials, refers to biopsy samples of a gingival mass from gorilla 'Upala' in Zoo d'Amnéville: 'Histology was consistent with a pyogenic granuloma/eruptive haemangioma'.

As stressed elsewhere in this book, there are many organs and tissues that appear not to have been fully studied in *Gorilla*. One example is the bone marrow. In human medicine there are whole books written specifically about bone marrow pathology (see, e.g., Islam, 2013) and the bone marrow of most species of domesticated and laboratory animals has been extensively covered in various publications, but very little has been published on the bone marrow of the genus *Gorilla*.

Techniques for the taking and examination of bone marrow are discussed in Chapter 7: Methods of Investigation − Sampling and Laboratory Tests. It must be remembered that bone marrow is made up of five elements − a highly vascular connective tissue, stroma supporting and protective bone spicules plus adipose, reticuloendothelial and haemopoietic elements. Recognition and interpretation of changes in the haemopoietic component is often not easy (a human medical, rather than a veterinary text, is recommended) and it is important to link findings with haematological examination of the peripheral blood. This same caveat is applicable to other haemopoietic organs such as spleen and haemo/haemic lymph nodes.

BLOOD PARASITES (BY IAN REDMOND AND JOHN E. COOPER)

There is surprisingly little published information about blood parasites of gorillas and few laboratory reports, even in Africa, include mention of whether 'examination of blood' included a specific search for such parasites or whether, for example, buffy coat preparations and routine examination of thick ('wet') blood smears formed part of the investigation.

Ruch (1959) reported *Plasmodium reichenowi* in a (western) lowland gorilla. In 1981 Dixson summarised the species of plasmodia recognised in *Homo* and the great

apes and listed *Plasmodium schwetzi* and *P. reichenowi* under malaria parasites, also in the lowland gorilla. No *Plasmodium* sp. had been reported in eastern lowland or mountain gorillas at that time − the latter probably because the high altitude mountain habitat is too cold for the *Anopheles* mosquito vector. Meder (1994) referred to 'various forms of malaria' diagnosed in lowland gorillas and said that 'the gorilla malaria species *P. reichenowi* and *P. schwetzi*, which are also found in chimpanzees, are apparently not dangerous for humans under natural conditions'.

Notwithstanding the earlier comment about altitude and *Anopheles*, malaria has apparently been suspected in mountain gorillas. The eight-year-old female 'Quince' died in 1978 and at necropsy it was concluded that malaria with complications was the cause of death (Fossey, 1983; Watts, 1983). Unfortunately the species of *Plasmodium* was not established.

The recent development of noninvasive methods has allowed reexploration of plasmodial diversity in African apes. The use of new DNA extraction kits has presented new strategies for sampling malaria parasites in apes' fresh faeces. More recent DNA analysis of 470 faecal samples from gorillas in Gabon, West Africa revealed the presence of *Plasmodium* spp. in 21.3% of the specimens (Boundenga et al., 2015). Earlier work by Liu et al. (2010) (see review by Holmes, 2010) involved the use of a single-genome amplification strategy to identify a *Plasmodium* spp. using DNA in faecal samples from free-living great apes. Of nearly 3000 specimens collected from field sites throughout Central Africa, the authors detected a *Plasmodium* sp. closely related to *Plasmodium falciparum* (human malaria) in western/lowland gorillas and chimpanzees but not in eastern/mountain gorillas or bonobos. Molecular analysis revealed that 99% of the parasites' DNA grouped within one of six host-specific lineages representing distinct *Plasmodium* species in the subgenus *Laverania* and one of these, from western gorillas, consisted of parasites that were nearly identical to *P. falciparum*. The authors concluded that *P. falciparum* has its origins in gorillas, not chimpanzees, bonobos or ancient humans.

Trypanosomiasis appears not to have been reported in gorillas but a number of different trypanosomes are known from primates in Africa, Asia and the New World. Again, mountain gorillas live at altitudes where tsetse flies (*Glossina* spp.) appear to be absent but some lowland gorillas may be exposed to − and presumably bitten by − these or other haematophagous Diptera. A serological survey would be of interest on the lines of that reported by Jeneby et al. (2002) who examined blood samples from three species of wild-caught monkeys in Kenya for circulating *Trypanosoma brucei* and for *T. brucei* antigen and antitrypanosome antibody. They detected titres of anti −

T. brucei antibodies and a field-orientated latex agglutination test demonstrated invariant *T. brucei* antigens in some samples. However, no trypanosomes were visible in blood smears, on wet blood films, or by buffy coat technique.

Meder (1994) reported, without a specific citation, that the threadworm *Tetrapetalonema vanhoofi* may be found in the lymphatic and the blood system of gorillas (see Chapter 11: Alimentary Tract and Associated Organs). Berghe and Chardome (1949), Berhe et al., 1964 and Bain et al. (1995) all reported filariae from free-living gorillas.

NEOPLASIA

The following is a brief résumé of some neoplasms reported in gorillas relating to the lymphoreticular and haemopoietic systems.

A T-cell lymphoma associated with immunological evidence of retrovirus infection in a lowland gorilla was reported by Prowten et al. (1985) (see Part II: A Catalogue of Preserved Materials). In 1986 Srivastava reported human T-cell leukaemia virus I provirus and antibodies in a captive gorilla with non-Hodgkin's lymphoma. Antibodies reactive against human T-cell leukaemia virus I (HTLV-I) (see also Chapter 3: Infectious Disease and Host Responses) were detected by indirect immunofluorescence assay, enzyme-linked immunosorbent assay (ELISA) and Western blot analysis in three sera (one collected in 1979) from a captive gorilla that had developed diffuse histiocytic lymphoma in 1983. The sera from four other healthy gorillas housed separately were HTLV-I antibody negative. These results indicate that the gorilla was infected with HTLV-I or a closely related simian virus several years before the development of lymphoma. Barrie et al. (1999) described the clinical management of a 6-month-old lowland gorilla that

presented with lethargy, pyrexia, abdominal distention and splenomegaly. Acute lymphocytic leukaemia was diagnosed on the basis of haematological findings and bone marrow cytology. Despite therapy based on human oncological practice, the leukaemia reappeared after 3 months and the gorilla was euthanased.

Barrie et al. (1996) reported an acute lymphocytic leukaemia in a captive infant lowland gorilla.

There is also, of course, interest in lymphoreticular neoplasia in free-living gorillas and other apes and its relationship to retroviruses (see Chapter 3: Infectious Disease and Host Responses). As an example, Nerrienet et al. (2004), related a serological survey for human T cell leukaemia virus (HTLV)/simian T cell leukaemia virus (STLV) antibodies in 61 wild-caught African apes, including five gorillas, originating from south Cameroon. One gorilla and a chimpanzee exhibited a pattern of complete HTLV-I seroreactivity; sequence comparison and phylogenetic analyses indicated the existence of two novel STLV-I strains, both HTLV-I/STLV-I molecular clade subtype B which is specific to central Africa. The authors argued that their findings reinforced the hypothesis of interspecies transmission of STLV-I to humans.

Part II, A Catalogue of Preserved Materials, refers to an animal at Buffalo Zoo that had a 'discharge that did not respond to antibiotics, increasing difficulty in swallowing and feeding, and progressive weight loss. Subsequent examination of biopsies of extensive nasopharyngeal and cervical tumours indicated a large-cell, histiocytic lymphoma'.

Reports continue to appear — some, like the paper above and the findings of Ikpatt et al. (2014) relating not just to gorillas. Lymphoreticular neoplasms have been only rarely reported in nonhuman hominins and it is important that both clinicians and pathologists remain alert to such diagnoses.

Chapter 13

Urinary and Reproductive Systems

John E. Cooper

To cure sometimes, to diagnose often, to investigate always.

B.M. Mount (1980)

INTRODUCTION

This chapter discusses the pathology of two organ systems, urinary and reproductive, that share similarities in respect of anatomical position and embryological development but which differ greatly in role and function.

URINARY SYSTEM

The kidneys are mesodermal derivatives, the bladder and urethrae endodermal.

The first detailed scientific descriptions of the anatomy of the urinary system of *Gorilla* were, respectively, the chapter on the abdominal viscera of the gorilla (kidneys) by Elftman and Atkinson and that on the regional anatomy of the gorilla (urinary bladder) by Raven and Hill, both in the book by Gregory (1950). Some years prior to that, Koch (1937) had included useful information on gorilla structure, including the urinary system, in his report on the famous gorilla 'Bobby' in the Berlin zoological gardens. General features of the structure and function of the primate urinary system were subsequently discussed in comparative terms by Scott (1992). The urinary system of the gorilla closely resembles that of humans, certainly insofar as its gross and histological appearances are concerned.

Schaller (1963) described urination patterns in the mountain gorilla.

Urinary Disease and Pathology

Reference to the human literature illustrates the broad spectrum of pathological changes that can affect the urinary tract of *Homo*. For example, amongst those discussed in the standard series by McGee et al. (1992) were congenital abnormalities, glomerulonephritis, pyelonephritis, interstitial nephritis, acute tubular necrosis, metabolic disorders (e.g., amyloidosis) and neoplasia. It seems likely that a comparable range of pathological changes occurs in *Gorilla* spp. – and, indeed, some are recognised and have been reported.

Unsubstantiated diagnoses or allusions to urinary diseases in gorillas appear in early 'popular' accounts and subsequent writings. For example, the cause of death of the famous circus gorilla 'Gargantua' (see Part II: A Catalogue of Preserved Materials) in 1949 was reported by Prince-Hughes (2001) to be 'double pneumonia compounded by kidney disease, tuberculosis, and cancer of the lip'. Part II, A Catalogue of Preserved Materials, includes entries with mention of urinary disease during an animal's life – for example, 'nephritis had caused the shrinkage of one kidney'; 'bilateral lobar pneumonia, complicated by a severe kidney disease'; 'nonfunction of her kidneys' and 'died of kidney failure' but confirmation of these diagnoses is either not supplied or not available.

Urinary disease and pathology in gorillas have been recorded by various zoological collections and some such cases have been published. O'Neill et al. (1978), for instance, described acute pyelonephritis and Robinson and Benirschke (1980) congestive heart failure and nephritis.

A 39-year-old male western lowland gorilla that developed renal failure following chronic interstitial nephritis with diffuse fibrosis of the kidneys was reported by Paixão et al. (2014). Paal et al. (1982) described a case of renal tubular nephrosis as a possible cause of death. In a brief account Burns (2006) recounted haemodialysis of a gorilla with fatal septicaemia and pyelonephritis that were secondary to urine stasis and a uterine leiomyosarcoma. Bradford et al. (2013), in their paper describing Turner syndrome in a gorilla, listed 'kidney abnormalities' as one possible feature of this abnormality. Abee et al. (2012) included a chapter on the urogenital system of (mainly laboratory) primates.

J. E. Cooper & G. Hull: Gorilla Pathology and Health. DOI: http://dx.doi.org/10.1016/B978-0-12-802039-5.00013-5

Fossey (1983) listed 14 postmortem reports (mainly, not entirely, of free-living animals). These are remarkably comprehensive and informative, in part probably because some of the gross, histological and microbiological examinations were performed by (human) medical practitioners or pathologists. Those reports include the following observations (slightly edited) that are of relevance to this chapter: kidney: all glomeruli show thickening of the Bowman's capsule; no interstitial reaction of fibrosis; peritonitis noted on dorsal side of large intestine in form of abnormal 'rubber-band' type of adhesions to pleura; intracellular accumulation of dark green-brown granules in the proximal tubule epithelium (11-month-old infant kidneys (right = 20.1 g; left = 25.3 g). In 'Pucker' was a transitional cell papilloma of the bladder.

In a personal communication, a brief update on 'mortality causes' in free-living *beringei*, Linda Lowenstine, Gorilla Doctors, made no mention of urinary conditions. However, in a recent paper concerning the comparative pathology of ageing great apes, Lowenstine et al. (2016) described chronic kidney disease and stated that (in contrast to orangutans) renal disease was considered to be the cause of death in only one adult female lowland gorilla and one free-living mountain gorilla in their series; both animals had pyelonephritis. Renal disease was not investigated as a comorbidity in either of these gorilla studies. Incidental ageing changes observed in captive and free-living gorilla kidneys included medullary interstitial hyalinisation, due to an increase in extracellular matrix positive for Masson's trichrome and periodic acid-Schiff, and mild multifocal cortical interstitial and periglomerular fibrosis. One male gorilla that died of aortic dissection showed hypertrophic changes in small renal arterioles that were suggestive of hypertension. Interestingly, but not surprisingly, Lowenstine et al. (2016) concluded that 'More detailed investigation of gorilla renal disease is needed'.

Quite apart from published accounts, clinical records from many zoos include confirmed or suspected urinary disease – for instance, pyelonephritis and 'possible kidney infection' in animals at Bristol Zoo Gardens, United Kingdom (see Appendix 6: Case Studies – Museums and Zoological Collections).

Likewise, many postmortem reports viewed by the author include mention of findings in the urinary system – almost entirely restricted to the kidneys. Foci of mononuclear cells in the renal cortex are the most frequently seen (Fig. 13.1), in some cases coupled with fibrosis and other changes suggestive of interstitial nephritis. One record from Bristol Zoo Gardens reported adhesion of the ventral aspect of the bladder to the peritoneum.

Sleeman and Mudakikwa (1998) – see below – argued that the establishment of urinary reference intervals would aid in the noninvasive detection of certain diseases in (mountain) gorillas and assist with

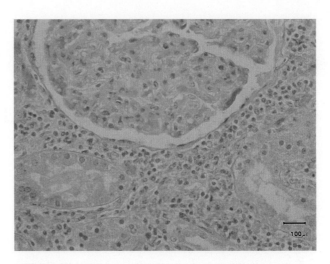

FIGURE 13.1 A focus of mononuclear cells in the interstitium of the renal cortex of an otherwise apparently healthy lowland gorilla.

FIGURE 13.2 Urine collection in a European zoo, using a syringe.

decision-making regarding the immobilisation and treatment of individual animals.

Sampling Techniques

Collection of clean, fresh, urine samples for examination is important. The ease with which this can be achieved depends upon many factors, not least whether the animal is captive or free-living. As far as the former is concerned, zoos have for decades generally made use of naturally voided samples on the cage floor (Fig. 13.2). Sunde and Sievert (1990) described how they had trained female lowland gorillas to urinate on request and this approach is now used in some collections.

As far as free-living gorillas are concerned, a number of approaches have been tried, almost exclusively by those working with free-living *beringei*. Animals are seen urinating but retrieval is hampered because often the sample soaks into the ground, especially if there is any delay in collection. Sometimes, however, a few drops of urine are trapped on the surface of bamboo leaves or other vegetation, especially on or near recently vacated night nests. Occasionally urine rains down from an animal in a tree. Such samples were collected opportunistically by the author (JEC) during the first 3 months of 1994, prior to the outbreak of genocide in Rwanda, and investigated in the laboratory of the VVC in Kinigi (see Chapter 5: Methods of Investigation – Observation, Clinical Examination and Health Monitoring and Appendix 3: Field Pathology).

A more rigorous technique was devised to collect urine samples voided by mountain gorillas by Robbins and Czekala (1997) and was subsequently applied to the reproductive monitoring (urinary hormone analysis) of free-living females by Czekala and Sicotte (2000). The approach was similar in principle to that used by the author and his colleagues referred to above. As soon as possible after seeing the animal urinate, a syringe was used to draw up the voided sample. Sleeman and Mudakikwa (1998) reported on the analysis of urine naturally voided by free-living mountain gorillas and produced a database of presumed normal values. Commercial dipsticks appeared to be unreliable for the measurement of specific gravity and leucocytes. Mountain gorillas were found to have a high urinary pH (8.45) and low specific gravity (1.013).

If a gorilla is anaesthetised/immobilised, urine may be passed during the procedure and can be used for laboratory tests. Preferably, for a fresh and uncontaminated sample, catheterisation or cystocentesis can be used.

Methods of Investigation

Standard urinalysis using commercial (human) dipsticks has long been a feature of routine health checks on captive gorillas but note should be taken of the cautionary comment by Sleeman and Mudakikwa (1998) above.

The samples collected by the author in Rwanda in early 1994 were largely used to prepare stained cytological preparations, as described by Brearley et al. (1986). Those slides were amongst most of the material lost (or irreparably damaged) in the ensuing genocide and its aftermath. Cytology and other laboratory investigations are discussed in more detail in Chapter 7, Methods of Investigation – Sampling and Laboratory Tests.

If a gorilla is anaesthetised/chemically immobilised, especially under zoo conditions, various investigative procedures are possible, ranging from non/minimally invasive imaging to the taking of bladder and renal biopsies. Animals with suspected renal disease may require a range of kidney function tests, including estimation of blood urea nitrogen.

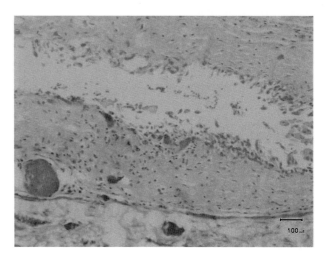

FIGURE 13.3 Section of otherwise normal bladder from a dead lowland gorilla, showing postmortem degeneration and sloughing of urothelium.

Specific Considerations

The gross and histological appearances of the urinary system of *Gorilla* closely resemble those of *Homo*.

Field material from free-living gorillas usually shows postmortem changes if there is an extended interval (hours) before the necropsy takes place. Pseudomelanosis, characterised by dark green or black discoloration of one or both kidneys, is frequently a feature, probably exacerbated by the pressure of the GIT on these organs. Under the same circumstances, or if postmortem examination of a captive animal is unduly delayed, the bladder wall often exhibits degenerative changes, including sloughing of urothelium (Fig. 13.3).

Iron deposition (haemosiderin) is sometimes seen in renal tubules – one notable case was Barcelona's albino gorilla 'Snowflake' (Marquez et al., 2008).

Subcapsular glomerulogenesis (a 'nephrogenic cap') is normal in the embryo and during the first few months of independent existence (Fig. 13.4). It is a measure of maturity; see later regarding intrauterine growth retardation.

REPRODUCTIVE SYSTEM

The gonads and genital ducts are intermediate mesodermal derivatives while the caudal vagina and vestibule are endodermal.

The human and nonhuman primate (NHP) reproductive tracts are anatomically very similar (Hill, 1953–1970); for example, the uterus is of the simple type seen in *Homo*.

The marked sexual dimorphism in body size and the male's pronounced canine teeth reflect a polygynous mating system in which each adult male has to defend his females against rival males.

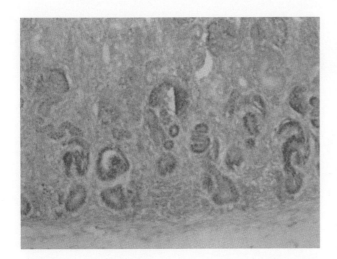

FIGURE 13.4 Even if the kidney shows autolysis, as here from a badly decomposed infant gorilla, subcapsular glomerulogenesis will be visible.

From the early days of the discovery of the gorilla there was interest in how its reproductive system compared with that of *Homo*. Owen (1859) described the penis of the gorilla as 'short and subconical; the prepuce is devoid of fraenum; the scrotum is broader and more sessile than in man'.

When Mivart published his *Man and Apes: An Exposition of Structural Resemblances and Differences Bearing upon Questions of Affinity and Origin* 150 years ago, he described primates, including humans, as being (amongst other things) 'placental mammals …. Penis pendulous; testes scrotal; always two pectoral mammae' (Mivart, 1873).

Excellent descriptions of the anatomy of the reproductive structures in *Gorilla* are to be found in chapters in Gregory (1950). Other early references include Koch (1937) and Steiner (1954). Dixson's (1981) book depicts external genitalia of the nonpregnant female gorilla.

Mammary Gland

One should always check mammary glands — of both sexes. A lactating female should have clearly bulging glands. Flat mammary tissue means low milk production (see Part II: A Catalogue of Preserved Materials — 'Souanké', Basel Zoo), a weak infant not sucking adequately and therefore not stimulating lactation, or an unwell mother with suppressed mammary function.

Dian Fossey set a good example insofar as checking the mammae of dead animals was concerned; in one necropsy report in her book (1983), she described the gorilla 'Old Goat' as having: 'Mammary glands in inactive phase'.

Even badly decomposed gorillas may have mammary glands that can still be examined to glean useful information. For example, in March 1995 the author (JEC) examined four gorillas that had been poached, and reported that 'the adult female in the group had active mammary tissue, suggesting that she had been suckling an infant'. Failure to locate the youngster supported the hypothesis that the four animals had been killed in order to obtain a live baby for a collector.

Obese captive gorillas of both sexes may appear to have pronounced mammary gland development. Postmortem, such animals will have large quantities of subcutaneous and intraabdominal fat (see Chapter 14: Musculoskeletal System).

Records from zoos — and unpublished observations on free-living animals — confirm that gorillas may develop mastitis. The pathogenesis is likely to be similar to that in humans.

Stewart (1988) drew attention to the important link between suckling and lactational anoestrus in gorillas. It should be noted that the term 'suckling' should be applied to the lactating female, not the infant — see Glossary.

Reproductive Physiology

Reproduction in gorillas and other great apes has attracted great interest over the past 35 years, largely prompted by a Symposium held in Gabon, West Africa, organised by R.V. Short and B.J. Weir and published as Proceedings (Short and Weir, 1980).

As Short and Weir pointed out, studies in zoological collections have done much to contribute to our understanding of reproduction in gorillas. Ryder (1989) added to this a strong argument for more genetic studies.

Attempts to propagate gorillas in captivity have yielded valuable information which has not only assisted in captive breeding but also in a greater understanding of reproductive behaviour in the wild (see, e.g., Mallinson et al., 1973) (Fig. 13.5).

Basic reproductive data on mountain gorillas were provided by Harcourt et al. (1980). They are summarised in Box 13.1.

Nadler and Collins (1984) reported on research on the reproductive biology of gorillas. Since the 1970s, zoos on both sides of the Atlantic have monitored and reported on their breeding programmes (Mallinson et al., 1973, 1976; Reichard et al., 1990).

Noback (1939) published the first account of the gorilla's menstrual cycle. Several decades later Loskutoff et al. (2004) reported on ovarian stimulation, transvaginal, ultrasound-guided oocyte retrieval and blastocyte production in the lowland gorilla, Atsalis et al. (2004) on sexual behaviour and hormonal oestrous cycles in aged animals and Atsalis and Margulis (2006) on sexual and hormonal cycles in geriatric gorillas.

FIGURE 13.5 An early photograph by the late Bob Campbell of a female mountain gorilla tending her baby.

BOX 13.1 Reproductive Data

Age of female at puberty (years)	7–8
Length of menstrual cycle (days)	28
Duration of oestrus (days)	2
Gestation (days)	255
Age at first birth (years)	10
Birth interval (years)	3–8
Average completed family size	2–5

Schaller (1963, 1964) and Fossey (1982, 1983) each published reproductive data based on their observations of free-living mountain gorillas. A preliminary investigation of urinary testosterone and cortisol levels in free-living males was carried out by Robbins and Czekala (1997). Habumuremyi et al. (2014) monitored ovarian cycles of the same species using progestogens in urine and faeces.

Copulation

Napier and Napier (1967), citing Lang, described copulation in captive western gorillas as 'both *more canum* and *more hominum*', mating is rarely observed in the wild. In nearly 2 years working with *beringei*, the author (JEC)

only experienced it once – and then only on the basis of the sounds made by the mating couple.

As Short and Weir (1980) pointed out, the testis:body-weight ratio in the gorilla is the lowest known for any primate. This low spermatogenic capacity is consistent with infrequent male encounters with an oestrous female, given an interbirth interval of 4 years. Likewise, overt signs of oestrus are few because, with a male present in the group all the time, it is scarcely needed.

Nadler (1995) investigated proximate and ultimate influences on the regulation of mating in the great apes. Semen collection in gorillas has been studied and reported by several authors including Brown (1998), Seager et al. (1982) and Platz et al. (1980).

Fossey (1983) observed homosexual mounting in the presence of heterosexual coupling, with this behaviour being more prevalent in the males.

Pregnancy

Gorillas were the last of the great apes to reproduce in captivity (Dixson, 1981).

Not only do female gorillas have very few outward signs of ovulation but they also exhibit scanty features of pregnancy. Schaller (1964) commented that this is probably the case in free-living females 'because the belly, like that of all gorillas, is distended from the vast bulk of greenery ... consumed daily'. In captivity signs such as swelling of the ankles may be seen in some females late in pregnancy.

Pregnancy testing of captive gorillas dates back 40 years and the monitoring of reproductive activity can now be considered part of health assessment of gorillas. It has developed hand-in-hand with the growth in understanding of reproductive physiology (see earlier) – from early studies on patterns of oestrogen during pregnancy (Seaton, 1978) and monitoring (Clift and Martin, 1978), through calculation of gestational age using ultrasonography (Yeager et al., 1981), to more recent studies on pregnancy and parturition (e.g., Randal et al., 2007).

Parturition

Early articles described the birth of individual gorillas in captivity (see, e.g., Bingham and Hahn, 1974; Fisher, 1972; Beck, 1984; Rahn, 1993). A paper from Canada, by a veterinarian and a human obstetrician (Randal et al., 2007), reported a pregnancy and parturition at Calgary Zoo, Canada; a complete necropsy was performed on the stillborn infant.

Births in the wild have been published by various authors including Stewart (1977) and Betunga (1997–98). The birth of twins is a rare event (Rosen, 1972) but is

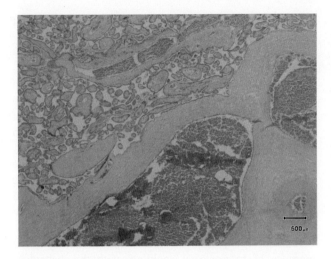

FIGURE 13.6 Histological appearance of normal placenta of a lowland gorilla.

reported from time to time – a mountain gorilla in the Virungas in May 2004, for instance.

The date of birth of a free-living gorilla can influence its health. For example, excessively cold or wet weather may increase vulnerability to respiratory disease (see Chapter 3: Infectious Disease and Host Responses).

Fetal and Placental Pathology

Interest in studying and preserving fetuses goes back nearly a century to the work of Coupin (1927) and Schultz (1927) (see also Part II: A Catalogue of Preserved Materials) and was continued by Cave (1959) and others. Research on the placentation of primates also dates back to the 1920s (Wislocki, 1929) with specific studies on *Gorilla* more recent (Ludwig, 1961). Gorillas have a haemochorial placenta: the vascular chorionic processes (villi) are bathed in crypts of maternal tissue.

Both clinician and pathologist need to be familiar with normal fetal and placental morphology and development (Fig. 13.6). Fetal pathology is a specialised subject and the veterinarian presented with a possibly abnormal fetus is well-advised to seek the assistance of a pathologist (veterinary or medical) with experience of such investigations. Useful reference texts are the book edited by Ernst et al. (2011) and the earlier publication (Royal College of Obstetricians and Gynaecologists & Royal College of Pathologists, 2001), about fetal and perinatal pathology.

Some lesions in fetus may only be detectable or decipherable histologically – for example, gonads that cannot be identified grossly, inflammatory changes in the umbilicus, lesions of hypoxia in the brain. Others will be apparent with the naked eye or lens but require microscopical and other tests in order to elucidate the pathology – for instance, hypoplastic adrenal glands.

There is a wealth of unpublished information and often material (see Part II: A Catalogue of Preserved Materials) in the world's zoos and laboratories. This needs proper collation and analysis as it could help elucidate the true prevalence and pathogenesis of perinatal disease in gorillas. For example, data kindly provided by Roberta Wallace, Milwaukee County Zoo (personal communication) includes a fetal death, a mummified abortus, a stillborn animal with umbilical cord around its neck at birth, and neonatal septicaemia in two separate 1-month-old neonates. Specimens held at Taronga Zoo, Australia (see Part II: A Catalogue of Preserved Materials) include material from an autolytic fetus.

A 'stillborn' infant examined by the author (JEC) at Durrell Wildlife Conservation Trust was reported as follows: 'Both hands oedematous. Left arm detached from body, no haemorrhage. Generalised brown discoloration of carcase. Brown discoloration of mucosae. Epidermis sloughs easily when carcase handled. No foul smell. Skull crushed latero-laterally, no haemorrhage associated. Generalised brown discoloration of all viscera. Generalised brown colour, including tongue and mucosae. Intestines thin walled, flaccid and filled with brown fluid. Brown coloured lungs, nonaerated. The tissues show the features of what paediatric pathologists call "macerated stillbirth" – autolysis when death occurs in utero. The dusky discoloration of the carcase described in the postmortem report is typical."

Examination of a fetal gorilla usually is on account of an abortion. It is an area of investigation to which many specialities can contribute – pathologists, clinicians, immunologists and geneticists. In any such studies it must be borne in mind that so-called 'spontaneous abortion' is a relatively common phenomenon in most mammals. The majority occur early in pregnancy and are not detected. The exact mechanism of abortion is uncertain in many cases, both in human and (domesticated) animals. Some may be due to a suboptimal maternal-conceptual hormonal balance, compromising the continuance of pregnancy, while others are likely to be related to such factors as abnormal site of implantation, incomplete cervix or uterine abnormalities.

Many pathological classifications of abortions have been proposed over the years in both human and veterinary medicine. A simple classification suitable for gorillas can be based on two criteria: (1) gross appearance and (2) placental histology. Space does not permit detailed elaboration but broad groupings (based on *Homo* and therefore probably applicable to *Gorilla*) are: (a) so-called 'blighted ova' where the conceptus shows various signs of pathological development such as a stunted embryo, an empty or ruptured sac or evidence of previous embryonic death and (b) macerated fetuses or embryos, the majority of which are anatomically normal and usually at a later stage of gestation than those in Group (a).

Review of unpublished zoo reports by the author has revealed many interesting, but sometimes unexplained, histological findings in fetuses of gorillas. One case that was fairly straightforward showed active haematopoiesis in the red pulp of the spleen and hepatic sinusoids and glomerulogenesis in the outer cortex of the kidney (all normal features), together with oedema and haemorrhage in the subcutaneous tissues and areas of haemorrhage in the thymus. The subcutaneous lesions are considered consistent with a difficult or prolonged birth. There was no gross or histological evidence of any underlying organic disease.

Another example, a near or full-term dead fetal male gorilla at Bristol Zoo Gardens, showed (inter alia) the following features: crown-rump length 14 cm, weight 450 g; large blood clot (80 g) present with the placenta (which weighed 160 g); no gross lesions in the placenta; skin over part of the head covered with tiny fawn-grey plaques; a few fine and coarse facial hairs − remainder of the body hairless; subcutaneous tissues slightly oedematous; lungs have been not inflated. No other abnormalities evident.

Placenta

Despite some interest in morphology (Pijnenborg et al., 2011) (Fig. 13.6), information about placental pathology in gorillas is very sparse. The scattered papers and abstracts include an account of diagnosis and treatment of choriocarcinoma by Cook et al. (1995).

There is useful background information about lesions in the placentae of rhesus macaque monkeys in the survey by Bunton (1986). The findings included subchorionic, perivillous and perilobular fibrin deposition, focal infarction, retroplacental haematoma and calcification. All lesions seen have also been reported in human placentae and are assumed to occur in *Gorilla*.

Zoo reports reviewed by the author suggest that placentae of gorillas are infrequently received or examined by pathologists and investigations are sometimes limited to gross observations. Foci of calcification are occasionally recorded − considered a normal finding in placentae.

The pathological features of fetal and placental lesions of gorillas clearly require further investigation, using standard techniques. It is interesting that some zoos (such as San Diego) have used dedicated submission and examination forms for placentae for many years. Specific mention of the placenta is also a feature of the Great Ape Tag Placental Examination (see Appendix 2: Protocols and Reports).

Neonatal Pathology and Health

'Neonatal' is defined here as 'newborn'. 'Perinatal' is taken to have a broader meaning, to include the embryonic, fetal and neonatal periods.

Birth and hand-rearing have been recounted by Haberle (1973), a zoo gorilla and Riddle et al. (1973), an animal at the Yerkes Regional Primate Research Center. Other reports have described individual problems such as septic abortion in a gorilla due to *Shigella flexneri* (Swenson and McClure, 1974), the neurosurgical and medical care of an infant (Rapley et al., 1982) and repair of an umbilical hernia (Graham-Jones, 2001).

In terms of neonatal postmortem examination, an early account was of a stillborn male in Jersey (Spencer, 1977). This was performed by a medical pathologist and remains an excellent example of how such a neonatal autopsy/necropsy should be performed. So also was the investigation of a baby gorilla that was performed for the City of Belfast Zoological Society in 1983 by medical pathologist Rachael Liebmann (personal communication). This was an interesting case in that there was diffuse fatty material within the air spaces of the lung, associated with moderate histiocyte, and slight neutrophil infiltration. These changes, combined with the evidence of squamous cells lying amongst the lipid, suggested that there had been significant aspiration, and that this was the cause of death. Extrapolating from humans, possible contributory causes might have been asphyxia, immaturity, underlying neurological or metabolic disease and/or congenital defects.

In their survey of gorilla deaths, Benirschke and Adams (1980) included as 'Primary cause of death' amongst fetuses and newborn animals (up to the age of 4 days) abortions, prematurity, abruptio placentae, subdural haematoma and rejection. In the same year Benirschke et al. (1980) produced a more general review of perinatal mortality in zoo animals. A recent, as yet unpublished, dissertation (Hassell, 2014) surveyed pathological causes of mortality and morbidity in infants of free-living mountain gorillas. Postmortem ($n = 112$) and histopathology ($n = 62$) reports for infants under the age of 3.7 years were collated. Trauma was found to be the most common cause of death (47%), followed by respiratory infections and aspiration (15%). Gastrointestinal parasitism (15%), lymphoid disease (suggestive of an EBV-like lymphocryptovirus on histology) (11%) and hepatic capillariasis (9%) accounted for most of the morbidity. All organ weights included in the study (with the exception of the brain) were strongly correlated with either crown-rump length or bodyweight, allowing them to be expressed as a generalised metric of body size for the assessment of infant development (see later).

There is a need for better analysis of neonatal/perinatal problems in *Gorilla*. Greater use could be made of the approach followed in *Homo*. Deaths during that period fall into three categories − those that occur (1) in utero, at parturition, or very soon after, (2) in the subsequent weeks or months, often due to infections or malformations and (3) long-term sequelae to earlier infectious or

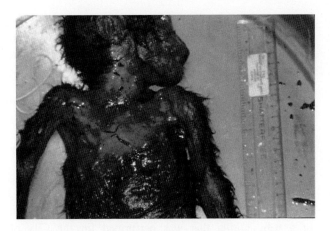

FIGURE 13.7 A dead baby free-living gorilla, finally dropped by its mother, shows hair loss and autolysis but still warrants a full necropsy.

BOX 13.2 Morphometrics – Neonates

Essential measurements:
- body length (crown-rump)
- head circumference

Essential weights:
- whole body
- placenta, if present
- liver
- brain
- thymus
- adrenals

---

noninfectious diseases, including (for example) spina bifida, hydrocephaly or cardiovascular defects.

The first requirement is to ensure that all neonates are submitted for examination, regardless of how autolysed or dismembered they are (Fig. 13.7). This is a particular problem in the wild as the female may continue to carry her baby long after it has died. Lowenstine (1990) drew attention to this when she wrote: 'Three neonates were severely autolyzed but maternal neglect was suspected in one, and aspiration during parturition in another'.

As stressed by Cooper and Cooper (2007), no carcase is 'too decomposed for a postmortem examination'. The remains of a gorilla may be only a liquefied 'soup' or a dry skeleton but these always warrant proper investigation. Nor should histological studies be dismissed, as findings of interest may still be detectable; the author routinely takes sets of tissues from *all* animals. Gorilla material is too valuable to waste.

The second requirement is to standardise the way dead fetuses and neonates are examined. Necropsy should be thorough and there must be a systematic assessment of the maturity of the animal. Recording the weight of a neonate and carrying out morphometrics are both *sine qua non* (see Box 13.2) and should be routine when an infant is either born dead or has to be handled alive for any reason – both in zoological work and in field studies.

The measurements of newborn gorilla infants, of both species, were reported by Schaller (1963), and birth weights by Napier and Napier (1967). Dixson (1981) published bodyweights of captive male and female gorillas from birth until 14 years of age. As far as neonatal anatomy is concerned, the monograph by Diogo et al. (2014) on the musculoskeletal structures, internal organs and skin of the 'baby gorilla', including CT scans, is recommended. Working with free-living mountain gorillas, Fossey (1979) described development in the first 36 months while Bolter and Zihlman (2002) compared growth and development in body tissues and proportions in the lowland gorilla and chimpanzee.

Important data that should be recorded include skin features and texture, the appearance of the genitalia, the shape of the ear (and whether pineal cartilage is present), characteristics of the nipples and the presence and extent of creases on the plantar surface of the feet.

During the postmortem examination of a neonate special attention needs to be paid to the appearance and size of internal organs. For a start, all organs should be weighed. They can provide useful clues as to the degree of maturity of the neonate, both in terms of when during pregnancy an abortion occurred or, if the gestation was full term, whether there was intrauterine growth retardation.

Intrauterine growth retardation produces a 'light-for-dates' (LFD) baby and in human paediatrics is defined as a neonate that is two standard deviations or more below the expected weight for that gestational age. Another term, that is not the same, is 'small for gestational age' which refers to fetuses that are smaller in size than would be normal for the gestational age, most commonly defined as a weight below the 10th percentile for the gestational age. The causes of LFD in humans are variable – physiological/genetic, fetal abnormalities, intrauterine infection and intrauterine malnutrition. LFD babies may die in utero but many are born alive, in which case they may exhibit immediate clinical signs, such as respiratory distress or hypoglycaemic fits.

Useful background to lesions in immature NHPs is to be found in the series on simian neonatology (Anver et al., 1973).

Forty years ago it was established experimentally that macaque monkeys could develop fetal and postnatal growth retardation if infected with certain viruses but the small number of neonatal gorillas that have been available for study means that little is known about intrauterine growth retardation in this genus. Nevertheless, awareness that newborn babies may exhibit a similar pattern to that seen in *Homo* is essential. Veterinarians need to be vigilant and, if LFD is suspected, ensure that relevant clinical and postmortem data are collected.

Orphans and Hand-Rearing

Orphans and hand-rearing are important issues, both in the wild and in captivity, and raise both legal and ethical dilemmas (see Chapter 18: Legal Considerations).

Infant gorillas in zoos attract public, as well as scientific, interest. As a result, there are 'popular' accounts of their rearing (e.g., Verrengia, 1997) as well as properly planned and executed studies of their growth and development.

Although sometimes free-living gorillas become separated from their mothers, solitary infants in the wild are almost invariably a product of poaching. It is estimated that each live orphan gorilla found in the markets of Congo-Brazzaville (CR) represents another 15−20 gorillas that will have been killed (consumed as bushmeat) while recovering the infant.

Sometimes a baby is found alive attached to its mother's dead body. Such babies are usually weak and often do not survive − 'Mutesi' for example (found in the Virungas in May 2010) − where the mother was thought to be a victim of extreme cold and rainy conditions on Mount Karisimbi (see Chapter 4: Noninfectious Disease and Host Responses), where the group was ranging. When such mortalities occur, even though the proximate cause of death of the infant may appear straightforward, detailed necropsies are essential.

Orphaned young gorillas present many challenges for the attending veterinarian, even in the environment of a modern zoo. In Africa there are often added complications such as dehydration, electrolyte imbalance and malnutrition. In both locations the pathologist has a role to play in diagnostic techniques and, if the animal dies, in elucidating the sequence of events. Many postmortem findings, gross and microscopical, reflect secondary changes rather than the initial cause of decline.

The question of hand-rearing baby gorillas has long attracted attention and public interest. The effects of such a practice on behavioural development were discussed by Bowen (1980) and Meder (1989). Maternal deprivation may have a detrimental impact on sexual behaviour, reproduction and nurturing (Ryan et al., 2002). Human imprinting can create psychological scarring (Cousins, 2015).

Reintroduction of hand-reared gorillas introduces new complexities and cannot be covered here (but see Chapter 17: Field Studies in Pathology and Health Monitoring).

Infanticide

See also Chapter 4: Noninfectious Disease and Host Responses.

Infanticide, defined here as the killing of a young animal by one of its own species, has been documented in many primate species (van Schaik and Kappeler, 1997). A useful review of this trait in gorillas was by Meder (1994), in her discussion of 'injuries and fatalities caused by conspecifics'. She wrote that 'at least 37% of all mountain gorillas in Rwanda which die within their first 3 years of life succumb to injuries caused by infanticide. In general, silverbacks are responsible for these infanticides, only rarely blackback males, and according to the observations which have been recorded so far, females never kill infants'.

Much has been written about infanticide in gorillas. Some of the research has noted the similarities to (human) child abuse (Nadler, 1983). Watts (1989) discussed infanticide in mountain gorillas and presented a 'reconsideration of the evidence'. He presented data on six cases of infanticide, one infanticide attempt and one suspected infanticide. These and previously reported instances in the same population were analysed. Watts related the circumstances and consequences of infanticide and argued that his findings support the sexual-selection hypothesis for the evolution of the trait of infanticide. Recently Yamagiwa et al. (2009) discussed infanticide and social flexibility in the genus *Gorilla*.

It cannot always be assumed that deaths in infant gorillas are due to infanticide, even if a male animal is in the vicinity. An instance of where premature or ill-informed statements were unhelpful was in 2014 when a family was found dead in the Aspinall Foundation's reserve in Gabon (Jane Hopper, personal communication). The British 'Daily Mail' carried an article claiming that the youngsters had been killed by a solitary male, 'Boumanga'. In fact, while Boumanga may have played a role in the deaths, he probably did not actually kill them. The youngsters did not appear to have wounds when found (although some bodies were decomposed, making detailed investigation difficult) and the veterinary opinion was that Boumanga caused a great deal of stress that may have led to infectious disease.

The instance above emphasises the need for a greater input into the study of infant deaths by veterinary pathologists. We still know so little about the sort of wounds that gorillas inflict upon one another or how these might be accurately differentiated from injuries caused by other animals or humans. In contrast to the situation in human paediatrics, we do not have categories of injuries or standard methods of recording and grading them. These are reasons why trauma is given so much prominence in Chapter 4: Noninfectious Disease and Host Responses and fractures in Chapter 14: Musculoskeletal System.

What can be interpreted as infanticide occurs sometimes in captivity too and has been the subject of study, again from a behavioural, not pathological, angle (see Enciso et al., 1999). Incidents are a cause of great concern to both the public and to zoo staff as, for example, at the

London Zoo in 2011 when the 7-month-old 'Tiny' died after suffering a broken arm and suspected internal injuries inflicted by a new silverback male 'Kesho'.

Developmental Abnormalities

These are discussed in detail in Chapter 8: Nonspecific Pathology (Fig. 13.8).

Bolton et al. (2012) provided useful background information on, and references to, the pathology of reproductive tract lesions in NHPs, including congenital abnormalities.

Female Reproductive Pathology

There are two aspects of female reproductive pathology: (1) that related to the animal's physiology (oestrus, cycling and pregnancy) and associated behavioural changes, and (2) the situation when a female gorilla has an infectious or noninfectious disease affecting her genital system.

As far as (1) is concerned, much has been published. In addition to papers cited earlier in this chapter, see Lasley et al. (1982), O'Grady et al. (1982), Böer (1983), Bahr et al. (2001) and Breuer et al. (2009).

Some techniques used to investigate infertility are also important tools in diagnosis — laparoscopy, for instance, pioneered in gorillas 35 years ago (Wildt et al., 1982), and hysterosalpingography (Ramsey et al., 1985).

Ultrasonographic assessment of reproductive diseases in captive gorillas and other great apes was the subject of a dissertation by Julia Braga Morais (2012). The work focused on the analysis of reproductive sonograms (which are beautifully reproduced) from 29 male and female captive great apes and was performed under the supervision of the Leibniz-Institut für Zoo-und Wildtierforschung (IZW), Berlin, Germany, reproduction management group (see Part II: A Catalogue of Preserved Materials). Of 22 female captive animals, 18 were diagnosed with reproductive tract lesions. Suggested diagnoses in gorillas included acute pelvic inflammatory disease, uterine leiomyomas, ovarian cysts, endometriosis, hydrosalpinges, cervical tumour and a paraovarian cyst.

Diseases affecting the female reproductive system do not feature prominently in surveys by (for example) Benirschke and Adams (1980) but Kaplan (1979) provided useful background on intrauterine infections in other NHPs. Individual reports over the years have included a description of endometriosis (Doré and Lagace, 1985) and a range of neoplasms — a unilateral ovarian adenocarcinoma (Huntress et al., 1988), metastatic endocervical adenocarcinoma (Olias et al., 2012), uterine leiomyoma (Mylniczenko et al., 2008) and uterine leiomyosarcoma (Burns, 2006). Individual, usually unpublished, reports sometimes include leiomyomas (Fig. 13.9); for example, Lydia Tong (personal communication) reported from Taronga, Australia, that gorilla 'Betsy' had two uterine leiomyomas in addition to a range of other lesions.

Surgical interventions have included various caesareans, obstetric management of protracted labour (Cole, 2000), hysterectomy (Fullerton et al., 1989) and operations on reproductive neoplasms (Stringer et al., 2016). Eclampsia is recognised in captive gorillas. A recent report from Bristol Zoo Gardens (Rowena Killick, personal communication) relates to an 11-year-old female lowland gorilla 'Kera' who 'was noticed to be "off colour" and lethargic' for 3 days approaching her parturition due date. Free-catch urine samples collected by keepers

FIGURE 13.8 Careful examination of a hand of a dead neonate in order to check for syndactyly or other digital abnormalities.

FIGURE 13.9 A uterine leiomyoma diagnosed clinically and on biopsy in a captive lowland gorilla.

showed increasing levels of protein so preeclampsia was suspected. Under anaesthesia, blood tests and an ultrasound scan were performed, and a urine sample obtained via catheterisation was tested. These all indicated preeclampsia as the most likely cause. An emergency caesarean was performed, with the help of human obstetricians, and the female baby was delivered alive but unresponsive. The baby required 2−3 hours of emergency treatment/resuscitation, which was probably at least partly due to receiving some of the ketamine given to the mother. The baby had a lot of fluid in her lungs, some of which she coughed up during the emergency treatment, and she was nebulised and treated with antibiotics for about a week after birth, but made a full recovery and has generally done well since. Kera's recovery from the caesarean has been long and difficult'.

The first report of possible eclampsia in a free-living gorilla was perhaps that by Cranfield (2001) on 'Flossie': 'pregnant (near-term 1.5-kg male fetus), acute heart failure, preeclampsia? (unfortunately fetus and placenta were lost)'.

The type of diseases and surgical interventions listed above are but the tip of the iceberg; in most instances the cases were published because they were dramatic examples and/or involved collaboration between the veterinary and medical professions. Of far more interest and significance are probably the less spectacular, but more common, conditions that feature regularly in records from zoos and field projects but which are not considered by some to warrant a scientific paper.

Such 'common' conditions include, for example, vaginitis. Sampling and diagnosis in such cases (usually live animals) requires care, especially in the field. Vaginal swabs may or may not yield significant organisms and may or may not aid diagnosis. In the author's experience, working with *beringei* in the field, mixed growths are obtained, including such organisms as *Citrobacter diversus*, assumed to be a commensal.

Foreign bodies are not uncommon in the vagina, sometimes (in captivity) inserted by the animal itself. For example, lowland gorilla 'Killakilla': examination with a speculum revealed pieces of straw in anterior vagina. These were removed with some difficulty. The vaginal mucosa was thickened and affected by a few small focal ulcers. Some straw and other foreign material was found in the vagina. Further dissection showed multiple purulent foci deep in the vaginal wall. Removal of most abdominal viscera revealed an extensive area of chronically inflamed and fibrosed soft tissue running along all the dorsal abdominal wall to the left of midline and involving the left abdominal wall itself. This continuous mass was very firm and cream-coloured and contained numerous interconnecting sinus tracts filled with creamy pus. 'A few focal large abscessated areas were also located in this

mass. Muscles and connective tissue elements were thickened and fibrosed and contained several pus-filled sinus tracts. These signs are consistent with an extensive purulent infection causing chronic and ongoing soft tissue lesions in the abdominal cavity and left leg. The aetiology is likely to be traumatic perforation of the rectal wall (possibly associated with 'self-inflicted' foreign body insertion) with consequent bacterial infection, peritonitis, sinus and fistulae formation. The left kidney shows marked chronic interstitial lesions with fibrosis. In retrospect I consider the vaginitis to have been the likely primary lesions with extension of infection into the pelvic and abdominal cavities, etc.'

Interpretation of findings in female reproductive disease is not easy. Standard veterinary texts are rarely relevant. An excellent book on human gynaecological pathology, applicable in part at least to gorillas, is by Mutter and Prat (2014).

In discussion of reproductive health, in any wild species, one should be aware of the Reproductive Health Surveillance Program (RSHP) (Moresco and Agnew, 2013), based in the United States. This collects, processes, evaluates and stores reproductive tracts from nondomesticated mammals submitted by zoos. The Program's archive has been used to document naturally occurring lesions and to detect those associated with exposure to certain contraceptives (see later). The RSHP requests that the whole reproductive tract, fixed in 10% formalin, is submitted. Submission forms and sampling protocols are available. The published paper by Moresco and Agnew includes an illustration of a cross-section of the ovaries and uterus of a lowland gorilla.

Male Reproductive Pathology

As in the female gorilla, male reproductive pathology is in part related to the animal's sexual activity and in part to infectious or noninfectious disease affecting the genitalia.

As far as the former is concerned, much has been published about fertility in the male, especially semen collection and examination (Brown, 1998; Seager et al., 1982; Platz et al., 1980; O'Brien et al., 2005) but also using other investigative methods (Schaffer et al., 1981; Breuer et al., 2010).

This low spermatogenic capacity of the male gorilla was mentioned earlier. Apparent testicular atrophy has interested research workers for some years. Wislocki (1942) published data on the size, weight and histology of the testes of *Gorilla*. A little over a decade later Steiner et al. (1955) described testicular atrophy in conjunction with neuropathy, cardiopathy and haemosiderosis in a captive lowland gorilla, which was followed by an account of testicular atrophy and pulmonary embolus by

Antonius et al. in 1971. Testicular atrophy in captive gorillas was discussed by Dixson et al. (1980) and it and Dixson's (1981) book included a plate demonstrating the microscopic features of such atrophic changes.

Dixson's work (1981) is worthy of more comment. He gave details of three captive male gorillas in which testicular atrophy had been detected postmortem. A lowland gorilla 'Oban' at the Yerkes Regional Primate Center (YRPRC) died of gastroenteritis and its testes were 'severely atrophic', with sloughing and degeneration of the germinal epithelium; mature spermatozoa could not be identified in either the seminiferous tubules or epididymes. A similar picture was seen in the testes of lowland gorilla 'Jojo' who died of pneumonia at Chester Zoo (UK). London Zoo's famous gorilla 'Guy' (see references to this animal elsewhere in this book and Appendix 5: Case Studies — Museums and Zoological Collections) had 'partially atrophic' testes. Dixson's description reads '... the testes were extremely small, measuring 27×18 mm (right) and 28×19 mm (left). The corresponding weights including the atrophic epididymis were only 6.85 g and 4.4 g, respectively. The entire genital tract of this male was removed, yet it was not possible to identify with certainty either the prostate or seminal vesicles'. Histologically the testes exhibited a diffuse nodular appearance, the seminiferous tubules were totally degenerate and fibrosed and none of the stages of spermatogenesis could be identified. Electron microscopy revealed that the Leydig cells, which normally produce testosterone, were nonfunctional.

A further word should be said about Dixson's first gorilla, 'Oban'. His seminiferous tubules were small and surrounded by large amounts of interstitial tissue. However, this is not necessarily a sign of pathology as both Wislocki (1942) and others have described these features in free-living (wild-shot) gorillas.

Studies on testicular morphology, with particular reference to spermatogenesis, have continued in recent years (Enomoto et al., 2004; Fujii-Hanamoto et al., 2011) but the causes of testicular atrophy remain unclear. Diet, infectious disease or psychological factors have been suggested, the last of these perhaps associated with husbandry practices.

Review by the author (JEC) of histological sections from male gorillas in British zoos has confirmed testicular changes in a number of cases. One of a series read at the University of Bristol showed much of the testis replaced by pale hyaline connective tissue — almost suggestive of prolonged ischaemia. In this same animal many small blood vessels had thickened eosinophilic walls due to amyloid deposition and in some there was thrombosis.

Other testicular pathology recounted in male gorillas has included neoplasms. In a survey of neoplasms at the Zoological Society of London, Wadsworth et al. (1985) reported an interstitial cell adenoma while Karesh et al. (1988) described a Leydig cell tumour. In her dissertation study Morais (2012) detected with ultrasonography three male gorillas with testicular lesions; all were suspicious of malignancy but only one was confirmed — a Leydig cell tumour.

Fixation of the testes may be best in modified Davidson's fixative as this preserves morphological features and tubule shrinkage is minimal (Latendresse et al., 2002).

Birth Control and Contraception

Breeding gorillas in captivity is not straightforward. It requires patience, expertise and an intuition that comes from working with these creatures for some years. Sometimes breeding success is low and warrants investigation of the animals by clinicians and pathologists, animal behaviourists and experienced zoo staff. Such a decline at Jersey Zoo/DWCT/Durrell prompted a request via the European Endangered Species Programmes for the transfer of one female animal to a breeding group in Germany and the male to a zoo in France. In return Durrell received a 12-year-old genetically important female from France.

In contrast to this, sometimes breeding in captivity needs to be controlled and contraceptive methods (in addition to changes in management) are warranted. There are health implications and the constant concern that chemical contraceptives may be associated with pathological changes — see earlier reference to reproductive health surveillance (Moresco and Agnew, 2013). Such worries have prompted study — for instance by Bolton et al. (2012) who investigated whether captive female great apes might experience a high incidence of lesions of the reproductive tract as an effect of contraception.

Another method of birth control that has been tried in captive gorillas in respect of the problem of surplus male animals is castration. This, however, is highly contentious. Cousins (2015) was particularly scathing, pointing out that it might appear to be a convenient and inexpensive solution while we still have populations of free-living apes but, in the event of catastrophic losses, potential breeding males could be very important. The controversy prompted a survey of castration by PASA and, at the time of writing (2015), a (totally separate) website https://www.change.org/p/eaza-eep-stop-castrating-gorillas that suggests that castrations are not necessary and can be avoided by various measures, including strict breeding restrictions, establishing all-male groups in more zoos and sending young males to be with lone gorillas. There is considerable interest in the whole question of nonbreeding/bachelor groups (Levréro et al., 2006).

Chapter 14

Musculoskeletal System

John E. Cooper

I warmly commend the physician who makes small mistakes: infallibility is rarely to be seen.

Hippocrates.

INTRODUCTION

This chapter is broad-based, discussing the pathology of those structures that contribute to locomotion — mainly bone, cartilage, connective tissues and muscle but some other tissues also (McGee et al., 1992). Nerves too are a component but these are primarily covered in Chapter 15: Nervous System and Special Senses.

Embryologically, the muscles, connective tissue, vertebrae and some bones are paraxial mesodermal derivatives while the limbs are derived from the parietal layer of mesoderm.

The first detailed scientific description of the anatomy of these structures is to be found in the chapter by Raven and Hill in Gregory (1950). More recently Gebo (2014) provided useful information (with an orientation towards adaptation for locomotor activity) about joints and muscles and their biology in nonhuman primates (NHPs). A series of monographs by Diogo and colleagues describe musculosketelal structures of gorillas, chimpanzees, hylobatids and orangutans (see, e.g., Diogo et al., 2011).

Scientific papers of relevance include the early work on skeletal development and variability in *Gorilla* by Randall (1943), the comparative study of a gorilla hind limb by Sakka et al. (1984) and recent research on ontogenetic changes in limb bone proportions in mountain gorillas by Ruff et al. (2013).

MOVEMENT DISORDERS

This section might alternatively be entitled 'locomotor pathology' as it is concerned with the pathogenesis of locomotor disorders. The subject has exercised comparative pathologists for many years; see, for example, Abee et al. (2012) which includes a comprehensive chapter on arthritis, muscles, adipose tissue and bone diseases of NHPs.

Pathological changes that affect the musculoskeletal system usually cause changes in gait and/or movement of body parts. A useful jingle suggested by Brian Livingstone (personal communication) is 'pain, limp, swelling, stiffness and deformity'. *Morphological* changes precipitate *functional* changes.

In order to analyse fully the pathogenesis of movement disorders, one needs to understand the locomotor biology of gorillas. These animals are quadrupedal and for much of the time move in a stooped position using knuckles to support part of their weight. This characteristic 'knuckle walking' means that their back and spine are held almost horizontal to the ground.

The use of correct terminology is fundamental to the precise analysis of gait and thereby the detection and diagnosis of locomotor disorders. As Birch et al. (2015) pointed out, the language used in human gait analysis is a mixture of clinical medicine and biomechanics. They stressed that a lack of consistency in terminology and definitions can lead to confusion. For example, 'gait' and 'walking' are not synonymous, nor are 'step' and 'stride'.

The detection and grading of lameness is well established in human and veterinary medicine (especially relating to horses and dogs); similar methods may be needed for great apes.

Numerous studies have been carried out on activity patterns and exercise in both free-living and captive gorillas — see, for example, Meder (1986). Changes in physical activity in the context of analysing pathology are discussed in Chapter 8, Nonspecific Pathology.

Locomotor disorders have long been recognised in gorillas and there were early reports of surgical interventions, such as the abstract by Ott-Joslin et al. (1987) recounting bilateral total hip replacement in a 26-year-old lowland gorilla.

J. E. Cooper & G. Hull: Gorilla Pathology and Health. DOI: http://dx.doi.org/10.1016/B978-0-12-802039-5.00014-7

THE PATHOLOGY OF BONE

There are some diseases of bone per se — inflammatory conditions, such as osteitis, fractures (see later), primary osteoporosis, and benign and malignant tumours — but secondary lesions occur frequently and the skeleton provides important clues to metabolic and other systemic disorders.

Bone shows only a limited repertoire of responses. Osseous tissue may be added (new bone growth), removed (causing porosity or thinning of cortices), become deformed (for instance, on account of nutritional disease) or become deformable (poor mineralisation). It is important to differentiate between osteolytic activity when it is thinning the cortex (e.g., in response to a change in the local biomechanical environment), and when cortical morphology remains unchanged, but bone is removed at the histological level (e.g., in osteoporosis).

The periosteum is a key feature of bone pathology. It is well innervated and is sensitive to insults and there are implications in terms of pain and welfare (see Chapter 8: Nonspecific Pathology). Periosteum responds to insults by intense osteoblastic activity. This may be beneficial, as in callus formation, or deleterious, unnecessary bone.

There are various terms used by medical and veterinary clinicians and pathologists to describe new bone deposits. These are not always the same as those employed by nonmedically trained scientists who examine (mainly dead) bone.

An *osteophyte* is new bone forming at a joint margin, typically in noninflammatory degenerative osteoarthrosis ('osteoarthritis'– see below).

An *exostosis* is, in humans and increasingly in domesticated animals, taken to mean a hamartoma (a benign, focal, tumour-like, proliferation of cells) arising from the epiphyseal cartilage plate, usually in a long bone. As the long bone continues to grow, the exostosis is 'left behind' attached to the shaft. It stops growing at skeletal maturity. Solitary exostoses are, microscopically, normal bone. Other examples of new bone formation, such as olecranon 'spurs' and exaggerated muscle attachment marks (see report on 'Guy' in Appendix 5: Case Studies — Museums and Zoological Collections) are best considered examples of *hyperostosis* although the term 'exostosis' or even 'osteophyte' is still sometimes erroneously used for these.

Trauma to the periosteum can damage it beyond repair in severe fractures, to the extent that such a fracture will not heal. More often, it results in new bone formation or *callus* which bridges the fragments and unites the fracture. This may remodel so well that in time the fracture site may be undetectable. More often some swelling and deformity will remain indefinitely.

A subperiosteal haematoma, formed without fracture, may ossify and remain as a permanent but asymptomatic swelling. Hyperostoses or 'spurs' at muscle attachments (see earlier) are possibly due to the traction on the periosteum but a constitutional tendency to new bone formation ('a bone-forming diathesis') is probably also necessary because otherwise such spurs would occur in almost all individuals. In humans, at least, this diathesis may manifest as 'diffuse idiopathic skeletal hyperostosis' (DISH), with new bone forming at multiple sites. DISH may occur in gorillas as well; work is in progress to confirm or refute this (Livingstone et al., 2016; personal communication).

Weston (2008), in a study of human long bones that showed periosteal new bone formation, used macroscopic and radiographic techniques to investigate lesions. The results indicated no qualitative or quantitative characteristics of the periosteal reactions that could be considered specific to certain diseases. The *progression* of a disease, rather than its type, was the most important determinant of the appearance of a periosteal lesion. A critical analysis by Weston of the bioarchaeological literature revealed that the varied pathogeneses of periosteal new bone formation have been largely ignored. Weston warned that assumptions regarding the infectious aetiology of periosteal lesions 'risk skewing the results of palaeoepidemiological studies' — a salutary warning for those working on gorillas.

Which lesions or specific diseases are recognisable in bones? The answer is 'a good number' but, as cautioned above, they need careful investigation before rushing into a diagnosis. Caution is essential.

Classification of skeletal lesions depends to some extent on the discipline that is carrying out the investigation. Pathologists (medical, dental and veterinary) use well-established terminology based essentially on aetiology and pathogenesis — in the case of the author (JEC), under the following five main headings:

- Developmental/congenital
- Physical
- Inflammatory (infectious and noninfectious)
- Neoplastic
- Other

This, though, is a rather different approach from that commonly employed by those who, while versed in osteology, do not have training in pathology. Thus, osteoarchaeologists tend to group 'diseases' seen in bones and teeth (the areas in which they mainly work) in broad categories, usually including the following:

- Arthropathies: joint disease such as osteoarthritis, due to age deterioration and activity; systemic disease affecting joint
- Tooth decay and loss, disease of the gums and supporting bone

- 'Stress indicators' (see later): considered to be signs of dietary deficiency or parasite infestation
- Changes due to activity and gait ('occupational changes'): changes in muscle markings and bone shape because of habitual movements
- Infection: specific (named diseases) or nonspecific (localised infections)
- Trauma: accidents, injuries due to interpersonal violence, surgery
- Epigenetic ('nonmetric') traits: small variants in bone features that have no clinical significance but can indicate relatedness
- Congenital and developmental disorders; changes in bone form occurring at the embryonic/fetal stage of development; metabolic diseases
- Neoplasia

The same applies to some areas of palaeontology. Terminology and categorisation of lesions used by palaeopathologists do not necessarily equate with those employed in diagnostic medicine. For example, 'arrested growth of long bones' (the Harris lines seen on radiographs) — and enamel hypoplasia of the teeth are regularly attributed by some to 'stress' without defining the term (see Chapter 16: Endocrinological and Associated Conditions). Rather loose categories of 'general disease' may also be used, as in the following examples:

- Anomaly
- Trauma — repairs
- Inflammatory/immune
- Circulatory (vascular)
- Metabolic
- Neuromechanical
- Neoplastic

Such divergences in terminology and approach are unfortunate and can cause confusion and scientific error. It can sometimes have repercussions insofar as work on great apes is concerned.

As was explained in Chapter 1, The genus *Gorilla* — Morphology, Anatomy and the Path to Pathology, over the past 150 years many studies on gorillas, by scientists from different disciplines, have been orientated towards attempting to elucidate evolutionary similarities and differences between the extant 'great apes', modern humans and fossil hominids. This continues to be the case — the recent paper by Joganic (2016), for example, detailing skeletal and dental development in a subadult lowland gorilla. It is clearly important that methods used and inferences drawn from such studies run in parallel with research by veterinarians and comparative pathologists. There is already some 'drift'; in her otherwise excellent study on the pathology of great apes, based on museum studies and referred to earlier, Lovell (1990a,b) used

terms for lesions that do not always correspond with current veterinary or (human) medical descriptions.

As Wood et al. (1992) pointed out, inferring prehistoric health from skeletal samples is an 'osteological paradox'. Frequently palaeontologists are trying to unravel complex interactions — such as the combined role of biology, genes and environment — in affecting human health (Weiss, 2015). Palaeontologists and archaeologists are often particularly interested in diet and nutritional disease in earlier societies. To support some of their deductions they regularly rely upon such radiological features as Harris lines (growth arrest lines) (see earlier). These are zones of increased bone density that represent the position of the growth plate at the time of an assumed 'insult'. There is not always a consensus over this, however; some researchers consider that Harris lines indicate normal growth spurts, not a period of malnutrition (Papageorgopoulou et al., 2011). In human clinical practice the radiological finding of a Harris line is not usually regarded as being of any real consequence (Brian Livingstone, personal communication).

Some palaeopathologists also lay weight on porotic hyperostosis (PH) (also known variously as osteoporosis symmetrica, cribracrani, hyperostosis spongiosa, and symmetrical osteoporosis), a change characterised by areas of porous bone in the cranial vault. The presence of PH is often interpreted as being indicative of anaemia but this again is challenged by some (Walker et al., 2009).

Writing about archaeology and domesticated animal disease, O'Connor (2000) discussed the gap that exists between (most) veterinarians and palaeopathologists and in his book provided a useful chapter on sickness and injury in animals, presenting material from both veterinary and zoo archaeological sources. O'Connor emphasised the need for closer collaboration if animal palaeopathology is to advance.

Notwithstanding the points made above, in skeletal pathology studies collaboration with those from other disciplines can often be mutually beneficial (Cooper and Cooper, 2007). Archaeologists, palaeontologists, anthropologists and others have for long been looking at bones of both humans and animals and as a result developed many skills and techniques that have applications in postmortem work on gorillas (see, for instance, Cox and Mays, 2000). The studies on 'Guy', summarised in Appendix 5: Case Studies — Museums and Zoological Collections, are an example of such interdisciplinary cooperation.

THE IMPORTANCE OF THE SKELETON

The investigation of the skeleton can reveal important information about antemortem health, including the detection of long-term physical and infectious insults. However, examination of bones has often in the past been neglected in routine post mortem work (Cooper and Cooper, 2008).

Skeletal examination is, however, now increasingly a feature of studies on wildlife disease and pathology. It can involve both radiographic investigation and the preparation for gross and microscopic examination of whole skeletons or selected bones (see Appendix CA2: Retrieval, Preparation, Preservation and Storage of Skeletal and Other Material).

Cooper and West (1988) described the use of routine radiography in the detection of traumatic and metabolic disease in endangered (Mascarene) species. Following their studies on the endangered Florida panther, Duckler and Van Valkenburgh (1998) declared 'the study of osteopathologies is a new tool for the conservation biologist'. The applicability of skeletal investigations to different taxa was well illustrated by the work of Brandwood et al. (1986) who examined the bones of birds and primates and the shells of molluscs and compared the incidence of naturally occurring fractures in these markedly different species.

The examination of skeletal remains of NHPs for evidence of pathological changes first attracted serious scientific attention as a result of the work of Schultz (1937, 1939) who published numerous reports of his findings in skeletons and teeth. Later studies threw more light on certain lesions (Lovell 1990a,b) — see later — and helped medical scientists to understand better the evolution and

pathogenesis of certain human diseases (Jurmain, 2000; Rothschild and Woods, 1992).

Lovell's (1990a,b) work is significant because it was the first proper survey of skeletal disease in great apes since the work on the skeletons and teeth of NHPs by Schultz (1937, 1939 and later).

Lovell's study was carried out at the Smithsonian Institution, United States, which has one of the largest collections of wild-shot 'great ape' skeletal material in North America. Some of the animals she examined had died of 'natural causes', others had been killed by poachers or had succumbed after becoming ensnared in traps (see Chapter 3: Infectious Disease and Host Responses). Although certain animals had been found shortly after death, others were located years later and were in relatively poor condition as a result of carnivore damage and post mortem change (see Appendix CA2: Retrieval, Preparation, Preservation and Storage of Skeletal and Other Material).

During the same period, a study of skeletal material from gorillas had started in Rwanda and then, after the outbreak of fighting there, extended to other countries in Africa and Europe. Cooper and Cooper (2008) described this and pointed out that there are very many bones of great apes available for investigation in the world's museums. These are often from animals that died, or were killed, decades ago and therefore they provide valuable reference ('baseline') data on the species, as well as opening up possibilities for DNA and other molecular studies. Modern imaging techniques such as computed tomography (CT) scanning offer new opportunities for research (Fig. 14.1).

SKULL

Skulls predominate in museum collections — see Part II, A Catalogue of Preserved Materials. Some include both the cranium (calvarium and facial bones) and the mandible (referred to as 'skull' in Part II: A Catalogue of Preserved Materials); others lack the mandible.

Skulls/crania have played a large part in osteological, palaeontological, pathological and odontological studies in gorillas. Osteometry (usually a series of cranial measurements) is important in taxonomic and diagnostic research (see studies on 'Guy' in Appendix 5: Case Studies — Postmortem Investigations). Geometric morphometrics has been employed extensively in primate studies, especially of the cranium and mandible.

Some specific pathology relating to the skull is discussed later.

VERTEBRAL COLUMN

The number of vertebrae in NHPs remains fairly constant within any one species but variations occur between

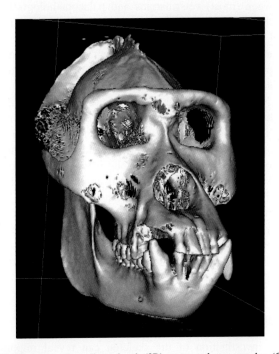

FIGURE 14.1 Three-dimensional (3D) computed tomography (CT) scan of skull of museum specimen (old silverback lowland gorilla).

species. The latter is exemplified by the lumbar vertebrae, where variation is connected to locomotion. Species with vertical, suspensory, postures have relatively short lumbar columns. Great apes generally have only four lumbar vertebrae while quadrupedal primates, most monkeys and prosimians, have a greater number. Napier and Napier (1967) provided basic data on the vertebrae of *Gorilla*. Cervical: 7 (3rd–7th spines very stout, nonbifid). Thoracic: 13. Lumbar: 3–4 ('commonly 17 thoracolumbar in lowland form and 16 in highland'). Sacral: 5–6.

Occasionally, in the author's experience, a gorilla has 12, not 13, thoracic vertebrae. Interestingly, a lumbar rib in humans is sometimes referred to as a 'gorilla rib'.

Vertebral diseases occur in gorillas and are discussed in detail later. Degenerative disc disease appears to be rare, probably because (1) their intervertebral discs are not subjected to as much downward force as is the bipedal human vertebral spine, and (2) their spinal musculature is greater than that in humans; this may reduce the pressure on the vertebrae and intervertebral discs but it could be argued that greater musculature would increase pressure across the joints.

However, Aryan et al. (2006) reported surgical removal of a lumbar disc from a human-habituated mountain gorilla. Spinal degenerative joint disease ('spondylosis') has been described in some species of primates (apparently not gorillas). Swales and Nystrom (2015) advocated a standardised approach to recording such cases.

JOINTS

The joints of *Gorilla* closely resemble those of *Homo* in structure and appearance and have attracted considerable attention by both clinicians and pathologists. Some of the 'diagnoses' made on the basis of gross examination alone of joints in osteological collections, often by scientists without human or veterinary medical training, were alluded to earlier.

Specific pathology relating to the joints is discussed later.

INVESTIGATION OF BONES

This section should be read in conjunction with the descriptions of the studies on 'Guy' at the NHM and of the three Osman Hill skeletons at the RCS (see Appendix 5: Case studies – Postmortem Investigations).

Box 14.1 shows what the investigation of skeletal material from gorillas encompasses.

As the Box indicates, investigation of skeletal disease may involve live gorillas or dead gorillas but the emphasis in the following section is on the latter, with particular reference to examination of assemblages of bones from

---

BOX 14.1 The Investigation of Skeletal Material From Gorillas

- Live animals
  - Clinical examination
  - Laboratory investigations including cytology and histology (bone biopsies)
  - Radiography, other imaging, etc.
- Dead animals
  - Preparation of bones
  - Gross examination
  - Radiography, other imaging
  - Laboratory investigations including histology
  - Ashing, etc.

---

museums or exhumed following burial (see Appendix CA2: Retrieval, Preparation, Preservation and Storage of Skeletal and Other Material). The latter also describes the preparation of bones for study.

Examination of skeletal material must be systematic and thorough, where appropriate (bearing in mind that the specimen is from an endangered species) following forensic procedures appropriate to wildlife (see Chapter 18: Legal Considerations) (Cooper and Cooper, 2013).

Basic Rules

1. Prepare equipment before starting to examine the bones. Check agreed protocol and explain it carefully to any assistants or new colleagues.
2. Have protective clothing available, even if (e.g., goggles) is not all immediately needed. Remind all present of the importance of health and safety and check the Risk Assessments. Always wear gloves (a) when the skeletal material is uncleaned or if obvious lesions are possible, and (b) in Africa – even if the gloves have to be shared (see Appendix 4: Hazards, Including Zoonoses)! In Africa special care should be taken, especially with unprepared ('dirty') bones, because of the risk of zoonoses (see Appendix CA2: Retrieval, Preparation, Preservation and Storage of Skeletal and Other Material; Appendix 4: Hazards, Including Zoonoses; and Chapter 18: Legal Considerations).
3. Handle bones carefully. They can be damaged easily (see later). Hold them over a table top or bench, preferably with two hands. The wearing of gloves helps to minimise trauma as well as reducing the risk of spread of pathogens, but they reduce dexterity. (see below). If forceps are used, pad them or employ those made of plastic; always use such forceps when handling metallic objects, such as bullets, that may form part of a forensic investigation (Cooper and Cooper, 2007).

4. Be particularly careful when moving or transporting bones, even over a short distance. Carry them on a tray or in a box. Wrap bones individually in cloth or soft paper (not newspaper – see Appendix CA1: Use of Collections and Handling of Biological Material) to reduce damage. If soft tissues are present, place them in jars or plastic bags to prevent or delay desiccation. When transporting bones by road, follow the advice given by NatSCA – 'choose flat, well-made roads where possible, go really slowly and avoid the road rage behind you'. This is very sound advice for countries such as the United Kingdom but in Africa, where many routes are on dirt tracks, can be totally impracticable.

5. If the bones are dirty, clean them with care, using a damp cloth or a soft brush. Do not apply hot water and do not scrub bone surfaces, especially of joints. Record on the record sheet if the bones have been cleaned. Remember that dirty bones are likely to present more of a health hazard than dry bones. Differences in colour of bones, even from the same animal, may be due to methods of preparation (see below and Appendix CA2: Retrieval, Preparation and Storage of Skeletal and Other Material) and/or their history – for instance, whether they were exposed to certain types of soil or vegetation.

6. When material is not properly labelled (see later), work on one skeleton at a time, so as to avoid mixing and possible transposition of bones. Ideally, lay out the bones in a skeletal arrangement on a table.

7. Use standard record sheets for your work, to avoid interobserver bias and to facilitate later analysis. Use clear, correct, unambiguous language – and be consistent in what you say. Write dates in full – for example, 11 September 2009, not '11/9/09' or '9/11/09'. When possible, supplement written records with photographs, drawings and tape recordings. The person compiling such records should be 'clean'. Ensure that the photographs and drawings are labelled with the animal's reference number and the date of the study (see above). In the field be prepared that assistants may not be confident at writing in English; be ready to have check sheets in other languages – in Central Africa, French and Swahili.

8. Make one or more photocopies or electronic scans of completed record sheets as soon as possible and keep in a safe place separate from the original. Retain all rough contemporaneous notes.

9. Do not remove bones for study overseas unless specifically authorised to do so, in which case both CITES permits and animal health documentation are likely to be needed (see Chapter 18: Legal Considerations).

Some additional points are discussed below.

The significance of proper, sensitive, handling of bones cannot be overemphasised. The words of Sir Arnold Theiler, the eminent South African veterinarian, when teaching osteology to students in the 1930s, are still pertinent: 'Gentlemen ... you may think it is only an old bone. You have to put your soul into that bone'. So too is the statement of Penelope Bodry-Sanders (1991), although it was written in a different context (following her finding the remains of Carl Akeley in Africa): 'I wanted to touch those bones myself, dig through the earth and hold them'.

Radiography of skeletal material should always be carried out when feasible, both before and during osteological examination. Other imaging techniques may also be appropriate (e.g., CT scanning – see discussion of postmortem scanning of carcases in Chapter 6: Methods of Investigation – Postmortem Examination).

Examination of what are often generally inaccessible features of bones, such as the interior surface of the cranium, can be facilitated if an endoscope or modified auriscope (otoscope) is employed (see Fig. 14.2). Researchers, especially in Europe and North America where resources are generally good, will ask 'why not just use imaging?' There are two reasons:

1. Colleagues working on gorillas in Africa are very likely to have access to auriscopes but not necessarily to imaging.
2. Endoscopy not only permits direct examination of the interior of the cranium, nasal cavity, sinuses and neural canal but also enables the operator to detect – and remove – material that can be missed on radiography. Plant and animal remains (especially invertebrates, such as puparia of Diptera), for instance, can provide invaluable clues to the origin of the bones and, sometimes, the circumstances of death.
3. The strong source of focused light can also be used (a) to facilitate the detailed examination of, e.g., minor

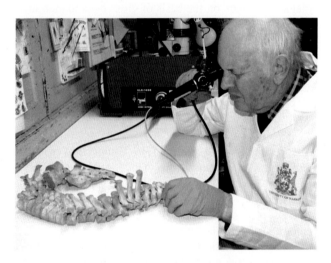

FIGURE 14.2 Use of a flexible endoscope permits examination of the spinal canal of a gorilla at the Booth Museum in Britain.

FIGURE 14.3 Transmitted light from an endoscope clearly shows the museum or collector's reference number on this gorilla bone.

dental damage, and (b) to help assess bone thickness/integrity, e.g., the presence of hairline fractures. This is achieved using transmitted light — placing the endoscope inside the calvarium. Reflected and transmitted light will sometimes also help to decipher faded writing on the bones — old museum reference numbers, for example (Fig. 14.3) — and thereby provide an extra clue to the origin of the specimen.

Meticulous examination of individual bones is important and is often best performed in a quiet place away from the hubbub that can characterise group activity around whole skeletons. For example, careful inspection of foramina is wise. Some are not present in all animals and can be mistaken for an abnormality. For instance, some male gorillas have a large obelionic foramen on the summit of the sagittal crest (Cave, 1994). Note should be taken of any foramina that appear to be unusual — that is, in an unexpected place, apparently enlarged or duplicated. Some such variations are normal (nonmetric, epigenetic) but on occasion they can be associated with disease and/or genetic factors.

Museum specimens have often not been measured and are rarely weighed. Such data are important. In some cases limited morphometrics have been carried out — see, for example, in Part II, A Catalogue of Preserved Materials, specimen A 12748 Skeleton (incomplete). 'The animal was very tall. Length of left humerus 535 mm; length of femur 450 mm'.

The use of appropriate equipment is vital, both in order to perform the work accurately and to protect the bones from harm. Repeated examination can cause damage. Gordon et al. (2013) warned against the overuse of great ape skeletal reference collections by researchers and, to help prevent unnecessary measurement-taking and

BOX 14.2 Equipment for Work on Gorilla Skeletons

- Trolley to transport skeleton or boxes of bones
- Protective sheets to cover skeleton or loose bones
- Paper to put under carcases
- Protective clothing — gloves (surgical and kitchen), gowns, masks, overshoes
- Warning signs for doors and corridors
- Camera, background and lighting equipment. Spare batteries and memory cards
- Tape recorder and spare tapes
- Labels with gorilla reference numbers, plus cm scale, for photographs
- Clipboard, pens, pencils and gorilla record forms
- Ruler or centimetre scale
- Tape measure/string for circumferential measurements/string
- Calipers
- Balance for weighing
- Torch (flashlight)/light source/auriscope (otoscope)
- Magnifying lens/dissecting loupe
- Scissors, scalpels, forceps (not rat-toothed, plus plastic)
- Probes, flexible and rigid
- Pots and plastic/paper bags for specimens, plus 10% formalin or 70% ethanol
- Swabs and transport medium, frosted glass slides, cover slips and saline
- Labels for pots and bags
- Disinfectants
- Clinical waste bags
- Marking pen/soft pencil for labelling bones
- Soft (artist's) brush and toothbrush for (gently) cleaning surfaces
- Books/dissection guide/anatomy charts/reference texts.
- Other special equipment, as appropriate, for example, endoscopes, imaging equipment

other routine procedures, proposed ways of standardising information about individual specimens, including making it available online.

A list of equipment that may be needed for work on gorilla skeletons is given in Box 14.2.

Plan any further tests carefully. They may range from noninvasive imaging to the drying of bones to a standard dry matter content for analysis. Check that the collection's rules permit this (see Appendix CA1: Use of Collections and Handling of Preserved Biological Material).

One can do dietary studies on bones — usually analysing nitrogen and carbon isotopes. Much of the research effort in this respect is led by archaeologists and others who are studying the feeding habitats of human ancestors.

Bone density studies can be used in both live and dead gorillas to assist in the detection of osteoporosis due to

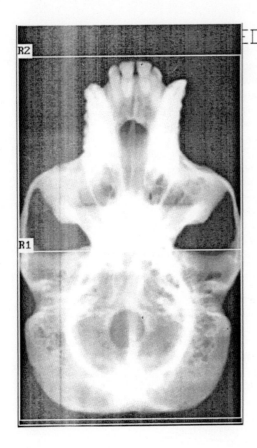

ED SCHOOL

Y0211990F   Thu 11.Feb.1999 11:11
Name: GORILLA NO1
Comment: 1 ST WITS
I.D.:   ZA 1312   Sex: M
S.S.#:   - -   Ethnic:
ZIPCode: Height: cm
Operator: ABG Weight: kg
BirthDate: / /   Age:
Physician: A. BISCARDI
Image not for diagnostic use

    C.F.   1.0300   0.9990   1.0000

Region   Area   BMC   BMD
          (cm2)   (grams) (gms/cm2)
_____ _____ _____ _____

GLOBAL 338.3076 338.0411   0.9992
    R1 197.1626 175.5557   0.8904
    R2 142.6095 164.2796   1.1520
NETAVG 338.1391 338.0131   0.9996

HOLOGIC

FIGURE 14.4 Bone density studies provide additional information — here, on the skull of a mountain gorilla.

dietary or metabolic disorders, osteomalacia, osteopoenia, and, in some cases, necrosis following infection.

In live animals bone density studies usually require sophisticated techniques, such as densitometry or X-ray absorption. Bone density studies on dead gorillas are also important, but care always has to be taken when skeletal material has been buried for some time as chemicals in the soil may have leached out some of the minerals. An example of research on the bones of dead gorillas was the detailed study in 1998 of four skulls and jaws of the mountain gorilla at the University of the Witwatersrand Medical School, Johannesburg, South Africa, reported in part by Cooper and Cooper (2007, 2013) using dual-energy X-ray absorption (Fig. 14.4). Values were obtained for bone mineral content (BMC) and bone mineral density (BMD). The findings provided BMD and BMC reference values for bone of the mountain gorilla, possibly the first time that this had been done, certainly a first for Africa — although it is worth mentioning that over 50 years ago, Galloway and Allbrook (1959) estimated the density of bone of one mountain gorilla in Uganda by simple displacement of water.

Histological studies can be very helpful when investigating the skeletons of gorillas. In the case of live animals bone biopsies, usually computer tomography-guided, can be taken and processed for histological and microbiological investigation, as in routine human and domesticated animal work. Taking bone samples from dead gorillas presents greater challenges, however, as such techniques are considered 'destructive' and few museums will give permission. From time to time small pieces of bone ('exostoses') become detached from a skeleton during the course of handling and it is sometimes possible to get authority to process these. Such was the case with 'Guy' at the NHM (see Appendix 5: Case Studies — Postmortem Investigations). Two such portions of bone were processed at Cambridge University Department of Veterinary Medicine with a view to ascertaining whether they were only basic osteoid or properly formed osseous tissue — and, if the latter, whether there was any evidence of inflammatory or other abnormality (Figs.14.5 and 14.6). The processing of these samples adhered to the following protocol (Scott Dillon, personal communication). Small flakes of bone were fixed in 10% neutral-buffered formalin for 24 hours before immersion in 8% formic acid for 48 hours, to facilitate decalcification. The samples were then placed into microbiopsy processing cassettes and submitted to routine histological tissue processing on an overnight cycle, allowing dehydration, clearing and infiltration of the tissue with paraffin wax. Tissues were oriented and embedded within paraffin wax blocks, before 3-μm sections were cut using a

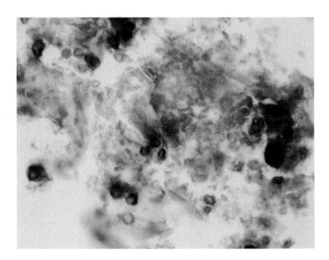

FIGURE 14.5 Excess 'bony' material from the skeleton of a gorilla proves on histological examination to be only osteoid.

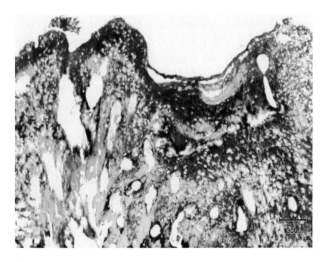

FIGURE 14.6 In contrast to above, this piece of excess 'bony' material is well-organised osseous tissue.

rotary microtome and mounted on to positively charged microscope slides. The slides were subsequently stained using routine H&E, along with Masson's Trichrome and Alcian Blue-PAS protocols.

SKELETAL DISEASE AND PATHOLOGY

An indication of the longstanding interest of scientists in skeletal diseases of NHPs is that the book *Pathology of Simian Primates* published 40 years ago (Fiennes, 1972) was able to devote nearly 80 pages to the subject. Of necessity, what follows here is only a summarised description of some findings in the skeletons of gorillas, together with a brief discussion of their relevance.

In general terms, findings in any collection of gorilla bones are likely to include various traumatic and chronic inflammatory lesions and smaller numbers of changes suggestive of metabolic and/or developmental abnormalities. Such was the case in the review by Lovell (1990b) of *beringei* skeletons in the Smithsonian Institution, and in concurrent studies by Cooper and Cooper (2008) of material from gorilla collections in Europe and Africa. In their study on ageing, Lowenstine et al. (2016) concluded that osteopoenia and arthritis are common to all great apes 'although gout is uniquely human'.

Asymmetry of the axial or appendicular skeleton may indicate infectious, inflammatory or genetic disease (see Part II: A Catalogue of Preserved Materials for examples). A common cause in monkeys is when sinuses are distended and the nasal cavity distorted. Such asymmetries are often suspected on initial examination but must be confirmed by measurement, making allowance for observer bias. Fluctuating asymmetry, which reflects small, random deviations from symmetry in otherwise bilaterally symmetrical characters, can be an indicator of stress (Parsons, 1992). Gosling (2004) warned of bias when carrying out assessments because most observers are themselves asymmetrical (left or right-handed).

An indication as to how detailed an investigation into asymmetry can be gleaned from Howell's (1925) detailed (seven-page) description of gorilla skulls in the US National Museum, which included a 'forensic' analysis of possible causes and the likely history. Examples of asymmetrical specimens are to be found in Part II, A Catalogue of Preserved Materials.

Developmental abnormalities affecting the skeleton have been reported on occasion – for example, Hill and Sabater (1971) described an anomaly of the hallux in a lowland gorilla.

Proliferative lesions are a common feature of osseous change in gorillas. The larger ones are often readily visible and lend themselves to hurried comment and 'diagnosis' without appropriate further proper investigation or comparison with sufficiently large numbers of other skeletons or bones. Proliferative changes in long bones may be metabolic in origin or reflect healing (see fractures, later), or be examples of an inflammatory lesion (osteitis or osteomyelitis) or – not apparently well recognised yet in gorillas – a primary or secondary bone neoplasm.

New bone formation close to joints seems to be a source of confusion. Researchers use the same terms to mean different lesions. In this book, following discussion with Brian Livingstone, FRCS, the following usages, referred to more briefly earlier, are recommended:

1. *Exostosis*. This should be reserved for a hamartomatous malformation of new bone derived from a growth plate (epiphyseal plate). That is the usage in human

pathology but, as far as the author (JEC) is aware, such lesions have not been definitively described in gorillas. Instead the term 'exostosis' has been used loosely and thus confusingly, to describe any area of bone proliferation, whatever its aetiology.

2. *Hyperostosis.* This is new bone forming at the site of a muscle, tendon or ligament attachment to bone. Such a site is called an enthesis. In some animals (including humans) production of the new bone appears to be stimulated by the constant traction force but there is probably a constitutional factor as well, otherwise they would form universally. Histologically hyperostoses look like normal bone. Differentiation from normal bony prominences may be subjective and comparison with the opposite limb (at least) or, preferably, another skeleton should be carried out before attributing any significance to these lesions.

3. *Osteophyte.* This term should be reserved for the new bone that forms at the articular margin of a bone at a synovial joint affected by degenerative change or 'osteoarthritis' (more correctly 'osteoarthrosis'). Such a joint should show evidence of degeneration which includes loss of cartilage, increased density of the bone surface that was under the cartilage (eburnation) and (if the bone is cut through) some cysts in the bone under the eburnated surface. Depression of some part of the articular bone surface and secondary deformity is common and, of course, osteophytes are seen.

Despite its name, osteoarthritis is not an inflammatory disorder; hence the preferred name of osteoarthrosis. Inflammatory arthritis (septic arthritis) and osteomyelitis *are* inflammatory and leave prominent skeletal features.

Autoimmune arthritis occurs in *Gorilla*, as in *Homo*; of which the best known form in humans (but not necessarily the most common) is rheumatoid arthritis.

Autoimmune disease occurs when the animal produces antibodies and lymphocytes that react with and damage its own tissues (see Chapter 12: Lymphoreticular and Haemopoietic Systems and Allergic Conditions). In arthritis this appears to be triggered by a response to certain bacteria in the GIT. Antibodies react with tissues of the joints such as the synovial membrane. It is of interest that in these disorders osteophyte formation is rare unless the original condition has been superseded by degenerative change. Also, the bone around the joint shows reduced calcification because of the increased blood flow associated with the inflammation.

For these forms of arthritis in *Gorilla*, the reader is referred to the excellent review by Rothschild and Rühli (2005) in which they are classified and described. That study is based on the examination of a very large series of skeletons, in both the United States and Europe, of free-living gorillas collected in the early part of the last century.

Given the genetic similarity between *Gorilla* and *Homo*, one would expect the two genera to have similar diseases although frequency may be affected by natural selection.

There was indeed a high incidence of inflammatory arthritis (assumed to be autoimmune) that could be demonstrated in these specimens, both in the lowland and the mountain species. Rothschild and Rühli recognised two patterns of peripheral joint damage, both associated with changes in the spine (i.e., axial skeleton), which they termed 'spondyloarthropathy'.

In the axial skeleton the changes were predominantly fusion of the intervertebral and sacroiliac joints. In the limbs the peripheral joints, by contrast, were eroded and destroyed but not usually fused. Two main patterns of peripheral joint involvement were seen. Just one or very few peripheral joints were affected in *beringei* while (when present) widespread peripheral joint involvement was noted in *gorilla* (either pattern is seen in humans). However, the incidence of spondyloarthropathy (with or without peripheral joint involvement) was the same in both species — at a surprisingly high rate of about 20%.

Many researchers seem to consider this spinal disorder as being the same as 'ankylosing spondylitis' in humans but there are some differences that may be significant enough to imply that distinct processes are causing the spinal joint fusion in these diseases. For example, the formation of large osteophytes[1] forming at the margins of the vertebral bodies and growing to bridge and thus fuse the joints between them but without destroying the intervertebral disc, appears to be illustrated (but not named) in some studies. This is a feature of the disorder known as 'disseminated idiopathic skeletal hyperostosis', or Forestier's disease (see Resnick et al., 1975, 1976) which is given the acronym DISH. Since this has a different prognosis from ankylosing spondylitis, the distinction may be important and further research is needed.

The reason for describing this in detail is to emphasise that it would be very valuable to record changes of peripheral arthritis and of fusion of spinal joints in any gorilla skeleton that is being examined. Most such skeletons do not appear to have been looked at with this distinction in mind. It may be that these bridging 'osteophytes' are in fact examples of hyperostosis as defined above. It should be noted that in a recent paper (Rothschild, 2015) an attempt was made to correlate gorilla intestinal flora (assessed by faecal samples) with the pattern of peripheral arthritis/spondyloarthropathy. Given the mechanism of the autoimmune reaction described above, this could clearly be important and the results do suggest that specific intestinal bacteria did correlate with more widespread distribution of the peripheral joint arthritis.

---

1. Rothschild calls these 'syndesmophytes' because, technically, the joint between two vertebral bodies and the interposed intervertebral disc is a syndesmosis rather than being a synovial joint.

Infectious Diseases of Bones

Bone and joint infections can leave a very characteristic appearance in the skeleton. An infected joint will be badly destroyed, often with involvement of the adjacent bone although the effects of florid autoimmune arthritis may be very similar. In the absence of other clues such as the clinical situation during life, the presence of multiple affected joints suggests autoimmune erosive arthritis while a solitary damaged joint is more likely to be due to septic arthritis.

Infection of bone may be confined to a subcutaneous surface, associated with ulceration of the overlying skin that allowed sepsis to penetrate the bone. By contrast, a blood-borne infection produces the very characteristic appearance of osteomyelitis. The blood-borne bacteria settle in the marrow and produce an abscess, the pus from which penetrates along the Haversian canals of the cortical bone destroying blood vessels and killing the bone. The pus emerges under the periosteum and raises it off the bone, further damaging the blood flow. The dead bone is the 'sequestrum' on which the bacteria can proliferate. The channels for pus drainage through the cortex enlarge and are called 'cloacae'. If the animal does not die from the infection, the condition can become chronic. The new bone produced by the elevated periosteum is called 'involucrum' and the cloacae extend through that also.

As noted, the final appearance is very characteristic with some dead cortical bone enveloped in involucrum that is perforated by the cloacae. Usually the skin breaks down over this to allow formation of a chronic discharging sinus. Such chronic osteomyelitis virtually never heals, even with antibiotics. All this produces chronic ill health which may be reflected in the other organs.

Different types of arthritis are recognised in *Homo*, some associated with (for example) conditions such as ulcerative colitis and Crohn's disease (see Chapter 11: Alimentary Tract and Associated Organs). A few of these types have been diagnosed in captive gorillas (Atkinson, 1996; Brand, 2009). In an early study Brown et al. (1980) investigated the possible role of mycoplasma–host interaction in pathogenesis and treatment. Janssen (1993) described 'osteoarthritis and rheumatoid arthritis in older gorillas' and suggested that the latter may be a result of immune complexes resulting from *Mycoplasma* infection.

Paixão et al. (2014) reported pathological findings of septic polyarthritis and chronic renal failure in a captive senile lowland gorilla.

There are diagnoses of 'arthritis' in museum specimens but it is not always clear as to who made these and on what basis. One example from Part II, A Catalogue of Preserved Materials, shows 'much evidence of osteoarthritis and rheumatoid arthritis' in the skeleton, but without defining how these very different diseases were distinguished, was by a PhD research associate at the museum.

So too are there many records in zoos and veterinary archives of arthritic and other conditions in great apes.

A recent (December 2015) student project by Dermot McInerney at the University of Bristol reviewed and digitised radiographs of NHPs at Bristol Zoo Gardens, including gorillas, and scored radiographs for osteoarthritis (OA, see Appendix 5: Case Studies – Museums and Zoological Collections).

Amongst a list of diseases of mountain gorillas recognised and listed by participants at the PHVA for this species (see Werikhe et al., 1997 and Chapter 3: Infectious Disease and Host Responses) were 'osteoarthritis (joint disease)', 'spondylitis/spondylosis (back disease)' healed fractures' and 'ankylosis (fusion of joints)' but these were not further defined or the diagnoses confirmed.

Little appears to have been published, even in the French literature, about joint disease in lowland gorillas but Tutin and Benirschke (1991) described a case of possible osteomyelitis of the skull that caused the death of one animal in the Lopé Reserve, Gabon. In his 12-year review of clinical conditions in the Howletts Zoo (United Kingdom) gorilla colony, Furley referred to various locomotor disorders including 'sprains', all in juvenile animals. Treponematoses, including yaws (*Treponema pallidum* subsp. *pertenue* infection) in humans affects skin, bones and joints as well as the integument. It is probable, but not yet apparently proven, that bone lesions similar to those of humans with treponematosis (Schultz, 1950; Lovell et al., 2000) are caused by the same disease. Skin lesions associated with yaws are discussed in Chapter 9, Skin and Integument.

Ian Redmond (KRC field notes, personal communication) reports that a female mountain gorilla 'with what appeared to be tertiary bone lesions of yaws, resulting in an irregular black hole in the middle of her face, was living an apparently normal life in Karisoke Group 6 on the eastern slopes of Mt Visoke, Rwanda, in the 1970s'.

Proliferative, usually bilaterally symmetrical, osseous outgrowths of the facial region are present in a number of skeletal collections. In Paris, for example, at the Museum National d'Histoire Naturelle, three gorilla skulls show such growths (Fig. 14.7) but in the Museum's catalogue only two include reference to the lesions ('voluminous malar exostoses'). They are assumed to be cases of 'yaws' but, as is so often the case with museum material, the 'diagnoses' to date have been made by nonmedical/nonveterinary personnel and are based only on the gross appearance of lesions. The skulls have been closely examined macroscopically (and using a lens) by the author (JEC). Both the aetiology and the pathogenesis of these and other similar specimens need proper investigation.

PCR can be used to detect treponemes in animal bone (Nascimento et al., 2015) and might be usefully applied to gorilla specimens.

A discussion of the possible causes of so-called 'Gundu' in primates, two in *Gorilla*, is to be found in Schultz and Starck (1977). Histology and radiography were used. The two cases showed symmetrical bony

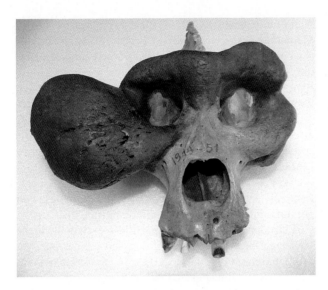

FIGURE 14.7 A skull from the Paris Natural History Museum, showing marked asymmetrical proliferative lesion – possibly caused by treponematosis.

'tumours' of the maxilla. The authors concluded that the origin is uncertain.

Other osseous lesions in bones may be predominantly destructive, as in the 'phalangeal osteolysis syndrome' in a male gorilla described by Brandt (1972).

Developmental Diseases of Bones

Developmental abnormalities often manifest themselves in the skeleton and require detection and proper description (see also Chapter 8: Nonspecific Pathology).

The suggested approach to investigation of possible skeletal developmental abnormalities is:

- Examine all bones carefully: describe, draw, photograph possible abnormalities, including asymmetries. Check neonates for 'common' (commonly seen) abnormalities, for example, cleft palate.
- Be prepared to carry out detailed investigation using (for example) radiography, and (in fetal bones) alizarine-staining.
- Try to link abnormalities with history of animal(s), especially any breeding records.

FRACTURES AND THEIR REPAIR

The natural history of a fractured bone is for it to heal. When the fracture is of a long bone, this is largely a function of the periosteum, from which new bone for healing is derived. This new bone is the callus that forms in the haematoma around the fracture, bridges the gap between the fragments and matures into calcified bone. Once united, remodelling can result in the fracture site

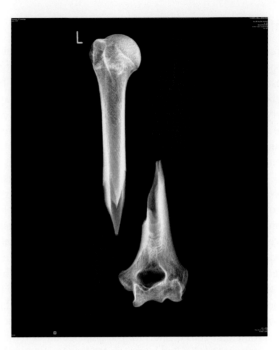

FIGURE 14.8 Routine radiography of a fractured humerus from a dead lowland gorilla helps confirm that it occurred at the time of death; there is no evidence of periosteal or endosteal callus.

becoming surprisingly inconspicuous, especially if the injury was to the immature skeleton. The final microscopic structure, with Haversian canals, looks no different from normal bone. However, most sites of fractures sustained by adults are still obvious postmortem as a deformed section of the shaft enveloped in new bone.

The factor that determines the rate and completeness of fracture union is the severity of the force of the injury that broke the bone. The greater this was, then the longer the fracture will take to heal. In free-living or captive gorillas most accidents will be low-energy events, such as falls or fights, often rarely enough to fracture the very strong bones of an adult. Nonetheless, long bone fractures do occur (Fig. 14.8) and in some museum skeletons show healing (see Part II: A Catalogue of Preserved Materials).

The pattern of fracturing is important to note because it gives a good clue to the mechanism of its production and the force required as to cause it. For example, a direct blow to the shaft of a femur or tibia will produce a more or less transverse fracture line. If the force was great enough there may be shattering into numerous fragments ('comminution'). The bone ends will have been forced apart and the surrounding soft tissues (which include the all-important periosteum and the blood supply) will be torn and severely damaged. Such a fracture may never heal.

On the other hand, the break may be due to a rotation injury of the limb which will produce a long spiral pattern of fracture. While this still takes a lot of force, it is mostly applied at some point away from the break. For example, the animal's foot may be trapped in a snare while the body is twisted to one side. The result is that the soft tissue around this fracture is much less damaged because it has not been bruised or torn directly. There is also some splinting of the fracture by the intact soft tissue envelope. Thus, even with a major bone, such a fracture may well heal by means of periosteum-produced callus in about 6–8 weeks (albeit with some deformity) and the injured animal might maintain enough mobility/limb use to survive that period.

Thus the examination of the fracture of a long bone can reveal the mechanism of production and allow a fair guess as to how it occurred. There are, of course, patterns that fall in between these types and one sometimes sees what must have been a combination of twisting and direct force that produces a short oblique fracture line in the broken shaft. This has an intermediate prognosis but is more likely to heal too slowly (like a transverse fracture) for the animal to survive. In captivity all these prognoses will be altered by treatment.

Flat bones include the wing of the ilium and those of the tarsus and carpus. Fractures of these are usually due to direct force resulting in their cancellous structure being crushed or the joints between them being pulled apart. The cancellous structure of bone trabeculae, with good blood supply, results in healing directly by the activity of the bone-forming cells that they contain (endosteal healing); the periosteum is less important. Very large force can produce shattering and compounding by skin damage, resulting in the same poor prognosis as for high-energy injuries to long bones. These comments also apply to the ends of long bones (which are expanded into cancellous structures such as the trochanteric region and the condyles of the femur or the head of the humerus).

Fractured vertebrae are due to twisting or forced bending rather than direct trauma although avulsion of a transverse process, a spinous process or fracturing at the vertebral attachment of the head end of a rib, could occur this way.

Fractured ribs are apparently common in gorillas and are due to direct blows (e.g., fighting or falls) but they are usually low-energy injuries and heal quickly by periosteal callus.

It is worth noting that, within limits, movement does not inhibit healing … otherwise no fractures would ever join in free-living animals.

If it is diseased, bone may fracture with much less applied force (or none at all). Metabolic disease or metastatic malignant neoplasms are the most common preexisting pathological factors in *Homo* and they often affect multiple areas of the skeleton. Immunosuppressive infections probably do not prevent healing of low-energy fractures but any overlying skin damage is more likely to become infected and fail to heal.

The same forces that may break a bone can pull joints apart instead (or as well). A trauma-induced dislocation might spontaneously relocate and leave no trace (except perhaps for some damage to the articular surfaces of the joint). However, if the capsule and ligaments are badly or completely torn they do not have much capacity to heal. Thus the dislocation may be still found post mortem, visible in the carcase as a deformity or detectable by noting limited mobility of the joint before dissection.

The structure of animal bones and their tensile strength was extensively studied by Alexander (1984). As discussed earlier, there has been much interest in traumatic injuries in free-living wildlife – in particular the incidence and significance of healed fractures – especially in birds (Goodman and Glynn, 1988; Roth et al., 2002). Fractures have also been investigated in free-living primates, including gorillas (see, e.g., Schultz, 1939; Lovell 1990a,b), using museum material.

As was emphasised in Chapter 4: Noninfectious Disease and Host Responses, detailed understanding of fractures and other physical insults can often be obtained from study of skeletal remains. Indeed, perhaps the earliest report of pathology related to injury in a gorilla was described in a skull, a then recent addition to the museum of the Odontological Society in London (Bennett, 1884).

Fractures are frequently reported, clinically and post-mortem, in gorillas, especially in the wild and to a lesser extent in captivity. Clinical management will not be discussed here other than in the context of healing. The proper assessment of fractures is important regardless of whether the animal is alive or dead. In view of this it is appropriate to mention how broken bones mend. As in humans, fracture repair in great apes involves mesenchymal tissues – fibrous, cartilaginous and bony – and can be considered to take place in six, often overlapping, stages:

- Immediate – tissue necrosis and haemorrhage
- Inflammatory response to the tissue damage
- Formation of provisional callus (osteoid)
- Formation of mature callus
- Remodelling of callus
- Final reconstruction

Cortical bone undergoes both modelling and remodelling, and these occur on both the endocortical and periosteal surfaces of the cortex, but within the cortex only (Haversian) remodelling takes place. Modelling is probably best defined as the uncoupled activity of bone cells at the periosteal and endosteal surfaces, resulting in an overall change in volume, dimensions or morphology. Remodelling is the coupled activity of bone cells within the cortex resulting in no net change, at least under nonpathological conditions.

Many factors – single and multiple – can impair healing of bones. They include:

- Infection – especially of wounds that communicate with the fracture ('compound fracture')

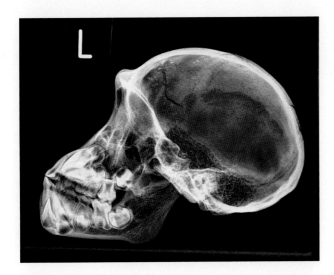

FIGURE 14.9 Radiography plays an important part in the examination of young gorillas, especially if traumatic injury is suspected. This infant's skull shows fractures in the calvarium.

- Antiinflammatory agents, for example, corticosteroids (also endogenous steroids – stress?)
- Nonsteroidal antiinflammatory agents, NSAIDs, for example, aspirin (inhibits production of prostaglandins)
- Immunosuppression – humoral and cellular: HIV/AIDS, probably because it makes any associated wounds more prone to infection and thus inhibits fracture healing
- Other immunosuppressive infections such as malaria: malnutrition (hypoproteinaemia)

Some of these may be relevant to free-living gorillas in the forest. There are various types of fracture but in the interests of space and because of their prevalence, particular attention will be paid here to skull fractures.

Damage to the bones of the head is not uncommon in gorillas, especially but not exclusively in adult males and in youngsters which are the victims of infanticide (Fig. 14.9). Fossey (1983) reported 'From 64 skeletal specimens collected throughout six Virunga volcanoes, I found that 74% of the silverback remains revealed signs of healed head wounds, and 80% had either missing or broken canines'.

While some such 'healed head wounds' may not have led to the animal's death, other skull injuries do prove fatal. Firearm injuries are one such cause; the other is infanticide, attacks on young animals by (usually) male gorillas (see Chapter 4: Noninfectious Disease and Host Responses).

Skull fractures in humans, especially in infants, have attracted much interest and research, in part for forensic reasons (Cooper and Cooper, 2007). Relevant references

go back several decades and include Ito et al. (1977), Zegers et al. (2003) and Graham and Glennarelli (2000). Techniques used to investigate such injuries in children include CT and MR imaging (for the role of imaging in necropsy studies on gorillas, see Chapter 6: Methods of Investigation – Postmortem Examination). Skull fractures are also sometimes a feature of 'battered-pet syndrome' – that is 'nonaccidental injury' in dogs and cats (Munro and Thrusfield, 2001).

Regrettably, in contrast to the situation in humans and household pets, the type and extent of skull fractures seen in gorillas have not been fully investigated and categorised, even in victims of infanticide.

SPECIFIC DIAGNOSIS OF DISEASES OF THE SKELETON

Investigation of any lesion involving bone, proliferative or degenerative, or a combination of both, needs always to be cautious and comprehensive and adhering to correct medical/veterinary medical terminology.

It has been said (Royal College of Pathologists) that 'Medicine is all about pattern recognition'. Diagnosis of bone disease depends upon detecting distinctive patterns of change, recognising the types of change and assessing the speed of development and any degree of healing. The best approach to studying skeletal lesions, macroscopical and microscopical, in both live and dead gorillas is to *observe, examine and describe*! Interpretation of the changes can then, with caution, follow.

Infectious Disease

This is not easy without recourse to a battery of laboratory investigations. Diagnoses based on gross examination alone must always be considered speculative. Regrettably, many nonmedical scientists and curators fall into the trap of observing lesions that superficially resemble 'confirmed' cases of disease A or B and assume that their specimen is similarly afflicted. An example of how meticulous differential diagnosis can be was provided by Hershkovitz et al. (1998) who investigated the specificity of osseous lysis in the differentiation of metastatic cancer, tuberculosis and fungal disease in humans. Osseous changes in the skeleton of a 47-year-old man, diagnosed in life as having blastomycosis, were characterised and compared with lytic lesions observed in 10 people with tuberculosis and 6 with metastatic cancer. Apparently distinguishing characteristics were identified. Blastomycosis was characterised by the presence of eroded areas with protruding, short, blunt, spicules of new bone surrounded by periosteal reaction, tuberculosis by zones of resorption and coalesced lesions with a smoothed marginal zone and 'space-occupied' appearance, and metastatic cancer by lytic, nonpermeative, lesions. Similar studies could and should be applied to hitherto undiagnosed skeletal lesions in gorillas (and other species) in the world's museums.

Noninfectious Disease

The accurate diagnosis of noninfectious skeletal disease also usually necessitates detailed technological and laboratory investigations, ranging from radiography to transmission electron microscopy. See earlier discussion. One additional example will be given.

Metabolic Bone Disease

Deficiencies of dietary calcium and/or vitamin D3 will lead to metabolic bone disease (MBD). Farrell et al. (2015) described osseous changes seen in skulls of previously captive NHPs characterised by general thickening with decreased bone density, affecting mainly the maxillofacial region and mandible. They performed CT scanning on a subsample of these skulls and suggested that a MBD was 'the causative agent' [*sic*], with osteomalacia the likely diagnosis. Farrell et al. related these findings to the inadequacy of zoo primate management during the late 19th and early 20th centuries.

Many papers over the years, in a wide range of different journals, have discussed differential diagnosis of noninfectious bone diseases of NHPs and some of these are relevant to studies on gorillas. For example, in a human medical journal, Fiennes (1974) analysed the problems of rickets in monkeys and apes while a publication suggesting a possible familial predisposition to MBD in rhesus macaques was directed towards laboratory animal scientists (Wolfensohn, 2003).

MUSCLE

Interest in the muscles of gorillas goes back for over a century; Kelly (1880) published a short paper on the sartorius muscle of the gorilla. A series of papers, mainly in German, in the 1960s, threw considerable light on the muscular biomechanics of gorillas and other apes (Preuschoft, 1961, 1965) and this focus continued (Preuschoft and Günther, 1994; Preuschoft et al., 2010) to the present (Zihlman et al., 2011; Diogo et al., 2011, 2014; Diogo and Wood, 2012).

Skeletal muscle is essential to physical function. As gorillas age, they, like humans, lose muscle mass (see age-related – Chapter 8: Nonspecific pathology). This is coupled with a reduction in strength and function – sarcopoenia. In captivity this can lead to frailty and cachexia (involuntary rapid loss of body weight and muscle mass). There are methods of assessing loss of muscle mass and strength and clinical measurements of

malnutrition and cachexia in Homo and these would be applicable to *Gorilla*. Loss of muscle mass can be due to other factors too, as in the female mountain gorilla 'Inyuma', examined post mortem by the author (JEC) in 1995 following chronic lameness; she had: 'left leg disuse atrophy. Healed fracture of rib (? 3 months)...'

Although specific diseases of muscles are not readily located in gorilla literature, the musculature is often involved in both trauma cases (including some snare injuries) and infectious conditions affecting the skin or (smooth muscle) the viscera. Investigation of muscle in the live gorilla usually involves biopsy. Useful information on technique and the interpretation of histological changes can be found in the book by Dubowitz et al. (2013).

CONNECTIVE TISSUE

Very little has been published on the connective tissues of gorillas per se; but, as in other species they play an important part in gorillas in supporting, binding and/or connecting organs and tissues and, of course, in providing the mechanical framework for the body, the skeleton, discussed above. They also contribute substantially to healing, both of soft tissues and bone (see above).

An entry in Part II: A Catalogue of Preserved Materials, refers to a tissue biopsy of a tendon. The histological diagnosis was a severe subacute fibrinosuppurative tenosynovitis, with multifocal collagen degeneration, perivascular neutrophil cuffing and fibrinoid change, and numerous intralesional mixed bacteria.

Occasionally the filarial worm *Loa loa* lives in the connective tissue of the abdominal cavity (Meder, 1994) as does *Mansonella perstans* (see Chapter 11: Alimentary Tract and Associated Organs).

ADIPOSE TISSUE

In gorillas, as in humans, adipose tissue is found in over a dozen discrete depots in different parts of the body. Little attempt has been made to categorise it, or interpret its real significance, in live or dead animals despite pleas over the years for such an approach (Pond, 2016). No evidence has been found that those necropsying gorillas have made any serious effort to weigh, measure or otherwise quantify the fat seen in different locations. Imprecise statements, such as saying that body fat was 'extensive' or 'minimal' have been used. Rarely is even the colour of the fat mentioned although there are exceptions, such as in the postmortem report on 'Colossus' at the Cincinnati Zoo (see Part II: A Catalogue of Preserved Materials): 'The epicardial surface contains a normal amount of glistening, yellow adipose tissue'. The yellow colour is typical of captive specimens, arising from the feeding of eggs and ruminant products.

Gorillas, like other mammals, use their adipose tissue (fat) as a reserve energy source. Like humans, gorillas (certainly infants) appear to have both white and brown adipose tissue but little seems to be known about the role of the latter. The adults, especially of mountain gorillas, may well have thermogenic 'beige' adipocytes (Caroline Pond, personal communication).

There is increasing evidence that adipocytes in the minor fat deposits that enclose lymph nodes may play a part in mobilising cells involved in immunity and thus assist in a prompt and effective response to pathogens (for an early review, see Pond, 2000).

Sex differences in the distribution and abundance of adipose tissue are well recognised in humans but appear not to have been studied in gorillas. Research on NHPs has largely been confined to *Macaca* monkeys and lemurs. Zihlman and McFarland (2000) computed body proportions and tissue composition in four adult captive lowland gorillas by dissection, and were able to provide data on the relative contributions of muscle, skin, bone and adipose tissue to total body mass.

Changes in the adipose tissue have often been reported in necropsy reports. Fossey (1983), for example said of 'Quince': 'Adipose tissue — Fat necrosis, accumulation of lipochromes (lipofuscin and ceroid) and infiltration of the interstitium with edema fluid, neutrophils, and macrophages'. Field necropsies in Rwanda have noted atrophy of fat, particularly in neonates. Some collections of gorilla material include samples of adipose tissue (see Part II: A Catalogue of Preserved Materials) and one includes a report of steatosis together with myocardial fibrosis.

Chapter 15

Nervous System and Special Senses

John E. Cooper

Of all types of animal I have observed naturalistically or used in experiments during almost 60 years as a psycho-biologist, none has stirred my curiosity and suggested so many questions as the gorilla.

Robert Yerkes (cited by Dixson, 1981)

INTRODUCTION

This chapter covers the nervous system — central (CNS) and peripheral (PNS), ear, nose and eye.

The nervous system is of ectodermal origin. The ear is derived from a mixture of embryological tissues — for instance, surface ectoderm (membranous labyrinth), neuroectoderm (spinal and vestibular ganglia), mesoderm (bony labyrinth) and endoderm (first pharyngeal pouch — lining of middle ear (tympanic cavity)). The ear, nasal septum and turbinates are derived from ectoderm, neural crest and mesoderm. The eye is also a mixture — neuroepithelium (e.g., retina and optic nerve), surface ectoderm (e.g., lens and corneal epithelium) and extracellular mesenchyme (e.g., sclera and vitreous humour).

The pathology of the nasal organs, nasal septum and paranasal sinuses is also reviewed in Chapter 10: Respiratory and Cardiovascular Systems.

Some general, but useful, data were provided by Dixson (1981), including discussion of the structure and function, senses and intelligence and the behaviour of gorillas and a table of cranial capacities of apes and humans.

PSYCHOLOGICAL STUDIES ON GORILLAS

As this chapter covers the brain, brief mention will also be made to the psychology of gorillas, a topic that is of some relevance to this book because it lies at the intersection of different disciplines, including medicine. The interaction between brain, mind and behaviour is important and, when this relationship is disturbed, the result may be clearly discernible 'behavioural pathology' (see Chapter 8: Non-specific Pathology) — syndromes that not only indicate a psychological disorder, with welfare considerations, but can also result in physical injury to the animal.

Psychological studies on gorillas and other great apes go back to the research of Yerkes (1928), quoted at the beginning of this chapter, who penned a series of papers on the 'mind of a gorilla'. Other early authors also wrote about the 'mentality' of apes — for example, Köhler (1926) — and interest in this topic continues to this day. In their book *The Mentalities of Gorillas and Orangutans: Comparative Perspectives* Parker et al. (1999) pointed out that, although research on the mental abilities of chimpanzees and bonobos has been extensively studied and applied to theories of human evolution, relatively little attention has been paid to the abilities of gorillas and orangutans. Those authors tried to rectify this with discussion on the kinds and levels of intelligence displayed by these primates compared with that shown by other great apes, with reference to such subjects as imitation, self-awareness, social communication and tool use. Interest in these activities is hardly new, however, as witnessed by the writings about the origin and progress of language by Lord Monboddo (1774).

Fifty years ago Morris and Morris (1966) reviewed the use of implements by great apes in the context of picture making. They pointed out that gorillas, chimpanzees, orangutans and even capuchin monkeys had been reported to create drawings or paintings; much has been described and analysed since then.

Some years later Byrne (1996) described the gorilla as a 'misunderstood ape' and discussed the cognitive skills of the species.

Mitchell (1999) provided a fascinating and well-researched review of scientific and popular conceptions of the psychology of great apes from the 1790s to the 1970s and appropriately entitled his paper 'Déjà vu all over again'.

J. E. Cooper & G. Hull: Gorilla Pathology and Health. DOI: http://dx.doi.org/10.1016/B978-0-12-802039-5.00015-9

STUDIES ON THE BRAIN OF GORILLAS

Since the first live gorillas came into captivity scientists and others have been intrigued by their apparent cognitive abilities, leading to extensive study of their nervous system, measurement of cranial capacities (see, e.g., Schultz, 1962) and, particularly, the collection of their brains (see, e.g., Von Bonin, 1946). Such was the fascination with brains that when the first live gorilla in the United States died, his brain was preserved (Kennedy and Whittaker, 1976). As long ago as 1916 Campbell described histological studies on the brain of the gorilla and interest in such microscopical investigations, often attempting to compare humans and other primates, continued for many decades – see, for example, Fulton (1938), Bianchi (1946), Hosokawa et al. (1965) and chapters in Noback and Montagna (1970).

Various publications in the second half of the last century explored the structure and development of the brain of gorillas, some of it of relevance to clinical medicine and pathological investigation of this taxon. As is so often the case, the research material was usually restricted to one animal. Thus, gross and fine motor development of an infant lowland gorilla formed the basis of a study by Thompson and Jankowski (1984) while Cave (1994) described venous drainage of the diploe (the cancellous tissue between the outer and inner tables of the skull) using anatomical material.

More recently Sherwood et al. (2004) compared the brain structure of the mountain gorilla with that of chimpanzees, western lowland gorillas and six orangutans, using magnetic resonance imaging (MRI) and their findings suggested possible functional differences among the taxa in terms of neural adaptations for ecological and locomotor capacities. This last piece of research, together with the work of Martin (1981, 1982, 1984, 1996), illustrates the move in primate brain studies from basic structural investigations to research that helps to throw light on whole-animal biology – for instance, metabolic rate, the maternal energy hypothesis, allometry and feeding strategies. Brain performance is important for fitness and survival and these in turn are associated with social acuity or the ability to interact with other animals in a group. There is now strong evidence that ecological, social and sexually selected pressures affect variation in brain size within different species (Dunbar and Schultz, 2007). The central nervous system (CNS) is, however, 'energetically expensive' and the expensive-tissue hypothesis predicts that increases in brain size will inevitably result in a decrease in the size of other metabolically costly tissues, such as the gastrointestinal tract (GIT) (Aiello and Wheeler, 1995). Sophisticated modern technology permits researchers to obtain information that would have been considered only fanciful even two decades ago. Such

research may be relevant to the health management of gorillas and to studies on their pathology. So, for example, Hopkins et al. (2015) who used structural neuroimaging (postmortem magnetic resonance) to examine neuroanatomical variation between gorilla species – 18 captive western lowland gorillas, 15 'wild' (=free-living) mountain gorillas and three Grauer's gorillas (both 'wild' and captive animals).

Stereological methods revealed that the volumes of the hippocampus and cerebellum were significantly larger in *Gorilla gorilla* than in *Gorilla beringei*, possibly relating to divergent ecological adaptations of the two species – *G. gorilla* is more arboreal and thus may rely more on cerebellar circuits and, because it tends to eat more fruit and has larger home ranges, may depend more on spatial mapping functions of the hippocampus than does *G. beringei*. Similar analytical work by McFarlin et al. (2012) (see also the Introduction to Part II: A Catalogue of Preserved Materials), using data derived from necropsy reports and endocranial volume measurements, found that in 'wild' (free-living) *beringei*, brain growth is completed at between 3 and 4 years of age, corresponding with the emergence of the first permanent molars. Their findings demonstrate that brain growth is completed earlier in mountain gorillas than in other great apes and suggested that this characteristic may be linked to 'life history characteristics' of the Virunga *beringei* population.

Another interesting paper, with applications to health and pathology, was by Barks et al. (2014) who explored variable temporoinsular cortex neuroanatomy (fusion between temporal and insular cortex). This is a relatively rare neuroanatomical feature that has become more common in eastern gorillas and Barks et al. (2014) postulated that this might be the result of a population bottleneck effect.

There are, of course, very many links between brain function and other aspects of health. Even the gut flora may be relevant – certainly in the case of the blood–brain barrier, the integrity of which is essential to proper brain function.

INVESTIGATION OF THE NERVOUS SYSTEM

Investigation of neurological disorders is not easy in non-human primates (NHPs). Clinical investigations cannot be discussed in any detail here but should follow methods used in human neurology.

In this context, in a recent paper in a (human) medical journal, O'Brien (2014) discussed the 'use and abuse of physical signs in neurology' and argued that the history is overwhelmingly the most important part of a neurological assessment and that signs (there should be a bias towards the use of motor signs) should be used to clarify the history.

Animals with central neurological signs may require a range of laboratory tests, including examination of cerebrospinal fluid (CSF). The latter cannot be discussed in detail here but it is worth mentioning that interpretation of laboratory findings often needs experience: if necessary, the assistance of a colleague should be sought. Important parameters include the appearance (colour and turbidity) of the CSF, the cell count, the cell type and the protein values.

A gorilla that suffers trauma may display clinical signs relating to nerve damage (see also later – 'Peripheral nerves' section). In the case of head injuries, which are common in gorillas, these can include:

- Partial or complete blindness (optic nerves)
- Inability to move ear and lip (facial nerve)
- Dysphagia (glossopharyngeal nerve)

Other effects, such as a loss or reduction of smell due to olfactory nerve damage, undoubtedly occur but are rarely diagnosed, even in conditions of captivity.

Neurological signs in gorillas can be due to metabolic disorders – for example, hypoglycaemia and electrolyte imbalances. It is unclear as to whether hepatic encephalopathy neurological manifestations, that can affect other species of animal that have liver dysfunction, occurs in gorillas.

Differential diagnoses in a gorilla showing CNS signs should include trauma, encephalitis, meningitis, abscessation, cerebral thrombosis, degenerative disorders (encephalomalacia) and neoplasms (see Box 15.1 – lesions affecting the nervous system).

Neuropathological investigation essentially requires a sound knowledge of neuroanatomy, an ability to remove and correctly trim brain and spinal cord (in cases of paralysis, careful dissection of the cranium and spinal cord may be necessary to detect localised damage – see later) and skill in detection and sampling of lesions for histological examination.

Fixation of nervous tissues presents particular challenges. Formalin may create artefacts. Glutaraldehyde has often been used for the CNS in the past but, even more than formaldehyde, can present health hazards.

Post mortem change occurs early in the CNS and has led in experimental pathology to the use of ante mortem perfusion (Scudamore, 2014). In work with gorillas this is rarely feasible because these animals are seldom euthanased and even when they are, full advantage has to be taken of the opportunity to sample tissues in various ways, not just to fix them for microscopical examination.

Freezing can be used for nervous tissue but can produce tissue artefacts if there is contact between the samples and dry ice/liquid nitrogen. Instead, immersion of the nervous tissue in cooled isopentane is advised, as used in toxicological pathology (Scudamore, 2014).

As in other species, artefactual changes are often seen in brain and spinal cord. Histopathological examination, using time-honoured H&E staining (Figs. 15.1–15.3) still provides the best information for diagnosis. However, rapid, careful, handling, sampling and fixation (see below)

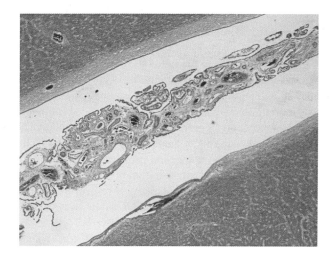

FIGURE 15.1 Part of normal choroid plexus of lowland gorilla.

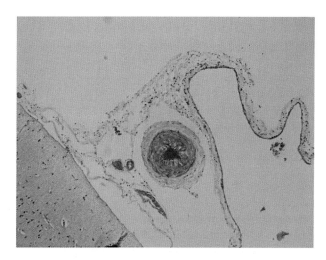

FIGURE 15.2 Part of normal meninges of lowland gorilla.

BOX 15.1 Lesions of the Nervous System

- Traumatic
- Inflammatory
- Vascular
- Degenerative
- Metabolic (storage diseases)/toxic

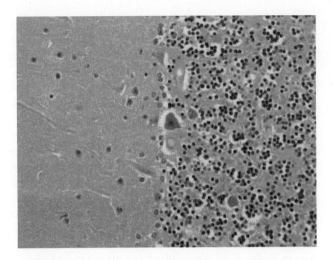

FIGURE 15.3 Part of normal cerebellum of lowland gorilla, showing Purkinje cells.

are an essential prerequisite for the correct interpretation of histological changes in nervous tissue. This can be a particular challenge in the field but is also important when necropsying a gorilla in captivity, especially if (for example) an animal dies on a Saturday evening and a thorough postmortem examination cannot be arranged until the Monday morning. Commonly seen artefacts include vacuolation, which can follow autolysis, rough handling of tissues, or prolonged immersion in alcohol (Scudamore, 2014). Differentiation of vacuolation that is associated with such factors from vacuolation that represents a pathological change is not easy as the latter is not always associated with gliosis or other cellular responses.

So-called 'dark neurons' are another artefactual change in the CNS and are characterised, as the name suggests, by the appearance of small, dark, basophilic neurons. They are usually associated with poor fixation or damage (pressure) on unfixed brain or spinal cord. Distinguishing dark neurons from necrotic neurons is not easy and requires experience. Scudamore (2014) pointed out that, in rodents at any rate, such differentiation is made more difficult by the fact that some published literature has reported dark neurons due to artefactual change as a pathological change, neurodegeneration.

Histological examination of the CNS of embryos requires particularly careful study and interpretation. Some features of normal development, such as the appearance and differentiation of glial cells, synapse apoptosis and the formation of neurons can be mistaken for pathological changes.

Nervous tissue can be fixed, like most other tissues, in neutral buffered formalin but this can cause artefactual changes — see earlier/later. For transmission electron microscopical examination of, for example, myelin sheaths, perfusion of glutaraldehyde — if feasible — has much to commend it.

Freezing can be used for nervous tissue but can produce tissue artefacts if there is contact between the samples and dry ice/liquid nitrogen. Instead, immersion of the nervous tissue in cooled isopentane is advised, as used in toxicological pathology (Scudamore, 2014).

Interpretation of findings needs recognition of major reactions to neural injury and common artefacts and familiarity with the principles of lesion classification and pattern recognition. Much of this can be learned from frequent examination of the CNS and PNS of other species, including monkeys.

Although as explained, the brain of the gorilla has attracted scientific attention for many decades, our knowledge of its diseases remains rudimentary and often still the CNS is not removed and examined during routine postmortem examinations. Many pathologists are daunted by the prospect of examining brain and spinal cord, thus perpetuating the 'myth of neuropathology' (Baumgartner, 2015). In fact, a morphological diagnosis is not always difficult: Baumgartner points out that most CNS structures can be identified in haematoxylin and eosin-stained sections, with assistance from a textbook on neuroanatomy and neuropathology.

As far as necropsy is concerned, a useful general reference text to aid in diagnosis is the *Atlas of Gross Neuropathology: A Practical Approach* (Kurian et al., 2014). There are also various journals that cover neurology and neurological pathology and an important one insofar as the clinical and laboratory investigation of traumatic brain and spinal cord injury is concerned, in both humans and animals, is the *Journal of Neurotrauma*.

PATHOLOGY AND DISEASES OF THE NERVOUS SYSTEM

Abee et al. (2012) include in their book a chapter on nervous system disorders of NHPs and research models.

McGee et al. (1992) pointed out that in humans important causes of damage to the nervous system are hypoxia, infections, cerebrovascular diseases, intracerebral and subarachnoid haemorrhage, metabolic diseases (toxic and storage), demyelination, degenerative disorders, tumours and developmental anomalies. It has traditionally been assumed that a very similar situation applies to gorillas, but published data on neuropathology, as opposed to neuroanatomy, of this taxon are remarkably few.

The sparse literature to date concerning neurological diseases of gorillas does, however, suggest that lesions are likely to fall within the categories in Box 15.1.

Regrettably, a large proportion of gorillas examined postmortem have not had their nervous system properly examined. The reasons for this are many but usually relate to (1) pressures of time; proficient removal of the brain (and especially of the spinal cord) is neither an easy nor a rapid task. In addition, however, there are often restrictions on how much can be done to the CNS because so often a museum requires the animal's body in as complete a state as possible. For whatever the reason, many free-living gorillas over the years have not had their nervous system examined and zoo postmortem records are littered with such comments as 'Nervous system: brain not examined, spinal cord not examined' or 'The brain is macroscopically normal. The spinal cord is not examined'.

So, what lesions and which diseases have been described in gorillas? The immediate answer is 'not many', as first pointed out by Scott (1992), who wrote (see also later) 'For reasons that are not clear, lesions of the CNS … seem not to be as frequent as in humans. For instance, no lesions of the CNS, resembling those in humans, were found in 288 animals seen in San Diego and other centres, between 1964 and 1983. Lesions were found in apes far more frequently than in monkeys, being reported most often in gorillas, orangutans, and chimpanzees'.

There follow brief notes and comments on conditions and findings in gorillas in which the nervous system was either primarily or secondarily affected.

Sixty years ago, Steiner et al. (1955) described neuropathological changes, in conjunction with cardiopathy, haemosiderosis and testicular atrophy, in a captive lowland gorilla. Ten years later poliomyelitis was reported in two gorillas in a laboratory by Allmond et al. (1965). A cerebral infarction associated with coarctation of the aorta was described in a captive lowland gorilla by Trupkiewicz et al. (1995).

Very little has been published on brain or spinal surgery in gorillas but Jandial et al. (2004) recounted a lumbar discectomy in a human-habituated lowland gorilla. The animal had developed progressive lower extremity weakness, which progressed until the gorilla was unable to walk. MRI showed a herniated disk at the L1−L2 level. Surgical laminectomy resulted in complete resolution of clinical signs.

Post mortem findings in 'an aged albino gorilla' were described by Márquez et al. (2008). At the time of examination the animal was aged (40 years old) and had suffered over the previous 2 years from progressive tetraparesis, nystagmus and dyskinesia of the arms, hands and neck, with accompanying abnormal behaviour. Post mortem examination revealed changes in the brain, some relating to ageing. There were numerous corpora amylacea, especially in the substantia nigra, and large numbers of axonal spheroids often linked with iron accumulation in the internal globus pallidus. Sequencing of the gorilla pantothenate kinase 2 (*PANK2*) gene failed to detect any mutation. The authors stated that the clinical, neuropathological and genetic findings in the gorilla suggested an age-related pallido-nigral degeneration that presented with PKAN (pantothenate kinase-associated neurodegeneration)-like neurological deficits. Pallidonigral spheroids associated with iron deposition have been observed in some aged clinically normal NHPs. In humans similar findings are seen in neurodegeneration associated with brain iron accumulation diseases which, in some cases, show mutations in the PANK2 gene that prompts the production of the enzyme pantothenate kinase.

Minter et al. (2012) gave a comprehensive account of Reye's or a Reye's-like syndrome in a captive female lowland gorilla. Reye's syndrome is a rare, but serious, condition that usually affects the liver and the brain of children and young people.

Neurological lesions are often seen incidentally on routine histology. For example, Lydia Tong (personal communication), Taronga Zoo, Australia, commented to the author (JEC) that tissues from gorilla 'Betsy' (metastatic carcinoma in the lung and liver, possibly from a pancreatic islet cell carcinoma), showed bilateral, locally extensive, leukomalacia of the corpus striatum.

Infectious causes of neurological disease in gorillas include those due to parasites. A very significant one is *Balamuthia mandrillaris*, the free-living amoeba that is the cause of granulomatous amoebic encephalitis in humans.

For example, data provided by Roberta Wallace, Milwaukee County Zoo (personal communication) included a male of 24 years that was euthanased following chronic progressive neurological signs for over a year. The final diagnosis was acute leukomyelomalacia of the cranial cervical spinal cord, superimposed on diffuse degeneration leukomyelopathy/malacia. It appeared to be an acute traumatic event resulting in lesions of the cranial cervical cord following chronic injury to the spinal cord, possibly secondary to degenerative vertebral disease.

Earlier, Wallace et al. (1997) had reported briefly on a case of leptomyxid amoebic meningoencephalitis in a captive lowland gorilla.

Rideout et al. (1997) presented a retrospective review of fatal infections associated with *B. mandrillaris* using the pathology database for the Zoological Society of San Diego over 30 years. They located five cases of amoebic meningoencephalitis, all in Old World primates and these included two western lowland gorillas. Two distinct patterns of pathology were identified. A gibbon, mandrill and the young (1-year-old) gorilla had an acute to subacute necrotising amoebic meningoencephalitis with a short clinical course while the adult gorilla and a colobus monkey had a granulomatous amoebic meningoencephalitis with extraneural fibrogranulomatous inflammatory lesions and a long clinical course. Indirect immunofluorescent staining of amoebae in brain sections with a *Balamuthia*-specific polyclonal antibody was positive in all five animals. Direct examination of water and soil samples from the gorilla and former mandrill enclosures revealed unidentified amoebae in 11/27 samples, but intraperitoneal inoculations in mice and attempts to isolate amoebae from frozen tissues from the adult male gorilla were unsuccessful.

A few years later Mätz-Rensing et al. (2011) gave an account of a fatal *B. mandrillaris* infection in a gorilla, described as 'the first case of balamuthiasis in Germany'. The 12-year-old female western lowland gorilla died in a zoological garden after exhibiting general neurological signs.

The animal was found to have a chronic, progressive, necrotising amoebic meningoencephalitis and *B. mandrillaris* was implicated as a result of indirect immunofluorescent staining of brain sections and PCR.

Gjeltema et al. (2016) reviewed amoebic meningoencephalitis and disseminated infection caused by *B. mandrillaris* and concluded that *B. mandrillaris* has been reported to account for 2.8% of captive gorilla deaths in North America over the past 19 years. They urged clinicians working with gorillas to have this condition on their list of differential diagnoses when evaluating animals with signs of centrally localised neurological disease. See also the fact sheet compiled by Manfred Brack, formerly of the German Primate Center, Göttingen/Germany (EAZWV Transmissible Disease Fact Sheet, 2008).

Other protozoa may also be involved. Anderson et al. (1986) recounted a case of pneumonia and meningoencephalitis associated with amoebae in a 371-day-old captive lowland gorilla. The affected gorilla died after a 19-day illness that had been characterised by a decline in CNS function, 'whimpering', lethargy and muscular weakness and terminating in coma. The animal presented with scabs on its ankles, wrists, near the left palpebra and ear, right ear, nape of the neck, and lower back. The right palpebrae were partially fused by crusted exudate. The pupils were moderately constricted and symmetrical. Clear, colourless, CSF was aspirated from the ventral aspect of the exposed occipitoatloid junction − see later.

On postmortem examination of the animal's brain three 1-cm-diameter foci, each composed of many petechiae, were seen in the left dorsal cerebrum. A soft focus of yellow discoloration 1−2 cm diameter in size was observed in the right dorsocaudal cerebrum. When this focus was incised for culture, a deeper 1-cm diameter area of numerous petechiae was found. Cross sections of the cerebrum revealed many red, tan or yellow foci situated mainly along sulci. The animal's cerebellum and medulla oblongata appeared superficially to be normal but the leptomeninges were oedematous and moderately hyperaemic.

Several grey-green areas of melmanosis (*sic* − presumably melanosis) were seen in leptomeninges of the thoracic and lumbar spinal cord. The lungs were oedematous and the right caudal lobe was firmer and had a darker red and blue coloration than did other portions of the lung. The right ventricle of the heart had 'slightly rounded contours'.

On histological examination large numbers of plasma cells, eosinophils, lymphocytes and macrophages were seen in all areas of the leptomeninges; they extended into the perivascular Virchow-Robin spaces of the cerebrum. Large irregular areas of malacia and diffuse gliosis were noted in the cortex, usually associated with large numbers of unicellular organisms that were later identified as *Acanthamoeba*, and with vascular necrosis, haemorrhage, infarction and early fibrous astrocytosis. Some lesions in the cortex were more diffuse and had large areas of mature fibrous astrocytosis. The lesions appeared to radiate from the vasculature. Numerous trophozoites and a few cysts containing amoebae were found in deep areas especially those adjacent to blood vessels. Small numbers of encysted amoebae were seen in chronic lesions. The leptomeningeal exudate of the cerebellum was similar to that over the cerebrum and numerous amoebic trophozoites were seen in some areas of gliosis. Extensive patches of gliosis and fibrosis were observed in folia of the cerebellum. Dorsal horn neurons of the thoracic and lumbar spinal cord showed central chromatolysis. There were no obvious changes in one enucleated eye, the nictitating membrane, the optic nerve, trigeminal ganglion, peripheral nerves or spinal root ganglia.

Many pulmonary alveoli were consolidated by haemorrhage, neutrophils, alveolar macrophages, serofibrinous exudate and numerous amoebic trophozoites. The nasopharynx had moderate aggregates of lymphocytes in the mucosa and submucosa; amoebae were absent.

No bacteria or fungi were cultured from brain specimens collected at necropsy. CSF aspirated 5 days prior to death, serum collected at necropsy and serum from the animal's dam, sire and a 2-year-old cagemate showed no indirect immunofluorescent antibody titre to five species of *Acanthamoeba*, two species of *Naegleria* or to *Hartmanella vermiformis*. The relatively large size of the amoebae and the presence of cyst forms within lesions suggested an *Acanthamoeba*-like rather than a *Naegleria*-like amoeba as the aetiological agent of the disease.

This case is described here in some detail not only because of the important part played in diagnosis by gross and histological investigations but also because of the environmental implications.

The collection's drains had been clogged on six occasions during the 3 months prior to the gorilla's demise. Three days after death, water samples were examined from locations within the zoo. Four water sources contained trace numbers of amoebic trophozoites but no attempt was made to identify the species of these amoebae.

The exact route of infection in this gorilla remained uncertain but the lack of lesions in the nasal and ethmoid areas, the absence of dermal reaction below scabs on the skin, the presence of amoebic pneumonia and the vascular distribution of amoebae in the brain led the investigators to postulate that the primary route of infection was respiratory, with embolic secondary infection of the brain. Encysted amoebae are hardy and can be harboured in moist soil and bodies of water; the gorilla may, therefore, have been exposed by contact with standing water or moist soil in the exhibit area.

Hygiene may be relevant to other diseases with a neurological component too. For example, tetanus is assumed to occur in gorillas and many zoos and wildlife projects include the administration of tetanus toxoid to animals that have suffered injuries where anaerobiosis is a feature — e.g., as a result of snare injury (see Chapter 3: Infectious Disease and Host Responses), as well as for staff working in gorilla habitat (see, e.g., Stoinski et al., 2008).

Rabies does not appear to have been definitively diagnosed in gorillas but the high prevalence of the disease in dogs in some parts of West and Central Africa, coupled with an increase in human–gorilla interaction and the fact that dogs are sometimes used to poach gorillas could change this. Other carnivores might also play a part. In a (now lost) letter to Joan and Alan Root on 28 July 1973, Dian Fossey said that a dog had been killed at Karisoke by Sandy Harcourt and that they 'had noted strange behavior in some gorillas'. She went on to express concern about the possible role of jackals in spreading rabies in the Virungas.

Various studies on captive NHPs of different species have demonstrated changes in the CNS associated with ageing and pathologists including the author have long noted that, as in many other animal species, there is a progressive increase in lipofuscin in the brain as a gorilla ages. There are more specific features in some NHPs too — for example, a decline in hippocampal neurogenesis (Gould et al., 1999). Psychological studies have been carried out on animals too — for instance, the research by Anderson et al. (2005) on relative numerousness judgment and summation in young and old lowland gorillas.

Hof et al. (2002) provided a complex, but valuable, overview of the comparative neuropathology of brain ageing in primates as understood then, nearly 15 years ago. The studies were performed in the context of the Great Ape Aging Project and involved specimens from lowland gorillas, common chimpanzees and orangutans. The brains of aged apes were compared with those from young specimens and also with brains obtained from elderly human controls. The results cannot be discussed here but included quantitative data on neuron (neurone) numbers and on volume rendition (using postmortem MRI scans).

More recently, in a comprehensive review of the comparative pathology of ageing great apes, including gorillas, Lowenstine et al. (2016) confirmed that there is

evidence that cognitive decline occurs in great apes. In aged chimpanzees and orangutans, only certain types of tasks were affected—such as social cognition, some motor skills, and, in the case of orangutans and gorillas, relative numerousness judgment (see also above). Pathological changes characteristic of brain ageing in *Homo*, such as the senile plaques of Alzheimer disease and cerebral vascular disease, have been reported in NHPs, including gorillas, chimpanzees and orangutans. Other lesions, such as dystrophic neurites and tau-neurofibrillary tangles, are less common, leading some authors to believe that true Alzheimer's disease is unique to the human ape.

Notwithstanding the comment above, Perez et al. (2013) described changes consistent with Alzheimer's disease in the neocortex and hippocampus of the lowland gorilla. They explained that the two major histopathological hallmarks of Alzheimer's disease in *Homo* are amyloid beta protein (Ab) plaques and neurofibrillary tangles (NFT). Ab pathology is a common feature in the aged NHP brain, whereas NFT are found almost exclusively in humans. In their study they used immunohistochemistry and histochemistry to search for Ab and tau-like lesions in the neocortex and hippocampus of aged male and female gorillas. They detected an age-related increase in Ab-immunoreactive plaques and vasculature; the plaques were more abundant in the neocortex and hippocampus of female animals, whereas Ab-positive blood vessels were more widespread in males. Ab plaques were less fibrillar (diffuse) in gorillas than in humans. Although phosphorylated neurofilament immunostaining revealed a few dystrophic neurites and neurons, choline acetyltransferase-immunoreactive fibres were not dystrophic. Neurons stained for the tau marker Alz50 were found in the neocortex and hippocampus of gorillas of all ages. Occasional Alz50-, MC1-, and AT8-immunoreactive astrocyte and oligodendrocyte coiled bodies and neuritic clusters were seen in the neocortex and hippocampus of the oldest gorillas.

The cellular aspects of the pathology of neurodegenerative disease is complex and cannot be discussed here. Most such conditions are characterised by an accumulation of mutant or toxic proteins. Autophagy aids cell survival by removing unwanted cellular organelle and protein aggregates.

Neoplasms affecting the nervous system of gorillas appear to be rare. Scott (1992) wrote 'For reasons that are not clear, lesions of the CNS, including tumours, seem not to be as frequent as in humans'. Therefore it was interesting to hear from Sam Young (personal communication) in Australia about lowland gorilla 'Mouila' who was diagnosed with a meningothelial meningioma (Fig. 15.4).

Part II: A Catalogue of Preserved Materials refers to a gorilla that was euthanased 'after a debilitating stroke caused by a large brain tumour' but no details are available. Other unsubstantiated reports include 'George', Pittsburgh Zoo, who 'died on 23 December 1978 of brain inflammation resulting from a dental abscess'.

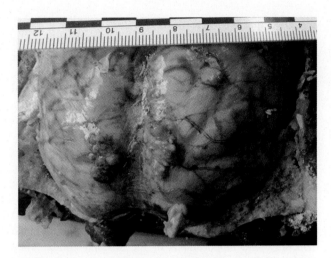

FIGURE 15.4 Gross appearance of brain of lowland gorilla, showing proliferative lesions on meninges, a meningothelial meningioma. *Image courtesy of Sam Young.*

Noninfectious causes of neurological disease in gorillas were mentioned briefly at the beginning of this section. A number of conditions can produce clinical signs suggestive of nervous disease but few detailed investigations appear to have been performed on the pathogenesis of such conditions. An exception was the account by Chatfield et al. (2010) of a case of severe idiopathic hypocalcaemia in a juvenile lowland gorilla which required veterinary investigation on account of clinical signs of tetany of both hands. The gorilla had had alternating periods of constipation, diarrhoea and bloating since birth. The diagnosis of idiopathic hypocalcaemia was based on the animal's markedly low blood calcium, a normal vitamin D value, the clinical response to therapy and eventual resolution of the condition.

Traumatic damage, often fatal, is an important cause of nervous tissue pathology in gorillas.

Damage to the skull is particularly prevalent in gorillas, especially in adult males and young, dependent, animals (infanticide) (see Chapter 14: Musculoskeletal System) (Fig. 14.9). Free-living gorillas in Africa may be exposed to blasts from ordnance in war-torn locations (see Table 4.1 in Chapter 4) and this can cause direct or indirect damage to the CNS, to the PNS and to organs of special sense.

Mental Disorders

Neuroses, defined here as mental illnesses that are not caused by organic pathology but often associated, certainly in humans, with symptoms and signs of depression, anxiety and obsessive behaviour, will only be covered briefly here.

It was Hebb (1947) who first postulated that nonhuman great apes might develop 'spontaneous neurosis'. He based this hypothesis on disturbing traits in chimpanzees in the Yerkes laboratories (United States), in particular those observed in an adult female chimpanzee ('Kambi') that exhibited signs that were 'almost certainly diagnostic for bipolar disorder (manic-depression)'. The animal gave the impression of being 'an unstable introvert' and post-adolescence her troughs of apparent depression deepened. Hebb described this as follows: 'She may appear to be fairly well-adjusted and not unhappy for several months at a time, and then to be in a profound depression without intermission for periods as long as 6−8 months'. During the long depressive episodes Kambi was extremely unresponsive to environmental events and showed a total lack of spontaneous activity. She would sit for hours with her back to the wall, hunched over and staring at the floor of the cage, not looking up when attendants passed. Researchers were not able to ascribe her condition to any physiological specific origin. Hebb stated that the various behavioural signs displayed by Kambi were all consistent with the human disease bipolar disorder. Kambi remained maladjusted until she died of dysentery.

Cousins (2015) described another syndrome, possibly a seasonal affective disorder, this time in a male gorilla, 'Kubie', who lived in Stone Park Zoo, Boston, United States, who developed an annual seasonal pattern of disorders, including vomiting and diarrhoea, as well as what was believed to be distress behaviour such as repeated nose-blowing, increased appetite and excessive regurgitation and reingestion (see Chapter 11: Alimentary Tract and Associated Organs). His father had also had a history of vomiting and had suffered epileptic seizures during infancy (Cousins, 1974). Kubie died in 1997 while undergoing root canal surgery.

Other disturbed behaviour in captive great apes, including gorillas, is discussed in Chapter 8: Nonspecific pathology.

Peripheral Nerves

Trauma is probably the commonest cause of peripheral nerve lesions in gorillas. Muscle and peripheral nerve injuries can also result from stretching or compression and this is a potential problem when anaesthetising gorillas because of their size and weight (Ludders et al., 1987). Gray (1965) described paraplegia in a male lowland gorilla. In their account of retroperitoneal abscess in two lowland gorillas Hahn et al. (2014) listed lameness and hindlimb paresis amongst the clinical signs displayed.

In concluding this section, the words of Scott (1992) nearly 25 years ago, would appear still to be valid: 'the scope of investigation of the primate nervous system leaves a certain amount to be desired compared with research into human diseases'.

THE SPECIAL SENSES

The following sections cover the eye, nose and ear, together with associated structures. The embryological origins of these were outlined earlier in this chapter. A useful general introduction to the pathology of the special senses was a chapter in the book edited by Fiennes (1962).

The Eye

According to the field observations of Schaller (1964) 'sight is the most important sense' in gorillas. The superficial appearance of the eyes of the genus *Gorilla* was described by Napier and Napier (1967) as 'deeply set under a prominent supra-orbital ridge'. An often overlooked paper in German by Rohen (1962) described the microscopic anatomy of the gorilla eye. There was a notable similarity to the human eye. Extensive pigmentation was found in the choroid only; the remaining areas of uvea and sclera contained only scattered chromatophores. On the surface of the iris crypts and lacunae were present. The retina of one eye had degenerated but the other was normal and resembled the retina of *Homo*. Interest in both the normal anatomy (Fig. 15.5) and the pathology of the eye of the gorilla continues − see below.

Fifty years later the comparative ocular anatomy of *Gorilla* was investigated by Knapp et al. (2007); amongst other things, their research encompassed measurements and studies on globe dimensions, limbal conjunctival epithelium, choroid, iris, posterior lens capsule and cornea. They concluded that, although there is some variation between *gorilla* and *beringei*, the gross and microscopical features are very similar. Knapp et al. emphasised that the eye of *Gorilla* 'appears remarkably similar to the human eye'.

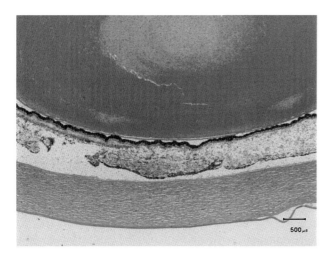

FIGURE 15.5 Normal histological appearance of part of the eye of the lowland gorilla. The lens is at the top of the image, with the cornea at the bottom and the iris between the two. This specimen was promptly removed and fixed postmortem; unfortunately, this is often not practicable under field conditions in Africa.

Vision appears to have played a very significant part in the origin and evolution of the primates and as a result much research has been directed towards elucidating the development of the eye. Relevant seminal references include Martin and Ross (2005) and Ross and Martin (2006).

The latter argued that interpreting the role of vision in primates requires an understanding of the integration of visual system function. Evolutionary changes to the visual apparatus were accompanied by improved abilities to localise sounds or movements, increased manual dexterity and changes to somatosensory and somatic motor systems that permitted greater control of visually guided reaching and grasping movements. These changes were manifest not only in the musculoskeletal periphery but also throughout the CNS, including the corticospinal tracts, the premotor areas that control limb movements, the visual cortex, and the primary and secondary sensorimotor areas.

Ocular diseases in gorillas of varying severity have been described or reported in publications. They range from mild conjunctivitis to cataracts, with an assortment of traumatic lesions.

Field observations usually provide little detailed information on the aetiology or pathogenesis of ocular diseases but are a useful indicator of the prevalence of such conditions. They date back many decades; thus, Schaller (1964) said of a female mountain gorilla 'Somehow she has injured her eye, blinding it, and now the lens was opaque . . .'.

At the PHVA for the mountain gorilla (Werikhe et al., 1997) two ocular conditions were recognised by those working in the field − 'eye infection' and 'strabismus (crossed eyes)'. The former was generally characterised by ocular discharge, often serous, and may well have been conjunctival or corneal damage, not necessarily an infection. Strabismus is mentioned later. In his book *The Magnificent Gorilla* Cousins (1990) mentioned *Dipetalonema vanhoofi* infection in captured gorillas and chimpanzees in Congo-Brazzaville (CR) and stated that some infected animals showed swelling of the eyelids.

Gorillas in zoos are also not infrequently reported as having 'conjunctivitis', which usually resolves spontaneously. More serious ocular conditions are also diagnosed or suspected; for example, possible macular degeneration described by Steinmetz et al. (2012) in a captive animal with osteoarthritis and severe periodontal disease that was noted to have problems with near, but not far, vision. Ophthalmological examination of this animal revealed depigmented spots in the maculae similar to optic disc Drusen bodies, leading to a diagnosis of the dry form of macular degeneration. In their review of the comparative pathology of ageing great apes, including gorillas, Lowenstine et al. (2016) − see earlier − were referring to the case above when they stated that 'NHPs have a macula and retinal anatomy similar to that of humans but there is only one published report of suspected macular degeneration in apes'. It is not clear whether this claim was based on a comprehensive, multilingual, literature search (see Introduction to References) or only publications in English.

Lowenstine et al. (2016) also recounted the case of retinal disease due to vascular lesions that was described in a 34-year-old female gorilla with a history of hypertension by Niemuth et al. (2014). This animal developed acute blindness due to bilateral retinal detachment and intraocular haemorrhage. She also had cataracts. Retinal vascular arteriolosclerosis characteristic of hypertension was found. Lemen et al. (1974) described marasmus in two infant gorillas with shigellosis.

Cataracts have been reported in gorillas on a number of occasions and Lowenstine et al. (2016) referred to them in both young gorillas and orangutans with reference to, amongst other works, Leiva et al. (2012) and Liang et al. (2005). As far as older animals are concerned (the focus of the paper by Lowenstine et al.), reports of cataracts in apes in zoos have included gorillas aged 37, 39, and 42 years. Lowenstine et al. also stated that 'cataracts have also been identified in wild [*sic*] mountain gorillas in the Mountain Gorilla Veterinary Project database (Gorilla Doctors, L.J.Lowenstein unpublished data)'. All these animals were said to be adults. In most cases, the lesions were mild, but one 24-year-old female had a mature mineralised cataract that would have impaired vision while another female, estimated to be 37 years old, had extensive bilateral cortical cataracts.

Lydia Tong (personal communication) reported bilateral cataracts, as well as atherosclerotic plaques, in the aorta and ulceration of the gastric mucosa in gorilla 'Buluman' (see Part II: A Catalogue of Preserved Materials). Published papers describing cataracts in gorillas usually discuss the surgical management of such cases rather than the aetiology and pathogenesis of the lesions. For instance, de Faber et al. (2004) reported that two juvenile, male, captive-born lowland gorillas with the same father but different mothers developed bilateral cataracts. The cataracts were surgically removed and foldable intraocular lenses were implanted. Intraocular lens implantation in a captive gorilla was also described by Clouser (2010).

Other cases have not been published but were reported in the media and on the internet; an example was 'Romina', a gorilla at Bristol Zoo Gardens, UK, that attracted a great deal of public interest in Britain in 2007, not least of all because the animal was operated upon by a joint veterinary/medical team (Uhlig, 2002).

Investigation of the eyes of gorillas, live or dead, does not differ greatly from that of humans or other primates. Liang et al. (2005) provided a useful account of clinical ophthalmological examination of the lowland gorilla and the equipment and techniques required. They presented some practical results including mean intraocular pressure, mean horizontal corneal diameter, mean vertical cornea diameter and mean lens thickness. The authors argued that their study suggested important similarities between the eyes of the western lowland gorilla and *Homo* and that these similarities may permit techniques, and equipment used for human eye surgery to be applied to ophthalmological work on gorillas.

The importance of meticulous and detailed examination of the eyes and associated structures, both in live and dead gorillas, cannot be overemphasised. In the context of postmortem work, this means that adequate time must be spent on the task, using a strong source of illumination and appropriate equipment — not always to hand when working in the field — before the necropsy commences. The following extract from a report on a 2-year-old female lowland gorilla at BZG, UK, illustrates this well: '*External examination*: The carcase of a small female gorilla. Bodyweight: 6.15 kg. The right eye is closed and both conjunctival sacs contain purulent material. A bulging 1 cm diameter abscess is present just below the right eye. There is a purulent discharge from the left nostril'.

Strabismus has been recorded in both free-living gorillas, of both species, and mention of it is to be found, usually as an incidental observation, in a number of papers dealing primarily with other topics. Strabismus is fairly easily noticed in habituated animals and was an incidental peculiarity ('a lazy eye') of the young mountain gorilla that features in a painting by the American poet and artist Charles Alexander.

The clinical management of a unilateral paralytic strabismus in a lowland gorilla, a captive animal, was described by Rapley et al. (1980). In humans, studies on twins and family members suggest that there may be a significant genetic component to the aetiology of strabismus; this may also apply to gorillas but research on the molecular genetics of the disorder is needed.

Other conditions that may be heritable and/or primarily genetic in aetiology have attracted interest because of the high inbreeding coefficient of some captive and free-living gorillas (see Chapter 17: Field Studies in Pathology and Health Monitoring). In the context of ocular disease, de Faber et al. (2004) — see earlier — urged that a possible genetic component in juvenile cataracts should be considered in breeding management programmes.

The famous white gorilla at Barcelona Zoo — see Chapter 9: Skin and Integument, and Jonch (1967) — displayed interesting ocular changes relating to depigmentation. Jonch wrote: 'The eye had a blueish sclera, a normal cornea, and a light blue iris which was very transparent to transillumination. The media were transparent and the fundus of the eye normal and totally depigmented. The choroidal vessels were perfectly visible and the pupil was normal. The animal displayed marked photophobia which caused it to close its eyes repeatedly when exposed to bright light. In diffuse light similar to that in its biotope, we calculated that it blinked on an average of 20 times a minute'.

A further case that may have had an underlying genetic aetiology and pathogenesis was the young gorilla with congenital obstruction of the nasolacrimal duct described by Nagashima et al. (1974).

Marked swelling in the area of the left lacrimal sac was first noted on the fifth day of life. There was slight epiphora but no discharge was seen. A large amount of transparent mucus with a little white pus was eliminated through the lacrimal puncta when digital pressure was applied over the lacrimal sac. A developmental anomaly at the lower end of the osseous nasolacrimal canal was detected on probing. Retrograde introduction of a polyethylene tube into the nasolacrimal duct and lacrimal sac through the inferior nasal meatus elicited full clinical recovery. It is worth noting in the context of the relevance of anatomical studies to clinical medicine (see Chapter 1: The genus *Gorilla* – Morphology, Anatomy and the Path to Pathology and Chapter 2: The Growth of Studies on Primate Pathology) that prior to attempting treatment of the condition Nagashima et al. acquired a dead young gorilla ('from a certain animal dealer'!) and perfected dissection of its lacrimal system.

The removal and correct fixation and sectioning of the eyes requires skills in dissection and processing. Fixation with formalin can result in tissue shrinkage and artefact, as pointed out by Knapp et al. (2007) – see earlier. Other fixatives may be preferable - glutaraldehyde (Yanoff et al., 1965) or Bouin's or Davidson's fluid (Latandresse et al., 2002).

The relative brevity of this section of the book that deals with the eye and other ocular structures speaks for itself. Lowenstine et al. (2016) put it concisely: 'Ophthalmologic pathology of the apes is yet another area needing further investigation'.

The Nose

Conspicuously flared nostrils are a characteristic of *Gorilla* (Napier and Napier, 1967).

Although Schaller (1964) described the sense of smell in mountain gorillas as 'roughly comparable to my own', the importance of it cannot be doubted, as evidenced by behaviour of these animals in the wild and captivity, but little research appears to have been performed on this. The relevance to pathology is that a gorilla with nasal damage, or perhaps a nasal infection, may be less able to respond to odour molecules that, through the olfactory bulb inform the animal of the presence of other creatures (including humans), food and possibly its location (spatial information). As in so many other species (Bhutta, 2007), the gorilla may use olfaction in pheromone communication.

The Ear

The ears are smaller in proportion than in humans, much smaller than in the chimpanzee; but the structure of the auricle is more like that of man. On a direct front view of the face, the ears are on the same parallel with the eyes.

The ears of *Gorilla* were described by Napier and Napier (1967) as 'small and set against the head'. Schaller (1964) said that 'Hearing, too, is well developed in gorillas'. The sounds made by gorillas and acoustic structure and variation have been studied by a number of researchers. The 'close call' vocalisation of mountain gorillas that serves the purpose of reassuring others in the group, was described by Norton (1990) as 'Mhemm, mhemm'.

Auditory disease and pathology are poorly represented in the literature but there are interesting observations in unpublished reports reviewed by the author (JEC). For example, one postmortem report from Bristol Zoo Gardens included: 'The left ear canal contains an elongated polyp which arises in the middle ear. The right ear appears healthy'.

The effect on gorillas of acoustic trauma, an increasing concern of city zoos, especially if their premises are used out of hours for social events such as concerts, does not appear to have been properly investigated.

Chapter 16

Endocrinological and Associated Conditions

John E. Cooper

The pituitary is the leader in the endocrine orchestra.

Sir Walter Langdon-Brown (1931)

For we are members one of the other.

Ephesians 4. 25

INTRODUCTION

This chapter is primarily concerned with diseases of the endocrine system together with discussion of various 'miscellaneous' conditions that relate in part to endocrine function, including stress and sleep.

The endocrine system is made up of various organs that produce hormones – pituitary, thyroid, parathyroid, adrenal, pancreas, pineal, ovary and testis. The last two are covered in Chapter 13, Urinary and Reproductive Systems.

THE ENDOCRINE SYSTEM

In appearance and function the endocrine organs appear to be very similar in *Gorilla* and *Homo*.

A useful starting point in consideration of endocrine pathology of nonhuman primates (NHPs) is the chapter by Spies and Clegg in Fiennes (1972). Prior to that, Rasmussen and Rasmussen (1952) had described the pituitary of 'Bushman', a gorilla at Lincoln Park Zoo, United States. In his overview of the primate endocrine system, Scott (1992) commented that 'There is a great dearth of information on endocrine disorders in primates'. That remains the case, especially in respect of gorillas – so this chapter is a short one!

This does not mean that pathological changes are rare in gorillas. On the contrary, lesions are not uncommonly noted in histological sections but they are generally relatively minor, often consisting of chronic changes such as fibrosis and mononuclear cell infiltration. Mention of

such lesions is to be found in many postmortem reports reviewed by the author (JEC), but rarely are they reported as such in publications.

It is possible that some diseases of endocrine origin are missed. The various endocrine organs, often working together, regulate metabolism, growth, development, tissue function, reproduction, sleep, and the body's response to stressors so clinical disease (and postmortem findings in particular) may or may not be linked to the appropriate organ. Endocrine pathology can be summarised as (1) abnormalities in synthesis and secretion of hormones, and (2) abnormalities affecting receptors (target cells) and postreceptor responses. The effects are summarised in Box 16.1.

As Scott (above) implied, more research is required on endocrine disorders in gorillas and other NHPs (see Box 16.2).

Each endocrine organ will now be discussed in turn.

Pituitary (Hypophysis)

Interest in the pituitary gland of the gorilla dates back nearly a century, when Keith (1926) analysed the then ongoing debate about human acromegaly and the apparent 'gorilla-like' characteristics shown by some affected patients. Keith was aware of the role of the pituitary in growth and development and wisely cautioned against drawing conclusions about apparent similarities in the

---

BOX 16.1 Effects of Endocrine Pathology

- The organ itself – change in:
  - size
  - shape
  - colour
- Target organs and the rest of the body

J. E. Cooper & G. Hull: Gorilla Pathology and Health. DOI: http://dx.doi.org/10.1016/B978-0-12-802039-5.00016-0

gland's possible action based solely on 'the crude use of scales or microscope'. Morphological investigation of the pituitary of *Gorilla* has been hampered by the need for meticulous and sometimes lengthy dissection. Removal of structures from the hypophysial fossa, the deepest part of the sella turcica, is never easy; as in humans, the pituitary is readily damaged. Gresham and Turner (1979) described in detail how to remove the brain (of *Homo*) and then inspect the inside of the base of the skull. They went on to explain how to open the various sinuses with a scalpel to check for thrombi before removing the pituitary in toto. Terio and Kinsel (2008) provided a useful, step-by-step, pictorial guide to removal of the pituitary but their publication lacks detailed text.

Pituitary-associated disease is probably more common in gorillas than the literature suggests.

Chatfield et al. (2006) reported on a 17-year-old lowland gorilla that presented with low fertility and hyperprolactinaemia. A magnetic resonance imaging (MRI) scan confirmed the presence of a pituitary mass. Following treatment with cabergoline (a potent dopamine receptor agonist), the mass decreased in size and serum prolactin returned to normal limits. The gorilla commenced breeding behaviour and became pregnant. Chatfield et al. pointed out that prolactin-secreting pituitary adenomas are one of the most common causes of infertility in women and are easily diagnosed and treated. They postulated that hyperprolactinaemia, secondary to a presumed small tumour, in this case a microprolactinoma, may be more common among breeding-age gorillas than is currently realised.

Thyroid

As far as is known, the thyroid glands serve the same role in *Gorilla* as in *Homo* − that is, the production of the hormones thyroxine (T4) and triiodothyronine (T3) that regulate growth rate of metabolism and other aspects of metabolism. Calcitonin, in humans and probably gorillas, has an undefined role, but is possibly involved in calcium homeostasis.

A detailed description of the thyroid of an adult mountain gorilla was published by Venable and Grafflin (1940) and of its histological appearance by Grafflin (1940).

A useful background to the pathology of the thyroid in NHPs is the account of findings in tumours of the gland by Ippen and Wildner (1984). This survey did not include gorillas but provided data on thyroid neoplasia from captive animals in various zoological gardens and circuses and also some from the wild, all over a 10-year period.

It has been shown (McLachlan et al., 2011) that there is a paucity of information regarding thyroid disease and pathology in great apes, especially regarding function tests and autoantibodies. In an 'extensive search' of the great ape literature for evidence of spontaneous Graves' disease (a form of thyroid autoimmunity condition causing hyperthyroidism), McLachlan et al. located five reports of non-congenital thyroid dysfunction, four with hypothyroidism and one with hyperthyroidism. The last of these, a gorilla 'Benno' in the Dresden Zoo, received antithyroidal treatment but subsequently died; unfortunately neither serum nor thyroid tissue from the animal had been retained.

The reports referred to above include four relating to gorillas − Lair et al. (1997, 1999, 2000) (diagnosis of hypothyroidism using human thyroid-stimulating hormone assay) and Schneider (1989) (hypothyroidism = hypothyreosis).

Studies on thyroid hormones, thyroid autoantibodies and thyroid histology are amongst data listed in *Publications on SSP Gorillas 2000−2008* (compiled by Elena Hoellein Less, Research Co-Advisor, Gorilla SSP): http://www.gorillassp.org/Forms/AZA_CaptiveGorillaResearch.pdf Omaha Zoo's 'Gorilla Thyroid Project' has a website: http://www.omahazoo.com/conservation/medicine/research/gorilla-thyroid-project/.

Thyroid disease may be uncommon in gorillas but lesions are not infrequently seen in histological sections of the gland from animals that have succumbed to other disease. For example, a 3-month-old female lowland gorilla died at Bristol Zoo Gardens in 1985 and a diagnosis of bronchopneumonia was made. At necropsy the thyroids were not positively identified but small cystic structures containing gelatinous material were found close to the larynx.

Histological examination showed the thyroid tissue to consist of lobules comprising colloid cysts of varying size (Figs. 16.1 and 16.2) but otherwise within normal limits. In the adjacent connective tissue there were also large cysts but these related to changes in the animal's thymus.

Parathyroid

The parathyroid glands are small endocrine glands that are usually positioned close to, but separate from, the thyroid gland. In both *Homo* and *Gorilla* there are two pairs.

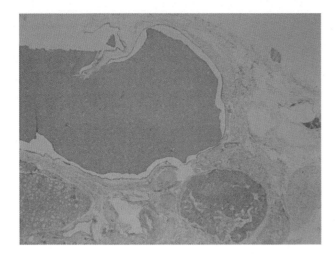

FIGURE 16.1 Thyroid of lowland gorilla, with adjacent parathyroid tissue, showing colloid cyst.

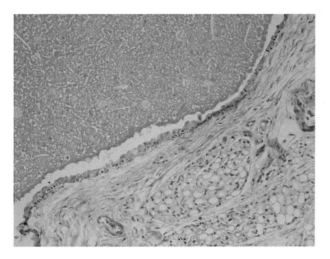

FIGURE 16.2 High power view of the above.

The superior pair is usually adjacent to or near the upper poles of the thyroid, while the position of the inferior pair is more variable; they sometimes migrate into the upper chest during growth. Their function is to secrete parathormone (parathyroid hormone) that by a complex mechanism controls the metabolism of calcium and phosphate.

The parathyroid glands were first identified by Richard Owen (see Chapter 1: The genus *Gorilla* – Morphology, Anatomy and the Path to Pathology) in 1852 in a rhinoceros that he dissected! As far as gorillas are concerned, the first thorough gross description of the parathyroids was in a paper detailing their detection and dissection in an adult mountain gorilla (Venable and Grafflin, 1940) and the histological appearance of the same two glands by Grafflin (1940).

In many species, including NHPs, parathyroid disorders range from those that are developmental (usually cysts) or degenerative (chief cells fuse to form syncytia) to hypoparathyroidism and hyperparathyroidism. In NHPs the last of these is – or certainly has been – the most prevalent. In *Gorilla* it is usually secondary, not primary, hyperparathyroidism and almost invariably is of nutritional origin (NSH = nutritional secondary hyperparathyroidism, often termed MBD (metabolic bone disease) or nutritional osteodystrophy) (see Chapter 14: Musculo-skeletal system). The cause is usually a diet that does not contain the correct Ca:P ratio (1.5:1.0 or 2.0:1.0) but the condition may be complicated, especially in primates, by low vitamin D3 causing bone resorption. NSH is one of the causes of the various 'bone diseases' that have been recognised in captive monkeys, especially but not exclusively New World species, for over a century (see, e.g., Fiennes, 1974).

In a recent paper Farrell et al. (2015) discussed metabolic disease in NHPs caused by dietary 'deficiency or inappropriate surroundings' with reference to NHP skulls in the Odontological Collection of the Royal College of Surgeons of England and a more recent assemblage at the National Museum of Scotland. They detected what they described as 'a recurring gross pathology' in 51 of these skulls, all apparently captive animals. Computed tomography (CT) scanning on a subsample of these skulls added weight to their suggestion that an MBD was 'the causative agent', with osteomalacia the likely diagnosis. The paper was written for a primatological, not a medical, readership and this is reflected in the terminology employed, especially that relating to pathology, but it provides a useful historical context and illustrates the inadequacy of zoo primate management during the late 19th and early 20th centuries. There is only one mention of gorillas – a reference to 'goundou' or yaws (see Chapter 14: Musculoskeletal System) but some of the text is helpful in describing calcium metabolism in primates and therefore emphasises the pivotal role played by the parathyroid glands.

A reference that is specific to gorillas, albeit only a single case history, is that by Chatfield et al. (2010), which describes severe idiopathic hypocalcaemia in a juvenile lowland gorilla. Outside peer-reviewed literature, in 2008 the British media publicised widely the involvement of a (human) consultant surgeon, in collaboration with veterinary staff, in surgery on a 22-year-old female lowland gorilla, 'Tambabi', at Howletts Animal Park, which had been diagnosed with hyperparathyroidism. The animal 'had suffered severe weight loss and required urgent neck surgery to remove two of the four parathyroid glands.'

These are benign tumours (adenomas), as described above, usually occurring in one, but occasionally more, of the parathyroid glands. (Previously these adenomas have been described as an inflammation or parathyroiditis but there seems to be little evidence for this in *Gorilla* and it is not the case in *Homo*). An adenomatous gland is no longer under the control of the normal feedback mechanism and autonomously secretes very high levels of parathormone. In humans and also possibly in gorillas, the clinical signs following decalcification of the bones can be so severe as to cause duodenal ulcers, kidney stones and, in severe cases, spontaneous fractures of the long bones due to 'osteitis fibrosis cystica' also called Von Recklinghausen disease — where whole lacunae of calcium have been reabsorbed so causing considerable bone weakness.

Other lesions are sometimes seen as 'incidental findings' in histological sections of the parathyroids of zoo animals, as the author (JEC) has observed when reviewing such reports. These are usually characterised by small clusters of mononuclear cells. Unfortunately, the parathyroid glands are often not specifically examined — a typical report from one British zoo read 'Pituitary and thyroids normal. Parathyroid glands not readily identified'. Others say 'The parathyroid glands appear usual in size and shape' — and, as a result are not submitted for histological investigation. Regrettably, they are also rarely weighed (see earlier).

The parathyroids are amongst the tissues that should be routinely stored as part of a reference or 'voucher' collection following the death and post mortem examination of a gorilla (see Part II: A Catalogue of Preserved Materials).

Adrenal

Although the literature contains a fair amount of information relating to the adrenals of gorillas, most of this refers to studies on 'stress' (see later). There is, apparently, very little published on the pathology of the gland or the various conditions that affect it in other species. In *Homo* the latter include hypoplasia, hyperplasia, traumatic injury (sometimes during surgery or endoscopy), adrenalitis and neoplasia. Even incidental findings, such as thrombosis, calcification and amyloidosis (seen in many animals) are rarely recorded — certainly in the reports from various zoos reviewed by the author. There are exceptions; necrosis of the adrenal is seen following infections and, particularly, toxaemia/septicaemia. Routine examination of the adrenals of a gorilla from one British zoo showed a degree of vacuolation and haemorrhage of the adrenal cortex (Fig. 16.3) but the significance was unclear. Information on findings in the adrenals of free-living animals, such as those mountain gorillas necropsied over the years by the MGVP, are unfortunately not freely available. However one female, 'Inyuma', examined postmortem by the author (JEC) in 1995 was reported many years

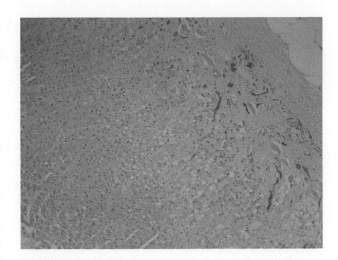

FIGURE 16.3 Vacuolation and haemorrhage of the adrenal cortex of a lowland gorilla.

later (Linda Lowenstine, personal communication) as showing (amongst other things) 'lipofuscinosis of the zona reticularis and mild medullary hyperplasia'.

Sometimes the paucity of information may be attributable to failure to examine the glands in detail. A report from one British zoo on a baby gorilla stated 'Unusually large adrenal glands but appear normal'; histological examination was not performed. In contrast, the report on 'Samson', the animal with an oesophageal leiomyoma (see Chapter 11: Alimentary Tract and Associated Organs), BZG, included the following descriptions: 'Gross. Both adrenal glands are abnormal. Their cortex is extremely narrow (0.2 cm thick) and surrounds a cavity lined by friable dark green brown tissue. Only small areas of apparently normal adrenal tissue are recognisable. Histology. Adrenals have an extraordinary appearance. In the areas lining the central cavity adrenal medulla, zona reticularis and the inner part of the zona fasciculate a are absent as though they had been torn out leaving a ragged margin of zona fasciculata but with no evidence of haemorrhage, necrosis or inflammation. The lack of any tissue response at the border of the central cavity suggests that either the cavity is normal in the adult gorilla or that this is a postmortem artefact. At the poles of both adrenals all zones are present and appear normal.'

It is probable that adrenal-associated disease is not recognised in some gorillas. Although adrenocortical pathology in mammals can include a drop in mineralocorticoids, causing retention of potassium (K) and loss of sodium (Na) — Addison's disease — it is the rise in glucocorticoids, often associated with 'stress' and capable of causing suppression of the immune response and probable retardation of healing that has attracted most interest (see Chapter 12: Lymphoreticular and Haemopoietic Systems and Allergic Conditions). In live gorillas ultrasonography can provide information on the size of the adrenals (and other visceral organs) and may help detect endocrinopathies.

Stress and stressors are discussed in a separate section below.

A paper that threw some light on adrenal development in gorillas was by Czekala et al. (1983). They collected and measured urinary oestrogen during the pregnancy of the gorilla and orangutan and compared the values with those in normal human pregnancies. They discovered that during pregnancy gorillas and chimpanzees excrete considerably less oestrogen than do orangutans and humans. Czekala et al. postulated that this reflects the smaller fetal adrenal in the former; they have both a reduced adrenal weight and an increased definitive to fetal zone ratio compared with the latter two species.

There is clearly a need for more work on the adrenals in *Gorilla* and elucidation of their pathology. The first step is to encourage all those performing necropsies to examine, describe and weigh the glands and to take appropriate material for laboratory investigation. Excellent advice on sampling is to be found in *Standardized Necropsy Report Form for Gorillas* (SSP, courtesy of Linda Lowenstine): 'Fix small adrenals whole and section larger ones … use a very sharp knife or new scalpel blade so as not to squash these very soft glands'.

Pancreas

The pancreas of African anthropoid apes was described by Stunkard and Goss (1950b).

Its examination and investigation are covered in Chapter 11: Alimentary Tract and Associated Organs.

Despite the long-term use of NHPs as experimental models for research on diabetes mellitus (Abee et al., 2012), reports of the disease in gorillas are rare. Historical records are of interest. Part II: A Catalogue of Preserved Materials refers to two cases – 'Djombo', who died at the Zoologischer Garten Hannover of 'arteriosclerosis and diabetes' (Christiane Schilling, 2016, personal communication with Gordon Hull) and 'Hobbit', who was euthanased at Pretoria Zoo on account of 'chronic diabetes mellitus' (interestingly, there were 'no histological lesions in pancreatic islets') (Kim Labuschagne and Emily Lane, 2014, personal communication with Gordon Hull). As Part II: A Catalogue of Preserved Materials, recounts, 'Hobbit' was found postmortem to have a range of lesions including glomerulosclerosis, nephrosis, arteriosclerosis, myocardial fibrosis, pleurisy and bronchopneumonia, with an equally spectacular span of histological changes.

Early descriptions of pancreatic disease associated with flukes in individual gorillas were by Prine (1968) and in the more lengthy report by Stunkard and Goss

(1950a). The latter gave details of a 'massive infection in a gorilla from the Congo' by the trematode *Eurytrema brumpti*. Important work was performed by McClure and Chandler (1982) who examined approximately 3000 histological sections of haematoxylin and eosin (H&E)-stained pancreases from 1000 NHPs including 100 great apes and reported pancreatic lesions of varied severity in 18.7% of the animals. The series did not include gorillas but provided valuable baseline data on types of pancreatic pathology in both simians and prosimians. Diabetes mellitus in the chimpanzee had been reported a year earlier by Rosenblum et al. (1981) and more recent studies on ageing chimps have revealed impaired glucose intolerance analogous to Type 2 diabetes in humans.

Reviews of the literature concerning the gross and histological changes associated with protozoal and metazoal parasitism of NHPs, in one case (1982) with particular reference to the alimentary tract and pancreas (including several references to gorillas), were presented by Toft (1982, 1986). All parasites, both pathogenic and apparently nonpathogenic, protozoan and metazoan, that had by that stage been reported in the pancreas of NHPs were tabulated.

What was confidently described as the 'first reported case of nesidioblastosis in a NHP' was presented 20 years later by Burns and Barrie (2002). The subject was a 10-year-old female lowland gorilla at the Columbus Zoo and Aquarium in the United States that exhibited episodic bouts of lethargy, trembling and stupor. Similar signs had been observed during the animal's first pregnancy and serum chemistry estimations on two separate occasions during that pregnancy had demonstrated hypoglycaemia. Episodic hypoglycaemia was suspected. Measurements of glucose, insulin, proinsulin and C-peptide on stored sera from this animal and from two other similarly aged, reportedly healthy, lowland gorillas gave results that were strongly suggestive of a diagnosis of insulinoma. A coeliac angiogram showed a small, abnormal contrast 'blush' in the region of the head of the pancreas while contrast-enhanced CT with arterial phase contrast revealed a contrast-enhanced mass in the region of the head of the pancreas. Partial pancreatectomy was performed, and the excised portion submitted for histopathological examination. No discrete neoplasm was detected but the pancreatic parenchyma had a multinodular to multilobular pattern with hyperplasia and increased numbers of islets, justifying a diagnosis of nesidioblastosis.

As stressed in Chapter 11, Alimentary Tract and Associated Organs, it is important to observe the pancreas, with the naked eye and with a lens or magnifying loupe, before handling it – gently. The colour and consistency should be recorded as accurately as possible.

Unpublished histopathological findings in the pancreas of gorillas appear to be widespread; as stressed elsewhere

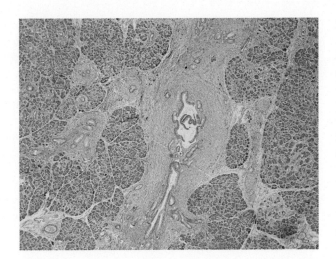

FIGURE 16.4 Pancreas of lowland gorilla, showing a moderate degree of fibrosis.

in the context of other organs, such changes need to be collated and published. In the author's (JEC) experience, varying degrees of pancreatic fibrosis are often seen incidentally (Fig. 16.4).

Lydia Tong (personal communication) reported from Taronga Zoo, Australia, that tissues from gorilla 'Betsy' showed metastatic carcinoma in the lung and liver, possibly from a pancreatic islet cell carcinoma. A female lowland gorialla from BZG that died of bronchopneumonia was also found on histology to have fibrosis and 'mucosis' of the pancreas (see Part II: A Catalogue of Preserved Materials). The pancreas was described grossly as being 'white in colour, triangular in shape and measured 3.5×0.7 cm' and the histopathological report stated: 'Lobules appear reduced in size, with an apparent increase in the size of individual acini. There is an increase in interstitial connective tissue and intralobular periductular connective tissue. The epithelium lining the ducts is tall columnar. There is a moderate amount of pink eosinophilic exudate in intra- and interlobular ducts, which is confirmed to be mucin using special stains. Mucin-containing cells are prominent within the duct epithelium'.

Pancreatic disease is a challenging area and the pathologist who is examining material from gorillas will need to refer to appropriate texts that focus on humans — for example, the book by Campbell and Verbeke (2013).

Pineal

The pineal, historically considered 'the seat of the soul' because it appeared to have no function, is a sac-like evagination from the diencephalon on the dorsal surface of the brain. It has an interesting background because it

has evolved from a sensory structure in fish to an endocrine gland in birds and mammals. There are no functioning photosensory cells in the pineal of mammals. Instead there are secretory cells called pinealocytes. It is assumed that in *Gorilla*, as in *Homo*, the pineal changes in size in response to light and that at night it produces melatonin. This hormone has a number of effects on body function and health, including controlling circadian rhythm ('sleep–wake cycles'), and probably playing a part in other aspects of metabolism such as ageing and immunity. Seasonal variation in the length of the menstrual cycle of captive gorillas may have been linked to pineal activity, as suggested by work at Barcelona Zoo (Abelló and Flamme, 2006).

See 'Sleep' section in this chapter.

The pineal is largely overlooked in gorilla necropsy reports but it is noteworthy that it is referred to in the *Standard Necropsy Report Form for Gorillas* (see earlier), with the instruction: 'If the pineal gland is evident, please submit for histology'. In humans and domesticated animals a number of conditions affecting the pineal are recognised — traumatic injuries, inflammatory (extension from meninges) and neoplasms (but these represent less than 1% of human brain tumours). It is likely that all these occur in gorillas — the pineal is certainly often badly damaged in cases of infanticide, for example — but records are lacking.

SLEEP

See 'Pineal' section in this chapter.

Both lowland and mountain gorillas have rest periods and at times sleep. Gorillas, as do older gorilla researchers, like to have a siesta during the day — in Central Africa, often in the afternoon. Norton (1990) described this 'midday siesta' and pointed out that it is a time when members of the group can socialise. 'Now in the warm sun, youngsters and adults like to lie close to each other, often reaching out to stroke or to groom another's fur'. After the morning bout of feeding the adults lie about and snooze during the middle of the day. Only the energetic youngsters are still active and this becomes a time of play'. Yawning in gorillas was described by Schaller (1963).

Various sleep disorders have been described in humans and in some other animals but there appear not yet to be any records of such conditions in *Gorilla*, although twitching during sleep ('hypnic jerks') — which is not pathological — is recognised in gorillas. If a sleep disorder is suspected, clinical examination should include neurological investigation and advanced imaging (MRI) and CSF analysis should be performed.

STRESS

See 'Adrenal' section in this chapter.

The term 'stress' is used loosely in many popular texts about gorillas. For example, Verrengia (1997), writing about Central Africa, stated 'With loss of habitat, gorilla groups shrunk as hunger and stress contribute to the premature deaths of adults and infants and the females' breeding cycles are interrupted'. The term is also often used loosely, without a clear definition, in scientific circles. Only 25 years ago, Hattingh (1992) stated that 'no universally accepted definition of stress exists. Stress as such can therefore not yet be quantified'. Fiennes (1964), in his Report at the Zoological Society of London's Pathologist for the year 1962 wrote of one mountain gorilla:

(A) Young male (DB 456/62) … The adrenal glands were completely exhausted, indicating a state of "shock" associated with this disease.

Our scientific understanding and interpretation of stress was originally based on Hans Selye's seminal work in the 1930s and later (Selye, 1950) and, although much has been researched and published since then, Selye's original concepts can provide a useful background to those who are unfamiliar with terminology.

According to Selye (1950) there is a 'general adaptation syndrome' which evolves in response to stress, the three stages of the syndrome being the 'alarm reaction' (AR), the 'stage of resistance' (SR) and the 'stage of exhaustion' (SE). All three stages can damage the body but the animal is usually able to overcome the AR; it is only when the stress is severe and prolonged that it may prove fatal. When an animal reaches the SE, profound and deleterious changes can take place, including the development of pathological lesions, such as gastric ulceration and alterations to the immune system. In a subsequent paper, Selye (1955) looked in more detail at the interaction between stress and disease. He stated that the normal, immediate, stress response — the release of adreno (gluco) corticoids — protects an animal but in the longer term, the elevated glucocorticoids are deleterious as they are linked with hyperglycaemia, and other changes, including suppression of both the immune and the reproductive systems.

Selye's concepts still provide a basis for an understanding of stress and are presented here for that purpose. However, it is argued by some that they are too imprecise (Broom and Johnson, 1993) and succeeding decades have seen changes in how stress is perceived; a useful summary of developments in the 50 years subsequent to Selye's book was provided by Moberg and Mench (2000). A landmark gathering insofar as this book is concerned was the 2-day meeting organised by the Primate Society of Great Britain (PSGB) in December 2009, entitled 'Primate Stress, Causes, Responses & Consequences'.

The presentations included papers on gorillas both in the wild and in captivity.

In practical terms, how does all the groundwork above relate to the pathology and health of gorillas?

Before considering this in any detail, it is essential that those working with these animals fully understand the difference between stressors (that impinge on the animal) and stress (physiological changes in the animal that can thereby result). This is depicted simply, graphically, in the figure below.

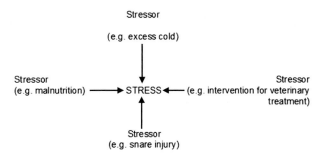

Some categories of stressors, all of which may affect gorillas, are listed in Box 16.3.

See also Chapter 8: Nonspecific pathology.

So, what effect might stress have on captive and free-living gorillas? The pathogenesis of stress is still poorly understood, especially in respect to how it applies to different species; but conventional wisdom (some of it based on research on monkeys, coupled with data on *Homo sapiens*) is that, in response to stressors, there is likely to be an enlargement of the adrenals (with increased corticosteroid production), involution of lymphoid tissues and (possibly only in mammals) gastrointestinal ulceration. The actual mechanism is, as explained above, a stimulation of the pituitary by stressors resulting in increased production of adrenocorticotropic hormone and a consequential rise in glucocorticoids. It is these glucocorticoid hormones that produce the pathological changes. Extrapolating from other NHPs, gross postmortem examination lesions may be slight but adrenal enlargement is seen (the adrenals need to be described, measured and

---

BOX 16.3 Examples (Some Categories) of Stressors in Gorillas

- Pain
- Distress
- Discomfort
- Fear
- Thirst/hunger
- Infectious disease
- Noninfectious disease

weighed), the gastrointestinal tract may show erosions or ulcers (careful examination is vital, including use of a lens and employing both reflected and transmitted light) and some degree of involution of the lymphoid tissues may be present (not easily detected; needs microscopical inspection too). Other secondary changes may be present, including amyloidosis.

Clinical signs and lesions that reflect this pathogenesis are certainly reported in NHPs. Research on baboons at the end of the last century by Robert Sapolsky showed that free-living baboons in Africa can develop GIT ulcers, have an elevated blood pressure, increase in serum cholesterol and display behavioural changes in a similar way to stressed humans (for a popular account, see Sapolsky, 1990).

Are such changes also seen in gorillas? It appears so. Lesions attributed to stress are often included in histological reports from zoos. One of many examples was 'Colossus' at Cincinnati Zoo, who died of cardiomegaly with fibrosing cardiomyopathy but whose tissues showed other interesting changes including the following: 'The gastric erosions are suspected stress related (associated with social status, environment, and/or tooth abscessation). Source: necropsy report (prosector Dr Pulley, pathologist Dr Monica P. Revelo).

Space does not permit detailed description of stress in either captive or free-living gorillas so only brief mention will be made of relevant published research.

Stress has long been recognised as an important, often dangerous, sequel to restraint and other procedures in captive animals, including primates (Fowler, 1993, 1999,

2003, 2007) and gorillas in particular (Bahr et al., 1998). In recent years there have been significant advances in our understanding of its role — and its relevance to zoo animal welfare; see, for example, the studies by Clark et al. (2012) on the relationship between behaviour, adrenal activity and environment in zoo-housed lowland gorillas. The association of stressors with captivity has been discussed by many authors — see, for example, Meder (1994) and in particular Cousins (2015) who suggested that 'gorillas repress and conceal their inner feelings, allowing stressful situations to become more damaging...'.

Impressive steps have also been taken to improve our ability to measure presumed stress in free-living gorillas (Minnis et al., 2004). A technique to measure urinary cortisol in such animals was developed by Robbins and Czekala (1997) (see Chapter 13: Urinary and Reproductive Systems). An easier method is the evaluation of glucocorticoids in gorilla faeces (Nizeyi, 2005; Nizeyi et al., 2011) or even saliva (Kuhar et al., 2005). More recent papers have tried to link studies on stress with such factors as habituation, research and 'ecotourism' (Shutt et al., 2014).

There are other possible 'markers' of stress, such as linear enamel hypoplasia and incremental lines of dental cementum (see Chapter 11: Alimentary Tract and Associated Organs).

The welfare of gorillas is linked to stress, as well as to many additional factors, especially but not exclusively in captivity. Welfare and other related topics, such as Pain, are discussed in Chapter 8: Nonspecific Pathology.

Chapter 17

Field Studies in Pathology and Health Monitoring

Jane Hopper, Gladys Kalema-Zikusoka, Celsus Sente and Jenny Jaffe

---

New Discoveries of the Earth discover new Diseases: for besides the common swarm, there are endemial and local Infirmities proper unto certain regions, which in the whole Earth make no small number: and if Asia, Africa, and America should bring in their List, Pandora's Box would swell, and there must be a strange Pathology.

From a 'Letter to a Friend' by Sir Thomas Browne (1690)

INTRODUCTION

John E. Cooper

This chapter consists of four short papers, presented as 'case studies' and written by four colleagues, two with experience of working in West Africa, the other two living in Uganda. The contributions demonstrate the wide range of activities in which a veterinary input is needed — and all four contributors are experienced veterinarians. Their contact details are given in the List of Contributors, at the beginning of the book.

The first paper, 'Pre-release health considerations for gorillas', is by Jane Hopper, who graduated from Cambridge University, United Kingdom, in 2002. She spent 4 years in British mixed practice and providing locum cover for zoological collections before becoming Head of Veterinary Services for the Aspinall Foundation (Howletts and Port Lympne Wild Animal Parks) in 2006. As part of this role, she oversees the health of animals in their conservation projects, including gorillas in their Congo and Gabon projects.

Gladys Kalema-Zikusoka's contribution is entitled 'Setting up long-term health-monitoring systems for mountain gorillas using a "One Health approach"'. Gladys is a Ugandan veterinary surgeon (veterinarian) who trained initially at the Royal Veterinary College in the United Kingdom and did postgraduate studies at North

Carolina State University and Zoological Park in the United States. In her work as a Director of Conservation Through Public Health (CTPH), based in the west of Uganda, she brings together the many strands that make up veterinary medicine, conservation and human health in order to create an integrated programme that benefits humans and livestock, as well as wild animals and their environment.

The third contribution is by Celsus Sente and entitled 'Recent work on helminths, cryptosporidia and *Giardia* in one group of gorillas in Mgahinga'. Celsus is a Ugandan veterinary surgeon (veterinarian), who gained his BVM degree from Makerere University and then went on to complete a Master's degree in Wild Animal Health. In addition to being an Assistant Lecturer, Department of Wildlife and Aquatic Animal Resources (WAA) at Makerere, Celsus is involved in various projects in East Africa. He has also studied and worked in Europe.

Jenny Jaffe is the author of the fourth account 'Working with great apes during an Ebola outbreak'. She graduated from Utrecht University, the Netherlands, in 2007. She has spent 3 years in British small animal practice. After wildlife veterinary jobs in Ecuador and Indonesia, she completed an MSc in Wild Animal Health at the RVC. She served at Tacugama Chimpanzee Sanctuary in Sierra Leone from September 2012 until April 2015. She currently works at the Institute of Zoology in London.

PRE-RELEASE HEALTH CONSIDERATIONS FOR GORILLAS

Jane Hopper

When considering returning gorillas to parts of their endemic range, it is essential to consider the effect that the gorillas can have on the environment into which they are being translocated and also the impact that this habitat

J. E. Cooper & G. Hull: Gorilla Pathology and Health. DOI: http://dx.doi.org/10.1016/B978-0-12-802039-5.00017-2

might have on the gorillas. More specifically, a thorough review and analysis must be performed in order to protect the gorillas that are being reintroduced, the existing population of gorillas, other wildlife and any humans in the location.

Infectious disease can affect the success of reintroductions in both direct and indirect ways. Direct effects are easy to envisage — death of a reintroduced animal or death of an animal in the recipient population. Indirect effects are less straightforward but amongst other things may be economic (if, for example, there is an effect on livestock) or related to human health or altered public perception of the project.

For these reasons, it is helpful to think of gorilla reintroduction in terms of a 'One Health' approach (see elsewhere in this book). This advocates the integration of human, animal and environmental health (Mazet et al., 2009).

Each reintroduction project should be accompanied by a health risk analysis aimed at summarising the possible dangers, either quantitatively (probabilities, percentages, etc.) or qualitatively (high, medium and low risk), to the humans and animals involved (Beck et al., 2007). The logical way of achieving this is to conduct a 'Disease Risk Analysis' (DRA). This assesses the disease threats posed by any reintroduced animals to other species, including humans, in the release area.

A thorough DRA is time-consuming and requires a lot of information about the species being released, the area where the release will take place and other species in the area. Once completed it provides a framework on which to work through disease risks and problems scientifically, and it will help assess which disease surveillance will be most appropriate (IUCN/SSC, 2013). Therefore, it is very important that veterinary surgeons with the appropriate experience are part of the management team from the outset.

The generally accepted method for creating a DRA has been adapted from risk analysis material in the IUCN Captive Breeding Specialist Group (CBSG) Disease Risk Assessment Manual (Armstrong et al., 2003). Each DRA can be broken down into four stages:

1. Hazard identification — identifying the list of hazards (pathogens).
2. Risk assessment — determining the likelihood of adverse health effects associated with hazard exposure.
3. Risk management — decreasing the likelihood of an adverse outcome, and reducing the consequences if it happens anyway.
4. Risk communication — communicating the identified risks to all interested parties.

Each DRA will be unique as it will be created for a *specific* group of animals being reintroduced to a *specific* location at a *specific* time. A DRA is a living and evolving document, and as such can be used as a framework for future releases. However, it will need to be updated to take into account changing health/disease information and altered circumstances of the reintroduction (Beck et al., 2007).

In this article the mite *Sarcoptes scabiei* and the disease that it can cause, sarcoptic mange, are used as an example.

Hazard Identification

Hazard identification is the first step of a DRA. An exhaustive list of hazards (pathogens) should be compiled. There are, of course, in addition to pathogens, other non-infectious hazards such as toxins, trauma and poaching. However, in a DRA only infectious diseases are usually considered.

The list of pathogens should include not only those that the newly released gorillas may be at risk from in their new environment, but also those that the gorillas may spread to their new environment and to other animals, including humans, living in that environment. To identify those pathogens it is essential to conduct a thorough literature review. There may also be health data available from animals already at the planned release site. These data may be pre-release data from animals that have already been released, or obtained from surveying wild populations (Levrero et al., 2007; Van Heuverswyn et al., 2006).

Risk Assessment

After the hazards have been identified, a risk assessment is carried out on each of the pathogens. This helps prioritise the list of diseases and it determines the likelihood of adverse health effects associated with hazard exposure (Armstrong et al., 2003).

The risk assessment should use a set of criteria specific to the situation and these should be ranked in a uniform way (e.g., ranking from 5 as the highest rank down to 1). This assessment identifies the potential severity of the pathogens, and the probability that they will affect the gorilla(s) being released.

Commonly used criteria are:

- *Likelihood of susceptibility.* What is the likelihood that the animal to be released will be susceptible to this disease? For example, if we consider *S. scabiei* in this and the following clauses, we can see how this exercise can apply to a situation. Here it ranks 3. We know that gorillas are susceptible to *S. scabiei* as there have been confirmed clinical cases (Kalema-Zikusoka et al., 2002; Thaddeus et al., 2001). This susceptibility is likely to increase during reintroduction on account of stress (Teixeira et al., 2007).

- *Likelihood of exposure.* What is the likelihood that the animal to be released will be exposed to the pathogen? Such exposure may occur at the site of origin of the animal, during transport to the release site, or during holding and preparation for release. It is important to consider what you know about the occurrence of the disease in other captive animals to which the released animals may be exposed, the prevalence of the disease in other species at the release site and how well the organism survives in the environment. Again using *S. scabiei*, it ranks 2 − low likelihood before release so long as there is regular observation and full clinical investigation of any apparent skin disease.

- *Likelihood of becoming infected.* What is the likelihood that the animal will become infected if exposed to the pathogen? Some diseases are very infectious; others such as tuberculosis require prolonged/repeat exposure. If the animals can be immunised (see Chapter 12: Lymphoreticular and Haemopoietic Systems and Allergic Conditions), they may be less susceptible. Again using *S. scabiei*, here it ranks 3. The mite needs prolonged skin-to-skin contact if it is to spread between animals but such contact is likely to occur because of the social structure and behaviour of gorillas.

- *Likelihood of transmitting to others.* What is the likelihood that the pathogen will be transmitted to other individuals? The social behaviour of the species is particularly significant here. A highly sociable species such as the gorilla is very likely to transmit pathogens that require direct contact. The likelihood also depends on the character of the pathogen and how long-lived it is in the environment. *S. scabiei* would rank 4 because gorilla behaviour involves close physical contact between members of the group.

- *Severity to the individual.* What is the likely severity of the disease if an individual in the wild population becomes clinically ill? *S. scabiei* ranks 3. The severity of disease in gorillas infected with *S. scabiei* seems to depend on age and health status. Young gorillas have been recorded as dying from sarcoptic mange whilst older gorillas may only suffer mild skin disease (Kalema-Zikusoka et al., 2002).

- *Severity for the population.* What is the severity for the population if it becomes infected with the disease? Organisms that are responsible for diseases that are severe would spread quickly through a population, e.g., *Ebolavirus* (Bermejo et al., 2003), or decrease the fitness of the overall population, e.g., *Brucella*-like organism (Niphuis, 2014). *S. scabiei* ranks 4 because the effects would vary throughout the population, depending on age and health status. They would range from death of some individuals to chronic pruritic skin disease in others, the latter probably decreasing the overall fitness of the whole population.

- *Probability of transmission.* The probability of transmission of the pathogen is then estimated. A different scale is often used for this stage: 0 (not transmissible) to 3 (highly transmissible). The probability of transmission between various groups should be considered:
  - *Probability of transmission from humans to apes. S. scabiei* rank 1. The probability that humans will infect gorillas with *Sarcoptes* is low due to limited contact.
  - *Probability of transmission between humans. S. scabiei* rank 2. The probability that humans will infect other humans with *Sarcoptes* is moderate as close physical contact is required.
  - *Probability of transmission between apes. S. scabiei* rank 3. The probability that gorillas will infect other gorillas with *Sarcoptes* is high because of close contact between animals.
  - *Probability of transmission from apes to humans. S. scabiei* rank 1. The probability that gorillas will infect humans with *Sarcoptes* is low due to limited contact.

The scores are then added up to give a 'significance to the programme' which allows each pathogen to be ranked in order of risk level. These significance values are used in the next step of risk management in order to create a health plan showing the scale of risk that enables a plan to be produced for reintroduction of the species to a new environment.

It can sometimes be difficult to carry out the risk assessment due to lack of data. Certain diseases have only recently been discovered or are poorly studied in gorillas. An example of the former is the *Brucella*-like organism (Niphuis, 2014). For other pathogens the infectivity or prevalence in the release site may not be known or may vary. It is always prudent to err on the side of caution when assigning risk. The ranking and significance of such diseases should be updated as new information becomes available (Beck et al., 2007).

When ranking, it is also important to bear in mind that the potential for pathogen transmission is increased during the reintroduction process because the gorillas to be released will be in close contact, under stressful conditions, both during transport and when arriving in a new environment (Table 17.1).

Risk Management

Once the risk analysis has been carried out, a veterinary protocol should be written as part of a risk management strategy (Beck et al., 2007). This aims both to decrease the likelihood of an adverse outcome (e.g., by testing pre-release) and to reduce the consequences should this

TABLE 17.1 Disease Risk Analysis for the Transfer of Three Infant Western Lowland Gorillas from the United Kingdom to Gabon in 2008

Disease	Likelihood of Susceptibility	Likelihood of Exposure	Likelihood of Becoming Infected	Likelihood of Transmitting to Other	Severity for the Individual if Clinical	Severity for the Population	Estimated Significance to the Programme	Prob. of Transmission From Humans to Apes	Prob. of Transmission Between Humans	Prob. of Transmission Between Apes	Prob. of Transmission from Apes to Humans	Updated Significance to the Programme
Ebola/Marburg	5	0	5	3	5	5	23	3	3	3	3	35
Shigella	5	3	4	4	3	3	22	3	3	3	3	34
Salmonella	5	3	4	4	3	3	22	3	3	3	3	34
Campylobacter	5	2	4	4	3	3	21	3	3	3	3	33
Enteropathogenic E. coli (EPEC)	5	2	4	4	3	3	21	3	3	3	3	33
Strongyloides	5	2	5	4	3	3	22	2	3	2	2	31
Bacillus coli	4	1	4	4	5	4	22	2	3	2	2	31
Hookworm	4	2	5	4	3	3	21	2	3	2	2	30
Pinworm	4	4	3	3	2	2	18	3	3	3	3	30
Streptococcus pneumoniae	4	2	4	4	4	4	22	2	3	2	1	30
Entamoeba histolytica	4	2	4	4	3	3	20	2	3	2	2	29
Yersinia	4	2	4	4	3	3	20	2	3	2	2	29
Whipworm	4	2	3	3	3	2	17	3	3	3	3	29
Oesophagostomum	4	1	5	4	3	3	20	2	2	2	1	27
Giardia	3	2	3	4	3	3	18	2	3	2	2	27
Tuberculosis	3	1	2	3	5	5	19	2	2	2	1	26
Dermatophytosis	4	1	4	4	3	3	19	1	2	3	1	26
Cryptosporidium	4	2	3	3	3	3	18	2	2	2	2	26
Klebsiella	4	1	3	3	3	3	17	2	2	2	2	25
Sarcoptes	3	2	3	3	4	3	18	1	2	3	1	25
Rabies	4	0	3	3	5	3	18	1	1	1	1	22
Malaria	3	0	3	2	3	2	13	1	1	1	1	17
Herpesvirus simplex	2	5	3	2	1	1	14	2	2	2	1	21
Measles	3	1	3	3	3	3	16	1	1	1	1	20
RSV	3	1	2	3	3	3	15	1	1	2	1	20
SIV/HIV	3	1	2	3	2	2	13	1	1	1	1	17
Influenza	2	2	3	2	2	2	13	1	3	1	1	19
Polio	2	2	2	3	3	2	14	1	1	1	1	18
Candida	3	3	2	2	2	2	14	1	1	1	1	18
Taenia	2	3	2	2	3	1	13	1	2	1	1	18
Yellow fever	1	2	2	2	3	3	13	1	2	1	1	18
Adenovirus	3	2	2	2	2	2	13	1	1	1	1	17
Hepatitis B	2	1	2	2	3	3	13	1	1	1	1	17
STLV	3	1	2	3	2	2	13	1	1	1	1	17
Hepatitis A	2	1	2	1	2	2	10	1	3	1	1	16
Parainfluenza	1	2	2	3	2	2	12	1	1	1	1	16
Hepatitis C	1	1	2	1	1	1	7	1	2	1	1	12

Source: Copyright Jane Hopper, Aspinall Foundation.

TABLE 17.2 Part of a Veterinary Protocol as Part of a Risk Management Strategy for the Release of Wild-Caught Orphan Western Lowland Gorillas in 2009

Disease Category	Disease/ Pathogenic Organism	Type of Diagnostic Test	Sample Details	How to Store Sample	Where Test Can Be Performed	Permits May Be Required (See Chapter 18: Legal Considerations)	Comments
Viral	Ebola	Serology	Serum (1 mL)	Freeze	CIRMF[a]	No	
	SIV	Serology	Serum (1 mL)	Freeze	CIRMF[a]	No	Request SIV specifically – do not want HIV
	STLV	Serology	Serum (1 mL)	Freeze	CIRMF[a]	No	Request STLV specifically – do not want HTLV
	Polio	Serology	Serum (1 mL)	Freeze	CIRMF[a]	No	Politically sensitive – liaise with health officials at CIRMF before submitting
	Measles	Serology	Serum (1 mL)	Freeze	CIRMF[a]	No	
	Herpes simplex	Serology	Serum (1 mL)	Freeze	CIRMF[a]	No	Submit for both herpes simplex I and II
	Yellow fever	Serology	Serum (1 mL)	Freeze	CIRMF[a]	No	
	Hepatitis A	Serology	Serum (1 mL)	Freeze	CIRMF[a]	No	
	Hepatitis B	Serology: HBsAg and anti-HBS	Serum (1 mL)	Freeze	CIRMF[a]	No	Request further tests if positive: HBeAg, anti-HBc and anti-HBe
	Hepatitis C	Serology	Serum (1 mL)	Freeze	CIRMF[a]	No	

[a]*Centre Internationale de Recherches medicales de Franceville, Gabon.*
Source: Copyright Jane Hopper, Aspinall Foundation.

happen anyway by outlining risk mitigation strategies. These strategies may involve testing for pathogens/disease or preventive healthcare.

The protocol is easiest to create in the form of a chart, outlining mitigation strategies for each disease, in order of the risk status determined by the risk assessment. An example is found in Table 17.2.

Testing for Pathogens and Disease

Testing as part of a risk management strategy is made more difficult by many factors, and it is essential that it be carefully planned (Fig. 17.1). This is also discussed under health monitoring; see Chapter 5: Methods of Investigation – Observation, Clinical Examination and Health Monitoring. The appropriate tests may or may not be available in the relevant country, and permits may be needed to send the samples to an appropriate laboratory in

FIGURE 17.1 Western lowland gorilla in The Republic of Congo (CR) undergoing health check as part of a risk management strategy. *Image courtesy and Copyright Tony King, Aspinall Foundation.*

another country (see Chapter 18: Legal Considerations). Some laboratories may decline to accept samples from NHPs, adding to the challenges. Even if the test is available, it may well not be validated in gorillas and there may be no diagnostic standards available for them. If there is a lack of knowledge of the sensitivity and specificity of the test in gorillas, it can be difficult to know how to interpret correctly the test results. It is important before the test is carried out to know how you intend to interpret the test result and the actions that you will take for both a positive and a negative result.

Some diseases prove difficult to manage because of political pressure. An example of this is poliomyelitis. This has been eradicated from many countries and a positive polio test in a gorilla would compromise the country's polio status as defined by the World Health Organization. Under such circumstances it is best to collaborate with the range country's public health authority (Beck et al., 2007).

Preventive Healthcare

Vaccination (Immunisation)

There appear to be no vaccines licensed for use in gorillas. There is also a paucity of published data about their efficacy and safety in these species. In some rehabilitation centres several inactivated human childhood vaccinations have been used, with gorillas being vaccinated against measles, rubella, poliomyelitis and tetanus (Hopper, 2009). Follow-up work is needed to determine whether these courses of immunisation have produced protective antibody titres in gorillas. It is recommended to use only inactivated vaccines in gorillas as there is a risk that live attenuated vaccines may revert to virulence (Hanley, 2011), especially in a species they have not been developed for (Bronson et al., 2003; Thomas-Baker, 1985). This could endanger the health of both the gorillas and the other species in the release environment.

Parasiticides

The release into the wild of parasite-free gorillas is not recommended as this may lead to acute parasitism and clinical disease if and when the animals become infected subsequently, perhaps during a stressful time. It is therefore helpful to know which parasites are 'normal' and endemic in the area during each season. This may involve screening members of the existing wild population. During their quarantine period gorillas should be free of both endoparasites and ectoparasites to ensure that nonendemic parasites are not introduced. They should then be allowed to build up gradually a low burden of endemic endoparasites during their prerelease

acclimatisation phase. Parasite numbers should be closely monitored.

A disease to consider if the gorillas are returning to their home range is malaria. *Plasmodium* infections of apes are very prevalent (Liu et al., 2010) but these infections appear to be subclinical unless the animal is immunosuppressed (see Chapter 12: Lymphoreticular and Haemopoietic Systems and Allergic Conditions). Indeed, *Plasmodium falciparum* appears to have originated from the western lowland gorilla (Liu et al., 2010). Gorillas that have been translocated may be at risk of developing clinical signs due to infection with *Plasmodium* spp. before they have developed natural immunity (Doolan et al., 2009), especially during the stressful acclimatisation period. Therefore, consideration should be given to giving prophylactic therapy to gorillas using antimalarial drugs for the quarantine period, allowing them to attain some immunity to the endemic species of *Plasmodium* (Hopper, 2009).

QUARANTINE

A quarantine period is essential for all gorillas prior to release. Quarantine allows the health assessment determined by the DRA to be carried out, which assists in the prevention of spread of pathogens. The quarantine period should be at least 90 days (Beck et al., 2007). As gorillas are highly sociable animals, it is often preferable to keep them as a group in a co-terminus quarantine (quarantine where more than one gorilla is housed in the same space at the same time). Some aspects of such quarantine are not ideal (spread of disease through the group, more difficult sample collection, etc.) but gorillas are known to be the most psychologically fragile of all the apes (King et al., 2009) and long periods alone can be detrimental to their health and welfare.

After a 90-day quarantine period it is advisable to provide an acclimatisation period of 3−6 months during which the gorillas can become accustomed to their new surroundings, habitat and food without being fully released and thereby having the added stress of interacting with the existing population. Such acclimatisation periods have been successfully carried out on large islands or areas of forest far away from existing gorilla populations.

Budgets

Financial restraints may mean that not every item on the risk management plan can be performed. If this is the case, the risk mitigation strategies will need to be prioritised. However, it is essential to produce a mitigation strategy for each hazard identified − after which the budget available can be discussed with the rest of the management team. A typical budget will need to cover such costs as health checks, analysis, storage and transportation of samples, permits (where necessary) and preventive healthcare.

Risk Communication

The final, and equally important, stage of the DRA is to communicate to all interested parties/stakeholders (staff, directors, other veterinarians, the Press, etc.) the risks identified, and to prompt discussion. Risk assessment and risk management strategies should be fully discussed with stakeholders so that they are satisfied that no unnecessary risks are being taken and that costs are a worthwhile insurance. It is also important to have a communication network that can be used to disseminate knowledge about the release programme.

SETTING-UP LONG-TERM HEALTH MONITORING SYSTEMS FOR MOUNTAIN GORILLAS USING A ONE HEALTH APPROACH

Gladys Kalema-Zikusoka

Mountain gorillas are critically endangered with an estimated 400 of the 880 individuals found in the Bwindi Impenetrable National Park, southwestern Uganda and the remainder in the Virungas in Rwanda, Democratic Republic of Congo (DRC) and a few gorillas in Mgahinga National Park in Uganda.

'One Health' is described as an approach that promotes human, animal and ecosystem health together, and thus encourages veterinarians, medical practitioners and conservationists to work together to address issues that require a multidisciplinary solution, such as zoonotic disease prevention and control and promotion of environmental health.

CTPH is a grassroots nongovernmental and nonprofit organisation set up in 2003 with the goal of preventing and controlling disease transmission at the human/wildlife/livestock interface while cultivating a winning attitude to conservation and public health in local communities. The mission of CTPH is to promote biodiversity conservation by enabling people, wildlife and livestock to coexist through improving their health and livelihoods in and around protected areas in Africa. CTPH envisages people, wildlife and livestock living together in balance, health and harmony with local communities acting as stewards of their environment. As such, CTPH has three main integrated programmes: wildlife health, community health and alternative livelihoods. This section will focus on the wildlife and

community health programmes, which have the following aims:

- Reduce threats to biodiversity conservation through proactive health monitoring
- Provide an early warning system for disease outbreaks
- Prevent cross-species disease transmission between people, gorillas and livestock (see also Fig. 3.2 in Chapter 3)
- Improve health monitoring and healthcare for gorillas, people and their livestock

Assessment of Disease Risks to Gorillas

In 1996 and 2000/01, a fatal scabies skin disease outbreak occurred in the critically endangered mountain gorillas. In 1996, this resulted in the death of an infant and morbidity in the rest of the Katendegyere gorilla group, which developed alopecia, pruritus and white scaly skin with crusty lesions. The severity of the disease was age- and size-related with the infant being the worst affected and the lead silverback gorilla only scratching with no hair loss. The 8-month-old gorilla infant died, and the rest of the group only recovered with ivermectin antiparasitic treatment. In 2000, another disease outbreak occurred affecting 17 gorillas, which developed the same clinical signs, and also only recovered after treatment with ivermectin. The source of the scabies was traced to the Bwindi local community who come into contact with gorillas, which range outside the park to forage on banana plants, and other attractive crops. Gorillas are curious and most likely touched dirty clothing placed on scarecrows to deter wildlife from eating crops.

Bwindi Impenetrable National Park is surrounded by some of the poorest people in Africa with less than adequate hygiene and sanitation and access to basic health services. Bwindi is also surrounded by a high human population density of 200–600 people per square kilometre, with typical family sizes of 10 above the national average of just below 7, placing Uganda as one of the countries in the world with the highest population growth rate of 3.2%. The large family sizes place great pressure on land for cultivation and have resulted in a hard edge with no buffer zone around most of the national park, exacerbating human, gorilla and other wildlife conflict (see Fig. 17.2).

Building a Gorilla Research Clinic for Analysis of Pathogens From Wildlife and Domestic Animals

The first activity in the wildlife health monitoring programme was building a field laboratory in 2005 to enable proactive gorilla health monitoring through regular

FIGURE 17.2 Stephen Rubanga of CTPH points out an area where gorillas regularly stray on to cultivated land.

analysis of normal gorilla faecal samples, and abnormal gorilla faecal samples from gorillas showing clinical signs. The Gorilla Research Clinic enables monthly laboratory analysis of faecal samples from each habituated gorilla group, gorillas showing clinical signs and unhabituated gorillas during the gorilla census conducted once every 4–5 years. Organisms that have been investigated to date include intestinal helminth parasites; protozoa including *Entamoeba*, *Giardia* and *Cryptosporidium*; and bacteria including *Salmonella* spp.

In the first year it was found that the Nkuringo gorilla group that regularly foraged most in community land had the highest parasite infection rate. This prompted UWA to train more community volunteers of the HUGO (Human and Gorilla Conflict Resolution) team from that sector. HUGO is made up of volunteers from the community who herd gorillas back to the forest when they forage in community land. A few years later the parasite infection rate reduced and was comparable to other gorilla groups.

Linking With Local Human Health Centres Through Comparative Disease Investigations

A survey using Faecal Antigen ELISA kits to test for *Giardia* and cryptosporidia established 40% prevalence *of Giardia*, a water-borne pathogen, in five human infants with diarrhoea, 7% prevalence in people who were asymptomatic, 16% prevalence in asymptomatic livestock, and 0% prevalence in asymptomatic gorillas. *Cryptosporidium* was found in all three: gorillas, humans and livestock. This prompted the local health centres to educate the infants' mothers on the importance of hygiene and sanitation by collecting water from protected water sources and utilising clean water storage containers, as well as boiling drinking water.

Linking With Local Human Health Centres Through Community-Based Health Promotion

CTPH's community health programme is centred around strengthening community-based healthcare working with established Ministry of Health structures of Village Health Teams. The programme started in two high human and gorilla conflict parishes, Mukono and Bujengwe in Kanungu District, where CTPH educated the local leaders about the programme and they selected trusted members of the community who could read and write, and were trained to conduct both health and conservation outreach. Thus CTPH links the community to local health centres through a model of Village Health and Conservation Teams (VHCTs) who are trained to promote hygiene and sanitation, infectious disease prevention and control, as well as refer ring people suspected to have scabies, diarrhoea, tuberculosis and HIV/AIDS to the nearest health centres for testing and treatment; family planning promotion including some who are trained to give the three-monthly interval Depo-Provera injection; and promote good nutrition and refer children with malnutrition to the nearest health centres. The VHCTs also assists gorilla and forest conservation and sustainable agriculture.

Ten years' work at the Gorilla Research Clinic has helped to identify some of the major causes of diarrhoea in the community, such as *Giardia*, which has led CTPH to encourage communities to better manage their water resources to reduce contamination with *Escherichia coli*, coliforms and other pathogens, including (1) teaching farmers to dig water troughs for their cattle to reduce defaecation in the streams where people collect water for home use, (2) collect water with clean containers from protected water sources and (3) boil the drinking water.

CTPH has built a permanent home by expanding the Gorilla Research Clinic into a larger Gorilla Health and Community Education Centre, with more rooms for laboratory work and a community resource meeting room. Through partnerships with UWA, local health centres and veterinary centres, the new One Health Centre, which is continuing to expand, will regularly analyse both normal and abnormal gorilla, human and livestock faecal samples (see Fig. 3.2 in Chapter 3).

Conclusion: Impact on Conservation

This programme has directly reduced threats to gorillas through improving the health and hygiene of homes where they range outside the park. There has been a 50% increase in hand-washing facilities outside the pit latrine from 10% to 60%; homes have also acquired pit latrines, drying racks and clean water storage containers. There has also been a significant increase in new users to modern family planning, which has increased the inter-birth interval from 20% to 60% of the women, above the national average of 28%. When a family has the children they can manage they are better able to provide good healthcare and education, thus reducing the threat of disease transmission between people and gorillas and helping to break the poverty cycle.

This programme has resulted in a higher quality of life for Bwindi local communities, which in turn has led to better attitudes towards conservation. Examples include the community protecting an aged silverback gorilla from Mubare gorilla group that could not keep up with the rest of the group (see Chapter 8: Nonspecific Pathology) and therefore chose to settle in community land rather than the forest; and the VHCT community volunteers, on their own, promoting sustainable agriculture, which utilises less land and good soil and water conservation methods resulting in reduced pressure on the gorillas' fragile habitat.

Future Recommendations for Strengthening the Programme

CTPH has established one of the first 'One Health' field programmes in the world, which has been expanded to another district surrounding Bwindi Impenetrable National Park, Kisoro District's Nteko and Rubuguri parishes, which experience considerable human and gorilla conflict. The programme is also being scaled-up to Virunga National Park in the DRC and there are plans to do so in other protected areas in Africa that have gorillas and chimpanzees by working with and/or training local partners, communities and the countries' wildlife authorities.

CTPH is also implementing this 'One Health' model in savannah ecosystems where the main threat to conservation is pathogen transmission between wildlife and livestock, which can also affect humans, including brucellosis, tuberculosis and Rift Valley Fever.

CTPH is grateful to many donors and partners who have supported this work, most notably Professor John Cooper, Margaret Cooper and Anthony Chadwick who raised funds through presentations on Webinar Vet, Uganda Wildlife Authority, Kanungu and Kisoro District Local Governments, Institut Congolais pour la Conservation de la Nature (ICCN), The Gorilla Organization, USAID, United States Fish and Wildlife Service, John D and Catherine T MacArthur Foundation, Development Cooperation Ireland, Whitley Fund for Nature, Tusk Trust, Global Development Network, Wildlife Conservation Network, Wildlife Direct, FHI 360, Intervet Netherlands/Merck Animal Health and Cornell University.

CRYPTOSPORIDIUM, GIARDIA AND HELMINTHS OF HABITUATED NYAKAGEZI MOUNTAIN GORILLA GROUP IN MGAHINGA GORILLA NATIONAL PARK

Celsus Sente

A cross-sectional study was carried out to determine the prevalence and shedding intensity of mountain gorilla faecal helminth eggs and protozoan oocysts (*Cryptosporidium* and *Giardia*), between January and February 2015, in Mgahinga Gorilla National Park, located in southwestern Uganda in Kisoro District. Mgahinga covers approximately 33.7 km^2 and is bordered by the slopes of Sabinyo ('Old Man's Teeth') and two peaks, Gahinga (3347 m) and Muhavura (4127 m). Mgahinga National Park harbours a variety of wildlife which includes giant forest hogs, elephants, buffalo, warthogs and, most importantly, is home to the endangered mountain gorillas. This provides enough ecological diversity for the proliferation of parasites in both the environment and host species.

Faecal samples were collected noninvasively from the habituated Nyakagezi mountain gorilla group, which consists of 10 individuals that originally roamed between Congo and Rwanda, but have recently been reported to be confined to Uganda. The mountain gorillas were tracked without making any effort to contact them, ensuring that sample collection remained nonintrusive and noninvasive. About 50 g of each sample were collected individually from the centre of faeces and clearly labelled. The faecal samples were then stored in an icebox at 4°C and transported immediately to Makerere University parasitology laboratory for analysis.

The samples were subjected to flotation, modified acid-fast stain (Ziehl-Neelsen's carbol-fuchsin solution) and McMaster technique for the recovery of helminth protozoan parasites. To maximise recovery of *Giardia* cysts, the samples were concentrated by centrifugation at $100 \times g$ for 5 minutes, before the normal wet preparation procedure. Ziehl-Neelsen's carbol-fuchsin solution was used to stain the faecal samples for 2 minutes and then rinsed with tap water for the recovery of *Cryptosporidium* oocysts. McMaster technique was used to determine the number of eggs/cysts per gram.

Identification of *Cryptosporidium, Giardia* and Helminths

Following staining with Ziehl-Neelsen's carbol-fuchsin solution, *Cryptosporidium* oocysts appeared as bright red circular granules on a blue-green background. On the other hand, in wet faecal smears *Giardia* cysts appeared as binucleate oval bodies, with two nuclei. *Ascaridia* and strongyle (Fig. 17.3), *Strongyloides, Trichuris* spp. and *Capillaria* spp. eggs were easily identified microscopically. Identification of the parasites was according to Kaufmann (1996) and Foreyt (1997).

Prevalence and Means

The prevalence of the parasites in all species was as follows (Table 17.3): *Cryptosporidium* spp. (50%), *Giardia* spp. (50%), strongyle (30%), *Strongyloides* (10%), *Ascaris* spp. (60%), *Trichuris* spp. (40%) and *Capillaria* spp. (50%). The shedding intensity/mean ($\pm$SE) was highest for strongyles (440 ± 76.3) followed by *Ascaris* (350 ± 67.1), and least for *Giardia* (170 ± 30) and *Capillaria* (130 ± 15.3). There was no significant difference ($p = 0.1$) in the prevalence of the parasites from the 10 different gorilla nests.

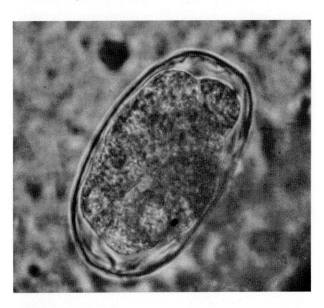

FIGURE 17.3 Strongyle egg.

Discussion

Prevalence of *Cryptosporidium* oocysts, *Giardia* cysts and helminth eggs was high except for *Strongyloides* that occurred in only one gorilla. In the present study, *Cryptosporidium* and *Giardia* prevalence, both at 50%, is higher than was reported by Arintereiho (2005). The high prevalence of the faecal parasites could be due to coprophagy among gorillas, with the older gorillas eating faeces of the young ones and likewise the young ones ingesting those faeces of the older ones. The high occurrence of faecal parasites could also be attributed to one or all of the following: increase in number of tourists and rangers that have access to the gorillas, interaction of the gorillas with other wild animals like buffalos, contact with domestic animals from neighbouring communities and contaminated water sources from which they drink.

Further research should be done to confirm the presence of similar species or genotypes in the different animal species and humans sampled in the study area, to provide information of possible cross infections between humans, nonhuman primates, domestic animals and other wild species.

WORKING WITH GREAT APES DURING AN EBOLA OUTBREAK

Jenny Jaffe

As a veterinarian working with nonhuman primates (NHPs), one should always be aware of the many possible zoonoses that can endanger the lives of staff dealing with the animals (Burgos-Rodriguez, 2011). As well, reverse zoonoses (also known as zooanthroponoses), diseases transmitted from man to animal, have caused high mortality rates in great apes in particular, with several documented cases in gorillas (Wallis, 1999). Basic techniques used in the field to avoid transmission to and from mountain

TABLE 17.3 Mountain Gorilla Faecal Parasites

	Mountain Gorilla ($n = 10$)			
Parasite	**Positive**	**% Prevalence**	**Mean Count (+SE)**	**Significance ($p < 0.05$)**
Cryptosporidium spp.	5	50	140 ± 22.1	0.1
Giardia spp.	5	50	170 ± 30	0.1
Strongyles	3	30	440 ± 76.3	0.1
Strongyloides	1	10	200 ± 47.1	0.1
Ascaris spp.	6	60	350 ± 67.1	0.1
Trichuris trichuris	4	40	230 ± 44.9	0.1
Capillaria spp.	5	50	130 ± 15.3	0.1

gorillas have been described (Cranfield, 2008; Wallis, 1999). Apart from B-virus (cercopithecine herpesvirus 1) in macaques, one of the most acutely fatal diseases that can be transmitted by NHPs is that caused by *Ebolavirus* (see also Chapter 3: Infectious Disease and Host Responses).

This chapter aims to give an overview of the role that NHPs have played in previous *Ebolavirus* disease (EVD) outbreaks, as well as the practicalities of running a great ape sanctuary in Sierra Leone during what was by far the biggest human outbreak to date.

Great Apes and Ebola

Great apes have been known to contract EVD. During the well-documented 1994 outbreak in Côte d'Ivoire, the mortality rate in the chimpanzee community was roughly 25%, whereas the case fatality rate was probably near 100% (Formenty et al., 1999). At that point it was not known that the cause was EVD, and one of the researchers performing an autopsy on a recently deceased chimp contracted Ebola. Thankfully, she survived, and did not start a human outbreak (Formenty et al., 1999). In February 1996, in Gabon, 18 people who had skinned and chopped up a chimpanzee cadaver that they found became ill with Ebola and did infect others in their community (Georges et al., 1999). More wild chimp — and gorilla — deaths due to Ebola were confirmed in Gabon and the Republic of Congo (CR) in 2001–02 (Leroy et al., 2004), leading to infections in hunters there. The start of every human outbreak is linked to the virus 'jumping' (being transmitted) from wildlife. Certain species of fruit bats are considered the most likely reservoir, and they can either infect humans directly or infect primates as intermediate host. Primates are not seen as a reservoir, as the virus is so deadly to them. They are either an intermediate host or a dead-end host.

Interestingly, antibodies against *Ebolavirus* have been detected in western lowland gorilla stools (faeces) in the Republic of Congo roughly 2 and 3 years after the outbreak in that region. The overall prevalence rate reached 10% on average; 8 of the 80 stool samples tested contained antibodies against *Ebolavirus*. This demonstrates that *Ebolavirus* exposure or infection is not uniformly lethal in this species (Reed, 2014).

Realities of Working at a Great Ape Sanctuary During a Major EVD Outbreak

The most recent outbreak in West Africa started in a small village in Guinea, where many of the villagers were known to hunt bats. After the initial crossover from, presumably, a bat to a young boy, all the subsequent cases were human-to-human (Bausch and Schwarz, 2014). A lot of the initial information regarding EVD prevention included a recommendation not to touch or eat bats, monkeys or apes. This is sound advice for preventing future outbreaks, but would probably per se make little difference once the West African pandemic was established.

Tacugama Chimpanzee Sanctuary in Freetown, Sierra Leone provides sanctuary for around 100 chimpanzees that were confiscated or handed in after being kept illegally. The majority came to the sanctuary as infants after the mothers were trapped or hunted and killed, most to be eaten as bushmeat, others because they were raiding farms that had replaced their former forest habitat. Apart from providing refuge for the orphaned chimps that were kept illegally, Tacugama is also very involved in public education and field research. Supporting the conservation of chimpanzees in the wild is a priority.

Though the disease itself was the primary concern, one of the side effects of the outbreak was a steep drop in visitors. The sanctuary had received a significant portion of its income from sightseers to the sanctuary and the eco-lodges on site. Tourism to Sierra Leone, a beautiful country, was on the rise. However, due to the EVD outbreak almost all expatriates left the country and visitors dropped to the lowest levels since the civil war, which ended in 2002. As well, several volunteers planning to come had to cancel their trips due to perceived risk and negative travel advice from their governments. To add to the financial troubles, the prices of supplies increased sharply due to transport restrictions. These factors combined created an important budget shortfall. Successful emergency appeals to private donors and funding organisations (e.g., International Primate Protection League (IPPL), Arcus and Humane Society amongst others) helped avert a major financial crisis.

Precautions at Work

Naturally, the main concern during the outbreak was that of staff becoming ill, as well as the potential of care staff transmitting the virus to the chimps that they were tending. Humans were the only known species to be affected in this specific outbreak after the initial spillover from bats. That is why they were seen as the only likely source of infection for the sanctuary's chimpanzees, rather than bats or free-living chimpanzees in the surrounding area. If one of the chimps had become infected with the virus, the end result could have been catastrophic. The set-up of the sanctuary, with large forest enclosures and interconnected night dens, meant that strict quarantine of many individuals would have been an immense challenge. Despite humans forming the largest risk, it was also decided not to accept any new chimpanzees to the sanctuary during the outbreak.

At the time of writing (July 2015), the outbreak is still ongoing, though it has diminished greatly. None of the staff has become ill. The precautions include taking the temperature of all personnel using an infrared thermometer during the daily morning meeting (Fig. 17.4). These thermometers have become common since the start of the outbreak. If well calibrated, they are accurate and very quick (Bindu et al., 2015). An important benefit is that nobody's contact occurs while taking the temperature. Another measure that the sanctuary took to prevent staff from contracting EVD was providing transport to and from the sanctuary, so they could avoid taking public transport. As well, anybody who arrived on site was obliged to wash their hands with soap and/or bleach solution before entering (Fig. 17.5) and during the height of the outbreak visitors were kept to an absolute minimum. Outreach and field research trips to the outlying provinces

FIGURE 17.4 The veterinarian takes the temperature of staff members with an infrared thermometer.

FIGURE 17.5 Sign at entrance of the sanctuary asking all people entering to wash their hands.

were cancelled and partially replaced by radio programmes addressing the importance of chimp conservation and biodiversity in general.

All staff were advised that if they felt ill with fever and vomiting or diarrhoea they must report to a health clinic rather than attend work. In principle, EVD is not infectious before the patient becomes symptomatic.

The daily morning meeting was also used as an opportunity to go through the latest reports on locations of new confirmed cases and for the provision of any relevant news concerning the outbreak. This included hearing from staff what they knew about any cases in their neighbourhoods. In one case, the niece of one of the staff members died of EVD and a special meeting was held to assess any resultant risks. The staff member involved, however, had not been to visit the niece for many months. She had been on her way to the funeral, but had been forewarned by phone not to attend, as the cause of death was likely to be EVD. In addition, she had not had contact with any family members who had seen the niece in the past month. It was decided, therefore, that she did not constitute an increased risk.

One other precaution involved the role of the veterinarian. This professional was traditionally the first port of call for staff with health issues. For instance, if any person suspected that they had malaria, the veterinarian had access to a dependable, quick test to diagnose the disease. This works with a drop of blood from the finger. Biosecurity measures (gloves used, careful disinfection of the site after testing) were tightened in this regard.

Precautions Outside the Sanctuary

Several measures were taken voluntarily or imposed by the government. Transport between provinces was severely restricted. All nightclubs were closed and gatherings of more than five people were prohibited. Football matches were cancelled. Stores and restaurants were closed on Sundays and after 1800 hours on all other days. Most establishments had buckets with a tap out front, filled with bleached water. There was normally someone present to make sure that all visitors washed their hands with the solution before entering the premises. In some hotels and supermarkets, body temperatures would be taken with an infrared thermometer before letting in staff and visitors. People quickly became familiar with these measures and, after a few weeks, hardly noticed the additional precautions.

In addition, the government organised several 'lock downs' or 'stay at home days' during which it was forbidden to travel on the roads without a special permit. These periods did add to the challenge of getting enough staff to and from work and also complicated planning the regular supply trips.

Challenges During the Ebolavirus Disease Outbreak

- Preventing EVD from infecting staff and animals
- Panic with regard to the role that great apes might play in the outbreak
- Drop in visitors meant a drop in income
- Supplies became scarce and/or increased in price
- Outreach and field research trips were impossible due to travel restrictions and raised risk of infection for staff

Precautions Taken to Prevent Ebolavirus Disease Affecting the Sanctuary

- Regularly assessing the specific risks posed by different participants in the outbreak (humans vs bats or free-living chimpanzees and monkeys)
- No longer accepting new chimpanzees

- Keeping visitors to an absolute minimum
- Training staff with regard to *Ebolavirus* transmission and preventive measures
- Increasing staff transport to prevent use of public transport
- Enforced washing of hands before entering the sanctuary
- Daily temperature check of all staff with infrared thermometer
- Daily discussion of news regarding EVD, new cases and preventive measures
- Careful consideration of individual cases, with assessment of perceived risk
- Increased vigilance when testing staff members for malaria (contact with body fluids prevented or minimised)
- No more outreach or field trips to the provinces (partially replaced by radio programmes)

Chapter 18

Legal Considerations

Margaret E. Cooper

Active law enforcement is what active conservation is all about.

Dian Fossey

INTRODUCTION

The gorilla was first described scientifically in the 19th century (see Chapter 1: The Genus Gorilla. Morphology – Anatomy and the Path to Pathology and Part II: A Catalogue of Preserved Materials). Once able to live relatively undisturbed in plentiful, remote and inaccessible habitat, gorillas have become increasingly exposed to diverse threats to their survival (see Chapter 19: Pathology, Health and Conservation: The Way Forward). The first national park in Africa, now known as the Virunga National Park, was created in 1925, primarily to protect gorillas and their habitat, and many protected areas have been gazetted since then (Fig. 18.1). Nevertheless, a majority of gorillas still live in areas that are not protected. The mountain gorilla (Fig. 18.2) is the only gorilla subspecies with every individual living in national parks (Ian Redmond, personal communication).

Loss of individual, or even groups of, gorillas caused by poaching, a long list of legal and illegal competing land uses or exploitation, not to mention disease, war and civil disturbance, corruption and the ever-closer proximity of humans (Fig. 18.3), has taken its toll of gorilla populations and their habitat, despite decades of efforts committed to their conservation (see Table 19.1 in Chapter 19: Pathology, Health and Conservation: The Way Forward). This has been widely documented by the numerous international, national, governmental and voluntary agencies that have been involved in gorilla conservation; see, for example, Beudels-Jamar et al. (2008), Stoinski et al. (2008) and the documents and agency websites listed in Table 18.1.

One of the essential elements of providing protection for gorillas and their habitat is legislation. This may not appear to be the most immediate tool for conservation and can often cause frustration on grounds of its inadequacy or ineffectuality when put to use. Legislation nevertheless establishes the authority and powers for those working in conservation and law enforcement. Legislation in conjunction with policy implementation and collaboration with stakeholders provides the framework for modern gorilla conservation. This is evident from the current Regional Action Plans and government policies for gorilla conservation in addition to other studies (Table 18.1). They emphasise the need to strengthen national conservation legislation and its enforcement. They also promote the benefits of harmonising such laws between countries and building capacity for cross-border and regional cooperation in law enforcement.

Since gorilla conservation laws and practices may have adverse effects on communities adjoining gorilla habitat it may be necessary to gain the cooperation of those who are affected (Nkurunungi and Ampumuza, 2013) (Fig. 18.4). It is also recommended by Egbe (2001), Nguiffo and Talla (2010), Jones (2008), Hakimzumwami (2000) and Kühl (2008) that local customary laws and traditions, where appropriate, should be integrated with national laws to enhance the participation of local communities in species and habitat conservation.

GENERAL

Definition of 'Wild'

It is important in any scientific or legal document that terms should be properly defined. One word that is widely applied without due clarity is the adjective 'wild'. It is frequently used to refer to 'free-living' animals (i.e., those living in the wild and not in captivity). However, it is also often loosely applied to captive specimens of species that occur in the wild in order to emphasise that they are not domesticated species (Cooper and Cooper, 2013) and is the sense that is used throughout this book.

The term 'free-living' is also referred to as 'free-ranging' in some countries (see Chapter 1: The

J. E. Cooper & G. Hull: Gorilla Pathology and Health. DOI: http://dx.doi.org/10.1016/B978-0-12-802039-5.00018-4

FIGURE 18.1 A view across the Bwindi Impenetrable National Park, Uganda, mountain gorilla habitat.

FIGURE 18.2 A silverback mountain gorilla in the Virungas.

FIGURE 18.3 The Volcanoes National Park boundary in 1994. Note the proximity of cultivation (foreground) to gorilla habitat.

Genus *Gorilla* – Morphology, Anatomy and the Path to Pathology and Appendix 1: Glossary of Terms). In this chapter reference to 'free-living' gorillas is to those that are living without restraint 'in situ' (i.e., in their own habitat). There may also be an intermediate stage when gorillas are on 'soft (partial) release' and kept in a restricted area of habitat. Gorillas under the control of a keeper (which might include an illegal captor) will be referred to as 'captive' or living 'ex situ' when in, for example, a zoo.

The use of the term 'wildlife' refers generally to those species that are commonly found free-living and also to their conservation. The context should make clear whether they are in the particular circumstances 'free-living' or 'captive'.

IUCN Red List of Threatened Species Status

Gorillas were designated by the International Union for Conservation of Nature (IUCN) as either 'endangered' or 'critically endangered' until 4 September 2016. This status has been revised; now all gorillas are critically endangered. The status recognises a species' risk of extinction according to criteria set by the IUCN (IUCN, 2012). While this is not a legal status, it may be used as a tool in assessing whether a species should be given a level of legal protection. The term 'endangered' is also used generically or with reference to specific designations in scientific or legal contexts.

Nomenclature

The International Commission on Zoological Nomenclature (ICZN) adjudicates on the correct scientific naming for animals and plants (see Appendix 6: Scientific Names of Species). Taxonomic names change frequently in this age of molecular biology (see Part II: A Catalogue of Preserved Materials) and are available on the ICZN website. The scientific names of species that are listed in legislation are not always up-to-date or, in the case of very old legal measures, only common names or even vernacular names may be used.

Legislation

Most gorillas are free-living although some, primarily western lowland gorillas, are to be found in captivity, mainly in zoological collections (Fig. 18.5), sanctuaries, rescue facilities or, in certain parts of the world and often illegally, in private hands. Most zoo-kept gorillas are captive-bred but occasionally a wild-caught gorilla may have been placed in a collection following confiscation as a result of illegal trade or smuggling.

There is extensive legislation relevant to gorillas, both free-living and captive. While there is some overlap, the two situations will be discussed separately.

TABLE 18.1 Action Plans and Other Documents as a Source of Information on Law and Enforcement Relating to Gorillas

Document	Gorilla Range States	References
Revised Regional Action Plan for the Conservation of the Cross River Gorilla (*Gorilla gorilla diehli*): 2014–19	Cameroon Nigeria	Dunn et al. (2014)
Regional Action Plan for the Conservation of the Cross River Gorilla (*Gorilla gorilla diehli*)	Cameroon Nigeria	Oates et al. (2007)
Regional Action Plan for the Conservation of Western Lowland Gorillas and Central Chimpanzees 2015–25	Angola Cameroon Central African Republic Equatorial Guinea Gabon Republic of Congo	IUCN (2014)
Grauer's Gorillas and Chimpanzees in Eastern Democratic Republic of Congo (Kahuzi-Biega, Maiko, Tayna and Itombwe Landscape): Conservation Action Plan 2012–22	Democratic Republic of the Congo	Maldonado et al. (2012) Plumptre et al. (2015)
Gorilla Agreement Action Plan – Cross River Gorilla 2009	Cameroon Nigeria	http://www.cms.int/sites/default/files/document/cross_river_ggd_AP_e_0.pdf
Gorilla Agreement Action Plan – Western Lowland Gorilla 2009	Angola, Cameroon Central African Republic Republic of Congo Democratic Republic of Congo Equatorial Guinea Gabon	http://www.cms.int/sites/default/files/document/GA_MOP2_Inf_7_1_AP_Ggg_postMoP1_E_0.pdf
Gorilla Agreement Action Plan – Eastern Lowland Gorilla 2009	Democratic Republic of Congo	http://cms.eaudeweb.ro/sites/default/files/document/GA_MOP2_Inf_7_3_AP_Gbg_ postMoP1_E_0.pdf
Mountain Gorilla (*Gorilla beringei beringei*) – Gorilla Agreement Action Plan 2009	Democratic Republic of Congo Rwanda Uganda	http://cms.eaudeweb.ro/sites/default/files/document/GA_MOP2_Inf_7_4_AP_Gbb_ postMoP1_E_0.pdf
Population and Habitat Viability Assessment Workshop for *Gorilla gorilla beringei*	Democratic Republic of Congo Rwanda Uganda	Cooper (1997) Werikhe et al. (1998)
Report on the Conservation Status of Gorillas. CMS Gorilla Concerted Action. CMS Technical Series Publication No. 17	All gorilla range states	Beudels-Jamar et al. (2008)
An IUCN Situation Analysis of Terrestrial and Freshwater Fauna in West and Central Africa	Cameroon, Central African Republic Equatorial Guinea, Gabon Nigeria Republic of the Congo (and other countries)	Mallon et al. (2015)
Empty Threat 2015: Does the law combat illegal wildlife trade? A review of legislative and judicial approaches in fifteen jurisdictions	Cameroon Democratic Republic of Congo Uganda	DLA Piper (2015)
Legislation and Policies relating to Protected Areas, Wildlife Conservation, and Community Rights to Natural Resources in countries being	Angola (and other countries)	Jones (2008)

(Continued)

TABLE 18.1 (Continued)

Document	Gorilla Range States	References
partner in the Kavango Zambezi Transfrontier Conservation Area		
The Wildlife Law as a Tool for Protecting Threatened Species in Cameroon	Cameroon	Djeukam (2012)
Stolen Apes — The Illicit Trade in Chimpanzees, Gorillas, Bonobos and Orangutans. A Rapid Response Assessment	Law enforcement and multinational networks	Stiles et al. (2013)
The Last Stand of the Gorilla — Environmental Crime and Conflict in the Congo Basin. A Rapid Response Assessment	All gorilla range states	Nellemann et al. (2010)
Country reports and, for the CBD, National Biodiversity Strategy and Action Plans prepared and adopted as government policy	Individual countries are required from time to time to submit 'Country Reports' to the United Nations or to the secretariat of treaties such as CITES, CBD or CMS. May contain accounts of current legislation	See Convention websites
Gorilla Journal — trilingual publication (German, French and English) produced by Berggorilla und Regenwald Direkthilfe	Coverage of all matters pertaining to gorillas. See numerous references in text	Biannually from No 9 (December 1994) http://www.berggorilla.org/en/journal/issues/journal-no-09/
FAOLEX ECOLEX US Library of Congress	Law databases	http://faolex.fao.org/ http://www.ecolex.org/start.php https://www.loc.gov/discover/
Great Apes Survival Partnership (GRASP). United Nations consortium of international and national bodies working for the survival of great apes and their habitat	Assisted several range states to prepare National Great Ape Survival Plans which were then adopted as policy documents by respective governments	http://www.un-grasp.org/ http://www.un-grasp.org/videos-resources/publications/
Pan African Sanctuaries Alliance (PASA). Organisation supporting primate sanctuaries, conservation and education in 12 African countries	Holds biannual workshops and produces guidance for sanctuary managers and veterinarians. Inspects and accredits sanctuaries based on its veterinary welfare and conservation standards	Ferrie et al. (2014) http://www.pasaprimates.org/about/

FIGURE 18.4 A singer from the Volcano Veterinary Centre dance team plays the *inanga*. Public performances facilitated communication between the project, the community and United Nations peacekeepers.

FIGURE 18.5 A western lowland gorilla at a zoo in Europe.

TABLE 18.2 Taxonomy and International Conservation and Legal Status of Gorillas

Taxonomy	Range States	Conservation Status Relative Risk of Extinction	Estimated Population
Various sources	As in CMS	IUCN Red List of Threatened Species (IUCN, 2016)	Various sources, including IUCN (2016), A.P.E.S. (undated)
Order: Primates			
Superfamily: Hominoidea			
Family: Hominidae			
Subfamily: Homininae			
Tribe: Gorillini			
Genus: *Gorilla*			
Species: *Gorilla gorilla* (Western gorilla)		Critically endangered	Maisels et al. (2016)
Subspecies			
Gorilla gorilla gorilla (Western lowland gorilla)	Angola	Critically endangered	100,000–150,000
	Cameroon		
	Central African Republic		
	Equatorial Guinea		
	Gabon		
	Republic of Congo		
	Democratic Republic of the Congo		
Gorilla gorilla diehli (Cross River gorilla)	Cameroon Nigeria	Critically endangered	250–300
Species: *Gorilla beringei* (Eastern gorilla)		Critically endangered (From 4 September 2016)	Plumptre et al. (2016)
Subspecies			
Gorilla beringei beringei (Mountain gorilla, Virunga gorilla (historic))	Democratic Republic of the Congo Rwanda Uganda	Critically endangered	880
Gorilla beringei graueri (Grauer's or Eastern lowland gorilla)	Democratic Republic of the Congo	Critically endangered (From 4 September 2016)	3800 Plumptre et al. (2015)
CMS and CITES Status and Terminology	All gorillas are included in:		
CMS Article 1	Appendix 1: species that are "endangered"; further defined in: CMS Res.11.33 as: "facing a very high risk of extinction in the wild in the near future"		
CITES Article 2	Appendix 1: "threatened with extinction which are or may be affected by trade"		

FREE-LIVING GORILLAS

General

There are two species and four subspecies of gorilla that are currently recognised taxonomically. They are set out in Table 18.2, together with their Red List status and the 10 countries in which they occur (usually described as 'range states') (IUCN, 2016; UNEP/WCMC, 2003).

A number of international conservation conventions (see Table 18.3) contribute directly or indirectly to gorilla conservation. These treaties, together with supplementary documentation, and information are readily found on the convention websites. National legislation (see Table 18.4)

TABLE 18.3 International Legislation Relating to Gorillas

Legislation	Details	Additional Points, Websites
Convention on International Trade in Endangered Species of Wild Fauna and Flora (CITES)	'Trade' = movement across national boundaries of species listed on Appendix 1 or 2. Any such movement must be authorised by an import and export permit. Permits are not usually given for primarily commercial movement of Appendix 1 species. Applies to live, dead animals and derivatives, including biological samples	All gorilla range states are parties. Illegal trade or failure to transpose CITES into national law can lead to a recommendation for suspension of a party https://cites.org/ Each member state manages imports and exports in national law. The European Union implements CITES as a bloc and imposes 'stricter measures' http://ec.europa.eu/environment/cites/legislation_en.htm
Convention on the Conservation of Migratory Species of Wild Animals (Bonn Convention) (CMS)	Requires parties to protect species that routinely migrate across national borders and are listed on Appendix 1(including all gorillas) and their habitats and reduce obstacles to migration	All gorilla range states are parties to CMS except the Central African Republic (signed not ratified) http://www.cms.int/
Agreement on the Conservation of Gorillas and their Habitats (Gorilla Agreement)	Action Plans for all gorilla subspecies, see Table 18.1	Angola and Cameroon not signed. Equatorial Guinea signed but not ratified. Other range states are parties http://www.cms.int/gorilla/en/documents/agreement-text *Gorilla Journal* 36, June 2008, p. 22
Greater Virunga Transboundary Collaboration	Conservation collaboration between the Democratic Republic of the Congo, Rwanda and Uganda with particular reference to gorillas	http://greatervirunga.org/
Convention Concerning the Protection of the World Cultural and Natural Heritage (World Heritage) (WH)	Protects important environmental and heritage sites. Parties designate protected areas as World Heritage sites. Six WH sites include gorilla habitat: Bwindi Impenetrable, Virunga, Kahuzi-Biega, National Parks, Sangha Trinational, Dja Faunal Reserve, Lopé-Okanda Ecosystem.	All gorilla range states are parties. http://whc.unesco.org/ See also Fung (2005) http://whc.unesco.org/document/135487
Convention on Wetlands of International Importance especially as Waterfowl Habitat (Ramsar)	Designation and protection of wetlands of international importance, some of which provide protected habitat for gorillas	All range states are parties except Angola, Republic of Congo and Uganda http://www.ramsar.org/
Convention on Biological Diversity (CBD)	Conservation of biodiversity, sustainable use of its components, equitable sharing of its benefits derived from genetic resources. Biological resources: includes genetic resources, and organisms with actual or potential use or value for humanity. 'Biological diversity' means the variability among living organisms from all sources including, inter alia, terrestrial, marine and other aquatic ecosystems and the ecological complexes of which they are part; this includes diversity within species, between species and of ecosystems.'(CBD Article 2)	All gorilla range states are parties. https://www.cbd.int/
Nagoya Protocol on Access to Genetic Resources and the Fair and Equitable Sharing of Benefits Arising from their Utilization	Provides a transparent legal framework for the access and utilisation of genetic resources and the fair and equitable sharing of benefits. Enforcement. 'Genetic resources' means genetic material of actual or potential value. (CBD Article 2)	Democratic Republic of the Congo, Gabon, Rwanda, Uganda are parties https://www.cbd.int/abs/ https://www.cbd.int/abs/about/default.shtml

(Continued)

TABLE 18.3 (Continued)

Legislation	Details	Additional Points, Websites
African Convention on the Conservation of Nature and Natural Resources (Revised Version 2003)	Requires parties to conserve species, habitat biodiversity and the environment Seven range states are parties; Angola, Republic of Congo and Rwanda have not yet ratified	http://faolex.fao.org/docs/pdf/mul45449.pdf Erinosho (2013)
Convention Relative to the Preservation of Fauna and Flora in their Natural State 1933	Protection for gorilla (Class A1 Gorilla all species) and other listed wildlife in Africa Superseded by African Convention above	http://www.ecolex.org/ecolex/ledge/view/RecordDetails?id=TRE-000069&index=treaties
Convention for the Preservation of Wild Animals, Birds and Fish in Africa (London Convention) 1900	Gorilla listed on Appendix 1, prohibition on hunting or destruction. Treaty did not come into force	http://iea.uoregon.edu/pages/view_treaty.php?t=1900-PreservationWildAnimalsBirdsFishAfrica.EN.txt
Kinshasa Declaration on Great Apes 2005	Nonbinding agreement negotiated under the UN Great Apes Survival Partnership. Followed by Gorilla Agreement	http://www.unesco.org/mab/doc/grasp/E_KinshasaDeclaration.pdf

TABLE 18.4 National Legislation Relating to Gorillas

Range States	Gorilla Subspecies	Legislation	Other Relevant Documents
Angola	*Gorilla gorilla gorilla* Western lowland gorilla	Environment Framework Act (No. 5/98 of 19 June 1998), which draws on articles 12 and 24 of the Angolan Constitutional Law (No. 23/92 of 16 September 1992) Decree No. 40.040 of 1955 Species and habitat protection. Protected species listed. Hunting is forbidden for gorilla in Cabinda (range area) Hunting Regulations 1957 Regulations for Forestry and National Parks Some laws repealed by Decree No. 43/77	Mention in one report of draft legislation in 2006 but drafts kept on file
Cameroon	*Gorilla gorilla gorilla* Western lowland gorilla and *Gorilla gorilla diehli* Cross River gorilla	Law No 94-01 of 20 January 1994 on Forestry and Wildlife Decree No 95-466 PM of 20 July 1995 on application of the Forest Law Gorilla has Class A (full) protection under Order No 0648/MINOF of 18 December 2006	National Anti-poaching Strategy (implementation) and Forest and Wildlife Control (strategy, enforcement, support of 1994 Law)
Central African Republic	*Gorilla gorilla gorilla* Western lowland gorilla	Forestry Code of 2008 Gorilla fully protected List A Annex 2 Ordinance 84.045 of 1984 Concerning the Protection of Wildlife and the Regulation of Hunting in the Central African Republic	
Democratic Republic of the Congo	*Gorilla gorilla gorilla* Western lowland gorilla and *Gorilla beringei graueri* Grauer's gorilla	Law No. 011-2002 of 29 August 2002 Forestry Code Loi No. 82-002 du 28 mai 1982 regulating hunting: Gorilla fully protected Arrêté No. 014-CAB-MIN-ENV-2004) 29 April 2004	Constitution Act 1969 'disturbing or hunting wildlife in any way' Exception for scientific research Virunga National Park declared 1925

(Continued)

TABLE 18.4 (Continued)

Range States	Gorilla Subspecies	Legislation	Other Relevant Documents
Equatorial Guinea	*Gorilla gorilla gorilla* Western lowland gorilla	Decree no 72/2007 banned hunting and consumption of apes and other primates Decree no 172/2005 implements CITES	
Gabon	*Gorilla gorilla gorilla* Western lowland gorilla	Decree No. 115/PR/MAEFDR (Décret no 115/PR/MAEFDR portant protection de la faune) of 3 February1981 Bans live capture of gorillas for export	Décret no 0164/PR/MEF Annexe 1 sont intégralement protégés: le gorille
Nigeria	*Gorilla gorilla diehli* Cross River gorilla	Nigeria, Cross River State Law No. 3 of 2010 S.71 and Schedule II Nigeria Federal law: National Environmental (Protection of Endangered Species in International Trade) Regulations 2011	Protection of Cross River gorillas in the Cross River State Protection from hunting and trade, including *Gorilla gorilla* [*sic*]
Republic of Congo	*Gorilla gorilla gorilla* Western lowland gorilla	Loi No 37-2008 du 28 novembre 2008 sur la faune et les aires protégées Arrêté no 6075 du 9 avril 2011 déterminant les espèces animales intégralement et partiellement protégées Gorilla has Class A (full) protection Wildlife Law No 48 of 1983 and Decree No 85/679 on wildlife and implementation of CITES	See Project for the Application of Law for Fauna Republic of Congo (PALF) http://palf-enforcement.org/wildlife-law/congo/
Rwanda	*Gorilla beringei beringei* Mountain gorilla	Article 417: Poaching, selling, injuring or killing a gorilla or other endangered animal species "Any person who poaches, sells, injures or kills a gorilla or any other protected endangered animal species shall be liable to a term of imprisonment of more than five (5) years to ten (10) years and a fine of five hundred thousand (500,000) to five million (5,000,000) Rwandan francs". Organic Law No 01/2012/OI of 02/05/2012 instituting the Penal Code	
Uganda	*Gorilla beringei beringei* Mountain gorilla	Uganda Wildlife Act, Cap 200 of 2000 Protects gorillas and their habitat Uganda Game (Preservation and Control) Act Cap 198 of 1959. Gorillas cannot be hunted or captured throughout Uganda without special permit (First Schedule, Part A)	1959 Act repealed except for Schedules by Uganda Wildlife Statute 14/1996 http://faolex.fao.org/

is more difficult to trace but recent Regional Action Plans and other studies listed in Table 18.1 have discussed the national conservation laws, so far as they can be ascertained, in the gorilla range states. Copies of many of these laws can be found on the ECOLEX (FAO/IUCN/UNEP, 2015) and FAOLEX (FAOLEX, undated) databases via the Internet. Local or customary laws are not readily available but are discussed in various studies, for example, Nguiffo and Talla (2010), Inogwabini (2014), Roe et al. (2016) and Sandbrook and Roe (2010).

International Legislation

Multilateral legislation to protect the rare species of Africa is not new. The first attempt, the Convention for the Preservation of Wild Animals, Birds and Fish in Africa (the London Convention) of 1900, never came into force. It was succeeded by the Convention relative to the Preservation of Fauna and Flora in their Natural State 1933 and then the African Convention on the Conservation of Nature and Natural Resources of 1968,

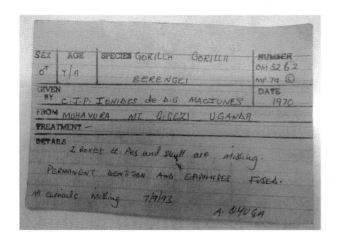

FIGURE 18.6 The record card of a gorilla collected by CJP Ionides for the National Museums of Kenya.

revised in 2003 (IUCN, 2004). From the outset the gorilla was listed as one of the species to be fully protected from killing and capture. Ionides (1965) wrote:

Permission to collect the mountain gorilla was obtained only with great difficulty. The animal ... is very carefully protected. But international agreement permits to hunt it can only be given for scientific purposes and mine was the first in 22 years. But it was not sufficient that I should just be able to supply the Coryndon Museum with a specimen, collected and mounted entirely at my own expense. I also had to turn the trip into a full-scale biological expedition (Fig. 18.6).

The modern international conventions that support the conservation of gorillas and their habitat are listed in Table 18.3. They impose a legal obligation on the countries that are parties to those conventions to implement them in their own national legislation and, where appropriate, to provide legal powers of enforcement.

Convention on Migratory Species

The Convention on Migratory Species (CMS) is important in that gorillas are listed as 'migratory species' on Appendix 1 of the Convention. As such they are recognised as being at 'very high risk of extinction throughout all or a significant part of their range'. Range states are expected to provide protection for the species and their habitat and to remove obstacles to their migration. In addition, CMS has instigated the Agreement on the Conservation of Gorillas and their Habitats (Gorilla Agreement) (Table 18.3) which requires the range states to review and improve their legislation and, where possible, to harmonise it so that the provisions and enforcement powers are consistent. The Action Plan in the Agreement includes 'implementation and enforcement of

gorilla conservation policies' but does not mention legislation despite the fact that strengthening the legal infrastructure is an obligation of the Agreement itself. This is essential for effective law enforcement and for dealing effectively with trans-boundary smuggling and is included in the Action Plans cited in Table 18.1.

The Mountain Gorilla Population and Habitat Viability Assessment (PHVA) (Cooper, 1997; Werikhe et al., 1998) (see also elsewhere in this book) recommended that the three mountain gorilla range states take note of the CMS as all gorillas are listed on its Appendix 1. Now, in 2016, all gorilla range states are parties except the Central African Republic and most have joined the CMS Gorilla Agreement UNEP/CMS (2011).

CITES

The Convention on International Trade in Endangered Species of Wild Fauna and Flora (CITES) regulates the cross-border movement of species (animal and plant) listed on the Appendices to the Convention for both trade and noncommercial (e.g., scientific) purposes. It applies to live animals, carcases and any parts or derivatives). Any such movement must be authorised by CITES import and export permits issued by the CITES Management Authority, on the advice of the Scientific Authority of the respective countries in accordance with the criteria set by the Convention. Permits for Appendix 1 species (considered by CITES to be threatened with extinction in the wild), including both gorilla species, can only be issued for movement that is 'not for primarily commercial purposes', for example, scientific, educational, captive-breeding or forensic purposes. Appendix 2 species (wild populations not at immediate risk of extinction but in need of monitoring) are allowed in trade under a CITES permit. Permits should not be issued retrospectively.

The Convention must be implemented in national laws together with provisions for its enforcement. Of the gorilla range, only Cameroon, Central African Republic, Democratic Republic of the Congo, Equatorial Guinea and Nigeria states have so far fully implemented CITES to the satisfaction of the CITES National Legislation Project (CITES, undated a). Some analysis of the progress of national CITES legislation in West African gorilla range states appears in the Last Great Ape Organisation (LAGA) Regional Library website (LAGA, undated). Hansard (2012) reports that the UK supported CITES technical legislation visitations to the gorilla range states of Cameroon, Gabon and Uganda in order to assist with this.

CITES is implemented in the European Union (EU) by Regulations that apply directly to all member states. Gorillas are listed on Annex A (the equivalent of CITES Appendix 1). The Regulations apply 'stricter measures' than the Convention; for example, a permit is required for the commercial use of Appendix 1 species, including exhibition to the public. On the other hand, once a CITES specimen can be shown to be legally within the EU it can be freely moved between the member states without further import and export permits. See EU (undated, ongoing) and EU/TRAFFIC (2015).

Further information on the working of CITES is to be found in Cooper and Cooper (2013), Cooper and Rosser (2002), Wijnstekers (2011) and on the websites of CITES, the EU and some national government websites.

CITES affects pathologists (and other veterinarians and scientists) working with gorillas in that samples and body parts taken from live animals or postmortem for health monitoring, diagnosis, forensic investigation or for research may have to be sent out of the country for processing or further study. In such situations CITES export and import permits are required not only for live gorillas but also carcases and 'any readily recognisable parts or derivatives'. This term includes not only trophies but any parts or samples taken from gorillas. Urine and faeces are not considered to be 'derivatives' and do not need CITES permits (CITES, undated b) but may need other (e.g., animal health) authorisation.

There are often difficulties and delays in the processing of paperwork on the part of both the sending and receiving countries and, in the meantime, the samples may have deteriorated or the results arrive too late for their original purpose (Cooper, 1993; Cooper and Cooper, 2013). It is difficult to obtain permits in advance because the type of sample cannot be predicted before the intervention.

The PHVA for the mountain gorilla and Cooper (1997) mentioned the need for 'fast-track' permits. These problems have been recognised by CITES in that Resolution Conf. 12.3 (Rev. CoP16) (CITES, undated c) provides for a 'simplified permit process' for 'urgent biological samples' whereby partially completed permits are issued in anticipation of the work to be undertaken. There are stringent conditions relating to the type of samples, their use and the permit procedure itself that are set out in CITES (undated c) and Cooper and Cooper (2013).

This process is not very well known so it may be necessary to show to the relevant CITES Management Authorities concerned the actual Resolution (or its equivalent in the EU, and the explanatory guide in EU/TRAFFIC, 2015).

Convention on Biological Diversity

This treaty applies to

- the conservation of biodiversity, that is, the variability of all living organisms, ecosystems and ecological complexes;
- the sustainable use of its components

- the fair and equitable sharing of the benefits arising from the utilisation (i.e., research and development) of genetic resources (or their biochemical composition) and any associated indigenous knowledge.

The Nagoya Protocol (implementing Article 15 of the CBD) requires its member states to provide national legislation for a transparent, legal framework regulating access to, utilisation of, and fair and equitable sharing of the benefits of genetic resources. It must also provide enforcement measures.

'Genetic resources' are defined as 'any material of plant, animal or microbial or other origin containing functional units of heredity'. Genetic resources are generally considered to be a subset of 'biological resources' as defined in Article 2 of the CBD (see Table 18.3). Allem (2000) urged that care should be taken when interpreting these definitions in biodiversity studies. Tvedt and Young (2007) also analysed the problems of providing a clear definition of the meaning of these and other terms used in the CBD.

It would appear that routine veterinary sampling or the preservation of body parts of gorillas for diagnosis, education, forensic evidence or reintroduction do not fall into the provisions for the utilisation (i.e., research and development) of genetic resources. However, if sampling (or using existing samples, possibly from a biobank) for research that could include or lead to development (e.g., of a vaccine), obtaining access to or using the material could be caught by the CBD provisions and national legislation implementing the Nagoya Protocol. No distinction is drawn by the CBD between commercial and noncommercial use (e.g., for taxonomy, gene banks or ecosystem research and development) although the Protocol suggests that a more flexible approach to non-commercial development might be considered (Correa, 2011; Greiber et al., 2012; IEEP, Ecologic and GHK, 2012).

In their paper Bartels and Kotze (2006) announced the setting up of a Biological Resource Bank in South Africa in accordance with the spirit of the CBD.

Other International Environmental Law

There are many other treaties that relate to the conservation of species and habitats and also to the wider environment that are broadly beneficial to gorillas. The Ramsar Convention on Wetlands, for example, lists a number of wetland sites that are important gorilla habitats, for example, Lac Tele in Congo and the Sangha River Ramsar Site in Central African Republic. The UNESCO World Heritage Convention lists World Heritage Sites 'of outstanding universal value', some of which are critical gorilla habitats such as the Virunga and Kahuzi-Biega National Parks in DRC. Similarly, the UNESCO Man and the Biosphere

Programme includes MAB Reserves such as the Volcanoes area in Rwanda and the Dja Reserve in Cameroon.

For a fuller understanding of this field of law see Birnie et al. (2009) and Sands and Peel (2012).

National Legislation

Most countries in the world have legislation on the conservation of free-living animals and their habitats (Cirelli, 2002). Countries with federal constitutions may delegate some wildlife law responsibilities to their constituent states; for example, in Nigeria and the Cross River State. However, the legislation may not be sufficiently robust to deal with the present-day threats to the animals and their habitat that it is intended to protect. Generally, the national legislation includes protection for the species found in that country and their habitats, plus the implementation of international conservation obligations. Some countries (such as Uganda and Gabon) legislate that indigenous wildlife is owned by the national or subnational government; there may also be local customary laws giving rights over wildlife or its utilisation (Cirelli, 2002). In much of Africa, protected areas, whatever their status, face the challenges presented by an unprecedented growth in human population and the consequential pressure on land (Fig. 18.7).

Examples of the national legislation of the range states that primarily relates to protection of the various subspecies of gorillas (so far as can be ascertained) are set out in Table 18.4.

These countries have a range of other conservation and environmental laws and subsidiary legislation that is discussed in the Regional Action Plans and other studies listed in Table 18.1 and in LAGA (undated), Morgera (2011) and The Law Library of Congress (2013). The general tenor of these sources is that the national legislation relating to gorillas needs to be rationalised, updated, implemented and enforced far more effectively

FIGURE 18.7 The boundary of a lush forest reserve in stark contrast to the unauthorised footpath and maize field.

(Rose, 2011). This is going ahead in west-central Africa (Djeukam, 2012; Mathot and Puit, 2008).

Access to Gorillas

Access to gorillas and their habitat, especially in protected areas, is usually tightly controlled; consequently, authorisation is normally required to visit them for any purpose. Even established research, conservation or veterinary projects may have limited access; visiting colleagues, friends or family have to be additionally approved or may be expected to pay gorilla tourist fees. Fees for commercial photography can be very high. Access to unprotected land may need the consent of the individual or community landowner and may take time to negotiate. Other considerations often affect access to gorilla habitat; for example, local administrative or community issues, civil unrest and guerrilla or military actions. This may involve additional authorisation, military and/or UN permits, armed or military escorts and enhanced measures for safety (Cooper and Cooper, 2001; Cooper, 2013; and see also Appendix 3: Field Pathology and Appendix 4: Hazards, Including Zoonoses).

Law Enforcement

Law enforcement is the responsibility of individual countries within their boundaries and is a matter of national legislation. It is now recognised in the Regional Action Plans and other studies in Table 18.1 that there is an urgent need to modernise and strengthen the national wildlife legislation in the gorilla range states, to increase penalties and to provide stronger powers for law enforcers such as police, wildlife officers, border controls, prosecuting authorities and the judiciary. The LAGA has been putting this into practice by providing training for magistrates and other law enforcement key stakeholders in Cameroon and within the subregion, leading efforts to combat wildlife crime and corruption in these areas. This is being replicated in other west-central African gorilla range states (Djeukam, 2012; LAGA, undated). The International Consortium on Combating Wildlife Crime (ICCWC) has carried out technical missions and training and is likely to be supported further by its Strategy Programme for 2016−2020 (see below). There is also a need for 'harmonisation' of the conservation law between different gorilla range states to facilitate better cross-border coordination when dealing with the smuggling of gorillas.

Interstate cooperation between the various agencies of the gorilla range states is a further step towards effective law enforcement. It is recommended by Nellemann et al. (2010), Stiles et al. (2013), the Gorilla Agreement (Table 18.3) and the Regional Action Plans for gorillas in Table 18.1. For example, range countries carry out joint patrols in the border areas of the mountain and Cross River

TABLE 18.5 International Organisations Concerned With Wildlife Law Enforcement

ICCWC	International Consortium on Combating Wildlife Crime	Parties: CITES, INTERPOL, UNODOC, World Bank, WCO (see below) Strategic Programme 2016 – 2020	https://cites.org/sites/default/files/eng/prog/iccwc/ICCWC_Strategic_Programme_2016-2020_final.pdf
CITES	Convention on International Trade in Endangered Species of Wild Fauna and Flora (CITES)	CITES Illegal Wildlife Trade – Recent high-level events and initiatives. CITES Secretariat, June 2014 COBRA I, II & III	https://www.cites.org/ https://cites.org/sites/default/files/eng/news/pr/CITES_Jun_2014_illegal_wildlife_trade.pdf
ICPO-INTERPOL	International Criminal Police Organization – INTERPOL	International cooperation between national police authorities	http://www.interpol.int/
UNODC	United Nations Office on Drugs and Crime	UN body for cooperation on illegal drugs including technical cooperation, research and strengthening of legislation	https://www.unodc.org/
WB	World Bank	With the Global Environment Fund runs the Global Partnership on Wildlife Conservation and Crime Prevention for Sustainable Development (Global Wildlife Programme). Will support projects in Cameroon, Gabon, Republic of Congo	http://www.worldbank.org/en/topic/environment/brief/global-wildlife-program
WCO	World Customs Organization	Collaboration between national customs authorities Declaration of the customs cooperation council on the illegal wildlife trade (June 2014)	http://www.wcoomd.org/en.aspx http://www.wcoomd.org/en/about-us/legal-instruments/~/media/BC96FE063BF848AD83E3ADB56B0A79BE.ashx
LATF	Lusaka Agreement Task Force Lusaka Agreement on Cooperative Enforcement Operations Directed at Illegal Trade in Wild Fauna and Flora	Member countries: the Republics of Congo (Brazzaville), Kenya, Liberia, Tanzania, Uganda, Zambia and the Kingdom of Lesotho Operations	http://lusakaagreement.org/
LAGA	Last Great Ape Organisation	Wildlife Law Enforcement	http://www.laga-enforcement.org

gorillas' habitat (Plumptre et al., 2015). Regional organisations, such as the Lusaka Agreement Task Force (LATF) (set-up in 1999), coordinate action on cross-border smuggling of wildlife and derivatives. LATF has organised multinational enforcement operations against, inter alia, the smugglers of great apes.

At the international level there are a number of multilateral organisations and initiatives for improving cross-border collaboration and law enforcement (Table 18.5).

The United Nations General Assembly Resolution 69/314 of 19 August 2015 and the Declaration of 13 February 2014 of the London Conference on the Illegal Use of Wildlife recognised the association of illegal wildlife and forest activities with serious and organised crime and regional insecurity. The ICCWC, formed in 2010, in its Strategic Programme for 2016–2020 (ICCWC, 2014) promised a concerted effort to solve the underlying problems of inadequate wildlife law enforcement by providing extensive technical support, training and other activities (see above).

The Strategic Programme includes the need for forensic evidence and training and capacity building in this field. In the main, their concern is for DNA techniques for tracing ivory and rhino horn and other artefacts of commercial value, such as gorilla trophies. The veterinary pathologist can answer questions relating to the cause, time and place of death, injuries or illness, make assessments of welfare and suffering, offer alternative methods of identification and provide forensic evidence to support prosecutions (see elsewhere in this book, and Cooper and Cooper, 2013).

GORILLAS IN CAPTIVITY

General

There is thought to be almost 1000 gorillas in zoos, sanctuaries and rescue centres according to the database of Gorillas Land (undated), Wilms (2011) and other sources. There may also be some gorillas in private collections or other facilities that are not recorded. Most in zoos are western lowland gorillas but a very few are *Gorilla gorilla graueri*.

A range of legislation applies to gorillas kept in captivity. This varies from country to country as does the extent and quality of the legislation and its enforcement. The relevant types of law will be outlined below to give an indication of the kind of regulation that one should look for when working in any given country.

Authorisation to Keep Gorillas

In their range states or neighbouring countries any captive gorillas (particularly young ones) are most likely to be kept in rescue facilities and sanctuaries following confiscation on account of illegal activities or due to injury or illness. These facilities are normally authorised by the wildlife authority, which may set standards for the care of the gorillas and inspect the premises from time to time. Animals kept or transported illegally are often kept in circumstances that do not meet their welfare and other needs (see Chapter 8: Non-specific Pathology).

Outside Africa, gorillas are to be found mainly in zoos although there may be others that are kept in sanctuaries or private collections. Since gorillas are endangered, large, and need specialised care and management in captivity, many countries regulate the keeping of gorillas. There are legal provisions for privately kept dangerous animals and also for zoos (i.e., collections that display to the public commercially). In either case the national or local government authority will normally require the licensing and inspection of the facilities where gorillas are held. It will probably be a condition of the licence that the owner or keeper must comply with guidance or standards for the animals' management, welfare and veterinary care and be insured against public liability in case anyone is harmed; see, for example, DEFRA (2012a,b). Compliance with other laws on, for example, health and safety, employment rights and public and animal health laws is also necessary. Many zoos belong to national, regional and global zoo associations and follow the guidelines, standards and ethical codes that are imposed as a condition of membership; for example, Dewar and Meijer (2015), EAZA (2013, 2014), Mellor et al. (2015). In addition, global and regional taxon advisory groups and species breeding programmes may provide general advice (IPS, 2007) or specific guidelines for keeping gorillas (Ogden and Wharton, 1997).

As gorillas are CITES-listed species the owner or keeper of gorillas should be able to demonstrate that they have been legally acquired through documentation of import, purchase, exchange, loan or gift and that the animals originated from a legal source. If the gorilla has been imported from another country, CITES permits should be kept to prove this. A permit to export an indigenous animal from its country of origin may be required under the national wildlife law (in addition to CITES permits). Other documentation relating to animal health, transportation and customs clearance should also be retained. Gorillas in well-regulated zoos are likely to be captive-bred and extensively documented by the zoo, the International Species Information System (see Chapter 7: Methods of Investigation — Sampling and Laboratory Tests), the stud books and breeding programmes of the various zoo associations.

Gorillas in Research

As will be apparent from other chapters in this book, a considerable amount of research has been carried out in respect of the biology and conservation of both free-living and captive gorillas. Field research relating to gorillas is noninvasive in that it is not intended to cause pain, distress after suffering or lasting harm (to use the definition from the UK legislation). Some data collection does not involve the presence of the gorillas — for example, the taking of hair and faecal samples from night nests (see Chapter 4: Noninfectious Disease and Host Responses). Invasive research on great apes in situ or ex situ is rare and generally not acceptable in the 21st century. It is forbidden in some countries (e.g., Austria, New Zealand, Spain and the Netherlands) although there is a narrow exemption for research in the EU Regulation on animal research that is also maintained by some EU member states (e.g., the United Kingdom). This permits research for the preservation of the species used or an unexpected pandemic affecting humans, providing that there is no alternative (EU, 2016). In practice, all research in gorillas requires some form of authorisation and usually, an ethical review; by for example the field project, by zoos, the research institutions concerned, grant-giving bodies and/or the wildlife authorities of the countries involved. A condition of a range state's research permit may require duplicates of samples or data to be left in the range country and the supply of scientific papers when they are published.

The IUCN Policy Statement on Research (IUCN, 1989) recommends compliance with 'all applicable laws, regulations and veterinary professional standards governing animal acquisition, health and welfare, and with all applicable agricultural and genetic resource laws involving species at risk of extinction'. This is supported by the Guidelines on the Implementation of the IUCN Policy together with other IUCN Guidelines such as those for re-introductions (see Baker, 2002; Beck et al., 2007).

Care should be taken to distinguish between procedures carried out for veterinary, for health monitoring or for research purposes and to ensure the appropriate authorisations are obtained (Bishop et al., 2013; Home Office, 2014).

An assurance that a research project complied with the law and any relevant ethical codes often has to be given to a journal when the results are submitted for publication. Ideally, a statement should be included in the paper of the relevant laws, a statement of compliance and details of any permits obtained, including CITES and animal health, if material is moved to another country. It may be difficult to publish the results of research that cannot be shown to be legally compliant and meeting generally accepted standards and this aspect may be questioned by readers. Project leaders and research supervisors should ensure that researchers understand the relevant laws, particularly about import/export requirements, and do not short-circuit the lengthy permit process.

Animal Welfare

Animals in captivity depend upon their keepers not only for food and water but their environmental and behavioural needs and for veterinary care when they are unwell or injured. In countries with well-developed legislation there is usually a specific law or laws that makes it an offence to treat an animal cruelly or to cause it unnecessary suffering. A number of countries, led by New Zealand (Wells, 2011), have modernised welfare legislation by including an additional obligation upon the person responsible for keeping the animal to take reasonable steps to provide for the animal's needs (Robertson, 2015). This can be enforced before the lack of proper care has reached the point of causing cruelty or unnecessary suffering.

Welfare is an essential concern in keeping gorillas in zoos (Fig. 18.8) or elsewhere (see Chapter 8: Nonspecific Pathology and contribution by Sophia Keen in Appendix 5: Case Studies — Museums and Zoological Collections). It is the keystone of the leading standards for zoos (cited above), both national and international, (Mellor et al., 2015; WAZA, 2013) and a recurring consideration in conservation interventions (IUCN/SSC, 2013; Pearson et al., 2007).

Many countries have little or no animal welfare provision and poor enforcement. Within the gorilla range states only Nigeria and Uganda retain legislation, based on the old British law, that makes it an offence to ill-treat, or cause unnecessary suffering to, domestic or captive

FIGURE 18.8 To meet the European zoo standards, gorillas are provided with spacious outdoor enclosures and environmental enrichment.

animals. In the range states, the standard of welfare given to a gorilla held in captivity is most likely to depend upon the person or organisation keeping it. When a primate is confiscated as a matter of successful law enforcement it is often difficult and costly for the enforcement authorities to accommodate and care for it properly and its future often depends on the resources and skills of NGOs. For those who wish to maintain a good quality of life for captive gorillas in such circumstances, it is possible to prepare voluntary guidelines for their care and management. These may be based on established zoo guidelines such as Abello (2006), Mellor et al. (2015) and Ogden et al. (1997) or on the 'Five Freedoms'.

The Five Freedoms (FAWC, 2009) are a distilled expression of the essential needs of an animal for its well-being. Many definitions of 'animal welfare' follow this concept. The freedoms were originally defined in respect of agricultural livestock but have since been applied to other species and circumstances such as zoos and animal research (e.g., DEFRA, 2012a). They are also a useful basis for voluntary guidelines that can be drawn up in situations where there is no animal welfare legislation or it is not enforced.

The Five Freedoms are:

1. Freedom from Hunger and Thirst — by ready access to fresh water and a diet to maintain full health and vigour.
2. Freedom from Discomfort — by providing an appropriate environment including shelter and a comfortable resting area.
3. Freedom from Pain, Injury or Disease — by prevention or rapid diagnosis and treatment.
4. Freedom to Express Normal Behaviour — by providing sufficient space, proper facilities and company of the animal's own kind.
5. Freedom from Fear and Distress — by ensuring conditions and treatment which avoid mental suffering.

Animal welfare concerns extend to the transport of animals with the usual caveat that in practice there is great variation in standards. An animal that is being sent to a country with high standards of welfare legislation should be prepared to a standard that meets its requirements, even if they are not required in the country of export, in case the law is enforced for violations once in the jurisdiction (including the airspace) of the receiving country. International transporters apply strict requirements to meet the standards of the Live Animals Regulations (IATA, 2016) for air transport and the CITES Guidelines for non-air transport (CITES, 2013).

Responsibility for Damage Caused by Captive Gorillas

Gorillas are capable of doing considerable damage, in the wild or in captivity. A person who owns or keeps gorillas is likely to be responsible for any death, injury or damage to property that they cause, depending upon the civil law in the jurisdiction concerned and may be liable to pay financial compensation (damages) to the person harmed. Insurance is advisable to cover such eventualities.

People employed to work with gorillas are at particular risk. An employer is normally also liable in civil law for harm that an employee suffers during work activities although the degree of liability will depend on the circumstances of each incident. This liability cannot usually be entirely excluded in an employment contract but this depends on the national law that governs the contract of employment. In addition, legislation usually requires an employer to hold workmen's compensation insurance.

OTHER RELEVANT LAW

Occupational Health and Safety

Some countries have strict occupational health and safety legislation that requires an organisation and its premises to be run with full regard for the safety of staff and visitors. Most health and safety management is based on risk assessment:

- Identify the hazards — anything that could cause harm
- Identify the risks — who might be harmed
- Identify how the above can be removed, reduced or controlled
- Record the findings — hazards, risks, preventive measures
- Review the risk assessment regularly to see whether it is still appropriate and is being complied with.
- See, for example, HSE, 2014(a) and 2014(b) and other website guidance

Hazards that are common in work with gorillas in the field or in captivity include:

- standard matters such as fire, heavy lifting, use of computers, driving vehicles and the safety of premises, working methods and environment. Every area of the workplace should be assessed, separately if necessary.
- specific hazards relevant to gorillas or pathology (Cooper, 2013; Lewis, 2003) such as:
  - veterinary work, dealing with biomaterials and drugs, dangerous substances including pathogens
  - laboratory and post-mortem equipment, disposal of waste and animal bodies; pathogens, zoonoses (BIAZA, 2014; DEFRA, 2012; Jakob-Hoff et al., 2014)
  - museum resources (e.g., feathers, dust, chemicals, obsolete preserving methods that may cause allergies and asthma, cuts and abrasions from bone fragments, instruments) (Hawks and Makos, 2000; Hawks et al., 2011; Irvin et al., 1972)
  - working near gorillas in captivity or in the field — physical injury, zoonoses
  - fieldwork — working alone, in dangerous conditions or in remote areas, lack of standard facilities (Fig. 18.9)
  - zoos (HSE, 2012; ZSL, undated) and other collections — gorilla handling, escapes and emergencies

A risk assessment is implemented by the use of protocols, codes of practice, standard operation procedures, data files, training, protective clothing and other strategies to reduce exposure to identified hazards in the workplace. Staff must be informed about them and given training. Allowance must be made for specific difficulties such as disability or lack of language comprehension and literacy (Fig. 18.10).

Risk assessment can be applied in any situation so may be useful for those employing staff in gorilla

FIGURE 18.9 A risk assessment and appropriate protective measures are needed when dealing with decomposing gorilla carcases in the field.

FIGURE 18.10 This sign caters to several languages and provides a non-literary warning too.

collections even where there is no legislation or it is not implemented. In these circumstances good working practices can be adapted from established guidelines and protocols in force in other countries — especially where a project or collection is funded from abroad. Taking such steps will ensure compliance with health and safety legislation where it exists and it will also help to reduce the risk of civil liability for accidents mentioned above.

Movement of Gorillas and Samples

Gorillas in captivity may from time to time be moved between countries, perhaps between zoos as part of a breeding or a stocking programme or a release programme. Samples and other derivatives taken from gorillas are frequently moved in-country and also overseas. In such circumstances a number of legal factors must be taken into consideration:

- In range states, national wildlife law permission to work with, keep or export gorillas
- Any in situ work should follow IUCN Guidelines such as Baker (2002), Beck et al. (2007), IUCN (2000), IUCN/SSC (2013)
- CITES permits. These apply to live gorillas, carcases and derivatives
- Animal health — import and export procedures including veterinary checks and quarantine; disposal of materials used during transport
- EU movement of gorillas within EU or into/out of EU. Simplified provisions for the movement of primates between premises approved under the 'Balai' Directive (BIAZA, 2013)
- Animal and human disease control measures, including notifiable diseases and zoonoses
- Samples and derivatives are treated similarly to live animals for CITES and animal health permits
- Samples must be packaged according to international standards (UN 3373) for biological samples

- Welfare — general welfare of live animals, transport requirements (see above)
- Carriage requirements and costs; import duties and clearance fees

See Cooper (2013), BIAZA (undated), Woodford (2000), and consult government and other websites on these topics as the requirements are frequently updated.

Law Regulating the Veterinary Profession

Veterinary legislation regulates the profession and the veterinarian's right to practise. Veterinary legislation normally defines:

- The scope of veterinary services that only an academically qualified and registered veterinarian is allowed to provide. This may vary from a narrow view of treatment and surgery in one country to a definition that includes all manner of veterinary work, including teaching, in another.
- The species to which the veterinary law applies.
- The requirements for a veterinary surgeon to be registered with the veterinary services board (VSB) that regulates the profession in the country or state in which the veterinarian is practising. Veterinary practice by unregistered persons is an offence in many countries, carrying the risk of a penalty (fine or imprisonment). In some gorilla range states government-employed veterinarians do not have to register and in one, the law states that owners may treat their own animals. In some countries such as Uganda, the law provides that free-living nondomesticated animals are the property of the national government.

The gorilla range states have a certain number of veterinarians, some in private practice or academia and others state-employed. Legislation regulates the veterinary profession in some countries such as Angola, Cameroon, Nigeria (Garba, 2014; GALVMED/Luseba, 2015). Rwanda and Uganda (EAC, 2015), most supervised by a veterinary board or council. The Democratic Republic of the Congo has no veterinary legislation and private owners may obtain drugs and treat their own animals.

The provision of veterinary care for gorillas (and other primates) in their range states attracts visiting veterinarians on account of the specialised expertise, equipment and training that they can offer — often as volunteers. In these situations, especially in an emergency or where there is a serious lack of veterinary skills in the area, the issue of registration or collaboration with the local veterinary profession may be ignored. In some countries there is little provision (or none at all) for registration but on the other hand a country like Kenya has become strict about registration and is carrying out awareness campaigns.

There is growing pressure on African countries to upgrade their veterinary professional legislation, to register veterinarians and to regulate the profession more closely. For example, the WOAH/OIE has revised its standard for veterinary legislation and developed a Veterinary Legislation Support Programme to strengthen national veterinary services and improve governance, animal health and disease control standards. There is also recognition of the converging interests of WOAH/OIE, CBD and CITES in animal health (improving controls and managing emerging diseases and zoonoses), the movement of animals and biological samples (CITES and WOAH/OIE, 2015); and likewise, with IUCN in the field of disease risk analysis (Jakob-Hoff et al., 2014). These are developments for which regular horizon scanning may be useful (Gusset et al., 2014; see Chapter 19: Pathology, Health and Conservation: The Way Forward).

In some parts of Africa, the veterinary services are becoming more organised, as the number of nationally trained vets and the range of their skills increase. In the growing private sector, the need to protect opportunities for employment and to exclude persons offering unqualified services has encouraged stricter enforcement of the veterinary law.

A point made by one African VSB is that once it starts enforcing the national legislation about registration and practice against unqualified or unregistered persons who are offering veterinary services (quite common in Africa in the long absence of proper regulation) the VSB cannot be seen to be lenient with unregistered foreign vets who fly in to deal with specialised veterinary treatment or surgery. Consideration should therefore be given to making an arrangement with the VSB for some form of temporary registration or authorisation for an external service before an incident arises that has the potential of carrying a fine or, at worst, time in prison.

Thus the visiting or foreign veterinarian should investigate registration in-country and liaise with the veterinary administration to build goodwill. It is also important to realise that a local gorilla NGO may not be aware of changing attitudes towards the governance of the veterinary profession within its own country. Maintaining good relations with the local veterinary services, both private and government, and within the wildlife authority will help to prevent regulatory problems.

In view of the fact that veterinary professional legislation is being revised in African countries under pressure from policies to strengthen the animal health services and control of disease, it is important to examine any proposed revisions to ensure that the proposed law (that is likely to be orientated towards domestic livestock) includes in its scope veterinary work with nondomesticated species in the wild and in captivity.

A non-national veterinary surgeon providing professional services is likely to require a work permit. This may be covered by a project's direct arrangement with the government's immigration department. Strictly speaking, a work permit is also normally required even for temporary and voluntary work but this is often ignored in many spheres of short-term charitable work. However, it is worth being aware of the vulnerability to investigation, detention or deportation if adverse attention is attracted. A gorilla veterinarian who applies for authorisation to work in the country should ascertain whether this extends to domesticated species since local project staff, friends and neighbours who keep livestock or other domesticated species and are likely to ask for veterinary help.

Medical, dental or other specialists in human medicine are at times called upon to advise, treat or operate on gorillas when techniques (e.g., for a caesarean section or necropsy or forensic investigation) can be adapted from human medicine (Allchurch, 2001). This is normally considered acceptable on account of the specialist expertise that they offer. However, the legality of such interventions will depend on the veterinary law of the country in question. In addition, formal consent of the owner should be obtained.

Veterinary Medicines

The import, manufacture, possession and supply of veterinary medicines are usually controlled by pharmaceutical legislation and should meet accepted standards (Health for Animals, 2015). In some parts of the world this is strictly enforced, but in the gorilla range states control is weak or minimal. Consequently, the quality and efficacy of medicines may not meet international standards. They may be counterfeit or adulterated or diverted from other countries. On the other hand, a person arriving in a country (even with poor overall enforcement) with veterinary drugs risks having them confiscated unless proper import documentation is presented on arrival. Duty may be payable on the imports unless the project has duty-free status; in this case it is wise to carry a supporting document or letter.

Correct and secure storage of veterinary medicines is important but may be made difficult by poor facilities and unreliable power supplies. Dangerous/controlled drugs, such as those used to immobilise gorillas, should be kept securely due to the possibility of misuse. Theft of any drugs is another high risk as they are readily sold. Taylor and Ingwersen (2016) indicate the importance of ketamine hydrochloride for veterinary anaesthesia in large animals, particularly in the field (just as it is in medical anaesthesia in under-resourced areas of the world). However, its use as a recreational drug makes it attractive to thieves.

Immobilising equipment is required to treat or manage either free-living or captive gorillas. In countries with strict regulation the equipment would normally be classed as a firearm and require a permit and the drugs used as dangerous/controlled drugs. This may not apply in the gorilla range states if laws are not strict or are unenforced but the same due regard to the hazards involved should be followed for reasons of safety. Additional ad hoc precautions are necessary in insecure areas or in times of war and civil unrest as vehicles are often stopped at checkpoints.

Permission, Reference Material, Data, Intellectual Property and Copyright

Academic and zoological institutions, museums, field and other projects hold gorilla material, generate research data, biological samples, veterinary and management records and other reference material. Questions of ownership and the right to use or reproduce such material often arise in research and veterinary work. Permission is often available for noncommercial purposes but should be requested and documented.

Veterinarians must consider legal professional requirements governing client confidentiality and the ownership of radiographic and other images, scans and clinical and post-mortem reports. When working with gorillas it is advisable to clarify in advance, with the relevant project or government authority, the ownership of records such as case notes, post-mortem reports, images and other records.

Access to research material depends upon the attitude of organisations towards sharing their resources. Some believe in open access but others strictly control information about, and use of, their raw data, biomaterial and other research assets. The would-be researcher may be required to complete a complex application (such as the US Association of Zoos and Aquariums Standard Research Proposal Form) to carry out research or even to request information on the material available. Thus, one zoo stated that 'even a listing of available biomaterials would require a full research committee proposal and approval process' and another collection advised that permission would have to be obtained from the government of the countries where the material was acquired. Whatever the ethics, applications for access to research materials or simply information about such resources are not for the faint-hearted and may be a deterrent to the young or disadvantaged scientist.

It is normally the host institution that sets the terms of use of its resources but any agreement should be carefully negotiated to cover all eventualities. Matters of joint authorship of papers, acknowledgement of the institution, staff or collaborators and the supply of eventual publications should be agreed. There may be restrictions on passing raw data to third parties, and major institutions may retain the ownership of intellectual rights to the proposed research in order to facilitate its future use in metadata research.

Written permission is often required to study, borrow or take samples from museum specimens (see Appendix CA1: Use of Collections and Handling of Biological Material). This may impose conditions on the use of material, especially if it is to be taken away from the institution or subjected to destructive sampling or analysis. The researcher may be required to return any unused material or preparations such as histological blocks (otherwise normally archived by the processing laboratory); copies of radiographs or other images may be required. The researcher may have to supply copies of publications of research based on the material. For the loan of valuable specimens, such as gorilla bones, a written agreement may include standards of care and insurance. It is not unknown for material to be alleged to have been removed from poorly curated museums so the level of trust may be low.

Field studies of free-living gorillas or of stored material such as bones will require authorisation by the range country. This may involve a permit or agreement, fees and the delivery of publications when completed; unfortunately, the last is not always honoured.

In publishing, permission forms are standard usage and are required for the reproduction of copyrighted images, text and other intellectual property in books and journals. Permission from large publishers is often available over the Internet. Payment or a copy of the publication may be requested.

Chapter 19

Pathology, Health and Conservation — The Way Forward

John E. Cooper and Gordon Hull

Probably no animal has fired the imagination of man to the same extent as has the gorilla.

George Schaller (1963)

INTRODUCTION

This book has explored the pathology of the genus *Gorilla* — how these animals respond to infectious and noninfectious agents — and how this in turn influences their health and welfare. The aim, as pointed out in the Preface, was to help in 'understanding disease' in these close relatives of *Homo sapiens*. So what are the salient points, the overall conclusions, that emerge from the preceding chapters?

We shall summarise them briefly.

KNOWLEDGE OF THE GENUS AND ITS BIOLOGY

Since their 'discovery' less than 2 centuries ago, gorillas have attracted much interest. Over the years, they have served as objects of human fantasy, public popularity (Fig. 19.1) and scientific curiosity. The stimulus to intensive research has been largely driven by the many similarities, anatomical, physiological and medical, between *Gorilla* and *Homo* — and, as the first two chapters describe, what this can teach us about the origins, evolution and success of our own species.

Despite this interest in *Gorilla*, it took nearly a hundred years for an authoritative textbook detailing the gross anatomy of gorillas to appear and even longer for anything substantive to be published about the gross and microscopic pathology of the genus. Only in the very recent past has scientific information begun to emerge

that defines how these animals respond to infectious and noninfectious insults.

THREATS TO SURVIVAL

Despite their popularity and scientific importance, gorillas face many threats to their survival, ranging from poaching to loss and deterioration of habitat, exemplified by the recent plans for oil exploitation in the Virunga massif (see Table 19.1).

THE IMPLICATIONS AND IMPORTANCE OF STUDIES ON PATHOLOGY AND HEALTH

A better understanding of diseases and their pathogenesis in *Gorilla* clearly serves to help gorillas themselves and thereby their welfare and conservation. In addition, however, such knowledge contributes to human health and well-being. The various facets of this relationship are discussed on numerous occasions throughout this book.

Alexander Pope (An Essay on Man: Epistle II) said 'Know then thyself, presume not God to scan; the proper study of mankind is man'. The same could be said for gorillas; the proper study is through the animal itself. However, in the absence of direct evidence, one must resort to extrapolation and here the many similarities between *Gorilla* and *Homo*, referred to above, clearly indicate that the best approach when investigating disease in a gorilla, alive or dead, is to 'think human'. A traditional veterinary approach, where the emphasis is on different domesticated animals, none of them primates, equips the veterinarian or research worker in terms of basic vertebrate pathology but little more.

In respect of studies on free-living gorillas, knowledge of (and preferably training in) tropical medicine is an

J. E. Cooper & G. Hull: Gorilla Pathology and Health. DOI: http://dx.doi.org/10.1016/B978-0-12-802039-5.00019-6

FIGURE 19.1 The statue of 'Jambo', the gorilla who sheltered a child who fell into his enclosure at Jersey Zoo (Durrell Wildlife Conservation Trust.). A much-loved national hero.

FIGURE 19.2 Routine gorilla sample collection for the Volcano Veterinary Centre, Rwanda, 1993.

FIGURE 19.3 The VVC (then a MAF project) veterinary surgeon and staff carry out a 'field' postmortem examination of a buffalo at Kinigi, Rwanda, 1994.

TABLE 19.1 Threats to Survival of Free-Living Gorillas

Threat	Notes
Habitat loss, forest encroachment	Alternative land uses; incursions into forest for cultivation, settlement, plantations (oil palm, monoculture trees) Mineral extraction, including oil exploration
Habitat destruction, degradation	Timber: logging, charcoal production, fuel wood, food gathering, hunting, honey gathering (fire), artisanal mining, warfare
Human−gorilla conflicts	Arising from intrusion in forest, gorillas invading smallholdings, crop raiding
Poaching	Hunting other species, damage from snares, disturbance by poachers
Poaching gorillas	Hunting gorillas for capture, trophies (e.g., hands, skulls), to undermine tourism, for bushmeat trade, local consumption
Pollution	See text
Climate change (external threats)	Changes in rainfall, drying of forests in sub-Saharan Africa predicted if average global temperatures rise by 4 degrees
War and its aftermath, civil disturbance	Refugees/population movements Invasion of gorilla habitat, urgent need for fuel to cook and food/ bushmeat; zoonoses and parasites

invaluable asset. Gorillas only occur naturally in Africa. They live in habitats where climatic factors have a great influence, where arthropod vectors flourish and where a whole range of 'exotic' infectious agents threaten humans, domesticated livestock and wildlife. A very different approach is needed on the part of both local and expatriate veterinarians from that which is taught and practised in more temperate climes.

ADVANCES IN HEALTH CARE

Developments in veterinary medicine and surgery have contributed much to the health and welfare of captive gorillas.

For those who worked two decades ago with the (free-living) mountain gorillas under the umbrella of MAF (Figs. 19.2 and 19.3), the advances have been outstanding and tribute must be paid to the MGVP for the way they have built on the initial achievements of veterinarians employed by MAF between 1986 and 1996. Other organisations have also done sterling work, such as CTPH (Fig. 19.4), which has led the way in community health 'on the spot' in Buhoma, Uganda.

FIGURE 19.4 Gladys Kalema-Zikusoka, Founder of CTPH, combines gorilla and human health in the proximity of the Bwindi gorillas. *Image courtesy of CTPH.*

FIGURE 19.5 PASA provides technical manuals and regular workshops for managers and veterinarians working at great ape sanctuaries in Africa.

Workshops run by European and North American zoos and by organisations such as the GPSG and PASA (Fig. 19.5) have done much to introduce modern veterinary practice and advanced pathology into work with mountain and lowland gorillas in different parts of Africa.

Pathology is fundamental to the clinical care of gorillas, in particular by helping to provide diagnoses and by guiding the treatment of diseases. Most important of all, however, it teaches us about issues such as host:parasite relations and the response of *Gorilla* to both infectious and noninfectious insults.

So, what are the messages for the future? What still needs to be done?

Some of the points made below are obvious because they are essentially a continuation and expansion of what

is already being achieved. Others have required a certain amount of 'horizon scanning' on the authors' part – that is, exploring what the future might look like, or, in simple terms, trying to discern 'new growing points'.

COLLABORATION AND SHARING OF INFORMATION

The need, especially in conservation programmes, for broader collaboration and less 'politics' and competitiveness has never been greater. It is salutary to read some of the recommendations on this that were put forward at the PHVA (Population and Habitat Viability Assessment) for the mountain gorilla held in Uganda nearly 20 years ago (Werikhe et al., 1998) and to ponder what might have ensued, had they been heeded.

In all studies, whether dealing with live gorillas, dead gorillas or their parts, interdisciplinary liaison pays dividends. For reasons explained above and illustrated with case studies in this book, the inclusion in the team of a 'human' pathologist, surgeon, paediatrician or odontologist usually proves synergistic in terms of results.

A GREATER ROLE FOR AFRICA

It is clear that involvement of the nationals of the range states, at all levels, can contribute substantially to our understanding of gorilla health and welfare. The participation of young people, such as veterinary students, in projects has already proved very helpful and has been a feature of teaching at Makerere University in Uganda – see later. As stressed throughout this book, African colleagues must be partners, not just employees or 'facilitators'.

There are so many areas in which Africans, of all races, can contribute uniquely to research on gorillas and wildlife. For a start, they have knowledge, since birth, of their locality and the people and animals that live there. They speak the language(s) and know the customs. Expatriates may attain some such knowledge but only in any real depth if they have shared similar experiences over a substantial period of time. This means not just visiting for short periods 'to do projects'. As Owen Rutter wrote in 1936: '... living abroad is a very different thing from travelling abroad'.

An example of a particular role that Africans can play in respect of free-living gorillas – and something that could help humans too – is in the field of ethnobotany, i.e., traditional local knowledge of medicinal plants. This has been the subject of research in many of the gorilla range states but the emphasis, understandably, has been on the treatment of humans. Evidence of self-medication by great apes (see text and Anon, 1997–98) suggests that applications to gorillas might be fertile ground for research.

In this respect, it is important that those working in situ with gorillas play a part in building stronger links throughout Africa, not just between the range states in the west and centre of the continent. The past and ongoing contributions to gorilla research elsewhere in Africa is emphasised in Chapter 2, The Growth of Studies on Primate Pathology. Kenya, for instance, has no free-living gorillas (or other great apes) of its own, but has for many years provided support through its world-renowned National Museums of Kenya (NMK) (Fig. 19.6) and Institute of Primate Research (IPR) (Fig. 19.7) to gorilla researchers and members of the Gorilla Pathology Study Group (GPSG). In 1994 NMK/IPR provided refuge to the senior author and his counterpart following evacuation from the fighting in Rwanda. Such history begets valuable scientific and personal links and serves as a reminder that, ultimately, gorillas belong to Africa.

FIGURE 19.6 Three veterinary pathologists examine the skull of a gorilla loaned by the NMK for a GPSG pathology workshop in Kenya.

FIGURE 19.7 Veterinarians study bones loaned by the NMK.

The sacrifices made by Africans who work with gorillas, so often in potentially dangerous parts of the world, are not always fully appreciated. When in 1999 guerillas captured members of a tour group in the Bwindi Impenetrable Forest, even the liberal British newspaper *The Guardian* emphasised that eight tourists had been killed and only at the bottom of column two did the report reveal that four Ugandan trackers/rangers had also been murdered. Eckhart and Lanjouw (2008) dedicated their beautifully illustrated book about the mountain gorilla to 'The Men and Women of the Protected Area Authority' and listed those who, at that stage, had been killed during the course of their duties in Rwanda, Uganda and the DRC.

Greater recognition, in their own right, of Africans who contribute to gorilla conservation continues to be needed. Several have been so acclaimed — for example, Edwin Sabuhoro (a graduate of DICE, University of Kent, UK) who in 2008 was named Young Conservationist of the Year by IUCN. Another, announced at the time of writing, was Inayom Imong, one of the 2015 Whitley Awardees, for working to save Cross River gorillas through community-based conservation in the Mbe Mountains, Nigeria. The list could go on.

INVOLVEMENT OF AFRICAN ACADEMIC INSTITUTIONS

Those working with gorillas in the field benefit greatly from close links with in-country universities and laboratories, not just those giving support from further afield — for example, in North America and Europe. Local institutions in Africa sometimes are sometimes poorly equipped and perhaps lack experience as far as gorillas are concerned but they often have other things to offer, ranging from readily accessible laboratories to knowledge of the country, terrain and populace where the gorilla work is being performed.

It was for this reason that the Coopers began so soon after the cessation of hostilities in Rwanda (July 1994) to forge links with Makerere University in Uganda, thereby playing a part in the establishment of the Department of Wildlife and Animal Resources Management (WARM) that was to play an important role in teaching and research, some of it relating to gorillas. That initial association between the VVC and Makerere has grown and now the MGVP boasts strong academic links with the University. It is to be hoped that in due course such a relationship will lead to a situation whereby Makerere routinely performs standard diagnostic procedures, such as the production and reading of histological sections, rather than having tissues sent thousands of kilometres by air to be processed in North America.

SCIENTIFIC ADVANCES

Advances in diagnostic methods in human and veterinary medicine have been prodigious in recent years and now include, for example, genetic techniques. There is a growing understanding of the importance of the microbiome, as discussed in several chapters in this book. Manual microscopy continues to be used for histopathology and other techniques but is increasingly being supplemented by the use of automated methods such as digital scoring and chemical analysers.

Laboratory services in sub-Saharan Africa are still often inadequate, even for human work. The organisation LabSkills Africa, supported by the (UK) Royal College of Pathology in partnership with the College of Pathologists of East, Central and Southern Africa and other bodies, seeks to improve the standards and quality of laboratory services through training, leadership development and mentoring. It has trained and supported 100 pathologists, laboratory technicians and biomedical scientists. This programme is at present restricted to human pathology but in this era of 'One Health' it should not simply be wishful thinking that the system might ultimately also involve veterinary and wildlife laboratories.

HUMAN HEALTH

Human health is of great significance to gorilla health and vice versa, as explained so often in the pages of this book. In captivity gorillas can infect humans and organisms can be acquired by them from keepers, veterinarians and visitors. In Africa, tourism and habituation (Fig. 19.8) have long been viewed with some concern (Russon and Wallis, 2014), particularly because of the opportunities such activities offer for the exchange of pathogens. In this age of *Ebolavirus* disease and other emerging infections the consequences are potentially very serious. Relatively cheap modern air travel has facilitated the rapid, almost

FIGURE 19.8 A tourist discusses the gorillas with the veterinarian.

immediate, spread of some human pathogens across the world. Until recently international disease surveillance was mainly based on medical reports following the investigation of infected patients. The approach is now changing to one of early detection of potential pathogens and potential risks such as the study involving the genome sequencing of organisms detected in the toilets of civil aircraft by the Technical University of Denmark.

ATTENTION TO WELFARE

The need to protect gorillas extends to those in captivity, with a particular emphasis on their welfare. There have been commendable improvements in how gorillas are kept in zoos since the days of the first gorillas in captivity and, as Chapter 4: Non-Infectious Disease and Host Responses, points out, noninvasive research is yielding valuable information about how these and other great apes respond to adverse stimuli and thrive best in captivity. The welfare of gorillas in the wild should also come under greater scrutiny, especially when animals are habituated and we become more responsible for them. The welfare of the animal should be an integral part of the cost:benefit analysis in interventions, treatment, study and viewing gorillas in the wild.

CONTINUED SEARCH FOR GORILLA MATERIAL

Commentary in the text of this book and, especially, Part II: A Catalogue of Preserved Materials, indicate the ongoing value of *Gorilla* specimens and collections. Nevertheless, there must be much gorilla material, skeletal or biological, lurking in unexpected locations that is not yet publically recorded, whatever its nature or source. For example, parts of gorillas have long been used for medical and religious purposes in Africa: some may still be in use; others are likely to be found in museums or cultural centres. Elsewhere in the world there must be gorilla hands and other 'trophies' in private homes, probably stored in trunks in attics. What happened to the 'massive collection of bones' brought by sea to London (The Royal Veterinary College, RVC) by Sir Arnold Theiler several decades ago? We do not know whether this included material from gorillas but the collection was amassed in South Africa and it is possible that it contained bones from East and Central African animals.

In the domain of the pathology laboratory, there are thousands of samples in different parts of the world, not all of which have found their way into the pages of the catalogue that accompanies this book. Much could be done with them. For a start, gorilla science could benefit from the rereading of histology slides, the cutting or recutting of paraffin blocks and reexamination of wet tissue using new procedures. Gorilla bones are another example; advanced study by pathologists and osteologists, working together, could be mutually beneficial to the two disciplines and to the advancement of primate pathology.

The challenge is there!

GREATER DISSEMINATION OF INFORMATION

There is a need for more publications about gorillas and their health, coupled with greater dissemination of information at all levels. Reporting and publishing should be part and parcel of all studies on the taxon, whether in the wild or in captivity.

The value of observational studies must not be underestimated. They were the basis of early knowledge about gorillas and observation remains the cornerstone of surveillance and health monitoring. There is a certain intellectual arrogance in some quarters that expresses itself in taking the view that research that is not based on a hypothesis, or 'asking real questions', is of little or no scientific value. Such an approach towards studies and publications on gorillas belies the value of those, at all levels, who regularly and conscientiously put together data on gorillas in the wild or captivity prompted only by a wish help to build a database/body of information on these endangered species. In a lecture in 2015 the primatologist Jane Goodall recounted her early struggles with the academic establishment to accept the observational methods that she used when studying chimpanzees in the Gombe Stream National Park in Tanzania. Things are changing and even the careful use of anthropomorphism is no longer considered a cardinal sin in scientific circles as it can often help provide an insight into the working of the minds of wild animals, including gorillas (Tudge, 1995).

The main necessity, however, as with so much relating to work with gorillas, is more openness and transparency with regard to findings, results and materials — and thereby making them readily available to all, in the interests of science, veterinary medicine and pathology and the gorillas themselves.

BROADENING THE HOLISTIC APPROACH

Over a hundred years ago, in a valedictory address, Mills (1896) told graduating veterinary students that 'comparative medicine is the medicine of the future and the sooner that is realised the better for man as well as for beast'. That philosophy has been reinforced in succeeding decades by, for example, Schwabe (1984), the 'father of modern epidemiology' who introduced the term 'one medicine'. He urged synergy between different health practitioners and scientists.

In recent years other terms relevant to work with gorillas have found favour, ranging from 'comparative medicine' to 'zoobiquity', each stressing, in its own way, the need for collaboration between different disciplines and urging a more integrated approach to the health of divergent species. The most favoured at present in veterinary circles is 'One Health' but others remain important — see below. The avowed aim of 'One Health' is to improve health and well-being through cooperative efforts at the interface between humans, animals and their environments.

A 'One Health' approach has been applied with enthusiasm by many in the field of gorilla research — see various chapters in this book. It is apt that this should be the case because of the strong medical/veterinary links that have long been integral to tropical medicine (Cooper, 1985). Recently Kamani et al. (2015) argued that Africa is the appropriate home for One Health, not least of all because people's lives there are intimately related to the health and productivity of livestock and the natural environment. As national medical and veterinary institutions are still maturing, African health professionals are presented with an opportunity to 'leapfrog' barriers that, in Europe particularly, have been imposed by more well-established and rigid systems. African scientists and African institutions have, they stated, the opportunity to become world leaders in One Health.

Notwithstanding this very valid point, the senior author (JEC) makes a strong case here for more attention to be paid to one of the other areas of collaborative medicine. This is 'Ecosystem Health', launched some years ago by Canada's International Development Research Centre (IDRC) (Lebel, 2003).

What is now known as the Ecohealth approach has application to work on gorillas and, indeed, other wildlife. It can, and should, be seen as a part of One Health. Its appeal to the senior author (JEC) is that it (the Ecohealth approach) brings in people from many and varied disciplines, not just 'medical professions'. It therefore, for example, provides a niche for field naturalists who, as this book emphasises, still have so much to offer to studies on gorillas. As Tewksbury et al. (2014) pointed out, many who worked professionally with wildlife in the past were strongly influenced by sustained contact with nature at an early age. Unfortunately, the growth of the human population and the relentless pace of urbanisation are fundamentally changing the way in which future generations will interact with the natural world.

Those working with, or recruiting staff for, projects concerning gorillas should be mindful of this. Veterinarians tending gorillas, especially in the wild, need to be sound naturalists, not just competent 'animal doctors'.

FIGURE 19.9 A mountain gorilla at ease in its natural habitat.

FIGURE 19.10 View across cultivated hills and valleys to the Virungas, mountain gorilla habitat in Rwanda.

CONCLUDING REMARKS

As we finalise this concluding chapter, on 22 April 2016, 180 nations are signing the treaty on climate change agreed last year in Paris. To get so many countries to agree on such a complex issue is a great achievement. But it is also a landmark. If it achieves what is hoped, it could ensure a future for humans, animals and plants on this planet. As such it is apposite to this book because of our overarching thesis that the health of gorillas is inextricably linked to the well-being of humans, other animals and the environment that they share. As Annette Lanjouw said in an interview in 2001, 'Welfare of people and conservation are one and the same'.

It is endorsement of the crucial importance to living things of Global Health, Ecosystem Health and One Health.

The closing chapter of Andrew Dixson's (1981) book was entitled *Conservation or extinction?* and that is plainly the most pressing question today, in the second decade of the 21st century, insofar as the genus *Gorilla* and its African haunts are concerned (Figs. 19.9 and 19.10).

The greatest need is knowledge. As the Senegalese writer and environmentalist Baba Dioum put it:

In the end we will conserve only what we love. We will love only what we understand. We will understand only what we are taught.

The authors of this book – and their fellow contributors – hope that it will play a part in fostering such teaching, love and conservation, not only in respect of gorillas, our magnificent and complex fellow primates, but also the wild, untamed places where they still survive.

Appendix 1

Glossary of Terms

John E. Cooper and Gordon Hull

INTRODUCTION

The list of words and definitions that follows is not comprehensive but explains the authors' use of various terms in this book. It is intended to encourage consistency in terminology and to assist those readers who are relatively unfamiliar with medical or biological terms or are not primarily English speakers. Certain other, especially lay (common parlance, colloquial), words and expressions, usually given in inverted commas, are also explained below.

Some definitions are given in the text.

LIST OF TERMS AND MEANINGS

Abrasion an injury to the skin involving the outer surface (epidermis) only. In common parlance (lay terminology), termed a 'scratch' or a 'graze'. Abrasions are caused by contact with, or against, a blunt object and are one of the triad of injuries caused by blunt force. (See also **Bruise** and **Laceration**). In pathology parlance, an erosion is a shallow abrasion (of any tissue), and an ulcer is a deeper abrasion that penetrates the surface layer

Abscess a focus of suppuration within a tissue, organ or region of a body

Acute describes diseases, clinical signs or host reactions that start abruptly, are usually characterised by marked intensity and then subside after a relatively short time. From Latin '*acutus*' – sharp

Aetiology (USA 'Etiology') the cause of a disease

Allergy altered bodily reactivity, as in a hypersensitivity, to an antigen, exaggerated or pathological immunological reaction to a substance

Alveolus (plural **Alveoli**, adjective **Alveolar**) empty cavities, as in the lung and tooth socket.

Anaerobic living only in the absence of free oxygen, as in 'anaerobic bacteria' in wounds

Anaphylaxis an acute allergic reaction, sometimes resulting in severe clinical signs or even death

Aneurysm an abnormal dilatation of a blood vessel, which may rupture

Anomaly marked deviation from normal

Anorexia absence of appetite (also often referred to as inappetance)

Anterior the front of the body, limb, organ, etc. (See also **Cranial**)

Antibody (immunoglobulin) serum protein produced in response to the presence of specific antigens. Detection of specific antibodies is useful in diagnosis

Antigen any substance that is capable of being recognised by the body as foreign, for example, invading organisms, toxins. Recognition gives rise to an immune response, such as antibody production

Antiseptic see **Disinfectant**

Apnoea cessation of breathing

Apoptosis 'programmed cell death', normal and controlled death of cells

Asphyxia lack of oxygen to the cells of the body caused by reduced oxygen in the inspired air. Often caused by obstruction of the external or internal airways (as in strangulation), inability of the blood to carry oxygen or of the cells and tissues to utilise oxygen

Asymptomatic an infection in which no signs of illness are present

Atrophy decrease in size of a tissue or organ

Autolysis postmortem breakdown of tissues, usually as a result of cell death, may involve the whole body or part. Can complicate postmortem examination but may assist in determination of time of death

Autophagy the process by which dysfunctional cellular components are degraded, within the cell, through the action of lysosomes

Autopsy postmortem examination, usually of a human. (See also **Necropsy**)

Bacteraemia the presence of bacteria in the blood

Basal length a commonly used measurement of cranium length. The distance from the anterior part of the foramen magnum to the prosthion along the underside of the skull

Basion the midpoint on the anterior margin of the foramen magnum

Behavioural flexibility behavioural responses to changing local conditions, reflecting solutions to ecological or social problems (sometimes referred to as 'behavioural adaptability')

Biocide a chemical substance, sometimes a microorganism, that can deter, render harmless, or exert a controlling effect on any harmful organism by chemical or biological means

Biological vector an animal, such as an insect, that transmits an infective agent from one host to another and in which the agent multiplies

Biomaterial any organic piece or derivative of a plant or animal

Bizygomatic breadth a commonly used measurement of cranium width. The distance between the zygomatic arches, measured from the outermost part of each zygoma

Bleeding leakage of blood caused by damage to blood vessels (see also **Haemorrhage**)

Blisters (vesicles) collection of fluid derived from blood below the outermost layer of the skin (epidermis)

Blowflies bluebottles, greenbottles and other sarcophagous flies in the order Diptera, class Insecta. Their larvae (maggots) and pupae can be used to assist in the determination of the time of death in cases of advanced decomposition

Brachycephaly the condition of having a broad, short head

Bregma the point on the top of the skull where the sutures between the frontal and parietal bones meet

Bruise see **Contusion**

Bulla ('blister') elevated, fluid-filled, lesion measuring more than 5 mm

Burns injury caused by either wet or dry heat. There are degrees of severity: first — outer surface of skin only, second — full thickness of skin, third — deeper tissues. Chemical and friction burns also occur

Bushmeat nondomesticated animals (wildlife) used as food

Caries demineralisation and loss of hard tooth tissue, leading to continued destruction of enamel and dentine, followed by cavitation of the tooth; may be caused by bacteria. A progressive acidic demineralisation of the inorganic matrix of dental tissues secondary to bacterial multiplication

Carrier an animal that harbours and may transmit the causative agent of a disease

Caudal relating to the cauda or tail. Also used for **Posterior**. (See later)

Cellulitis a spreading infection so that the exudate travels between cells and dissects through the cleavage planes of the interstitial and tissue spaces

Chlorinated hydrocarbons synthetic organic insecticides that block transmission of nerve impulses, have relatively low acute toxicity, but are persistent in the environment

Choking accidental or deliberate obstruction of the upper airways leading to asphyxiation

Chronic describes diseases or clinical signs that may arise slowly and persist for a relatively long time. From Greek '*chronos*' — time

CITES Convention on International Trade in Endangered Species of Wild Fauna and Flora

Clinical signs the features of a disease that can be observed, for instance, lameness, diarrhoea, etc. (See also **Symptoms**)

Co-existing species species that occur at the same time period and in the same place and thus can potentially interact (also known as 'sympatric species')

Co-occurring species species that occur at the same time, but not in the same location (also known as 'synchronic species')

Commensal an organism which, although not a parasite, lives in, on or with another (and is nourished by the same food)

Comparative forensic medicine that branch of forensic science relating to different species of animals, including humans, and its application to the judicial and other processes

Concussion lay term for the effects of a head injury. No generally agreed medical definition exists. However, in humans a group of symptoms and clinical signs including amnesia, confusion and altered consciousness are referred to as Post Concussional Syndrome

Condyle rounded bony prominence that provides a point for articulation

Condylobasal length a commonly used measurement of cranium length. The distance between the posterior part of the occipital condyle to the prosthion along the underside of the skull

Congenital present from birth; congenital defects may be immediately apparent or may not become manifest themselves until later in life

Congestion excessive accumulation of blood in a tissue or organ. This is not the same as 'bleeding' because the blood is still in the blood vessels

Contusion leakage of blood from a damaged blood vessel, most commonly the smallest vessels — capillaries

Cotype an additional type specimen from the same brood as the original type specimen

Cranial relating to the cranium or skull. Also used for '**Anterior**' (See earlier)

Cranial length a measurement of the length of the braincase. The distance from the inion to the glabella

Cranium skull, not including the lower jaw

'Cut' in popular parlance, a break in the surface of the skin. Correct terminology depends on the cause of the skin injury; blunt trauma causes lacerations, sharp trauma causes incisions

Cyanosis blueness of the skin caused by reduced oxygen in the blood

Death the cessation of life. Usually ascertained by the absence of heartbeat and respiration

Decomposition decay of body after death. (See also **Autolysis**). Very variable and exact type and speed of decomposition will depend on body type, environmental temperature, availability of water, etc.

Dehydration lack of water in the body. Can occur in starvation, neglect, hyperthermia, etc.

Demographics study of factors that affect a population, such as birth and death rates

Dental calculus (plural 'calculi') (Tartar) mineralised plaque, usually a yellowish film formed of calcium phosphate and calcium carbonate, food particles and other organic matter, deposited on teeth by saliva

Diarrhoea loose (sometimes watery) faeces

'Discharge' an excretion or substance leaving the body. (See also **Exudate**)

Disease an unhealthy condition of the body or a part thereof, disordered state of an organism or organ. Anything that causes impairment of normal function. Diseases can be caused by infectious agents (e.g., viruses) or noninfectious factors (e.g., burns)

Disinfectant an antimicrobial agent that is applied to nonliving objects to destroy microorganisms. Disinfection does not usually kill all microorganisms, in contrast to **sterilisation**, which can be physical (e.g., use of heat) or chemical. Disinfectants and sterilants act by destroying the cell wall of microorganisms or by interfering with their metabolism. An **antiseptic** is used to destroy microorganisms on living tissue. Disinfectants are different from **biocides** — the latter are intended to destroy all forms of life, not just microorganisms

Disseminated intravascular coagulation (DIC) a complex disease process in which there is uncontrolled clotting of the blood inside blood vessels. It can be caused by many factors including septicaemia, trauma, etc.

Distal situated away from the midline of the body or from the point of attachment

Domesticated a species of animal that has for long lived in association with human beings for purposes such as food production or companionship. In biological parlance, a 'domesticated' animal is one that breeds under human control, provides a product or service useful to humans, is tame, and has been selected away from the wild type. There are no domesticated gorillas; all are 'wild animals', even if in captivity

Drowning death caused by immersion in a fluid, usually water. Pathological features may be absent if death was rapid and they differ for fresh and salt water drowning

Dysentery blood in faeces, may be fresh (red) or partly digested (dark and tarry)

Dyskeratosis abnormal keratinisation

Dysphagia difficulty in swallowing

Dyspnoea difficulty in breathing

Ecology the study of the interrelationships of organisms and their environment

Ecosystem a dynamic complex of plant, animal and microorganism communities and their nonliving environment that serve as a functional unit

Effusion escape of fluid into a body cavity

Emaciation excessive leanness; a wasted condition of the body

Embolism (plural emboli) blockage of a blood vessel, usually an artery, by an extraneous object or material

Enzootic the continual presence of a disease in an animal community

Epidemiology the study of disease in a population, including its prevalence/incidence, distribution and spread

Epizootic a disease suddenly affecting a large number of animals in a given region ('epidemic' in humans)

Erosion focal loss of epidermis (see also **Ulcer**)

Ethology the study of animal behaviour

Euthanasia humane killing (Greek: 'good death')

Excoriation linear, traumatic, lesion resulting in epidermal damage

Exostosis (plural 'exostoses') see text. Sometimes used in the broad sense of any new osseous tissue on the surface of a bone (see also **Osteophyte**)

Exhumation recovery of a body from a grave or burial place

Exudate a fluid, usually of inflammatory origin, with high protein content, a specific gravity of more than 1.020 and many cells. It can be 'serous' (watery), 'mucous' (containing mucus) or 'purulent' (containing pus). Exudation is characterised by an escape of such fluid, into a body cavity or into interstitial tissue

Facial length The distance from the glabella to the prosthion

Faeces droppings, excreta, 'stool' (colloquial)

Fat embolism blockage of blood vessels, particularly of the lungs and kidneys, by portions of fat. Seen following trauma, particularly fractures

Fetopathy disease or disorder of a fetus

Flatulence (adjective 'flatulent') gas in the lower intestine. A flatulent gorilla may pass gas through the anus

Flotation test the placing of tissues or samples in water to see if they float or sink. Sometimes used to determine if a baby or young animal was born alive: inflated lung floats but this is not a totally reliable test. Pneumonic lung sometimes sinks, sometimes floats, depending upon the changes present. Liver that floats may be fatty or contain air/gas

Foramen magnum large hole in the occipital bone that allows the spinal cord to enter the skull

Forensic veterinary medicine the application of veterinary knowledge to the purpose of the law

Friable crumbly or powdery

Frontal bone the bone that forms the forehead

Gangrene necrosis (death) of tissue, due to a failure of the blood supply, disease or direct injury. Such necrosis of tissue may be followed by bacterial invasion leading to putrefaction. 'Dry gangrene' is necrosis without subsequent infection, the tissues becoming dry and shrivelled

Genetic material material of plant, animal, microbial or other origin that contains functional units of heredity (see Chapter 18)

Genetic resources genetic material of actual or potential value

Gingivitis inflammation of the gums

Glabella the supraorbital region located between the orbits and just above the highest point at which the nasal bones meet the frontal bone

Haematemesis vomiting of blood

Haematoma used to describe a significant collection of blood outside blood vessels and in the tissues of the body

Haemoglobin pigment in red cells responsible for carrying oxygen

Haemoptysis coughing-up of blood

Haemorrhage bleeding

Health a state of physical and mental well-being of the body (not merely the absence of disease). 'A harmonious participation in the resources of the environment, which allows individuals the full play of their functions and aptitudes' (Lebel, 2003)

Health monitoring ongoing evaluation of health status; may involve clinical examination, haematology, parasitology, etc. See also **Screening**

Histology microscopic study of the composition of tissues and organs

Holotype a single type specimen upon which the description and name of a new species is based

Homeostasis constancy in the internal environment (the '*milieu interieur*' of the French physiologist, Claude Bernard) of the body, naturally maintained by adaptive responses that ensure healthy survival

Hyperaemia increase in blood supply to an organ or tissue

Hyperkeratosis excess keratinisation

Hyperpnoea increase in respiratory rate

Hyperostosis an unusual growth or thickening of bone; see text

Hyperthermia an increase in body temperature

Hypertrophy increase in size of a tissue or organ

Hypoglycaemia low blood sugar

Hypostasis (adjective: hypostatic) the red staining of the skin of a carcase caused by the settling of blood under the influence of gravity. Areas of the body that are in contact with surfaces will be unaffected and will remain white (if the skin is not heavily pigmented)

Hypothermia decreased body temperature

Hypoxia low levels of oxygen in the blood and tissues

Iatrogeny (noun; adjective **Iatrogenic**) an adverse reaction that results from treatment, induced by human intervention

Icterus (Jaundice) yellowing of the skin or mucous membranes due to bile pigment

In situ in situ conservation may be defined as that which takes place within the natural range of the species involved, as opposed to ex situ efforts, usually involving captive breeding, which may be far removed from that area, often in a different country

Incidence the number of new cases of a particular disease during a stated period of time

Incision break in the continuity of the skin caused by an object with a sharp edge

Incubation period the time between acquisition of infection and onset of clinical signs

Infarct an area of tissue that has been deprived of a blood supply

Infection the entry of an organism into a susceptible host (human or animal) in which it may persist, but detectable clinical or pathological effects may or may not be apparent. The means by which a pathogenic organism becomes transferred to a new host. The invasion of a host by organisms, with or without manifest disease. The pathological state caused by organisms

Infectious disease a disease caused by the actions of a living organism (virus, bacterium, etc.), as opposed to (for example) physical injuries, endocrinological disorders or genetic abnormalities

Inferior (adjective) situated below or directed downward; also refers to the lower surface of a structure

Inflammation a local response characterised by swelling, redness, heat and pain and provoked by physical, chemical and/or biological agents

Inion the most posterior point of the midline of the cranium, where the occipital and parietal bones meet

Injury tissue damage. (See also **Wound**)

Insult any adverse stimulus, infectious or noninfectious, to the body, an organ or a tissue. An action that causes damage to a tissue or organ

Invasiveness the ability of a microorganism to gain a footing and to spread within the tissue of the host

Jaundice see **Icterus**

Laceration break in the continuity of the skin or organ caused by application of blunt force. One of the triad of injuries associated with this type of force. Lacerations are usually characterised by jagged edges. They are not caused by sharp objects which produce incisions

Latent infection an inapparent infection in which the pathogen persists within a host, but may be activated to produce clinical disease by such factors as stressors or impaired host resistance

Lectotype a biological specimen or other element that is selected as the type specimen when a holotype was not originally designated

Lesion any abnormality, usually a well-defined alteration, in an organ or tissue caused by disease; usually it is characterised by changes in appearance, for example, a raised nodule on the skin. From Latin 'laesio' − injury

Ligature mark a form of friction abrasion due to the effect of a ligature

Lumen cavity of a hollow organ, for example, the lumen of the stomach or uterus

Maceration destruction of tissues or a body caused by a combination of trauma and decomposition

Macule flat, circumscribed, lesion distinguished from surrounding skin by its coloration

Malocclusion malposition of teeth, resulting in faulty meeting of teeth or jaws

Mandible the bone of the lower jaw

Mastoid a bony process of the temporalis just behind the ear region on the lower part of the cranium, which is an area of attachment for postural musculature

Maxilla bone of the upper jaw holding the canine, premolar and molar teeth and forming the sides of the nasal passage

Mechanical vector an animal such as an insect which transmits an infective agent from one host to another but is not essential to the life cycle of the organism

Melanocytes cells that produce melanin pigment

Microbiology the study of microscopic or ultramicroscopic organisms such as viruses and bacteria

Monitoring the routine collection of information on disease, productivity and other characteristics.

Morbidity rate the proportion of clinical cases during a given time

Mortality rate the proportion of deaths during a given time

Mucous membrane the thin layer of mucus-secreting epithelium lining such areas as the mouth, nose, eyes, vagina and rectum

Mummification a form of decomposition that is associated with warm dry conditions whereby the body dehydrates. Often results in remarkably good preservation. Usually involves the whole body but may involve only part, most commonly the extremities

Nasal bones the pair of facial bones that form the upper part of the nasal passage (the 'bridge of the nose' in humans)

Necropsy autopsy, postmortem examination (of an animal)

Necrosis death of a cell/tissue/organ

Neoteny (noun; adjective **Neotenous** or **Neotenic**) the retention of juvenile characteristics in adulthood

Nodule elevated, solid, lesion measuring more than 5 mm in diameter

Nuchal crest a bony ridge on the occipital bone that provides an area for attachment of the muscles of the neck

Nuchal height the distance from the midline posterior margin of the foramen magnum to the inion

Objective synonym each of two or more different names applied to one and the same taxon based on the same type

Occipital bone the bone that forms the back and base of the skull and includes the foramen magnum

Oedema (United States: 'edema') abnormal accumulation of fluid under the skin or elsewhere (e.g., pulmonary oedema)

Opportunist an organism usually incapable of inducing disease in a healthy host, but able to do so in a less resistant or injured host

Orbit the bony cavity containing the eyeball and its associated muscles, vessels, and nerves

Organophosphates synthetic organic insecticides that inhibit the function of cholinesterase; acutely toxic but generally not persistent in the environment

Osteoma (plural 'osteomata') a benign, proliferative, lesion of bone

Osteomyelitis inflammation of bone and bone marrow

Osteophyte see text. Sometimes used in the broad sense of any osseous (bony) projection (see also **Exostosis**)

Outbreak the confirmed presence of disease, clinically expressed or not, in at least one animal in a defined location during a specified period of time

Palate the roof of the mouth, consisting of an anterior hard portion and a posterior fleshy portion (soft palate), that separate the oral cavity from the nasal cavity

Palaeontology the study of fossil animals and plants

Papule elevated, solid, lesion measuring 5 mm or less in diameter

Parakeratosis abnormal retention of nuclei in stratum corneum

Paralectotype any additional specimen from among a set of syntypes, after a lectotype has been designated from among them

Parasitaemia parasites present in the blood

Parasites organisms that live in or on another and benefit from it. Nowadays, especially by biologists, divided into 'macroparasites', such as worms and ticks, and 'microparasites', such as bacteria and protozoa

Paratype a specimen other than a type specimen that is used for the original description of a taxonomic group and specifically stated to be the one on which the original description of the taxon was based

Parietals the pair of bones that join to form the sides and roof of the cranium

Plastination a technique for preserving bodies, organs and other biological tissues that involves replacing the water and fat with a polymer, such as silicone or polyester, in order to produce a dry and durable specimen for display or anatomical study

Plastotype an artificial specimen cast or moulded directly from a type specimen

Pathogen any living organism or other agent that is the cause of disease. Can range from a microscopical bacterium to a worm or a flea (see text)

Pathogenesis the *mechanism* of a disease, how a disease develops

Pathological (**United States: 'Pathologic'**). Of or relating to pathology

Pathology strictly (from its Greek roots), the science of the study of disease. This usually includes its causes, its development and its appearance (see text). In common parlance 'pathology' tends to be linked with death, carcases and postmortem examination and not, as it should, to the investigation also of clinical disease and application of laboratory tests. The word 'pathologies' has gained currency in certain disciplines, but is considered by conventionalists, especially academics, to be at variance with its Greek roots

Penetrating injuries injuries that pass through the skin

Peracute a disease with an exceedingly sudden onset and exceptionally short course

Periodontitis inflammation of the periodontium

Periodontium tissues investing and supporting the teeth

Peritonitis inflammation of the lining of the abdominal cavity and surfaces of the abdominal organs

Petechiae (singular Petechia; adjective Petechial) pinpoint haemorrhages, due to rupture of small venules

Pharyngitis inflammation of the pharynx

Poison (Toxicant) (Toxin) a substance which can harm the body if taken in sufficient quantity. It is the dose that determines whether or not the substance is poisonous

Poisoning (Intoxication) disease produced by a poison (See also **Toxicosis**)

Polychlorinated biphenyls (PCBs) a group of synthetic industrial compounds that are toxic to animals and are persistent in the environment

Posterior the back of the body, limb, organ, etc. (See also **Caudal**)

Postmortem after death

Postural musculature the muscles of the neck and back that control body posture

Premortem before death

Prevalence the total number of cases of a particular disease at a given moment of time

Primary infection first infection; may not be evident and thus should be considered cause

Primatology the study of primates, including monkeys, apes and humans

Prognathic (adjective; noun **Prognathism**) having jaws that project forward to a marked degree

Prognosis forecast of the probable course and outcome of a disease

Prosection programmed dissection of a cadaver or carcase for demonstration and/or teaching

Prosthion the most anterior point on the maxillary alveolar process, midway between the central incisor teeth

Proximal situated nearer to the midline of the body or the point of attachment

Pruritus (adjective 'pruritic') itching

Purulent containing, consisting of or forming pus

Pus an exudate rich in neutrophils, usually viscous; may be creamy, white, yellow or green in colour, depending upon the pathogens involved

Pustule discrete, pus-filled, raised area

Putrefaction the enzymic decomposition or of organic matter, especially proteins, by anaerobic microorganisms

Pyogenic producing pus

Quality control (QC) procedures that provide for a continual check on the accuracy of testing methods, equipment, reagents and working practices

Ramus height the straight line distance from the top of the mandibular condyle to the corresponding point on the lower margin of the mandible

Regurgitation see **Vomiting**

Reservoir the host or hosts in which a pathogen is maintained in nature, often as a subclinical infection

Rigor mortis stiffening of the muscles after death. Very variable process which is affected by many factors and generally cannot be used to give an accurate time of death but may provide some clue as to antemortem history

Rugose A surface that is rough, wrinkled or ridged

Sample in statistics is a subset of a population — and there are many types of statistical sample — but in common parlance, and in much of biomedical science, a sample is a representative specimen or small quantity of something

Scald a type of burn due to hot liquid, usually water

Scale dry, plate-like, excrescence on the skin

Screening a form of monitoring of a group or population of animals, using sampling rather than examination of each individual. The term can also be used when an individual is monitored but only a limited number of tests or samples are possible

Secondary infection an infection which only follows if the host is altered by the primary infection

Semeion a category of clinical information, third in human medicine after Symptoms and Clinical signs. A semeion is what is written down in notes

Sepsis see **Septicaemia** and also see text

Septicaemia multiplication of organisms in the blood, usually with pathological effects on organs

Serology the study of antigen/antibody reactions, especially the detection of antibodies in blood

Shock a physiological response to trauma, haemorrhage, infection, toxaemia and stressors, characterised by inadequate blood flow to the body's tissues (largely a deterioration of capillary perfusion) and sometimes life-threatening cellular dysfunction

Speciation either (1) the evolutionary process by which new biological species arise, or (2) the first description of a new species. Sometimes incorrectly used to mean the determination (identification) of an animal or plant in (for example) forensic evidence

Specimen the whole animal or a part thereof. In laboratory work can also refer to a sample of, say, water or soil

Squamous epithelial cells (keratinocytes) these are the majority of epidermal cells and the major mechanical barrier; also a source of cytokines, which regulate the cutaneous environment

Stab a penetrating injury caused by a sharp object that is deeper than it is wide

Starvation illness or death usually due to lack of food intake

Stillbirth birth of a dead human baby after a 24-week gestation period or of an animal (mammal) late in gestation

Stomatitis inflammation of the buccal cavity

Strangulation application of pressure to the neck resulting in obstruction of the airways with or without obstruction of the blood vessels

'Subclinical' versus 'asymptomatic' some note that a 'subclinical infection' is the same as 'an asymptomatic infection'. However, there is probably a subtle difference. While an asymptomatic infection has no effect at all on the host (the carriage of bacteria, including potential pathogens, in the intestine, for example), a subclinical infection may, as the term suggests, produce changes that cannot be detected during clinical examination but are revealed if extra, nonclinical, tests are performed (e.g., a change in blood values)

Suckle to give (a baby or young animal) milk from the breast. A word that should be applied to the lactating female, not the infant; a baby gorilla 'sucks', its mother 'suckles'

Sudden death no specific definition. Usually death without premonitory signs. Some medical authorities state that the time period should be less than a few minutes. Must not be confused with 'unexpected death' where the animal may have been unwell for some time

Suffocation asphyxiation caused by obstruction of the external airways or lack of oxygen in the inspired air

Supraorbital region of the frontal bone situated above the orbit

Superior (adjective) situated above or directed upward; also sometimes still used to refer to the upper surface of an organ, limb, etc.

Suppuration the formation of pus

Symbiosis a mutually beneficial relationship between different organisms

Symptoms the features of a disease that are experienced and can be recounted by a human patient, for instance, giddiness, abdominal pain. The term is not used for animals — see **Clinical signs**

Syndrome a group of symptoms, clinical signs or pathological changes that consistently occur together (Greek — 'to run together')

Syndromic surveillance the tracking of syndromes, groups of symptoms or signs of disease, rather than the tracking of specific diseases

Synonym a taxonomic name which has the same application as another, especially one which has been superseded and is no longer valid

Systemic disease affecting the whole body (cf. local disease)

Syntype (1) One of two or more biological specimens or other elements used for the original published description of a species or subspecies when no holotype was designated. (2) One of two or more biological specimens or other elements simultaneously designated as type specimens in the original published description of a species or subspecies

Tachypnoea rapid breathing

Taphonomy from Greek '*taphos*' (grave); the postmortem fate of biological remains. Forensic taphonomy is the application of such studies to assist legal investigations

Tartar see **Dental calculus**

Temporal bone either of a pair of bones forming part of the side of the cranium, including the inner and middle ear and the posterior portion of the zygomatic arch

Tooth row length measurement from anterior alveolus of the first premolar to the posterior alveolus of the third molar

Topotype a biological specimen that is of the same species or subspecies as a type specimen and has been collected from the same location

Type series a group of representatives of a taxon (as a subspecies or species) selected to demonstrate the extent of variation of that unit

Type specimen the specimen, or each of a set of specimens, on which the name of a new species is based

Thrombosis blockage of a blood vessel by a blood clot in a live animal (distinguish from postmortem blood-clotting)

Toxaemia the presence of a toxin in the blood. Usually caused by soluble bacterial toxins

Toxicity the amount of poison needed to produce an adverse effect

Toxicological pathology the pathology of poisoning. The study of the pathogenesis of poisoning — the structural and functional changes in cells, tissues and organs that are induced by toxins

Toxicosis any disease due to poisoning

Translocation deliberate movement of living organisms from one area for free release into another

Transudate a fluid with low protein content and a specific gravity of less than 1.012 and containing few cells

Trauma injury

Triage the assignment of degrees of urgency to decide the order of treatment of wounds, illnesses, etc.

Tumour a neoplasm, growth of tissue which is uncontrolled or abnormal

Ulcer focal *complete* loss of epidermis: may include dermis and subcutaneous fat

Vagal inhibition slowing or stoppage of the heart caused by stimulation of the vagus nerve, can be caused by pressure on the neck. Sometimes sudden death, in the absence of specific signs, is attributed to vagal inhibition

Vector a carrier of a disease or parasite, often an arthropod

Vesicle ('blister') elevated, fluid-filled, lesion measuring 5 mm or less in diameter

Viraemia the presence of a virus in the blood

Virulence The capacity of any organism to produce disease in any stated host, part of a host/parasite relationship

Vomiting forcible ejection of stomach contents through the mouth, generally some time after eating. In British English often colloquially referred to as 'being sick'. Regurgitation usually implies immediate ejection after ingestion

Voucher specimen is any specimen (see **Specimen**), often but not always a cadaver, that serves as a basis of study and is retained as a reference. (A voucher should be in an accessible collection; however, even if it is not, it remains a voucher.)

Wheal erythematous, elevated, area resulting from dermal oedema, usually pruritic

Wound a breach of the integrity of the skin or other surrounding structure (See also **Injury**)

Zoonosis (adjective 'zoonotic') an infection or disease that is naturally transmitted between vertebrate animals and humans

Zoopharmacognosy behaviour whereby animals apparently self-medicate by selecting and ingesting or topically applying plants, soils, insects and psychoactive agents

Appendix 2

Protocols and Reports

John E. Cooper

Therefore if you see frequently in monkeys the position and size of each tendon and nerve, you will remember them accurately ... you will discover each thing as you have observed it; but if you are entirely without practice you will profit not at all from such an opportunity. ...

Galen, in Mattern (2013)

INTRODUCTION

Not only techniques, but also protocols and proper formats for submission and report forms are essential if postmortem studies on gorillas are to be of maximal value. This Appendix provides some examples, but should be read in conjunction with Chapter 6: Methods of Investigation – Postmortem Examination and Appendix CA2: Retrieval, Preparation and Storage of Skeletal and Other Material.

The forms reproduced here or accessible electronically are as follows:

Submission form:
Forensic cases . This is based on the format used by the author (JEC) for any possible legal case, not just for gorillas.

Report forms:
Necropsy report form for great apes and other primates
Great Ape TAG form for placental examination
Worksheet for fetus/neonate/infant post-mortem examination
Form for post-mortem examination of the air sacs
Guide to dissection of the heart
Skeletal examination form

Five of the documents above are replicated with permission. The author (JEC) is grateful to Linda Lowenstine and Karen A. Terio for providing the most recent versions of the General Ape Necropsy Form, the GAHP Recommended Cardiac Necropsy (Prosection) Guide and the GAHP Guide for Pathologists. Those using these documents are advised to refer to the originals for coloured illustrations of prosection techniques.

The North American (USA) spelling and word usage in original documents have been retained. So also, in order to preserve their integrity and veracity, are any typographical errors or questionable spellings, such as 'spinal chord' and 'clavical'.

Other documents below are based on formats used by the author (JEC) and include some general guidance for veterinarians who are relatively new to work with gorillas and the special challenges that they present.

POSTMORTEM EXAMINATION OF GORILLAS – SOME BASIC RULES

The following Course Notes were issued to participants at the PASA Veterinary Workshop in Entebbe, Uganda, on 15 November 2011 at a session organised by the GPSG. They were intended for African colleagues who might have to examine a dead gorilla, perhaps with limited equipment and support.

1. Always practise strict hygiene. Be aware of zoonotic infections and protect yourself and others. As a routine, devise and adhere to a **Risk Assessment**, following the five steps below:
   - Identify the hazards
   - Assess who might be harmed and how
   - Evaluate the risks and decide precautions
   - Record the findings
   - Review your assessment and update it regularly.

2. Try to ensure that the carcase is as fresh as possible by carrying out the necropsy promptly or by chilling/insulating the dead body.

3. Have appropriate equipment available. When doing field necropsies, be prepared to improvise and to use lightweight, portable, instruments.

4. Before starting the examination, do your best to weigh the carcase. Always carry out a full range of morphometrics (standard measurements). Radiography, if an X-ray machine is available, is of great value, especially if subsequent examination of the skeleton is not possible.

5. Before opening the carcase, carefully check the skin for ectoparasites and lesions: remember that some parasites may already have left the host and might be found in any wrappings or in the environment around the carcase. Animals that have been dead for some time may appear to be dehydrated and other artefactual skin (and deeper tissue) changes may be present.

6. After opening the carcase, carry out a standard examination of thoracic and abdominal cavities. Observe before touching. Try to familiarise yourself with the size, shape and appearance of organs and tissues by examining normal animals, whether primate or not, whenever this is possible. Weigh and measure body organs, if feasible.

7. Make wet preparations of gastro-intestinal contents for microscopy and search for internal parasites. Collect any parasites that can be found.

8. Prepare touch preparations of tissues for cytological examination.

9. Take a wide range of samples for bacteriological and histopathological investigation and be prepared to process these whenever facilities and finances permit. As a basis, consider culturing heart-blood and taking at the very least lung, liver and kidney for histopathology. If microbiological examination is to be performed, be sure to sterilise instruments between samples.

10. Record all findings. A voice-activated tape-recorder facilitates data collection. Take photographs. Do drawings if necessary. Use a scoring system (e.g., 1−3) to record changes that are seen, e.g., pulmonary congestion, splenic enlargement. Measure lesions, such as skin wounds.

11. Do not discard the carcase or samples prematurely. Additional tests may be necessary later. Store tissues for histology, virology or serology at a later date. Consider establishing a reference collection of gorilla material for teaching and research purposes.

Submission Forms

Gorillas and their derivatives submitted for necropsy or laboratory investigation must be accompanied by a submission form. Different zoos and projects still often use their own forms but there is increased standardisation. In the case of samples submitted to a laboratory for tests, the submission form can either be a general document, on which the person submitting material inserts the nature of the specimen (e.g., gut contents for parasitological examination, a placenta for bacteriological culture) or it may comprise a specific, different, sheet for each type of sample. The author prefers the latter as this is generally easier to complete and focuses the mind of both the person doing the submission and the veterinarian/technician who is to carry out the laboratory tests.

Until relatively recently generalised postmortem forms, often designed for domesticated animals, were used for gorillas in zoos and, to an extent, in the field. They are not recommended, however. Necropsy paperwork should relate specifically to the genus *Gorilla* and include relevant features of the taxon such as the laryngeal sacs.

Paper necropsy and submission forms still have much to commend them. The latter can actually accompany the sample to the laboratory, enclosed in a separate plastic bag. Paper forms should be backed up with electronic versions, photographs and tape-recordings (when working alone).

Forensic Cases

Submission form:
Forensic cases

SUBMISSION AND REPORT FORMS

SUBMISSION FOR *POST-MORTEM* EXAMINATION

Submitting Client/Agency……………………….   Reference No ……………………….

Lab Ref……………………………………….   Date (in full)………………………….

Address ………………………………………………………………………………….

……………………………………………Post (Zip) Code………………………..

Tel ……………………… Fax……………………………. Email…………………………..

Veterinary Surgeon (Veterinarian) (where applicable)………………………………………………..

Address and contact details……………………………………………………………..

Other relevant persons, e.g. Police Officer, Wildlife Inspector…………………………………

Species of animal (English and scientific name)………………………………………………

Local name…………………….………. Breed/variety………………………………………….

Colour/markings……………………………….

Age………… Sex …………….…Name……………………………………

Ownership of animal or sample(s)………………………………………………………………

Number/Ring (Band)/Tattoo/Microchip/Other methods of identification

………………………………………………………………………………………

Background to the case/history………………………………………………………………

………………………………………………………………………

Sample submitted……………………………………………………………………

Sent by hand/courier/post/other……………………………….

Method and description of packing ....

Tag No............ Prior storage condition...

Signature of courier .... Date............................

Carcase..

Organs (state)..

Other tissues..

Parasites..

Blood (and details of how presented)..

Swabs...

Other..

Comment on chain of custody (attach relevant paperwork) ....

....

Questions being asked (expand as necessary)

- Why did this animal die? ....

- When did this animal die? ....

- How did this animal die? ....

- Did this animal suffer pain or distress? ....

- What species is this? ....

- What is its provenance/parentage, etc? ....

- What is this material? Species? Sex?..

- Other questions ....

Investigations required ……………………………………………………………………

Post-mortem examination ………………………………………………………………

Laboratory tests:
     Toxicology ………… ………………………………………………………
     Bacteriology………………………………………………………………………
     Histology ……………………………………………………………………….
     Other ………………………………………………………………………

Other tests as necessary to answer the questions above ……………………………………..

……………………………………………………………………………………

……………………………………………………………………………………

Results to be sent to ………………………………………………………………….

……………………………………………………………………………………

Special instructions regarding the storage/transfer/disposal of samples/wrappings/carcase

……………………………………………………………………………………

……………………………………………………………………………………

……………………………………………………………………………………

……………………………………………………………………………………

Other comments, e.g. cruelty case, civil action, insurance claim, professional malpractice hearing, etc.

……………………………………………………………………………………

……………………………………………………………………………………

Signed by recipient at (location) …………………………………………………………….

Date ……………………………..   Time …………………………………

Reproduced and modified, with permission, from J. E. Cooper and M. E Cooper (2007*). Introduction to Veterinary and Comparative Forensic Medicine.* Blackwell, Oxford.

……………………………………………………………………………………

Report Forms

Necropsy report forms for great apes and other primates
Great Ape TAG form for placental examination
Worksheet for fetus/neonate/infant postmortem examination
Form for post-mortem examination of the air sacs

STANDARDIZED NECROPSY REPORT FOR GREAT APES AND OTHER PRIMATES
(reproduced, with permission, from the website of the American Association of Zoo Veterinarians).

STANDARDIZED NECROPSY REPORT FOR NON-HUMAN PRIMATES WORK SHEET

Pathology # _____Species_____ Date_____
Animal #/Name _____ Sex _____ Age(DOB)_____
Date of Death/Euthanasia _____ Time _____ (am/pm)
Method of euthanasia _____
Time and date of necropsy _____ Duration of necropsy _____
Post mortem state _____ Nutritional state _____
Pathologist or prosector and
institution:_____

Gross diagnoses:

Abstract of clinical history:

Please check tissues submitted for histopathology.
External Examination (note evidence of trauma, exudates, diarrhea etc):
_____Hair coat:
_____Skin:
_____Scent glands:
_____Mammary glands and nipples:
_____Umbilicus (see neonatal/fetal protocol):
_____Subcutis (note: fat, edema, hemorrhage, parasites):
_____Mucous membranes (note: color, exudates):
Ocular or nasal exudate?:
_____Eyes and ears:
____ External genitalia:
_____Oral cavity, cheek pouches and pharynx: 19

Dentition (see attached dental form):
_____Tongue:
Musculoskeletal System:
Note fractures, dislocations, malformations?:
____ Bone growth plate (rib, distal femur, sternabra)
_____Muscles:
_____Bone marrow (femur):
_____Joints (note any exudates or arthritis):
_____Spinal column (examine ventral aspect when viscera removed)
Examination of the neck region:
_____Larynx:
_____Laryngeal air sac (see protocol for great apes):
____Mandibular and parotid salivary glands: 20

_____Thyroids and parathyroids:

_____Cervical/cranial lymph nodes:

_____Esophagus:

Thoracic Cavity:

Note any effusions, adhesions, or hemorrhage:

Note amount, color and any lesions in mediastinal and coronary fat:

_____Thymus:

_____Heart (see attached protocol):

_____Great vessels:

_____Trachea and bronchi:

_____Lungs:

_____Esophagus: 21

_____Lymph nodes:

Abdominal Cavity:

Note any effusions, adhesions, or hemorrhage?:

Note amount, color or lesions in omental, mesenteric and perirenal fat:

_____Liver and gall bladder:

_____Stomach:

_____Pancreas:

_____Duodenum:

_____Jejunum:

_____Ileum:

_____Cecum and (in apes) appendix:

_____Colon and rectum:

_____Lymph nodes: 22

_____Kidneys and ureters:

_____Adrenals:

_____Gonads:

_____Uterus:

_____Bladder and urethra:

_____Male accessory sex glands (prostate and seminal vesicles):

_____Umbilical vessels, round ligaments of bladder in neonates:

_____Abdominal aorta and caudal vena cava:

Nervous System:

_____Meninges:

_____Brain:

_____Pituitary: 23

_____Trigeminal (gasserian) ganglia:

_____Spinal cord (please note to which lumbar segment the cord extends):

_____Brachial plexus and sciatic nerves:

Is there an identifiable pineal gland?

WEIGHTS AND MEASUREMENTS (in grams, kilograms, and cm, please):

Body weight:_____

Lymphoid tissue:

R. axillary LN _____ L. axillary LN _____

R. inguinal LN _____ L. inguinal LN _____

Jejunal LN _____

Spleen _____ Thymus _____

Abdominal Organs:

Liver _____

R. kidney _____ L. kidney _____

R. adrenal _____ L. adrenal _____

R. ovary _____ L. ovary _____

uterus _____

placenta (weigh and measure disc(s)): 24

Thoracic Organs:

Heart wt._____ Thymus (above)

Height:_____ Circumference at coronary groove:_____

Left Vent.thickness_____ Rt. vent.thickness_____

Septum_____

Lt. AV valve circ._____ Rt. AV valve circ._____

Aortic valve circ._____ Pulmonary v. circ._____

R. lung _____ L. lung _____

Other:

Brain _____ Tumors? _____

R. testes (wt.)_____ L. Testes _____

Length x dia._____ _____

Penis (length x diameter) _____

STANDARDIZED BODY MEASUREMENTS FOR NONHUMAN PRIMATES INCLUDING APES:
crown rump length (linear)_____
crown rump length (curvalinear)_____
cranial circumference (above brow ridge)_____
Length of head (tip of jaw to top of crest)_____
width of brow ridge_____
chest circumference (at nipples)_____
abdominal circumference (at umbilicus)_____
Left arm: Shoulder-elbow:_____ 25 elbow-
wrist:_____
wrist-tip of middle finger:_____
pollex:_____
Right arm: Shoulder elbow:_____
elbow- wrist:_____
wrist-tip of middle finger:_____
pollex:_____
Left leg: hip-knee:_____
knee-ankle:_____
ankle-tip of big toe:_____
heel-tip of big toe:_____
hallux:_____
Right leg: hip-knee:_____
knee-ankle:_____
ankle-tip of big toe:_____
heel-tip of big toe:_____
hallux:_____

ANCILLARY DIAGNOSTICS (CHECK IF PERFORMED, GIVE RESULTS IF AVAILABLE, NOTE LOCATION IF STORED, OR TO WHOM SENT):
Cultures:
bacterial:
fungal:
viral:
Heart blood:
serum:
filter paper blot:
Parasitology:
feces:
direct smears:
parasites:
Tissues fixed in 10% formalin (list tissues or specific lesions other than those checked above):
Tissue fixed for EM:_____ Tissue frozen:_____ 27

Impression smears:_____
Comments (interpretation of gross findings):

NONHUMAN PRIMATE POSTMORTEM EXAMINATION

Collection of Tissues

Tissues to be fixed in 10% neutral buffered formalin should be less than 0.5 cm thick to ensure penetration of formalin for fixation.

Initial fixation should be in a volume of fixative 10 times the volume of the tissues. Agitation of the tissues during the first 24 hours is helpful to prevent pieces from sticking together and inhibiting fixation. Once fixed tissues may be transferred to a smaller volume for shipment.

Labeling of Specimens

The formalin container should be labeled with the animal's name or number, the age and sex, the date and location, and the name of the prosector.

Tissues to Be Preserved

From the skin submit at least one piece without lesions, a nipple and mammary gland tissue, scent gland, any lesions and subcutaneous or ectoparasites.

Axillary and or inguinal lymph nodes may be submitted whole from small animals and should be sectioned transversely through the hilus in large primates.

Mandibular and/or parotid salivary glands should be sectioned to include lymph node with the former and ear canal with the latter.

Thyroids, if it is a small primate, may be left attached to the larynx and submitted with the base of tongue, pharynx, esophagus as a block. In larger primates, take sections transversely through the thyroids trying to incorporate the parathyroids in the section.

Trachea and esophagus and laryngeal air sac sections may be submitted as a block.

Cervical lymph nodes may be submitted whole if small or sectioned transversely.

A single sternebra should be preserved as a source of bone marrow. A marrow touch imprint may be made from the cut sternebra and air dried for marrow cytology.

Section of thymus or anterior pericardium should be taken perpendicular to the front of the heart.

Heart: Weigh and measure heart after opening but before sectioning. Please fix longitudinal sections of left and right ventricles with attached valves and atria in large animals and the whole heart opened and cleaned of blood clots in smaller animals. In tiny animals the heart may be fixed whole after cutting the tip off the apex.

Lungs: If possible inflate at least one lobe by instilling clean buffered formalin into the bronchus under slight pressure. Fix at least one lobe from each side and preferably samples from all lobes. In little animals the entire "pluck" may be fixed after perfusion and sampling for etiologic agents.

Gastrointestinal track: Take sections of all levels of the GI track including: gastric cardia, fundus and pylorus (or presaccus, saccus, tubular stomach and pylorus in colobines); duodenum at the level of the bile duct with pancreas attached; anterior, middle and distal jejunum; ileum; ileocecocolic junction with attached nodes; cecum and (in apes) appendix; ascending, transverse and descending colon. Open loops of bowel to allow exposure of the mucosa and allow serosa to adhere momentarily to a piece of paper before placing both bowel section and paper in formalin; or gently inject formalin into closed loops.

Liver: Take sections from at least two lobes, one of which should include bile ducts and gall bladder.

Spleen: Make sure sections of spleen are very thin if the spleen is congested; formalin does not penetrate as far in very bloody tissues.

Mesenteric (jejunal) nodes: section transversely; colonic nodes may be left with colon sections.

Kidneys: Take sections from each kidney: Cut the left one longitudinally and the right one transversely so they will be identifiable (or label). Please make sure the sections extend from the capsule to the renal pelvis. 29

Adrenals: Small adrenals may be fixed whole but larger ones should be sectioned (left longitudinal and right transversely) making sure to use a very sharp knife or new scalpel blade so as not to squash these very soft glands.

Bladder: Sections should include fundus and trigone. Please make sure to include round ligaments (umbilical arteries) in neonates.

Male gonads and accessory sex glands: Section the prostate with the urethra and seminal vesicles transversely. Section testes transversely. If testes are being collected perimortem for sperm retrieval, try to arrange to take small sections before the gonads are manipulated.

Female reproductive organs: Fix the vulva, vagina, cervix, uterus and ovaries from small and medium sized primates as a block (after making a longitudinal slit to allow penetration of formalin). Rectum and bladder (opened) can also be included in this block. In somewhat larger animals make a longitudinal section through the entire track. In great apes make transverse sections of each part of the track and the ovaries. (See reproductive track protocols from the contraception advisory group if animals are to be included in their database.)

Gravid females: Weigh and measure placenta and fetus. Perform a post mortem examination of the fetus. Take sections of placental disc(s) from periphery and center and from extraplacental fetal membranes. Take sections of major organs and tissues of fetus (see fetal protocol).

Nervous system: The brain should be fixed whole, or, if too large for containers, may be cut in half longitudinally (preferred) or transversely through the midbrain. It should be allowed to fix for at least a week before sectioning transversely (coronally) into 0.5−1.0 cm slabs to look for lesions. Submit the entire brain if possible and let the pathologist do the sectioning, otherwise submit slabs from medulla, pons and cerebellum, midbrain, thalamus and hypothalamus, prefrontal, frontal, parietal and occipital cortex including hippocampus and lateral ventricles with choroid plexus. (Note: Limited sectioning is advised if the brain is to go to the great ape brain project).

Pituitary and pineal gland: Fix the pituitary whole. Put pituitary in an embedding bag if it is small. If the pineal gland is identifiable, fix it whole. Also remove and fix the Gasserian (trigeminal) ganglia.

Spinal chord − if clinical signs warrant, remove the cord intact and preserve it whole or in anatomic segments (e.g., cervical, anterior thoracic, etc.) (Please note to which lumbar vertebra the cord extends)

Bone marrow: Take bone marrow by splitting or sawing across the femur, to get a cylinder and then make parallel longitudinal cuts to the marrow. Try to fix complete cross sections or hemisections of the marrow. 30

Additional sections for fixation: Take sections of any and all lesions, putting them in embedding bags if they need special labeling.

NOTE: It is better to save "too many" tissues than to risk missing essential lesions or details.

THANK YOU for taking the time to perform this necropsy.

CARDIAC EXAMINATION FOR GREAT APES (AND OTHER PRIMATES IN WHICH CARDIAC DISEASE IS PRESENT)

Examine heart in situ. Check for position, pericardial effusions or adhesions. Collect for culture or fluid analysis if present.

Remove heart and entire thoracic aorta with 'pluck'.

Examine heart again. Check the ligamentum (ductus) arteriosus for patency. Check position of great vessels. Open pulmonary arteries to check for thrombi.

Remove heart and thoracic aorta from the rest of the 'pluck'. Examine for presence of coronary fat.

Examine external surfaces especially coronary vessels. Note relative filling of atria and state of contracture (diastole or systole at death) and general morphology. (The apex should be fairly pointed.)

Measure length from apex to top of atria. Measure circumference at base of atria (around coronary groove).

Open the heart:

Begin at the tip of the right auricle and open the atrium parallel to the coronary groove continuing into the vena cava. Remove blood clot and examine the AV valves. Cut into the right ventricle following the caudal aspect of the septum and continuing around the apex to the anterior side and out the pulmonary artery. Remove postmortem clots and examine inner surface. Open left atrium beginning at the auricle and continuing out the pulmonary vein. Remove any clots and examine valves. Open the left ventricle starting on the caudal aspect and following the septum as for the right ventricle. When you reach the anterior aspect, clear the lumen of blood and identify the aortic outflow. Continue the incision around the front of the heart and into the aorta, taking care to cut between the pulmonary artery and the auricle. Open the entire length of the thoracic aorta. Remove all postmortem clots. You may gently wash the heart in cool water or dilute formalin to better visualize the internal structures and valves. Sever the thoracic aorta from the heart just behind the brachiocephalic arteries. Examine intima and adventitia and section aorta for formalin. Sever the pulmonary vessel and vena cava close to the heart. 31

Weigh and measure the heart and record.

Measure thickness of right and left ventricular free walls and the septum. (On the left side, do not measure directly through a papillary muscle.)

Measure the circumference of the right and left AV valves and the aortic and pulmonary valves using a pliable measuring tape (or use a piece of string and measure the string on a straight ruler).

Take sections for histopathology:

Sections should include:

longitudinal sections of left and right ventricles AV valves and atria.

Sections of myocardium from left and right ventricles including coronary vessels. Sections of papillary muscles. Sections from the septum at the vase of the AV valves (area of conduction system).

In small animals like callitrichids, you may fix the heart whole.

Fix the entire heart, if possible by immersion in 10% buffered formalin for more detailed examination by a cardiac pathologist.

Other vessels:

Make sure to examine the abdominal aorta, iliac arteries and popliteal arteries (frequent sites of aneurysms in humans).

Note the location and severity of fibrous plaques, fatty streaks and atherosclerotic plaques and presence of mineralization or thrombosis.

POSTMORTEM EXAMINATION OF NONHUMAN PRIMATE FETUSES AND NEONATES

External examination of the fetus:

Weigh the fetus and make body measurements.

Measure the placental disc(s) and weigh the placenta. Note umbilical length and vascular patterns on the placenta.

Note presence of hair, freshness of the carcase (if dam is dead, is the decomposition of the fetus consistent with that of the dam) and any evidence of meconium staining. 32

Internal examination of the fetus:

Follow the general nonhuman primate necropsy protocol.

Make sure to note whether ductus arteriosus and foramen ovale are patent. Note also whether the lungs are aerated and to what extent.

Note dentition/erupted teeth.

Identify umbilical vein and arteries and check for inflammation. Make sure to save umbilicus and round ligaments of the bladder (umbilical arteries) for histology.

Mae sure to save a growth plate (e.g., costochondral junction or distal femur) in formalin.

Cultures:

Take as many of the following as possible: Stomach content or swab of the mucosa; lung; spleen or liver; placental disc and extra-placental membranes. Do both aerobic and anaerobic cultures if possible.

POSTMORTEM EXAMINATION OF THE AIR SACS OF ORANGUTANS AND OTHER NONHUMAN PRIMATES

Examine the skin over the air sac for signs of fistulae or scars. Note thickness of the skin and presence of fat or muscle overlying the air sac.

Incise the air sac through the skin on the anterior aspect.

Note color and texture of air sac lining.

Note presence or absence of exudate.

Note presence or absence of compartmentalization by connective tissue and presence of diverticulae.

Note extent of air sacs (e.g., under clavicle, into axilla, etc.)

Identify and describe the opening(s) from the larynx into the air sac (e.g., single slit-like opening, paired oval openings, etc.). Note any exudate.

Note the location, size and shape of the opening in the larynx (e.g., from lateral saccules or centrally at the base of the epiglottis). Note length of any connecting channel between larynx and air sac and direction a probe must take to go from inside the larynx to the air sac. 33

Cultures:

Please culture several different sites within the air sacs (we need data to determine normal flora and if infections are "homogeneous" or compartmentalized).

STANDARDIZED NECROPSY REPORT FORM FOR GORILLAS

**Please send completed path report to the SSP Veterinary and Pathology Advisors
(ljlowenstine@ucdavis.edu, Pathology Advisor)
(TOM.MEEHAN@CZS.org, Veterinary Advisor)
(pmd@clevelandmetroparks.com, Veterinary Advisor)
(hmurphy@zoonewengland.com, Veterinary Advisor)**

Please see attached pages for description of specific methods for examination of organs of special interest

WORK SHEET
-
Pathology # gross exam _____Pathology # histologic exam_____
_____
Animal   #/Name   _____   Stud   Book   Number_____   Sex   _____
Age(DOB)_____
Date of Death/Euthanasia _____ Time _____ (am/pm)
Method of euthanasia _____
Time and date of necropsy _____Duration of necropsy_____
_____
Post mortem state _____ Nutritional state subjective or body score_____
_____
Pathologist   or   prosector/
institution:_____
Institution   to   which   histology
sent_____
_____
____
Gross diagnoses:
_____
__
Abstract of clinical history:
_____
__
Please check tissues submitted for histopathology.
External Examination (note evidence of trauma, exudates, diarrhea):
_____Hair coat:
_____Skin:
_____Scent glands (axillary organ in gorillas):
_____Mammary glands and nipples:
_____Umbilicus (see neonatal/fetal protocol):
_____Subcutis (note: fat, edema, hemorrhage, parasites):
_____Mucous membranes (note: color, exudates):

(reproduced, with permission, from the website of the SSP).

_____Ocular or nasal exudate:
_____Eyes and ears:
_____External genitalia:
_____Oral cavity and pharynx
_____Pharyngeal and lingual tonsils:
_____Dentition
_____Tongue:

Musculoskeletal System:
Are there any fractures, degenerative lesions or malformations?:
_____Muscles:
_____Bone marrow (femur):
_____Joints (note effusions, synovial proliferation, arthritis):
_____ Growth plate in young animals (distal femur, costochondral junction or sternabra)
_____Spinal column (examine ventral aspect for spondylosis when viscera removed)

Examination of the neck region:
_____Larynx:
_____Laryngeal air sac (see attached protocol):
_____Mandibular and parotid salivary glands:
_____Thyroids and parathyroids:
_____Cervical/cranial lymph nodes:
_____Esophagus (please take sections from proximal, mid and distal esophagus to map striated and smooth muscle distribution):

Thoracic Cavity:
Are there any effusions, adhesions, or hemorrhage?:
_____Mediastinal and coronary fat (color and relative abundance):
_____Thymus (are there cervical portions as well as anterior mediastinal?):
_____Heart (see attached protocol):

_____Great vessels (see attached protocol):
_____Trachea and bronchi:
_____Lungs:
_____Esophagus:
_____Lymph nodes:

Abdominal Cavity:
_____Effusions, adhesions, or hemorrhage?:
_____Omental, mesenteric and perirenal fat:
_____Liver and gall bladder:
_____Stomach:
_____Pancreas:
_____Duodenum:
_____Jejunum:
_____Ileum:
_____Cecum and (in apes) appendix:
_____Colon and rectum:
_____Lymph nodes:

(Continued)

_____Kidneys and ureters:

_____Adrenals:

_____Gonads:

_____Uterus:

_____Bladder and urethra:

_____Male accessory sex glands (prostate and seminal vesicles):

_____Umbilical vessels, round ligaments of bladder in neonates:

_____Abdominal aorta (note evidence of atherosclerosis) and caudal vena cava:

Nervous System:

_____Meninges:

_____Brain:

_____Pituitary:

_____Gasserian (trigeminal) ganglia:

Is there an identifiable pineal gland?

_____Spinal cord (not necessary in every case, but if removed, please note to what level the spinal cord extends):

_____Brachial plexus and sciatic nerves:

Note: Great Ape Aging Project is accepting brains from gorillas.

To submit an ape brain, contact

Patrick R. Hof

Department of Neuroscience

Mount Sinai School of Medicine

1425 Madison Avenue, 9th Floor, Rm 9-02

New York, NY 10029

212-659-5904

e-mail: patrick.hof@mssm.edu or mraghant@kent.edu

(Continued)

WEIGHTS AND MEASUREMENTS (in grams, kilograms, and cm, please):

Body weight:_____

Lymphoid tissue:

      R. axillary LN _____ L. axillary LN _____

R. inguinal LN _____ L. inguinal LN _____

Jejunal LN _____

Spleen _____ Thymus _____

Abdominal Organs:

      Liver _____

R. kidney _____ L. kidney _____

R. adrenal _____ L. adrenal _____

R. ovary _____ L. ovary _____

uterus _____

placenta (weigh in toto and measure disc(s):

Thoracic Organs:

      Heart _____ Thymus (above)

               Height _____ Circumference at coronary groove _____

Left Vent._____ Rt. vent._____ Septum_____

Lt. AV valve_____ Rt. AV valve_____

Aortic valve_____ Pulmonary valve._____

             –

R. lung _____ L. lung _____

      Other:

      Brain _____ Pituitary _____

Thyroids (wt) Left _____ Right _____

Thyroids (3 dimensions) Left_____ Right _____

        Testes (wt.) Left _____ Right _____

        Testes Length x dia. Left _____ Right_____

        Penis (length x diameter) _____

Tumors(?) Measurements (3 dimensions) _____ Weight _____

STANDARDIZED BODY MEASUREMENTS FOR NONHUMAN PRIMATES INCLUDING APES:

crown rump length (linear)_____

crown rump length (curvalinear)_____

cranial circumference (above brow ridge)_____

Length of head (tip of jaw to top of crest)_____

width of brow ridge_____

chest circumference (at nipples)_____

abdominal circumference (at umbilicus)_____

Left arm: Shoulder-elbow:_____

elbow-wrist:_____

wrist-tip of middle finger:_____

pollex:_____

Right arm: Shoulder elbow:_____

elbow- wrist:_____

wrist-tip of middle finger:_____

pollex:_____

Left leg: hip-knee:_____

knee-ankle:_____

ankle-tip of big toe:_____

heel-tip of big toe:_____

hallux:_____

Right leg: hip-knee:_____

knee-ankle:_____

ankle-tip of big toe:_____

heel-tip of big toe:_____

hallux:_____

ANCILLARY DIAGNOSTICS (CHECK IF PERFORMED, GIVE RESULTS IF AVAILABLE, NOTE LOCATION IF STORED, OR TO WHOM SENT):

Cultures:

bacterial:

fungal:

viral:

Heart blood:

serum:

filter paper blot:

Parasitology:

feces:

direct smears:

          parasites:

Tissues fixed in 10% formalin (list tissues or specific lesions other than those checked above):

Tissue fixed for EM:_____ Tissue frozen:_____

Impression smears:_____

Comments (interpretation of gross findings):

GORILLA POSTMORTEM EXAMINATION

Collection of Tissues

Tissues to be fixed in 10% neutral buffered formalin should be less than 0.5 cm thick to allow for adequate penetration of formalin for fixation.

Initial fixation should be in a volume of fixative 10 times the volume of the tissues. Agitation of the tissues during the first 24 hours is helpful to prevent pieces from sticking together and inhibiting fixation.

Labeling of Specimens

If pieces are small or not readily recognizable (e.g., individual lymph nodes) they can be fixed in cassettes or embedding bags or wrapped in tissue paper labeled with pencil or indelible ink. Another alternative is to submit lymph nodes with attached identifiable tissue, e.g., axillary with brachial plexus, inguinal with skin, bronchial with bronchus, etc.

Sections from hollow viscera or skin can be stretched flat on paper (serosal side down) and allowed to adhere momentarily before being placed in formalin with the piece of paper. The paper can be labeled with the location from which the tissue came.

The formalin container should be labeled with the animals name or number, the age and sex, the date and location and the name of the prosector.

Tissues to Be Preserved

From the skin submit at least one piece without lesions, a nipple and mammary gland tissue, scent gland, and any lesions and subcutaneous or ectoparasites.

Axillary and or inguinal lymph nodes may be submitted whole from small animals and should be sectioned transversely through the hilus in large primates.

Mandibular, and/or parotid salivary glands should be sectioned to include lymph node with the former and ear canal with the latter.

Thyroids, if it is a small primate, may be left attached to the larynx and submitted with the base of tongue, pharynx, esophagus as a block. In larger primates, take sections transversely through the thyroids trying to incorporate the parathyroids in the section.

Trachea and esophagus and laryngeal air sac sections may be submitted as a block.

Cervical lymph nodes may be submitted whole if small or sectioned transversely.

Marrow: A single sternebra can be preserved as a source of bone marrow. A marrow touch imprint may be made from the cut sternebra and air dried for marrow cytology.

Thymus: Section of thymus or anterior pericardium should be taken perpendicular to the front of the heart.

Heart: Weigh and measure heart after opening but before sectioning. Longitudinal sections of left and right ventricles with attached valves and atria in large animals and the whole heart opened and cleaned of blood clots in smaller animals. In tiny animals the heart may be fixed whole after cutting the tip off the apex. Please see attached Cardiovascular Protocol.

Lungs: If possible inflate at least one lobe by instilling clean buffered formalin into the bronchus under slight pressure. Fix at least one lobe from each side and preferably samples from all lobes. In little animals the entire "pluck" may be fixed after perfusion.

GI tract: Take sections of all levels of the GI tract including: upper, mid and distal esophagus; gastric cardia, fundus and pylorus; duodenum at the level of the bile duct with pancreas attached; anterior, middle and distal jejunum; ileum; ileocecocolic junction with attached nodes; cecum and (in apes) appendix; ascending, transverse and descending colon. Open loops of bowel to allow exposure of the mucosa and allow serosa to adhere momentarily to a piece of paper before placing both bowel section and paper in formalin; or gently inject formalin into closed loops.

Liver: One section should include bile ducts and gall bladder and take sections from at least one other lobe.

Spleen: Make sure sections of spleen are very thin if the spleen is congested; formalin does not penetrate as far in very bloody tissues.

Mesenteric (jejunal) nodes should be sectioned transversely; colonic nodes may be left with colon sections.

Kidneys: cut the left one longitudinally and the right one transversely so they will be identifiable, submit sections from both.

Adrenals: Fix small adrenals whole and section larger ones (left-longitudinal and right transversely) making sure to use a very sharp knife or new scalpel blade so as not to squash these very soft glands.

Bladder: sections should include fundus and trigone. Make sure to include round ligaments (umbilical arteries) in neonates.

Accessory sex glands: Section the prostate with the urethra and seminal vesicles transversely. Section testes transversely.

Female reproductive tract: In small females fix the vulva, vagina, cervix, uterus and ovaries as a block after making a longitudinal slit to allow penetration of formalin. Rectum and bladder (opened) can also be included in this block. In somewhat larger animals make a longitudinal section through the entire tract. In large primates make transverse sections of each part of the tract and the ovaries.

Gravid uterus: Measure placenta and fetus. Perform a post mortem examination of the fetus. Take sections of disc from periphery and center and from extraplacental fetal membranes. Take sections of major organs and tissues of fetus (see protocol)

Nervous system: The brain should be fixed whole, or, if too large for containers, may be cut in half longitudinally (preferred) or transversely through the midbrain. It should be allowed to fix for at least a week before sectioning transversely (coronally) into 0.5–1.0 cm slabs to look for lesions. Submit the entire brain if possible and let the pathologist do the sectioning, otherwise submit slabs from medulla, pons and cerebellum, midbrain, thalamus and hypothalamus, prefrontal. **Note: Institutions may elect to send brains to the Great Ape Aging Project.**

Pituitary: Fix whole. Put pituitary in an embedding bag or cassette if it is small. Also remove and fix the Gasserian (trigeminal) ganglia. If the pineal gland is evident, please submit for histology.

Spinal cord - if clinical signs warrant, remove the cord intact and preserve it whole or in anatomic segments (e.g., cervical, anterior thoracic, etc.). Please note to which spinal column segment the cord extends.

Bone marrow: Split or saw across the femur, to get a cylinder and then make parallel longitudinal cuts to the marrow. Fix complete cross sections or hemisections of the marrow. Alternatively make a transverse cut and scoop out marrow.

Take sections of any and all lesions, putting them in embedding bags or cassettes if they need special labeling. Remember, it's better to save "too many" tissues than to risk missing essential lesions or details.

This represents a lot of work on the part of the prosector, often under less than comfortable conditions. But the effort expended at the time of the gross post mortem is much appreciated by the histopathologist, and is crucial to our investigations of the causes of morbidity and mortality of gorillas

THANK YOU!!!!!

CARDIAC EXAMINATION FOR APES AND OTHER PRIMATES

Examine heart in situ. Check for position, pericardial effusions or adhesions. Collect for culture or fluid analysis if present.

Remove heart and entire thoracic aorta with "pluck".

Examine heart again. Check the ligamentum (ductus) arteriosus for patency. Check position of great vessels. Open pulmonary arteries to check for thrombi.

Remove heart and thoracic aorta from the rest of the "pluck".

Examine for presence of coronary fat. Examine external surfaces especially coronary vessels and palpate for evidence of plaque or mineralization. Note relative filling of atria and state of contraction (diastole or systole at death) and general morphology. (The apex should be fairly sharp.)

Measure length from apex to top of atria. **Measure circumference** at base of atria (around coronary groove).

Open the heart:

Begin at the tip of the right auricle and open the atrium parallel to the coronary groove continuing into the vena cava. Remove blood clot and examine the AV valves and foramen ovale. Cut into the right ventricle following the caudal aspect of the septum and continuing around the apex to the anterior side and out the pulmonary artery. Remove postmortem clots and examine inner surface.

Open left atrium beginning at the auricle and continuing out the pulmonary vein. Remove any clots and examine valves. Open the left ventricle starting on the caudal aspect and following the septum as for the right ventricle. When you reach the anterior aspect, clear the lumen of blood and identify the aortic outflow. Continue the incision around the front of the heart and into the aorta, taking care to cut between the pulmonary artery and the atrium. Open the entire length of the thoracic aorta.

Remove all postmortem clots. You may gently wash the heart in cool water or dilute formalin to better visualize the internal structures and valves. Examine the foramen ovale for patency.

Sever the thoracic aorta from the heart just behind the brachiocephalic arteries. Examine intima and adventitia and section aorta for formalin. Sever the pulmonary vessel and vena cava close to the heart.

Weigh and measure the heart and record (please see work sheet).

Weigh the heart after it has been opened, cleaned of luminal clots and the vessels removed (see above)

Measure height of heart and circumference.

Measure thickness of right and left ventricles and septum.

Measure the circumference of the right and left AV valves and the aortic and pulmonary valves.

Take sections for histopathology:

Sections should include:

Longitudinal sections of left and right ventricles AV valves and atria.

Sections of myocardium from left and right ventricles including coronary vessels.

Sections of papillary muscles.

Sections from the septum at the vase of the AV valves (area of conduction system).

Section of the ascending aorta just above the valves (the most common site of dissecting aneurysms in great apes) as well as sections of descending thoracic aorta and abdominal aorta.

Sections from any lesions noted.

Alternatively, fix the entire heart after opening by emersion in 10% buffered formalin for more detailed examination by the SSP pathologist or a cardiac pathologist.

Other vessels;

Make sure to open and examine the entire aorta, iliac arteries and popliteal arteries (frequent sites of aneurysms in humans)

Note the location and severity of fibrous or fatty streaks and overt atherosclerosis. (see diagram)

Plese note location of atherosclerosis, aneurysms, dissections or other abnormalities (mark directly on diagram)

POSTMORTEM EXAMINATION OF PRIMATE FETUSES, NEONATES AND PLACENTAS

Follow the general primate necropsy protocol.

Note presence of hair, freshness of the carcase (if dam is dead, is the decomposition of the fetus consistent with that of the dam) and any evidence of meconium staining.

Make sure to **weigh** the fetus (without placenta) and **make morphologic measurements.**

In addition, **measure the placental disc(s) and weigh the placenta.**

Describe the placental discs and membranes and the vascular pattern.

Measure umbilical length and diameter and note degree of twisting. If possible, please photograph the placenta.

Culture placenta:

Fix sections from margins of discs, extra placental membranes, and from any areas of discoloration.

Internal examination:

Note dentition/ erupted teeth and carefully examine the palate.

Identify umbilical vein and arteries and check for inflammation. Make sure to save umbilicus and round ligaments of the bladder (umbilical arteries) for histology.

Make sure to save a growth plate (e.g., costochondral junction or distal femur) in formalin.

Before removing the heart from the pluck, open the pulmonary artery to check for patency of ductus arteriosus. Open the lateral side of the right atrium and examine the foramen ovale for patency.

Cultures:

Culture as many of the following as possible (both aerobic and anaerobic cultures if possible):

Stomach content or swab of the mucosa;

lung;

spleen or liver;

placental disc and extra-placental membranes.

This is important as there have been several cases of prenatal pneumonia in apes

Tissues for histology (see general necropsy protocol)

POSTMORTEM EXAMINATION OF THE AIR SACS OF GORILLAS AND OTHER APES

Examine the skin over the air sac for signs of fistulae or scars. Note thickness of the skin and presence of fat.

Incise the air sac through the skin on the anterior (ventral) aspect.

Note color and texture of air sac lining.

Note presence of absence of exudates, and character of exudate.

Note presence or absence of compartmentalization by connective tissue.

Note extent of air sacs (e.g., under clavical, into axilla, etc.)

Is there a central compartment?

Are the lateral sacs symmetrical (they may vary in size in chimpanzees and bonobos)

Identify and describe the opening(s) from the larynx into the air sac (e.g., single slit-like opening or paired oval openings). Are the openings parallel or perpendicular to the long axis of the larynx and trachea? Note any exudate.

Note the location, size and shape of the opening in the larynx (e.g., from lateral saccules or centrally at the base of the epiglottis).

Cultures: Please culture several different sites within the air sacs (we need data to determine if infections are "homogeneous" or compartmentalized).

Diagrams of air sacs to aid in measurements and descriptions.

··

STANDARDIZED NECROPSY REPORT FOR GREAT APES AND OTHER PRIMATES

**this is a fillable form – if printing / hand writing please circle or highlight need information

Pathology #   Necropsy Date
Species   If other, then specify species
Name   ISIS/ID   SB#
Age/DOB   DOD
Euthanized ☐Yes? ☐No?   Post-mortem condition of carcass

Institution

Contact   Contact Email
Prosector (if different from contact)

Abstract of clinical history:

Gross Diagnoses:

External examination:

Measurements:
Body weight   (kg)   Crown-rump length (sitting height)   (cm)
Chest circumference (level of nipples)   (cm)
Width across the back at level of axilla   (cm)
Abdominal circumference (level of umbilicus)   (cm)
Skin fold thickness at dorsum/ level of lower ribs   (cm)
Depth of abdominal fat   (cm)   Depth of fat over throat sac   (cm)

If not examined, please enter "NE" in description. Sections of all tissues should be saved in formalin but not all tissues need to be saved frozen or photographed. See "Tissue collection guide" at the end of the worksheets for recommended frozen tissue collection. Please check whether tissues were saved in formalin, frozen and whether a gross photo was taken of lesions.

Site	Description	Formalin	Frozen	Photo
Eyes (fix whole- do not incise)	Ocular discharge?			
Ears				
Nose	Nasal discharge?			
Mammary gland (incl nipples)				
Skin/Hair				
Umbilicus (neonates only)				
External genitalia				
Scent glands				
Subcutis	(note fat, edema, hemorrhage, parasites)			

Head & Neck Region:

Site	Description	Formalin	Frozen	Photo
Oral Cavity (gingiva, lips, cheek)				
Dentition				
Larynx/ Pharynx				
Tongue				
Tonsils				
Laryngeal Air Sacs	Are they symmetrical yes☐ no☐?			
See appendix for air sac examination	Are there septa yes☐ no☐?			
Salivary glands				
Thyroids	Combined weight (g)			
Parathyroids				
Lymph Nodes (cervical)				
Esophagus				

Other Notes:
Thoracic Cavity:

Site	Description	Formalin	Frozen	Photo
Cavity	Note effusions/hemorrhage: volume (ml) Pleural Adhesions yes☐ no☐? Mediastinal adipose yes☐ no☐?			
Thymus	Weight (g) Size x x (cm)			
Pericardium	Effusion yes☐ no☐? Volume (ml) Pericardial fat yes☐ no☐?			
Great Vessels				
Heart * (See appendix for requested photographs and measurements)	Weight (g) Circumference at groove (cm)			*
Trachea/Bronchi				
Lungs	Weight: Left (g) Right (g)			
Lymph Nodes (tracheobronchial)				
Diaphragm				

Other Notes:
Abdominal Cavity:
For sections of gastrointestinal tract, remember to note contents

Site	Description	Formalin	Frozen	Photo
Cavity	Effusion yes☐ no☐? Volume (ml) Adipose yes☐ no☐? Amount: Adhesions yes☐ no☐? If yes, severity:			
Liver	Weight (g)			
Gall Bladder				
Stomach				
Duodenum				
Pancreas				
Jejunum				
Ileum				
Cecum/Appendix				
Colon				
Rectum				
Lymph nodes (mesenteric)				
Spleen				

Site	Description	Formalin	Frozen	Photo
Abdominal Aorta (open past bifurcation)				
Adrenals (weigh/measure L and R)	L Weight (g) Size x x (cm)			
	R Weight (g) Size x x (cm)			
Kidneys (weigh/measure L and R)	L Weight (g) Size x x (cm)			
	R Weight (g) Size x x (cm)			
Ureters				
Urinary bladder				
Gonads (ovaries/testes) (weigh/measure L and R)	L Weight (g) Size x x (cm)			
	R Weight (g) Size x x (cm)			
Uterus/Cervix				
Prostate/Penis/Seminal vesicles				

Other Notes:

CNS/MUSCULOSKELETAL/OTHER

Site	Description	Formalin	Frozen	Photo
Skeletal Muscle				
Joints				
Spinal Column				
Bone Marrow				
(femur or rib)				
Brain (describe also meninges)	Weight (g)			
Pituitary	Weight (g)			
Trigeminal ganglia				
Spinal cord				
Peripheral nerve (brachial plexus & sciatic)				
Lymph Nodes	Specify other submitted sites:			

Other Notes:

<u>Cardiac Worksheet *see GAHP Recommended Cardiac Necropsy Protocol for details:</u>

Whole Heart Submission

<u>Photographs:</u>

☐ *In situ* ☐ Heart base ☐ 4 Views: Anterior, Right, Posterior, Left

<u>Measurements:</u>

Heart weight   (g)   Heart circumference   (cm)

<u>Fixed in Formalin to Submit:</u>

☐Entire heart

Selected Section Submission

<u>Photographs:</u>

☐ *In situ*   ☐ 4 Views: Anterior, Right, Posterior, Left

☐ Heart base   ☐ 3 (or 4 for gorillas) slab sections from apex

☐ R AV valve   ☐ Pulmonic valve

☐ L AV valve   ☐ Aortic valve

<u>Measurements:</u>

Heart weight   (g)   Heart circumference   (cm)

R AV valve   (cm)   Pulmonic valve   (cm)

L AV valve   (cm)   Aortic valve   (cm)

<u>Fixed in Formalin to Submit:</u>

☐3 or 4 cm slab cross-section

☐ R Atrium-Ventricle with R AV valve

☐ Interventricular septum w/ aortic valve

☐ L Atrium-Ventricle with L AV valve

☐ Aorta

☐ Conduction System (if submitting for detailed protocol)

GUIDE TO THE NONHUMAN PRIMATE POSTMORTEM EXAMINATION TIPS FOR TISSUE COLLECTION DURING THE NECROPSY EXAMINATION

Collection of Tissues

Tissues to be fixed in 10% neutral buffered formalin should be less than 0.5 cm thick to (exception is brain, see below) allow for adequate penetration of formalin for fixation.

Initial fixation should be in a volume of fixative 10 times the volume of the tissues. Agitation of the tissues during the first 24 hours is helpful to prevent pieces from sticking together and inhibiting fixation.

Labeling of Specimens

If pieces are small or not readily recognizable (e.g., individual lymph nodes) they can be fixed in cassettes or embedding bags or wrapped in tissue paper labeled with pencil or indelible ink. Another alternative is to submit lymph nodes with attached identifiable tissue, e.g., axillary with brachial plexus, inguinal with skin, bronchial with bronchus, etc.

Sections from hollow viscera or skin can be stretched flat on paper (serosal side down) and allowed to adhere momentarily before being placed in formalin with the piece of paper. The paper can be labeled with the location from which the tissue came.

The formalin container should be labeled with the animals name or number, the age and sex, the date and location, and the name of the prosector.

Tissues to be Frozen

Archiving or biobanking is an important component of a thorough post mortem examination. Frozen tissues can provide a resource for pathogen discovery, toxicology, nutritional analysis, and genetic studies. Freezing at refrigerator freezer temperatures (about 0°F or −18−20°C) is adequate for toxicology and most nutritional studies, while ultralow temperatures (about −80°C or colder) are better for genetic studies and pathogen discovery.

Recommended tissues:

Samples to be held at −20°C include 5−10 g of liver, kidney, fat, stomach content, lower GI content.

Samples to be held at −80°C include 1−2 g lung, liver, kidney spleen, brain, and any specific lesions for which you can envision wanting pathogen discovery.

Additional samples

Swabs

Serum retrieved from chicken fat clots by centrifugation

Containers for freezing:

For -80 wrap small samples individually in foil and put together in a freezer safe baggie.

For -20 place tissues in individual freezer safe baggies such as WhirlPak.

Liquids can be frozen in freezer-safe cryotubes

Tissues to be Preserved (10% Neutral Buffered Formalin)

From the skin submit at least one piece without lesions, a nipple and mammary gland tissue, scent gland, and any lesions and subcutaneous or ectoparasites.

Axillary and or inguinal lymph nodes may be submitted whole from small animals and should be sectioned transversely through the hilus in large primates.

Mandibular, and/or parotid salivary glands should be sectioned to include lymph node with the former and ear canal with the latter.

Thyroids, if it is a small primate, may be left attached to the larynx and submitted with the base of tongue, pharynx, esophagus as a block. In larger primates, take sections transversely through the thyroids trying to incorporate the parathyroids in the section.

Trachea and esophagus and laryngeal air sac sections may be submitted as a block.

Cervical lymph nodes may be submitted whole if small or sectioned transversely.

Rib or femur can be used as a source of bone marrow. A marrow touch imprint may be made and air dried for marrow cytology.

Section of thymus or anterior pericardium should be taken perpendicular to the front of the heart.

Heart: See cardiac necropsy protocol for recommended measurements, photos and prosection guidelines.

Lungs: if possible inflate at least one lobe by instilling clean buffered formalin into the bronchus under slight pressure. Fix at least one lobe from each side and preferably samples from all lobes. In little animals the entire "pluck" may be fixed after perfusion.

Take sections of all levels of the GI tract including: gastric cardia, fundus and pylorus; duodenum at the level of the bile duct with pancreas attached; anterior, middle and distal jejunum; ileum; ileocecocolic junction with attached nodes; cecum and (in apes) appendix; ascending, transverse and descending colon. Open loops of bowel to allow exposure of the mucosa and allow serosa to adhere momentarily to a piece of paper before placing both bowel section and paper in formalin; or gently inject formalin into closed loops.

Liver: One section should include bile ducts and gall bladder and take sections from at least one other lobe.

Make sure sections of spleen are very thin if the spleen is congested; formalin does not penetrate as far in very bloody tissues.

Mesenteric (jejunal) nodes should be sectioned transversely; colonic nodes may be left with colon sections. Take sections from each kidney: cut the left one longitudinally and the right one transversely so they will be identifiable.

Fix small adrenals whole and section larger ones (left -longitudinal and right transversely) making sure to use a very sharp knife or new scalpel blade so as not to squash these very soft glands.

Bladder sections should include fundus and trigone. Make sure to include round ligaments (umbilical arteries) in neonates.

Section the prostate with the urethra and seminal vesicles transversely. Section testes transversely.

In small females fix the vulva, vagina, cervix, uterus and ovaries as a block after making a longitudinal slit to allow penetration of formalin. Rectum and bladder (opened) can also be included in this block. In somewhat larger animals make a longitudinal section through the entire track. In large primates make transverse sections of each part of the track and the ovaries.

If gravid: weigh and measure placenta and fetus. Perform a post mortem examination of the fetus. Take sections of disc from periphery and center and from extraplacental fetal membranes. Take sections of major organs and tissues of fetus.

The brain should be fixed whole, or, if too large for containers, may be cut in half longitudinally (preferred) or transversely through the midbrain. It should be allowed to fix for at least a week before sectioning transversely (coronally) into 0.5−1.0 cm slabs to look for lesions. Submit the entire brain if possible and let the pathologist do the sectioning, otherwise submit slabs from medulla, pons and cerebellum, midbrain, thalamus and hypothalamus, prefrontal, frontal, parietal and occipital cortex including hippocampus and lateral ventricles with choroid plexus. In older apes it is especially important to examine prefrontal and frontal cortex and hippocampus for senile plaques and vascular changes.

Instead, institutions may elect to send brains to the Great Ape Aging Project (separate protocol). This is a research project which does not perform diagnostic histopathology (as of Jan 2015). If histopathology is desired, the prosecting pathologist may need to modify diagnostic tissue handling/ selection. Contact the Great Ape Aging Project PIs for more information.

Fix the pituitary whole. Put pituitary in an embedding bag if it is small. Also remove and fix the Gasserian (trigeminal) ganglia.

Spinal cord − if clinical signs warrant, remove the cord intact and preserve it whole or in anatomic segments (e.g., cervical, anterior thoracic, etc.)

Take bone marrow by splitting or sawing across the femur, to get a cylinder and then make parallel longitudinal cuts to the marrow. Try to fix complete cross sections or hemi-sections of the marrow. Take sections of any and all lesions, putting them in embedding bags if they need special labeling. Remember, it's better to save "too many" tissues than to risk missing essential lesions or details.

This represents a lot of work on the part of the prosector, often under less than comfortable conditions. But the effort expended at the time of the gross post mortem is much appreciated by the histopathologist, and is crucial to our investigations of the causes of morbidity and mortality of free-living nonhuman primate

THANK YOU !!!!!

GREAT APE TAG PLACENTAL EXAMINATION

GREAT APE TAG PLACENTAL EXAMINATION

WORKSHEET

GREAT APE TAG PLACENTAL EXAMINATION

Dam name _____Stud book #_____

Infant/fetus weight _____gm, Infant crown-rump length_____cm Sex: M F U

Status of infant (circle all that apply): term, preterm, alive healthy, alive weak or ill, dead, singleton, twin, vaginal birth, C-section, other _____

PLEASE INSERT PHOTOGRAPHS OF BOTH SIDES OF THE PLACENTA WITH ATTACHED CORD AND MEMBRANESDESCRIPTIONS AND MORPHOMETRICS OF THE PLACENTA AND CORD

PLACENTA (Circle all that apply): complete, partial, disc, membranes, cord, fresh, desiccated, clean, contaminated, meconium, hemorrhage.

Describe other_____

Umbilical cord:

        Cord length_____cm, Cord diameter_____cm, Twists: N= _____

        Cord cut surface: number of arteries_____vein(s)_____ other structures?_____,

        Warthin's jelly_____ Desication?_____

        Cord color (white, tan, brown, green, red. Other_____),

        Lesions: hematomas, exudate, edema, knots (N=_____),

        other_____

        Cord insertion: central, marginal, on disc, within membranes

Fetal membranes

        Insertion; percent_____ location: marginal, circumvallate, circummarginal

        Color_____ Exudates?_____Hemorrhage?_____

Trimmed placental weight (minus membranes and cord)_____ gm

Placental disc greatest diameter_____cm x thickness_____cm

Fetal surface (photograph): WNL, smooth, rough, vessels, thrombi, hemorrhage, percent surface affected _____other_____

Maternal surface (photograph): WNL, complete, disrupted, excessively nodular or masses, hematomas, pallor, fibrin, percent surface affected _____, other_____

Parenchyma cut surface: normal (= meaty, spongy, red); lesions: marginal, central, dark, pallor, exudative, percent disc affected_____ Other_____

Samples taken:

        Histology/formalin – location and number of samples

        Culture: (bacterial, fungal viral)

        Frozen (refrigerator freezer, ultralow freezer, Liquid nitrogen, dry ice)

Adapted from: http://www.uptodate.com/contents/gross-examination-of-the-placenta#

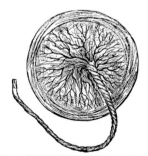

usf.edu/clipart/26100/26112/placenta_26112.htm
You may use this diagram to note extent and location of lesions

<u>WORKSHEET GREAT APE TAG FETUS/NEONATE/INFANT POSTMORTEM EXAMINATION</u>

WORKSHEET

GREAT APE TAG FETUS/NEONATE/INFANT POSTMORTEM EXAMINATION

Identification number or name_____ Stud Book #_____

Age _____ days. Weight _____gm. Crown-rump length _____cm.

Placenta available? Yes No , eaten by dam, eaten by other group members, fate unknown, other?_____

Post mortem condition (fresh, mild, moderate or severe autolysis)
Post mortem interval (death to examination):_____ hours, days
Other body measurements: head circumference _____cm, thoracic circumference _____cm, abdominal circumference _____cm

EXTERNAL EXAMINATION (circle all that apply or describe)

Nutritional status: emaciated, thin, adequate, abundant fat, other_____
Muscle development: hypoplastic or atrophic, well-muscled, pink, dark red
Umbilicus: Fresh desiccated, color_____(stump length _____cm),

Hair coat: naked, sparse, luxuriant, head only, epilates easily,
other_____

Skin: hemorrhage, other discoloration, lacerations, maceration, other_____,
location of lesions _____

Peripheral lymph nodes: indicate nodes examined and any abnormalities _____
_____

ORAL CAVITY:

Dentition: erupted teeth _____, enamel (pigmented, pitted, linear erosions, on
_____ teeth,

Palate: intact, cleft (photo or describe _____)

Tongue: pigmented, coated, erosions, plaques, ulcerations, other_____
Lips and buccal surfaces: NSL, lacerations, hemorrhage, other _____

EYES: NSL, micro-ophthalmia, cornea cloudy, lens opaque, hemorrhage, other_____

EARS: pinna: hemorrhage, lacerations, other _____

NECK REGION:

Thyroid(s): flat, visible follicles, nodular, red, amber, tan, parathyroids visible? ☐Yes ☐ No
Retropharyngeal and mandibular lymph nodes: small, enlarged, other_____
Esophagus: empty, full, dilated, thickened, erosions, ulcerations, plaques, other

INTERNAL EXAMINATION:

Viscera position (photo viscera in situ): normal, situs inversus, individual organs displaced_____

THORAX: negative pressure? Yes No. Effusion? none, clear, serosanguinous, blood, pus, fibrin, other
_____ amount _____ cc

THYMUS: mediastinal only, mediastinal and cervical, size: _____cm x____cm x ____cm; _____gm

HEART: Pericardial effusion? ☐ Yes ☐ No , character_____, epicardial fat: none, little, moderate, abundant, moderate, little, serous atrophy; epicardial fibrosis? ☐Yes ☐ No. Please photo (all 4 sides if possible).

 DUCTUS ARTERIOSUS: open, probe patent, closed, length _____cm

Mid-Ventricular transverse section _____ cm from apex: right ventricular free wall
_____cm, left ventricular free wall _____cm, septum _____cm;

Open along lines of flow: Foramen ovale (closed, open, probe patent, dye patent)

Myocardium: NSL, pale streaking, masses, other_____

Right AV valve circumference _____cm, Left AV valve circumference_____ cm, pulmonic valve circumference _____cm, aortic valve circumference _____cm

Valves (RtAV): normal number of leaflets, abnormal number leaflets (photo or describe), smooth nodules, rough nodules, adherent thrombi? Other:

_____

LAV valve: normal number of leaflets, abnormal number of leaflets (photo or describe), smooth nodules, rough nodules, adherent thrombi? Other:

Pulmonic valve: normal number of leaflets, abnormal number of leaflets (photo or describe), smooth nodules, rough nodules, adherent thrombi? Other:

Aortic valve: normal number of leaflets, abnormal number of leaflets (Photo or describe), smooth nodules, rough nodules, adherent thrombi? Other:

Coronary ostia: number and location (photo if possible):

LUNGS:
Color: light pink, red, purple, other _____; Atelectasis: none, partial, diffuse) weight

left_____gm, right _____gm,

Lobation left N= _____ , right N= _____

Cut surface: aerated, dry, oozes fluid: clear, foamy, tan, pink, red, other_____.

Trachea and Bronchi: clear, foam, thin fluid, bloody fluid, mucus, pus

Hilar (tracheobronchial) lymph nodes; small, enlarged, cut surface: dry, oozes lymph, exudate?

Other _____

ABDOMINAL CAVITY: Effusions? ☐Yes ☐ No. Type: clear, serosanguinous, blood, pus, fibrin.

Adhesions Yes, No. Character: fibrinous, fibrous, easily broken down, firm, other _____

Diaphragm: intact, hernia, other _____

Omental and mesenteric fat: none, sparse, moderate, abundant. Color: white, off white, yellow, orange

LIVER: extends beyond sternum? ☐ Yes ☐No _____cm); Weight _____gm

Color (tan, brown, red-brown, dark red/purple, green tinged, other _____),

Gall bladder: empty, full, opaque, translucent. Bile: yellow, green, brown, red, watery, chunks or flakes.

Other?_____

Umbilical vein/falciform ligament: NSL, thickened, rough surface, discolored, other _____

SPLEEN: Size _____cm x _____cm x _____cm. Weight:_____gm;

Color: pale, dark red, purple, other _____ cut surface (dry, oozes blood, exudate, nodules, visible white pulp), Other_____

KIDNEYS:
   Left kidney _____cm x_____cm x_____cm. Weight _____gm
   Right kidney _____cm x _____cm x _____cm. Weight _____gm
Capsules: smooth, pitted, undulating, other_____; capsule peels easily, with difficulty, other, Cut surface: pale streaks (fibrosis), wedge shaped foci pale or red (infarcts), exudates, Other _____

_____

ADRENALS: Left _____cm x _____cm x _____cm Weight_____

                   Right _____cm x _____cm X_____cm. Weight_____

                   Cortex_____, medulla_____ lesions?

BLADDER: empty, full _____cc, color: clear, yellow, red, opaque, granular, other _____,
Round ligaments (umbilical arteries) (NSL, hemorrhage, fibrin, exudates)

STOMACH: empty, full, distended. Content: water, mucus, curdled milk, other_____.
Mucosa: NSL, multifocal erosison, red or black spots, ulceration, discoloration, describe or other

_____

DUODENUM: content: empty, scant, abundant, mucus, curdled milk, other _____
Mucosa: Tan, green, brown, red, other _____

PANCREAS: Size: NSL, abundant, scant, Other _____
Color/appearance: cream-colored, tan, brown, hemorrhagic, edematous, other _____

JEJUNUM: content: empty, scant, abundant, mucus, color of content _____other

_____
Color of content and mucosa: tan, green, brown, red, other _____

ILEUM: content: empty, scant, abundant, mucus, other _____
Color of content and mucosa: tan, green, brown, red, other _____

CECUM: content: empty, scant, abundant, mucus, feces, other _____
Color of content and mucosa: tan, green, brown, red, other _____

APPENDIX: _____ cm long x _____cm diameter. Content: empty, scant,
abundant, mucus, feces, other _____
Color of content and mucosa: tan, green, brown, red, other _____

COLON: empty, scant, abundant, mucus, feces, other _____
Color of content and mucosa: tan, green, brown, red, other _____

RECTUM: Content: empty, distended, liquid feces, pasty feces, formed normal feces, hard dry feces
Color of content and mucosa: tan, green, brown, red, other _____

MESENTERIC, ILEOCECAL and COLONIC LYMPH NODES: small, enlarged, cut surface edematous, bulging
cortex, Other _____

PERIAORTIC and INTERNAL ILIAC LYMPH NODES:

SKULL: sutures (open, closed) PHOTO; Anterior Fontanelle closed, open _cm x __cm; posterior fontanelle closed, open _____cm x ___cm. (Photo if possible)

BRAIN: meninges (wet, dry, congestion, edema, exudates, hemorrhage, other_____)
Weight _____gm

SPINE:

Spinal column: NSL, spinal bifida, scoliosis, kyphosis, other defects

_____

Spinal cord: not examined, NSL, hemorrhage, exudates other_____

APPENDICULAR SKELETON:

Growth plates and costochrondral junctions: NSL, wide, flared, inflamed, other:

_____

Ration of cortices to medullary cavity: _____

ANCILLARY DIAGNOSTICS:

      Cultures _____

      Tissues frozen _____

      Cytogenetics _____

PLEASE SUMMARIZE YOUR IMPRESSION OF THIS CASE:

POSTMORTEM EXAMINATION OF THE AIR SACS OF APES

Form for Postmortem Examination of the Air Sacs

Information on air sac anatomy is especially important for bonobos, chimpanzees and orangutans as there are no definitive papers on their air sac anatomy

Examine the skin over the air sac for signs of fistulae or scars. Note thickness of the skin and presence/amount of fat.

Incise the air sac through the skin on the anterior (ventral) aspect.

Note color and texture of air sac lining.

Note presence of absence of exudates, and character of exudate.

Note presence or absence of compartmentalization by connective tissue.

Note extent of air sacs (e.g., under clavical, into axilla, etc.)

Is there a central compartment?

Are the lateral sacs symmetrical (they may vary in size in chimpanzees and bonobos) Identify and describe the opening(s) from the larynx into the air sac (e.g., single slit-like opening or paired oval openings). Are the openings parallel or perpendicular to the long axis of the larynx and trachea. Note any exudate within the ostia.

Note the location, size and shape of the opening in the larynx (e.g., from lateral saccules or centrally at the base of the epiglottis).

Cultures: Please culture several different sites within the air sacs (we need data to determine if infections are "homogeneous" or compartmentalized).

Diagrams of air sacs to aid in measurements and descriptions.

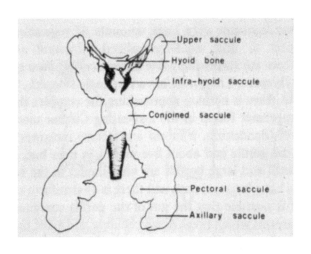

Gorilla air sacs (From Dixon)

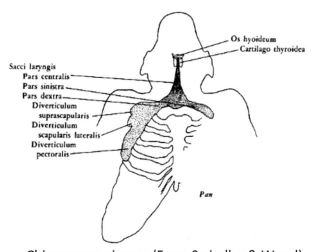

Chimpanzee air sacs (From Swindler & Wood)

GAHP RECOMMENDED CARDIAC NECROPSY PROSECTION GUIDE

The goal of this guide is to assist the prosector in following the recommended protocol for evaluation of ape hearts at necropsy. Before beginning the necropsy, the prosector should choose which protocol for histologic evaluation they wish to pursue as this will impact how the heart is sectioned and fixed at the time of necropsy (see Decision Point below). The "Basic Cardiac Protocol" protocol provides an evaluation of the myocardium, valves and a limited evaluation of coronary vessels. In contrast, the "Detailed Cardiac Protocol" protocol provides the same evaluation as the Basic but evaluates the conduction system, a more extensive evaluation of the myocardium and coronary vessels and can evaluate foramen ovale patency. In cases with known conduction system abnormalities, it is highly recommended that the "detailed" protocol be followed as this will provide the most information possible.

Guide to Dissection of the Heart

Either the whole heart or sections of the heart can be submitted for evaluation. Submission of heart sections are adequate for the basic protocol. If the whole heart is submitted, then either the "Basic" or "Detailed" protocol can be followed. Sectioning of the conduction system is not covered in this guide. Therefore, if the prosector is uncomfortable obtaining sections of the conduction system and the "Detailed" protocol is requested, it is recommended that either the heart be submitted whole or the 3 (or 4) cm slab and the entire top of the heart be submitted to the pathologist for evaluation.

In this guide, where it mentions to take a photograph with a ruler, it is critical that the ruler is visible so that measurements can be made off of the digital images.

Please note − in this guide images are provided for general guidance and do NOT represent a "Normal" ape heart.

What You Need:

- Digital Camera
  - Blue or black solid background
  - Small ruler or size marker for photograph
- String or twine
- Scale for weighing the heart
- Larger (1 foot/30 cm) ruler for measuring
- Necropsy knife or scalpel and scissors
- Either surgical staple, or surgical needle/thread, to mark the myocardial slices

Questions? Feel Free to Contact:

Dr. Rita McManamon (Orangutan SSP and GAHP lead Pathologist) ritamcm@uga.edu

Dr. Linda Lowenstine (Bonobo, Gorilla and Orangutan SSP and Ape TAG Pathologist) ljlowenstine@ucdavis.edu

Dr. Karen Terio (Chimpanzee SSP Pathologist) kterio@illinois.edu

APE CARDIAC NECROPSY PROTOCOL

1. Open the chest and take a photograph of the heart *in situ*
2. Separate the heart and weigh
   Cut the great vessels close to the lungs, flush out clots and weigh the whole heart. Great vessels should be cut approximately 5 cm from the base of the heart.
3. Photograph whole heart with a small ruler/size marker taking 4 images
   Start with the anterior side facing up (right auricle is on the left; left auricle is on the right in the photograph). For each photograph, rotate the heart 90o from the previous image
   Anterior Right Posterior Left
4. Photograph heart base
5. Measure circumference
   Take string or twine and wrap it around the heart at the level of the coronary groove then lay the length of the string that outlines the circumference along the larger ruler to measure.
6. Ape Cardiac Necropsy Worksheet
   Heart weight
   Photographs
   In situ

DECISION POINT

Whole Heart Vs Selected Section Submission − See Above

If submitting whole heart:

1. Single transverse cut 3 cm from apex (bonobos, chimpanzees, and orangutans) or 4 cm from the apex (gorillas), flush out clots and cradle or suspend heart in large volume of formalin. After complete fixation (>48 hour), ship in smaller volume of formalin.

If submitting heart sections:

1. Make 1 cm slices up to 3 (or 4) cm from the apex. Photograph with ruler & fix final slice (3 or 4 cm) in formalin.

   Make parallel slices perpendicular to the long axis. Final slice should be 3 cm from the apex in bonobos, chimpanzees and orangutans but 4 cm from the apex in gorillas (thus for a gorilla you would have 4 slices in the image). Mark the posterior side with a removable staple or thread to assist the pathologist in sectioning.

2. Open right side of heart along lines of flow. Measure R AV valve and pulmonic valve circumference. Photograph both valves. Section R atrium and ventricle and fix in formalin.

   Open the atrium from posterior vena cava to the auricle. Cut from the back (posterior side) of the right atrium into the right ventricle and out the pulmonary artery. Use string to measure the right atrioventricular (tricuspid) valve circumference and photograph. Take a long axis section of the right atrium and ventricle with valve and fix in formalin. Using string, measure the circumference of the pulmonic valve and photograph.

3. Open the left side of the heart. Measure the L AV valve circumference and photograph. Measure aortic valve circumference and photograph. Section L atrium, AV valve and ventricular free wall as well as the interventricular septum with the aorta and fix in formalin.

   Open the left atrium from pulmonary vein to auricle and then make a single longitudinal cut perpendicular to this through the middle of the left ventricular free wall. Measure the left atrioventricular valve circumference using the string method. Photograph the inside of the left side of the heart with a ruler alongside the heart. Take a longitudinal section through the left atrium, atrioventricular valve and ventricle and fix in formalin. Cut through the left AV valve along the septum and into the aorta to open the aorta. Measure the aortic valve circumference using the string method. Take a longitudinal section of the septum from the aorta into the left ventricle and fix in formalin.

4. Take a cross section of the aorta at the arch and fix in formalin.

5. Submit all photographs and measurements along with tissues to the pathologist to be incorporated into the final electronic necropsy report.

GAHP RECOMMENDED CARDIAC TRIMMING PROTOCOL FOR PATHOLOGISTS

Protocol for Trimming the Heart

We are recommending one of two trimming protocols. The "basic" protocol provides an evaluation of the myocardium, valves and a limited evaluation of coronary vessels. In contrast, the "detailed" protocol provides the same evaluation as the basic but evaluates the conduction system as well as a more extensive evaluation of the myocardium and coronary vessels. Additionally, foramen ovale patency can be assessed. In cases with known conduction system abnormalities, it is highly recommended that the "detailed" protocol be followed as this will provide the most information possible. An excellent reference if assistance is needed is Sheppard MN. 2012. Approach to Cardiac Autopsy. J Clin Pathol 65: 484-95.

For inclusion in the GAHP database, please submit to gahpinfo@gmail.com the final electronic necropsy report, gross photos (see below) and a key so we can determine which slide number corresponds to which specific myocardial section (either as text within necropsy report, attached table or by labeling a photo of slab in sections similar to pg 2 of this guide).

Either the whole heart or sections of the heart may be submitted for evaluation. If the whole heart is submitted, then please follow the steps on page 4 for the following additional requested digital images and measurements be taken during prosection (see also check sheet):

Photographs:

☐ 4 Views: Anterior, Right, Posterior, Left **Photograph w/ ruler or size marker

☐ Heart base ☐ 3 (or 4 for gorillas) slab sections from apex

☐ R AV valve ☐ Pulmonic valve

☐ L AV valve ☐ Aortic valve

Measurements:

Heart weight _____ (g) Heart circumference _____ (cm)

R AV valve _____ (cm) Pulmonic valve _____ (cm)

L AV valve _____ (cm) Aortic valve _____ (cm)

Questions? Feel Free to Contact:

Dr. Rita McManamon (Orangutan SSP and GAHP lead Pathologist) ritamcm@uga.edu

Dr. Linda Lowenstine (Gorilla SSP and Ape TAG Pathologist) ljlowenstine@ucdavis.edu

Dr. Karen Terio (Chimpanzee SSP Pathologist) kterio@illinois.edu

Basic Protocol:

Note: Cassette numbers are approximate and will vary based on the size of the heart, other cassettes trimmed for the case, and any lesions noted. All slides should be stained with HE. Additional specials may include: Masson's trichrome on sections of myocardium from slab section, PAS and Elastin on sections of aorta.

- Cassette 1: section right atrium-ventricle with R AV valve
- Cassettes 2−6: Section the myocardium from the entire 3 cm (bonobos, chimpanzees and orangutans) or 4 cm (gorillas) slab section, please sample the following areas:

  Example of "slab" sectioning from two hearts. The number of cassettes varies based on heart size but it is important to make sure to sample the entire slab "face" (from endocardium to epicardium). Do not put the entire 1 cm thickness into cassettes but trim to appropriate thickness for normal cassettes. Use the 3 or 4 cm face of the slab (not the 2 cm side) for histology and place this side "down" in the cassettes. In other words, we should be able to put the slides together to recreate this gross image − just with an HE stain on it.

- Cassette 7: Left atrium-ventricle with L AV valve
- Cassette 8: Interventricular septum with aortic valve
- Cassette 9: Aorta
- Cassette 10-?: Other lesions

Detailed Protocol:

  Follow all of the same steps as the basic protocol but also include:

- Cassette 10: AV node
- Cassette 11: SA node
- Cassette 12: Additional sections of descending coronary arteries

For Whole Submitted Hearts:

1. Make 1 cm slices up to 3 (or 4) cm from the apex. Photograph with ruler & section the final slice (3 or 4 cm) along the cut surface (the 3 or 4 cm cut surface − see above).

   Make parallel slices perpendicular to the long axis. Final slice should be 3 cm from the apex in bonobos, chimpanzees and orangutans but 4 cm from the apex in gorillas (thus for a gorilla you would have 4 slices in the image).

2. Open right side of heart along lines of flow. Measure R AV valve and pulmonic valve circumference. Photograph both valves. Section R atrium and ventricle.

   Open the atrium from posterior vena cava to the auricle. Cut from the back (posterior side) of the right atrium into the right ventricle and out the pulmonary artery. Use string to measure the right atrioventricular (tricuspid) valve circumference and photograph. Take a long axis section of the right atrium and ventricle with valve and fix in formalin. Using string, measure the circumference of the pulmonic valve and photograph.

3. Open the left side of the heart. Measure the L AV valve circumference and photograph. Measure aortic valve circumference and photograph. Section L atrium, L AV valve and ventricular free wall as well as the interventricular septum with the aorta.

   Open the left atrium from pulmonary vein to auricle and then make a single longitudinal cut perpendicular to this through the middle of the left ventricular free wall. Measure the left atrioventricular valve circumference using the string method. Photograph the inside of the left side of the heart with a ruler alongside the heart. Take a longitudinal section through the left atrium, atrioventricular valve and ventricle and fix in formalin. Cut through the left AV valve along the septum and into the aorta to open the aorta. Measure the aortic valve circumference using the string method. Take a longitudinal section of the septum from the aorta into the left ventricle and fix in formalin.

4. Take a cross section of the aorta at the arch (additional sections if there is evidence of an aneurysm or aortic dissection).

5. Section SA Node (if detailed protocol requested)

6. Section AV Node (if detailed protocol requested)

GAHP CARDIAC NECROPSY CHECK SHEET (FILLABLE PDF) − PATHOLOGISTS: PLEASE SUBMIT WITH REPORT, SLIDE KEY AND IMAGES

Whole Heart Submission

Photographs:

☐ *In situ* ☐ 4 Views: Anterior, Right, Posterior, Left
☐ Heart base ☐ 3 (or 4 for gorillas) slab sections from apex
☐ R AV valve ☐ Pulmonic valve
☐ L AV valve ☐ Aortic valve

Measurements:

Heart weight _____ (g) Heart circumference _____ (cm)
R AV valve _____ (cm) Pulmonic valve _____ (cm)
L AV valve _____ (cm) Aortic valve _____ (cm)

Slide Key:

Please attach key so we can determine which slide number corresponds to which specific myocardial section (either as text within necropsy report, attached table or by labeling a photo of slab in sections similar to pg 2 of this guide).

SKELETAL EXAMINATION FORM

Professor John E. Cooper Consultant Veterinary Pathologist

GORILLA SKELETAL STUDY

SHEET 1

Study Ref:

* DATE OF EXAMINATION _____ PERSONNEL INVOLVED _____

* CURRENT LOCATION OF MATERIAL _____ CONTACT PERSON _____

INSTITUTION & ADDRESS _____

SPECIMEN NUMBER & OTHER IDENTIFICATION _____

PUBLISHED REFERENCES ON SPECIMEN/OTHER RECORDS _____

ORIGIN OF ANIMAL/BONES _____

* METHOD OF PRESERVATION/PREPARARATION _____ STATE OF PRESERVATION _____

AGE: INFANT/JUVENILE/YOUNG ADULT/OLD ADULT   SEX: MALE/FEMALE/UNCERTAIN _____

MATERIAL AVAILABLE/EXAMINED – WITH GENERAL COMMENTS:

CRANIUM (SKULL)	MANDIBLE (JAW)	POST-CRANIAL SKELETON	OTHER (eg SKIN)

SUMMARY OF PATHOLOGICAL FINDINGS:

CRANIUM (SKULL)	
MANDIBLE	
POST-CRANIAL SKELETON	

RADIOGRAPHY OR OTHER TESTS PERFORMED _____ PHOTOS: Y/N

DATE OF COMPLETION OF FORM _____   SIGNATURE OF: 1) PERSON COMPLETING THIS FORM _____

2) PERSON PERFORMING EXAMINATION _____

FOLLOW-UP NEEDED _____

* See subsequent sheets

SHEET 1

Study Ref.

Inventory of bones present, with comments as appropriate

cranium

mandible

sternebrae

Vertebrae:

cervical

| C1 | C2 | C3 | C4 | C5 | C6 | C7 |

Comment

thoracic

| T1 | T2 | T3 | T4 | T5 | T6 | T7 | T8 | T9 | T10 | T11 |

thoracic/lumbar

| T12 | (T13) |

Comment

| L1 | L2 | L3 | L4 | L5 |

Comment

Comment

Sacrum

	right			left			Comment
	present	weight	length	present	weight	length	
clavicles							
scapulae							
ribs							
humerus							
ulna							
radius							

SHEET 3

Study Ref.

right hand

sca	lun	tri	pis

trm	trd	cap	ham

mc1	mc2	mc3	mc4	mc5

pp1	pp2	pp3	pp4	pp5

mp2	mp3	mp4	mp5

dp1	dp2	dp3	dp4	dp5

left hand

pis	tri	lun	sca

ham	cap	trd	trm

mc5	mc4	mc3	mc2	mc1

pp5	pp4	pp3	pp2	pp1

mp5	mp4	mp3	mp3

dp5	dp4	dp3	dp2	dp1

Comments

Weights and measurements of selected bones:			
Bone	Weight	Measurements(s)	Comments

SHEET 3

Study Ref.

	right			left			Comment
	present	weight	length	present	weight	length	
innom							
femur							
patella							
tibia							
fibula							
pelivs							

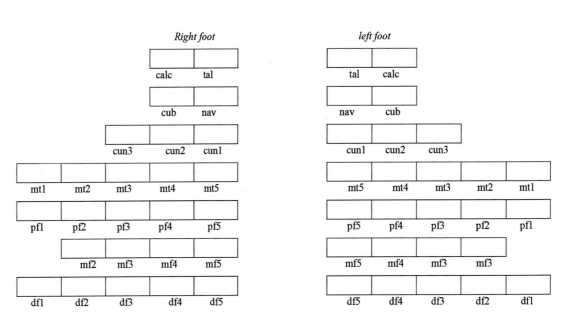

Right foot

| | |
| calc | tal |

| | |
| cub | nav |

| | | |
| cun3 | cun2 | cun1 |

| | | | | |
| mt1 | mt2 | mt3 | mt4 | mt5 |

| | | | | |
| pf1 | pf2 | pf3 | pf4 | pf5 |

| | | | |
| mf2 | mf3 | mf4 | mf5 |

| | | | | |
| df1 | df2 | df3 | df4 | df5 |

left foot

| | |
| tal | calc |

| | |
| nav | cub |

| | | |
| cun1 | cun2 | cun3 |

| | | | | |
| mt5 | mt4 | mt3 | mt2 | mt1 |

| | | | | |
| pf5 | pf4 | pf3 | pf2 | pf1 |

| | | | |
| mf5 | mf4 | mf3 | mf3 |

| | | | | |
| df5 | df4 | df3 | df2 | df1 |

Comments

Sexing

Sex | Determined from |

Comments |

Professor John E Cooper, Consultant Veterinary Pathologist

Study Ref.

<u>Ageing</u>

Infant	Determined from

Juvenile	Determined from

Young adult	Determined from

Old adult	Determined from

(See Lovell, 1990)

SHEET 5

Study Ref.

Dental record   **UPPER (maxilla)**

R																	L
other																	
malocclusion																	
hypoplasia																	
abscesses																	
caries																	
recession																	
calculus																	
attrition																	
presence																	
Permanent	8	7	6	5	4	3	2	1	1	2	3	4	5	6	7	8	
R Decudyiys				e	d	c	b	a	a	b	c	d	e				**L**

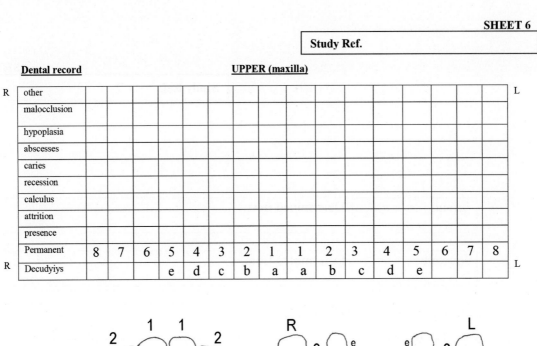

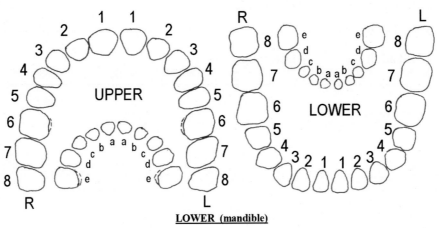

LOWER (mandible)

R																	L
Deciduous				e	d	c	b	a	a	b	c	d	e				
Permanent	8	7	6	5	4	3	2	1	1	2	3	4	5	6	7	8	
presence																	
attrition																	
calculus																	
recession																	
caries																	
abscesses																	
hypoplasia																	
malocclusion																	
R other																	**L**

Key: N/A = not applicable   * = severe   ✓ = present   x = absent

COMMENTS ON DENTAL PATHOLOGY:

Study Ref.

Weight of skull: **Weight of mandible:**

Selected measurements:

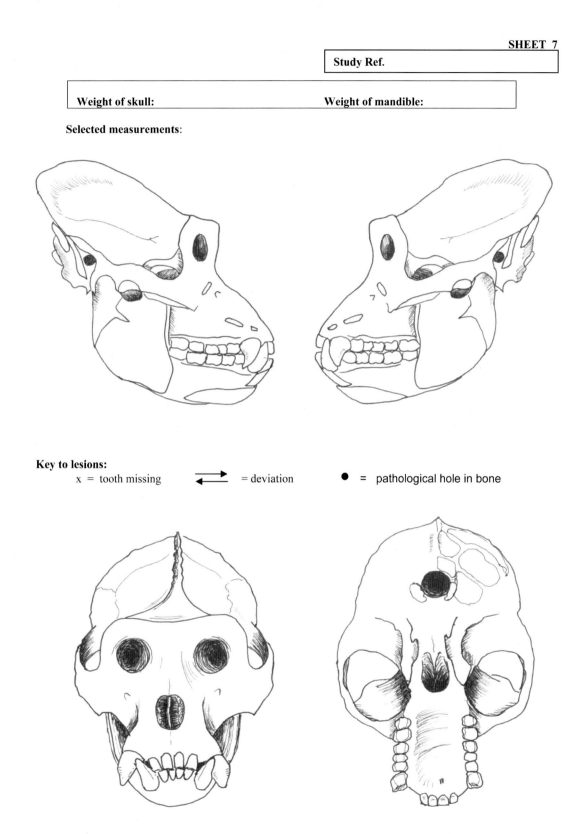

Key to lesions:

x = tooth missing   ⟶ = deviation   ● = pathological hole in bone

COMMENTS ON PATHOLOGY:

Appendix 3

Field Pathology

John E. Cooper and Margaret E. Cooper

Society is like a lawn, where every roughness is smoothed, every bramble eradicated, and where the eye is delighted by the smiling verdure of a velvet surface; he, however, who would study nature in its wildness and variety, must plunge into the forest, must explore the glen, must stem the torrent, and dare the precipice.

Washington Irving (1783–1859)

INTRODUCTION

This Appendix looks at the question of 'fieldwork' especially as it relates to the clinical and postmortem examination of free-living gorillas and the investigation of samples derived from them. However, fieldwork in a veterinary or medical context can refer to any procedures that are performed away from the usual place of work, be it clinic, hospital, dispensary or laboratory (Fig. A3.1).

Fieldwork can vary from a domiciliary visit to a patient to providing veterinary services or carrying out research in remote locations (Cork and Halliwell, 2002). In the context of gorillas, the latter is the usual scenario — working with free-living animals in the mountains or forests of Africa — but occasionally 'field' techniques are needed when giving attention to captive gorillas in a zoo.

As is clear from the description above, fieldwork necessitates assessment of the conditions in which one is to work and making any special provisions or adaptations that are required to operate successfully in nonstandard conditions (Cooper, 2004).

Efficient fieldwork, like veterinary medicine, is a combination of art and science. Aptitude is important. Some people perform well in the field while others constantly yearn for a comfortable clinic or laboratory with home comforts. Traditionally, Africans 'adapted' best to

working in the field because they came from rural villages where improvisation and ingenuity were already part of everyday life, but that is changing. Africans who have grown up in urban areas and received 'Western' education sometimes find it difficult to forego such luxuries.

Both the basics and the specifics of fieldwork in respect of 'exotic' animals and wildlife were covered in the Special Issue of the *Journal of Exotic and Pet Medicine* (JEPM) edited by Cooper (2013a,b). Several contributions to that volume are of direct relevance to those working with gorillas — legal, ethical and practical considerations (Cooper, 2013), lessons learned from the CTPH Gorilla Research Clinic in Uganda (Kalema-Zikusoka and Rubanga, 2013), communications and data collection (Ojigo and Daborn, 2013) and field techniques in zoo and wildlife work (Goodman et al., 2013). The paper on fieldwork in a cold climate (Haigh, 2013) is an important *aide memoire* when working with *Gorilla beringei* high in mountain terrain. Under such conditions 'windchill' increases heat loss, which can affect humans if the skin is exposed and wet, and gorillas if their body temperature drops — when chemically immobilised, for example.

The challenges of fieldwork are many and various. The veterinarian may have to operate without or with unreliable mains electricity, running water or refrigeration; there may be little or no road access. The veterinarian may therefore have to walk or climb for hours, in order to see a particular gorilla, to collect diagnostic samples such as faeces, or to search for a carcase. If equipment is needed, this all has to be carried with the aid of local people. Traversing the forest in Central Africa is certainly an adventure — one sees butterflies, frogs, ferns, mosses, giant lobelias and *Hagenia* trees draped in lichens — but there are also nettles that sting, brambles that scratch, and thick stems and roots that catch the feet as one struggles to keep up with the porters — all a far cry from life in the western world of the

FIGURE A3.1 Necropsy of an infant gorilla soon after the cessation of hostilities in Rwanda. Although performed in a plot, adjacent to human habitation, it can be considered a field procedure because of the limited facilities, equipment and assistance available.

small-animal veterinarian in a modern practice or the large-animal practitioner visiting farms or stables.

Under such conditions equipment may have to be carried and therefore needs to be portable and robust. Portable items may range from lightweight, disposable, spatulae and plastic Coplin staining jars to pocket refractometers and hand-operated centrifuges. Likewise, 'point-of-care', usually handheld, equipment for measuring blood gases, as well as for more specialised diagnostic purposes, such as ketone meters, can be used. Coupled with this new equipment, portable diagnostic kits that provide results within a few minutes are now being marketed — for instance, battery-operated palm-sized thermocyclers that permit fast and sensitive nucleic acid amplification.

Despite the advent of molecular techniques such as rapid diagnostic tests, microscopy remains the 'gold standard' for the verification of malaria and other blood-borne diseases in Africa. In capable hands microscopy is able to diagnose a wide range of tropical diseases at far less cost and provides information on other matters such as haematological parameters. Microscopes (and some other items) for use in the field will need to be run on rechargeable batteries or solar power (sometimes just an 'old-fashioned' mirror). An example of a modern, battery-operated, field microscope is The Newton (Fig. A3.2), http://newtonmicroscopes.com/about.html that was launched at WHO in Geneva in 2013.

The Newton microscope, like its predecessors, is monocular, which some people find difficult after using binocular instruments; but those who are dismissive of this feature

FIGURE A3.2 The Newton microscope in use in the field.

Topic	General Comment	Further Information
Planning	Plan carefully every aspect of a fieldwork project Assess risks and possible threats – write protocols, plans, record decisions Allow for the unexpected and prepare for emergencies Review and amend assessments and plans in the light of experience Insurance Maps, GPS, travel notes; route planning Check travel advice from diplomatic missions and reliable sources. Safety of field and laboratory workers and others	See Cooper (2013a,b) If new to working overseas, read the guidance given by the British Veterinary Association (undated)
Authorisation	Obtain all authorisations and permits; file and prepare copies to carry Store carefully all personal documents – identity card, passport, visa/work permit, contract of employment, academic and professional credentials (possibly notarised copies to protect originals); driving licence Make copies and keep them separately from originals	See Cooper (1996, 2013)
Equipment	For use in the field, the ideal equipment may be modern and sophisticated but sometimes older, sturdy, more basic, items prove very suitable to remote locations where both facilities and skills are limited Project may have to operate without, or with unreliable, mains electricity, running water refrigeration, poor road access In either case training to use equipment properly and safely is essential for accurate results Regular maintenance of equipment is crucial; may dictate the use of simpler equipment	See Frye et al. (2001) and Cooper and Samour (1997) Communications, phones and equipment must be appropriate to conditions and location and include back-up kit, spare batteries and chargers Computers should be lightweight with long-life batteries and spares and nonreflective screens Electrical plug adaptors for equipment from different countries; voltage transformers; surge protection; cables Choose equipment that is easy to repair. Invest in essential repair equipment and spares. Ingenuity and ability may be needed to adapt and repair equipment, vehicles
Transport and vehicles	Be prepared to go on foot, to use bicycle, motorbike, tuk-tuk, public transport or local 'taxi' Generator and fuel may be needed Vehicle must be reliable and relatively safe: may need to carry a sick or injured person or even a gorilla (Fig. A3.3)	See Cooper J.E. (2013a, b); Cooper M.E., (2013)
Phones for data collection and other uses	Communications are dependent on a signal A smartphone can be used to collect and transmit data Carry data; with adaptor, record images from a microscope Data recording depends on battery and storage space May be able to obtain diagnosis and advice based on images – see Chapter 7: Methods of Investigation – Sampling and Laboratory Tests	See Ojigo and Daborn (2013)
Medicines/drugs	Some are dependent on cold storage; may need bottled gas, solar power, battery-insulated containers for travel Will require secure, locked, storage and special safety measures	See also Appendix 4: Hazards, Including Zoonoses
Leadership and coordination	Proper chain of command Management of people, activities Planning and delegation Unexpected and unusual situations Awareness of legal requirements and ethical issues Locally-employed personnel may have, or be able to advise on, access to in-country officials and public offices	See Cooper (1996) A sense of humour is essential, as is 'aequanimitas' – see Chapter 6: Methods of Investigation – Postmortem Examination

Topic	General Comment	Further Information
Community relations	Access to land, outreach Building good relations with local people — requests for/provision of advice, education, healthcare	See Cooper (1996) Fig. A3.4 Fig. A3.5
Authorisation	Vehicle, firearms and other licences Project business, premises, or other licences Payment for utilities — water, electricity, property tax Insurance: employees, buildings, vehicles, other property, pensions Insurance: health, hospital, air evacuation Permits — research, protected areas Letters of authorisation/introduction	See Cooper (1996)
Personal	Documents see above Appropriate clothing (see Appendix 5: Case Studies — Museums and Zoological Collections); Spare essentials such as spectacles and repair kits Essential medication and extra for other people Personal hygiene Essential currency; payment methods	See British Veterinary Association (undated), Johnson et al. (2015) and Werner et al. (2015)
Occupational health and safety	Various	See Chapter 18: Legal Considerations and Appendix 4: Hazards, Including Zoonoses
Use of local languages	Difficulty in dealing with local people, staff relations and emergencies	See also Appendix 4: Hazards, Including Zoonoses
Cultural sensitivity	Cultural mistakes. Causing offence or adverse action relating to customs, religion, other sensitivities, dress Care over photography, especially of local people who may be onlookers, or their homes	See Cooper (2013a,b)
Knowledge of field conditions	Knowledge improves fieldwork and makes it safer. Advantages accrue to those who have previous experience of/training in working/surviving under field conditions; other benefits if interest/experience in natural history, biodiversity, environment	See Cooper (2013a,b)
Clinical, postmortem and laboratory work under field conditions	Clinical and postmortem work require the correct equipment and carefully planned protocols Live and dead gorillas may need to be carried to a safer or more convenient location (but not if a forensic case — 'crime scene') (Fig. A3.3) Field laboratory work may be temporary, in situ, or permanent, with fixed building and facilities (Fig. A3.7). The correct reagents will be needed for (for example) cytology (Cooper, 1995) Field kits will be needed for any clinical work, postmortem examinations, sample collection or laboratory tests performed in situ. See considerations below Rigid, plastic boxes for kits; for example, toolbox; lightweight and compact; suitable for porterage, vehicle or bicycle Pouched, padded or insulated, waterproofed bags Padding, insulation, protection from spillage, vibration rain/rivers Good water and dust-proofing The transportation of samples to a local laboratory or further afield (Lowenstine, 1990) is an important consideration (Cooper, 2013b). The use of transport media is desirable as is keeping samples cool by employing a solar-operated refrigerator or more basic equipment such as vacuum flasks and ice. In an emergency, leaves, soil and running stream water can be used to chill specimens and containers (Cooper and Cooper, 2013)	For general advice see Frye et al. (2001) and for establishing a field laboratory Kalema-Zikusoka and Rubanga (2013) Fig. A3.6 For field forensic work see Cooper and Cooper (2013) and Chapter 6: Methods of Investigation — Postmortem Examination

FIGURE A3.3 An injured, still anaesthetised, mountain gorilla is put on the back of a vehicle to be transported across the volcanic lava to its group.

FIGURE A3.6 A dead gorilla is carried from the waterlogged valley where it was found, to a drier (and safer) location for a field necropsy.

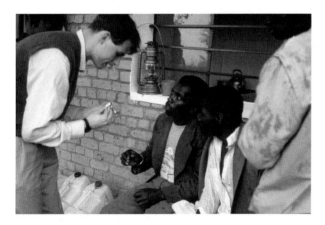

FIGURE A3.4 Requests for help with health matters are commonplace in rural Africa. Here a medical student provides assistance, mainly in the form of simple medicines and advice, at a veterinary centre in Central Africa.

FIGURE A3.7 The CTPH permanent field laboratory in Buhoma, Uganda. Situated in the forest, it uses solar panels for power.

FIGURE A3.5 Zoonosis awareness in a community close to gorilla habitat in Uganda.

forget that it has the advantage, especially (but not only) in the absence of a camera, in that what is seen can also be sketched. They should also be mindful of the words of the famous surgeon Denis Burkitt, who as a child lost his right eye in an accident but went on to make major contributions to tropical medicine: 'when I eventually reached Africa, God in his mercy enabled me with my one eye to see things which my predecessors have missed with two'.

Whatever equipment is used, or techniques employed, in the field, a carefully planned programme of quality control (QC) is important.

Versatility and adaptability are essential when working in the field in Africa. This often means improvisation. Anderson (1996) described the many veterinary uses of a Swiss Army knife. Plastic or paper bags can be used in the place of gloves or to cover feet. An umbrella is very useful for shelter and as a temporary container. In the absence of formalin, alchoholic drinks can be used to fix tissue samples (Cyrus, 2005; Spratt, 1993).

Appendix 4

Hazards, Including Zoonoses

John E. Cooper and Margaret E. Cooper

We must evacuate in 45 minutes. The rain pours all the way down the mountain. At the bottom we pack 11 people and luggage into one car. The road is deserted. We head for the veterinary clinic run by the Morris Animal Foundation.

Amy Bevis (her escape from Rwanda, April 8, 1994)

INTRODUCTION

This Appendix covers, mainly in tabular form, some of the hazards associated with working with gorillas — live or dead, captive or free-living — or material from them. Such hazards can be noninfectious or infectious (see Chapter 18: Legal Considerations).

There is strong emphasis on zoonoses here because the senior author (JEC) is very aware of their dangers and conscious that, on at least three occasions, he has been fortunate to escape infection. The first incident was long before he worked with either free-living or captive gorillas. In 1965, while still a veterinary student, he offered to climb a tree in Bristol in order to help rescue and capture a vervet monkey that had escaped from a laboratory. This plan was vetoed by the late Professor C.W. Ottaway. The Laboratory Director, Professor K.R. Hall, instead attempted to rescue the animal.

He was bitten and within 8 weeks had died following an ascending encephalomyelitis, never definitively diagnosed (Fiennes, 1967) but assumed to be due to a neurotropic virus. Two years later, 1967, when working in Tanzania, JEC was fortuitously not on duty at Arusha Airport when a consignment of vervet monkeys arrived from Uganda; if he had been, he would, as always, have given the animals a comprehensive health examination, to include close contact, sometimes handling. The vervets received a fairly superficial check by a colleague, after which they were flown on to Europe. There they were the source of three simultaneous outbreaks of Marburg disease in Marburg, Frankfurt and Belgrade. This killed seven people and another 24 developed clinical illness. In the 1970s and 1980s JEC was regularly in contact with recently imported macaque monkeys, many of them seropositive for *Herpesvirus simiae*, and amongst other things sustained a splash of monkey saliva in his eye while administering medication *per os*. This had no adverse sequelae, other than causing understandable anxiety, but served as a further warning to him that, while primates are fascinating creatures that warrant study and concern for their welfare and conservation, they are also potentially dangerous, especially in respect of pathogens.

Table of Hazards

Hazards	Risks and Considerations	Action or Comments
General		
In all circumstances a risk assessment should be made, hazards recognised, risks analysed and measures taken to remove, reduce or provide protection from the hazards Written risk assessments and protocols are essential Applies to all aspects of employment, systems of work and the workplace, equipment and vehicles, illness, injury and death Protective measures, equipment and training should be implemented Regular review and updating is essential Consult existing literature in countries where health and safety legislation is well established and strictly enforced (Cooper, 2013)	Sickness, accidents, injury, death Physical fitness Fire and other emergencies Natural occurrences, storm, flood, earthquake Use of equipment, office, laboratory, surgery, workshop, vehicles Dangerous substances	Risk assessment and implementation Protocols, working practices, notices, training, protective clothing and equipment First aid training, handbook and equipment Fire alarms and evacuation Handwashing and other facilities for hygiene Emergency contacts and evacuation plan Insurance: personal, accident, health, property, personal liability, medical evacuation Project: property, car, employer's liability/workers' compensation Legal expenses
Premises	Condition of buildings, facilities, services (electricity, etc.), lighting Slipping, tripping, falling	Regularly check premises for hazards Prompt and regular repairs Warning notices
Location	Altitude Harsh, unstable terrain Remote areas Getting lost	Appropriate clothing, equipment, first aid Communications: GPS, maps, local guide, whistles, flares Plan and train for emergencies; rescue plan
Weather	Heat, cold, fog, rain/lack of rain. Wind chill factor. Exposure to extremes Sudden or unexpected events	Appropriate vehicle and equipment Appropriate clothing; shelter Weather forecasts, local advice Planning for emergencies
Natural phenomena Fire, flooding, earthquake	Injury, death; displacement, damage to buildings, personal property Lack of communication, security Loss of supplies, equipment, personnel	Plan and practise for known or possible events. Consult local support staff and use local knowledge
Animals Gorilla behaviour Other species in location, for example, buffalo, snakes Plants Stings, poisons, allergies	General: injuries, death Infection, zoonoses, toxicity Proximity, aggression, bites, scratches Attacks, bites, stings, allergic reaction Free-living gorillas - during observation, interventions Injuries when gorilla enters a village Captive gorillas — restrictions on access and handling; also plan for emergencies and escapes	Plan and practise for known or possible events. Consult local support staff and utilise local knowledge. Have access to specialist advice on hand (for example, snakebite)
Carrying heavy loads, e.g., in fieldwork Immobilisation equipment, medicines and veterinary/scientific equipment	Overloading; lifting Damage to equipment Exhaustion, injury	Use lightweight equipment Need to be fit and versatile (Fig. A4.1) Plan safe weights, lifting; spare porters Consider terrain and activity involved See Appendix 3, Field Pathology
Personal safety during fieldwork	Hygiene, health Accidents, illness and injuries Getting lost	Avoid lone working. Have an adequate number of reliable personnel Communications. Appropriate clothing for protection from elements and accidents (Fig. A4.2) Emergency contacts Rescue plan See Appendix 3, Field Pathology
Lone working in field or zoo/rehabilitation centre	Needing help for any hazards, accidents, conflict, weather event	Prepare note of planned work, route, location, timing and keep to it. Leave message with reliable person if this information changes Have means of communication, if possible, arrange regular checks See Appendix 3: Field Pathology

(Continued)

Hazards	Risks and Considerations	Action or Comments
General		
Security – day to day	Personal – assault, burglary, accidents Property – theft, damage Instability (see below)	Secure buildings, property, equipment, veterinary drugs, firearms Guards and/or security service Dogs, geese, snakes provide a deterrent, where acceptable Plan and train for emergencies In the field – awareness, appropriate precautions
Security – instability	Local or national instability or insecurity, demonstrations, civil/military disturbance, civil or military forces/controls, counter-insurgence, spies, firearms, landmines, roadblocks. War. Genocide Insurance may be invalidated in conflict situations	Emergency plans and training Intelligence gathering and sharing Need for guide, armed escort (Fig. A4.3) Alternative routes, means of travel, communication Evacuation plans; triage and packs of minimal essentials for different scenarios In the context of gorillas see Cooper (1997a) and MacFie (1992)
Communications		Phone, radio and alternatives and supporting equipment Contacts including medical and diplomatic mission Emergency numbers on speed dial; ICE ('in case of emergency') numbers Emergency plan Information-sharing with other people and organisations
Vehicles	Night driving, off-road driving; overloading Vehicle breakdown Giving lifts Accidents; stopping for accidents	Safe driving skills training for drivers Seat belts fitted and strict rules on use Rules on night driving, lifts, armed passengers May be safer not to stop for an accident but to report it later.
Medical	Accidents, attacks, illness, injury, death Diseases: rabies, polio, smallpox, typhoid, paratyphoid, tuberculosis, tetanus meningitis, for which vaccination should be carried out. Allergies	Emergency plan Insurance/money for air/medical evacuation and treatment Records of medical conditions, blood group, allergies, vaccinations. Store securely; carry spare copies. If confidentiality is assured, provide copy to appointed local medical service. First aid/medical/survival kit; medicines Contact numbers; local and main hospitals
Languages	Difficulty in dealing with authorities Cultural mistakes. Dangers if there is an emergency. Understanding situations, especially in the field. Staff relations Courtesy to host country	Speak one or more national languages and a little in local vernaculars. Always learn local greetings and courteous phrases Carry pocket dictionary or list of key words Interpreter for important matters
Cultural issues	Causing offence or adverse action relating to customs, religion, other sensitivities, dress, photography Offence caused by foreign manners, behaviour, way of life, freedom of expression and personal conduct Courtesy to host country	Awareness of/sensitivity to local culture, customs, religions. Heed local advice Treat older people with respect. Use formal introductions and business cards. Learn official or local languages – see above Dress modestly (well-covered) in keeping with locality (male and female). Have formal clothes to meet officials and academics (men often wear ties, suits and formal shoes, especially in government, business, education and diplomacy) Respect access to property and locations Photography: obtain permission – especially people, livestock, official buildings and strategic structures (e.g., bridges, airports)

Hazards	Risks and Considerations	Action or Comments
General		
Field postmortem examinations	Observers, intruders Assistants Equipment Poor facilities, poor hygiene Limited equipment, drugs Special precautions for nonhuman primates; zoonoses Disposal of postmortem material	Demarcate area/guards if people are watching. monitor property. Make fire and boil water, if possible. Use table or large leaves on a natural surface. Assign assistants and areas as 'clean'/'dirty' (i.e., contaminated) Protection regarding postmortem equipment, material and effluent Burn or bury (secure from scavengers) all remains Dispose of syringes, needles and other sharps to prevent reuse See Appendix 3, Field Pathology

Animal collections	Gorilla behaviour
	Attacks, injuries
	Zoonoses
	Working practices:
	   Handling
	   Cleaning
	   General management
	Handling and rearing of infant gorillas
		In zoos: strict working practices limiting direct access to adult gorillas unless sedated, or anaesthetised. Barriers and warning signs around enclosures and in front of cages to restrict public access (Fig. A4.4)
		In the field: strict limits on proximity other than during veterinary care
		Adhere to appropriate national standards, regulations or laws. See DEFRA (2012) and HSE (2012)

Veterinary pathology and laboratory work	Procedures Equipment Animals, live and dead, wild and domesticated Dangerous substances/pathogens, biological material, drugs, medicines, chemicals Storage — secure, refrigerated Personal safety — working practices; declaration of pregnancy and medical conditions that require special precautions Injury, infection	Risk assessment, protocols, protection, training Ensure correct, safe, use of equipment, protection training supervision Follow established guidance even if there is no national legislation If not possible, use simple precautions (e.g., colour coding) and simple barriers (red line on floor to regulate entry) Labelling of containers, especially if reusing old ones (e.g., poisons). Add coloured dye to colourless fluids to avoid their being mistaken for water by children or local people Employees to declare pregnancy and medical conditions (balance with privacy) Animal handling/restraint protocols —live or dead See Lehn (2008) regarding biomaterials Consider using respiratory protection/hood with full face when working on tissues. Mask appears to be inadequate when electric saw is used. See Posthaus et al. (2011) General recommendation — disease awareness, prompt diagnosis, staff disease surveillance, correct waste disposal. Radiography: well-established installation, precautions, protection, standards, training Refer to bodies with appropriate knowledge, for example, The National Consortium for Zoonosis Research, hosted by the University of Liverpool, UK, but NOTE: most information is not on open website
Pathology standards	Taking, preserving, storage and transportation of samples/pathological material	General recommendation — disease awareness, prompt diagnosis, staff disease surveillance, correct waste disposal See Appendix 3, Field Pathology

Hazards	Risks and Considerations	Action or Comments
General		
Equipment	'Sharps' (scalpel blades, needles, glass, etc.) Pipetting Electric saws, especially in tuberculosis cases (see earlier) Gas/inhalation anaesthesia — risks of exposure, especially with large animals/long procedures. Radiography — exposure hazards. Radiography of dead material — no legal or ethical issues but health risk if excessive Firearms and immobilisation equipment	Protocols. Colour codes for receptacles — for example, red for hazard, contaminated. Secure storage, restricted access Safety cabinet (see below) (Fig. A4.7 and Fig. A4.8) Monitoring of anaesthetic agents See below
Dangerous substances hazardous to health	Chemicals, for example, formaldehyde, formalin, glutaraldehyde Drugs Biological agents Pathogens Zoonoses	Laboratory design and equipment for safe use of dangerous substances Write guidelines for safe storage, use and disposal Follow established guidelines if available Keep data sheets Be aware of ketamine misuse for 'recreation' Secure cabinet for dangerous drugs Anaesthetic gases — scavenging systems (risk of over exposure) Chemicals and poisons — careful labelling, storage, handling and training Colourless chemicals can be dyed to prevent confusion with water Biological agents categorised in four hazard groups Levels of containment according to hazard group. See, for example, for UK: HSE/ADCP (2013, 2005) Other countries have similar provisions
Medicines Dangerous/controlled drugs	Dosage, abuse, theft	Follow data sheets for use, safety and storage Locked storage; restrict access Note risk of theft/misuse of ketamine hydrochloride for recreational purposes or because it is in demand in human medicine
Work in museums	Manual handling, heavy lifting, sharp objects, knife blades (training). Biological hazards by contact or inhalation — chemicals (some used decades ago, e.g., arsenious compounds), frass, invertebrates. Infectious diseases. Allergies	See Chapter 18, Legal considerations and Irvin et al. (1972) Fig. A4.5
Zoonoses — defined here as 'those diseases and infections that are naturally transmitted between vertebrate animals and humans' (see Chapter 3: Infectious Disease and Host Responses)	Zoonoses — those at risk are: • Working with live/dead animals • Working with pathogens • Working with affected persons • Inadvertently in contact with animals, invertebrate vectors or pathogens Additional risk to humans who are immunosuppressed/immunocompromised (e.g., patients with AIDS, malaria or receiving chemotherapy) Endemic zoonoses are a particular problem in the tropics (Halliday et al., 2015) and are often being tackled using 'One Health' measures, an initiative led by Africa	Standard procedures for prevention and control of zoonoses: 1. Awareness, education and training (Fig. A4.6) 2. Minimising unnecessary contact with live/dead animals or their products, vectors and pathogens 3. Specific protection and monitoring of those at risk by (for example) immunisation, health checks. A health surveillance programme for staff in contact with/in proximity to gorillas or their tissues

(Continued)

Hazards	Risks and Considerations	Action or Comments
General		
	(Kamani et al., 2015). Ebolavirus disease (EVD) is currently a particular concern in Africa There is also the question of protection of gorillas from zoonoses acquired from humans	4. Closer liaison between medical and veterinary professions and other disciplines 5. Correct, safe, handling of animals and their products 6. High standards of hygiene and use of barriers/colour codes 7. Maintenance and monitoring of health of animals The above can be supplemented with adherence to standards, protocols and production of information sheets, notices and training literature Simple visual procedures and warnings (allow for literacy, learning/language difficulties and/or limited education) Protective clothing and equipment – must be cleaned regularly, properly and safely Prompt reporting of illness or injury in humans or gorillas, with prompt treatment and containment When working with gorilla tissues, especially if a zoonotic infection is suspected, precautions must be taken. If available, an appropriate safety cabinet should be used (Figs. A4.7 and A4.8). The person(s) using the cabinet must be adequately trained and experienced in such work. Often safety cabinets are not immediately available, especially in Africa, but also in the context of zoos in poorer or less well-equipped countries. Even a basic plastic or glass hood or chamber, in which the material is handled, can reduce splash See various references and discussion in text and Appendix 3: Field Pathology Guidance on handling or avoidance of carcases or sick gorillas found in situ if EVD is a possible diagnosis
Waste	Waste disposal Hazardous used equipment Animal byproducts Cage material Effluent and sewage Infection of people and animals Contamination Vermin, such as rodents and cockroaches	Follow laws on waste disposal, for example, separation of hazardous waste such as sharps and medical waste from domestic or office waste; distinct containers If no legislation: adopt guidelines Bury or incinerate material safely to prevent access by scavengers, human or animal. See Appendix 3, Field Pathology Low-cost, self-built, medical waste incinerator Manage effluent and sewage safely including drains and eventual outlet Control vermin

FIGURE A4.1 An injured gorilla (or person) may have to be transported on foot in the field. This presents a number of risks.

FIGURE A4.4 Barriers are important. Warning notices in zoos or elsewhere should be in appropriate languages.

FIGURE A4.2 Visibility, in this case by the provision of conspicuous jackets to veterinary staff, can be both an advantage and a danger in the event of civil unrest or war. Although essential at times for safety, the colour startles the gorillas and makes the wearer very visible.

FIGURE A4.5 Old museum specimens, such as these skins, may present dangers from toxic chemicals that were originally used in preservation.

FIGURE A4.3 Local or national instability brings its own dangers. The arrival of blue-helmeted Ghanaian soldiers in Rwanda in 1994 meant a degree of security for local people as well as for the gorilla projects.

FIGURE A4.6 Safety on training courses in Africa may require ingenuity. Here the demonstrators, from Tanzania and Uganda, respectively, had to share the only available pair of intact reinforced gloves.

FIGURE A4.7 Safety cabinets, however basic, should be used whenever possible when dealing with gorilla material or, as here, trimming tissues for histology.

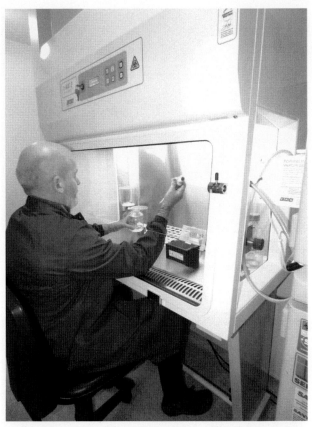

FIGURE A4.8 A Class 2 safety cabinet at Cambridge University Veterinary School suitable for working with gorilla tissues that may harbour zoonotic pathogens.

Appendix 5

Case Studies — Museums and Zoological Collections

INTRODUCTION

John E. Cooper

The following are reports of studies that are still in progress but which illustrate how much research still needs to be carried out on undescribed or inadequately described skeletal material in institutional collections, taking account of unpublished zoo records where available. A more detailed account of the findings in each of the four studies will be published in full in due course, after publication of this book.

The first two reports concern skeletal material in museums. The first describes ongoing studies on 'Guy', the western lowland gorilla who lived in London Zoo from 1947 to 1978, and whose skeletal remains are now housed in the Natural History Museum (NHM), London. The second outlines current investigations on three skeletons in the Osman Hill Collection, preserved in the Royal College of Surgeons of England (RCS), also in London. The last two reports are of two student (veterinary undergraduate) projects that were successfully completed during the summer and autumn of 2015.

One reason for including these four museum specimens is to illustrate the value of retrospective studies in throwing light on an animal's health status in life and the pathological processes leading to its death. This is not only of relevance where well-known captive specimens are concerned, but can also be applied to the numerous skeletons and bones of *Gorilla* in collections around the world. An important element in such retrospective investigation is the utilisation of modern techniques and equipment by specialists from different disciplines. In the case of Guy, these included a veterinary pathologist, an orthopaedic surgeon, a veterinary professor of orthopaedic science and an osteologist.

STUDIES ON THE SKELETON OF 'GUY' AT THE NHM

Brian N. Livingstone, John E. Cooper, Paolo Viscardi, Keith Maybury, Allen Goodship, Gordon Hull and Roberto P. Miguez

BACKGROUND (GORDON HULL)

'Guy' is the first gorilla to have reached adulthood in London Zoo. His life history is summarised in Part II: A Catalogue of Preserved Materials (see entry for the NHM, London). His developmental years were spent in the Monkey House (opened in 1927), and the inadequacies of this building and its enclosures were well understood by the Zoological Society of London (ZSL) (Brambell and Mathews, 1976; Toovey and Brambell, 1976). The Michael Sobell Pavilions, built to replace the Monkey House, were opened in 1972, and Guy was given the benefit of a larger enclosure and the companionship of a female gorilla, 'Lomie'. He died in June 1978 at the age of 32 years (his estimated date of birth was May 1946). The Sobell Pavilions were replaced in 2007 by a new exhibit, Gorilla Kingdom, which houses a group of western lowland gorillas.

MATERIALS AND METHODS (JOHN E. COOPER)

The studies on Guy to date comprise the following:

1. Preliminary visual and gross assessment of the skeletal remains (by John E. Cooper and Gordon Hull), together with written and tape-recorded descriptions

and initial photographs. On one visit a battery-operated endoscope was used to investigate the internal appearance of hollow bones and hyperostotic lesions.

2. Follow-up visits with specialists in different disciplines (Brian N. Livingstone, Paolo Viscardi, Keith Maybury, Allen Goodship, Gordon Hull and Roberto P. Miguez—see above) to obtain opinions on specific aspects of the skeleton and in order to start unravelling, as far as practicable, Guy's clinical history.

NOTES ON THE CRANIAL MORPHOLOGY (PAOLO VISCARDI)

The skull of Guy (NHMUK.1978.1226) is at the smaller end of the size range of the adult male gorillas examined and, when compared against male gorilla specimens collected from the wild (visual inspection $n = 25$, measured $n = 8$), the cranium and mandible show differences in morphology.

Guy has broad orbits, a proportionally high cranium, and a concave and short facial profile compared with other gorillas (Fig. A.5.1). This difference in cranial profile appears to be caused by an upward deflection of the premaxilla, retraction of the maxilla and forward deflection of the supraorbital region of the frontals, which may also influence orbit breadth. Overall, this gives Guy's skull a brachycephalic and less prognathic appearance than is usual which, combined with the broad orbits, offers a slight neotenous cast to Guy's facial area. Guy also has an unusually thick-bordered nuchal crest and robust and somewhat rugose mastoid portion of the temporalis, suggesting some differences in the development of the postural musculature between Guy and free-living gorillas in the wild.

The nature of the bone of Guy's skull is somewhat different from that of free-living gorillas, in terms of texture and development. The texture is rather friable and waxy, which may reflect postmortem preparation, or could indicate a premortem issue with bone growth relating to nutrition or disease. Lending support to a premortem cause is the uneven and overdeveloped configuration of the supraorbital ridge and glabella, in association with the unusually fine configuration of the upper region of the nasals (Figs. A.5.2 and A.5.3; Table A.5.1).

EXAMINATION OF THE SKELETON (BRIAN N. LIVINGSTONE)

This report is based on my examination of Guy's disarticulated skeleton stored at the NHM, London. There is no soft tissue still attached.

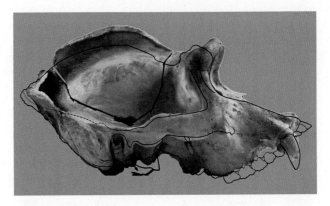

FIGURE A.5.1 Lateral view of *Gorilla gorilla* specimen NHMUK.1978.1226, Guy, with outline of more typical free-living gorilla specimen superimposed for comparison. *Original photo taken by Paolo Viscardi.* © *Natural History Museum, London.*

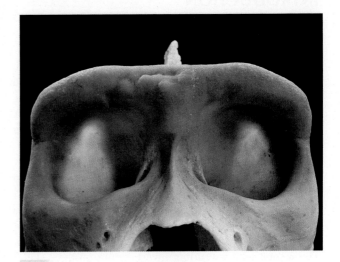

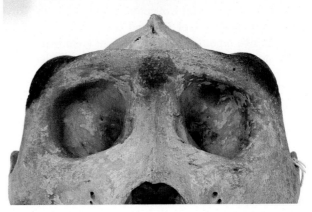

FIGURES A.5.2 AND A.5.3 These figures show the uneven and overdeveloped configuration of the supraorbital ridge and glabella in association with the unusually fine configuration of the upper region of the nasals in Guy's skull (top), when compared with the 'normal' skull of a non-captive male gorilla LDHRM-NH.H2 from the Horniman Museum (bottom). *Original photos taken by Paolo Viscardi.* © *Natural History Museum, London and Horniman Museum and Gardens, respectively.*

TABLE A.5.1 Table of Measurements (mm)

Condylobasal length (occipital condyle to prosthion)	200.5
Basal length (anterior of foramen magnum to prosthion)	176.8333
Bicanine breadth	72.32
Upper canine length (largest)	26.57
Lower canine length (largest)	24.77
Bizygomatic breadth	160.5
Cranial length (inion to glabella)	194.8333
Facial length (glabella to prosthion)	102.8333
Maxillary toothrow length (mean of both sides)	69.08667
Mandibular toothrow length (from anterior alveolus of p1 to posterior alveolus of m3)	81.2
Cranial height (basion to bregma)	125.6667
Orbit breadth (mean of both sides)	43.155
Orbit height left (mean of both sides)	38.93333
Palate length	102.71
Supraorbital ridge width	130.88
Nuchal height	89.24667
Mandible height	111.6833
Ramus height	100.62
Mandible length (horizontal distance from prosthion to midpoint between posterior-most position of ascending rami)	160.3333
Mandible length (prosthion to condyle)	176.8333

The Skeleton

The vertebrae are threaded together. In *Gorilla gorilla gorilla* there are commonly 17 thoracolumbar vertebrae. While this is the same total as for humans, for *Gorilla*, the spine is usually considered to be made up of 13 thoracic vertebrae with ribs and four lumbar (instead of the 12 + 5 for humans). This '13 + 4' is the terminology used in this report.

Ribs were not articulated and were stored separately, with only the pair of ribs attached to the 13th thoracic vertebra still attached. The pelvis is preserved as the ring with the sacrum and 4th lumbar vertebra included. The sternum is preserved as a single piece with manubrium articulated to the body of the sternum. Hands and feet are stored as loose bones, as are the long bones of the limbs.

The Spine

There is new bone or hyperostosis that encloses the 13th thoracic and the 4th lumbar vertebral bodies over the front and sides. This shows a 'flowing' appearance (Fig. A.5.4). The skeleton appears to have been separated between the fused 3rd and 4th lumbar vertebrae leaving the 4th lumbar, which is fused to the sacrum, as part of the pelvis described below.

Some gaps exist in this encasing bone and they appear to connect with the spaces between the vertebrae. In life, the spaces would have been occupied by the intervertebral discs. Higher up the spine, several vertebral bodies exhibit lesser degrees of hyperostosis arising from the lower margins of the vertebral bodies. This is not extensive enough to bridge over to the adjacent vertebra. The uppermost affected bone is the 5th thoracic vertebra. Above that the thoracic spine and the cervical spine appear normal.

The joints between the posterior elements of the vertebrae (facet joints) are unaffected. The external surfaces of the vertebral bodies preserve their normal concavity with no squaring off in their shape to suggest the rheumatic disorder of ankylosing spondylitis.

As noted, the 13th thoracic vertebra has the lumbar ribs attached. The costovertebral and costotransverse joints of this rib are encased by hyperostosis (Fig. A.5.4).

The Pelvis

Viewed from the back, the joints between the iliac bones of the pelvis and either side of the sacrum (sacroiliac joints) are covered with new bone that give a 'flowing' appearance. However, from the internal aspect of the pelvis, the lines of these joints can easily be made out, which suggests that the sacroiliac joints are preserved under the hyperostosis.

In a gorilla the 4th lumbar vertebra (and sometimes the 3rd also) are deeply set between the posterior aspect of the iliac wings. That is the case with the 4th lumbar in this skeleton. The iliolumbar ligaments appear to be ossified, but there is no sign that other pelvic ligaments such as sacrotuberous were affected by hyperostosis.

Finally, I note a partial fusion of the pubic symphysis on superficial and deep aspects. The hip joint socket (acetabulum) does not show any abnormality and, in particular, there are no signs of osteoarthritis or inflammatory arthritis (Fig. A.5.5).

The Sternum

This is formed from an upper piece, the manubrium, which articulates with the body of the sternum. That body section has four segments that are often fused as the skeleton matures and ages. However a synovial joint persists between the manubrium and the body, which is important in the mechanism of breathing. In Guy's skeleton the manubriosternal joint is bridged and splinted by hyperostosis front and back so that no movement can occur there.

However the joints between the medial ends of the collar bones and the manubrium (sternoclavicular joints)

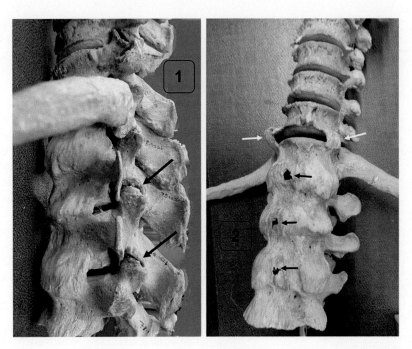

FIGURE A.5.4 The spine: the thoracolumbar section. (1) Left lateral view of fused, thoracolumbar vertebrae. The 13th thoracic vertebra bears the lowest ribs. *Dark arrows* indicate the unfused posterior facet joints. (2) Left anterolateral view of thoracolumbar vertebra. *Dark arrows* mark gaps in the encasing hyperostosis and those are over the spaces that, in life, would have been occupied by intervertebral disc. These spaces appear to have been well preserved with no evidence of spondylosis (degenerative disc disease). *Light arrows* point to hyperostosis that was forming in the ligamentous tissue between the 12th and 13th thoracic vertebrae but which had not extended to fully bridge between them. Note the 'flowing' appearance of this new bone. *Original photos taken by Keith Maybury FRCS FLS.* © *Natural History Museum, London.*

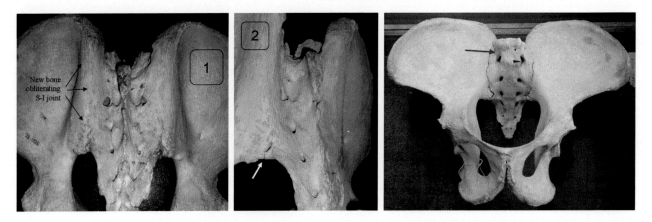

FIGURES A.5.5 The pelvis and sacroiliac joints. (1) From the back. New bone (hyperostosis) obliterates the posterior surfaces of the sacroiliac joints on both sides X marks spinous process of 4th lumbar vertebra. (2) As in (1), but oblique from left. The new bone has a streaked or flowing appearance. The lower end of left sacroiliac joint is still visible (*arrow*) with new bone 'flowing' towards it from above and below. (3) Frontal view. The lines of the sacroiliac joints are clearly seen. Note new bone partly bridges the pubic symphysis. The 4th lumbar vertebra is fused to the top of the sacrum (*red arrow*). *Original photos taken by Keith Maybury FRCS FLS.* © *Natural History Museum, London.*

are intact, with no evidence of any new bone bridging over them or of any arthritis. There are pronounced spikes of bone extending upwards from the lower corners of the manubrium, and I think these are hyperostosis tracking into the ligament that connects this bone to the lower margin of the first rib on each side.

The Limb Bones

There is apparent hyperostosis at various points of muscle attachment to the limb bones. Particular sites are the tuberosities at the upper end of the humerus in both shoulders, the attachments of triceps to the olecranon process of the ulnae,

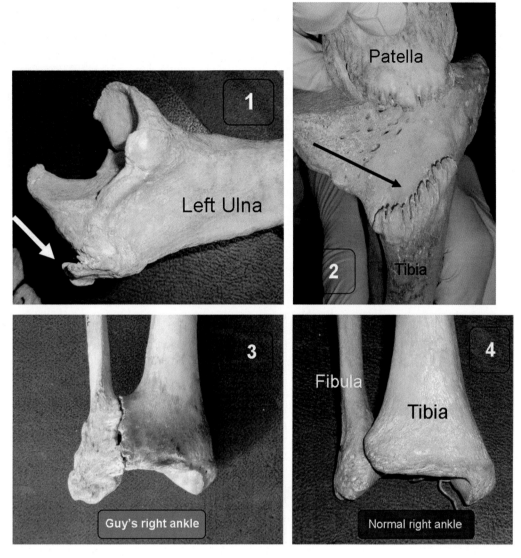

FIGURE A.5.6   Hyperostosis: Elbow, knee and ankle (these appearances are present at multiple sites bilaterally; see text). (1) Upper end of the left ulna showing prominent hyperostosis spur at the attachment of triceps to the olecranon (*white arrow*). (2) Upper end of right tibia showing hyperostosis of patella ligament attachment to tibial tuberosity (*dark arrow*). (3&4) Guy's right ankle with a normal ankle for comparison. Hyperostosis is affecting the inferior tibiofibular joint and the subcutaneous surface of the lateral malleolus (fibula). *Original photos taken by Keith Maybury FRCS FLS.* © *Natural History Museum, London.*

the upper and lower poles of the patella and tibial tuberosities in the knees and on the calcanea at the attachments of Achilles' tendons and plantar fascia (Fig. A.5.6). Particularly striking are the inferior tibiofibular joints bilaterally, where the anterior ligament of the joint has been ossified leaving very prominent flanges of hyperostosis on both bones.

Sites of muscular attachments to the long bones are raised and prominent, more so than would be considered normal.

The Tarsus

I am advised that there was a dissection of the left foot performed shortly after Guy's death and that this reported

the finding of a partially fused accessory ossicle between the tarsal navicular bone and the lateral cuneiform. I have examined this area. Given that the previous examiner would have only seen this foot and would not necessarily have been aware of the widespread hyperostosis and joint fusions elsewhere in the skeleton, that diagnosis was reasonable.

However, with the whole prepared skeleton available, it is apparent that a similar appearance is present in the right foot, and in my opinion this is not an accessory ossicle but hyperostosis affecting the ligament between the anterolateral corner of the navicular bone and the lateral cuneiform.

The Skull

This is described separately by Paolo Viscardi (see above). I would simply mention that only the four canine teeth remain. Nowhere on the skull could I see any hyperostosis except for, possibly, a little prominence and roughening on the lower border of the zygomatic arches corresponding to the attachments of the masseter muscles on both sides.

The NHM holds other gorilla skeletons and 16 of these were examined for comparison with that of Guy. One other skeleton showed fusion of the lumbar spine very similar to that on Guy but not as extensive. Areas of hyperostosis were seen in some of the other skeletons but none was as widespread or extensive as exhibited in Guy's skeleton.

Radiography

This has been performed by Roberto P. Miguez. Anteroposterior and lateral images of the fused thoracolumbar section of the spine are illustrated in Fig. A.5.7. The large 13th rib obscures the thoracolumbar junction on the lateral view. Taken together, the images confirm the impression that the hyperostosis that has caused the fusion is all extraarticular. The spaces for the intervertebral discs are well preserved with no evidence of any bone damage due to erosive arthritis.

The sacroiliac joints have not been imaged, and at the time of writing CT scanning is being arranged. Thus, at this stage a radiological assessment is not complete.

Comment

The skeleton shows widespread and extensive hyperostosis with fusion of the lower thoracolumbar vertebrae to each other and to the sacrum. The sacroiliac joints are fused (at least posteriorly). The joints of the sternum are also fused, while the inferior tibiofibular joints are almost fused. There is widespread hyperostosis affecting the peripheral skeleton causing the appearance of spurs at muscle attachments. In other reports of the appearances of *Gorilla* skeletons (Rothschild and Ruhli, 2005; Lovell, 1990), this appearance is named 'spondyloarthropathy' and in some cases appears to have been regarded as equivalent to the autoimmune rheumatic disease 'ankylosing spondylitis' that affects humans. Similarly spur formation has been described in *Gorilla* skeletons, with the new bone often described as 'osteophyte' or 'exostosis' formation. Since these terms have other specific pathological definitions in human (often also veterinary) pathology (see text of book), they are not used in this report. The term hyperostosis (excessive new bone formation) is preferred.

Peripheral joint arthropathy is not seen in Guy's skeleton, at least in so far as the actual articular surfaces are

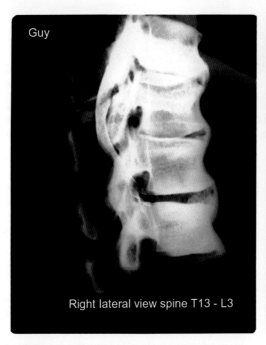
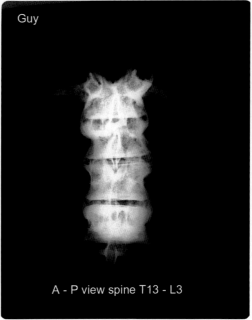

FIGURE A.5.7 Radiographs of the fused spinal segment in Guy's spine. On the lateral view the 13th rib obscures the thoracolumbar junction, but taking the views together it is clear that the spaces for the intervertebral discs are well preserved with no arthritic destruction of the bone of the vertebral bodies. *X-ray images courtesy of Mr Roberto P. Miguez.* © *Natural History Museum, London.*

not eroded or damaged. The probability must be considered that there is a combination of an autoimmune arthritis affecting the spine and sacroiliac joints, together with widespread and extensive hyperostosis in the peripheral skeleton. However, I am advised that, in life, Guy never showed clinical signs to suggest that he was in pain or having difficulty moving.

The human disorder of disseminated idiopathic skeletal hyperostosis or 'DISH' (Resnick et al., 1975; Resnick and Niwayama, 1976) — see elsewhere in text — can produce appearances similar to those exhibited by Guy, with extra-articular hyperostosis encasing and fusing joints and this can be distinguished from the type of joint fusion seen in ankylosing spondylitis. At time of writing this possibility is being explored and further investigation is in progress with a view to reporting Guy's case in more detail.

A REVIEW OF THE CLINICAL RECORDS (BRIAN LIVINGSTONE AND ALLEN GOODSHIP)

1. We have been provided with copies of veterinary records from the ZSL that cover the later period of Guy's life from March 1971 up to his death in June 1978.
2. We have reviewed a copy of the postmortem report.
3. Guy arrived at ZSL in 1947 as a 10 kg infant and is said to have been in rather poor physical condition. For about 25 years he was the solitary gorilla at the Zoo and we understand it was common for visitors to feed him in his cage in the early years.
4. By current standards it is probably fair to state that his enclosure was small.
5. As far as we can tell, he never had or required any radiographs of his skeleton while he was alive.
6. The veterinary records start in 1971 and they almost entirely refer to dental caries and the need either for antibiotics, dental extractions or both. Blood tests in 1971 showed haemoglobin of 10 g/100 mL, which was considered low and he was prescribed 'haliborange'. By 1976, haemoglobin level was 12 g/100 mL, but red cells still had an appearance that, in humans, would indicate mild iron deficiency.
7. We do not find any records or estimates of his weight at maturity but postmortem he was 240 kg (including weight of net).
8. If his urine was ever collected and examined, it is not recorded.

    NB: noting the absence of such data is not to imply criticism of care but simply to indicate what assumptions about his health can or cannot be justified.
9. In 1976 he had anaesthesia for dental extraction and is noted to have recovered uneventfully.
10. His final illness was in June 1978 when he again required anaesthesia for dental extraction.

    There is no preoperative record of clinical signs to suggest any disorder such as heart disease. He failed to recover from the anaesthesia. The copy of the postmortem records that we have is not fully legible, but we note the following findings.
    a. There was a large quantity of subcutaneous fat.
    b. There was what was described as a massive congestion of both lungs and a large quantity of froth in the trachea and major bronchioles. We take the term pulmonary congestion to mean fluid accumulation in the lungs or pulmonary oedema which in turn implies he had heart failure. The myocardium had a quantity of surface fat. There is no record about the coronary arteries but there were raised plaques at the origin of the aorta. This last record could be referring to plaques of atheroma but the note does not specify.
    c. The cause of death written in the record is 'cardiac arrest following massive pulmonary congestion'.

Comment

It has to be said that the records do not provide very robust data for making a diagnosis, but it does appear that the cause of death was due to previously asymptomatic heart disease. The postmortem examination does seem to have excluded any mechanical cardiac disorder such as valvular disease. Experience from human medicine is relevant given the close genetic relationship between the species. On that basis, the most likely cause of the heart disease will have been coronary artery atheroma, but that cannot be diagnosed for certain and see Chapter 10: Respiratory and Cardiovascular Systems. However, it would be consistent with his early lifestyle, which featured relatively low activity level and an uncertain diet. Since examination of the skeleton raises the possibility that Guy was suffering with DISH, this is very relevant. In humans, at least, DISH is associated with obesity and diabetes which, in turn, would be compatible with Guy's early lifestyle. This topic is the subject of research in progress.

FURTHER WORK (JOHN E. COOPER)

The skeletal remains of Guy are proving to be of great interest but much remains to be done. In addition to more in-depth studies on the skeleton, outlined above, the team will:

a) attempt to obtain more information about Guy's early life at the ZSL, including any records prior to 1971, and

b) search for stored soft tissues and/or paraffin blocks that might still exist in order to supplement the small collection of histological sections of Guy's (1978) tissues that have been read by John E. Cooper but not reported here.

The results of these studies so far amply vindicate the decision to make this a multidisciplinary investigation.

STUDIES ON THREE SKELETONS IN THE OSMAN HILL COLLECTION

Carina Phillips, Martyn Cooke, Jaimie Morris, Paul Budgen, Gordon Hull, John E. Cooper

BACKGROUND

William Charles Osman Hill (1901–75) was a renowned anatomist and primatologist. In 1950, following time in academia in Ceylon (Sri Lanka) and Edinburgh, Scotland, he became Prosector at the Zoological Society of London (ZSL) (Anon, 1975). There he prepared specimens for his extensive collection of primate material from all over the world and published the first three volumes of his seminal work on primate anatomy and taxonomy (Hill, 1953–1970).

Much of Osman Hill's collection was bequeathed to the Royal College of Surgeons of England (RCS). It comprises an estimated 1250 specimens and includes skeletal preparations, fluid-preserved specimens, a small number of embalmed remains and some casts. Of the 81 macroscopic gorilla specimens currently in the RCS museums (see Part II: A Catalogue of Preserved Materials), 7 came from the Osman Hill collection.

The three gorilla skeletons that are the subject of this study comprise the semiarticulated skeletons of an adult female, an adult male and a juvenile (Table A.5.2). Unfortunately there is little surviving information about these specimens.

These three gorilla skeletons are currently being examined in detail. The background to the study and some preliminary findings are given below.

THE RESEARCH PROPOSAL

The proposal to examine these skeletons was in the name of the Gorilla Pathology Study Group (GPSG) (see Preface and elsewhere in the text of the book).

This skeletal material had only basic catalogue entries. The GPSG team considered that it held potentially important information and data of direct relevance to research areas encouraged by the RCS, including many biomedical, zoological and other scientific possibilities.

In addition, it was considered that a multidisciplinary approach to examining the skeletons could yield invaluable detail and contribute to other fields of research, including information about the life and work of Dr Osman Hill, together with the collation of morphological and taxonomic data of endangered species, particularly *Gorilla* spp.

During the course of initial meetings, held between members of GPSG and RCS staff, it was agreed that members of the GPSG should approach the investigation in three stages:

1. The initial stage: a preliminary visual assessment of the condition of the specimens, from which it was hoped that recommendations could be proposed for the preservation and presentation of the material so that it might be of optimal value for future research.
2. The second stage: to build on the preliminary findings, but still using only noninvasive methodology.
3. The third stage, for which a research application was successfully made to the RCS Library, Museums and Archives Committee (LMAC) to carry out the following:
   a. Detailed study of joints, ligaments and tendons
   b. The taking of very small pieces of some of the remaining, desiccated, tissues for (1) DNA studies and (2) rehydration for in-depth histological study and possibly electron-microscopical investigation.
4. A further, fourth, stage of investigation, which might include non-invasive imaging of the skeletons using radiography, CT scanning etc, will be considered in due course.

Following the initial meetings, risk assessments were compiled which were cognisant of the relevant legislation,

TABLE A.5.2 Details of Three Gorilla Skeletons from the Osman Hill Collection at the RCS

RCS Reference no.	Osman Hill no.	Sex	Age	Notes
RCSOM/OH/118	PA 61	Male	Adult (possibly old)	
RCSOM/OH/119	PA 63	Unknown	Juvenile	Metal tag-90. Possibly acquired from the collections of F.G. Merfield (Series III)
RCSOM/OH/398	PA 62	Female	Adult	

RCS policy, facilities available and equipment required to conduct the investigation in order to achieve the aims set for each stage.

INITIAL OBSERVATIONS ON THE SPECIMENS

The three gorilla skeletons exhibit changes associated with previous, not current, invertebrate infestation, lipid migration causing discoloration, and greasiness. It is apparent that the three carcases were eviscerated and roughly defleshed. Marks consistent with defleshing were observed during examination. Often a quantity of previously soft tissue was left adhered to the bone, particularly at the joints, and this flesh had subsequently become desiccated. Some individual hairs had also survived, attached to the tissue.

While it was generally considered that there was minimal risk from biohazard exposure during the examinations, appropriate health and safety procedures were strictly followed throughout. The procedures were considered to be important, not just per se, but because one of the aims of the study was to develop techniques that might be applied to gorilla work in Africa, where hazards were likely to be far more significant.

Visual examination verified the sex of the two adults and suggested that the adult male (RCSOM/OH/118) was possibly an older animal. The adult female (RCSOM/OH/398) is almost complete, with only some hand and foot bones missing. The adult male is missing its skull, but the postcranial skeleton is largely complete, again lacking only some of the small hand/foot bones. The juvenile (RCSOM/OH/119) is missing its lower right limb. The female adult exhibits thickening of the bone on the distal half of the right ulna, suggesting a possible healed fracture.

SAMPLING

Each procedure was carefully assessed and agreed by the team before tissue sampling was undertaken (Figs. A.5.8 and A.5.9). Due to the extreme desiccation and toughness of the material, great difficulty was experienced when removing dry tissues. Scalpels, Stanley knives, scissors, bone cutters and hacksaws were all used.

Following removal the samples were placed in prelabelled, resealable polythene bags along with a Resistall label (Preservation Equipment Ltd) marked with a special Rotring pen.

Hair samples were also collected and retained as above.

Debris, including various sloughed skins from insects and pupal cases, were found loose in the storage container and were also randomly sampled.

FIGURE A.5.8 The appearance of one of the skeletons on examination. Note the attached, now very dry and desiccated, soft tissues and the articulated joints.

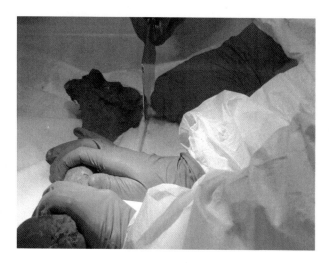

FIGURE A.5.9 A sample is taken for histology and DNA. Because it is so hard and dry, a saw has to be used.

PROCESSING THE SAMPLES FOR HISTOLOGICAL EXAMINATION

Full personal protective equipment (PPE) kit was worn while sampling (see Fig. A.5.10).

A random choice of tissue was made based on the size of sample (the biggest piece of tissue that could be sectioned into multiple parts) for carrying out a tissue rehydration trial.

It is extremely difficult to obtain satisfactory material for histological investigation when tissue is hard, desiccated and shrunken. The cellular structure is severely distorted, making analysis very difficult (Blaney and Johnson, 1989). A process of softening, rehydrating and controlled expansion of the tissue is therefore

FIGURE A.5.10 Personal protective equipment worn for tissue sampling.

desirable. There are several long established methods available (Wilder, 1904), including those used on desiccated Egyptian mummies (Sandison, 1955), but none is ideal. This part of the project was designed to compare some traditional tissue softening methods with more recent approaches and thereby be able to select the most suitable procedure for use on the remainder of the gorilla material.

A total of eight small pieces of tissue from a single sample obtained from specimen RCSOM/OH/398 was placed into individual screw-capped glass universal containers, each uniquely labelled on the jar and internally, using a Resistall label marked with a Rotring pen.

The following test samples were used as a direct comparison, with sample 1 being a distilled water control.

1. Distilled water (control)
2. 3% Sodium hydroxide
3. 50% Polyethylene glycol (PEG)
4. 3% Trichloroacetic acid
5. 4% Buffered formaldehyde
6. 5% Sodium carbonate
7. Fairy fabric conditioner (Procter and Gamble)
8. Comfort fabric conditioner (Lever Brothers)

Each sample was photographed, measured and weighed before being put into the respective test fluid. Following rehydration, the samples were again assessed, photographed, measured and weighed. Each was then rinsed in running water and fixed in 4% buffered formaldehyde. Routine paraffin wax processing and haematoxylin and eosin-stained sections were prepared for histological evaluation.

PROCESSING OF SAMPLES FOR DNA TESTING

Full PPE was worn when sampling. It was paramount that measures were taken to avoid cross-contamination during collection of tissues as this would lead to misleading results.

Before sampling took place, the sampling area and all tools/instruments (forceps, scalpel, scissors, etc.) were disinfected with ethanol. The instruments were also held in a flame for a few seconds to supplement this. Soft tissue samples were placed in 2-mL microcentrifuge tubes and clearly labelled. Between each specimen, all tools and instruments were immersed in ethanol and flamed, and gloves were changed, as was the protective covering on the examination table.

RESULTS

The study is still underway, but the indications are that the techniques employed were both successful and safe in respect of obtaining scientific information from these three important gorilla specimens.

The appearance of the remains suggests that Osman Hill partially processed them, either deliberately to retain the soft tissue or because he planned to work on them further at a later date.

The sloughed skins from insects and pupal cases are currently being examined and identified.

It is too soon to make firm recommendations as to the most effective solution for rehydration as a histological evaluation has yet to be made to assess the cellular structure, but both the fabric conditioners have shown favourable preliminary results.

To date, DNA extractions have been performed using the QIAGEN DNeasy Blood & Tissue Kit and initial analysis showed that DNA is present in the Osman Hill specimens. Two of the samples (RCSOM/OH/118 and RCSOM/OH/119) have produced more encouraging results in comparison with the third sample (RCSOM/OH/398), based on agarose gel electrophoresis with SYBR safe (Invitrogen) staining and visualisation under UV light. Further analyses involve quantifying the amount of DNA present in the samples, to be performed using a Qubit fluorometer (Life Technologies), DNA sequencing and microsatellite genotyping.

It is hoped that when the work is completed and the results have been published, the methods used will prove helpful to others, including colleagues in Africa, who are confronted with similar material and comparable challenges.

ACKNOWLEDGEMENTS

We are grateful to the Royal College of Surgeons of England, The RCS Library, Museums and Archives Committee and Dr Sam Alberti for permitting this study.

STUDENT PROJECTS, BRISTOL, UNITED KINGDOM, 2015

INTRODUCTION

Two student (veterinary undergraduate) projects were successfully completed during the summer and autumn of 2015. They are reported briefly here and will be published in due course. These projects were supported financially by the Dick Smith Memorial Scholarship (established in memory of Dr RN Smith FRCVS after his death), carried out under the supervision of staff at the University of Bristol (Professor Kate Robson-Brown) and Bristol Zoological Society (BZS) (Dr Michelle Barrows, Dr Christoph Schwitzer) and overseen by Professors Allen Goodship and John E. Cooper. The CEO of BZS, Dr Bryan Carroll, kindly gave permission for the two projects to proceed: we are grateful to him and his staff for the welcome and encouragement that they afforded the two students and for their willingness to provide and to discuss zoo records.

PROJECT 1: A REVIEW OF THE DATA RELATING TO THE ANIMAL MANAGEMENT AND WELFARE OF GORILLAS AT BRISTOL ZOO GARDENS FROM 1930 TO 2015

Sophia Keen

Summarised Report

Background

Available archival material held at Bristol Zoo Gardens (BZG) was collated and analysed to compile three separate documents: BZG Gorilla List (Keen, 2015a), BZG Gorilla Tables (Keen, 2015b) and this review. The resources used included: gorilla keepers' diaries, handwritten medical cards, gorilla folders, veterinary files, online Animal Record Keeping System (ARKS) records and Zoological Information Management System (ZIMS) records as well as some postmortem reports held at the Department of Pathology of the University of Bristol's School of Veterinary Sciences — see elsewhere in text.

This review aimed to use the archival information in order to track the changes in husbandry, care, management, health and causes of death of the gorillas held at BZG between 1930 and 2015. A more extensive list of relevant health records and postmortem examination reports is included in the BZG Gorilla List (Keen, 2015a). The review and described accompanying documents aimed to act as a resource for future research covering areas such as enclosure design, animal management, veterinary medicine, animal health and welfare.

The 85 years of gorilla history at BZG were split into five distinct periods under the headings of: Enclosures, Animal Management and Veterinary Medicine, Diet, and Welfare.

The time periods used in the review were:

1. 1930—48 ('Alfred')
2. 1954—66 ('Congo' and 'Josephine')
3. 1967—75 (expanding large group)
4. 1976—95 (problematic new enclosure)
5. 1998—2015 (new group of 'Claus', 'Undi' and 'Salome')

In this summarised report attention is focused on only two aspects: Animal Management and Health.

Animal Management

Our understanding of housing, feeding and management of zoo animals has advanced greatly since BZG first commenced keeping gorillas. It is important to identify the husbandry factors that have changed and investigate how they may have affected gorilla pathology and welfare.

Health

Preventive and evidence-based medicine has increasingly replaced reactionary medicine in zoo veterinary care in recent years (Fig. A.5.11). Few early records were found detailing the treatment that gorillas

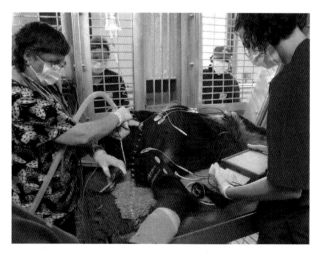

FIGURE A.5.11 A health check is carried out on a captive lowland gorilla, under anaesthesia, at Bristol Zoo Gardens.

received at BZG, although the postmortem reports were relatively detailed and often pinpointed bacteria and parasites considered responsible for ill health or death. General health trends are described below, with focused sections on health problems in the 1976 Ape House and referring briefly to the outbreak of tuberculosis in gorillas in the 2000s. A full list of available records is obtainable from the BZG Gorilla List (Keen, 2015a).

1. 1930—48

'Alfred' (see Part II: A Catalogue of Preserved Materials)

Apart from the record of Alfred's death due to a year-long tuberculosis infection, few health records exist for this animal to illustrate his routine veterinary care. He contracted what was described as whooping cough as a "youngster", from a visitor, and in 1946 was diagnosed with a thyroid deficiency [*sic*] but no treatment was recorded (Bristol Museums & Art Gallery, 1999).

2. 1954—66

'Congo' and 'Josephine'

The health records for Congo and Josephine were relatively sparse, but a few trends emerged. Congo suffered from thyroid deficiency [*sic*] like Alfred before him (see Chapter 16: Endocrinological and Associated Conditions). There was a recurrence of parasites and diarrhoea with *Entamoeba histolytica* cysts and later ascarids detected soon after his arrival in 1954, and a record of mixed bacteria in his loose faeces, mainly *Klebsiella* sp. in 1968. Records describe his death as due to a dissecting aneurysm of the aorta and accompanying haemorrhage.

Josephine suffered from 'loose bowels' in 1965; she was treated with only chloramphenicol and was fed on milk. Her postmortem report identified the cause of her death as viral pneumonia, probably influenza, accompanied by a streptococcal infection of the lungs. It was noted that extreme obesity probably contributed to her death (see Chapter 11: Alimentary Tract and Associated Organs).

3. 1967—75

Various gorillas

Records from 1968 onwards related the introduction of a quarantine procedure for new gorilla arrivals to BZG, for example 'Caroline'. Loose faeces were a common complaint, not normally restricted to one animal. Routine faecal sampling, treatment for worms (recorded at least from the 1970s onwards) and annual visual and/or general anaesthetic health checks were recorded (Fig. A.5.11).

4. 1976—95

Table A.5.3 summarises the causes of death of 16 gorillas between 1976 and 1995.

Three other adult gorillas died in this enclosure during the period of 1976—95. 'Susie' (1984), aged 21 years, died of a possible respiratory infection; 'Gogal' (1989) aged 27 years, of diarrhoea, haematochezia and weakness; and 'Jason' (1994) aged 31, of a suspected chest/lung infection, recorded as 'inconclusive sudden death'.

The project review goes on to discuss an outbreak of tuberculosis in the gorillas at BZG and the measures taken to investigate its origins and to limit disease spread and duration. A review of the relevant diagnostic investigations and epidemiological/epizootiological investigations was presented by Redrobe (2003). DNA studies revealed that a new strain (subspecies) of *Mycobacterium tuberculosis*, affecting several different species of animal, was involved.

Later in the project review the welfare of animals in zoos is discussed, with particular emphasis on the 'Five Freedoms' (see Chapter 18: Legal Considerations) and the application of these, coupled with other guidelines, including the Secretary of State's Standards (under the Zoo Licensing Act), in promoting higher standards. BZG's many achievements in this area, especially in respect of gorillas, are reflected by the greatly improved health, longevity and reproductive performance of these animals.

PROJECT 2: THE DEVELOPMENT OF OSTEOARTHRITIS IN CAPTIVE PRIMATES

Dermot McInerney

Summarised Report

Background

This report presents the findings of a research project undertaken with the University of Bristol and BZG, studying the development of osteoarthritis (OA), degenerative joint disease, in several of the primate species held at BZG between 1999 and 2009.

OA is a degenerative condition associated with the destruction of articular cartilage, leading to the formation of bony spurs (osteophytes) and subchondral sclerosis of the underlying bone (see also definitions in Chapter 14: Musculoskeletal System). The joint space or disc space may also be seen to be narrowed and can be accompanied by the formation of a subchondral cyst

TABLE A.5.3 Table of 16 Gorilla Deaths 1976–95

Gorilla	Year of Death	Age at Time of Death (Months, Days)	Cause of Death (see also discussion of some of the cases elsewhere in text)
'Rebecca'	1976	17, 0	Bronchopneumonia; balantidial colitis
'Benjamin'	1980	6, 16	Interstitial pneumonia; catarrhal enteritis
'Zachary'	1980	5, 3	Terminal bronchopneumonia; liver abscess (umbilical infection); ulcerative colitis; healing fractures of ribs (rough behaviour of adults)
Unnamed 1 Female	1981	0, 1	Rough handling from Delilah; crushed liver
Unnamed 2 Female	1982	2, 19	Viral pneumonitis; collapse and congestion of lungs
'Leah'	1983	13, 16	*Shigella flexneri* infection; oesophageal moniliasis
'Deborah'	1983	23, 16	Multiple abscessation; mucopurulent rhinitis; acute exudative (broncho) pneumonia; keeper in contact recently ill with measles/German measles
Unnamed 3 Female	1983	0,3	Very premature birth
Unnamed 4 Male	1984	3, 7	Diarrhoea/enteritis (mother had diarrhoea)
Unnamed 5 Male	1984	1, 28	Born small; acute pneumonia of bacterial origin
Unnamed 6 Female	1984	0, 1	Sudden death: respiratory distress; abdominal haemorrhage
'Dorcas'	1985	3, 12	Pneumonia; fibrosis pancreas; cystic thymus
'Abigail'	1986	1, 7	Sudden infant death syndrome; rough handling by mother?
Unnamed 7 Male	1988	0, 0	Dystocia: brain trauma
Unnamed 8 Female	1990	0, 0	Acute purulent bronchopneumonia with *Klebsiella oxytoca* infection
'Nadia'	1994	19, 12	Failure to recover from general anaesthesia (reason for anaesthetic unclear); bacterial infection?

(Altman and Gold, 2007; DeRousseau, 1985; Videan et al., 2011). In western 'developed countries', evidence of OA can be seen in the majority of people by the age of 65 (Arden and Nevitt, 2006) and around 8.75 million people in the United Kingdom are treated for OA each year (*Arthritisresearchuk.org*, 2015). However, despite the widespread nature of this disease, its pathogenesis is poorly understood.

The study examined four species of primate, one of them the western lowland gorilla. Only that species is included in this summarised report.

The development of OA was documented using a database of radiographs, provided by BZG, by scoring the joints of individual animals as they progressed through life.

None of the radiographs used in this study was taken for the purpose of the investigation, but many of the animals in the database had had repeated radiography, performed for various reasons, over a number of years. All of the radiographs studied were taken before the introduction of the digital system now present at BZG, so each X-ray film had first to be photographed, digitised and filed before being inspected. The digitising process provided an opportunity to take some long-exposure photographs and apply filters, thereby often producing better quality images than the original radiographs.

Results

A total of 345 radiographs was examined, with the number available for each animal ranging between 1 and 77.

One of the OA cases diagnosed was in the hip of a gorilla. This animal was found to have a subchondral cyst in the femoral head of the hip. The progression of OA could be seen as the researcher moved through the radiographs chronologically.

Another incidental, but interesting, finding in a gorilla was a fracture of a metacarpal bone (Fig. A.5.12).

Although much work needs to be done, examination of the database of radiographs provided by BZG uncovered some interesting information and helped to shed more light on the development of OA among some less-studied species of nonhuman primate.

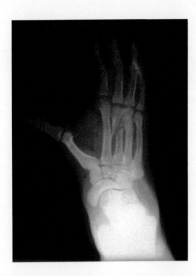

FIGURE A.5.12   An early radiograph of a gorilla's hand at Bristol Zoo Gardens reveals a fractured phalanx and marked soft tissue reaction.

ACKNOWLEDGEMENTS

In addition to the tributes earlier (see Introduction), Sophia Keen and Dermot McInerney are grateful to those who supervised their projects – Dr Michelle Barrows at the Bristol Zoo Gardens and Professor Kate Robson-Brown at the University of Bristol – and to Professors Allen Goodship and John E. Cooper for arranging funding from the Dick Smith (RN Smith) Memorial Scholarship. In addition, Dermot McInerney thanks Mr B. Skerrett for his help during the early stages of this project and Miss H. Beattie and Mr D. Topham for their assistance in editing his preliminary report.

Appendix 6

Scientific Names of Species and Taxa Mentioned in Text

John E. Cooper and Gordon Hull

What's the use of their having names," the Gnat said, "if they won't answer to them?

No use to them," said Alice, "but it's useful to the people that name them, I suppose.

Lewis Carroll, Alice through the Looking-Glass

The English or other vernacular names below are generally given as they appear in the text, with additions or changes where this may assist the reader.

Mammals

African elephant	*Loxodonta africana*
African green (vervet) monkey	*Chlorocebus aethiops*
Baboons	*Papio* spp.
Black rhinoceros	*Diceros bicornis*
Bonobo	*Pan paniscus*
African Buffalo	*Syncerus caffer*
Chimpanzee	*Pan troglodytes*
Colobus monkey	*Colobus* spp.
Cynomolgus monkey	*Macaca fascicularis*
De Brazza's guenon	*Cercopithecus neglectus*
Domestic cat	*Felis catus*
Domestic dog	*Canis familiaris*
Domestic goat	*Capra hircus*
Domestic ox	*Bos taurus/indicus*
Domestic pig	*Sus scrofa*
Domestic sheep	*Ovis aries*
Donkey	*Equus asinus*

Duikers	Subfamily Cephalophinae, several species
Drill	*Mandrillus leucophaeus*
Florida panther	*Puma concolor coryi*
Fruit bats	*Pteropus* spp.
Gazelles	Family Bovidae, many species
Giant anteater	*Myrmecophaga tridactyla*
Giant rat	*Cricetomys gambianus*
Guinea pig	*Cavia porcellus*
Hamster, golden	*Mesocricetus auratus*
Howler monkey	*Alouatta* sp.
Human	*Homo sapiens*
Leopard	*Panthera pardus*
Lowland gorilla	*Gorilla gorilla*
Macaques	*Macaca* spp.
Macaque, Barbary	*Macaca sylvanus*
Macaque, Rhesus	*Macaca mulatta*
Mandrill	*Mandrillus sphinx*
Marmoset	*Callithrix* spp.
Marmoset, common	*Callithrix jacchus*
Mongooses	Family Herpestidae
Mountain gorilla	*Gorilla beringei*
Mouse, house, laboratory	*Mus musculus*
Orangutan	*Pongo* spp.
Rat, brown, laboratory	*Rattus norvegicus*

Red-tailed guenon	*Cercopithecus ascanius*
Rhinoceros	Family Rhinocerotidae
Sooty mangabey monkey	*Cercocebus atys*
Squirrel monkey	*Saimiri* spp.

Birds

Domestic duck	*Anas platyrhynchos domesticus*
Domestic fowl	*Gallus domesticus*
Domestic turkey	*Meleagris gallopavo*
Guinea fowl	*Numida meleagris*
Waterfowl	Order Anseriformes

Reptiles

Nile crocodile	*Crocodylus niloticus*
Snakes	Class Reptilia, suborder Serpentes
Tortoise, leopard	*Geochelone (Stigmochelys) pardalis*

Amphibians

Toad, African clawed	*Bufo bufo*

Fish

	A very large group; three classes, many species
Tilapia	*Tilapia* spp.

Invertebrates (see also text — Parasites)

Ant, safari, *siafu*	*Dorylus* spp.
Bee, honey	*Apis mellifera*
Carrion beetles	*Silpha* spp.
Leeches	Class Hirudinea
Scorpions	Class Arachnida
Ticks	Order Acarina

References and Further Reading

John E. Cooper and Gordon Hull

Look deep into nature, and then you will understand everything better.

—Albert Einstein

Access to published and unpublished literature is a vital prerequisite to the study of the biology of any species, but awareness of the existence of such material is essential too. In his excellent monograph on 'great ape psychology', Mitchell (1999) explored scientific evidence and popular commentary relating to the subject from the 1790s to the 1970s. Although it is primarily concerned with psychology, Mitchell's thesis is a rich source of information about gorillas and other apes and he makes the very pertinent comment: 'It is an unfortunate fact that scientists often appear unaware of much relevant earlier research when publishing and publicising their own'.

As is obvious from the pages of this book, there is substantial published information about gorillas, not least data relevant to their pathology and health. The problem is that such material is widely scattered, extremely time-consuming to locate, laborious to digest, and indicative of what has been described as 'information overload' – but this appears not to be unique to modern times. Blair (2010) pointed out that the invention of the printing press and the ensuing abundance of books provoked 16th- and 17th-century European scholars to register complaints, very similar to our own, about the surfeit of published work – and that was long before gorillas were known to science!

This list of References and Further Reading is intended not only to provide original sources but also to point the reader in the direction of additional information.

The References include publications in languages other than English. Some of these are important historically and others represent seminal studies. In a few cases no readily available or reliable, detailed, translations exist but where comprehensive abstracts in English are provided, either in the original or by CABI, the publication has been quoted in English.

The word 'science' comes from the Latin for knowledge and knowledge should embrace all available sources of information. Those studying the genus *Gorilla* should always be aware of the existence of literature, published and not published, in different languages and be prepared (at the very least) to seek a translation of relevant portions. To depend *solely* upon material in English (or perhaps French) that has been published or put on the Internet is professionally irresponsible (Cooper, 2010a,b,c).

At the time of writing (2015), there are over 200 veterinary journals available to the reader: this figure is greatly amplified if one adds medical and dental literature and specific disciplines such as pathology, pharmacology and paediatrics. For those interested in locating papers about diseases of gorillas, a literature search containing the word 'gorilla' is clearly the first step, made easier by the fact that the scientific name of the taxon is identical to the English name. However, when looking for relevant literature, it is important not to rely solely on searches for the word 'gorilla'. Similar, but different terms are used in other languages (e.g., *gorille*, French) and various local (African) names, such as *ingagi, ngagi, engagi, n'gi, ingheena, gina, ngina* are to be found in in-country reports, especially those produced 50 or more years ago (most still not available electronically) where the spellings reflect who first wrote the word down – often a local French/Belgian/British missionary or a colonial administrator.

Unpublished reports and protocols may, strictly, count as 'grey literature' but can, nevertheless, provide useful information and should not be ignored. Such is the case with, for example, the early records of the Karisoke Research Center in Rwanda in the years before the establishment of the CVV/VVC. With a few exceptions, such records were not compiled by veterinarians even though some relate to sick or injured gorillas or the findings postmortem.

The reader should also be aware that some of the papers listed below do not mention gorillas as such but are included because they deal with diseases or diagnostic

317

procedures that are relevant, or can be applied, to the genus *Gorilla*.

All this means is that no one person searching the literature is likely to have the time to locate everything that is relevant to his/her subject. The answer, it may be suggested, is a 'literature review'. This, however, also has its drawbacks. For example, it has been recognised in human medicine for two decades that experts in a particular subject are less likely to produce a consistent and reliable literature review on their subject of expertise than are nonexperts with training in research methodology (Oxman and Guyatt, 1993). Where there is a vast amount of possibly relevant published literature it may be necessary to resort to 'text mining'. Text mining, also called 'text data mining', is a process whereby high-quality information is obtained by recognising patterns and trends in the various written texts and putting these into a database for evaluation and interpretation. It is increasingly used by life science companies involved in the development and testing of drugs.

Textbooks are sometimes referred to as 'an endangered species' and, indeed, many young people rely almost entirely on the Internet. The latter has, indeed, much to commend it and to some extent has revolutionised access to information, but caution must be exercised when seeking reliable data. It is a fallacy to assume that the Internet is sufficient on its own; some material is not accessible electronically (unpublished field notes, for instance — see earlier) and little-known journals may not be located in a search. Internet information may also not be reliable, or confusing. An important point to remember when using search engines is that the search term queried is matched by a computer that has organised and sifted website information to return what it believes to be the most relevant website data. This information may well be skewed if, for example, a website is better managed or updated daily. In other words, key pieces of research may not be returned in a search query as the search engine has been unable to correctly index the information. For this reason — at the very least — online research should delve far deeper than the first page of Google. The popular press has suggested that academics and researchers generally oppose the use of *Wikipedia*, the online, user-edited encyclopaedia, because of its inaccuracies and impermanence. Interestingly, though, some scientists endorse it enthusiastically — for example, Bond (2011) evaluated the value of *Wikipedia* in the context of ornithological information studies and, while accepting that some of its coverage of material relating to birds was very poor, advocated that ornithologists should 'embrace and contribute to' it (*Wikipedia*).

A few Internet sources essential to gorilla pathology are listed below but others are given in the text. Various organisations catalogue gorilla-related publications on their websites. Examples are:

Mountain Gorilla Veterinary Project/Gorilla Doctors, http://www.gorilladoctors.org/
The Great Ape Heart Project, http://greatapeheartproject.org/about/
Berggorilla, http://www.berggorilla.org/en/extras/literature/

The Pan African Sanctuary Alliance (PASA) pasaprimates.org. Operations manual, veterinary manual and education manual are available at http://pasaprimates.org/about/resources/. These resources form part of a Primate Healthcare Manual and comprise:

- Common Parasitic Diseases—Diagnosis and Control
- Diagnostic Sampling Procedures
- Necropsy Procedures
- Preventative Health Programme Overview
- Best Practices for Great Ape Conservation-IUCN/SSC Primate Specialist Group

In addition, the Primate Specialist Group of IUCN has produced a series of guidelines that address critical issues in great ape conservation. Six titles in the series have been published and are available in PDF format:

- Best Practice Guidelines for Great Ape Tourism
- Best Practice Guidelines for Heath Monitoring and Disease Control in Great Ape Populations
- Best Practice Guidelines for Preventing and Mitigating Human-Great Ape Conflict
- Best Practice Guidelines for Surveys and Monitoring of Great Ape Populations
- Best Practice Guidelines for the Re-introduction of Great Apes
- Best Practice Guidelines to Reduce the Impacts of Logging
- Great Apes and FSC

These guidelines are currently being translated into French and Bahasa Indonesia.

Online archives such as PubMed provide useful data. Other websites have information relevant to Part II: A Catalogue of Preserved Materials, in this book. For instance, the online catalogue Surgicat http://surgicat.rcseng.ac.uk/ details all the odontological specimens in the museums of the Royal College of Surgeons of England (see Chapter 2: The Growth of Studies on Primate Pathology).

It is hoped that this book will assist those working in Africa, particularly in the (gorilla) range states for whom access to the Internet is difficult and costly. At the time of writing, there is strong pressure within the European Commission that all companies that publish journals should provide 'open access' for readers in poorer countries to their scientific, peer-reviewed, publications. This is to be welcomed.

REFERENCES

Abee, C.R., Mansfield, K., Tardif, S., Morris, T. (Eds.), 2012. Nonhuman Primates in Biomedical Research, vol. 2: Diseases. Second ed. Academic Press, Elsevier.

Abelin, 1912. Observations sur un squelette de gorille offert à la Société d'Histoire Naturelle de Toulon. Ann. Soc. Hist. Nat. Toulon, 3, 156–177.

Abelló, M.T., Flamme, G., 2006. Monitoring ovulatory cycles in western lowland gorillas *Gorilla gorilla gorilla* at Barcelona Zoo. Int. Zoo Yearb. 39 (1), 185–191.

Abelló, M.T., Bemment, N., Rietkerk, F., 2006. Gorilla EEP Husbandry Guidelines (Revision 2005). EAZA Great Ape Taxon Advisory Group.

Adenusi, A.A., Adewoga, T.O.S., 2013. Studies on the potential and public health importance of non-biting synanthropic flies in the mechanical transmission of human enterohelminths. Trans. R. Soc. Trop. Med. Hyg. 107, 812–818.

Advisory Committee on Dangerous Pathogens Health and Safety Executive, UK. <http://www.hse.gov.uk/biosafety/biologagents.pdf>.

Agerer, R., Ammirati, J., Blanz, P., Courtecuise, R., Desjardin, D.E., Gams, W., et al., 2000. Always deposit vouchers. Mycol. Res. 104, 642–644.

Aguirre, A.A. (Ed.), 2002. Conservation Medicine: Ecological Health in Practice. Oxford University Press, Inc, New York, NY, 432 pp.

Aiello, L.C., Wheeler, P., 1995. The expensive-tissue hypothesis—the brain and the digestive system in human and primate evolution. Curr. Anthropol. 36, 199–221.

Ainsworth, G.C., 1955. Host-parasite relationships. J. Gen. Microbiol. 12, 352–355.

Akeley, C.E., 1920. In Brightest Africa. Garden City Publishing, New York, NY.

Akeley, M.L.J., 1929. Carl Akeley's Africa. The Account of the Akeley-Eastman-Pomeroy Africa Hall Expedition of the American Museum of Natural History. Dodd, Mead & Co, New York, NY.

Akeley, M.L.J., 1941. The Wilderness Lives Again. Carl Akeley and the Great Adventure. Dodd, Mead & Co, New York, NY.

Akeley, M.L.J., 1951. Congo Eden: A Comprehensive Portrayal of the Historical Background and Scientific Aspects of the Great Game Sanctuaries of the Belgian Congo with the Story of a Six Months Pilgrimage throughout the most primitive Region in the Heart of the African Continent. Victor Gollancz Ltd, London.

Akers, J., Schildkraut, D., 1985. Regurgitation/reingestion and coprophagy in captive gorillas. Zoo Biology. 4, 99–109.

Albrecht, G.H., 1982. Collections of nonhuman primate skeletal materials in the United States and Canada. Am. J. Phys. Anthropol. 57, 77–97.

Albrecht, G.H., Gelvin, B.R., Miller, J.M.A., 2003. The hierarchy of intraspecific craniometric variation in gorillas: a population-thinking approach with implications for fossil species recognition studies. In: Taylor, A.B., Goldsmith, M.L. (Eds.), Gorilla Biology: A Multidisciplinary Perspective. Cambridge University Press, New York, NY, pp. 62–103.

Alderton, D., 2011. Animal Grief – How Animals Mourn. Hubble & Hattie, Dorchester, UK.

Alexander, R. Mc.N., 1984. Optimum strength for bones liable to fatigue and accidental damage. J. Theor. Biol. 109, 621–636.

Alix, E., Bouvier, A., 1877. Sur un nouvel anthropoide (*Gorilla mayéma*). Bull. Soc. Zool. France. 2, 488–490.

Allchurch, A.F., 1993. Sudden death and cardiovascular disease in the lowland gorilla. Dodo J. Jersey Wildl. Preserv. Trust. 29, 172–178.

Allchurch, A.F., 2001. Multidisciplinary medical management of zoo animals. Dodo J. Jersey Wildl. Preserv. Trust. 37, 88–95.

Allchurch, A.F., Dutton, C.J., 2000. Suspected toxic epidermal necrolysis or Stevens-Johnson syndrome in a captive lowland gorilla *Gorilla gorilla gorilla*. Dodo J. Jersey Wildl. Preserv. Trust. 36, 96–97.

Allem, A.C., 2000. The terms genetic resource, biological resource, and biodiversity examined. Environmentalist. 20, 3135–341, Kluwer Academic Publications, Doordrecht.

Allen, E., Cooper, J.E., 1983. Zoography: the use of animal terms in medicine. Br. Med. J. 286, 14.

Allender, M.C., McCain, S.L., Ramsay, E.C., Schumacher, J., Ilha, M.R., 2009. Cervical necrotizing fasciitis and myositis in a western lowland gorilla (*Gorilla gorilla gorilla*). J. Med. Primatol. 38 (3), 156–159.

Alley, M., 2000. What is disease? In: Proceedings of the Seminar Wildlife Health in Conservation, 11–13 July 2000. Massey University, New Zealand. Publication No. 204. Department of Conservation/Te Papa Atawhai, New Zealand.

Allmond Jr, B.W., Froeschle, J.E., Guilloud, N.B., 1965. Paralytic poliomyelitis in large laboratory primates: virologic investigation and reports on the use of oral poliomyelitis virus (OPV) vaccine. Am. J. Epidemiol. 85, 229–239.

Altman, R., Gold, G., 2007. Atlas of individual radiographic features in osteoarthritis, revised. Osteoarthr. and Cartil. 15, A1–A56.

Anderson, J., 1881. Catalogue of Mammalia in the Indian Museum, Calcutta. Part 1. Primates, Prosimiae, Chiroptera, and Insectivora. Indian Museum.

Anderson, L., 1996. Veterinary applications of the Swiss army knife. In. Pract. 18 (7), 337–339.

Anderson, M.P., Oosterhuis, J.E., Kennedy, S., Benirschke, K., 1986. Pneumonia and meningoencephalitis due to *Amoeba* in a lowland gorilla. J. Zoo Animal Med. 17 (3), 87–91.

Anderson, U.S., Stoinski, T.S., Bloomsmith, M.A., et al., 2005. Relative numerousness judgment and summation in young and old western lowland gorillas. J. Comp. Psychol. 119 (3), 285–295.

Angst, R., Storch, G., 1967. Bemerkungen über den Schädel des Gorilla Abraham aus dem Frankfurter Zoologischen Garten. Nat. Mus. 97, 417–420.

Anonymous, 1975. Obituary. William Osman-Hill. Nature 253, 667.

Anonymous, 1997. Lessons in self-medication from great apes? Afr. Primates. 3 (1-2), 38–39.

Anonymous, 2000. Ape tourism and human diseases. Gorilla J. 20, 19–21.

Anthony, N.M., Johnson-Bawe, M., Jeffery, K., Clifford, S.L., Abernethy, K.A., Tutin, C.E., et al., 2007. The role of Pleistocene refugia and rivers in shaping gorilla genetic diversity in Central Africa. Proc. Natl. Acad. Sci. U.S.A. 104, 20432–20436.

Antoine, S.E., Child, A.M., Nicholson, R.A., Pollard, A.M., 1992. Biochemistry and microbiology of buried human bone, in relation to dietary reconstruction. Circaea. 9 (2), 65–79.

Antonation, K.S., Grützmacher, K., Dupke, S., Mabon, P., Zimmermann, F., Lankester, F., et al., 2016. *Bacillus cereus* Biovar Anthracis causing anthrax in sub-saharan Africa—chromosomal monophyly and broad geographic distribution. PLoS. Negl. Trop. Dis. 10 (9), e0004923.

Antonius, J.I., Ferrier, S.A., Dillingham, L.A., 1971. Pulmonary embolus and testicular atrophy in a gorilla. Folia Primatol. 15, 277–292.

Anver, M.R., Hunt, R.D., Price, R.A., 1973. Simian Neonatology. II Neonatal pathology. Vet. Pathol. 10, 16–36.

A.P.E.S., undated. Ape Populations Environments Surveys Portal. <http://apesportal.eva.mpg.de/about>.

Arden, N., Nevitt, M., 2006. Osteoarthritis: epidemiology. Best. Pract. Res. Clin. Rheumatol. 20 (1), 3–25.

Arlian, L.G., Bruner, R.H., Stuhlman, R.A., Ahmed, M., Vyszenski-Moher, D.L., 1990. Histopathology in hosts parasitized by *Sarcoptes scabiei*. J. Parasitol. 76 (6), 889–894.

Armstrong, B., 1996. The Columbus Zoo husbandry program. Gorilla Gaz. 10 (2), 8.

Armstrong, B., White, S., 1996. Update on the surrogacy program at the Columbus Zoo. Gorilla Gaz. 10 (1), 10–11.

Armstrong, D., Jakob-Hoff, R., Seal, U.S., 2003. Conservation Breeding Specialist Group (SSC/IUCN). Animal Movements and Disease Risk: A Workbook. CBSG, Austin, Minnesota.

Aryan, H.E., Jandial, R., Nakaji, P., et al., 2006. Lumbar diskectomy in a human-habituated mountain gorilla (Gorilla beringei beringei). Clin. Neurol. Neurosurg. 108 (2), 205–210.

Ashford, R.W., et al., 1990. The intestinal faunas of man and mountain gorillas in a shared habitat. Ann. Trop. Med. Parasitol. 84, 337–340.

Ashford, R.W., et al., 1996. Patterns of intestinal parasitism in the mountain gorilla Gorilla gorilla in the Bwindi Impenetrable forest, Uganda. J. Zool. 239 (3), 507–514.

Ashford, R., Crewe, W., 1998. Parasites of Homo sapiens: An Annotated Checklist of the Protozoa, Helminths and Arthropods for which we are Home. Liverpool School of Tropical Medicine, Liverpool, UK.

Ashton, E.H., Zuckerman, S., 1956a. Age changes in the position of the foramen magnum in hominids. Proc. Zool. Soc. Lond. 126, 315–325.

Ashton, E.H., Zuckerman, S., 1956b. Cranial crests in the Anthropoidea. Proc. Zool. Soc. Lond. 126, 581–634.

Atkinson, J.P., 1996. A remembrance of Fred, the lowland gorilla. Arthritis. Rheum. 39 (6), 891–893.

Atsalis, S., Margulis, S., 2006. Sexual and hormonal cycles in geriatric Gorilla gorilla gorilla. Int. J. Primatol. 27, 1663–1687.

Atsalis, S., Margulis, S., Bellem, A., Wielebnowski, N., 2004. Sexual behavior and hormonal estrus cycles in aged lowland gorillas (Gorilla gorilla). Am. J. Primatol. 62 (2), 123–132.

Attwater, M., 1990. Thoughts on the reintroduction of lowland gorillas. Gorilla Gaz. 4 (2), 13–15.

Attwater, M., 1994a. Congo project update. Gorilla Gaz. 8, 4.

Attwater, M., 1994b. The Congo Gorilla protection project—1993. Gorilla Conserv. News. 8 (1), 12.

Attwater, M., 1999. My Gorilla Journey. Sidgwick and Jackson, London.

Attwater, M., 2001. Challenging developments in primate rehabilitation programs, Africa. Re-Introduction News. 20, 12–13.

Attwater, M., Hudson, H., Blake, S., 1992. Project de Protection de Gorilles, Brazzaville, 1991. Gorilla Conserv. 6, 677.

Auerbach, M., 1913. Kameruner Säugetiere des Grossh. Naturalienkabinetts zu Karlsruhe. I. Teil. Verh. Nat. Ver. Karls. 25, 3–28.

Aufderheide, A.C., 2011. The Scientific Study of Mummies. Cambridge University Press, Cambridge.

Ayoub, T., Chow, J., 2008. The conventional autopsy in modern medicine. J. R. Soc. Med. 101 (4), 177–181.

Bahr, N.I., Pryce, C.R., Döbeli, M., Martin, R.D., 1998. Evidence from urinary cortisol that maternal behavior is related to stress in gorillas. Physiol. Behav. 64 (4), 429–437.

Bahr, N.I., Martin, R.D., Pryce, C.R., 2001. Peripartum sex steroid profiles and endocrine correlates of postpartum maternal behavior in captive gorillas (Gorilla gorilla gorilla). Horm. Behav. 40, 533–541.

Bahuchet, S., 1989. Les Pygmées Aka et Baka (Ph.D. thesis). Université René Descartes Paris 5, Paris.

Bain, O., Molsson, P., Huerre, T.M., et al., 1995. Filariae from a wild gorilla in Gabon with a description of a new species of Mansonella. Parasite 2, 315–322.

Baird, J.K., 1987. False parasites in tissue sections. Parasitol. Today 3 (9), 273–276.

Baitchman, E.J., Goldberger, M.H., Nanna, M. et al, 2003. Evaluation of blood lipid profiles as a part of a complete cardiac assessment in western lowland gorillas (Gorilla gorilla gorilla). In: Proceedings of the American Association Zoo Veterinarians, pp. 72–74.

Baker, B.W., DeHaan, J.D., Hegdahl, D., Chamberlain, J.A., Espinoza, E.O., 2009. Taphonomic effects of burning on ivory: preliminary results. Curr. Res. Pleistocene 26, 129–131.

Baker, R.J., 1994. Some thoughts on conservation, biodiversity, museums, molecular characters, systematics, and basic research. J. Mammal. 75, 277–287.

Baker, L, (Ed). 2002. Guidelines for Nonhuman Primate Re-introductions. Re-introduction Specialist Group of The World Conservation Union Species Survival Commission. In: Edited by Lynne R. Baker Group of the World Conservation Union. 48 pp. Occasional Paper of the IUCN Species Survival Commission No. 36.

Baloch, Z.W., Elsheikh, T.M., Faquin, W.C., Vielh, P. (Eds.), 2014. Head and Neck Cytohistology. Cambridge University Press, Cambridge.

Banish, L.D., Sims, R., Bush, M., Sack, D., Montali, R.J., 1993. Clearance of Shigella flexneri carriers in a zoologic collection of primates. J. Am. Vet. Med. Assoc. 203 (1), 133–136.

Barbot, J., Casseneuve, J., 1732. A voyage to Congo-River. an abstract of a voyage to Congo River, or the Zair, and to Cabinde, in the year 1700, A Collection of Voyages and Travels, Some Now First Printed from Original Manuscripts, Others Now First Published in English, vol. 5. Messrs Churchill, London.

Barks, S.K., Bauernfeind, A.L., Bonar, C.J., Cranfield, M.R., de Sousa, A.A., Erwin, J.M., et al., 2014. Variable temporoinsular cortex neuroanatomy in primates suggests a bottleneck effect in eastern gorillas. J. Comp. Neurol. 522, 844–860.

Barnard, C.J., Behnke, J.M., 1990. Parasitism and Host Behaviour. Taylor & Francis, London.

Barnes, K.M., 2013. Forensic entomology. In: Cooper, J.E., Cooper, M.E. (Eds.), Wildlife Forensic Investigation: Principles and Practice. Taylor & Francis/CRC Press, Boca Raton, FL.

Barnett, C.H., Harrison, R.J., Tomlinson, J.D.W., 1958. Variations in the venous systems of mammals. Biol. Rev. 33 (4), 442–487.

Barns, T.A., 1922. The Wonderland of the Eastern Congo: The Region of the Snow-Crowned Volcanoes, the Pygmies, the Giant Gorilla, and the Okapi. G.P. Putnam's Sons, London.

Barrie, L.D., Backuse, K.A., Grunow, J., Nitschke, R., 1996. Acute lymphocytic leukemia in an infant western lowland gorilla (Gorilla gorilla gorilla). In: Proceedings of the American Association Zoo Veterinary Annual Conference, 1996, p. 404.

Barrie, M.T., Backues, K.A., Grunow, J., Nitschke, R., 1999. Acute lymphocytic leukemia in a six month-old western lowland gorilla. J. Zoo Wildl. Med. 30 (2), 268–272.

Bartels, P., Kotze, A., 2006. Wildlife biomaterial banking in Africa for now and the future. J. Environ. Monit. 8, 779–781.

Bartlett, A.D., 1899. Wild Animals in Captivity. Chapman & Hall, London.

Bartlett, E., 1900. Bartlett's Life Among Wild Beasts in the Zoo. Chapman and Hall, London.

Bartosiewicz, L., 2013. Shuffling Nags, Lame Ducks. The Archaeology of Animal Disease. Oxbow Books, Oxford and Oakville.

Bastawrous, A., Armstrong, M.J., 2013. Mobile health use in low- and high-income countries: an overview of the peer-reviewed literature. J. R. Soc. Med. 106 (4), 130–142.

Baumgärtel, W., 1976. Up Among the Mountain Gorillas. E P Dutton, UK.

Baumgartner, R., Jörger, K., 1988. Funktionelles megakolon bei einem jungen flachlandgorilla (Gorilla g. gorilla) im Zürcher Zoo. (Functional megacolon in a young lowland gorilla in Zurich Zoo). Erkrankungen der Zootiere Verhandlungsbericht des 30 Internatinalen Symposiums über die Erkrankgen der Zoo-und Wildtiere, Sofia 1988. Akademie-Verlag, Berlin.

Beaune, D., 2015. What would happen to the trees and lianas if apes disappeared? Oryx 49 (3), 442–446.

Beck, B., 1984. The birth of a lowland gorilla in captivity. Primates. 25, 378–383.

Beck, B., Walkup, K., Rodrigues, M., Unwin, S., Travis, D., Stoinski, T., 2007. Best Practice Guidelines for the Re-Introduction of Great Apes. IUCN/SSC Primate Specialist Group of the World Conservation Union, Gland, 48 pp.

Beddard, F.E., 1906. Prosector to the Society 62 (June 19, 1906).... By W. Woodland, F.Z.S., Demonstrator of Zoology, King's College, London December 11, 1906. The Secretary. Exhibition of a sketch of a young Gorilla.

Behrensmeyer, A.K., Dechant, D.E., 1980. The recent bones of Amboseli Park, Kenya in relation to East African palaeoecology. In: Behrensmeyer, A.K., Hill, A.P. (Eds.), Fossils in the Making: Vertebrate Taphonomy and Paleoecology. University of Chicago Press, Chicago, IL, pp. 72−93.

Benavides, J.A., Godreuil, S., Bodenham, R., et al., 2012. No evidence for transmission of antibiotic-resistant *Escherichia coli* strains from humans to wild western lowland gorillas in Lope National Park, Gabon. Appl. Environ. Microbiol. 78 (12), 4281−4287.

Benirschke, K., Adams, F.D., 1980. Gorilla diseases and causes of death. J. Reprod. Fertil. (Suppl. 28), 139−148.

Benirschke, K., Adams, F.D., Black, K.L., Gluck, L., 1980. Perinatal mortality in zoo animals. In: Montali, R.J., Migaki, G. (Eds.), Comparative Pathology of Zoo Animals. Smithsonian Institution, Washington, DC.

Bennett, S., 1884. On some recent additions to the museum of the Odontological Society. Trans. Odontol. Soc.187−194.

Bennett, S., 1885. Casual communication (questioner). Trans. Odontol. Soc. 17, 82.

Bennett, S., 1886. On some recent additions to the Museum of the Odontological Society. Trans. Odontol. Soc. 18, 193−194 (and 19: 226).

Berghe, L. van den, Chardome, M., 1949. Une microfilaire du gorille, *Microfilaria gorillae*. Ann. Soc. Belg. Med. Trop. 29, 495−499.

Berghe, L. van den, Chardome, M., Peel, E., 1964. The filarial parasites of the eastern gorilla in the Congo. J. Helminthol. 38 (3−4), 349−368.

Bergl, R.A., Vigilant, L., 2007. Genetic analysis reveals population structure and recent migration within the highly fragmented range of the Cross River gorilla (*Gorilla gorilla diehli*). Mol. Ecol. 16, 501−516.

Bernstein, I., 1969. A comparison of nesting patterns among the three great apes. In: Bourne, G. (Ed.), The Chimpanzee, vol. 1. Karger, Basel/New York, pp. 393−402.

Bernstein, J.J., 1972. An epizootic of hydatid disease in captive apes. J. Zoo Anim. Med. 3, 16−20.

Betunga, W., 1997. Birth of a wild gorilla. Afr. Primates 3 (1−2), 40−41.

Beudels-Jamar, R.C., Lafontaine, R.-M., Devillers, P., Redmond, I., Devos, C., Beudels, M.-O., et al., 2008. Report on the conservation status of Gorillas. CMS Gorilla Concerted Action. CMS Technical Series Publication No 17.

Bhutta, M.F., 2007. Sex and the nose: human pheromonal responses. J. R. Soc. Med. 100, 268−274.

BIAZA, undated. Animal Movement Protocol Work Book. Produced for BIAZA by Matt Hartley and Zoo and Wildlife Solutions. < http://www.biaza.org.uk/uploads/legislation/Animal%20Movements%20Workbook%20(2).pdf > .

BIAZA, 2013. Primates and the Balai Directive EC92/65. < http://www.biaza.org.uk/uploads/legislation/balai/BIAZA_BalaiPrimate_InfoNote%202015.pdf > .

BIAZA, 2014. Guidance on Disease Risk Protocols for New Stock. British and Irish Association of Zoos and Aquariums, London, < http://www.biaza.org.uk/uploads/governance/Guidance%20on%20Disease%20Risk%20Protocols%20for%20New%20Stock.pdf > .

Bieniasz, P.D., Rethwin, A., Pitman, R., et al., 1995. A comparative study of higher primate foamy viruses, including a new virus from a gorilla. Virology 207, 217−228.

Bingham, L., Hahn, T., 1974. Observations on the birth of a lowland gorilla in captivity. Int. Zoo Yearb. 14, 113−115.

Birch, I., Vernon, W., Walker, J., Young, M., 2015. Terminology and forensic gait analysis. Sci. Just. 55 (4), 279−284.

Birnie, P., Boyle, A., Redgwell, C., 2009. International Law and the Environment. Oxford University Press, Oxford.

Bisschop, deD., 1956. Un cas de saturnisme chez un gorile. Soc. R. Zool. d'Anvers. An unpublished communication to the International Union of Zoological Societies, Directors, Chicago, USA. Referred to by Zook (1971).

Bishop, I., Hosey, G., Plowman, A. (Eds.), 2013. Handbook of Zoo Research Guidelines for Conducting Research in Zoos. British and Irish Association of Zoos and Aquariums, London.

Bittle, J.M., 1993. Use of vaccines in exotic animals. J. Zoo Wildl. Med. 24 (3), 352−356.

Blainville, H. M. D. de, 1839−1864. Ostéographie ou description iconographique comparée du squelette et du système dentaire des mammifères récents et fossiles pour servir de base à la zoologie et à la geologie. Atlas Vol. 1 (1849) Text Vol. 4 (1855). J. B. Baillière et Fils, Paris.

Blair, A.M., 2010. Too Much to Know: Managing Scholarly Information Before the Modern Age. Yale University Press, New Haven, CT.

Blancou, L., 1951. Notes sur les mammifères de l'Equateur africain français. Le gorille. Mammalia 15, 143−151.

Blaney, S.P., 1986. An allometric study of the frontal sinus in *Gorilla, Pan* and *Pongo*. Folia. Primatol. (Basel). 47 (2-3), 81−96.

Blaney, S., Johnson, B., 1989. Techniques for reconstituting fixed cadaveric tissue. Anat. Rec. 224, 550−551.

Blann, A., Ahmed, N., 2014. Blood Sciences: Principles and pathology. Wiley-Blackwell, Oxford.

Blasier, M., 2001. The risk of measles and the efficacy of the measles vaccine in gorillas. Report of Dr Scholl Summer Fellow 2001. Lincoln Park Zoo, IL.

Blasier, M.W., Travis, D.A., 2005. Retrospective evaluation of measles antibody titers in vaccinated captive gorillas (*Gorilla gorilla gorilla*). J. Zoo Wildl. Med. 36 (2), 198−203.

Blom, A., Cipolletta, C., Brunsting, A.M.H., Prins, H.H.T., 2004. Behavioral responses of gorillas to habituation in the Dzanga-Ndoki National Park, Central African Republic. Int. J. Primatol. 25 (1), 179−196.

Blomqvist, J. (Ed.), 1979. The Date and Origin of the Greek Version of Hanno's periplus: with an Edition of the Text and a Translation. Scripta Minora Regiae Societatis Humaniorum Litterarum Lundensis. Liber Läromede/Gleerup, Lund.

Bloomsmith, M., 1989. Feeding enrichment for captive great apes. In: Segal, E. (Ed.), Housing, Care and Psychological Wellbeing of Captive and Laboratory Primates. Noyes Publications, Park Ridge, NJ, pp. 336−356.

Blumenthal, S.A., Chritz, K.L., Rothman, J.M., Cerling, T.E., 2012. Detecting intra-annual dietary variability in wild mountain gorillas by stable isotope analysis of feces. Proc. Natl. Acad. Sci. U.S.A. 109, 21277−21282.

Bodrey-Sanders, P., 1991. Carl Akeley: Africa's Collector, Africa's Savior. Paragon House, New York, NY.

Boekhorst, I., van Oorschot, I., 1990. Social structure of a group of captive lowland gorillas (*Gorilla gorilla gorilla*). Acta. Zool. Pathol. Antverp. 81, 17−30.

Böer, M., 1983. Several examinations on the reproductive status of lowland gorillas (*Gorilla g. gorilla*) at Hannover Zoo. Zoo Biol. 2 (4), 267−280.

Bogusz, M.J., 2011. Quality Assurance in the Pathology Laboratory: Forensic, Technical And Ethical Aspects. CRC Press (Taylor & Francis), Boca Raton, FL.

Bollen, K., Novak, M., 2000. A survey of abnormal behaviors in captive zoo primates. Abstract. Am. J. Primatol. (Suppl. 51), 47.

Bolter, D.R., Zihlman, A.L., 2002. Growth and development in body tissues and proportions in African apes (*Gorilla gorilla* and *Pan troglodytes*): a preliminary report. Am. J. Phys. Anthropol. Suppl. 34, 46.

Bolton, R.L., Masters, N.J., Milham, P., Lea, R.G., 2012. Environment and reproductive dysfunction in captive female great apes (Hominidae). Vet. Rec. 170, 676.

Bond, A.L., 2011. Why ornithologists should embrace and contribute to Wikipedia. Ibis. 153, 640−641.

Bookhout, T.A. (Ed.), 1996. Research and Management Techniques for Wildlife and Habitats. Fifth ed. revised. The Wildlife Society, Bethesda, MD (Chapters 4, 5, 6, 11, 12, 13, 15).

Booth, T.J., 2015. An investigation into the relationship between funerary treatment and bacterial bioerosion in European archaeological human bone. Archaeometry. Available from: http://dx.doi.org/10.1111/arcm.12190.

Bornstein, S., 1995. *Sarcoptes scabiei* Infections of the Domestic Dog, Red Fox and Pig (Dissertation). University of Uppsala, Sweden.

Bornstein, S., Wallgran, P., 1997. Serodiagnosis of sarcoptic mange in pigs. Vet. Rec. 141, 8–12.

Borriello, S.P., Murray, P.R., Funke, G., 2005. Topley and Wilson's Microbiology and Microbial Infections, (tenth edition) Bacteriology, vol. I. Hodder Arnold, London.

Boundenga, L., Ollomo, B., Rougeron, V., Mouele, L.Y., Mve-Ondo, B., Delicat-Loembet, L.M., et al., 2015. Diversity of malaria parasites in great apes in Gabon. Malar. J. 14, 111.

Bourne, D.C., 2009. Monkeypox in Great Apes. <http://wildpro.twycrosszoo.org/S/00dis/viral/Monkeypox.htm> (accessed 13 February 2016).

Bowditch, T.E., 1819. Mission from Cape Coast Castle to Ashantee, with a Statistical Account of that Kingdom, and Geographical Notices of Other Parts of the Interior of Africa. John Murray, London.

Bowen, R., 1980. The behaviour of three hand-reared gorillas, *Gorilla g. gorilla*, with emphasis on the response to a change in accommodation. Dodo J. Jersey Wildl. Preserv. Trust. 17, 63–79.

Bowen, R., 1981. Social integration in lowland gorillas (*Gorilla gorilla gorilla*) at the Jersey Wildlife Preservation Trust. Dodo J. Jersey Wildl. Preserv. Trust. 18, 51–59.

Boylston, A., 2014. John Haygarth's 18th-century 'rules of prevention' for eradicating smallpox. J. R. Soc. Med. 107 (12), 494–499.

Brack, M., 1987. Agents Transmissible from Simians to Man. Springer Verlag, Berlin, Germany.

Brack, M., 1993. Virus infections in nonhuman primates: survey. Ver. Int. Symp. Erkankungen Zoo. 35, 5–38.

Brack, M., Göltenboth, R., Rietschel, W., 1995. Diseases of zoo and wild animals. Primates. In: Göltenboth, R. (Ed.), Krankheiten Der Zoo-Und Wildtiere. Blackwell Wissenschafts Verlag, Berlin, pp. 25–66.

Bradford, C.M., Tupa, L., Wiese, D., Hurley, T.J., Zimmerman, R., 2013. Unusual Turner syndrome mosaic with a triple x cell line (47,X/49, XXX) in a western lowland gorilla (*Gorilla gorilla gorilla*). J. Zoo Wildl. Med. 44 (4), 1055–1058.

Bradley, R.D., Bradley, L.C., Garner, H.J., Baker, R.J., 2014. Assessing the value of natural history collections and addressing issues regarding long-term growth and care. Bioscience. 64, 1150–1158.

Brambell, M.R., Mathews, S.J., 1976. Primates and carnivores at Regent's Park. Symp. Zool. Soc. London. 40, 147–165. Academic Press, London.

Brand, R.A., 2009. 50 years ago in CORR: osteoarthritis of the hip in a gorilla: report of a third case. Robert M. Stecher CORR 1958.12, 307–314. Clin. Orthop. Relat. Res. 467 (1), 305–307.

Brandt, H.P., 1972. A phalangeal osteolysis syndrome in a male gorilla (*Gorilla g. gorilla*). In: Ippen, R., Schroder, H.D. (Eds.), Proc. 14th Int. Symp. Erkrankungen Zootiere, Wroclaw. Akademic Verlag, Berlin, pp. 299–300.

Brandwood, A., Jayes, A.S., Alexander, R.M., 1986. Incidence of healed fractures in the skeletons of birds, molluscs and primates. J. Zool. Lond. 208, 55–62.

Brearley, M.J., Cooper, J.E., Wedderburn, N., 1986. Urinary cytology as an adjunct to bacteriology in the laboratory investigation of canine bladder disease. J. Small Anim. Pract. 27 (7), 425–432.

Breuer, T., Ndoundou-Hockemba, M., Fishlock, V., 2005. First observation of tool use in wild gorillas. PLoS Biol. 3 (11), e380.

Breuer, T., Breuer-Ndoundou Hockemba, M., Olejniczak, C., Parnell, R.J., Stokes, E.J., 2009. Physical maturation, life-history classes and age estimates of free-ranging western gorillas—insights from Mbeli Bai, Republic of Congo. Am. J. Primatol. 71, 106–119.

Breuer, T., Robbins, A.M., Olejniczak, C., Parnell, R.J., Stokes, E.J., Robbins, M.M., 2010. Variance in the male reproductive success of western gorillas: acquiring females is just the beginning. Behav. Ecol. Sociobiol. 64, 515.

Bridson, T.L., Govan, B.L., Norton, R.E., Schofield, L., Ketheesan, N., 2014. The double burden; a new-age pandemic meets an ancient infection. Trans. R. Soc. Trop. Med. Hyg. 108, 676–678.

Brinkworth, J.F., Pechenkina, K. (Eds.), 2013. Primates, Pathogens, and Evolution. Springer, New York, NY.

Bristol Museums & Art Gallery, 1999. Alfred the gorilla: the dictator of Bristol. Bristol City Council. Archives, BZG.

British Museum, 1862. Catalogue of the Bones of Mammalia in the Collection of the British Museum. Dept. of Zoology [compiled by Edward Gerrard], London.

British Veterinary Association (undated). Veterinary work overseas. <http://www.bva.co.uk/Professional-development/Careers/Working-overseas/>.

Broom, D.M., 1996. Animal welfare defined in terms of attempts to cope with the environment. Acta Agric. Scand. A. Anim. Sci.(27), 22–28.

Broom, D.M., Johnson, K.G., 1993. Stress and Animal Welfare. Chapman & Hall, London.

Brown, C., 1998. A training program for noninvasive semen collection in captive western lowland gorilla (*Gorilla gorilla gorilla*). Zoo Biol. 17, 143–151.

Brown, L.E., 2012. *Pseudacris streckeri illinoensis* (Illinois chorus frog): large wound in holotype. Herpetol. Bull. 121, 43–44.

Brown, D.L., Gold, K.C., 1997. Effects of straw bedding on non-social and abnormal behavior of captive lowland gorillas (*Gorilla gorilla gorilla*). In: Holst, B. (Ed.), Proceedings of the 2nd International Conference on Environmental Enrichment. Copenhagen Zoo, Frederiksberg, pp. 27–35.

Brown, S., Wagster, M., 1986. Socialization processes in a female lowland gorilla. Zoo Biol. 5, 269–279.

Brown, T.M., Clark, H.W., Bailey, J.S., 1980. Rheumatoid arthritis in the gorilla: a study of mycoplasma-host interaction in pathogenesis and treatment. In: Montali, R.J., Migaki, G. (Eds.), The Comparative Pathology of Zoo Animals. Smithsonian Institution Press, Washington, DC, pp. 259–266.

Brown, S., Dunlap, W., Maple, T., 1982. Notes on water-contact by a captive male lowland gorilla. Zoo Biol. 1, 243–249.

Browne, T., 1690. Letter to a Friend.

Bruner, E., Gippoliti, S. (Eds.), 2006. Le collezione primatologiche italiane. Istituto di Antropologia, Roma.

Bruner, G., 1989. From bars to boughs: design and construction of Zoo Atlanta's great ape exhibits. In: AAZPA Regional Conference Proceedings.

Bunton, T., 1986. Incidental lesions in nonhuman primate placentae. Vet. Pathol. 23 (4), 431–438.

Burbridge, B., 1928. Gorilla: Tracking and Capturing the Ape-Man of Africa. The Century Company, New York, NY.

Burbridge, J., 2013. Blood samples from Mt Tshiaberimu gorillas. Gorilla J. 46, 3–4.

Burks, K., 2000. Bachelor gorilla introductions: using empirical data in decision making. Paper presented at "The Apes: Challenges for the 21st Century." Brookfield Zoo, Chicago, IL.

Burks, K., Maple, T., 1995. Isolate adult male gorilla (*Gorilla gorilla gorilla*) socialization: scientific management of aggression. In: AZA Annual Conference Proceedings, pp. 62–67.

Burns, R., 2006. Hemodialysis of a western lowland gorilla (*Gorilla gorilla gorilla*) with fatal septicemia and pyelonephritis secondary to urine stasis and uterine leiomyosarcoma. In: Proceedings of the American Association Zoo Veterinarians, pp. 154–155.

Burns, R.P., Barrie, M.T., 2002. Nesidioblastosis in a western lowland gorilla (*Gorilla gorilla gorilla*). In: Proceedings of the American Association Zoo Veterinarians, p. 140.

Burton, R.F., 1876. Two Trips to Gorilla Land and the Cataracts of the Congo. Sampson Low, London.

Butynski, T.M., Kingdon, J., Kalina, J., 2013. Mammals of Africa; an Introduction and Guide. vol. II. Primates. Bloomsbury, London.

Byers, A.C., Hastings, B., 1991. Mountain gorilla mortality and climatic factors in the Parc National des Volcans, Ruhengeri Prefecture, Rwanda, 1988. Possible contributing climatic factors. Mt. Res. Dev. 11 (2), 145–151.

Byrne, R., 1996. The misunderstood ape: cognitive skills of the gorilla. In: Russon, A., Bard, K., Parker, S. (Eds.), Reaching Into Thought: The Minds of the Great Apes. University of Cambridge, Cambridge, pp. 111–130.

Byrne, R.W., Stokes, E., 2002. Effects of manual disability on feeding skills in gorillas and chimpanzees. Int. J. Primatol. 23, 539–554.

Caillaud, D., Ndagijimana, F., Giarrusso, A.J., Vecellio, V., Stoinski, T.S., 2014. Mountain gorilla ranging patterns. Am. J. Primatol. 76 (8), 730–746.

Caldecott, J., Miles, L. (Eds.), 2005. World Atlas of Great Apes and their Conservation. University of California Press, Berkeley and Los Angeles, CA.

Callanan, J.J., 2012. Pathology, a path well worn. J. Small Anim. Pract. 53 (12), 671–672.

Campbell, F., Verbeke, C.S., 2013. Pathology of the Pancreas. Springer, New York, NY.

Cancelliere, E., DeAngelis, N., Raubenheimer, D., Nkurunungi, J.B., Rothman, J.M., 2014. Mineral nutrients in the foods eaten by mountain gorillas (Gorilla beringei). PLoS ONE ie112117.

Canfield, D., Brignolo, L., Peterson, P.E., Hendrickx, A.G., 2000. Conjoined twins in a rhesus monkey (Macaca mulatta). J. Med. Primatol. 29, 427–430.

Carder, G., Semple, S., 2008. Visitor effects on anxiety in two captive groups of western lowland gorillas. Appl. Anim. Behav. Sci. 115 (3–4), 211–220.

Caro, T.M., 1976. Observations of the ranging behaviour and daily activity of lone silverback mountain gorillas (Gorilla gorilla beringei). Anim. Behav. 24, 889–897.

Carter, F.S., 1974. Treatment of acute dehydration in a 283 day old lowland gorilla, Gorilla g. gorilla at Jersey Zoological Park. In: JWPT 11th Annual Report, pp. 60–61.

Casas-Marce, M., Revilla, E., Fernandes, M., Rodríguez, A., Delibes, M., Godoy, J.A., 2012. The value of hidden scientific resources: preserved animal specimens from private collections and small museums. Bioscience. 62, 1077–1082.

Caswell, J.L., Callanan, J.J., 2014. The intriguing pathology of infectious diseases. Vet. Pathol. 51 (2), 313–314.

Cato, P.S., 1991. The value of natural history collections in Latin American conservation. In: Mares, M.A., Schmidly, D.J. (Eds.), Latin American Mammalogy: History, Biodiversity, and Conservation. University of Oklahoma Press, Norman, pp. 416–430.

Cattley, R.C., 2012. Autopsy. Vet. Pathol. 49 (5), 879.

Cave, A.J.E., 1959. The nasal fossa of a foetal gorilla. Proc. Zool. Soc. Lond. 133, 73–77.

Cave, A.J.E., 1961. The frontal sinus of the gorilla. Proc. Zool. Soc. Lond. 136, 359–374.

Cave, A.J.E., 1994. Note on the venous drainage of the gorilla (Gorilla gorilla) diploë. J. Zool. Lond. 233, 37–43.

Cave, A.J.E., Jones, T.S., 1992. Canine tooth fracture in two Congolese gorillas. J. Zool. 227 (4), 685–690.

Cerveny, S., Sleeman, J.M., 2014. Great ape anesthesia. In: West, G., Heard, D., Caulkett, N. (Eds.), Zoo Animal and Wildlife Immobilization and Anesthesia, second ed. Wiley Blackwell.

Cerveny, S.N., D'Agostino, J.J., Davis, M.R., et al., 2012. Comparison of laryngeal mask airway use with endotracheal intubation during anesthesia of western lowland gorillas (Gorilla gorilla gorilla). J. Zoo Wildl. Med. 43 (4), 759–767.

Chamanza, R., Parry, N.M.A., Rogerson, P., Nicol, J.R., Bradley, A.E., 2006. Spontaneous lesions of the cardiovascular system in purpose-bred laboratory nonhuman primates. Toxicol. Pathol. 34, 357–363.

Chang, S., Moatamed, N.A., Christina, K.Y., Salami, N., Apple, S.K., 2014. The sheep in wolf's clothing: vegetable and fruit particles mimicking cells and microorganisms in cytology specimens. J. Cytol. Histol. 5 (1), 1–11.

Chapman, H.C., 1892. Observations upon the brain of the gorilla. Proc. Acad. Nat. Sci. Philadelphia. 44, 203–212.

Chatfield, J., 1990a. Notes on the introduction of an aggressive male gorilla at the Los Angeles Zoo. In: Columbus Zoo Gorilla Workshop Proceedings, pp. 2–4.

Chatfield, J., 1990b. Use of an experienced female as a role model in a non-reproductive group of gorillas. In: Columbus Zoo Gorilla Workshop Proceedings, pp. 5–7.

Chatfield, J., Zhang, L., Ramey, J., et al., 2006. Resolution of a hyperprolactinemia in a western lowland gorilla (Gorilla gorilla gorilla). J. Zoo Wildl. Med. 37 (4), 565–566.

Chatfield, J., Stones, G., Jalil, T., 2010. Severe idiopathic hypocalcemia in a juvenile western lowland gorilla, Gorilla gorilla gorilla. J. Zoo Wildl. Med. 43 (1), 171–173.

Chaubaud, A.G., Rousselot, R., 1956. Un nouveau spiruridae du gorille Chitwoodspirura gorille Chitwoodspirura wehri. Bull. Soc. Pathol. Exot.467–472.

Chelluri, G.I., Ross, S.R., Wagner, K.E., 2012. Behavioral correlates and welfare implications of informal interactions between caretakers and zoo-housed chimpanzees and gorillas. Appl. Anim. Behav. Sci. 147, 306–315.

Cipolletta, C., 2003. Ranging patterns of a western gorilla group (Gorilla gorilla gorilla) during habituation to humans in the Dzanga-Ndoki National Park, Central African Republic. Int. J. Primatol. 24 (6), 1207–1226.

Cipolletta, C., 2004. Effects of group dynamics and diet on the ranging patterns of a western gorilla group (Gorilla gorilla gorilla) at Bai Hokou, Central African Republic. Am. J. Primatol. 64, 193–205.

Cipolletta, C., Spagnoletti, N., Todd, A.F., Robbins, M.M., Cohen, H., Pacyna, S., 2007. Termite feeding by western lowland gorillas (Gorilla gorilla gorilla) at Bai Hokou, Central African Republic. Int. J. Primatol. 28 (2), 457–476.

Cipriano, A., 2002. Cold stress in captive great apes recorded in incremental lines of dental cementum. Folia Primatol. 73, 21–31.

Cirelli, M.T., 2002. Legal Trends in Wildlife Management. FAO Legislative Study, 74. Development Law Service FAO Legal Office. Food and Agriculture Organization of the United Nations, Rome, <http://www.fao.org/docrep/005/y3844e/y3844e00.htm#Contents>.

CITES, undated a. National laws for implementing the Convention. <https://cites.org/eng/legislation>.

CITES, undated b. Resolution Conf. 9.6 (Rev. CoP16) Trade in readily recognizable parts and derivatives. <https://cites.org/eng/res/09/09-06R16.php>.

CITES, undated c. Resolution Conf. 12.3 (Rev. CoP16). Permits and certificates. <https://www.cites.org/eng/res/12/12-03R16.php>.

CITES, 2013. CITES Guidelines for non-air transport. <https://cites.org/eng/resources/transport/index.php>.

CITES and WOAH/OIE, 2015. Cooperation Agreement between CITES Secretariat and WOAH(OIE) for Dec 2015. <https://cites.org/sites/default/files/eng/disc/sec/Cooperation_Agreement_CITES_and_OIE_dec_15.pdf>.

Clapp, N.K., McArthur, A.H., Carson, R.L., Henke, M.A., Peck, O.C., Wood, J.D., 1987. Visualization and biopsy of the colon in tamarins and marmosets by endoscopy. Lab. Anim. Sci.217–219, April, 1987.

Clark, F.E., Fitzpatrick, M., Hartley, A., King, A.J., Lee, T., Routh, A., et al., 2012. Relationship between behavior, adrenal activity, and environment in zoo-housed western lowland gorillas (Gorilla gorilla gorilla). Zoo Biol. 31 (3), 306–321.

Clarke, A., Juno, C., Maple, T., 1982. Behavioral effects of a change in the physical environment: a pilot study of captive chimpanzees. Zoo Biol. 1, 371–380.

Clay, P., Maurer, C., Elia, R., Sass, P., Meyers, E., 1989. 72 hour birth protocol at Lincoln Park. Gorilla Gaz. 3 (1), 1–3.

Clift, J., Martin, R., 1978. Monitoring of pregnancy and postnatal behavior in a lowland gorilla. Int. Zoo Yearb. 18, 165–173.

Clouser, S., 2010. Cataract surgery with intraocular lens implantation on the captive western lowland gorilla. Insight. 35 (1), 13–16.

Clutterbuck, A., Woods, E.J., Knottenbelt, D.C., Clegg, P.D., Cochrane, C.A., Percival, S.L., 2007. Biofilms and their relevance to veterinary medicine. Vet. Microbiol. 121, 1–17.

Cobb, W.M., 1933. Human materials in American institutions available for anthropological study. Am. J. Phys. Anthropol. 17 (4), Supplement: iv+45pp.

Codner, M., Nadler, R., 1984. Mother-infant separation and reunion in the great apes. Primates 25, 204–217.

Coe, J., 1983. A greensward for gorillas: adventures in zoo horticulture. In: AAZPA Annual Conference Proceedings, pp. 177−185.

Coe, J., 1985. Design and perception: making the zoo experience real. Zoo Biol. 4, 197−208.

Coe, J., 1992. Advances in facility design for great apes in zoological gardens. In: Erwin, J., Landon, J. (Eds.), Chimpanzee Conservation and Public Health: Environments for the Future. Diagnon/Bioqual, Inc, Rockville, MD, pp. 103−111.

Coe, J., 1999. Increasing affiliative behavior between zoo animals and zoo visitors. In: AZA Annual Conference Proceedings, pp. 216−220.

Coe, J., Lee, G., 1996. One hundred years of evolution in great ape facilities in American zoos. In: AZA Regional Conference Proceedings, pp. 489-498.

Coe, J., Maple, T., 1984. Approaching Eden: a behavioral basis for great ape exhibits. In: AAZPA Annual Conference Proceedings, pp. 117−128.

Coetzee, M., 2014. How important are Dipteran vectors of disease in Africa? Trans. R. Soc. Trop. Med. Hyg. 108, 179−180.

Coffey, P., Pook, 1974. Breeding, hand-rearing and development of the third lowland gorilla, *Gorilla g. gorilla* at the Jersey Zoological Park. In: JWPT 11th Annual Report, pp. 45−52.

Cole, B.C., Ward, J.R., Golightly−Rowland, L., Graham, C.E., 1970. Characterization of mycoplasmas isolated from the great apes. Can. J. Microbiol. 16, 1331−1339.

Cole, M., 1949. A History of Comparative Anatomy from Aristotle to the Eighteenth Century. Macmillan & Co, London.

Cole, M., 1990. The effects of separation from and subsequent reintroduction to an established group on a primiparous female lowland gorilla. In: Columbus Zoo Gorilla Workshop Proceedings, pp. 16-26.

Cole, M., Ervine, L., 1983. Maternal behavior and infant development of the lowland gorillas at Metro Toronto Zoo. AAZK Forum. 10 (2), 387−391.

Cole, R.E., 2000. Obstetric management of a protracted labor in a captive western lowland gorilla. Am. J. Obstet. Gynecol. 182 (6), 1306−1311.

Colyer, J.F., 1915. Injuries of the jaws and teeth in animals. Dent. Rec. 35, 4−6.

Colyer, J.F., 1931. Abnormal Conditions of the Teeth of Animals in Their Relationship to Similar Conditions in Man. Dental Board of the UK, London.

Colyer, J.F., 1936. Variations and Diseases of the Teeth of Animals. John Bale, Sons and Danielsson, Ltd., London.

Conn, P.M. (Ed.), 2013. Animal Models for the Study of Human Disease. Academic Press, London.

Conniff, R., 2008. The missionary and the gorilla. A nineteenth-century tale of disease, perseverance, scientific infighting, and a landmark of natural history. Yale Alumni Mag. 72 (1), New Haven.

Connolly, C.J., 1933. The brain of a mountain gorilla, Okero (*G. beringei*). Am. J. Phys. Anthropol. 17, 291−307.

Cook, J.A., Edwards, S.V., Lacey, E.A., Guralnick, R.P., Soltis, P.S., Soltis, D.E., et al., 2014. Natural history collections as emerging resources for innovative education. Bioscience. 64, 725−734.

Cook, R.A., Clarke, D.A., 1985. The use of isoflurane as a general anesthetic in the western lowland gorilla (*Gorilla gorilla gorilla*). J. Zoo Anim. Med. 16, 122−124.

Cook, R.J., Calle, P., Mangold, B. et al., 1995. Choriocarcinoma in a young adult gorilla (*Gorilla g. gorilla*): diagnosis, treatment and outcome. In: Proc. Joint Conf. Am. Assoc. Zoo. Vet./Wildlife Dis. Assoc./Am. Assoc. Wildl. Vet., pp. 329−330.

Coolidge, H.J., 1929. A revision of the genus *Gorilla*. Mem. Mus. Comp. Zool., Harv. 50, 291−381.

Coolidge, H.J., 1936. Zoological results of the George Vanderbilt African Expedition of 1934. Part IV, Notes on four gorillas from the Sanga River region. Proc. Acad. Nat. Sci. Philadel. 88, 479−501.

Cooper, J.E., 1985. Medico-veterinary collaboration. A review of its importance and relevance, especially in the tropics. Trop. Doct. 15, 187−191.

Cooper, J.E., 1989. Care, cure or conservation: developments and dilemmas in wildlife conservation. In: Harris, S., Thomas, T. (Eds.), Proceedings of the Inaugural Symposium of the British Wildlife Rehabilitation Council, 19 November 1988, London. British Wildlife Rehabilitation Council, c/o RSPCA, Horsham, UK, pp. 14−23.

Cooper, J.E., 1994. Biopsy techniques. Semin. Avian Exot. Pet Med. 3 (3), 161−165.

Cooper, J.E., 1995a. Are we guilty of neocolonism in our work on wildlife diseases? Wildl. Dis. Newsl. (Supplement to the Journal of Wildlife Diseases). 31 (4), October 1995.

Cooper, J.E., 1995b. Permit problems. New Sci. 149, 53.

Cooper, J.E., 1995c. The role of cytology in the diagnosis and investigation of disease. In: Zwart, P., Matz, G. (Eds.), 5th International Colloquium on the Pathology of Reptiles and Amphibians, Alphenaan den Rijn, Netherlands.

Cooper, J.E., 1996a. Protocol for collection and examination of faecal samples from the mountain gorilla (*Gorilla gorilla beringei*). Afr. Primates. 2, 24−25.

Cooper, J.E., 1996b. Physical injury. In: Fairbrother, A., Locke, L.N., Hoff, G.L. (Eds.), Non-infectious Diseases of Wildlife, second ed. Iowa State University Press, Ames, IO.

Cooper, J.E., 1996c. Parasites and pathogens of non-human primates. Vet. Rec. 139, 48.

Cooper, J.E., 1997a. Gorillas in the Midst of War, Challenges to Conservation in Central Africa. Salim Ali Memorial Lecture. Centre of Wildlife and Ornithology, Aligarh Muslim University, India, February 1996.

Cooper, J.E., 1997b. Medical terminology—mountain gorillas. Newsl. World Assoc. Wildl. Vet. (WAWV). 28, 1−2.

Cooper, J.E., 1997c. Skin diseases of gorillas. Proceedings of the Autumn Meeting of the BVDSG (British Veterinary Dermatological Study Group) October 1997. Stratford-upon-Avon, England, UK, pp. 41−48.

Cooper, J.E., 1998. Minimally invasive health-monitoring of wildlife. Anim. Welfare. 7, 35−44.

Cooper, J.E., 2002. Diagnostic pathology of selected diseases in wildlife. Sci. Tech. 21 (1), 77−89.

Cooper, J.E., 2003a. Pathology of exotic species. J. Caribbean Vet. Med. Assoc. 3 (1), 7−12.

Cooper, J.E., 2003b. Principles of clinical pathology and post-mortem examinations. In: Mullineaux, E., Best, D., Cooper, J.E. (Eds.), Manual of Wildlife Casualties. British Small Animal Veterinary Association, Gloucester, UK.

Cooper, J.E., 2003c. Jiggers and sticktights: can sessile fleas help us in our understanding of host:parasite responses? In: Perez Jiminez, J.M. (Ed.), Memoriam of Prof. Dr. Isidoro Ruiz Martinez. University of Jaen, Spain.

Cooper, J.E., 2004a. Searching the literature. Vet. Rec. 155, 375.

Cooper, J.E., 2004b. The need for portable field kits in improved veterinary diagnosis in the Neotropics. Proceedings of Third Seminar on Appropriate Medical Technology for Developing Countries. IEE, London, UK.

Cooper, J.E., 2004c. The need for portable field kits in improved veterinary diagnosis in the Neotropics. Newsl. World Assoc. Wildl. Vet. 14 (52), 1−3.

Cooper, J.E., 2005. The microbiology of animal bites. Bull. R. Coll. Pathol. 130, 67.

Cooper, J.E., 2009a. In-practice and field techniques for the investigation of parasitic infections. J. Exot. Pet Med. 18 (4), 280−290.

Cooper, J.E., 2009b. Amyloid, apples and analogies. Bull. R. Coll. Pathol. 146, 176−177.

Cooper, J.E., 2010a. References, reliability and readership (editorial). J. Small Anim. Pract. 51, 359−360.

Cooper, J.E., 2010b. Heat stress, climate change and animal welfare. Vet. Rec. 166, 729.

Cooper, J.E., 2010c. Doctor! The monkeys are dying.... Bull. R. Coll. Pathol. 149, 13−15.

Cooper, J.E., 2011. Organ and tissue responses in animals. Symposium on the Pathology of Abuse in Animals and Humans. The Royal College of Pathologists, London.

Cooper, J.E., 2012. What is comparative pathology? Bull. R. Coll. Pathol. 160, 295–296.

Cooper, J.E., 2013a. What is fieldwork? In Field techniques in exotic animal medicine. J. Exot. Pet Med. 22 (1), 7–16.

Cooper, J.E., 2013b. Field techniques in exotic animal medicine. J. Exotic Pet Med. 22 (1).

Cooper, J.E., 2015. Ebola, primates and people. Vet. Rec.152, 7 February.

Cooper, J.E., Anwar, M.A., 2001. Blood parasites of birds: a plea for more cautious terminology. Ibis. 143, 149–150.

Cooper, J.E., Baskerville, A., 1976. An outbreak of epistaxis in cynomolgus monkeys (*Macaca fascicularis*). Vet. Rec. 99, 438–439.

Cooper, J.E., Cooper, M.E., 1994. A trip to the mountain gorillas: a naturalist's perspective. Bull. E. Afr. Nat. Hist. Soc. 24 (2), 3l–32, June 1994.

Cooper, J.E., Cooper, M.E., 1996. Mountain gorillas—a 1995 update. Afr. Primates. 2 (1), 30–31.

Cooper, J.E., Cooper, M.E., 2001a. Rueben Rwanzagire, Ugandan gorilla tracker, is remembered. Afr. Primates. 5 (1&2), 56–57.

Cooper, J.E., Cooper, M.E., 2001b. Legal and ethical aspects of working with wildlife, with particular reference to Africa. ANZCCART News. 14 (4), 4–7.

Cooper, J.E., Cooper, M.E., 2004. African veterinarians and conservation work. Vet. Times. 39, 15 March 2004.

Cooper, J.E., Cooper, M.E., 2006. Ethical and legal implications of treating casualty wild animals. In. Pract. 28, 2–6.

Cooper, J.E., Cooper, M.E., 2007. Introduction to Veterinary and Comparative Forensic Medicine. Blackwell, Oxford.

Cooper, J.E., Cooper, M.E., 2013. Wildlife Forensic Investigation: Principles and Practice. Taylor & Francis/CRC Press, Boca Raton, FL.

Cooper, J.E., Cooper, M.E., 2015. The increasing importance of laboratory tests. In: Proceedings of the Manchester Veterinary Conference (VetsNorth 2015), Manchester, UK, 25–26 June 2015.

Cooper, J.E., Holt, P.E., 1975. Veterinary work at a primate centre in Kenya. Trop. Anim. Health. Prod. 7, 219–228.

Cooper, J.E., Hull, G., 2015. Studies on the pathology and health of gorillas. Vet. Rec. 177 (14), 374.

Cooper, J.E., Jones, C.G., 1986. A reference collection of endangered Mascarene specimens. Linnean 2 (3), 32–37.

Cooper, J.E., Samour, J.H., 1997. Portable and field equipment for avian veterinary work. In: Proceedings of European Committee of the Association of Avian Veterinarians, London, England, 19–24 May 1997.

Cooper, J.E., Sylvanose, C., 2016. Cytology. In: Samour, J.S. (Ed.), Avian Medicine. Elsevier, London.

Cooper, J.E., West, C.D., 1988. Radiological studies on endangered Mascarene fauna. Oryx. 22, 18–24.

Cooper, J.E., Slavin, G., Kreel, L., Keymer, I.F., 1978. A survey of lung lesions in non-human primates using correlative histological and X-ray techniques. In: Chivers, D.C. (Ed.), Recent Advances in Primatology; vol. 4: Medicine. Academic Press, London, pp. 79–82.

Cooper, J.E., Dutton, C.J., Allchurch, A.F., 1998. Reference collections: their importance and relevance to modern zoo management and conservation biology. Dodo J. Jersey Wildl. Preserv. Trust. 34, 159–166.

Cooper, J.E., Cooper, M.E., 2013. Wildlife Forensic Investigation: Principles and Practice. CRC Press, Taylor and Francis/CRC Press, Boca Raton, Florida, USA.

Cooper, M.E., Rosser, A.M., 2002. International regulation of wildlife trade: relevant legislation and organisations. Rev. Sci. Tech. Off. Int. Epiz. 21 (1), 103–123.

Cooper, M., Cooper, S., 2014. Resisting decomposition as evidence of bodily strength: funerary practice and psychological well-being in rural Guinea Bissau. Leeds Afr. Stud. Bull. 76, 67–69.

Cooper, M.E., 1996. Community responsibility and legal issues. Semin. Avian Exot. Pet Med. 5 (1), 3745.

Cooper M.E., 1997. Unpublished report to the Universities Federation for Animal Welfare on the Population and Habitat Viability Assessment Workshop for the mountain gorilla (*Gorilla gorilla beringei*). 17 December 1997.

Cooper, M.E., 2013. Legal, ethical and practical considerations of working in the field. In Field techniques in exotic animal medicine. J. Exot. Pet Med. 22 (1), 17–33.

Corbet, G.B., 1967. Nomenclature of the "Eastern Lowland Gorilla". Nature. 215, 1171–1172.

Cork, S.C., Halliwell, R.W., 2002. The Veterinary Laboratory and Field Manual. Nottingham University Press, Nottingham, UK.

Correa, C., 2011. Implications for BioTrade of the Nagoya Protocol on Access to Genetic Resources and the Fair and Equitable Sharing of Benefits Arising from their Utilization. UNCTAD BioTrade Initiative. UNCTAD/DITC/TED/2011/9. United Nations, New York and Geneva, < http://www.unctad.org/biotrade >.

Corrin, B., Nicholson, A.G., 2011. Pathology of the Lungs. Third ed. Churchill Livingstone, Elsevier, London, UK.

Coupin, F., 1927. L'indice de valeur cerebral au cours de l'enfance chez les Anthropoides. C. R. Hebd. Seances Acad. Sci. 184, 396–398.

Courage, A., 2001. Projet Protection des Gorilles: Lesio-Louna Reserve, Republic of Congo. The John Aspinall Foundation. The Conservationist, November.

Cousins, D., 1972. Diseases and injuries in wild and captive gorillas, *Gorilla gorilla*. Int. Zoo Yearb. 12, 211–218.

Cousins, D., 1974. A review of some complaints suffered by captive gorillas with notes on some causes of death in wild gorillas (*Gorilla gorilla*). Der Zool. Garten (NF) Jena. 44, 201–210.

Cousins, D., 1978. Notes on bacterial and fungal infections in the gorilla. Int. Zoo News 151, 19–24.

Cousins, D., 1982. Dental diseases in the gorilla. Int. Zoo News 29 (1), 10–18.

Cousins, D., 1983. Mortality factors in captive gorillas. Int. Zoo News 30 (3), 5–17.

Cousins, D., 1984. Notes on the occurrence of skin infections in gorillas (*Gorilla gorilla*). Der Zool. Garten (NF) Jena. 54, 333–338.

Cousins, D., 1990. The Magnificent Gorilla. The Life History of a Great Ape. The Book Guild, Sussex, UK.

Cousins, D., 2008. Possible goundou in gorillas. Gorilla J. 37, 22–24.

Cousins, D., 2015. Ruminations on disturbed behaviours in captive great apes. Int. Zoo News. 62 (1), 5–28.

Cousins, D., Huffman, M.A., 2002. Medicinal properties in the diet of gorillas: an ethno-pharmacological evaluation. Afr. Study. Monogr. 23 (2), 65–89.

Cox, C., DuBois, T., Renzetti, V., 2002. Effects of supplemental feeding enrichment on gorilla (*Gorilla gorilla gorilla*) activity at the Los Angeles zoo. Am. J. Primatol. 57, 83.

Cox, M., Mays, S., 2000. Human Osteology in Archaeology and Forensic Science. Greenwich Medical Media, London, UK.

Cranfield, M., 2002. Utilizing a biological resource center of stored mountain gorilla samples to measure changes in gorilla flora biodiversity. In: Proceedings of the American Association of Zoo Veterinarians, 2002, p. 277.

Cranfield, M., Minnis, R., 2007. An integrated health approach to the conservation of mountain gorillas *Gorilla beringei beringei*. Int. Zoo Yearb. 41 (1), 110–121.

Cranfield, M., Gaffikin, L., Cameron, K., 2001. Conservation medicine as it applies to the mountain gorilla (*Gorilla gorilla beringei*). In: The Apes: Challenges for 21st Century Conference Proceedings, Brookfield, IL, 10–13 May 2001, pp. 238–240.

Cranfield, M.R., Gaffikin, L., Sleeman, J., Rooney, B.A., 2002. The mountain gorilla and conservation medicine. In: Aguirre, A.A., Ostfeld, R.S., Tabor, G.M., House, C., Pearl, M.C. (Eds.), Conservation Medicine: Ecological Health in Practice. Oxford University Press.

Cranfield, M.R., 2001. A Brief Review of the MGVC Pathology Data From 1998–2000. Mountain Gorilla Veterinary Project, Rwanda, Document provided to NGOs. Made available by the author. Received 6th February 2001.

Cranfield, M.R., 2004. Links between human and mountain gorilla health: an example from the mountain gorilla (*Gorilla beringei beringei*) veterinary project in Rwanda. In: Proceedings AAZV/AAWV/WDA Joint Conference, 2004, pp. 200–201.

Cranfield, M.R., 2008. Mountain gorilla research: the risk of disease transmission relative to the benefit from the perspective of ecosystem health. Am. J. Primatol. 70 (8), 751–754.

Cray, C., Zaias, J., Altman, N.H., 2009. Acute phase response in animals: a review. Comp. Med. 59, 517–526.

Crispin, S, 2007. Gorillas in our midst. RCVS News.

Crockett, C., 1996. Data collection in the zoo setting emphasizing behavior. In: Kleiman, D., Allen, M., Thompson, K., Lumpkin, S. (Eds.), Wild Mammals in Captivity. University of Chicago Press, Chicago, IL, pp. 545–565.

Crockett Wilson, C., 1978. Methods of observational research in the zoo setting. In: Crockett, C., Hutchins, M. (Eds.), Applied Behavioral Research at the Woodland Park Zoological Gardens. Pika Press, Seattle, WA, pp. 51–73.

Crosby, J., Lukas, K., 2000. A longitudinal study of maternal behavior and group cohesion in a gorilla group at Lincoln Park Zoo. Paper presented at "The Apes: Challenges for the 21st Century." Brookfield Zoo, Chicago, IL.

Crowe, D.C., 2003. An appeal for better automated medical records. J. Am. Vet. Med. Assoc. 223 (11), 1566–1567.

Cunningham, A., 1921. A gorilla's life in civilization. Bull. N.Y. Zool. Soc. 24, 118–124.

Cunningham, A.A., 1996. Disease risks of wildlife translocations. Conserv. Biol. 10, 349–353.

Cyrus, C., 2005. The Dr Cecil Cyrus Museum: public attitudes to tissue donation for display in St. Vincent. Bull. R. Coll. Pathol. 131, 44–47.

Czekala, N.M., Benirschke, K., McClure, H., Lasley, B.L., 1983. Urinary estrogen excretion during pregnancy in the gorilla (*Gorilla gorilla*), orangutan (*Pongo pygmaeus*) and the human (*Homo sapiens*). Biol. Reprod. 28 (2), 289–294.

Dallmann, R., Steinlechner, S., von Hörsten, S., Karl, T., 2006. Stress-induced hyperthermia in the rat: comparison of classical and novel recording methods. Lab. Anim. 40, 186–193.

Dalton, R., 2003. Natural history collections in crisis as funding is slashed. Nature. 423, 575.

D'Ambrosio, M.V., Bakalar, M., Bennuru, S., Reber, C., et al., 2015. Point-of-care quantification of blood-borne filarial parasites with a mobile phone microscope. Sci. Transl. Med. 7 (286), 286.

Danese, S., et al., 2010. The protein C pathway in tissue inflammation and injury: pathogenic role and therapeutic implications. Blood. 115, 1121–1130.

Dart, R.A., 1925. *Australopithecus africanus*: the man-ape of South Africa. Nature. 115, 195–199.

Darwin, C., 1859. On the Origin of Species by Means of Natural Selection, or the Preservation of Favoured Races in the Struggle for Life. John Murray, London.

Darwin, C., 1872. The Expression of the Emotions in Man and Animals. John Murray, London.

Daszak, P., Cunningham, A.A., Hyatt, A.D., 2000. Emerging infectious diseases of wildlife—threats to biodiversity and human health. Science. 287, 443–449.

Davenport, R.K., 1979. Some behavioural disturbances of great apes in captivity. In: Hamburg, D.A., Mccown, E.R. (Eds.), The Great Apes: Perspectives on Human Evolution, vol. 5. Benjamin/Cummings, Menlo Park, CA, pp. 341–357.

Davis, P.G., 1997. The bioerosion of bird bones. Int. J. Osteoarchaeol. 7, 388–401.

Day, M.J., Schultz, R.D., 2010. Veterinary Immunology: Principles and Practice. CRC Press, Boca Raton, FL.

Dayan, A.D., Morgan, R.J.I., Trefty, B.T., 1978. Naturally occurring diatomaceous pneumoconiosis in sub-human primates. J. Comp. Pathol. 88, 321–325.

Debenham, J.J., Atencia, R., 2014. Homologous whole blood transfusion during treatment of severe anemia in a chimpanzee (*Pan troglodytes*). J. Zoo. Wildl. Med. 45 (3), 654–657.

de Boer, H.H., Maat, G.J.R., 2012. The histology of human dry bone (a review). Cuadernos Prehist. Arqueol. Univ. Granada. 22, 49–65.

Debyser, I.W.J., 1995. Jeugdsterfte bij primaten in gevangenschap: een epidemiologische en pathologische studie in Nederlandse en Belgische dierentuinen en in het Duitse primatencentrum (meteen samenvatting in het Nederlands). (Juvenile Mortality in Captive Populations of Primates: An Epidemiological and Pathological Study in Dutch and Belgian Zoological Gardens and a German Primate Centre). Proefschrift University of Utrecht, The Netherlands.

Decision Tree Writing Group, 2006. Clinical response decision tree for the mountain gorilla (*Gorilla beringei*) as a model for great apes. Am. J. Primatol. 68, 909–927.

de Faber, J.T., Pameijer, J.H., Schaftenaar, W., 2004. Cataract surgery with foldable intraocular lens implants in captive lowland gorillas (*Gorilla gorilla gorilla*). J. Zoo. Wildl. Med. 35 (4), 520–524.

DEFRA, 2012a. Secretary of State's Standards of Modern Zoo Practice. Department for Environment, Food and Rural Affairs, Bristol, UK.

DEFRA, 2012b. Zoos Expert Committee Handbook. Department for Environment, Food and Rural Affairs, Bristol, < https://www.gov.uk/government/uploads/system/uploads/attachment_data/file/69611/pb13815-zoos-expert-committee-handbook1.pdf > .

de Graaf, M., Osterhaus, A.D., Fouchier, R.A., Holmes, E.C., 2008. Evolutionary dynamics of human and avian metapneumoviruses. J. Gen. Virol. 89, 2933–2942.

Dekeyser, P.L., 1955. Les mammifères de l'Afrique noire francaise. Inst. Francais d'Afrique Noire, Dakar.

Delahay, R., Smith, G., Hutchings, M., 2009. Management of Disease in Wild Mammals. Springer, New York, NY.

de Lahunta, A., 2012. Nomenclature. Vet. Pathol. 49 (4), 735.

de Lahunta, A., 2014. Correct use of nomenclature. Vet. Pathol. 51 (2), 540.

Demanche, C., Petit, T., Moisson, P., Ollivet, F., Rigoulet, J., Chermette, R., et al., 2003. Assessment of *Pneumocystis* species carriage in captive primates. Vet. Rec. 152, 811–813.

Denis, A., 1963. On Safari. Collins, London.

Dennis, M.J., 1986. The effects of temperature and humidity on some animal diseases—a review. Br. Vet. J. 142, 472–485.

de Queiroz, K., 2007. Species concepts and species delimitation. Syst. Biol. 56, 879–886.

DeRousseau, C., 1985. Aging in the musculoskeletal system of rhesus monkeys: II. Degenerative joint disease. Am. J. Phys. Anthropol. 67 (3), 177–184.

Dessauer, H.C., Hafner, M.S. (Eds.), 1984. Collections of Frozen Tissues — Value, Management, Field and Laboratory Procedures, and Directory of Existing Collections. Association of Systematics Collections, Lawrence, Kansas.

Dessauer, H.C., Cole, C.J., Hafner, M.S., 1996. Collection and storage of tissues. In: Hillis, D.M., Moritz, C., Mable, B.K. (Eds.), Molecular Systematics, second ed. Sinauer Associates, Inc, Sunderland, pp. 29–47.

Dewar, D.J., Meijer, K., 2015. EU Zoos Directive Good Practices Document. European Union, Luxemburg, < http://ec.europa.eu/environment/nature/pdf/EU_Zoos_Directive_Good_Practices.pdf > .

Dillner, J., Rabin, H., Letvin, N., et al., 1987. Nuclear DNA-binding proteins determined by the Epstein-Barr virus-related simian lymphotropic herpesviruses H. gorilla, H. pan, H. pongo and H. papio. J. Gen. Virol. 68, 1587–1596.

DiMaio, V.J.M., DiMaio, D.J., 2001. Forensic Pathology. second ed. CRC Press, FLUSA/London, UK.

Diogo, R., Wood, B.A., 2012. Comparative Anatomy and Phylogeny of Primate Muscles and Human Evolution. CRC Press, Boca Raton, FL.

Diogo, R., Barbosa, M., Bello, G., de Paz, F.J., Ferrero, E.M., Pastor, J.F., et al., 2011. Photographic and Descriptive Musculoskeletal Atlas of Gorilla: With notes on the Attachments, Variations, Innervation, Synonymy and Weight of the Muscles. CRC Press, Boca Raton, FL.

Diogo, R., Pastor, J.F., Hartstone-Rose, A., Muchlinski, M.N., 2014. Baby Gorilla: Photographic and Descriptive Atlas of Skeleton, Muscles and Internal Organs Including CT scans and comparison with adult gorillas, humans and other primates. CRC Press, Boca Raton, FL.

Dixson, A., Moore, H.D.M., Holt, W.V., 2009. Testicular atrophy in captive gorillas (*Gorilla g. gorilla*). J. Zool. 191, 315−322.

Dixson, A.F., 1981. The Natural History of the Gorilla. Weidenfeld and Nicholson, London.

Djeukam, R, 2012 The Wildlife Law as a Tool For Protecting Threatened Species In Cameroon. Ministry of Forestry and Wildlife, Department of Wildlife and Protected Areas and the Last Great Ape Organisation (LAGA). < http://www.laga-enforcement.org/Portals/0/Documents/ >, < http://www.laga-enforcement.org/Portals/0/Documents/Legal%20 documents/l%C3%A9gislation%20faunique_Protection_esp%C3%A8ce_menac%C3%A9es-_En.pdf >.

DLA Piper, 2015. Empty threat: does the law combat wildlife trade? An eleven-country review of legislative and judicial approaches. < http://www.dlapiperprobono.com/export/sites/pro-bono/downloads/pdfs/Illegal-Wildlife-Trade-Report-2014.pdf >.

Doherty, J. 1999. Animal management in the Bronx Zoo's Congo Gorilla Forest. In: AZA Annual Conference Proceedings, pp. 222−226.

Dohoo, I., Martin, W., Stryhn, H., 2010. Veterinary Epidemiologic Research. Second ed. UPEI, Canada.

Dolensek, E.P., Napolitano, R.L., Kazimiroff, J., 1977. Gastrointestinal geotrichosis in six adult gorillas. J. Am. Vet. Med. Assoc. 171, 975−976.

Dolphin, R.E., McNally, J.D., Schnoor, J.W., Backer, B.G., 1976. Dental prophylaxis for a lowland gorilla. Vet. Med. Small Anim. Clin. 71, 1732−1735.

Doolittle, R.L., Grand, T.I., 1995. Benefits of the zoological park to the teaching of comparative vertebrate anatomy. Zoo Biol. 14 (5), 453−462.

Doran, D.M., McNeilage, A., 2001. Subspecific variation in gorilla behavior: the influence of ecological and social factors. In: Robbins, M.M., Sicotte, P., Stewart, K.J. (Eds.), Mountain Gorillas: Three Decades of Research at Karisoke. Cambridge University Press, Cambridge, UK, pp. 123−149.

Doré, M., Lagacé, A., 1985. Spontaneous external endometriosis in a gorilla (*Gorilla gorilla*). Can. Vet. J. 26 (11), 347−349.

Dorobisz, A., Gucwinski, A., Opyrchal, A., Rybrinezyk, L., 1983. Personal circulation disorders in lowland gorillas. Sonderdruck aus Verhandlungsbericht des 25 Internationalen Symposiums über die Erkrankungen der Zootiere. Vienna 1983. Akademie-Verlag, Berlin.

Doski, J.J., Heiman, H.S., Solenberger, R.I., Stefko, R.M., Kuivila, T., Rozanski, T.A., et al., 1997. Successful separation of ischiopagus tripus conjoined twins with comparative analysis of methods for abdominal wall closure and use of the tripus limb. J. Pediatr. Surg. 32 (12), 1761−1766.

Douglas, T.A., 1977. Standard international units. Vet. Rec. 100, 28−29.

Downman, M., 2000a. Formation of a bachelor group: gorillas at Loro Parque. Gorilla Gaz. 14 (1), 17−19.

Downman, M., 2000b. Introducing gorillas to a naturalistic environment. Gorilla Gaz. 14 (1), 20−24.

Dubowitz, V., Sewry, C., Oldfors, A., 2013. Muscle Biopsy: A practical approach. Fourth ed. Elsevier Inc, San Diego.

Du Chaillu, P., 1871. Adventures in the Great Forest of Equatorial Africa and the Country of the Dwarfs. Harper & Brothers, New York, NY.

Duckler, G.L., Van Valkenburgh, B., 1998. Osteological corroboration of pathological stress in a population of endangered Florida pumas (*Puma concolor* coryi). Anim. Conserv. 1, 39−46.

Duckworth, W.L.H., 1915. Morphology and Anthropology. Univ. Press, Cambridge, UK.

Duncan, I.J.H., 1996. Animal welfare defined in terms of feelings. Acta Agric. Scand., A. 27, 29−35.

Duncan, J., 2015. Making the most of in-clinic and external laboratory testing. In: Villiers, E., Blackwood, L. (Eds.), BSAVA Manual of Canine and Feline Clinical Pathology, second ed. BSAVA, Gloucester, UK.

Dunn, A., Nicholas, A., Takang, E., Omeni, F., Imong, I., Fotso, R., et al., 2014. Revised Regional Action Plan for the Conservation of the Cross River Gorilla (*Gorilla gorilla diehli*): 2014−2019. IUCN/SSC Primate Specialist Group and Wildlife Conservation Society, New York, NY, USA.

Durette-Desset, M.-C., et al., 1992. Two new species of the Trichostrongylidae (Nematoda: Trichostrongylidae) parasitic in *Gorilla gorilla beringei* in Uganda. Syst. Parasitol. 23, 159−166.

Durrell, G., 1989. Introduction. In: A Jubilee Bibliography 1963-1988, Jersey Wildlife Preservation Trust, Jersey, Channel Isles (Christopher M. Perkins, author).

Dwight Davis, D., 1951. The baculum of the gorilla. Fieldiana Zool. 31 (54), 645−647.

EAC, 2015. Veterinary Services in the EAC. East African Community, Arusha, < http://www.rr-africa.oie.int/docspdf/en/2015/VS_EAC.pdf >.

EAZA, 2013. The Modern Zoo: Foundations for Management and Development. European Association of Zoos and Aquaria, Amsterdam, < http://eaza.portal.isis.org/activities/technical-assistance/Documents/Zoo%20Management%20Manual.pdf >.

EAZA, 2014. Standards for the Accommodation and Care of Animals in Zoos and Aquaria. European Association of Zoos and Aquaria, < http://www.eaza.net/assets/Uploads/Standards-and-policies/Standards-for-the-Accommodation-and-Care-of-Animals-2014.pdf >.

EAZWV Transmissible Disease Fact Sheet, 2008. <http://eaza.portal.isis.org/activities/tdfactsheets/100%20Acanthamoebiasis.doc.pdf>.

Eckhart, G., Lanjouw, A., 2008. Mountain Gorillas: Biology, Conservation, and Coexistence. The Johns Hopkins University Press, Baltimore, MD.

Egbe, S.E., 2001. The Law, Communities and Wildlife Management in Cameroon. Rural Development Forestry Network paper 25 July 2001. Overseas Development Institute, London.

Eichler, W., 1942. Die Entfaltungsregel und andere Gesetzmäßigkeiten in den parasitogenetischen Beziehungen der Mallophagen und anderer ständiger Parasiten zu ihren Wirten. Zool. Anz. 137, 77−83.

Elgart, A.A., 2010. Are the gorillas in Bwindi Impenetrable National Park "true" mountain gorillas? Am. J. Phys. Anthropol. 141 (4), 561−570.

Elliot, D.G., 1913. A review of the primates. Am. Mus. Nat. Hist. Monogr. 3 (No. 1), 224−226.

Elliott, R.C., 1976. Observations on a small group of mountain gorillas (*Gorilla gorilla beringei*). Folio Primatol. 25, 12−24.

Ellis, R.A., Montagna, W., 1962. The skin of the primates VI. The skin of the gorilla (*Gorilla gorilla*). Am. J. Phys. Anthropol. 20, 79−94.

Elsner, R., 2000a. An overview of training at Lincoln Park Zoo's Great Ape House. In: AAZK National Conference Proceedings, pp. 226−235.

Elsner, R., 2000b. Examining the effects of positive reinforcement training on the abnormal behaviors of regurgitation and reingestion and ear-covering in a female western lowland gorilla (*Gorilla gorilla gorilla*) at Lincoln Park Zoo's Lester E.Fisher Great Ape House. Lincoln Park Zoo, Chicago, IL.

Embury, A., 1992. Gorilla rainforest at Melbourne Zoo. Int. Zoo Yearb. 31, 203−211.

Enciso, E., Calcagno, J., Gold, K., 1999. Social interactions between captive adult male and infant lowland gorillas: implications regarding kin selection and zoo management. Zoo Biol. 18, 53−62.

Enomoto, T., Matsubayashi, K., Nakano, M., et al., 2004. Testicular histological examination of spermatogenetic activity in captive gorillas (*Gorilla gorilla*). Am. J. Primatol. 63 (4), 183−199.

Eriksen, A.M., Simonsen, K.P., Rasmussen, A.R., 2013. Conservation of mitochondrial DNA in fast enzyme-macerated skeletal material. Int. J. Conserv. Sci. 4, 127−132.

Erinosho, B.T., 2013. The Revised African Convention on the Conservation of Nature and Natural Resources: Prospects for a Comprehensive Treaty for the Management of Africa's Natural Resources. Afr. J. Int. Comp. Law. 21 (3), 378−397, Oct 2013.

Ernst, L.M., Ruchelli, E.D., Huff, D.S. (Eds.), 2011. Color Atlas of Fetal and Neonatal Histology. Springer, New York, NY.

Erwin, J., Deni, R., 1979. Strangers in a strange land: abnormal behaviors or abnormal environments? In: Erwin, J., Maple, T.L., Mitchell, G. (Eds.), Captivity and Behavior. Van Nostrand Reinhold, New York, NY, pp. 239−272.

Erwin, J.M., Hof, P.R. (Eds.), 2002. Aging in Nonhuman Primates. Interdisciplinary Topics in Gerontology, 31. Karger, Basel and New York, 240 pp.

Etienne, L., Locatelli, S., Ayouba, A., et al., 2012. Noninvasive follow-up of simian immunodeficiency virus infection in wild-living nonhabituated western lowland gorillas in Cameroon. J. Virol. 86 (18), 9760−9772.

EU, 2016. Legislation for the protection of animals used for scientific purposes. EU Commission, < http://ec.europa.eu/environment/chemicals/lab_animals/legislation_en.htm > .

EU, undated, ongoing. The European Union and Trade in Wild Fauna and Flora. < http://ec.europa.eu/environment/cites/legislation_en.htm > .

EU/TRAFFIC, 2015. Reference guide to the European Union Wildlife Trade Regulations. < http://ec.europa.eu/environment/cites/pdf/referenceguide_en.pdf > .

Evans, T.S., Gilardi, K.V.K., Barry, P., Ssebide, B.J., Kinani, J.F., Nizeyimana, F., et al., 2016. Detection of viruses using discarded plants from wild mountain gorillas and golden monkeys. American Journal of Primatology.DOI: 10.1002/ajp.22576.

Fain, A., 1957. L'Acariase pulmonaire chez le Chimpanzé et le Gorille par des Acariens du genre *Pneumonyssus* Banks. Rev. Zool. Bot. Afr. LVI, 3−4.

Fain, A., 1962. *Pangorillages pani* g.n., sp.n. Acarien psorique du chimpanzé (Psoralgidae: Sarcoptiformes). Rev. Zool. Bot. Afr. 66, 290−293.

Fairbrother, A., Locke, L.N., Hoff, G.L. (Eds.), 1996. Noninfectious Diseases of Wildlife. Second ed Iowa State University Press, Ames, IO.

Falconer, T., 1797. The Voyage of Hanno Translated, and Accompanied With the Greek text; Explained From the Accounts of Modern Travellers; Defended Against the Objections of Mr Dodwell, and Other Writers; and Illustrated by Maps From Ptolemy, d'Anville, and Bougainville. Sold by T. Cadell Jnr and Davies, in the Strand, London.

FAO/IUCN /UNEP, 2015. Ecolex: the gateway to environmental law. < http://www.ecolex.org/ > .

FAOLEX, undated. FAO Legal Office FAOLEX. < http://faolex.fao.org/faolex/ > .

Farmer, K.H., Courage, A., 2008. Sanctuaries and reintroduction: a role in gorilla conservation? In: Stoinski, T.S., Steklis, H.D., Mehlman, P. (Eds.), Conservation in the 21st Century: Gorillas as a Case Study. Springer, New York, NY.

Farnsworth, R.J.R.E., Findleton, J.F., 1979. Conservative treatment of an intestinal foreign body in a gorilla. J. Zoo Anim. Med. 10, 56−57.

Farrell, M., 2010. News from the Royal College of Surgeons: a new habitat for the odontological collection of primates. NatSCA News.(Issue 20), 25−27.

Farrell, M., 2011. Hidden treasures: natural history exhibits at the Royal College of Surgeons. NatSCA News.(Issue 21), 72−74.

Farrell, M., Rando, C., Garrod, B., 2015. Lessons from the past: metabolic bone disease in historical captive primates. Int. J. Primatol. 36 (2), 398−411.

Farrington, S., 2013. Hanno, Periplus. In: Bagnall, R.S., Brodersen, K., Champion, C.B., Erskine, A., Huebner, S.R. (Eds.), The Encyclopedia of Ancient History. Blackwell Publishing Ltd, pp. 3058−3060.

FAWC, 2009. Five Freedoms. Farm Animal welfare Council. National Archive < http://webarchive.nationalarchives.gov.uk/20121007104210/ > , < http://www.fawc.org.uk/freedoms.htm > . Crown Copyright published under Open government Licence v2.0. < http://www.nationalarchives.gov.uk/doc/open-government-licence/version/2/ > .

Fehlman, P., Hauser, B., Isenbügel, E., 1983. Strongyloidose bei Einem Jungen Gorilla. Sonderdruck aus Verhandlungsbericht des 25 Internationalen Symposiums über die Erkrankungen der Zootiere. Wien. Akademie-Verlag, Berlin.

Ferreira, A.J.D.L., Athayde, A., De Màgàlhães, H., 1945. Gorilas do Maiombe Português. Memórias Ministério das Colónias, Junta das Missões Geográficas e de Investigações Coloniais, Série Zoológica I. Lisboa. 155 pp.

Ferreira, A.L., 1938. Un cas de lésions déformantes polyarticulaires chez un "Gorilla gina" d'Angola. An. Fac. Ciênc. Univ. Porto. 23, 157−174.

Ferrer, N., 1997. Shipping and regulations. In: Ogden, J., Wharton, D. (Eds.), The Gorilla Husbandry Manual. Fulton County Zoo, Atlanta, GA, pp. 181−185.

Ferrie, G.M., Farmer, K.H., Kuhar, C.W., Grand, A.P., Sherman, J., Bettinger, T.L., 2014. The social, economic, and environmental contributions of Pan African Sanctuary Alliance primate sanctuaries in Africa. Biodivers. Conserv. January 2014. 23 (1), 187−201.

Field, A.S., Geddie, W.R. (Eds.), 2014. Lymph Node and Spleen Cytohistology. Cambridge University Press, Cambridge, UK.

Fiennes, R.N.T.-W., 1962. The skin and appendages, and special senses. In: Fiennes, R.N.T.-W. (Ed.), Pathology of Simian Primates, Part I. Karger, Basel, Switzerland.

Fiennes, R.N.T-W., 1966. The Zoological Society of London Report of the Society's Pathologist for the year 1963. J. Zool. 148, 341−362.

Fiennes, R., 1967. Zoonoses of primates. The Epidemiology and Ecology and Simian Diseases in Relation to Man. Cornell University Press, Ithaca, NY.

Fiennes, R.N.T.-W. (Ed.), 1972. Pathology of Simian Primates. Karger, Basel, Switzerland.

Fiennes, R.N., 1974. Problems of rickets in monkeys and apes. J. R. Soc. Med. 67, 309−314.

Finlay, T., James, L., Maple, T., 1988. Zoo environments influence people's perceptions of animals. Environ. Behav. 20 (4), 508−525.

Finlayson, R., 1965. Spontaneous arterial disease in exotic animals. J. Zool. 147, 239−343.

Finley, R.B., 1987. The value of research collections. Bioscience. 37, 92.

Fischer, E., Hinkel, H., 1992. Natur Ruandas: Einfuhrung in die Flora und Fauna Ruandas. Ministerium des Inne ren und fiir Sport Rheinland-Pfalz, Mainz.

Fischer, F., 1989. Training program to enhance maternal behavior. Gorilla Gaz. 3 (1), 5−6.

Fischer, R., 1983. Maternal subgrouping in lowland gorillas. Behav. Processes. 8, 301−306.

Fischer, R., 1984. Observations of group introductions in lowland gorillas. Behav. Process. 9, 293−296.

Fischer, R., Nadler, R., 1977. Status interactions of captive female lowland gorillas. Folia Primatol. 28, 122−133.

Fischer, R., Nadler, R., 1978. Affiliative, playful, and homosexual interactions of adult female lowland gorillas. Primates. 19 (4), 657−664.

Fisher, L., 1972. The birth of a lowland gorilla at Lincoln Park Zoo, Chicago. Int. Zoo Yearb. 12, 106−108.

Fisher, L.E., 1954. Lead poisoning in a gorilla. J. Am. Vet. Med. Assoc. 125, 478−479.

Fitzgibbons, J.F., Simmons, L., 1975. Autopsy findings of a seven-year-old male lowland gorilla. J. Zoo. Anim. Med. 6, 25−30.

Fitzgibbons, J.F., Thomas, W., Simmons, L., 1971. Autopsy findings of two jungle-born female lowland gorillas. J. Zoo. Anim. Med. 2, 11−23.

Flahou, B., Modry, D., Pomajbikova, K., et al., 2014. Diversity of zoonotic enterohepatic *Helicobacter* species and detection of a putative novel gastric *Helicobacter* species in wild and wild-born captive chimpanzees and western lowland gorillas. Vet. Microbiol. 174, 186−194.

Flanagan, K.L., 2015. Vaccines have sex differential non-targeted heterologous effects: a new dawn in vaccine research. Trans. R. Soc. Trop. Med. Hyg. 109, 1−2.

Fleming, K., 2013. Pathology is global: developing our international agenda. Bull. R. Coll. Pathol. 164, 261.

Flower, W.H., Garson, J.G., 1884. Catalogue of the Specimens Illustrating the Osteology and Dentition of Vertebrated Animals, Recent and Extinct, Contained in the Museum of the Royal College of Surgeons of England. Part II: Class Mammalia, Other Than Man. Taylor and Francis, London.

Fogel, A.T., 2015. The gut microbiome of wild lemurs: a comparison of sympatric *Lemur catta* and *Propithecus verreauxi*. Folia Primatol. 86, 85–95.

Fontana, L., Partridge, L., Longo, V.D., 2010. Extending healthy life span—from yeast to humans. Science 328 (5976), 321–326.

Fontenot, D., Terrell, S., Miller, M., Robbins, P.K., Stetter, M., Weber, M., 2005. *Clostridium septicum* myositis in a western lowland gorilla (*Gorilla gorilla gorilla*). J. Zoo Wildl. Med. 36 (3), 509–511.

Fossey, D., 1979. Development of the mountain gorilla (*Gorilla gorilla beringei*): the first thirty-six months. In: Hamburg, D., McCown, E. (Eds.), *The Great Apes (Perspectives on Human Evolution, vol. 5)*. Benjamin/Cummings, Menlo Park, CA, pp. 139–184.

Fossey, D., 1982. Reproduction among free-living mountain gorillas. Am. J. Primatol. (Suppl. 1), 97–104.

Fossey, D., 1983. Gorillas in the Mist. Houghton Mifflin Company, Boston, MA.

Foster, J., 1992. Mountain gorilla conservation: a study in human values. J. Am. Vet. Med. Assoc. 200 (5), 629–633.

Foster, J.W., 1993. Health plan for the mountain gorillas of Rwanda. In: Fowler, M.E. (Ed.), Zoo and Wild Animal Medicine: Current Therapy 3. W.B. Saunders, Philadelphia, PA, pp. 331–333.

Fotso, R., et al., 2002. Distribution and conservation status of gorilla population in the forests around Belabo, Eastern Province, Cameroon. Unpublished report, Cameroon Oil Transportation Company (COTO).

Fowler, M.E. (Ed.), 1993, 1999, 2003, 2007. Zoo and Wild Animal Medicine. Third, fourth, fifth and sixth editions. W.B. Saunders, Philadelphia, PA.

Fowler, M.E. (Ed.), 1993. Zoo and Wild Animal Medicine: Current Therapy 3. W.B. Saunders, Philadelphia, PA.

Fowler, M.E., 1999. Murray: Hummingbirds to Elephants and Other Tales: Autobiography of Murray E. Fowler, D.V.M. Clay Press, CA.

Fox, J., 1923. Diseases in Captive Wild Mammals and Birds. Lippincott, Philadelphia, PA.

Frechkop, S., 1943. Mammifères. Exploration du Parc National Albert. Mission S. Frechkop (1937–8). Fascicule 1, pp. 1–186. Institut des Parcs Nationaux du Congo Belge, Bruxelles.

Frechkop, S., 1953. Animaux protégés au Congo belge et dans le territoire sous mandat du Ruanda-Urundi ainsi que les espèces dont la protection est assurée en Afrique. Institut des Parcs Nationaux du Congo Belge, Bruxelles.

Freeman, A.S., Kinsella, J.M., Cipolletta, C., Deem, S.L., Karesh, W.B., 2004. Endoparasites of western lowland gorillas (*Gorilla gorilla gorilla*) at Bai Hokou, Central African Republic. J. Wildl. Dis. 40 (4), 775–781.

Freeman, S., 1998. Oldest living gorilla tells all. Atlanta, 36–43, 132–133 (June issue).

Frey, J.C., Rothman, J.M., Pell, A.N., Bosco-Nizeyi, J., Cranfield, M.R., Angert, E.R., 2006. Fecal bacterial diversity in a wild gorilla. Appl. Environ. Microbiol. 72 (5), 3788–3792.

Friend, M., 2006. Disease Emergence and Resurgence: The Wildlife-Human Connection. US Geological Survey.

Fritze, A., 1911–1912. Kleinere Mitteilungen. Hannover Jahrbuch des Provinzial-Museums: 113 and plate 8.

Frye, F.L., Cooper, J.E., Keymer, I.F., 2001. Outfitting and employing a compact field laboratory. ZooMed, Bull. Br. Vet. Zool. Soc. 1, 28–36.

Fujii-Hanamoto, H., Matsubayashi, K., Nakano, M., et al., 2011. A comparative study on testicular microstructure and relative sperm production in gorillas, chimpanzees, and orangutans. Am. J. Primatol. 73 (6), 570–577.

Fukui, K., Kato, N., Kato, H., Watanabe, K., Tatematsu, N., 1999. Incidence of *Prevotella intermedia* and *Prevotella nigrescens* carriage among family members with subclinical periodontal disease. J. Clin. Microbiol. 37 (10), 3141–3145.

Fullerton, D.S., Kit, R.D., Dingle, H.R., Purcell-Jones, G., Allchurch, A.F., Blampied, N. Le. Q., 1989. Uterine rupture and hysterectomy in a western lowland gorilla *Gorilla g. gorilla* at the Jersey Wildlife Preservation Trust. Dodo J. Jersey Wildl. Preserv. Trust. 26, 87–93.

Fulton, J.F., 1938. Cytoarchitecture of the gorilla brain. Science 88, 426–427.

Fung, M, 2005. A World Heritage Species Case Study: The Virunga Mountain Gorillas, < https://www.lclark.edu/live/files/192 >.

Funk, V., 2004. 100 uses for an herbarium (well at least 72). Am. Soc. Plant Taxonomists Newsl. 17 (2), 17–19.

Furley, C.W., 1997. Clinical conditions encountered in the Howletts gorilla colony. A twelve year review. Proceedings of British Veterinary Zoological Society (BVZS) Meeting. 14–15 June 1997. Howletts and Port Lympne Wild Animal Parks, UK, pp. 13–21.

Furuichi, T., Inagaki, H., Angoue-Ovono, S., 1997. Population density of chimpanzees and gorillas in the Petit Loango Reserve, Gabon: employing a new method to distinguish between nests of the two species. Int. J. Primatol. 18 (6), 1029–1046.

Gaffikin, P., 1949. *Gorilla gorilla beringei* post mortem report. East. Afr. Med. J. 26, 1–4.

Galbany, J., Imanizabayo, O., Romero, A., Vecellio, V., Glowacka, H., Cranfield, M., et al., 2016. Tooth wear and feeding ecology in mountain gorillas from Volcanoes National Park, Rwanda. Am. J. Phys. Anthropol. 159 (3), 457–465.

Galloway, A., Allbrook, D., 1959. The study of *Gorilla gorilla berengei*. With a post-mortem report by A. M. M. Wilson. S. Afr. J. Sci. 55, 205–209.

GALVMED/Luseba, D, 2015. Review of the Policy, Regulatory and Administrative Framework for Delivery of Livestock Health Products and services in West and Central Africa. GALVMED, Edinburgh.

Gamble, K.C., et al., 2011. Blood groups in the Species Survival Plan, European Endangered Species Program, and managed in situ populations of bonobo (*Pan paniscus*), common chimpanzee (*Pan troglodytes*), gorilla (*Gorilla* ssp.) and orangutan (*Pongo pygmaeus* ssp.). Zoo Biol. 30, 427–444.

Gao, Z., Tseng, C.-h, Pei, Z., Blaser, M.J., 2007. Molecular analysis of human forearm superficial skin bacterial biota. Proc. Natl. Acad. Sci. U.S.A. 104 (8), 2927–2932.

Garba, A., Danbirni, S., Ahmed, A., Ambursa, A.U., Suleiman, A., Mohammed, M.N., et al., 2014. Veterinary Laws And Administration In Nigeria: Historical And Current Perspectives. Zariya Vet. 8 (2), 21.

Gardner-Roberts, D., Lowenstine, L.J., Spelman, L., 2007. An outbreak of stomatitis in a group of orphan eastern gorillas (*Gorilla beringei* spp.) in an interim quarantine facility, Rwanda. In: Proceedings of British Veterinary Zoological Society (BVZS), November 2007, pp. 48–49.

Garner, K.J., Ryder, O.A., 1996. Mitochondrial DNA diversity in gorillas. Mol. Phylogenet. Evol. 6, 39–48.

Garrison, D.H., Hast, M.H., 2014. Vesalius, A.; The Fabric of the Human Body. An annotated translation of the 1543 and 1555 editions of "De Humani Corporis Fabrica Libri Septem". Karger Medical Publishers, Basel, Switzerland.

Gaud, J., Till, W., 1957. Analgesoidea ectoparasites des Singes et de Lémuriens. Ann. Parasitol. Hum. Comp. XXXI (1-2), 136–144.

Gebo, D.L., 2014. Primate Comparative Anatomy. John Hopkins University Press, Baltimore, MD.

Genton, C., Cristescu, R., Gatti, S., et al., 2012. Recovery potential of a western lowland gorilla population following a major Ebola outbreak: results from a ten year study. PLoS ONE. 7 (5), pe37106.

Geoffroy Saint-Hilaire, I., 1852. Sur le. gorille, C. R. Acad. Sci., Paris. 34 pp. 81–84.

Geoffroy Saint-Hilaire, I., 1853. Sur les rapports naturels du gorille; remarques faites à la suite de la lecture de M. Duvernoy, C. R. Acad. Sci., Paris. 36, pp. 933–936.

Geoffroy Saint-Hilaire, I., 1858. Description des mammifères nouveaux ou imparfaitement connus de la collection du Muséum d'Histoire Naturelle et remarques sur la classification et les caractères des mammifères. Quatrième Mémoire. Famille des Singes. Second Supplément. Arch. Mus. Natur. X, 1–102, Paris.

Giebel, C.G., 1855. Die Säugethiere in Zoologischer, Anatomischer und Palaeontologischer Beziehung. Ambrosius Abel, Leipzig, pp. 1083–1084.

Gilardi, K.V., Gillespie, T.R., Leendertz, F.H., Macfie, E.J., Travis, D.A., Whittier, C.A., et al., 2015. Best Practice Guidelines for Health Monitoring and Disease Control in Great Ape Populations. IUCN SSC Primate Specialist Group, Gland, Switzerland, 56 pp.

Gilardi, K.V.K., Oxford, K.L., Gardner-Roberts, D., Kinani, J.-F., Spelman, L., Barry, P.A., et al., 2014. Human Herpes Simplex Virus Type 1 in confiscated Gorilla. Emerg. Infect. Dis. 20 (11), .

Gillespie, T., 2006. Noninvasive assessment of gastrointestinal parasite infections in free-ranging primates. Int. J. Primatol. 27, 1129–1143.

Gippoliti, S., 2006. Applied primatology in zoos: history and prospects in the field of wildlife conservation, public awareness and animal welfare. Primate Rep. 73, 57–71.

Gippoliti, S., Kitchener, A.C., 2007. The Italian zoological gardens and their role in mammal systematic studies, conservation biology and museum collections. Hystrix (n.s.) 18, 173–184.

Gjeltema, J.L., Troan, B., Muehlenbachs, A., Liu, L., Da Silva, A.J., Qvarnstrom, Y., et al., 2016. Amoebic meningoencephalitis and disseminated infection caused by *Balamuthia mandrillaris* in a Western lowland gorilla (*Gorilla gorilla gorilla*). J. Am. Vet. Med. Assoc. 248 (3), 315–321.

Glick, C., Swart, G., Woolf, A., 1979. Dental caries, periodontal abscesses, and extensive cranial osteitis in a captive lowland gorilla (*Gorilla gorilla gorilla*). J. Zoo Anim. Med. 10, 94–97.

Goerke, B., Fleming, L., Creel, M., 1987. Behavioral changes of a juvenile gorilla after a transfer to a more naturalistic environment. Zoo Biol. 6, 283–295.

Gold, K., 1992. Nonsocial behavior of captive infant gorillas. Am. J. Primatol. 26, 65–72.

Gold, K., 1997a. Social structure—overview. In: Ogden, J., Wharton, D. (Eds.), The Management of Gorillas in Captivity. Atlanta/Fulton County Zoo, Atlanta, GA, p. 12.

Gold, K., 1997b. The conservation role of primate exhibits in the zoo. In: Wallis, J. (Ed.), Primate Conservation: The Role of Zoological Parks. American Society of Primatologists, pp. 42–61.

Gold, K., Benveniste, M., 1995. Visitor attitudes and behavior toward great apes. In: AZA Annual Conference Proceedings, pp. 152–158.

Gold, K., Maple, T., 1994. Personality assessment in the gorilla and its utility as a management tool. Zoo Biol. 13, 509–522.

Göltenboth, R., 1982. Special section: diseases of zoo animals. Nonhuman primates (apes, monkeys, prosimians). In: Klöss, H.G., Lang, E.M. (Eds.), Handbook of Zoo Medicine. Van Nostrand Reinhold Company, New York, NY, pp. 46–85.

Gomez, A., Petrzelkova, K., Carl, J., et al., 2015. Gut microbiome composition and metabolomic profiles of wild western lowland gorillas (*Gorilla gorilla gorilla*) reflect host ecology. Mol. Ecol. 24, 2551–2565.

Gomez, J., 1999. Development of sensorimotor intelligence in infant gorillas: The manipulation of objects in problem solving and exploration. In: Parker, S., Mitchell, R., Miles, H. (Eds.), The Mentalities of Gorillas and Orangutans: Comparative Perspectives. Cambridge University Press, Cambridge, pp. 160–178.

Goodall, A., 1979. The Wandering Gorillas. Collins, London.

Goodall, J., 1964. Tool-using and aimed throwing in a community of free-living chimpanzees. Nature. 201, 1264–1266.

Goodman, G., Hedley, J., Meredith, A., 2013. Field techniques in zoo and wildlife conservation work. In Field Techniques in Exotic Animal Medicine. J. Exot. Pet Med. 22 (1), 58–64.

Gordon, A.D., Marcus, E., Wood, B., 2013. Great ape skeletal collections: making the most of scarce and irreplaceable resources in the digital age. Am. J. Phys. Anthropol. 152 (Suppl. 57), 2–32.

Gorillas Land, undated. Captive Population. < http://gorillasland.yolasite.com/captive-population.php > .

Gould, E., Bres, M., 1986a. Regurgitation in gorillas: possible model for human eating disorders (rumination/bulimia). Dev. Behav. Pediatr. 7 (5), 314–319.

Gould, E., Bres, M., 1986b. Regurgitation and reingestion in captive gorillas: description and intervention. Zoo Biol. 5, 241–250.

Gould, E., Reeves, A.J., Graziano, M.S.A., Gross, C.G., 1999. Neurogenesis in the neocortex of adult primates. Science. 286 (5439), 548–552.

Gould, J., 1990a. Conspecific introduction, socialization, and attempts to breed a solitary-raised, silverbacked male gorilla. In: Columbus Zoo Gorilla Workshop Proceedings, pp. 56–79.

Gould, J., 1990b. Enriched-environment adaptation of a deprived, captive, silverbacked male gorilla. In: Proceedings of the Columbus Zoo Gorilla Workshop, pp. 42–55.

Graczyk, T.K., Lowenstine, L.J., Cranfield, M.R., 2000. *Capillaria hepatica* (Nematoda) infections in human-habituated mountain gorillas (*Gorilla gorilla beringei*) of the Parc National de Volcans, Rwanda. J. Parasitol. 85 (6), 1168–1170.

Graczyk, T.K., Mudakikwa, A.B., Cranfield, M.R., et al., 2001a. Hyperkeratotic mange caused by *Sarcoptes scabiei* (Acariform Sarcoptidae) in juvenile human-habituated mountain gorillas (*Gorilla gorilla beringei*). Parasitol. Res. 87, 1024–1028.

Graczyk, T.K., DaSilva, A.J., Cranfield, M.R., Bosco-Nizeyi, J., Kalema, G.A., Pieniazek, N.J., 2001b. *Cryptosporidium parvum* genotype 2 infections in free-ranging mountain gorillas (*Gorilla gorilla beringei*) of the Bwindi Impenetrable National Park, Uganda. Parasitol. Res. 87 (5), 368–370.

Graczyk, T.K., Bosco-Nizeyi, J., da Silva, A.J., Moura, I.N., Pieniazek, N.J., Cranfield, M.R., et al., 2002a. A single genotype of *Encephalitozoon intestinalis* infects free-ranging gorillas and people sharing their habitats in Uganda. Parasitol. Res. 88 (10), 926–931.

Graczyk, T.K., Nizeyi, J.B., Ssebide, B., Thompson, R.C.A., Read, A.C., Cranfield, M.R., 2002b. Anthropozoonotic *Giardia duodenalis* genotype (assemblage) a infections in habitats of free-ranging human-habituated gorillas, Uganda. J. Parasitol. 88 (5), 905–909.

Grafflin, A.L., 1940. Histology of the thyroid and parathyroid glands in the mountain gorilla, with observations upon autofluorescence, fat and pigment. J. Morphol. 67 (3), 455–470.

Graham, D.I., Glennarelli, T.A., 2000. Pathology of brain damage after injury. In: Cooper, P.R., Golfinos, J.G. (Eds.), Head Injury, fourth ed. McGraw-Hill, New York, NY.

Graham, J.M., 1978. *Clostridium botulinum* type C and its toxin in fly larvae. Vet. Rec. 102, 242–243.

Graham-Jones, O., 2001. Zoo Tails. Bantam Press, London.

Gray, C.W., 1965. Paraplegia in a male lowland gorilla, *Gorilla gorilla gorilla* at the National Zoological Park. Int. Zoo Yearb. 5, 186–189.

Gray, J.E., 1870. Catalogue of Monkeys, Lemurs, and Fruit-Eating Bats in the Collection of the British Museum. British Museum Trustees, London.

Gray, R., O'Neal, R.M., Jordan, F.B., 1981. Sudden death associated with atherosclerosis in a gorilla. J. Am. Vet. Med. Assoc. 179 (11), 1306–1307.

Greenwood, E.J.D., 2014. The Outcome of Simian Immunodeficiency Virus Infection in Two African Primate Species (Ph.D. thesis). University of Cambridge, Cambridge, UK.

Greenwood, E.J.D., et al., 2014. Simian immunodeficiency virus infection of chimpanzees (*Pan troglodytes*). Natural hosts of SIV. Implication in AIDS. Elsevier, London, pp. 85–101.

Greenwood, E.J.D., Schmidt, F., Kondova, I., Niphuis, H., Hodara, V., Clissold, L., et al., 2015. Simian immunodeficiency virus infection of chimpanzees (*Pan troglodytes*) shares features of both pathogenic and nonpathogenic lentiviral infections. PLoS. Pathog. 11 (9), e1005146, 2015 Sep 11.

Gregory, W.K. (Ed.), 1950. The Anatomy of the Gorilla. Columbia University Press, New York, NY.

Greiber, T, Moreno, SP, Åhrén, A, Carrasco, JN, Kamau, EC, Medaglia, JC, et al., O. F., in cooperation with Ali, N. and Williams, C. 2012. An Explanatory Guide to the Nagoya Protocol on Access and Benefit-sharing. IUCN, Gland, Switzerland. pp. xviii + 372 < https://cmsdata.iucn. org/downloads/an_explanatory_guide_to_the_nagoya_protocol.pdf > .

Gresham, G.A., Jennings, A.R., 1962. An Introduction to Comparative Pathology. Academic Press, London and New York.

Gresham, G.A., Turner, A.F., 1979. Post-Mortem Procedures (An Illustrated Textbook). Wolfe Medical Publications, London.

Griffith, B., Scott, M., Carpenter, J.W., Reed, C., 1993. Animal translocations and potential disease transmission. J. Zoo Wildl. Med. 24, 231–236.

Griffiths, C.S., Bates, J.M., 2002. Morphology, genetics and the value of voucher specimens: an example with *Cathartes* vultures. J. Raptor Res. 36 (3), 183–187.

Grogan, E.S., Sharp, A.H., 1900. From the Cape to Cairo. Hurst and Blackett, London.

Gropp, R.E., 2003. Are university natural science collections going extinct? Bioscience. 53, 550.

Groves, C.P., 1967. Ecology and taxonomy of the gorilla. Nature 213, 890–893.

Groves, C.P., 1970a. Gorillas. Barker, London.

Groves, C.P., 1970b. Population systematics of the gorilla. J. Zool., London. 161, 287–300.

Groves, C.P., 1986. Systematics of the great apes. In: Swindler, D.R., Erwin, J. (Eds.), Comparative Primate Biology, vol. 1, Systematics, Evolution and Anatomy. Alan R. Liss, New York, NY, pp. 187–217.

Groves, C.P., 2000. What, if anything, is taxonomy? Gorilla J. 21, 12–15.

Groves, C.P., 2001. Primate Taxonomy. Smithsonian Institution Press, Washington, DC.

Groves, C.P., 2003. A history of gorilla taxonomy. In: Taylor, A.B., Goldsmith, M.L. (Eds.), Gorilla Biology: A Multidisciplinary Perspective. Cambridge University Press, Cambridge, UK.

Groves, C.P., 2008. Extended Family: Long Lost Cousins. Conservation International, Arlington, CA.

Groves, C., Meder, A., 2001. A model of gorilla life history. Aust. Primatol. 15, 2–15.

Groves, C.P., Napier, J.R., 1966. Skulls and skeletons of *Gorilla* in British collections. J. Zool. 148, 153–161.

Groves, C.P., Stott, K.W., 1979. Systematic relationships of gorillas from Kahuzi, Tshiaberimu and Kayonza. Folia Primatol. 32, 161–179.

Grubb, P., Butynski, T.M., Oates, J.F., Bearder, S.K., Disotell, T.R., Groves, C.P., et al., 2003. Assessment of the diversity of African primates. Int. J. Primatol. 24, 1301–1357.

Gruber-Thalmann, G., 1931. Gorillaschädel vom Likouala. Ann. Natur. Mus. Wien. 46, 165–183.

Grzimek, B., 1956. Maße und Gewichte von Flachland-Gorillas. Z. Säugetierk. 21, 192–194.

Grzimek, B., 1957. Blinddarmentzündung als Todesursache bei Gorilla-Kleinkind. (Appendicitis as cause of death in a Gorilla toddler). Zool. Garten (N.S.) 23, 249.

Gucwinski, A., Gucwinska, H., Ippen, R., Wajton, A., 1975. Uber Haltungsfragen und Erkrankungen bei jungen Flachlandgorillas im Zoo Wroclaw. (Aspects of keeping and diseases in young lowland gorillas at Wroclaw Zoo). In: Ippen, R., Schroder, H.D. (Eds.), Proc. 17th Int. Symp. Erkrankungen Zootiere, Tunis. Akademie Verlag, Berlin, pp. 35–41.

Guerrera, W., Sleeman, J.M., Ssebide, B.J., Pace, L., Ichinose, T.Y., Reif, J. S., 2003. Medical survey of the local human population to determine possible health risks to the mountain gorillas of Bwindi Impenetrable Forest National Park, Uganda. Int. J. Primatol. 24 (1), 197–207.

Guillot, J., Vermeulen, B., Lafosse, S., Chaffour, S., Cibot, M., Narat, V.S., et al., 2011. Nematodes of the genus *Oesophagostomum*: an emergent risk for humans and apes in Africa? Bull. Acad. Natl. Méd. 195 (8), 1955–1963.

Guilloud, N.B., FitzGerald, F.L., 1967. Spontaneous hyperthermia in the gorilla. Folia Primatol. 6, 177–179.

Gusset, M., Fa, J.E., Sutherland, W.J., the Horizon Scanners for Zoos and Aquariums, World Association of Zoos and Aquariums, 2014. A horizon scan for species conservation by zoos and aquariums. Zoo Biol. 33 (5), 375–380.

Gyldenstolpe, N., 1928. Zoological results of the Swedish expeditions to Central Africa 1921. Vertebrata 5. Mammals from the Birunga Volcanoes, north of Lake Kivu. Ark. Zool. 20A, 1–76.

Haberle, A., 1973. Notes on the birth and hand-rearing of a gorilla. J. Zoo An. Med. 4 (4), 20–21.

Habumuremyi, S., Robbins, M.M., Fawcett, K.A., Deschner, T., 2014. Monitoring ovarian cycle activity via progestagens in urine and feces of female mountain gorillas: a comparison of EIA and LC-MS measurements. Am. J. Primatol. 76, 180–191.

Hackett, C.J., 1981. Microscopical focal destruction (tunnels) in exhumed human bones. Med. Sci. Law. 21 (4), 243–265.

Haddow, A.J., Ross, R.W., 1951. A critical review of Coolidge's measurements of gorilla skulls. Proc. Zool. Soc. London. 121, 43–54.

Haeckel, E., 1903. Anthropogenie. Fifth ed. Wilhelm Engelmann, Leipzig, pp. 426–430., Part I.

Haggquist, 1933. Zeitschrift für Wissenschaftliche Mikroskopie und für Mikroskopische Technik, vol. 50. Schwetschke, p. 77.

Hahn, A., D'Agostino, J., Cole, G.A., Raines, J., 2014. Retroperitoneal abscess in two western lowland gorillas (*Gorilla gorilla gorilla*). J. Zoo Wildl. Med. 45 (1), 179–183.

Haigh, J., 2013. Fieldwork in a cold climate. In Field techniques in exotic animal medicine. J. Exot. Pet Med. 22 (1), 51–57.

Haime, J., 1851. Recherches sur le gorille. Ann. Sci. Nat. 16, 158–161.

Hakimzumwami, E., 2000. Community Wildlife Management In Central Africa A Regional Review Submitted to IIED August 1998, published February 2000 Evaluating Eden Series Discussion Paper No.10. International Institute for Environment and Development, London.

Hall, E.R., Russell, W.C., 1933. *Dermestes* beetles as an aid in cleaning bones. J. Mammal. 14, 372–374.

Hall, J., 1995. Recensement de gorilles dans le Parc National du Kahuzi-Biega au Zaire. Afr. Primates. 1 (1), 10–12.

Hall-Craggs, E.C.B., 1961a. The skeleton of an adolescent gorilla (*Gorilla gorilla beringei*). S. Afr. J. Sci. 57 (11), 299–302.

Hall-Craggs, E.C.B., 1961b. The blood vessels of the heart of *Gorilla gorilla beringei*. Am. J. Phys. Anthropol. 19 (4), 373–377.

Halliday, J.E., Allan, K.J., Ekwem, D., Cleaveland, S., Kazwala, R.R., Crump, J.A., 2015. Endemic zoonoses in the tropics: a public health problem hiding in plain sight. Vet. Rec. 176, 220–225.

Hansard, 2012 Primates: Conservation. House of Commons Debates. Written Questions and Answers 17-December 2012, Column 512W. Hansard, London.

Hamada, Y., Udono, T., Teramoto, M., Sugawara, T., 1996. The growth pattern of chimpanzees: somatic growth and reproductive maturation in *Pan troglodytes*. Primates. 37, 279–295.

Hamerton, A.E., 1929. Report on the deaths occurring in the Society's Gardens during the year 1928. Proc. Zool. Soc. Lond. 49–59.

Hamerton, A.E., 1935. Report on the deaths in the Society's Gardens during 1934. Proc. Zool. Soc. Lond. 105, 443–474.

Hamerton, A.E., 1938. Report on the deaths in the Society's Gardens during 1937. Proc. Zool. Soc. Lond. 108, 489–526.

Hamerton, A.E., 1939a. Review of mortality rates and report on the deaths occurring in the Society's Gardens during the year 1938. Proc. Zool. Soc. Lond. 109, 281–287.

Hamerton, A.E., 1939b. Report on the deaths occurring in the Society's Gardens during the year 1939-1940. Proc. Zool. Soc. Lond. 111, 151–187.

Hamerton, A.E., 1943. Report on the deaths occurring in the Society's Gardens during the year 1941-1942. Proc. Zool. Soc. Lond. 112, 151-137.

Harcourt, A., 1987. Behaviour of wild gorillas (*Gorilla gorilla*) and their management in captivity. Int. Zoo Yearb. 26, 248–255.

Harcourt, A., Stewart, K., Fossey, D., 1981. Gorilla reproduction in the wild. In: Graham, C. (Ed.), Reproductive Biology of the Great Apes: Comparative and Biomedical Perspectives. Academic Press, New York, NY, pp. 265−279.

Harcourt, A.H., 1978. Strategies of emigration and transfer by primates, with particular reference to gorillas. Z. Tierpsychol. 48, 401−420.

Harcourt, A.H., 1979. Contrasts between male relationships in wild gorilla groups. Behav. Ecol. Sociobiol. 5, 39−49.

Harcourt, A.H., 1988. Bachelor groups of gorillas in captivity: the situation in the wild. Dodo J. Jersey Wildl. Preserv. Trust. 25, 54−61.

Harcourt, A.H., 1989. Release of gorillas to the wild. Gorilla Conserv. 3, 18−23.

Harcourt, A.H., 1990. Adult male gorillas cohabit in the wild. In: Columbus Zoo Gorilla Workshop Proceedings, pp. 80−81.

Harcourt, A.H., Stewart, K.J., 1978. Coprophagy by wild mountain gorillas. E. Afr. Wildl. J. 16, 223−225.

Harcourt, A.H., Stewart, K.J., 1981. Gorilla male relationships: can differences during immaturity lead to contrasting reproductive tactics in adulthood? Anim. Behav. 29, 206−210.

Harcourt, A.H., Stewart, K.J., 1987. The influence of help in contests on dominance rank in primates: hints from gorillas. Anim. Behav. 35, 182−190.

Harcourt, A.H., Fossey, D., Sabater, P.J., 1981. Demography of Gorilla gorilla. J. Zool. 195, 215−233.

Harcourt, A.H., Stewart, K.J., Harcourt, D.E., 1986. Vocalisations and social relationships of wild gorillas: a preliminary analysis. In: Taub, D.M., King, F.A. (Eds.), Current Perspectives in Primate Social Dynamics. Van Nostrand Reinhold Co, New York, NY, pp. 346−356.

Harden, D., 1962. The Phoenicians. Thames and Hudson, London.

Harris, H.A., 1926. Endocranial form of gorilla skulls, with special reference to the existence of dolichocephaly as a normal feature of certain primates. Am. J. Phys. Anthropol.157−172.

Hart, T.B., Hart, J.A., Dechamps, R., Fournier, M., Ataholo, M., 1996. Changes in forest composition over the last 4000 years in the Ituri basin, Zaire. In: van der Maesen, L.J.G., et al., (Eds.), The Biodiversity of African Plants. Kluwer Academic Publishers, Dordrecht, the Netherlands, pp. 545−563.

Harvey, R.G., Lloyd, D.H., 1995. The distribution of bacteria (other than staphylococci and Propionibacterium acnes) on the hair, at the skin surface and within the hair follicles of dogs. Vet. Dermatol. 6 (2), 79−84.

Hassell, J., 2014. Mortality, Morbidity and Development of Infant Mountain Gorillas (Gorilla beringei beringei): A Retrospective Analysis of 46 Years' Worth of Necropsy Data (M.Sc. dissertation). University of London, London, UK.

Hassell, J.M., Blake, D.P., Cranfield, M.R., Ramer, J., Hogan, J.N., Noheli, J. B., et al., 2013. Occurrence and molecular analysis of Balantidium coli in mountain gorilla (Gorilla beringei beringei) in the Volcanoes National Park, Rwanda. J. Wildl. Dis. 49 (4), 1063−1065.

Hastings, B., 1988. Clinical signs of disease in wild mountain gorilla. In: Proceedings of the American Association of Zoo Veterinarians Annual Conference, 1988, p. 107.

Hastings, B., 1991. The veterinary management of a laryngeal air sac infection in a free-ranging mountain gorilla. J. Med. Primatol. 20, 361−364.

Hastings, B., Kenny, D., Lowenstine, L., Foster, J., 1991. Mountain gorillas and measles: ontogeny of a wildlife vaccination program. In: Proceedings of the American Association of Zoo Veterinarians, pp. 198−205.

Hastings, B.E., Gibbons, L.M., Williams, J.E., 1992. Parasites of free-ranging mountain gorillas: survey and epidemiological factors. In: Proceedings of the Joint American Association Zoo Veterinarians and AAWV, Oakland, CA, USA, pp. 301−302.

Hattingh, J., 1992. Stress in measurement in animals. In: Ebedes, H. (Ed.), The Use of Tranquillizers in Wildlife. Department of Agriculture Development, Pretoria, South Africa.

Hawkins, P. (Ed.), 2004. Husbandry refinements for rats, mice, dogs and non-human primates used in telemetry procedures. In: Seventh Report of the BVA AWF/FRAME/RSPCA/UFAW Joint Working Group on Refinement, Part B. Laboratory Animals 38, pp. 1−10.

Hawks, C., McCann, M., Makos, C., Goldberg, L., Hinkamp, D., Ertel, D., et al., 2011. Health and Safety for Museum Professionals. Society for the Preservation of Natural History Collections, New York.

Hawks, C.A., Makos, K.A., 2000. Inherent and Acquired Hazards in Museum Objects. Cultural Resour. Manage. 23 (5), 31−37.

Hayman, D.T., King, T., Cameron, K., 2010. Successful treatment of acute systemic anaphylaxis in a western lowland gorilla (Gorilla gorilla gorilla). J. Zoo Wildl. Med. 41 (3), 522−525.

Health for Animals, 2015. Global Principles and Perspectives on the Responsible Use of Medicines in Animals. < http://healthforanimals.org/wp-content/files_mf/1441700586GPPRUMA.positionppaper.July2015FINAL.pdf > .

Hebb, D.O., 1947. Spontaneous neurosis in chimpanzees. Psychiatr. Med. 10, 3−19.

Hebert, P., Courtois, M., 1994. Twenty-five years of behavioral research on great apes: trends between 1967 and 1991. J. Comp. Psychol. 106 (4), 373−380.

Heckel, J.O., Rietschel, W., Hufert, F.T., 2001. Prevalence of hepatitis B virus infections in nonhuman primates. J. Med. Primatol. 30 (1), 14−19.

Hedeen, S., 1982. Utilization of space by captive groups of lowland gorillas (Gorilla g. gorilla). Ohio J. Sci. 82 (1), 27−30.

Hediger, H., 1950. Wild Animals in Captivity (G. Sircom, Trans.). Butterworths Scientific Publications, London.

Heminway, J., 1972. A walk with the gorillas. Africana 4 (11), 22−23, 26.

Hemphill, J., McGrew, W., 1998. Environmental enrichment thwarted: Food accessibility and activity levels in captive western lowland gorillas (Gorilla gorilla gorilla). Zool. Garten. 68, 381−394.

Hendrickx, A.G., Binkerd, P.E., 1993. Congenital malformations in nonhuman primates. Nonhuman primates I. Monographs on Pathology of Laboratory Animals. Springer, Berlin, Heidelberg, pp. 170−180.

Hepper, P.G., Wells, D.L., 2010a. Individually identifiable body odors are produced by the gorilla and discriminated by humans. Chem. Senses. 35 (4), 263−268.

Hepper, P.G., Wells, D.L., 2010b. Olfactory discrimination in the western lowland gorilla, Gorilla gorilla gorilla. Primates 53 (2), 121−126.

Herbers, J.M., 1981. Time resources and laziness in animals. Oekologia. 49, 252−262.

Herrmann, E., Wobber, V., Call, J., 2008. Great apes' (Pan troglodytes, Pan paniscus, Gorilla gorilla, Pongo pygmaeus) understanding of tool functional properties after limited experience. J. Comp. Psychol. 122 (2), 220−230.

Hershkovitz, I., Rothschild, B.M., Dutour, O., Greenwald, C., 1998. Clues to recognition of fungal origin of lytic skeletal lesions. Am. J. Phys. Anthropol. 106 (1), 47−60.

Hetherington, C.M., Cooper, J.E., Dawson, P., 1975. A case of syndactyly in the white-lipped tamarin Saguinus nigricollis. Folia Primatol. 24, 24−28.

Heuschele, W.P., 1960. Varicella (Chicken pox) in three young anthropoid apes. J. Am. Vet. Med. Assoc. 136, 256−257.

Hewer, H.R., 1965. The Length and Breadth of Zoology. Inaugural lecture, Imperial College of Science and Technology, London, UK, pp. 205−219.

Heymann, E.W., 1999. Primate behavioural ecology and diseases—some perspectives for a future primatology. Primate Rep. 55, 53−65.

Hill, C.A., 1966. Coprophagy in apes. Int. Zoo Yearb. 6, 251−257.

Hill, S.P., 2009. Do gorillas regurgitate potentially-injurious stomach acid during 'regurgitation and reingestion'? Anim. Welfare. 18, 123−127.

Hill, W.C.O., 1953−1970. Primate Comparative Anatomy and Taxonomy, vols. 1−8. Edinburgh University Press, Edinburgh.

Hill, W.C.O., Sabater Pi, J., 1971. Anomaly of the hallux in a lowland gorilla (Gorilla g. gorilla). Folia Primat. 14, 252−255.

Hime, J.M., Keymer, I.F., Appleby, E.C., 1972. Hypertrophic pulmonary osteoarthropathy in an orang-utan (Pongo pygmaeus). Vet. Rec. 91 (14), 334−337.

HMPC (Human Microbiome Project Consortium), 2012. Structure, function and diversity of the healthy human microbiome. Nature. 486, 207–214.

Hockings, K.J., 2015. Leaps and boundaries. How great apes are adapting to human impact. Pulse.(Issue 27), September 2015, 4–5.

Hockings, K.J., Humle, T., 2009. Best Practice Guidelines for the Prevention and Mitigation of Conflict Between Humans and Great Apes. IUCN/SSC Primate Specialist Group, Gland, Switzerland.

Hockings, K.J., McLennan, M.R., Carvalho, S., Ancrenaz, M., Bobe, R., Byrne, R., et al., 2015. Apes in the Anthropocene: flexibility and survival. Trends Ecol. Evol. 30, 215–222.

Hodgkinson, C., 2009. Tourists, Gorillas and Guns: Integrating Conservation and Development in the Central African Republic (Ph.D. thesis). University College, London.

Hodgkinson, C., Cipolletta, C., 2009. Western lowland gorilla tourism: Impact on gorilla behaviour. Gorilla J. 38, 29–31.

Hof, P.R., Gilissen, E.P., Sherwoodg, C.C., Duana, H., Lee, P.W.H., Delman, B.N., et al., 2002. Comparative neuropathology of brain aging in primates. In: Erwin, J.M., Hof, P.R. (Eds.), Aging in Nonhuman Primates. Interdisciplinary Topics in Gerontology, vol. 31. Karger, Basel, pp. 130–154.

Hoff, M., Maple, T., 1995. Post-occupancy modification of a lowland gorilla. Int. Zoo Yearb. 34, 153–160.

Hoff, M., Nadler, R., Maple, T., 1980. The development of infant play in a captive group of lowland gorillas (*Gorilla gorilla gorilla*). Am. J. Primatol. 1, 65–72.

Hoff, M., Nadler, R., Maple, T., 1981. Development of infant independence in a captive group of lowland gorillas. Dev. Psychobiol. 14 (3), 251–265.

Hoff, M., Nadler, R., Maple, T., 1982. Control role of an adult male in a captive group of lowland gorillas. Folia Primatol. 38, 72–85.

Hoff, M., Forthman, D., Maple, T., 1994. Dyadic interactions of infant lowland gorillas in an outdoor exhibit compared to an indoor holding area. Zoo Biol. 13, 245–256.

Hoff, M., Hoff, K., Horton, C., Maple, T., 1996. Behavioral effects of changing group membership among captive lowland gorillas. Zoo Biol. 15, 383–393.

Hoff, M., Burks, K., Maple, T., 1997. Abnormal behaviors in captive gorillas. In: Ogden, J., Wharton, D. (Eds.), The Management of Gorillas in Captivity. Atlanta/Fulton County Zoo, Atlanta, GA, pp. 26–34.

Hoff, M., Hoff, K., Maple, T., 1998. Behavioural response of a western lowland gorilla (*Gorilla gorilla gorilla*) group to the loss of the silverback male at Zoo Atlanta. Int. Zoo Yearb. 36, 90–96.

Hoffmann, A.R., Proctor, L.M., Surette, M.G., Suchodolski, J.S., 2016. The microbiome: the trillions of microorganisms that maintain health and cause disease in humans and companion animals. Vet. Pathol. 53 (1), 10–21.

Hoffmann, M., Hawkins, C.E., Walsh, P.D., 2008. Action needed to prevent extinctions caused by disease. Nature. 454, 159.

Hofreiter, M., Siedel, H., Van Neer, W., Vigilant, L., 2003. Mitochondrial DNA sequence from an enigmatic gorilla population (*Gorilla gorilla uellensis*). Am. J. Phys. Anthropol. 121, 361–368.

Hog, M., Schindera, I., 1989. Neurodermitis bei Primaten: Fallschilderung eines Gorillaweibchens. (Neurodermatitis in primates: a case report of a female gorilla). Der Hautarzt. 40, 150–152.

Hogan, J.N., Miller, W.A., Cranfield, M.R., Ramer, J., Hassell, J., Bosco Noheri, J., et al., 2014. *Giardia* in mountain gorillas (*Gorilla beringei beringei*), forest buffalo (*Syncerus caffer*), and domestic cattle in Volcanoes National Park, Rwanda. J. Wildl. Dis. 50 (1), 21–30.

Hohmann, G., Robbins, M.M., Boesch, C., 2006. Feeding Ecology in Apes and Other Primates. Ecological, Physical, and Behavioral Aspects. Cambridge University Press, Cambridge, UK.

Holland, W.J., 1924. Account of a skeleton of a gorilla remarkable because showing recovery from gunshot wounds. Ann. Carnegie Mus. 15 (2–3), 293–299.

Home Office, 2014. Guidance on the Operation of the Animals (Scientific Procedures) Act 1986. Presented to Parliament pursuant to Section 21 (5) of the Animals (Scientific Procedures) Act 1986. < https://www.gov.uk/government/uploads/system/uploads/attachment_data/file/291350/Guidance_on_the_Operation_of_ASPA.pdf > .

Homsy, J., 1999. Ape Tourism and Human Diseases: How Close Should We Get? A Critical Review of the Rules and Regulations Governing Park Management & Tourism for the Wild Mountain Gorilla *Gorilla gorilla beringei*. International Gorilla Conservation Program, Kampala, Uganda.

Honess, P., Gimpel, J., Wolfensohn, S., Mason, G., 2005. Alopecia scoring: the quantitative assessment of hair loss in captive macaques. Altern. Lab. Anim. 33 (3), 193–206.

Honess, P.E., Johnson, P.J., Wolfensohn, S.E., 2003. A study of behavioural responses of non-human primates to air transport and re-housing. Lab. Anim. 38, 119–132.

Hopkins, K., Cranfield, M.R., Mudakikwa, A., Stoinski, T.S., Patterson, F.G., Erwin, J.M., et al., 2015. Brain organization of gorillas reflects species differences in ecology. Am. J. Phys. Anthropol. 156, 252–262.

Hopkins, W.D., Russell, J.L., Cantalupo, C., Freeman, H., Schapiro, S.J., 2005. Factors influencing the prevalence and handedness for throwing in captive chimpanzees (*Pan troglodytes*). J. Comp. Psychol. 119, 363–370.

Hopkins, W.D., Phillips, K.A., Bania, A., Calcutt, S.E., Gardner, M., Russell, J., et al., 2011. Hand preferences for coordinated bimanual actions in 777 great apes: implications for the evolution of handedness in hominins. J. Hum. Evol. 60, 605–611.

Hopkins, W.D., Russell, J.L., Schaeffer, J.A., 2012. The neural and cognitive correlates of aimed throwing in chimpanzees: a magnetic resonance image and behavioural study on a unique form of social tool use. Philos. Trans. R. Soc. B. Biol. Sci. 367 (No. 1585), 37–47.

Hopper, J.S., 2009. Veterinary aspects of rehabilitation and eventual release of 3 infant gorillas born in the UK to Bateke Plateau Reserve, Gabon. In: BVZS Conservation Medicine 25/04/09–26/04/09, Jersey, Channel Isles.

Hornaday, W.T., 1922. The Minds and Manners of Wild Animals: A Book of Personal Observations. Charles Scribner's Sons, New York, pp. 93–100.

Horne, W.A., Norton, T.M., Loomis, M.R. 1997. Cardiopulmonary effects of medetomidine-ketamine-isoflurane anesthesia in the gorilla (*Gorilla gorilla*) and chimpanzee (*Pan troglodytes*). In: Proceedings of the American Association of Zoo Veterinarians, pp. 140–142.

Hosey, G., Druck, P., 1987. The influence of zoo visitors on the behaviour of captive primates. Appl. Anim. Behav. Sci. 18, 19–29.

Hosokawa, H., Kamiya, T., 1961. Anatomical sketches of visceral organs of the mountain gorilla (*Gorilla gorilla beringei*). Primates. 3, 1–28.

Hosokawa, H., Kamiya, T., Hirosawa, K., 1965. The brain of the mountain gorilla (*Gorilla gorilla beringei*). Primates. 6, 419–449.

Howell, A.B., 1925. Asymmetry in the skulls of mammals. Proc. US Nat. Mus. 67, 1–18 (article 27, 2599).

Hoyt, A.M., 1941. Toto and I. A Gorilla in the Family. J. B. Lippincott Co, Philadelphia, PA.

HSE, 2012. Managing Health and Safety in Zoos. Health and Safety Executive, London, UK, < www.hse.gov.uk/pubns/books/hsg219.htm > .

HSE, 2014a. Risk Assessment. A Brief Guide to Controlling Risks in the Work Place. Health and Safety Executive, Sudbury, < http://www.hse.gov.uk/pubns/indg163.pdf > .

HSE, 2014b. Health and Safety Made Easy. The Basics for Your Business. Health and Safety Executive, Sudbury, < http://www.hse.gov.uk/pubns/indg449.pdf > .

HSE, 2015. Managing Health and Safety in Zoos. Health and Safety Executive HSG219. HSE Books Ltd, < http://www.hse.gov.uk/pubns/priced/hsg219.pdf > (Revised version pending).

Hudson, P., Rizzoli, A., Grenfell, B., Hesterbeek, H., Dobson, A., 2002. The Ecology of Wildlife Diseases. Oxford University Press, Oxford, UK.

Huff, J.F., 2010. Surgical extractions for periodontal disease in a western lowland gorilla. J. Vet. Dent. 27 (1), 24–32.

Huffman, M.A., Chapman, C.A. (Eds.), 2009. Primate Parasite Ecology: The Dynamics and Study of Host-Parasite Relationships. Cambridge University Press, Cambridge, UK.

Hunter, J.A., 1794. Treatise on the Blood, Inflammation and Gun Shot Wounds. G. Nicholl, London.

Huntress, S.L., Luskutoff, N.M., Raphael, B.L., 1988. Unilateral ovarian adenocarcinoma and in-vitro fertilization in the gorilla. In: Proceedings of the American Association of Zoo Veterinarians, pp. 168–169.

Hurst, L., 2006. Disease and Endangered Species: A Case Study Investigation into Endoparasite Levels of the Western Lowland Gorilla (*Gorilla gorilla gorilla*) (M.Sc. thesis). University of Kent, Kent, UK.

Hutchinson, J.E., Fletcher, A.W., 2010. Using behavior to determine immature life-stages in captive western gorillas. Am. J. Primatol. 72 (6), 492–501.

Huxley, A., 1998. Analysis of shrinkage in human fetal diaphyseal lengths from fresh to dry bone using Petersohn and Kohler's data. J. Forensic Sci. 43 (2), 423–426.

Huxley, T.H., 1863. Evidence as to Man's Place in Nature. Williams & Norgate, London.

Huxley, T.H., 1879. The Crayfish: An Introduction to the Study of Zoology. C. Kegan Paul, London.

Hyde, W.W., 1947. Ancient Greek Mariners. Oxford University Press, Oxford, UK.

IATA, 2016. Live Animals Regulations. Forty-second ed. < https://www.iata.org/publications/pages/live-animals.aspx > .

ICCWC, 2014. International Consortium on Combating Wildlife Crime Strategic Mission 2014 – 2016. International Consortium on Combating Wildlife Crime, < https://cites.org/eng/prog/iccwc.php/Strategy > .

IEEP, Ecologic and GHK, 2012. Study to analyse legal and economic aspects of implementing the Nagoya Protocol on ABS in the European Union. Final report for the European Commission, DG Environment. Institute for European Environmental Policy, Brussels and London, < http://www.ieep.eu/assets/1227/ABS_FINAL_REPORT.PDF > .

Ikpatt, O.F., Reavill, D., Chatfield, J., Clubb, S., Rosenblatt, J.D., Fonte, G., et al., 2014. Diagnosis and treatment of diffuse large b-cell lymphoma in an orangutan (*Pongo pygmaeus*). J. Zoo Wildl. Med. 45 (4), 935–940.

Innes, J.R.M., Hull, W.B., 1972. Endoparasites-lung mites. In: Fiennes, R.N. (Ed.), Pathology of Simian Primates. Part II: Infectious and Parasitic Diseases. Karger, Basel, pp. 177–193.

Inogwabini, B-I, 2014. Conserving biodiversity in the Democratic Republic of Congo: a brief history, current trends and insights for the future *PARKS* 2014 Vol 20.2.

Inoue, M., Hayama, S., 1961. Histopathological studies on two mountain gorilla specimens (*Gorilla gorilla beringei*) "Munidi" and "Emmy". Primates. 3, 29–46.

IOM, 1992. Emerging Infections: Microbial Threats to Health in the United States. Institute of Medicine, Washington, DC.

Ionides, C.J.P., 1965. Mambas and Man-Eaters: A Hunter's Story. Holt, Rinehart and Winston, New York, NY.

Ippen, R., Wildner, G.P., 1984. Comparative pathological investigations of thyroid tumors of animals in zoos and in the wild. In: Oliver, A., Ryder, Mary, L., Byrd (Eds.), One Medicine. Springer, Berlin Heidelberg, pp. 280–295.

IPS, 2007. IPS International Guidelines for the acquisition, care and breeding of nonhuman primates. Second ed. International Primatological Society, < http://www.internationalprimatologicalsociety.org/docs/ips_international_guidelines_for_the_acquisition_care_and_breeding_of_nonhuman_primates_second_edition_2007.pdf > .

Irvin, A.D., Cooper, J.E., Hedges, S.R., 1972. Possible health hazards associated with the collection and handling of post-mortem zoological material. Mamm. Rev. 2, 43–54.

Islam, A., 2013. Manual of Bone Marrow Examination. Trafford Publishing.

Islam, M.S., Siddique, A.K.M., Salam, A., Akram, K., Majumdar, R.N., Zaman, K., et al., 1995. Microbiological investigation of diarrhoea epidemics among Rwandan refugees in Zaire. Trans. R. Soc. Trop. Med. Hyg. 89 (5), 506.

IUCN, 1989. IUCN Policy Statement on Research. International Union for Conservation of Nature, Gland, Switzerland.

IUCN, 2000. Guidelines for the Placement of Confiscated Animals. < https://portals.iucn.org/library/efiles/edocs/2002-004.pdf > .

IUCN, 2004. An Introduction to the African Convention on the Conservation of Nature and Natural Resources. IUCN, Gland, Switzerland and Cambridge, UK, < http://www.ecolex.org/server2.php/libcat/docs/LI/MON-072455.pdf > .

IUCN, 2012. IUCN Red List Categories and Criteria: Version 3.1. Second ed. IUCN, Gland, Switzerland and Cambridge, UK, pp. iv + 32.

IUCN/SSC, 2013. Guidelines for Reintroduction and Other Conservation Translocations. Version 1.0. IUCN Species Survival Commission, Gland, Switzerland.

IUCN, 2014. Regional Action Plan for the Conservation of Western Lowland Gorillas and Central Chimpanzees 2015–2025. Gland, Switzerland: IUCN SSC Primate Specialist Group.

IUCN, 2016. The IUCN Red List of Threatened Species. Version 2016-2. < http://www.iucnredlist.org >.

Iverson, W.O., Popp, J.A., 1978. Meningo-encephalitis secondary to otitis in a gorilla. J. Am. Vet. Med. Assoc. 173, 1134–1136.

Jaffe, J., 2012. Anthropozoonotic Diseases in Mountain Gorillas (Review Paper) (Unpublished MSc in Wild Animal Health). University of London, UK.

Jaffe, J.E., Zimmerman, D., Alibhai, H., Cranfield, M., 2012. Anesthesia with Medetomidine-Ketamine and Dexmedetomidine-Ketamine in Mountain Gorillas (*Gorilla beringei beringei*) (Unpublished Master's Thesis by Jenny Jaffe). Royal Veterinary College, London, UK.

Jaffery, M.S., Chaudhry, M.A., Khan, N.A., Khokhar, M.A., 1976. *Candida albicans* infection in gorillas at the Lahore zoological gardens. Pakistan J. Anim. Sci. 18, 11–16.

Jakob-Hoff, R.M., MacDiarmid, S.C., Lees, C., Miller, P.S., Travis, D., Kock, R., 2014. Manual of procedures for wildlife disease risk analysis. In: World Organisation for Animal Health, Paris, France, Published in association with the International Union for Conservation of Nature and the Species Survival Commission.

Janssen, D.L., 1993. Diseases of great apes. In: Fowler, M.E. (Ed.), Zoo and Wild Animal Medicine: Current Therapy 3. W.B. Saunders, Philadelphia, PA, pp. 334–338.

Janssen, D.L., Bush, R.M., 1990. Review of medical literature of great apes in the 1980s. Zoo Biol. 9, 123–134.

Jarofke, D., Klös, H.-G., 1975. Über einige Hauterkrankungen bei Zootiere des Zoologischen Garten Berlin. Sonderdruck aus Verhandlungsbericht des 17. Internationale Symposiums über die Erkrankungen der Zootiere. Tunis, 1975. Akademie-Verlag, Berlin.

Jarrell, H., 2011. Associations Between Skeletal Fractures and Locomotor Behavior, Habitiat Use, and Body Mass in Nonhuman Primates (Ph.D. dissertation). The Ohio State University.

Jendry, C., 1996. Utilization of surrogates to integrate hand-reared gorilla infants into an age/sex diversified group of conspecifics. Appl. Anim. Behav. Sci. 48, 173–186.

Jendry, C., Absi, A., 1989. Gorilla introductions. Gorilla Gaz. 3 (3), 5–6.

Jeneby, M.M., Suleman, M.A., Gichuki, C., 2002. Sero-epizootiologic survey of *Trypanosoma brucei* in Kenyan nonhuman primates. J. Zoo Wildl. Med. 33 (4), 337–341.

Jenkins, P.D., 1990. Catalogue of Primates in the British Museum (Natural History) and Elsewhere in the British Isles. Part V: The Apes, Superfamily Hominoidea. Natural History Museum Publications, London.

Jensen, K.J., Ndure, J., Plebanski, M., Flanagan, K.L., 2015. Heterologous and sex differential effects of administering vitamin A supplementation with vaccines. Trans. R. Soc. Trop. Med. Hyg. 109, 36–45.

Jirků, M., Pomajbíková, K., Petrželková, K.J., Hůzová, Z., Modrý, D., Lukeš, J., 2012. Detection of *Plasmodium* spp. in human feces. Emerg. Infect. Dis. 18 (4), 634, Vol. 33, No. 4 (Dec., 2002), pp. 337-341.

Jobbins, S.E., Alexander, K.A., 2015. Whence they came—antibiotic-resistant *Escherichia coli* in African wildlife. J. Wildl. Dis. 51 (4), 1—10.

Joganic, J.L., 2016. Skeletal and dental development in a sub-adult western lowland gorilla (*Gorilla gorilla gorilla*). Am. J. Phys. Anthropol. 159, 174—181.

Johnson, C., Anderson, S.R., JonDallimore, J., Winser, S., Warrell, D., Imray, C., Moore, J. (Eds.), 2015. Oxford Handbook of Expedition and Wilderness Medicine. Third ed. Oxford University Press Medical Handbooks.

Johnstone-Scott, R., 1984. Integration and management of a group of lowland gorillas (*Gorilla gorilla gorilla*) at the Jersey Wildlife Preservation Trust. Dodo J. Jersey Wildl. Preserv. Trust. 21, 67—79.

Johnstone-Scott, R., 1988. The potential for establishing bachelor groups of western lowland gorillas (*Gorilla g. gorilla*). Dodo J. Jersey Wildl. Preserv. Trust. 25, 61—66.

Johnstone-Scott, R., 1995. Jambo: A Gorilla's Story. Macmillan, London.

Johnstone-Scott, R., 1998. The effects of a change in leadership in a breeding group of western lowland gorillas *Gorilla g. gorilla* at the Jersey Wildlife Preservation Trust. Dodo J. Jersey Wildl. Preserv. Trust. 34, 42—65.

Joines, S., 1977. A training programme designed to induce maternal behavior in a multiparous female lowland gorilla, *Gorilla g. gorilla*, at the San Diego Wild Animal Park. Int. Zoo Yearb.185—188.

Jonch, A., 1967. The white gorilla (*Gorilla g. gorilla*) at Barcelona Zoo. Int. Zoo Yearb. 13, p. 196.

Jones, D.M., Dixson, A.F., Wadsworth, P.F., 1980. Interstitial cell tumour of the testis in a western lowland gorilla (*Gorilla gorilla gorilla*). J. Med. Primatol. 9 (5), 319—322.

Jones, E.E., Alford, P.L., Reingold, A.L., Russell, H., Keeling, M.E., Broome, C.V., 1984. Predisposition to invasive pneumococcal illness following parainfluenza type 3 virus infection in chimpanzees. J. Am. Med. Vet. Assoc. 185, 1351—1353.

Jones, F.W., 1938. The so-called maxillary antrum of the gorilla. J. Anat. 73 (Pt 1), 116—119.

Jones, T.S., Cave, A.J.E., 1960. Diet, longevity and dental disease in the Sierra Leone chimpanzee. Proc. Zool. Soc. Lond. 135, 147—155.

Jones-Engel, L., Engel, G., 2006. Disease risk analysis: a paradigm for using health-based data to inform primate conservation and public health. Am. J. Primatol. 68, 851—854.

Jones-Engel, L., Engel, G.A., Schillaci, M.A., Lee, B., Heidrich, J., Chalise, M., et al., 2006. Considering human—primate transmission of measles virus through the prism of risk analysis. Am. J. Primatol. 68, 868—879.

Jones, B.T.B., 2008. Legislation and Policies relating to Protected Areas, Wildlife Conservation, and Community Rights to Natural Resources in countries being partner in the Kavango Zambezi Transfrontier Conservation Area. Brian T. B. Jones Windhoek December 2008 A. < http://www.tbpa.net/docs/KAZAPolicy > .

Jonsson, N.N., Reid, S.W.J., 2000. Global climate change and vector borne diseases. Vet. J. 160, 87—89.

Jordan, D.S. (Ed.), 1910. Leading American Men of Science. Henry Holt and Company, New York, NY.

Jortner, B.S., 2012. Pathologic lesions? Vet. Pathol. 49 (5), 880.

Jurmain, R., 2000. Degenerative joint disease in African great apes: an evolutionary perspective. J. Hum. Evol. 39 (2), 185—203.

Kaandorp, 2010. EAZWV Transmissible Diseases Handbook. European Association of Zoo and Wildlife Veterinarians (EAZWV). Autumn, cited by < http://wildpro.twycrosszoo.org/S/00dis/Parasitic/lung_mites_bonobo.htm > .

Kalema, G., 1995. Epidemiology of the intestinal parasite burden of mountain gorillas *Gorilla gorilla beringei* in Bwindi Impenetrable National Park, South-west Uganda. BVZS Newsl. Autumn, 19—34.

Kalema, G., Kock, R.A., Macfie, E., 1998. An outbreak of sarcoptic mange in free-ranging mountain gorillas (*Gorilla gorilla berengei*) in Bwindi

Impenetrable National Park, South Western Uganda. In: Proceedings of the American Association of Zoo Veterinarians/American Association of Wildlife Veterinarians Joint Conference, Omaha, Nebraska, 17—22 October 1998, p. 438.

Kalema-Zikusoka, G., 2012. Ruhondeza laid to rest. Br. Vet. Assoc. Overseas Newsl. August p. 6.

Kalema-Zikusoka, G., 2013. Mountain gorilla disease: implications for conservation. In: Cooper, J.E., Cooper, M.E. (Eds.), Wildlife Forensic Investigation: Principles and Practice. Taylor & Francis/CRC Press, Boca Raton, FL.

Kalema-Zikusoka, G., Lowenstine, L., 2001. Rectal prolapse in a free-ranging mountain gorilla (*Gorilla beringei beringei*): clinical presentation and surgical management. J. Zoo. Wildl. Med. 32 (4), 509—513.

Kalema-Zikusoka, G., Rubanga, S.V., 2013. The establishment and use of field laboratories: lesion from the CTPH Gorilla Research Clinic, Uganda. J. Exot. Pet Med. 22, 34—38.

Kalema-Zikusoka, G., Kock, R.A., Macfie, E.J., 2002. Scabies in free ranging mountain gorillas (*Gorilla beringei beringei*) in Bwindi Impenetrable National Park, Uganda. Vet. Rec. 150, 12—15.

Kalema-Zikusoka, G., Rothman, J.M., Fox, M.T., 2005. Intestinal parasites and bacteria of mountain gorillas (*Gorilla beringei beringei*) in Bwindi Impenetrable National Park, Uganda. Primates. 46, 59—63.

Kalina, J., Butynski, T.M., 1995. Close encounters between people and gorillas. Afr. Primates. 1 (1), 20.

Kalter, S.S., 1980. Infectious diseases of the great apes of Africa. J. Reprod. Fertil. 28 (Suppl.), 149—159.

Kamani, T.M., Kazwala, R., Mfinanga, S., Haydon, D., Keyyu, J., Lankester, F., et al., 2015. One health: a concept led by Africa, with global benefits. Vet. Rec. 176 (19), 496—497.

Kambale, E.S., Ramer, J.C., Gilardi, K., et al., 2014. Cardiovascular and hepatic disease in wild eastern lowland gorillas (*Gorilla beringei graueri*). Proc. Am. Assoc. Zoo Vet.

Kaplan, C.G., 1979. Intrauterine infections in nonhuman primates. J. Med. Primatol. 8, 233—243.

Kaplan, H., Hill, K., Lancaster, J., Hurtado, A.M., 2000. A theory of human life history evolution: diet, intelligence and longevity. Evol. Anthropol. 9, 156—184.

Kapoor, A., Mehta, N., Esper, F., et al., 2010. Identification and characterization of a new bocavirus species in gorillas. PLoS ONE (United States). 5 (7), pe11948.

Karesh, W.B., 1993. Cost evaluation of infectious disease monitoring and screening programs for wildlife translocation and reintroduction. J. Zoo Wildl. Med. 24, 291—295.

Karesh, W.B., Cook, R.A., 1995. Application of veterinary medicine to *in situ* conservation efforts. Oryx. 29, 244—252.

Karesh, W.B., et al., 1988. Leydig cell tumor in a western lowland gorilla. J. Zoo Anim. Med. 19 (1-2), 51—54.

Karisoke Archives, 1981—1987. Karisoke Research Center, Ruhengeri, Rwanda.

Kaufmann, M.H., 2004. The embryology of conjoined twins. Child's Nerv. Syst. 20, 508—525.

Kaur, T., Singh, J., Tong, S., Humphrey, C., Clevenger, D., Tan, W., et al., 2008. Descriptive epidemiology of fatal respiratory outbreaks and detection of a human-related metapneumovirus in wild chimpanzees (*Pan troglodytes*) at Mahale Mountains National Park, Western Tanzania. Am. J. Primatol. 70, 755—765.

Kawashima, T., Sato, F., 2012. Detailed comparative anatomy of the extrinsic cardiac nerve plexus and postnatal reorganization of the cardiac position and innervation in the great apes: orangutans, gorillas, and chimpanzees. Anat. Rec. 295 (3), 438—453.

Keay, J.M., Singh, J., Gaunt, M.C., Kaur, T., 2006. Fecal glucocorticoids and their metabolites as indicators of stress in varius mammalian species: a literature review. J. Zoo Wildl. Med. 37 (3), 234—244.

Keen, S., 2015. A Review of the Data Relating to the Animal Management and Welfare of Gorillas at Bristol Zoo Gardens From 1930—2015. Student Project. University of Bristol and Bristol Zoo Gardens, UK.

Keita, M.B., Hamad, I., Bittar, F., 2014. Looking in apes as a source of human pathogens. Microb. Pathog. 77, 149—154.

Keith, A., 1896. An introduction to the study of anthropoid apes. 1 The Gorilla. Nat. Sci. 9, 26—57.

Keith, A., 1899. On the chimpanzees and their relationship to the gorilla. In: Proceedings of the General Meetings for Scientific Business of the Zoological Society of London, 7 March 1899, pp. 296—312.

Keith, A., 1926. The gorilla and man as contrasted forms. Lancet. 207 (5349), 490—492.

Keizer, F., 1989. Apenheul gorilla diet. Gorilla Gaz. 3 (2), 8—9.

Keizer, F., 1990. Intragroup mother-rearing and intergroup transfers at Apenheul. In: Columbus Zoo Gorilla Workshop Proceedings, pp. 109—116.

Keizer, F., Keizer, M., 2004. The gorillas of "Petit Evengue". Gorilla J. 29, 28—30.

Kelley, J., Schwartz, G.T., 2010. Dental development and life history in living African and Asian apes. Proc. Natl. Acad. Sci. U.S.A. 107, 1035—1040.

Kelly, D.F., 1994. Veterinary pathology: the benefits and the cost. J. Small Anim. Pract. 35 (2), 65—67.

Kelly, H.A., 1880. Sartorius muscle of the gorilla. Proc. Acad. Nat. Sci. Philadelphia. 32, 128.

Kennard, M.A., Willner, M.D., 1941. Findings at autopsies of seventy anthropoid apes. Endocrinology. 28, 967—976.

Kennedy-Stoskopf, S., 1997. In: Fowler, M.E., Miller, E. (Eds.), Fowler's Zoo and Wild Animal Medicine Current Therapy, Vol. 7. Saunders.

Kenny, D.E., Cambre, R.C., Alvarado, T.P., et al., 1994. Aortic dissection: an important cardiovascular disease in captive gorillas (Gorilla gorilla gorilla). J. Zoo. Wildl. Med. 25 (4), 561—568.

Kerbis, J.P., 1990. The Role of Leopards, Hyaena and Porcupines in Ungulate Carcass Dispersal: Implications for Paleoanthropology (Ph.D. thesis). University of Chicago, Chicago.

Khabbaz, R.F., Heneine, W., George, J.R., Parekh, B., et al., 1994. Infection of a laboratory worker with simian immunodeficiency virus. N. Engl. J. Med. 330, 172—177.

Kilbourn, A., Froment, J.-M., Rouguet, P., Leroy. E., Karesh, W., 2002. Disease trends in Central Africa: the implications on great ape health and conservation. In: American Association of Zoo Veterinarians Annual Proceedings, 2002. pp. 380—381.

Killough, M.L., Hunt, D., Eriksen, A.B., et al., 2015. Skeletal pathology in individually documented wild Virunga mountain gorillas. Am. J. Phys. Anthropol. 156, 186—187.

Kim, K.C., Emerson, K.C., 1968. Descriptions of two species of Pediculidae (Anoplura) from great apes (Primates, Pongidae). J. Parasitol. 54 (4), 690—695.

Kimura, T., Hamada, Y., 1996. Growth of wild and laboratory born chimpanzees. Primates. 37, 237—251.

King, J.M., Roth-Johnson, L., Newson, M.E., 2007. The Necropsy Book. Fifth ed. C.L. Davis DVM Foundation, Gurnee, IL.

King, T., 2004. Reintroduced western gorillas reproduce for the first time. Oryx. 38 (3), 251—252.

King, T., 2005. Gorilla re-introduction program, Republic of Congo. Gorilla Gaz. 18 (1), 28—31.

King, T., Chamberlan, C., 2007. Orphan gorilla management and reintroduction: progress and perspectives. Gorilla J. 34, 21—25.

King, T., Courage, A., 2007. Reintroduced western gorillas reproduce again. Oryx. 41 (1), 14.

King, T., Boyen, E., Muilerman, S., 2003. Variation in reliability of measuring behaviours of reintroduced orphan gorillas. Int. Zoo News. 50 (5), 288—297.

King, T., Chamberlan, C., Courage, A., 2005a. Rehabilitation of orphan gorillas and bonobos in the Congo. Int. Zoo News. 52 (4), 198—209.

King, T., Chamberlan, C., Courage, A., 2005b. Reintroduced gorillas: reproduction, ranging and unresolved issues. Gorilla J. 30, 30—32.

Kingdon, J., 1988. Island Africa. The Evolution of Africa's Rare Animals and Plants. Collins, London.

Kingdon, J., et al., 2013. Primates. Mammals of Africa. Vol II. Bloomsbury.

Kirk, H., 1940. The London Zoological Society and the veterinary profession. Vet. Rec. 52 (3), 54.

Kirkwood, J.K., 1983. Influence of body size in animals on health and disease. Vet. Rec. 113, 287—290.

Kirkwood, J.K., 1992. Wild animal welfare. In: Ryder, R.R. (Ed.), Animal Welfare and the Environment. G. Duckworth & Co, London, pp. 139—145.

Kirkwood, J.K., 2000. Interventions for the conservation or welfare of wild animals. In: Legood, G. (Ed.), Veterinary Ethics: an Introduction. Continuum, London, pp. 121—128.

Kirkwood, J.K., 2003. Introduction: wildlife casualties and the veterinary surgeon. In: Mullineaux, Cooper, J.E. (Eds.), Manual of Wildlife Casualties. BSAVA, Gloucester.

Kitchener, A.C., 1997. The role of museums and zoos in conservation biology. Int. Zoo Yearb. 35, 325—336.

Klailova, M., 2011. Interunit, Environmental and Interspecific Influences on Silverback-Group Dynamics in Western Lowland Gorillas (Gorilla gorilla gorilla) (Ph.D. thesis). Department of Psychology, University of Stirling, Stirling, UK.

Klailova, M., Hodgkinson, C., Lee, P., 2010a. Behavioral responses of one western lowland gorilla (Gorilla gorilla gorilla) group at Bai Hokou, Central African Republic, to tourists, researchers and trackers. Am. J. Primatol. 72 (10), 897—906, Special Issue on Ethnoprimatology.

Klailova, M., Hodgkinson, C., Lee, P., 2010b. Human impact on western lowland gorilla behaviour. Gorilla J. 40, 19—22.

Klailova, M., Casanova, C., Henschel, P., Lee, P., Rovero, F., Todd, A., 2012. Non-human predator interactions with wild great apes in Africa: a review focusing on large felid predation and a discussion of novel methods for assessing predator-prey relationships. Folia Primatol. 83, 312—328, Special Issue on Primate-Predators.

Klein, R.G., Cruz-Uribe, K., 1984. The Analysis of Animal Bones from Archeological Sites. University of Chicago Press, Chicago, IL.

Kleinnijenhuis, J., van Crevel, J., Netea, M., 2015. Trained in consequences for the heterologous effects of BCG vaccination. Trans. R. Soc. Trop. Med. Hyg. 109, 29—35.

Knapp, S., James, P., McCulley, J.P., Alvarado, T.P., Hogan, R.N., 2007. Comparative ocular anatomy of the western lowland gorilla. Vet. Ophthalmol. 10 (6), 357—362.

Knauf, S., Liu, H., Harper, K.N., 2013. Treponemal infection in nonhuman primates as possible reservoir for human yaws. Emerg. Infect. Dis. 19 (12), 2059—2060.

Knight, B., 1986. The cause of death. J. R. Soc. Med. 79, 191—192.

Knight, M., 1968a. How to Keep a Gorilla. Wolfe Publishing, London.

Knight, M., 1968b. Be a Nature Detective. Frederick Warne, London.

Knott, C.D., Thompson, M.E., Wich, S.A., 2009. The ecology of female reproduction in wild orangutans. In: Wich, S.A., Utami Atmoko, S.S., Mitra Setia, T., van Schaik, C.P. (Eds.), Orangutans: Geographic Variation in Behavioral Ecology and Conservation. Oxford University Press, Oxford, UK, pp. 171—188.

Knox, A.G., Walters, M.P., 1994. Extinct and endangered birds in the collections of the Natural History Museum. British Ornithologists' Club Occasional Publications, No. 1, Maidenhead, UK. p. 292.

Koch, W., 1937. Bericht über das Ergebnis des Obduktion des Gorilla Bobby des Zoologischen Gartens zu Berlin: ein Beitrag zur verleichenden Konstitutionspathologie. Ver. Kunst Wehrpathol. 9, 1—36.

Köhler, W., 1926. The Mentality of Apes. Harcourt, Brace & Company, Inc., New York, NY.

Kompanje, E.J.O., Hermans, J.J., 2008. Cephalopagus conjoined twins in a leopard cat (*Prionailurus bengalensis*). J. Wildl. Dis. 44 (1), 177–180.

Köndgen, S., Kühl, H., N'Goran, P.K., Walsh, P.D., Schenk, S., Ernst, N., et al., 2008. Pandemic human viruses cause decline of endangered great apes. Curr. Biol. 18 (4), 260–264.

Kondo, H., Wda, Y., Bando, G., et al., 1996. Alveolar hydatidosis in a gorilla and in a ring-tailed lemur in Japan. J. Vet. Med. Sci. 58, 447–449.

Kopff, H., Mager, W., 1990. Further developments in the breeding population of lowland gorillas at Apenheul. Zoo Biol. 9, 165–170.

Krapp, F., Lampel, G., 1973. Zahnanomalien bei Altweltaffen (Catarrhina). Rev. Suis. Zool. 80, 83–150.

Krief, S., Jamart, A., Hladik, C.M., 2004. On the possible adaptive value of coprophagy in free-ranging chimpanzees. Primates. 45, 141–145.

Krogman, W.M., Schultz, A.H., 1938. Anthropoid ape materials in American collections. Am. J. Phys. Anthropol. 24, 199–234.

Krueger, G.R.F., 1979. Coco und Pucker, die Berg-gorillas (*Gorilla g. beringei*) des Kolner Zoo – ein Epilog. Zeitschrift des Kolner Zoo. Cologne Zoo, Germany.

Krueger, G.R.F., Neumann, E.P., Kullmann, E., 1980. Cystic thymic dysplasia in two mountain gorillas with lethal gram-negative septicemia. In: Montali, R.J., Migaki, G. (Eds.), The Comparative Pathology of Zoo Animals. Smithsonian Institution Press, Washington, DC.

Kuehl, H.S., Todd, A.F., Boesch, C., Walsh, P.D., 2007. Manipulating decay time for efficient large-mammal density estimation: gorillas and dung height. Ecol. Appl. 17 (8), 2403–2414.

Kuhar, C.W., Bettinger, T., Laudenslager, M., 2005. Salivary cortisol and behaviour in an all-male group of western lowland gorillas (*Gorilla g. gorilla*). Anim. Welfare. 14 (3), 187–193.

Kühl, H., Maisels, F., Ancrenaz, M., Williamson, E.A., 2008. Best Practice Guidelines for Surveys and Monitoring of Great Ape Populations. IUCN, Gland, Switzerland.

Kuiken, T., van den Hoogen, B.G., van Riel, D.A., Laman, J.D., van Amerongen, G., Sprong, L., et al., 2004. Experimental human metapneumovirus infection of cynomolgus macaques (*Macaca fascicularis*) results in virus replication in ciliated epithelial cells and pneumocytes with associated lesions throughout the respiratory tract. Am. J. Pathol. 164, 1893–1900.

Kukavica-Ibrulj, I., Hamelin, M.E., Prince, G.A., Gagnon, C., Bergeron, Y., Bergeron, M.G., et al., 2009. Infection with human metapneumovirus predisposes mice to severe pneumococcal pneumonia. J. Virol. 83, 1341–1349.

Kurian, K.M., Moss, T.H., Camelo-Piragua, S., 2014. Atlas of Gross Neuropathology: A Practical Approach. Cambridge University Press, Cambridge, UK.

Kurth, R., Norbert Bannert, N., 2010. Retroviruses. Caister Academic, Wymondham, UK.

Laber-Laird, K.E., Jokinen, M.P., Lehner, N.D.M., 1987. Fatal fatty liver-kidney syndrome in obese monkeys. Lab. Anim. Sci., 205–209.

Lacépède, B.G.E. de la V., 1799. Tableau des divisions, sous-divisions, ordres et genres des mammifères. In: Discours de l'ouverture et de clôture du cours d'histoire naturelle donné dans le Muséum National d'Histoire Naturelle, l'an VII de la République, et tableaux méthodiques des mammifères et des oiseaux. Plassan, Paris.

Lacroix, W.F.G., 1998. Africa in antiquity: a linguistic and toponymic analysis of Ptolemy's map of Africa, together with a discussion of Ophir, Punt and Hanno's voyage. Nijmegen Studies in Development and Cultural Change, 28. Verlag für Entwicklungspolitik Saarbrücken GmbH, Germany, xi + 416 pp.

LAGA, undated. < http://www.laga-enforcement.org/Resources/LegalRegional Library/tabid/176/language/en-US/Default.aspx > .

Lair, S., Crawshaw, G.J., Mehren, K.G., Pare, J., 1997. Clinical investigation of hypothyroidism in a western lowland gorilla (*Gorilla gorilla gorilla*). In: American Association of Zoo Veterinarians Annual Conference Proceedings, Lawrence, KS.

Lair, S., Crawshaw, G.J., Mehren, K.G., Perrone, M.A., 1999. Diagnosis of hypothyroidism in a western lowland gorilla (*Gorilla gorilla gorilla*) using human thyroid-stimulating hormone assay. J. Zoo. Wildl. Med. 30, 537–540.

Lair, S., Crawshaw, G.J., Mehren, K.G., Perrone, M.A., 2000. Evaluation of a human immunometric assay for the determination of thyroid-stimulating hormone in nonhuman primates. J. Zoo. Wildl. Med. 31 (2), 267–268.

Landsoud-Soukate, J., Tutin, C.E.G., Fernandez, M., 1995. Intestinal parasites of sympatric gorillas and chimpanzees in the Lope Reserve, Gabon. Ann. Trop. Med. Parasitol. 89, 73–79.

Landsteiner, K., Miller, C.P., 1925. Serological studies on the blood of primates. II. The blood groups in anthropoid apes. J. Exp. Med. 42 (6), 853–862.

Lange, B., 2005. Die Allianz von Naturwissenschaft, Kunst und Kommerz in Inszenierungen des Gorillas nach 1900. In: Zimmermann, A., Sichtbarkeit und Medium. Austausch, Verknüpfung und Differenz naturwissenschaftlicher und ästhetischer Bildstrategien: 183–210. Hamburg University Press.

Lange, B., 2006. Echt, Unecht, Lebensecht. Menschenbilder im Umlauf. Kadmos Kulturverlag, Berlin.

Langer, S., Jurczynski, K., Gessler, A., Kaup, F.-J., Bleyer, M., Mätz-Rensing, K., 2013. Ischiopagus tripus conjoined twins in a western lowland gorilla (*Gorilla gorilla*). J. Comp. Pathol. 150, 469–473.

Lanjouw, A., 2002. Mountain gorillas—a challenge for modern conservation. Fauna Flora. 2, 6–14, April 2002.

Lankester, F., Lankester, F., Mätz-Rensing, K., Kiyang, J., Jensen, S.A., Weiss, S., et al., 2008. Fatal ulcerative colitis in a western lowland gorilla (*Gorilla gorilla gorilla*). J. Med. Primatol. 37 (6), 297–302.

Lankester, F., Kiyang, J.A., Bailey, W., et al., 2010. *Dientamoeba fragilis*: initial evidence of pathogenicity in the western lowland gorilla (*Gorilla gorilla gorilla*). J. Zoo Wildl. Med. 41 (2), 350–352.

Lanyon, L., 2014. Evidence-based veterinary medicine: a clear and present challenge. Vet. Rec. 174 (7), 173–175.

Larkin, N.R., 2016a. Cleaning, packing and moving a 115 year old taxidermied adult male orang-utan, stuck in a very fragile old nest of leaves in a tree with other nests. J. Nat. Sci. Collect. 3, 38–42.

Larkin, N.R., 2016b. Japanese tissue paper and its uses in osteological conservation. J. Nat. Sci. Collect. 3, 62–67.

Lash, N., Ogden, J., Meller, L., Wall, V., 1997. Design. In: Ogden, J., Wharton, D. (Eds.), The Gorilla Husbandry Manual. Fulton County Zoo, Atlanta, GA, pp. 217–266.

Lasley, B., Czekala, N., Presley, S., 1982. A practical approach to evaluation of fertility in the female gorilla. Am. J. Primatol.(Suppl. 1), 45–50.

Latendresse, J.R., Warbrittion, A.R., Jonassen, H., Creasy, D.M., 2002. Fixation of testes and eyes using a modified Davidson's fluid: comparison with Bouin's fluid and conventional Davidson's fluid. Toxicol. Pathol. 30, 524–533.

Laudati, A., 2010. Ecotourism: the modern predator? Implications of gorilla tourism on local livelihoods in Bwindi Impenetrable National Park, Uganda. Environ. Plann. D –Soc. Space. 28, 726–743.

Laule, G., 1993. The use of behavioral management techniques to reduce or eliminate abnormal behavior. Anim. Welfare Inf. Cent. Bull. 4 (4), 1–3, 8-11.

Laule, G., Whittaker, M., 2000. A behavioral management approach to caring for great apes. Paper Presented at The Apes: Challenges for the 21st Century. Brookfield Zoo, Chicago, IL.

Lavoie, C., 2012. Biological collections in an ever changing world: herbaria as tools for biogeographical and environmental studies. Perspect. Plant Ecol. Evol. Syst. 15, 68–76.

Leadbetter, S., 2011. Organ and tissue responses in humans. Symposium on the Pathology of Abuse in Animals and Humans. The Royal College of Pathologists, London.

Leakey, L.S.B., 1970. The relationship of African apes, man, and Old World monkeys. Proc. Natl. Acad. Sci. U.S.A. 67 (2), 746–748.

Lebel, J., 2003. Health. An Ecosystem Approach. International Development Research Centre (IDRC), Ottawa, Canada, Issued also in French under the title *La santé, une approche écosystémique*.

Lee, R.V., Orlick, A.O., Dolensek, E.P., Doherty, J.G., 1981. The electrocardiogram of the lowland gorilla (*Gorilla gorilla*). J. Zoo Anim. Med. 12, 73–80.

Leendertz, F.H., Ellerbrok, H., Boesch, C., et al., 2004. Anthrax kills wild chimpanzees in a tropical rainforest. Nature. 430 (6998), 451–452.

Leendertz, F.H., Lankester, F., Guislain, P., Neel, C., Drori, O., Dupain, J., et al., 2006a. Anthrax in western and central African great apes. Am. J. Primatol. 68, 928—933.

Leendertz, F.H., Pauli, G., Maetz-Rensing, K., Boardman, W., Nunn, C., Ellerbrok, H., et al., 2006b. Pathogens as drivers of population declines: the importance of systematic monitoring in great apes and other threatened mammals. Biol. Conserv. 131, 325—337.

Lehn, C., 2008. Biomaterials in gorilla research and conservation. In: Stoinski, T.S., Steklis, H.D., Mehlman, P.T. (Eds.), Conservation in the 21st Century: Gorillas as a Case Study. Springer Science, pp. 253—268.

Leiva, M., Peña, T., Bayón, A., de León, M., Morales, I., 2012. Phacoemulsification considerations in nonhuman primates. J. Med. Primatol. 41 (5), 317—324.

Lemen, R., Lemen, S., Morrish, R., Tooley, W., 1974. Marasmus and shigellosis in two infant gorillas. J. Med. Primatol. 3, 365—369.

Lenz, A., Franklin, G.A., Cheadle, W.G., 2007. Systemic inflammation after trauma. Injury 38, 1336—1345.

Less, E.H., Lukas, K.E., Kuhar, C.W., et al., 2010. Behavioral response of captive western lowland gorillas (*Gorilla gorilla gorilla*) to the death of silverbacks in multi-male groups. Zoo Biol. 29 (1), 16—29.

Less, E.H., Kuhar, C.W., Dennis, P.M., Lukas, K.E., 2012. Assessing inactivity in zoo gorillas using keeper ratings and behavioral data. Appl. Anim. Behav. Sci. 137, 74—79.

Levréro, F., Gatti, S., Ménard, N., Petit, E., Caillaud, D., Gautier-Hion, A., 2006. Living in nonbreeding groups: an alternative strategy for maturing gorillas. Am. J. Primatol. 68, 275.

Levréro, F., Gatti, S., Gautier-Hion, A., Ménard, N., 2007. Yaws disease in a wild gorilla population and its impact on the reproductive status of males. Am. J. Phys. Anthropol. 132 (4), 568—575.

Lewis, J.C.M., 2003. Preventive Health Measures for Primates and Keeping Staff in British and Irish Zoological Collections. International Zoo Veterinary Group. A report to the British and Irish Primate Taxon Advisory Group (B&I PTAG), London, Federation of Zoos, < http://www.2ndchance.info/vaccination-Lewis2003.pdf > .

Liang, D., Alvarado, T.P., Oral, D., Vargas, J.M., Denena, M.M., McCulley, J.P., 2005. Ophthalmic examination of the captive western lowland gorilla (*Gorilla gorilla gorilla*). J. Zoo Wildl. Med. 36 (3), 430—433.

Lilly, A.A., et al., 2002. Intestinal parasites in gorillas, chimpanzees, and humans at Mondika Research Site, Dzanga-Ndoki National Park, Central African Republic. Int. J. Primatol. 23, 555—573.

Linnemann, C.C., Kramer, L.W., Askey, P.A., 1984. Familial clustering of hepatitis B infections in gorillas. Am. J. Epidemiol. 119, 424—430.

Lintz, G.D., 1942. Animals Are My Hobby. R.M. McBride & Co, New York.

Litchfield, C., 1997. Treading Lightly: Responsible Tourism With the African Great Apes. Travellers' Medical & Vaccination Centre (TMVC), Chatswood, NSW, Australia.

Liu, W., Li, Y., Learn, G.H., Rudicell, R.S., Robertson, J.D., Ndjango, J.B. N., et al., 2010. Origin of the human malaria parasite *Plasmodium falciparum* in western gorillas (*Gorilla gorilla*). Nature. 467, 420—442.

Liversidge, H., 2003. Variation in modern human dental development. In: Krovitz, G.E., Nelson, A.J., Thompson, J.L. (Eds.), Patterns of Growth and Development in the Genus *Homo*, Cambridge University Press, Cambridge, pp. 73—113.

Lokai-Owens, A., 1999. Browse for primates. Gorilla Gaz. 13 (1), 30—31.

Lombardi, D., 1987. Nursery rearing of infant gorillas. Gorilla Gaz. 1 (2), 3.

Lönnberg, E., 1917. Mammals collected in Central Africa by Captain E. Arrhenius. Kungl. Sven. Vetensk. akad. Handl. 58, 3—110.

Loomis, M.R. 1990. Update of vaccination recommendations for nonhuman primates. In: American Association of Zoo Veterinarians Annual Proceedings, pp. 257—260.

Loskutoff, N., Bowsher, T., Chatfield, J., Stones, G., Ramey, J.W., Zhang, L., et al., 2004. Ovarian stimulation, transvaginal, ultrasound-guided oocyte retrieval, ICSI and blastocyte production in sequential media in the western lowland gorilla (*Gorilla gorilla gorilla*). Reprod. Fertil. Dev. 16 (2), 225.

Lovell, N.C., 1990a. Skeletal and dental pathology of free-ranging mountain gorillas. Am. J. Phys. Anthropol. 81, 399—412.

Lovell, N.C., 1990b. Patterns of Injury and Illness in Great Apes. A Skeletal Analysis. Smithsonian Institution Press, Washington and London.

Lovell, N.C., 1991. An evolutionary framework for assessing illness and injury in nonhuman primates. Am. J. Phys. Anthropol. Suppl.: Yearb. Phys. Anthropol. 34 (Issue Suppl. S13), 117—155.

Lovell, N.C., et al., 2000. Skeletal evidence of probable treponemal infection in free-ranging African apes. Primates. 41, 275—290.

Lowenstine, L.J., 1990. Long-distance pathology, or—will a mountain gorilla fit in the diplomatic pouch? In: American Association of Zoo Veterinarians Annual Proceedings, p. 178.

Lowenstine, L.J., 2003. A primer of primate pathology: lesions and non-lesions. Toxicol. Pathol. 32 (Suppl.), 92—102.

Lowenstine, L.J., Lerche, N.W., 1988. Retrovirus infections in non-human primates: a review. J. Zoo Wildl. Med. 19, 168—187.

Lowenstine, L.J., Lerche, N.W., 1993. Nonhuman primate retroviruses and simian acquired immunodeficiency syndrome. In: Fowler, M.E. (Ed.), Zoo and Wild Animal Medicine: Current Therapy, 3. W.B. Saunders, Philadelphia, PA, pp. 373—377.

Lowenstine, L.J., McManamon, R., Terio, K.A., 2016. Comparative pathology of aging great apes: bonobos, chimpanzees, gorillas, and orangutans. Vet. Pathol. 53 (2), 250—276.

Lowenstine, L.J., Montali, R.J., 2016. Historical perspective and future directions in training of veterinary pathologists with an emphasis on zoo and wildlife species. J. Vet. Med. Educ. 43 (3), 338—345.

Loxdale, H.D., 2013. Gorillas were his neighbours.... and lots of large tropical insects too. Antenna. 37 (3), 116—123.

Ludders, J.W., Sedgwick, C.J., Manley, S.V., Haskins, S.C., 1982. Anesthesia for restraint and transportation of five lowland gorillas (*Gorilla gorilla*). J. Zoo Anim. Med. 13, 78—81.

Ludwig, K.S., 1961. Beitrag zum Bau der Gorilla placenta. Acta Amal. 45, 110—123.

Lukas, K., 1999. A review of nutritional and motivational factors contributing to the performance of regurgitation and reingestion in captive lowland gorillas (*Gorilla gorilla gorilla*). Appl. Anim. Behav. Sci. 63, 237—249.

Lukas, K., 2003. Gorilla behavior in response to systematic alternation between zoo enclosures. Appl. Anim. Behav. Sci. 81 (4), 367—386.

Lukas, K., Hoff, M., Stoinski, T., Maple, T., 1997a. A preliminary analysis of R/R in gorillas at Zoo Atlanta. Abstract (w/supplemental notes). In: Pittsburgh Zoo Gorilla Workshop Program, p. 23.

Lukas, K., Stoinski, T., Maple, T., 1997b. All male gorilla groups: planning for the future. Abstract (w/supplemental notes). In: Pittsburgh Zoo Gorilla Workshop Program, p. 24.

Lukas, K., Forthman, D., Bloomsmith, M., Mar, M., Blanchard-Fields, F., Maple, T., 2000. An inter-institutional study of individual and nutritional factors associated with regurgitation and reingestion in captive gorillas. Paper Presented at "The Apes: Challenges for the 21st Century." Brookfield Zoo, Chicago, IL.

Luke, S., Gandi, S., Verma, R.S., 1995. Conservation of the Down syndrome critical region in humans and great apes. Gene. 161, 283—285.

Lyon, L., 1975. The saving of the gorilla. Africana. 5 (9), 11—13, 23.

MacArthur, R.H., Wilson, E.O., 1967. The Theory of Island Biogeography. Princeton University Press, Princeton, NJ.

Mace, G., 1990. Birth sex ratio and infant mortality rates in captive western lowland gorillas. Folia Primatol. 55, 156—165.

Mace, G.M., 1988. The genetic and demographic status of the Western lowland gorilla (*Gorilla g. gorilla*) in captivity. J. Zool. 216, 629—654.

MacFie, E.J., 1992. An update on current medical management program for Rwanda's mountain gorillas: veterinarians as population managers, and the effects of war. In: Proceedings of the American Association of Zoo Veterinarians/AAWV, pp. 45—47.

Macfie, E.J., Williamson, E.A., 2010. Best Practice Guidelines for Great Ape Tourism. IUCN SSC Primate Specialist Group, Gland, Switzerland.

Macfie, L., 1996. Bericht von einer Krätze-Infektion bei den Bwindi-Gorillas. Case report on scabies infection in Bwindi gorillas. Gorilla J.13.

Machado, C., Mihm, F.G., Noe, C., 1989. Diagnosis of ventricular septal defect in a gorilla using in vivo oximetry. J. Zoo. Wildl. Med. 20, 199−202.

MacLarnon, A.M., Martin, R.D., Chivers, D.J., Hladik, C.M., 1986a. Some aspects of gastrointestinal allometry in primates and other mammals. In: Sakka, M. (Ed.), Définition et Origines de l'Homme. Editions du CNRS, Paris, pp. 293−302.

MacLarnon, A.M., Chivers, D.J., Martin, R.D., 1986b. Gastro-intestinal allometry in primates and other mammals including new species. In: Else, J.G., Lee, P.C. (Eds.), Primate Ecology and Conservation. Cambridge University Press, Cambridge, UK, pp. 75−85.

Mahé, S., 2006. Reintroduction in Gabon of 7 Captive-Bred Gorillas From UK: Progress Report 2003 to 2005. PPG-Gabon/The Aspinall Foundation, Franceville, Gabon, 48 pp.

Mainka, S.A., 1990. *Balantidium coli* bei einem Flachlandgorilla. Verh. ber. Erkrg. Zootiere. 32, 19−22.

Maisels, F., Bergl, R.A., Williamson, E.A., 2016. *Gorilla gorilla*. The IUCN Red List of Threatened Species 2016. e.T9404A17963949.

Makuwa, M., Souquière, S., Telfer, P., Bourry, O., Rouquet, P., Kazanji, M., et al., 2006. Hepatitis viruses in non-human primates. J. Med. Primatol. 35 (6), 384−387.

Maldonado, O., Aveling, C., Cox, D., Nixon, S., Nishuli, R., Merlo, D., et al., 2012. Grauer's Gorillas and Chimpanzees in Eastern Democratic Republic of Congo (Kahuzi-Biega, Maiko, Tayna and Itombwe Landscape): Conservation Action Plan 2012−2022. IUCN/SSC Primate Specialist Group, Ministry of Environment, Nature Conservation & Tourism, Institut Congolais pour la Conservation de la Nature & the Jane Goodall Institute, Gland, Switzerland, 66 pp. < http://www.cms.int/sites/default/files/document/IUCN%20Grauer%E2%80%99s%20Gorillas%20and > .

Mallinson, J.J.C., Coffey, P., Usher-Smith, J., 1973. Maintenance, breeding and handrearing of lowland gorilla, *Gorilla g. gorilla*, (Savage & Wyman, 1847) at the Jersey Zoological Park. In: JWPT 10th Annual Report, pp. 5−28.

Mallinson, J.J.C., Coffey, P., Usher-Smith, J., 1976. Breeding and handrearing of lowland gorillas at the Jersey Zoo. Int. Zoo Yearb. 16, 189−194.

Mallon, D.P., Hoffmann, M., Grainger, M.J., Hibert, F., van Vliet, N., McGowan, P.J.K., 2015. An IUCN situation analysis of terrestrial and freshwater fauna in West and Central Africa. Occasional Paper of the IUCN Species Survival Commission No. 54. IUCN, Gland, Switzerland and Cambridge, UK, x + 162 pp. < https://portals.iucn.org/library/sites/library/files/documents/SSC-OP-054.pdf > .

Maloney, M.A., Leighty, K.A., Kuhar, C.W., et al., 2011. Behavioral responses of silverback gorillas (*Gorilla gorilla gorilla*) to videos. J. Appl. Anim. Welf. Sci. 14 (2), 96−108.

Maly, J., 1939. Kostra raněné gorily. Anthropologie 17 (1−4), 21−36.

Manning, D.P., Cooper, J.E., Stirling, I., Jones, C.M., Bruce, M., McCausland, P.C., 1985. Studies on the footpads of the polar bear (*Ursus maritimus*) and their possible relevance to accident prevention. J. Hand Surg. 10-13, 303−307.

Maple, T., 1979. Great apes in captivity: The good, the bad, and the ugly. In: Erwin, J., Maple, T.L., Mitchell, G. (Eds.), Captivity and Behavior. Van Nostrand Reinhold, New York, NY, pp. 239−272.

Maple, T., Finlay, T., 1986. Evaluating the environments of captive nonhuman primates. In: Bernischke, K. (Ed.), Primates: The Road to Self Sustaining Populations. Springer-Verlag, New York, NY, pp. 479−488.

Maple, T., Finlay, T., 1987. Post-occupancy evaluation in the zoo. Appl. Anim. Behav. Sci. 18, 5−18.

Maple, T., Finlay, T., 1989. Applied primatology in the modern zoo. Zoo Biol. (Suppl. 1), 101−116.

Maple, T., Hoff, M., 1982. Gorilla Behavior. Van Nostrand Reinhold Company, New York, NY.

Maple, T., Perkins, L., 1996. Enclosure furnishings and structural environmental enrichment. In: Kleiman, D., Allen, M., Thompson, K., Lumpkin, S. (Eds.), Wild Mammals in Captivity. University of Chicago Press, Chicago, IL, pp. 212−222.

Maple, T., Stine, W., 1982. Environmental variables and great ape husbandry. Am. J. Primatol.(Suppl. 1), 67−76.

Maple, T., Warren-Leubecker, A., 1983. Variability in the parental conduct of captive great apes and some generalizations to humankind. In: Reite, M., Caine, N. (Eds.), Child Abuse: The Nonhuman Primate Data. Alan R. Liss Inc., New York, NY, pp. 119−137.

Marennikova, S.S., Maltseva, N.N., Shelukhina, E.M., Chenkman, L.S., Korneeva, V.I., 1973. A generalised herpetic infection simulating smallpox in a gorilla. Intervirology. 2, 280−286.

Mares, M.A., 2002. A Desert Calling: Life in a Forbidding Landscape. Harvard University Press, Cambridge, MA.

Margulis, S., 2000. Social interactions among female gorillas before and after the introduction of a new silverback. Paper Presented at "The Apes: Challenges for the 21st Century.". Brookfield Zoo, Chicago, IL.

Margulis, S.W., Atsalis, S., Bellem, A., Wielebnowski, N., 2007. Assessment of reproductive behavior and hormonal cycles in geriatric western lowland gorillas. Zoo Biol. 26 (2), 117−139.

Marks, M., Mitjà, O., Vestergaard, L.S., Pillay, A., Knauf, S., Chen, C.-Y., et al., 2015. Challenges and key research questions for yaws eradication. Lancet. Infect. Dis. 15 (10), 1220−1225.

Marks, M., 2016. Yaws: towards the WHO eradication target. Trans. R. Soc. Trop. Med. Hyg. 110, 319−320.

Márquez, M., Serafin, A., Fernández-Bellon, H., Serrat, S., Ferrer-Admetlla, A., Bertranpetit, J., et al., 2008. Neuropathologic findings in an aged albino gorilla. Vet. Pathol. 45 (4), 531−537.

Marriner, L., Drickamer, L., 1994. Factors influencing stereotyped behavior of primates in a zoo. Zoo Biol. 13, 267−275.

Martin, R.D., 1981. Relative brain size and metabolic rate in terrestrial vertebrates. Nature. 293, 57−60.

Martin, R.D., 1982. Allometric approaches to the evolution of the primate nervous system. In: Armstrong, E., Falk, D. (Eds.), Primate Brain Evolution: Methods and Concepts. Plenum Press, New York, NY, pp. 39−56.

Martin, R.D., 1984. Body size, brain size and feeding strategies in primates. In: Chivers, D.J. (Ed.), Food Acquisition and Processing in Primates. Plenum Press, London, UK.

Martin, R.D., 1996. Scaling of the mammalian brain: the maternal energy hypothesis. News Physiol. Sci. 11, 149−156.

Martin, R.D., Ross, C.F., 2005. The evolutionary and ecological context of primate vision. In: Kremers, J. (Ed.), The Primate Visual System: A Comparative Approach. John Wiley, New York, NY.

Martin, R.D., Seaton, B., Lusty, J., 1975. Application of urinary hormone determinations in the management of gorillas. In: Jersey Wildlife Preservation Trust Annual Report, vol. 12, pp. 61−70.

Martin, R.D., Chivers, D.J., MacLarnon, A.M., Hladik, C.M., 1985. Gastrointestinal allometry in primates and other mammals. In: Jungers, W.L. (Ed.), Size and Scaling in Primate Biology. Plenum Press, New York, NY, pp. 61−89.

Martin, W.C.L., 1841. A general introduction to the natural history of mammiferous animals, with a particular view of the physical history of man, and the more closely allied genera of the order Quadrumana, or monkeys. Wright and Co, London.

Masi, S., 2003. Uso del substrato arboreo di un gruppo di gorilla di pianura occidentale nel Parco Nazionale di Dzanga-Ndoki, Repubblica Centraficana. [tesi di laurea − master equivalent]. University of Rome La Sapienza, Italy [Tree use by a western lowland gorilla group (*Gorilla gorilla gorilla*)].

Masi, S., 2008. Seasonal Influence on Foraging Strategies, Activity and Energy Budgets of Western Lowland Gorillas (*Gorilla gorilla gorilla*) in Bai-Hokou, Central African Republic (Ph.D. thesis). University of Rome, Roma, Italy.

Masi, S., 2009. Habituation, ecotourism and research for conservation of western gorillas in Central African Republic—Bai Hokou. Gorilla Gaz. 21 (1), 31—34.

Masi, S., 2010. Western gorilla conservation and research in Bai Hokou. Gorilla J. 40, 19—22.

Masi, S., 2011. Differences in gorilla nettle-feeding between captivity and the wild: local traditions, species typical behaviors or merely the result of nutritional deficiencies? Anim. Cogn. 14 (6), 921—925.

Masi, S., Cipolletta, C., Robbins, M.M., 2009. Western lowland gorillas (Gorilla gorilla gorilla) change their activity in response to frugivory. Am. J. Primatol. 71 (2), 91—100.

Masi, S., Narat, V., Todd, A., Krief, S., 2011. How do great apes acquire information on unusual feeding behaviors? Role of sociality and physiology on learning process to understand origins of self-medication in humans. J. Biol. Res. 1 (LXXXIV), 293—296.

Masi, S., Chauffour, S., Bain, O., Todd, A., Guillot, J., Krief, S., 2012a. Seasonal effects on great ape health: a case study of wild chimpanzees and western gorillas. PLoS ONE 7 (12), e49805.

Masi, S., Gustafsson, E., Jalme, M.S., Narat, V., Todd, A., Bomsel, M.C., et al., 2012b. Unusual feeding behaviour in great apes, a window to understand origins of self-medication in humans: role of sociality and physiology on learning process. Physiol. Behav. 105 (2), 337—349.

Masi, S., Mundry, R., Ortmann, S., Cipolletta, C., Boitani, L., Robbins, M. M., 2012c. The influence of seasonal frugivory on nutrient and energy intake in wild western gorillas. PLoS ONE. 10 (7), e0129254.

Mason, J.K., Purdue, B.N., 2000. The Pathology of Trauma. Third ed. Arnold, London, UK.

Masters, N., Niphuis, H., Verschoor, E., et al., 2010. Debilitating clinical disease in a wild-born captive western lowland gorilla (Gorilla gorilla gorilla) co-infected with varicella zoster virus (VZV) and simian T-lymphotropic virus (STLV). J. Zoo Wildl. Med. 41 (4), 713—716.

Mathot, L, Puit M, 2008 Educational Activities in the Republic of Congo Gorilla Journal, Issue 36, pp 20—22. Berggorilla & Regenwald Direkthilfe. <http://www.berggorilla.org/fileadmin/user_upload/pdf/journal/journal-en/gorilla-journal-36-english.pdf>.

Matschie, P., 1903. Über einen Gorilla aus Deutsch-Ostafrika. Sitzungsber. Ges. Naturf. Freunde Berlin, 253—259.

Matschie, P., 1904. Bemerkungen über die Gattung Gorilla. Sitzungsber. Ges. Naturf. Freunde Berlin, 45—53.

Matschie, P., 1905. Merkwürdige Gorilla-Schädel aus Kamerun. Sitzungsber. Ges. Naturf. Freunde Berlin, 279—283.

Matschie, P., 1914. Neue Affen aus Mittelafrika. Sitzungsber. Ges. Naturf. Freunde. Berlin, 324—325.

Mattern, S.P., 2013. The Prince of Medicine: Galen in the Roman Empire. Oxford University Press, New York, NY.

Mätz-Rensing, K., Kunze, M., Zoller, M., et al., 2011. Fatal Balamuthia mandrillaris infection in a gorilla—first case of balamuthiasis in Germany. J. Med. Primatol. 40 (6), 437—440.

Mayden, R.L., 1997. A hierarchy of species concepts: the denouement in the saga of the species problem. In: Claridge, M.F., Dawah, H.A., Wilson, M.R. (Eds.), Species: the Units of Biodiversity. Chapman and Hall, London, pp. 381—424.

Mayer, A.F.J.C., 1856. Zur Anatomie des Orang-Utang und des Chimpanse. Arch. Nat. 22, 281—304, Berlin.

Mayigane, L.N., 2013. Evaluation of the clinical response decision tree for respiratory diseases in mountain gorillas (Gorilla beringeii beringeii) in the Virunga massif. Critical Links Between Human and Animal Health. University of Guelph, Guelph, Ontario, pp. 13—37.

Mayr, E., 1942. Systematics and the Origin of Species. Columbia University Press, New York, NY.

Mayr, E., 1970. Populations, Species, and Evolution. Harvard University Press, Cambridge, MA.

Mayr, E., 1996. What is a species, and what is not? Philos. Sci. 63, 262—277.

Mayr, E., Ashlock, P.D., 1991. Principles of Systematic Zoology. McGraw-Hill, New York, NY.

Mays, S., Taylor, G.M., 2002. Osteological and biomolecular study of two possible cases of hypertrophic osteoarthropathy from mediaeval England. J. Archaeol. Sci. 29, 1267—1276.

McCabe, C.M., Reader, S.M., Nunn, C.L., 2014. Infectious disease, behavioural flexibility and the evolution of culture in primates. Proc. R. Soc. B Biol. Sci. 282 (1799), 20140862.

McCann, C., Rothman, J., 1999. Changes in nearest-neighbor associations in a captive group of western lowland gorillas after the introduction of five hand-reared infants. Zoo Biol. 18, 261—278.

McClure, H.M., Chandler, F.W., 1982. A survey of pancreatic lesions in nonhuman primates. Vet. Pathol. I9 (Suppl. 7), 193—209.

McClure, H.M., Keeling, M.E., Guilloud, N.B., 1972. Hematologic and blood chemistry data for the gorilla (Gorilla gorilla). Folia Primatol. 19, 300—316.

McClure, H.M., Swenson, R.B., Kalter, S.S., Lester, T.L., 1980. Natural genital 'Herpes hominis' infection in chimpanzees (Pan troglodytes and Pan paniscus). Lab. Anim. Sci. 30 (5), 895—901.

McConnell, E.E., Basson, P.A., Vos, V., de Myers, B.J., Kuntz, R.E., 1974. A survey of diseases among 100 free-ranging baboons (Papio ursinus) from the Kruger National Park. Onderstepoort. J. Vet. Res. 41 (3), 97—168.

McFarlin, S.C., Bromage, T.G., Lilly, A.A., Cranfield, M.R., Nawrocki, S.P., Eriksen, A., et al., 2009. Recovery and preservation of a mountain gorilla skeletal resource in Rwanda. Am. J. Phys. Anthropol. (Suppl. 48), 187—188.

McFarlin, S.C., Barks, S.K., Tocheri, W.M., Massey, J.S., Eriksen, A.B., Fawcett, K.A., et al., 2012. Early brain growth cessation in wild Virunga mountain gorillas (Gorilla beringei beringei). Am. J. Primatol. 75 (5), 450—463.

McFarlin, S.C., Reid, D.J., Arbenz-Smith, K., Cranfield, M.R., Nutter, F., Stoinski, T.S., et al., 2014. Histological examination of dental development in a juvenile mountain gorilla from Volcanoes National Park, Rwanda. Bull. Int. Assoc. Paleodontol. 8 (1), 149.

McGee, J. O'D., Isaacson, P.G., Wright, N.A., 1992. Oxford Textbook of Pathology, vols. I, 2a and 2b. Oxford University Press, Oxford,UK.

McGrath, K.D., Guatelli-Steinberg, K., Arbenz-Smith, et al., 2015. Linear enamel hypoplasia prevalence in wild Virunga mountain gorillas from Rwanda. Am. J. Phys. Anthropol. 156, 221.

McInnes, E.F., 2012. Background Lesions in Laboratory Animals: A Color Atlas. Saunders Elsevier, Philadelphia, PA.

McKenney, E.A., Williamson, L., Yoder, A.D., Rawls, J.F., Bilbo, S.D., Parker, W., 2015. Alteration of the rat cecal microbiome during colonization with the helminth Hymenolepis diminuta. Gut Microbes. 6, 182—193.

McKenney, F.D., Traum, J., Bonestell, A.E., 1944. Acute coccidioidomycosis in a mountain gorilla (Gorilla beringei) with anatomical notes. J. Am. Vet. Med Assoc. 104, 136—140.

McKenzie, A.A., 1993. The Capture and Care Manual. Wildlife Decision Support Service and the South African Veterinary Foundation, South Africa.

McLachlan, S.M., Alpi, K., Rapoport, B., 2011. Review and hypothesis: does Graves' disease develop in non-human great apes? Thyroid. 21 (12), 1359—1366.

McLaren, S.B., Schlitter, D.A., Genoways, H.H., 1984. Catalog of the Recent Scandentia and Primates in the Carnegie Museum of Natural History. Ann. Carnegie Mus. 53, 463—525.

McManamon, R., Lowenstine, L., 2012. Cardiovascular disease in great apes. In: Miller, R.E., Fowler, M.E. (Eds.), Fowler's Zoo and Wild Animal Medicine Current Therapy. Elsevier.

McManamon, R., et al., 1989. Localized periodontitis in an adult male lowland gorilla. In: American Association of Zoo Veterinarians Annual Proceedings, p. 185.

McMillan, N., 1996. Robert Bruce Napoleon Walker, F.R.G.S., F.A.S., F.G.S., C.M.Z.S. (1832–1901), West African trader, explorer and collector of zoological specimens. Arch. Nat. Hist. 23, 125–141.

McRae, M., Nichols, M., 2000. Central Africa's orphan gorillas: will they survive in the wild? Natl. Geogr. Mag. 197 (2), 84–97.

Meder, A., 1985. Integration of hand-reared gorilla infants in a group. Zoo Biol. 4, 1–12.

Meder, A., 1986. Physical and activity changes associated with pregnancy in captive lowland gorillas (*Gorilla gorilla gorilla*). Am. J. Primatol. 11, 111–116.

Meder, A., 1989. Effects of hand-rearing on the behavioral development of infant and juvenile gorillas (*Gorilla g. gorilla*). Dev. Psychobiol. 224, 357–376.

Meder, A., 1990. Integration of hand-reared gorillas into breeding groups. Zoo Biol. 9, 157–164.

Meder, A., 1992. Effects of the zoo environment on the behaviour of lowland gorillas in zoos. Primate Rep. 32, 167–183.

Meder, A., 1993. Gorillas. Ökologie und Verhalten. Springer-Verlag, Berlin.

Meder, A., 1994. Causes of death and diseases of gorillas in the wild. Gorilla J. 2, 19–20.

Meder, A., 1995. Some effects on the reproductive success of captive gorillas. Gorilla Gaz. 9 (1), 5–6.

Meder, A., 1996. Should we consider the translocation of gorilla populations? Gorilla J. 13, 21.

Meder, A., Groves, C.P., 2005. Where are the gorillas? Gorilla J. 30, 21–28.

Medina, R.G., Ponssa, M.L., Guerra, C., Aráoz, E., 2013. Amphibian abnormalities: historical records of a museum collection in Tucuman Province, Argentina. Herpetol. J. 23, 193–202.

Medjo Mvé, P., 1997. Essai sur la Phonologie Panchronique des Parlers Fang du Gabon et ses Implications Historiques (Ph.D. thesis). Université Lumière-Lyon 2, Lyon.

Meehan, T., 1997a. Disease concerns in lowland gorillas. In: Ogden, J., Wharton, D. (Eds.), Management of Gorillas in Captivity Gorilla Species Survival Plan. Fulton County Zoo, Atlanta, GA, pp. 153–159.

Meehan, T., 1997b. Disease concerns in lowland gorillas. In: Ogden, J., Wharton, D. (Eds.), The Gorilla Husbandry Manual. Fulton County Zoo, Atlanta, GA, pp. 143–149.

Meehan, T., 1997c. Immobilization and shipping. In: Ogden, J., Wharton, D. (Eds.), Management of Gorillas in Captivity. Gorilla Species Survival Plan. Fulton County Zoo, Atlanta, GA, pp. 197–201.

Meehan, T., 1997d. Immobilization and shipping. In: Ogden, J., Wharton, D. (Eds.), The Gorilla Husbandry Manual. Fulton County Zoo, Atlanta, GA, pp. 168–171.

Meehan, T., Zdziarski, J., 1997. Zoonotic diseases. In: Ogden, J., Wharton, D. (Eds.), The Gorilla Husbandry Manual. Gorilla Species Survival Plan. Fulton County Zoo, Atlanta, GA, pp. 163–167, 191–196.

Meehan, T., Lowenstine, L., 1997. Causes of mortality in captive lowland gorillas: a survey of the SSP population. In: Ogden, J., Wharton, D. (Eds.), Gorilla Husbandry Manual. Fulton County Zoo, Atlanta, GA, pp. 150–152.

Mellor, D.J., Hunt, S., Gusset, M. (Eds.), 2015. Caring for Wildlife: The World Zoo and Aquarium Animal Welfare Strategy. WAZA Executive Office, Gland, 87 pp. < http://www.waza.org/files/webcontent/ 1.public_site/5.conservation/animal_welfare/WAZA%20Animal%20Welfare %20Strategy%202015_Portrait.pdf >.

Merfield, F.G., Miller, H., 1956. Gorillas Were My Neighbours. Longmans, Green and Co, London, UK.

Mesa, D.P., 2005. Protocolos para el manejo y preservación de las colecciones científicas. Bol. Cient. Mus. Hist. Nat. 10, 117–148.

Miles, A.E.W., Grigson, C., 1990. Colyer's Variations and Diseases of the Teeth of Animals. Cambridge University Press, Cambridge, UK.

Miller, C.L., Schwartz, A.M., Barnhart Jr., J.S., Bell, M.D., 1999. Chronic hypertension with subsequent congestive heart failure in a western lowland gorilla (*Gorilla gorilla gorilla*). J. Zoo Wildl. Med. 30 (2), 262–267.

Miller, R.A., 1941. The laryngeal sacs of an infant and an adult gorilla. Am. J. Anat. 69 (1), 1–17.

Miller, R.E., Fowler, M., 2012. Fowler's Zoo and Wild Animal Medicine, vol. 7. Saunders, Philadelphia, PA.

Miller-Schroeder, P., Paterson, J.D., 1989. Environmental influences on reproductive and maternal behavior in captive gorillas: results of a survey. In: Segal, E. (Ed.), Housing, Care and Psychological Well-Being of Captive and Laboratory Primates. Noyes, Park Ridge, NJ, pp. 389–415.

Mills, W., 1896. Valedictory address to graduates in comparative medicine and veterinary medicine and clinical research. J. Comp. Med. Vet. Arch. 17, 25.

Mims, C.A., Nash, A., Stephen, J., 2000. Mims' Pathogenesis of Infectious Disease. Fifth ed. Elsevier, Amsterdam.

Minnis, R.B., Cranfield, M.R., Ssebide, B., Rwego, I., Whittier, C.A., Travis, D., et al., 2004. Adrenal gland response of three individuals in Mubale Mountain Gorilla family to specific stressors in Bwindi Impenetrable National Park, South-Western Uganda. Abstract for the Conference on Diseases—The Third Major Threat for Wild Great Apes? Leipzig, Germany.

Minter, L.J., Cullen, J.M., Loomis, M.R., 2012. Reye's or Reye's-Like syndrome in western lowland gorilla (*Gorilla gorilla gorilla*). J. Med. Primatol. 41 (5), 329–331.

Mitani, M., 1992. Preliminary results of the studies on wild western lowland gorillas and other sympatric diurnal primates in the Ndoki Forest, Northern Congo. In: Itoigawa, N., Sugiyama, Y., Sackett, G., Thompson, R. (Eds.), Topics in Primatology, vol. 2: Behavior, Ecology, and Conservation. University of Tokyo Press, Tokyo, pp. 215–224.

Mitchell, G., Tromborg, C., Kaufman, J., Bargabus, S., Simoni, R., Geissler, V., 1992. More on the 'influence' of zoo visitors on the behaviour of captive primates. Appl. Anim. Behav. Sci. 35, 189–198.

Mitchell, R., 1989. Functions and social consequences of infant-adult male interaction in a captive group of lowland gorillas (*Gorilla gorilla gorilla*). Zoo Biol. 8, 125–137.

Mitchell, R.W., 1999. Scientific and popular conceptions of the psychology of great apes from the 1790s to the 1970s: Déjà vu all over again. Primate Rep. 53, 1–118.

Mitchell, W.R., Luskutoff, N.M., Czekala, N.M., Lasley, B.L., 1982. Abnormal menstrual cycles in the female gorilla (*Gorilla gorilla*). J. Zoo Anim. Med. 13, 143–147.

Mitra, M., Mahanta, S.K., Sen, S., Ghosh, C., Hate, A.K., 1995. Transmission of *Sarcoptes scabiei* from animal to man and its control. J. Indian Med. Assoc. 93, 142–143.

Mittermeier, R.A., Rylands, A.B., Wilson, D.E., 2013. Handbook of the Mammals of the World, vol. 3, Primates. Edicions in association with Conservation International and IUCN.

Moberg, G.P., Mench, J.A., 2000. The Biology of Animal Stress. CABI, Wallingford, UK.

Moggi-Cecchi, J., Crovella, S., 1991. Occurrence of enamel hypoplasia in the dentitions of simian primates. Folia Primatol. 57, 106–110.

Molina-Lopez, R.A., Valverdú, N., Martin, M., Mateu, E., Cummings, P.M., Trelka, D.P., et al., 2010. Atlas of Forensic Histopathology. Cambridge University Press, Cambridge, UK.

Molteni, A., Scarpelli, D.G., Sparberg, M., Maschgan, E.R., 1980. Chronic idiopathic diarrhea with enterocolitis and malabsorption in a captive, lowland gorilla—a case report. In: Montali, R.J., Migaki, G. (Eds.), The

Comparative Pathology of Zoo Animals. Smithsonian Institution Press, Washington, DC, pp. 105–112.

Monard, A., 1951. Résultats de la mission zoologique suisse au Cameroun. In: Mémoires de l'Institut français d'Afrique noire, centre du Cameroun: série sciences naturelles no. 1, Paris, p. 57.

Monboddo, J. B. [Lord], 1774. Of the Origin and Progress of Language, vol. I, second ed., revised. J. Balfour, Edinburgh, pp. 270–361.

Moñino, Y., 1988. Lexique Comparatif des Langues Oubanguiennes. Geuthner, Paris.

Montali, R.J., 1988. Comparative pathology of inflammation in the higher vertebrates (reptiles, birds and mammals). J. Comp. Pathol. 99, 1–25.

Montali, R.J., Migaki, G., 1978. The Comparative Pathology of Zoo Animals. Proceedings of a Symposium Held at the National Zoological Park, Smithsonian Institution, October 2–4. Smithsonian Institution Press, Washington, DC, p. 684.

Montgomery, S., 1991. Walking with Great Apes. Houghton Mifflin Company, Boston, MA.

Moresco, A., Agnew, D.W., 2013. Reproductive health surveillance in zoo and wildlife medicine. J. Zoo Wildl. Med. 44 (4 Suppl.), S26–S33.

Morgera, E., 2011. An IUCN Situation Analysis of Terrestrial and Freshwater Fauna in West and Central Africa Law and the Empowerment of the Poor. FAO Legislative Study 103 Part III. Development Law Service FAO Legal Office Food and Agriculture Organization of the United Nations, Rome.

Morgan, B.J., Sunderland-Groves, J., 2006. The Cross-Sanaga gorillas: the northernmost gorilla populations. Gorilla J.32.

Morgan, B.J., et al., 2003. Newly discovered gorilla population in the Ebo forest, Littoral Province, Cameroon. Int. J. Primatol. 24 (5), 1129–1137.

Morris, D., 1958. The behaviour of higher primates in captivity. In: Zool. Soc. London, XVth International Congress of Zoology, Sect 1, Paper 34.

Morris, D., Morris, R., 1966. Men and Apes. Hutchinson, London, UK.

Morris, D.O., Dunstan, R.W., 1996. A histomorphological study of sarcoptic acariasis in the dog: 19 cases. J. Am. Anim. Hosp. Assoc. 32, 119–124.

Morris, P.A., 2010. A History of Taxidermy: Art, Science and Bad Taste. MPM Publishing, Ascot.

Morton, B.F., Todd, A.F., Lee, P., Masi, S., 2013. Observational monitoring of clinical signs during the last stage of habituation in a wild western gorilla group at Bai Hokou, Central African Republic. Folia Primatol. 84, 118–133.

Mowat, F., 1988. Woman in the Mists. Warner Books Ltd, New York, NY.

Mudakikwa, A., 2001. An outbreak of mange hits the Bwindi gorillas. Gorilla J. 22, 24.

Mudakikwa, A., 2002. Ubuzima, a 13-month-old re-introduced to her group. Gorilla J. 25, 8.

Mudakikwa, A., Cranfield, M., Sleeman, J., Eilenberger, U., 2001. Clinical medicine, preventive health care and research on mountain gorillas (*Gorilla gorilla beringei*) in the Virunga Volcanoes region. In: Robbins, M., Sicotte, P., Stewart, K.J. (Eds.), Mountain Gorillas: Three Decades of Research at Karisoke. Cambridge University Press, Cambridge, UK, pp. 341–360.

Mudakikwa, A.B., Sleeman, J.M., 1997 Analysis of urine from free-ranging mountain gorillas (*Gorilla gorilla beringei*) for normal physiological values. In: American Association of Zoo Veterinarians Annual Proceedings, Houston, TX, October 1997, p. 278.

Mudakikwa, A.B., Sleeman, J.M., Foster, J.W., Meader, L.S., Patton, S., 1998. An indicator of human impact: gastrointestinal parasites of mountain gorillas (*Gorilla gorilla beringei*) from the Virunga Volcanoes Region, Central Africa. In: American Association of Zoo Veterinarians and AAWV Joint Conference, Omaha, Nebraska, pp. 436–437.

Muhangi, D., 2009. Pathological Lesions of the Gastrointestinal Tract in the Free Ranging Mountain Gorilla (Ph.D. thesis). Makerere University, Kampala, Uganda.

Mulhearne, D., 2013. Facing extinction. Museums J. 113, 24–29.

Mundy, N.I., Ancrenaz, M., Wickings, E.J., Lunn, P.G., 1998. Protein deficiency in a colony of western lowland gorillas (*Gorilla g. Gorilla*). J. Zoo Wildl. Med. 29 (3), 261–268.

Munro, H.M.C., Thrusfield, M.V., 2001. "Battered pets": non-accidental physical injuries found in dogs and cats. J. Small Anim. Pract. 42, 279–290.

Munro, R., Munro, H.M.C., 2012. Some challenges in forensic veterinary pathology: a review. J. Comp. Pathol. 149 (1), 1–17.

Munson, L., 1999. Necropsy Procedures for Wild Animals. Wildlife Health Center, School of Veterinary Medicine, University of California, Davis.

Munson, L.L., 1990. Future directions of zoological pathology. J. Zoo. Wildl. Med. 21, 385–390.

Murata, H., Shimada, N., Yoshioka, M., 2004. Current research on acute phase proteins in veterinary diagnosis: an overview. Vet. J. 168, 28–40.

Murphy, H.W., Dennis, P., Devlin, W., et al., 2011. Echocardiographic parameters of captive western lowland gorillas (*Gorilla gorilla gorilla*). J. Zoo Wildl. Med. 42 (4), 572–579.

Mutter, G.L., Prat, J., 2014. Pathology of the Female Reproductive Tract. Third ed. Churchill Livingstone, Edinburgh.

Mwanza, N., Yamagiwa, J., Yumoto, T., Maruhashi, T., 1992. Distribution and range utilization of eastern lowland gorillas. In: Itoigawa, N., Sugiyama, Y., Sackett, G., Thompson, R. (Eds.), Topics in Primatology, vol. 2: Behavior, Ecology, and Conservation. University of Tokyo Press, Tokyo, pp. 283–300.

Mwebe, R., 1998. A survey on *Cryptosporidium* prevalence within the mountain gorilla population (*Gorilla gorilla berengei*) in the Bwindi Impenetrable National Park in South Western Uganda. Research Project Submitted in Partial Fulfilment for the Award of the Degree of Bachelor of Veterinary Medicine. Makerere University, Kampala, Uganda.

Myers, M.G., Kramer, L.W., Stanberry, L.R., 1987. Varicella in a gorilla. J. Med. Virol. 23, 317–322.

Mylniczenko, N.D., Murrey, S.S., Smith, S., Sewall, L.W., Facchini, F. 2008. Management of a uterine leiomyoma in a western lowland gorilla (*Gorilla gorilla gorilla*). Presented at American Association of Zoo Veterinarians Conference, Los Angeles, CA, USA.

Nadler, R., 1974. Periparturitional behavior of a primiparous lowland gorilla. Primates 15, 55–73.

Nadler, R., 1983. Experimental influences on infant abuse of gorillas and some other nonhuman primates. In: Reite, M., Caine, N. (Eds.), Child Abuse: The Nonhuman Primate Data. Alan R. Liss, New York, NY, pp. 139–149.

Nadler, R., 1989. The psychological well-being of captive gorillas. In: Segal, E. (Ed.), Housing, Care and Psychological Wellbeing of Captive and Laboratory Primates. Noyes Publications, Park Ridge, NJ, pp. 416–420.

Nadler, R., 1995. Proximate and ultimate influences on the regulation of mating in the great apes. Am. J. Primatol. 37, 93–102.

Nadler, R., Collins, D., 1984. Research on the reproductive biology of gorillas. Zoo Biol. 3, 13–25.

Nagashima, K., Takizawa, A., Yasui, K., Nishiyama, M., Ikai, A., Husegawa, T., 1974. Congenital obstruction of the nasolacrimal duct in a lowland gorilla. J. Zoo Anim. Med. 5, 9-6.

Nagel, M., Dischinger, J., Turck, M., et al., 2013. Human-associated *Staphylococcus aureus* strains within great ape populations in Central Africa (Gabon). Clin. Microbiol. Infect. 19 (11), 1072–1077.

Nall, J.D., Lindsey, J.R., Baker, H.J., Truett, F.R., 1972. Hepatitis in a gorilla. J. Zoo Anim. Med. 3, 27–28.

Napier, J.R., Napier, P.H., 1967. A Handbook of Living Primates. Academic Press, London.

Nascimento, L.V., Mauerwerk, M.T., dos Santos, C.L., Filho, I.R., de Barros, Júnior, E.H.B., Sotomaior, C.S., et al., 2015. Treponemes detected in digital dermatitis lesions in Brazilian dairy cattle and possible host reservoirs of infection. J. Clin. Microbiol. 53 (6), 1935–1937.

National Research Council of the National Academies, USA, 2003. Occupational Health and Safety in the Care and Use of Non-Human Primates. Report from the Committee on Occupational Health and Safety in the Care and Use of Non-Human Primates. The National Academies Press, Washington, DC.

Neder, A., 1993. Gorillas—Ökologie und Verhalten. Springer-Verlag, Berlin.

Neel, C., Etienne, L., Li, Y.Y., Takehisa, J., Rudicell, R.S., Bass, I.N., et al., 2010. Molecular epidemiology of simian immunodeficiency virus infection in wild-living gorillas. J. Virol. 84 (3), 1464—1476.

Neiffer, D.L., Rothschild, B.M., Marks, S.K., Urvater, J.A., Watkins, D.I., 2001. Reactive arthritis in a juvenile gorilla (Gorilla gorilla gorilla): effective management with long-term sulfasalazine therapy. J. Zoo Wildl. Med. 32 (4), 539—551.

Nellemann, C., Redmond, I., Refisch, J. (Eds.), 2010. The Last Stand of the Gorilla — Environmental Crime and Conflict in the Congo Basin. A Rapid Response Assessment. United Nations Environment Programme, GRID-Arendal, < www.grida.no >.

Nerrienet, E., Meertens, L., Kfutwah, A., Foupouapouognigni, Y., Ayouba, A., Gessain, A., 2004. Simian T cell leukaemia virus type I subtype B in a wild-caught gorilla (Gorilla gorilla gorilla) and chimpanzee (Pan troglodytes vellerosus) from Cameroon. J. Gen. Virol. 85 (Pt 1), 25—29.

Nguiffo, S., Talla, M., 2010. Cameroon's Wildlife Legislation: Local Custom Versus Legal Conception Forests, People and Wildlife. Unasylva No. 236, Vol. 61. Food and Agriculture Organization of the United Nations, Rome, 2010/3.

Niemuth, J.N., De Voe, R.S., Jennings, S.H., Loomis, M.R., Troan, B.V., 2014. Malignant hypertension and retinopathy in a western lowland gorilla (Gorilla gorilla gorilla). J. Med. Primatol. 43, 276—279.

Nishida, T., McGrew, W.C., Marchant, L.F., 2012. Wild chimpanzees at Mahale are not manually lateralised for throwing. Pan Afr. News 19 (No. 2), December.

Nizeyi, J.B., 2005. Non-invasive Monitoring of Adrenocortical Activity of Free-Ranging Mountain Gorillas in Bwindi Impenetrable National Park, South-Western Uganda (Ph.D. thesis). Makerere University, Kampala, Uganda.

Nizeyi, J.B., Mwebe, R., Nanteza, A., et al., 1999. Cryptosporidium sp. and Giardia sp. infections in mountain gorillas (Gorilla gorilla beringei) of the Bwindi Impenetrable National Park, Uganda. J. Parasitol. 85 (6), 1085—1088.

Nizeyi, J.B., Innocent, R.B., Erume, J., et al., 2001. Campylobacteriosis, salmonellosis, and shigellosis in free-ranging human-habituated mountain gorillas in Uganda. J. Wildl. Dis. 37, 239—244.

Nizeyi, J.B., Cranfield, M.R., Graczyk, T.K., 2002a. Cattle near the Bwindi Impenetrable National Park, Uganda, as a reservoir of Cryptosporidium parvum and Giardia duodenalis for local community and free-ranging gorillas. Parasitol. Res. 88, 380—385.

Nizeyi, J.B., Sebunya, D., Dasilva, A.J., Cranfield, M.R., Pieniazek, N.J., Graczyk, T.K., 2002b. Cryptosporidiosis in people sharing habitats with free-ranging mountain gorillas (Gorilla gorilla beringei), Uganda. Am. J. Trop. Med. Hyg. 66 (4), 442—444.

Nizeyi, J.B., Czekala, N., Monfort, S.L., Taha, N., Cranfield, M., Linda, P., et al., 2011. Detecting adreno-cortical activity in gorillas: a comparison of faecal glucocorticoid measures using RIA versus EIA. Int. J. Anim. Vet. Adv. 3 (2), 103—115.

Nkurunungi, J.B., Ampumuza, C., 2013. Assessment of the Impacts of Mountain Gorilla Habituation and Tourism on their Sustainable Conservation. Report of a consultancy for the International Gorilla Conservation Programme. Summary Gorilla J. (48), 10—13, English, June 2014 < http://www.berggorilla.org/fileadmin/user_upload/pdf/journal/journal-en/gorilla-journal-48-english.pdf >.

Noback, C.R., 1939. The changes in the vaginal smears and associated cyclic phenomena in the lowland gorilla. Anat. Rec. 73, 209—225.

Noback, C.R., Montagna, W., 1970. The Primate Brain. Appleton Century-Crofts, New York, NY.

Norton, B., 1990. The Mountain Gorilla. Swan Hill Press, Shrewsbury, UK.

Noss, R.F., 1996. The naturalists are dying off. Conserv. Biol. 10, 1—3.

Nottidge, H.O., Omobowale, T.O., Olopade, J.O., Oladiran, O.O., Ajala, O. O., 2007. A case of craniothoracopagus (Monocephalus thoracopagus tetrabrachius) in a dog. Anat. Histol. Embryol. 36, 179—181.

Novilla, M.N., Jackson, M.K., Reim, D.A., Jacobson, S.B., Nagata, R.A., 2014. Occurrence of multinucleated hepatocytes in cynomolgus monkeys (Macaca fascicularis) from different geographical regions. Vet. Pathol. 51 (6), 1183—1186.

Nowell, A., Fletcher, A., 2007. Development of independence from the mother in Gorilla gorilla gorilla. Int. J. Primatol. 28, 441—455.

Nunn, C.L., Altizer, S., 2006. Infectious Diseases in Primates: Behavior, Ecology and Evolution. Oxford University Press, Oxford.

Nunn, C.L., Heymann, E.W., 2005. Malaria infection and host behavior: a comparative study of neotropical primates. Behav. Ecol. Sociobiol. 59, 30—37.

Nutter, F.B., Whittier, C.A., Cranfield, M.R., Lowenstine, L.J. 2005a. Causes of death for mountain gorillas (Gorilla beringei beringei and G b undecided) from 1968-2004: an aid to conservation programs. In: Proceedings of the Wildlife Disease Association International Conference. Cairns, Queensland, Australia, pp. 200—201.

Nutter, F.B., Whittier, C.A., Lowenstine, L.J., 2005b. Cleft palate in a neonatal Virunga mountain gorilla (Gorilla beringei beringei). In: Proceedings of the 54th Annual Wildlife Disease Association Conference. Abstract, p. 62.

Oates, J.F., McFarland, K.L., Groves, J.L., Bergl, R.A., Linder, J.M., Disotell, T.R., 2003. The Cross River gorilla: natural history and status of a neglected and critically endangered subspecies. In: Taylor, A.B., Goldsmith, M.L. (Eds.), Gorilla Biology: A Multidisciplinary Perspective. Cambridge University Press, Cambridge, UK, pp. 472—497.

Oates, J, et. al. 2007. Regional Action Plan for the Conservation of the Cross River Gorilla (Gorilla gorilla diehli). IUCN/SSC Primate Specialist Group and Conservation International, Arlington, VA, USA < https://portals.iucn.org/library/sites/library/files/documents/2007-012.pdf >.

O'Brien, J.K., Stojanov, T., Crichton, E.G., Evans, K.M., Leigh, D., Maxwell, W.M., et al., 2005. Flow cytometric sorting of fresh and frozen-thawed spermatozoa in the western lowland gorilla (Gorilla gorilla gorilla). Am. J. Primatol. 66, 297—315.

O'Connor, T., 2004. The Archaeology of Animal Bones. Sutton Publishing Limited, Stroud, UK.

Ogden, J., Wharton, D., 1997. Management of Gorillas in Captivity. Gorilla Species Survival Plan. Fulton County Zoo, Atlanta, GA.

Ogden, J., Horton, C., Perkins, L., 1987. New facilities open at Zoo Atlanta. Gorilla Gaz.12—17.

Ogden, J., Finlay, T., Jackson, D., Maple, T., 1989. Gorillas of Cameroon: a post-occupancy evaluation. AAZPA Regional Conference Proceedings, pp. 557—563.

Ogden, J., Finlay, T., Maple, T., 1990. Gorilla adaptations to naturalistic environments. Zoo Biol. 9, 107—121.

Ogden, J., Bruner, G., Maple, T., 1992. A survey of the use of electric fencing with captive great apes. Int. Zoo Yearb. 31, 229—236.

Ogden, J., Lindburg, D., Maple, T., 1993. Preference for structural environmental features in captive lowland gorillas (Gorilla gorilla gorilla). Zoo Biol. 12, 381—395.

Ogden, J., Porton, I., Gold, K., Jendry, C., Snowden, S., Sevenich, M., 1997. Introductions and socializations in captive gorillas. In: Ogden, J., Wharton, D. (Eds.), The Management of Gorillas in Captivity. Atlanta/Fulton County Zoo, Atlanta, GA, pp. 124—130.

O'Grady, J.P., et al., 1978. Cesarean delivery in a gorilla. J. Am. Vet. Med. Assoc. 173 (9), 1137—1140.

O'Grady, J.P., et al., 1982a. Ultrasonic evaluation of echinococcosis in four lowland gorillas. J. Am. Vet. Med. Assoc. 181 (11), 1348—1350.

O'Grady, J.P., Esra, G.N., Yeager, C.H., Thomas, W.D., 1982b. Evaluation of secondary infertility in the gorilla. Zoo Biol. 1, 135–140.

Ojigo, D.O., Daborn, C.J., 2013. Communications and data collection: lessons from studies in the extensive livestock production areas of Kenya. J. Exotic Pet Med. 22, 39–45.

Olejniczak, C., 2001. The 21st century gorilla: progress or perish? In: The Apes: Challenges for the 21st Century, Conference Proceedings, May 10–13, 2000, Brookfield Zoo, Brookfield, IL, pp. 36–42.

Olias, P., Schulz, E., Ehlers, B., et al., 2012. Metastatic endocervical adenocarcinoma in a western lowland gorilla (*Gorilla g. gorilla*): no evidence of virus-induced carcinogenesis. J. Med. Primatol. 41 (2), 142–146.

Olson, F., Gold, K., 1985. Behavioral differences in mother-reared and hand-reared infant lowland gorillas (*Gorilla g. gorilla*). In: AAZPA Annual Conference Proceedings, pp. 143–152.

O'Neill, W.M., Hammer, S.P., German, R., Bloomer, H.A., Moore, J.G., 1978. Acute pyelonephritis in an adult gorilla. Lab. Anim. Sci. 28, 100–101.

Ostenrath, F., 1978. Therapeutic measures against a pyoderma in young *Gorilla g. gorilla*. Verh. Ber. Erkrg. Zootiere. 20, 285–287.

Ott-Joslin, J., Turner, T., Galante, J., Torgerson, B., 1987. Bilateral total hip replacement in a 26-year old lowland gorilla (*Gorilla gorilla*). Am. Assoc. Zoo Vet. Annu. Proc., 516–518.

Ott-Joslin, J.E., 1993. Zoonotic diseases of nonhuman primates. In: Fowler, M.E. (Ed.), Zoo and Wild Animal Medicine: Current Therapy, 3. W.B. Saunders, Philadelphia, PA, pp. 358–373.

Owen, R., 1848a. On a new species of chimpanzee. Proc. Zool. Soc. London, Part XVI, 27–35.

Owen, R., 1848b. Supplementary note on the great chimpanzee (*Troglodytes gorilla*, Savage, *Troglodytes savagei*, Owen), Proc. Zool. Soc. London, Part XVI, pp. 53–56.

Owen, R., 1849. Osteological contributions to the natural history of the chimpanzee (*Troglodytes*, Geoffroy), including the description of the skull of a large species (*Troglodytes gorilla*, Savage), discovered by Thomas S. Savage, M.D., in the Gaboon country, West Africa. Trans. Zool. Soc. London, 3 (6), 381–422.

Owen, R., 1853. Descriptive Catalogue of the Osteological Series Contained in the Museum of the Royal College of Surgeons of England. Vol. II. Mammalia. Placentalia. Taylor and Francis, London, pp. 782–805.

Owen, R., 1859a. On the Extinction and Transmutation of Species. John W. Parker and Son, West Strand, London, UK.

Owen, R., 1859b. On the Gorilla (*Troglodytes gorilla*, Sav.). Proc. Zool. Soc. London 27, p. 21.

Owen, R., 1861. The Gorilla and the Negro. Lecture at the Royal Institution, Tuesday 19th of March. Athenaeum, 1769, p. 373.

Owen, R., 1862. Osteological contributions to the natural history of the chimpanzee (*Troglodytes*) and orangs (*Pithecus*). No. IV. Description of the cranium of an adult male gorilla from the River Danger, west coast of Africa, indicative of a variety of the great chimpanzee (*Troglodytes gorilla*), with remarks on the capacity of the cranium and other characters shown by sections of the skull, in the orangs (*Pithecus*), chimpanzees (*Troglodytes*), and in different varieties of the human race. Trans. Zool. Soc. London. 4 (3), 75–88.

Owen, R., 1865a. Memoir on the Gorilla (*Troglodytes gorilla*, Savage). Taylor and Francis, London, UK.

Owen, R., 1865b. Contributions to the natural history of the anthropoid apes. No. VIII. On the external characters of the gorilla (*Troglodytes Gorilla*, Sav). Trans. Zool. Soc. London, 5, 243–284.

Owen, R., 1866. Osteological contributions to the natural history of the anthropoid apes. No. VII. Comparison of the bones of the limbs of the *Troglodytes gorilla*, *Troglodytes niger*, and of different varieties of the human race; and on the general characters of the skeleton of the gorilla. Trans. Zool. Soc. London, V, 1–32.

Oxman, A.D., Guyatt, G.H., 1993. The science of reviewing research. N. Y. Acad. Sci. 703, 125–133.

Paal, S., Kasza, L., Boros, G., Fabian, L., 1982. Tubulonephrose als mögliche Todesursache bei einem Gorilla. (Tubular nephrosis-possible cause of death of a gorilla). Verh. Ber. Erkrg. Zootiere. 24, 205–208.

Paixão, T.A., Coura, F.M., Malta, M.C.C., Tinoco, H.P., Pessanha, A.T., Pereira, F.L., Leal, C.A.G., Heinemann, M.B., Figueiredo, H.C.P., Santos, R.L., 2014. Draft genome sequences of two *Salmonella enterica* serotype Infantis strains isolated from a captive western lowland gorilla (*Gorilla gorilla gorilla*) and a cohabitant black and white tegu (*Tupinambis merianae*) in Brazil. Genome Announc. 4 (1), e01590–15.

Paixão, T.A., Tinoco, H.P., de Campos Cordeiro Malta, M., Teixeira da Costa, M.E.L., Soave, S.A., Pessanha, A.T., Silva, A.P.C., Santos, R.L., 2014. Pathological findings in a captive senile western lowland gorilla (*Gorilla gorilla gorilla*) with chronic renal failure and septic polyarthritis. Braz. J. Vet. Pathol. 7 (1), 29–34.

Palacios, G., Lowenstine, L.J., Cranfield, M.R., Gilardi, K.V.K., Spelman, L., Lukasik-Braum, M., et al., 2011. Human metapneumovirus infection in wild mountain gorillas, Rwanda. Emerg. Infect. Dis. 17 (4), 711–713.

Papageorgopoulou, C.S.K., Suter, F.J., Rühli, Siegmund, F., 2011. Harris lines revisited: prevalence, co-morbidities and possible etiologies. Am. J. Hum. Biol. 23 (3), S.381–391.

Paredes, J., Sandrine, M., Paerson, L., McCort, R., Polderman, A.M., 2004. Oesophagostomiasis (*Oesophagostomum stephanostomum*) during the reintroduction of a group of western lowland gorillas in Gabon. In: Joint BVZS/WAWV/RVC Conference, RVC, London, UK.

Parker, S., Mitchell, R., Miles, H. (Eds.), 1999. The Mentalities of Gorillas and Orangutans: Comparative Perspectives. Cambridge University Press, Cambridge, UK.

Parnell, R., 2001. Will bachelors party? Perspectives from the wild. Presentation given at "Bachelor Gorilla 2000." Disney's Animal Kingdom, Lake Buena Vista, FL.

Parrott, T.Y., 1997. An introduction to diseases of nonhuman primates. In: Rosenthal, K.L. (Ed.), Practical Exotic Animal Medicine. Veterinary Learning Systems, Trenton, NJ, pp. 258–263.

Patton, L., Ogden, J., Czezcala-Gruber, N., 1997a. Estrous cycle and copulation. In: Ogden, J., Wharton, D. (Eds.), The Management of Gorillas in Captivity. Atlanta/Fulton County Zoo, Atlanta, GA, pp. 33–38.

Patton, L., Ogden, J., Czezcala-Gruber, N., Loskutoff, N., 1997b. Gestation and parturition. In: Ogden, J., Wharton, D. (Eds.), The Management of Gorillas in Captivity. Atlanta/Fulton County Zoo, Atlanta, GA, pp. 39–44.

Patton, S., Sleeman, J.M., Mudakikwa, A.B., Meader, L.L., Foster, J.W., 1998. Parasites of Mountain Gorillas (*Gorilla gorilla beringei*) from the Virunga Volcanoes Region, Central Africa, and Implications of Human Contact. Southeastern Society of Parasitologists, Clemson, South Carolina.

Paul-Murphy, J., 1993. Bacterial enterocolitis in nonhuman primates. In: Fowler, M.E. (Ed.), Zoo and Wild Animal Medicine: Current Therapy 3. W.B. Saunders, Philadelphia, PA, pp. 344–351.

Pavlovsky, E.N., 1966. Natural Nidality of Transmissible Diseases with Special Reference to the Landscape Epidemiology of Zooanthroponoses (English translation). University of Illinois Press, Urbana and London.

Pearson, L., Aczel, P., Mahé, S., Courage, A., King, T., 2007. Gorilla reintroduction to the Batéké Plateau National Park, Gabon: an analysis of the preparations and initial results with reference to the IUCN guidelines for the re-introduction of Great Apes. In: Projet Protection des Gorilles, BP 583, Franceville, Gabon. < http://643world.com/publications/ PPG_2007_gorilla_reintoduction_in_Gabon_report.pdf >.

Peden, C., 1960. Report on post-mortem performed on adult mountain gorilla. Mimeographed report by the Department of Veterinary Services and Animal Husbandry, Uganda.

Peeters, M., Delaporte, E., 2012. Simian retroviruses in African apes. Clin. Microbiol. Infect. 18, 514–520.

Peeters, M., Courgnaud, V., Abela, B., Auzel, P., Pourrut, X., Bibollet-Ruche, F., et al., 2002. Risk to human health from a plethora of simian immunodeficiency viruses in primate bushmeat. Emerg. Infect. Dis. 8 (5), 451–457.

Pellis, S., Iwaniuk, A., 1999. The problem of adult play fighting: a comparative analysis of play and courtship in primates. Ethology 105, 783–806.

Penner, L.R., 1981. Concerning threadworm (*Strongyloides stercoralis*) in great apes – lowland gorillas (*Gorilla Gorilla*) and chimpanzees (*Pan troglodytes*). J. Zoo Anim. Med. 12 (4), 128–131.

Pennington, D.D., Simpson, G.L., McConnell, M.S., Fair, J.M., Baker, R.J., 2013. Transdisciplinary research, transformative learning, and transformative science. Bioscience 63, 564–573.

Pereira, M.E., Pond, C.M., 1995. Organization of white adipose tissue in Lemuridae. Am. J. Primatol. 35, 1–13.

Pérez, J.M., 2009. Parasites, pests and pets in a global world: new perspectives and Challenges. J. Exotic Pet Med. 18 (4), 248–253.

Perez, S.E., Raghanti, M.A., Hof, P.R., Kramer, L., Ikonomovic, M.D., Lacor, P.N., et al., 2013. Alzheimer's disease pathology in the neocortex and hippocampus of the western lowland gorilla (*Gorilla gorilla gorilla*). J. Comp.Neurol. 521, 4318–4338.

Petersen, H.H., Nielsen, J.P., Heegard, P.M., 2004. Application of acute phase protein measurements in veterinary clinical chemistry. Vet. Res. 35, 163–187.

Petiniot, C., 1990. A successful introduction: behavior enrichment and training at the Toledo Zoo. In: Columbus Zoo Gorilla Workshop Proceedings, pp. 129–132.

Petiniot, C., 1995. Operant conditioning and great ape management at the Toledo Zoo. In: AZA Regional Conference Proceedings, pp. 116–119.

Petiniot, C., Favata, G., 1997. Successful introduction of orphaned seven-and-one-half-month old female gorilla to surrogate mother gorilla using operant conditioning. Abstract. In: Pittsburgh Zoo Gorilla Workshop Program, p. 14.

Pfitzer, P., Schulte, H. D., 1972. The nuclear DNA content of myocardial cells of monkeys as a model for the polyploidization in the human heart. In: Medical Primatology 1972. Proc. 3rd Conf. Exp. Med. Surg. Primates, Lyon 1972, part I: 379–389. S. Karger, Basel.

Philipp, C., 1995. Operant conditioning with the great apes. In: AAZK National Conference Proceedings, pp. 156–163.

Philipp, C., 1997a. Developing a positive reinforcement program with primates at the Memphis Zoo. In: AZA Regional Conference Proceedings, pp. 102–106.

Philipp, C., 1997b. Developing a positive reinforcement program with primates at the Memphis Zoo. Abstract (w/supplemental notes). In: Pittsburgh Zoo Gorilla Workshop Program, p. 13.

Philipp, C., Breder, C., MacPhee, M., 2000. Training as a management tool for facilitating maternal care and infant behaviors in gorillas. Paper Presented at The Apes: Challenges for the 21st Century. Brookfield Zoo, Chicago, IL.

Phillips, R.L., 1997. Preventive medical program for captive gorillas. In: Ogden, J., Wharton, D. (Eds.), Gorilla Husbandry Manual. Fulton County Zoo, Atlanta, GA, pp. 153–162.

Pijnenborg, R., Vercruysse, L., Carter, A.M., 2011. Deep trophoblast invasion and spiral artery remodelling in the placental bed of the lowland gorilla. Placenta 32 (8), 586–591.

Pilbrow, V., 2010. Dental and phylogeographic patterns of variation in gorillas. J. Hum. Evol. 59 (1), 16–34.

Pires De Lima, J.A., 1933. Fracture de l'avant bras chez un 'Gorilla gina'. Folia Anat. Univ. Conimbr. 8 (19).

Pitman, C., 1931. A Game Warden Among His Charges. Nisbet, London, UK.

Pitman, C., 1942. A Game Warden Takes Stock. Nisbet, London, UK.

Pizzi, R., 2009. Veterinarians and taxonomic chauvinism: the dilemma of parasite conservation. J. Exotic Pet Med. 18 (4), 279–282.

Platz, C.C.J., et al., 1980. Electroejaculation and semen analysis in a male lowland gorilla, *Gorilla gorilla gorilla*. Primates 21 (1), 130–132.

Plumptre, A.J., 1995. The chemical composition of montane plants and its influence on the diet of the large mammalian herbivores in the Parc National des Volcans, Rwanda. J. Zool. 235, 323–337.

Plumptre, A.J., Nixon, S., Critchlow, R., Vieilledent, G., Nishuli, R., Kirkby, A., et al., 2015. Status of Grauer's Gorilla and Chimpanzees in Eastern Democratic Republic of Congo: Historical and Current Distribution and Abundance. Wildlife Conservation Society, Fauna & Flora International and Institut Congolais pour la Conservation de la Nature, New York.

Plumptre, A.J., Robbins, M., Williamson, E.A., 2016. *Gorilla beringei*. The IUCN Red List of Threatened Species 2016: e.T39994A17964126.

Plowden, G., 1972. Gargantua: Circus Star of the Century. Bonanza Books, New York.

Pomajbíková, K., Petrželková, K.J., Profousová, I., Petrášová, J., Modrý, D., 2010. Discrepancies in the occurrence of *Balantidium coli* between wild and captive African great apes. J. Parasitol. 96 (6), 1139–1144.

Pomajbíková, K., Oborník, M., Horak, A., Petrželková, K.J., Norman Grim Bruno, J., Levecke, B., et al., 2013. Novel insights into the diversity of *Balantidium*-like ciliates in primates based on the conservative and hypervariable nuclear gene markers. PLoS. Negl. Trop. Dis. 7 (3), e2140.

Pond, C.M., 2000. Adipose tissue: quartermaster to the lymph node garrisons. Biologist 47 (3), 147-5.

Pond, C.M., 2005. Adipose tissue and the immune system. Prostaglandins. Leukot. Essent. Fatty. Acids. 73, 17–30.

Pond, C.M., 2016. The evolution of mammalian adipose tissues. In: Symonds, M.E. (Ed.), Adipose Tissue Biology. Springer Science, Heidelberg and New York.

Poole, T., 1988. Normal and abnormal behaviour in captive primates. Primate Rep. 22, 3–12.

Popilskis, S.J., Kohn, D.F., 1997. Anesthesia and analgesia in nonhuman primates. In: Kohn, D.F., Wixson, S.K., White, W.J., Benson, G.J. (Eds.), Anesthesia and Analgesia in Laboratory Animals. Academic Press, San Diego, CA, pp. 233–255.

Popovich, D., Dierenfield, E., 1997. Nutrition. In: Ogden, J., Wharton, D. (Eds.), Gorilla Husbandry Manual. Fulton County Zoo, Atlanta, GA, pp. 134–138.

Porton, I., 1997a. Birth control options for gorillas. In: Ogden, J., Wharton, D. (Eds.), Gorilla Husbandry Manual. Fulton County Zoo, Atlanta, GA, pp. 71–78.

Porton, I., 1997b. Birth management and hand-rearing of captive gorillas. In: Ogden, J., Wharton, D. (Eds.), The Management of Gorillas in Captivity: Husbandry Manual of the Gorilla Species Survival Plan, AZA Gorilla SSP. Atlanta/Fulton County Zoo, Atlanta, GA, pp. 111–123.

Porton, I., White, M., 1996. Managing an all-male group of gorillas: eight years of experience at the St. Louis Zoological Park. In: AZA Regional Conference Proceedings, pp. 720–728.

Prado-Martinez, J., Hernando-Herraez, I., Lorente-Galdos, B., Dabad, M., Ramirez, O., Baeza-Delgado, C., et al., 2013. The genome sequencing of an albino Western lowland gorilla reveals inbreeding in the wild. BMC Genomics. 14, 363.

Prendini, L., Hanner, R., DeSalle, R., 2002. Obtaining, storing and archiving specimens for molecular genetic research. In: DeSalle, R., Giribet, G., Wheeler, W. (Eds.), Techniques in Molecular Systematics and Evolution. Birkhauser, Basel, pp. 176–248.

Preuschoft, H., 1961. Muskeln und gelenke der hinterextremität des gorillas (*Gorilla gorilla* Savage et Wyman, 1847). Gegenbaurs. Morphol. Jahrb. 101, 432–540.

Preuschoft, H., 1965. Muskeln und Gelenke der Vorderextremität des Gorilla. Gegenbaurs. Morphol. Jahrb. 107, 99–183.

Preuschoft, H., Günther, M.M., 1994. Biomechanics and body shape in primates and horses. Z. Morphol. Anthropol. 80, 149–165.

Preuschoft, H., Hohn, B., Scherf, H., Schmidt, M., Krause, C., Witzel, U., 2010. Functional analysis of the primate shoulder. Int. J. Primatol. 31 (Issue 2), 301–320.

Prince-Hughes, D., 2001. Gorillas Among Us. A Primate Ethnographer's Book of Days. University of Arizona Press, Tucson, AZ.

Prine, J.R., 1968. Pancreatic flukes and amoebic colitis in a gorilla. Abstract 50. In: 19th Annual Meeting, Association Laboratory Animal Science, Las Vegas, USA.

Prowten, A.W., Lee, R.V., Krishnamsetty, R.M., Satchidanand, S.K., Srivastava, B.I., 1985. T-cell lymphoma associated with immunologic evidence of retrovirus infection in a lowland gorilla. J. Am. Vet. Med. Assoc. 187 (11), 1280–1282.

Proyart, Abbé. 1776. History of Loango, Kakongo and N'Goyo. Paris.

Purchas, S., 1625. Hakluytus posthumus, or, Purchas His Pilgrimes. Contayning a history of the World, in Sea Voyages, and Lande-Travells, by Englishmen and Others. In Four Parts, Each Containing Five Bookes. William Stansby for Henrie Fetherstone, London.

Pyke, G.H., Ehrlich, P.R., 2010. Biological collections and ecological/environmental research: a review, some observations and a look to the future. Biol. Rev. 85, 247–266.

Pyle, R.M., 2001. The rise and fall of natural history, Orion. Autumn 2001, pp. 16–23.

Rabeeah, A.A., 2006. Conjoined twins—past, present, and future. J. Pediatr. Surg. 41, 1000–1004.

Rademacher, A., 2000. The Little Rock Zoo social gorilla groups. Gorilla Gaz. 14 (1), 1–3.

Rahn, P., 1993. Zwei Geburten und Adoptionen bei Westlichen Flachland gorillas (Gorilla g. gorilla) im Zoologischen Garten Berlin. Bongo 21, 19–24.

Ralls, K., Brugger, K., Ballou, J., 1979. Inbreeding and juvenile mortality in small populations of ungulates. Science 206, 1101–1103.

Ramsey, E.C., et al., 1985. Hysterosalphingography in gorilla. J. Zoo Anim. Med. 16 (3), 85–89.

Randal, P., Taylor, P., Banks, D.R., 2007. Pregnancy and stillbirth in a lowland gorilla (Gorilla gorilla gorilla). Int. Zoo Yearb. 23 (1), 183–185.

Randall, F.E., 1943. The skeletal and dental development and variability of the gorilla. Hum. Biol. 15, 236–254.

Rapley, W.A. et al, 1979. The clinical management of a unilateral paralytic strabismus in a lowland gorilla. In: American Association of Zoo Veterniarians Annual Meeting, pp. 91–92.

Rapley, W.A., Mehren, K.G., Prober, C.G., 1982. Neurosurgery and medical care of an infant lowland gorilla (Gorilla gorilla gorilla). In: Am. Assoc. Zool. Parks Aquariums. Regional Conf. Proc.

Rasmussen, A.T., Rasmussen, T., 1952. The hypophysis cerebri of Bushman, the gorilla of Lincoln Park Zoo, Chicago. Anat. Rec. 113 (3), 325–347.

Ratcliffe, H.L., 1961. Rep. Penrose Res. Lab., Zool. Soc. Philad. 13, 1961.

Ratcliffe, H.L., Cronin, M.T.I., 1958. Changing frequency of arteriosclerosis in mammals and birds at the Philadelphia Zoological Garden: review of autopsy records. Circulation 18, 41–52.

Rau, J., 2005. Biodiversidad y colecciones científicas. Rev. Chil. Hist. Nat. 78, 341–342.

Reade, W.W., 1864. Savage Africa: Being the Narrative of a Tour in Equatorial, Southwestern, and Northwestern Africa; With Notes on the Habits of the Gorilla; on the Existence of Unicorns and Tailed Men; on the Slave-Trade; on the Origin, Character, and Capabilities of the Negro, and on the Future Civilization of Western Africa. Harper and Brothers, New York, NY.

Redmond, I., 1978a. Post mortem examination of male Gorilla g. beringei juvenile, known as Kweli. 26th October 1978. Karisoke Research

Centre Records, summarised in Fossey, D. (1983) Gorillas in the Mist, Appendix F.

Redmond, I., 1978b. Post mortem examination of female Gorilla g. beringei infant, known as Mwelu. 6th December 1978. Karisoke Research Centre Records, summarised in Fossey, D. (1983) Gorillas in the Mist, Appendix F.

Redmond, I., 1983. Summary of parasitology research, November 1976 to April 1978. In: Fossey, D. (Ed.), Gorillas in the Mist, Appendix G. Houghton Mifflin, Boston, MA, pp. 271–286.

Redrobe, S., 2003. Diagnostic and epidemiology of tuberculosis at a British zoo — multispecies infection with a new form of Mycobacterium, preliminary overview. In: Proceedings Conference of American Association Zoo Veterinarians, 2003.

Redrobe, S., 2008. Neuroleptics in great apes, with specific reference to modification of aggressive behavior in a male gorilla. In: Fowler, M., Miller, E. (Eds.), Zoo and Wild Animal Medicine Current Therapy, vol. 6. Saunders Elsevier, Philadelphia, PA.

Redshaw, M.E., Mallinson, J.J.C., 1991. Stimulation of natural patterns of behaviour: studies with golden lion tamarins and gorillas. In: Box, H.O. (Ed.), Primate Responses to Environmental Change. Chapman and Hall, London, UK, pp. 217–238.

Reel, M., 2013. Between Man and Beast: An Unlikely Explorer, the Evolution Debates, and the African Adventure that Took the Victorian World by Storm. Doubleday, New York, NY.

Reichard, T.A., Favata, G.F., Petiniot, C., Gheres, A., Czekala, N.M., Lasley, B.L., 1990. Reproductive management plan of Toledo Zoo lowland gorillas over the past 25 years. Zoo Biol. 9, 149–156.

Renaud, F., de Meeüs, T., 1991. A simple model of host-parasite evolutionary relationships. Parasitism: compromise or conflict? J. Theor. Biol. 152, 319–327.

Rennick, S.L., Fenton, T.W., Foran, D.R., 2005. The effects of skeletal preparation techniques on DNA from human and non- human bone. J. Forensic Sci. 50, 1016–1019.

Resnick, D., Shaul, S.R., Robins, J.M., 1975. Diffuse idiopathic skeletal hyperostosis (DISH): Forestier's disease with extraspinal manifestations. Radiology 115, 513–524.

Resnick, D., Niwayama, G., 1976. Radiographic and pathologic features of spinal involvement in diffuse idiopathic skeletal hyperostosis (DISH). Radiology 119, 559.

Reynolds, M.G., Emerson, G.L., Pukuta, E., Karhemere, S., Muyembe, J.J., Bikindou, A., et al., 2013. Detection of human monkeypox in the Republic of the Congo following intensive community education. Am. J. Trop. Med. Hyg. 88 (5), 982–985.

Reynolds, V., Lloyd, A.W., English, C.J., Lyons, P., Dodd, H., Hobaiter, C., et al., 2015. Mineral acquisition from clay by Budongo forest chimpanzees. PLoS One 10 (7), e0134075.

Rich, J., 2011. Missing Links: The African and American Worlds of R. L. Garner, Primate Collector. The University of Georgia Press, Athens.

Riddle, K.E., Keeling, M.E., Roberts, J., 1973. Birth of a lowland gorilla (Gorilla gorilla gorilla) at the Yerkes Regional Primate Research Center. J. Zoo An. Med. 4 (4), 22–27.

Rideout, B.A., Gardiner, C.H., Stalis, I.H., Zuba, J.R., Hadfield, T., Visvesvara, G.S., 1997. Fatal infections with Balamuthia mandrillaris (a free-living amoeba) in gorillas and other Old World primates. Vet. Pathol. 34 (1), 15–22.

Ridges, P., 1999. Forming bachelor groups at Port Lympne. Gorilla Gaz. 13 (1), 8–9.

Riopelle, A.J., 1967. Snowflake the world's first white gorilla. Natl. Geogr. Mag. 131, 442–448, March.

Robbins, M., Sicotte, P., Stewart, K. (Eds.), 2001. Mountain Gorillas: Three Decades of Research at Karisoke. Cambridge University Press, Cambridge, UK.

Robbins, M.M., Czekala, N.M., 1997. A preliminary investigation of urinary testosterone and cortisol levels in wild male mountain gorillas. Am. J. Primatol. 43, 51–64.

Robbins, M.M., Bermejo, M., Cipolletta, C., Magliocca, F., Parnell, R.J., Stokes, E., 2004. Social structure and life-history patterns in western gorillas (*Gorilla gorilla gorilla*). Am. J. Primatol. 64, 145–159.

Robbins, M.M., Steklis, D., Steklis, N., Robbins, A.M., 2007. Sociological influences on the reproductive success of female mountain gorillas (*Gorilla beringei beringei*). Behav. Ecol. Sociobiol. 61, 919–931.

Roberts, J.A., 1995. Occupational health concerns with nonhuman primates in zoological gardens. J. Zoo Wildl. Med. 26 (1), 10–23.

Robertson, I.A., 2015. Animals, Welfare and the Law; Fundamental Principles for Critical Assessment. Routledge, Oxford.

Robinson, P.T., Benirschke, K., 1980. Congestive heart failure and nephritis in an adult gorilla. J. Am. Vet. Med. Assoc. 177 (9), 937–938.

Robinson, P.T., Lambert, D., 1986. A review of 226 chemical restraint procedures in great apes at the San Diego Zoo. In: Proceedings of American Association of Zoo Veterinarians, p. 183.

Robson, S.L., Wood, B., 2008. Hominin life history: reconstruction and evolution. J. Anat. 212, 294–425.

Rode, P., Solas, L., 1946. Un crâne de gorille à 36 dents. L'Odontologie, Paris.

Roe, D., Biggs, D., Dublin, H., Cooney, R., 2016. Engaging Communities To Combat Illegal Wildlife Trade: a Theory of Change. IIED Briefing, February 2016. International Institute for Environment and Development, London, < http://pubs.iied.org/pdfs/17348IIED.pdf >.

Rogers, M.E., Abernethy, K., Bermejo, M., Cipolletta, C., Doran, D., Mcfarland, K., et al., 2004. Western gorilla diet: a synthesis from six sites. Am. J. Primatol. 64, 173–192.

Rogers, R., Eastham-Anderson, J., DeVoss, J., Lesch, J., Yan, D., Xu, M., et al., 2016. Image analysis-based approaches for scoring mouse models of colitis. Vet. Pathol. 53, 200–210.

Rohen, J.W., 1962. Beitrag zur mikroskopischen Anatomie des Gorilla-Auges (Contribution to the microscopic anatomy of the gorilla eye). Albrecht von Graefes Arch. Ophthalmol. 164 (4), 374–385.

Ronco, C., McCullough, P., Anker, S.D., et al., 2010. Cardio-renal syndromes: report from the consensus conference of the acute dialysis initiative. Euro Heart J. 31, 703–711.

Rooney, M., Sleeman, J., 1998. Effects of selected behavioral enrichment devices on behavior of western lowland gorillas (*Gorilla gorilla gorilla*). J. Appl. Anim. Welf. Sci. 1 (4), 339–351.

Roscoe, N., 2007. Dissection: a dying art? Biologist 54 (1), 6.

Rose, G.L., 2011. Gaps in the Implementation of Environmental Law at the National, Regional and Global Level. World Congress on Justice, Governance and Law for Environmental Sustainability12 - 13 October 2011 - Kuala Lumpur, Malaysia. < http://www.unep.org/delc/Portals/24151/FormatedGapsEL.pdf >.

Rose, P.E., Croft, D.P., 2015. The potential of social network analysis as a tool for the management of zoo animals. Anim. Welf. 24, 123–138.

Rosen, S.I., 1972. Twin gorilla fetuses. Folia Primatol. 17, 132–141.

Rosenbaum, S., Hirwa, J.P., Silk, J.B., Stoinski, T.S., 2016. Relationships between adult male and maturing mountain gorillas (*Gorilla beringei beringei*) persist across developmental stages and social upheaval. Ethology 122 (2), 134–150.

Rosenblum, I.Y., Barbolt, T.A., Howard, C.F., 1981. Diabetes mellitus in the chimpanzee (*Pan troglodytes*). J. Med. Primatol. 10, 93–101.

Ross, C.F., Martin, R.D., 2006. The role of vision in the origin and evolution of primates. In: Kaas, J.H. (Ed.), Evolution of Nervous Systems, vol. IV: Primates. Academic Press, Oxford, pp. 59–78.

Ross, R., Glomset, J., Harker, L., 1977. Response to injury and atherogenesis. Am. J. Pathol. 86 (3), 675–684.

Ross, S.R., Holmes, A.N., Lonsdorf, E.V., 2009. Interactions between zoo-housed great apes and local wildlife. Am. J. Primatol. 71 (6), 458–465.

Rothman, J.M., Bowman, D.D., Eberhard, M.L., et al., 2002. Intestinal parasites in the research group of mountain gorillas in Bwindi Impenetrable National Park, Uganda: preliminary results. Ann. N. Y. Acad. Sci. 969, 346–349.

Rothman, J.M., Bowman, D.D., Kalema-Zikusoka, G., et al., 2006a. The parasites of the gorillas in Bwindi Impenetrable National Park, Uganda. In: Paterson, J., Reynolds, V., Notman, H., Newton-Fisher, N. (Eds.), Primates of Western Uganda. Springer Science and Business Media, LLC, New York, NY, pp. 171–192.

Rothman, J.M., Van Soest, P.J., Pell, A.N., 2006b. Decaying wood is a sodium source for mountain gorillas. Biol. Lett. 2, 321–324.

Rothman, J.M., Plumptre, A.J., Dierenfeld, E.S., Pell, A.N., 2007. Nutritional composition of the diet of the gorilla (*Gorilla beringei*): a comparison between two montane habitats. J. Trop. Ecol. 23, 673–682.

Rothman, J.M., Dierenfeld, E.S., Hintz, H.F., Pell, A.N., 2008. Nutritional quality of gorilla diets: consequences of age, sex, and season. Oecologia. 155 (1), 111–122.

Rothschild, B.M., 2015. Intestinal flora modification of arthritis pattern in spondyloarthropathy. J. Clin. Rheumatol. 21, 296–299.

Rothschild, B.M., Rühli, F.J., 2005. Comparison of arthritis characteristics in lowland *Gorilla gorilla* and mountain *Gorilla beringei*. Am. J. Primatol. 66, 205–218.

Rothschild, B.M., Woods, R.J., 1992. Erosive arthritis and spondyloarthropathy in Old World primates. Am. J. Phys. Anthropol. 88 (3), 389–400.

Rothschild, W., 1904. Notes on anthropoid apes. Proc. Zool. Soc. London 2, 413–440.

Rothschild, W., 1908. Note on *Gorilla gorilla diehli* Matschie. Novit. Zool. 15, 391–392.

Rothschild, W., 1923. Remarks on a Mountain Gorilla from near Lake Kivu. Proc. Zool. Soc. London 93 (1), 176–177.

Rothschild, W., 1927a. On a new race of bongo and of gorilla. Ann. Mag. Nat. Hist. 19, 271.

Rothschild, W., 1927b. On the skull of *Gorilla gorilla halli*, Rothsch. Ann. Mag. Nat. Hist. 19, 512.

Rotstein, D.S., 2008. How to perform a necropsy if a toxin is suspected. J. Exotic Pet Med. 17 (1), 39–43.

Rouquet, P., et al., 2005. Wild animal mortality monitoring and human ebola outbreaks, Gabon and Republic of Congo, 2001-2003. Emerg. Infect. Dis. 11, 283–290.

Rousselot, R., Pellissier, A., 1952. Pathologie du gorille. III. Esophagostomose nodulaire à *Esophagostomum stephanostomum* Pathologie du gorille et du chimpanzee. Bull. Soc. Pathol. Exot. 9, 569–574.

Routh, A., Sleeman, J.M. A preliminary survey of syndactyly in the mountain gorilla (*Gorilla gorilla beringei*). In: Proceedings of British Veterinary Zoological Society (BVZS) Meeting, 14–15 June 1997. Howletts and Port Lympne Wild Animal Parks, UK.

Royal College of Obstetricians and Gynaecologists & Royal College of Pathologists, 2001. Fetal and Perinatal Pathology. Royal College of Pathologists, London, UK.

Ruch, T., 1959. Diseases of Laboratory Primates. W. B. Saunders, Philadelphia, PA.

Ruedi, D., et al., 1978. Herpes virus-induced sepsis in a 13 day-old gorilla in the Zoological Garden of Basel. Verh. Ber. Erkrg. Zootiere. 20, 279–282.

Ruempler, U., 1992. The Cologne Zoo diet for lowland gorillas (*Gorilla gorilla gorilla*) to eliminate regurgitation and reingestion. Int. Zoo Yearb. 31, 225–229.

Ruff, C.B., Burgess, L.M., Bromage, T.G., Mudakikwa, A., McFarlin, S.C., 2013. Ontogenetic changes in limb bone structural proportions in mountain gorillas (*Gorilla beringei beringei*). J. Hum. Evol. 65 (6), 693–703.

Rush, E.M., Ogburn, A.L., Monroe, D., 2010a. Clinical management of a western lowland gorilla (*Gorilla gorilla gorilla*) with a cardiac resynchronization therapy device. J. Zoo Wildl. Med. 42 (2), 263–276.

Rush, E.M., Ogburn, A.L., Hall, J., et al., 2010b. Surgical implantation of a cardiac resynchronization therapy device in a western lowland gorilla (*Gorilla gorilla gorilla*) with fibrosing cardiomyopathy. J. Zoo Wildl. Med. 41 (3), 395–403.

Russell, W.C., 1947. Biology of the dermestid beetle with reference to skull cleaning. J. Mammal. 28, 284–287.

Russon, A.E., Wallis, J. (Eds.), 2014. Primate Tourism—A Tool For Conservation? Cambridge University Press, Cambridge, UK.

Rutty, G.N., 2004. The pathology of shock versus post-mortem change. In: Rutty, G.N. (Ed.), Essentials of Autopsy Practice. Recent Advances, Topics and Developments. Springer-Verlag, London, UK.

Ruvolo, M., Pan, D., Zehr, S., Goldberg, T., Disotell, T.R., von Dornum, M., 1994. Gene trees and hominoid phylogeny. Proc. Natl. Acad. Sci. USA 91, 8900–8904.

Rwego, I.B., et al., 2008. Gastrointestinal bacterial transmission among humans, mountain gorillas, and livestock in Bwindi Impenetrable National Park, Uganda. Conserv. Biol. 22, 1600–1607.

Ryan, A.S., 1981. Anterior dental microwear and its relationships to diet and feeding behavior in three African primates (*Pan troglodytes troglodytes, Gorilla gorilla gorilla* and *Papio hamadryas*). Primates 22 (4), 533–550.

Ryan, E.B., Proudfoot, K.L., Fraser, D., 2012. The effect of feeding enrichment methods on the behavior of captive Western lowland gorillas. Zoo Biol. 31 (2), 235–241.

Ryan, S., Roth, A., Thompson, S., 2000. Effects of hand-rearing on the reproductive success of western lowland gorillas in North America. Paper Presented at "The Apes: Challenges for the 21st Century." Brookfield Zoo, Chicago, IL.

Ryan, S., Thomson, S.D., Roth, A.D., Gold, K.C., 2002. Effects of hand-rearing on the reproductive success of western lowland gorillas in North America. Zoo Biol. 21, 389–401.

Ryan, S.J., Walsh, P.D., 2011. Consequences of non-intervention for infectious disease in African great apes. PLoS ONE 6 (12), e29030.

Ryder, O.A., 1989. Genetic studies in support of conservation of gorillas. In: International Union of Directors of Zoological Gardens, Scientific Session of the 44th Annual Conference, San Antonio, CA, USA, pp. 54–58.

Sabater Pi, J., 1966. Gorilla attacks against humans in Rio Muni, West Africa. J. Mammal. 47, 123–124.

Sabater Pi, J., 1967. An albino lowland gorilla from Rio Muni, West Africa and notes on its adaptation to captivity. Folia Primatol. 7, 155–160.

Saez, H., Chauvien, G., 1977. Champignons kératinophiles décelés dans le pelage de primates sans lésions mycosiques du revêtement cutané. (Keratinophilic fungi detected in the coat of primates not showing any mycotic injury of the skin tissue). Sci. Tech. Anim. Lab. 2 (2), 83–87.

Sainsbury, A.W., Eaton, B.D., Cooper, J.E., 1989. Restraint and anaesthesia of primates. Vet. Rec. 125, 640–643.

Sak, B., Petrželková, K.J., Květoňová, D., Mynářová, A., Shutt, K.A., Pomajbíková, K., et al., 2013. Long-term monitoring of microsporidia, *Cryptosporidium* and *Giardia* infections in western lowland gorillas (*Gorilla gorilla gorilla*) at different stages of habituation in Dzanga Sangha Protected Areas, Central African Republic. PLoS ONE 8 (8), e71840.

Sakka, M., Trouilloud, P., Binnert, D., 1984. Etude radiotomodensitométrique comparée d'un membre pelvien de gorille (*Pan gorilla*). Applications morpho-fonctionnelles à l'origine de l'homme. Mammalia 48 (3), 437–460.

Samuel, W.M., Pybus, M.J., Kocan, A.A. (Eds.), 2001. Parasitic Diseases of Wild Mammals. Second ed. ISU Press, Ames, IO.

Sandbrook, C., Roe, D., 2010. Linking Conservation and Poverty Alleviation: the case of Great Apes. An overview of current policy and practice in Africa. Arcus Foundation, < http://pubs.iied.org/pdfs/G02770.pdf >.

Sandbrook, C., Semple, S., 2006. The rules and the reality of mountain gorilla *Gorilla beringei beringei* tracking: how close do tourists get? Oryx 40, 428–433.

Sandground, J.H., 1928. Some new cestode and nematode parasites from Tanganyika territory. Proc. Boston Soc. Nat. Hist. 39 (No. 41928).

Sandison, A.T., 1955. The histological examination of mummified material. Stain. Technol. 30, 277–283.

Sands, P., Peel, J., 2012. Principles of International Environmental Law. Third ed. Cambridge University Press, Cambridge.

Sapolsky, R.M., 1990. Stress in the wild. Sci. Am. 262, 106–113.

Sapolsky, R.M., Share, L.J., 1998. Darting terrestrial primates in the wild: a primer. Am. J. Primatol. 44, 155–167.

Sarfaty, A., Margulis, S.W., Atsalis, S., 2012. Effects of combination birth control on estrous behavior in captive western lowland gorillas, *Gorilla gorilla gorilla*. Zoo Biol. 31 (3), 350–361.

Sarmiento, E.E., 2003. Distribution, taxonomy, genetics, ecology, and causal links of gorilla survival: the need to develop practical knowledge for gorilla conservation. In: Taylor, A.B., Goldsmith, M.L. (Eds.), Gorilla Biology: A Multidisciplinary Perspective. Cambridge University Press, Cambridge, pp. 432–471.

Sarmiento, E.E., Butynski, T.M., 1996. Present problems in gorilla taxonomy. Gorilla J. 12, 5–7.

Sarmiento, E.E., Oates, F., 2000. The Cross River gorilla: a distinct subspecies *Gorilla gorilla diehli* Matschie 1904. Am. Mus. Novit. 3304, 1–55.

Sasseville, V.G., Mankowski, J.L., Baldessar, A., et al., 2013. Viral infections and nonhuman primate case reports. Meeting report: emerging respiratory viral infections and nonhuman primate case reports. Vet. Pathol. 50 (6), 1145–1153.

Saudargas, R., Drummer, L., 1996. Single subject (small N) research designs and zoo research. Zoo Biol. 15, 173–181.

Savage, T.S., Wyman, J., 1843. Observations on the external characters and habits of *Troglodytes niger*, Geoff. and on its organization. Boston J. Nat. Hist. 4, 362–376, 377–386.

Savage, T.S., Wyman, J., 1847. Notice of the external characters and habits of *Troglodytes gorilla*, a new species of orang from the Gaboon River; Osteology of the same. Boston J. Nat. Hist. 5, 417–442.

Saxena, C.B., Rai, P., Shrivastava, V.P., 1998. Veterinary Post-mortem Examination – A Laboratory Manual. Vikas Publishing House, New Delhi, India.

Scally, A., Dutheil, J.Y., Hillier, L.W., Jordan, G.E., Goodhead, I., Herrero, J., et al., 2012. Insights into hominid evolution from the gorilla genome sequence. Nature 483 (7388), 169–175.

Schaffer, N., Seager, S.W., Platz, C.C., Wildt, D.E., 1981. Fertility evaluation in male gorillas (*Gorilla gorilla gorilla*). Am. Assoc. Zoo Vet. Annu. Proc.133–135.

Schaller, G.B., 1963. The Mountain Gorilla: Ecology and Behavior. University of Chicago Press, Chicago, IL.

Schaller, G.B., 1964. The Year of the Gorilla. University of Chicago Press, Chicago, IL.

Schattner, A., 2015. Curiosity. Are you curious enough to read on? J. R. Soc. Med. 108 (5), 160–164.

Schaumburg, F., Alabi, A.S., Kock, R.A., Mellmann, P.G., Kremsner, C., et al., 2012a. Highly divergent *Staphylococcus aureus* isolates from African non-human primates. Environ. Microbiol. Rep. 4 (1), 141–146.

Schaumburg, F., Mugisha, L., Peck, B., Becker, K., Gillespie, T.R., Peters, G., et al., 2012b. Drug-resistant human *Staphylococcus aureus* in sanctuary apes pose a threat to endangered wild ape populations. Am. J. Primatol. 74 (12), 1071–1075.

Schaumburg, F., Mugisha, L.P., Fichtel, C., Köck, R., Köndgen, S., et al., 2013. Evaluation of non-invasive biological samples to monitor *Staphylococcus aureus* colonization in great apes and lemurs. PLoS ONE 8 (10), e78046.

Scheels, J.L., 1989. Gorilla dental case reports 1982–1988. In: Proceedings of the 1989 Exotic Animal Dentistry Conference, pp. 39–52.

Schenker, N.M., Buxhoeveden, D.P., Blackmon, W.L., Amunts, K., Zilles, K., Semendeferi, K., 2008. A comparative quantitative analysis of cytoarchitecture and minicolumnar organization in Broca's area in humans and great apes. J. Comp. Neurol. 510, 117–128.

Schlegel, H 1876. Muséum d'Histoire Naturelle des Pays-Bas. Revue méthodique et critique des collections déposées dans cet établissement. Tome VII. Monographie 40, Simiae. Leiden, pp. 7–8.

Schmidley, D.J., 2005. What it means to be a naturalist and the future of natural history at American universities. J. Mammal. 86, 449–456.

Schmidt, D.A., Ellersieck, M.R., Cranfield, M.R., Karesh, W.B., 2006. Cholesterol values in free-ranging gorillas (Gorilla gorilla gorilla and Gorilla beringei) and Bornean orangutans (Pongo pygmaeus). J. Zoo Wildl. Med. 37 (3), 292–300.

Schmidt, R.E., Hubbard, G.B., 1987. Atlas of Zoo Animal Pathology, vol. 1. Mammals. CRC Press, Boca Raton, FL.

Schneider, H.-E., 1989. Hyperthyreose bei einem männlichen Flachland gorilla (Gorilla gorilla gorilla) (Hyperthyreosis in a male lowland gorilla (Gorilla gorilla gorilla). Verh. ber. Erkrg. Zootiere. 31, 243–244.

Schoffner, T., 1997. Introduction of juvenile male gorillas to a blackback bachelor group. Abstract (w/ supplemental notes). In: Pittsburgh Zoo Gorilla Workshop Program, p. 26.

Schouteden, H., 1927. Note in: Schwarz, E (1927). Un gorille nouveau de la forêt de l'Ituri Gorilla gorilla rex-pygmaeorum subsp. n. Rev. Zool. Afr. 14, 333–336.

Schulman, F.Y., Farb, A., Virmani, R., Montali, R.J., 1995. Fibrosing cardiomyopathy in captive western lowland gorillas (Gorilla gorilla gorilla) in the United States: a retrospective study. J. Zoo Wildl. Med. 26 (1), 43–51.

Schultz, A.H., 1927. Studies on the growth of gorilla and of other higher primates with special reference to a foetus of gorilla, preserved in the Carnegie Museum. Mem. Carnegie Mus. 11, 1–86.

Schultz, A.H., 1930. The skeleton of the trunk and limbs of higher primates. Hum. Biol. 2, 303–438.

Schultz, A.H., 1934. Some distinguishing characters of the mountain gorilla. J. Mammal. 15 (1), 51–61.

Schultz, A.H., 1935. Eruption and decay of the permanent teeth in primates. Am. J. Phys. Anthropol. 19, 489–581.

Schultz, A.H., 1937. Proportions, variability, and asymmetries of the long bones of the limbs and the clavicles in man and apes. Hum. Biol. 9, 281–328.

Schultz, A.H., 1939. Notes on diseases and healed fractures of wild apes and their bearing on the antiquity of pathological conditions in man. Bull. Hist. Med. 7, 571–582.

Schultz, A.H., 1950. Morphological observations on gorillas. In: Gregory, W. K. (Ed.), The Anatomy of the Gorilla. Columbia University Press, New York, NY, pp. 227–251.

Schultz, A.H., 1956. The occurrence and frequency of pathological and teratological conditions and of twinning among non-human primates. Primatologia, 1, 965–1014.

Schultz, A.H., 1962. Die Schädelkapazität männlicher Gorillas und ihr Höchstwert (The cranial capacity of male gorillas and its maximum value). Anthropol. Anz. 25, 197–203.

Schultz, A.H., 1964. A gorilla with exceptionally large teeth and supernumerary premolars. Folia Primatol. 2, 149–160.

Schultz, M., Starck, D., 1977. Neue Beobachtungen und Überlegungen zur Pathologie des Primatenschädels Ein Beitrag zur «Gundu»-Frage. Folia Primatol. 28, 81–108.

Schwabe, C.W., 1984. Veterinary Medicine and Human Health. Williams & Wilkins, Baltimore, MD.

Schwartz, G.T., Reid, D.J., Dean, M.C., Zihlman, A.L., 2006. A faithful record of stressful life events preserved in the dental development record of a juvenile gorilla. Int. J. Primatol. 27, 1201–1219.

Schwartz, J.H., 1983. Premaxillary-maxillary suture asymmetry in a juvenile gorilla. Implications for understanding dentofacial growth and development. Folia Primatol. 40 (1-2), 69–82.

Schwarz, E., 1927. Un gorille nouveau de la forêt de l'Ituri Gorilla gorilla rex-pygmaeorum subsp. n. Rev. Zool. Afr. 14, 333–336.

Schwarz, E., 1928. Die Sammlung afrikanischer Affen im Congo-Museum. Rev. Zool. Bot. Afr. 16, 105–152.

Scott, G.B.D., 1977. The value of comparative pathology. Editorial. Proc. R. Soc. Med. 70, 372–374.

Scott, G.B.D., 1979. The comparative pathology of the primate colon. J. Pathol. 127 (2), 65–72.

Scott, G.B.D., 1980. The primate caecum and appendix vermiformis: a comparative study. J. Anat. 131 (3), 549–563.

Scott, G.B.D., 1992. Comparative Primate Pathology. Oxford University Press, Oxford.

Scott, G.B.D., Keymer, I.F., 1975. Ulcerative colitis in apes: a comparison with the human disease. J. Pathol. 115 (4), 241–244.

Scott, G.B.D., Keymer, I.F., 1976. Mucosal herniations in the colons of non-human primates. J. Pathol. 120 (3), 177–181.

Scott, J., Lockard, J., 1999. Female dominance relationships among captive western lowland gorillas: comparisons with the wild. Behaviour 136, 1283–1310.

Scott, P., 2013. Ultrasonographic findings in adult cattle with chronic suppurative pneumonia. In Pract. 35, 460–469.

Scott, S., 2015. No laughing matter. Biologist 62 (1), 27–29.

Scudamore, C.L.C., 2014. A Practical Guide to the Histology of the Mouse. Wiley-Blackwell, New York, NY.

Seager, S.W.J., et al., 1982. Semen collection and evaluation in Gorilla gorilla gorilla. Am. J. Primatol. 3 (S1), p. 13.

Seaton, B., 1978. Patterns of oestrogen and testosterone excretion during pregnancy in a gorilla (Gorilla gorilla gorilla). Reprod. Fertil. 53 (2), 231–236.

Seimon, T.A., Olson, S.H., Lee, K.J., et al., 2015. Adenovirus and herpesvirus diversity in free-ranging great apes in the Sangha Region of the Republic of Congo. PLoS ONE 10 (3), e0118543.

Selbitz, H.-J., Elze, K., Seifert, S., 1980a. Staphylococcus aureus als Erreger einer Pyodermic beim Gorilla (Gorilla gorilla) im Zoologischen Garten Leipzig. (Staphylococcus aureus as a cause of pyoderma in a gorilla (Gorilla gorilla) in the Leipzig Zoo). In: Sonderfruck aus Verhandlungsbericht des 22 Internationalen Symposiums über die Erkrankunken des Berlin.

Selenka, E., 1899. Studien ueber Entwickelungsgeschichte der Tiere, vi Heft (Menschenaffen). C.W. Kreidel's Verlag, Wiesbaden, Germany.

Selye, H., 1950. The Physiology and Pathology of Exposure to Stress. Acta Montreal, Canada.

Selye, H., 1955. Stress and disease. Science 122, 625–631.

Semendeferi, K., Armstrong, E., Schleicher, A., Zilles, K., Van Hoesen, G. W., 1998. Limbic frontal cortex in hominoids: a comparative study of area 13. Am. J. Phys. Anthropol. 106, 129–155.

Semendeferi, K., Armstrong, E., Schleicher, A., Zilles, K., Van Hoesen, G. W., 2001. Prefrontal cortex in humans and apes: a comparative study of area 10. Am. J. Phys. Anthropol. 114, 224–241.

Semple, S., Cowlishaw, G., Bennett, P.M., 2002. Immune system evolution among anthropoid primates: parasites, injuries and predators. Proc. Biol. Sci. 269 (1495), 1031–1037.

Shalukoma, C., 2000. Attempt to re-introduce a young gorilla to the Kahuzi-Biega forest. Gorilla J. 21, 3–4.

Shapiro, A.E., Tukahebwa, E.M., Kasten, J., Clarke, S.E., Magnussen, P., Olsen, A., et al., 2005. Epidemiology of helminth infections and their relationship to clinical malaria in southwest Uganda. Trans. R. Soc. Trop. Med. Hyg. 99, 18–24.

Shave, R., Oxborough, D., Somauroo, J., Feltrer, Y., Strike, T., Routh, A., et al., 2014. Echocardiographic assessment of cardiac structure and function in great apes: a practical guide. Int. Zoo Yearb. 48, 218–233.

Shellabarger, W., 1990. Animal training techniques at the Toledo Zoo in different species to aid in introductions, movement, and for behavioral enrichment. Am. Assoc. Zoo Vet. Annu. Proc.309–311.

Shepherd, N.A., Warren, B.F., Williams, G.T., Greenson, J.K., Lauwers, G. Y., Novelli, M.R. (Eds.), 2013. Morson and Dawson's Gastrointestinal Pathology. Fifth ed. Wiley-Blackwell.

Sheppard, M.N., 2012. Approach to cardiac autopsy. J. Clin. Pathol. 65, 484–495.

Sherwood, C.C., Holloway, R.L., Gannon, P.J., Semendeferi, K., Erwin, J.M., Zilles, K., et al., 2003. Neuroanatomical basis of facial expression in monkeys, apes, and humans. Ann. N.Y. Acad. Sci. 1000, 99–103.

Sherwood, C.C., Holloway, R.L., Erwin, J.M., Schleicher, A., Zilles, K., Hof, P.R., 2004. Cortical orofacial motor representation in Old World monkeys, great apes, and humans. I. Quantitative analysis of cytoarchitecture. Brain Behav. Evol. 63, 61–81.

Sherwood, C.T., Cranfield, M.R., Mehlman, P.T., Lilly, A.A., Garbe, J.A.L., Whittier, C.A., et al., 2004. Brain structure variation in great apes, with attention to the mountain gorilla (Gorilla beringei beringei). Am. J. Primatol. 63 (3), 149–164.

Sholley, C.R., 1989. Mountain gorilla update. Oryx 23, 57–58.

Short, R.V., Weir, B.J. (Eds.), 1980. The Great Apes of Africa: Proceedings of a Symposium Held in Gabon, West Africa. Society for Reproduction and Fertility Supplement No. 28. Society for Reproduction and Fertility Ltd, Cambridge, UK.

Shumaker, R., 1997. Gorilla enrichment. In: Ogden, J., Wharton, D. (Eds.), The Management of Gorillas in Captivity. Atlanta/Fulton County Zoo, Atlanta, GA, pp. 102–110.

Shumaker, R.W., Wich, S.A., Perkins, L., 2008. Reproductive life history traits of female orangutans (Pongo spp.). Interdiscip. Top. Gerontol. 36, 147–161.

Shutt, K., Setchell, J.M., Heistermann, M., 2010. Non-invasive monitoring of physiological stress in the western lowland gorilla (Gorilla gorilla gorilla): validation of a fecal glucocorticoid assay and methods for practical application in the field. Gen. Comp. Endocrinol. 179 (2), 167–177.

Shutt, K., Heistermann, M., Kasim, A., Todd, A., Kalousova, B., Profosouva, I., et al., 2014. Effects of habituation, research and ecotourism on faecal glucocorticoid metabolites in wild western lowland gorillas: implications for conservation management. Biol. Conserv. 172, 72–79.

Sicotte, P., 1993. Inter-group encounters and female transfer in mountain gorillas: influence of group composition on male behavior. Am. J. Primatol. 30, 21–36.

Sikarskie, J.G., 2000. The role of veterinary medicine in wildlife rehabilitation. J. Zoo Wildl. Med. 23, 397–400.

Sikes, S.K., 1969. Habitat and cardiovascular disease: observations made on elephants (Loxodonta africana) and other free-living animals in East Africa. Trans. Zool. Soc. London 32, 1–104.

Sill, F.G., Reyes, J.P., et al., 1996. Treatment of balantidiasis in lowland gorillas and a case complicated with salmonellosis and lead poisoning. In: Proceedings of American Association of Zoo Veterinarians Annual Conference 1996, pp. 410–417.

Simmons, L., 1972a. Case Report. Observations of three gastro-intestinal problems encountered in baby gorillas. In: Proceedings of American Association of Zoo Veterinarians Annual Conference 1972, pp. 273-277.

Simmons, L., 1972b. Fatality in an infant gorilla due to gastric dilation. In: Proceedings of American Association of Zoo Veterinarians Annual Conference 1972, pp. 278–279.

Simmons, L.G., 1972/73. Fatal shigellosis in a two-year-old gorilla. In: Proceedings of American Association of Zoo Veterinarians, pp. 280-281.

Simons, N.D., Wagner, R.S., Lorenz, J.G., 2012. Genetic diversity of North American captive-born gorillas (Gorilla gorilla gorilla). Ecol. Evol. 3 (1), 80–88.

Slack, J.H., 1862. [Untitled note on Gorilla castaneiceps]. Proc. Acad. Nat. Sci. Philadel. 14, 159–160.

Sleeman, J.M., 1998. Preventive medicine programme for the mountain gorillas (Gorilla gorilla beringei) of Rwanda: a model for other endangered primate populations. In: European Association of Zoo and Wildlife Veterinarians 2nd Scientific Meeting, Chester, UK, 21-24 May, 1998, pp. 127–132.

Sleeman, J.M., 2004. The role of Ebola virus in the decline of great ape populations. Oryx 38 (2), 136 (Letter to the Editor).

Sleeman, J.M., 2005. Disease risk assessment in African great apes using geographic information systems. EcoHealth 2 (3), 222–227.

Sleeman, J.M., Mudakikwa, A.B., 1998. Analysis of urine from free-ranging mountain gorillas (Gorilla gorilla beringei) for normal physiologic values. J. Zoo. Wildl. Med. 29 (4), 432–434.

Sleeman, J.M., Mudakikwa, A.B., Foster, J.W., 1996. Human activities affecting the health of the mountain gorillas (Gorilla gorilla beringei). In: Proceedings of the 2nd European Conference of the Wildlife Diseases Association, Wroclaw, Poland.

Sleeman, J.M., Cameron, K., Mudakikwa, A.B., Anderson, S., Cooper, J.E., Hastings, B., et al., 1998. Field anesthesia of free-ranging mountain gorillas (Gorilla gorilla beringei) from the Virunga Volcano region, Central Africa. In: Joint Proceedings of the American Association of Zoo Veterinarians and American Association of Wildlife Veterinarians, Omaha, NE, USA, October, 1998, pp. 1–4.

Sleeman, J.M., Cameron, K.K., Mudakikwa, A.B., Nizeyi, J.B., Anderson, S.S., Cooper, J.E., et al., 2000a. Field anesthesia of free-living mountain gorillas (Gorilla gorilla beringei) from the Virunga Volcano region, Central Africa. J. Zoo Wildl. Med. 31 (1), 9–14.

Sleeman, J.M., Meader, L.L., Mudakikwa, A.B., Foster, J.W., Patton, S., 2000b. Gastrointestinal parasites of mountain gorillas (Gorilla gorilla beringei) in the Parc National des Volcans, Rwanda. J. Zoo Wildl. Med. 31 (3), 322–328.

Sleeman, J.M., Guerrera W., Ssebide, B.J., Pace, L., Ichinose, T.Y., Reif, J. S., 2002c. Medical survey of the local human population to determine possible health risks to the mountain gorillas of Bwindi Impenetrable Forest National Park, Uganda. In: Proceedings of the Wildlife Diseases Association Annual Meeting, Arcata, CA, USA, July–August 2002.

Smiley, T., Spelman, L., Lukasik-Braum, M., Mukherjee, J., Kaufman, G., Akiyoshi, D.E., et al., 2010. Noninvasive saliva collection techniques for free-ranging mountain gorillas and captive eastern gorillas. J. Zoo Wildl. Med. 41 (2), 201–209.

Smiley Evans, T., Barry, P.A., Gilardi, K.V., Goldstein, T., Deere, J.D., Fike, J., et al., 2015. Optimization of a novel non-invasive oral sampling technique for zoonotic pathogen surveillance in nonhuman primates. PLoS Negl. Trop. Dis. 9 (6), e0003813.

Smith, A.J., 1995. Vaccination recommendations for nonhuman primates. Proceedings of the North American Veterinary Conference. Eastern State Veterinary Association, Gainesville, FL, pp. 714–715.

Smith, B.H., Crummet, T.L., Brandt, K.L., 1994. Age of eruption of primate teeth: a compendium for aging individuals and comparing life histories. Yearb. Phys. Anthropol. 37, 177–231.

Smith, R.J., Gannon, P.J., Smith, B.H., 1995. Ontogeny of australopithecines and early Homo: evidence from cranial capacity and dental eruption. J. Hum. Evol. 29, 155–168.

Smith, R.M., Giannoudis, P.V., 1998. Trauma and the immune response. J. R. Soc. Med. 91 (8), 417–420.

Smith, W., 1744. A New Voyage to Guinea: Describing The Customs, Manners, Soil, Climate, Habits, Buildings, Education, Manual Arts, Agriculture, Trade, Employments, Languages, Ranks of Distinction, Habitations, Diversions, Marriages, and Whatever Else Is Memorable Among the Inhabitants. Likewise, An Account of their Animals, Minerals, &c. John Nourse, London.

Snowden, S., Sevenich, M., 1997. Implementation of positive reinforcement training. In: Ogden, J., Wharton, D. (Eds.), The Management of Gorillas in Captivity. Atlanta/Fulton County Zoo, Atlanta, GA, pp. 92–101.

Soifer, F., 1971. A post Shigella infection electrolyte imbalance in a gorilla. J. Zoo Anim. Med. 2, 6–7.

Sonntag, C.F., 1924. The Morphology and Evolution of the Apes and Man. John Bale, Sons & Danielson, London, UK.

Spalding, M.G., Forrester, D.J., 1993. Disease monitoring of free-ranging and released wildlife. J. Zoo Wildl. Med. 24, 271–280.

Spelman, L.H., Gilardi, K.V.K., Lukasik-Braum, M., Kinani, J.-F., Nyirakaragire, E., Lowenstine, L.J., et al., 2013. Respiratory disease in mountain gorillas (Gorilla beringei beringei) in Rwanda, 1990-2010: outbreaks, clinical course, and medical management. J. Zoo Wildl. Med. 44 (4), 1027–1035.

Spencer, D., 1977. A post mortem examination carried out on a stillborn male lowland gorilla *Gorilla g. gorilla* at the Jersey Zoological Park. Dodo J. Jersey Wildl. Preserv. Trust. 14, 96–98.

Spencer, R., 2000. Theoretical and analytical embryology of conjoined twins: Part I: embryogenesis. Clin. Anat. 13 (1), 36–53.

Spies, H.G., Clegg, M.T., 1962. Diseases of the endocrine system. In: Fiennes, R.N.T.-W. (Ed.), Pathology of Simian Primates, Part I. Karger, Basel, Switzerland.

Spikins, P., 2015. How Compassion Made Us Human: The Evolutionary Origins of Tenderness, Trust and Morality. Pen & Sword Archaeology, Barnsley, UK.

Spratt, D., 1993. Gin' ll fix it. Inst. Med. Lab. Sci. Gaz. 29 (1), 26–27.

Srivastava, B.I.S., Wong-Staal, F., Getchell, J.P., 1986. Human T-cell leukemia virus I provirus and antibodies in a captive gorilla with non-Hodgkin's lymphoma. Cancer. Res. 46, 4756.

Stehbens, W.E., 1963. Cerebral aneurysms of animals other than man. J. Path. Bact. 86, 161–168.

Steedle, A., 1997. Better living through training. Abstract. In: Pittsburgh Zoo Gorilla Workshop Program, p. 13.

Steele, S., Fried, J., Bennet, J., 1993. The effects of a silverback lowland gorilla on the behavior and spacing of females in a naturalistic enclosure. In: AZA Annual Conference Proceedings, pp. 312–318.

Stehbens, W.E., 1963. Cerebral aneurysms of animals other than man. J. Path. Bact. 86, 161–168.

Steiner, P.E., 1954. Anatomical observations in a *Gorilla gorilla*. Am. J. Phys. Anthropol. 12 NS, 45–179.

Steiner, P.E., Rasmussen, T., Fisher, L., 1955. Neuropathy, cardiopathy, hemosiderosis, and testicular atrophy in a *Gorilla gorilla*. Am. Med. Assoc. Arch. Pathol. 59, 5–25.

Steinmetz, A., Bernhard, A., Sahr, S., et al., 2012. Suspected macular degeneration in a captive western lowland gorilla (*Gorilla gorilla gorilla*). Vet. Ophthalmol. 15 (Suppl. 2), 139–141.

Sterling, E.J., Bynum, N., Blair, M.E. (Eds.), 2013. Primate Ecology and Conservation: A Handbook of Techniques. Oxford University Press, Oxford, UK.

Stetter, M.D., Cook, R.A., Calle, P.P., Raphael, B., Shayegani, M., 1993. Epidemiology of *Shigella flexneri* in captive lowland gorillas. In: Proceedings of the 1993 Annual Conference of the American Association of Zoo Veterinarians, pp. 61–62.

Stetter, M.D., Cook, R.A., Calle, P.P., Shayegani, M., Raphael, B.L., 1995. Shigellosis in captive western lowland gorillas (*Gorilla gorilla gorilla*). J. Zoo Wildl. Med. 26 (1), 52–60.

Stevens, A., 1990. Gorilla husbandry/enrichment at the Dallas Zoo. Poster abstract. In: Proceedings from the Columbus Zoo Gorilla Workshop, Appendix A-5.

Stevenson, M., 1983. The captive environment: its effect on exploratory and related behavioural responses in wild animals. In: Archer, J., Birke, L. (Eds.), Exploration in Animals and Humans. Van Nostrand Reinhold (UK) Co. Ltd, Berkshire, England, pp. 176–197.

Stewart, K., 1977. The birth of a wild mountain gorilla (*Gorilla gorilla beringei*). Primates 18, 965–976.

Stewart, K., 1990. Reproductive strategies of wild gorillas. In: Columbus Zoo Gorilla Workshop Proceedings, pp. 141–150.

Stewart, K., Harcourt, A., 1987. Gorillas: variations in female relationships. In: Smuts, B., Cheney, D., Seyfarth, R., Wrangham, R., Struhsaker, T. (Eds.), Primate Societies. University of Chicago Press, Chicago, IL, pp. 155–164.

Stiles, D., Redmond, I., Cress, D., Nellemann, C., Formo, R.K. (Eds.), 2013. Stolen Apes – The Illicit Trade in Chimpanzees, Gorillas, Bonobos and Orangutans. A Rapid Response Assessment. United Nations Environment Programme, GRID-Arendal.

Stöber, M., Serrano, H.S., 1974. Gross findings in bovine faeces. Vet. Med. Rev. 4, 361–379.

Stoinski, T.S., Steklis, H.D., Mehlman, P.T. (Eds.), 2008. Conservation in the 21st Century: Gorillas as a Case Study. Springer Science, New York, NY.

Stoinski, T.S., Jaicks, H.F., Drayton, L.A., 2012. Visitor effects on the behavior of captive western lowland gorillas: the importance of individual differences in examining welfare. Zoo Biol. 31 (5), 586–599.

Stoinski, T.S., Perdue, B., Breuer, T., Hoff, M.P., 2013. Variability in the developmental life history of the genus *Gorilla*. Am. J. Phys. Anthropol. 152, 165–172.

Stolpe, H.-J., 1986. Chronische Enteropathie bei einem weiblichen Gorilla mit letalem Ausgang. (Chronic enteropathy in a female gorilla with lethal outcome). Verh. Ber. Erkrg. Zootiere. 28, 275–282.

Stone, A., 2003. IPBIR update for AAPA. Phys. Anthropol. 4 (1), 2.

Stork, N.E., Lyal, C.H.C., 1993. Extinction or 'co-extinction' rates. Nature 366, 307.

Stout, C., Ogilvie, P.W., Lemmon, W.B., 1969. The correlation of behaviour and pathology in zoo animals. Int. Zoo Yearb. 9, 197–198.

Straus, W.L., 1950. The microscopic anatomy of the skin of the gorilla. In: The Anatomy of the Gorilla (W. K. Gregory, Ed.). Columbia University Press, New York, NY, pp. 213–226.

Stringer, C., 2011. The Origin of our Species. Allen Lane, London.

Stringer, E.M., De Voe, R.S., Valea, F., et al., 2010. Medical and surgical management of reproductive neoplasia in two western lowland gorillas (*Gorilla gorilla gorilla*). J. Med. Primatol. 39 (5), 328–335.

Stuart, M.D., Strier, K.B., 1995. Primates and parasites: a case for a multidisciplinary approach. Int. J. Primatol. 16, 577–593.

Stunkard, H.W., Goss, L.J., 1950a. *Eurytrema brumpti* Railliet, Henry and Joyeux, 1912 (Trematoda: Dicrocoeliidae), from the pancreas and liver of African anthropoid apes. J. Parasitol. 36, 1–8.

Stunkard, H.W., Goss, L.J., 1950b. The pancreas and liver of African anthropoid apes. J. Parasitol. 36, 574–581.

Suárez, A.V., Tsutsui, N.D., 2004. The value of museum collections for research and society. BioScience 54, 66–74.

Sugimoto, Y., Ohkubo, F., Ohtoh, H., Honjo, S., 1986. Changes of hematologic values for 11 months after birth in the cynomolgus monkeys. Exp. Anim. S35 (4), 449–454.

Summers, B.A., 2009. Climate change and animal disease. Vet. Pathol. 46, 1185–1186.

Sunde, V., Sievert, J., 1990. Training female lowland gorillas to urinate on request. In: Poster presentation. Columbus Zoo Gorilla Workshop Proceedings, Appendix A-6.

Sunderland-Groves, J.L., 2008. Population, Distribution and Conservation Status of the Cross River Gorilla (*Gorilla gorilla diehli*) in Cameroon (Master of Philosophy thesis). University of Sussex, Sussex, UK.

Sutherland, R., 1990. The effects of temporary space reduction on lowland gorilla socialization. In: Columbus Zoo Gorilla Workshop Proceedings, pp. 151–160.

Sutherland, W.J., Barlow, R., Clements, A., Harper, M., Herkenrath, P., Margerison, C., et al., 2011. What are the forthcoming legislative issues of interest to ecologists and conservationists in 2011? BES Bulletin. 42 (1), 26–31.

Sutherland, W.J., Allison, H., Aveling, R., Bainbridge, I.P., Bennun, L., Bullock, D.J., et al., 2012. Enhancing the value of horizon scanning through collaborative review. Oryx 46, 368–374.

Swales, D.M., Nystrom, P., 2015. Recording primate spinal degenerative joint disease using a standardised approach. In: Karina, G.-R., McSweeney, K. (Eds.), Proceedings of the British Association for Biological Anthropology and Osteoarchaeology (BABAO). Oxbow Books, pp. 10–20.

Swayne, D.E., 2012. Pathologic lesions? Vet. Pathol. 49 (5), 881.

Swenson, R.B., 1993. Protozoal parasites of great apes. In: Fowler, M.E. (Ed.), Zoo and Wild Animal Medicine: Current Therapy 3. W.B. Saunders, Philadelphia, PA, pp. 352–355.

Swenson, R.B., McClure, H.M., 1974. Septic abortion in a gorilla due to *Shigella flexneri*. Proceedings of American Association Zoo Veterinarians Annual Conference, 194–196.

Swindler, D.R., 2002. Primate Dentition: An Introduction to the Teeth of Non-human Primates. Cambridge University Press, Cambridge, UK.

Swindler, D.R., Wood, C.D., 1982. An Atlas of Primate Gross Anatomy: Baboon, Chimpanzee, and Man. Krieger, Malabar, FL.

Switzer, W.M., Wolfe, N.D., Burke, D.S., Mpoudi-Ngole, E.I., Folks, T.M., Heneine, W., 2004. Frequent and widespread simian retrovirus infection in persons exposed to nonhuman primates. In: Proceedings of American Association Zoo Veterinarians Annual, AAWV, WDA Joint Conference.

Szabo, K.T., 1989. Congenital malformations in Laboratory and Farm Animals. Academic Press, San Diego, CA; London, UK.

Szentiks, C.A., Kondgen, S., Silinski, S., Speck, S., Leendertz, F.H., 2009. Lethal pneumonia in a captive juvenile chimpanzee (*Pan troglodytes*) due to human-transmitted human respiratory syncytial virus (HRSV) and infection with *Streptococcus pneumoniae*. J. Med. Primatol. 38, 236–240.

Taibo, A., 2014. Veterinary Medical Terminology. Wiley Blackwell, Oxford.

Taichman, N.S., Simpson, D.L., Sakurada, S., Cranfield, M.R., DiRienzo, J., Slots, J., 1987. Comparative studies on the biology of *Actinobacillus actinomycetemcomitans* leukotoxin in primates. Oral Microbiol. Immunol. 2 (3), 97–104.

Tanner, J., Byrne, R., 1999. The development of spontaneous gestural communication in a group of zoo-living lowland gorillas. In: Parker, S., Mitchell, R., Miles, H. (Eds.), The Mentalities of Gorillas and Orangutans: Comparative Perspectives. Cambridge University Press, Cambridge, pp. 210–239.

Taylor, A.B., Goldsmith, M.L. (Eds.), 2003. Gorilla Biology: A Multidisciplinary Perspective. Cambridge University Press, Cambridge.

Taylor, D., 1991. Vet on the Wild Side: Further Adventures of a Wildlife Vet. Arrow (Random Century Group).

Taylor, P., Ingwersen, W., 2016. Veterinary medicines: importance of ketamine (letter). Veterinary Rec. 178 (11), 271.

Teare, J.A., Loomis, M.R., 1982. Epizootic of balantidiasis in lowland gorillas. J. Am. Vet. Med. Assoc. 181 (11), 1345–1347.

Teixeira, C.P., De Azevedo, C.S., Mendl, M., Cipreste, C.F., Young, R.J., 2007. Revisiting translocation and reintroduction programmes: the importance of considering stress. Anim. Behav. 73, 1–13.

Terio, K., Kinsel, M., 2008. Primate Necropsy Manual. Primate Necropsy Workshop, Kigoma, Tanzania, 2007. Also in Swahili – Mwongozo wa Kupasua Mizoga ya Wanyama Jamii (with Catherine Vallance and Titus Mlengeya).

Tewksbury, J.J., Anderson, J.G.T., Bakker, J.D., Billo, T.J., Dunwiddie, P. W., Groom, M.J., et al., 2014. Natural history's place in science and society. BioScience 64, 300–310.

Thalmann, O., Fischer, A., Lankester, F., Pääbo, S., Vigilant, L., 2007. The complex evolutionary history of gorillas: insights from genomic data. Mol. Biol. Evol. 24, 146–158.

The Law Library of Congress, 2013. Wildlife Trafficking and Poaching. Botswana Central African The Law Library South Africa Tanzania. The Law Library of Congress, Global Legal Research Center, < http://www.law.gov > .

Thoemmes, M.S., Fergus, D.J., Urban, J., Trautwein, M., Dunn, R.R., 2014. Ubiquity and diversity of human-associated *Demodex* mites. PLoS ONE 9 (8), e106265.

Thoen, C.O., Steele, J.H., Kaneene, J.B., 2014. Zoonotic Tuberculosis: *Mycobacterium bovis* and Other Pathogenic *Mycobacteria*. John Wiley & Sons, Chichester, UK.

Thomas, C., Anderson, D., Grigg-Damberger, M., et al., 1996. Polyclonal lymphoid tumor of the choroid plexus presenting as an intraventricular mass in a young gorilla. Acta Neuropathol. 92, 621–624.

Thomas, N.J., Atkinson, C.T., Hunter, D.B., 2007. Infectious and Parasitic Diseases of Wild Birds. Blackwell Publishing, Oxford.

Thompson, R.C., Allam, A.H., Lombardi, G.P., Wann, L.S., Sutherland, M. L., Sutherland, J.D., et al., 2013. Atherosclerosis across 4000 years of human history: the Horus study of four ancient populations. Lancet 381 (9873), 1211–1222.

Thompson, S.W., Luna, L.G., 1978. An Atlas of Artifacts Encountered in the Preparation of Microscopic Tissue Sections. Charles Louis Davis, DVM, Foundation, Gurnee, IL.

Thomson, R.G., 1978. General Veterinary Pathology. WB Saunders, Philadelphia, PA; Toronto, Canada.

Tobias, P.V.T., 1961. The work of the Gorilla Research Unit in Uganda. South Afr. J. Sci. 57 (11), 297–298.

Tobias, P.V.T., Cooper, J.E., 2003. The mountain gorilla: a little known chapter of pioneering studies. Trans. R. Soc. South Afr. 58 (1), 75–77.

Tocheri, M.W., Dommain, R., McFarlin, S.C., Burnett, S.E., Troy Case, D., Orr, C.M., et al., 2016. The evolutionary origin and population history of the grauer gorilla. Am. J. Phys. Anthropol. 159 (Suppl. 61), S4–S18.

Todd, A.F., 2008. First observation of the birth of a western gorilla in the wild. Gorilla J. 36, 16–17.

Todd, A.F., Kuehl, H.S., Cipolletta, C., Walsh, P.D., 2008. Using dung to estimate gorilla density: modeling dung deposition rate. Int. J. Primatol. 29, 549–563.

Toft, J.D., 1982. The pathoparasitology of the alimentary tract and pancreas of nonhuman primates: a review. Vet. Pathol. 19 (7 Suppl.), 44–92.

Toft, J.D., 1986. The pathoparasitology of nonhuman primates: a review. In: Benirschke, K. (Ed.), Primates. The Road to Self-Sustaining Populations. Springer, New York, NY, pp. 571–679.

Tokuyama, N., Emikey, B., Bafike, B., Isolumbo, B., Iyokango, B., Mulavwa, M.N., et al., 2012. Bonobos apparently search for a lost member injured by a snare. Primates 53, 215–219.

Tomasello, M., Call, J., 1997. Primate Cognition. Oxford University Press, Oxford.

Tomson, F.N., Schulte, J.M., 1978. Root canal therapy for a fractured canine tooth in a gorilla. J. Zoo Anim. Med. 9, 101–102.

Toovey, J., Brambell, M., 1976. The Michael Sobell Pavilions for apes and monkeys. Int. Zoo Yearb. 16, 212–217.

Torondel, B., Gyekye-Aboagye, Y., Routray, P., Boisson, S., Schimdt, W., Clasen, T., 2015. Laboratory development and field testing of sentinel toys to assess environmental faecal exposure of young children in rural India. Trans. R. Soc. Trop. Med. Hyg. 109 (6), 386–392.

Tranquilli, S., et al., 2011. Lack of conservation efforts leads to rapid great ape extinction. Conserv. Lett. 5 (1), 48–55.

Travis, D.A., Hungerford, L., Engel, G.A., Jones-Engel, L., 2006. Disease risk analysis: a tool for primate conservation planning and decision making. Am. J. Primatol. 68, 855–867.

Trupkiewicz, J.G., McNamara, T.S., Weidenheim, K.M., Cook, R.A., Grenell, S.L., Factor, S.M., 1995. Cerebral infarction associated with coarctation of the aorta in a lowland gorilla (*Gorilla gorilla gorilla*). J. Zoo. Wildl. Med. 26, 123–131.

Tsuchida, S., Kitahara, M., Nguema, P.P.M., et al., 2014. *Lactobacillus gorillae* sp nov., isolated from the faeces of captive and wild western lowland gorillas (*Gorilla gorilla gorilla*). Int. J. Syst. Evol. Microbiol. 64, 4001–4006.

Tsuchiya, Y., Isshiki, O., Yamada, H., 1970. Generalized cytomegalovirus infection in a gorilla. Jpn. J. Med. Sci. Biol. 23, 71–73.

Tudge, C., 1995. In praise of anthropomorphism. Biologist. 42 (1), 48.

Tulp, N., 1641. Observationum Medicarum, Libre Tres. Amsterdam, The Netherlands.

Turk, J.L., Allen, E., Cooper, J.E., 2000. The legacy of John Hunter, pioneer in comparative pathology. Eur. J. Vet. Pathol. 6 (1), 11–18.

Turner, P.S. Gorilla Doctors: Saving Endangered Great Apes. Houghton Mifflin Harcourt, Boston, MA.

Tutin, C., Fernandez, M., Rogers, M., Williamson, E., 1992. A preliminary analysis of the social structure of lowland gorillas in the Lope Reserve, Gabon. In: Itoigawa, N., Sugiyama, Y., Sackett, G., Thompson, R. (Eds.), Topics in Primatology, vol. 2: Behavior, Ecology, and Conservation. University of Tokyo Press, Tokyo, pp. 245–253.

Tutin, C.E.G., Benirschke, K., 1991. Possible osteomyelitis of skull causes death of a wild lowland gorilla in the Lopé Reserve, Gabon. J. Med. Primatol. 20, 357–360.

Tuyisingize, D., Kerbis Peterhans, J.C., Bronner, G.N., Stoinski, T., 2013. Small mammal community composition in the Volcanoes National Park, Rwanda. Bonn Zool. Bull. 62 (2), 177–185.

Tvedt, MW, Young, T, 2007. Beyond Access: Exploring Implementation of the Fair and Equitable Sharing Commitment in the CBD. IUCN Environmental Policy and Law Paper No. 67/2IUCN, Gland, Switzerland, xx + 148 pp. < http://www.ecolex.org/server2.php/libcat/docs/LI/MON-080368.pdf > .

Tyson, E., 1699. Orang-outang, sive Homo sylvestris: or, The Anatomy of a Pygmie Compared With That of a Monkey, an Ape, and a Man. Printed for Tomas Bennet and Daniel Brown, London.

Udono, T., Hamada, Y., Okayasu, N., Yamagiwa, J., 1997. Veterinary clinical findings in orphans of gorillas and bonobos in Congo. Reichorui Kenkyu/Primate Res. 13 (3), 264.

Uhlig, R., 2002. Eye surgeon brings a gorilla out of the mist. The Telegraph.<http://www.telegraph.co.uk/news/uknews/1390557/Eye-surgeon-brings-a-gorilla-out-of-the-mist.html>.

UNEP/CMS, 2011. Gorilla Agreement Action Plan Mountain Gorilla (*Gorilla beringei beringei*). < http://www.cms.int/sites/default/files/document/GA_MOP2_Inf_7_4_AP_Gbb_postMoP1_E_0.pdf > .

UNEP/WCMC, 2003. *Report* on the status and conservation of the Mountain Gorilla *Gorilla gorilla beringei*. CMS/ScC12/Doc.5 Attach 4. < http://www.cms.int/sharks/sites/default/files/document/Doc_05_Attach4_MountainGorilla_E_0.pdf > .

Unwin, S., Robinson, I., Schmidt, V., Colin, C., Ford, L., Humle, T., 2012. Does confirmed pathogen transfer between sanctuary workers and great apes mean that reintroduction should not occur? Commentary on drug-resistant human *Staphylococcus aureus* findings in sanctuary apes and its threat to wild ape populations. Am. J. Primatol. 74 (12), 1076–1083.

Unwin, S., Chatterton, J., Chantrey, J., 2013. Management of severe respiratory tract disease caused by human respiratory syncytial virus and *Streptococcus pneumoniae* in captive chimpanzees (*Pan troglodytes*). J. Zoo Wildl. Med. 44 (1), 105–115.

Van Kruiningen, H.J., Dobbins, W.O., John, G., 1991. Bacterial histiocytic colitis in a lowland gorilla (*Gorilla gorilla gorilla*). Vet. Pathol. 28 (6), 544–546.

Van Kuppeveldt, T.H., Robbesom, A.A., Versteeg, E.M., et al., 2000. Restoration by vacuum inflation of original alveolar dimensions in small human lung specimens. Eur. Respir. J. 15, 771–777.

van Schaik, C.P., Kappeler, P.M., 1997. Infanticide risk and the evolution of male-female association in primates. Proc. R. Soc. London, B, Biol. Sci. 264 (1388), 1687–1694.

van Zijll Langhout, M., Reed, P., Fox, M., 2010. Validation of multiple diagnostic techniques to detect *Cryptosporidium* sp. and *Giardia* sp. in free-ranging western lowland gorillas (*Gorilla gorilla gorilla*) and observations on the prevalence of these protozoan infections in two populations in Gabon. J. Zoo Wildl. Med. 41 (2), 210–217.

Varki, A., 2010. Uniquely human evolution of sialic acid genetics and biology. Proc. Natl. Acad. Sci. U.S.A. 107 (Suppl. 2), 8939–8946.

Venable, J.H., Grafflin, A.L., 1940. Dissection of the thyroid and parathyroid glands in an adult mountain gorilla. J. Mammal. 21 (1), 71–73.

Vercammen, F., Bauwens L., De Deken R., Brandt, J., 2007. Entomophthoromycosis in an Eastern lowland gorilla (*Gorilla gorilla graueri*). In: Proceedings of the American Association of Zoo Veterinarians, 20–25 October 2007, Knoxville, USA, pp. 244–247.

Vercruysse, J., Mortelmans, J., 1978. The chemical restraint of apes and monkeys by means of phencyclidine or ketamine. Acta Zool. Pathol. Antverp. 70, 211–220.

Verrengia, J., 1997. Cenzoo: The Story of a Baby Gorilla. Roberts Rinehart, Dublin, Ireland.

Videan, E., Lammey, M., Lee, D., 2011. Diagnosis and treatment of degenerative joint disease in a captive male chimpanzee (*Pan troglodytes*). J. Am. Assoc. Lab. Anim. Sci. 50 (2), 263–266.

Viggers, K., Lindenmayer, D., Spratt, D., 1993. The importance of disease in reintroduction programmes. Wildl. Res. 20 (5), 687–698.

Vigilant, L., Bradley, B.J., 2004. Genetic variation in gorillas. Am. J. Primatol. 64 (2), 161–172.

Viscardi, P., 2013. A survival strategy for natural science collections: the role of advocacy. J. Nat. Sci. Collect. 1, 4–7.

Vlčková, K., 2010. Description of Microflora of Gastrointestinal Tract of Western Lowland Gorillas (*Gorilla gorilla gorilla*). (Bachelor thesis). Masaryk University.

Vlckova, K., Mrazek, J., Kopecny, J., et al., 2012. Evaluation of different storage methods to characterize the fecal bacterial communities of captive western lowland gorillas (*Gorilla gorilla gorilla*). J. Microbiol. Methods. 91 (1), 45–51.

Voevodin, A.F., Marx, P.A., 2009. Simian Virology. Wiley-Blackwell, Ames, IO.

Von Beringe, A., 2002. On the trail of the man who discovered the mountain gorilla. Gorilla J. 24, 9–11.

Von Beringe, O., 1903. Bericht des Hauptmanns von Beringe über seine Expedition nach Ruanda. Deutsch. Kolonialbl. 234–235, 264–266, 296–298, 317–319.

Wagner, J.A., 1855. Die Säugethiere in Abbildungen nach der Natur mit Beschreibungen. In: von Schreber, J.C.D. (Ed.), Fünfte Abtheilung: Die Affen, Zahnlücker, Beutelthiere, Hufthiere, Insektenfresser und Handflügler, Supplementband 1. T. O. Weigel, Leipzig. pp. 8–11.

Walker, A., Shipman, P., 1997. The Wisdom of Bones: In Search of Human Origins. Phoenix.

Walker, J.B., Keirans, J.E., Horak, I.G., 2000. The Genus *Rhipicephalus* (Acari, Ixodidae): A Guide to the Brown Ticks of the World. Cambridge University Press, Cambridge, UK.

Walker, P.L., Bathurst, R.R., Richman, R., Gjerdrum, T., Andrushko, V.A., 2009. The causes of porotic hyperostosis and cribra orbitalia: a reappraisal of the iron-deficiency-anemia hypothesis. Am. J. Phys. Anthropol. 139 (2), 109–125.

Wallace, R.S.A., Gendron–Fitzpatrick, A., Teare, J.A., Morris, G., 1997. Leptomyxid amoebic meningoencephalitis in a western lowland gorilla (*Gorilla gorilla gorilla*). In: Am. Assoc. Zoo Vet. Annu. Conf. Proc. 1997, pp. 59–61.

Wallach, J.D., 1968. Angioedema associated with strawberry ingestion by a gorilla. J. Am. Vet. Med. Assoc. 153, 879–880.

Waller, B.M., Cherry, L., 2012. Facilitating play through communication: significance of teeth exposure in the gorilla play face. Am. J. Primatol. 74 (2), 157–164.

Wallis, J., Lee, D.R., 1999. Primate conservation: the prevention of disease transmission. Int. J. Primatol. 20 (6), 803–826.

Walsh, P.D., et al., 2003. Catastrophic ape decline in western equatorial Africa. Nature 422, 611–614.

Walsh, P.D., Biek, R., Real, L.A., 2005. Wave-like spread of Ebola Zaire. PLoS Biol. 11, 1946–1953.

Ward, J.M., Youssef, S.A., Treuting, P.M., 2016. Why animals die: an introduction to the pathology of aging. Vet. Pathol. 53 (2), 229–232.

Warmington, B.H., 1960. Carthage. Robert Hale, London.

Washington, H., Baillie, J., Waterman, C., Milner-Gulland, E.J., 2015. A framework for evaluating the effectiveness of conservation attention at the species level. Oryx 49, 481–491.

Watts, D., 1984. Composition and variability of mountain gorilla diets in the central Virungas. Am. J. Primatol. 7, 323–356.

Watts, D., 1989a. Ant eating behavior of mountain gorillas. Primates 30, 121.

Watts, D., 1989b. Infanticide in mountain gorillas: new cases and a reconsideration of the evidence. Ethology 81, 1–18.

Watts, D., 1990. Mountain gorilla life histories, reproductive competition, and sociosexual behavior and some implications for captive husbandry. Zoo Biol. 9, 185–200.

Watts, D., 1995. Post-conflict social events in wild mountain gorillas (Mammalia, Hominoidea). I. Social interactions between opponents. Ethology 100, 139—157.

Watts, D.P., 1983. Foraging Strategy and Socioecology of Mountain Gorillas (*Pan gorilla beringei*). (Ph.D. thesis). University of Chicago, Chicago, IL.

Watts, D.P., 1988. Environmental influences on mountain gorilla time budgets. Am. J. Primatol. 15, 195—211.

Watts, E., Meder, A., 1996. Introduction and socialization techniques for primates. In: Kleiman, D., Allen, M., Thompson, K., Lumpkin, S. (Eds.), Wild Mammals in Captivity. University of Chicago Press, Chicago, IL, pp. 67—77.

WAZA, 2013. WAZA Code of Ethics and Animal Welfare. World Association of Zoos and Aquariums, < http://www.waza.org/files/webcontent/1.public_ site/5.conservation/code_of_ethics_and_animal_welfare/Code%20of% 20Ethics_EN.pdf >.

Webb, T., 1997. A training plan to induce nursing in a female western lowland gorilla (*Gorilla gorilla gorilla*). Abstract (w/supplemental notes). In: Pittsburgh Zoo Gorilla Workshop Program, p. 14.

Weber, B., 2000. Boyz in the hood: Establishing a bachelor troop at Disney's Animal Kingdom. Paper Presented at "The Apes: Challenges for the 21st Century." Brookfield Zoo, Chicago, IL.

Webster, I., 1963. Asbestosis in non-experimental animals in South Africa. Nature 197 (4866), 506.

Webster, J., 2005. Animal Welfare: Limping Towards Eden. Wiley-Blackwell, Oxford.

Weiman, F.D., 1927. Observations on skin conditions. Report of the Laboratory of Comparative Pathology, Philadelphia, PA. pp. 36—38.

Weinberg, M., 1908. Un cas d'appendicite chez le gorille. Bull. Soc. Pathol. Exot. I, 556—560.

Weiss, D., Hampshire, V., 2015. Primate wellness exam. Lab. Anim. Euro. 15 (10), 30—33.

Weiss, E., 2015. Paleopathology in Perspective: Bone Health and Disease Through Time. AltaMira Press, USA.

Wells, D.L., 2005. A note on the influence of visitors on the behaviour and welfare of zoo-housed gorillas. Appl. Anim. Behav. Sci. 93, 13—17.

Wells, N., 2011. Animal Law in New Zealand. Thomson Reuters/Brookers Ltd, Wellington.

Weninger, M., 1948. Zur Reduktion der Prämolaren. (Drei Prämolaren im Oberkiefer eines Gorilla). Z. Stomatol. 45, 223—231.

Werikhe, S., Macfie, L., Rosen, N., Miller, P. (Eds.), 1998. Can the Mountain Gorilla Survive? Population and Habitat Viability Assessment for *Gorilla gorilla beringei*. Kampala, Uganda. 8—12 December, 1997. Conservation Breeding Specialist Group (CBSG), Minnesota, MN.

Werner, D., Thuman, C., Maxwell, J., 2015. Where There Is No Doctor. Hesperian health guides, Berkeley, CA, <http://theboatgalley.com/ where-there-is-no-doctor-free-download>.

West, G., Heard, D., Caulkett, N. (Eds.), 2007. Zoo Animal and Wildlife Immobilization and Anesthesia. Blackwell Publishing, Ames, IO.

Weston, D.A., 2008. Investigating the specificity of periosteal reactions in pathology museum specimens. Am. J. Phys. Anthropol. 137, 48—59.

Wevers, D., Leendertz, F.H., Scuda, N., et al., 2010. A novel adenovirus of western lowland gorillas (*Gorilla gorilla gorilla*). Virol. J. 7, 303.

Wevers, D., Metzger, S., Babweteera, F., et al., 2011. Novel adenoviruses in wild primates: a high level of genetic diversity and evidence of zoonotic transmissions. J. Virol. 85 (20), 10774—10784.

Wharton, D., 2000. Gorilla management for the 21st century. In: AZA Annual Conference Proceedings, pp. 323—327.

White, M., 1999. An overview of the St. Louis Bachelor Group(s). Gorilla Gaz. 13 (1), 34—36.

White, M., Armstrong, B., Shumaker, R., Sunde, V., Sitherland, R., 1997. Caregiver and gorilla relationship. In: Ogden, J., Wharton, D. (Eds.), Gorilla Husbandry Manual. Fulton County Zoo, Atlanta, GA, pp. 80—84.

White, R.J., Simmons, L., Wilson, R.B., 1972. Chickenpox in young anthropoid apes: clinical and laboratory findings. J. Am. Vet. Med. Assoc. 1616, 690—692.

White, T.D., Folkens, P.A., 1991. Human Osteology. Academic Press, Burlington, CA.

White, T.D., Folkens, P.A., 2005. The Human Bone Manual. Elsevier, Burlington, CA.

Whiten, A., 1999. Parental encouragement in *Gorilla* in comparative perspective: implications for social cognition and the evolution of teaching. In: Parker, S., Mitchell, R., Miles, H. (Eds.), The Mentalities of Gorillas and Orangutans: Comparative Perspectives. Cambridge University Press, Cambridge, pp. 342—366.

Whittier, C, 2001. Leadership battle rages amongst silverbacks. Digit News (DFGF). Issue 21, Autumn, p. 3.

Whittier, C., 2004. Mountain gorillas and other primate orphans of Rwanda. Gorilla Gaz. 17 (1), 6—7.

Whittier, C., 2006. Application of the RSG Guidelines in the case of confiscated mountain gorillas, Virunga Massif: Rwanda, Uganda & DRC. Reintroduction News 25, 40—41.

Whittier, C., Nutter, F.B., Cranfield, M.R., 2005. Seroprevalence of infectious agents in free-living mountain gorillas (*Gorilla beringei* spp). In: Proceedings of the Wildlife Disease Association International Conference 2005, Cairns, Australia. Abstract 174, p. 291.

Whittier, C.A., Nutter, F.B., Stoskofp, M.K., 2001. Zoonotic disease concerns in primate field settings. In: The Apes: Challenges for the 21st Century, Conference Proceedings, 10—13 May 2000. Brookfield Zoo, Brookfield, IL, USA, pp. 232—237.

Whittier, C.A., Horne, W., Slenning, B., Loomis, M., Stoskopf, M.K., 2004. Comparison of storage methods for reverse-transcriptase PCR amplification of rotavirus RNA from gorilla (*Gorilla g. gorilla*) fecal samples. J. Virol. Methods 116 (1), 11—17.

Whittier, C.A., Cranfield, M.R., Stoskopf, M.K., 2010a. Real-time PCR detection of *Campylobacter* spp. in free-ranging mountain gorillas (*Gorilla beringei beringei*). J. Wildl. Dis. 46 (3), 791—802.

Whittier, C.A., Milligan, L.A., Nutter, F.B., Cranfield, M.R., Power, M.L., 2010b. Proximate composition of milk from free-ranging mountain gorillas (*Gorilla beringei beringei*). Zoo Biol. 30 (3), 308—317.

Wiard, J., 1992. Reduction of regurgitation and reingestion (R&R) in lowland gorillas at the Oklahoma City Zoo. Gorilla Gaz. 6 (3), 6—7.

Wich, S.A., Utami-Atmoko, S.S., Mitra Setia, T., Rijksen, H.D., Schürmann, C., van Hooff, J.A.R.A.M., et al., 2004. Life history of wild Sumatran orangutans (*Pongo abelii*). J. Hum. Evol. 47, 385—398.

Wijnstekers, W., 2011. The Evolution of CITES. Ninth ed. International Council for Game and Wildlife Conservation, Budapest.

Wilbush, J., 1984. Clinical information—signs, semeions and symptoms: discussion paper. J. R. Soc. Med. 77, 766—773.

Wilder, H.H., 1904. The restoration of dried tissues with especial reference to human remains. Am. Anthropol. 6, 1—17.

Wildt, D.E., Chakraborty, P.K., Cambre, R.C., Howard, J.G., Bush, M., 1982. Laparoscopic evaluation of the reproductive organs and abdominal cavity content of the lowland gorilla. Am. J. Primatol. 2 (1), 29—42.

Williams, E.S., Barker, I.K. (Eds.), 2001. Infectious Diseases of Wild Mammals. Third ed. ISU Press, Ames, IO.

Williams, F.L., Orban, R., 2007. Ontogeny and phylogeny of the pelvis in *Gorilla, Pongo, Pan, Australopithecus* and *Homo*. Folia Primatol. 78 (2), 99—117.

Williams, K. 1997. Gorillas: the enrichment experience. Abstract (w/ supplemental notes). In: Pittsburgh Zoo Gorilla Workshop Program, p. 23.

Wilms, T., 2011. International Studbook for the Western Lowland Gorilla (*Gorilla g. gorilla*; Wyman, 1847) - 2010. Frankfurt Zoo, Frankfurt am Main.

Wilson, A.M., 1958. Notes on a Gorilla Eviscerated at Kabale, August 1958. Laboratory report on faeces and stomach contents. Animal Health Research Centre, Entebbe, Uganda.

Wilson, S., 1982. Environmental influences on the activity of captive apes. Zoo Biol. 1, 201—209.

Winker, K., 1996. The crumbling infrastructure of biodiversity: the avian example. Conserv. Biol. 10, 703—707.

Winker, K., 2004. Natural history museums in a postbiodiversity era. Bioscience 54, 455–459.

Winker, K., Fall, B.A., Klicka, J.T., Parmelee, D.F., Tordoff, H.B., 1991. The importance of avian collections and the need for continued collecting. Loon 63, 238–246.

Winslow, S., Ogden, J., Maple, T., 1992. Socialization of an adult male lowland gorilla (*Gorilla gorilla gorilla*). Int. Zoo Yearb. 31, 221–225.

Wislocki, G.B., 1929. On the placentation of primates, with a consideration of the phylogeny of the placenta. Contrib. Embryol. 20, 51–80.

Wislocki, G.B., 1942. Size, weight, and histology of the testes in the gorilla. J. Mammal. 23 (3), 281–287.

Withington, E.T., 1894. Medical history from the earliest times, a popular history of the healing art. The Scientific Press, Limited, London, UK.

Wittiger, L., Sunderland-Groves, J.L., 2007. Tool use during display behavior in wild Cross River gorillas. Am. J. Primatol. 69 (11), 1307–1311.

Wobeser, G.A., 2010. Disease in Wild Animals: Investigation and Management. Springer-Verlag, Berlin, Germany.

Wolf, T.M., Sreevatsan, S., Travis, D., Mugisha, L., Singer, R.S., 2014. The risk of tuberculosis transmission to free-ranging great apes. Am. J. Primatol. 76 (1), 2–13.

Wolfe, N.D., Escalante, A.A., Karesh, W.B., Kilbourn, A., Spielman, A., Lal, A.A., 1998. Wild primate populations in emerging infectious disease research: the missing link? Emerg. Infect. Dis. 4 (2), 149–158.

Wolfensohn, S.E., 2003. Case report of a possible familial predisposition to metabolic bone disease in juvenile rhesus macaques. Lab. Anim. 37 (2), 139–144.

Wolff, P.L., Seal, U.S., 1993. Implications of infectious disease for captive propagation and reintroduction of threatened species. J. Zoo Wildl. Med. 24 (3), 229–230.

Wolfheim, J.H., 1983. Primates of the World: Distribution, Abundance, and Conservation. University of Washington Press, Seattle, WA.

Wood, J.G., 1893. The New Illustrated Natural History. George Routledge and Sons, Manchester and New York, NY.

Wood, V.J.W., Milner, G.R., Harpending, H., Weiss, K.M., 1992. The osteological paradox. Problems of inferring prehistoric health from skeletal samples. Curr. Anthropol. 33, 343–370.

Woodford, M.H. (Ed.), 2000. Quarantine and Health Screening Protocols for Wildlife prior to Translocation and Release into the Wild. IUCN Species Survival Commission's Veterinary Specialist Group, Gland, Switzerland, the Office International des Epizooties (OIE), Paris, France, Care for the Wild, U.K., and the European Association of Zoo and Wildlife Veterinarians, Switzerland.

Woodford, M.H., 2001. Quarantine and Health Screening Protocols for Wildlife Prior to Translocation and Release into the Wild. Office International des Epizooties, the IUCN – The World Conservation Union (IUCN), Species Survival Commission (SSC) Veterinary Specialist Group, Care for the Wild International, Geraldine R. Dodge Foundation, and European Association of Zoo and Wildlife Veterinarians.

Woodford, M.H., Kock, R.A., 1991. Veterinary considerations in re-introduction and translocation projects. In: Gipps, J.H.W. (Ed.), Beyond Captive Breeding: Re-introducing Endangered Mammals to the Wild. In: Symposium of the Zoological Society of London, vol. 62. Clarendon Press, Oxford, pp. 101–110.

Woodford, M.H., Rossiter, P.B., 1994. Disease risks associated with wildlife translocation projects. In: Olney, P.J.S., Mace, G.M., Feistner, A.T.C. (Eds.), Creative Conservation: Interactive Management of Wild and Captive Animals. Chapman and Hall, London, UK, pp. 178–200.

Woodford, M.H., Keet, D.F., Bengis, R.G., 2000. Postmortem Procedures for Wildlife Veterinarians and Field Biologists. Office International des Epizooties, Care for the Wild International and IUCN/SSC Veterinary Specialist Group.

Woodford, M.H., Butynski, T.M., Karesh, W.B., 2002. Habituating the great apes: the disease risks. Oryx 36 (2), 153–160.

Woods, S. 1990. Behavioral enrichment of a group of captive gorillas: a quantitative study. In: Proceedings of the Columbus Zoo Gorilla Workshop, pp. 205–209.

Woods, S., 1995. Ear-covering by captive great apes: a shared behavior related to stress. In: AAZK National Conference Proceedings, p. 196.

Woods, S. 2000. Stress-related ear covering by captive great apes: a second look. In: Paper Presented at The Apes: Challenges for the 21st Century. Brookfield Zoo, Chicago, IL.

World Organisation for Animal Health (OIE), International Union for the Conservation of Nature (IUCN), 2014. Guidelines for Wildlife Disease Risk Analysis. OIE, Paris, France (Published in association with the IUCN Species Survival Commission).

Worthington, H.M., 2006. Endpiece. The name of the illness. Br. Med. J. 332, 1070.

Wright, B.M., Slavin, G., Kreel, L., Callan, K., Sandin, B., 1974. Postmortem inflation and fixation of human lungs. Thorax 29, 189–194.

Wu, H.-J., Ivanov, I.I., Darce, J., Hattori, K., Shima, T., Umesaki, Y., et al., 2010. Gut-residing segmented filamentous bacteria drive autoimmune arthritis via T helper 17 cells. Immunity 32 (6), 815–827.

Wyman, J., 1847. A communication from Dr Thomas S. Savage, describing the external character and habits of a new species of *Troglodytes* (*T. gorilla*, Savage,) recently discovered by Dr S. in Empongwe, near the river Gaboon, Africa. Proc. Boston Soc. Nat. Hist. 2, 245–247.

Wyse, C.A., et al., 2004. Current and future uses of breath analysis as a diagnostic tool. Vet. Rec. 154 (12), 353–360.

Xue, Y., Prado-Martinez, J., Sudmant, P.H., Narasimhan, V., Ayub, Q., Szpak, M., et al., 2015. Mountain gorilla genomes reveal the impact of long-term population decline and inbreeding. Science 348, 242–245.

Yamagiwa, J., 1986. Activity rhythm and the ranging of a solitary male mountain gorilla (*Gorilla gorilla beringei*). Primates 27, 273–282.

Yamagiwa, J., 1987. Intra- and inter-group interactions of an all-male group of Virunga mountain gorillas (*Gorilla gorilla berengei*). Primates 28, 1–30.

Yamagiwa, J., Goodall, A., 1992. Comparative socio-ecology and conservation of gorillas. In: Itoigawa, N., Sugiyama, Y., Sackett, G., Thompson, R. (Eds.), Topics in Primatology, vol 2: Behavior, Ecology, and Conservation. University of Tokyo Press, Tokyo, pp. 209–213.

Yamagiwa, J., Kahekwa, J., Basabose, A.K., 2009. Infanticide and social flexibility in the genus *Gorilla*. Primates 50, 293.

Yanoff, M., Zimmerman, L.E., Fine, B.S., 1965. Glutaraldehyde fixation of whole eyes. Am. J. Clin. Pathol. 44, 167–171.

Yasui, K., Takizawa, A., 1975. A case of treatment of myositis purulenta in a lowland gorilla. J. Jap. Ass. Zoos Gardens Aqua. 17, 69–72.

Yeager, C.H., et al., 1981. Ultrasonic estimation of gestational age in the Lowland Gorilla: a biparietal diameter growth curve. J. Am. Vet. Med. Assoc. 179 (11), 1309–1310.

Yerkes, R.M., 1927a. The mind of a gorilla. Genet. Psychol. Monogr. 2, 1–193.

Yerkes, R.M., 1927b. The mind of a gorilla. Part II. Mental development. Genet. Psychol. Monogr. 2, 375–551.

Yerkes, R.M., 1928. The mind of a gorilla: Part III. Memory. Compar. Psychol. Monogr. 5, 1–92.

Yerkes, R.M., Yerkes, A.W., 1929. The Great Apes: A Study of Anthropoid Life. Yale University Press, New Haven, CT.

Yoshida, H., 2000. Who likes where? How to use the outdoor enclosure of captive western lowland gorillas in the Columbus Zoo. In: Presentation Given at the AAZK National Conference.

Zabo, A., 2015. Telepathology in daily routine histological diagnostics: experiences from Sweden. Bull. R. Coll. Pathol. 169, 39.

Zihlman, A., Bolter, D., Boesch, C., 2004. Wild chimpanzee dentition and its implications for assessing life history in immature hominin fossils. Proc. Natl. Acad. Sci. U.S.A. 101, 10541−10543.

Zihlman, A.L., McFarland, R.K., 2000. Body mass in lowland gorillas: a quantitative analysis. Am. J. Phys. Anthropol. 113 (1), 61−78.

Zihlman, A.L., Mcfarland, R.K., Underwood, C.E., 2011. Functional anatomy and adaptation of male gorillas (*Gorilla gorilla gorilla*) with comparison to male orangutans (*Pongo pygmaeus*). Anat. Rec. 294, 1842−1855.

Zook, B.C., 1971. Lead poisoning in nonhuman primates. Comp. Pathol. Bull. III (1), 3−4.

Zorich, D., Hoagland, K.E., 1995. Status, Resources, and Needs of Systematic Collections. The Association of Systematic Collections, Washington, DC.

ZSL, undated. Health and safety at the Zoo. < https://www.zsl.org/sites/default/files/document/2014-02/health-and-safety-activity-map-ks4-and-5-during-visit-1768.pdf > .

Zuckerman, Professor Lord and Staff of the Publications Department, 1976. The Zoological Society of London 1826-1976 and Beyond. Proceedings of a Symposium, 25th-26th March, 1976. Academic Press, London.

Zuckerman, S., 1998. The Ape in Myth & Art. Verdigris Press, Scotland.

Part II

A Catalogue of Preserved Materials

Introduction to the Catalogue

Gordon Hull

SCOPE AND METHOD OF COMPILATION

A well-established natural history collection is likely to include many rarities. There are numerous attestations to the intrinsic and enduring value of such collections, and the need to maintain and extend them via continued financial support and ongoing scientific collecting (see, e.g., Finley, 1987; Winker et al., 1991; Baker, 1994; Zorich and Hoagland, 1995; Winker, 1996; Kitchener, 1997; Cooper et al., 1998; Suárez and Tsutsui, 2004; Winker, 2004; Mesa, 2005; Rau, 2005; Pyke and Ehrlich, 2010; Medina et al., 2013). Biological reference collections vary widely, of course, depending on when, why, by whom and for whom they were accumulated, and since no two collections are ever the same, potentially all of them can be utilised by scientists working in many different fields of basic and applied research. Nowadays, historical collections are proving their worth in ecological and environmental studies, as well as the more traditional ones. For example, using 1651 specimens from a single collection, Medina et al. (2013) were able to analyse long-term patterns of abnormalities in South American anurans. Such studies are important because abnormalities found in amphibians could be a warning of mutagenic agents that might affect vertebrates as a whole, and the authors' investigation demonstrated how collections can provide a valuable historical perspective on temporal biological changes in extant and extinct populations. Baker (1994) made the important point that, because polymerase chain reaction can amplify DNA sequences from small biological samples, it enhances the value of museum skins, bones and tissues, while at the same time creating a dilemma, because the technique necessitates the partial destruction of specimens; therefore, as much tissue as possible should be saved in frozen form whenever material from a rare or endangered species becomes available. DNA from small amounts of tissue can be cloned into molecular libraries or used for PCR amplification without any damage to the voucher specimen, and frozen samples will be available for future studies without any impact on the remaining living individuals, thus helping to preserve biodiversity. In view of the crisis we face in the conservation of biodiversity, Baker called for cooperation among museums, so that frozen tissues from endangered species can be deposited in an increasing number of accredited, systematically arranged and computerised museum collections, beyond those already noted by Dessauer and Hafner (1984).

Gippoliti and Kitchener (2007) contested the strong belief, still held by many in the museum world, that specimens derived from captive animals are in some way inferior, or unimportant for research. They demonstrated how, when applied to diverse scientific scenarios, this is a misconception, and they advocated a future stronger cooperation between zoos, museums and universities, so that the scientific, educational and conservation value of living and preserved mammal collections can be maximised. While their recommendation was aimed primarily at Italian collections, it applies equally to institutions elsewhere around the world.

An informative table of the uses of collections in biological research can be found in Cato (1991).

Unfortunately, traditional natural history collections now face multiple threats to their continued existence. The problems are summarised in the following two paragraphs, written especially for this Introduction by Paolo Viscardi:

Tyranny of the test-tube. With the decline in academic interest in whole organism biology in favour of mathematical and molecular techniques over the past 50 or more years (Noss, 1996; Pyle, 2001; Schmidley, 2005; Tewksbury et al., 2014), there has been a corresponding decrease in the perceived value of natural history collections (Dalton, 2003; Gropp, 2003; Funk, 2004; Tewksbury et al., 2014). Ironically, the development of fast and inexpensive gene sequencing means that the data locked in collections are now more accessible than ever, providing a unique insight into the genetic diversity, distribution and even viral exposure of past and present populations (Suárez and Tsutsui, 2004; Pyke and Ehrlich, 2010; Casas-Marce et al., 2012; Lavoie, 2012; Pennington et al., 2013; Bradley et al., 2014; Tewksbury et al., 2014). Unfortunately, the current generation of biological scientists seem largely unaware of this potential, having been trained in the sterile environs of a laboratory and experiencing limited (if any) specimen-based research (Noss, 1996; Mares, 2002; Schmidley, 2005; Cook et al., 2014; Tewksbury et al., 2014).

This situation has been exacerbated by the global economic downturn, which has seen collections and expertise in natural history preferentially cut from universities in response to reduced funding (Dalton, 2003; Gropp, 2003; Suárez and Tsutsui, 2004; Schmidley, 2005; Viscardi, 2013; Bradley et al., 2014; Tewksbury et al., 2014). A similar pattern has been seen in many local authority and regional museums in countries with a well-established history in collecting, such as the UK (Viscardi, 2013), where there has been a greater than 35% loss of natural science curators in the last 10 years (Mulhearne, 2013). Overall, this equates to fewer skilled staff making collections available for use and contributes to a vicious circle of decline. Academic exercises to document repositories of particular collection types can make a valuable contribution to increasing awareness of the research resources available, and help with the important task of reconnecting researchers with collections.

Lehn (2008) reviewed the numerous ways in which gorilla biological samples of different types can be utilised, and reminded us of the urgency for collecting and archiving these resources, since there may be a limited amount of time left to acquire them, while the future holds as yet unknown research possibilities. The curation of tissues and gametes is relatively new and requires specific curatorial methods (which would seem to fall more within the province of zoos and wildlife parks than of natural history museums), but nonetheless, tissue collections have now been established in many parts of the world (Dessauer et al., 1996; Prendini et al., 2002).

While the immense usefulness of museum collections and the educational worth of their public exhibitions are beyond doubt, the need for comprehensive inventories of those assemblages is not so readily understood. Yet no collection can ever really attain its full potential as an educational and research resource if it has not been carefully inventoried, and the results published, or at least made available in one form or another to all interested parties. Certainly, a detailed catalogue may be time-consuming and expensive to produce initially, but once it is available in printed or computerised form, it will provide a valuable ongoing service to students and researchers who may need to examine scarce materials that are stored away out of their sight, yet could be made available to them if only they knew where to find them!

Gordon et al. (2013) sounded a cautionary note, however, by raising concerns about the over-use of certain great ape skeletal reference collections, whereby repeated examination by numerous researchers over time causes irreparable damage to the specimens concerned.

To help prevent unnecessary measurement-taking and other repeat procedures, the authors discussed ways of standardising individual specimen information across skeletal collections and making it available in an online database. Before publication, one of the authors (ADG) conducted a systematic inventory of all *Gorilla* and *Pan* skeletal elements in the Powell-Cotton Museum, UK. The information is freely available as specimen data sheets accessible through the Human Origins Database (www.humanoriginsdatabase.org).

A small number of surveys of human and nonhuman primate materials have been published so far (e.g., Anderson, 1881; Cobb, 1933; Krogman and Schultz, 1938; Groves and Napier, 1966; Albrecht, 1982; Jenkins, 1990; Bruner and Gippoliti, 2006), and each of these has been a welcome addition to the sparse literature on the subject. This catalogue differs from all the previous surveys because it is unrestricted as far as particular collections or geographical areas are concerned, being international in its scope, but it *is* restricted in the sense that it deals with only one primate genus, comprising just two species according to current taxonomic thinking. This approach allowed the data to be gathered in a more thorough and painstaking way than might otherwise have been the case. The project was begun as a pastime, with no intention of publication, but as the data gradually accumulated it became clear that the end result should be made as widely available as possible.

The information in this catalogue has been obtained mainly by means of correspondence with the curators, collection managers, heads of departments and various other staff members of the institutions concerned, and occasionally with private individuals who were able to provide some valuable leads or additional facts. For the original survey, carried out mainly in the 1980s, letters in English were sent to approximately 1500 institutions in 62 countries. Of those that replied, 860 indicated that they had no *Gorilla* material, while a further 416 collections (in 45 countries) indicated that they did have *Gorilla* material, amounting to more than 4500 specimens, which form the basis of this catalogue.

Prior to publication, efforts have been made to update the information as far as possible, and to widen the scope to include biological materials typically archived in living collections (zoos, wildlife parks, etc.). Unfortunately, there are many problems where living collections are concerned. There was some concern from zoos on how they would prioritise tissue and sample requests once they were listed as a world resource in this book. Many zoos do not keep a tissue repository (samples collected from deceased animals are submitted to the appropriate histopathology service and then disposed of; carcases are usually cremated). Where zoos maintain just a small number of samples, they would be unable to provide outside researchers with access to these, as they need to retain them for their own studies/health assessments retrospectively. There may also be reservations about providing a list of samples primarily because many of these are associated with strict permit restrictions on how they can be used, and an institution will not want to give the impression that they are an easily accessible resource when in fact their uses are restricted. Furthermore, it is often the case that the government authority under which samples have been imported does not allow for indefinite archiving, and eventually those samples have to be destroyed. On the other hand, some collections which have extensive clinical and pathology data do not ordinarily share that information in its raw form or without analysis and interpretation. It has been suggested that an electronic search table on a website may be more useful and user-friendly, but concerns remain about how the resource will be maintained if the samples are depleted. These obstacles have inevitably reduced the number of living collections that could be included here.

Whenever possible, the entries in the catalogue have been augmented by data from the published literature, summarised by the author (GH) and duly referenced. It is hoped that by combining and summarising scientific and (occasionally anecdotal) historical information in this way, the catalogue will be a useful tool for researchers in various academic disciplines.

Despite every effort having been made to maintain accuracy, it should be emphasised that a survey such as this will inevitably remain incomplete and out of date, because some organisations do not, or cannot, respond to requests for information, and others may have acquired new specimens or disposed of old ones since the survey was last carried out. In any instance where previous information could not be updated, the organisation concerned is marked with an asterisk *, so that the reader will immediately understand that the published data are based on historical information alone.

Thus, the information in the catalogue is provided on a strictly bona fide basis, with the caveat that its accuracy should not be relied upon by any researcher, but should instead be treated merely as a guide to what might be available for study and research purposes.

Unless stated otherwise, measurements and details of pathological changes and diagnoses have been supplied by the respondents and have not been confirmed by the authors (GH or JEC) in person, and consequently are not propounded to be definitive as far as any individual specimen is concerned.

LAYOUT

Institutions are listed alphabetically by country, and then by city or town. The name and postal address or physical location of each institution is given.

The names of the persons who have kindly provided information are given in italics under the name and address of each institution, and elsewhere in the text (in brackets) at appropriate points. These people (some of whom are now retired and/or deceased) should not be regarded as a point of contact for their institutions. However, in most cases the institution, and its current staff, can be traced on the Internet, and contact can be made via email or letter.

It should be emphasised that the catalogue entries are not confined solely to the data available from a particular institution's records. In numerous instances, these data have been augmented by information obtained from other reliable sources, including correspondence with individuals based elsewhere, such as, for example, in zoos where the animals were kept when alive (a brief summary of an animal's life history may be relevant for researchers who are studying its remains). One consequence of combining additional information in this way is that it can become inextricable from the rest of the text, and precise indication of its source impracticable. While this is regrettable, it is thought to be preferable to the alternative of omitting the information altogether.

The entry for each specimen includes the following details, insofar as they are available:

Accession Number

This is usually prefixed by the organisation's abbreviation or acronym, which helps to make it unique. Accession numbers may be from modern, computerised records but often relate to original, handwritten catalogues or card indexes. The accession number will usually also be written on the specimen itself or on an attached label. It is a sad fact that when a computer database is created, the original records are sometimes lost or destroyed at the time or soon afterwards, and in the interests of streamlining, valuable concomitant data are deliberately or inadvertently discarded. Original labels should always be retained (see Appendix CA1).

Nature of Specimen

With a few exceptions, the materials in the catalogue have not been examined personally by the authors (GH and JEC), but care has been taken when requesting information, to ensure as far as possible that the descriptions are standardised.

Thus the term 'skull' can be taken to mean the cranium and mandible combined. 'Cranium' is used when the mandible is missing, but the calvaria and facial bones are intact. 'Skeleton' means the skull plus the postcranial skeleton; any missing bones may be noted. 'Mounted skeleton' signifies that the skull and postcranial skeleton are articulated and held by supporting rods in a manner suitable for display. 'Articulated skeleton' indicates that the overall supports are absent. 'Skin' means an unmounted hide, which may be salted or tanned. 'Mounted skin' is used for a hide that has been tanned and prepared for display.

Apart from skeletal remains and skins, many other types of material (except modern casts) are also listed.

Age Category and Sex

The age category assigned to each specimen may be precise in some cases, but only approximate in others, depending on how assiduously the specimen has been documented or studied, and so is best taken only as a guide.

In general terms, 'fetus' means a prenatal specimen with unerupted dentition; an 'infant' will have mainly deciduous teeth; a 'juvenile' will have a mixed but predominantly permanent dentition, and the postcranial skeleton will be immature, with the long bones unfused; a 'subadult' will have a fully erupted permanent dentition, but the spheno-occipital synchondrosis and the epiphysis of the proximal humerus will be unfused; and an 'adult' will have a fully erupted permanent dentition, probably showing a degree of attrition, and the spheno-occipital synchondrosis and the proximal humerus will be fused. In their study of *Pan* and *Gorilla* skeletal materials, Gordon et al. (2013) found that cranial suture fusion had occurred after dental maturity and postcranial fusion in all cases for which complete skeletons were available. They further found that, in both taxa, approximately one-quarter to one-third of dentally mature individuals were not skeletally mature, and therefore were unlikely to have reached their full adult size. Assessing the age at death of a gorilla from its skin alone is less certain still; size may be taken as a rough guide, and also an adult male may display a 'saddle' of grey or silver hairs on the back: only the male develops the secondary sexual characteristic of a grey or silver-back (see discussion in Part 1).

There is striking sexual dimorphism in the gorilla, the adult male being on average considerably heavier and taller (crown to heel length) than the adult female. The adult male skull is usually (but not always) furnished with a well-developed sagittal crest, a feature lacking in the adult female skull, and the canine teeth are much larger and project more prominently than in the female. The age categories (or ageing of growth stages) recognised for living animals vary slightly, depending on the subspecies being considered and the authority being consulted, but the following is a rough guide: infant 0–3 years; juvenile 3–6 years; subadult female 6–8 years; subadult (blackback) male 6–13 years; adult female 8+ years; adult (silverback) male 13+ years.

Numerous studies have been made of the life history traits (rate of maturation and reproduction, age at primiparity, gestation length, age at weaning, interbirth interval, lifespan, etc.) and the ways they differ between species in primates and other animals. Factors thought to contribute to differences include body size, relative brain size, predation risks and diet. Some of these are referred to in different parts of the book (Part I: Gorilla Pathology and Health), especially in Chapter 8: Nonspecific Pathology, in the section about age-related changes in gorillas. Gorillas are useful subjects for study because both the eastern and western populations live in a wide range of habitats and under different ecological conditions.

Groves and Meder (2001) considered it inescapable that chimpanzees survive, on average, to greater ages in the wild and also in zoos than do gorillas, and that, for reasons not entirely understood, gorillas may simply age more quickly. With regard to reproductive parameters, they also thought it possible that 'in Mountain Gorillas and Eastern Lowland Gorillas, females begin breeding later than in Western Gorillas (despite apparently the same age at menarche); Mountain Gorillas have shorter but Eastern Lowland have longer interbirth intervals; Mountain Gorillas have shorter cycles. Male chimpanzees begin breeding earlier in the wild than male gorillas, but females begin later (Kaplan et al., 2000), in line with their much later menarche; their gestation is shorter, their interbirth intervals and cycles are notably longer. Orangutans seem more to resemble chimpanzees in age at maturity and length of interbirth interval; they have an early menarche like gorillas; and their gestation is of intermediate length'. The authors concluded that 'Gorillas grow faster and breed more rapidly than do other hominids. Adults have a relatively short life expectancy; silverback males, in particular, seem to have a hard life and to die young. Gorillas are, in the terminology of MacArthur and Wilson (1967), more *r*-selected'. This is alluded to in the context of gerontological studies in Chapter 8: Nonspecific Pathology.

Other authors (Kelley and Schwartz, 2010; Robson and Wood, 2008) have also described gorillas as having faster life histories than chimpanzees and orangutans. In a more recent paper, however, Stoinski et al. (2013) asserted that this is not true at the genus level, and the likelihood is that western gorillas' developmental life history patterns are quite similar to those of chimpanzees. They presented and analysed data on the interbirth intervals in captive and free-living western lowland gorillas, and the weaning age of captive western lowland gorillas, compared with similar datasets for free-living mountain gorillas. Their comparisons showed 'earlier weaning and faster resumption of reproduction in mountain gorillas and captive western gorillas as compared to wild western lowland gorillas', and they believed that 'energetic risk' strongly influences gorilla life history. They noted that mountain gorillas are specialised folivores (see Chapter 11: Alimentary Tract and Associated Organs), whose primary diet of herbaceous vegetation remains abundant throughout the year and of constant high quality (Watts, 1984, 1988), whereas western lowland gorillas are frugivorous, and have to rely on herbaceous vegetation when fruits are not available. As pointed out in Chapter 11, Alimentary Tract and Associated Organs, such ecological differences are likely to be responsible for several variations in social organisation, behaviour, brain development and life history traits observed between the species (Doran and McNeilage, 2001; Robbins et al., 2004; Nowell and Fletcher, 2007; Breuer et al., 2009; McFarlin et al., 2012).

Kelley and Schwartz (2010) contended that, among extant primates, the timing of many life history events is correlated with the age at which the first permanent molar (M1) erupts, and this criterion can be used to infer the life histories of fossil species. They reported M1 emergence ages of 4.6 years for the orangutan (*Pongo pygmaeus pygmaeus*) and 3.8 years for the gorilla (*Gorilla gorilla gorilla*), obtained from the dental histology of wild-shot individuals in museum collections. For comparison, they mentioned one previously reported age at M1 emergence in a free-living chimpanzee (*Pan troglodytes verus*) of approximately 4.0 years (Zihlman et al., 2004). Only free-living individuals were considered, because of evidence suggesting that captive animals develop faster, with earlier dental emergence (Hamada et al., 1996; Kimura and Hamada, 1996; Schwartz et al., 2006). Calculations also need to take account of the fact that, in primates, the mandibular M1 tends to emerge somewhat in advance of the maxillary M1 (Smith et al., 1994). The late age at M1 emergence in *Pongo* is consistent with findings that free-living orangutans have more protracted life histories (Wich et al., 2004; Shumaker et al., 2008; Knott et al., 2009). The average age at M1 emergence in modern humans is 5.8 years (Smith et al., 1995; Liversidge, 2003).

The study by McFarlin et al. (2013) found that in both male and female free-living Virunga mountain gorillas, brain growth is completed between 3 and 4 years of age, corresponding with the emergence of the first permanent molars. By comparison, chimpanzees reportedly attain adult brain size at 4−5 years (despite having smaller brains than gorillas). Modern humans attain adult brain mass only slightly later than chimpanzees, at 5−7 years.

In this catalogue, where the age or sex of a specimen has not been recorded or inferred, it has been omitted and has not been replaced by a category such as 'unknown' or 'indeterminate'. Inferred or otherwise uncertain age and/or sex categories are qualified by a question mark.

Species and Subspecies

The vast majority of specimens listed in the catalogue belong to the nominate subspecies *G. gorilla gorilla*, the western lowland gorilla. To save space, specimens belonging to this subspecies are not noted as such; only specimens that belong to a different species or subspecies are noted accordingly. However, if the original classification of a particular specimen is an obsolete synonym of *Gorilla gorilla* it is noted, both for the sake of historical accuracy, and because it may well be the only available guide to the specimen's approximate antiquity and provenance. See also the notes below on *Gorilla* taxonomy and synonymy.

Locality

Throughout recent history, the names and borders of all the African countries that comprise the gorilla's geographical range have been changed from time to time, usually for political reasons. In this catalogue, the locality names as originally recorded for each specimen are given, for the sake of historical fidelity, and no attempt has been made to bring them up to date. However, where there is an obvious, simple misspelling of a place name in the original records, it has been corrected. Summaries of gorilla distribution can be found in, for example, Wolfheim (1983) and Caldecott and Miles (2005). A convenient, tabulated history of the names of African nations in which great apes are found is given in Gordon et al. (2013).

Name of Collector and Date of Collection

Apart from the above, the collector's field notes, measurement lists, etc., are included if available. Measurements are always given as originally recorded, whether in imperial or metric units, so as to avoid any slight inaccuracies that could result from conversion.

Method and Date of Acquisition

This information is given wherever it is available. Specimens are normally acquired as a result of a purchase or a donation, but in the past some have been procured through specially arranged museum collecting expeditions.

Other Information

This may include the identity of the preparator, taxidermic measurements and references to relevant publications. If the animal died in captivity, its life history may be summarised. It must be emphasised that any diagnoses or observations on pathology have been repeated from institutional records alone, and are not those of the authors, unless stated otherwise.

GORILLA HISTORY, TAXONOMY AND SYNONYMY

From around the 3rd millennium BCE onwards, the ancients gradually became acquainted with various kinds of monkey, and it is possible that at least one kind of anthropoid ape − the chimpanzee − was known as far back as the 6th century BCE (Groves, 2008). Also in the 6th century BCE, Hanno, a Carthaginian Admiral, led the earliest known circumnavigation of the coast of West Africa, for the dual purpose of colonisation and exploration. At the end of their voyage the explorers encountered some hairy 'savage people', whom their interpreters called *gorillai*, according to a Greek translation of the original Punic text. The true identity of these *gorillai* is uncertain, but it is possible that they were chimpanzees or gorillas (see further discussion below about the possible identity of Hanno's *gorillai*).

From that time onwards, up until the 17th century, virtually no new information about what we might confidently suppose to be anthropoid apes came to light. To make matters worse, throughout the 17th, 18th and early 19th centuries there was much confusion surrounding the scientific and vernacular names of these creatures, with the chimpanzee and orangutan both being referred to as orangutans, the black variety coming from Africa and the red variety from Asia. The name *Simia*, which originated with Linnaeus, also caused confusion as it was applied to all three genera of great ape by different authors at different times. It was finally suppressed as a generic name by the International Commission on Zoological Nomenclature (ICZN) in 1955.

As an example of the confused situation, we may cite the case of the young chimpanzee brought to England from Angola in 1698, which, dying soon after its arrival, was the subject of what is generally considered to be the first known dissection of an ape, undertaken by Edward Tyson. Tyson (1699) regarded his subject as being intermediate between man and 'ape', and decided to assign to it the status of 'pygmie', apparently believing it to be representative of the tiny creatures first mentioned by Homer in the *Iliad*; it should be stressed that he was not referring to the small-statured human groups to which this name is sometimes applied today. The mounted skeleton of Tyson's chimpanzee is preserved in the Natural History Museum in London.

Intimations of the existence of a giant, manlike ape first began to filter out of Africa in the 17th century. Andrew Battell, an English sailor from Leigh in Essex, had been taken prisoner by the Portuguese in Angola in 1559, but during his 18 years in captivity he was afforded enough freedom to be able to visit several neighbouring countries, and to observe his surroundings, including both the people and the wildlife. He either saw, or was told about, two kinds of 'monster' common in the woods, the larger called 'Pongo' and the smaller called 'Engeco'. He did not describe the Engeco, which is assumed to be the chimpanzee, but his description of the Pongo leaves little doubt that he was referring to the gorilla. The name Pongo may have been derived from the name of the Mpongwe people of Gabon, or from 'mpungu', a local name for the gorilla in what was then the Kingdom of Loango (modern-day Republic of the Congo, CR). In any event, it was adopted by Lacépède (1799) as the generic name of the Asian ape, which he called *Pongo borneo* (equivalent to the *Simia pygmaeus* of Linnaeus, 1760). *Pongo* remains valid today as the generic name of the orangutan. When Battell returned to England he recounted his adventures to a local vicar, Samuel Purchas, who subsequently included the narrative in a publication entitled *Hakluytus Posthumus, or Purchas his Pilgrimes* (1625).

In the 18th century, the only unmistakable reference to the gorilla seems to be the one provided by James Burnett, Lord Monboddo, who in the second edition of his essay, *Of the Origin and Progress of Language* (1774), while writing of the 'orang outang', inserted a letter from a former 'captain of a ship trading with the slave coast of Africa' in which the gorilla is mentioned as the first and largest of three 'classes or species' of African apes, 'by the natives of Loango, Malemba, Cabenda and Congo, called or named Impungu' (Yerkes and Yerkes, 1929). The letter was reproduced in full by Reade (1864). A probable reference also occurs in a work by the missionary Proyart (1776), who wrote about 'baboons four feet high' living in the forests of what is present-day Angola, Cabinda and Republic of the Congo. Barbot and Casseneuve (1732) mentioned 'a sort of baboon' seen by John Casseneuve at Cabinda, which had been brought there from inland. Casseneuve reported that there were many such apes in the woods of the Kingdom of Angola or Dongo, and in Guinea, and that they were called 'Quojas Morrow'. It is clear from his description that he was referring to the chimpanzee, but he confusedly believed it was the same creature known to the Indians as the 'Orang autang'.

Bowditch (1819) wrote of the 'African ourang outang', which he saw for himself, and was called by the local people 'Inchego', and the similar, but much larger 'Ingena', of which he had only heard rumours. These were undoubtedly the chimpanzee and the gorilla. Bowditch hoped that his brief description of the Ingena would provide the stimulus for a proper investigation into its existence, but unfortunately it did not.

Halfway through the 19th century, however, the gorilla finally emerged from obscurity and into the spotlight of scientific scrutiny (see also Chapter 1: The Genus *Gorilla* — Morphology, Anatomy and the Path to Pathology). Thomas S. Savage (1843/44), an American medical missionary to West Africa, published a paper with Professor Jeffries Wyman on the structure and habits of the chimpanzee. Then, in 1847, he sent some skulls and bones, which he believed were from a new species of chimpanzee, from Gabon to Professor Wyman in Boston, United States. Wyman realised that the animal in question was indeed new to science, and the outcome was a joint memoir, published in the *Boston Journal of Natural History* that same year (Savage and Wyman, 1847). They gave it the scientific name *Troglodytes gorilla*; *Troglodytes* because they mistakenly believed it was a new species of 'orang' (i.e., chimpanzee, then known as *Troglodytes niger*), and *gorilla* because they thought the creature mentioned by Hanno was 'probably one of the species of the Orang' (see below).

Historically, the classification of the anthropoid apes, especially at the lower hierarchical levels, has been fraught with confusion and instability. In the case of the gorilla, as with the orangutan and chimpanzee, a number of forms have been described since its scientific discovery, but most have proved to be spurious and have had to be discarded or reassigned as our knowledge of the animals, together with our understanding of the species/subspecies concept, improved. The following is a summary of the species, subspecies and synonyms, given in the chronological order they were first described.

Troglodytes gorilla

This is the type species. See Savage and Wyman (1847). Dr Thomas Staughton Savage (1804—80) was a missionary of the Protestant Episcopalian establishment in New York, United States. While on the voyage home from Cape Palmas he was unexpectedly detained on the Gaboon River, at latitude 0°15′ N, and spent the month of April 1847 at the house of the Revd John Leighton Wilson, Senior Missionary of the American Board of Commissioners for Foreign Missions to West Africa, who showed him a skull represented by the local people to be that of a large, ferocious, monkey-like creature. Savage believed that it might belong to a new species of chimpanzee, and with Wilson's help he acquired four skulls (two male and two female) and other odd bones, which he forwarded along with an explanatory letter to Jeffries Wyman (1814—74), Hersey Professor of Anatomy at Harvard University, Cambridge, United States. According to Conniff (2008), Savage paid a Mpongwe chieftain $25.00 for the four skulls, a male and female pelvis, and assorted ribs, vertebrae and limb bones, and sent the material to Wyman on 16 July 1847 from New York City, after he got back from West Africa. Apparently, Wyman reimbursed him in full, and also suggested to Savage that he should write to the *Annals and Magazine of Natural History* in London to establish priority for his discovery, which he duly did. A placeholder notice appeared in the issue for October 1847; it was the first printed reference to the gorilla, under the heading 'New Orang-Outang', and announced that a description would appear shortly in the *Boston Journal of Natural History* (it appeared in December, 1847, in volume V, no. IV). In the meantime, Wyman had read a communication from Savage to the Boston Society of Natural History on 18 August 1847, which was subsequently published in the *Proceedings of the Boston Society of Natural History* in 1848. Wyman recognised that the animal in question was distinguishable from the previously described African ape, the chimpanzee (then known scientifically as *T. niger*), and therefore gave it the specific name *gorilla*. This name is the singular version of the word *gorillai*, used in the Greek account of Hanno's voyage to describe a tribe of savage, hairy people (see at the end for further notes on Hanno's *gorillai*). Who chose the gorilla's specific name: Savage or Wyman? Certainly, either one could have been familiar with the story of Hanno's voyage. The answer can be found in Jordan (1910), wherein Wyman's handwritten account of the early stages of the discovery, signed and dated 'Cambridge, June 18, 1866', is reproduced in full. The account was bound up in Wyman's privately printed version of the 1847 paper published in the *Boston Journal of Natural History*, to which Burt G. Wilder (Wyman's biographer in Jordan's book) had been given access by Wyman's family. Wyman stated:

The credit of the discovery clearly belongs to Drs Wilson and Savage, chiefly to the latter, who first became convinced of the fact that the species was new and who first brought it to the notice of naturalists. The species therefore stands recorded Troglodytes gorilla, *Savage.*

In the following account, the notice of the external characters and habits was prepared by Dr Savage. The introductory portion and the description of the crania and bones, and also the determination of the differential characters on which the establishment of the species rests, was prepared by me. In view of this last fact Dr Savage thought, as will be seen in letter, that the species should stand in my name; but this I declined.

In a conversation I had with Dr A. A. Gould with regard to a suitable name, when I informed him that Hanno stated that the natives called the wild men of Africa Gorillæ, he at once suggested the specific name gorilla, which was adopted.

Although the type locality is 'Empongwe, near the mouth of the Gaboon River, Gaboon, West Africa', Meder and Groves (2005) pointed out that Empongwe (also given as Mpongwe) is not a town, but the name of a people living close to the southern bank of the Gaboon River (about 0°4′ N, 9°39′ E). Conniff (2008) went on to say that no-one knows what happened to two of the four skulls Savage brought home: in a letter to Wyman, he had asked that they be set aside for J. L. Wilson. The two remaining skulls (one male and one female) are the cotypes and are housed in the Museum of Comparative Zoology at Harvard University.

Troglodytes savagei

This is a synonym of *G. gorilla*. See Owen (1848a,b). On 24 April 1847 Dr Savage wrote a letter from the Protestant Mission House on the Gaboon River to Professor Richard Owen (1804−92) in London, enclosing a sketch of a gorilla cranium, drawn by the wife of Dr Prince, the English Baptist missionary at Fernando Pó. Savage asked Owen whether he could identify the skull with that of any animal already known to him. Owen realised that the skull resembled that of the chimpanzee, but was unable to decide whether it belonged to a distinct species, or was merely an old adult male *T. niger*. At an evening meeting of the Zoological Society of London, held on 22 February 1848, Owen read a paper 'on a new species of chimpanzee' and proposed to call it *Troglodytes savagei* in honour of its discoverer, but acknowledged that this was provisional, and would be a synonym of any name already put forward. Having received the publication by Savage and Wyman in the meantime, Owen read a 'supplementary note on the great chimpanzee' at an evening meeting of the Zoological Society of London, held on 11 April 1848, in which he afforded priority to the name bestowed by Savage. (See also Chapter 1: The Genus *Gorilla* − Morphology, Anatomy and the Path to Pathology).

Pithecus gesilla

This is a synonym of *G. gorilla*. See Blainville (1849). This name is a misprint for *Pithecus gorilla*, and appears on two lithographic plates (I *bis* and V *bis*) in an atlas published in 1849 as a fascicule of *Ostéographie* (the whole work appeared in instalments between 1839 and 1864). The typographic errors were corrected in a later explanation of the plates, contained in Vol. IV, pp. 5−6, published posthumously in 1855 by Arthus Bertrand, Paris. Plate I *bis* depicts an adult female skeleton presented to the Muséum national d'histoire naturelle, Paris, in April 1849 by Monsieur Gauthier-Laboulaye, a naval surgeon serving aboard the hospital-corvette 'l'Aube'. The animal had been killed on 20 September 1848, and its body obtained from Walker, an American missionary in Gaboon. Plate V *bis* illustrates two adult skulls, one male and one female, which were also brought back from Gaboon by Gauthier-Laboulaye and donated to the Paris museum along with the skeleton. For further use of the name *Pithecus gorilla*, see Giebel (1855).

Chimpanza gorilla

This is a synonym of *G. gorilla*. See Haime (1851). In a footnote (on p. 160) to his analysis and translated extracts of Richard Owen's published works on the gorilla, Jules Haime noted that the generic name *Troglodytes*, proposed for the chimpanzee by Étienne Geoffroy Saint-Hilaire in 1812, could not be used to designate a mammal, because Louis Jean Pierre Vieillot had already given it to a genus of birds in 1806. Haime suggested adopting the name chimpanzee, employed by Georges Cuvier, in a generic sense. Thus, *Chimpanza gorilla* and *Chimpanza troglodytes* would replace *T. gorilla* and *T. niger*, respectively.

Gorilla gina

This is a synonym of *G. gorilla*. See Geoffroy Saint-Hilaire (1852, 1853b). In 1851 Dr Franquet, of the French navy, returned from Gaboon with a complete carcase of an adult male gorilla, preserved in a tun of tafia (a type of rum). The carcase was examined by Professors Duvernoy and Geoffroy Saint-Hilaire of the Muséum national d'histoire naturelle in Paris, whilst the skin was mounted by Monsieur Poortmann, the head of the taxidermy laboratory. In his subsequent publications, Isidore Geoffroy Saint-Hilaire expressed dissatisfaction with the scientific name *T. gorilla*, and proposed that the specific name should be used to create a distinct genus, which he regarded as being intermediate between *Troglodytes* and *Simia*. He adapted the new specific name from 'N'gina', the indigenous name for the animal in Gaboon.

Gorilla savagei

This is a synonym of *G. gorilla*. See Geoffroy Saint-Hilaire (1853a). The February 1853 edition of a scientific magazine edited by Henri Aucapitaine included an article based on Isidore Geoffroy Saint-Hilaire's lecture on the genus *Gorilla*. The description below plate 2, a sketch of a gorilla, reads simply '*T. gorilla*, Savage. An *Gorilla savagei*?' The name is also used by Gray (1870).

Simia gorilla

This is a synonym of *G. gorilla*. See Wagner (1855). Johann Andreas Wagner employed the genus *Simia*, originated by Carl Linnaeus in his *Systema Naturae* ed. 12, i, p. 34 (1766). The name is also used by Schlegel (1876).

Satyrus adrotes

This is a synonym of *G. gorilla*. See Mayer (1856). In his paper, Professor Mayer stated that the name *Troglodytes*, which Linnaeus gave to his fabulous *Homo nocturnus*, had been taken over incorrectly by Johann Friedrich Blumenbach for the chimpanzee, and that anyway, no ape is a cave dweller. He then suggested that the name *Satyrus* (the one already used for apes by the Roman poet Horace) should be used for the genus, comprising three species: *Satyras* [*sic*] *knekias*, the orangutan; *Satyrus adrotes*, the gorilla; and *Satyrus lagaros*, the chimpanzee and tschego (=the 'Inchego' or black-faced chimpanzee).

Gorilla castaneiceps

This is a synonym of *G. gorilla*. See Slack (1862). On 8 April 1862 Dr Slack exhibited a coloured cast of the head of a gorilla to members of the Academy of Natural Sciences of Philadelphia, United States, which he considered to be a new species. According to Slack, the original specimen came from Kamma, Fernan Vaz, Congo and the principal external specific character is a circular patch of reddish hairs upon the top of the head, while 'the skull presents important differences from that of the ordinary gorilla'. The whereabouts of the holotype is unknown, but the cast is kept in the Academy of Natural Sciences of Philadelphia. Groves (2003) mentioned that when he saw the cast, ironically it had been painted completely black. He also pointed out that the colour of the crown is polymorphic in gorilla populations, and so has no validity as a distinguishing feature.

Gorilla mayêma

This is a synonym of *G. gorilla*. See Alix and Bouvier (1877). The authors described the skeleton and skin of an old adult female gorilla which had been collected by Dr Lucan and Monsieur Petit during their hunting expedition on the banks of the Quilo River, 4°35′ S, in Portuguese Congo. The specimen was considered to be sufficiently different to warrant the creation of a new species, named after Maniéma, chief of the village of Conde, near to which it had been killed.

Anthropopithecus gorilla

This is a synonym of *G. gorilla*. See Anderson (1881). John Anderson applied the name to casts of the skulls of an adult male, adult female and a young female gorilla, all of which were presented by Edward Blyth to the Indian Museum, Calcutta, in May 1864. The type locality is 'Gaboon, West Africa'. The generic name had been given first to the chimpanzee, by Blainville, in 1838.

Gorilla gigas

This is a synonym of *G. gorilla*. See Haeckel (1903). This form is based solely on the 'giant gorilla' shot by H. Paschen in the region of Yaunde, Kamerun, in 1900. According to Ernst Haeckel, its unusual size and curious skull formation distinguished it from the normal species, *Gorilla gina*. The Hon. Lionel Walter Rothschild (1868–1937) purchased the specimen for his Zoological Museum in Tring, UK: the original registration number was AD 15. The museum building and Lord Rothschild's collection were bequeathed to the British Museum (Natural History) in 1938, which led to the syntypes being given new accession numbers: 1939.3405 for the mounted skin and 1939.3406 for the mounted skeleton. These are still on public display in the Tring museum. See also the entry for *Gorilla gorilla matschiei*.

Gorilla beringei

See Matschie (1903). A number of explorers travelled within, or close to gorilla habitat in Central Africa in the 19th century, and Ewart Scott Grogan (Grogan and Sharp, 1900) came across the skeleton of a 'gigantic ape' in the bamboo forest north of Mount Mikeno, although he never saw a live specimen. According to Schaller (1963), had Grogan sent the skeleton home, he would be known today as the discoverer of the mountain gorilla. As it is, that honour belongs to Captain Friedrich Robert von Beringe (1865–1940), an officer in the Imperial Colonial Army for German East Africa. In August 1902, whilst Commander of the Military District of Usumbura, he travelled northward through Ruanda-Urundi to visit Sultan Msinga, and then several German field posts. The main purpose of the trip was to impress the local chiefs and the Belgian border posts with the military might of the

German empire, especially as Germany disputed the exact location of the eastern Congo boundary with the Belgians (Schaller, 1963). On 17 October 1902 von Beringe, accompanied by Dr Engeland, five *askaris* and additional porters, attempted to climb Mount Sabinio (Sabinyo) in the Virunga Volcanoes, north of Lake Kivu. At an altitude of 3100 m they set up camp, from which they spotted a group of large black apes ascending the highest point of the volcano. They managed to shoot two of them, which fell inconveniently into a ravine, necessitating 5 hours of hard work just to get one animal (a large male weighing over 200 lb) up on a rope. This event is described in a German colonial newspaper (von Beringe, 1903), but the author's first name is given, apparently erroneously, as Oscar. This mistake has been repeated over the years, and indeed the same error appears on a commemorative plaque at the entrance to the Virunga Conservation Area (von Beringe, 2002). Von Beringe sent the remains of the ape to the Zoologisches Museum in Berlin for identification, although unfortunately its skin and one of its hands had been eaten by a hyaena on the way back to Usumbura (von Beringe, 2002). Professor Paul Matschie (1861–1926) studied the skeleton and decided that it represented a new species. On page 257 of his paper Matschie writes: 'Ich nenne ihn *Gorilla beringeri* nach seinem Entdecker'. This is an obvious misspelling of the specific name, which is correctly spelt as *Gorilla beringei* on page 259 of the same paper. The holotype was registered under the number 13254 in the Berlin museum. (See also Fig. 1.5 in Chapter 1: The Genus *Gorilla* – Morphology, Anatomy and the Path to Pathology, and Fig. CA1.4 in Appendix CA1: Use of Collections and Handling of Biological Material).

Gorilla diehli

This is a subspecies of *G. gorilla*. See Matschie (1904). In his paper, Matschie recognised four species: *Gorilla gorilla*, *Gorilla castaneiceps*, *Gorilla beringeri* (repeating his original misspelling), and a new species, *Gorilla diehli*. The holotype is an adult male skull, originally numbered 12789 in the Zoologisches Museum, Berlin. The animal was killed by local people, and collected by Herrn S. Diehl on a hilltop near Dakbe, SW Kamerun, approx. 9°20′ E, 6°6′ N. This location is on the Cross River, and is written nowadays as 'Takpe' or 'Nfakwe'. It lies at the southern edge of the Takamanda Forest Reserve (Meder and Groves, 2005).

Gorilla gorilla matschiei

This is a synonym of *G. gorilla* and an objective synonym of *Gorilla gigas*. See Rothschild (1904). The holotype is the adult male skin and skeleton which The Hon. Walter Rothschild purchased in 1901 for his museum in Tring, and which came originally from near Yaunde, southern Kamerun. Rothschild was apparently unaware that Haeckel had already used it as the type of *G. gigas*. In his paper, Rothschild created the new subspecies for all southern Kamerun gorillas, to distinguish them from the typical *G. gorilla* of Gabon, and reduced *G. diehli* to the status of a subspecies, *Gorilla gorilla diehli*.

Gorilla jacobi

This is a synonym of *G. gorilla*. See Matschie (1905). The holotype is an adult male skull, original number 28051 (now number 83558) in the Museum für Naturkunde, Berlin. It was collected by Herrn Leutnant Jacob at the station at the mouth of the Lobo River, near the influx of the Njong River, within the Dscha [Dja] River system, West Central Kamerun. The distinguishing characters are simple individual variation (Groves, 2003), but this is the largest gorilla skull seen by Colin Groves, having a total length of 340 mm (Meder and Groves, 2005).

Gorilla gorilla manyema

This is a synonym of *Gorilla graueri*. See Rothschild (1908). Rothschild stated that he had received specimens of *Gorilla manyema* [sic] and saw that it was not a chimpanzee, as he had originally believed, but a gorilla of the 'South Congo region': by inference, the province of Manyema, Congo-Kinshasa (Groves, 1970). Groves (1967) cited one of the Rothschild specimens – an adult male skull, number ZD. 1939.945 in the Natural History Museum, London – as the possible holotype. The words '*Gorilla gorilla manyema* Upper Congo' are written on the skull, which assorts as a *G. graueri*. Groves (1967) originally used the name *manyema* for the eastern lowland gorilla, but Corbet (1967) pointed out that Rothschild clearly attributed this name to Alix and Bouvier, who in 1877 described *Gorilla mayema* from the River Quilo, near Landana, lower Congo, which is in the range of the western lowland gorilla, *G. gorilla gorilla*. Therefore, *manyema* is clearly an inadvertent error for *mayema* and consequently has no taxonomic validity.

Gorilla gorilla schwarzi

This is a synonym of *G. gorilla*. See Fritze (1912). The holotype is an adult male mounted skeleton and dermoplastic, collected by F. Knauss in the vicinity of Sogemafam, on the Dja River, South Kamerun. The skin and skeleton were acquired by the Provinzial-Museums, Karlsruhe, Germany in 1911, and mounted by the museum taxidermist, Herrn Schelenz. Here, Prof. Dr Auerbach examined the specimen and later gave a brief description (1913), in which he referred to it as *Gorilla gorilla schwarzi* Matschie, based on a notification given to him in a letter from Prof. Matschie, who was to have published a full description shortly afterwards, naming it *Gorilla schwarzi*. Auerbach recorded the maximum length of some of the bones, as follows: clavicle 171 mm; sternum 207 mm; humerus 476 mm; ulna 391 mm; radius 369 mm; femur 395 mm. The animal's height from top to toe was reportedly 188 cm, and its arm span, fingertip to fingertip, 336 cm. It is believed that this type specimen was destroyed during the Second World War.

Pseudogorilla mayema

This is a synonym of *G. gorilla.* See Elliot (1913). Daniel Giraud Elliot's description is based on 3 mounted specimens in the Senckenberg Museum, Frankfurt, Germany: an adult male, an old adult female and a baby, which 'differ in size and colour from all known species'. The skeleton of the male and the skull of the female are also in the museum's collection. The type locality was given by Elliot as Upper Congo, but the labels on the specimens say that they are from Fernan Vaz (which is on the southern Gabon coast), as indeed correctly reported by Frechkop (1943). Elliot considered that these specimens exhibited affinities to both the gorilla and chimpanzee, possessing characteristics of each, and thus represented a distinct, interconnecting genus. He thought that they closely resembled the *G. mayema* of Alix and Bouvier, but as he had been unable to locate and examine their holotype, he could not be certain that they were the same, so he placed a question mark after the name — *Pseudogorilla mayema?* He acknowledged that his form would have to be renamed if it were eventually proved that the true *G. mayema* was a different species.

Gorilla graueri

This is a subspecies of *G. beringei.* See Matschie (1914). The holotype is an adult male, collected by Rudolf Grauer (1870−1927) 80 km northwest of Boko, above the northwestern shore of Lake Tanganyika, Belgian Congo, in the summer of 1908. The skin and skull, original number A. 48, 09, 1 (now numbered 31618 and 31619) are in the Museum für Naturkunde (formerly the Zoologisches Museum), Berlin, Germany. The skull is rather small, being only 306 mm in length (Meder and Groves, 2005), and the inscription on it says also 'Nähe des Nutamba-Flusses', which is possibly the Mutambala River. Meder and Groves pointed out that the extreme south of the Itombwe Massif lies 80 km northwest of Boko. See also Chapter 18: Legal Considerations.

Gorilla hansmeyeri

This is a synonym of *G. gorilla.* See Matschie (1914). The holotype is an adult male, collected by Sergeant Peter on the road from Assobam, between Mensima and Bimba, south of the Dume River, west of Mokbe, Kamerun, on 27 January 1907. The skin and skeleton were presented by privy councillor Prof. Dr Hans Meyer, of Leipzig, to the Zoologisches Museum (now the Museum für Naturkunde) in Berlin, and are numbered 17961 and 17960, respectively. According to Meder and Groves (2005), the skeleton is now missing, but the skull, 333 mm long, is one of the largest gorilla skulls in any museum. They also stated that Mensima is correctly spelt Mesima.

Gorilla zenkeri

This is a synonym of *G. gorilla.* See Matschie (1914). The holotype is a young adult male, collected by G. Zenker near Mbiawe, on the Lokundje River, 6 hours downstream from Bipindi, Kamerun, in January 1908. The skeleton and mounted skin, original number A. 15, 09, 1 (now 30260 and 30261) are in the Museum für Naturkunde, Berlin, Germany. According to Meder and Groves (2005), the type skull (number 30261) is only 299 mm long, a small specimen even though it is not completely mature.

Gorilla beringei mikenensis

This is a synonym of *G. beringei.* See Lönnberg (1917). The cotypes are an old adult male, a semi-adult male, an adult female and several juveniles, collected by Captain Elias Arrhenius in the bamboo forest on Mount Mikeno, Virunga Volcanoes, about 30 km northeast of Kissenji, Belgian Congo. All are in the Naturhistoriska riksmuseet, Stockholm, Sweden.

Gorilla gorilla halli

This is a synonym of *G. gorilla.* See Rothschild (1927a, b). The holotype was collected by Dr Leon Fadhi at Punta Mbouda, Spanish Guinea, and was named by The Hon. Walter Rothschild after its owner, Mr John Hall of Charnes Hall, Stafford. Rothschild subsequently purchased the mounted skin for his private museum in Tring. Its original registration number was G. 15; it is now number 1939.3415. According to Jenkins (1990) the skeleton number 1986.757 almost certainly comes from the same individual. The correct spelling of the type locality is Punta Mbonda, which lies at 2°6′ N, 9°46′ E in the north of Río Muni, Equatorial Guinea (Meder and Groves, 2005).

Gorilla gorilla rex-pygmaeorum

This is a synonym of *G. graueri*. See Schwarz (1927). The holotype is an adult male collected by Buxant at Luofu, west of Lake Edward, Belgian Congo. According to Ernst Schwarz, its skin, skull and partial skeleton are in the Koninklijk Museum voor Midden-Afrika (Royal Museum of Central Africa), Tervuren, Belgium, the registration number being 8187.

Gorilla uellensis

This is a synonym of *G. gorilla*. See Schouteden (1927). The cotypes are three skulls, numbered 101, 102 and 103, in the Koninklijk Museum voor Midden-Afrika, Tervuren, Belgium. They were collected by Lemarinel in the Bondo region of Uele, Belgian Congo, and were briefly described by Ernst Schwarz. In a short footnote to page 335 of Schwarz's paper, Henri Schouteden stated that Matschie had proposed the name *Gorilla uellensis* for these skulls. However, it is not clear whether, or when, he proposed that exact name. The doubt arises because, on 1 February 1905, Emile Coart, the director of the Musée du Congo, sent a number of primate specimens, including gorillas no. 63 and nos. 100−103, to Prof. Matschie of the Königlichen Zoologisches Museum zu Berlin for a determination (Wim Wendelen, personal communication). In his reply dated 18 February 1905, Matschie assigned skull no. 63 to *Gorilla gorilla mayêma*, and 100−103 to 'Gorilla gorilla uellensis Matschie subsp. nov.' In the same letter, he also used the name *Gorilla gorilla beringei*, rather than *G. beringei*. So at that time at least, Matschie clearly favoured creating gorilla subspecies, rather than full species. In any event, since the name *G. uellensis* had not been previously published, Schouteden must be regarded as the valid author. According to Meder and Groves (2005), there are in fact four specimens in the Tervuren museum which are labelled as syntypes of *Gorilla uellensis*. Number 100 is the smoke-blackened cranium of an adult male, and number 101 is the similarly blackened cranium of a subadult male. These are both from Djabbir, Bondo. Number 102 is an adult female skull, not smoke-blackened, and number 103 is an adult male mandible, also not smoke-blackened (which happens to fit exactly with number 100), both labelled as coming from Mbili, Itimbiri. Researchers from the Max Planck Institute for Evolutionary Anthropology in Leipzig have determined a mitochondrial DNA sequence from one of these specimens, and compared it to sequences from other gorillas. Contrary to expectations, they found that the sequence obtained did not exhibit the phylogenetic distinctiveness typical of a representative of a peripheral isolated population (as the skulls were collected in an area where gorillas do not

occur naturally today). The researchers believed that the results suggested a scenario in which the museum specimens did not originally derive from the northern Congo, but were brought from the area of current distribution of western gorillas to that location, and their subsequent discovery and collection there gave rise to the false inference that a local gorilla population on the Uele River had survived at least until the end of the 19th century. See Hofreiter et al. (2003). See also Schwarz (1928).

Gorilla (Pseudogorilla) ellioti

This is a synonym of *G. gorilla*. See Frechkop (1943). Serge Frechkop proposed the name *Pseudogorilla ellioti* for the specimens described by Elliot (1913) under the name *P. mayema*, because he believed that they were probably not the same as Alix and Bouvier's *Gorilla mayêma*. The type locality is given as the Rembo Nkomi delta, south of Fernan Vaz, Gabon.

The chaotic state of *Gorilla* taxonomy, as demonstrated by the above compilation, prompted the zoologist Harold Jefferson Coolidge, Jr (1904−85) to review the evidence for the plethora of recognised forms and try to bring some order to the situation. He measured 213 adult male skulls, including examples of all the previously described forms, and came to the conclusion that all gorillas belong to a single species. He recognised only two subspecies, *G. gorilla gorilla* from West Africa and *G. gorilla beringei* from Central Africa. See Coolidge (1929).

Haddow and Ross (1951) thoroughly reexamined Coolidge's statistical methods and found discrepancies. Although unable to accept his computations, they nonetheless agreed with his taxonomic conclusions.

With access to additional material and novel biometric techniques, notably multivariate analysis, Groves (1967, 1970) reviewed the systematics of the gorilla. A careful metrical analysis of 747 adult gorilla skulls (469 male and 278 female) led him to recognise three subspecies. He confirmed the taxonomic integrity of *G. gorilla gorilla* (which he referred to as the 'western lowland gorilla'), but divided the East-Central African gorillas into two subspecies. In his classification, *G. gorilla beringei* (the 'mountain gorilla') is confined to the Virunga Volcanoes and the Mount Kahuzi area (including Mounts Nakalongi and Biega); the remaining populations (Utu lowlands of eastern Congo, Itombwe and Tshiaberimu highlands and Kayonza Forest) are considered to be *Gorilla gorilla graueri* (the 'eastern lowland gorilla'). Subsequently (Groves and Stott, 1979; Groves, 1986) this arrangement was modified in the light of some new information, by transferring the Kahuzi population to *graueri* and including the Kayonza Forest population with *beringei* (the Kayonza Forest is now usually referred to as the Bwindi

Impenetrable Forest). Later still, Groves (2000, 2001, 2003) came to the conclusion that, phenotypically, eastern and western gorillas are very different and do, after all, rate as distinct species: *G. gorilla* and *G. beringei*. He also concluded that the latter encompasses two valid subspecies, *Gorilla beringei beringei* and *Gorilla beringei graueri*, but the Bwindi population may yet turn out to be distinct from *beringei*, and one or more of the Utu, Tshiaberimu and Kahuzi populations may be separable from *graueri*. The name *rex-pygmaeorum*, given to Mount Tshiaberimu gorillas, is a synonym of *graueri*, but it will have to be resurrected if it is ever found that the Mount Tshiaberimu gorillas are a different subspecies from those of Itombwe (the type locality of *G. graueri*).

Just as Coolidge (1929) found the skulls from Bondo (the cotypes of *G. uellensis*) indistinguishable from *G. gorilla gorilla*, Groves also perceived no differences at all from those of western gorillas, and so concluded that, despite their geographically isolated locality, they are probably examples of *G. gorilla gorilla*. Sarmiento (2003) has also studied these skulls, and concurs, believing they indicate a previous continuous distribution of gorillas across tropical Africa in forests north of the Congo and Oubangui Rivers. Today, longer dry seasons and less rain at latitudes farther from the equator prevent the growth of continuous forest and thus the spread of gorillas beyond latitudes 6°30′ N and 5°30′ S (Groves, 1970; Sarmiento and Oates, 2000; Sarmiento, 2003). The end result is that the whole Congo basin effectively separates gorillas into western and eastern groups that are approximately 650 miles (1046 km) apart (Coolidge, 1929; Schaller, 1963). Based on evidence from various sources, Schaller hypothesised that at some time during the late Pleistocene, during an as yet undated pluvial period, there was a continuous belt of rainforest suitable for gorilla habitation extending from West to Central Africa, which disappeared and was replaced by savannah during a succeeding drier period. Thus, gorillas may well at one time have inhabited the northern bank of the Uele River and penetrated southward toward the present range of eastern forms via the headwaters of that river. Colin Groves (personal communication) suspects that the Oubangui is of recent origin; that the Uele, its main tributary, was at one time (until the early Holocene, perhaps?) part of the Chari-Logone system, but then was captured by the Congo, forming the Oubangui. Because most of the forest in the strip between the Middle Congo and the Uele is not typical rainforest, there is a gap in distribution of many mammals in that region today. Those that can inhabit it go right across the Oubangui as if it does not exist, the major taxonomic boundary in that region being not the Oubangui but the Itimbiri. The recency of the Oubangui might facilitate the intermittent genetic exchange between eastern and western gorillas that Thalmann et al. (2007)

deduced. When rainforest was more extensive, one or both populations would have spread and come into contact; then when the rainforest contracted to its present extent, their ranges would have contracted with it. What seems to keep eastern gorillas confined at the moment is the wide *Gilbertiodendron* monodominant belt of the Ituri forest and westward (as suggested by Groves, 1967); John Hart's excavations (1996) showed that at least some areas of the Ituri were mixed forest just a few hundred years ago, and clearly this should be taken into account, but we do not know what climatic regime permits monodominant forest to spread.

For reasons that are not understood, gorillas have not subsequently spread back into the central Congo basin. The results of recent genetic studies are consistent with an east-west divergence of gorilla populations within the Pleistocene, though not during the last glacial maximum. It is estimated that the eastern and western populations became separated at least two million years ago, whereas eastern lowland gorillas have probably been isolated from mountain gorillas for 400,000 years: see Ruvolo et al. (1994); Garner and Ryder (1996); Vigilant and Bradley (2004). As one might expect from this time difference, the divergence between eastern and western populations is far greater than that observed between mountain and eastern lowland gorillas, and there are high levels of substructuring present in western gorillas (Anthony et al., 2007).

Phylogenetic analyses support two regionally defined, major haplogroups within western gorillas. The first extends from the Cross River area in Nigeria to southeastern Cameroon and comprises two subgroups, one of which is distributed across the entire region, while the other is limited to Dja in central Cameroon, Minkébé in northern Gabon, and the west bank of the Ivindo/Ayala River in central Gabon. The second haplogroup extends from coastal Gabon eastwards to Congo and the southern tip of the Central African Republic, and comprises three subgroups: in the montane regions of Equatorial Guinea and the adjacent Monts de Cristal in northwestern Gabon; in the Dzanga-Sangha region of the Central African Republic; and across much of Gabon and east to Lossi, Congo. Anthony et al. considered that the strong patterns of regional differentiation are consistent with forest refugia and rivers playing a role in the partitioning of gorilla genetic diversity.

Tocheri et al. (2016) report that two rare morphological traits are present in the hands and feet of eastern gorillas at much higher frequencies than are found in western gorillas, and the intrageneric distribution of these rare traits suggests that they became common among eastern gorillas after diverging from their western relatives during the early to middle Pleistocene. According to the authors, the particularly high frequencies observed among Grauer gorillas imply that Grauers originated relatively recently from a small founding population of eastern gorillas, and a small group probably dispersed westward from the Virungas into present-day Grauer range, either immediately before or directly after the Younger Dryas stadial. The authors propose that a rapid expansion of the lowland forests of central Africa during the early Holocene caused a connection with the expanding highland forests along the Albertine Rift, thus enabling the descendants of this small founding group to significantly expand their geographic range and population numbers relative to the gorillas of the Virunga Volcanoes and the Bwindi Impenetrable Forest, ultimately resulting in the Grauer gorilla subspecies recognised today.

Some authorities have advocated retaining all gorillas in a single species. For example, Albrecht et al. (2003) believed the morphological and genetic evidence for recognising western and eastern subspecies as separate species to be equivocal, and suggested that they may be semispecies (borderline cases between species and subspecies that have not acquired all of the attributes of species rank). They followed Mayr and Ashlock (1991) in preferring to treat allopatric populations of doubtful rank as subspecies, because the use of trinomial nomenclature conveys closest relationship and allopatry. Sarmiento (2003) also cautioned that a two-species taxonomy is not without its problems, since eastern and western gorillas are allopatric populations, as are the two currently recognised species of chimpanzee (*P. troglodytes* and *Pan paniscus*), meaning that 'there is no objectively determined species within great apes to establish a bench-mark as to the magnitude of phenotypic distinctiveness corresponding to species differences'. Sarmiento (personal communication) sought in the quoted passage to remind the reader that, in great apes, species differences are subjectively and not objectively decided, and must therefore for the sake of practicality strive for consistency. He was the first to suggest that eastern and western gorillas cannot be referred to the same species if pygmy and common chimpanzees are referred to different species (see Sarmiento and Butynski, 1996). Albrecht et al. and Sarmiento were writing when the Biological Species Concept (BSC) of Mayr (1942, 1970, 1996) was still widely adopted. However, Sarmiento (personal communication) does not defend either the Biological Species Concept or the Phylogenetic Species Concept (PSC), and has only described conceptual problems in both.

There has been a widespread rejection of the BSC over the past 20 years or so (Colin Groves, personal communication). The main reasons for this are that (1) it is simply inapplicable to allopatric populations and (2) DNA work has shown that there has been wide interbreeding between different species throughout time. It took DNA to show that interbreeding can occur without altering species' separateness. Several authors — Mayden (1997), Groves (2001) and de Queiroz (2007) — have argued that a species is an evolutionary lineage, and that the way to recognise an evolutionary species is above all that it is consistently different from its sister species.

De Queiroz was one of the most extreme proponents of the school of thought striving to make all taxonomy (especially species) conform to phylogeny (Esteban Sarmiento, personal communication). However, while species have comparatively short life spans, the evolutionary lineages phylogeny describes extend over long time spans. Moreover, when it comes to primate evolutionary lineages, these may comprise multiple species even during the same time span. (We must also remember that the binomial species classification of life-forms predates Darwinism and is thus not designed to account for evolution). As such taxonomy and phylogeny can never fully conform, and de Queiroz's systems have proved to be impractical for the classification of life-forms, as are strict interpretations of the phylogenetic species concept (i.e., one species, one evolutionary lineage).

Just as the relationships among eastern gorilla populations have proved to be more complex than previously appreciated, so with western gorilla populations, the status of the Cross River gorillas has been recently reassessed, and it is now generally accepted that the Cross River gorilla (*G. gorilla diehli*) is morphologically distinct (Sarmiento and Oates, 2000; Groves, 2001; Grubb et al., 2003). The Cross River gorillas have the most northern and western distribution of all gorilla populations, and are isolated from other western gorillas by a distance of about 200 km (Oates et al., 2003). *G. gorilla diehli* is classified as critically endangered, with a total population of fewer than 300 individuals, at first thought to be split into at least 10 subpopulations, straddling the Nigeria-Cameroon border (Morgan and Sunderland-Groves, 2006), but now known to inhabit 11 discrete localities. Recent genetic studies suggest that gorillas at 10 of these localities constitute one population, divided into three subpopulations, which still occasionally exchange individuals (Bergl and Vigilant, 2007).

There are two other main gorilla populations north of the Sanaga River. One is restricted to three forest blocks in the Bélabo region, about 320 km northeast of the Ebo Forest (Fotso et al., 2002). This population was surveyed by Eno Nku and Jacqueline L. Sunderland-Groves in 2001–2002. The other population is in the Ebo Forest itself, less than 100 km north of the Sanaga River (Morgan et al., 2003). The one known cranium from this locality was subjected to canonical discriminant analysis, the results of which suggest that the Ebo population does not fit into any of the known demes of western gorillas, but is instead a relict of a formerly more widespread population living north of the Sanaga (Groves, 2005).

Although there is no absolute consensus on the specific and subspecific classification of *Gorilla*, the current taxonomic situation may be summarised thus:

Western Lowland Gorilla, *Gorilla gorilla gorilla* (Savage, 1847)

Found in West-Central Africa, i.e., southern Cameroon, southwestern Central African Republic, Equatorial Guinea, Gabon, northern Republic of Congo, Cabinda province (Angola) and extreme southwestern Democratic Republic of Congo (Djabbir region). NB: Groves (2003) reminded us that if a species was originally described in one genus and later transferred to another, the original author's name is placed in parentheses, as in *G. gorilla* (Savage and Wyman). He went on to explain that, under the *International Code of Zoological Nomenclature* (4th edition) rules, the author of the original name, *T. gorilla*, is Savage, not 'Savage and Wyman' as commonly cited. This is because the *Proceedings of the Boston Society of Natural History* (1847) states that 'Dr J. Wyman read a communication from Dr Thomas S. Savage, describing the external character and habits of a new species of *Troglodytes* (*T. gorilla*, Savage) recently discovered by Dr Savage in Empongwe, near the river Gaboon, Africa'. However, Savage himself looked upon Wyman as the sole describer, and left the choice of scientific name to him (Conniff, 2008).

Cross River Gorilla, *Gorilla gorilla diehli* (Matschie, 1904)

Found on the border of southeastern Nigeria and southwestern Cameroon, West-Central Africa.

Grauer's Gorilla (or Eastern Lowland Gorilla), *Gorilla beringei graueri* (Matschie, 1914)

Found in eastern Democratic Republic of Congo, East-Central Africa.

Mountain Gorilla, *Gorilla beringei beringei* (Matschie, 1903)

Found in the Virunga Volcanoes region on the international borders of Rwanda, Uganda and Democratic Republic of Congo and the Bwindi Impenetrable National Park in southwestern Uganda, East-Central Africa.

HANNO'S *GORILLAI*

The Voyage and Related Geographical Interpretations

Hanno's voyage took place in antiquity, probably in the first half of the 6th century BCE. Somewhere around that time, the Carthaginian senate decreed that Hanno, a high-ranking official, should lead a voyage beyond the Pillars of Hercules (the Strait of Gibraltar) and found Phoenician colonies in Libya (in classical times, the term Libya meant the whole of Africa). This he did with considerable success, so much so that upon his return to Carthage an account of the voyage was inscribed on a stone tablet which was consecrated and hung up in the temple of Ba'al Hammon (known as Kronos to the Greeks and Saturn to the Romans). Unfortunately, the stone tablet no longer exists, but an abridged version of the account has come down to us in the form of a Greek translation, entitled *Hannonis periplus*, preserved within a 9th century manuscript known as the *Codex Palatinus Graecus* 398, held by the Universitätsbibliothek Heidelberg. It was first published by the Czech philologist Sigismond Gelenius in Basel in 1533. Other translations followed, including into English by Thomas Falconer in 1797. Hanno sailed with 60 pentaconters (50-oared ships) and 30,000 men and women on board, but we cannot be sure of exactly how far they travelled around the African coast, because many of the places mentioned in the text are difficult to identify with any modern location, and consequently various interpretations of the geography have been proffered. The general consensus is that the expedition reached at least as far as Senegal, but scholarly opinions as to where to locate the furthest limit differ: Sierra Leone, Cameroon, Corisco Bay in Equatorial Guinea or Gabon. It would appear, though, that this voyage went further than any other the Carthaginians had undertaken. The first colonies they mentioned might already have existed, but as they sailed on, they clearly entered uncharted lands. Indeed, they sailed past their ability to communicate, and had to take on translators along the way (Scott Farrington, personal communication). According to Falconer (1797), Hanno procured his interpreters from a shepherd tribe living on the banks of the Lixus river (generally identified as the Draa River in Morocco). Presumably their native tongue was Berber, but they must have been familiar with one or more languages widely used for trade in West Africa at the time.

Despite problems with the toponyms, and gaps of uncertain length and other discrepancies in the Greek translation, the account is generally considered to be authentic rather than fabulous, and thus its historical importance as the earliest known account of West Africa (and perhaps of the chimpanzee or gorilla) has to be acknowledged. The conclusion of the account is arguably the most intriguing part. Late into the voyage, three days after passing a large fiery hill called the Chariot of the Gods, they arrived at a bay called the Southern Horn, 'at the bottom of which lay an island like the former, having a lake, and in this lake another island, full of savage people, the greater part of whom were women, whose bodies were hairy and whom our interpreters called gorillai. Though we pursued the men, we could not seize any of them; but all fled from us, escaping over the precipices, and defending themselves with stones. Three women were however taken; but they attacked their conductors with their teeth and hands, and could not be prevailed on to accompany us. Having killed them, we flayed them and brought their skins with us to Carthage. We did not sail further on, our provisions failing us'.

The true identity of the 'gorillai' remains indeterminable. They may have been baboons, chimpanzees, gorillas, or a race of humans where both sexes are hairy, although the last-mentioned is unlikely: no race of humans is particularly hirsute, but all races have very thin skins, and if they were human it would have been difficult to flay them, preserve and transport their skins back to Carthage and then display them for any length of time. According to Pliny the Elder and Gaius Julius Solinus, the skins of the hairy women were placed in the temple of Tanit Pne Ba'al (the wife of Ba'al Hammon), known as Juno to the Romans, and the name 'gorillas' was changed to 'gorgones'. Two of the skins were still there when Carthage was destroyed by the Romans in 146 BCE. The Carthaginians clearly knew dark-skinned humans, and it seems impossible that the sight of dark-skinned people would have been so exotic that it would have inspired them to try to preserve their skins, which is why most classicists conclude that the *gorillai* must be a species of primate (Scott Farrington, personal communication).

If the *gorillai* were chimpanzees or gorillas, one would expect the encounter to have taken place somewhere on the African mainland, since anthropoid apes do not live on islands. That said, it is reasonable to suppose that populations of apes lived on what was the mainland during the low sea levels of the last ice age, and became isolated on newly-formed islands on higher levels of ground as postglacial sea levels rose 11,000−7000 years ago. In that case, they may still have been present on islands in the shallower seas off West or Central Africa in Hanno's time (Jonathan Miles Adams, personal communication). For large animals, such island populations are usually very susceptible to extinction, especially where dense human populations are around. The prima facie candidate in the area is Bioko Island (formerly known as Fernando Pó), which lies about 32 km off the coast of Cameroon and is the largest island in the Gulf of Guinea. Until it was cut off by rising sea levels, it constituted the extremity of a peninsula attached to the mainland. There

is no evidence that apes ever occurred on Bioko, but even so, gorillas and chimpanzees might have been there, and disappeared, perhaps several times (Thomas Butynski, personal communication). As several species of monkeys and prosimians reached Bioko, it is likely that the more mobile gorilla and chimpanzee got there as well, but high rainfall, humidity and volcanic activity are likely to have destroyed any evidence in the meantime. All this would have occurred well before the 6th century BCE. Furthermore, Bioko has no sizeable lakes, and does not come close to meeting the description given in Hanno's account.

Bioko does have a subspecies of drill, *Mandrillus leucophaeus poensis*, and although the name 'drill' is assumed to have an African etymology, any link with 'gorilla' seems implausible. The word 'mandrill' appears to be simply the English 'man' plus 'drill', so it would seem that the comparison between that species and people is not uncommon (Scott Farrington, personal communication). The earliest report of the name is to be found in Smith (1744, p. 51): 'I shall next describe a strange sort of animal, call'd by the White Men in this country, a Mandrill, but why it is so called I know not, nor did I ever hear of the Name before, neither can those who call them so tell, except it be for their near Resemblance of a human Creature, though nothing at all like an Ape'. However, it is clear that William Smith was in fact describing the chimpanzee. Martin (1841, footnote to p. 372) said: 'It is very probable that the word Gorilloi [*sic*] may be identical with the terms Drill and Mandrill — old African names still given in some districts to the Chimpanzee, and, therefore, belonging to it, rather than to the Baboons — so called by European naturalists'. Writing about the chimpanzee, Martin (1841, p. 371) said: 'Little is known of the habits and manners of the Chimpanzee; it is an animal of comparatively recent discovery, though, probably, alluded to in a work of great antiquity — naturalists are indebted to Mr Ogilby for directing their attention to a passage in the *Periplus Hannonis*, respecting the discovery, in an island near the African coast, of wild men who were, in all probability, animals of this species'. Real mandrills would not seem to be credible candidates, because the unique appearance of males would probably have caught the attention of the crew and thus would have appeared in the narrative. Also, there is no reference to tails in the description of the *gorillai* (James L. Newman, personal communication).

It is feasible that what is being described geographically is one of the western Central African river deltas, some of which are massive, with large lagoons, estuaries and islands; very complicated places where one would find islands within lagoons, and sizeable lakes within those islands. The Ogooué Delta, Gabon, would be such a candidate. Perhaps the Niger Delta held gorillas at that time — but probably not — and they are not there today (Thomas Butynski, personal communication). The 'fiery hill' in Hanno's account is thought by some to be Mount Cameroon, an active volcano close to the Gulf of Guinea coast, which is easily visible from both land and sea. Following on naturally from this conclusion, it has been suggested that the Southern Horn is the Gabon estuary, and the expedition fled the effects of the volcanic eruption they were witnessing by sailing up the Ogooué as far as the Lambaréné region, where the river expands into a great lake containing numerous small islands (see http://www.metrum.org/mapping/hanno.htm). But this is an extremely long way from Mount Cameroon (much more than three days' sailing), and there would be no need to go far at all from the base of Mount Cameroon to flee its lava flows or ash clouds (Thomas Butynski, personal communication).

Geographically, 'horn' usually means either a branch of a stream forming a delta, or a cape/peninsula (James L. Newman, personal communication). A conservative view of Hanno's progress would suggest the latter, and Cape Palmas in particular. As for which mountain is recorded, Sierra Leone seems a likely location. Given the flat nature of the coast until this point, it would have looked impressive enough to be the 'Chariot of the Gods'. How much further would three days have taken them? Maybe to the vicinity of where Abidjan sits today? A key in identifying the so-called island are the comments about the males climbing steep hills and there being a lake in the middle. But was it really a lake? For here and elsewhere, such as Mount Cameroon and Corisco Bay, an examination of detailed topographic maps might provide clues.

For more detailed discussion of the historical and geographical aspects of Hanno's Periplus, see for example Hyde, 1947; Warmington, 1960; Harden, 1962; Blomqvist, 1979; Lacroix, 1998; and Farrington, 2013.

ETYMOLOGICAL CONSIDERATIONS

The word 'gorillai' as preserved for us is a Hellenisation of a Punic word that was in turn borrowed from a native West African language, and the etymon has not been established with any certainty so far. According to Du Chaillu (1871), the Mbondemo word for gorilla is *nguyla*, the equivalent of *ngina* in the Mpongwe language. At a stretch of the imagination, *nguyla* (especially if pronounced something like 'ing-guy-lah') could have sounded like 'gorilla' to Hanno's men. However, there appears to be no such word in use in Gabon. Many languages of Gabon have the form *ngila*, *ngiya*, or *ngina* (Jacky Maniacky, personal communication). In linguistics, there is agreement about a possible link with the word 'taboo', and 'lion' as well! In the dialects of the Fang language, *ndji*, *ndzi*, *ngi* and *nji* are found. In Mpongwe (a dialect of the Myene language) 'gorilla' is *njina*. Unfortunately, 'ngiya'-type words cannot be linked to the origin of the word 'gorilla'; it is still too difficult to solve.

In several Bantu (e.g., Mkako) and Ubangian (e.g., Gbaya and the Baka pygmies) languages of the southeastern Cameroon area, *ngile* (pronounced 'ing-gee-leh') is the term for 'gorilla', and cognate forms of this term are likely to be found in other neighbouring languages (Philip Burnham, personal communication).

The old root word, which the regular sound laws of the Bantu languages would reconstruct as *-gida, has been preserved (Christopher Ehret, personal communication). [NB: in historical linguistics, the asterisk is used to indicate that the root is a hypothetical reconstruction, based on phonological rules]. This means that it occurs in languages of both of the two lines of descent from proto-Bantu (the common ancestor) leading down (via complicated further divergences) to all the Bantu languages of today. Coastlands Bantu, which consists of just Mpongwe and its close relatives, forms one of the two primary branches. This root for 'gorilla' occurs in the languages of this group. Regular reflexes of this root word occur in at least two of the several primary branches into which the Nyong-Lomami branch subsequently split: in Benga; in Bulu; and in Njem. It also occurs farther south as a loanword in the Teke languages, which are spoken just south of the areas in which gorillas occur, so presumably the Teke people adopted their word in more recent eras from their neighbours to the north in Gabon. The form *njina* is a regular reflex of this root in the Mpongwe group of the lower Ogowe region of Gabon. In this region, proto-Bantu *d not uncommonly yields an l. In the Bulu language the pronunciation is in fact *ngilo*. In the Gabon area, *d may have the following reflexes: l, n, or Ø (the last often pronounced as [y]), according to the local languages. In Geviya, for example, it is *ŋgia* ([ŋgiya] (Lolke van der Veen, personal communication). The n reflex found in Myene has a special status: it is due to nasal harmony (*d > n) triggered by word-initial nasal consonants.

Du Chaillu's recording is presumably from a language which had l. (In a majority of the languages in which reflexes of the root are found, the *d had been lost entirely between vowels; in the Mpongwe group, *d became n when it occurred between vowels, due to a process of nasal harmony triggered by the word-initial prenasal consonant). Nevertheless, it is doubtful that 'gorilla' derives from a mispronunciation of *ngila*. When people borrow words from languages they find hard to pronounce, they tend to contract the word rather than lengthen it by adding additional syllables. If the Bantu languages had a word similar to *ngolila*, it would not be a surprise to see an outcome such as *gola* or *gila* in the Hanno story. The reverse seems much less probable, but cannot be wholly ruled out (Christopher Ehret, personal communication).

The root *-gida is also widespread in the Congo Basin (Bahuchet, 1989). For instance, one can find it in the Gbaya languages spoken in Central African Republic, according to

Moñino (1988). Moñino considers that it may even come from proto-Gbaya. However, another etymon is used in the western part of the Congo Basin: *bobo (Pither Medjo Mvé, personal communication). One can find it in some Ubangian languages and also in the extreme south of Gabon (in Mbede, Ndumu and Teke). The Aka and Baka pygmies also use the word *bobo* as the name of the gorilla species, but they call a solitary male *ngile*. In some languages spoken in this area the name which designates the species can be quite different from the one dedicated to the 'solitary male' (as in the case of Fang, Ewondo, Ngom, Sangu and Punu). Fang, for instance, uses *ngi* for the species and the compound word *ntore-ngi* for the 'solitary' animal which is often excluded from the group, sometimes perhaps for health reasons. In the Bata variety of Fang situated in the coastal part of Equatorial Guinea, one finds surprisingly the word *engore* both for the chimpanzee and gorilla. However, in this language the gorilla is also called *nji*, as is the case of most Fang dialects (Medjo Mvé, 1997). When writing his PhD dissertation, Medjo Mvé simply considered that the word 'engore' may be a loanword, probably borrowed from a Latin language (Spanish, Portuguese, or maybe French). However, this interpretation may not be a good one in light of the hypothesis of a direct African connection (Pither Medjo Mvé, personal communication).

The regional reconstruction (Northwestern Bantu) for 'gorilla' is *-gida n 9/10 LL, and as there appears to be no possible connection with the form 'gorilla', an affinity with West African languages looks more plausible (Lolke van der Veen, personal communication). However, one should never forget that animals (as well as people and other entities) may be referred to by several names (simple nouns or compounds); or that the names collected do not necessarily match the ones that are (or were) used locally; and also that names may change (lexical replacement) or may have a different referent nowadays. The possible biases are numerous, which makes things rather complicated.

However, the word 'gorilla' may not come from the Bantu-speaking area at all: according to a respected French etymological dictionary (available online at http://atilf.atilf.fr/) the root is more likely to be *gor*, *kor*, or *gur*, meaning 'man' in several current Senegalese languages. In Wolof, for instance, it is *góor* (but 'monkey' is *golo*). The act of comparing an ape with a human in the Bantu-speaking area employs words such as *sokomuntu*, used in eastern Democratic Republic of Congo, and meaning literally 'human-like' (Jacky Maniacky, personal communication).

If we assume that there could have been gorillas in the forests of western tropical Africa, there must have been a word (or several) for it among the people there. That this word (or one of these words) came to mean 'man', and look like *gor is of course not impossible, but highly speculative. And the present-day languages of the western coast of Africa do not show any derivational suffix that would explain the change from *gor to gorilla. But these languages are so distant from each other today that if a proto-language is to be posited, it would surely be quite different from today's languages. It seems that no language in this area has a word for 'gorilla', although some have words for 'rhinoceros' or 'giraffe', also absent from the region (Guillaume Segerer, personal communication). However, although giraffes and rhinoceroses are absent today from the west coast of Africa, giraffes were certainly found as far west as Senegal, and rhinos as far west as Benin, in the 19th century (Colin Groves, personal communication). And who knows what other large mammals might have lived there, in this most densely populated area of Africa.

In the Atlantic languages, all the forms for 'man' that could be related to this root are as follows: *o-gude* (Bijogo); *góor* (Wolof); *gor-ko* (Fula) and *goor* (Sereer). Even if we do not accept a connection with the Bijogo form, Fula-Sereer-Wolof separation is very ancient, so it is clear that this word existed at least 3000 years ago (Konstantin Pozdniakov, personal communication).

BEHAVIOURAL CONSIDERATIONS

Du Chaillu explored the lower Ogooué, including the Crystal Mountains to the north, and the Chaillu Massif to the south, which was later so-named in his honour. He referred to both granite ranges as the 'Sierra del Crystal'. In his book (1871), Du Chaillu is noncommittal about 'the passage in the *Periplus*, or voyage of Hanno, in which it is supposed he alludes to the animal now known as the gorilla'; but one of Du Chaillu's own passages does evoke a semblance of similarity to Hanno's account: 'As I saw the gorillas running – on their hind legs – they looked fearfully like hairy men: their heads down, their bodies inclined forward, their whole appearance was like men running for their lives.' Elsewhere though, he concedes that 'The common walk of the gorilla is not on his hind legs, but on all-fours. In this posture, the arms are so long that the head and breast are raised considerably, and, as it runs, the hind legs are brought far beneath the body. The leg and arm on the same side move together, which gives the beast a curious waddle'. This last statement is, of course, incorrect.

Sunderland-Groves (2008) observed that, of the 99 villages in the region of Cameroon encompassing Takamanda, Mone and Mbulu, in at least 8 (located around the base of the Kagwene Mountain) the inhabitants do not hunt or eat gorillas, because they believe that the gorilla is really human. This leaves open the small possibility that the indigenous people encountered by Hanno had a similar belief, and looked upon the *gorillai* as genuine hairy 'people', which they transmitted in the word they gave to Hanno's interpreters. Hanno's report that the majority of the 'savage people' were females, accords well with the usual formation of gorilla troops and drill groups; but less so with chimpanzee communities, which have a more evenly balanced male-to-female sex ratio.

If it is genuine, the Hanno document includes the first known report of tool use (in the form of throwing) in hominins. The use of objects for aimed throwing is more commonly reported for captive primates than for wild ones, but in any event, throwing in non-humans is comparatively rare, and has been observed most frequently in free-living and captive chimpanzees (Hopkins et al., 2012). Jane Goodall saw aimed aggressive throwing for the first time towards the end of her early studies of the eastern chimpanzee (*Pan troglodytes schweinfurthii*) in the Gombe Stream Chimpanzee Reserve, Tanganyika (now Tanzania), East Africa (Goodall, 1964). The results of a study of manual laterality in throwing by an eastern chimpanzee population in the Mahale Mountains of western Tanzania (Nishida et al., 2012) showed that, overall, there was no clear bias, in contrast to the studies by Hopkins et al. (2005, 2011) which found a species-level right-sided laterality in captive chimpanzees.

The reported behaviour of climbing steep inclines and throwing stones may be indicative more of humans or chimpanzees than gorillas. Interestingly, however, during a three-year ecological study of Cross River gorillas on the Kagwene Mountain in Cameroon, researchers observed three cases of this kind of tool use, which may have been learned through interactions with humans (Wittiger and Sunderland-Groves, 2007). In the first instance, the silverback in a nonhabituated group of gorillas bluff-charged the research team, whilst three nonsilverback individuals tore up fistfuls of muddy grass and threw them toward the humans, using an underarm motion. In the second encounter, a subadult male in a resting group picked up a detached branch and threw it, underarm and somewhat casually, in the direction of the researchers. The silverback bluff-charged several times but the group as a whole did not appear to be unduly concerned by the presence of humans. The third encounter occurred between a resting focal group and a local man, who tried to frighten the gorillas away from the path he was on by banging his machete on the ground loudly, then throwing stones and small rocks at them. The gorillas reacted by tearing up bundles of grass and throwing them in the man's direction, again in an underarm fashion. This encounter lasted for over an hour, before the gorillas calmly left the area. The authors speculated that, since similar observations have not been reported for other western lowland gorilla groups in the wild, the throwing behaviour of the Kagwene mountain Cross River gorillas may be unique to that particular subpopulation. The Kagwene gorillas in general are not hunted, and consequently have less fear of humans than other groups throughout the Cross River gorilla range. The authors believed it possible that the gorillas had learned object throwing from humans trying to chase them away from nearby farmland. This idea was supported by the authors' observations of captive gorillas at the Limbe Wildlife

Centre in Cameroon throwing stones and grit toward visitors, something they may have learned from the chimpanzees housed next door. Conversely, observations by the English hunter Fred Merfield suggest that the behaviour is innate, not learned. In a letter received at Quex Park (home to the Powell-Cotton Museum) on 29 December 1933, Merfield described a hunt for gorillas with the Mendjim Mey tribe in Cameroun, which took place on the morning of 16 November 1933. After spending a year in the local forest, this was the first time that Merfield had been allowed to accompany the Mendjims on one of their gorilla hunts (since they thought that he would bring them bad luck). The hunters were each armed with a long spear, a crossbow and a quiver containing about 50 darts. An hour after leaving the village of Arteck they came across six new gorilla beds in an abandoned plantation and at once took up the fresh tracks leading into dense forest. In 20 minutes they heard the grunts and barks of the beasts and Merfield was told to wait, while the hunters went forward, armed only with spears. After they had surrounded three gorillas up a tree, and having scared off the old male as well as some others, they allowed Merfield to join them. While the men shot at the gorillas with their crossbows, Merfield placed himself next to Besalla (a sub-chief of Arteck) and remained focused on one particular animal, because he wanted to see exactly what happened when it was hit. Many shots missed, but when hit, the gorilla broke off the darts with its hands or teeth. Merfield wrote: 'By this time the animals were infuriated, tearing at branches with their teeth, but I did not see them beat their chests. I could hardly believe my eyes when I saw one of the beasts break off a big piece of dead branch and hurl it down at us, it was certainly a deliberate throw. I had heard of this from natives, but thought it was the movement of the beasts that dislodged pieces of dead wood, but I now know that this is not the case. I distinctly saw the beast throw the branch, after looking down to see where we were. Besalla told me that both chimps and gorillas do this and that some people have had some nasty cracks on the head. One of the beasts hanging on with its hands and with its feet on a lower branch was bouncing up and down in the effort, so the hunters told me, of breaking off the branch so that it would fall on them'. Two gorillas were procured during the hunt: an adult female (series 2, no. 716 in Merfield's list) and a three-quarters-grown male (series 2, no. 717). The third gorilla of the trio crawled along to a fork in the tree and died there, so that they were obliged to leave it.

Captive gorillas' propensity for throwing objects has not only been frequently observed, but also purposefully encouraged at the Jersey Zoo/Durrell Wildlife Conservation Trust/ 'Durrell'. The subject was touched on briefly by Johnstone-Scott (1995), who described how the adult male 'Jambo' was conditioned to throw back objects accidentally dropped into his enclosure by members of the public in return for a small reward. Even delicate items such as spectacles were returned undamaged in this way by the silverback. 'Jambo's' successor, 'Ya Kwanza', was likewise trained to hand back items through the bars, which he found no problem, but he became impatient when having to negotiate the height of the surrounding wall of his outside enclosure in order to return objects by throwing, and at times would lose interest and wander off. However, on a number of occasions he performed as 'Jambo' had done, by successfully returning a tearful child's lunch box (after emptying!), a pair of pliers, a drill bit and a couple of weighty clamps (Richard Johnstone-Scott, personal communication). 'Ya Kwanza' also developed the habit of throwing clods of earth, sticks and even pieces of vegetables left over from midday feeds at certain noisy visitors who chose to tease him, and because he was frequently accurate with his aim, notices had to be put up informing the public that 'The gorillas are likely to throw things'. The females 'G-Ann', 'Julia' and 'Sakina' also developed this behaviour following 'Ya Kwanza's' arrival, and 'G-Ann' in particular was extremely accurate. She would also spend time searching for a missile rather than just snatching up the handiest twig or pulling up a clump of grass while displaying in a customary strutting run.

In 1981 Richard Johnstone-Scott (personal communication) spent 6 weeks up on the southern slopes of Mount Visoke (Bisoke), Rwanda, courtesy of Drs Sandy Harcourt and Kelly Stewart, and witnessed when in close proximity, 'Icarus', one of two silverbacks in group 5 at that time, tear up quantities of vegetation and hurl it at villagers working on their crops close to the park boundary. Some of the younger members of the group (then 14- strong) followed Icarus's example. This display, accompanied by cough grunts and barks, lasted only for a minute or so before the group turned back into the security of the dense undergrowth.

Chimpanzees both at Jersey and Howletts Wild Animal Park in the United Kingdom (where Mr Johnstone-Scott worked previously) would often throw things at visitors, usually those who antagonised them. The Jersey male 'Chumley' would hurl anything handy, which for certain unfortunate individuals included faecal matter, whilst 'Bustah' at Howletts would also resort to filling his mouth from a water feeder and showering people if a suitable missile was not immediately available. The hurling of objects by both gorillas and chimpanzees at members of the visiting public was in most cases an act of aggressive retaliation. The throwing technique, whether in aggressive displays or when returning objects for reward was always a low underarm movement with a strong wrist flick to propel the missile with remarkable accuracy.

'Massa', a long-lived male gorilla at the Philadelphia Zoo, was known for throwing his faeces at the public, and did so when the public stared at him and made a lot of noise (Esteban Sarmiento, personal communication). For this reason, he was moved from a cage with bars to an enclosure with a protective glass screen. During his first year there, the glass was smeared with faeces. (See also Chapter 8: Nonspecific Pathology).

CORRESPONDENTS

Jonathan Miles Adams, College of Natural Sciences, Seoul National University, Seoul, South Korea.
Philip Burnham, Department of Anthropology, University College London, United Kingdom.
Thomas M. Butynski, Sustainability Centre Eastern Africa, Lolldaiga Hills, Laikipia, Kenya.
Christopher Ehret, History Faculty, University of California Los Angeles, CA, United States.
Scott Farrington, Department of Classics, University of Miami, Coral Gables, FL, United States.
Colin P. Groves, School of Archaeology and Anthropology, Australian National University, Canberra, Australia.
Richard Johnstone-Scott, Jersey, British Channel Islands.
Jacky Maniacky, Service for Linguistics, Koninklijk Museum voor Midden-Afrika, Tervuren, Belgium.
Pither Medjo Mvé, Faculté des Lettres et Sciences Humaines, Département des Sciences du Langage, Université Omar Bongo, Libreville, Gabon.
James L. Newman, Syracuse University, NY, United States.
Konstantin Pozdniakov, Institut National des Langues et Civilisations Orientales (INALCO); Institut Universitaire de France (IUF); Le Laboratoire Langage, Langues et Cultures d'Afrique Noire (LLACAN), Paris, France.
Esteban E. Sarmiento, Human Evolution Foundation, East Brunswick, NJ, United States.
Guillaume Segerer, Centre National de la Recherche Scientifique (CNRS); INALCO; LLACAN, Paris, France.
Lolke van der Veen, Laboratoire Dynamique du Langage (UMR 5596), Institut des Sciences de l'Homme, Lyon, France.
Paolo Viscardi, National Museum of Ireland — Natural History, Dublin, Ireland.
Wim Wendelen, Vertebrate Section, Koninklijk Museum voor Midden-Afrika, Tervuren, Belgium.

Catalogue of Preserved Gorilla Materials

ARGENTINE REPUBLIC

Buenos Aires

Museo Argentino de Ciencias Naturales 'Bernardino Rivadavia', Avenida Angel Gallardo 470, Casilla de Correo 220, Sucursal 5, 1405 Buenos Aires*

Martha Piantanida

10.33 Mounted skeleton. Adult male. *Gorilla jacobi.* Purchased from Carl Hagenbeck on 2 December 1910. This skeleton, together with that of an adult female gorilla, was exhibited at the World Fair in Buenos Aires. Their mounted skins were also exhibited, along with those of a subadult male and two juveniles. The family group was displayed with chimpanzees, antelopes and other animals in a large African diorama within a case measuring 20 m long by 6 m wide by 6 m high. All the specimens had been supplied by the taxidermy firm of Umlauff, in Hamburg, to Carl Hagenbeck on 24 March 1910.

La Plata

Universidad Nacional de La Plata, Facultad de Ciencias Naturales y Museo, Paseo del Bosque, 1900 La Plata

Jorge Luis Frangi

go 963 Mounted skeleton. Adult male. Gaboon. Received in exchange from the British Museum (Natural History), London, in 1896.

go 18.IX.84.1 Skull. Adult male. Cameroon. Donated by the Jardin Zoologico de La Plata. It arrived at the Jardin Zoologico on 30 June 1970, aged about 3 years and was named 'Negro'.

AUSTRALIA

Adelaide

South Australian Museum, North Terrace, Adelaide, South Australia 5000

Catherine Kemper

M 1426 Mounted skin. Male. *Gorilla savagei.*
M 1427 Mounted skin. Female. *Gorilla savagei.*
M 1436 Mounted skeleton. Male. *Gorilla savagei.*
M 1437 Mounted skeleton. Female. *Gorilla savagei.*

The above 4 specimens were all purchased, and all were registered on 7 August 1922. The only locality noted for them is 'West Africa'.

M 5347 Skull (cast). *Gorilla savagei.* Registered in 1945.

University of Adelaide, Faculty of Dentistry, G.P.O. Box 498, Adelaide, South Australia 5001*

Roger J. Smales

Skull. Adult male.

Canberra

Australian Institute of Anatomy, Canberra, ACT 2601

Mrs N.W. Keith

NB: This Institute is defunct, having closed down c.1985. Four of the specimens listed here are now in the Department of Prehistory and Anthropology, Australian National University (see below). The current whereabouts of the other specimens is unknown.

2606 Skull.
2696 Femur.

2701 Brain (cast).

2706 Femur (cast).

2787 Skeleton.

2813 Skull. Female.

2814 Skull. Male.

2863 Skull (cast).

2864 Skull (cast). Female.

Australian National University, School of Archaeology and Anthropology, Department of Prehistory and Anthropology, G.P.O. Box 4, Canberra, ACT 2601

Colin P. Groves

Skull (bisected). Adult male.

Skull. Adult female.

Skull (cast). Adult female.

Mounted skeleton. Adult female. 'Young male' is written on the skull.

The above specimens were acquired from the Australian Institute of Anatomy, Canberra (see above) when it closed.

Skull (most teeth missing); both humeri; both ulnae; one radius; both femora; both tibiae; one fibula; one calcaneus; one ilium; sundry ribs, vertebrae, metapodials and phalanges. Adult male. *Gorilla beringei beringei*. These remains were collected (picked up) at the foot of Mount Karisimbi, near the Karisoke Research Center, mainly by a Rwandan guide employed at Karisoke, and the residue by Colin P. Groves and Dian Fossey, in 1971. The cranium and mandible are grossly asymmetrical, due to malformation of the left mandibular condyle, perhaps caused by some birth or early infancy trauma, which may have resulted in underdevelopment of the whole masticatory apparatus on the left side. The animal may have been the victim of attempted infanticide. There is also what appears to be some spinal osteoarthritis.

Launceston

Queen Victoria Museum and Art Gallery, 2 Invermay Road, Inveresk, Launceston, Tasmania 7248

Tammy Gordon

Skull (cast). Adult female.

Hand (cast).

Foot (cast).

Melbourne

Museum Victoria, Division of Natural History and Anthropology, 285–321 Russell Street, Melbourne, Victoria 3000

Joan M. Dixon; Kevin C. Rowe

NMV 17752 Mounted skin. Adult male. Received on 20 August 1865. The top of the head has a bright chestnut hue.

NMV 17753 Mounted skin. Adult female. Received on 20 August 1865.

NMV 17754 Mounted skin. Infant. Received on 20 August 1865. This infant was being nursed by the above female, no. 17753, when they were shot.

NMV 18613 Mounted skeleton. Adult male. Received on 20 August 1865. It is from the same animal as no. 17752.

NMV 18614 Mounted skeleton. Adult female. Received on 20 August 1865. It is from the same animal as no. 17753.

The above specimens were collected in West Africa by Paul Belloni Du Chaillu (1831–1903) and obtained under the auspices of Dr J.E. Gray of the British Museum. They were preserved in salt and sent to the taxidermist Edward Gerrard Jr, 31 College Place, Camden Town, London. Gerrard informed Prof. Frederick McCoy, the museum director, that he had the specimens, in a letter dated 25 April 1864 (McCoy was in correspondence with Gerrard about the possibility of obtaining some of Du Chaillu's gorillas as early as 24 August 1861). The specimens were prepared in December 1864 and supplied at a total cost of £150. They were sent to Australia on board the 'Holmsdale' on 28 February 1865. The skins were mounted as a family group atop an artificial rock outcrop, the male with its arm raised aloft, giving the exhibit a total height of about 9 ft.

NMV C31059 Skull (cast). Adult male. From the largest gorilla in the Paris Museum. Purchased from Maison Verreaux. Registered on 6 October 1865.

NMV C31096 Skull (cast). Infant.

NMV C31097 Skull (cast). Male. Received from the British Museum (Du Chaillu specimen). Registered on 6 October 1865. Previous number 18616.

NMV C31100 Skull (cast). Male. Made in 1980 from old cast C31102 from the Paris Museum.

NMV C31102 Skull (cast). Male. Received from the Paris Museum. Registered on 6 October 1865. Previous number 18615. Cast 31100 was made from this one in 1980.

NMV C31354 Skull (cast). Adult male. This is a polyurethane plastic reproduction taken from the skull of 'Buluman', who was born in Cameroon c.1958 and arrived at the Taronga Zoo, Mosman, on 18 October 1961, along with another male ('Little John') and two females ('Annabelle' and 'Wild One'). All were purchased from Dr Deets Pickett, Trans World Animal, Inc., Kansas City, Missouri, USA (and Yaoundé, Cameroon). 'Buluman' was transferred on loan to Royal Melbourne Zoological Gardens, Parkville, Victoria 3052, on 15 October 1980. He was euthanased there on 30 March

1998 due to poor health and an ulcerated/infected foot. Last recorded weight 184 kg on 20 February 1997. Previous number 12332. [Some additional information added by GH from correspondence with Taronga Zoo personnel.]

Mogo

Mogo Zoo, 222 Tomakin Road, Mogo, NSW 2536

Samantha Young

B30039 Serum (frozen) × 3. Male. These are from 'Mahali', who was born on 18 August 2008 in Taronga Zoo, Sydney; transferred to Mogo Zoo in August 2013, then (with males 'Fataki' and 'Fuzu') to Orana Wildlife Park, Christchurch, New Zealand, in July 2015.

B30042 (1) Tissue samples — in 10% neutral buffered formalin, sent to an external laboratory for histopathology — of heart, lung, liver, spleen, kidney, stomach, small intestine, large intestine, pancreas, bladder, nictitating membrane, brain, spinal cord, lymph node and aorta. (2) Tissue samples — duplicate set retained at Mogo Zoo in 10% neutral buffered formalin — of heart, lung, liver, spleen, kidney, stomach, small intestine, large intestine, pancreas, bladder, nictitating membrane, spinal cord, lymph node, aorta; plus ovary and uterus. (3) Tissue samples — frozen at Mogo Zoo — of heart, lung, liver, spleen, kidney, stomach, small intestine, large intestine, pancreas, bladder, nictitating membrane, spinal cord, lymph node and aorta. Female. These are from 'Mouila', who was wild-caught in Cameroon, and arrived at the Stichting Apenheul, Apeldoorn, Netherlands, on 22 November 1974, when aged about 1 year. She had been purchased, with a male named 'Bongo', from G. van den Brink. She was transferred to Sydney Zoo on 6 December 1996 and to Mogo Zoo in August 2013. She died in November 2015 of an intracranial neoplasia (see Chapter 15: Nervous System and Special Senses in Part I: Gorilla Pathology and Health). [Some additional information added by GH from correspondence with Apenheul personnel.]

Perth

Western Australian Museum, Francis Street, Perth, Western Australia 6000

D.J. Kitchener

NB: The Western Australian Museum is currently closed for renovation, and is scheduled to reopen in 2020, as the New Museum for Western Australia. For further information, please visit http://museum.wa.gov.au.

Mounted skeleton. It is in the Mammal Display Gallery.

South Brisbane

Queensland Museum, Corner of Grey and Melbourne Streets, South Bank, South Brisbane, Queensland 4101

Andrew P. Amey; Heather Janetzki; S. van Dyck

QM J4435 Skull (plaster cast). Juvenile.

QM J4436 Skull (plaster cast). Adult female or sub-adult male.

QM J4437 Skull (plaster cast). Adult male.

QM J 4438 Skull (plaster cast). Adult male.

The above casts were purchased (with QM J4433, an orangutan skull plaster cast, and QM J4434, a chimpanzee skull plaster cast) for 176 francs from Les Fils d'Émile Deyrolle, 48, rue du Bac, Paris 7e, and were registered on 16 January 1926.

QM J4845 Mounted skin. Adult female. *Gorilla beringei beringei*. Virunga Volcanoes, near Lake Kivu, Belgian Congo. Collected by Thomas Alexander Barns (1881–1930). Purchased for £150 from Rowland Ward, 'The Jungle', 167 Piccadilly, London W1, and despatched to Australia on the 'ss Fordadale'. Registered on 1 February 1929. Collector's measurements: chest girth 47½ in.; squatting height 31 in.; arm span from tips of fingers across back 88 in.

QMJM 13563 Brain (plaster cast). Donated by Prof. J.L. Shellshear, Faculty of Medicine, Hong Kong University, on 21 November 1935. It was originally registered as QM E2508 in the anthropological collection but was de-registered in 1967 and moved to the mammal collection.

Sydney

Australian Museum, 6–8 College Street, Sydney, New South Wales 2000

Sandy Ingleby; B.J. Marlow

A.9828.001 Skull (cast). Received from the National Museum, Melbourne.

A.16562.001 Skull. West Africa. Received from Gerrard, London.

A.16562.002 Skeleton. West Africa. Received from Gerrard, London.

A.16568 Skin.

B.2669.001 Mounted skin. Adult male. Gabon. Received from E. Gerrard Jr.

B.2685 Skeleton. Juvenile. Gaboon. Purchased in 1884.

M.1920 Mounted skin. Adult female. Registered on 25 October 1907.

M.1920.001 Mounted skin. Registered on 25 October 1907.

M.10855 Skeleton (incomplete). Adult female. Ex captive. Registered on 17 August 1978.

M.36146 Part skin (head). Registered on 15 November 2001.

M.36147 Part skin (manus). Registered on 15 November 2001.

M.36148 Skull. Registered on 15 November 2001.

M.36149 Skull. Registered on 15 November 2001.

M.37431 Skull. Adult female. Ex captive (zoo). Registered on 18 August 2004.

M.37431.001 Skull. Adult female. Ex captive (zoo). Registered on 18 August 2004.

M.38612.001 Mounted skin. Adult male. Ex Taronga Zoo. Registered on 13 February 2006.

M.38630.001 Mounted skin. Adult female. Ex Taronga Zoo. Registered on 15 February 2006.

M.40075.001 Part skin (2 pes). Registered on 26 June 2008.

M.43538.001 Skull. Adult male. Ex captive (zoo). Registered on 17 November 2011.

M.45154.001 Part skin (2 forelimbs). Registered on 30 August 2012.

M.45209.001 Brain (cast). Registered on 4 October 2012.

M.45607.001 Skull (cast). Registered on 6 December 2012.

M.45607.002 Skull (cast). Registered on 6 December 2012.

Postcranial skeleton. Adult male. Received from the Taronga Zoo, Mosman, NSW, on 30 October 1968. This is 'King Kong', who arrived at the Taronga Zoo on 7 June 1959, aged about 8 years, having been purchased from Marine Enterprises, Inc., Ocean Park, California, USA. He died of nephritis, chronic enteritis and chronic pancreatitis on 29 October 1968. The skull was incorporated into a whole mount of the head and face which was prepared by wax impregnation techniques, as were also the hands and feet. [Some additional information added by GH from correspondence with Taronga Zoo personnel.]

Taronga Zoo, P.O. Box 20, Mosman, New South Wales 2088

Karrie Rose

Fixed tissues in microscope slides and in paraffin blocks are held from each of the cases below:

TARZ-765/1 Neonate female. Dam: 'Frala'. Necropsy/biopsy date 26 June 1999.

TARZ-7043/1 Autolytic fetus biopsy. Adult female 'Kriba'. Biopsy date 25 June 2009.

TARZ-7043/2 Probable reproductive tract tissue biopsy. Adult female 'Kriba'. Biopsy date 6 October 2009.

TARZ-9456/1 Normal skin biopsy. Male 'Kibali'. Biopsy date 3 September 2013.

University of Sydney, Department of Anatomy and Histology, J.L. Shellshear Museum of Physical Anthropology and Comparative Anatomy, Anderson Stuart Building F 13, Sydney, New South Wales 2006

Denise Donlon

Skeleton (partially disarticulated). Male.

20 Skull. Adult male.

22 Skull (papier mâché).

21 Skull. Male.

56 Endocranial cast.

107 Endocranial cast.

2701 Endocranial cast.

CA 555 Left femur (cast). Male. Yaunde, Kamerun. No. 43 in Kgl. Zool. Mus. Berlin. From S. Tenker. No. 16 of series.

CA 556 Tibia (cast). Male. Yaunde, Kamerun. No. 17 of series.

CA 557 Right humerus (cast). Male. East of Balaga, S. Kamerun. No. 11609 in Kgl. Zool. Mus. Berlin. From Oberleutnant Schwarz. No. 18 of series.

CA 558 Radius (cast). Male. East of Balaga, S. Kamerun. No. 19 of series.

CA 559 Tibia (cast). Male. East of Balaga, S. Kamerun. No. 20 of series.

CA 560 Right humerus (cast). Male. Dume, Kamerun. No. 12809 in Kgl. Zool. Mus. Berlin. From Oberleutnant Schipper. No. 21 in series.

CA 561 Ulna (cast). Male. Dume, Kamerun. No. 22 of series.

CA 562 Radius (cast). Male. Dume, Kamerun. No. 23 of series.

CA 563 Right femur (cast). Male. Dume, Kamerun. No. 24 of series.

CA 564 Tibia (cast). Male. Dume, Kamerun. No. 25 of series.

CA 565 Right humerus (cast). Male. Baraka, Tanganyika. Large specimen. No. A 48091 in Kgl. Zool. Mus. Berlin. From S. Grauer. No. 26 of series.

CA 566 Right femur (cast). Male. Baraka, Tanganyika. Large specimen. No. 27 of series.

CA 567 Tibia (cast). Male. Baraka, Tanganyika. No. 28 of series.

CA 568 Ulna (cast). Female. Baraka, Tanganyika. Small specimen. No. A 4809 in Kgl. Zool. Mus. Berlin. From S. Grauer. No. 29 of series.

The maker of the above 14 casts, CA 555—CA 568 inclusive, was Krantz of Bonn.

AUSTRIA

Admont

Naturhistorischen Museums, Benediktinerstift Admont, 8911 Admont 1*

Gunter Morge

Life-size plaster bust (head and right shoulder), left hand (plaster) and left foot (plaster). *Gorilla gina*.

Received from the Königliches Zoologisches Museum (now Staatliches Museum für Tierkunde), Dresden, in February 1898.

Skull (plaster cast). Purchased for 7.5 marks from Firma Schlüter, Halle. This specimen is now lost.

Graz

Steiermärkisches Landesmuseum Joanneum, Abteilung für Zoologie, Raubergasse 10, 8010 Graz

Erich Kreissl; Ursula Stockinger

29438/1 Skull. Adult male. *Gorilla castaneiceps*. Ngomo, on the Ogooué River, Gabon. Collected by the missionary Haug c.1900. Acquired on 9 May 1942. This specimen is located in Hall 7.

29438/2 Skull. Juvenile male. *Gorilla castaneiceps*. Ngomo, on the Ogooué River, Gabon.

29440 Skull. Subadult female. *Gorilla beringei graueri*. Lake Kivu district, Ruanda-Urundi, German East Africa. This may be an adult male chimpanzee.

29464 Skull. Subadult male. *Gorilla castaneiceps*. French Congo. Purchased from Dr H. Strohmeyer, Marburg, on 19 October 1942. This may be an adult male chimpanzee.

Innsbruck

Medizinische Universität Innsbruck, Sektion für Klinisch-Funktionelle Anatomie, Müllerstrasse 59, 6020 Innsbruck

Erich Brenner; W. Platzer

2745 Skull. Adult male. Mounted on a wooden base; mandible articulated. Obtained by Prof. Dr Sieglbauer in 1920. The upper right canine is possibly broken and has been covered by a yellowish wax.

3449 Mounted skeleton. Juvenile female. Obtained by Prof. Dr Sieglbauer in 1920. Mounted by preparator Kurz. Renovated by Romed Hörmann in 2014. Overall height 116 cm. The upper right canine is missing.

Leopold-Franzens-Universität Innsbruck, Institut für Zoologie und Limnologie, Technikerstrasse 25, 6020 Innsbruck*

Reinhard Rieger; Wolfgang Schedl

Skull. Adult male. Lower reaches of the Ogooué River, in the vicinity of Enionga village, Gabon. The upper left lateral incisor is missing.

Skull. Adult? female? One canine and 1 incisor are missing from the upper jaw.

Salzburg

Haus der Natur, Museumplatz 5, 5020 Salzburg

Max Kobler; Robert Lindner

HNS-Mam-D-0218 Mounted skin. Adult male. Little Noamali, Yaoundé, Cameroun. Collected on 27 November 1924 (attributed to Rudolf Grauer, but Grauer's last African journey dates to 1910–11). Donated in 1927. Prepared by Franz Wald.

HNS-Mam-D-0230 Mounted skin. Subadult male. *Gorilla beringei beringei*. Kirum, Rwanda – Burundi. It died of tuberculosis in the Zoologischer Garten Hannover in 1964, aged approximately 6–7 years. Purchased. Prepared by Alfred Höller. It is from the same animal as the skeleton HNS-Mam-S-0550.

HNS-Mam-D-0242 Mounted skin. Subadult female. This is 'Eseka', who arrived at the Zoologischer Garten Berlin on 3 July 1936, when she was about 4 years old (weight 40 kg). She died of general debility on 17 July 1940 (weight 224 lb). Donated by Dr Lutz Heck, Zoologischer Garten Berlin. Accession date 1 September 1941. Prepared by Gerhard Schröder. [Some additional information added by GH from correspondence with Berlin Zoo personnel.]

HNS-Mam-D-0255 Skull and mounted skin. Juvenile. *Gorilla beringei beringei*. The animal was approximately 5 years old.

HNS-Mam-D-0258 Mounted skin. Infant. Cameroun. Presented by Heinrich Demmer in 1960. The animal was approximately 2 years old. Prepared by Leopold Wald.

HNS-Mam-D-0271 Mummified specimen. Infant female. Cameroun. She had arrived at the Munchener Tierpark on 2 November 1937, and was known as 'Molly'. She died there on 9 June 1938. Presented by Dr Lutz Heck in 1938. [Some additional information added by GH from correspondence with Berlin Zoo personnel.]

HNS-Mam-D-0284 Mounted skin. Infant male. Cameroun. The animal was approximately 1½–2 years old. Collected pre-1927. Prepared by Franz Wald?

HNS-Mam-D-0847 Mounted skin. Subadult male. Cameroun. The animal was approximately 8 or 9 years old. Donated by Dr Lutz Heck in 1938. From the same animal as the mounted skeleton HNS-Mam-S-0662.

HNS-Mam-S-0073 Skull. Adult male. Cameroun? Donated by Heinrich Demmer in 1959?

HNS-Mam-S-0074 Skull. Adult male. Cameroun. Donated by Dr Lutz Heck in 1938.

HNS-Mam-S-0075 Skull. Subadult. Cameroun. Received from Dr Schlüter and Dr Mass, Naturwissenschaftliche Anstalt, Halle (Saale).

HNS-Mam-S-0076 Skull. Adult male. Cameroun. Donated by Dr Lutz Heck in 1938.

HNS-Mam-S-0077 Skull. Subadult female. From the same animal as the mounted skin HNS-Mam-D-0242. Donated by Dr Lutz Heck. Prepared by Gerhard Schröder.

HNS-Mam-S-0078 Cranium. Juvenile female. Cameroun. Collected by Obermaier in 1940; donated by him.

HNS-Mam-S-0519 Skull. Juvenile male. Cameroun. Donated by Dr Lutz Heck in 1938.

HNS-Mam-S-0521 Skull (cast). Adult male. Cameroun. Received from Dr F. Krantz, Bonn a. Rh.

HNS-Mam-S-0522 Skull (cast). Juvenile male. Cameroun. Received from Dr F. Krantz, Bonn a. Rh.

HNS-Mam-S-0523 Skull (cast). Adult male. Cameroun. Received from Dr F. Krantz, Bonn a. Rh.

HNS-Mam-S-0524 Skull. Adult male. Cameroun. Donated by Dr Lutz Heck in 1938.

HNS-Mam-S-0525 Skull. Subadult male.

HNS-Mam-S-0550 Articulated postcranial skeleton. Subadult male. *Gorilla beringei beringei*. It is from the same individual as the mounted skin HNS-Mam-D-0230. Some vertebrae are missing. Prepared by Alfred Höller.

HNS-Mam-S-0628 Skull (cast). Adult female. Cameroun. Received from Dr F. Krantz, Bonn a. Rh.

HNS-Mam-S-0633 Face (plaster cast). Adult male. This is the captive specimen 'Bobby'. Donated by Dr Lutz Heck in 1938.

HNS-Mam-S-0634 Left hand (plaster cast). Adult male ('Bobby'). Donated by Dr Lutz Heck in 1938.

HNS-Mam-S-0635 Right foot (plaster cast). Adult male ('Bobby'). Donated by Dr Lutz Heck in 1938.

HNS-Mam-S-0662 Mounted skeleton. Subadult male. Donated by Dr Lutz Heck in 1938.

HNS-Mam-S-0742 Skull. Juvenile. *Gorilla* sp. Cameroun. Donated by Dr Lutz Heck in 1938.

Skin and mounted skeleton. Infant female. Purchased from the Salzburger Tiergarten Hellbrunn, Anif, where it had arrived in 1972 and died through accidental self-strangulation (hanging) in March 1973. It was approximately 1 year old. This specimen appears to be missing.

Vienna

Naturhistorisches Museum, Burgring 7, Postfach 417, 1014 Wien

Anthropologische Abteilung*

21. 495 Skull. Adult male. Cameroun. Collected by Ernst Alexander Zwilling (1904−90) in 1933. Transferred from the Säugetiersammlung, where its inventory number was 5289.

Erste Zoologische Abteilung (Wirbeltiere)

Kurt Bauer; Frank E. Zachos

NMW 792/st664 Dermoplastic and skeleton. Adult male. Gabon. Purchased from Jean Baptiste Édouard Verreaux (1810−68), Paris, in 1856.

NMW 807 Skull (plaster cast). Adult female. *Gorilla savagei/dyina* (ind. *gina*). Gabon.

NMW 808 Skull (plaster cast). Adult male. *Gorilla djego* (Slack). Gabon.

NMW 809 Skull (plaster cast). Adult male. *Gorilla savagei/dyina* (ind. *gina*).

The above 3 casts were donated by E. Verreaux, Paris, in 1863.

NMW 1779 Skeleton. Adult male. *Gorilla beringei graueri*. Sibatwa, Kivu, Belgian Congo. Collected by Rudolf Grauer in October 1908.

NMW 3083/st668 Dermoplastic and skeleton. Adult female. Ogooué, Gabon. Collected in 1906. Purchased from G. Schneider in 1909.

NMW 3111/st665 Dermoplastic and skeleton. Adult male. *Gorilla beringei graueri*. Sibatwa, Kivu, Belgian Congo. Collected by Rudolf Grauer in March 1910.

NMW 3112/st1649 Dermoplastic and skeleton. Adult female. *Gorilla beringei graueri*. Locality and collection details same as for 3111/st665 above.

NMW 3113 Skull (sagittally sectioned). Adult male.

NMW 3114 Skull. Adult male. It has 3 premolars in both maxillae, and is described and illustrated in Weninger (1948).

NMW 3115 Skull. Adult male.

NMW 3116 Skull. Adult male.

NMW 3117 Skull. Adult male.

NMW 3118 Skull. Adult male.

NMW 3119 Skull. Adult female.

NMW 3120 Skull. Adult female.

The above series of skulls (3113−3120) are described in Gruber-Thalmann (1931). In this paper, no. 3113 is figured (plate VIII) before it had been bisected.

NMW 6090/st669 Dermoplastic and postcranial skeleton. Juvenile male.

The above 9 specimens (3113−6090/st669) were collected by A. Weidholz at Mabili, Likouala-Mossaka, Moyen-Congo, in 1928, and donated by him.

NMW 7136 Skull. Adult male. Cameroun. Collected by Ernst A. Zwilling in 1936 and donated by him.

NMW 7777/st1710 Dermoplastic and skeleton. Juvenile male. He had been captured alive in July 1952, when he was about 2½ years old, by a Ntumu man who sold him to Ernst Zwilling in the village of Njabessang, South-East Cameroun. Zwilling named him 'August' and took him to the Tiergarten Schönbrunn in Vienna, where he died on 6 July 1953.

NMW 15249 Skull. Adult male. Nkom River, Gabon. Collected by Ernst A. Zwilling in October 1950 and sold to Prof. Dr Fritz Kincel in 1952. Received in exchange (for the skull of a female orangutan) from Prof. Dr Kincel on 24 June 1971. Dentition very worn. Damage to the sagittal crest and left supraorbital torus has been repaired by Prof. Dr Kincel, using a mixture of plaster and Russian glue. Prof. Dr Kincel (personal communication) recorded the following measurements: gnathion−inion 305 mm; prosthion−inion 292 mm; condylobasal length 215 mm; glabella−inion 191 mm; bizygomatic width 176.5 mm; prosthion−foramen lacrymale 103.7 mm; rostral width over canines 78.3 mm; rostral width over P3 78.7 mm; orbital width 43 mm; orbital height 39 mm; interorbital width 25.5 mm; cranial capacity 514 cc.

NMW 24942 Skull. Adult female. Bongo, Ogooué-Maritime, Gabon. Killed by local hunters: the skull was purchased from them by A. Radda in 1975.

NMW 28707 Skin and skeleton. Juvenile female. Received in exchange from the Institut für Pathologie, Veterinärmedizinischen Universität Wien, on 12 June 1975.

NMW 32843 Skin and skeleton. Juvenile female. Acquired from the Veterinärmedizinischen Universität Wien on 12 November 1974. It came originally from an animal dealer. Head-body length 500 mm; hind foot length 150 mm; ear length 35 mm. The calotte has been detached.

NMW 41460 Mounted skeleton. Adult male. Near the confluence of the Dja and Lobo Rivers, 3°15′N, 12°25′E, approximately SW of Bengbis, Kamerun. Collected and donated by Oberleutnant von Heigelin (Schutztruppe). Acquired in November 1906.

Grauer II Mandible and partial postcranial skeleton. Adult male. *Gorilla beringei graueri*. Sibatwa, Kivu, Belgian Congo. Collected by Rudolf Grauer in March 1910.

NMW Grauer IV/st666 Dermoplastic and postcranial skeleton. Adult female. *Gorilla beringei graueri*. Locality and collector details same as for Grauer II above.

B 6236 Skin and skeleton. Juvenile female. Imported alive in May 1954. It died in the Tiergarten Schönbrunn on 5 September 1954.

Naturhistorisches Museum Wien, Pathologisch-anatomische Sammlung im Narrenturm, Spitalgasse 2, 1090 Wien
Eduard Winter

MN.24.924/186/2 Mounted skeleton. Adult male. Ntem River, 80−100 km west of Ebebiyín, on the border of Spanish Guinea and Cameroun. Collected by Erwein Graf von Schönborn-Buchheim (1912−98) on 14 January 1939. Acquired from him by Prof. Dr Fritz Kincel (1904−88) on 29 April 1963. Bequeathed by Prof. Dr Kincel (his collection no. 186/2). Dentition very worn. The upper left canine projects forwards, and has been worn through to the pulp by its antagonist. Prof. Dr Kincel recorded the following measurements: gnathion−inion 334 mm; prosthion−inion 311 mm; condylobasal length 223 mm; glabella−inion 206 mm; bizygomatic width 193.7 mm; orbital width 48 mm; interorbital width 34.6 mm; orbital height 47.4 mm; rostral width over canines 80 mm; rostral width over P3 85 mm; facial height gnathion−supraorbital torus 220 mm; cranial capacity 585 cc; standing height calcaneus−sagittal crest 1630 mm; coccyx−sagittal crest 860 mm; ischial tuberosity−sagittal crest 1000 mm; shoulder width between acromia 460 mm; arm length 1050 mm; humerus length 450 mm; radius length 370 mm; ulna length 390 mm; hand length 280 mm; leg length trochanter−calcaneus 750 mm; femur length 400 mm; tibia length 330 mm; fibula length 300 mm; foot length 280 mm; iliac width 430 mm. Received in 1988.

MN.24.924/186/3 Skull. Adult male. Purchased by Prof. Dr Fritz Kincel from the dealing firm Fischer, of Vienna, in 1971. It came from the estate of Stefan Staufer in Vienna, who owned a large collection of primate skulls. Bequeathed by Prof. Dr Kincel (his collection no. 186/3). Dentition shows very moderate wear. Prof. Dr Kincel recorded the following measurements: gnathion−inion 326 mm; prosthion−inion 320 mm; condylobasal length 230 mm; glabella−inion 199.5 mm; bizygomatic width 185.8 mm; interorbital width 35.5 mm; rostral width over canines 77.5 mm; rostral width over P3 72 mm; cranial capacity 565 cc. Received in 1988. [Some additional information above added by GH from correspondence with Prof. Dr Fritz Kincel and Erwein Graf von Schönborn-Buchheim.]

MN.32.497/3/96-456 Brain (serially sectioned in the horizontal plane). The preparations previously formed part of the comparative neuroanatomical collection of histologic serial brain sections belonging to the Universität Wien, Neurologisches Institut (Obersteiner-Institut), Schwarzspanierstrasse 17, 1090 Wien, and may have been made c.1905. The brain was alcohol-fixed and embedded in celloidin (personal communication, Prof. Dr Franz Seitelberger, Neurologisches Institut, Wien). NB: The historic collection of Prof. Obersteiner was transferred to the Pathologisch-anatomisches Bundesmuseum (now the Pathologisch-anatomische Sammlung im Narrenturm) in 1999 (personal communication, Prof. Dr Johannes Hainfellner, Obersteiner Institut, Wien).

MN.32.801 Skin of face. Juvenile. Zaïre. It was probably acquired in the 1970s.

Veterinärmedizinischen Universität Wien, Institut für Anatomie, Veterinärplatz 1, 1210 Wien

Gerhard Forstenpointer; Wolfgang Künzel

Skull. Adult female. Abong-Mbang, Cameroun.

BELGIUM

Brussels

Institut Royal des Sciences Naturelles de Belgique, Rue Vautier 29, 1000 Bruxelles

NB: General Inventory numbers are in parentheses.

869 (4275) Mounted skin. Juvenile. Gabon. Died in captivity. Purchased from the Royal Zoological Society, Antwerp (Antwerp Zoo). Registered on 25 February 1879.

869b (2271) Skull. Adult male. Purchased from Pierre Joseph van Beneden. Registered on 20 December 1859.

869d (4275) Skeleton (incomplete). Died in captivity. Purchased from the Royal Zoological Society, Antwerp. Registered on 25 February 1879.

869e (4275) Skeleton (incomplete). Adult. Other details same as for 869d.

869y (4769) Mounted skeleton. Juvenile. Purchased from Edward Gerrard, London. Registered on 9 December 1881.

869z (4275) Skeleton (incomplete). Fetal. Other details same as for 869d.

870 (4275) Skull. Adult. Other details same as for 869d.

870b (2643) Skull (cast). Adult male. Purchased from Leven. Registered on 29 February 1868.

870d (2643) Skull (cast). Adult female. Other details same as for 870b.

870e (2290) Skull (cast). Adult male. Received from Poelman. Registered on 10 March 1860.

870y (2029) Skull (cast). Adult male. Purchased from Verreaux. Registered on 25 May 1858.

871 Skeleton and mounted skin. Adult male. *Gorilla mayêma*. Lândana, Portuguese Congo. Purchased for 1500 Fr B from Dr Lucan. Registered on 18 December 1882.

1583 (12132) Skull. Male. *Gorilla beringei graueri*. Maniema Forest, Belgian Congo. Collected in 1937. Donated.

1583b (12132) Skull. Male. *Gorilla beringei graueri*. Details same as for 1583.

1583d (12132) Skull; bones of one hand and one leg; skin of both hands and both feet. Male. *Gorilla beringei graueri*. Other details same as for 1583.

1583e (12132) Skull. Female. *Gorilla beringei graueri*. Other details same as for 1583.

1583y (12132) Skull. Male. *Gorilla beringei graueri*. Other details same as for 1583.

1583z (12132) Skull. Juvenile. *Gorilla beringei graueri*. Other details same as for 1583.

4394 (13688) Skull. *Gorilla beringei graueri*. Belgian Congo. Donated by Auriol. Registered on 4 September 1942.

4398 (11605) Cranium. *Gorilla beringei graueri*. Donsale region, Kabare territory, Kivu, Belgian Congo. Donated by Dr Dukeyser. Registered on 20 May 1938.

4414 (14144) Cranium. Male. *Gorilla beringei graueri*. Region of Parc National Albert, Kivu, Belgian Congo. Purchased from Flament. Registered on 7 September 1943.

4906 (14277) Skeleton and mounted skin. Adult male. *Gorilla beringei graueri*. Alimbongo, region of Parc National Albert, Kivu, Belgian Congo, at an approximate altitude of 2250 m. Collected by Commandant E. Hubert (a member of the Serge Frechkop expedition) on 16 May 1938. Donated by the Institut des Parcs Nationaux du Congo Belge. This animal was the leader of a band of about 7 individuals. Measurements: height (top of head to soles of feet) 195 cm; sitting height 121 cm; arm length 98 cm; hand length 25 cm; leg length (approximate) 75 cm; foot length 32.5 cm; chest circumference 168 cm; chest width 74 cm; shoulder width (neck to beginning of arm) 22 cm; neck circumference 74 cm; arm circumference (at biceps) 66 cm; arm span (tip of third finger of one hand to tip of corresponding finger of the other hand) 270 cm.

5485 (14656) Skull. Male. *Gorilla beringei graueri*. Maniema region, Belgian Congo. Donated by V. Karcher. Registered on 2 June 1945.

5486 (14656) Skull. Male. *Gorilla beringei graueri*. Other details same as for 5485.

5487 (14656) Skull. Female. *Gorilla beringei graueri*. Other details same as for 5485.

5488 (14656) Skull. Female. *Gorilla beringei graueri*. Other details same as for 5485.

6869 (15855) Skull and tanned skin. Juvenile male. French Equatorial Africa. Deposited at the Lèopoldville Zoo by Lt Col. P.P.M. Offermann. Registered on 25 November 1947.

6870 (15856) Skull. Juvenile male. *Gorilla beringei graueri*. Received from F. Hendrickx. Registered on 25 November 1947.

7503 (15414) Skeleton. Male. *Gorilla beringei graueri*. Lubutu, east of Kirundu, between Ponthierville and Kindu, Belgian Congo. Collected in 1942. Donated by Maquet, Lèopoldville.

8667 (18002) Skull. Adult male. *Gorilla beringei graueri*. Lubutu, Goma, Lower Belgian Congo. Collected in 1949. Donated by Mme Van Audenard.

9904 (19602) Skeleton. Male. *Gorilla beringei graueri*. Rumangabo, between Goma and Rutshuru, Parc National Albert, Kivu, Belgian Congo. Collected by Hoier. Donated by the Institut des Parcs Nationaux du Congo Belge. Registered in 1946. It is missing one rib and the ungues.

12769 (19605) Skin (tanned). Male. *Gorilla beringei graueri*. Pouvinzargue, Parc National Albert, Kivu, Belgian Congo. Collected by R. Germain in February 1945. It is in bad condition.

13094 (21806) Skeleton and tanned skin. Juvenile female. French Equatorial Africa. Purchased from the Royal Zoological Society of Antwerp. Registered on 13 August 1959. She had arrived at the Antwerp Zoo from Marseilles, France, on 16 June 1958, aged about 1 year, and was named 'Lea'. She died of general cattharal enteritis and gastritis on 8 August 1959.

15043 (17833) 17 maxillae. *Gorilla beringei graueri*.

15044 (17833) 13 mandibles. *Gorilla beringei graueri*.

15045 (17833) 45 scapulae. *Gorilla beringei graueri*.

15046 (17833) 41 humeri. *Gorilla beringei graueri*.

15047 (17833) 38 ulnae. *Gorilla beringei graueri*.

15048 (17833) 34 radii. *Gorilla beringei graueri*.

15049 (17833) 48 femora. *Gorilla beringei graueri*.

15050 (17833) 38 tibiae. *Gorilla beringei graueri*.

15051 (17833) 32 fibulae. *Gorilla beringei graueri*.

15052 (17833) 33 clavicles. *Gorilla beringei graueri*.

15053 (17833) 401 ribs. *Gorilla beringei graueri*.

15054 (17833) 435 vertebrae. *Gorilla beringei graueri*.

15055 (17833) 22 sacra or coccyges. *Gorilla beringei graueri*.

15056 (17833) Series of sternal plates. *Gorilla beringei graueri*.

15057 (17833) 3 hands and 1 foot (incomplete). *Gorilla beringei graueri*.

15058 (17833) Series of phalanges. *Gorilla beringei graueri*.

15059 (17833) Series of carpal and tarsal bones. *Gorilla beringei graueri*.

17750 (17833) 50 ilia. *Gorilla beringei graueri*.

The above 18 specimens, 15043–17750, came from a total of 32 gorillas collected by a capture group in Angumu, Bafwasende territory, Belgian Congo, February 1948–January 1949. Donated.

19929 (24617) Skeleton and tanned skin. Female. *Gorilla beringei graueri*. Busiangwa, north section of Parc National Albert, Kivu, Belgian Congo. Collected by

the G.F. de Witte expedition on 27 March 1954. Donated by the Institut des Parcs Nationaux du Congo Belge.

19930 (24617) Skeleton and tanned skin. Male. *Gorilla beringei graueri*. Angumu, Bafwasende territory, Belgian Congo. Collected by the G.F. de Witte expedition on 12 April 1957. Donated by the Institut des Parcs Nationaux du Congo Belge.

39812 (31387) Skull. Female. *Gorilla beringei graueri*. Lubutu, Kalima, Belgian Congo, 2°34'S, 26°37'E.

39813 (31387) Skull. Female. *Gorilla beringei graueri*. Lubutu, Kalima, Belgian Congo, 2°34'S, 26°37'E.

39814 (31387) Skull. Male. *Gorilla beringei graueri*. Lubutu, Kalima, Belgian Congo, 2°34'S, 26°37'E.

39815 (31387) Skull. Male. *Gorilla beringei graueri*. Lubutu, Kalima, Belgian Congo, 2°34'S, 26°37'E.

39816 (31387) Skull. Male. *Gorilla beringei graueri*. Lubutu, Kalima, Belgian Congo, 2°34'S, 26°37'E.

39817 (31387) Skull. Male. *Gorilla beringei graueri*. Lubutu, Kalima, Belgian Congo, 2°34'S, 26°37'E.

39818 (31387) Skull. Female. *Gorilla beringei graueri*. Lubutu, Kalima, Belgian Congo, 2°34'S, 26°37'E.

39819 (31387) Skull. Male. *Gorilla beringei graueri*. Lubutu, Kalima, Belgian Congo, 2°34'S, 26°37'E.

39820 (31387) Skull. Female. *Gorilla beringei graueri*. Lubutu, Kalima, Belgian Congo, 2°34'S, 26°37'E.

39821 (31387) Skull. Male. *Gorilla beringei graueri*. Lubutu, Kalima, Belgian Congo, 2°34'S, 26°37'E.

39822 (31387) Skull. Female. *Gorilla beringei graueri*. Lubutu, Kalima, Belgian Congo, 2°34'S, 26°37'E.

The above 11 specimens, 39812–39822 inclusive, were collected by Villageois during 1975–77.

39152 (30781) Skull. *Gorilla beringei graueri*. Kivu, Mt Kahuzi, west of Lake Kivu, Belgian Congo. Collected by G. Parent in 1959.

Université Libre de Bruxelles, Musée de Zoologie Auguste Lameere, Faculté des Sciences, Avenue F.-D. Roosevelt 50, 1050 Bruxelles*

G. Lenglet; Robert Peuchot

1283 Mounted skeleton. Male. Purchased for 600 Fr B from the Royal Zoological Society of Antwerp in January 1897. Height of skeleton 123 cm; upper canine length 3 cm. The upper and lower 3rd molars are unerupted. The sagittal and nuchal crests are absent.

2042 Skull. Male. Salo, a post on the Sangha River, French Equatorial Africa. Apparently donated by Dr Georges Marchal (surgeon, gynaecologist and obstetrician), Rue Elise 69, Bruxelles, probably c.1925. The animal was 158 cm tall and weighed 105 kg.

2043 Skull. Female. Details as for 2042. The animal was 147 cm tall and weighed 68 kg.

2044 Mounted skin. Male. Believed to be from the same animal as the skeleton 1283.

Ghent

Universiteit Gent, Zoölogisch Museum, K.L. Ledeganckstraat 35, 9000 Gent

Dominique Adriaens; Dominick Verschelde

UGMD 50010 Mounted skeleton. Male. *Gorilla gorilla manyema*. Majumba (=Mayumba), French Congo. Collected in May 1885. Purchased for 1040 Fr B from Tramond, Paris, in 1890. The original skull was stolen in 1994; the skull used to complete this skeleton is UGMD 55250.

UGMD 55246 Skull (plaster cast). Female. Purchased from Verreaux in 1859. Canine broken off/missing from maxilla; left I1 and right I2 and I2 broken off/missing from mandible.

UGMD 55247 Skull (plaster cast). Male. Upper incisors broken off/missing; upper right canine missing; lower right canine broken off/missing. Received in exchange from the Musée Royale d'Histoire Naturelle, Brussels, Belgium.

UGMD 55248 Skull (plaster cast). Male. Purchased in March 1859. The mandible is broken and repaired.

UGMD 55250 Skull. Male. It was placed on the postcranial skeleton UGMD 50010 to replace the original skull, which was stolen in 1994. The lower right incisor is split.

Schoolmuseum Michel Thiery, Sint-Pietersplein 14, 9000 Gent*

A. Van Gijseghem

Mounted skin. Female. Purchased from A. Schlüter KG, Winnenden, Germany, in 1981. Schlüter had acquired it from Jac Bouten, Venlo, Netherlands, in 1975. It was a zoo specimen, 2 years old. The skeleton of the same animal was supplied to the Karlsruhe Natural History Museum, Germany, in 1977.

Leuven

Katholieke Universiteit Leuven, Departement Biologie, Verdeling van Ecologie, Evolutie en Biodiversiteit, Laboratorium voor Biodiversiteit en Evolutionaire Genomica, Charles Deberiostraat 32, 3000 Leuven

Willem Van Neer; Jos Snoeks

ZLV 122 Skull. Adult male. *Gorilla beringei graueri*. Kivu region, Belgian Congo. Collected by Prof. Vandebroek c.1955. It is missing the upper right lateral and central incisors, lower right lateral and central incisors, lower right 3rd molar and lower left lateral incisor.

Louvain-La-Neuve

Université Catholique de Louvain, Faculté des Sciences et l'Institut des Sciences de la Vie, Groupe de Recherche de Biochimie Nutritionnelle, Croix-du-Sud 5, 1348 Louvain-la-Neuve

Elisa de Jacquier; Jean-François Rees

ZLV 124 Cranium. Adult male. *Gorilla beringei graueri*. Kima, Lubutu, Belgian Congo. From the old collection of the U.C.L.; its original number was 1709.

ZLV 128 Skull. Adult female. From the old collection of the U.C.L.

ZLV 133 Skull. Adult female. *Gorilla beringei graueri*. Utu, Belgian Congo. Collected by Roberti (Vandebroek expedition) in 1955.

ZLV 138 Skull. Adult male. *Gorilla beringei graueri*. Utu, Belgian Congo. Collected by the Vandebroek expedition in 1955. Collector's number 52. The animal weighed 162 kg.

ZLV 139 Skull. Adult female. *Gorilla beringei graueri*. Mwenga, Belgian Congo. Collection same as for ZLV 138; collector's number 47. The animal weighed 80 kg.

ZLV 140 Skull. Female. *Gorilla beringei graueri*. Details same as for ZLV 139; collector's number 48. The animal weighed 13.4 kg. Deciduous teeth present, M1 erupting.

ZLV 141 Skull and some bones. Adult male. *Gorilla beringei graueri*. Details same as for ZLV 139; collector's number 45.

ZLV 147 Skull. Adult male. Nola, Haute Sangha, Congo-Brazzaville.

ZLV 152 Skull. Female. *Gorilla beringei graueri*. Collection details same as for ZLV 139. Collector's number 46. The animal weighed 30 kg. Deciduous teeth present, M1 erupting.

ZLV 153 Skull. Adult male. *Gorilla beringei graueri*. Utu, Belgian Congo. Collection details same as for ZLV 138; collector's number 55. The animal weighed 124.5 kg.

Tervuren

Koninklijk Museum voor Midden-Afrika, Leuvensesteenweg 13, 3080 Tervuren

Emmanuel Gilissen; Wim Wendelen

RMCA 63 Skeleton. Adult male. Mayombe forest, Gabon. Received in 1897.

RMCA 100 Cranium. Adult male (M3 erupted). *Gorilla uellensis* (holotype). Djabbir, Haut-Uele.

RMCA 101 Cranium. Subadult male (M2 erupted). *Gorilla uellensis* (cotype). Djabbir.

RMCA 102 Skull. Adult female? (M3 erupted). *Gorilla uellensis* (cotype). Mobele, Itimbiri.

RMCA 103 Mandible. Adult male (M3 erupted). *Gorilla uellensis* (cotype). Uele, Itimbiri. It probably belongs to RMCA 100.

The above 4 specimens, RMCA 100−RMCA 103, were collected by Le Marinel and were received in September 1898.

RMCA 804 Skull and mounted skin. Adult male (M3 erupted). *Gorilla beringei graueri*.

RMCA 805 Mounted skin. Juvenile male. *Gorilla beringei graueri*.

RMCA 806 Skeleton and mounted skin. Adult female. *Gorilla beringei graueri*.

RMCA 807 Skeleton and mounted skin. Adult female (M3 erupted). *Gorilla beringei graueri*.

RMCA 808 Skeleton and mounted skin. Juvenile female (M1 erupted). *Gorilla beringei graueri*.

RMCA 809 Skeleton and mounted skin. Juvenile female (M1 erupted). *Gorilla beringei graueri*.

RMCA 810 Skeleton and mounted skin. Juvenile female (M1 erupted). *Gorilla beringei graueri*.

RMCA 811 Skeleton and mounted skin. Juvenile female (M2 erupted). *Gorilla beringei graueri*.

RMCA 812 Skeleton and mounted skin. Adult female (M3 erupted). *Gorilla beringei graueri*.

RMCA 813 Skeleton and mounted skin. Adult female (M3 erupted). *Gorilla beringei graueri*.

The above 10 specimens, RMCA 804−RMCA 813, were collected by Pauwels at Baraka, Sibatwa forest (altitude 2100 m), Belgian Congo, and were received on 10 February 1910.

RMCA 833 Skeleton. Adult female (M3 erupted). *Gorilla beringei graueri*. West of Baraka, Sibatwa forest (altitude 2100 m), Belgian Congo. Collected by Pauwels. Received on 15 May 1910.

RMCA 834 Skeleton and skin. Female (M2 erupted). *Gorilla beringei graueri*. West of Baraka, Sibatwa forest (altitude 2100 m), Belgian Congo. Collected by Pauwels. Received on 15 May 1910.

RMCA 881 Skeleton and skin. Female (M2 erupted). *Gorilla beringei graueri*. West of Baraka, Sibatwa forest, Belgian Congo. Collected by Pauwels. Received on 25 July 1910.

RMCA 994 Skeleton and skin. Adult male (M3 erupted). *Gorilla beringei graueri*.

RMCA 995 Skeleton and skin. Adult female (M3 erupted). *Gorilla beringei graueri*.

RMCA 996 Skeleton and skin. Juvenile female (M1 erupted). *Gorilla beringei graueri*.

RMCA 997 Skeleton and skin. Juvenile male (M1 erupted). *Gorilla beringei graueri*.

RMCA 998 Skeleton and skin. Adult male (M3 erupted). *Gorilla beringei graueri*.

RMCA 999 Skeleton and skin. Adult male (M3 erupted). *Gorilla beringei graueri*.

RMCA 1000 Skeleton and skin. Adult male (M3 erupted). *Gorilla beringei graueri*. Collected on 5 January 1912.

RMCA 1001 Skeleton and skin. Adult male (M3 erupted). *Gorilla beringei graueri*.

RMCA 1002 Skeleton and skin. Juvenile male (M2 erupted). *Gorilla beringei graueri*. Collected on 5 January 1912.

The above 9 specimens, RMCA 994−RMCA 1002, were collected west of Baraka, Sibatwa forest, by De l'Épine, and were received on 2 February 1912.

RMCA 1099 Cranium. Adult (M3 erupted). N'Counda, right bank of the lower Congo, above the mouth of the Alima River. Collected by Dusseljé. Received on 11 July 1912.

RMCA 1100 Cranium. Adult (M3 erupted). Details same as for RMCA 1099.

RMCA 1223 Cranium. Adult (M3 erupted). Ganda Sundi. Collected by C. Claessens. Received on 25 November 1912.

RMCA 1432 Cranium. Adult (M3 erupted).

RMCA 1433 Skull. Juvenile (PD4 erupted).

RMCA 1434 Cranium. Adult (M3 erupted).

RMCA 1435 Cranium. Adult (M3 erupted).

RMCA 1436 Cranium. (M3 erupting).

RMCA 1437 Cranium. Adult (M3 erupted).

RMCA 1438 Cranium. (M3 erupting).

RMCA 1439 Cranium. (M3 erupting).

RMCA 1440 Cranium. (M2 erupted).

RMCA 1441 Cranium. Adult (M3 erupted).

RMCA 1442 Cranium. Adult (M3 erupted).

The above 11 specimens, RMCA 1432−RMCA 1442, were collected by Dusseljé, Ngoko, Alima River, Congo (French Equatorial Africa), and were received on 24 February 1913.

RMCA 2253 Skull (cast of type). Adult (M3 erupted). *Gorilla gorilla matschiei*. Southern Cameroons. Received on 2 October 1913.

RMCA 2254 Cranium (cast of type). Adult (M3 erupted). *Gorilla gorilla diehli*. Region of the Mun-Aya (tributary of the Cross River), near Dakbe, Nigeria. Received on 2 October 1913.

RMCA 2255 Skull (cast of type). Adult (M3 erupted). *Gorilla gorilla jakobi*. Lobo mouth station, junction of the Dja and the Njong, Cameroun. Received on 2 October 1913.

RMCA 2257 Skeleton and skin. Adult male (M3 erupted). *Gorilla beringei beringei*.

RMCA 2258 Skeleton and skin. Adult female (M3 erupted). *Gorilla beringei beringei*.

RMCA 2259 Skeleton and skin. Juvenile (PD4 erupting). *Gorilla beringei beringei*.

RMCA 2260 Skeleton. Adult male (M3 erupted). *Gorilla beringei beringei.*

RMCA 2261 Skeleton and skin. Juvenile female (M2 erupting). *Gorilla beringei beringei.*

RMCA 2262 Skeleton and skin. Juvenile female (M1 erupted). *Gorilla beringei beringei.*

RMCA 2263 Skeleton and skin. Adult female (M3 erupted). *Gorilla beringei beringei.*

The above 7 specimens, RMCA 2257−RMCA 2263, were collected in the Virunga Volcanoes, northwest of Mt Mikeno (altitude 2300−2400 m), by Pauwels, and were received on 13 October 1913.

RMCA 5484 Skull and skin. Juvenile (M1 erupted). *Gorilla beringei beringei.* Mt Mikeno, Virunga Volcanoes. Collected by Thomas Alexander Barns. Received on 14 March 1922.

RMCA 5485 Skull and skin. Juvenile (M1 erupted). *Gorilla beringei beringei.* Virunga Volcanoes. Collected by Thomas Alexander Barns. Received on 14 March 1922.

RMCA 5585 Skull. Adult male (M3 erupted). Ganga, road from Yaunde to Dume, Kamerun. Collected by A. Bal in 1914. Received via Houssiaux in 1922.

RMCA 6142 Skeleton and skin. Juvenile male (M2 erupted). *Gorilla gorilla rex-pygmaeorum* (paratype). North of Kilimamensa, Ituri Forest. Collected by A. Pilette. Received on 21 June 1923.

RMCA 7059 Cranium. (M1 erupting). *Gorilla beringei graueri.* South of Lake Edward, Kivu. Collected by Flamand. Received on 10 January 1924.

RMCA 8187 Skeleton. Adult male (M3 erupted). *Gorilla gorilla rex-pygmaeorum* (holotype). Luofu, altitude 2000 m., Lubero territory, Belgian Congo. Collected by J. Buxant. Donated by the Cercle Zoologique Congolais. Received on 19 September 1925.

RMCA 8607 Skeleton. Adult female (M3 erupted). *Gorilla beringei beringei.* Mount Karisimbi, Virunga Volcanoes. Collected by Burbridge. Received in 1926.

RMCA 8608 Skeleton. Adult female (M3 erupted). *Gorilla beringei beringei.* Details same as for RMCA 8607.

RMCA 9220 Skull. Adult male (M3 erupted). *Gorilla gorilla rex-pygmaeorum.* Upper Lowa River, north of Walikale, Manyema, Belgian Congo. Collected by M. Dorsinfang. Received in 1927.

RMCA 9291 Skeleton. Adult male (M3 erupted). Upper Sangha River, Congo. Collected by Carlier. Donated by Houssiaux. Received on 30 September 1927.

RMCA 9405 Skeleton (on display). Adult male. *Gorilla beringei graueri.* West of Tshibinda, near Katana, West Kivu. Collected by Henrard, Commandant of the Kivu district. Received on 27 March 1928.

RMCA 9406 Skull. Adult (M3 erupted). Ogooué, Gabon. Collected by Blaehe. Received on 2 April 1928. In fragments, no label.

RMCA 9424 Skull. Adult male (M3 erupted). *Gorilla beringei graueri.* Shabunda, Utega territory, upper Ulindi River, Manyema, Belgian Congo. Collected by Schoumaker. Received on 16 May 1928.

RMCA 9564 Cranium. Adult male (M3 erupted). *Gorilla beringei graueri.* Shabunda forest. Collected by Douce. Received on 2 October 1928.

RMCA 9581 Skeleton. Adult male (M3 erupted). *Gorilla beringei graueri.* Kilimamensa, Ituri, Belgian Congo. Collected by Flamand. Received on 1 March 1929. No mandible.

RMCA 10051 Skeleton. Adult male (M3 erupted). *Gorilla beringei graueri.* Collected 10 km from Pinga, between Peuty and Penekendi, Belgian Congo, by Moulaert. Received on 28 January 1930.

RMCA 10151 Skeleton. Juvenile (M1 erupted). *Gorilla beringei graueri.* Tshibinda forest, west of Lake Kivu, Belgian Congo. Collected by Pescatore. Received on 10 February 1930.

RMCA 11724 Skull. Juvenile male (M1 erupted) *Gorilla beringei graueri.*

RMCA 11725 Skull. Adult female (M3 erupted). *Gorilla beringei graueri.*

RMCA 11726 Skull. Juvenile male (M1 erupted). *Gorilla beringei graueri.*

The above 3 specimens, RMCA 11724−RMCA 11726, were collected in the bamboo forest 25 km south of Lubero, Belgian Congo, by Moriamé (Lady Buxton expedition). They were received on 24 June 1932.

RMCA 12282 Skull. Juvenile (PD4 erupting). *Gorilla beringei graueri.*

RMCA 12283 Skull. Juvenile (PD4 erupted). *Gorilla beringei graueri.*

RMCA 12284 Skull. Juvenile (PD4 erupted). *Gorilla beringei graueri.*

RMCA 12285 Skull. Juvenile (M1 erupting). *Gorilla beringei graueri.*

The above 4 specimens, RMCA 12282−RMCA 12285, were collected in Babembi territory by the commandant of Costermansville province. They were received on 25 May 1934.

RMCA 12675 Skeleton. Juvenile (PD4 erupted). *Gorilla beringei graueri.* Lubero region, Belgian Congo. Collected by Van Saceghem (it died in a laboratory in Kisenyi). Received on 15 February 1935.

RMCA 13243 Cranium. Adult male (M3 erupted). *Gorilla beringei graueri.* Niolo, 15 km from Tshela, Belgian Congo. Collected by Nauwelaert. Donated by Tordeur. Received on 10 March 1936.

RMCA 14304 Skull. Adult male (M3 erupted). *Gorilla beringei graueri*. Elila river, NNE Fizi territory, Belgian Congo. Collected by Spitaels. Received on 21 February 1938.

RMCA 14305 Skull. Juvenile (PD4 erupted). *Gorilla beringei graueri*. Details same as for RMCA 14304.

RMCA 14577 Skull and fragments of skeleton. *Gorilla beringei graueri*. Forest to the west of Matale. Collected by D. Hautmann. It died in October 1937. Received on 18 May 1938.

RMCA 14615 Skull. Adult female (M3 erupted). *Gorilla beringei graueri*. Lubero, Belgian Congo. Collected by the territorial administrator. Received on 27 June 1938.

RMCA 14616 Skull. Juvenile (M1 erupting). *Gorilla beringei graueri*. Details same as for RMCA 14615.

RMCA 14768 Skull. Adult female (M2 erupted). *Gorilla beringei graueri*.

RMCA 14769 Skull. Adult female (M3 erupted). *Gorilla beringei graueri*.

RMCA 14770 Skull. Adult female (M2 erupted). *Gorilla beringei graueri*.

The above 3 specimens, RMCA 14768—RMCA 14770, were collected in Shabunda region, Belgian Congo, by A. Braun. They were received on 26 August 1938.

RMCA 14912 Skull. Adult female (M3 erupted). *Gorilla beringei graueri*. Elila river, NNE Fizi territory, Belgian Congo. Collected by Spitaels. Received on 21 November 1938.

RMCA 15232 Cranium. Adult (M3 erupted). Loango Mission, Niari-Kwilu region, Congo (French Equatorial Africa). Collected by E. Dartevelle. Received on 4 March 1939.

RMCA 15234 Skull. Adult male (M3 erupted). *Gorilla beringei graueri*. Warega chiefdom, Lubongola region, Belgian Congo. Collected by D. Hautmann. Received on 6 March 1939.

RMCA 15236 Cranium (M1 erupted). *Gorilla beringei graueri*. Bukavu region, Belgian Congo. Collected by D. Hautmann. Received on 6 March 1939.

RMCA 15237 Mandible. Adult (M3 erupted). *Gorilla beringei graueri*. Details same as for RMCA 15236.

RMCA 15290 Skeleton and brain (in alcohol). Adult female (M3 erupted). *Gorilla beringei beringei*. Nbayo, Mt Kahuzi. Collected by J. Vandelannoite. Received on 28 April 1939.

RMCA 15291 Skeleton and skin. Juvenile male (M2 erupted). *Gorilla beringei graueri*. Collected by Duncan M. Hodgson (McGill University expedition) in November 1938, 30 km south of Lubero, Costermansville territory, Belgian Congo. Donated by Hodgson: received on 3 May 1939.

RMCA 15351 Skeleton. Juvenile (PD4 erupted). *Gorilla beringei graueri*. Kitunda, Shabunda territory, Belgian Congo. Collected by A. Braun. Received on 13 June 1939.

RMCA 15352 Skull. Adult female (M3 erupted). *Gorilla beringei graueri*.

RMCA 15353 Skull. Adult female (M3 erupted). *Gorilla beringei graueri*.

RMCA 15354 Skull. Adult female (M3 erupted). *Gorilla beringei graueri*.

RMCA 15355 Skull. Adult female (M3 erupted). *Gorilla beringei graueri*.

RMCA 15356 Skull. Adult female (M3 erupted). *Gorilla beringei graueri*.

RMCA 15357 Skull. Adult female (M3 erupted). *Gorilla beringei graueri*.

RMCA 15358 Skull. Juvenile (M1 erupting). *Gorilla beringei graueri*.

RMCA 15359 Cranium. Juvenile (PD4 erupted). *Gorilla beringei graueri*.

RMCA 15360 Cranium. Juvenile (M1 erupting). *Gorilla beringei graueri*.

RMCA 15361 Cranium. Juvenile (PD4 erupted). *Gorilla beringei graueri*.

RMCA 15362 Cranium. Adult female (M2 erupted). *Gorilla beringei graueri*.

RMCA 15363 Cranium. Adult female (M3 erupted). *Gorilla beringei graueri*.

RMCA 15364 Cranium. Juvenile male (PD4 erupted). *Gorilla beringei graueri*.

RMCA 15365 Cranium. Adult female (M2 erupted). *Gorilla beringei graueri*.

RMCA 15366 Cranium. Juvenile (PD4 erupted). *Gorilla beringei graueri*.

RMCA 15367 Cranium. Juvenile (PD4 erupted). *Gorilla beringei graueri*.

RMCA 15368 Skeleton (of at least 3 specimens from the series RMCA 15352—RMCA 15367). *Gorilla beringei graueri*.

The locality for the above 17 specimens, RMCA 15352—RMCA 15368, is Ibatsero. They were collected by the Administrator of Lubero territory on 20 December 1938. Received on 27 June 1939.

RMCA 15573 Cranium. Adult (M3 erupted). *Gorilla beringei graueri*. Bibugwa, between Matale and Lubongola, Shabunda territory. Collected by D. Hautmann. Received on 8 August 1939.

RMCA 15574 Skull. Adult male (M3 erupted). *Gorilla beringei graueri*. Details same as for RMCA 15573.

RMCA 15580 Skull. Juvenile (PD4 erupted). *Gorilla beringei graueri*. Tshibinda, Kahuzi, Belgian Congo. Collected by D. Hautmann in December 1938. Received on 8 August 1939.

RMCA 15648 Cranium. Adult (M3 erupted). *Gorilla beringei graueri*. Itombwe region, Fizi territory. Collected by D. Hautmann. Received on 5 October 1939.

RMCA 15724 Skinned carcase (in alcohol). Juvenile. *Gorilla beringei graueri*. Mbayo, Kahuzi, Belgian Congo. Collected by Guyaux; captured on 9 May 1939. Received on 6 November 1939.

RMCA 15856 Skull. Juvenile (M1 erupted). *Gorilla beringei graueri*. Kabare, southwest of Lake Kivu, Belgian Congo. Collected by J. Schwetz between 10 August and 5 September 1939. Received on 18 January 1940.

RMCA 16400 Brain (in alcohol). Received from Zoo Antwerpen on 21 October 1925.

RMCA 16401 Brain (in alcohol). Juvenile female. *Gorilla beringei graueri*. Kivu, Belgian Congo. Received from Zoo Antwerpen.

RMCA 16408 Viscera (in alcohol). From the same specimen as RMCA 16401.

RMCA 16409 Viscera (in alcohol). Received from Zoo Antwerpen on 21 October 1925. From the same specimen as RMCA 16400.

RMCA 16433 Carcase (in alcohol). Received from Zoo Antwerpen on 21 October 1925. From the same specimen as RMCA 16400. Skull damaged.

RMCA 17136 Brain (in alcohol). Received from the Administrator of Lubero territory, Belgian Congo, on 8 March 1946.

RMCA 17161 Skull. Adult male (M3 erupted). *Gorilla beringei beringei*. Virunga Volcanoes. Collected by Guyaux. Received on 23 July 1946.

RMCA 17202 Skeleton. Adult female (M3 erupted). Gabon, French Equatorial Africa. Captured in the wild, died in Zoo Léopoldville in July 1941. Sent to the museum by M. Caroll, Société de Botanique et de Zoologie Congolaise. Received on 23 July 1946.

RMCA 17663 Skin. Collected by Guyaux. Received on 23 July 1946.

RMCA 17770 Skull. Adult male (M3 erupted). *Gorilla beringei graueri*. Lubutu, Belgian Congo. Collected by J. Maquet. Received on 23 July 1946.

RMCA 17785 Skeleton. Adult male (M3 erupted). *Gorilla beringei graueri*. Received from Lepersonne, Léopoldville, Belgian Congo, on 29 October 1947.

RMCA 18191 Skull. Adult male (M3 erupted). *Gorilla beringei graueri*. Wazimu chefferie, Shabunda territory, Belgian Congo. Collected by Guyaux in 1947. Received on 29 October 1947.

RMCA 18739 Skeleton and skin. Adult male (M3 erupted). *Gorilla beringei graueri*. Lubero territory, Belgian Congo. Collected by Bormans in 1948. Received on 18 May 1949.

RMCA 20085 Skeleton. Adult male (M3 erupted). *Gorilla beringei graueri*. Butembo, Belgian Congo. Collected by A. Prigogine in 1950. Received on 16 October 1950.

RMCA 20086 Postcranial skeleton. Adult male. *Gorilla beringei graueri*. Details same as for RMCA 20085.

RMCA 21532 Skull. Adult male (M3 erupted). *Gorilla beringei graueri*.

RMCA 21533 Skull. Adult female (M3 erupted). *Gorilla beringei graueri*.

RMCA 21534 Skull. Adult female (M3 erupted). *Gorilla beringei graueri*.

RMCA 21535 Skull. Adult female (M3 erupted). *Gorilla beringei graueri*. It is incomplete.

RMCA 21536 Postcranial skeleton (bones of specimens from the series RMCA 21352−RMCA 21535). *Gorilla beringei graueri*.

The above 5 specimens, RMCA 21532−RMCA 21536, were collected by Guyaux near Lubero, Northeast Kivu, Belgian Congo, in 1950. They were received on 13 March 1950.

RMCA 21537 Skull. Adult female (M3 erupted). *Gorilla beringei graueri*. Near Lubero, Northeast Kivu, Belgian Congo. Its original number was RMCA 21532; it was given its current number at the end of 1952.

RMCA 21538 Cranium. Adult male (M3 erupted). *Gorilla beringei graueri*. Near Lubero, Northeast Kivu, Belgian Congo.

RMCA 22761 Skeleton. Adult male (M3 erupted). *Gorilla beringei graueri*. Chibimbi (Katana), Kabare territory, north of Mt Kahuzi, Mitumba mountains, Belgian Congo. Collected by the Colonial Agriculture Service, Bukavu, on 28 August 1954. Received on 3 January 1955.

RMCA 22762 Skull. Adult male (M3 erupted). *Gorilla beringei graueri*. Shabunda territory, Belgian Congo. Collected by the Colonial Agriculture Service, Bukavu, in 1954. Received on 3 January 1955.

RMCA 22763 Skull. Adult male (M3 erupted). *Gorilla beringei graueri*. Shabunda territory. Collection details and date of receipt same as for RMCA 22762.

RMCA 22903 Skeleton. Adult male (M3 erupted). *Gorilla beringei graueri*.

RMCA 22904 Skull. Adult female (M3 erupted). *Gorilla beringei graueri*.

RMCA 22905 Skeleton. Male (M3 erupted). *Gorilla beringei graueri*. The skull is incomplete.

The above 3 specimens, RMCA 22903−RMCA 22905, were collected in Lubero, Belgian Congo, by the Section Chasse et Pêche, Léopoldville. They were received on 25 March 1955.

RMCA 22924 Skeleton. Adult male (M3 erupted). *Gorilla beringei graueri*. Mwabi, Mwenga territory, Belgian Congo. Collected by J. Roberti (Colonial Agriculture Service). Received on 19 April 1955.

RMCA 22932 Postcranial skeleton. Male. Maduda region, Mayumbe. Collected by Hacquart. Received on 3 June 1955.

RMCA 23136 Skeleton. Adult female (M3 erupted). Tshela region. Collected by the Colonial Agriculture Society, Mayumbe. Received on 4 November 1955.

RMCA 23138 Postcranial skeleton (not from one animal: mix of *Pan* and *Gorilla*). Details same as for RMCA 23136.

RMCA 23958 Skull. Adult male (M3 erupted). *Gorilla beringei graueri*. Mt Biéga, Kabare territory, Belgian Congo. Collected by the Agriculture and Forests Service, Bukavu. Received on 15 October 1956.

RMCA 24001 Skull. Adult male (M3 erupted). *Gorilla beringei graueri*. Kima, Kabare territory, Belgian Congo. Collected by Charles Mauchant in 1931. Weight 225 kg. Received on 14 December 1956.

RMCA 26206 Skeleton. Adult male (M3 erupted). *Gorilla beringei graueri*. Bukavu, Belgian Congo. Collected by the Agriculture and Forests Service. Received on 3 May 1957.

RMCA 27108 Skull. Adult male (M3 erupted). *Gorilla beringei graueri*. Saulia, north of Punia, Belgian Congo. Collected by A. Prigogine. Received on 22 April 1958.

RMCA 27222 Postcranial skeleton. Adult male. *Gorilla beringei graueri*. Belgian Congo. Collected by the Agriculture Service, Bukavu.

RMCA 27495 Skull. Adult male (M3 erupted). *Gorilla beringei graueri*. Between Shabunda and Walikale, Belgian Congo. Collected by F. Wilmet. Received on 18 November 1958.

RMCA 27669 Carcase (2 arms, 2 legs, 1 head) and viscera (in alcohol, 2 jars). Female. Mayumbe region. Collected by the Colonial Agriculture Society, Mayumbe. Received on 12 December 1958.

RMCA 27755 Skeleton. Adult female (M3 erupted). *Gorilla beringei graueri*. Twenty-two kilometres from the Itebero−Utu road, Walikale territory, Belgian Congo. Collected by Charles Cordier (Provincial Agriculture Service, Bukavu), 17−18 September 1958. Received on 31 January 1959.

RMCA 27756 Skeleton. Adult female (M3 erupted). *Gorilla beringei graueri*. All details same as for RMCA 27755.

RMCA 27759 Skull. Adult male (M3 erupted). *Gorilla beringei graueri*. Road Katshungu−Cobelmin, 48 km, 6 km northeast of Lulingu, Kato, Bamuguba N. chiefdom, Shabunda territory, Belgian Congo. Collected by R. Gheysen on 7 September 1958. Received on 6 February 1959.

RMCA 27839 Skeleton and skin. Adult female (M3 erupted). *Gorilla beringei graueri*. Chamaka, road Itebero−Kasese, Belgian Congo. Collected by Charles Cordier (Provincial Agriculture Service, Bukavu) on 18 November 1958. Received on 25 April 1959.

RMCA 27840 Skeleton and skin. Adult female (M3 erupted). *Gorilla beringei graueri*. All details same as for RMCA 27839. Weight 62 kg.

RMCA 27841 Skull. Adult male (M3 erupted). *Gorilla beringei graueri*. Right bank of Ulindi River, Kamituga region, Belgian Congo. Collected by J. Feltes. Received on 21 May 1959.

RMCA 28511 Carcase (in alcohol). Juvenile male. Received from the Royal Zoological Society, Antwerp (Zoo Antwerpen). Received on 8 April 1960.

RMCA 29098 Skull. Adult male. *Gorilla beringei graueri*. Utu, Belgian Congo. Collected by G. Vandebroek. Received on 28 November 1960. Specimen not located; it may be in the mounted skin RMCA b2.018-M-0297.

RMCA 29099 Cranium. Adult male (M3 erupted). *Gorilla beringei graueri*. Utu, Belgian Congo. Collected by G. Vandebroek. Received on 28 November 1960.

RMCA 29100 Skull. Adult male (M3 erupted). *Gorilla beringei graueri*. Details same as for RMCA 29099.

RMCA 29101 Skull. Adult male (M3 erupted). *Gorilla beringei graueri*. Details same as for RMCA 29099.

RMCA 29102 Cranium. Adult female (M3 erupted). *Gorilla beringei graueri*. Details same as for RMCA 29099.

RMCA 29103 Cranium. Adult male (M3 erupted). *Gorilla beringei graueri*. Kima, Lubutu territory, Belgian Congo. Collected by Vandebroek. Received on 28 November 1960.

RMCA 29104 Cranium. Adult female (M3 erupted). *Gorilla beringei graueri*. Details same as for RMCA 29103.

RMCA 29317 Skull. Adult male (M3 erupted). *Gorilla beringei graueri*. Lulingu, Shabunda territory. Collected by H. De Maere in June 1960. Received on 9 February 1961.

RMCA 29318 Skull. Juvenile female (M2 erupting). *Gorilla beringei graueri*. Details same as for RMCA 29317.

RMCA 29538 Skull. Adult female (M3 erupted). *Gorilla beringei graueri*. Kivu Province, Belgian Congo. Collected by J. van Mol. Received on 29 June 1961.

RMCA 30659 Carcase (in alcohol). Juvenile. Received from the Royal Zoological Society, Antwerp (Zoo Antwerpen), on 29 June 1961.

RMCA 31131 Skull. Adult male (M3 erupted). *Gorilla beringei graueri*. Lulingu. Collected by A. Prigogine in January 1962. Received on 29 May 1962.

RMCA 31132 Skull. Adult female (M3 erupted). *Gorilla beringei graueri*. Lulingu. Collected by A. Prigogine in February 1962. Received on 29 May 1962.

RMCA 31133 Skull. Adult female (M3 erupted). *Gorilla beringei graueri*. Details same as for RMCA 31132.

RMCA 31543 Cranium. Adult male (M3 erupted). Bomo village (114 km from Zanaga?), Letili district, Gabon. Collected (bought) by Timmermans on 4 August 1963. Received on 25 October 1963.

RMCA 31544 Cranium. Adult male (M3 erupted). Near Ndongo, Gabon. Collected (bought) by Timmermans, July/August 1963.

RMCA 31545 Cranium. Adult male (M3 erupted). Details same as for RMCA 31544.

RMCA 31546 Cranium. Adult female (M3 erupted). Details same as for RMCA 31544.

RMCA 33161 Carcase (in alcohol). Lomié region, Cameroon. Collected by E. Hancq in 1964. Received on 2 February 1965.

RMCA 33162 Carcase (in alcohol). Details same as for RMCA 33161.

RMCA 33466 Skull. Adult (M3 erupted). Lomié region, Cameroon. Collected by E. Hancq in 1964. Received on 2 February 1965.

RMCA 33467 Skull. Adult (M3 erupted). Djoum region, Cameroon. Collected by E. Hancq in 1964. Received on 2 February 1965.

RMCA 33468 Mummy. Juvenile. Sangmélima region, Cameroon. Collected by E. Hancq in 1964. Received on 2 February 1965.

RMCA 33481 Cranium. Adult (M3 erupted). Akoundoum, Cameroon. Collected by Dirk Thys van den Audenaerde in 1964. Received on 2 February 1965.

RMCA 34476 Skull. Adult female (M3 erupted). *Gorilla beringei graueri*. Kabare territory, Democratic Republic of Congo. Collected by Van Vijve in 1966. Received on 4 January 1967.

RMCA 35274 Skeleton and skin. Juvenile male (PD4 erupted). Democratic Republic of Congo. Received from Zoo Antwerpen, where it had died on 9 April 1968.

RMCA 73.016-M-0151 Skull. Adult (M3 erupted). Eight kilometres from Onkimoyos, Cameroon. Collected by Puylaert and Elsen on 20 September 1971. Received in 1973.

RMCA 73.018-M-0001 Skull. Adult female (M3 erupted). Mvam, Cameroon.

RMCA 73.018-M-0002 Cranium. Juvenile (PD4 erupting). Mvam, Cameroon.

RMCA 73.018-M-0003 Skull. Adult (M3 erupted). Akoabas, Cameroon.

RMCA 73.018-M-0004 Cranium. Adult (M3 erupted). Akoabas, Cameroon.

RMCA 73.018-M-0006 Skull. Adult (M3 erupted). Alen, Cameroon.

RMCA 73.018-M-0007 Cranium. Adult (M3 erupted). Cameroon.

RMCA 73.018-M-0008 Skull. Adult male (M3 erupted). Meyo Nkoulou, Cameroon.

RMCA 73.018-M-0009 Skull. Adult male (M3 erupted). Meyo Nkoulou, Cameroon.

RMCA 73.018-M-0010 Skull. Adult male (M3 erupted). Messea region, Cameroon. It does not belong to the 73.018 collection, but will remain with it for reasons of continuity.

RMCA 73.018-M-0012 Skull (M2 erupted). Meyo Nkoulou, Cameroon.

The above 10 specimens, RMCA 73.018-M-0001−RMCA 73.018-M-0012, were collected by Dirk Thys van den Audenaerde in March−April 1973.

RMCA 73.018-M-0013 Cranium. Adult (M3 erupted). Ngoazik, Cameroon. Collected by Dirk Thys van den Audenaerde in 1975. It does not belong to the 73.018 collection, but will remain with it for reasons of continuity.

RMCA 73.029-M-0002 Cranium. Adult (M3 erupted). Bertoua, Cameroon. Collected by Dirk Thys van den Audenaerde and Opdenbosch in 1969. Received in 1973.

RMCA 73.029-M-0003 Skull. Adult female (M3 erupted). Ebogo, Cameroon. Collected by Dirk Thys van den Audenaerde and Opdenbosch in 1970. Received in 1973.

RMCA 75.051-M-0002 Skull (cast). Río Muni. Received from the Wenner-Gren Foundation in 1975. On loan to the Katholieke Universiteit Leuven.

RMCA 75.051-M-0003 Skull (cast). All details same as for RMCA 75.051-M-0002 (including loan to KUL).

RMCA 75.056-M-0001 Cranium. Adult (M3 erupted). Mang, Cameroon. Collected by Dirk Thys van den Audenaerde and Van Der Veken in February−March 1975.

RMCA 75.056-M-0002 Skull. Adult female (M3 erupted). Djaposten, Cameroon. Collected on 7 February 1975.

RMCA 75.056-M-0003 Cranium. Adult (M3 erupted). Masins, Cameroon. Collected on 15 February 1975.

RMCA 75.056-M-0004 Skull. Adult male (M3 erupted). Messea region, Cameroon. Collected in 1975.

RMCA 75.056-M-0005 Skull. Adult male (M3 erupted). Messea region, Cameroon. Collected in 1975.

RMCA 75.056-M-0006 Skull. Adult male (M3 erupted). Djaposten, Cameroon. Collected on 7 February 1975.

RMCA 75.056-M-0007 Skull. Adult male (M3 erupted). Mpane, Cameroon. Collected in February−March 1975.

RMCA 75.056-M-0008 Cranium. Adult (M3 erupted). Mang, Cameroon. Collected in February−March 1975.

RMCA 75.056-M-0009 Skull. Adult male (M3 erupted). Mang, Cameroon. Collected in February−March 1975.

RMCA 75.056-M-0010 Skull. Adult female (M3 erupted). Messea region, Cameroon. Collected in February−March 1975.

RMCA 75.056-M-0011 Cranium. Adult (M3 erupted). Essengbot, Cameroon. Collected in February−March 1975.

RMCA 75.056-M-0012 Skull. Adult male (M3 erupted). Essengbot, Cameroon. Collected in February−March 1975.

RMCA 75.056-M-0013 Skull. Adult male (M3 erupted). Essengbot, Cameroon. Collected in February−March 1975.

RMCA 75.056-M-0014 Skull. Adult male (M3 erupted). Zoulabot, Cameroon. Collected on 8 February 1975.

RMCA 75.056-M-0015 Skull. Juvenile (M2 erupted). Djaposten, Cameroon. Collected on 7 February 1975.

RMCA 75.056-M-0016 Cranium. Adult (M3 erupted). Mang, Cameroon. Collected in February−March 1975.

RMCA 75.056-M-0017 Skull. Adult female (M3 erupted). Masins, Cameroon. Collected on 15 February 1975.

RMCA 75.056-M-0018 Cranium. Adult (M3 erupted). Mbang, Cameroon. Collected in February−March 1975.

RMCA 75.056-M-0019 Skull. Juvenile (M1 erupted). Messea region, Cameroon. Collected in February−March 1975.

RMCA 75.056-M-0020 Skull. Juvenile (M1 erupted). Mang, Cameroon. Collected in February−March 1975.

RMCA 75.056-M-0021 Skull. Adult female (M3 erupted). Ntimbe, Cameroon. Collected on 7 February 1975; obtained from a pygmy camp.

The above 21 specimens, RMCA 75.056-M-0001−RMCA 75.056-M-0021, were collected by Dirk Thys van den Audenaerde and Van Der Veken.

RMCA 76.064-M-0008 Fetus (in alcohol). *Gorilla beringei graueri*. Received from the Royal Zoological Society, Antwerp (Zoo Antwerpen), on 28 December 1976.

RMCA 77.032-M-0001 Skull. Adult male (M3 erupted). Nemeyong, Dja region, Cameroon. Collected in March/May 1977.

RMCA 77.032-M-0002 Skull. Adult male (M3 erupted). Mimbo Mimbo, Cameroon. Collected on 25 March 1977.

RMCA 77.032-M-0003 Skull. Adult female (M3 erupted). Masins, Cameroon. Collected on 8 March 1977.

RMCA 77.032-M-0004 Cranium. Adult (M3 erupted). Moboe, Dja region, Cameroon. Collected on 25 March 1977.

RMCA 77.032-M-0005 Skull. Adult male (M3 erupted). Moboe, Dja region, Cameroon. Collected on 25 March 1977.

RMCA 77.032-M-0006 Skull. Adult male (M3 erupted). Mimbo Mimbo, Cameroon. Collected on 25 March 1977.

RMCA 77.032-M-0007 Skull. Adult male (M3 erupted). Mimbo Mimbo, Dja region, Cameroon. Collected on 25 March 1977.

RMCA 77.032-M-0008 Skull. Adult male (M3 erupted). Mimbo Mimbo, Dja region, Cameroon. Collected on 25 March 1977.

RMCA 77.032-M-0009 Skull. Adult male (M3 erupted). Long, Cameroon. Collected on 7 March 1977.

RMCA 77.032-M-0010 Skull. Adult female (M3 erupted). Dja region, Cameroon. Collected on 7 March 1977.

RMCA 77.032-M-0011 Cranium (M2 erupted). Messea region, Cameroon. Collected March−May 1977.

RMCA 77.032-M-0012 Skull. Juvenile (M1 erupted). Dja region, Cameroon. Collected March−May 1977.

RMCA 77.032-M-0013 Skull. Adult female (M3 erupted). Dja region, Cameroon. Collected March−May 1977.

RMCA 77.032-M-0014 Skull. Juvenile (M1 erupted). Nemeyong, Cameroon. Collected on 14 March 1977.

RMCA 77.032-M-0015 Skull. Adult female (M3 erupted). Moboe, Dja region, Cameroon. Collected on 25 March 1977.

RMCA 77.032-M-0016 Skull. Juvenile (M2 erupted). Moboe, Dja region, Cameroon. Collected on 25 March 1977.

RMCA 77.032-M-0017 Cranium. Adult (M3 erupted). Moboe, Dja region, Cameroon. Collected on 25 March 1977.

RMCA 77.032-M-0018 Skull. Adult male (M3 erupted). Moboe, Dja region, Cameroon. Collected on 25 March 1977.

RMCA 77.032-M-0019 Cranium. (M1 erupted). Moboe, Dja region, Cameroon. Collected on 25 March 1977.

RMCA 77.032-M-0020 Skull. Adult (M3 erupted). Moboe, Dja region, Cameroon. Collected on 25 March 1977.

RMCA 77.032-M-0021 Skull. Adult (M3 erupted). Moboe, Dja region, Cameroon. Collected on 25 March 1977.

RMCA 77.032-M-0022 Skull (M2 erupting). Dja region, Cameroon. Collected March−May 1977.

RMCA 77.032-M-0023 Skull (M2 erupting). Dja region, Cameroon. Collected March−May 1977. It was classified originally as *Pan troglodytes troglodytes*.

The above 23 specimens, RMCA 77.032-M-0001−RMCA 77.032-M-0023, were collected by G. Philippart de Foy.

RMCA 83.001-M-0005 Carcase (in alcohol). *Gorilla beringei graueri*. Received from the Royal Zoological Society, Antwerp (Zoo Antwerpen), on 14 January 1983.

RMCA 83.001-M-0006 Skin (untanned). *Gorilla beringei graueri*. Received from the Royal Zoological Society, Antwerp (Zoo Antwerpen), on 14 January 1983.

RMCA 86.044-M-0001 Skull. Adult male (M3 erupted). *Gorilla beringei graueri*. Amisi, Zaïre.

RMCA 86.044-M-0002 Cranium. Adult male (M3 erupted). *Gorilla beringei graueri*. Obomongo, Zaïre.

RMCA 86.044-M-0003 Cranium. Adult male (M3 erupted). *Gorilla beringei graueri*. Bitule, Zaïre.

RMCA 86.044-M-0004 Cranium. Adult male (M3 erupted). *Gorilla beringei graueri*. Obomongo, Zaïre.

RMCA 86.044-M-0005 Skull. Adult male (M3 erupted). *Gorilla beringei graueri*. Mungele, Zaïre.

RMCA 86.044-M-0006 Cranium. Adult male (M3 erupted). *Gorilla beringei graueri*. Mungele, Zaïre.

RMCA 86.044-M-0007 Skull. Adult female (M3 erupted). *Gorilla beringei graueri*. Mungele, Zaïre.

RMCA 86.044-M-0008 Skull. Adult female (M3 erupted). *Gorilla beringei graueri*. Mungele, Zaïre.

RMCA 86.044-M-0009 Skull. Adult female (M3 erupted). *Gorilla beringei graueri*. Mungele, Zaïre.

RMCA 86.044-M-0010 Skull. Adult male (M3 erupted). *Gorilla beringei graueri*. Kasese, Zaïre.

RMCA 86.044-M-0011 Skull. Adult male (M3 erupted). *Gorilla beringei graueri*. Kasese, Zaïre.

RMCA 86.044-M-0012 Skull. Adult male (M3 erupted). *Gorilla beringei graueri*. Kasese, Zaïre.

RMCA 86.044-M-0013 Skull. Adult male (M3 erupted). *Gorilla beringei graueri*. Kasese, Zaïre.

RMCA 86.044-M-0014 Skull. Adult male (M3 erupted). *Gorilla beringei graueri*. Kasese, Zaïre.

RMCA 86.044-M-0015 Skull. Adult male (M3 erupted). *Gorilla beringei graueri*. Kasese, Zaïre.

RMCA 86.044-M-0017 Skull. Adult male (M3 erupted). *Gorilla beringei graueri*. Kasese, Zaïre.

RMCA 86.044-M-0018 Skull. Adult male (M3 erupted). *Gorilla beringei graueri*. Kikunda, Zaïre.

RMCA 86.044-M-0019 Skull. Adult female (M3 erupted). *Gorilla beringei graueri*. Kasese, Zaïre.

RMCA 86.044-M-0020 Skull. Adult female (M3 erupted). *Gorilla beringei graueri*. Kikunda, Zaïre.

RMCA 86.044-M-0021 Skull. Adult female (M3 erupted). *Gorilla beringei graueri*. Kabangola, Zaïre.

RMCA 86.044-M-0022 Skull. Adult female (M3 erupted). *Gorilla beringei graueri*. Kibireketa, Zaïre.

RMCA 86.044-M-0023 Skull. Adult female (M3 erupted). *Gorilla beringei graueri*. Kasese, Zaïre.

RMCA 86.044-M-0024 Skull. Adult female (M3 erupted). *Gorilla beringei graueri*. Mutuma, Zaïre.

RMCA 86.044-M-0025 Skull. Adult female (M3 erupted). *Gorilla beringei graueri*. Katchokolo, Zaïre.

RMCA 86.044-M-0026 Skull. Juvenile (M1 erupted). *Gorilla beringei graueri*. Zaïre.

The above 26 specimens, RMCA 86.044-M-0001–RMCA 86.044-M-0026, were collected in 1980–84 by the Faculty of Sciences, University of Kisangani, and received in 1986.

RMCA 90.061-M-0001 Skull. Adult (M3 erupted). Region of Onkassa, Republic of the Congo. Purchased by John Buytaert in 1975.

RMCA a3.040-M-0009 Arm bones. Received from Zoo Antwerpen on 25 September 2003.

RMCA a3.040-M-0010 Arm and foot bones. Received from Zoo Antwerpen on 25 September 2003.

RMCA a6.031-M-0003 Skull (PD4 erupted). It was found among the chimpanzees.

RMCA b2.018-M-0297 Mounted skin (on display). Adult male. *Gorilla beringei graueri*, Utu, Walikale, Belgian Congo. Live weight 162 kg. Field no. 52.

RMCA b2.018-M-0298 Mounted skin (on display). Adult female. *Gorilla beringei graueri*, Mwana, Mwenza. Live weight 80 kg. Field no. 47.

RMCA b2.018-M-0299 Mounted skin (on display). Juvenile. *Gorilla beringei graueri*, Mwenza. Live weight 13 kg. Field no. 48.

The above 3 specimens were mounted in 1956. The collector is probably Vandebroek.

Tournai

*Musée d'Histoire Naturelle et Vivarium, Rue Saint-Martin 52, 7500 Tournai**

Mrs P. Simon
Skin. Male. Gabon. Acquired in 1872.

BRAZIL

Rio de Janeiro

*Universidade Federal do Rio de Janeiro, Laboratório de Anatomia Comparata, Departamento de Anatomia, Instituto de Ciências Biomédicas, Ilha do Fundão, Cidade Universitária, Rio de Janeiro**

Espedito Cordeiro da Silva Jr
AC No. 098 Mounted skeleton. Adult male. *Gorilla gina*. Purchased in Paris, France, and donated by Dr Olímpio da Fonseca on an unknown date (but before 1956). It is in need of some restoration.

BULGARIA

Plovdiv

Prirodonauchen muzei, Ch. G. Danov 34, 4000 Plovdiv*

Tzeno Petrov

Skull. Adult male. Some teeth are missing.

Sofia

Natsionalen prirodonauchen muzei kăm BAN, Blvd Tsar Osvoboditel 1, 1000 Sofia

Assen Ignatov; Nikolai Spassov

Mounted skin. Juvenile female. Collected in 1939.

CANADA

Calgary

University of Calgary, Museum of Zoology, Department of Biology, Faculty of Science, 2500 University Drive NW, Calgary, AB T2N 1N4*

Anthony P. Russell

UCMZ (M) 1975.76 Skeleton. Adult female. Acquired from the Calgary Zoological Society in the autumn of 1970. This specimen is 'Toni', who died of a fracture of the skull on 13 October 1970, aged 8 years and weighing 155 lb. She had been received from the Tiergarten Nürnberg, Germany, on 23 August 1963 in exchange for a male named 'Ruff', who had been sold to Calgary, with another male named 'Tuff', by the animal dealer Karl Krag. Since Calgary wanted a true pair, Krag arranged the swap with Nürnberg. The incisors are somewhat broken and the braincase was removed with an autopsy saw at the time of the postmortem examination. The top of the braincase is preserved with the rest of the disarticulated skeleton. A skull fracture extends across both parietals, but is interrupted by the sagittal crest. The fracture crosses the parietals transversely and is located about three-tenths of the distance between the frontoparietal suture lines (very faint) and the occipital crest. The fracture line is more extensive on the left side. This is a teaching specimen. [Some additional information added by GH from correspondence with Calgary Zoo personnel.]

Skeleton and muscles preserved in 70% ethanol, skin removed. All internal organs and brain removed. Neonate. Acquired from the Calgary Zoological Society in April 1980. Captive bred, died at birth, not sexed. [This may be the male which was stillborn to 'Caroline' and 'Tuff' on 12 April 1980. It was about 2 weeks premature, and weighed 1276 g. The cause of death was hypoxia and cranial trauma. GH]

Edmonton

Royal Alberta Museum, 12845 102nd Avenue NW, Edmonton, AB T5N 0M6

Hugh C. Smith; Bill Weimann

85.12.1 Skeleton and skin. Juvenile female. Obtained from the Calgary Zoological Society in 1985. This specimen is 'Natasha', who died on 26 January 1985 of terminal heart failure caused by hypovolemic shock. This condition occurred as a consequence of severe enteritis associated with a *Balantidium coli* infestation which was diagnosed after a colectoctomy had been performed. She had arrived on 26 August 1984 on loan (along with another female named 'Tabitha') from Metropolitan Toronto Zoo, where she had been born on 4 November 1980. Her mother was 'Samantha', but the sire's identity is uncertain: it may have been 'Charlie'. She was hand-reared from 20 days. The eviscerated carcase weighed approximately 29.5 kg. Total length 790 mm; foot length 250 mm; ear (from notch) 50 mm. [Some additional information added by GH from correspondence with Calgary Zoo and Toronto Zoo personnel.]

University of Alberta, Department of Anthropology, 13–15H.M. Tory Building, Edmonton, AB T6G 2H4

Julie L. Cormack; Pamela Mayne Correia; Nancy C. Lovell

999.2.19a/b Skull. Juvenile female. Received from the University of Alberta, Division of Anatomy, Faculty of Medicine and Dentistry, 5 -01 Medical Sciences Building, Edmonton, Alberta T6G 2H7. The specimen was 3 years old. The dentition has minor defects only. Upper 1st and 2nd molars erupted, 3rd unerupted and deep; 2 premolars erupted; deciduous canine on left in situ, right deciduous canine lost and adult canine visible but unerupted; 1st and 2nd incisors erupted. Mandible: 1st and 2nd adult molars, 3rd unerupted; 2 premolars erupted; deciduous canines in situ; adult incisors erupted. The skull is in a very deteriorated condition, apparently due to overprocessing, but has been stabilized by the osteology technologist in the Anthropology Department.

University of Alberta, Museum of Zoology, Department of Zoology, Faculty of Sciences, CW-312 Biological Sciences Building, Edmonton, AB T6G 2E9

Andrew E. Derocher; W.E. Roberts

UAMZ 3715 Skull. Received from W. Rowan.

UAMZ 4644 Skull.

M4937 Skull (plaster cast). Adult male. Received from W. Rowan.

London

University of Western Ontario, Department of Biology, Faculty of Science, Biological and Geological Sciences Building, London, ON N6A 5B7

Sandra Johnson; David M. Ogilvie; Nina M. Zitani

UWO T 265 Skull. Adult male. Purchased from a biological supply company c.1965. All teeth are present; canines are cracked and broken.

Montréal

McGill University, Redpath Museum, 859 Sherbrooke Street West, Montréal, QC H3A 2K6

Delise Alison; Anthony Wayne Howell

RM 2390 Mounted skelton. Adult male. Donated by the Free Museum, Liverpool, England, in the 1880s. There are bullet holes through the sternum and two ribs.

RM 2391 Mounted skin (standing position). Adult male. *Gorilla beringei graueri.* Belgian Congo: 22 km south of Lubero, Costermansville Province, at an elevation of 7300 ft. (2225.04 m). Collected by Duncan M. Hodgson in 1938. Received in 1939. Total length 168 cm; chest 137 cm; arm spread 253 cm; weight 204 kg.

RM 5021 Incomplete skeleton, partially articulated. Adult male. It appears that this was once a complete mounted skeleton, but it is now missing the skull, the left arm and the right leg. The torso is articulated; separated from the torso are a fully articulated right arm (hand, ulna, radius and humerus) and a fully articulated left leg (foot, tibia, fibula and femur). There is a fracture in the tibia and fibula of the left leg, with evidence of healing; the following measurements were recorded when both limbs were present: right tibia 264 mm; right fibula 245 mm; left tibia 182 mm; left fibula 293 mm. The skeleton had been painted with cream enamel, much of which still remains.

RM 7305 Mounted skeleton. It was donated by McGill University Department of Anatomy, 3640 University Street, Montréal, in July 2011. It had been in the departmental museum collection probably since the turn of the 19th–20th century. Its overall height is 5 ft. [Dennis G. Osmond].

Saint-Hyacinthe

Université de Montréal, Département de Biomédecine Vétérinaire, Faculté de Médecine Vétérinaire, 3200 Rue Sicotte, Saint-Hyacinthe, QC J2S 2M2

André Bisaillon; Christopher A. Price

69-09-19B Mounted skeleton. Adult female. Received from the Société Zoologique de Granby Inc., 347 rue Bourget, Granby, Québec, J2G 1E8.

Saskatoon

University of Saskatchewan, Department of Archaeology and Anthropology, 55 Campus Drive, Saskatoon, SK S7N 5B1

Gary R. Bortolotti; Angela R. Lieverse

Skeleton. Subadult. Female?

Toronto

Royal Ontario Museum, 100 Queen's Park, Toronto, ON M5S 2C6

Department of Mammalogy:

Nancy S. Grepe; Jacqueline R. Miller; Susan M. Woodward

95.8.13.1a Mounted skin. Adult male. Thought to be a topotype of *Gorilla castaneiceps* (Slack), probably the only one in North America. Lake Fernan Vaz basin, Gabon. Collected and presented by Richard Lynch Garner (1848–1920). This specimen is 'No. 6' in his book *Gorillas and Chimpanzees* (Osgood, McIlvaine & Co., London, 1896): see pp. 191 and 198–204. It is associated with the skeleton 95.8.13.4.

95.8.13.1b Mounted skin. Juvenile female. Omboué, Lac Fernan Vaz basin, West Gabon. Collected and presented by Richard L. Garner. This specimen is 'No. 2' in his book: see pp. 203–204. According to Garner, it was nearly 4 years old. It is from the same animal as 95.8.13.2.

95.8.13.1c Mounted skin. Juvenile male. Collected and presented by Richard L. Garner. This specimen is 'No. 1' in his book: see pp. 203 and 233–238. Garner had the animal alive, naming it 'Othello', but it soon died of a gastric complaint. The suddenness of its illness, and death within hours, led Garner to believe that it had been poisoned by his boy assistant.

95.8.13.2 Postcranial skeleton (partial). Juvenile female. Received from Gerrard, London, England. It is associated with the mounted skin 95.8.13.1b.

95.8.13.3 Skeleton. Juvenile male. Fingers and toes absent, some teeth wanting. From the same animal as the skin no. 95.8.13.1c. It was about 1½ years old.

95.8.13.4 Skeleton (formerly mounted). Adult male. From the same individual as the mounted skin 95.8.13.1a. Garner reports that it had a facial wound, the result of a severe blow in early life, but the bone fragments had knitted together. Abnormal dentition: right maxillary abscess with facial disruption and asymmetry; right upper M1 advanced decay; lower right M2 and M3 impaction with lateral eruption of extra molar; mandibular arc distortion. Two incisors are absent. It was killed by a local person. Its head was surmounted by a red crown. 'Othello' had the same mark.

95.8.13.5 Skull. Adult male. South of Lambaréné, near Lake Ezanga, Gabon. Collected and presented by Richard L. Garner. This specimen is 'No. 8' in his book: see pp. 199 and 204. It had been given to Garner by James A. Deemin, an English trader, in February 1893. It has been missing since at least March 1985.

27931 Skeleton. Infant male. Donated on 29 July 1957 by the Riverdale Zoo, Toronto (now defunct), where it had died on 26 July 1957. It was reputedly 1½ years old and weighed 13 lb (eviscerated).

46982 Skeleton. Juvenile male. Donated by the Riverdale Zoo, Toronto, on 4 July 1968. This specimen is 'Leonard', who arrived at the Riverdale Zoo on 1 June 1965 from the National Zoological Park, Washington DC, where he had been born on 10 January 1964. The dam was 'Moka' and the sire was 'Nikumba'. He weighed 5 lb 7¾ oz and was hand-reared. He died of pesticide poisoning on 24 June 1968. Total length 980 mm; foot length 245 mm; ear 43 mm; chest girth 738 mm. [Some additional information added by GH from correspondence with Toronto Zoo personnel.]

94424 Skeleton. Infant male. Received from the Metro Toronto Zoo; entered 23 November 1987. Foot length 105 mm; ear 35 mm; weight 1580 g.

120498 Skull and partial skeleton. Adult female. Damage from necropsy to cranial vault. Some cervical vertebrae (seem to be C2−C4) with skull, with hyperostosis on the anteromedial surface of the vertebral bodies (probably arthritic). There are some dental and associated pathological changes: right lower P2 and M1 missing, complete alveolar filling with some remodelling; left lower M2 out of socket (bagged), socket empty, not refilled; left upper P2 out (bagged) with some refill; some maxillary thinning around molar roots evident. This is 'Samantha', who arrived at the Metro Toronto Zoo, West Hill, Ontario, on 24 September 1974, when about 16 months old. She arrived with a male named 'Charlie' (aged about 20 months). Both came originally from Gabon, and were purchased from Interfauna. She suffered a stroke and was euthanased on 16 August 2010. Length 1030 mm; foot 235 mm; ear 58 mm; forearm 360 mm; weight 103 kg. Donated on 19 August 2010. [Some additional information added by GH from correspondence with Toronto Zoo personnel.]

Department of Vertebrate Paleontology (Comparative Osteology Collection):

2323 Skeleton. Infant female (8 days old). Received from the Metro Toronto Zoo.

8118 Skull. Adult. Received from the Metro Toronto Zoo.

University of Toronto, J.C.B. Grant Museum, Division of Anatomy, Department of Surgery, Faculty of Medicine, Medical Sciences Building, Toronto, ON M5S 1A8

Ian M. Taylor; Michael J. Wiley

Mounted skeleton. Adult male. Purchased from Adam, Rouilly & Co., 18 Fitzroy Street, London W.1, England, some time before 1943.

Partial skull (mounted). Adult female. Ouesso, Moyen-Congo. Collected in 1928. Purchased as above. It includes the facial bones and mandible.

Mandible. Purchased from Adam, Rouilly & Co. The number 89 is written in ink on the lingual side.

University of Toronto, Faculty of Dentistry, 124 Edward Street, Toronto, ON M5G 1G6

John T. Mayhall; Susan Mazza

Skull. Adult male. It is in the oral anatomy teaching collection.

Vancouver

University of British Columbia, Beaty Biodiversity Museum, 2212 Main Mall, Vancouver, BC V6T 1Z4

Christine Adkins; Richard J. Cannings; Elaine Marr; Christopher M. Stinson

M003483 Skull. Adult male. Belgian Congo. Collected by J. Estanove in 1950. The tips of all 4 canines are broken off, and the lower left canine is detached from its socket. It is part of the Cowan Tetrapod Collection (formerly the Cowan Vertebrate Museum).

CENTRAL AFRICAN REPUBLIC

The WWF Primate Habituation Programme (PHP) was established by Allard Blom in 1997 under Dzanga-Sangha's Ecotourism Programme. Its aim is to habituate gorillas for tourism and research. Further information can be found at www.dzanga-sangha.org. An in situ wildlife health laboratory was set up in 2012 and it is likely that sampling will increase and become extensive over the coming years.

Bai Hokou Base Camp, Dzanga-Ndoki National Park

Shelly Masi; Angelique Todd

Cranium. Subadult female. It was found c.2004 near the group being habituated at the time (although it is not known if it came from that group, or another group). The animal had a neck wound, but it was uncertain whether this was a result of scavenging or an attack by a leopard. Missing all teeth except the molars.

Cranium. Juvenile male? Found by Chloe Cipolletta c.2003 in the range of the gorillas being habituated. Its identity as a gorilla is uncertain. Missing all teeth except the molars.

The above crania sit in the open camp and have some termite damage.

Tourist Welcome Centre, Dzanga-Sangha Protected Areas

Angelique Todd

Cranium. Adult male. Teeth missing.

Cranium. Adult male. Teeth missing, postmortem damage to supraorbital torus and maxilla.

Cranium. Adult female? Some teeth remain. Its identity as a gorilla is uncertain (may be chimpanzee).

Cranium. Adult female? Some teeth remain. Its identity as a gorilla is uncertain (may be chimpanzee).

Cranium. Adult female? Some teeth remain. Its identity as a gorilla is uncertain (may be chimpanzee).

CHANNEL ISLANDS

Jersey

Durrell Wildlife Conservation Trust, Les Augrès Manor, La Profonde Rue, Trinity, Jersey JE3 5BP

Ann L. Thomasson

The listed blood samples are in the cryopreservation bank.

Female 'Kishka' (date of birth 9 November 1978, Howletts Wild Animal Park). Collected 24 April 2001: 4 × serum; 4 × serum, clear. Coll. 6 July 2007: 4 × plasma (heparin, lithium), clear; 2 × plasma (heparin, lithium), haemolysis; 1 × plasma (heparin, lithium), clear; 3 × serum, clear; 2 × serum, haemolysis.

Female 'Sakina' (date of birth 14 July 1986). Collected 6 July 2007: 5 × plasma (heparin, lithium), clear; 3 × serum, clear.

Female 'Hlala Kahilli' (date of birth 23 January 1988). Collected 10 August 2011: 6 × plasma (heparin, lithium), clear.

Male 'Asato' (date of birth 20 October 1991). Collected 31 March 1999: 1 × serum, clear. Coll. 7 May 1999: 2 × plasma (heparin, lithium), haemolysis; 1 × plasma (heparin, lithium), clear; 2 × serum, clear.

Male 'Ya Kwanza' (date of birth 3 June 1984, Royal Melbourne Zoological Gardens). Collected 19 August 2006: 2 × serum, clear; 2 × plasma (heparin, lithium), clear. Coll. 6 November 2006: 3 × plasma (heparin, lithium), clear; 2 × serum, clear; 3 × serum, icteric. Coll. 3 June 2007: red blood cells. Coll. 17 June 2007: 5 × plasma (heparin, lithium), clear; 2 × serum, clear. Coll. 15 October 2009: 1 × serum, clear; 1 × plasma (heparin, lithium), clear.

Male 'Mapema' (date of birth 30 April 1996). Collected 1 December 2003: 13 × plasma (heparin, lithium), clear; 3 × plasma (heparin, lithium), cultured with primagam antigen re-TB testing, clear.

Female 'Bahasha' (date of birth 23 April 1994). Collected 19 August 2006: 1 × plasma (heparin, sodium), clear; 1 × plasma (heparin, lithium), slight haemolysis; 1 × serum, slight haemolysis. Coll. 18 June 2008: 4 × plasma (heparin, lithium), clear; 10 × serum, clear.

Female 'Ya Pili' (date of birth 22 July 2003). Collected 11 October 2007: 3 × serum, clear. Coll. 12 October 2007: 3 × postmortem heparinised plasma (heparin, lithium), clear.

M1 Tissues in formalin: gall bladder, epiglottis, trachea, 3 × lung, 2 × liver, urinary bladder, 2 × heart, coronary arteries, 2 × endometrium, spleen, 2 × kidney, adrenal, 2 × pancreas, stomach, 3 × sacroiliac, lumbar 1. Adult female. These are from 'N'Pongo', who arrived on 22 November 1959, aged about 1½ years (weight 18 lb). She came originally from Cameroun, and was purchased from Tyseley Pet Stores Ltd, 771 Warwick Road, Birmingham 11. She died on 24 September 1999. Crown to coccyx 980 mm; head circumference 577 mm; shoulder to elbow 400 mm; elbow to wrist 300 mm; hand length 202 mm; hip to knee 320 mm; knee to ankle 380 mm; foot length 230 mm. [Some additional information added by GH from correspondence with Jeremy J.C. Mallinson.]

M2639 Frozen tissues: skin and muscle, carcase, spleen, lung, kidney, liver, placenta, gastrointestinal tract. Tissues in formalin: spleen, lung, heart, 2 × liver, kidney, 4 × placenta. Female fetus, stillborn on 19 September 2000. Weight 318 g (obtained within 12 hours of death).

M2819 Blocks taken from brain: left frontal lobe, right frontal lobe, left hippocampus, right hippocampus, left internal/external capsule, right internal/external capsule, anterior corpus callosum, posterior corpus callosum, left parietal lobe, right parietal lobe, left temporal pole, right temporal pole, left occipital pole, right occipital pole, cerebellum, pons, 2 × medulla oblongata. Female. These are from 'Ya Pili', who died on 12 October 2007 (weight 40 kg). Brain weight 520 g.

M3227 Frozen tissues: liver, heart, spleen, lung, kidney, brain, skin, muscle. Tissues in formalin: liver, heart, spleen, lung, kidney, brain. Female. Born prematurely on 8 April 2013, died on 14 April 2013. Weight 890 g (obtained more than 12 hours after death).

CHILE

Santiago

Museo Nacional de Historia Natural, Interior Parque Quinta Normal, Casilla 787, Santiago

Jhoann Luis Canto Hernandez; José Yáñez

MNHNCL/Mamex 012 Mounted skin. Adult male. It had 3 bullet impacts in the thorax.

MNHNCL/Mamex 013 Mounted skin. Adult female.

MNHNCL/Mamex 014 Mounted skin. Juvenile.

MNHNCL/Mamex 015 Mounted skeleton. Adult male. Gabon.

All the above specimens entered the collection before 1912.

CUBA

Havana

Museo Nacional de Historia Natural, Obispo No. 61, Plaza de Armas, La Habana 10 100*

Yazmin Peraza Diez; Mariana Sáker Labrada

13−61 Mounted skin. Adult male. Height 145 cm.

13−62 Mounted skin. Adult female. Height 131 cm.

13−63 Mounted skin. Juvenile male. Height 64 cm.

The above specimens were acquired by Dr Mario Sánchez Roig, a private collector, in 1915 in Germany, where they formed part of an old collection. He kept them in his home until 1934, when he sold them to the Institute of Secondary Education, Santiago de las Vegas, Habana Province, where they remained as a study collection until 1962. In that year they were transferred to the museum and placed in a diorama representing their natural environment.

CZECH REPUBLIC

Prague

Národní Muzeum v Praze, Václavské Náměstí 68, 115 79 Praha 1

Petr Benda; Ivan Horáň

NMP 9589 Mounted skin. Adult male. French Congo. Purchased from Rowland Ward, London, in 1906.

NMP 9590 Mounted skeleton. Adult female. French Congo. Purchased for 13 Guldens (=26 Koruna) from Václav Frič, Prague, in 1896. Total height 150 cm.

NMP 9603 Mounted skeleton. Juvenile. Purchased by Prof. Antonín Frič as an alcoholic preparation in London in 1866. Total height 47 cm. A mounted skin, perhaps from the same specimen (but accessioned under a different number) was destroyed due to war events in 1945.

NMP 10769 Skull. Adult male. Collected by Mr Škulina in 1953. Acquired as an inheritance from Prof. Dr Med. Babor, Brno, in 1956. The rostral part of the left mandible shows pathological change (inflammation).

NMP 10770 Skull. Adult female.

NMP 46409 Skull and skin. Adult male. Donated by the Zoologická zahrada Praha (Prague Zoo) in 1990. This specimen is 'Jimmy', who died of actinomycosis on 21 March 1990. He had arrived at the Prague Zoo on 26 June 1975, when about 1½ years old and weighing 8 kg. He came originally from Sangmélima, Cameroon, and was purchased from the animal dealer G. van den Brink. [Some additional information added by GH from correspondence with Prague Zoo personnel.]

NMP 46815 Bones. Female. From a captive specimen imported by van den Brink on 21 June 1985.

NMP 46816 Skeleton. Received from Dvůr Králové Zoo.

NMP 50432 Skull. Adult male. *Gorilla diehli*. This is 'Pong', who was wild-born in Cameroon in 1993 and was deposited in the Al Wabara Farm, Qatar, in 1994. He was transferred to Prague Zoo on 12 June 2001, and died there during a flood, on 13 August 2002.

Univerzity Karlovy, Hrdlička Museum, Katedra Antropologie, Prirodovědecké Fakulty, Viničná 7, 128 44 Praha 2

Helena Malá; Marco Stella

II.2 Skull (defective). Purchased from A. Frič for 100 crowns in 1913. It now appears to be missing.

4068 Right foot (plaster cast). *Gorilla beringei beringei*.

4069 Left hand (plaster cast). *Gorilla beringei beringei*.

4070 Foot (plaster cast), with uncovered muscles and bones.

4071 Foot (plaster cast), with uncovered muscles and bones.

30001 Mounted skeleton. Adult female. Purchased from V. Frič for 1840 Czechoslovak crowns in 1919.

30002 Mounted skeleton. Adult male. Kadei-Nola region, Haute Sangha, French Congo. Acquired from N. Rouppert, Paris, France, and presented by Dr Aleš Hrdlička in 1928. Height (calcaneus to top of sagittal crest) 158.5 cm. For a fully detailed description of this specimen (from which the following summary is taken), see Maly (1939).

> There are expressive pathological changes in the skull, spinal column, pelvis, and lower limb bones. An extensive healed wound is indicated in the pelvic and lumbar regions of the right side, directly affecting the hip bone, lower ribs, and lumbar vertebrae. The greater part of the iliac crest, together with the proximal part of the ala ossis ilii, are displaced. This wound, and the consequent changes caused by the traction of the respective muscles, have resulted in the ilium changing shape, so that the whole bone is protracted in the direction of a line from the anterior superior iliac spine to the posterior superior iliac spine, and is also very much diminished in height. The change of shape is restricted to the ilium; the ischium and pubis are unaffected. The spinal column is deformed scoliotically and lordotically between the eighth thoracic vertebra and the fourth lumbar vertebra. The lordosis is a direct result of the original trauma, whereas the scoliosis is a secondary complication. The last three ribs on the right are broken. The gorilla was also injured in the face: the front part of the upper jaw, as far as the lower edge of the piriform aperture, together with all the incisor teeth, both canines, and the first bicuspids, are missing. The lower jaw is quite normal. Besides the changes resulting from the wound, the skeleton also has remarkably developed traces, in the regions of the left knee joint and the right tarsometatarsal joint, of a past chronic inflammation of the bones of these joints. The changes in the pelvis and spine show that the gorilla was wounded in its youth, during the period when intense reparation was possible. Also, the incisors, canines, and first bicuspids in the lower jaw are but slightly abraded, which proves that the loss of part of the upper jaw occurred in the gorilla's youth, and that it really was a traumatic event and not some atrophic process. In spite of all its problems, the animal lived to quite an old age and most probably died a natural death. Vestiges of soil in the mental foramen and the nasal cavity, as well as the dark brown colour of the whole skeleton, prove that it was skeletonized in the field and not in the preparator's laboratory.

30003 Left hand (in alcohol). Probably early 1930s.

30004 Right foot (in alcohol). Probably early 1930s.

30005 Skull. Adult male. Fernan Vaz lagoon, Gabon. Donated by Dr Aleš Hrdlička in 1929.

30006 Skull. Adult male. Fernan Vaz lagoon, Gabon. Donated by Dr Aleš Hrdlička in 1929.

30007 Skull. Adult male. *Gorilla beringei*? There is a massive perimortem injury of the neurocranium.

30008 Skull. Adult female. Purchased from Rouppert, Paris, for 1070 Czechoslovak crowns in 1925.

30009 Skull. Infant. *Troglodytes gorilla*. Purchased from V. Frič for 35 crowns in 1917.

DENMARK

Aarhus

Naturhistorisk Museum Aarhus, Bygning 210, Universitetsparken, 8000 Aarhus C

Birger Jensen; Christina Vedel-Smith

NHMA11059 Skull. Adult male. Cameroun. Purchased from Dr Curt Schlüter, Halle, Germany, on 27 March 1922.

NHMA11058 Mounted skeleton. Purchased from J. Umlauff, Hamburg, Germany, on 20 September 1922.

NHMA11060 Skull. Purchased from J. Umlauff, Hamburg, Germany, on 20 September 1922.

NHMA3174 Skull, skin, death mask and entire carcase in formol. Juvenile female. Received from Aalborg Zoologiske Have (Aalborg Zoo), Molleparkvej 63−65, 9000 Aalborg, on 3 September 1965.

NHMA3175 Skin, skeleton and death mask. Juvenile male. Received from Aalborg Zoologiske Have on 14 January 1965.

Copenhagen

København Zoo, Roskildevej 32, 2000 Frederiksberg

Christina Hvilsom

Copenhagen Zoo is housing the official great ape biobank within the European Association of Zoos and Aquaria (EAZA). The vision is to store biomaterial (blood, serum, tissue, DNA, hair, etc.) collected from the captive great apes in Europe, as well as other regions, and from the wild. The biomaterial is stored and can be made available for research projects improving the management and conservation of great apes. However, the usage right is restricted and a researcher seeking permission to access sample(s) will be requested to provide a detailed project description and to complete a sample agreement form, which is to be approved by the institution owning the sample or the EAZA Great Ape Taxon Advisory Group (GATAG) if rights have been transferred.

Københavns Universitet, Statens Naturhistoriske Museum, Universitetsparken 15, 2100 København

Kim Aaris-Sørensen; Hans J. Baagøe; Daniel K. Johansson

CN 1 Mounted skeleton. Female. Gabon. Received from the University Physics Museum. Registered on 24 July 1865.

CN 181 Cranium. Male. Gabon. Received from the Institut Linnaea. Registered on 15 October 1889.

CN 182 Cranium. Female. Gabon. Received from the Institut Linnaea. Registered on 7 February 1890.

CN 210 Cranium. Juvenile. Purchased (as a *Simia troglodytes*) from Frank. Registered on 25 July 1893.

CN 295 Carcase (minus guts, external sex organs and ears; in alcohol). Juvenile. Registered on 17 October 1907.

CN 543 Skeleton, mounted skin and brain (in alcohol). Adult male. *Gorilla beringei graueri*. Forests west of Wimbi, Belgian Congo. Collected by Dr Bøje Benzon, University Central African Expedition, on 19 December 1946. Registered on 1 July 1948.

CN 638 Skeleton and skin (defective, in alcohol). Female. Received from the animal dealer Carl Krag. It died of tuberculosis on 30 June 1962, aged 5 months. Registered on 10 November 1962.

M 746 Skeleton and skin. Adult male. Cameroon? This is 'Samson', who arrived at the Copenhagen Zoo on 14 July 1973, and was transferred to Givskud Zoo on 24 November 1998. He died on 22 July 2015.

M 747 Skeleton and skin. Adult female. *Gorilla diehli*? This is 'Nille', who arrived at the Aalborg Zoo in April 1966, aged about 1½ years. She was captured by Gunnar Børglum of Monrovia, Liberia. She was sent on loan to Copenhagen Zoo on 3 January 1985, and to Givskud Zoo on 24 November 1998. She died in 2015. [Some additional information added by GH from correspondence with Søren Rasmussen, Aalborg Zoo.]

EGYPT

Cairo

Cairo University, Faculty of Oral and Dental Medicine, Department of Oral Biology, 11 El-Saraya Street, Manial, Cairo*

Mohamed Abdul Aziz Attia

22 Skull. Adult male. Purchased in Sudan, 1930. Length basion−nasion 160 mm; bizygomatic width 150 mm; cranial width 100 mm. The skull contains a full set of permanent teeth.

29 Skull. Juvenile female. Purchased in Sudan, 1930. Length basion−nasion 110 mm; bizygomatic width 90 mm. The skull contains a full set of mixed dentition.

FINLAND

Helsinki

Helsingin Yliopisto, Luonnontieteellinen keskusmuseo Eläinmuseo, Pohjoinen Rautatiekatu 13, 00100 Helsinki 10

Ann Forsten; Eirik Granqvist; Martti O. Hildén

MZH 535 Mounted skin. Juvenile male. Possibly received from the dealer G.A. Frank, Amsterdam?

MZH 536 Mounted skin. Juvenile female. Possibly received from G.A. Frank?

MZH 2315 Skull. Adult female. *Gorilla gina*. Zolowarf, South Kamerun. It was part of a skeletal collection donated by the Anatomy Department, Helsinki University, in 1960. The skull was originally acquired in January 1911.

MZH 1416 Partial skeleton and mounted skin. Female. *Gorilla beringei beringei*. Received in exchange from the Zoologisches Forschungsinstitut und Museum Alexander König, Bonn, Germany, in 1982. This specimen is 'Pucker', who arrived at the Zoologischer Garten Köln, Riehler Strasse 173, Köln, Germany, on 4 May 1969, along with another female named 'Coco'. Both animals were captured in the Parc National des Volcans, Rwanda, and were purchased from the Government of Rwanda. 'Pucker' weighed 22.2 kg and was about 5½ years old on arrival in Köln. She died on 1 April 1978 (weight 61.1 kg) of Gram-negative septicaemia and shock, and severe cystic dysplasia of the thymus. 'Coco' died of the same illness on 1 June 1978. See Chapter 12, Lymphoreticular and Haemopoietic Systems and Allergic Conditions in Part I, Gorilla Pathology and Health. [Some additional information added by GH from correspondence with Cologne Zoo personnel.]

MZH 1720 Partial skin. Received in exchange from the Zoologisches Forschungsinstitut und Museum Alexander König, Bonn. Originally from Zoologischer Garten Köln. It may be from 'Coco' (see MZH 1416 above).

FRANCE

Auxerre

Musée d'Histoire Naturelle, 5 Boulevard Vauban, 89000 Auxerre

F. Pavy; Sophie Rajaofera

79-2524 Skull. Adult male. Gabon. Donated in 1874 by Monsieur Drillon. There is a bullet hole behind the supraorbital ridge. The upper left lateral incisor is missing.

Avignon

Muséum Requien, 67 Rue Joseph Vernet, 84000 Avignon

Evelyne Crégut

MR 1996-181 Skull. Adult male. It is part of the Sylvain Gagnière collection, donated in 1987. Length, prosthion−acrocranion 378 mm; length basion−prosthion

219 mm; toothrow length, canine (mesial)−molar (distal) 88.2 mm; length premolar−molar 68.5 mm; width zygion−zygion 188.5 mm.

MR 2008-262 Cranium. Adult female. Donated in 2008 by the Office national de la Chasse et de la Faune sauvage. Length, prosthion−acrocranion 287 mm; length basion−prosthion unavailable (no condyles, basioccipital broken); toothrow length, canine (mesial) − molar (distal) 87 mm; length premolar−molar 67 mm; width zygion−zygion 187 mm.

Besançon

Université de Franche-Comté, Laboratoire de Zoologie et Embryologie, Faculté des Sciences et des Techniques de Besançon, Place Maréchal Leclerc, 25030 Besançon*

C.R. Marchand
Skull. Male. The upper left central incisor is unerupted; both upper right incisors and the lower right central incisor are missing.

Bourges

Muséum d'Histoire Naturelle, Allée René Ménard, 18000 Bourges

Ludovic Besson; Michèle Lemaire
BOUM 11.Pon.1 Skull. Adult male. Purchased in 1966 from Nérée Boubée & Cie.
BOUM 11.Pon.3 Skull. Adult male. Collected before 1950. Donated on 4 July 2007.
BOUM 11.Pon.5 Skull. Adult male. Near Etéké, Gabon. Collected between 1948 and 1952.

Clermont-Ferrand

Muséum Henri-Lecoq, 15 Rue Bardoux, 63000 Clermont-Ferrand

Marie-Françoise Faure; Mme A.M. Vivat
MHLCLEA328 Skull. Juvenile male. Gabon. Donated by Monsieur Dauzat, a naval surgeon.
MHLCLFEA329 Skull. Adult male. Donated by Monsieur Gérard Sabatier (collection of Monsieur Pierre Sabatier, Curator of the Museum of Riom).
MHLCLFEA330 Skull. Adult male. Gabon. Donated by Monsieur Dauzat, a naval surgeon. The mandible is broken.

Dijon

Musée d'Histoire Naturelle, Parc de l'Arquebuse, 1 Avenue Albert-1er, 21000 Dijon*

Christian Morizot
210 98 MA 0000 35 Skull. Male.
210 98 MA 0000 38 Skull. Female.
210 99 MA 0000 72 Skull. Male?

Université de Bourgogne, L'Unité de Formation et de Recherche des Sciences de la Terre et de l'Environnement, 6 Boulevard Gabriel, 21000 Dijon*

Jean Chaline
Skull. Adult male. Vicinity of Lambaréné, Gabon. Donated by Sir G. de Jesse, Dijon.
Pelvis (plaster cast). It originated from the Museum of Natural History in Paris.

Elbeuf

Musée Municipal, Place Aristide Briand, 76500 Elbeuf*

Georges Hazet; J. Tabouelle
Os. 42. ME Cranium. Male. *Gorilla gina*. Gabon. It lacks 2 incisors.
Os. 43. ME Skull. Female. *Gorilla gina*. Gabon. It lacks 2 canines and the incisors.
Both the above specimens were acquired in 1894 by Pierre Noury, who was the founder and first curator of the museum.

Grenoble

Muséum d'Histoire Naturelle, 1 Rue Dolomieu, 38000 Grenoble

Philippe Candegabe; Michèle Dunand; Armand Fayard
MHNGr. MA. 981 Head (mounted). Adult. Donated by Jean-Claude Mestrallet in 1984.
MHNGr. MA. 1287 Mounted skin. Juvenile male. *Gorilla gina*. French Congo. It was purchased (already mounted) for 300 francs from the Jardin Zoologique de Marseille. Received on 8 June 1913. It was cleaned in February/March 2007 with a mixture of acetone and toluene.
MHNGr. OS. 362 Skull. Adult male. *Gorilla gina*. Vicinity of Odzala, Moyen-Congo. Donated by Madame Boilly in 1963. The animal had an arm span of 264 cm.
MHNGr. OS. 363 Skull. Adult.

MHNGr. OS. 364 Head (plaster mould). Juvenile. This is from the same individual as the mounted skin MHNGr. MA. 1287. It was cast by Mr Siepi. Purchased from the Jardin Zoologique de Marseille and received on 8 June 1913.

MHNGr. OS. 365 Skull. Male. *Gorilla gina*. Donated by Lieutenant de vaisseau Jules Carradot (1851–89).

MHNGr. OS. 366 Skull. Female. Donated by Lieutenant de vaisseau Jules Carradot.

MHNGr. OS. 2005 Skull. Male. It was a seizure of Grenoble Customs. Received in 1993.

MHNGr. OS. 2006. Skull. Adult male. It was a seizure of Grenoble Customs. Received in 1993.

MHNGr. OS. 2007 Skull. Male. It was a seizure of Grenoble Customs. Received in 1993.

Ile d'aix

Musée Africain, Rue Napoléon, 17123 Ile d'Aix*

Etienne Féau

MAIA Z-1 Mounted skin, Male. Cameroun. Height 136 cm.

MAIA Z-2 Mounted skin. Female. Cameroun. Height 123 cm.

MAIA Z-3 Mounted skin. Juvenile. Cameroun. Height 62 cm.

The above specimens were sold, in 1936, by Rowland Ward Ltd, 'The Jungle', 166 Piccadilly, London W. 1, to Baron Gourgaud, founder of the Musée Napoléonien and the Musée Africain, which were bequeathed to the French State in 1952.

La Rochelle

Muséum d'Histoire Naturelle, 28 Rue Albert 1er, 17000 La Rochelle

Guillaume Baron; R. Duguy; Mme J. Péré; R. Rosoux

MHNLR M.151 Calvarium. Adult male. Gabon. The donor is probably Georges-Louis-Marie-Félicien Jousset de Bellesme, who came to La Rochelle in 1892 and 1894.

MHNLR M.153 Skull. Adult female. Gabon. Collected on 4 September 1892.

MHNLR M.163 Mounted skeleton. Adult male. Sembé region, Ngoko basin, Moyen-Congo.

MHNLR M.164 Skeleton. Adult male. Ngoko basin, Moyen-Congo.

MHNLR M.165 Mounted skeleton. Juvenile male. Sembé region, Ngoko basin, Moyen-Congo.

MHNLR M.169 Skeleton. Adult female. Sembé region, Ngoko basin, Moyen-Congo. This is the mother of the fetus M. 170.

MHNLR M.170 Fetus (fluid-preserved in flask). Sembé region, Ngoko basin, Moyen-Congo.

The above 5 specimens, MHNLR M.163– MHNLR M.170, were donated in 1925 by Monsieur A. Petit Renaud.

MHNLR M.208 Skull. Male. Denis village, Gabon. Purchased from Ward's, Rochester, NY, USA, in 1919.

MHNLR M.224 Skull. Adult male. Bokiba, Likouala region, French Congo. Donated in 1920 by Monsieur A. Petit Renaud.

MHNLR M.225 Skull. Adult male. Near Ntem, Nyas Cham region, Cameroun. Received in exchange in 1938.

MHNLR M.260 Mounted skin. Male. Gabon. Donated by the Muséum National d'Histoire Naturelle, Paris, in 1930.

MHNLR M.261 Mounted skin. Adult female. Gabon. Collected by Alfred de Sennal on 1 April 1860. Donated by the Muséum National d'Histoire Naturelle, Paris, in 1930.

MHNLR M.262 Mounted skin. Juvenile female. Gabon. Collected by Aubry Lecomte in 1856. Donated by the Muséum National d'Histoire Naturelle, Paris, in 1930. This specimen has grey hair.

MHNLR M.294 Skull. Adult male. Bayanga, Kadéi, Haute Sangha, French Congo. Collected in 1929. Donated by Lieutenant Ecarlat in 1933.

MHNLR M.295 Skull. Adult female. Bayanga, Kadéi, Haute Sangha, French Congo. Collected in 1929. Donated by Lieutenant Ecarlat in 1933.

MHNLR M.420 Skull. Adult male. Moloundou, Cameroun. Donated by Monsieur A. Petit Renaud in 1934. Total length 318 mm; condylobasal length 287 mm; zygomatic width 189 mm; weight 1507 g; endocranial capacity 600 cc. This skull has 36 teeth, since each half of the upper and lower jaws includes a 4th molar. It is described in Rode and Solas (1946).

MHNLR M.586 Skull. Adult male. Bania, Shanga, Moyen-Congo. Collected in February, 1928. Donated by Monsieur L. Rouaud. Condylobasal length 289 mm; zygomatic width 204 mm. This skull exhibits an osseous exostosis of the right maxilla.

MHNLR M.664 Fetus (fluid-preserved in flask). Moloundou, Cameroun. Collected in 1930. Donated by Monsieur A. Petit Renaud in 1934.

MHNLR M.666 Skeleton. Juvenile. Moloundou, Cameroun. Donated by Monsieur A. Petit Renaud in 1934.

MHNLR M.667 Brain (fluid-preserved in flask). Moloundou, Cameroun. Collected in 1930. Donated by Monsieur A. Petit Renaud in 1934.

MHNLR M.836 Skeleton (incomplete). Adult male. Moyen-Congo. Collected by Dr R. Boisseau. Donated between 1954 and 1957. Condylobasal length 249 mm; zygomatic width 155 mm. The skull lacks a sagittal crest.

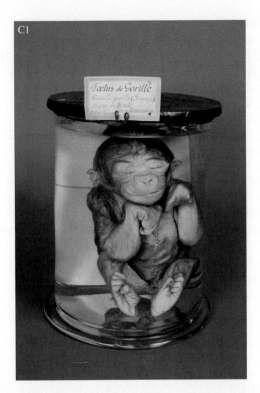

FIGURE C-1 Fluid-preserved gorilla fetus (no. MHNLR M.170). This specimen is in an old-style museum 'pot' which has many advantages in terms of ease of handling and safety but makes further study on the specimen difficult. *Photograph by Guillaume Baron, courtesy of Muséum d'Histoire Naturelle, La Rochelle.*

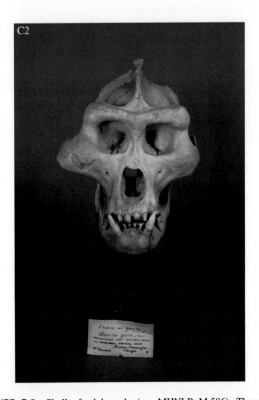

FIGURE C-2 Skull of adult male (no. MHNLR M.586). There is a clear asymmetry affecting the zygomatic arch and the bone on the animal's right hand side appears to be porous. Histological investigations would be necessary to elucidate the cause. *Photograph by Guillaume Baron, courtesy of Muséum d'Histoire Naturelle, La Rochelle.*

Le Havre

Muséum d'Histoire Naturelle, Place du Vieux-Marché, 76600 Le Havre

Juliette Galpin; Thierry Vincent

2012.20.27 Skull. Subadult male. Purchased from a stall in a second-hand market in Le Havre in June 1987. It is missing the upper central incisors.

2015.5.13 Cranium. Adult male. It has been in the museum collections since before 1960. The central and left incisors are missing.

In addition to the above specimens, four skulls (two adult males and two adult females) have been donated by French Customs, following confiscation from luggage under CITES regulations. These skulls have not yet been given individual accession numbers.

Lille

Université Catholique de Lille, Laboratoire de Craniologie Comparée, Faculté de Médecine et de Maïeutique, 56 Rue du Port, 59046 Lille

R. Fenart; Patrick Hautecoeur; A. Remane

Cranium. Adult male.

Cranium. Juvenile.

Université Catholique de Lille, Faculté des Sciences, 60 Boulevard Vauban, 50916 Lille

Benoît L.M. Hubert; J.P. Panent

Skeleton (plaster cast). Adult male. Acquired from Maison Tramond, Paris. It probably dates back to before 1919.

Skull. Male. North Cameroun. It was purchased at a market in its place of origin, then given to the University. The incisor teeth are missing, and the foramen magnum is damaged on the right side.

Calvarium. Female. It was purchased at a market in North Cameroun, then given to the University. Two incisor teeth are broken, and the other teeth are worn.

Musée d'Histoire Naturelle, 19 Rue de Bruxelles, 59000 Lille

R. Defretin; Muriel Lecouvez

ZOO 4963 Mounted skin. Adult male. Purchased from Frank, London, in 1890. Height 173 cm.

ZOO 7375 Mounted skin. Adult female. *Gorilla beringei graueri.* Purchased from Antwerp Zoo in 1990.

ZOO 7811 Skeleton. Adult female. *Gorilla beringei graueri.* Purchased from Antwerp Zoo in 1990. It is from the same animal as ZOO 7375.

ZOO 7812 Mounted skeleton. Adult male. Purchased from Frank, London, in 1885. Height 155 cm.

Lyon

Musée des Confluences, 86 Quai Perrache, 69002 Lyon

Didier Berthet; Joël Clary; Michel Philippe

50001751 Skeleton. Adult female. *Gorilla mayêma*. Donated by Bouvier in 1884.

50001752 Skeleton. Adult female. Gabon. Collected by Maurice Halley. Donated in 1935.

50001761 Mounted skeleton. Juvenile male. *Gorilla gina*. Gabon.

50001762 Skeleton. Adult male. Gabon. Collected and donated by Maurice Halley in1932.

50001763 Skeleton (missing cranium). Adult male. Donated by Bouvier in 1884.

50001764 Skeleton. Adult male. *Gorilla gina*. Gabon. Donated by Association Lyonnaise des Amis des Sciences Naturelles in 1892.

50001765 Mounted skeleton. Adult female. Donated by Bouvier in 1884.

50001772 Skeleton. Adult male. Gabon.

50001773 Skull. Adult male. Haute-Sangha, Cameroun. Donated by Crétin in 1922.

50001774 Skull. Adult male. Gabon. Donated by Association Lyonnaise des Amis des Sciences Naturelles in 1892.

50001775 Skull. Adult male. Gabon. Acquired in 1890.

50001776 Skull. Adult female. Haute-Sangha, Cameroun. Donated by Crétin in 1922.

50001777 Skull. Adult female. Cameroun. Donated by Janin in 1935.

50001778 Skull. Adult male. Gabon. Donated in 1885. There is a pathological area of the left frontal bone.

50001779 Skull. Adult male. Gabon.

50001780 Skull. Adult male. Gabon. Donated by Maurice Halley in 1946.

50001781 Skull. Adult male. Cameroun. Donated by Janin in 1935.

50001782 Skull. Adult male. Gabon. Donated by Maurice Halley in 1946.

50001783 Skull. Adult male. Cameroun. Donated by Janin in 1935.

50001784 Skull. Adult male. Donated by Janin in 1935.

50001785 Skull. Adult male. Cameroun. Donated by Janin in 1935.

50001786 Skull. Juvenile male. Cameroun. Donated by Janin in 1935.

50001787 Cranium. Adult male. Guinea. Donated by Bouvier.

50001788 Skull. Adult female. Haute-Sangha, Cameroun. Donated by Janin in 1935.

50001789 Skull. Adult female. Haute-Sangha, Cameroun. Acquired in 1922.

50001790 Skull. Adult female. Cameroun. Donated by Petit in 1919.

50001804 Skull. Adult male. Cameroun. Donated by Maurice Halley in 1933.

50001964 Mounted skeleton. Adult female. *Gorilla gina*. Gabon. Donated by Gravier in 1885.

50001965 Mounted skeleton. Adult male. Donated by Bouvier in 1884.

50001966 Mounted skeleton. Adult male. *Gorilla gina*. French Congo. Donated by Bouvier in 1884.

50001967 Mounted skeleton. Adult male. Donated by Bouvier in 1884.

50001971 Mounted skeleton. Adult female. *Gorilla mayêma*.

50001977 Mounted skeleton. Adult female. *Gorilla mayêma*. Donated by Bouvier in 1884.

50001994 Skeleton. Juvenile. *Gorilla mayêma*.

50001995 Skull. Juvenile male. *Gorilla mayêma*.

50001996 Skeleton. Juvenile male. Haute-Sangha, Cameroun. Donated by Janin in 1935.

50002604 Skeleton. Juvenile male. *Gorilla mayêma*.

50002625 Cranium. Adult male. Donated by French Customs of Léman in 1998.

50003307 Skull. Adult male. Epéna, French Congo. Collected in 1945. Donated by Auguste Jeannet in 1990.

40000036 Mounted skin. Juvenile female. *Gorilla mayêma*. Congo.

40000037 Mounted skin. Adult female. *Gorilla mayêma*. Congo.

40000038 Mounted skin. Adult female. *Gorilla mayêma*. Congo. It is in poor condition.

40000039 Mounted skin. Juvenile. *Gorilla mayêma*. Congo. It is in poor condition.

40002919 Fetus (in fluid). Gabon. Donated by Maurice Halley in 1935.

40003636 Brain (cast).

40003655 Hand (cast). *Gorilla gina*.

Université Claude Bernard (Lyon I), Faculté d'Odontologie, 11 Rue Guillaume Paradin, 69372 Lyon*

J. Poulard

84 Skull. Adult male. Donated by Dr Gondrand in July 1949. The lower left canine is broken.

119 Skull. Adult male. Acquired on 4 September 1950. The incisors and canines are false, being made of bone. The first upper left premolar is missing. The base of the cranium is broken.

135 Skull. Adult male. Gabon. Donated by Prof. Verchere in September 1983. The lower right canine and

the upper right incisor are missing. The lower left canine is broken.

Université Claude Bernard (Lyon I), Laboratoire d'Anatomie, Faculté de Médecine Lyon Est, 8 Avenue Rockefeller, 69373 Lyon*

A. Morin
   Skeleton.

Vet Diagnostics, 14 Avenue Rockefeller, 69008 Lyon

Karin Lemberger

All tissue samples are embedded in paraffin blocks. This private laboratory's policy is that access to all material is possible in principle, but only after obtaining authorisation from the referring veterinarian of the submitting institution.

Zoo de la Palmyre. Referring veterinarian Dr Thierry Petit.

V08-01390 Necropsy samples. Tissue samples of liver, lung, kidney, brain, heart, spleen, stomach, intestine, skeletal muscle, skin, pancreas, oesophagus. Male #4635. Died on 26 March 2008 (12 days old). Histology revealed interstitial pneumonia with intralesional amniotic squames, consistent with aspiration pneumonia during difficult parturition.

V11-00227 Biopsy samples. Gorilla #8, no age or sex indicated. Tissue sample of skin from the nose. Histology revealed superficial bacterial dermatitis; no evidence of associated viral infection.

V13-05877 Necropsy samples. Tissue samples of liver, lung, kidney, brain, heart, spleen, stomach, intestine, skeletal muscle, pancreas, lymph node. Female #2552. Died on 21 September 2013 (21 years old). Cause of death undetermined. Histology revealed hepatic and splenic haemosiderosis, membranous glomerulonephritis and interstitial pneumonia with emphysema, all relatively nonspecific findings attributed to age and of unknown (though suspected limited) clinical significance.

V14-06673 Necropsy samples. Tissue samples of aorta, liver, lung, kidney, heart, spleen. Male #8 'Mike' (25 years old). Died on 31 October 2013 of a ruptured aortic aneurysm. Histology confirmed the dissecting aneurysm of the aorta with underlying atherosclerosis. Hepatic and splenic haemosiderosis were also present.

ZooParc de Beauval. Referring veterinarian Dr Baptiste Mulot.

V13-06972 Necropsy samples. Tissue samples of skin, gingiva, tongue, liver, lung, kidney, heart, spleen, brain, lymph nodes, stomach, intestine, urinary bladder, skeletal muscle, diaphragm. Male 'Kivu'. Born on 21 January 2013 (to dam 'Inge' and sire 'Asato'). Died on 4 December 2013 of Varicella-Zoster virus infection.

Histology revealed vesicular dermatitis, glossitis and stomatitis with intranuclear inclusions and syncitia and fibrinonecrotic pneumonia with intranuclear inclusions and syncitia and concomitant bacterial infection, considered the proximal cause of death. Also noted: fibrinonecrotic hepatitis and lymphadenitis of presumed similar origin.

V14-01893 Biopsy samples. Tissue samples of stomach, duodenum, colon. Female 'Tamarilla'. Born in Howletts Zoo Park, Bekesbourne, on 29 July 1986 (to dam 'Killa Killa' and sire 'Bitam'); transferred to Beauval on 29 June 2005. Histology revealed gastrointestinal lesions consistent with chronic inflammatory bowel disease.

Zoo d'Amnéville. Referring veterinarian Dr Alexis Maillot.

V14-05212 Biopsy samples. Tissue samples of a gingival mass. Male 'Upala'. Born in Zoo Heidelberg on 27 May 2008 (to dam 'Chuma' and sire 'Bobo'). Histology was consistent with a pyogenic granuloma/eruptive haemangioma.

V14-07494 Recurrence of previous case.

Marseilles

Muséum d'Histoire Naturelle, Palais Longchamp, Rue Espérandieu, 13004 Marseille*

Robert Jullien

519 Mounted skeleton. Female. Donated by Alphonse Baux in 1880.

8342 Skull. Female. French Congo. Donated by Baudon on 26 November 1926.

8466 Postcranial skeleton (incomplete, not cleaned). Donated by Mlle Guyard, Aix-en-Provence, in October 1929.

8840 Cranium; female skull. French Congo. They were donated by the Faculty of Science, Université d'Aix-Marseille, upon the dispersal of the Colonial Museum in November 1961.

1974-275 Skull. Male. Sendara, Ngounié, tributary of Ogooué, northeast of Lambaréné, Gabon. Donated by A. G. Lions in March 1927.

Mounted skin. Juvenile female. Donated by HSH Prince Rainier of Monaco on 11 February 1971.

Long bones, 'probably *Gorilla*'.

Aix-Marseille Université, Faculté de Médecine–Secteur Nord, Laboratoire d'Anthropologie, 51 Boulevard Pierre Dramond, 13916 Marseille*

Skull. Adult male. Complete dentition.

Skull. Adult female. Both upper M3 are missing, as is one lower M3. One lower canine is broken.

Skull. Juvenile. The basioccipital is missing. One lower milk canine and the four milk molars are present, and the first permanent molars are forming in their crypts.

Nancy

Muséum-Aquarium de Nancy, 34 Rue Sainte-Cathérine, 54000 Nancy

Sandra Delaunay; Alain Philippot

2685 Left hand (plaster cast).

7298 Skull. Adult (male?).

8756 Mounted skeleton. Juvenile male. Congo. Purchased in 1914.

9403 Skeleton (cast). Adult male. Purchased from Rouppert.

9536 Skull and mounted skin. Adult male. Mission Catholique Sainte Anne, Gabon. Collected in 1932. Purchased from naturaliste Conrad, Strasbourg, in 1934.

10853 Skull. Adult male. Lambaréné, Gabon. Donated by Mercier.

Nantes

Muséum d'Histoire Naturelle, 12 Rue Voltaire, 44000 Nantes

Mme Baudouin; Marie-Laure Guérin

MHNN.Z.19071 Mounted skin. Juvenile male. *Gorilla mayêma.* Congo. Collected by Monsieur Petit and donated by Dr Lucan in 1882.

MHNN.Z.19862 Skull. Adult female. Ogowai, Gabon. Collected by A. Eutrope in June 1874.

MHNN.Z.19864 Mounted skeleton. Juvenile male. Congo. From the same animal as the skin MHNN.Z.19071.

MHNN.Z.25850 Skull. Adult male.

MHNN.Z.25851 Skull. Adult male. The teeth are missing.

MHNN.Z.25852 Skull. Adult male. The teeth are missing.

MHNN.Z.25854 Skull. Adult female.

MHNN.Z.25857 Skull. Adult male. The teeth are missing.

MHNN.Z.57260 Skull. Adult male. Cameroun. Collected by Abacucio before 1970. Received on 7 July 2007.

MHNN.Z.57292 Skull. Adult male. Collected by Gioux. Received in 1989. It lacks the upper right canine.

Nice

Muséum d'Histoire Naturelle, 60 bis, Boulevard Risso, 06300 Nice

Olivier Gerriet; Ewald Philippe

MHNNice-2012.0.10373 Skull. Adult male. Donated by Blondin on 8 March 1920.

MHNNice-2012.0.10374 Skull. Juvenile male? Donated by Blondin on 8 March 1920.

NB: The above skulls apparently had been obtained by Blondin in Dahomey (=Benin), but their place of origin must have been different, since gorillas do not inhabit this area.

MHNNice-2968/1 Skeleton. Adult male. Cameroun (littoral region). Donated by Combaud on 10 June 1982.

MHNNice-3990/1 Skull. Adult male. Democratic Republic of Congo (Bas-Congo). Donated by Eugène Clerissi on 5 January 1995.

Nîmes

Muséum d'Histoire Naturelle et Préhistoire, 13 bis, Boulevard Amiral Courbet, 30000 Nîmes*

Gérard Gory

Mounted skeleton. Adult. Height 160 cm.

Mounted skeleton (papier mâché cast). The original specimen came from Gabon.

Head (plaster cast).

Paris

Muséum National d'Histoire Naturelle, Laboratoire d'Anatomie Comparée, 55 Rue de Buffon, 75005 Paris

Céline Bens; Jacques Cuisin; Christine Lefèvre

NB: The registration numbers given first are from the 'Journal d'Anatomie comparée' (JAC): these comprise the year of inscription in the catalogue and the entry number. The numbers in brackets correspond to the 'Catalogue de la Galerie d'Anatomie comparée' (CAC). When the gallery was opened in 1898, the specimens on display were registered in this catalogue. The Mammifères et Oiseaux (MO) collection has one 'Catalogue général' (CG) with the same year and entry number as the JAC. In 2004, the AC and MO collections were grouped, which is why the prefix ZM (for 'Zoologie Mammifères') had to be added to the numbers. CAG numbers correspond to the 'Catalogue des Anciennes Galeries', which means the gallery opened by Cuvier at the beginning of the 19th century. 'Removed taxidermy mount' means that the skin was in such poor condition that it has been destroyed and only the bones are now in the collection. Sometimes, there is only the skull, sometimes the skull, hands and feet.

ZM-AC-1854-113 (A 504) Skull. Juvenile male. *Gorilla gina.* Makokou, Gabon. Donated by Monsieur Aubry-Lecomte. The secondary dentition is in eruption. It is exhibited in the gallery.

ZM-AC-1850-459 (A 8078) Cast. Female.

ZM-AC-1854-113 (A 504) Juvenile male. Makokou, Setté Cama, Gabon. Collected by Aubry-Lecomte. CAG no. I-992.

ZM-AC-1856-27 (A 8870) Skeleton. Juvenile female. A note on the inventory says that all the bones look like 'male'. The skull is on display in the 'Galerie d'Anatomie comparée'; the postcranial skeleton is in the Laboratoire d'Anatomie comparée.

ZM-AC-1856-67 (A 10656 & A 12762) Postcranial skeleton. Adult female. *Gorilla gina*. Makokou, Setté Cama, Gabon. Received from Monsieur Bouët, through the agency of the First Lord of the Admiralty.

ZM-AC-1866-92 (A 507 & A 8872) Skeleton (incomplete). Adult male. Makokou, Setté Cama, Gabon. Prepared by Dr Méry and donated by Rear-Admiral Laffon de Ladebat. It lacks the vertebral column. One radius, one ulna and one humerus are sawn into several segments. The skull, with the calvarium detached, is exhibited in the gallery.

ZM-AC-1867-20 (A 8076) Cast (no other details available).

ZM-AC-1869-142 (A 506) Cranium. Adult male. Gabon. Donated by Dr Méry of the Imperial Navy. It was sent to the Laboratoire d'Anthropologie on 3 February 1868, and returned by Monsieur de Quatrefages on 30 April 1869.

ZM-AC-1870-150 Calvarium. Juvenile female. Gabon.

ZM-AC-1870-153 (A 526) Skull. Adult male. Makokou, Setté Cama, Gabon.

ZM-AC-1877-759 (A 510) Skull. Adult female. Ogooué, Gabon. Collected during the voyage of Monsieur Marcke.

ZM-AC-1878-452 (A 8077) Cast. *Gorilla gina*.

ZM-AC-1880-1083 (A 615) Removed taxidermy mount. Male. *Gorilla gina*. Makokou, Setté Cama, Gabon. Collected by de Brazza.

ZM-AC-1883-2028 (A 12764) Postcranial skeleton (incomplete). Adult male. Donated by Monsieur Masson, Governor of Gabon.

ZM-AC-1883-2030 (A 8782) Postcranial skeleton (incomplete). Juvenile. Makokou, Setté Cama, Gabon. Donated by Monsieur Masson, Governor of Gabon.

ZM-AC-1884-1180 Postcranial skeleton. Infant. Gabon. Donated by navy surgeon Dr E. Franquet in 1852.

ZM-AC-1885-482 (A 12746) Skull (incomplete). Adult male. Nzela village, Haut Benito, Equatorial Guinea. Donated by Monsieur Guiral. It lacks the calvarium and mandibular rami.

ZM-AC-1885-483 (A 10666). Cranium. Adult female. Mbonda village, Haut Benito, Equatorial Guinea. Donated by Monsieur Guiral.

ZM-AC-1885-668 Cranium (sagittally bisected). Adult male. French West African Mission, Cap Lopez station, Gabon.

ZM-AC-1885-707 Skull (incomplete). Adult male. Received from Monsieur Guiral, Mbonda, Haut Benito, Equatorial Guinea. It lacks the mandibular rami.

ZM-AC-1885-709 (A 10662) Removed taxidermy mount and skull. Adult female. Collected by Guiral.

ZM-AC-1885-710 (A 10660) Skin and skeleton. Female. Collected by Guiral.

ZM-AC-1886-295 Skeleton (incomplete). Adult male. Received from the de Brazza expedition. It lacks one tibia, both ulnae, one fibula and the truncal bones. The cranium and the articular region of the mandible are damaged.

ZM-AC-1891-1142 (A 8501) Cranium. Juvenile male. *Gorilla gina*. Collected by Lienard. Received from H. Neuville.

ZM-AC-1894-144 (A 8871) Skin and skeleton. Male. Mbila, French Congo.

ZM-AC-1894-145 (A 8861 & A 12750) Postcranial skeleton. Juvenile female. Mbila, French Congo. Donated by Monsieur J. Dybowski.

ZM-AC-1894-154 (A 8891) Skin and skeleton. Juvenile female. *Gorilla gina*. Bagamée Forest, Togodo, 176 km north of Arecho, Togo, Great Lakes region. Collected by J. Dybowski.

ZM-AC-1894-351 Skeleton.

ZM-AC-1897-276 Skeleton (incomplete). Female. Fernan Vaz, Gabon. Donated by Monsieur Auguste Forêt, Colonial Administrator. It lacks the mandible and the right hemicalvarium. Many of the bones are fragmented.

ZM-AC-1898-162 (A 12747) Postcranial skeleton (incomplete). Adult male. Gabon estuary. Donated by Monsieur Durand.

ZM-AC-1898-316 (A 12763) Skeleton (incomplete). Subadult male. Donated by Monsieur Brandon, a merchant in Libreville, Gabon.

ZM-AC-1899-16 Postcranial skeleton (incomplete). Adult male. Originally from the Saint Anne Catholic Mission, Cap Lopez, Gabon. Received from Monsieur Georges Bichet.

ZM-AC-1901-660 Cranium (sagittally sectioned). Female? Donated by Henri Loustau, Colonial Administrator of Mayumba, French Congo.

ZM-AC-1903-108 (A 14178) Removed taxidermy mount and skeleton. Female.

ZM-AC-1906-440 (A 13922) Skull. Adult male. Souanké, Ivindo basin, Gabon. Collected by Dr J. Pavot in April 1906. Received from Monsieur Cottes, South Cameroun Mission. It is exhibited in the gallery.

ZM-AC-1906-441 (A 13923) Skull. Adult male. Souanké, Ivindo basin, Gabon. Collected by Dr J. Pavot in April 1906. Received from Monsieur Cottes, South Cameroun Mission.

ZM-AC-1906-444 (A 13924) Skull (incomplete). Ivindo River basin, right tributary of Ogooué, Gabon. Collected by Cottes.

ZM-AC-1907-102 Cranium (sagittally sectioned). Adult female. Haut Mbeï. Collected by Monsieur Foula, J.M. Bel Congo Expedition, in 1906.

ZM-AC-1908-13 Cranium. Adult male. Tema, left bank of Ogooué basin, Gabon. It was collected in September 1893 and sent to Milne-Edwards by Alexandre Le Roy, Bishop of Alinda at Libreville. It became part of the Filhol collection and was sent by Monsieur Boule to the Laboratoire d'Anatomie Comparée at the time of acquisition of this collection. This specimen has voluminous malar exostoses. For a discussion of the possible causes, see Schultz and Starck (1977).

ZM-AC-1909-358 Skull. Female.

ZM-AC-1912-489 (A 14167) Skeleton. Fetus. Bokiba, French Congo. Collected by G. Demange.

ZM-AC-1919-50 Skeleton. Juvenile female. Cameroun. Donated by Lucien Fourneau, Republican Commissioner of Cameroun, through the agency of Prof. Lacroix.

ZM-AC-1919-51 Skull. Adult male. Cameroun. Donated by Lucien Fourneau, Republican Commissioner of Cameroun, through the agency of Prof. Lacroix. This specimen has a voluminous exostosis of the right malar bone. For a discussion of the possible causes, see Schultz and Starck (1977).

ZM-AC-1924-383 (A 18078) Skeleton. Female.

ZM-AC-1925-187 Skull. Subadult male. Evouzok, Cameroun. Donated by Monsieur Azey.

ZM-AC-1926-19 Skull. Adult male. Cameroun. Donated by Monsieur Bang. The cranium is broken in two obliquely through the middle of the face.

ZM-AC-1926-20 Skeleton (incomplete). Juvenile. Cameroun. Donated by Monsieur Bang. It lacks the face, upper jaw, occiput, mandible, several long bones and the truncal bones.

ZM-AC-1926-22 Skeleton (incomplete). Infant. Cameroun. Donated by Monsieur Bang. It lacks the occiput and most of the long bones.

ZM-AC-1926-114 (A 14488) Skeleton. Infant male. Age 3 years.

ZM-AC-1926-116 (A 14493) Skeleton. Infant female. Belgian Congo. Collected by Hindzé. Age c. 2 years.

ZM-AC-1928-272 Skeleton (incomplete). Adult male. From the collection of the Prince d'Orléans.

ZM-AC-1929-129 Removed taxidermy mount and skull.

ZM-AC-1931-16 (A 14544) Skull (incomplete). Juvenile.

ZM-AC-1931-601 Skeleton (incomplete). Juvenile female. A zoo specimen.

ZM-AC-1931-605 Skeleton. Juvenile female. A zoo specimen.

ZM-AC-1931-608 Skeleton (incomplete). Juvenile male. A zoo specimen, 1½ years old.

ZM-AC-1931-748 Skeleton. Juvenile male. A zoo specimen.

ZM-AC-1932-25 Skeleton. Juvenile female. A zoo specimen.

ZM-AC-1933-74 Skeleton. Adult male. Donated by Governor Bonamy, of the Commissariat of the Mandated Territories, Gabon and Cameroun. It was received from the Colonial Exhibition of 1931.

ZM-AC-1933-75 Skeleton (incomplete). Adult female. Donated by Governor Bonamy, of the Commissariat of the Mandated Territories, Gabon and Cameroun. It was received from the Colonial Exhibition of 1931. The skin of this specimen is preserved in the Laboratoire des Mammifères et Oiseaux (ZM-MO-1985-228).

ZM-AC-1938-195 Skull. Adult male. Cameroun. Donated by Monsieur and Madame Maury, 7 rue Voltaire, Paris 11e.

ZM-AC-1940-393 Cranium. Adult male. Gabon. Donated by Madame Veuve Péqueur, Libreville. It was received from the Paris Exposition of 1889.

ZM-AC-1940-398 Skull. Adult male. Gabon. Donated by Madame Veuve Péqueur, Libreville. It was received from the Paris Exposition of 1889.

ZM-AC-1946-123 Skull. Juvenile female. Dakar, Senegal (arrived dead at the zoo).

ZM-AC-1947-77 Skeleton. Infant female. Cameroun. It was collected by the Carayon expedition. The animal was approximately 1½ years old.

ZM-AC-1952-168 Skeleton. Male. A zoo specimen (having been collected in the wild).

ZM-AC-1954-332 Skeleton. Adult female. The animal was 25 years old, having lived in a zoo for 23 years.

ZM-AC-1962-268 Skeleton. Female. A zoo specimen.

ZM-AC-1990-49 Skeleton. Male. Received from the Zoo de la Palmyre.

ZM-AC-1995-151 Removed taxidermy mount. Mount Visoke, Rwanda. Collected by Cagnon Reine.

ZM-AC-2004-281 Removed taxidermy mount and skull. Adult.

ZM-AC-A509 Removed taxidermy mount. Adult female. *Troglodytes gina*. Makokou, Setté Cama, Gabon. CAG no. III-11.

ZM-AC-A512 Removed taxidermy mount. Juvenile male? *Gorilla gina*. Bagamée Forest, Togodo, 176 km north of Arecho, Togo, Great Lakes region. CAG no. I-1.

ZM-AC-A513 Skull. Juvenile female? *Gorilla gina*. Probably from Makokou, Setté Cama, Gabon. The braincase is sawn. It is exhibited in the gallery. CAG no. I-2.

ZM-AC-A10718 Mounted skeleton. Adult male. *Gorilla gina*. Makokou, Setté Cama, Gabon. It is exhibited in the gallery. CAG no. III-8.

ZM-AC-A10719 Mounted skeleton. Female. *Gorilla gina*. Makokou, Setté Cama, Gabon. It is exhibited in the gallery. CAG no. III-9.

A 12748 Skeleton (incomplete). Adult male. Donated by Monsieur Métayer, Deputy Administrator of Produce of the Sangha region, French Congo. It lacks the skull and other bones. The animal was very tall. Length of left humerus 535 mm; length of femur 450 mm.

A 12770 Skeleton (incomplete). Subadult female. Donated by Monsieur Métayer, Deputy Administrator of Produce of the Sangha region, French Congo. It lacks the left mandibular ramus and the vertebral column. The posterior part of the left upper jaw is damaged.

ZM-AC-A12771 Skin and skeleton. Male. *Gorilla gina*.

Muséum National d'Histoire Naturelle, Laboratoire des Mammifères et Oiseaux, 55 Rue de Buffon, 75005 Paris

Céline Bens; Jacques Cuisin; Christine Lefèvre

36.1 Mounted skin. Adult male. Gabon. Donated by Dr E. Franquet, an Imperial Navy surgeon, who had purchased it from local people who had, apparently, found it already dead. It had several old, healed wounds, including a badly mutilated upper left jaw. Franquet placed the carcase in a 367-litre cask of alcohol, and it was brought back to France on board the steam frigate 'Eldorado'. It arrived at the museum on 16 January 1852. The skin, although defective because of putrefaction prior to its acquisition by Franquet, was mounted by Monsieur Poortmann. The following measurements were taken by Dr Franquet in Africa: height 167 cm; neck circumference 75 cm; chest circumference 135 cm; arm span 218 cm. It is stored in the hangar de taxidermie.

37.6 Mounted skin. Juvenile male. Gabon. It is stored in the hangar de taxidermie. This specimen had been purchased alive, with a young chimpanzee, and taken on board the steam frigate 'Eldorado'. Both animals succumbed during the passage to France, at the end of 1851. They were eviscerated and preserved entire in alcohol by Dr Franquet. They arrived at the museum, with the above-mentioned adult male, on 16 January 1852. All three were mounted by Monsieur Poortmann for exhibition in the zoology gallery, and they were also shown at the Universal Exposition of 1855. Donated by Captain (later Admiral) Charles Penaud, commander of the 'Eldorado'. For further information, see Saint-Hilaire (1858).

38.5a (ZM-MO-1877-1354) Mounted skin. Juvenile. Ogooué, Gabon. Collected by Marche in 1877. It is stored in the hangar de taxidermie.

39.8 (ZM-MO-1856-113) Mounted skin. Adult female. Gabon. Donated by Monsieur Gaillard, a navy commissioner in Gabon, through the agency of his superior, Captain Bouët, in December 1856. It is in very bad condition, and is stored in the hangar de taxidermie. The postcranial skeleton of this specimen is preserved in the Laboratoire d'Anatomie Comparée (ZM-AC-1856-67).

40.1 (ZM-MO-1899-713) Mounted skin. Male. French Congo. Collected by Brandon in 1899. It is on display in the Grande Galerie de l'Evolution, 36, rue Geoffroy-Saint-Hilaire, Paris 75005.

41.1 (ZM-MO-1899-714) Mounted skin. Male. French Congo. Collected by Brandon in 1899. It is on display in the Grande Galerie de l'Evolution.

42.1 (ZM-MO-1898-1572) Mounted skin. Female. Eschiras country. Collected by Buléon in 1898. It is on display in the Grande Galerie de l'Evolution.

43.2 Left hand (cast).

44.3 Right hand (cast). From the same animal as 43 2.

45.4 Foot (cast).

46.5 Head (cast).

47.7b Right and left feet (casts). Female.

48.7c Head (cast). Juvenile male. This animal arrived at the Mènagerie du Jardin des Plantes on 17 January 1884, aged about 3 years. It had been purchased from the animal dealer Cross, Liverpool. It died on 21 April 1884.

The above 6 casts are stored in the Zoothèque.

ZM-MO-1884-2387 Cranium. Juvenile. Collected by Groos.

ZM-MO-1929-503 Skull and skin. Adult male. *Gorilla beringei graueri*. Nakalongi, west of Katanga, Lake Kivu, Belgian Congo. Collected by the Guy Babault expedition in 1929.

ZM-MO-1931-832 Skull, limb bones and skin. Juvenile male. Cameroun. Collected alive by the Berthollet expedition, it arrived at the Mènagerie du Jardin des Plantes on 20 September 1931. It died on 24 September 1931.

ZM-MO-1931-833 Skull, limb bones and skin. Subadult female. Cameroun. Collected alive by the Berthollet expedition, it arrived at the Mènagerie du Jardin des Plantes on 20 September 1931. It died on 15 December 1931.

ZM-MO-1936-1982 Skull. Adult male. Loumalou, Babelay canton, subdivision of Mimongo, Gabon. Donated by Monsieur Alain Maclatchy in September 1934.

ZM-MO-1936-1983 Skull. Adult male. Lidiembo, Bakelai canton, subdivision of Mimongo, Gabon. Donated by Monsieur Alain Maclatchy in August 1934.

ZM-MO-1936-1984 Cranium. Adult male. Northeast of Mimongo, Gabon. Donated by Monsieur Alain Maclatchy.

ZM-MO-1936-1985 Skull. Adult female. Lidiembo, Bakelai canton, subdivision of Mimongo, Gabon. Donated by Monsieur Alain Maclatchy in August 1934.

ZM-MO-1947-225 Skull. Subadult female. Haute Sangha? Central African Republic. Donated by Dr René Malbrant in 1946.

ZM-MO-1947-277 Skull. Juvenile male. Dimako, subdivision of Doumé, Cameroun. Collected by Charles

Kieffer on 4 July 1946. It is stored in the Laboratoire d'Anatomie comparée.

ZM-MO-1947-278 Skull. Juvenile female. Dimako, subdivision of Doumé, Cameroun. Collected by Charles Kieffer on 4 July 1946. It is stored in the Laboratoire d'Anatomie comparée.

ZM-MO-1947-279 Skull. Subadult female. Dimako, subdivision of Doumé, Cameroun. Collected by Charles Kieffer on 4 July 1946.

ZM-MO-1948-639 Skull. Adult male. Mossendjo, Moyen-Congo. Donated by Monsieur Paix in 1941.

ZM-MO-1949-332 Skull and skin. Juvenile male. Makak, subdivision of Eséka, Cameroun. Donated by Monsieur Jean-Marie Collet. It was captured on 10 May 1949 and died on 31 May 1949.

ZM-MO-1949-523 Skull. Adult male. Batouala, 96 km east of Makokou, Gabon. Donated by Madame Paul Rode.

ZM-MO-1949-524 Skull. Adult male. Batouala, 96 km east of Makokou, Gabon. Donated by Madame Paul Rode and Dr René Malbrant.

ZM-MO-1962-1489 Skull. Adult male. Dimonika, Moyen-Congo. Donated by Dr Georges Guillon.

ZM-MO-1962-1490 Skull. Adult male.

ZM-MO-1962-1491 Skull. Adult male. Collected by Raymond Rollinat.

ZM-MO-1962-1492 Skull. Adult male.

ZM-MO-1962-1493 Skull. Adult male. Sangatanga, Libreville, Gabon. Donated by Monsieur Durand-Dassier.

ZM-MO-1962-1494 Skull. Adult male. Mayombé, Gabon. Donated by Dr Vétérinaire Alfred Prunier.

ZM-MO-1962-1495 Skull, limb bones and skin. Adult female. Cameroun. Collected by Georges Bertholet in June, 1931.

ZM-MO-1962-1496 Skull. Adult female. Batouala, 96 km east of Makokou, Gabon. Collected by Monsieur R. Thallon.

ZM-MO-1962-1497 Skull. Adult female.

ZM-MO-1962-1498 Skull and limb bones. Adult male. Cameroun. Collected by Philippe Bertholet in June 1931.

ZM-MO-1962-1499 Skull. Adult male.

ZM-MO-1962-1500 Skull. Adult male.

ZM-MO-1962-1501 Skull. Adult male. Fernan Vaz, Gabon. Donated by Captain Léon Modeste.

ZM-MO-1962-1502 Skull. Adult male.

ZM-MO-1962-1503 Skull. Adult female.

ZM-MO-1962-1504 Skull. Adult male. Collected by H. Terry.

ZM-MO-1962-1505 Skull. Subadult female. Cameroun. Collected by R. Bertholet in June 1931.

ZM-MO-1962-1506 Skull. Juvenile. French Congo. Collected by Dr C. Maclaud.

ZM-MO-1963-275 Skull. Adult male. Mayombé, Gabon. Purchased at Pointe Noire, Moyen-Congo, in October 1962. Donated by Francis Petter.

ZM-MO 1963-276 Skull. Adult female. Mayombé, Gabon. Purchased at Pointe Noire, Moyen-Congo, in September 1962. Donated by Francis Petter.

ZM-MO-1964-230 Skull. Adult male. Moyen-Congo. Collected by Charles Kieffer in 1950.

ZM-MO-1964-1539 Skull and limb bones. Adult male. Bélinga, Gabon. Collected by Gérard Dubost in 1961. Donated by Prof. Pierre-Paul Grassé.

ZM-MO-1964-1540 Skull and limb bones. Adult male. Makokou, Gabon. Collected by Gérard Dubost in 1961. Donated by Prof. Pierre-Paul Grassé.

ZM-MO-1966-222 Skull. Adult male. Zanaga forest, French Congo. Collected by Dr René Taufflieb. It has bilateral osseous proliferation, similar in extent but not necessarily with the same aetiology as ZM-AC-1908-13 and ZM-AC-1919-51 [JEC, author's personal observation].

ZM-MO-1974-313 Skull. Adult male. Collected by Pierre-Médard.

ZM-MO-1985-228 Skin. Adult female. Cameroun. Received from the Colonial Exposition of 1931. The incomplete skeleton of this specimen is preserved in the Laboratoire d'Anatomie Comparée (ZM-AC-1933-75).

ZM-MO-1987-164 Skin and skull. Adult male.

ZM-MO-1987-250 Skull. Adult male. Mbigou, Gabon. Collected by Laurent Mbumb in 1978.

ZM-MO-1990-426 Skull. Adult female. Batouala, 96 km east of Makokou, Gabon.

ZM-MO-1990-427 Skull. Adult male. Batouala, 96 km east of Makokou, Gabon. Collected by S. Fiorina.

ZM-MO-1991-726 Skull. Adult male. Batouala, 96 km east of Makokou, Gabon. Collected by Raymond Bosc.

ZM-MO-1994-1115 Mounted skin. Adult male. *Gorilla beringei graueri*. Lake Kivu region, Belgian Congo. This specimen was acquired from the Musée du Duc d'Orléans, 45 bis, rue de Buffon, 75005 Paris, upon its closure in 1960 (the building was demolished in 1962). The skin was mounted by Rowland Ward-Burlace, London. Height 180 cm; chest circumference 150 cm; arm span 300 cm. It is on display in the Grande Galerie de l'Evolution (Salle des Espèces Disparues).

ZM-MO-1998-1960 Skull. Adult male. Cameroun. Collected by Lapeyre.

ZM-MO-2000-152 Removed taxidermy mount. Juvenile male. Batouala, 96 km east of Makokou, Gabon. Collected by Petit Aîné. Catalogue of mounted specimens no. 36A.

ZM-MO-2000-667 Removed taxidermy mount and skull. Adult female. Catalogue of mounted specimens no. 2000-22.

ZM-MO-2001-1647 Removed taxidermy mount and skeleton. Adult male. Dachang assay station, Cameroun.

ZM-MO-2001-2139 Skull. Adult male. Cameroun. Collected by Gaëtan Mollez.

ZM-MO-2001-2140 Skull. Male. Ebolowa, Cameroun. Collected by Gaëtan Mollez.

ZM-MO-2001-2141 Skull. Male. Cameroun. Collected by Gaëtan Mollez.

ZM-MO-2001-2142 Skull. Female. Ebolowa, Cameroun. Collected by Gaëtan Mollez.

ZM-MO-2001-2143 Skull. Female. Cameroun. Collected by Gaëtan Mollez.

ZM-MO-2001-2144 Skull. Female. Cameroun. Collected by Gaëtan Mollez.

ZM-MO-2001-2145 Skull. Female. Cameroun. Collected by Gaëtan Mollez.

ZM-MO-2001-2146 Skull. Female. Cameroun. Collected by Gaëtan Mollez.

ZM-MO-2001-2147 Skull. Male. Cameroun. Collected by Gaëtan Mollez.

ZM-MO-2001-2148 Skull. Male. Cameroun. Collected by Gaëtan Mollez.

ZM-MO-2001-2149 Skull. Female. Dimonika, French Congo. Collected by Gaëtan Mollez.

ZM-MO-2001-2150 Skull. Male. Dimonika, French Congo. Collected by Gaëtan Mollez.

ZM-MO-2001-2151 Skull. Male. Dimonika, French Congo. Collected by Gaëtan Mollez.

ZM-MO-2001-2152 Skull. Male. Dimonika, French Congo. Collected by Gaëtan Mollez.

ZM-MO-2001-2153 Skull. Male. Dimonika, French Congo. Collected by Gaëtan Mollez.

ZM-MO-2001-2154 Skull. Male. Dimonika, French Congo. Collected by Gaëtan Mollez.

ZM-MO-2001-2155 Skull. Male. Dimonika, French Congo. Collected by Gaëtan Mollez.

ZM-MO-2001-2156 Skull. Male. Dimonika, French Congo. Collected by Gaëtan Mollez.

ZM-MO-2001-2157 Skull. Male. Brazzaville, French Congo. Collected by Gaëtan Mollez.

Muséum National d'Histoire Naturelle, Laboratoire de Préhistoire, Institut de Paléontologie Humaine, 1 Rue René Panhard, 75013 Paris

Henry de Lumley; Amelie Vialet

1916-1 Skeleton. Juvenile male. Sernanni village, Fernan-Vaz lagoon, Gabon. Purchased from A. Defaye, Fernan-Vaz. Date of entry 10 November 1916.

1920-1 Cranium. Adult male. Gabon. Donated by Prof. Breuil. Date of entry 15 January 1920.

1920-2 Mandible. Female. Gabon. Donated by Prof. Breuil. Date of entry 15 January 1920.

1943-13 Skull. Adult male. Gabon. Donated by Dr Vallois (purchased from Fenasse, 1931). Date of entry 1943.

1943-14 Skull. Adult male. Haut-Ogoué, Gabon. Collected by Dr Castex. Donated by Dr Vallois. Date of entry 1943.

1943-15 Skull. Adult male. Haut-Ogoué, Gabon. Collected by Dr Castex. Donated by Dr Vallois. Date of entry 1943. It appears to belong to the skeleton no. 1943-40.

1943-16 Skull. Adult male. Collected by Dr Florence. Donated by Dr Vallois. It belongs to the skeleton no. 1943-41. Date of entry 1943.

1943-17 Skull and right humerus. Subadult male. Procured by Prof. Delmas. Donated by Dr Vallois. Date of entry 1943.

1943-18 Cranium. Adult female. Gabon. Donated by Dr Vallois (purchased from Fenasse, 1931). Date of entry 1943.

1943-19 Cranium. Adult male. Haut-Ogoué, Gabon. Collected by Dr Castex. Donated by Dr Vallois. Date of entry 1943. It is very damaged.

1943-20 Skull. Juvenile. Haut-Ogoué, Gabon. Collected by Dr Castex. Donated by Dr Vallois. Date of entry 1943.

1943-21 Cranium. Adult female. Procured by Dr Bergés. Donated by Dr Vallois. Date of entry 1943.

1943-22 Cranium. Juvenile. Haut-Ogoué, Gabon. Collected by Dr Castex. Donated by Dr Vallois. Date of entry 1943.

1943-23 Cranium. Juvenile. Cameroun. Procured by Dr Calvet. Donated by Dr Vallois. Date of entry 1943.

1943-24 Cranium. Juvenile. Gabon. Donated by Dr Vallois (purchased from Fenasse, 1931). Date of entry 1943.

1943-25 Cranium. Adult female. Gabon. Donated by Dr Vallois (purchased from Fenasse, 1931). Date of entry 1943.

1943-26 Cranium. Haut-Ogoué, Gabon. Collected by Dr Castex. Donated by Dr Vallois. Date of entry 1943.

1943-32 Mandible. Adult male. It previously belonged to the Laboratoire d'Anthropologie de l'École des Hautes Études. Date of entry 1943.

1943-33 Mandible. Adult. It previously belonged to the Laboratoire d'Anthropologie de l'École des Hautes Études. Date of entry 1943.

1943-34 Skull. Adult female. It previously belonged to the Laboratoire d'Anthropologie de l'École des Hautes Études. Date of entry 1943.

1943-40 Postcranial skeleton. Adult male. Haut Ogoué, Gabon. Donated by Dr Vallois. Date of entry 1943. The skull appears to be no. 1943-15.

1943-41 Postcranial skeleton. Adult male. Donated by Dr Vallois. Date of entry 1943. The skull is no. 1943-16.

1943-43 Skull. Juvenile. Donated by Dr Vallois. Date of entry 1943.

1951-4 Skull. Adult male. Cameroun. Donated by Prof. Piedelièvre.

Muséum National d'Histoire Naturelle, Collection d'Anthropologie, Musée de l'Homme, 17 Place du Trocadéro, 75116 Paris

Véronique Laborde

Limb bones (articulated). Female. From the Broca collection. On deposit from the Société d'Anthropologie, Paris.

Mounted skeleton. Adult male. On deposit from the Société d'Anthropologie, Paris (their no. 226).

Mounted skeleton. Adult male. On deposit from the Société d'Anthropologie, Paris (their no. 227).

MNHN-HA-19439 Postcranial skeletal elements (complete vertebral column, sacrum, 2 innominate bones, ribs, 2 scapulae, 2 clavicles, 2 humeri, 2 radii, 2 ulnae, bones of hands and feet). Adult male. Gabon. Donated by Halley, Ngouo, via Port Gentil, in 1934.

MNHN-HA-20497 Skeletal elements (skull, complete vertebral column, sacrum, right innominate bone, ribs, 2 scapulae, 2 clavicles, 2 humeri, 2 radii, 2 ulnae, 2 femora, 2 tibiae, 2 fibulae, bones of hands and feet, fragments of sternum). Adult male. Oubangui-Chari. Acquired from Madame P. Bouvard, Rochefort sur Mer, in 1939.

MNHN-HA-20498 Skeletal elements (skull, 2 scapulae, 1 clavicle, 2 humeri, 2 radii, 2 ulnae, 2 innominate bones, 2 femora, 2 tibiae, 2 fibulae, 2 patellae, ribs, 1 vertebra, sternum, bones of hands and feet). Adult male. Oubangui-Chari. Acquired from Mme P. Bouvard in 1939.

MNHN-HA-20526 Skeletal elements (skull, complete vertebral column, sacrum, 2 innominate bones, 2 femora, 2 tibiae, 2 fibulae, 2 clavicles, 2 humeri, 2 radii, 2 ulnae, 1 scapula, ribs, bones of hands and feet). Adult male. Haut-Ogooué, Gabon. Donated by Trézenen in 1939.

MNHN-HA-20527 Skeletal elements (skull, complete vertebral column, sacrum, 2 innominate bones, 2 scapulae, 2 clavicles, 2 humeri, 2 radii, 2 ulnae, left femur, 2 tibiae, 2 patellae, 2 fibulae, fragments of sternum, 24 ribs, bones of hands and feet). Adult female. Haut-Ogooué, Gabon. Donated by Trézenen in 1939.

MNHN-HA-20528 Skeleton (articulated). Juvenile. Haut-Ogooué, Gabon. Donated by Trézenen in 1939. This specimen is the offspring of the above female, MNHN-HA-20527.

MNHN-HA-20607 Mounted skeleton. Adult male. Purchased from Boubée in 1941.

MNHN-HA-20841 Skeletal elements (2 scapulae, 2 humeri, 2 radii, 2 ulnae, left femur, 2 tibiae, 1 fibula, sternum, 7 ribs, pelvis, 1 calcaneus). Adult male. Received from the École d'Anthropologie in 1944.

MNHN-HA-20842 One innominate bone. Juvenile. Received from the École d'Anthropologie in 1944.

MNHN-HA-20868 Skull (bisected in sagittal plane). Male. Kribi, Cameroun. Donated by Fontana.

MNHN-HA-24401 Cranium. Adult female. Cameroun. Donated by Durand-Ferté in 1957.

MNHN-HA-24402 Skull. Adult male. Cameroun. Donated by Durand-Ferté in 1957.

MNHN-HA-24873 Skull. Adult male. Donated by Prof. Urbain in 1963. It is now missing.

MNHN-HA-26159 Skin of hand and foot (preserved in a glass jar). Donated by Anatomie Comparée in 1974.

MNHN-HA-27024 Cranium (cast). Adult female.

Perpignan

Muséum d'Histoire Naturelle, 12 Rue Fontaine Neuve, 66000 Perpignan

Amy Benadiba; Robert Bourgat

2008.0.271 Skull. Adult male. Purchased in 1950. Bizygomatic width 175 mm; length of sagittal crest (inion–bregma) 118 mm; condylobasal length (from the middle of the border of the intercondylar notch to the anterior border of the incisive suture: alveolar-basilar line in man) 192 mm; length of mandible 171.5 mm.

Poitiers

Université de Poitiers, Laboratoire de Biologie Animale, Faculté des Sciences, 40 Avenue du Recteur Pineau, 86022 Poitiers*

Michel Brunet

Mounted skeleton. Adult male.

Cranium (plaster cast). Adult male.

11 Skull. Adult male.

Université de Poitiers, Laboratoire de Paléontologie des Vertébrés et Paléontologie Humaine, Faculté des Sciences, 40 Avenue du Recteur Pineau, 86022 Poitiers*

P.M. 319 Skull. Adult male.

P.M. 320 Skull. Adult female.

P.M. 321 Cranium. Subadult female.

P.M. 322 Cranium. Subadult female.

P.M. 323 Cranium. Adult female.

P.M. 324 Skull. Adult male.

P.M. 325 Skull. Infant. The teeth are deciduous.

P.M. 326 Skull. Infant. The teeth are deciduous.

P.M. 327 Skull. Infant. The teeth are deciduous.

8311 Skull. Subadult female.

Rouen

Muséum d'Histoire Naturelle, 198 Rue Beauvoisine, 76000 Rouen

Mlle Monique Fouray; Thierry Kermanach

6237 Skull. Juvenile. Donated by Delavoipière in April 1891.

8602 Skull. Adult male. Congo. Donated by Dr Leplé on 2 May 1900.

NB: Three skulls (an adult male, adult female and juvenile) are figured in the *Catalogue des Mammifères-Actes du Muséum de Rouen*, 1905, fascicule 9, p. 12. It is not clear whether 6237 and 8602 listed above are two of these skulls; and the disposition of the figured adult female skull is unknown.

9936 Assorted bones. Female. Purchased from Tramond on 24 January 1907.

12479 Skull. Female. Purchased at a sale in Rouen on 6 April 1943.

12646 b Skull. Juvenile. Ivory Coast. Donated by L. Hedin on 2 May 1970.

MAM.2009.0.869.MHN (old number 287) Mounted skin. Adult male. Region of Lac Oguémoué, Gabon. Collected by A. Marche and the Marquis de Compiègne in 1871. It was shot dead whilst raiding a plantation of banana trees at night, by the Welshman who owned the plantation, and who was giving shelter and hospitality to the hunters, who had been following the animal's tracks that day. Purchased from Bouvier in May 1875.

MAM.2009.0.870.MHN (old number 10298) Mounted skin. Adult female. Salto, Haute Sangha, Moyen-Congo. Collected by Monsieur Poirries in July, 1910. Received on 11 April 1911.

MAM.2009.0.871.MHN (old number 10246) Mounted skin. Juvenile. Purchased from Lipschitz. Acquired on 14 June 1910. It is from the same individual as MAM.2009.0.1019.MHN.

MAM.2009.0.1011.MHN (old number 12646 a) Skull. Adult male. Ivory Coast. Donated by L. Hedin on 2 May 1970.

MAM.2009.0.1013.MHN (old number 12529) Skull. Male. Donated by Madame Lancesseur in 1947.

MAM.2009.0.1019 Skull. Juvenile. Purchased from Lipschitz. Acquired on 14 June 1910. It is from the same individual as MAM.2009.0.871.MHN. Noted on the skull: 'skin mounted, skeleton defective'.

MAM.2009.0.1293.MHN (old number 10010) Mounted skeleton. Male. Purchased from Monsieur de Limur at Vannes (Morbihan) on 21 November 1907.

MAM.2009.0.1294.MHN (old number 10052) Mounted skeleton. Adult female. Received as loose bones on 28 February 1906. Mounted in May 1908.

MAM.2009.0.1310.MHN Appendix (in fluid). Juvenile.

Strasbourg

Université de Strasbourg, Institut d'Anatomie, Faculté de Médecine, 4 Rue Kirschleger, 67085 Strasbourg

J.G. Koritké; Jean-Marie Le Minor; Henri Sick

X – Zool. 197 (1923/5686 – C 1669) Skull. Male. *Gorilla gina*. Received between 1857 and 1870.

1885-86/3-Zool. 228 (1923/5824 – C 1807) Skull (sectioned). Female. *Gorilla gina*. Purchased from Frank, London, in 1885 for the sum of 150 marks.

1908-09/26 – Zool. 1445 (1923/5824 – C 1807) Skull. Male. Cameroun. Presented by Prof. Dr Haberer, Griesbach, Germany, in 1908.

X – 1923/5687 – C 1670 Skull. Received before 1918. The mandible is partially broken.

Université de Strasbourg, Musée Zoologique, 29 Boulevard de la Victoire, 67000 Strasbourg

J. Devidts; Marie-Dominique Wandhammer

MZS Mam00238 Mounted skin. Adult male. This is from the same specimen as the mounted skeleton, MZS Mam01286. It was prepared by Preuschen and purchased from Rolle, Berlin, in 1916.

MZS Mam00839 Skull. Réserve de la Biosphère de Dimonika, Mayombe forest, Republic of the Congo. Collected c.1972. Received from Roger Miesch in 2006.

MZS Mam01286 Mounted skeleton. Adult male. Ebamina, Cameroun. Collected by Chamder on 10 March 1913. Purchased from Franz Hermann Rolle, Berlin, in 1916. Mounted in 1960.

MZS Mam01808 Postcranial skeleton. Adult male. Spanish Guinea. Collected on 26 May 1908. Received from Edelmann.

MZS Mam02618 Skull. It is from the same specimen as MZS Mam01808.

MZS Mam02619 Skull. Adult female. Gabon. Collected in 1852. Received from G. Conrad.

MZS Mam02871 Mounted skeleton. Adult female. Gabon. Collected in 1860.

MZS Mam02872 Mounted skeleton. Adult male. Received from Poehl.

MZS Mam06971 Skull. Cameroun. Received from Police Nationale on 11 March 2015.

Toulon

Muséum d'Histoire Naturelle de Toulon et du Var, 737 Chemin du Jonquet, Jardin du Las, 83200 Toulon*

Philippe Orsini; J.P. Riste Rucci

4/H12 GOR GOR (1) Mounted skeleton. Adult male. Gabon. It was brought back by Lieutenant-Commander (later Commandant) Colin in about 1884 and was donated by Dr Abelin in 1912. Full details of this specimen are to be found in Abelin (1912). The following details are taken from this article:

The skeleton was partly covered by mummified tissue, which served to hold some of the bones in their natural articulation. Some small bones are missing, and there are various old lesions. The right maxilla is injured above the first premolar, which is missing, and the second premolar, which is fractured. A fissure extends from the injury and diverges either side of the infraorbital foramen. This lesion is probably the result of a bullet wound. There is a comminuted fracture of the left mandibular ramus, just behind the last molar; the ascending ramus is uniformly splintered and a number of fragments are missing. This was probably one of the wounds which led to the animal's death. All the lower incisors are carious. The lower right canine is broken. The incisors and molars are moderately worn. The vertebral column has 14 thoracic vertebrae and 3 lumbar vertebrae, in contrast to the normal arrangement of 13 thoracic and 4 lumbar. The first thoracic vertebra is missing, but all 14 pairs of ribs are present; the 14th pair are probably a supernumerary lumbar pair. An abnormal, elongated aperture about 11 mm long by 2 mm wide is visible just above the third posterior sacral foramen, and is in simultaneous contact with the foramen and the sacral canal. The 4th left rib has an incomplete, comminuted fracture 14 cm from its chondrosternal articulation, whilst the 4th right rib has a complete fracture 3 cm from its chondrosternal articulation. Both these fractures were probably caused by the same projectile. The left scapula has been perforated at the rim of the deep groove below the glenoid fossa, probably by a bullet. This injury may be related to the rib fractures mentioned above. Both ulnae are moderately curved anteriorly, but the curvature of the left ulna is the more accentuated. This exaggerated curvature of the ulnae may be the consequence of old fractures of both radii. The medial surface of the right ulna has an oval depression which articulates precisely with the rounded end of the inferior fragment of the radius. The left ulna is enlarged at a point approximately one third of the distance from its inferior extremity. This may be the callus of an old fracture, or a hyperostosis following an inflammation of the bone. The lesions of the radii are much more important than those of the ulnae. The right radius is in three pieces. The left radius presents traces of an old fracture, healed by a voluminous, flattened callus. Its width at this point is 55 mm. In spite of their deformities, both radii are the same length overall, 350 mm. Almost all the medial and distal phalanges of the right hand are missing. In the left hand the carpus is complete, but the 4th and 5th metacarpals are missing, as are 6 phalanges, although the thumb, forefinger, and little finger are complete. The left fibula has a voluminous fusiform exostosis, the callus of an old fracture which healed without causing any notable distortion or contraction of the bone. In this subject, the comparative lengths of the radii are notably less than normal for the gorilla, because of

contraction following their fracture. The right foot is nearly complete, lacking only the 3rd and 4th digits. The left foot is more incomplete, lacking the lateral cuneiform, the 5th metatarsal, the distal phalange of the great toe, and the medial and distal phalanges of the other digits.

5/H12 GOR GOR (3) Skull. Adult male. It is missing both upper central incisors and the lower left central incisor. The remaining teeth have enamel hypoplasia.

663/H12 GOR GOR (2) Skull. Adult male. It is missing the upper right central incisor and the lower right lateral incisor.

Skull. Adult male. Gabon. It was collected c.1946−56, and received from a private collection in 2001.

Toulouse

Muséum d'Histoire Naturelle de Toulouse, 35 Allées Jules-Guesde, 31000 Toulouse

Henri Cap; Mlle C. Sudre

MHNT PRI 1 Mounted skin. *Gorilla mayêma.* Juvenile male. Conde, near Lândana, Portuguese Congo. Purchased from Petit on 22 April 1884.

MHNT PRI 2 Mounted skin. Adult male. *Gorilla mayêma.* Conde, near Lândana, Portuguese Congo. Purchased from Petit on 22 October 1884.

MHNT PRI 3 Mounted skin. Adult female. *Gorilla mayêma.* Conde, near Lândana, Portuguese Congo. Purchased from Petit on 22 August 1884. The following measurements were taken in the flesh by Petit: circumference of head at level of cheeks and lower lip 90 cm; neck 75 cm; chest 140 cm; abdomen 135 cm; lower abdomen 130 cm; upper arm 43 cm; forearm 34 cm; wrist 28 cm; length of middle finger 12 cm; circumference of hand below fingers 29 cm; thigh 68 cm; calf 38 cm; ankle 35 cm.

MHNT OST 1996.160 Mounted skeleton. Male.

MHNT OST 2002.65 Mounted skeleton. Juvenile. *Gorilla mayêma.* Portuguese Congo. Donated by Petit in 1884.

MHNT OST 2002.67 Mounted skeleton. Adult male. Gabon. Donated by Lazorthes.

MHNT OST AC 342 Clavicle. Cameroun.

MHNT OST AC 343 Clavicle. Gabon.

MHNT ZOO 2011.0.2 Skull. Juvenile. Gabon. Donated by Janin in 1939.

MHNT ZOO 2011.0.18.1 Skull. Adult female. Gabon. Brought back by Bestion; purchased from Constantin on 18 May 1880.

MHNT ZOO 2011.0.18.2 Skull. Male. Donated by Ganin and Lazorthes in 1923.

MHNT ZOO 2011.0.18.3 Skull. Female. Gabon. Donated by Moura on 20 July 1886.

MHNT ZOO 2011.0.18.4 Skull. Female. Gabon. Donated by de Blandinière.

MHNT ZOO 2011.0.18.5 Skull. Juvenile. Gabon. Donated by de Blandinière.

MHNT ZOO 2011.0.18.6 Skull.

MHNT ZOO 2011.0.18.7 Skull. Male. Chakes region, Haut Ogooué, Gabon. Purchased from Dubairgh in January 1919.

MHNT ZOO 2011.0.18.8 Skull. Adult male. Donated by Gayraud in 1998.

MHNT ZOO 2011.0.18.9 Skull. Adult male. Congo. Purchased in 1939.

MHNT ZOO 2011.0.18.10 Skull. Male. Gabon. Donated by Laumagne.

MHNT ZOO 2011.0.18.11 Skull. Male. Gabon. Brought back by Bestion; purchased from Constantin on 18 May 1880.

MHNT ZOO 2011.0.18.12 Skull. Male. Gabon. Donated by de Blandinière.

MHNT ZOO 2011.0.18.13 Skull. Male. Gabon. Donated by Lazorthes.

MHNT ZOO 2011.0.18.14 Skull. Female. Donated by Ganin and Lazorthes in 1923.

MHNT ZOO 2011.0.18.18 Skull. Juvenile. Donated by Ganin and Lazorthes in 1923.

MHNT ZOO 2011.0.18.19 Cranium. Female. Donated by Moura on 20 July 1886.

MHNT ZOO 2011.0.20.2 Postcranial skeleton. Adult male.

MHNT ZOO 2011.0.20.3 Skeleton. Adult female.

FIGURES C-3—C-5 Mounted skin of adult female (no. MHNT PRI 3). There is a raised, skin-covered structure in the supraorbital region, but the aetiology and pathogenesis would require detailed investigation. *Photographs by Didier Descouens, courtesy of Muséum d'Histoire Naturelle de Toulouse.*

Troyes

Musée d'Histoire Naturelle, Ancienne Abbaye Saint-Loup, 1 Rue Chrestien de Troyes, 10000 Troyes

Ghislain Grégoire; Céline Nadal

SM 9 Skull. It is missing the lower left canine and both lower left incisors.

SM 10 Skull (plaster cast). Adult male. It is damaged.

Villeneuve d'Ascq

Université des Sciences et Techniques de Lille Flandres Artois, U.F.R. de Biologie, Bâtiment SN3, 59655 Villeneuve d'Ascq*

Andre Dhainaut

817 Mounted skeleton. Adult male. *Gorilla gina.* Donated. Height 150 cm.

818 Mounted skeleton. Adult female. *Gorilla gina*. Donated. Height 120 cm. The skull is missing.

GERMANY

Alfeld

Museum der Stadt Alfeld (Leine), Am Kirchhof 4/5, 31061 Alfeld (Leine)

Ina Gravenkamp; Gerhard Kraus

Mounted skin. Adult male. Congo.

Mounted skin. Adult female. Congo.

Mounted skin. Juvenile female. Congo.

The above specimens are displayed in an African diorama. They were purchased from a local animal dealer named Ruhe, having died in quarantine in Alfeld. They were prepared in Alfeld either by the präparator Carl Bartels or the teacher and founder of the Tiermuseums Alfeld, Alois Brandmüller (1867–1939), between c. 1900 and 1933, when the museum first opened.

Bamberg

Naturkunde-Museum Bamberg, Fleischstrasse 2, 96047 Bamberg

Matthias Mäuser

NKMB-Z-5155 Mounted skeleton. Adult female. Purchased from Hoppe, Hamburg, in 1896. The skull is exhibited in the 'Hall of Birds'; the postcranial skeleton is in a store room. There is a button osteoma on the body of the left mandible, and alveolar resorption at the upper left 1st molar.

Berlin

Freie Universität Berlin, Institut für Biologie, Fachbereich Biologie, Chemie, Pharmazie, Königin-Luise-Strasse 1/3, 14195 Berlin

Dieter Jung; Jens Rolff; Heidi Schindler

Mounted skeleton. Infant male. Received from the Zoologischer Garten und Aquarium Berlin, Hardenbergplatz 8, 10787 Berlin, after its death in December 1956, aged about 6–8 months. Mounted by Steinmetzler. Total height 54 cm.

Leibniz-Institut für Zoo-und Wildtierforschung, Alfred-Kowalke-Strasse 17, 10315 Berlin

Heribert Hofer

Access to the Institute's samples is possible in principle but restricted in the sense that they usually work out a Memorandum of Understanding prior to access, to sort out what this access entails, how the research will be published and how the contribution to the research will be shared.

Department of Reproductive Management:

Roland Frey; Thomas B. Hildebrandt

Digitalised video sequences of examinations by ultrasound of 18 gorillas (i.e., 12 adult females, 1 juvenile female and 5 adult males).

Left hand. Adult female. Donated by John E. Cooper on 21 July 1998. The skin of the dorsum has been removed over an area of 12×9 cm, with damage to the underlying muscles and tendons. On the palm, a piece of skin has been taken from an area of 11×3 cm at the distal end of the metacarpals, likewise with damage to the underlying muscles and tendons. Pieces of skin 2 cm wide and 0.5 to 1 cm long have been excised from the dorsal area of the distal phalanx of the 2nd, 3rd and 4th digits, with damage to the underlying layers of tissue. This hand came from the female 'G-Ann', who was born in the Oklahoma City Zoo, USA, on 8 June 1979. The dam was 'Fern' and the sire was 'Moemba'. She was nursery-reared, and weighed 3½ lb. at 2 days old. She was sent to Jersey Zoo (Durrell Wildlife Park), in exchange for the male 'Tatu', on 14 August 1983. She lost a hand (amputated, surgically) due to an unanticipated adverse reaction to a neuroleptic (antipsychotic) agent prior to her move to Melbourne Zoo, where she arrived on 7 December 1997 (along with another female from Jersey, named 'Julia'). [Some additional information added by GH from correspondence with Oklahoma City Zoo personnel.]

Department of Wildlife Diseases:

Gudrun Wibbelt

NB: All tissue samples are embedded in paraffin blocks.

376/71 Tissue samples of liver, lung, kidney, brain, heart, spleen and stomach/intestine. Female. This is 'Kama', who arrived at the Tierpark Berlin-Friedrichsfelde GmbH on 9 July 1965 (weight c. 10 kg). She was caught in the wild, possibly in Cameroon. She died on 25 May 1971 (weight 55.2 kg) of cardiac paralysis during anaesthesia after tooth extraction.

766/74 Tissue samples of liver, lung, kidney, brain, heart, spleen, stomach/intestine and adrenal gland. Infant female (1½ years old). Captive specimen in Wrocław Zoo, Poland; wild-caught in Cameroon. Date of necropsy 27 November 1974 (weight 8.045 kg). Cause of death: unspecified parasitosis.

437/75 Tissue samples of liver, lung, kidney, brain, heart, spleen, stomach/intestine and lymph node. Subadult female. This is 'Fanta', who was wild-caught in Cameroon, and arrived at the Wrocław Zoo on 28 August 1974, when aged about 3 years (weight 15 kg). She was purchased, with a male named 'Tadao', from G. van den Brink, Soest, Netherlands. The pair were owned by Dvůr Králové Zoo, Czech Republic. 'Fanta' died of purulent/gangrenous pneumonia on 17 July 1975 (weight 12.2 kg). [Some additional information added by GH from correspondence with Wrocław Zoo Zoo personnel.]

798/78 Tissue samples of liver, lung, kidney, brain, heart, spleen, stomach/intestine and thyroid. Adult female. This is 'Sani', who was wild-caught in Cameroon, and arrived at the Wrocław Zoo on 29 June 1974 when aged about 2½ years (weight 8 kg; very thin). She was purchased, with a male named 'Soko', from G. van den Brink. She died on 13 December 1978 of haemorrhagic colitis; salmonellosis; and dilatation and atrophy of the cardiac muscle. [Some additional information added by GH from correspondence with Wrocław Zoo Zoo personnel.]

132/79 Tissue samples of liver, lung, kidney, heart, spleen, stomach/intestine, skin, adrenal gland and thyroid. Adult female (8 years old). Date of necropsy: 23 February 1979 (weight 66.2 kg). Captive specimen in Wrocław Zoo, Poland; wild-caught in Cameroon. Cause of death: pleuropneumonia.

434/79 Tissue samples of liver, lung, kidney, brain, heart, spleen, stomach/intestine, lymph node and thyroid. Adult male. This is 'Willy', who arrived at the Wrocław Zoo on 17 November 1970, aged about 3 years (weight 16.5 kg). He was wild-caught in Cameroon and was purchased from G. van den Brink. He died of pneumonia on 12 July 1979. [Some additional information added by GH from correspondence with Wrocław Zoo Zoo personnel.]

300/85 Tissue samples of liver, lung, kidney, brain, heart, spleen and stomach/intestine. Adult male (16 years old). Date of necropsy: 26 April 1985 (weight 70 kg). Wild-caught in Cameroon; captive specimen in Wrocław Zoo. Cause of death: nephritis.

218/07 Tissue sample of rectum. Adult female. Received on 28 November 2007. This is from 'Dufte', who was born in the Zoologischer Garten Berlin on 30 October 1974, after a gestation period of 244 days. The dam was 'Fatou' and the sire was 'Knorke II'. She was mother-reared. She died of chronic pyometra with uterus rupture and subsequent peritonitis on 13 December 2001. [Some additional information added by GH from correspondence with Berlin Zoo Zoo personnel.]

Museum für Naturkunde, Invalidenstrasse 43, 10115 Berlin

Christiane Funk

Skulls × 6 (no numbers or available data).

Study skin (stuffed, unmounted). Juvenile.

Skull. Collected by Escherich.

Study skin (stuffed, unmounted).

Skin. Female. Bipindi, Kamerun. Collected by Georg A. Zenker.

ZMB_Mam_6046 Partial skeleton (no mandible, legs, arms), partly wired. Adult female. Gabon. Acquired from Gerrard.

ZMB_Mam_6962 Skull. Adult male. Gabon. Collected by Handmann.

ZMB_Mam_6964 Skull. Subadult. Gabon. Collected by Handmann.

ZMB_Mam_6965 Skull. Female. Ogowe, Gabon. Collected by B.S. Lenz.

ZMB_Mam_6966 Skull. Male. Kamerun. Collected by Kohrs on 2 March 1905.

ZMB_Mam_6968 Skull. Adult female. Gabon. Collected by Brehmer.

ZMB_Mam_6969 Skull. Gabon. Collected by Handmann.

ZMB_Mam_6980 Skull. Ogowe, Gabon. Collected by Lenz.

ZMB_Mam_6982 Skull. Juvenile. Gabon. Collected by Handmann.

ZMB_Mam_6989 Skull. Gabon. Collected by Buchholz.

ZMB_Mam_7157 Skull. Subadult male. Samakïta, Ogowe, Gabon. Collected by Schmidt.

ZMB_Mam_7793 Skull. Juvenile female. Gabon. Collected by Buchholz.

ZMB_Mam_7907 Skeleton (with ligaments); parts in alcohol. Juvenile female. Gabon. Collected on 6 November 1868. Acquired from Gerrard.

ZMB_Mam_10493 Skeleton. Subadult male. Kakamöeka, Mayombe, French Congo. Collected by Falkenstein.

ZMB_Mam_11642 Partial skeleton (no hind limbs). Adult male. Yaunde, Kamerun. Collected by Georg A. Zenker.

ZMB_Mam_11643 Skull. Bipindi, Kamerun. Collected by Georg A. Zenker.

ZMB_Mam_11644 Skeleton (partly wired). Juvenile female. Collected by Garraud.

ZMB_Mam_11645 Skull. Male. Bipindi, Kamerun. Collected by Georg A. Zenker.

ZMB_Mam_11646 Skull. Bipindi, Kamerun. Collected by Georg A. Zenker.

ZMB_Mam_11652 Skull. Male. Sanaga, Semikore, Kamerun. Collected by Scheunemann.

ZMB_Mam_11653 Skull. Male. Kamerun. Collected by von Stein.

ZMB_Mam_11654 Skull. Juvenile male. Acquired from Gerrard.

ZMB_Mam_11683 Skeleton (partly wired). Subadult male. Gabon. Collected by Brehmer.

ZMB_Mam_12205 Skull. Female. Yaunde, Campo, Kamerun. Collected by Paschen on 13 March 1905.

ZMB_Mam_12206 Partial skeleton (no hind limbs). Female. Melem, Kamerun. Collected by Georg A. Zenker on 16 March 1905.

ZMB_Mam_12789 Skull. Adult male. *Gorilla diehli* (holotype). Adult male. Dakbe, Cross River area, Kamerun. Collected by Diehl.

ZMB_Mam_12790 Skull. Adult male. *Gorilla diehli* (paratype). Forest at Oboni, Kamerun. Collected by Diehl.

ZMB_Mam_12791 Skeleton (incomplete: no mandible; cranium black). Adult male. *Gorilla diehli* (paratype). Oboni, Kamerun. Collected by Diehl.

ZMB_Mam_12792 Cranium. Adult male. *Gorilla diehli* (paratype). Dakbe, Kamerun. Collected by Diehl.

ZMB_Mam_12793 Cranium. Adult female. *Gorilla diehli* (paratype). Basho, Kamerun. Collected by Diehl.

ZMB_Mam_12794 Cranium (black). Adult female. *Gorilla diehli* (paratype). Basho, Kamerun. Collected by Diehl.

ZMB_Mam_12795 Cranium. Male. *Gorilla diehli* (paratype). Basho, Kamerun. Collected by Diehl.

ZMB_Mam_12796 Cranium. Adult female. *Gorilla diehli* (paratype). Oboni, Kamerun. Collected by Diehl.

ZMB_Mam_12799 Skull. Female. Basho, Kamerun. Collected by Diehl on 21 March 1905.

ZMB_Mam_13254 Partial skeleton (no hands). Adult male. *Gorilla beringei* (holotype). Kirunga ya Sabinyo, Rwanda. Collected by Friedrich Robert von Beringe and Engeland in 1902.

ZMB_Mam_14644 Skull. Male. Loango, Angola. Collected by Güssfeldt.

ZMB_Mam_14645 Skull. Male. Campo, Kamerun. Collected by Lenz.

ZMB_Mam_14647 Skull. Female. Campo, Kamerun. Collected by Lenz.

ZMB_Mam_16058 Skin. Juvenile male. Received from Berliner Aquarium.

ZMB_Mam_17658 Skeleton (incomplete). Adult male. North of Momie, Kamerun. Collected by von Ramsey.

ZMB_Mam_17802 Skull. Male. Bush near Abong-Mbang, Kamerun. Collected by Antelmann on 27 March 1905.

ZMB_Mam_17960 Skeleton. Male. *Gorilla hansmeyeri* (holotype). Between Mensina and Bimba, south of Dume River, Kamerun. Collected by Peter on 27 January 1907. Received from Hans Meyer. It is now missing.

ZMB_Mam_17961 Skin. Male. *Gorilla hansmeyeri* (holotype). Between Mensina and Bimba, south of Dume River, Kamerun. Collected by Peter on 27 January 1907. Received from Hans Meyer.

ZMB_Mam_17963 Skull. Male. Nola, Kamerun. Collected by Escherich.

ZMB_Mam_18515 Partial skeleton (no hind limbs). Adult male. Edea region, Kamerun. Collected by Göpfert.

ZMB_Mam_18516 Partial skeleton (no hind limbs). Adult male. Edea region, Kamerun. Collected by Göpfert.

ZMB_Mam_18519 Skin. Juvenile. Edea region, Kamerun. Collected by Göpfert, Stein and Herbst on 2 April 1905.

ZMB_Mam_18643 Skull. Male. Basho, Kamerun. Collected by Adametz.

ZMB_Mam_20305 Skull. Male. Area of Nginda, 20 km N. of Molundu, Kamerun. Collected by R. von Stetten.

ZMB_Mam_20306 Skull. Area of Nginda, N. of Molundu, Kamerun. Collected by R. von Stetten.

ZMB_Mam_20307 Skull. Male. Area of Nginda, Molundu, Kamerun. Collected by R. von Stetten.

ZMB_Mam_20318 Skull. Area of Nginda, Molundu, Kamerun. Collected by R. von Stetten.

ZMB_Mam_24835 Partial skeleton (no legs, but feet). Adult female. *Gorilla beringei beringei*. Mount Sabinio, altitude 2800 m, Belgian Congo. Collected by Stemmermann.

ZMB_Mam_24836 Partial skeleton (no hands and feet; not prepared, 'mummified'). Infant male. *Gorilla beringei beringei*. Mount Sabinio, Belgian Congo. Collected by Stemmermann on 7 August 1916.

ZMB_Mam_27516 Scalp (skin of). Male. Hinterland of Fernan Vaz, Gabon. Collected by B. Schneider.

ZMB_Mam_30260 Postcranial skeleton and skin. *Gorilla zenkeri* (holotype). Bipindi, Mbiawe, Lokundje, Kamerun. Collected by Georg A. Zenker. Both are now missing.

ZMB_Mam_30261 Skull. Adult male. *Gorilla zenkeri* (holotype). Mbiawe, Lokundje, 6 hours downstream from Bipindi. Collected by Georg A. Zenker on 22 March 1905.

ZMB_Mam_30306 Skeleton. Juvenile. Lodibonga at Edea, Kamerun. Collected by Stein; acquired from Herbst.

ZMB_Mam_30352 Skin. Area of Edea, Kamerun. Collected by Göpfert, Stein and Herbst on 2 April 1905.

ZMB_Mam_30890 Skeleton (not prepared), skin and organs (liver, spleen, intestine) in alcohol. Juvenile male. Ajos, Akonolinga, Kamerun. Collected by Reichenow on 3 April 1905.

ZMB_Mam_30891 Skull and skin. Male. Ajos, Akonolinga, Kamerun. Collected by Reichenow on 3 April 1905.

ZMB_Mam_30892 Skeleton (not prepared). Subadult male. Ajos, Akonolinga, Kamerun. Collected by Reichenow on 3 April 1905.

ZMB_Mam_30893 Skeleton, skin and intestines (in alcohol). Juvenile male. Ajos, Akonolinga, Kamerun. Collected by Reichenow on 3 April 1905.

ZMB_Mam_30938 Partial skeleton (no legs, but feet; partly wired). Male. Bipindi, Kamerun. Collected by Georg A. Zenker.

ZMB_Mam_30939 Skeleton (complete with ligaments). Juvenile male. Received from von Hansemann.

ZMB_Mam_30940 Skull. Male. Kamerun.

ZMB_Mam_30941 Skull. Male. Maka, French Congo.

ZMB_Mam_30942 Skull. Acquired from Umlauff.

ZMB_Mam_30943 Skull. Female. Acquired from Umlauff.

ZMB_Mam_31229 Skeleton. Juvenile female. Between Dume and Kadei, Kamerun. Collected by Bluhm; received from Zukowsky.

ZMB_Mam_31275 Skull. Mbusu, Ogowe, Gabon. Collected by Buchholz.

ZMB_Mam_31276 Skull. Juvenile male. Gabon. Collected by Buchholz in 1875.

ZMB_Mam_31277 Skull. Male. Gabon. Collected by Buchholz.

ZMB_Mam_31432 Skin. Juvenile. Ajos, Akonolinga, Kamerun. Collected by Reichenow on 3 April 1905.

ZMB_Mam_31433 Skeleton. Juvenile male. Ajos, Akonolinga, Kamerun. Collected by Reichenow on 3 April 1905.

ZMB_Mam_31434 Skin. Ajos, Akonolinga, Kamerun. Collected by Reichenow on 3 April 1905.

ZMB_Mam_31435 Partial skeleton (missing right leg). Adult female. Ajos, Akonolinga, Kamerun. Collected by Reichenow on 3 April 1905.

ZMB_Mam_31436 Skin. Ajos, Akonolinga, Kamerun. Collected by Reichenow on 3 April 1905.

ZMB_Mam_31437 Partial skeleton (right hand and feet missing; left hand mummified). Adult female. Ajos, Akonolinga, Kamerun. Collected by Reichenow on 3 April 1905.

ZMB_Mam_31616 Skin. Juvenile male. *Gorilla beringei beringei*. Mount Sabinio, Belgian Congo. Collected by Wintgens on 12 June 1911.

ZMB_Mam_31617 Skeleton. Juvenile male. *Gorilla beringei beringei*. Mount Sabinio, Belgian Congo. Collected by Wintgens on 12 June 1911.

ZMB_Mam_31618 Skin. Adult male. *Gorilla graueri* (holotype). Wabembe, Belgian Congo. Collected by Rudolf Grauer in June 1908.

ZMB_Mam_31619 Skeleton. Adult male. *Gorilla graueri* (holotype). Wabembe, 80 km northwest of Boko, on the West bank of Lake Tanganjika, 2000–3000 m, Belgian Congo. Collected by Rudolf Grauer in 1908.

ZMB_Mam_31621 Cranium. Adult male. *Gorilla beringei graueri*. Wabembe, 80 km northwest of Boko, near Mutambale, 2200–2600 m, Belgian Congo.

ZMB_Mam_31622 Skull. Adult male. *Gorilla beringei graueri*. Wabembe, 80 km northwest of Boko, near Mutambale, 2200–2600 m, Belgian Congo.

ZMB_Mam_31624 Partial skeleton (no legs, but feet) and skin. Subadult male. *Gorilla beringei graueri*. Wabembe, 80 km northwest of Boko, near Mutambale, Belgian Congo. Collected by Rudolf Grauer.

ZMB_Mam_31626 Partial skeleton (no ribs). Adult female. *Gorilla beringei graueri*. Wabembe, 80 km northwest of Boko, near Mutambale, Belgian Congo.

ZMB_Mam_31627 Skull. Juvenile female. *Gorilla beringei graueri*. Wabembe, 80 km northwest of Boko, near Mutambale, Belgian Congo.

ZMB_Mam_31628 Cranium. Juvenile female. *Gorilla beringei graueri*. Wabembe, 80 km northwest of Boko, near Mutambale, Belgian Congo.

ZMB_Mam_31630 Skull. Juvenile female. *Gorilla beringei graueri*. Wabembe, 80 km northwest of Boko, near Mutambale, Belgian Congo.

ZMB_Mam_31632 Skeleton (not prepared). Infant female. *Gorilla beringei graueri*. Wabembe, 80 km northwest of Boko, near Mutambale, Belgian Congo.

ZMB_Mam_31655 Skeleton (with ligaments). Juvenile male. Gabon. Supplied originally by Frank, received from Anatomisch Sammlung.

ZMB_Mam_31656 Skull. Juvenile male. Dume, Kamerun. Collected by Freyer.

ZMB_Mam_31702 Skull and skin. Juvenile female. Bipindi, Kamerun. Collected by Georg A. Zenker on 23 March 1905.

ZMB_Mam_31806 Partial skeleton (no cranium or hind limbs). Female. Kamerun. Collected by O. Nagy.

ZMB_Mam_32393 Skeleton (badly prepared). Infant. Received from Anat. Biol. Institut.

ZMB_Mam_33490 Skull. Male. French Congo. Collected by Bier.

ZMB_Mam_33669 Skull. Male. French Congo. Received in exchange from O. Ruge.

ZMB_Mam_37523 Skull and skin. Juvenile female. *Gorilla beringei graueri*. Edge of the mountains of the north shore of Lake Tanganyika, Burundi. Collected on 6 July 1908.

ZMB_Mam_41660 Specimen in alcohol. Collected on 16 December 1892.

ZMB_Mam_44480 Skull. Female. Besam, South Kamerun. Collected by C.W.H. Koch. Received in exchange from O. Ruge.

ZMB_Mam_44550 Skull. Female. Río Muni, Equatorial Guinea. Collected by O. Krohnert. Received from Umlauff.

ZMB_Mam_47484 Specimen in alcohol. Male. Received from Prof. Dr Gustav Brandes (1862–1941), Zoo Dresden.

ZMB_Mam_47526 Skeleton, mounted skin, death mask, and hand and organs in alcohol. Male. Probably from Sembé, southwest of Molundu, Cameroun. Received from Zoologischer Garten Berlin in 1935. This is 'Bobby', who lived in Zoo Berlin from 1928 until his death on 1 August 1935. NB: The death mask should not be confused with the cast made from the face after the mount was prepared. Copies of the latter were sold and given away as presents.

ZMB_Mam_48170 Skull. Male. Tinto, Ossidinge, Kamerun. Collected by von Oertzen.

ZMB_Mam_48171 Skull. Male. Akoafim, Ebolowa, Kamerun. Collected by von Oertzen in 1907.

ZMB_Mam_48172 Skull. Lomié, Lomié District, Kamerun. Collected by von Oertzen.

ZMB_Mam_48173 Cranium. Male. Ossidinge District, Kamerun. Collected by von Oertzen.

ZMB_Mam_74708 Skull. Male.

ZMB_Mam_78193 Skin. South of Molundu, Kamerun. Collected by von der Marwitz on 10 April 1912.

ZMB_Mam_78194 Skin.

ZMB_Mam_78195 Skin.

ZMB_Mam_83519 Skull. Ossidinge, Kamerun. Collected by Mansfeld, 1913/1914.

ZMB_Mam_83520 Skull. Male. Balaga, between Kadel and Bange, eastern border of Kamerun. Collected by Schwarz.

ZMB_Mam_83521 Skeleton and testes (in alcohol). Subadult male. Molundu, Kamerun. Collected by von der Marwitz on 26 March 1905.

ZMB_Mam_83522 Skull. Male. Ossidinge, Kamerun. Collected by Mansfeld, 1913/1914.

ZMB_Mam_83524 Skull. Juvenile. Yoko, Kamerun. Collected by von Oertzen.

ZMB_Mam_83525 Cranium. Basho, Kamerun. Collected by Adametz.

ZMB_Mam_83526 Skull. Ossidinge, Kamerun. Collected by Mansfeld, 1913/1914.

ZMB_Mam_83527 Cranium. Male. Anjany, Basho, Kamerun. Collected by Adametz.

ZMB_Mam_83528 Cranium. Male. Mele Mtun, Kamerun. Collected by Tessmann.

ZMB_Mam_83529 Skull. Male. Noja District. Collected by Escherich.

ZMB_Mam_83530 Partial skeleton (no hind limbs). Adult male. Dume, Kamerun. Collected by Schipper on 24 March 1905.

ZMB_Mam_83531 Skull.

ZMB_Mam_83532 Skull.

ZMB_Mam_83533 Skull. Female.

ZMB_Mam_83534 Skull.

ZMB_Mam_83535 Skull.

ZMB_Mam_83536 Cranium.

The above 6 specimens are from Ossidinge, Kamerun, and were collected by Mansfeld, 1913/1914.

ZMB_Mam_83537 Cranium. Male. Mouth of the Lobo river, Kamerun. Collected by Jacob on 19 March 1905.

ZMB_Mam_83538 Skull.

ZMB_Mam_83539 Skull.

ZMB_Mam_83540 Skull.

ZMB_Mam_83541 Skull.

ZMB_Mam_83542 Skull.

The above 5 specimens are from Ossidinge, Kamerun, and were collected by Mansfeld, 1913/1914.

ZMB_Mam_83543 Skull. Male. Lomie or mouth of Dume river, Kamerun. Collected by Reuter.

ZMB_Mam_83544 Cranium. Male. Ossidinge, Kamerun. Collected by Mansfeld, 1913/1914.

ZMB_Mam_83545 Partial skeleton (no hind limbs). Adult male. Kamerun. Collected by Schwarz.

ZMB_Mam_83546 Skull. Male. Between Lomié and Malen, Kamerun. Collected by Eichelberger; received from Rolle.

ZMB_Mam_83549 Cranium. Ossidinge, Kamerun. Collected by Mansfeld, 1913/1914.

ZMB_Mam_83550 Skull. Male. Lolodorf, Kamerun. Collected by Jacob on 2 April 1905.

ZMB_Mam_83551 Skull. Female. Mouth of Lobo river, Kamerun. Collected by Jacob.

ZMB_Mam_83552 Skull. Male. Ossidinge, Kamerun. Collected by Mansfeld, 1913/1914.

ZMB_Mam_83553 Skull. Male. Basho. Collected by Rohde.

ZMB_Mam_83554 Skull. Ossidinge, Kamerun. Collected by Mansfeld, 1913/1914.

ZMB_Mam_83555 Skull. Ossidinge, Kamerun. Collected by Mansfeld, 1913/1914.

ZMB_Mam_83556 Skull. Male. Mouth of Lobo river, Kamerun. Collected by Jacob.

ZMB_Mam_83557 Skull. Ossidinge, Kamerun. Collected by Mansfeld, 1913/1914.

ZMB_Mam_83558 Cranium. Adult male. *Gorilla jacobi* (holotype). Mouth of Lobo river, South Kamerun. Collected by Gerhard Jacob on 19 March 1905.

ZMB_Mam_83559 Skull. Female. Ossidinge, Kamerun. Collected by Mansfeld, 1913/1914.

ZMB_Mam_83560 Skull. Male. Collected by O. Eichelberger; received from Rolle.

ZMB_Mam_83561 Partial skeleton (no spine, hands or hind limbs) and skin. Adult male. Lolodorf, Kamerun. Collected by Jacob.

ZMB_Mam_83562 Skull. Juvenile male. Mouth of Lobo river, Kamerun. Collected by Jacob on 19 March 1905.

ZMB_Mam_83565 Skull. Ossidinge, Kamerun. Collected by Mansfeld, 1913/1914.

ZMB_Mam_83566 Cranium. Male. Mouth of Lobo river, Kamerun. Collected by Jacob on 19 March 1905.

ZMB_Mam_83568 Skull. Male. Surroundings of Nginda, Molundu, Kamerun. Collected by von Stetten.

ZMB_Mam_83569 Skull. Male.

ZMB_Mam_83570 Skull. Male.

ZMB_Mam_83571 Skull.

ZMB_Mam_83572 Skull. Male.

ZMB_Mam_83573 Skull.

ZMB_Mam_83574 Skull. Juvenile male.

ZMB_Mam_83575 Skull. Female.

ZMB_Mam_83576 Skull. Male.

ZMB_Mam_83577 Cranium. Female.

ZMB_Mam_83578 Skull. Male.

ZMB_Mam_83579 Skull. Male.

The above 11 specimens are from Ossidinge, Kamerun, and were collected by Mansfeld, 1913/1914.

ZMB_Mam_83580 Skull. Juvenile. Northwest of Molundu, between Lomié and Malen, Kamerun. Collected by von der Marwitz on 24 March 1905.

ZMB_Mam_83581 Skull and skin. Lolodorf, Kamerun. Collected by Jacob on 21 March 1905.

ZMB_Mam_83582 Skull. Male. Eastern border of Kamerun. Collected by Schwarz on 24 March 1905.

ZMB_Mam_83583 Cranium. Female. Ossidinge, Kamerun. Collected by Mansfeld, 1913/1914.

ZMB_Mam_83584 Skeleton. Juvenile female. Bipindi, Kamerun. Collected by Georg A. Zenker.

ZMB_Mam_83585 Partial skeleton (no spine, hind limbs or hands). Ogowe, Gabon. Collected by Lenz.

ZMB_Mam_83586 Partial skeleton (no spine, hind limbs, or ribs). Bipindi, Kamerun. Collected by Georg A. Zenker on 22 March 1905.

ZMB_Mam_83588 Skeleton and skin. Infant. Molundu, Kamerun. Collected by Reichenow.

ZMB_Mam_83590 Partial skeleton (hands and feet incomplete). Juvenile. Ogowe, Gabon. Collected by Lenz.

ZMB_Mam_83688 Skeleton. Infant. Lolodorf, Kamerun. Collected by Jacob on 2 April 1905.

ZMB_Mam_83689 Skull. Juvenile. Ajos, Akonolinga, Kamerun. Collected by Reichenow on 3 April 1905.

ZMB_Mam_83807 Skin. Male. Yokaduma, Kamerun. Collected by Koch on 2 April 1905.

ZMB_Mam_83808 Skin. Kamerun. Collected by Schwarz on 24 March 1905.

ZMB_Mam_83809 Skin. Molundu, Kamerun. Collected by von Stetten on 25 March 1905.

ZMB_Mam_83810 Skin. Kamerun. Collected by Schwarz on 24 March 1905.

ZMB_Mam_83811 Skin. Nzimuland, between Ngoko and Sanga. Collected by Förster on 18 March 1905.

ZMB_Mam_83812 Skin. Juvenile. Lolodorf, Kamerun. Collected by Jacob.

ZMB_Mam_83813 Skin. Juvenile. Lolodorf, Kamerun. Collected by Jacob on 2 April 1905.

ZMB_Mam_83828 Skin. South Kamerun. Collected by Förster.

ZMB_Mam_83862 Skull. Female. *Gorilla jacobi* (paratype). Mouth of Lobo river, South Kamerun. Collected by Gerhard Jacob on 10 July 1905.

ZMB_Mam_85827 Skull. Male. Kamerun. Collected by Mansfeld, 1913/1914.

ZMB_Mam_85828 Skull. Female. Kamerun. Collected by Mansfeld, 1913/1914.

ZMB_Mam_85829 Ossidinge, Kamerun. Collected by Mansfeld, 1913/1914.

ZMB_Mam_85830 Skull. Juvenile.

ZMB_Mam_85832 Skull. Juvenile. Acquired from Umlauff.

ZMB_Mam_85833Skull. Male. Ossidinge, Kamerun. Collected by Mansfeld, 1913/1914.

ZMB_Mam_85834 Mandible and skin. Infant. Molundu, Kamerun. Collected by von der Marwitz.

ZMB_Mam_85835 Skull. Female. Noja basin, southern Spanish Guinea. Collected by Escherich.

ZMB_Mam_85836 Skull. Ossidinge, Kamerun. Collected by Mansfeld, 1913/1914.

ZMB_Mam_85837 Skull. Ossidinge, Kamerun. Collected by Mansfeld, 1913/1914.

ZMB_Mam_85838 Skull. Female. Ossidinge, Kamerun. Collected by Mansfeld, 1913/1914.

ZMB_Mam_85839 Skull. Juvenile. Nehring collection, received from Umlauff.

ZMB_Mam_85840 Skull. Male. Ossidinge, Kamerun. Collected by Mansfeld, 1913/1914.

ZMB_Mam_85841 Skull. Male. Ossidinge, Kamerun. Collected by Mansfeld, 1913/1914.

ZMB_Mam_85842 Skull. Male. Ossidinge, Kamerun. Collected by Mansfeld, 1913/1914.

ZMB_Mam_85843 Skull. Male. Collected by Noack.

ZMB_Mam_85844 Partial skeleton (no hind or fore limbs, hands incomplete). Adult male. Collected by Lenz.

ZMB_Mam_85845 Mandible. Infant. Received from Museum Hannover.

ZMB_Mam_88245 Skin. Lambaréné, Gabon. Received from Museum Basel.

ZMB_Mam_88247 Skin. Juvenile. Collected on 22 January 1910.

ZMB_Mam_88248 Skin.

ZMB_Mam_88795 Skull. Adult male.

ZMB_Mam_90290 Baculum. Male.

ZMB_Mam_102662 Specimen in alcohol. Male. Bipindi, Kamerun. Collected by Georg A. Zenker on 20 October 1913.

ZMB_Mam_102663 Specimen in alcohol. *Gorilla hansmeyeri*. Collected by Meyers.

ZMB_Mam_102664 Specimen in alcohol. Collected by Reichenow.

ZMB_Mam_102665 Specimen in alcohol. Kamerun. Collected by Knuth on 8 March 1912.

ZMB_Mam_102666 Specimen in alcohol. Kamerun. Collected by Knuth on 8 March 1912.

ZMB_Mam_102667 Specimen in alcohol. Collected by Reichenow.

ZMB_Mam_102668 Specimen in alcohol. Molundu, Kamerun. Collected by Poll.

ZMB_Mam_102669 Specimen in alcohol.

ZMB_Mam_102670 Specimen in alcohol. Juvenile. Collected by Grauer.

ZMB_Mam_102671 Specimen in alcohol. Collected by Marwitz on 11 April 1912.

ZMB_Mam_102672 Specimen in alcohol. Bipindi, Kamerun. Collected by Georg A. Zenker on 29 June 1912.

ZMB_Mam_102673 Specimen in alcohol.

ZMB_Mam_102674 Specimen in alcohol. Kamerun. Collected by Reichenow.

ZMB_Mam_102675 Specimen in alcohol. Kamerun. Collected by Reichenow.

ZMB_Mam_102676 Specimen in alcohol. Kamerun. Collected by Reichenow.

ZMB_Mam_102677 Specimen in alcohol. Kamerun. Collected by Schwarz.

ZMB_Mam_102678 Specimen in alcohol. Female. Kamerun. Collected by Jacob on 1 December 1906.

ZMB_Mam_102679 Specimen in alcohol. Kamerun. Collected by Reichenow.

ZMB_Mam_102680 Specimen in alcohol. Kamerun. Collected by Reichenow.

ZMB_Mam_102681 Specimen in alcohol. Infant male. Ajos, Akonolinga, Kamerun. Collected by Reichenow on 17 February 1920.

ZMB_Mam_102682 Specimen in alcohol. Bipindi, Kamerun. Collected by Georg A. Zenker.

Bochum

Ruhr-Universität Bochum, Institut für Anatomie, Abteilung für Funktionelle Morphologie, Universtitätsstrasse 150, 44801 Bochum*

Holger Preuschoft

Mounted skeleton (incomplete). Subadult male. The skull has been cut in the medial sagittal plane, and one-half of the mandible into smaller pieces. The postcranial skeleton consists mainly of the vertebral column and ribs, pelvis and parts of the fore- and hindlimbs. The skull and postcranial bones have been combined to make a whole skeleton, but it is not known whether the skull belongs with the other parts. Pig teeth have been

inserted into the jaws in place of the 3rd molars. The material was obtained in the early 1970s from the Anthropological Institute in Tübingen, having been in Germany since 1950 at least.

Bonn

Zoologisches Forschungsinstitut und Museum Alexander Koenig, Adenauerallee 160, 53113 Bonn

Rainer Hutterer

ZFMK 38.135 Skull. Adult male. Received from E.A. Böttcher in April 1938.

ZFMK 57.3 Skull. Adult male. Abong-Abong (=Abong-Mbang), Kamerun. Purchased from Umlauff, Hamburg.

ZFMK 57.4 Skull. Adult male. Abong-Abong, Kamerun. Purchased from Umlauff, Hamburg.

ZFMK 58.72 Skull. Female. Yaoundé, Cameroun. Collected by Sohl in 1957. It is currently missing.

ZFMK 60.124 Skull; skin; casts of head/shoulder, right hand and left foot. Juvenile male. Received from the Zoologischer Garten Köln on 22 June 1960.

ZFMK 63.660 Skeleton and skin. Adult male. Received from the Zoologischer Garten und Aquarium Berlin on 4 September 1963. This may be 'Knorke', who arrived at the Zoo Berlin on 7 June 1957, aged about 2 years. He died of peritonitis on 30 August 1963, weighing 55 kg. The specimen is currently missing. [Some additional information added by GH from correspondence with Berlin Zoo personnel.]

ZFMK 65.500 Skeleton and skin. Juvenile female. Received from Zoo Duisburg AG, Mülheimer Strasse 273, 4100 Duisburg, on 7 January 1965.

ZFMK 97.076 Skeleton. Adult female. *Gorilla graueri* (paratype?). Sibatwa, 3°50′S, 28°55′E, SE Itombwe massif, Congo Free State. Collected in 1907. A hand-written label by E. Schwarz states: 'Sibatoa's, W Baraka, Ubemba, NW Tanganyika, Belg. Kongo. From the original series.'

ZFMK 97.087 Mounted skin. Adult female. *Gorilla beringei graueri*. Mountain forests at Moira, altitude 1100 m, 0°39′N, 29°30′E, NW Lake Tanganyika, Congo Free State. Collected by Rudolf Grauer in 1907. It is currently missing.

ZFMK 97.088 Mounted skin. Adult male. Dume River, S Dume [Doumé], 4°13′N, 13°30′E, SW Kamerun. Collected by Wichers in 1911 (field no. 183). Purchased from J. Umlauff, Hamburg.

ZFMK 97.089 Ankle bone. Female. Received from Kerz.

ZFMK 99.933 Death mask. Adult male. Donated by I. Riemer. This is from 'Bobby', who lived in the Zoo Berlin from 1928 until his death on 1 August 1935.

ZFMK 2001.302 Skeleton and skin. Adult male. Received in 2001 from Zoologischer Garten Wuppertal. This is 'Nangai', who was born in Zoo Basel on 8 October 1990 (to dam 'Faddama' and sire 'Pepe'). He was transferred to Wuppertal on 30 April 1997, and died there on 4 November 2001. Head and body length 1220 mm; foot length 330 mm. [Some additional information added by GH from International Studbook for the Western Lowland Gorilla, 2010.]

ZFMK 2008.326 Carcase in fluid. Infant male. This is 'Wim', who was born in Kölner Zoo on 4 July 1999 (to dam 'N'Datwa' and sire 'Kim'). He was killed by conspecifics on 15 August 1999. [Some additional information added by GH from the International Studbook for the Western Lowland Gorilla, 2010.]

Bremen

Übersee-Museum, Bahnhofsplatz 13, 28195 Bremen*

2283 Mounted skeleton.

2299 Skull and skin. Juvenile male. Donated by Frau Köhne, Kamerun, on 21 June 1937.

2962 Skull (sectioned longitudinally). The occipital region is damaged.

3077 Skull. The mandible is broken and parts are missing.

3168 Skeleton (raw) and mounted skin. Juvenile male. Donated by the Tierpark Bremen on 10 June 1966. The two lower left incisors are abnormal.

3169 Skeleton (raw) and skin. Juvenile female. Donated by the Tierpark Bremen on 13 June 1966. Note: the Tierpark Bremen was founded in 1966 and closed in 1973. It was owned by George Munro Jr, an animal dealer based in Calcutta, India.

4019 Skull. Purchased from Fritsche in 1956.

4052 Cranium. Kamerun. Received from the Landesmuseum für Kunst- und Kulturgeschichte (Focke Museum), Bremen, in 1964. It had been purchased by the Focke Museum from the collection of August Wulff, Bremen, in 1953. It was originally collected by engineer Rogge on 1 November 1941 and was received by Wulff from Dr Kirchhoff, Richtweg. All the teeth are missing.

4518 Skull. Purchased from Konrad Müller, Knechting bei Bocholt, Westfalia.

Brunswick

Staatliches Naturhistorisches Museum, Pockelsstrasse 10, 38106 Braunschweig*

Kirstin Kuczius; O. v. Frisch

Mounted skin (dermoplastic). Adult male. Prepared in the almost upright position, its height from the foot to the top of the head is 160 cm.

Mounted skin (dermoplastic). Infant female. Age c. 9 months—1 year. Standing height 74 cm.

Mounted skin (dermoplastic). Infant female. Kamerun. Age c. 9 months—1 year. Standing height 58 cm.

Cranium. Female.

Chemnitz

Museum für Naturkunde, Moritzstrasse 20, 09111 Chemnitz

Ronny Rössler

I 01846A Skull. Adult male. It is on exhibition in the Historical Cabinet.

I 00046M Brain (fluid-preserved in ethanol and formaldehyde). It is in K17/1 (Archive).

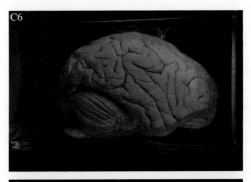

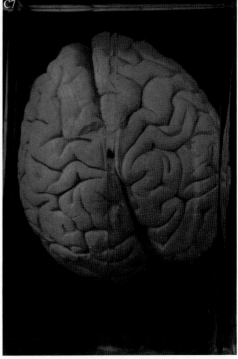

FIGURES C-6—C-8 Fluid-preserved gorilla brain (no. I 00046M). Right lateral (C-6), superior (C-7) and inferior (C-8) aspects. *Photographs by Ronny Rössler, courtesy of Museum für Naturkunde, Chemnitz.*

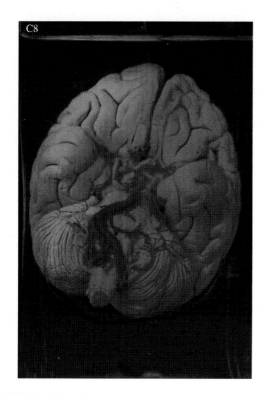

C8

FIGURES C-6—C-8 (Continued)

Cologne

Universität zu Köln, Medizinische Fakultät, Institut I für Anatomie, Joseph-Stelzmann-Strasse 9, 50931 Köln*

XII-t F-2 Skeleton. Juvenile.

XII-t F-3 Skeleton. Subadult male.

XII-t F-4 Skeleton. Male. The occipital region of the skull is damaged.

XII-t F-5 Skull. Juvenile female.

XII-t F-6 Skull. Juvenile male.

XII-t F-7 Skull. Adult male. The mandible is damaged.

XII-t F-8 Skull. Male

XII-t F-9 Skull. Male.

XII-t F-10 Skull. Subadult male.

XII-t F-11 Skull. Male.

XII-t F-12 Skull. Female.

Xii-t F-13 Skull. Male. The occipital region is damaged.

XII-t F-14 Skull. Male.

XII-t F-15 Skull. Male. There is osteomyelitis of the left mandible.

XII-t F-16 Skull. Male.

XII-t F-17 Skull. Male.

XII-t F-18 Skull. Female. It is in a damaged condition.

XII-t F-19 Skull. Male.

XII-t F-20 Cranium. Male. *Gorilla beringei beringei.*

XII-t F-21 Cranium. Male. *Gorilla beringei beringei.* It is in a damaged condition.

XII-t F-22 Cranium. Male. *Gorilla beringei beringei.* It is in a damaged condition.

XII-t F-24 Skull. Adult male.

XII-t F-25 Skull. Juvenile male?

XII-t F-27 Skull. Juvenile.

XII-t F-28 Skull. Juvenile.

XII-t F-30 Mandible.

XII-t F-31 Skeleton. Adult male.

Universität zu Köln, Medizinische Fakultät, Institut II für Anatomie, Kerpener Strasse 62, Gebäude 35, 50937 Köln

Kirsten Brockes

Skeleton.

Skeleton. Adult male.

Skull.

Skull.

Universität zu Köln, Zoologisches Institut, Biozentrum Köln, Gebaude 304, Zülpicher Strasse 476, 50674 Köln*

H. Engländer

Skull. Adult male. Sangmélima, Cameroun. Purchased from Mr Reichert in the early 1950s.

Skull. Adult female. Abong-Mbang, Cameroun. Purchased from Mr Reichert in the early 1950s.

Darmstadt

Hessisches Landesmuseum, Friedensplatz 1, 64283 Darmstadt

Hanns Feustel; Francisco Hita-Garcia

G/82 Mounted skeleton. Adult female. Documented on 2 March 1871.

HMLD-M-1389 Mounted skeleton. Adult male. Guinea. Purchased from G.A. Frank, 9 Haverstock Hill, London NW, England, in 1893.

G/77, 1907 Skull and mounted skin (dermoplastic). Adult male. Purchased from G.A. Frank, 9 Haverstock Hill, London NW, England, in 1896. The dermoplastic was prepared by the taxidermist Karl Küsthardt. The skull is damaged.

HLMD-M-1394 Postcranial skeleton (incomplete) of the above individual.

G/81, 1909 Postcranial skeleton (incomplete). Juvenile female. Kamerun. Received from Captain Dominik in 1907. Arm, leg, vertebrae and ribs only.

HLMD-M-1320 Skull fragment with upper jaw and incomplete skeleton. Kamerun. Received from Captain Dominik in 1909.

Detmold

Lippisches Landesmuseum Detmold, Ameide 4, 32756 Detmold

Rainer Springhorn; Michael Zelle

Sk. Slg. 86 Cranium. Adult male. Kamerun. From the estate of August Ferdinand Wilhelm Kirchhof (1876–1926). Kirchhof worked for the German State Colonial Service and served as a magistrate in several areas of Kamerun (Buea, Duala and Jaunde) between 1903 and the outbreak of World War I. He donated a small part of his ethnological and zoological collection to the Lippisches Landesmuseum in 1920 and 1924. His family transferred further material in 1944. Finally, in 1950, the museum added ethnographic and zoological objects from Kamerun from an executor auction. Both zygomatic arches have the maxillary portions broken off. The front section of the skull base around the foramen magnum is irregularly broken. I1–I2 right and left are missing.

Sk. Slg. 208 Mounted skeleton. Adult male. Kamerun.

Dresden

Senckenberg Naturhistorische Sammlungen Dresden, Museum für Tierkunde, Königsbrücker Landstrasse 159, 01109 Dresden

Siegfried Eck; Alfred Feiler; Clara Stefen

B 55 Skeleton. Male. Ogowé, Gaboon. Collected by Hugo von Koppenfels in 1875.

B 56 Mandible. Male. Ogowé, Gaboon. Collected by von Koppenfels in 1875.

B 57 Skull. Juvenile male. Ogowé, Gaboon. Collected by von Koppenfels in 1875.

B 132 Skull (plaster cast). Purchased from Schmidt in 1875. Original in Vienna Museum.

B 254 Skull (plaster cast). Male. Purchased from Gerrard in 1876. Original in the Liverpool Museum.

B 255 Skull (plaster cast). Male. Purchased from Gerrard in 1876. Original in the British Museum.

B 260 Skull. Juvenile. Elobey islands area, Río Muni. Purchased from Gerrard in 1876.

B 281 Skeleton. Juvenile. Ogowé, West Africa. Received from Bouvier in 1876.

B 285 Skull (plaster cast). Juvenile female. Purchased from Horst Lehmann in 1876. Original in the Berlin Museum.

B 286 Cranium. Adult male. Purchased from Gerrard in 1876.

B 287 Cranium. Male. Ogowé, Gabon. Received from Bouvier in 1876.

B 288 Cranium. Adult male. Cap Lopez, W Gabon. Received from Bouvier in 1876.

B 289 Cranium. Male. Ogowé, Gabon. Received from Bouvier in 1876.

B 1271 Skull. Purchased from Gerrard in 1879.

B 1293 Skeleton. Juvenile male. Purchased from Gerrard in 1880.

B 1306 Skull (plaster cast). Purchased from the Free Museum in 1880. Original in the Liverpool Museum.

B 1383 Skull. Juvenile. Elobe, West Africa. Purchased from Gerrard in 1881.

B 1384 Skull. Juvenile. Elobe, West Africa. Purchased from Gerrard in 1881.

B 1442 Skull. Juvenile male. Purchased from Gerrard in 1882.

B 4127 Piece of cardboard with several samples of hair affixed.

B 5528 Mounted skin (dermoplastic). Adult male. Jaunde, Kamerun. Collected by Erich Moye in 1914. Purchased from Umlauff in 1920.

B 5530 Skin. Female. Jaunde Province, Kamerun. Purchased from Umlauff in 1920.

B 5532 Skin. Dascha Bogen, Jaunde Province, Kamerun. Collected by Erich Moye. Purchased from Umlauff in 1921.

B 5534 Skin. Juvenile. Jaunde and Ebolowa Province, Kamerun. Purchased from Umlauff in 1920.

B 5535 Skin. Juvenile. Jaunde, Kamerun. Collected by Zenker. Purchased from Umlauff in 1920.

B 6950 Skin. Juvenile male. Received from the Zoologischer Garten Dresden on 18 March 1932. This may be 'Sonny Boy', who arrived at the Zoologischer Garten Dresden on 15 August 1930 (weight 8.25 kg), and died there in 1932. [Some additional information added by GH from correspondence with Dresden Zoo personnel.]

B 11746 Skin. Juvenile. Río Muni. Collected by O. Krohnert. It was sold by Umlauff, Hamburg, to Julius Riemer in 1933. Received from Frau Charlotte Riemer in 1976.

B 12034 Skull and skin. Adult male. Ngato Bush, Moloundou-Yokadouma, SE Cameroun. Collected by Ernst Alexander Zwilling (1904–90) on 4 November 1933. Purchased by Julius Riemer in June 1939. Received from Frau Charlotte Riemer in 1977.

B 12062 Skull. Male? Kamerun? Collected before 1900. Received from von Haase, Dresden, in 1977 (his grandfather was in Africa).

B 12099 Skeleton and skin. Adult female. Río Muni. Collected by O. Krohnert. Sold by Umlauff, Hamburg, to Julius Riemer in 1933. Received from Frau Charlotte Riemer in 1977.

B 12100 Skin. Adult female. Río Muni. Collected by O. Krohnert. Sold by Umlauff, Hamburg, to Julius Riemer in 1933. Received from Frau Charlotte Riemer in 1977.

B 12101 Skin. Adult male. Río Muni. Collected by O. Krohnert. Sold by Umlauff, Hamburg, to Julius Riemer in 1933. Received from Frau Charlotte Riemer in 1977.

B 12177 Skin. Female. Río Muni. Collected by O. Krohnert. Sold by Umlauff, Hamburg, to Julius Riemer in 1933. Received from Frau Charlotte Riemer in 1977.

B 12178 Skin. Adult female. Río Muni. Collected by O. Krohnert. Sold by Umlauff, Hamburg, to Julius Riemer in 1933. Received from Frau Charlotte Riemer in 1977.

B 13141 Skull. Adult male. Río Muni. Collected by O. Krohnert. Sold by Umlauff, Hamburg, to Julius Riemer in 1933. Received from Frau Charlotte Riemer in 1979.

B 13142 Skull. Female. Njiba, Cameroun. Collected by Aseaan Roosen on 10 April 1936. Received from Frau Charlotte Riemer in 1979.

B 13669 Skull. Adult male. Received from the Zoologisches Institut der Karl-Marx-Universität, Leipzig, in 1973. It was originally from the Zoologischer Garten Leipzig.

B 14704 Skull. Male. Gabon? Received from the Zoological Institute, University of Leipzig, before 1970. (In the old catalogue it is B14704, but the skull is marked incorrectly as B 14705).

B 15807 Skin.

B 15808 Skull. Received from the Zoological Institute, University of Leipzig, before 1970. The teeth are missing.

B 15809 Skull. Received from the Zoological Institute, University of Leipzig, before 1970. The teeth are missing.

B 15810 Skull. Received from the Zoological Institute, University of Leipzig, before 1970. The cranium and mandible are damaged.

B 15811 Skeleton. Female. Ogowé, Gaboon. Collected by von Koppenfels in 1878. Received from the Zoological Institute, University of Leipzig.

B 17807 Mounted skin and some skeletal elements. Adult female. This is 'Dima', who arrived at the Zoologischer Garten Dresden, Tiergartenstrasse 1, 8020 Dresden, on 10 January 1963, aged about 2½ years (weight 17 kg). She was purchased from L. Ruhe, Hannover, who had acquired her in November 1962. She was euthanased on 11 March 1993 because of severe dysplasia and luxation of both coxae, chronic obstipation, obesity and siderosis in the liver and spleen. Received on 12 March 1993. [Some additional information added by GH from correspondence with Dresden Zoo personnel.]

B 27526 Cranium. From the Kleinschmidt collection, Wittenberg. Received in 2010.

B 27531 Skull. From the Kleinschmidt collection. Received in 2011.

Skull (plaster cast). From the Kleinschmidt collection. Received in 2010.

Düsseldorf

Löbbecke-Museum und Aquarium, Kaiserswerther Strasse 380 im Nordpark, 40200 Düsseldorf

Joseph Boscheinen; Silke Stoll

LMA 89/292 Skull. Subadult female. Received as part of the sculptor Josef Pallenberg's collection in 1952. It was originally a zoo specimen, named 'Assabam', and had been acquired by Pallenberg between 1900 and 1940. Since Pallenberg (1882−1946) lived in Düsseldorf, presumably it came from the Zoologischer Garten Düsseldorf, which was founded in 1876 and was completely destroyed by an air-raid on 2 November 1944 (however, Pallenberg also had close connections with other zoological gardens, for example, Hagenbeck's in Hamburg). The second incisors of the left side are lost. The canines are deciduous, the premolars mixed deciduous/permanent (PM1 and 2 right, PM2 left). All four 3rd molars are in eruption. Total length 22 cm; height 17 cm.

LMA 89/293 Skull. Adult male. The acquisition details are the same as for LMA 89/292 above; it was a zoo specimen, named 'Njium'. The lower right canine is broken and partly missing, and both canines of the left side are damaged, all premortem. Total length 28 cm; height 22 cm.

LMA 89/295 Cranium. Adult male. It was collected by Horst Sieloff in 1936−38, possibly in Cameroun and was probably received in 1938. It is subfossil, with fragmentary teeth. The crista sagittalis is slightly twisted. Length 25 cm; height 18 cm.

Death mask. Adult male. From the collection of Josef Pallenberg, Düsseldorf. It was probably a captive specimen. Size 25 cm × 18 cm.

Heinrich-Heine-Universität Düsseldorf, Universitätsklinikum, Cécile und Oskar Vogt-Institut für Hirnforschung, Moorenstrasse 5, 40225 Düsseldorf

Katrin Amunts; A. Hopf

A 128 Brain. Adult male. It is cut in frontal serial sections (without the lower brain stem) and is stained with cresyl violet and for myelin. It came from 'Bobby', who lived in the Zoologischer Garten Berlin from 30 March 1928 until his death 1 August 1935. The fresh brain weight was 610 g.

YN-82-140 Brain. Adult female. Sectioned coronally in 20-μm thick sections, stained with Merker cell body staining. Fresh brain weight 376 g. This is from 'Jini', who arrived at the Yerkes Regional Primate Research Center, Emory University, Atlanta, GA 30322, on 20 November 1964 (weight 5.02 kg), having been purchased from the Far East Animal Co., Tilburg. She died on 29

September 1982 (weight 84.7 kg). [Some additional information added by GH from correspondence with Yerkes RPRC personnel.]

The brains have been mentioned in the following publications: Semendeferi et al. (1998, 2001); Sherwood et al. (2003, 2004); Schenker et al. (2008); Barks et al. (2014).

The following 3 specimens were formerly in the Max-Planck-Institut für Hirnforschung, Deutschordenstrasse 46, 60528 Frankfurt am Main:

A 374 Brain (serially sectioned, sagittally). Adult male. Bilota, Belgian Congo. Hunted by gun, collected on 25 April 1957. Brain weight (with dura mater) 542 g.

A 375 Brain (serially sectioned, frontally). Juvenile male. Bilota, Belgian Congo. Hunted by gun, collected on 29 April 1957. Brain weight 471 g (with dura mater); 450 g (without dura mater).

1001 Brain (serially sectioned, frontally). Adult. Utu, Belgian Congo. Collected in 1957. Prepared by IRSAC (Institut pour la Recherche Scientifique en Afrique Centrale).

Heinrich-Heine-Universität Düsseldorf, Abteilung für Cytopathologie, Zentrums für Pathologie und Biophysik, Moorenstrasse 5, 40225 Düsseldorf

Stefan Biesterfeld; Peter Pfitzer

NB: The histological sections listed below (and any associated paraffin blocks) no longer exist, but are mentioned purely as a matter of historical interest.

Heart (histological slides of the left and right ventricles). From a 4-year-old male which died of unknown causes at IRSAC, Bukavu, Belgian Congo. The material was examined by Prof. Dr Peter Pfitzer between November 1971 and February 1972, as part of an investigation of the DNA content of the cell nuclei of primates. The heart of the gorilla and also the hearts of three chimpanzees were supplied by Dr Peter Kunkel, who was the director of IRSAC at that time. See Pfitzer and Schulte (1972).

Erlangen

Institut für Anatomie II, Friedrich-Alexander Universität Erlangen-Nürnberg, Universitätsstrasse 19, Erlangen 91054

Friedrich Paulsen

Mounted skeleton. Adult male. Purchased between 1885 and 1910 from Johannes Umlauff taxidermy firm, Hamburg. It was purchased for the Institute of Zoology in Erlangen, which was opened in 1885. In the early 1980s, the Institute of Zoology was moved to a new building, and most of its collection was dispersed, but the gorilla skeleton remained in the building and became part of the Anatomy II collection. Height 169 cm.

Frankfurt Am Main

Senckenberg Forschungsinstitut und Naturmuseum , Senckenberganlage 25, 60325 Frankfurt am Main

Heinz Felten; Jens L. Franzen; D. Kock; Katrin Krohmann; Ottmar Kullmer; Gerhard Storch

Abteilung Paläoanthropologie und Messelforschung, Sektion Tertiäre Säugetiere:

SMF 1130 Skull. Male. Donated by Freifrau Louise von Rothschild in 1892.

SMF 1131 Skull. Female. Donated by Freifrau Louise von Rothschild in 1892.

SMF 1132 Skull. Male. Cap Lopez, Gabon.

SMF 1133 Skull. Male.

SMF 1134 Skull. Female. Purchased from Herr Freckmann in 1884.

SMF 1135 Skeleton and dermoplastic. Subadult male. Lectotype of *Pseudogorilla mayema* Elliot 1913 and *Gorilla (Pseudogorilla) ellioti* Frechkop 1943. Rembo Nkomi delta, southern Fernan Vaz, Gabon. Donated by Dr A. von Weinberg on 2 June 1907.

SMF 1136 Skeleton and dermoplastic. Adult female. Paralectotype of *Pseudogorilla mayema* Elliot 1913 and lectotype of *Gorilla (Pseudogorilla) ellioti* Frechkop 1943. Rembo Nkomi delta, southern Fernan Vaz, Gabon. Donated by Dr A. von Weinberg on 2 June 1907.

SMF 1137 Skeleton and dermoplastic. Juvenile male. Paralectotype of *Pseudogorilla mayema* Elliot 1913 and lectotype of *Gorilla (Pseudogorilla) ellioti* Frechkop 1943. Rembo Nkomi delta, southern Fernan Vaz, Gabon. Donated by Dr A. von Weinberg on 2 June 1907.

The above type-series probably formed one family party. It seems very likely that the skull with the male skeleton SMF 1135 actually belongs to the female, SMF 1136, and vice versa. The interchange of skulls becomes apparent when comparing them with the photographs in Elliot (1913).

SMF 1580 Skull.

SMF 1590 Mounted skeleton.

SMF 3254 Mounted skeleton. Male. It probably belongs to the uncatalogued mounted skin of a male from Molundu, Kamerun, which is on public exhibition.

SMF 4108 Skull. Kumilla, north of Molundu, SE Kamerun. Collected by the Deutsche Zentral-Afrika-Expedition, led by Herzog A.F. zu Mecklenburg, in 1910−11. The mandible is in the public exhibition.

SMF 4109 Skull. Male. Nginda, north of Molundu, SE Kamerun. Collected by the Deutsche Zentral-Afrika-Expedition, led by Herzog A.F. zu Mecklenburg, in 1910−11. Shot by A. Schultze on 31 May 1911.

SMF 5277 Skeleton and skin. Adult female. Abong Mbang, Nyong, SE Kamerun. Collected by Holtmann, and purchased from him on 29 August 1913. The skeleton has been missing since 1977. This female was pregnant: the fetus is SMF 16709.

SMF 16180 Skull and skin. Juvenile. Equatorial Guinea. Received from the Zoologischer Garten der Stadt Frankfurt am Main, Alfred-Brehm-Platz 16, 60316 Frankfurt am Main, where it died on 19 June 1956.

SMF16709 Fetus (fluid-preserved). It belongs to the female SMF 5277. It was housed in the Comparative Anatomy Section, with the number SMF 90.5908 (the original number was SMF 5276). Comparative Anatomy is no longer a separate Section, but is overseen by Mammalogy.

SMF 17826 Mounted skeleton and skin. Male. Micomeseng, Spanish Guinea. Received from the Zoologischer Garten der Stadt Frankfurt am Main on 25 October 1957. This specimen was captured in the spring of 1953, aged approximately 5½ years and arrived at the zoo on 2 July 1953, along with 2 other males ('Thomas' and 'Carlo'). Named 'Rafiqi', he died on 24 October 1957. [Some additional information added by GH from correspondence with Frankfurt Zoo personnel.]

SMF 25143 Mounted skeleton and skin. Juvenile female. SW of Ebolowa, near Méyo Centre, Cameroon. Received from the Zoologischer Garten der Stadt Frankfurt am Main in December 1965. This specimen was estimated to have been born in the spring of 1961, since she weighed 3 kg when caught in July, 1961. She arrived at the zoo on 22 April 1964 and was named 'Licorice Drop' ('Lic'). She died on 11 December 1965. [Some additional information added by GH from correspondence with Frankfurt Zoo personnel.]

SMF 26103 Skeleton and mounted skin. Adult male. Received from the Zoologischer Garten der Stadt Frankfurt am Main on 6 January 1957. He arrived at the zoo on 14 December 1957, weighing 21.3 kg and was named 'Abraham'. He died on 5 January 1967. A skull modification due to captivity is described by Angst and Storch (1967) [Some additional information added by GH from correspondence with Frankfurt Zoo personnel.]

SMF 28993 Skin. Male. Received from the Zoologischer Garten der Stadt Frankfurt am Main on 9 October 1932.

SMF 45713 Skull. Near Bélabo, Cameroon. Collected by R. Tanglmayer in February, 1972; purchased from him on 7 April 1972.

SMF 59158 Skull (sectioned medially). Juvenile male. Received from Herr Stephan, Max-Planck-Institut für Hirnforschung, Frankfurt am Main, on 8 December 1980. It died in the Zoo Duisburg AG, Mülheimer Strasse 273, 4100 Duisburg, on 14 April 1965, aged 11–13 months, and was known there as 'Massa'.

SMF 59159 Skull (sectioned medially). Received with the skull SMF 59158 above.

SMF 63976 Skull, postcranial skeleton (partial) and skin (damaged). *Gorilla beringei beringei.* Uganda. Received from the CITES authorities of the Federal Republic of Germany on 14 February 1983. It had been killed illegally in 1981–82, and was subsequently confiscated.

SMF 91512 Skin. Juvenile female. Equatorial Guinea. It was captive in the Zoologischer Garten der Stadt Frankfurt am Main and died in 1956. Acquired from D. Starck, Johann Wolfgang Goethe-Universität, Frankfurt am Main, in 2000.

SMF 92766 Carcase (in alcohol). Infant male. It was acquired in 2004 from the Zoologischer Garten der Stadt Frankfurt am Main, where it was born on 28 November 2002 and died on 16 August 2003.

SMF 94796 Skull. Adult male. *Gorilla beringei beringei.* Rwanda. Collected c.1974. Donated by Dr Roland Plesker, Paul-Ehrlich-Institut, Paul-Ehrlich-Strasse 51–59, 63225 Langen, on 19 July 2012. It was originally a gift in 1974 from 'mountain pygmies' of the Virunga volcanoes to a person working in a development project in Rwanda. When he finished his work, he brought the skull with him to Germany, and in 1993, he gave it to Dr Plesker for his primate collection. It lacks both zygomatic arches. The following teeth are missing from the maxilla: M3 and I2 and 1 (both incisors) on the right side; M3, ½ M1 (split in the middle), P2 and P1 (only the roots are present), C, I2 (lateral incisor) on the left side. [Roland Plesker].

Zentrum der Morphologie (Dr Senckenbergische Anatomie), Klinikum der Johann Wolfgang Goethe-Universität, Theodor-Stern-Kai 7, 60590 Frankfurt am Main*

F. Huckinghaus

48.50 Assorted bones (including 1 femur, 2 tibiae, 2 patellae and 1 tarsus). Subadult male. Received from the collection of Prof. Starck.

Vg 53.1 Mounted skeleton. Adult male. Purchased from Frohlich and Umlauff, Hamburg, in 1953.

56.27 Postcranial skeleton. Infant female. Spanish Guinea. It died in the Zoologischer Garten der Stadt Frankfurt am Main when aged 6 months. Received from the collection of Prof. Starck.

56.36 Postcranial skeleton; hand; leg; genitals; spleen; skin of hand and foot. Juvenile male. Spanish Guinea. He had arrived at the Zoologischer Gartren der Stadt Frankfurt am Main on 2 July 1953, along with 2 other males ('Carlo' and 'Rafiqi') and was named 'Thomas'. He died on 6 September 1956, aged about 5 years. Received from the collection of Prof. Starck. [Some additional information added by GH from correspondence with Frankfurt Zoo personnel.]

Vg 57.52 Mandible. Female. Bata, Spanish Guinea. Collected in 1957. Purchased from Fritsche, Bremerhaven, on 18 May 1957.

Vg 57.82 Mounted skeleton. Adult male. Bata, Spanish Guinea.

Vg 58.2 Mounted skeleton. Adult male. Bata, Spanish Guinea. Collected in 1957.

Vg 58.41 Mounted skeleton. Juvenile. Bata, Spanish Guinea.

Vg 58.43 Mounted skeleton. Juvenile. Bata, Spanish Guinea.

Vg 58.44 Skull. Juvenile. Bata, Spanish Guinea.

59.118 Postcranial skeleton. Adult male. South Cameroon. Collected by J. Sabater. Received from the collection of Prof. Starck.

59.119 Skull. Adult male. Spanish Guinea. Received from the collection of Prof. Starck.

59.145 Skeleton. Juvenile male. Bata, Spanish Guinea. Collected by J. Sabater.

59.146 Skeleton. Juvenile female. Bata, Spanish Guinea. Collected by J. Sabater.

59.148 Skull. Subadult male. Bata, Spanish Guinea. Collected by J. Sabater.

Vg 60.3 Skull. Adult male. South Cameroun. Collected by Sohl in 1959.

Vg 60.4 Skull. Female. South Cameroun. Collected by Sohl in 1959.

60.5 Diverse bones. Infant. South Cameroun. Collected by Sohl in 1959. From the collection of Prof. Starck.

60.62 Postcranial skeleton. Adult male. Spanish Guinea. Collected by J. Sabater.

60.63 Postcranial skeleton. Adult female. Spanish Guinea. Collected by J. Sabater.

69.28 Skull and dissected body (in formol). Adult male. This specimen arrived at the Zoologischer Garten der Stadt Frankfurt am Main on 20 April 1967, having been purchased from Oklahoma City Zoological Park, USA. Known as 'Solomon', he had arrived at the Oklahoma City Zoo on 30 September 1960, weighing 18 lb, along with a female named 'Sheba' (who died on 20 April 1967). The pair were purchased from F.J. Zeehandelaar, Inc. 'Solomon' died of peritonitis and gastroenteritis on 19 April 1969. [Some additional information added by GH from correspondence with Oklahoma City Zoo personnel.]

69.43 Skull and dissected body (in formol). Juvenile female. This specimen was named 'Alice' and was a twin to 'Ellen'. They were born in the Zoologischer Garten der Stadt Frankfurt am Main on 3 May 1967. The dam was 'Makulla' and the sire was 'Abraham'. Both infants were hand reared; 'Alice' weighed 1700 g and 'Ellen' weighed 1900 g. 'Alice' died on 7 June 1969. [Some additional information added by GH from correspondence with Frankfurt Zoo personnel.]

69.69 Dissected body (in formol) and diverse organs (in formol). Male. This specimen lived in the Zoologischer Garten der Stadt Frankfurt am Main and died in April 1969, aged about 10 years.

Freiburg Im Breisgau

Albert-Ludwigs-Universität, Anatomisches Institut, Lehrstuhl Anatomie III, Albertstrasse 17, 79001 Freiburg im Breisgau*

Skull. Adult. It probably came from the University's Institut für Humangenetik und Anthropologie (this Institute's collection of animal remains was dissolved in the 1950s). The jaw is articulated. There is slight damage, sustained postmortem, to the roof of the braincase on the right side. The maxilla holds 18 teeth in all, 9 per side, there being supernumerary (3rd) premolars.

Albert-Ludwigs-Universität, Institut für Biologie I (Zoologie), Hauptstrasse 1, 79104 Freiburg im Breisgau*

Odwin Hoffrichter

1643 Skull. Adult female? It was transferred from the Institut für Humangenetik und Anthropologie of Freiburg University in 1947. It is in a rather bad condition, not prepared very finely, with dry brown remnants of tissue.

Görlitz

Senckenberg Museum für Naturkunde, Am Museum 1, 02826 Görlitz

Hermann Ansorge; W. Dunger

M.9723 Skull. Adult female. This is believed to be 'Copina', who arrived at the Zoologischer Garten Rostock, Rennbahnallee 21, 18059 Rostock, along with a male named 'Copo', on 24 October 1977. They were purchased from the Zoologischer Garten Zürich, Switzerland, where they had arrived on 9 April 1963. 'Copina' weighed about 11 kg and was aged about 2 years upon arrival in Zürich. She came originally from Equatorial Guinea, and had been purchased by the animal dealer L. Ruhe from Barcelona Zoo. At the time of departure for Rostock she weighed 76 kg. She died on 11 April 1985. [Some additional information added by GH from correspondence with Zürich Zoo personnel.]

Z 62/60 Mounted skeleton and dermoplastic. Adult male. Ndognjok, altitude 300 m, Edea district, Kamerun. It was collected by Dr Hans Schäfer in 1914, and donated in the same year.

Z 62/153 Skull. Adult male. Kamerun? Donated by Putzler?

Z 85/019 Skull and parts of skeleton. Juvenile male. This is 'Carlo', who arrived at the Zoologischer Garten

Rostock on 23 November 1975, and died on 7 November 1976.

Göttingen

Deutsches Primatenzentrum GmbH, Leibniz-Institut für Primatenforschung, Kellnerweg 4, 37077 Göttingen

Robert Teepe; Stefan Treue

Access to the Institute's samples is possible in principle but restricted in the sense that they usually work out a Memorandum of Understanding prior to access, to sort out what this access entails, how the research will be published, and how the contribution to the research will be shared.

Gene Bank of Primates, Primate Genetics Laboratory:

Christian Roos

Blood sample. Obtained from Zoologischer Garten Berlin on 9 March 1995. Female 'Dufte', captive-born on 30 October 1974. Sample available: DNA.

Blood sample. Obtained from Tierpark München on 13 July 1995. Male 'Polepole', captive-born on 28 December 1989. Sample available: DNA, serum.

Blood sample. Obtained from Tierpark München on 26 November 1995. Female 'Diane', captive-born on 4 January 1989. Sample available: DNA, serum.

Blood sample. Obtained from Stichting Apenheul on 4 April 1996. Male 'Dibo', captive-born on 14 June 1980. Sample available: DNA, serum.

Blood sample. Obtained from Allwetterzoo Münster on 12 October 1996. Male 'Kimba', wild-born in 1972. Sample available: DNA, serum.

Blood sample. Obtained from Stichting Apenheul on 15 November 1996. Female 'Pabebo', captive-born on 18 March 1987. Sample available: DNA, serum.

Blood sample. Obtained from Zoologischer Garten Leipzig on 21 December 1996. Female 'Virunga', wild-born in 1971. Sample available: DNA, serum.

Blood sample. Obtained from Tiergarten Nürnberg on 10 April 1997. Female 'Lena', wild-born in 1977. Sample available: DNA, serum.

Blood sample. Obtained from Tiergarten Nürnberg on 19 April 1997. Female 'Bianca', wild-born in 1973. Sample available: DNA, serum.

Blood sample. Obtained from Zoologischer Garten Basel on 3 May 1997. Male 'Nangai', captive-born on 8 October 1990. Sample available: DNA, serum.

Blood sample. Obtained from Zoologischer Garten Basel on 3 May 1997. Female 'Muna', captive-born on 6 March 1989. Sample available: DNA, serum.

Blood sample. Obtained from Zoologischer Garten Basel on 3 May 1997. Female 'Tamtam', captive-born on 2 May 1971. Sample available: DNA, serum.

Blood sample. Obtained from Tiergarten Nürnberg on 30 July 1997. Male 'Fritz', wild-born in 1963. Sample available: DNA, serum.

Blood sample. Obtained from Tiergarten Nürnberg on 30 July 1997. Female 'Delphi', wild-born in 1962. Sample available: DNA, serum.

Blood sample. Obtained from Stichting Apenheul on 16 October 1997. Female 'Sheila', captive-born on 18 February 1991. Sample available: DNA, serum.

Blood sample. Obtained from Tiergarten Nürnberg on 28 May 1998. Male 'Yanude', captive-born on 21 November 1983. Sample available: DNA, serum.

Blood sample. Obtained from Zoologischer Garten Leipzig on 17 October 1998. Male 'Gaidi', wild-born in 1971. Sample available: DNA, serum.

Blood sample. Obtained from Zoologischer Garten Leipzig on 17 October 1998. Female 'Babsy', wild-born in 1971. Sample available: DNA, serum.

Blood sample. Obtained from Zoologischer Garten Leipzig on 25 January 2001. Male 'Vinoto', captive-born on 11 March 1995. Sample available: DNA.

Blood sample. Obtained from Zoologischer Garten Leipzig on 25 January 2001. Female 'Vizuri', captive-born on 18 May 1995. Sample available: DNA.

Blood sample. Obtained from Zoologischer Garten Leipzig on 25 January 2001. Female 'Viringika', captive-born on 23 March 1995. Sample available: DNA.

Blood sample. Obtained from Zoo Duisburg on 5 June 2002. Captive-born, sex and birth date unknown. Sample available: DNA.

Blood sample. Obtained from Zoo Duisburg on 5 June 2002. Captive-born, sex and birth date unknown. Sample available: DNA.

2741 Blood sample. Obtained from Bristol Zoo Gardens on 16 June 2008. Captive-born, sex and birth date unknown. Sample available: DNA.

7925 Blood sample. Obtained from Bristol Zoo Gardens on 16 June 2008. Captive-born, sex and birth date unknown. Sample available: DNA.

9018 Blood sample. Obtained from Bristol Zoo Gardens on 16 June 2008. Captive-born, sex and birth date unknown. Sample available: DNA.

9582 Blood sample. Obtained from Bristol Zoo Gardens on 16 June 2008. Captive-born, sex and birth date unknown. Sample available: DNA.

Blood sample. Obtained from Zoologischer Garten Basel on 17 October 2008. Male 'Viatu', captive-born on 20 December 1998. Sample available: DNA.

Blood sample. Obtained from La Vallée des Singes on 22 October 2009. Female 'Kwanza', captive-born on 1 April 2002. Sample available: DNA.

Blood sample. Obtained from Zoologischer Garten Basel on 17 March 2010. Female 'Wima', captive-born on 9 January 1999. Sample available: DNA.

Blood sample. Obtained from Zoologischer Garten Basel on 17 March 2010. Female 'Chelewa', captive-born on 31 December 2005. Sample available: DNA.

Blood sample. Obtained from La Vallée des Singes on 30 March 2010. Male 'Yaounde', captive-born on 21 November 1983. Sample available: DNA.

Blood sample. Obtained from Zoologischer Garten Basel on 19 April 2010. Female 'Quarta', captive-born on 17 July 1968. Sample available: DNA.

Blood sample. Obtained from Zoologischer Garten Basel on 19 April 2010. Male 'Kisoro', captive-born on 14 March 1989. Sample available: DNA.

Blood sample. Obtained from Zoologischer Garten Basel on 19 April 2010. Male 'Zungu', captive-born on 4 August 2002. Sample available: DNA.

Blood sample. Obtained from Zoologischer Garten Basel on 19 April 2010. Female 'Goma', captive-born on 23 September 1959. Sample available: DNA.

Blood sample. Obtained from Zoologischer Garten Basel on 19 April 2010. Female 'Faddama', captive-born on 2 February 1983. Sample available: DNA.

Blood sample. Obtained from Zoologischer Garten Basel on 19 April 2010. Female 'Joas', captive-born on 6 July 1989. Sample available: DNA.

Blood sample. Obtained from La Vallée des Singes on 30 April 2010. Female 'Miliki', captive-born on 22 February 2006. Sample available: DNA.

Blood sample. Obtained from La Vallée des Singes on 22 September 2010. Male 'Badongo', captive-born on 27 September 1999. Sample available: DNA.

Blood sample. Obtained from Zoo Hannover on 21 September 2011. Female 'Mambele', captive-born on 21 February 1999. Sample available: DNA.

Department of Infectious Pathology:

Kerstin Mätz-Rensing

All tissue samples are embedded in paraffin blocks or frozen at −80°C; frozen tissue is partly available via Dr Christian Roos, Gene Bank of Primates.

7870 Tissue samples of liver, lung, kidney, brain, heart, spleen, stomach/intestine, reproductive system and lymph nodes. Infant male. This is 'Claudio', who was born in Allwetterzoo Münster on 9 May 2008 (weight 1725 g) to dam 'Gana' and sire 'Nkwango', and died on 16 August 2008 (carried by mother until 22 August 2008). Suspected cause of death: enteritis.

8053 Tissue samples of liver, lung, kidney, brain, heart, spleen, stomach/intestine, adrenal gland, reproductive system and lymph nodes in formalin, paraffin blocks and frozen at −80°C; exhibit of skull and taxidermy preparation of right hand. Adult male. This is 'Tamtam', who was born in Zoologischer Garten Basel on 2 May

1971 to dam 'Goma' and sire 'Jambo'? and was mother-reared. He was sent to Zoologischer Garten Wuppertal on 30 April 1997, and died there on 23 July 2009 (weight 140 kg). Cause of death: pericardial tamponade after anaesthesia for tooth extraction.

8158 Tissue samples of liver, lung, kidney, brain, heart, spleen, stomach/intestine, adrenal gland, reproductive system and lymph nodes in formalin, paraffin blocks and frozen at −80°C. Adult female. This is 'Gana', who was born in Aktiengesellschaft Zoologischer Garten Köln on 18 January 1997 to dam 'Gina' and sire 'Kim'. She was sent on loan to Allwetterzoo Münster on 1 June 2004, developed neurological signs in January 2010 and died of meningoencephalitis induced by *Balamuthia mandrillaris* on 17 January 2010 (weight 70 kg). See Chapter 15, Nervous System and Special Senses in Part I, Gorilla Pathology and Health.

8367 Tissue samples of liver, lung, kidney, brain, heart, spleen, stomach/intestine, reproductive system, lymph nodes in formalin, paraffin blocks and frozen at −80°C. Neonate female. Died of asphyxia during birth at Zoologischer Garten Wuppertal on 20 January 2011 (weight 2353 g). The dam was 'Vimoto' and the sire was 'Ukiwa'.

8797 Skeletal elements (frozen). Neonate female conjoined twins. Born in Zoo Duisburg on 2 March 2013 (to dam 'Safiri' and sire 'Mapema'). They were born alive and lived for a short time, but were already dead when first observed by the keepers. They were fused ventrocaudally and had 3 legs. See Langer et al. (2014).

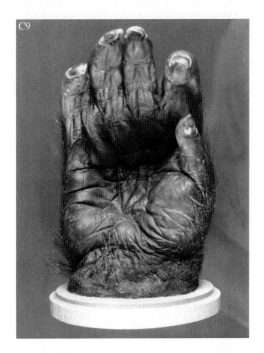

FIGURE C-9 Mounted right hand of gorilla (no. 8053), palmar aspect. This is apparently normal. It illustrates the deep pigmentation and thickening with patterning of the epidermis (dermatoglyphics). (See Chapter 9: Skin and Integument in Part I: Gorilla Pathology and Health). *Photograph by Kerstin Mätz-Rensing, courtesy of Deutsches Primatenzentrum GmbH, Göttingen.*

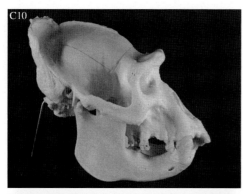

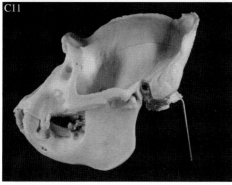

FIGURES C-10–C-11   Skull of adult male (no. 8053). Right lateral (C-10) and left lateral (C-11) views. There is marked dental pathology and apparent osseous changes. This skull is probably brachycephalic but confirmation of this would require osteometric studies. *Photographs by Kerstin Mätz-Rensing, courtesy of Deutsches Primatenzentrum GmbH, Göttingen.*

Georg-August-Universität Göttingen, Anatomisches Institut, Fachbereichs Medizin, Kreuzbergring 36, 3400 Göttingen*

Hans-Jürg Kuhn

Mounted skeleton. Adult male. It is complete except for some casts of phalanges and metacarpals. There is a healed fracture of the left radius.

Skull. Adult male. Southern Cameroun. Collected by a merchant in the late? 1950s. It is included in the personal collection of Prof. Dr Hans-Jürg Kuhn.

Skull. Adult female. Southern Cameroun. Collected by a merchant in the late? 1950s. It is included in the personal collection of Prof. Dr Hans-Jürg Kuhn.

Georg-August-Universität Göttingen, Johann-Friedrich-Blumenbach-Institut für Zoologie und Anthropologie, Abteilung Historische Anthropologie und Humanökologie, Bürgerstrasse 50, 3400 Göttingen

Birgit Grosskopf; C. Vogel

Postcranial skeleton (25 costae, 22 vertebrae, autopodia mounted, some distal phalanges recreated in wood).

3/5f Skull. Adult female. 3rd molar erupting.

688 Mounted skeleton. Adult female.

826 Pelvis (mounted). Adult male?

209000b Skull. Adult male.

209001b Skull. Adult female.

209002b Skull. Juvenile.

209003b Skull (cranium sagittally sectioned). Deciduous teeth; 1st molar erupting.

209003b Skull (cranium sagittally sectioned). 3rd molar erupting.

209003b Skull (cranium sagittally sectioned). Adult female?

209003b Skull (cranium sagittally sectioned). Adult male.

209003b Skull (cranium sagittally sectioned). Adult male. Only left side of cranium present.

209005b Mounted skeleton. Infant female.

209067 Skeleton (incomplete; cranium sagittally sectioned). Neonate.

209067 Postcranial skeleton (incomplete: 20 vertebrae, 23 costae, no pelvis, no fibulae, autopodia not examined). Juvenile (epiphyseal plates clearly recognisable).

209068 Long bones of lower extremity and left humerus. No epiphyseal plates recognisable; note says 'juvenile'.

Georg-August-Universität Göttingen, Johann-Friedrich-Blumenbach-Institut für Zoologie und Anthropologie, Zoologisches Museum, Berliner Strasse 28, 37073 Göttingen*

Gert Tröster

Skeleton. Male.

Greifswald

Ernst Moritz Arndt Universität Greifswald, Institut für Anatomie und Zellbiologie, Friedrich-Loeffler-Strasse 23c, 17487 Greifswald

Jochen Fanghänel; Thomas Koppe

Mounted skeleton. Adult male. There is an arrowhead still embedded in the nuchal crest (left side), and the skeleton has a number of 'pathological peculiarities'.

Mounted skeleton. Adult female.

15.2.8 Skull. Adult female. Gabon.

15.3.2 Skull. Adult male. *Gorilla gorilla matschiei.* Molundu, Kamerun.

VA0170 Skull. Adult female. *Gorilla gorilla matschiei.* Molundu, Kamerun. From the collection of Richard N. Wegner (1884–1967), who was Head of the Anatomy Department from 1946 to 1959.

VA0171 Skull. Adult female. Gabon. The squama of the occipital bone is partially destroyed.

VA0172 Skull. Adult male. *Gorilla gorilla matschiei.* Bangandu, northern Molundu, South Kamerun. The left nuchal crest is partially destroyed.

VA0173 Skull. Adult male. Gabon. R.N. Wegner collection. The left cheek bone has signs of a healed bony fracture.

VA0174 Skull. Adult male. Molundu, Kamerun. R.N. Wegner collection.

VA0176 Cranium (sagittal section). Adult female. *Gorilla gina.* French Congo coast. Collected in 1880. The right temporal bone and zygomatic arch are partially removed.

VA0177 Cranium (right side). Adult male. Kamerun. Partially removed facial wall with view inside the maxillary sinus.

VA0179 Skull. Juvenile. Njong, Kamerun. R.N. Wegner collection.

VA0180 Skull (frontally sectioned 3 times). Adult male. Gabon. R.N. Wegner collection.

VA0181 Skull. Adult male. Gabon. R.N. Wegner collection. The right ala major ossis sphenoidalis and right orbital cavity are partially destroyed.

VA0182 Skull. Adult male. Gabon. The mandible and occipital bone are partially destroyed.

VA0183 Skull. Adult male. Gabon. R.N. Wegner collection.

VA0184 Cranium. Juvenile. Gabon. R.N. Wegner collection. The teeth (molars and premolars) on the right side are missing antemortem.

VA0185 Skull. Adult female. Gabon. R.N. Wegner collection.

VA0186 Skull (horizontally sectioned 3 times). Adult male. Bangandu area, northern Molundu, Kamerun. R.N. Wegner collection.

VA0187 Skull. Adult male. *Gorilla gorilla matschiei.* Molundu, Kamerun.

VA0188 Skull. Adult male. Lichtenberg, Kamerun. R.N. Wegner collection. There is a bony formation at the upper left supraorbital margin.

VA0189 Skull. Adult male. *Gorilla gorilla matschiei.* Molundu, Kamerun. R.N. Wegner collection.

VA0190 Os frontale, os nasale and maxilla. Female. *Gorilla gorilla matschiei.* R.N. Wegner collection. The canine and M3 are not yet erupted.

VA0192 Mandible. Juvenile.

VA0193 Skull. Adult male. *Gorilla gorilla matschiei.* Molundu, Kamerun. R.N. Wegner collection. The left canine projects laterally. There is a hole in the left nuchal crest.

VA0194 Death mask. Adult male. *Gorilla gorilla matschiei.* Kamerun.

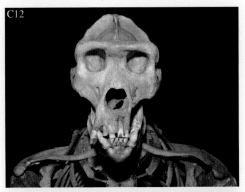

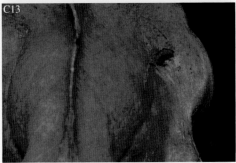

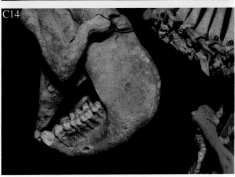

FIGURES C-12–C-17 Adult male mounted skeleton (no number): No detailed history available. C-12: Frontal view of skull. There appears to be resorption of bone and alveoli of teeth are exposed. C-13: Close-up of arrowhead embedded in nuchal crest. Presumed contributory to death as there is no evidence of regrowth of bone. C-14: Left side of skull. The maxilla is completely edentulous, apparently due to severe bone resorption. The mandibular canine is blunted; all post-canine teeth are present and remarkably free of wear but all have exposed roots, also apparently due to bone resorption. C-15: Right side of skull. The changes are less severe than those on the left but there is clear evidence of deviation of teeth and apparent porosity of the maxilla. C-16: Left elbow joint. There is discoloration of the joint surfaces, apparent lipping and new bone growth affecting the radius and ulna. C-17: Left elbow joint. Both radius and ulna show new bone growth but it is difficult to determine the cause without detailed, possibly multidisciplinary, investigation. *Photographs by courtesy of Institut für Anatomie und Zellbiologie, Greifswald.*

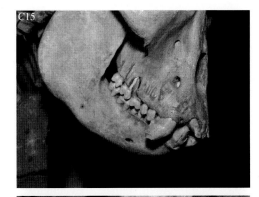

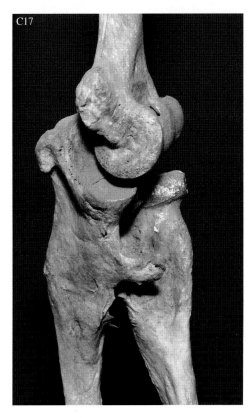

FIGURE C-12–C-17   (Continued)

ZIMG I/4120 Skull. Juvenile female. Mbusi, West Africa. Collected by Reinhold Buchholz in July 1875, and donated by him.

ZIMG I/4121 Cranium. Adult female. Mbusi, West Africa. Collected by Reinhold Buchholz in July 1875, and donated by him. The right side is missing.

Hamburg

Zoologisches Institut und Zoologisches Museum der Universität Hamburg, Martin-Luther-King-Platz 3, 20146 Hamburg

Thomas M. Kaiser; Harald Schliemann

S 2709 Skull. Adult male.

S 2710 Skull. Adult.

S 2711 Skull. Adult male.

S 2712 Skull. Adult female.

S 2713 Skull. Subadult female.

S 2753 Mounted skeleton. Adult female.

S 2754 Mounted skeleton. Adult male. The upper jaw lacks the left canine and 1 incisor.

The above 7 specimens were purchased from Th. Umlauff in 1943.

S 3225 Skull. Male. The upper jaw lacks all the incisors.

FIGURE C-12–C-17   (Continued)

Ernst-Moritz-Arndt-Universität, Zoologisches Museum, Loitzer Strasse 26, 17489 Greifswald

Peter Michalik

ZIMG I/4116 Skull. Adult male. Mbusi, West Africa. Collected by Reinhold Buchholz in July 1875, and donated by him. The left side of the cranium is missing.

ZIMG I/4119 Skull. Adult male. Mbusi, West Africa. Collected by Reinhold Buchholz in July 1875, and donated by him.

S 3226 Skull. Adult male. The lower jaw lacks the 2nd molar.

S 3227 Skull. Adult. The upper jaw lacks 2 incisors.

S 3228 Skull. Adult male.

S 3229 Skull. Adult male. The lower jaw lacks 1 incisor.

S 3230 Skull. Adult.

S 3231 Skull. Adult male. The upper jaw lacks the left canine.

S 3232 Skull. Adult. The upper jaw lacks the right canine.

S 3233 Skull. Adult male. The upper jaw lacks 2 incisors; the lower jaw lacks 1 incisor.

S 3234 Skull. Adult male. The right zygomatic arch is broken.

S 3235 Cranium. Adult male.

S 3291 Mounted skeleton. Adult male. The upper jaw lacks 1 incisor.

S 3292 Mounted skeleton. Adult male.

S 3293 Mounted skeleton. Adult male.

S 3294 Skeleton. Juvenile.

S 3414 Skull. Adult male. The lower jaw lacks 1 molar.

S 3415 Skull. Adult male.

S 3425 Skull. Adult male. The upper jaw lacks the right canine; the right side of the lower jaw is damaged.

S 3432 Bones of 1 foot. Adult.

S 3433 Bones of 1 hand. Juvenile.

S 3434 Bones of 1 hand. Adult.

S 3435 Bones of 1 hand. Adult.

S 3436 Bones of 1 foot. Adult.

The above 23 specimens were purchased from Reinbek on 10 January 1957.

S 6168 Skeleton, skin and brain. Juvenile male. Received from C. Hagenbeck on 1 February 1965. This is probably 'Jim', who arrived at the Carl Hagenbeck Tierpark GmbH, Stellingen, on 23 July 1960, weighing 20 kg. He had been imported from Lac Cachimba, Mayumba district, Gabon. He died of narcosis on 1 February 1965 (weight 84 kg). [Some additional information added by GH from correspondence with Carl Hagenbeck Tierpark personnel.]

S 6992 Skull. Adult male. Bayakka territory, Kanngula, near Kinshasa (Léopoldville), Republic of the Congo. Collected in 1962 or 1963. Received from K.O. Flemming on 2 December 1971.

S 6993 Skull. Subadult male. Kelle, Anall district, near Brazzaville, Republic of the Congo. Collected in 1962 or 1963. Received from K.O. Flemming on 2 December 1971. The upper jaw lacks the 3rd molars on both sides; the lower jaw lacks the 2nd and 3rd molars on both sides.

S 7113 Skull. Adult male. *Gorilla gina*. Upper Sanga, near Brazzaville, French Congo. Received from K.O. Flemming on 2 July 1953.

S 7887 Skull. Adult male. Received from Customs at Hamburg-Veddel on 10 March 1983.

S 7900 Skull. Adult male. Makokou, Gabon. Collected in January 1968. Received from H.L. Blonk, Netherlands, on 17 April 1985.

S 7901 Skull. Adult female. Makokou, Gabon. Collected in January 1968. Received from H.L. Blonk, Netherlands, on 17 April 1985.

S 7902 Skull. Adult female. Evonmitok, Gabon. Collected in January 1968. Received from H.L. Blonk, Netherlands, on 17 April 1985.

S 8024 Skull. Adult female. Jaunde, Kamerun. Collected early in the 20th century. Received from H.L. Blonk, Netherlands.

S 8025 Skull. Adult female. Surroundings of Brazzaville, French Congo. Collected c.1950. Received from H.L. Blonk, Netherlands. The upper jaw lacks 2 molars; the lower jaw lacks 4 molars and 2 premolars.

S 8026 Skull. Adult male. Surroundings of Brazzaville.

S 8192 Skull. Adult male. Jaunde, Kamerun. Received from J. Jungbauer on 10 March 1983.

S 8354 Skeleton and brain. Adult male. The upper jaw lacks 1 molar. Received from Zoologischer Garten Berlin in 1992.

S 9606 Skull (cast). Adult.

S 9874 Skull (cast). Adult. Original came from Ogowe-Kelle region, near Brazzaville, Congo.

S 9875 Skull (cast). Adult.

S 10319 Dermoplastinate. Adult male. Received from Tierpark Hagenbeck, Hamburg.

S 10483 Cranium. Subadult female. Received in exchange from Frau B. Hedwig, Museum der Arbeit, Hamburg, on 30 April 2009. It lacks the incisors and canines.

Hanover

Niedersächsisches Landesmuseum Hannover, Willy-Brandt-Allee 5, 30169 Hannover

G. Boenigk; Annina Böhme; Christiane Schilling

Left hand and right foot (casts). Male. From the captive specimen 'Bobby', Zoo Berlin. Received from Zoologisches Museum Berlin on 5 February 1938.

NLMH 956 Dermoplastic. Adult male. Purchased for DM500 from the Zoologischer Garten Hannover on 10 September 1986. This animal was born in Cameroon and was imported by Müller, Krechting, who sold it to L. Ruhe, Alfeld*. Originally named 'Djombo', he arrived at the Ruhr-Zoo Gelsenkirchen, Bleckstrasse 64, 4650

Gelsenkirchen, on 12 July 1966, weighing 6.5 kg and aged about 10 months. He was sent to the Zoologischer Garten Hannover (with a female named 'Chacha') on 27 December 1971, and was known there as 'Jambo'. He died of arteriosclerosis and diabetes on 9 September 1986 (weight 155 kg). His maximum recorded weight was 195 kg in the spring of 1986. The Museum taxidermists were G. Pucka and C. Wortmann. [Some additional information added by GH from correspondence with Ruhr-Zoo Gelsenkirchen personnel.]

*The Firma Ruhe was started in 1860 by Ludwig Ruhe. His son Hermann Ruhe (1895–1978) leased Hannover Zoo (near to Alfeld) from the town council between 1931 and 1971, using it as a transit station for his animals, and keeping them on public display until they were sold on to various other zoos. Ruhe went into liquidation in 1993.

NLMH O.956 Mounted skeleton of the above specimen.

NLMH O.2537 Skull. Male. Dscha River, South Kamerun. Purchased for 125 marks from Johannes Umlauff (1874–1951), Naturalien-Handlung, Hamburg, in 1914.

NLMH O.2538 Skull. Male. Purchased from Ruhe, Alfeld, in 1911. It was a captive specimen, imported from West Africa. The skull-cap is detachable. Some teeth are missing.

NLMH O.2539 Skull. Juvenile female. Congo. Purchased for 32 marks from Ruhe, Alfeld, in May 1906. The animal died on 17 May 1906. The taxidermist was Th. Till, Bad Kösen.

NLMH O.2540 Skull. Female. Kamerun. Purchased from Johannes Umlauff, Naturalien-Handlung, Hamburg.

Tierärztliche Hochschule Hannover, Institut für Zoologie, Bünteweg 17, 30559 Hannover*

Thomas Geissmann

2/66 Skeleton. Female. Received from the Zoologischer Garten Hannover on 22 July 1966. It was 4½ years old. Body weight 20.6 kg; brain weight 392 g.

25/66 Skeleton. Male. Received from the Zoologischer Garten Hannover on 14 November 1966. It was 2 years old. Body weight 5.9 kg.

28/66 Skeleton. Male. Received from the Zoologischer Garten Hannover on 18 November 1966. It was 3 years old. Body weight 9.2 kg; brain weight 470 g.

79/67 Skeleton. Male. Received from the Zoologischer Garten Hannover on 6 June 1967. It was 2½ – 3 years old. Body weight 10 kg; brain weight 415 g.

137/67 Skeleton. Male. Received from the Zoologischer Garten Hannover on 30 November 1967. It

was approximately 2 years old. Body weight 6 kg; brain weight 355 g.

464/70 Skeleton. Male. Received from the Zoologischer Garten Hannover on 10 November 1969. It was 3 years old. This is probably 'Ebobo', who arrived at the Zoo Hannover on 22 July 1968, having been acquired from Herrn Steinfurth, Río Muni, and died there on 8 November 1969. Body weight 14 kg; brain weight 457 g.

Heidelberg

Institut für Plastination e.K., Im Bosseldorn 17, 69126 Heidelberg

Christiane Casott

Carcase (plastinated). Adult male. Donated via the Zoo Hannover. The specimen has been dissected to display both the superficial and intermediate musculature, together with associated structures such as tendons, aponeuroses, fasciae, etc., as well as neurovascular bundles and cutaneous nerves. The body was fixed and later plastinated to demonstrate mainly the muscles in a static posture. Inner organs are represented as a whole block separately. This is 'Artis', who was born in the Stichting Koninklijke Rotterdamse Diergaarde, Van Aerssenlaan 49, 3000 AM Rotterdam, Netherlands, on 6 November 1983 to dam 'Salomé' and sire 'Ernst'. He was reared by hand. He was sent on loan to the Zoologischer Garten Hannover, Adenauerallee 3, 3000 Hannover 1, where he arrived on 27 October 1993. He drowned in the moat surrounding his enclosure on 14 May 2000. NB: This specimen is included in Körperwelten (Body Worlds) touring exhibitions, and does not have a fixed location at the time of writing. [Some additional information added by GH from International Studbook for the Western Lowland Gorilla, 2010.]

Zoologisches Museum, Zoologisches Institut der Universität Heidelberg, Im Neuenheimer Feld 230, 69120 Heidelberg

Bernhard Glass; E. Mielenz

Mounted skeleton and dermoplastic. Adult female. Donated by the Tiergarten der Stadt Nürnberg, Am Tiergarten 30, 90480 Nürnberg, in 1977. This is 'Liane', who arrived at the Nuremberg Zoo on 29 April 1964. Originally from Congo, she was purchased from L. Ruhe, Ruhr-Zoo Gelsenkirchen. Her weight was 15.7 kg on 4 May 1964. She died on 19 June 1977. She gave birth to a male (named 'Schorschla') on 3 March 1972, which was hand-reared. The sire was 'Fritz', who came originally from Ebolowa, Cameroon, and arrived from the Munchener Tierpark Hellabrunn AG on 2 December 1970 (weight 66 kg). [Some additional information added by GH from correspondence with Nürnberg Zoo personnel.]

Dermoplastic. Infant male. It was donated by the privately owned zoo in Neuwied, and arrived preserved in alcohol. It was aged about 6 months at death.

Jena

Friedrich-Schiller-Universität, Phyletisches Museum, vordem Neutor 1, 07743 Jena

Rolf Georg Beutel; D. v. Knorre

Mam. 11 Mounted skin. Adult female. Dume River, Kamerun. Collected by Hans Paschen in 1910.

Mam. 136 Mounted skin. Adult female. Preparator Umlauff.

Mam. 469 Cranium. Adult male. Saale River, Thuringia; origin unknown.

Mam. 477 Skeleton. Adult female. Received from the Anatomical Museum, Jena.

Mam. 485 Skeleton. Adult male.

Mam. 486 Cranium. Juvenile. The teeth are missing.

Mam. 487 Skull. Adult male. Kamerun. Collected by Hans Paschen in 1910. The teeth are missing.

Mam. 488 Skull. Adult male. Kamerun. Collected by Hans Paschen in 1910.

Mam. 489 Skull. Adult male. Kamerun. Collected by Hans Paschen in 1914. The teeth are missing.

Mam. 490 Cranium. Adult male. Kamerun. Collected by Hans Paschen in 1911. The teeth are missing.

Mam. 491 Cranium. Juvenile male. Kamerun. The teeth are missing.

Mam. 492 Cranium. Adult female. Délélé province, Kamerun. Collected by Hans Paschen in August 1910.

Mam. 493 Skull. Adult female.

Mam. 494 Cranium. Adult male. Kamerun. Collected by Hans Paschen in August 1910.

Mam. 495 Skull (cast).

Mam. 496 Skeleton (incomplete). Adult. Kamerun. Collected by Hans Paschen.

Mam. 516 Skeleton. Adult female.

Mam. 735 Cranium. Juvenile female. Received from the Anatomical Museum, Jena.

Mam. 1252 Skull (cast). Made in Weimar.

Mam. 1293 Skull. Adult female.

Mam. 1303 Skull. Adult female.

Mam. 1389 Arm bones.

Mam. 1390 Leg bones.

Mam. 1403 Skeleton. Adult male.

Mam. 1441 Postcranial skeleton. Adult female. Kamerun. Collected by Hans Paschen in November 1910.

Mam. 1442 Skeleton. Adult male.

Mam. 1443 Postcranial skeleton. Adult female. Kamerun. Collected by Hans Paschen in November 1910.

Mam. 1444 Foot bones.

Mam. 1830 Skull. Adult female.

Mam. 1881 Mounted skin (habitat preparation). Adult male. Kamerun. Collected by Meyer in 1908. Preparator Umlauff.

Mam. 1908 Skull (cast). Male.

Mam. 2139 Skeleton. Male. Received from the Tierpark Berlin.

Mam. 2172 Vertebrae of the neck.

Mam. 2434 Skull (cast). From the collection of the Institut für Anthropologie (their number A 11578).

Mam. 2435 Skull (cast). From the collection of the Institut für Anthropologie (their number A 11577).

Mam. 2436 Skull (cast). From the collection of the Institut für Anthropologie (their number A 11579).

Mam. 2541 Jar with 6 eyes. Kamerun. Collected by Hans Paschen in 1900.

Mam. 3229 Head (embedded in paraffin). Juvenile. From the collection of D. Starck.

Mam. 3230 Brain (in ethanol) and cast of head. Female. Imported on 13 April 1960. From the collection of D. Starck (his number 6403).

Mam. 3231 Death mask. Male. Zoo Frankfurt am Main 9 August 1956. From the collection of D. Starck. This may be 'Carlo', who arrived at the Zoo Frankfurt with two other males ('Thomas' and 'Rafiqui') on 2 July 1953. All three came from Spanish Guinea. 'Carlo' died on 8 August 1956. [Some additional information added by GH from correspondence with Frankfurt Zoo personnel.]

Mam. 3232 Death mask. Male. Zoo Frankfurt am Main 9 August 1956. From the collection of D. Starck.

Mam. 3255 Head (cast). Juvenile male. Spanish Guinea. From the collection of D. Starck.

Mam. 4091. Hair sample. From the collection of D. Starck.

Mam. 4444 Brain (in ethanol). Male, 15 years old, Zoo Frankfurt 5 January 1965. From the collection of D. Starck (his number 246).

Mam. 4531 Brain and hypophysis cerebri (in ethanol). Male, 4−5 years old, Zoo Frankfurt 7 September 1956. From the collection of D. Starck (his number 5636). This may be 'Thomas' (see also Mam. 3231 above), who died in Zoo Frankfurt on 6 September 1956.

Mam. 4811 Brain (in formaldehyde). Male, Zoo Frankfurt am Main, 22 April 1969. Originally from Oklahoma Zoo. From the collection of D. Starck (his number 6928). This is probably 'Solomon', who weighed 18 lb when he arrived at the Oklahoma City Zoo, with a female named 'Sheba', on 30 September 1960. Both had been purchased from F.J. Zeehandelaar, Inc. 'Solomon' was sold to Frankfurt Zoo and arrived there on 20 April 1967. He died of peritonitis and gastro-enteritis on 19 April 1969. [Some additional information added by GH from correspondence with Oklahoma City Zoo personnel.]

Karlsruhe

Staatliches Museum für Naturkunde Karlsruhe, Erbprinzenstrasse 13, 76133 Karlsruhe

Ralf Angst; Albrecht Manegold

SMNK-MAM 381 Skull (cast). Adult male. Gaboon. Purchased.

SMNK-MAM 382 Skull (cast). Adult female. Gaboon. Purchased.

SMNK-MAM 2194 Scapula. Adult. Purchased from M. Schoch, Munich, on 16 June 1975.

SMNK-MAM 2245 Skull. Adult male. Río Muni. Collected, through local people by Dr Clyde Jones. Donated by Prof. Dr H.O. Hofer, Covington, Louisiana, USA, on 8 November 1975.

SMNK-MAM 2246 Skull. Adult male. Río Muni. Collected, through local people by Dr Clyde Jones. Donated by Prof. Dr H.O. Hofer, Covington, Louisiana, USA, on 8 November 1975.

SMNK-MAM 2247 Skull. Adult male. Río Muni. Collected, through local people by Dr Clyde Jones. Donated by Prof. Dr H.O. Hofer, Covington, Louisiana, USA, on 8 November 1975. It has no sagittal crest.

SMNK-MAM 2406 Mounted skeleton. Infant male. Río Muni? Purchased from Prof. Dr H.O. Hofer, Covington, Louisiana, USA, on 29 July 1976.

SMNK-MAM 2506 Skull (cast). Adult male. Purchased on 5 October 1973.

SMNK-MAM 2516 Postcranial skeleton. Juvenile. Purchased from Dr H. Himmelheber, Heidelberg.

SMNK-MAM 2517 Postcranial skeleton. Juvenile. Purchased from Dr H. Himmelheber, Heidelberg.

SMNK-MAM 2518 Postcranial skeleton. Juvenile. Purchased from Dr H. Himmelheber, Heidelberg.

SMNK-MAM 2519 Postcranial skeleton. Juvenile. Purchased from Dr H. Himmelheber, Heidelberg.

The above 4 hominid postcranial skeletons are all fragmentary, as found in the forest. They are labelled as 'Pongidae gen. et spec. indet'. Possibly, some of the bones are from chimpanzees. These skeletons may belong to the skulls SMNK-MAM 2520, SMNK-MAM 3179 and SMNK-MAM 3180 listed below.

SMNK-MAM 2520 Skull. Juvenile. Cameroun. From the collection of Pieter Sohl, Heidelberg. Purchased from Dr Hans Himmelheber, Heidelberg, on 5 August 1974.

SMNK-MAM 3179 Skull. Juvenile. Cameroun. From the collection of Pieter Sohl, Heidelberg. Purchased from Dr H. Himmelheber, Heidelberg, on 5 August 1974.

SMNK-MAM 3180 Skull. Juvenile. Cameroun. From the collection of Pieter Sohl, Heidelberg. Purchased from Dr H. Himmelheber, Heidelberg, on 5 August 1974.

The above 3 skulls are all slightly defective, as found in the forest.

SMNK-MAM 3786 Skull. Juvenile female.

SMNK-MAM 3787 Skull. Adult female.

SMNK-MAM 3788 Skull. Adult female.

SMNK-MAM 3789 Skull. Juvenile female.

SMNK-MAM 3790 Skull. Adult female.

SMNK-MAM 3791 Skull. Adult female.

SMNK-MAM 3792 Skull. Adult female.

SMNK-MAM 3841 Cranium. Adult female.

SMNK-MAM 3842 Cranium. Subadult female.

The above 9 specimens, SMNK-MAM 3786−SMNK-MAM 3842 and SMNK-MAM 3966 (listed below), were all originally collected long ago by the princes of Mannheim. They were stored in a coal cellar for many years after the court of Mannheim ceased to exist, and eventually were given to the museum by the government of Baden-Württemberg in 1977.

SMNK-MAM 3870 Skeleton. Infant female. Purchased from Achim Schlüter, Winnenden, on 24 November 1977. It had been supplied to Schlüter by the taxidermist Jac Bouten, Venlo, Netherlands, in 1975. The mounted skin of this specimen is in the Schoolmuseum Michel Thiery, Ghent, Belgium.

SMNK-MAM 3966 Skull. Adult female.

SMNK-MAM 10978 Skull. Adult male. South Cameroun, 300 km east of Kribi. Purchased from K. Kellermann on 23 March 1981.

SMNK-MAM 10979 Skull. Adult female. Cambo, Cameroun. Purchased from K. Kellermann on 23 March 1981.

SMNK-MAM 11691 Humerus, radius, ulna, both scapulae, pelvis and bones of both legs. Adult male. Donated by W. John, Goethegymnasium Karlsruhe, on 9 September 1982.

SMNK-MAM 13448 Death mask. Adult male. Captive specimen. Donated in 1987.

SMNK-MAM 26023 Skull (cast). Adult male. Received on permanent loan in 1977.

SMNK-MAM 26030 Skull (cast). Juvenile female. Received on permanent loan in 1977.

SMNK-MAM 26031 Skull (cast). Adult male. Received on permanent loan in 1977.

SMNK-MAM 26032 Cadaver (partially dissected). Neonate male. Captive specimen. Donated on 27 March 1990.

Kassel

Naturkundemuseum im Ottoneum, Steinweg 2, 34117 Kassel

Jürgen Fichter; Franz Malec; Peter Mansfeld

NMOK 51 MAM 0051 Mounted skeleton. Adult male.

NMOK 51 MAM 0166 Cranium (sectioned sagittally). Adult male. *Gorilla gina*.

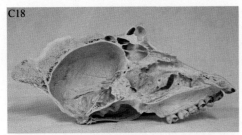

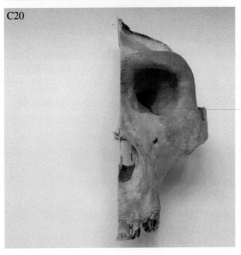

FIGURES C-18–C-20 Cranium of adult male (no. NMOK 51 MAM 0166), midsagittal section. *Photographs by Peter Mansfeld, courtesy of Naturkundemuseum im Ottoneum, Kassel.*

Leipzig

Naturkundemuseum Leipzig, Lortzingstrasse 3, 04105 Leipzig*

Michael Meyer

Ls Mam. 89 Mounted skin. Juvenile male. Gabon. Received from the Zoologisches Institut der Karl-Marx-Universität in 1968. This is probably the specimen which arrived at the Zoologischer Garten Leipzig on 29 September 1905 (purchased from Hagenbeck?) and died on 16 January 1906. The height, as mounted, is about 65 cm.

Magdeburg

Museum für Naturkunde, Otto-von-Guericke-Strasse 68–73, 39104 Magdeburg

Hans Pellmann

Z 1443 Mounted skeleton. Female.

Munich

Ludwig-Maximilians-Universität, Medizinische Fakultät, Institut für Anatomie und Zellbiologie, Pettenkoferstrasse 11, 80336 München

Jens Waschke

Skull. Adult male.
Skull. Adult female.
Mounted skeleton. Adult male.

Staatssammlung für Anthropologie und Paläoanatomie, Karolinenplatz 2a, 80333 München

Joachim Boessneck; Joris Peters

NB: The specimens below were part of the reference collection of recent mammal skeletons transferred from the Institut für Palaeoanatomie, Domestikationsforschung und Geschichte der Tiermedizin der Universität München, Schellingstrasse 10, München, in 2000.

IPM 1 Skeleton. Adult female. Received from the Munchener Tierpark Hellabrunn AG, Tierparkstrasse 30, München. This specimen is 'Bess', who came originally from Ebolowa, Cameroon, and arrived at the zoo on 19 August 1965, aged about 2 years. She arrived with a male named 'Porgy', who was about 3 years old, and also came from Ebolowa. 'Bess' died of purulent peritonitis on 24 August 1986, and was pregnant at the time of death. The skeleton of the unborn young has also been prepared. [Some additional information added by GH from correspondence with Munich Zoo personnel.]

IPM 2 Skeleton. Juvenile male. Received from the Tierpark Hellabrunn, where he was born on 17 August 1984 and was named 'Tarzan'. He died of Streptococcal septicaemia on 6 November 1986. The dam was 'Wilma', who was received at Munich on 26 July 1977 from the Tiergarten Nürnberg, having been born there (to dam 'Delphi' and sire 'Fritz') on 9 September 1975. The sire was 'Rututu', who came originally from Kribi, Cameroon, and arrived at the Tierpark Hellabrunn on 26 March 1974, aged about 8 months and weighing 4.6 kg. [Some additional information added by GH from correspondence with Munich Zoo personnel.]

Zoologische Staatssammlung, Münchhausenstrasse 21, 8124 München

Michael Hiermeier; Richard Kraft; Olav Röhrer-Ertl

NB: The Anthropological State Collection and the Zoological State Collection in Munich lost the greater part of their collections in 1944, owing to bombing, and the original catalogues were also burnt. Secondary catalogues were started in 1947 (ASM) and 1948 (ZSM). The specimens listed below are housed by the Zoologische Staatssammlung on behalf of the Primatology Section of the Staatssammlung für Anthropologie und Paläoanatomie, Abteilung Anthropologie, Karolinenplatz 2a, 80333 München.

ZSM 1907/706 Skull. Adult male. SW Kamerun. Collected c.1900. Received from the collection of Dr Emil Selenka.

ZSM 1907/708 Mandible. Juvenile male. SW Kamerun. Collected c.1900. Received from the collection of Dr Emil Selenka.

ZSM 1908/36a Skull. Adult male. Epolobingen, upper Nyong, SW Kamerun. Collected by Oberleutnant Bertram in 1908. The top of the skull is broken.

ZSM 1911/2397 Postcranial skeleton with lower jaw. Adult male. SW Kamerun. Collected and donated by von Oertzen. The cranium is missing.

ZSM 1913/1162 Skeleton. Adult male. Nola, Sangha, Congo. Collected and donated by Capitaine Michell in 1913.

ZSM 1913/1163 Skull. Infant. Nola, Sangha, Congo. Collected and donated by Capitaine Michell in 1913.

ZSM 1913/1164 Skull. Adult male. Ouesso, Sangha, Congo. Collected and donated by Capitaine Michell in 1913.

ZSM 1913/1165 Skull. Adult male. Probably Ouesso, Sangha, Congo. Collected and donated by Capitaine Michell in 1913.

ZSM 1913/1168 Skin. Male. Near the border of Kamerun and Spanish Guinea. Collected and donated by Capitaine Michell in 1913.

ZSM 1950/90 Skull. Adult male. SW Kamerun. Collected by R. Apitz of Meissen, from whom it was purchased for DM150.

ZSM 1951/269 Skull (cast). Adult male. Purchased from Dr Uthmöller on 22 August 1951.

ZSM 1954/201 Skeleton. Adult female. Bangomo, Congo, French Equatorial Africa. Collected by Konsul Dr H. Meissner in July 1954, and donated by him on 13 September 1954.

ZSM 1962/331 Skull. Juvenile male. Spanish Guinea, near the border with Cameroon. Collected in 1960.

ZSM 1962/332 Skull. Juvenile female. Spanish Guinea. Received from the Parque Zoologico de Barcelona, Spain, where it had arrived in October 1960. It was about 6 years old at the time of its death.

ZSM 1962/333 Skeleton. Adult male. Spanish Guinea, near the border with Cameroon. Received in exchange from Palaus.

ZSM 1963/83 Skull. Adult male. Spanish Guinea. Purchased from the Reitter family on 21 June 1963.

ZSM 1963/106 Cranium. Adult male. Yokadouma, SE Cameroun. Purchased from Prof. Lutz Heck in 1938.

ZSM 1963/107 Cranium. Adult female. Yokadouma, SE Cameroun. Purchased from Prof. Lutz Heck in 1938.

ZSM 1963/108 Skull. Juvenile female. Yokadouma, SE Cameroun. Purchased from Prof. Lutz Heck in 1938. The animal weighed 10 kg.

ZSM 1963/109 Skull. Adult male. Yokadouma, SE Cameroun. Purchased from Prof. Lutz Heck in 1938.

ZSM 1963/110 Skull. Adult male. Yokadouma, SE Cameroun. Purchased from Prof. Lutz Heck in 1938.

ZSM 1963/111 Skull. Adult male. Yokadouma, SE Cameroun. Purchased from Prof. Lutz Heck in 1938.

ZSM 1963/112 Skull. Adult male. Yokadouma, SE Cameroun. Purchased from Prof. Lutz Heck in 1938.

ZSM 1963/113 Skull. Juvenile female. Yokadouma, SE Cameroun. Purchased from Prof. Lutz Heck in 1938.

ZSM 1963/235 Carcase (in alcohol). Juvenile male. Donated by G. Blonk in 1963. It had died of influenza 6 days after arrival in the Netherlands.

ZSM 1965/203 Skeleton and skin. Adult female. Donated by the Zoologischer Garten Hannover on 26 May 1965. This animal arrived in Hannover on 4 May 1964, having been received in exchange from the Lincoln Park Zoo, Chicago, Illinois, USA, via F.J. Zeehandelaar, Inc. Known as 'Rajah', she had arrived at the Lincoln Park Zoo on 1 October 1948 (weight 18 lb), along with three other gorillas (the males 'Irving Young' and 'Sinbad', and the female 'Lotus'). All were originally from Cameroun and were donated by Mr Irvin B. Young of Chicago. 'Rajah' died on 18 May 1965, while being transported to the Amsterdam Zoo, Netherlands. [Some additional information added by GH from correspondence with Lincoln Park Zoo personnel.]

ZSM 1965/263 Skeleton. Infant male. Donated by Zoo Hannover on 30 November 1965. It was aged about 1½ years at the time of death.

ZSM 1971/543 Skull (cast). Adult male. Donated by Markus Sommer, Coburg, in May 1971. It is a cast of skull 1963/83 in the collection. Further casts have been sold commercially by Sommer, as SOMSO model no. ZoS 50.

ZSM 1971/544 Skull (cast). Juvenile male. Donated by Markus Sommer, Coburg, in May 1971. It is a cast of skull 1965/263 in the collection, sold commercially by Sommer as SOMSO model no. ZoS 50/1.

ZSM 1971/545 Skull (cast). Adult female. Donated by Markus Sommer, Coburg, in May 1971. It is a cast of skull 1965/203 in the collection. Further casts have been sold commercially by Sommer, as SOMSO model no. ZoS 51.

ZSM 1974/58 Skull. Adult male. Bélabo, 300 km from Yaoundé, SE Cameroon. Collected by Richard Tanglmayer in February 1972. Purchased in March 1974.

ZSM 1981/463 Skull. Female. Mbalmayo, Nyong, Kamerun. Received from the Anthropologische Staatssammlung, München, on 31 March 1981. Its original number in that collection was AS-P-443 (294).

ZSM 1981/464 Skull. Female. Sango Belika, Kamerun. Received from the Anthropologische Staatssammlung, München, on 31 March 1981. Its original number was AS-P-444 (295).

ZSM 1981/465 Skull. Female. Río Muní. Received from the Anthropologische Staatssammlung, München, on 31 March 1981. Its original number was AS-P-445 (296).

ZSM 1981/466 Skull. Adult female. Lomié, South Kamerun. Received from the Anthropologische Staatssammlung, München, on 31 March 1981. Its original number was AS-P-446.

ZSM 1981/467 Cranium. Adult male. Kamerun? Received from the Anthropologische Staatssammlung, München, on 31 March 1981. Its original number was AS-P-447.

ZSM 1981/468 Cranium. Adult male. SW Kamerun. Collected by Haberer in 1907. It was originally part of Dr Selenka's collection. Received from the Anthropologische Staatssammlung, München, on 31 March 1981. Its original number was AS-P-448 (4031).

ZSM 1981/469 Cranium. Adult male. Kamerun. Collected by Haberer in 1907. It was originally part of Dr Selenka's collection. Received from the Anthropologische Staatssammlung, München, on 31 March 1981. Its original number was AS-P-449 (4032).

ZSM 1981/470 Cranium. Juvenile female. SW Kamerun. Collected by Haberer in 1906. Received from the Anthropologische Staatssammlung, München, on 31 March 1981. Its original number was AS-P-450 (4033).

ZSM 1981/471 Cranium. Juvenile male. Kamerun. Collected by Haberer in 1906. Received from the Anthropologische Staatssammlung, München, on 31 March 1981. Its original number was AS-P-451 (4034).

ZSM 1981/472 Skull. Adult female. SW Kamerun. Collected by Haberer in 1907. Received from the Anthropologische Staatssammlung, München, on 31 March 1981. Its original number was AS-P-452 (4037).

ZSM 1981/473 Cranium. Adult male. Kamerun. Collected by Haberer in 1906. Received from the Anthropologische Staatssammlung, München, on 31 March 1981. Its original number was AS-P-453 (8996).

ZSM 1981/474 Skull. Adult male. Lomié, South Kamerun. Collected c.1914. Received from the Anthropologische Staatssammlung, München, on 31 March 1981. Its original number was AS-P-454 (9000).

ZSM 1981/475 Skull. Adult female. SW Kamerun. Collected by Haberer. It belonged originally to Dr Selenka's collection. Received from the Anthropologische Staatssammlung, München, on 31 March 1981. Its original number was AS-P-560 (9001).

ZSM 2006/0004 Skull. Adult male. From the collection of Prof. Rupprecht Bernbeck (1916–2003); donated by Dr Fehringer. It may have come originally from Zoo Hellabrunn.

ZSM 2006/0004 Skull. Adult female. From the collection of Prof. Rupprecht Bernbeck; donated by Dr Fehringer. It may have come originally from Zoo Hellabrunn.

Nuremberg

Tiergarten Nürnberg, Am Tiergarten 30, 90480 Nürnberg*

Hans Lichei

NB: The specimens listed below belong to the Naturhistorische Gesellschaft Nürnberg e.V., Marientorgraben 8, 90402 Nürnberg. Parts of this Society's collection (including *Gorilla* specimens) were taken over by the Tiergarten der Stadt Nürnberg in 1988/89 on permanent loan. Previously, they had been held on loan at the Zoological Museum in the old Zoological Institute of the University of Erlangen [Ronald Heissler].

Mounted skeleton. Male. *Gorilla castaneiceps*. Brazzaville area, French Congo.

K138 8124[1] Skull. Adult male.

K501 7497[2] Skull. Adult male. Acquired in 1966.

K502 7497[3] Skull. Adult female.

Dermoplastic. Adult male. This animal arrived at the Tiergarten Nürnberg on 28 September 1963 (weight 27 kg), having been purchased from Karl Krag, and was named 'Schorsch'. Krag had originally delivered him (with another male, 'Tuff') to the Calgary Zoo, Canada, on 14 February 1962. Known then as 'Ruff', he was transferred back to Krag in August 1963, in exchange for a female ('Toni'). He died on 4 August 1970 (weight 114.9 kg) of liver cirrhosis, uraemia and self-mutilation. Nephritis had caused the shrinkage of one kidney. [Some additional information added by GH from correspondence with Calgary Zoo personnel.]

Reutlingen

Naturkundemuseum, Weibermarkt 4, 72764 Reutlingen*

Jan Brinkmann; Barbara Karwatzki

Z 8 Dermoplastic. Subadult male. Cameroon. Purchased for DM2500 in 1974 from Scharf, Wilhelma Zoo, Stuttgart. It is in quite bad condition.

Z 9 Mounted skeleton. Subadult male. Prepared by the taxidermist Frank Döring in 1978 at a cost of DM2485. It is from the same individual as the above dermoplastic. Known as 'Ebebin', this animal lived in a private zoo in Ludwigsburg (near Stuttgart). One day he broke out of his cage and was shot with a submachine gun.

Z 12 Mounted skeleton. Adult female. French Congo. Purchased for DM760 in 1953 from the taxidermist Schmucker.

Stuttgart

Staatliches Museum für Naturkunde, Zoologische Abteilung, Schloss-Rosenstein 1, 70191 Stuttgart

F. Dieterlen; Stefan Merker; Reinhard Ziegler

SMNS 1731 Skull and mounted skin. Adult male. Gabon. Collected in 1879. Purchased from Gerrard in December 1879.

SMNS 1784 Skeleton and mounted skin. Subadult female. Eliva Comi, Gaboon. Collected by von Koppenfels in 1881. Received in 1881.

SMNS 2646 Mounted skin. South Kamerun. Collected by H. Preuss in 1905.

SMNS 2674 Skull and mounted skin. Subadult female. South Kamerun. Collected by von Heigelin in 1906. Received in 1906.

SMNS 5633 Cranium. Adult male. Jaünde, South Kamerun. Collected by Achenbach. Received from the Linden-Museum Stuttgart (Staatliches Museum für Völkerkunde), Hegelplatz 1, Stuttgart, on 23 December 1958. It is in bad condition, and lacks the canines, both 3rd molars and 3 incisors.

SMNS 5634 Cranium. Adult male. Bascho, South Kamerun. Collected by Diehl? Received from the Linden-Museum Stuttgart (Staatliches Museum für Völkerkunde), Hegelplatz 1, Stuttgart, on 23 December 1958. The cranium lacks 2 incisors and the occiput.

SMNS 5636 Left femur. Adult male. Bascho, South Kamerun. From the Diehl collection. Received from the Linden-Museum Stuttgart (Staatliches Museum für Völkerkunde), Hegelplatz 1, Stuttgart, on 23 December 1958.

SMNS 6872 Mounted skeleton. Adult male.

SMNS 6873 Skull and partial skeleton. Juvenile. Kribi, Cameroun. Collected by L. Obst in 1954. Donated by L. Obst. It is in bad condition.

SMNS 6874 Skull and partial skeleton. Juvenile. Kribi, Cameroun. Collected by L. Obst in 1954. Donated by L. Obst. It is in bad condition.

SMNS 7464 Skull and partial skeleton. Subadult male. Gabon. Collected in 1870. Received from the Paris Museum in 1870.

SMNS 7465 Skull. Adult female. Gaboon. Collected by, and received from, von Koppenfels in 1883. The upper jaw lacks the right canine and all the incisors; the lower jaw lacks all the incisors.

SMNS 7466 Skull. Adult male. Gaboon. Collected by, and received from, von Koppenfels in 1880. The upper jaw lacks 3 incisors; the lower jaw lacks 1 incisor and the 3rd molar. It is now missing (noticed in 2003).

SMNS 7467 Cranium. Adult male. *Gorilla gorilla diehli*. Kamerun. Collected by Preuss in 1905. Received from the Württembergischer Verein für Handelsgeographie und Förderung Deutscher Interessen im Ausland e.V. in 1905. All the teeth are missing.

SMNS 7468 Cranium. Adult male. *Gorilla gorilla diehli*. Kamerun. Collected by Oerzen in 1905. Received from the Württembergischer Verein für Handelsgeographie und Förderung Deutscher Interessen im Ausland e.V. in 1905. It lacks the right canine, right premolars, right 1st molar and all the incisors.

SMNS 7469 Skull. Juvenile. South Kamerun. Collected by von Heigelin in 1909. Received in 1909. The upper jaw lacks 1 incisor; the lower jaw lacks 2 incisors.

SMNS 7470 Skull. Juvenile. South Kamerun. Collected by von Heigelin in 1909. The upper jaw lacks 1 incisor.

SMNS 7533 Skull. Juvenile. South Kamerun? Collected by von Heigelin in 1909?

SMNS 7588 Two vertebrae (atlas and axis). Juvenile. Collected by Witz. Received in 1966.

SMNS 19158 Skull and partial skin. Juvenile. *Gorilla beringei graueri*. Kahuzi, Zaïre. Killed by poachers February/March 1967. Collected by Dr F. Dieterlen. The skin lacks the head, arms and legs.

SMNS 32010 Skull. Male. South Kamerun. Collected by Preuss in 1905. Received from the Württembergischer Verein für Handelsgeographie und Förderung Deutscher Interessen im Ausland e.V. in 1905.

SMNS 38230 Skeleton and mounted skin. Adult female. Received from Wilhelma Zoologisch-Botanischer Garten, Stuttgart, in August 2006. This is 'Mora', who was born in Wilhelma Zoo on 12 December 1973 to dam 'Mimi' and sire 'Tschamba' and hand-reared. She died on 11 May 1984 (leukaemia, euthanased).

SMNS 38971 Skull. Adult male. Collected by Preuss in 1905. Received from the Württembergischer Verein für Handelsgeographie und Förderung Deutscher Interessen im Ausland e.V. in 1905.

SMNS 38972 Skull. Adult female. Kamerun. Collected by von Heigelin in 1909.

SMNS 46001 Skull. Male. Collected before 1930.

SMNS 46002 Skull. Male. Collected before 1930.

SMNS 46766 Skull. Juvenile. Received from G. Pilleri in 2006. Cranium opened for cast of braincase.

SMNS 49058 Skull, skin and death mask. Adult male. Cameroon. This is 'Banjo', who arrived at the Wilhelma

Zoo on 27 April 1966 (weight 6 kg). He died on 21 August 2001.

SMNS 51672 Carcase (in ethanol). Infant male. Received from Wilhelma Zoo in May 2003. It was born on 12 May 2003 to dam 'Mutasi' and sire 'Kibo', and died on 18 May 2003. Head-and-body length 410 mm.

Zoologisch-Botanischer Garten Wilhelma, Wilhelmaplatz 13/Neckartalstrasse, 70376 Stuttgart

Tobias Knauf-Witzens
Refrigerated serum:
Male 'Tano' (date of birth 8 November 2011): 17 March 15; 7 April 15 (× 4).
Male 'Okanda' (d.o.b. 17 April 2011): 3 December 14 (× 3); 7 April 15 (× 4).
Female 'Vana' (d.o.b. 16 March 2012): 4 April 14 (× 9).
Female 'Undi' (estimated d.o.b. 1970): 12 March 13; 30 April 13.
Male 'Meru' (d.o.b. 4 February 2004): 30 May 11.
Male 'Kimbali' (d.o.b. 22 December 2009): 30 April 13 (× 2).
Male 'Monza' (d.o.b. 2 September 2007): 30 May 11.
Male 'Upala' (d.o.b. 27 May 2008): 30 May 11.
Female 'Claudia' (d.o.b. 2 August 2009): 4 July 12.
Female 'Mimi' (est. d.o.b. 1964): 30 April 13 (× 2).
Female 'Shira' (d.o.b. 28 August 2004): 20 September 11 (× 4).
Female 'Kolo' (d.o.b. 16 October 1986): 30 April 13 (× 3).
Female 'Tuana' (d.o.b. 25 February 2005): 6 May 13 (plasma; serum × 2).
Female 'Mutasi' (d.o.b. 12 June 1994): 30 April 13.
Male 'Lengai' (d.o.b. 19 April 2003): 30 May 11.

Tübingen

Eberhard Karls Universität Tübingen, Institut und Museum für Geologie und Paläontologie, Sigwartstrasse 10, 72076 Tübingen*

Alexander Liebau
Skull. Female. It is exhibited in the 'Primates' showcase.

Eberhard Karls Universität Tübingen, Museum der Universität Tübingen, Zoologische Sammlung, Sigwartstrasse 3, 72076 Tübingen

G. Mickoleit; Erich Weber
SZ Mamm Dermoplastic. Female. Kamerun.

SZ Mamm 4 Skeleton. Juvenile female. Gabon. Purchased from the Berlin Aquarium, Unter den Linden 68a, Berlin, in 1896.

SZ Mamm 814 Skull. Female. Lower Guinea, between Niger and Congo rivers. Received in 1885.

SZ Mamm 816 Skull. Male. Lower Guinea, between Niger and Congo rivers. Received in 1885.

Ulm

Naturkundliches Bildungszentrum, Kornhausgasse 3, 89073 Ulm*

Peter Jankov
Skeleton. Juvenile. About 70 km southeast of Ebolowa, South Cameroon. Preserved in 1970 by Frank Döring, Birkach (Stuttgart). Received on 5 August 1970 from Herr Scharpf, Wilhelma Zoo, Stuttgart Bad-Canstatt. The animal was about 3–4 years old.

Wiesbaden

Museum Wiesbaden, Friedrich-Ebert-Allee 2, 65185 Wiesbaden

F. Geller-Grimm; G. Heinrich; Hannes Lerp; R. Mentzel
MWNH-MA-SK-110 Skull. Adult male. Purchased from A. Frank, Amsterdam, c.1900. It is damaged.

MWNH-MA-SK-482 Skull. Adult. Yaoundé, Cameroun. Received from Captain Jakoby.

MWNH-MA-SK-496 Mounted skeleton. Adult. Cameroun. Purchased from J. Umlauff, Hamburg, on 5 May 1927.

MWNH-MA-PR-1016 Mounted skin. Adult male. Yaoundé, Cameroun. Collected by Hans Zenker in 1922. Purchased from J. Umlauff, Hamburg, on 17 February 1926.

HUNGARY

Budapest

Természettudományi Múzeum Allattára, Baross utca 13, 1088 Budapest*

A. Demeter; Mihály Gasparik
1572 a/b Skeleton and mounted skin. Male. Gabon. Purchased from Frank, London, in 1887. The left tibia, right radius and right ulna are missing from the skeleton.

2767 Skull. Adult female. Kamerun. Donated by Harry W. Róna in 1910.

64.27.1 Skull. Adult male. Donated by the University of Veterinary Science, Budapest.

64.28.1 Skull. Subadult female. Donated by the University of Veterinary Science, Budapest. The occipital region has been inexpertly removed, possibly for ritualistic purposes whilst the skull was still in Africa.

64.31.1 Mandible. Donated by the University of Veterinary Science, Budapest.

66.198.1 Skeleton. Male. Donated by the Department of Zoology, Eötvös University, Budapest.

66.202.1 Mounted skin. Male.

74.57.1 Skull. Male. Donated by the Department of Zoology, Eötvös University, Budapest.

INDIA

Kolkata

Indian Museum, 27 Jawaharlal Nehru Road, Kolkata, West Bengal 700016*

P.K. Das; A.P. Kapur

11687 Mounted skin. Male.

11688 Mounted skin. Female.

11689 Mounted skeleton. Male.

11690 Mounted skeleton. Male.

New Delhi

University of Delhi, Department of Anthropology, New Delhi 110007*

K. Guha

Mounted skeleton. Male. Height 154.5 cm.

Skull. Male. The mandible is articulated.

Skull. Female. The mandible is articulated.

Cranium (plaster cast). Male.

IRELAND

Cork

University College Cork, Department of Archaeology, Connolly Building, Dyke Parade, Cork T12 CY82

William O'Brien; Barra O'Donnabhain; Catryn Power

Cranium. Subadult female. It is missing the canines, incisors and left 3rd molar. The basioccipital suture is unfused. The foramen magnum has been artificially enlarged.

Dublin

National Museum of Ireland, Merrion Street, Dublin 2

Nigel T. Monaghan; C.E.O. O'Riordan

NMINH: 1878.18.1 Skeleton. Male. Purchased for £100 from Monsieur A. Bouvier on 12 April 1878. Label from 1961: 'Dentition examined by Prof. S. Friel on 28 March 1961 in connection with forthcoming catalogue of British and Irish pongid material by Dr J.R. Napier, Royal Free Hospital, School of Medicine, University of London. The 2nd and 3rd molars, right and left, of upper jaw are those of an ox; left upper canine is plaster; all upper incisors have been modified by filing; they may have been a horse?'

NMINH:1883.370.1 Skull. Female. Gabon. Purchased for £2 10s 0d from Edward Gerrard, London, in 1883.

NMINH:1911.400.1 Mounted skin. Adult male. Mboug tributary of Doumé River, Cameroons hinterland. Shot by Hartherna. Mounted by Knutte, Schwerin, Germany. Purchased from H. Rolle, Speyererstrasse 8, Berlin, in November 1911 (total of £350 paid for 2 mounted skins and 2 mounted skeletons). It had been identified by Prof. Matschie of Berlin as a *Gorilla graueri* [*sic*].

NMINH:1911.401.1 Mounted skin. Adult female. Wabembeland, northwest of Lake Tanganyika. Collected by the Duke of Mecklenburg. Mounted by Knutte, Schwerin. Purchased from H. Rolle, Speyererstrasse 8, Berlin, in November 1911. It had been identified by Prof. Matschie as a *Gorilla beringei*.

NMINH:1911.402.1 Mounted skeleton. Adult male. It is from the same specimen as the mounted skin NMINH:1911.400.1, and was purchased with it. It was set up by Knutte, but later was sent to Edward Gerrard in London for remounting of the bones of the hands and feet at a cost of 5 shillings each, and was returned to Dublin on 22 July 1913. The hands and feet of the skeletons mounted by Knutte were found to be very faulty, the bones being mixed, and some even belonging to other specimens and species.

NMINH:1911.403.1 Mounted skeleton. Adult female. It is from the same specimen as the mounted skin NMINH:1911.401.1, and was purchased with it. It was set up by Knutte, but later was sent to Edward Gerrard in London for remounting of the bones of the hands and feet at a cost of 5 shillings each; returned to Dublin on 22 July 1913.

NMINH:2008.85.95 Skull (plaster cast).

NMINH:2008.85.765 Cranium (plaster cast), bisected sagittally.

Galway

Zoology and Marine Biology Museum, National University of Ireland, University Road, Galway

James S. Fairley; Eoin MacLoughlin

NUIGZ.PRI.33 Skull. Adult male.

NUIGZ.PRI.34 Skull. Adult female?

Both the above specimens were acquired before 1911.

NUIGZ.PRI.35 Skull.

ISRAEL

Jerusalem

The Hebrew University of Jerusalem, Institute of Life Sciences, Department of Zoology, Givat Ram, 91904 Jerusalem*

Goggy Davidovitz; Hefzibah Eyal-Giladi
    Skull. Female. Belgian Congo. Collected by Dr Weisberg.

Tel Aviv

Tel Aviv University, Sackler School of Medicine, Department of Anatomy and Anthropology, Ramat Aviv, 69978 Tel Aviv

Baruch Arensburg
    Cranium. Adult male. Acquired from a private donor c.1981. It has a healed fracture of the brow.

ITALY

Catania

Università di Catania, Istituto di Anatomia Umana Normale, Via Biblioteca 4, Catania*

Adalberto Passaponti
    Skull. Adult male. Purchased for 120 lire in 1884. The upper canines and incisors are missing, the maxillary alveolar bone is damaged and the lower right lateral and central incisors are missing. Maximum length, glabella—opisthocranion 192 mm; length glabella—inion 161 mm; length metopon—opisthocranion 150 mm; length metopon—inion 134 mm; maximum width, between euryons 98 mm; biauricular width 144 mm; bimastoid width 150 mm; skull height, basion—vertex (top of sagittal crest) 120 mm; skull height, basion—vertex (base of sagittal crest) 100 mm; basi-bregmatic height 107 mm; auricular height, porion—bregma 111 mm.
    Skull. Adult female. Purchased as above. The upper central incisors and upper right canine are missing. Maximum length, glabella—opisthocranion 151 mm; length glabella—inion 144 mm; length metopon—opisthocranion 127 mm; length metopon—inion 122 mm; maximum width, between euryons 100 mm; biauricular width 125 mm; bimastoid width 131 mm; skull height, basion—vertex 90 mm; auricular height, porion—bregma 91 mm.

Ferrara

Civico Museo di Storia Naturale, Via de 'Pisis 24, 44121 Ferrara

Fausto Pesarini

    227 Skull. Adult male. Purchased from N. Boubée, Paris, before 1967.

Florence

Università di Firenze, Museo di Storia Naturale, Sezione di Antropologia ed Etnologia, Via del Proconsolo 12, 50122 Firenze

Pier Virgilio Arrigoni; B. Lanza; P. Messeri; Jacopo Moggi-Cecchi; Caterina Scarsini
    3631 Skull. Adult female. Donated by Signora Giorgina Hamming, c.1886.
    5248 Skull. Juvenile male. Belgian Congo. Donated by Cavalier E. Brissoni on 19 November 1921. The occipital bone is missing, apart from a small part of the asterion and two Wormian bones along the left lambdoid suture.
    5361 Skull. Juvenile female. *Gorilla gina.* Collected in 1911.
    5362 Skull. Juvenile male. *Gorilla gina.*
    5363 Skull. Adult male. *Gorilla gina.* Kamerun.
    5364 Skull. Adult male. *Gorilla beringei?*
    5365 Skull. Adult male. *Gorilla gina.* Kamerun.
    5366 Skull. Adult male. *Gorilla gina.* Kamerun.
    The above 6 specimens were received from Dr Riccardo Folli on 12 July 1926.
    5700 Skeleton. Adult male. *Gorilla beringei graueri.* Tchibinda Forest, Bukavu area, Belgian Congo. Collected by Attilio Gatti on 13 December 1930. Acquired in 1932. It is incomplete, consisting of the skull, right scapula, right clavicle, both humeri, both radii, left ulna, pelvis, left femur, right tibia, right fibula, 17 vertebrae and 23 ribs. The tibia and fibula present a healed, irregular fracture. All the teeth are present, but I2 and and C1 were broken at the gum margin after death. According to the method of Martin and Saller (1957), *Lehrbuch der Anthropologie*, p. 414, tooth wear is: C, P1, P2, M1, M2 and M3 grade 2; I1, I2, grade 3. Skull length, glabella—opisthocranion 180 mm; length glabella—lambda 150 mm; length nasion—basion 143 mm; basion—prosthion 228 mm; bizygomatic width 185 mm; basi-bregmatic height 101 mm; nasal width 38 mm; nasal height, nasion—nasospinal 99 mm; cranial capacity 563 cc; maximum length of left humerus 444 mm; maximum length of left radius 375 mm; maximum length of left femur, to great trochanter 388 mm. The skin of this specimen is in the Museo Zoologico 'La Specola', Florence.
    5701 Death mask (plaster cast) of the above specimen (no. 5700).
    5702 Brain, genital organs and eyes (in formalin) of the above specimen (no. 5700).

Università di Firenze, Museo Zoologico 'La Specola', Via Romana 17, 50125 Firenze

Paolo Agnelli; Maria Luisa Azzaroli Puccetti
Skull (plaster cast).

MZUF-22 Mounted skin. Adult male. *Gorilla savagei.* Gabon. Collected in September 1888. Received in 1890.

MZUF-23 Mounted skin. Adult female. *Gorilla savagei.* Gabon. Collected in September 1888. Received in 1890.

MZUF-24 Mounted skeleton. Adult male. Gabon. Purchased from Edward Gerrard, London, in 1887.

MZUF-25 Skeleton. Adult female. Gabon. Acquired in 1874.

MZUF-27 Cranium. Adult male. Lower Congo. Collected in June 1885. Purchased from Umlauff, Hamburg, in 1887.

MZUF-7536 Mounted skeleton. Male. *Gorilla beringei graueri.* Belgian Congo. Collected by Vittorio Emanuele di Savoia-Aosta, Count of Turin (1870–1946), 1909–10.

MZUF-7537 Mounted skeleton. Female. *Gorilla beringei graueri.* Belgian Congo. Collected as above.

MZUF-7544 Skin. Juvenile. *Gorilla beringei graueri.* Belgian Congo.

MZUF-7545 Skin. *Gorilla beringei graueri.* Belgian Congo.

MZUF-8591 Skin and head mounted on canvas. Male. *Gorilla beringei graueri.* Belgian Congo.

The above 5 specimens were donated by Duchesse Irene di Savoie-Aosta (1904–74) in May 1955.

MZUF-21780 Skin. Male. *Gorilla beringei graueri.* Tchibinda Forest, west of Lake Kivu, Belgian Congo. Collected by Attilio Gatti in 1928. Received in 1932. For further information, see Ducci et al. (2015).

Genoa

Museo Civico di Storia Naturale 'Giacomo Doria', Via Brigata Liguria 9, 16121 Genova*

Gianna Arbocco
90 Skull. Adult female. *Gorilla gina.* Gabon. Received in 1869.

91 Skull. Adult male. Details same as for no. 90.

92 Skull. Adult male. Details same as for no.90.

93 Skull. Adult male. *Gorilla gina.* Lac Onangué, Lambaréné, Gabon. Collected by Leonardo Fea (1852–1903) on 14 September 1902.

94 Skull. Adult male. *Gorilla gina.* Gabon. Received in 1867.

119 Mounted skeleton. *Gorilla gina.* Catholic Mission, Fernan-Vaz, Gabon. Collected by Leonardo Fea on 21 August 1902.

42902 Mounted skin. Juvenile male. Congo. Purchased from G.J. Erritzoe, Bockum-Hövel, Germany on 26 July 1971.

42903 Skull. Juvenile male. From the same specimen as the mounted skin above.

Gorla Minore

Collegio Rotondi, Via San Maurizio 4, 21055 Gorla Minore*

Cranium. Adult male. *Gorilla beringei beringei.* Katanga, Belgian Congo. Found by Gino Imeri in 1950.

Milan

Museo Civico di Storia Naturale, Corso Venezia 55, 20121 Milano

Giorgio Bardelli; Luigi Cagnolaro
MSNM Ma132 Mounted skeleton. Adult male. Spanish Guinea. It died in captivity in Barcelona Zoo. Purchased from Sig. Mario Dugone in 1954. The right fibula has a callus, probably from a fracture.

MSNM Ma133 Mounted skeleton. Adult female. Molundu, Cameroun.

MSNM Ma192 Mounted skin. Adult male. Died in captivity in Germany. The skin was purchased in 1970 from the taxidermist Arno Belger, Leutenbach, Germany. It was mounted by Sig. Giangaleazzo Giuliano, the museum's deputy head taxidermist, in 1973. The mount is about 170 cm high, and the chest circumference is 140 cm.

MSNM Ma4274 Partial skeleton (plus a right humerus from another, smaller gorilla). Adult male. Cameroun/Guinea. Collected in 1958. The bones were purchased from Sig. Mario Dugone in 1959. Some bones (mandible, rib and pelvis) are broken by gunfire. A big lead bullet is still lodged in the pelvis.

MSNM Ma4597 Skull. Adult female. Spanish Guinea. Collected in August 1953. Purchased from Sig. Mario Dugone in 1954.

MSNM Ma4598 Skull. Adult female. Spanish Guinea. Collected in 1949. Purchased from Sig. Mario Dugone in 1954.

MSNM Ma4600 Skull. Adult male. Gabon. Collected in 1935. Purchased from Sig. A. Grasselli in 1961.

MSNM Ma4601 Skull. Adult male. Well preserved but some teeth seem anomalous.

MSNM Ma4602 Skull. Adult male. Near Ntem river, Nyabesam, Cameroun. Collected in 1938.

MSNM Ma7840 Skull. Adult male. Sexo, Mayombe, Kouilou, Republic of the Congo. Collected in 1976. Donated by Sig. Luigi Pezzoli on 15 December 2011.

MSNM Ma7841 Skull. Adult female. Sexo, Mayombe, Kouilou, Republic of the Congo. Collected in 1976. Donated by Sig. Luigi Pezzoli on 15 December 2011.

Naples

Università degli Studi, Istituto e Museo di Zoologia, Via Mezzocannone 8, 80134 Napoli*

G. Tomasetta
Cranium. Acquired in 1907.
Cranium. Adult male. Acquired in 1907.
Mounted skin. Female. Gabon. Acquired in 1914. It measures 85 cm from top of head to base of spine.

Padua

Università degli Studi di Padova, Museo di Antropologia, Palazzo Cavalli, Via Matteotti 30, 35121 Padova

Nicola Carrara
Inv. (not readable) Mounted skeleton. Fetus/infant. Received in 1874?
Inv.91 Skull. Adult male. Received in 1905.
Inv.9241 Mounted skeleton. Adult male. Received in 1913.

Università degli Studi di Padova, Museo di Zoologia, Via G. Jappelli 1/a, 35121 Padova

Nicola Carrara
M39 Skull. Adult female. Received in 1874.

Palermo

Università degli Studi, Dipartimento di Scienze e Tecnologie Biologiche Chimiche e Farmaceutiche, Museo di Zoologia Pietro Doderlein, Via Archirafi 16, 90123 Palermo

Luca Sineo
An 1247 Skull (plaster cast). Adult male. *Gorilla gorilla* (*graueri* ?). It is of poor quality.

Università degli Studi, Dipartimento di Scienze e Tecnologie Biologiche Chimiche e Farmaceutiche, Laboratorio di Antropologia, Via Archirafi 18, 90123 Palermo

Luca Sineo
Right humerus (plaster cast). Male. Labelled F. Kranz.
Right ulna (plaster cast). Female. Catalogue no. 199 (ex Istituto di Antropologia).
Right radius (plaster cast). Male.
Right tibia (plaster cast). Male.

Genomic DNA in solution. Male and female. Held in the Laboratory DNA bank.

Università degli Studi, Dipartimento di Scienze e Tecnologie Biologiche Chimiche e Farmaceutiche, Raccolta di Anatomia Comparata, Via Archirafi 18, 90123 Palermo

Skull (plaster cast). Adult male.
Skull (plaster cast). Adult female. Cameroun.
Skull (plaster cast). Adult female.
NB: All the plaster casts in the University collections are presumed to have derived from a larger collection assembled during the period 1930−70.

Pavia

Università di Pavia, Museo Zoologico 'Lazzaro Spallanzani', Piazza Botta 9, 27100 Pavia

Daniele Formenti; Paolo Galeotti; Stefano Maretti
Mamm. 3.4 Skull and mounted skin. Adult male. Congo. Purchased for 1800 lira by Prof. Pavesi from Dr Ludovico Eger in Vienna, Austria in June 1887. The current whereabouts of the skull is unknown.
2540 Mounted skeleton. *Troglodytes gorilla*. From the banks of the Gabon River. Local people had killed it and skinned it, and sold the bones in a small village. Purchased in 1879 from the preparator, Dr Leopold Eger, Vienna, with funds donated by the Board of Directors of the Collegio Ghislieri di Pavia.
2947 Skeleton. Juvenile. The hands are missing.
3082 Skull. Infant. Acquired in 1891.
3087 Mounted skeleton. *Troglodytes gorilla*. Acquired in 1892?
3371 Mounted skeleton. Male. *Troglodytes gorilla*. Acquired in 1897.
3692 Skull. Male. Acquired in 1907. Deformed (asymmetrical) jaw; microdontia; upper right canine in abnormal position.
4139 Skull. Adult male. Gabon. Collected by Prof. Edoardo Zavattari. Acquired in 1925.
4203 Skull. Adult male. Gabon. Acquired in 1926.
6262/1 Skull. *Troglodytes gorilla*. Received in 1970. Originally from the collection of the Museo Civico di Storia Naturale di Pavia, which was later absorbed by, and located at, the Istituto Tecnico 'A. Bordoni'.

Pisa

Università di Pisa, Museo di Storia Naturale e del Territorio, Via Roma 79, 56011 Calci (Pisa)

Carla Nocchi; Luca Ragaini; Marco Zuffi

132 Skeletal material. Identification uncertain; cannot be found.

349 Skeletal material. Male. Identification uncertain; cannot be found.

1546 Skull (cast). Adult male. It is on display. Old number 3157.

1565 Skull (plaster cast). Adult male. *Gorilla savagei*. Old number 3155.

1569 Skull (cast). Adult female. It is on display.

1732 Mounted skeleton. Juvenile female. *Gorilla savagei*. Gabon. Old number 2699. The skull is missing.

1777 Mounted skeleton (papier mâché). Adult male. *Gorilla savagei*. Old number 3387.

2251 Brain (wet preserved). Juvenile. *Gorilla savagei*. Old number 3372. It belongs to skull no. 3388 (not found in the collection). It is in poor condition.

2256 Brain (wet preserved). Juvenile male. *Gorilla savagei*. Old number 3166. It is in poor condition.

3457 Skull (papier mâché). Adult male.

Rome

Museo Civico di Zoologia, Via Ulisse Aldrovandi 18, 00197 Roma

Francesco Baschieri-Salvadori

00005 Mounted skin. Adult male. Donated by Dr Aurelio Rossi in 1929.

00006 Mounted skin. Female. Donated as above.

00007 Mounted skin. Male. Arrived at the Giardino Zoologico, Viale del Giardino Zoologico 20, Roma on 15 May 1935, as a gift from the Menagerie du Jardin des Plantes, 57 rue Cuvier, Paris. He had arrived in Paris on 12 June 1934, aged about 3 years and was known as 'Jacques'. In Rome he was known as 'Jacco'. He died of hepatitis on 26 October 1936.

643 Mounted skin. Juvenile. Donated by Dr Aurelio Rossi in 1929.

6352 Brain. From the captive male 'Jacco'.

6353 Tongue, glottis, etc. From a captive specimen in the Giardino Zoologico, Roma.

6354 Skull. Adult male.

6358 Mounted skeleton. Male. Received from the Università di Roma. From the same specimen as 00007 ('Jacco'). Height 120 cm.

7572 Skin. Male. *Gorilla beringei graueri*. Captive specimen, arrived at the Giardino Zoologico, Roma, on 11 August 1952, aged about 1 year and weighing 25 lb. He had been purchased from the Department of the French Congo [*sic*] and was named 'Bongo'. He died of enteritis on 27 February 1953. [Some additional information added by GH from correspondence with Rome Zoo personnel.]

9470 Encephalon. Female. Captive specimen, arrived at the Giardino Zoologico, Roma, on 11 August 1952, having been purchased with the male 'Bongo', and given the name 'Zuma'. She was about 14 months old at the time and weighed 32 lb. She died of enteritis on 11 May 1955. [Some additional information added by GH from correspondence with Rome Zoo personnel.]

9483 Skeleton. Juvenile male. *Gorilla beringei graueri*. From the captive specimen 'Bongo' (no. 7572).

20920 Skull. Adult male. Donated by Dr Aurelio Rossi on 18 October 1960.

Skin. Female. From the captive specimen 'Zuma' (no. 9470).

Mounted skin. Adult male. This is 'Bongo II', who arrived at the Giardino Zoologico, Roma, on 3 February 1966 (weight 13 kg). He came originally from Spanish Guinea, and was acquired from Turin Zoo. He died on 30 August 1995. [Some additional information added by GH from correspondence with Rome Zoo personnel.]

Sapienza Università di Roma, Museo di Anatomia Comparata 'Battista Grassi', Via Borelli 50, 00161 Roma

Ernesto Capanna

AC.P.18 Skull. Adult male. Purchased for 250 lire in 1884. Condylobasal length (gnathion−condylion) 194 mm; maximum braincase width, between euryons 98 mm; biauricular width 144 mm; bimastoid width 150 mm; skull height, basion−vertex (top of sagittal crest) 115 mm. It is missing the upper right central incisor. Department inventory no. 111-341.

Sapienza Università di Roma, Museo di Antropologia 'Giuseppe Sergi', Piazzale Aldo Moro 5, 00185 Roma

Fabio Di Vincenzo

Skull. Adult male. Central African Republic?

673 Cranium. Subadult female. Democratic Republic of Congo?

2893 Skull. Adult male.

Siena

Museo Zoologico dell'Accademia della Scienze di Siena detta de 'Fisiocritici, Piazza Sant 'Agostino 4, 53100 Siena

Baccio Baccetti; Giuseppe Manganelli

MUSNAF/ZOO-MAM 125 Cranium. Juvenile. Collected after 1880. It was formerly in the collection of the Istituto di Anatomia Comparata e Zoologia, Università di Siena (their catalogue no. 199).

MUSNAF/ZOO-MAM 126 Skull. Adult male. Gabon. Collected in April 1868. Donated by Marchesa Strozzi to the Museo Zoologico 'La Specola', Università di Firenze, in 1875 (their catalogue no. 26). It was probably sent in exchange in 1879 to the zoological collection formerly held by Siena University, which has since been transferred in its entirety to the Siena Academy of Science.

Treviso

Museo Zoologico 'G. Scarpa', Seminario Vescovile, Piazzetta Benedetto XI, 2, 31100 Treviso.*

Giordano Gabris
Skull. Adult male. Probably acquired before 1914.

Trieste

Museo Civico di Storia Naturale, Via Dei Tominz 4, 34139 Trieste

Deborah Arbulla; Renato Mezzena
M973 Mounted skin. Adult female. *Gorilla gina.* South Cameroun.
M1916 Mounted skeleton. Adult female.

Turin

Museo Regionale di Scienze Naturali, Via G. Giolitti 36, 10123 Torino

Franco Andreone; Maria Stella Siori, Dipartimento di Scienze della Vitae Biologia dei Sistemi, Via Accademia Albertina 13, 10123 Torino
NB: The collection is the Museo di Zoologia dell'Università di Torino (MZUT), which also incorporates specimens from the Museo di Anatomia Comparata dell'Università di Torino (MACUT).
MZUT T 323 (MACUT 6) Mounted skeleton. Adult male.
MZUT T 347 (MACUT 3895) Cranium. Adult female. Sanga River, Congo Free State. Donated in 1897 by Ing. Gariazzo.

Università degli Studi di Torino, Museo di Anatomia Umana 'Luigi Rolando', Corso Massimo D'Azeglio 52, 10126 Torino

Giacomo Giacobini
Skull. Juvenile.

Venice

Museo di Storia Naturale di Venezia, Santa Croce 1730, 30135 Venezia

Mauro Bon; Giampaolo Rallo
MSNVe 4706 Skin. Adult male. Congo. Received from the de Reali collection in 1925.
MSNVe 4908 Skeleton. Adult male. From the same specimen as the above skin, and received with it.

Verona

Museo Civico di Storia Naturale, Lungadige Porta Vittoria 9, 37129 Verona

G.B. Osella; Roberta Salmaso; Lorenzo Sorbini
CO462 Skull. Adult male. Purchased in Paris by Mr C. Zanella on 3 November 1967, and donated by him. All teeth present. Gnathion−inion 220 mm; glabella−inion 150 mm; glabella−lambda 162 mm; height of sagittal crest 36 mm; bicondylar width 132 mm. A cast is on display in the museum.

JAPAN

Inuyama

Japan Monkey Centre, Kanrin 26, Inuyama, Aichi 484-0081

Yuta Shintaku
JMC-Pr0789 Mounted skin; mounted skeleton; fluid-preserved organs and brain. Adult male. *Gorilla beringei graueri.* This is the captive specimen 'Munidi'.
JMC-Pr0790 Mounted skin, mounted skeleton; fluid-preserved organs and brain. Adult female. *Gorilla beringei graueri.* This is the captive specimen 'Emmy'.
NB: For further information about the above specimens, see the entry for the University of Tokyo.
JMC-Pr3554 Mounted skin; parts of skeleton; fluid-preserved organs and brain. Infant, aged 4−5 months. Arrived at the Japan Monkey Centre on 7 June 1980. Died on 2 July 1980 (weight 2.95 kg).
JMC-Pr3711 Mounted skin and postcranial skeleton. Juvenile female, aged c. 3 years. This is 'Olive', who was wild-born, and arrived at the Miyazaki City Phoenix Zoo, 3083-42 Shioji, Miyazaki, Miyazaki Prefecture 880-0122, on 13 March 1980. She died of acute pneumonia on 9 September 1981 (weight 14.2 kg). Donated (having been necropsied at the JMC). [Some additional information added by GH from International Studbook for the Western Lowland Gorilla, 2010.]
JMC-Pr3958 Mounted skin; skeleton; fluid-preserved organs and brain. Adult female, aged c. 12−13 years. Kept privately since 1972, died on 5 December 1983. Donated.

JMC-Pr4297 Mounted skin; skeleton; fluid-preserved organs and brain. Adult male. This is 'Fujio', who was born in the Ritsurin Park Zoo, 1-20-2 Ritsurin-cho, Takamatsu, Kagawa 760, Japan, on 2 February 1971, to sire 'Rikky' and dam 'Rinko'; mother-reared. He arrived at the Japan Monkey Centre on 20 March 1973, and died of pneumonia on 27 May 1987 (weight 186.5 kg). [Some additional information added by GH from International Studbook for the Western Lowland Gorilla, 2010.]

JMC-Pr6025 Fluid-preserved organs and brain. Adult female, aged c. 42 years. Born in the wild (estimated 1967), she arrived at the Kamine Zoological Park, 5-2-22 Miyata-cho, Hitachi, Japan, on 20 December 1969, where she was named 'Yuki'. She was moved to the Japan Monkey Centre on 6 November 1980, where she was renamed 'Hanako'. She died on 25 December 2009 (estimated weight 140 kg). She had megalocardia, but the exact cause of death is unclear. [Some additional information added by GH from International Studbook for the Western Lowland Gorilla, 2010.]

JMC-NP0396 Articulated postcranial skeleton.

JMC-P-S-0168 Mounted skin. Donated by the Hirakata Park Monkey Land, Hirakata, Japan.

JMC-P-S-0171 Mounted skin of hand. Adult male, 10 years old. Kept at the Kyoto City Zoo. Donated.

Kyoto University, Primate Research Institute, Inuyama, Aichi 484-8506

Eishi Hirasaki

PRI 5375 Skull. Some teeth are missing.

PRI 7219 Skull. Adult male. Gabon. Acquired on 12 February 2002. Some teeth are missing.

PRI 7220 Cranium. Adult male. Gabon. Acquired on 6 January 2001. Some incisors are missing premortem.

PRI 7221 Mandible. Female. Gabon. Acquired on 5 December 2001. The canines and incisors are missing. The 3rd molars are in eruption (visible in their sockets). The ascending rami are broken and missing.

PRI 7902 Carcase (in formalin solution). Juvenile.

PRI 8155 Skeleton. Juvenile. Acquired (in 1975?) from Kamine Zoo, Hitachi, Japan.

PRI 8532 Brain (in formalin solution).

PRI 9253 Oesophagus, ears and eyes (in formalin solution). Juvenile? female.

PRI 9573 Postcranial skeleton. Juvenile female.

Niigata

Nippon Dental University, Department of Oral Anatomy, 1—8 Hamaura-cho, Niigata 951*

Keiichi Takahashi

NDN 06097 Skull. Adult female. Maximum length 192 mm; bizygomatic width 175 mm; greatest width of mandible 137 mm; height of mandible 97 mm.

Sapporo

Hokkaido University, Graduate School of Veterinary Medicine, N18 W9, Sapporo 060-0818

Mutsumi Inaba; Yasuhiro Kon; Makoto Sugimura

Brain (siliconised). Juvenile male. Originally it weighed 400 g and was preserved whole in formalin. It is cut off between the brain stem and the cerebrum. The gorilla had been kept in the Japan Animal Zoo, a privately owned travelling exhibition, and died on 17 June 1957, aged 4 years, whilst the zoo was staying in Sapporo. It was examined in the Dept of Veterinary Anatomy, Hokkaido University, where the causes of death were found to be chronic enterocolitis, acute hepatitis and pneumonia. The brain is exhibited in the Faculty museum. Parts of some other organs, such as the liver, spleen, kidney, etc., were preserved in formalin as histological samples, but these are now thought to have been lost when the Veterinary School and its museum were reformed and reconstructed.

Tokyo

National Museum of Nature and Science, Department of Zoology, 7-20 Ueno Park, Taito-ku, Tokyo 110-8718

Shin-ichiro Kawada

NSMT-M 28246 Mounted skin. Male. Cameroon. Received from Tobu Zoo, Saitama. Date of death 23 January 1988 (weight 43.3 kg).

NSMT-M 29618 Skeleton. Adult female. Received from Tama Zoological Park, 7-1-1 Hodokubo, Hino-shi, Tokyo 191-0042 on 16 July 1990. This is 'Kiki', who arrived at theTama Zoo on 13 September 1971. She died of pancreatitis on 1 June 1990.

NSMT-M 31594 Skeleton. Adult male. Received from Ueno Zoological Gardens, 9-83 Ueno Park, Taito-ku, Tokyo 110-8711. This is 'Biju', who was born in Howletts Wild Animal Park, UK, on 28 July 1987 to dam 'Ju-Ju' and sire 'Bitam' (mother-reared). He was transferred to Ueno Zoo on 3 December 1997 and died there on 28 October 1999. [Some additional information added by GH from International Studbook for the Western Lowland Gorilla, 2010.]

NSMT-M 31595 Postcranial skeleton. Adult male. Received from Ueno Zoo. This is 'Drum', who arrived at the Phoenix Zoo, 3083-42 Hamayama Shioji, Miyazaki 880-0122 on 13 March 1980, aged about 3 years. He was purchased from Aritake Animals and Birds Trading Co. Ltd. He was sent on loan to Ueno Zoo on 21 November 1993, and died there on 20 July 1998. [Some additional information added by GH from International Studbook for the Western Lowland Gorilla, 2010.]

NSMT-M 31823 Skeleton. Adult male. Received from Ueno Zoo. This is 'Bul-Bul', who arrived at the Ueno Zoo on 17 November 1957 (weight 15 kg), having been purchased from Phillip J. Carroll. He died on 1 November 1997. [Some additional information added by GH from correspondence with Ueno Zoo personnel.]

NSMT-M 31824 Skeleton. Adult male. Received from Kyoto Municipal Zoo, Okazaki, Hoshoji-cho, Sakyo-ku, Kyoto 606-8333. This is 'Sultan', who arrived at the Tama Zoo on 13 September 1971, aged about 2 years. He was sent on loan to Ueno Zoo on 5 October 1993, and transferred to Kyoto Zoo on 15 December 1997, where he died on 23 March 1998. [Some additional information added by GH from International Studbook for the Western Lowland Gorilla, 2010.]

NSMT-M 35965 Skull, parts of postcranial skeleton, DNA sample and most of the internal organs in fluid. Female. Received from Ueno Zoo.

NSMT-M 42664 Mounted skin. Juvenile. It was donated by the local museum in Yubari City, Hokkaido.

University of Tokyo, Faculty of Medicine, University Museum, Hongo 7-3-1, Bunkyo-ku, Tokyo 113-0033*

Toshiro Kamiya

107 Brain stem serial sections. Adult male. *Gorilla beringei beringei*. Received from the Japan Monkey Centre, Inuyama, Aichi 484-0081, in 1961. The animal was captured by Charles Cordier at Utu, Belgian Congo, in 1956, and kept in a stockade at Tshibati until purchased by the Japan Monkey Centre. Named 'Munidi', he arrived at the JMC on 30 May 1961, together with a female named 'Emmy', who had the same origin. 'Munidi' died of caseous [sic] pneumonia to a severe degree, suggestive of tuberculosis, and septicaemia, on 2 June 1961 (weight 130 kg). 'Emmy' died of catarrhal pneumonia, cardiac insufficiency and gastrointestinal catarrh [sic] on 6 June 1961 (weight 68 kg). Both animals arrived at the JMC in poor condition after a long journey by air from Nairobi to Tokyo Haneda airport. For further information, see Inoue and Hayama (1961). Brain weight 510 g.

KENYA

Nairobi

National Museums of Kenya, P.O. Box 40658, Nairobi 00100

R.H. Carcasson; Nina Mudida; Ogeto Mwebi

OM 3264 Skeleton and mounted skin. Adult male. *Gorilla beringei beringei*. Mount Muhavura, Virunga volcanoes, Kigezi District, British East Africa. Collected by Constantine John Philip Ionides (1901–68) on 17 September 1946, using a double-barrelled Evans .470 rifle and solid bullets. The skin was mounted by Messrs Zimmermann of Nairobi. [Additional information by GH via correspondence with C.J.P. Ionides.]

Department of Osteology:

OM 3259 Skull and long bones; foot bones incomplete. Neonate. Collected on 6 June 1967. The skeleton was previously fixed in formalin, and the preservation is poor.

OM 3260 Skull, long bones, pelvic sacrum and some foot bones. Adult female. Rwanda. Collected by Dr Dian Fossey.

OM 3261 Scapula, pelvic bones and a few long bones. Adult male. Rwanda. Collected by Dr Dian Fossey. The lumbar vertebrae are fused. The specimen has been missing since at least 2004.

OM 3262 Skull, pelvic bones and long bones. Adult male. Mount Muhavura, Virunga volcanoes, Kigezi district, Uganda. Collected by Johides Macjunes in 1970.

OM 3263 Postcranial skeleton. Adult female. Collected in 1970.

OM 3542 Skull. Adult male. Collected in 1970.

OM 3543 Skull. Adult male. Collected in 1970.

OM 3544 Skull. Adult female. Collected in 1970.

OM 3545 Skull. Juvenile.

OM 7174 Cranium. Adult male. Democratic Republic of Congo. Collected in 1965. NB: This is a *Gorilla gorilla gorilla*.

KUWAIT

Safat

Ministry of Education, Science and Natural History Museum, Abdallah Mubarak Street, Safat*

Hamad M. Al-Ateeqi

1/6/64 Mounted skin. Adult female. Received from F. G. Bode Co., West Germany, on 2 November 1983. The animal was 17 years old, and came from the Berlin Zoo in 1982. [GH — this is probably 'Cocotte', who arrived at the Berlin Zoo on 13 March 1968, aged about 3 years and died of pneumonia on 5 September 1982.]

Luxembourg

Musée National d'Histoire Naturelle, 25 rue Münster, 2160 Luxembourg

Edmée Engel; J.-M. Guinet; Marc Meyer

MNHNL18824 Skull. Adult male. Donated on 5 August 1998 by Mrs Roeder, Luxembourg, whose husband shot the specimen in Africa.

MNHNL18827 Skull. Adult female.

MNHNL18828 Skull. Adult male.

Monaco

Musée d'Anthropologie Préhistorique de Monaco, 56 bis, Boulevard du Jardin Exotique, 98000 Principauté de Monaco

Elena Rossoni-Notter; Mme S. Simone

61 Cranium. Adult male. Both canines and both central incisors are missing. The cranium is slightly asymmetrical.

cc-f-0001 Skull. Adult male. A small piece of bone is missing from the right zygomatic arch. Both upper central incisors are missing.

MOROCCO

Rabat

Université Mohammed V, Institut Scientifique, Département de Zoologie et Écologie Animale, Ibn Batota, B.P. 703, Rabat*

Michel Thevenot

179001 Skull. Adult male. Nola, Haute Sangha, Oubangui-Chari, Central African Republic. Collected in 1944. Donated by Dr Roland Choumara of the World Health Organisation.

NETHERLANDS

Amsterdam

Universiteit van Amsterdam, Academisch Medisch Centrum, Anatomisch-Embryologisch Laboratorium en Museum Vrolik, Meibergdreef 15, 1105 AZ Amsterdam*

J.A. Los

AEL 1907-91 Skull. Male.

AEL 1908-54 Skull. Male.

AEL 1911-1 Skull. Male. Cameroun.

AEL 1911-12 Skull. Male. Cameroun.

AEL 1912-53 Skull. Female.

AEL 1912-111 Skull. Female. French Congo. Acquired from Maison Tramond—N. Rouppert Suc., Paris.

AEL 1912-115 Skull. Male. French Congo.

AEL 1912-116 Skull. Male. French Congo. Acquired from Maison Tramond—N. Rouppert Suc., Paris.

AEL 1912-117 Skull. Male. Acquired from Maison Tramond—N. Rouppert Suc., Paris.

AEL 1912-118 Skull (on stand). Male. French Congo.

AEL 1912-139 Skull. Cameroun.

AEL 1913-38 Skull. Male. French Congo.

AEL 1913-40 Skull. Male. French Congo.

AEL 1913-263 Skull. Female.

AEL 1913-269 Skull. Female. French Congo. Acquired from Maison Tramond—N. Rouppert Suc., Paris.

AEL 1913-303 Skull. Male.

AEL 1914-1 Skull. Female. Cameroun.

AEL 1917-2 Skull. Male. Cameroun.

AEL 1919-39 Skull. Male. French Congo. Acquired from Maison Tramond—N. Rouppert Suc., Paris.

AEL 1919-40 Skull. Male.

AEL 1919-42 Skull. Male. Acquired from Maison Tramond—N. Rouppert Suc., Paris.

AEL 1919-43 Skull. Female.

AEL 1919-44 Skull (on stand). Female. Acquired from Maison Tramond—N. Rouppert Suc., Paris.

AEL 1919-46 Skull (on stand). Female. Acquired from Maison Tramond—N. Rouppert Suc., Paris.

AEL 1919-48 Skull (on stand). Male. Acquired from Maison Tramond—N. Rouppert Suc., Paris.

AEL 1919-49 Skull. Male. Acquired from Maison Tramond—N. Rouppert Suc., Paris.

AEL 1919-50 Skull. Male. Acquired from Maison Tramond—N. Rouppert Suc., Paris.

AEL 1919-51 Skull. Male.

AEL 1919-52 Skull. Male.

The above 11 skulls, AEL 1919-39 to AEL 1919-52 inclusive, were presented by Prof. J.A. Barge of Leyden University to Prof. L. Bolk of the Anatomical and Embryological Laboratory, University of Amsterdam.

AEL 1927-13 Skull. Female. 'No. 16'.

AEL 1927-14 Skull. Female.

AEL 1927-15 Skull. Female. 'No. 36'.

AEL 1927-17 Skull. Male.

AEL 1927-19 Skull. Female.

AEL 1928-4 Skull (on stand). Male. The roots of the upper and lower teeth are laid bare.

AEL 1928-9 Skull. Female.

AEL no number-a Skull. Female.

AEL no number-b Skull. Female. Acquired from Maison Tramond—N. Rouppert Suc., Paris.

AEL no number-c Skull. Female. 'No. 35'.

Groningen

Rijksuniversiteit Groningen, Universiteitsmuseum, Oude Kijk in 't Jatstraat 7a, 9712 EA Groningen

John Le Grand; P.F.A. Martinez Martinez

NB: The specimens below were originally kept in the Rijksuniversiteit Groningen, Laboratorium voor Anatomie en Embryologie, Anatomisch Museum, Oostersingel 69, 9713 EZ Groningen. In 2003 the building was demolished, and the Anatomisch Museum merged with the Universiteitsmuseum.

A-1393 Mounted skeleton. Adult male. Dume River, South Kamerun. It was purchased for 803 marks from Johannes Umlauff, Hamburg, in 1915. The animal had been felled by a shot to the left pelvis.

A-1180 Femur (plaster cast). Adult male. Dume, Kamerun.

A-1182 Femur (plaster cast). Adult female. Baraka, Tanganyika.

Leyden

Naturalis Biodiversity Center, Darwinweg 2, 2300 RA Leiden

Pepijn Kamminga; C. Smeenk; P.J.H. van Bree; Steven van der Mije; J. de Vos

NB: The Naturalis Biodiversity Center originated from the merger of the Rijksmuseum van Natuurlijke Historie (RMNH) and the Rijksmuseum van Geologie en Mineralogie (RGM) in 1984.

Cat.Ost.a Mounted skeleton. Adult male. Congo. Donated by the West African Commercial Society, Rotterdam, in 1871.

Cat.Ost.b Mounted skeleton. Juvenile female. *Simia gorilla*. Mayombo, Loanga loop, Chilango, Congo. Received from H.P. Braam, Rotterdam, in 1878.

Cat.Ost.c Skull (bisected longitudinally). Adult male. Congo. Donated by the West African Commercial Society, Rotterdam, in 1871.

Cat.Ost.d Cranium. Subadult. Gabon. Acquired in 1866. It is now missing.

Cat.Ost.e Cranium. Juvenile. Gabon. Acquired in 1866. It is now missing.

The four specimens Cat.Ost.a, c, d and e above were described by H. Schlegel (1876) under the scientific name *Simia gorilla*. He described the skeleton as a very old female.

Cat.Ost.f Mounted skeleton. Adult male. Purchased from G.A. Frank, London, in 1891. It is from the same animal as the mounted skin, Cat.Syst.a.

Cat.Syst.a Mounted skin. Adult male. Purchased from G.A. Frank, London, in 1891 (see Cat.Ost.f above).

Cat.Syst.b Mounted skin. Adult female. Purchased from G.A. Frank, London, in 1891.

Cat.Syst.c Mounted skin. Juvenile male. Purchased from G.A. Frank, London, in 1891.

RMNH.MAM.1802 Mounted skin. Adult male. Udongo, Ugoka-Sangha, French Congo. Collected on 12 June 1929.

RMNH.MAM.1802 Mounted skin. Adult female. Ouesso, French Congo.

RMNH.MAM.1802 Mounted skin. Adult female. Udongo, Ugoka-Sangha, French Congo. Collected on 10 June 1929.

RMNH.MAM.1802 Mounted skeleton. Adult male. Ouesso, French Congo. The skull is missing.

RMNH.MAM.1802 Mounted skeleton. Adult female. Ouesso, French Congo.

RMNH.MAM.1802 Skull. Male. Ouesso, French Congo.

RMNH.MAM.1802 Skull. Male. Ouesso, French Congo.

RMNH.MAM.1802 Skull. Male. Ouesso, French Congo.

RMNH.MAM.1802 Skull. Male. Ouesso, French Congo.

RMNH.MAM.1802 Skull. Male. Ouesso, French Congo.

RMNH.MAM.1802 Skull. Male. Ouesso, French Congo.

RMNH.MAM.1802 Skull. Male. Ouesso, French Congo.

RMNH.MAM.1802 Skull. Female. Ouesso, French Congo.

RMNH.MAM.1802 Skull. Female. Ouesso, French Congo.

The above 15 specimens were purchased from Dr A. Rossi, Rome, on 28 December 1929.

RMNH.MAM.11948 Skin. Juvenile male. Brazzaville, Moyen-Congo. It arrived at the Rotterdam Zoo (Stichting Koninklijke Rotterdamse Diergaarde, Van Aerssenlaan 49, 3000 AM Rotterdam) in 1953, when aged about 2 years, and died there on 20 August 1953.

RMNH.MAM.11948 Skull and limb bones of the above individual.

RMNH.MAM.18001 Mounted skin. Female. Kassai, Congo. Donated by Groenhuizen v. Oldenbarnevelt, Utrecht.

Mandible. *Simia gorilla*. Possible origin Amsterdam Anatomical Laboratory (from Dubois collection).

NB: The following specimens were previously kept in the Universiteit van Amsterdam, Instituut voor Taxonomische Zoölogie, Zoölogisch Museum, 1090 GT Amsterdam. In 2011 the collection was merged with that of the Naturalis Biodiversity Center.

ZMA.MAM.21 Skull. Male. The mandible is attached to the cranium with metal springs.

ZMA.MAM.22 Skull. Male.

ZMA.MAM.23 Skull (plaster cast). Female. Received from the Museum Vrolik (original no. 272).

ZMA.MAM.24 Mounted skeleton. Male.

ZMA.MAM.55 Skull. *Gorilla castaneiceps*. French Congo/Gabon. Donated by Bernhard Freiherr von Friesen.

ZMA.MAM.868 Skull (plaster cast).

ZMA.MAM.1024 Skull. Mayomba, 100 km inland from Pointe-Noire, French Equatorial Africa. It was collected and donated by J.F.V. Huese between 1945 and 1955, via J.F.O. Huese.

ZMA.MAM.1656 Skull. *Gorilla castaneiceps*. Gabon. Collected on 25 September 1918. Donated by Bernhard Freiherr von Friesen.

ZMA.MAM.1657 Cranium. Banana, Congo. Collected in 1890. The I1 is missing on both sides.

ZMA.MAM.1658 Skull. Juvenile. Loanda, Angola. Collected on 30 April 1910. Donated by Jacob de Groot. Several teeth are missing.

ZMA.MAM.2539 Skull; and mounted skin on exhibition. Female. Mékambo, NE Gabon. This is 'Moussa', who was captured on 15 March 1955, at the age of about 9½ months. She arrived at the Amsterdam Zoo on 26 April 1955, having been purchased from the Brazzaville Zoo. She died of heart dilatation and *Klebsiella* pneumonia on 7 September 1957.

ZMA.MAM.2720 Skeleton and flat skin. Juvenile male. Lokiamba district, 150 km inland, next to the railway at Douala, Cameroun. This is 'Jaap', who arrived at the Amsterdam Zoo on 6 March 1936, when aged between 1 and 1½ years, having been purchased from Mr F.J. de Vries, a Dutch steward aboard the SS *Reggestroom*. 'Jaap' died of an intestinal circulation problem owing to acute torsio mesenterialis on 28 October 1943. [Some additional information added by GH from correspondence with Amsterdam Zoo personnel.]

ZMA.MAM.6546 no. 285 (M.B.G.) Skull. Male? Purchased at Borassier, near Bélinga, NE Gabon, on 17 February 1964.

ZMA.MAM.6546a no. 339 (M.B.G.) Cranium. Male? Purchased between Makokou and Mékambo, NE Gabon, on 7 March 1964.

ZMA.MAM.6547 no. 340 (M.B.G.) Cranium. Male? Purchased between Makokou and Mékambo, NE Gabon, on 7 March 1964.

ZMA.MAM.6548 no. 286 (M.B.G.) Cranium. Male? Purchased at Mvadi, near Bélinga, NE Gabon, on 17 February 1964.

ZMA.MAM.6549 no. 288 (M.B.G.) Cranium. Male? Purchased at Mvadi, near Bélinga, NE Gabon, on 17 February 1964.

ZMA.MAM.6550 no. 287 (M.B.G.) Cranium. Male? Purchased at Ebà, near Bélinga, NE Gabon, on 17 February 1964.

ZMA.MAM.6551 no. 335 (M.B.G.) Skull. Purchased at Ebanda, between Makokou and Bélinga, NE Gabon, on 3 March 1964.

The above 7 specimens, ZMA.MAM.6546 no. 285 to ZMA.MAM.6551 no. 335 inclusive, were collected and donated by Dr P.J.H. van Bree, of the Mission Biologique au Gabon (M.B.G.).

ZMA.MAM.6552 Mounted hand and foot. Adult. Vicinity of Makokou, NE Gabon. Collected in 1963. Donated by Prof. P.P. Grassé (M.B.G.).

ZMA.MAM.6553 no. 334 (M.B.G.) Cranium. Juvenile male. Purchased at Ebankak, between Makokou and Bélinga, NE Gabon, on 3 March 1964.

ZMA.MAM.6597 no. 336 (M.B.G.) Skull. Female? Purchased at Ebankak, between Makokou and Bélinga, NE Gabon, on 3 March 1964.

ZMA.MAM.6598 no. 290 (M.B.G.) Cranium. Purchased at Ebà, near Bélinga, NE Gabon, on 17 February 1964. It is in bad condition.

ZMA.MAM.6599 no. 274 (M.B.G.) Skull. Female. Purchased at Bélinga, Route de Massa, NE Gabon, on 1 February 1964.

ZMA.MAM.6995 Skull. Male. Purchased on the road between Makokou and Mékambo, NE Gabon, on 7 March 1964. No M.B.G. number.

The above 5 specimens, ZMA.MAM.6553 no. 334 to ZMA.MAM.6995 inclusive, were collected and donated by Dr P.J.H. van Bree (M.B.G.).

ZMA.MAM.7275 no. 341 (M.B.G.) Cranium. Male. Purchased on the road between Makokou and Mékambo, NE Gabon, on 7 March 1964. Collected and donated by R.P. Peters.

ZMA.MAM.7276 no. 342 (M.B.G.) Cranium. Male. Purchased in the area between Makokou and Mékambo, NE Gabon, on 7 March 1964. Collected and donated by R.P. Peters.

ZMA.MAM.7277 no. 343 (M.B.G.) Cranium. Male. Shot in the area between Makokou and Mékambo, NE Gabon, on 9 March 1964. Donated by Dr Raynal, Makokou.

ZMA.MAM.14831 Skull. Male. Probably from the hinterland of Pointe-Noire, Congo. Collected in 1971. Donated by Pierre L.H. Opic.

ZMA.MAM.14832 Skull. Female? Probably from the hinterland of Pointe-Noire, Congo. Collected in 1971. Donated by Pierre L.H. Opic.

ZMA.MAM.14833 Skull. Female? Probably from the hinterland of Pointe-Noire, Congo. Collected in 1971. Donated by Pierre L.H. Opic.

ZMA.MAM.14834 Skull. Female? Probably from the hinterland of Pointe-Noire, Congo. Collected in 1971. Donated by Pierre L.H. Opic.

ZMA.MAM.15674 Skull. Male? Vicinity of Moukitsa, Cotovindo (c. 11°55′E, 3°45′S), People's Republic of the Congo. Purchased from hunters at Moukitsa on 5 December 1972. Collected and donated by W. Bergmans.

ZMA.MAM.15675 Skull. Female? Vicinity of Moukitsa, Cotovindo (c. 11°55′E, 3°45′S), People's Republic of the Congo. Purchased from hunters at

Moukitsa on 5 December 1972. Collected and donated by W. Bergmans.

ZMA.MAM.16865 Skeleton. Male. This is 'Babar', who was captured at Likoma, Haute Likouala, French Congo, in August 1952, when aged about 6 months. He arrived at the Amsterdam Zoo (Stichting Koninklijk Zoologisch Genootschap Natura Artis Magistra, Plantage Kerklaan 38–40, 1018 CZ Amsterdam C) on 30 May 1953 (weight 8.86 kg), having been purchased from the Brazzaville Zoo. He died in Arnhem Zoo (Burgers' Zoo and Safari, Schelmseweg 85, Arnhem) on 30 June 1974. The pelvis is broken. [Some additional information added by GH from correspondence with Amsterdam Zoo personnel.]

ZMA.MAM.16968 Skull and skin. Infant male. It came from the animal trade in August 1974. Donated by J. Docters van Leeuwen. Total length 355 mm; foot length 106 mm; ear 39 mm; weight 2941 g.

ZMA.MAM.22923 Skull. Female. This is 'Poupee', who arrived at the Wrocław Zoo (Miejski Ogrod Zoologiczny, ul. Wróblewskiego 1–5, 51-618 Wrocław, Poland) on 17 November 1970, aged about 1½ years and weighing 8 kg. She had been purchased (with a 3-year-old male named 'Willy') from Gabria van den Brink, Soest, Netherlands. She died in the Amsterdam Zoo on 22 September 1985. [Some additional information added by GH from correspondence with Wrocław Zoo personnel.]

ZMA.MAM.22924 Skull. Female. This is 'Tonky', who arrived at the Wrocław Zoo on 18 January 1974, when aged about 3 years and weighing 20 kg. She had been purchased from G. van den Brink, Soest, along with 3 other females ('Sarah', 'Messy' and 'Minouche'). She died in Amsterdam Zoo on 19 September 1985. [Some additional information added by GH from correspondence with Wrocław Zoo personnel.]

ZMA.MAM.23047 Skull. Male. Purchased from indigenous hunters between Abong Mbang and Lomié, East Cameroon, in 1984. Donated by H.P.L. Beijersbergen van Henegouwen in September 1986.

ZMA.MAM.23048 Skull. Male. Purchased from indigenous hunters between Abong Mbang and Lomié, East Cameroon, between 1980 and 1983. Donated by H.P.L. Beijersbergen van Henegouwen in September 1986.

ZMA.MAM.23880 Skeleton (almost complete) and mounted skin (on exhibition). Male. This is 'François', who arrived at Arnhem Zoo with 6 other gorillas ('Kéké', male; 'Ndiki', female; 'Laika', female; 'Bébé', female; 'Sophie', female and 'Iris', female) on 6 June 1984, aged about 6½ years and weighing 60 kg. These 7 gorillas came from M and Mme Robert Roy, Sangmélima, Republic of Cameroon. The International Primate Protection League thwarted efforts by 3 US zoos to import the gorillas. Although housed at the Burgers' Zoo, ownership and responsibility for exchanges with other zoos and participation in breeding programmes and 'survival plans' was bound to the Foundation for the Protection of Gorillas and other Threatened Species of Animals, realised under IUCN Netherlands. 'François' died of intestinal strangulation in Burgers' Zoo on 23 December 1989. A plaster death mask is in Burgers' Zoo. Total length 136 cm; foot length 28 cm; ear 58 mm. [Some additional information added by GH from correspondence with Arnhem Zoo personnel.]

ZMA.MAM.24211 Skull. Male. Djaposten (3°25′N, 13°32′E), Cameroon. It was shot in the autumn of 1990. Donated in 1990 by A.P.M. van der Zon.

ZMA.MAM.24698 Skull. Neonate male. Received from Natura Artis Magistra (Amsterdam Zoo) on 18 June 1992. It was stillborn on 17 June 1992 to dam 'Shinda' and sire 'Yaounde'.

ZMA.MAM.24974 Skull. Seized under CITES regulations at Schiphol Airport. Received in 1995.

ZMA.MAM.25608 Skull. Male. Seized under CITES regulations at Schiphol Airport. Received on 25 August 1970.

Leids Universitair Medisch Centrum, Anatomisch Museum, Hippocratespad 21, 2333 ZD Leiden

Pancras C.W. Hogendoorn; Dries van Dam; Mrs Reina J. van Velzen; Bas Wielaard

Ba 2931 Skull. Adult male. Congo. Presented by F. de La Fontaine Verweij in 1890. There is a bullet hole above the nasal aperture.

Ba 2933 Pelvis and 2 femora. Subadult female. Acquired from the Department of Zoology, University of Leiden.

Cf.Prim.Eus.Ant.11 Skull. Adult female. Acquired in 1925. Mandible articulated; nasal conchae partly missing.

Cf.Prim.Eus.Ant.12 Skull. Adult male. Acquired in 1925. Mandible articulated; nasal conchae partly missing.

Two pairs of arms and legs, one with a shoulder still intact. Acquired from the Department of Zoology, University of Leiden. These preparations were kept in formalin and were partly dissected. They were probably used for research and later cremated, since they are no longer present in the collection.

Leids Universitair Medisch Centrum, Fysiologisch Laboratorium, Albinusdreef 2, 2300 RC Leiden

Enrico Marani

NB: The specimens listed below have since been passed to the Nederlands Herseninstituut, Meibergdreef 47, 1105 BA Amsterdam, which in addition preserves 5 PFA-fixed gorilla brains in its Primate Brain Bank (www. primatebrainbank.org).

Series H 5945 Brain. Juvenile female. Originally from Utrecht: no. D 63/189. The animal was aged 1½−2 years. The brain is fixed in 4% formalin, sectioned transversely, and stained with iron-haematoxylin using the method of Haggquist (1933): Zeitschrift für Wissenschaftliche Mikroskopie und für Mikroskopische Technik 50, p. 77.

Series H 5965 Brain. Received from Utrecht. It is preserved and prepared as above.

Rotterdam

Natuurhistorisch Museum Rotterdam, Westzeedijk 345 (Museumpark), 3015 AA Rotterdam

Henry van der Es

NMR9990-2659 Penis (in alcohol). Adult male. Donated by G. Th. de Vries on 31 December 2001. It is a zoo specimen.

NMR9990-2989 Skull and partial skeleton. Neonate. Donated by H.L. Blonk on 9 August 2010. Originally obtained from Aart Walen. It is a zoo specimen.

Steyl

Missiemuseum Steyl, St Michaëlstraat 7, 5935 BL Steyl*

Bert van der Houwen

Mounted skin. Adult male. Kamerun. Mounted by the Präparatorium E. Ultsch, Halle, Germany.

The Hague

Museon, Stadhouderslaan 41, 2517 HV Den Haag

Arno van Berge Henegouwen; Frits van Rhijn

53010 Skull. Adult male. Purchased from Mr Van der Linde and received on 5 December 1946. The upper jaw lacks both canines and both central incisors; the lower jaw lacks all the incisors.

54863 Skull. Male. Donated by Dr (Miss) Wilhelmina Smidt; received on 17 August 1956. The incisors are missing from the lower jaw.

58923 Skull. Adult female. Zaïre. Purchased from Mr A. van Steyn; received on 15 August 1978. There is a hole measuring 6 cm × 8 cm around the foramen magnum; according to the donor, this was made so that the brain could be eaten.

72632 Mounted skin. Infant female? A captive specimen, from Dierenpark Wassenaar, Rikksstraatweg 667, 2245 CB Wassenaar (this zoo closed in 1985). Purchased from Mr A. Louwman; received on 1 July 1980.

72664 Mounted skin. Infant. A captive specimen, from Dierenpark Wassenaar. Purchased from Mr A. Louwman; received on 1 July 1980.

90123 Mounted skin. Adult female. Purchased from Jac Bouten & Zn b.v. Received on 12 November 1985.

99913 Cranium. Received in exchange from Dr H.L. Blonk.

214265 Skeleton. Infant male. Donated by Diergaarde Blijdorp, Rotterdam on 3 March 1989. This is 'Salem', who was born on 30 November 1988 (dam 'Salomé', sire 'Ernst') and was mother-reared. He died of catarrhal enteritis on 6 February 1989.

218580 Cast of body (black rubber). Infant male ('Salem'). Received as above.

220581 Mounted skin. Infant male. Donated by Diergaarde Blijdorp, Rotterdam; received on 16 July 2010.

Utrecht

Rijksuniversiteit Utrecht, Faculteit Diergeneeskunde, Departement van Pathobiology, Afdeling Anatomie en Fysiologie, Yalelaan 1, 3584 CL Utrecht

Claudia F. Wolschrijn

Skeleton. Adult female.

Rijksuniversiteit Utrecht, Universiteitsmuseum, Lange Nieuwstraat 106, 3512 PN Utrecht

Steven de Clercq; Paul H. Lambers; D. van den Tooren

UZ-58 Skull. Adult female. Nigeria-Cameroon border. From the old collection of the University's Zoological Museum.

UZ-60 Skull. Adult male. Nigeria-Cameroon border. From the old collection of the University's Zoological Museum.

UZ-192 Skull. Adult male. Acquired by the Anatomical Institute of the State University at Utrecht in 1912 (original no. 141).

UZ-294 Mounted skeleton. Adult male. Mounted by Maison Tramond−N. Rouppert successeur, 9 rue de l'École de Médecine, Paris. Acquired by the Anatomical Institute of the State University at Utrecht in 1912 (original no. 139). Height 176 cm.

UZ-1476 Mounted skin. Juvenile male. Sangmélima, Cameroon Republic. It was captured in about September 1966 and arrived at the animal dealer G. van den Brink, Den Blieklaan 52a, Soest, in October 1966, where it died shortly afterwards. Donated by G. van den Brink on 7 March 1967. Total length 448 mm; foot length 140 mm; ear length 43 mm; total ear length 49 mm; weight (without organs) 5750 g. Preparator: P. v.d. Brand.

UZ-1555.001/UZ-1555.002 Two ilia. Acquired in 1915?

UZ-3271 Left and right scapula. Male. Acquired in 1915?

UZ-7443 Plaster cast of left hand. Thumb and part of the middle finger are missing.

UZ-7444 Plaster cast of left foot.

UZ-7992 Skull. Adult male. Ouesso, French Congo. Donated by the Rijksmuseum van Natuurlijke Historie at Leyden to the Zoölogical Laboratory at Utrecht on 28 October 1953. It had been purchased from Dr A. Rossi, Rome, on 28 December 1929 (original no. RMNH 1802). Abscess in lower left jaw, right jugal bone damaged, damage around foramen magnum, tooth damage.

A-144 Skull. It probably originates from the Van Loon collection, dealing with comparative dental anatomy, dating back to c.1875–1910. An incisor is missing from the lower jaw. A molar and premolar in the lower right jaw are damaged.

The following 6 skulls were part of the collection of the Instituut voor Antropobiologie, Rijksuniversiteit Utrecht. Probably all were collected in 1938.

UP-1214 Skull. Male

UP-1215 Skull. Adult male. Spanish Guinea.

UP-1216 Skull. Adult male. Kana, French Congo. The animal weighed 210 kg.

UP-1217 Skull. Female.

UP-1218 Skull. Female. Spanish Guinea.

UP-1219 Skull. Female? Animal aged about 7 years.

NEW ZEALAND

Christchurch

Canterbury Museum, Rolleston Avenue, Christchurch 8001

Paul Scofield; G.A. Tunnicliffe

F. Ma. 214 Mounted skeleton. Adult male. It was acquired from Sir William Henry Flower (1831–99) in 1872. It is missing some of the phalanges of the right foot; also, the lower right canine is missing through natural causes (i.e., lost before the animal died). Old catalogue number: M.S. 95.

F.Ma.185 Skin. Male. It was acquired from Sir William Henry Flower in 1874. It was disposed of in the 1950s due to insect damage.

F. Ma. 3616 Skull (cast).

Dunedin

Otago Museum, 419 Great King Street, Dunedin 9030

Emma Burns; John T. Darby; Davina Hunt

OMNZ VT734 Skull (left sagittal section, mounted on wood). Female. Gabon. Collected in 1881. Purchased in January 1882. It is missing 2 teeth in the mandible and 1 tooth in the maxilla.

OMNZ VT2421 Mounted skin. Female. Bapmule (= Bapouli?) district, Cameroon. It was found in the forest. Purchased from Rowland Ward Ltd in May 1920. The hair has a predominantly medium-brown coloration.

OMNZ VT2468 Skull. Adult male. Purchased, probably in 1882. It is missing 2 teeth in the mandible and 3 teeth in the maxilla (all replaced with wooden replicas). A portion of the occipital bone around the foramen magnum is missing (broken rather than cut).

OMNZ VT3269 Skull (right sagittal section). It is the opposite side of OMNZ VT734. It is missing 4 teeth from the mandible and 3 teeth from the maxilla. Each bone of the cranium is painted a different colour, since it was part of a comparative skull anatomy teaching set. Donated in 2005 by the University of Otago Zoology Department.

C21

FIGURES C-21–C-23 Mounted skin of female (no. OMNZ VT2421) with unusual brown coloration. Specialist tests of the hair would be needed to determine whether the variations in colour intensity are original, or due to fading by sunlight over time. *Photographs courtesy of Otago Museum.*

C22

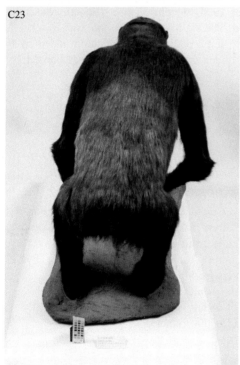

C23

FIGURES C-21–C-23   (Continued)

NORWAY

Bergen

Universitetsmuseet i Bergen, Muséplass 3, 5007 Bergen

Ingvar Byrkjedal

ZMUB 110 Mounted skin. Male. Guinea. Received in exchange from Prof. van Beneden, c.1860–80.

ZMUB 243 Skull. Adult male. Guinea. Received c.1860–80. It is now missing.

ZMUB 244 Skull. Juvenile. Guinea. Purchased from Du Chaillu c.1860–80. It is now missing.

ZMUB 245 Skull. Juvenile. Guinea. Purchased from Du Chaillu c.1860–80.

ZMUB 653 Carcase (preserved whole in alcohol). Juvenile. Received c.1860–80.

ZMUB 2904 Mounted skin. Adult male. Njong River, Kamerun. Purchased from Umlauff on 30 June 1917.

ZMUB 2905 Mounted skin. Adult female. Jaunde, Kamerun. Purchased from Umlauff on 30 June 1917.

ZMUB 2906 Mounted skin. Juvenile. Jaunde, Kamerun. Purchased from Umlauff on 30 June 1917.

ZMUB 2914 Mounted skeleton. Adult male. From the same animal as the mounted skin no. 2904.

ZMUB 3370 Skull. Adult male. Kamerun. Purchased from Oberprediger Bärthold.

ZMUB 3371 Skull. Male. Kamerun. Purchased from Oberprediger Bärthold.

Oslo

Universitetet i Oslo, Naturhistorisk Museum, Sars gate 1, 0562 Oslo

Øystein Wiig

14133 Skull. Adult female. It was collected by Ole Eigil Ommundsen in 1987, some tens of kilometres northeast of Pointe Noire, Congo-Brazzaville. It was bought from local people for 1000 CFA. It was imported legally to Norway in 1988. The collector had grown up in the area as the son of missionaries. It was given to the Natural History Museum in 2005.

22120 Mounted skeleton and mounted skin. Infant male. Democratic Republic of Congo. Its mother was shot in the autumn of 1966, and it was captured alive, aged about 1½ years. Purchased from Copenhagen Zoo, Denmark, in 1966. Received in January 2016 from the University of Oslo, Department of Bioscience, Kristine Bonnevies hus, Blindernveien 31, 0371 Oslo. [Additional data from: Eivind Østbye; Nils-Jarle Vtreberg, University Division of Zoology.]

M 4794 Skull. Adult male. Received from Bremen Museum.

M 7771 Skull. Juvenile. Belgian Congo. Collected in 1932. Received through an exchange on 14 October 1939. The preparator was Bartels.

Stavanger

Stavanger Museum, Muségata 16, 4005 Stavanger*

Olav J. Runde

Skull. Congo. Donated by K. Knudson in 1883.

Mounted skin. Adult female. Bibundi, British Cameroon. Donated by E. Berentsen, C. Middelthon and J. Dreyer in 1919.

Mounted skin. Juvenile. Bibundi, British Cameroon. Donated by E. Berentsen, C. Middelthon and J. Dreyer in 1919.

Trondheim

NTNU Norwegian University of Science and Technology, University Museum, Erling Skakkes gate 47b, 7491 Trondheim

Reidar Andersen; Torkild Bakken; Torleif Holthe; John W. Jensen; Gustav Thingstad

3327 Mounted skin. Juvenile male. Acquired in 1894. It was in rather poor condition, but has recently gone through a restoration process.

9833 Mounted skeleton. Adult male. Acquired in 1896. Height 136 cm. It has a low cranial crest (<2 cm).

POLAND

Cracow

Uniwersytet Jagielloński, Instytut Zoologii, Muzeum Antropologiczne, ul. Mieczyslawa Karasia 6, 30-060 Kraków*

Henryk Głąb

BI a2 Skull. Adult female. Gabon. Received from the Izydov Kopernicki collection in 1912.

BI a3 Skull. Adult female. Gabon. Received from the Izydov Kopernicki collection in 1912.

BI a4 Skull. Subadult male? Gabon. Received from the Izydov Kopernicki collection in 1912.

BI b4 Skull. Adult female. Gabon. Received from the Izydov Kopernicki collection in 1912.

BII 67 Skull. Adult male.

Wroclaw

Uniwersytet Wrocławski im Bolesława Bieruta, Instytut Zoologiczny, Zaklad Antropologii, ulica Kuźnicza 35, 50-951 Wrocław*

Skull. Adult male.
Skull. Adult female.
Cranium. Adult female.
Skull. Juvenile male?
Calotte. Infant.

Uniwersytet Wrocławski im Bolesława Bieruta, Muzeum Przyrodnicze, ulica Sienkiewicza 21, 50-335 Wrocław*

Louis Tomiałojć

Mounted skin. Female. *Gorilla castaneiceps*. This is 'Pussi', who arrived at the Breslau Zoo on 3 September 1897, having been purchased from the animal dealer Cross, in Liverpool, England. She weighed 15.75 kg upon arrival, and was said to be 4 years old (if true, she was unusually small). She died of chronic nephritis on 6 October 1904 (weight 33 kg). The skin was mounted by Gerrard. The skull was sent to the Zoölogisches Museum, Berlin. [Some additional information added by GH from correspondence with Wrocław Zoo personnel.]

PORTUGAL

Coimbra

Universidade de Coimbra, Instituto de Antropologia, Museu e Laboratório Antropológico, 3000-056 Coimbra*

Eugénia Cunha; Maria Augusta A. Tavares da Rocha

Skeleton. Adult male. *Gorilla gina*. Purchased for 1000 francs from E. Deyrolle, Paris, in 1889.

Skeleton. Adult female. *Gorilla gina*. Purchased for 800 francs from E. Deyrolle, Paris, in 1889.

Skull. Adult male.
Skull. Adult male.
Skull. Adult female.

Lisbon

Instituto de Investigação Científica Tropical, Trav Conde da Ribeira, no. 9, 1300-142 Lisboa

Rui Figueira

CZ000001149 Skin. Juvenile male. Cabinda, Angola. Received from Lisbon Zoo in 1959.

CZ000001542 Skull. Adult female. Received on 25 August 1969. This may be 'Conchita', who arrived at the Lisbon Zoo on 28 October 1959, aged about 2½ years. She was captured in the jungles of Maiombe-Cabinda, and donated by Lt-Col. Silvério Marques, Governor-General of Angola. She died of cirrhosis of the liver on 24 August 1969. [Some additional information added by GH from correspondence with Lisbon Zoo personnel.]

Universidade de Lisboa, Museu Nacional de História Natural e da Ciência, Rua da Escola Politécnica 56, 1250-102 Lisboa

Maria Judite Alves; Cristiane Bastos-Silveira; Graça Ramalhinho

NB: Nearly all of the museum collections were lost in an extensive fire on 18 March 1979. The only *Gorilla* specimens that survive are those listed below.

MB01-006010 Cranium. Adult male. Mayombe, Cabinda, Angola. Collected by José de Anchieta in 1865. It has a supernumerary (4th) right molar.

MB01-006011 Cranium. Juvenile male. The 3rd molars are unerupted.

MB01-006012 Skull. Juvenile male. Only the 1st permanent molars are erupted.

MB01-006013 Skull. Adult female. The mandibular 3rd molars are slightly procumbent.

MB01-006014 Cranium. Female.

A47 Mandible. Gorilla? Subadult?

Oporto

Universidade do Porto, Faculdade de Medicina, Instituto de Anatomia do Prof. J.A. Pires de Lima, Alameda Hernâni Monteiro, 4200 Porto*

J. Pinto Machado; Manuel M. Paula-Barbosa

Skeleton (incomplete). Adult male. Candambaco, Maiombe forest, Cabinda enclave, Angola. Presented by Dr Manuel J. dos Santos in December 1929. It comprises the cranium, both scapulae, both clavicles, sternum, 12 pairs of ribs, both humeri, both cubits, both radii, right femur, right tibia, and some of the cervical and thoracic vertebrae. There is an irregularly healed fracture of the left radius, which measures 350 mm in total length, and also a consolidated, incomplete fracture of the left ulna,

which measures 380 mm in total length. This specimen is described and illustrated in Pires De Lima (1933).

Universidade do Porto, Museu de História Natural e da Ciência, Praça Gomes Teixeira, 4099-002 Porto

Maria João Guimarães Fonseca; Rita Gaspar; Maria Teresa Madureira; Elsa Oliveira

UP-MHNFCP-080682 Skull (cast).

UP-MHNFCP-154865 Skull. Male.

UP-MHNFCP-154867 Skin (without face). Male.

UP-MHNFCP-154868 Skin (upper part only, with face and the arms).

UP-MHNFCP-154869 Skin (part only, without face, cut longitudinally).

The following specimens were collected in Maiombe, Cabinda territory, Portuguese West Africa, by António J. de Liz Ferreira during the period 1934–37. For measurements, see: Ferreira et al. (1945).

UP-MHNFCP-154866 Cranium. Adult male [*sic*]. No. 6 in Ferreira et al. (1945). The teeth are slightly worn.

UP-MHNFCP-154870 Mounted skeleton. Adult female. No. 5 in Ferreira et al. (1945). This animal was in the last stage of pregnancy, and was carrying a full-term fetus.

UP-MHNFCP-154871 Mounted skeleton. Adult male. No. 3 in Ferreira et al. (1945).

UP-MHNFCP-154872 Mounted skeleton. Adult male. No. 2 in Ferreira et al. (1945). The teeth are very worn. This skeleton exhibits polyarticular osteoarthrosis. The inferior epiphysis of the left humerus is deformed by disease. The remarkable difference in length between the right and left clavicles may be due to the deformations observed on the left side of this skeleton. For further information, see Ferreira (1938).

UP-MHNFCP-154873 Skeleton. Adult male. No. 1 in Ferreira et al. (1945). The museum has the skull, both innominate bones, both scapulae, 12 ribs, both radii, 81 hand and foot bones, both femora, both humeri, both tibiae, both fibulae, sacrum, sternum, right clavicle and 20 vertebrae of this specimen. The teeth are very worn.

UP-MHNFCP-154874 Skeleton. Subadult female (epiphyses of the humeri, femora and tibiae incompletely fused). No. 7 in Ferreira et al. (1945). The museum has only the skull, both innominate bones, 8 ribs, both scapulae, left femur, sternum, 24 vertebrae and 31 hand and foot bones of this specimen.

UP-MHNFCP-080792 Mounted skeleton. Adult male. No. 4 in Ferreira et al. (1945).

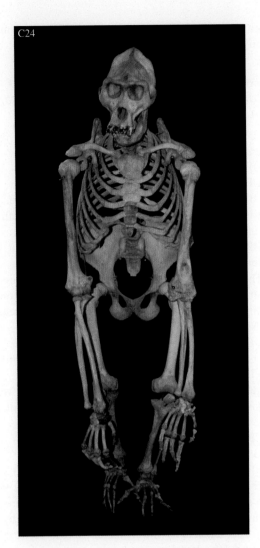

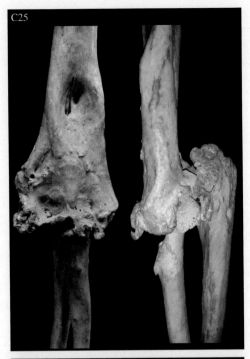

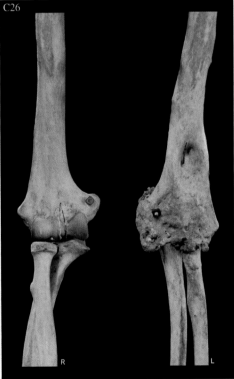

FIGURES C-24–C-26 Skeleton of adult male gorilla (no. UP-MHNFCP-154872). The skeleton appears to be complete. C-25: Left elbow joint, frontal and lateral (proximal surface) views. C-26: Right (normal) and left (diseased) elbow joints. The three images indicate that the right elbow joint is within normal limits with apparently intact articular surfaces. The left joint, however, shows excessive new bone growth engulfing much of the distal humerus with some involvement of the proximal ulna. This was clearly a chronic progressive process. The cause was possibly trauma or an infection, with reactive bone deposition. *Photographs courtesy of Museu de História Natural e da Ciênca da Universidade do Porto (MHNC-UP).*

FIGURES C-24–C-26   (Continued)

ROMANIA

Bucharest

Muzeul Naţional de Istorie Naturală 'Grigore Antipa', Şoseaua Kisselef 1, 79744 Bucureşti*

Dumitru Murariu

Skeleton and mounted skin. Adult male. Kamerun.

Skeleton and mounted skin. Adult female. Kamerun.

Skeleton and mounted skin. Juvenile. Kamerun.

Skeleton and mounted skin. Juvenile. Kamerun.

Mounted skin. Adult male. Kamerun.

Mounted skin. Adult female. Kamerun.

All the above specimens were purchased by Dr Grigore Antipa from J.F. Umlauff, Hamburg, in 1907.

RUSSIAN FEDERATION

Moscow

State Darwin Museum, Vavilova ul. 57, 117292 Moskva

Igor Fadeev; B.H. Ignatyeva; V. Ignatyeva; Dmitry Miloserdov

1651 Mounted skin. Adult male. Purchased from Wilhelm Schlüter, Ludwig-Wuchererstrasse 9, Halle a.S., Germany, in 1912.

1652 Mounted skin. Adult male. Purchased from Johannes Umlauff, Eckernförderstrasse 85, Hamburg 4, Germany, in 1913.

OF 1653 Mounted skin. Subadult male. Purchased from Schlüter.

OF 1654 Mounted skin. Juvenile. Purchased from Schlüter.

1656 Mounted skeleton. Adult male. Purchased from Schlüter in 1912.

1657 Mounted skeleton. Adult male. Purchased from Umlauff in 1913.

OF 1658 Mounted skeleton. Adult male. Esumba-Mbeck, Kamerun. Collected in 1916. Purchased from Umlauff. The skull is abnormal, deformed as the result of illness or injury (the animal was blind in one eye). Condylobasal length 22 cm.

OF 1659 Mounted skeleton. Adult male. Purchased from Oskar Fritsche, Taucha, bei Leipzig, Germany. There is a hole in the skull, made by a sharp object. Condylobasal length 22 cm.

OF 1660 Mounted skeleton. Adult female. Purchased from Schlüter. The cranium is numbered NVF 3792.

OF 1669 Mounted skeleton. Juvenile. Purchased from Schlüter.

OF 1673 Mounted skeleton. Juvenile. Purchased from Schlüter.

OF 1677 Skull. Adult male.

1678 Skull. Adult male. Purchased from V. Frič, Wladislawsgasse 21a, Prague.

1679 Skull. Adult female.

1681 Skull. Subadult.

1682 Skull. Juvenile. Purchased in 1909.

OF 4128 Mounted skin. Adult female. Purchased from Schlüter in 1912.

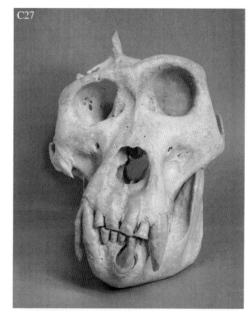

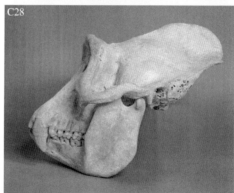

FIGURES C-27–C-31 Skull of adult male (no. OF 1658). C-27: Frontal view. C-28: Left lateral view. C-29: Right lateral view. C-30: Basal view of cranium. C-31: Superior view of mandible. There is marked asymmetry in both the cranium and mandible. The causes of asymmetry are discussed in Chapter 14, Musculoskeletal System in Part I, Gorilla Pathology and Health. The right temporomandibular joint is damaged. The dentition is complete, but the lower left canine has erupted ectopically, inferior to the lateral incisor. *Photographs courtesy of the State Darwin Museum, Moscow.*

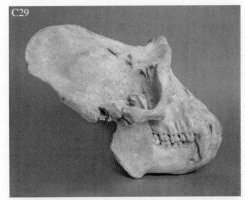

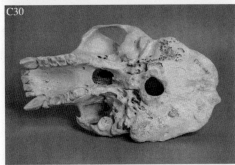

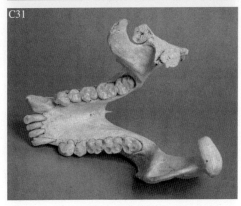

FIGURES C-27—C-31 (Continued)

Moscow State Lomonosov University, Zoological Museum, ul. B. Nikitskaya 6, 125009 Moskva

Andrey A. Lissovsky; Olga L. Rossolimo

S-4 Skull. Male. *Gorilla gina.*

S-5 Skull. Female. *Gorilla gina.*

S-151313 Skeleton and mounted skin. Adult male. Received from Moscow Zoo in 1990. This is 'Batanga', who came originally from Gabon, and arrived in Ruhr-Zoo Gelsenkirchen, Germany, on 31 July 1971. He was sold to Moscow Zoo, and arrived there on 16 June 1972, aged about 3 years (weight 18 kg). He died on 3 July 1990. [Some additional information added by GH from correspondence with Moscow Zoo personnel.]

S-152705 Mounted skeleton.

S-152706 Mounted skeleton. Juvenile.

St Petersburg

Russian Academy of Sciences, Zoological Institute, Vasilevski ostrov, Universitetskaya nab. 1, 199034 St Petersburg

Alexei V. Abramov; A.N. Tychonov

ZIN O.1 Skull. Adult male. Gabon. Received from Bouvier, Paris, in 1879.

ZIN O.1654 Skull. Adult male. Gabon. Received from Johann Friedrich Gustav Umlauff (1833—89), Hamburg, in 1883.

ZIN O.1666 Skull. Adult female. Gabon. Received from J.F.G. Umlauff in 1883.

ZIN O.2598 Skull. Adult male. Received from J.F.G. Umlauff in 1883.

ZIN O.2612 Skeleton. Adult female. Received from Frank in 1884. It is missing the right 1st metatarsal.

ZIN O.2703 Skeleton. Adult male. Received from Frank in 1884.

ZIN O.2835 Skull. Adult female. Gabon. Received from J.F.G. Umlauff in 1883.

ZIN O.3008 Skull. Adult male. Received in 1883.

ZIN O.3033 Skull. Adult male. Received from Wilhelm Schlüter, Halle, in 1884. The condyles are broken.

ZIN O.3034 Skull. Adult male. Received from Wilhelm Schlüter, Halle, in 1884.

ZIN O.6231 Skull. Adult male. Gabon. Received from Staudinger in 1891.

ZIN O.6232 Skull. Adult female. Gabon. Received from Staudinger in 1891.

ZIN C.19962 Mounted skin. Adult male. Received from the Kaliningrad Zoo in 1954.

ZIN C.28656 Skin (minus head). Received from the Museum of Anthropology and Ethnography, Academy of Sciences, Leningrad.

ZIN O.35617 Skull. Adult female. Received from Leningrad Zoo, 1 Aleksandrovsky Park, 197198 St Petersburg, in 1992. It had died at the age of 23 years.

ZIN C.71852 Parts of skin from face and hands. Adult female. Received from Kaliningrad Zoo in 1986. This is 'Leama', who died on 13 July 1986.

Skull (plaster cast). Male. South Cameroon. Received from J.F.G. Umlauff in 1907. It is of poor quality.

Skull (plaster cast). Male. Libreville, Gabon. Received from Capt. A. Termain in 1908.

RWANDA

Musanze (formerly called Ruhengeri)

John and Margaret Cooper

Rwanda has for over a century been intimately associated with the mountain gorilla *Gorilla beringei*. Over the years material from both dead and live animals has regularly been collected by individuals and institutions and either stored in Rwanda or sent overseas to museums, private collections and universities. Interest in the importance of such specimens dates back to the sighting in 1900 by Ewart Scott Grogan of a skeleton of a 'gigantic ape' in the forest near Mount Mikeno (Grogan and Sharp, 1900) followed soon after by the shooting in 1902 of two "black apes" by Friedrich Robert von Beringe (von Beringe, 2002). The carcase of one of these was sent to Berlin, where it still remains (see Part I, Fig 1.5). Over subsequent years, research by George Schaller (1963) and Dian Fossey (1983) included the collection of skeletal and other material from mountain gorillas, some of which remained in Rwanda while many specimens were sent to the United States and elsewhere. In 1994 a study of skeletal remains started at the Morris Animal Foundation Volcano Veterinary Center at Kinigi (see Fig CA2.1) but, during the fighting in April—July, some was damaged or lost (see Part I and Cooper and Cooper, 2007). At the time of writing (August 2017), only one formal collection appears to be properly documented within Rwanda. This is situated at the Karisoke Research Center in Musanze, and it is based on the work of the Mountain Gorilla Skeletal Project, Rwanda, the history of which is recounted in Appendix CA2. The collection in Musanze comprises more than 140 specimens of *G. beringei*, as well as other mammalian and avian species, and two- and three-dimensional image datasets (dental radiographs, CT scans) — see: https://cashp.columbian.gwu.edu/databases. The Karisoke Research Center is open to the public on weekdays. Scientists and others wishing to study the material should contact the Center for the Advanced Study of Human Paleobiology, Washington, DC 20052, USA — see: https://cashp.columbian.gwu.edu/hard-tissue-biology-field-projects. In addition to the Musanze collection, material from *G. beringei beringei* and *G. beringei graueri* forms part of centres maintained by Gorilla Doctors, formerly MGVP, in Kigali and elsewhere (see under "Davis, USA" in this catalogue). The centres contain material such as wet tissues, blocks and histology slides, blood smears, parasites and hair samples. Although Gorilla Doctors manages the resource, all samples are owned by the governments of the countries of origin.

SLOVENIA

Ljubljana

Prirodoslovni Muzej Slovenije, Prešernova 20, 1000 Ljubljana

Boris Kryštufek

PMS 22028 Cranium. Adult male. Central African Republic; probably Bayanga. It was discovered in the ground on a building site by F. Šetinc in 1978. Condylobasal length 296 mm; zygomatic width 178 mm; rostrum width 70 mm.

SOUTH AFRICA

Cape Town

South African Museum, P.O. Box 61, Cape Town 8000

Denise Hamerton; Erika Mias

ZM-001877 Skull and mounted skin. Purchased from E. Gerrard. Accessioned on 28 December 1896.

ZM-001878 Skull and mounted skin. Juvenile male. Purchased from E. Gerrard through P.L. Sclater. Accessioned on 28 December 1896.

ZM-033680 Skull. Juvenile. Cranial capacity 455 cc.

ZM-037009 Skull. Adult male?

ZM-037010 Mounted skeleton. Adult male?

ZM-037011 Skull. Juvenile. Cranial capacity 335 cc.

ZM-037012 Skull. Juvenile. Cranial capacity 362 cc.

University of Cape Town, Medical School, Department of Human Biology, Observatory, Cape 7925

Alan G. Morris; Caroline Powrie

z-199 Skull. Adult male. The teeth are very worn. Cranial capacity 565 cc.

z-200 Skull. Adult female. The teeth are moderately worn. Cranial capacity 525 cc.

Both the above specimens were purchased from N. Rouppert, Paris.

Johannesburg

University of the Witwatersrand, School of Animal, Plant, and Environmental Sciences, Life Sciences Museum, Private Bag 3, Wits 2050

James du G. Harrison; Caroline Robinson

336 Skull, both ulnae, one radius, parts of a humerus, both tibiae, part of sternum, sacrum, cervical vertebrae

and cast of the brain. Adult male. Gabon. It was collected by an Italian expedition in 1934.

University of the Witwatersrand, Medical School, School of Anatomical Sciences, 7 York Road, Parktown, Johannesburg 2193

Brendon Billings; Kevin L. Kuykendall; Phillip V. Tobias

C 10 Bust. Juvenile male. This is a cast of 'John Daniel', who died in captivity in the USA in 1921, aged about 5 years. The death mask has been retouched. Received from the American Museum of Natural History, New York, NY, USA. Not found in 2014.

C 69 Hand and foot (casts). Adult male. *Gorilla beringei graueri.* From the same individual as the skeleton no. Za 95. Not found in 2014.

C 70 Foot (cast). Juvenile male. From the same individual as the bust, no. C 10. Received from the American Museum of Natural History, New York, NY, USA.

C 87 Endocranial cast. Adult male? From a skull in the Zoology Department, University of the Witwatersrand. Volume 572 cc. Not found in 2014.

Ce 73 Endocranial cast. From a specimen in the South African Museum, Cape Town, in which the 1st molar is erupting. Presented by Solly Zuckerman (1904−93). Volume 450 cc.

Ce 80 Endocranial cast. From a specimen in the South African Museum, Cape Town, with a complete milk dentition. Presented by Solly Zuckerman. Volume 361 cc.

Ce 88 Endocranial cast. Adult male? From the skull no. C 6 in the Anatomy Department, Medical School, University of the Witwatersrand. Volume 548 cc.

Ce 93 Endocranial cast. Adult male. From the skull of the skeleton no. Za 95. Volume 548 cc. Not found in 2014.

Ce 96 Endocranial cast. Adult. From a skull in the Dental Hospital Museum, University of the Witwatersrand. Volume 573 cc.

Ce 97 Endocranial cast. Adult. From a specimen in the (sometime) possession of Prof. Grafton Elliot Smith (1871−1937), University College London. Volume 572 cc.

Cs 971 Skull (cast). Adult male. *Gorilla beringei beringei.* Received from the Wenner-Gren Foundation, New York, NY, USA (their catalog no. F-KR3).

M 215 Mounted skeleton (papier-mâché model). Adult male. *Gorilla beringei beringei.* Purchased from Anatomie Clastique du Dr Auzoux, Paris, in 1920.

MS 365 Skull. Adult male. It is part of the John Cecil Middleton Shaw collection.

Za 95 Partial skeleton (upper limbs, vertebrae, 26 ribs, and skull). Adult male. *Gorilla beringei graueri.* Mountains on the NW side of Lake Kivu (Belgian Scientific Centre). This was the second of two adult male

gorillas collected by Attilio Gatti's Italian Scientific Expedition to Africa of 1930−31 (the skeleton of the first gorilla is in the Institute of Anthropology, University of Florence, Italy). Donated by Attilio Gatti (1896−1969) and Prof. Raymond Arthur Dart (1893−1988). The teeth are well worn. There is a bullet hole in the left ilium. The measurements in the flesh were: top of head to sole of foot 65 in.; bottom of feet to tips of uplifted fingers 90 in.; circumference of abdomen 70 in.; right armpit to tips of fingers 43 in.; circumference of biceps (stretched) 17 in.; circumference of wrist 10¼ in. This specimen has been transferred to the Phillip V. Tobias Fossil Primate and Hominid Laboratory, Evolutionary Studies Institute, University of the Witwatersrand, 1 Jan Smuts Avenue, Braamfontein 2000.

Za 1290 Skull. Adult male. *Gorilla beringei beringei.* Kisoro, SW Uganda. It was killed by a leopard in 1959. Presented by Max Walter Baumgartel (1902−97) to Prof. Phillip Vallentine Tobias (1925−2012), who in turn presented it to the Department of Anatomy in 1969. This specimen has been transferred to the Phillip V. Tobias Fossil Primate and Hominid Laboratory. See Tobias and Cooper (2003).

Za 1311 Both humeri, both tibiae, both femora, both fibulae, both scapulae, sacrum, both ulnae, left radius, sternum (1 piece), 19 ribs, 21 vertebrae, pelvis (left and right), both calcanei, 1 talus and 1 metatarsal. Adult male. *Gorilla beringei beringei.* Kisoro, SW Uganda. Presented by Prof. A. Galloway in 1964. The skull is in the Phillip V. Tobias Fossil Primate and Hominid Laboratory.

Za 1312 Both scapulae, both clavicles, 25 ribs, both patellae, both ulnae, both tibiae, both fibulae, right humerus, right radius, sternum (4 pieces), 24 vertebrae, 1 foot and 1 hand. Adult male. *Gorilla beringei beringei.* Kisoro? Uganda. Presented by A. Galloway to Prof. R.A. Dart. The skull is in the Phillip V. Tobias Fossil Primate and Hominid Laboratory.

Pietermaritzburg

University of Kwazulu-Natal, Department of Zoology and Entomology, P.O. Box 375, Pietermaritzburg 3200*

G.L. Maclean

NUZ 2030 Mounted skeleton. Adult female. Kamerun. Collected by F. Izzard of Cambridge, UK. Purchased from W. Schlüter, Halle, Germany.

KwaZulu-Natal Museum, 237 Jabu Ndlovu Street, Pietermaritzburg 3201

Peter F. Tempest

Skull. Subadult.

934 Mounted skin. Adult female. Cameroon. Purchased from Rowland Ward in 1931.

934 Mounted skin. Juvenile. Cameroon. Purchased from Rowland Ward in 1931.

Mounted skin. Subadult male. *Gorilla beringei graueri*. Belgian Congo. Purchased at Christie's auction house, South Kensington, London, UK, on 29 September 1979. When purchased, the skin was in poor condition, having been extensively patched with parts from a second animal, and the head was poorly mounted and misshapen. The museum taxidermist subsequently remodelled and remounted the head. Apparently, the rolled-up skin had been smuggled out of the Congo during the civil war of the early 1960s.

Port Elizabeth

Port Elizabeth Museum, Corner of Beach Road and Brookes Hill Drive, Humewood, Port Elizabeth

Gillian Watson

Skeleton. *Troglodytes gorilla*. Purchased. This specimen is mentioned in an entry in the museum's Annual Report of 31 December 1897. There appears to be no reference to it in the accession records, and it is not on display or in storage.

PEM No. 1322 Mounted skeleton. Adult female? *Troglodytes savagei*. Central Africa. Collected in 1912. Purchased.

Pretoria

Ditsong National Museum of Natural History, Paul Kruger Street, Pretoria 0001

Shaw Badenhorst; Elizabeth M. Jones; I.L. Rautenbach; Miss Karin Scott

Mounted skeleton (on display). Adult male. This may be 'Wodan', who came originally from Cameroon, and arrived at the National Zoological Gardens of South Africa, Pretoria, on 5 June 1967, having been purchased, with another male name 'Kaiser', from G. van den Brink. Both were about 3 years old. Both were eventually found to be sterile, and were sold to the Hartbeespoort Dam Snake and Animal Park on 31 March 1983. 'Wodan' died there in May 1986. [Some additional information added by GH from correspondence with Pretoria Zoo personnel.]

A 2712 Hand bones.

AZ 1137 Bones from 1 pes; carpals. The 2nd and 5th digits are incomplete.

AZ 1152 Skeleton (incomplete). Adult female. Cameroon. This is 'Sylvia', who arrived at the Johannesburg Zoological Gardens, Jan Smuts Avenue, Parkview, Johannesburg 2193, on 24 June 1972, weighing

12 kg. She had been purchased, with a male named 'Maxi', from Heini Demmer, Mahlerstrasse 7, Wien I, Austria. She died of toxaemia due to *Actionomyces viscosus* on 24 May 1988. Donated by Johannesburg Zoo in August 1988. [Some additional information added by GH from correspondence with Johannesburg Zoo personnel.]

AZ 1162 Postcranial skeleton (incomplete). Adult female. This is 'Bobbie', who arrived at the National Zoological Gardens of South Africa, Pretoria, on 4 October 1972, when aged about 2½ years. She had been purchased, with a male named 'Frank', from G. van den Brink. She died of toxic shock due to suspected fungal necrotic enteritis on 7 September 1988. Donated by Pretoria Zoo. The zoo kept the skull. [Some additional information added by GH from correspondence with Pretoria Zoo personnel.]

AZ 1803 Skeleton. Adult male. Cameroon. This is 'Kaiser', who arrived at the National Zoological Gardens of South Africa, Pretoria, on 5 June 1967. He had been purchased with another male, 'Wodan'. He died of brain cancer in the Hartbeespoort Dam Snake and Animal Park on 13 April 1987. Donated by Hartbeespoort Dam Snake and Animal Park on 17 October 1990. [Some additional information added by GH from correspondence with Pretoria Zoo personnel.]

TM 16737 Mounted skeleton. Adult male. It is in the Mammal Department.

National Zoological Gardens of South Africa, 232 Boom Street, Pretoria 0001

Kim Labuschagne; Emily Lane

[Some additional information below added by GH from correspondence with Zoo personnel in Johannesburg and Stuttgart, and from the International Studbook for the Western Lowland Gorilla, 2010.]

NB: The National Zoological Gardens of South Africa/National Research Foundation Biobank is a facility that is accessible for research purposes, but any request for biomaterials must be accompanied by a research proposal, and permission must be obtained from the NZG Research and Ethics Scientific Committee before the samples can be released and/or subsampled.

4361 2 × blood: serum; 2 × blood: plasma EDTA (ethylenediamine tetraacetic acid); blood: plasma heparin; 2 × blood: white blood cells EDTA; blood: white blood cells heparin; hair; skin; organ: formalin. Male. This is 'Max' (also called 'Maxi'), who arrived at Johannesburg Zoological Gardens, Jan Smuts Avenue, Parkview, Johannesburg 2193, on 24 June 1972 (weight 15 kg). He was purchased, with a female named 'Sylvia', from Heini Demmer, Mahlerstrasse 7, Wien I, Austria. He died on 5 May 2004.

5535 Organs: frozen. From the same animal as 4361 above ('Max').

6420 Blood: serum; blood: plasma EDTA; blood: red blood cells EDTA; blood: whole blood; hair; skin; organ: frozen. Female. This is 'Liza', who came originally from Cameroon, and arrived at Rostov-on-Don Zoo, Zoologicheskaya ulica 3, Rostov-na-Donu, 344039 Russia, on 8 July 1975, when aged about 4 years. She was sent to Moscow Zoo on 20 July 1982, in exchange for the female 'Metti'. She was then sent on loan to Johannesburg Zoo, where she arrived on 22 June 1991. She died on 6 April 2006 while under anaesthetic, undergoing surgery for the examination of a lump next to her breast.

6563 8 × blood: serum; 7 × blood: plasma EDTA; 2 × blood: plasma heparin; 6 × blood: red blood cells EDTA; 2 × blood: red blood cells heparin; 14 × blood: whole blood; 3 × blood: white blood cells EDTA; 4 × hair; organ: frozen; FFPE (formalin-fixed, paraffin-embedded) block; faecal sample: frozen. Male. This is 'Izan', who arrived at Zoo Taiping, Perak, Malaysia, on 18 January 2002. He was transferred to Pretoria Zoo on 14 April 2004, then to Limbe Wildlife Centre, Cameroon, on 30 November 2007, where he died on 27 December 2008. History while in Pretoria: focal alopecia, depigmentation, mild pruritis. Pathology: skin biopsy showed nonspecific dermatitis. Diagnosis: suspected irritant or hypersensitivity skin disease.

6564 8 × blood: serum; 5 × blood: plasma EDTA; 2 × blood: plasma heparin; 4 × blood: red blood cells EDTA; 3 × blood: red blood cells heparin; 10 × blood: whole blood; 3 × blood: white blood cells EDTA; 3 × hair; 2 × organs: frozen; faecal sample: frozen. Female.This is 'Abbey', who arrived at Taiping Zoo on 18 January 2002. She was transferred to Pretoria Zoo on 14 April 2004, then to Limbe Wildlife Centre, Cameroon, on 30 November 2007.

6565 6 × blood: serum; 5 × blood: plasma EDTA; 3 × blood: red blood cells EDTA; 8 × blood: whole blood; 3 × blood: white blood cells EDTA; 2 × hair; organ: formalin; 2 × organs: frozen; FFPE block; faecal sample: frozen. Female. This is 'Oyin', who arrived at Taiping Zoo on 18 January 2002. She was transferred to Pretoria Zoo on 14 April 2004, then to Limbe Wildlife Centre, Cameroon, on 30 November 2007, where she died of intestinal pathogens on 2 June 2008. History while in Pretoria: head, shoulders, chest alopecia; pruritis since March 2007. Pathology: skin biopsy showed nonspecific dermatitis. Diagnosis: possible behavioural alopecia (cold, dry weather).

6566 Organs: frozen; faecal sample: frozen. Male. This is 'Makokou', who was born in Wilhelma Zoologisch-Botanischer Garten, Stuttgart, Germany, on 9 July 1985 (sire 'Banjo', dam 'Mimi'). He was reared by hand from the second day (weight 2360 g). He was sent on loan to Johannesburg Zoo on 27 November 2004.

6567 Blood: serum; blood: plasma EDTA; blood: red blood cells EDTA; 4 × blood: whole blood; blood: white blood cells EDTA; hair; skin; 2 × urine; organ: formalin; 2 × organs: frozen; faecal sample: frozen. Male. This is 'Hobbit', who was born in Stuttgart Zoo on 2 March 1979 (sire 'Banjo', dam 'Undi'), and was reared by hand from the 11th day. He was sent on loan to Pretoria Zoo on 3 June 1988, where he was euthanased due to chronic diabetes mellitus on 11 December 2009. Pathology: glomerulosclerosis; oxalate nephrosis; mesenteric and choroid arteriosclerosis; coronary artery mineralisation with myocardial fibrosis; purulent and granulomatous pleuritis and bronchopneumonia; mediastinal lymph node granulomas and amyloidosis; ocular lens vacuolisation; retinal degeneration; anterior synechiae; no histological lesions in pancreatic islets; hepatic hydropic degeneration and glycogenosis; adrenocortical atrophy; thyroid atrophy and lipofuscinosis; splenic amyloidosis; hydrothorax; hydropericardium; left atrioventricular valvular endocardiosis; testicular atrophy. Other tests: *Mycobacterium tuberculosis* and *Streptococcus pneumoniae* isolated from the lung. Diagnosis: chronic diabetes mellitus with secondary pneumonia.

6568 7 × blood: serum; 6 × blood: plasma EDTA; 2 × blood: plasma heparin; 4 × blood: white blood cells EDTA; 2 × blood: white blood cells heparin; 2 × hair; 2 × organs: frozen. Female. This is 'Tinu', who arrived at Taiping Zoo on 18 January 2002. She was transferred to Pretoria Zoo on 14 April 2004, and then to Limbe Wildlife Centre on 30 November 2007.

15251 4 × blood: serum; 2 × blood: plasma EDTA; 4 × blood: red blood cells EDTA; 7 × blood: whole blood. Male. This is 'Binga', who was born in Zoologischer Garten Zürich on 11 December 2001 (sire 'Ngola', dam 'Mamitu'), and was mother-reared. He was sent on loan to Pretoria Zoo on 20 January 2010.

15254 3 × blood: serum; 3 × blood: plasma EDTA; 5 × blood: red blood cells EDTA; 6 × blood: whole blood. Male. This is 'Bonsenga', who was born in Zoo Zürich on 9 September 2001 (sire 'Ngola', dam 'Nache') and was mother-reared. He was sent on loan to Pretoria Zoo on 20 January 2010.

15967 blood: serum; 4 × blood: red blood cells EDTA; blood: whole blood. Male. This is 'Louie', who was born in the Zoological Center Tel Aviv-Ramat Gan, Israel, on 9 April 2004 (sire 'Lukas', dam 'Lia') and was mother-reared. He was sent on loan to Pretoria Zoo on 1 March 2010.

15968 blood: serum; blood: plasma EDTA; blood: red blood cells EDTA; blood: whole blood. Male. This is 'Asali', who was born in the Zoological Center Tel Aviv-Ramat Gan, Israel, on 17 November 2003 (sire 'Lukas', dam 'Anya'), and was mother-reared. He was sent on loan to Pretoria Zoo on 1 March 2010.

SPAIN

Barcelona

Museu de Cièncias Naturals de Barcelona, Passeig Picasso s/n, 08003 Barcelona, Catalonia, Spain

Eulàlia Garcia Franquesa; Javier Quesada Lara; Francesc Uribe

Skeleton, muscle tissue and skin sample. Adult male. This is 'Snowflake' (known as 'Copito de Nieve' in Spanish, and 'Floquet de Neu' in Catalan), who arrived at Barcelona Zoo, Parc de la Ciutadella, s/n, 08003 Barcelona, on 1 November 1966 (weight 9 kg). He was imported from the Centro de Ikunde, Spanish Guinea, where he had been received on 5 October 1966 (weight 8.75 kg), when aged about 2 years. He had been captured on 1 October 1966 by Benito Mañé, an indigenous Ntumu Fang tribesman and collector for the Centro, in the Nkò forest, Ntem-Bongola river region, 2°8′N, 10°11′E, near the border with Cameroon. He is the first albino gorilla known to science, exhibiting a white pelage and skin, pink face and blue eyes. His official name was 'Nfumu', meaning 'white' in the Fang language. He was euthanased on 24 November 2003, having been suffering from skin cancer. By agreement with the donor, the skeleton may only be used for scientific study, and not for exhibition. See Márquez et al. (2008) and Prado-Martínez et al. (2013). [Some additional data added by GH from correspondence with Barcelona Zoo personnel, A. Jonch and J. Sabater Pí.]

MZB 82-0543 Skull. Female.

MZB 82-0544 Skull and axial skeleton (no arms or legs). Infant.

MZB 82-0545 Skull. Subadult male.

MZB 82-0548 Mounted skeleton. Juvenile female.

MZB 82-0561 Cranium. Male.

MZB 82-0562 Cranium. Female.

MZB 82-7153 Mandible. Subadult male. Equatorial Guinea? Received from Parc Zoològic de Barcelona (Barcelona Zoo) in 1970.

MZB 82-7170 Mandible. Subadult male. Equatorial Guinea? Received from Barcelona Zoo in 1974.

MZB 82-7200 Skeleton (half complete, no skull). Juvenile. Received from Barcelona Zoo in 1973.

MZB 82-7225 Skull. Adult male.

MZB 82-7626 Mounted skin. It was discarded in 1987, because it was extremely damaged.

MZB 82-7699 Mounted skin. Juvenile female. Donated. It is believed to have been removed from the collection before 1995, due to poor conservation.

MZB 82-7799 Postcranial skeleton. Male. Received from Barcelona Zoo in 1971.

MZB 89-0149 Recorded sound of adult female. Barcelona Zoo, 13 July 1989.

MZB 93-0008 Skeleton. Adult female. It died in Barcelona Zoo on 19 July 1989. Donated on 21 February 1992.

MZB 94-0468 Postcranial skeleton. Juvenile male. Donated by Barcelona Zoo on 18 April 1994.

MZB 94-0469 Postcranial skeleton. Juvenile female. Donated by Barcelona Zoo on 18 April 1994.

MZB 94-0642 Skeleton. Subadult male. It died in Barcelona Zoo on 6 November 1985. Donated by Barcelona Zoo on 18 April 1994.

MZB 94-1075 Sundry bones (femur and humerus). Adult male. Donated by Barcelona Zoo on 18 April 1994.

MZB 96-0862 Skeleton. Equatorial Guinea. Captive specimen. Died on 9 September 1996.

MZB 97-0053 Skeleton. Adult female. Mbini, Equatorial Guinea. Captive specimen. Died 6 February 1997.

MZB 97-0053-B Skeleton. Replica of MZB 97-0053.

MZB 99-1309 Mounted skeleton. Adult male. Donated on 13 September 1999 by Casp private school.

MZB 2001-0681 Skeleton. Infant male. Captive specimen. Died on 16 August 2001.

MZB 2001-0681-T Muscle tissue. From the above specimen.

MZB 2002-1261 Flat skin. Male.

MZB 2003-0544 Mounted skin. Adult male. Captive specimen. This is 'Urko', who was born in Barcelona Zoo on 22 February 1978 to sire 'Snowflake' and dam 'Ndengue'. He was hand-reared. 'Urko' sired one infant, a female stillborn on 5 July 1988 (dam 'Ntao'), but later became infertile due to testicular atrophy. He died on 12 April 2003, during surgery to treat peritonitis. Received on 13 April 2003.

MZB 2003-0544-B Skeleton. From the above specimen.

MZB 2003-0544-T Muscle tissue. From the above specimen.

MZB 2003-1582 Skeleton. Male. Captive specimen.

MZB 2003-1582-T Muscle tissue; skin sample. From the above specimen.

MZB 2010-1163 Skeleton. Female. Captive specimen.

MZB 2010-1163-B Flat skin. From the above specimen.

MZB 2010-1163-T Muscle tissue. From the above specimen.

Banc Biològic, Parc Zoològic de Barcelona, Parc de la Ciutadella, 08003 Barcelona; & Banc de Teixits Animals de Catalunya, Departement de Medicina i Cirurgia Animals, Facultat de Veterinària, Universitat Autònoma de Barcelona, Bellaterra, 08193 Barcelona

Hugo Fernández Bellon; Annaïs Carbajal Brossa

Tissue samples:

3 × umbilical cord. *Gorilla beringei beringei*. Date 27 March 2015. All are of poor quality.

Placenta. Date 4 September 2015. Poor quality.

Spleen, liver, kidney, adrenal gland, ileocaecal lymph node, duodenum, jejunum, ileum, proximal colon, distal colon. Date 31 January 2012. All are of good quality.

Histology:

Uterus and oviduct. Female. Date 3 September 2015.

Ovaries and uterus. Female. Date 14 June 2012.

Testicles. Male. Date 19 September 2013.

Testicles. Male. Date 22 October 2013.

Testicles. Male. Date 11 February 2014.

Other samples:

Hair. Female ('Makena', date of birth 15 March 2004, Munich). Date 3 November 2014.

Faeces extracts processed for hormone analysis. Male ('Xebo', d.o.b. 6 October 1985, Rotterdam). Date 17 October 2013–17 November 2013.

Faeces extracts processed for hormone analysis. 'Xebo' group. Date 17 October 2013–17 November 2013.

Serum:

Male. Collected 2 February 1999: mucus; 2 × serum, clear; 3 × plasma (citrate), clear; 3 × plasma (heparin), clear.

Female. Coll. 5 June 2006: 4 × serum, clear; 3 × blood, whole.

Female. Coll. 26 August 2006: plasma (sodium EDTA); serum, clear.

Female. Coll. 31 August 2006: serum, clear.

Female. Coll. 21 November 2007: serum, clear; 3 × plasma (lithium heparin), lipaemic and slight haemolysis; 2 × red blood cells (sodium heparin); plasma (sodium heparin); rbc (sodium EDTA).

Female. Coll. 27 July 2009: 2 × serum, lipaemic and slight haemolysis; rbc (EDTA); plasma (EDTA), clear.

Sex unrecorded. Coll. 23 September 2009: 2 × serum, clear; 2 × serum, haemolysis; 2 × blood, whole (EDTA).

Female. Coll. 21 October 2009: 2 × serum, clear; 2 × plasma (EDTA), clear; 3 × blood, whole (EDTA).

Female. Coll. 20 March 2011: 2 × plasma (EDTA), haemolysis; 2 × rbc (EDTA); serum, haemolysis.

Female. Coll. 20 September 2011: 3 × serum, clear; 2 × plasma (EDTA), clear; rbc (EDTA).

Female. Coll. 19 January 2012: 2 × plasma (EDTA), haemolysis; pellet (EDTA).

Female. Coll. 26 January 2012: 2 × plasma (EDTA), clear; blood, whole (EDTA); serum, clear; 2 × serum, lipaemic and slight haemolysis.

Female. Coll. 28 January 2012: pellet (EDTA); serum; plasma (EDTA), clear.

Male. Coll. 28 January 2012: serum, clear; serum, slight haemolysis; 2 × plasma (EDTA), clear; plasma (sodium heparin ACD solution), clear; 2 × plasma (citrate), clear; 3 × pellet (EDTA).

Male. Coll. 2 July 2012: serum, clear; plasma (EDTA), clear; plasma (EDTA), slight haemolysis; 4 × pellet (EDTA); serum, slight haemolysis; 2 × plasma (EDTA), clear.

Male. Coll. 17 February 2012: plasma (EDTA), clear; 2 × serum, haemolysis; 2 × plasma (EDTA), slight haemolysis; 2 × pellet (EDTA).

Female. Coll. 9 March 2012: serum, clear; pellet (EDTA); plasma (EDTA), clear.

Female. Coll. 6 November 2012: 2 × serum, clear.

Female. Coll. 3 January 2013: serum, haemolysis; 3 × plasma (EDTA), haemolysis; plasma, slight haemolysis; 3 × pellet (EDTA).

Female. Coll. 12 February 2013: serum, slight haemolysis; pellet (EDTA); plasma (EDTA), slight haemolysis.

Female. Coll. 9 April 2013: pellet (EDTA); plasma (EDTA), slight haemolysis.

Female. Coll. 10 March 2014: 3 × pellet (EDTA); 5 × plasma (EDTA), clear; 3 × plasma (heparin), clear; 3 × serum, clear; serum, slight haemolysis.

Female. Coll. 13 March 2014: plasma (EDTA), slight haemolysis.

Madrid

Museo Nacional de Antropología, Departamento de Antropología Física, Calle de Alfonso XII, 68, 28014 Madrid

Luis Pérez Armiño; Fernando Sáez Lara

CE1263 Bust (plaster of paris). Male.

CE10190 Cranium. Male.

CE10191 Cranium. Male.

CE10192 Cranium. Male.

CE10194 Cranium. Female.

CE10195 Cranium. Female.

CE10199 Cranium. Adult male.

The above six crania came from the former Spanish colony of Equatorial Guinea and were collected during an expedition led by Amado Ossorio in 1884–86. The expedition was financed by the Sociedad Española de Africanistas y Colonistas; it was not a hunting expedition, and the skulls were probably collected as cultural artefacts. According to the documentation, they were used as fetishes by the Pamues. Ossorio's collection was donated to the museum in 1915.

Museo Nacional de Ciencias Naturales, J. Gutierrez Abascal 2, 28006 Madrid

Josefina Barreiro; Antonio García-Valdecasas; Ángel Luis Garvía Rodriguez

MNCN-M2104 Mounted skin. Adult male. *Gorilla gorilla matschiei*. Equatorial Guinea.

MNCN-M2105 Mounted skin. Adult female. *Gorilla gorilla matschiei*. Equatorial Guinea.

MNCN-M2106 Mounted skin. Juvenile male. *Gorilla gorilla matschiei*. Equatorial Guinea.

MNCN-M2107 Mounted skin. Juvenile male. *Gorilla gorilla matschiei*. Equatorial Guinea.

MNCN-M2108 Mounted skeleton. Adult male. *Gorilla gorilla matschiei*. Akon, Micomeseng, Río Muni. Collected by O. Krohnert. Donated by J. González Gallarza; registered on 24 July 1948.

MNCN-M2109 Skull. Adult male. *Gorilla gorilla matschiei*. Akon, Micomeseng, Río Muni. Collected by O. Krohnert. Donated by J. González Gallarza; registered on 24 July 1948.

MNCN-M2110 Skull. Adult female. *Gorilla gorilla matschiei*. Akon, Micomeseng, Río Muni. Collected by O. Krohnert. Donated by J. González Gallarza; registered on 24 July 1948.

MNCN-M2187 Skull. Adult male. Upper Benito, Equatorial Guinea. Collected by J. Alonso Martínez. Received on 31 March 1924.

MNCN-M2219 Mandible. Female.

MNCN-M2253 Skeleton. Female. Received from Zoo de Madrid. Prepared by L. Castelo and F. Yagüe.

MNCN-M2273 Skull. Male.

MNCN-M2274 Skull. Male.

MNCN-M2275 Cranium (broken). Female.

MNCN-M2278 Cranium (broken).

MNCN-M2282 Mounted head.

MNCN-M 2284 Mandible (broken).

MNCN-M 5442 Skeleton. Adult female. Received from Zoo de Madrid on 6 June 1991. This is 'Afri', who arrived at Madrid Zoo (with another female named 'Gadi') on 1 July 1976, when she was about 4 years old. Both animals came originally from Gabon. She died of tuberculosis on 4 June 1991. Prepared by L. Castelo and F. Yagüe.

MNCN-M 18972 Mandible. Río Muni. Collected by L. Sorela.

MNCN-M 18973. Mandible. Missing all teeth.

Universidad Complutense, Facultad de Medicina, Cátedra de Anatomia, Madrid 3*

Juan Jiménez Collado

Mounted skeleton. Male. Guinea. Donated in 1937.

Mounted skeleton. Male. Guinea. Donated in 1937.

Universidad Complutense, Facultad de Medicina, Escuela de Medicina Legal, Ciudad Universitaria, Pabellón núm. 7, 28003 Madrid*

Francisco Gómez Bellard

Cranium. Adult male.

Cranium. Adult male.

The above crania are kept in the Forensic Anthropology Laboratories. They probably belonged to the old collection of the Museo Etnológico, and were brought to this Faculty just after the Spanish Civil War of 1936–39.

Universidad Complutense, Facultad de Veterinaria, Museo Veterinario Complutense, Avenida Puerta de Hierro, 28040 Madrid

Álvaro Mezcua

MV-001482 Skull, thorax and pelvis (mounted). Adult female. It was received from the Museo de Anatomía Javier Puerta, Departamento de Anatomía Humana y Embryología I, Facultad de Medicina, Universidad Complutense, on 24 April 2009. Height 60 cm; width 33 cm; depth 40 cm.

Santiago De Compostela

Universidad de Santiago de Compostela, Museo de Historia Natural, Parque de Vista Alegre, s/n, 15782 Santiago de Compostela

Xan Guerra

Mounted skeleton. Adult male. Equatorial Guinea. It was donated in 1998 by the Departamento de Ciencias Morfologicas, Anatomia y Anatomia Patalogica Comparadas, Facultad de Medicina, Universidad de Santiago de Compostela, where it had originally been received as a donation in 1957 [Javier Jorge Barreiro].

Seville

Consejo Superior de Investigaciones Cientificas, Estación Biológica de Doñana, Departamento de Colecciones, c/Américo Vespucio s/n, 41092 Sevilla

Teresa García Díez; José Cabot Nieves

EBD 6855 Skull. Male. Oreng, Acurenam, Equatorial Guinea. Found in the forest on 10 September 1984. Castelo and Cabezas no. B 2106.

EBD 14104 Skull. Male. Utonde, Bata, Equatorial Guinea. Collected on 15 May 1986. Juste-Castelo no. B 4512.

EBD 14105 Skull. Male. Moka, Evinayong, Bata, Equatorial Guinea. Collected on 30 June 1986. Juste-Castelo no. B 4626.

EBD 14106 Skull. Male. Moka, Evinayong, Bata, Equatorial Guinea. Collected on 26 May 1986. Juste-Castelo no. B 4510.

EBD 14309 Skin and postcranial skeleton. Female. Moca, Evinayong-Niefang, Bata, Equatorial Guinea. Collected on 27 June 1986. Juste-Castelo-Jimenez no. B 4616. Ears 56 mm/74 mm; foot 230 mm; arm span 1787 mm; head to heel 970 mm. Greasy condition.

EBD 14349 Skull. Juvenile female. Moca, Evinayong, Bata, Equatorial Guinea. Collected on 30 June 1986. Juste-Castelo-Jimenez no. B 4625.

EBD 14483 Skull. Female. Esong, Evinayong, Equatorial Guinea. Collected on 30 June 1986. Castelo-Juste-Jimenez no. B 4623.

EBD 14484 Skull. Female. Moca, Moga, Evinayong, Equatorial Guinea. Collected on 30 June 1986. Castelo-Juste-Jimenez no. B 4623.

EBD 15334 Skeleton. Male. It died on 8 May 1974 in the Jardin Zoologico y Botanico 'Alberto Duran', Jerez de la Frontera, Cadiz. Acquired on 25 May 1974. Head to base of spine 1010 mm; arm span 2160 mm; ears 42 mm/45 mm; foot 284 mm; head length 380 mm; head width 193 mm.

EBD 15773 Skull. Female. Equatorial Guinea. J. Juste-R. Castelo- A. Jimenez no. B 5342.

EBD 15777 Skull. Male. Mbia, Equatorial Guinea. Collected in 1986. J. Juste-R. Castelo- A. Jimenez no. B 5343.

EBD 15778 Skull. Female. Equatorial Guinea. J. Juste-R. Castelo- A. Jimenez no. B 5339.

EBD 15779 Skull. Female. Equatorial Guinea. J. Juste-R. Castelo- A. Jimenez no. B 5338.

EBD 15780 Skull. Male. Equatorial Guinea. J. Juste-R. Castelo- A. Jimenez no. B 5340.

EBD 15781 Skull. Male. Niefang, Equatorial Guinea. Collected in 1985. Juste-Castelo-Jimenez B no. 5341.

EBD 17445 Skull. Mikan, 32 km Bata-Niefang. Collected in August 1987. Castelo-Juste-Jimenez B no. 5712.

EBD 17960 Skull. San Joaquin de Ndyacong, Equatorial Guinea. Collected in 1988. Castelo-Juste-Jimenez no. B 7342.

EBD 17961 Skull. Nfaman, Niefang, Equatorial Guinea. Collected on 28 January 1988. Castelo-Juste-Jimenez no. B 6395.

EBD 22252 Bones of hands and feet. Adult. Equatorial Guinea. Collected in 1964. J.A. Valverde no. 2076.

EBD 24777 Skull and some small bones (probably from hands and feet). Juvenile female. Equatorial Guinea. Collected by J.A. Valverde in 1957. Died in captivity.

EBD 25152 Skeleton. Juvenile. Equatorial Guinea. Collected by J.A. Valverde in 1964.

EBD 28118 Skeleton. Juvenile. Equatorial Guinea. Collected by J.A. Valverde in 1964. Donated by Rosa Albacete.

EBD 28129 Skeleton. Male. Equatorial Guinea. Collected by J.A. Valverde in 1964. Donated by Rosa Albacete.

EBD 29954 Skull. Gabon. Collected by J.A. Valverde and J. Castroviejo.

EBD 29955 Skull. Adult. Equatorial Guinea. Collected in 1964. J.A. Valverde no. JAV 2076.

EBD 29956 Cranium. Equatorial Guinea. Collected 1960–70.

EBD 29957 Skull.

EBD 29958 Skull.

EBD 29960 Skull. Izaguirre to Sembin, Río Muni. Collected in May 1964. J.A. Valverde no. JAV 700. The mandible is broken and the left half is missing.

EBD 29961 Skull. Río Muni. Collected on 6 July 1964. J.A. Valverde no. JAV 706.

Valladolid

Universidad de Valladolid, Facultad de Medicina, Departamento de Anatomia Humana, Museo Anatómico, Ramon y Cajal 7, 47005 Valladolid

Juan Francisco Pastor Vázquez

NB: Specimens no. 1835, 5730 and 5731 are the subjects used in Diogo et al. (2011).

Specimen no. 6600 is the subject used in Diogo et al. (2015).

303 Skull. Juvenile female. Equatorial Guinea? Received in 1900.

723 Skull. Adult female. Equatorial Guinea? Received in 1900.

855 Cranium. Juvenile. Equatorial Guinea? Received in 1900.

914 Skull. Adult male. Equatorial Guinea? Received in 1900.

915 Skull. Adult male. Equatorial Guinea? Received in 1900.

1243 Skull. Adult male.

1767 Skull. Adult female. Equatorial Guinea? Received in 1900.

1835 Skeleton. Adult female. Received from Zoo de la Casa de Campo, Madrid, in 1997. This is 'Gadi', who arrived at Madrid Zoo on 1 July 1976, with another female named 'Afri'. Both were aged about 4 years, and came originally from Gabon. 'Gadi' died on 16 January 1997. [Some additional information added by GH from correspondence with Madrid Zoo personnel.]

2059 Skeleton. Adult male. Received from Loro Parque, Punta Brava, Puerto de la Cruz, Tenerife, Canary Islands, in 1998.

2334 Skeleton. Neonate female (2 days old). Received from Parc Zoològic de Barcelona in 2000. Date of birth: 30 January 2000 (to dam 'Machinda' and sire 'Xebo'); died 31 January 2000. The tongue is preserved in formol.

3043 Skeleton. Adult male. Received from Zoo Madrid in 2002. This is 'Bioko', who came originally from Equatorial Guinea and arrived at Madrid Zoo on 1 August 1981, when aged about 1 year. He died on 12 July 2002. [Some additional information added by GH from correspondence with Madrid Zoo personnel.]

4030 Skull. Adult male. Equatorial Guinea? Received in 1900.

5730 Skeleton. Adult female. Received from Bioparc Fuengirola in 2009. Also histological sections; organs in formol.

5731 Skeleton. Adult male. Received from Bioparc Valencia in 2007. Also histological sections.

5792 Skeleton. Adult female. Received from Bioparc Valencia in 2009. Also histological sections; organs in formol. This is 'Dorle', who was born in Zoologischer Garten der Stadt Frankfurt am Main on 27 May 1971 to dam 'Makulla' and sire 'Matze'. She died in Valencia on 9 December 2009. [Some additional information added by GH from International Studbook for the Western Lowland Gorilla, 2010.]

6600 Skeleton. Neonate male (1 day old). Received from Bioparc Valencia in 2012. Date of birth: 3 April 2012 (to dam 'Fossey' and sire 'Mambi'); died same day. Also tongue in formol; CT scans.

SWEDEN

Gothenburg

Naturhistoriska Museet, Slottskogen, Box 7283, 402 35 Göteborg

Sven Mathiasson; Göran Nilson

GNM Coll. an. 2277 Mounted skeleton. Male. *Gorilla mayema*. French Congo.

GNM Coll. an. 2278 Skull and mounted skin. Male. *Gorilla castaneiceps*. Gabon.

The above specimens were purchased for £250 altogether from Rowland Ward, London, in 1906.

Lund

Lunds Universitet, Biologiska Museet, Sölvegatan 37, 223 62 Lund

Lennart Cederholm; Jonas Ekström

NB: The specimens below were originally in the Zoologiska Museet, Helgonavägen 3, 223 62 Lund. The Zoologiska Museet merged with the Botaniska Museet in 2002, and all the zoological collections were moved to new facilities in 2012. Biologiska Museet is the new name for this museum.

L 863/3195 Mounted skeleton. Subadult male. *Gorilla gina*. Vicinity of Gabon River, Lower Guinea. Purchased

from 'Preparateur anatomiste' Guérin in 1863. The left scapula and the back of the skull have remains of old shot, overgrown by new tissue.

L 882/3044 Skeleton (partly mounted). Adult female. Purchased from G.A. Frank, London, in 1882.

Lxxx/5933 Cranium. Belolo, French Congo. It was preserved by a missionary, Mr David Johansson, and later purchased by Mr Jonas Lundberg, Töreboda, Sweden, in 1939.

Cranium. Adult female. Purchased from Frank, Amsterdam. It now appears to be missing from the collection.

Malmo

Malmö Museer, Malmöhusvägen 6, 201 24 Malmö

Per-Göran Bentz; Kristina Berggren; Yvonne Magnusson

Mam.ex.1.1 Mounted skin. Adult male. Loniie (= Lomié?), Kamerun. Collected by colonial Captain Schöneman in 1913. Stuffed by O. Bock, Berlin. Purchased in 1922.

Mam.ex.2.419 Mounted skin. Juvenile. It was bought by a sailor in 1956. Purchased from the firm Sct. Pauls Naturaliekabinett, Copenhagen, on 10 September 1969. The animal was approximately 9 months old.

Mam.ex.3.420 Mounted skin. Juvenile. It was bought by a sailor in 1956. Purchased from the firm Sct. Pauls Naturaliekabinett, Copenhagen, on 10 September 1969. The animal was approximately 9 months old.

Mam.os.1.217 Mounted skeleton. Adult male. Uganda. Purchased from W. Schlüter, Halle, Germany, in February 1920.

Mam.os.2.221 Cranium. Adult male. Jaunde, Kamerun. Purchased from Katzenstein, Berlin.

Mam.os.3.115 Cranium (cast). Acquired in 1912?

Mam.os.4.1431 Cranium. Juvenile. From the same individual as the mounted skin Mam.ex.2.419.

Mam.os.5.1432 Cranium. Juvenile. From the same individual as the mounted skin Mam.ex.3.420.

Mam.os.6.1276 Trunk skeleton (not mounted). Female. Donated by the taxidermist J.G. Erritzoe. Imported to Sweden as a carcase, flesh destroyed. The animal was 1 year 9 months old. Journal: 18 October 1963.

Solna

Stockholms Universitet, Osteologiska Forskningslaboratoriet, Ulriksdal Kungsgård, Ulriksdal, 17171 Solna*

Torstein Sjøvold

Mounted postcranial skeleton. Male? Donated by the Veterinary School when it moved to another town and disposed of some skeletons which were no longer required.

Stockholm

Naturhistoriska Riksmuseet, Vertebrate Zoology Section, Frescativägen 40, 114 18 Stockholm

Bo Fernholm; Olavi Grönwall; Daniela C. Kalthoff; Bengt-Olov Stolt

NRM 600433 Skull. Adult. Gabon.

NRM 601015 Skull. Kamerun. Collected by G. Waldau in 1905.

NRM 610941 Skull. Adult male. *Gorilla beringei beringei*. South of Mount Mikeno, Virunga volcanoes. Collected by Nils Gyldenstolpe on 20 March 1921. NB: The details of tooth eruption and skull measurements given below for the specimens 611540, 611639, 611930, 631164, 631168, 641165 and 641166 are taken from Lönnberg (1917b).

NRM 611540 Skeleton and skin. Juvenile male. *Gorilla beringei mikenensis* (cotype). Bamboo forest at the foot of Mount Mikeno, Virunga volcanoes. Collected by Karl Johan Ludvig Elias Arrhenius (1883–1923) on 27 December 1913. The 2nd molar is developed.

NRM 611639 Skeleton and skin. Juvenile female. *Gorilla beringei mikenensis* (cotype). Bamboo forest at the foot of Mount Mikeno, Virunga volcanoes. Collected by Elias Arrhenius on 27 December 1913. The 2nd molar is just cutting the gum.

NRM 611930 Skeleton and mounted skin. Adult female. *Gorilla beringei mikenensis* (cotype). Mount Mikeno, Virunga volcanoes. Collected by Elias Arrhenius on 14 May 1914. The skull has been badly damaged by the shot, so that almost the whole facial region has been torn away, but the interorbital region is left. Zygomatic width 147.3 mm; width of braincase 107 mm; inner height of orbit 38.8 mm; inner width of orbit 35.8 mm; least interorbital width 23.5 mm; width of planum nuchale 134.5 mm; height of planum nuchale from foramen magnum 64 mm; length of upper molar series 67 mm; length of M1 15 mm; width of M1 15.5 mm; length of M2 15.5 mm; width of M2 16.2 mm; length of M3 15 mm; width of M3 15.5 mm; length of lower molar series 75 mm.

NRM 612018 Skeleton and skin. Adult male. *Gorilla beringei graueri*. Lobero, 75 miles NE of Stanleyville, Belgian Congo. Collected by Lorens Marmstedt (1908–66) in March 1956. The specimen was collected during the making of the Swedish film 'Gorilla', which was co-directed and co-produced by Marmstedt, and was

released in Sweden on 27 August 1956. It was also released in several other countries, including the United States, where it was titled 'Gorilla Safari'. The cranium is broken in two; the left side of the cranium and the left posterior portion of the mandible are missing.

NRM 612395 Mounted skeleton and mounted skin. Adult male. South Kamerun. Prepared by Umlauff, and donated by Mrs Anna Broms in 1906.

NRM 612396 Mounted skeleton and mounted skin. Juvenile female. South Kamerun. Prepared by Umlauff, and donated by Mrs Anna Broms in 1906.

NRM 612397 Mounted skeleton and mounted skin. Juvenile male. French Congo. Prepared by Umlauff, and donated by Mrs Anna Broms in 1906.

NRM 612398 Mounted skeleton and mounted skin. Juvenile female. French Congo. Prepared by Umlauff, and donated by Mrs Anna Broms in 1906.

NRM 612581 Skin. Juvenile male. *Gorilla beringei beringei*. Mount Mikeno, Virunga volcanoes. Collected by Elias Arrhenius on 14 May 1914.

NRM 613159 Skull. Adult male. Collected at Masité, Belgian Congo, by Elias Arrhenius in 1919.

NRM 613160 Skull. Kamerun. Collected by Georg Waldau in 1905.

NRM 613183 Skull. Juvenile female. Kamerun. Collected by Georg Waldau in 1905.

NRM 613384 Skull. Spanish Guinea. Collected by Willborg in 1960.

NRM 613385 Skull. Subadult male. Spanish Guinea. Collected by Willborg in 1959.

NRM 613386 Skull. Subadult. Spanish Guinea. Collected by Willborg in 1959.

NRM 613883 Skull. Adult male. Collected by L. Ruben in 1906.

NRM 613884 Skull. Adult female. Collected by L. Ruben in 1906.

NRM 613916 Skeleton. Juvenile female. French Congo. Collected by L. Ruben in 1906.

NRM 613925 Skull. Adult female. Collected by L. Ruben in 1906.

NRM 615050 Mounted skeleton.

NRM 616106 Skull. Adult male. Kamerun. Collected by G. Waldau in 1905.

NRM 616106 Skull. Adult male. Between Pandu and Yokaduma, South Kamerun. Collected by G. Pape in October 1912.

NRM 616135 Skull. Adult male. Evinayong territory, Spanish Guinea. Collected by Willborg in 1965.

NRM 616136 Skull. Adult male. Micomeseng territory, Esaquaq, Nsomo, Spanish Guinea. Collected by Willborg in 1964.

NRM 616137 Skull. Adult male. Valladolid de los Bimbiles, Spanish Guinea. Collected by Willborg in January 1966.

NRM 616138 Skull. Subadult. Valladolid de los Bimbiles, Nkin Esaban, Spanish Guinea. Collected by Willborg in 1964.

NRM 616139 Skull. Adult. Micomeseng. Esafuman, Mbeayap, Spanish Guinea. Collected by Willborg in April 1965.

NRM 616501 Skull. Adult male. Spanish Guinea: 60 km from Bimbiles. Collected by Willborg on 7 March 1964.

NRM 616502 Skull. Adult male. Acam, Micomeseng. Spanish Guinea. Collected by Willborg in March 1964.

NRM 616503 Skull. Adult female. Nkin, Bimbiles, Spanish Guinea. Collected by Willborg in March 1964.

NRM 616504 Skull. Adult female. Cameroon, near the border with Spanish Guinea. Collected by Willborg in February 1964.

NRM 616505 Skull. Adult female. Micomeseng, Mbeayap, Spanish Guinea. Collected by Willborg in March 1964.

NRM 616506 Cranium. Adult. Micomeseng, Mbeayap, Spanish Guinea. Collected by Willborg in March 1964.

NRM 616507 Skull and partial skeleton. Adult. Acam, Micomeseng, Spanish Guinea. Collected by Willborg in March 1964.

NRM 616508 Cranium. Adult. Acam, Micomeseng, Spanish Guinea. Collected by Willborg in March 1964.

NRM 616509 Skull. Adult male. Nkin, Bimbiles, Spanish Guinea. Collected by Willborg in March 1964.

NRM 616510 Skull. Adult male. Ebebiyin, 140 km from Bata, Spanish Guinea. Collected by Willborg in March 1964.

NRM 618330 Skeleton. Adult male. Spanish Guinea. Collected by Willborg in 1959.

NRM 618494 Skeleton. Subadult male. Nsoc, Ebebiyin, Spanish Guinea. Collected by Jordi Sabater Pí (1922−2009) on 25 May 1956.

NRM 618495 Skeleton. Adult male. Nsoc, Ebebiyin, Spanish Guinea. Collected by J. Sabater Pí on 25 May 1956.

NRM 618496 Skeleton. Adult female. Nsoc, Ebebiyin, Spanish Guinea. Collected by J. Sabater Pí on 25 May 1956.

NRM 618498 Skull. Spanish Guinea. Collected by Willborg in 1962.

NRM 618499 Skull. Juvenile. Micomeseng, Spanish Guinea. Collected by Willborg in 1962.

NRM 620431 Skeleton and skin. Adult male. *Gorilla beringei beringei*. Mount Sabinio, Virunga volcanoes. Collected by Nils Gyldenstolpe on 18 February 1921.

NRM 620537 Skeleton and skin. Adult male. *Gorilla beringei beringei*. Northwest slopes of Mount Mikeno,

Virunga volcanoes. Collected by Prins Wilhelm on 10 March 1921. All teeth heavily worn. The skull is strongly asymmetrical towards the right. Cranial capacity 607 cc.

NRM 621229 Skeleton and skin. Juvenile female. *Gorilla beringei beringei*. Mount Sabinio, Virunga volcanoes. Collected by the Swedish Zoological Expedition (led by Prins Wilhelm) on 18 February 1921.

NRM 621346 Skeleton and skin. Adult male. *Gorilla beringei beringei*. Eastern slopes of Mount Mikeno, Virunga volcanoes. Collected by Kenneth Charles Carr (1888−1981) on 25 March 1921.

NRM 621413 Skin and skeleton. Female.

NRM 621444 Skin. Adult.

NRM 621445 Skull. Adult male. *Gorilla beringei beringei*. Northeast slopes of Mount Karisimbi, Virunga volcanoes. Collected by Nils Gyldenstolpe on 24 March 1921.

NRM 622170 Skull. Adult female. Obtained from Gerrard.

NRM 622279 Skin. Adult male. *Gorilla beringei beringei*. Mount Mikeno, Virunga volcanoes. Collected by Elias Arrhenius on 14 May 1914.

NRM 622330 Mounted skin. Juvenile male. Gabon. Collected by Theorin in 1882.

NRM 622580 Skin. Adult female. *Gorilla beringei beringei*. Mount Mikeno, Virunga volcanoes. Collected by Elias Arrhenius on 14 May 1914.

NRM 630739 Skeleton and skin. Adult female. *Gorilla beringei beringei*. Northwest slopes of Mount Mikeno, Virunga volcanoes. Collected by Nils Gyldenstolpe on 11 March 1921.

NRM 630840 Skeleton and skin. Adult male. *Gorilla beringei beringei*. South of Mount Mikeno, Virunga volcanoes. Collected by Nils Gyldenstolpe on 20 March 1921.

NRM 631028 Skeleton and skin. Adult female. *Gorilla beringei beringei*. Mount Sabinio, Virunga volcanoes. Collected by the Swedish Zoological Expedition on 18 February 1921.

NRM 631030 Skull, partial skeleton and skin. Juvenile male. *Gorilla beringei beringei*. Mount Sabinio, Virunga volcanoes. Collected by the Swedish Zoological Expedition on 18 February 1921.

NRM 631042 Skeleton and skin. Adult female. *Gorilla beringei beringei*. Southern slopes of Mount Mikeno, Virunga volcanoes. Collected by Nils Gyldenstolpe on 20 March 1921.

NRM 631143 Skull, partial skeleton and skin. Juvenile male. *Gorilla beringei beringei*. South of Mount Mikeno, Virunga volcanoes. Collected by Nils Gyldenstolpe on 20 March 1921.

NRM 631164 Skeleton and mounted skin. Adult male. *Gorilla beringei mikenensis* (cotype). Bamboo forest, Mount Mikeno, Virunga volcanoes. Collected by Elias

Arrhenius on 14 May 1914. An old male with worn teeth. The lower incisors and canines are more worn than the upper. The upper left canine has been broken off and the lower right one is completely broken out and the jaw bone has healed so that the alveolus has filled up. Condylobasal length 255 mm; zygomatic width 184.5 mm; braincase width 110.5 mm; width of planum nuchale 165 mm; height of planum nuchale from foramen magnum 102.5 mm; length of upper molar series 72.5 mm; length of M1 15 mm; width of M1 17 mm; length of M2 17.2 mm; width of M2 18.1 mm; length of M3 17.2 mm; width of M3 17.8 mm; length of lower molar series 83.3 mm; inner height of orbit 40 mm; inner width of orbit 40.5 mm; least interorbital width 31.9 mm; cranial capacity 585 cc.

NRM 631168 Skeleton and mounted skin. Juvenile female. *Gorilla beringei mikenensis* (cotype). Mount Mikeno, Virunga volcanoes. Collected by Elias Arrhenius on 14 May 1914. It has the milk dentition and the 1st molar developed.

NRM 631244 Skeleton and skin. Juvenile female. *Gorilla beringei beringei*. Southern slopes of Mount Mikeno, Virunga volcanoes. Collected by Nils Gyldenstolpe on 20 March 1921.

NRM 632582 Skin. Juvenile. *Gorilla beringei beringei*. Mount Mikeno, Virunga volcanoes. Collected by Elias Arrhenius on 14 May 1914.

NRM 640638 Skeleton and skin. Adult male. *Gorilla beringei beringei*. Northwest slopes of Mount Mikeno, Virunga volcanoes. Collected by Nils Gyldenstolpe on 11 March 1921. Cranium fragmented into several parts; mandible intact.

NB: For further details of specimens A 61 0941, A62 0431, A62 0537, A62 1229, A62 1346, A62 1445, A63 0739, A63 0840, A63 1028, A63 1030, A63 1042, A63 1143, A63 1244, and A64 0638 above, see Gyldenstolpe (1928). A documentary film of the expedition was produced by Pathé Frères and released in 1922 under the title 'Med prins Wilhelm på afrikanska jaktstigar'. In English-speaking countries it was known as 'The Cradle of the World'.

NRM 641165 Skeleton and mounted skin. Subadult male. *Gorilla beringei mikenensis* (cotype). Bamboo forest, Mount Mikeno, Virunga volcanoes. Collected by Elias Arrhenius on 14 May 1914. The occipital and sagittal crests are not developed. The upper 3rd molars are not quite fully developed. Condylobasal length 212 mm; zygomatic width 168.5 mm; width of braincase 119 mm; cranial capacity 597 cc.

NRM 641166 Skeleton and skin. Juvenile male. *Gorilla beringei mikenensis* (cotype). Mount Mikeno, Virunga volcanoes. Collected by Elias Arrhenius on 14 May 1914. It has the milk dentition and the 1st molar developed.

Stockholms Universitet, Institutionen för Biologisk Grundutbildning, 106 91 Stockholm

Lars-Olof Hagelin; Bengt Karlsson; Birgitta Tullberg

1765 Skull. Adult male. *Gorilla gina*. Kamerun. Collected by Theorin.

2689 Mounted skeleton. Adult male. *Gorilla gina*. Purchased from Frank, London, in 1898.

5193 Skull. Juvenile female. *Gorilla gina*. Molundu, Kamerun. Purchased from Umlauff in 1915.

Uppsala

Uppsala Universitet, Evolutionsmuseet, Norbyvägen 16, 752 36 Uppsala

Erica Mejlon; Lars Wallin

UPSZMC 162848 Skull. Adult male. Micomeseng territory, Esaguaq, Spanish Guinea. Received in 1964. The upper canines and incisors do not belong to this skull.

UPSZMC 162849 Mounted skeleton. Adult female. It is missing the terminal phalanges, and the skull is defective (the right zygomatic arch is broken off and missing; the right mandibular ramus is broken off and missing; a large area of the left maxilla is broken off and missing; most of the teeth are missing, some of these at least have been lost pre-mortem).

UPSZMC 162850 Mounted skeleton. Adult male.

UPSZMC 162851 Cranium. Female.

SWITZERLAND

Basle

Naturhistorisches Museum Basel, Augustinergasse 2, 4001 Basel

G. Stocker; Urs Wüest

Osteology Department:

Loïc Costeur

NMB-colyn 660 (provisional number only) Skull. Male. *Gorilla beringei graueri*. Lubutu, Zaïre. Collected in 1984.

NMB-colyn Z 5135 (provisional number only) Skull. Female. *Gorilla beringei graueri*. Kasese, Zaïre. Collected in March 1985.

NMB-colyn Z 5216 (provisional number only) Skull. Male. *Gorilla beringei graueri*. Kasese, Zaïre. Collected in July 1985.

NMB-colyn Z 5217 (provisional number only) Skull. Male. *Gorilla beringei graueri*. Kasese, Zaïre. Collected in July 1985.

NMB-1346 Skull (plaster cast). Male. Made and supplied by J. Leven, München, 1867.

NMB-1347 Skull (plaster cast). Female. Made and supplied by J. Leven, München, 1867.

NMB-1348 Foot (plaster cast). Male. Made and supplied by J. Leven, München.

NMB-1349 Hand (plaster cast). Male. Made and supplied by J. Leven, München.

NMB-2713 Skull. Male. Purchased from Gerrard in 1890.

NMB-2891 Mounted skeleton. Male. Donated. The sternum and left clavicle are made of wood. The upper left canine and central incisor are missing.

NMB-3072 Calvarium. Adult male. Cap Lopez, French Congo. It is defective.

NMB-3073 Skull. Adult male. Kumba, British Cameroon.

NMB-3074 Skull. Adult male. Fernan Vaz, French Congo. Purchased in 1948. The cranium is defective.

NMB-3075 Skull. Adult male. French Congo. Purchased from Heini Hediger (1908–1992). The upper right 1st incisor is missing, and the upper left canine is defective.

NMB-3076 Skull. Adult male. Gabon. Both upper 1st incisors and the lower left 2nd incisor are defective. There is a spear wound? to the right cheek. The animal's height was 180 cm.

NMB-3077 Skull. Adult female. French Congo. The lower jaw is fixed to the cranium.

NMB-3078 Upper and lower jaws (plaster cast). Infant. From a specimen which died in 1951 in Zoologischer Garten Basel.

NMB-3079 Plaster cast of P2−M3 sin. and P1−M3 dex. Adult.

NMB-3080 Cranium. Female. Kamerun.

NMB-3765 Skull. Male. Gabon. Received from Missionary Hermann in 1905.

NMB-3955 Skull (plaster cast). Juvenile. Received from the Naturhistorisches Museum, Lübeck, in 1909.

NMB-3956 Skull (plaster cast). Male. Received from the Naturhistorisches Museum, Lübeck, in 1909.

NMB-6294 Skeleton. Male. *Gorilla castaneiceps.* Hinterland of Fernan-Vaz, Gabon. Purchased from Gustav Schneider in 1920. It is from the same individual as the mounted skin NMB-MAMM.3110 in the Zoology Department.

NMB-8025 Skeleton. Female. *Gorilla castaneiceps.* Hinterland of Fernan-Vaz, Gabon. Purchased from Gustav Schneider in 1928. It is missing the proximal end of the left radius, the distal phalange of the left 3rd metatarsal, both patellae, the atlas and the coccyx. It is from the same individual as the mounted skin NMB-MAMM.4119 in the Zoology Department.

NMB-8026 Skeleton. Juvenile female. *Gorilla castaneiceps.* Hinterland of Fernan-Vaz, Gabon. Purchased from Gustav Schneider in 1928. It is missing the 1st and 2nd cervical vertebrae, the distal phalange of the left 5th metatarsal, and both patellae. It is from the same individual as the mounted skin NMB-MAMM. 4120 in the Zoology Department.

NMB-8197 Skull. La Gandja, Haute Sangha, Moyen Congo. Received from Paul Schmid in 1931. There is a supernumerary upper molar. A cast of this skull is also numbered 8197.

NMB-8198 Skull. La Gandja, Haute Sangha, Moyen Congo. Received from Paul Schmid in 1931. A supernumerary upper molar is missing.

NMB-9035 Skull. Juvenile male. Gabon. Donated by Dr Fritz Sarasin (1859–1942) in 1937.

NMB-9036 Skull. Juvenile. Gold Coast (Accra?). Donated by Dr Fritz Sarasin.

NMB-10143 Skeleton. Juvenile. Cameroun. Donated by Zoologischer Garten Basel in 1951.

NMB-10246 Skeleton. Female. Ogooué, Gabon. Purchased from Gustav Schneider in 1952.

NMB-10280 Postcranial skeleton. Male. French Congo. From the estate of preparator Gustav Schneider; obtained from E. Matter, Bern.

NMB-10429 Postcranial skeleton. Male. *Gorilla castaneiceps.*

NMB-10544 Skeleton. Juvenile male. French Congo, 40 km WSW of Souanké, Sangha. Purchased from Zoologischer Garten Basel; donated by the Voluntary Museum Society, Basel, in 1962. It is from the same individual as the skin NMB-MAMM.6044 in the Zoology Department.

Zoology Department:

Denis Vallan

NMB-MAMM.3110 Mounted skin. Male. *Gorilla castaneiceps.* Purchased from Gustav Schneider in 1920; mounted by Ter Meer, Leipzig, in 1922. It is the property of the Voluntary Museum Society, Basel. It is from the same animal as the skeleton NMB-6294 in the Osteology Department.

NMB-MAMM.4119 Mounted skin. Female. *Gorilla castaneiceps.* Purchased by the Friends of the Museum from Gustav Schneider in 1928 (the money collecting was put into effect by the supplier). Mounted by Ter Meer, Leipzig, in 1928. It is from the same animal as the skeleton NMB-8025 in the Osteology Department.

NMB-MAMM.4120 Mounted skin. Juvenile female. *Gorilla castaneiceps.* It was acquired with the above skin, no. 4119, in 1928. It is from the same animal as the skeleton NMB-8026 in the Osteology Department.

NMB-MAMM.5623 Skin. Juvenile female. Yaoundé, Cameroun. Donated by Zoologischer Garten Basel on 11 November 1951. It was aged about 14 months. This is probably 'Bütschi', who arrived at Zoologischer Garten Basel on 17 July 1951, weighing 5.6 kg. She was originally from Libamba Uakak, Cameroun. She died on 26 October 1951.

NMB-MAMM.5678 Skin. Female. Fernan-Vaz, Gabon. Originally from the zoological business of Gustav Schneider, Basel, who sold it to Dr M. Vogel. Donated by E. Matter, Bern, in January 1952.

NMB-MAMM.6044 Skin. Juvenile male. Purchased from Zoologischer Garten Basel on 17 February 1962; donated by the Voluntary Museum Society, Basel. It is from the same animal as the skeleton no. 10544 in the Osteology Department. This is 'Cäsar', who arrived at Zoologischer Garten Basel on 10 February 1962, weighing 10.28 kg, having been acquired from Tissot, Brazzaville. He was captured 40 km WSW of Souanké, Congo, and was acquired by Tissot at Souanké on 9 May 1961. He died of massive inflammation of the small intestine on 17 February 1962. [Some additional information added by GH from correspondence with Basel Zoo personnel.]

NMB-MAMM.8033 Carcase (in alcohol). Infant female. Donated by Zoologischer Garten Basel on 28 June 1984. This is 'Souanké', who was born in Basel Zoo on 23 November 1970 and died of malnutrition (insufficient milk supply) on 29 November 1970. The dam was 'Käti' and the sire was 'Jambo'. 'Käti' had the same origin and acquisition circumstances as 'Cäsar', and weighed 8.56 kg on arrival. 'Jambo' was himself born in Basel Zoo, on 17 April 1961, and was raised by his mother, 'Achilla'. His father was 'Stefi'. [Some additional information added by GH from correspondence with Basel Zoo personnel.]

NMB-MAMM.M08-159 Brain (in formalin). Juvenile female. Received from Zoologischer Garten Basel in 1962. The animal was 1½ years old.

NMB-MAMM.M08-160 Brain (in formalin). Juvenile female. Received from Zoologischer Garten Basel in 1951.

Zoo Basel, Binningerstrasse 40, 4054 Basel

Christian Wenker

NB: The gorilla samples available in Zoo Basel are exclusively EDTA, plasma and serum samples which are stored in their blood bank at −20°C. They do not provide samples for commercial purposes, and each request for scientific studies is carefully evaluated. Serum, plasma: about 1.0 mL per sample; EDTA-blood: about 2.0 mL per sample.

Female 'Goma' (born 23 September 1959 to dam 'Achilla' and sire 'Stefi'): 10 × serum; 6 × EDTA-blood.

Male 'Nangai' (born 8 October 1990 to dam 'Faddama' and sire 'Pépé' or 'Tam-Tam'; sent to Wuppertal Zoo on 30 April 1997; died there on 4 November 2001): 2 × EDTA-plasma.

Female 'Muna' (born 6 March 1989 to dam 'Quarta' and sire 'Pépé' or 'Tam-Tam'; sent to Krefeld Zoo on 30 April 1997): serum.

Female 'Faddama' (born 2 February 1983 to 'Quarta' and 'Pépé'): 28 × serum; 8 × EDTA-blood.

Male 'Vizuri' (born 17 February 1998 to 'Quarta' and 'Pépé'; sent to Moscow Zoo on 11 July 2006): 2 × serum.

Male 'Zungu' (born 4 August 2002 to dam 'Joas' and sire 'Kisoro'): 25 × serum; 10 × EDTA-blood.

Male 'Viatu' (born 20 December 1998 to dam 'Faddama' and sire 'Pépé'; sent to Frankfurt Zoo on 14 October 2008): 3 × serum; EDTA-blood.

Female 'Wima' (born 9 January 1999 to dam 'Joas' and sire 'Kisoro'; sent to Cabárceno Nature Park, Cantabria, Spain, on 16 March 2010; died on 3 September 2010): 3 × serum; 4 × EDTA-blood.

Female 'Chelewa' (born 31 December 2005 to dam 'Wima' and sire 'Viatu'; sent to Cabárceno Nature Park, Cantabria, Spain, on 16 March 2010): 2 × serum; 4 × EDTA-blood.

Female 'Joas' (born 6 July 1989 at Stichting Apenheul, Apeldoorn, Netherlands, to dam 'Kriba' and sire 'Kibabu'; received in Basel Zoo on 6 December 1996): 6 × serum; 5 × EDTA-blood.

Female 'Quarta' (born 17 July 1968 to dam 'Achilla' and sire 'Stefi'): 5 × serum; 5 × EDTA-blood.

Male 'Kisoro' (born 14 March 1989 in Krefeld Zoo, Germany, to dam 'Boma' and sire 'Massa'; arrived at Basel Zoo on 6 December 1996; died on 24 May 2014): 25 × serum; 6 × EDTA-blood.

Female 'Enea' (born 29 November 2004 in Zoologischer Garten Zürich to dam 'Mamitu' and sire 'Ngola'; euthanased on 26 November 2012): 10 × serum; 8 × EDTA-blood; EDTA-plasma.

Male 'M'Tonge' (born 23 February 1999 in Stichting Apenheul, Apeldoorn, Netherlands, to dam 'Tsimi' and sire 'Bongo'; received in Basel Zoo on 18 August 2014): 4 × serum.

Bern

Naturhistorisches Museum der Burgergemeinde, Bernastrasse 15, 3005 Bern

Beatrice Blöchlinger; Peter Lüps; Marc Nussbaumer; H. Sägesser

NMBE 1018418 Skull. Lambaréné, Gabon. Donated by Dr M. Lauterburg on 24 January 1929.

NMBE 1018419 Skull. Juvenile. Lambaréné, Gabon. Donated by Dr M. Lauterburg in January 1929.

NMBE 1018420 Skull. Lambaréné, Gabon. Donated by Dr M. Lauterburg in January 1929.

NMBE 1018421 Skull. Adult male. Donated by Klopfenstein, Thun, on 20 October 1940.

NMBE 1018422 Skull. Donated by G. von Burg on 23 June 1920.

NMBE 1018423 Skeleton and mounted skin (on exhibition). Adult male. *Gorilla beringei graueri.* Tshibinda district, Kivu region, Belgian Congo. Collected and donated by Eric Miville, 1927−29. Chest circumference 180 cm; arm span 280 cm.

NMBE 1018424 Mounted skeleton and skin. Juvenile female. Donated by Zoologischer Garten Basel on 30 October 1980. This is 'Akaba', who was born in Basel Zoo on 23 June 1978 and was reared by her mother, 'Quarta' (herself born in the zoo on 17 July 1968). The sire was either 'Pépé' or 'Migger'. She died on 11 June 1980, weighing 8.5 kg. [Some additional information added by GH from correspondence with Basel Zoo personnel.]

NMBE 1053181 Mounted skeleton. Adult male. Donated in 2002 by the Institut für Anatomie, Universität Bern (their no. A II.a.112). It had originally been purchased from Gustav Schneider, Basel, in July 1877, and was later restored by G. Schneider in 1942. During more recent restoration work, the museum taxidermist found some metal (steel) shrapnel in one of the thoracic vertebrae, and a damaged rib. Also, several pieces of lead were implanted in the skull bones. A specialist who was consulted came to the assumption that the gorilla was first wounded by an explosive device (steel is never used in guns, as it damages the barrel) and then finished off by gunshot (the lead particles in the skull). See Chapter 4, Noninfectious Disease and Host Responses in Part I, Gorilla Pathology and Health. NB: The Institute of Anatomy donated their complete collection of comparative anatomy, comprising approximately 300 animal skeletons and 518 individual bones, to the Natural History Museum as a 'permanent loan' in 2002. This historic collection is displayed in an exhibit that was inaugurated in 2005.

Skull (plaster cast). Adult male. Donated in 2002 by the Institut für Anatomie, Universität Bern (their no. A II. b.66). It was originally purchased from Guérin, Paris.

Skull (plaster cast). Adult female. Donated in 2002 by the Institut für Anatomie, Universität Bern (their no. A II. b.107). It was originally purchased from Leven, München.

Fribourg

Musée d'Histoire Naturelle, Chemin du Musée 6, 1700 Fribourg

André Fasel; Peter Wandeler

1872-1895 Skeleton. Male. Gabon. Purchased for 1300 francs from G.A. Frank, London.

1943-1897 Skeleton. Female. Gabon. Purchased for 675 francs.

7229-1951 Mounted skin. Female. *Gorilla castaneiceps.* Fernan-Vaz, Gabon. Purchased for 1500 francs from

E. Matter, Bern. Mounted by Schlier in 1951, at a cost of 730 francs.

15715-2015 Skull. Adult male. Date of acquisition 2010 or shortly thereafter. It was received from the Département de Biologie, Université de Fribourg, after the Department was restructured and its zoological collection dissolved (the accession number there was Gg 1).

Université de Fribourg, Département de Biologie, Chemin du Musée 10, 1700 Fribourg

Gerolf Lampel; Simon Sprecher

NB: The specimens listed below were originally in the Institute of Comparative Zoological Anatomy, Faculty of Natural Sciences, University of Fribourg. The Institute no longer exists, and its zoological collection, which was transferred to the Department of Biology, was dissolved in a subsequent reorganisation. Parts of the collection were transferred to the Museum of Natural History, Fribourg. However, only 1 gorilla skull (Gg 1) was transferred there; the fate of the remaining 5 skulls remains unknown. They are listed here purely as a matter of historical and scientific interest.

Gg 2 Skull. Adult male. Río Muni. Collected by Dr J. Sabater Pí, and purchased from him in 1963.

Gg 3 Skull. Adult Female. Río Muni. Collected by Dr J. Sabater Pí, and purchased from him in 1963. Both P4 are irregularly positioned, in the lingual direction.

Gg 4 Skull. Adult Female. Mobumuom region, Río Muni. Collected by Dr J. Sabater Pí, and purchased from him in 1963. The I1 on the left side is rotated 45 degrees from normal, and both M1 bear abscesses.

Gg 5 Skull. Juvenile. Río Muni. Collected by Dr J. Sabater Pí, and purchased from him in 1963.

Gg 6 Skull. Adult male. Purchased from Boubée, Paris. It has supernumerary disto-molars (M4).

The dental anomalies of skulls Gg 3, Gg 4, and Gg 6 are described in Krapp and Lampel (1973).

Geneva

Muséum d'Histoire Naturelle, 1 Route de Malagnou, 1208 Genève

Francois J. Baud; Louis de Roguin; Manuel Ruedi

MHNG-MAM-581.018 Skeleton. Adult female. Gabon. Collected by the expedition of A. Marche and de Compiègne in 1873.

MHNG-MAM-618.071 Mounted skin. Juvenile male. Gabon. Collected c.1886. Purchased from Bouvier. It is probably from the same animal as the mounted skeleton 727.053.

MHNG-MAM-653.050 Skull. Male. Gabon. Collected c.1898. Purchased from Tramond, Paris.

MHNG-MAM-727.048 Mounted skeleton. Adult male. Gabon. Collected c.1908. Donated by Société Auxiliaire du Muséum.

MHNG-MAM-727.053 Mounted skeleton. Juvenile male. Collected c.1886. It is most probably from the same animal as the mounted skin MHNG-MAM-618.071.

MHNG-MAM-738.031 Articulated skeleton and mounted skin. Adult female. *Gorilla beringei graueri.* Kigoma, Wabembeland, NW Tanganyika. Collected by Duke Adolf Friedrich of Mecklenburg's German Central African Expedition of 1908. Donated in 1909 by Société Auxiliaire du Muséum.

MHNG-MAM-797.049 Skull. Male. Purchased from Rouppert in 1919. It is now missing.

MHNG-MAM-797.050 Skull. Adult female. Purchased from Rouppert in 1919.

MHNG-MAM-850.014 Humerus and radius. Gabon. Donated by the Mission F. Grébert c.1931.

MHNG-MAM-851.066 Skull. Adult male. Ovan (=Collioura), Ivindo River, Gabon. Purchased by Mr Galley, a missionary in Gabon, during the winter of 1932–33. Height 175 cm; weight (without entrails) 123.6 kg; weight of head 18.5 kg; length of arm 111 cm; weight of arm 14.2 kg; length of leg 72 cm; weight of leg 11.7 kg.

MHNG-MAM-854.088 Articulated skeleton and mounted skin. Adult male. Bipindi, Cameroun. Collected on 16 September 1932.

MHNG-MAM-854.089 Skull, a few other bones, and mounted skin. Adult female. Lelo, 4¼°N, 14¼°E, Batouri district, Cameroun. Collected on 15 February 1933.

MHNG-MAM-854.090 Skull and mounted skin. Juvenile male. Obala, 3¾°N, 14¼°E, Batouri district, Cameroun. Collected on 14 October 1932. The skull is missing. The skin is mounted together with the female MHNG-MAM-854.089, sitting on its back.

The above 3 specimens, MHNG-MAM-854.088–MHNG-MAM-854.090 inclusive, were brought to Rowland Ward in London in December 1934. Purchased from Société Auxiliaire du Muséum.

MHNG-MAM-909.099 Skull. Adult female. Ebolowa, South Cameroun. Collected on 18 April 1952. Obtained from local people by Jean-Luc Perret. Donated.

MHNG-MAM-1495.001 Skull. Adult. Congo. Collected c.1920. Donated by Mrs Haissly in 1977.

Université de Genève, Département de Génétique et Évolution, Unité d'Anthropologie, 12 Rue Gustave-Revilliod, 1211 Genève 4

Mathias Currat; Alicia Sanchez-Mazas; Christian Simon

PO15_2 Skull. Adult male. Gabon.
PO15_3 Skull. Adult male. Gabon.
PO15_4 Skull. Adult female.
PO15_5 Skull. Adult female.
PO15_6 Skull. Adult female. The base of the cranium has been destroyed.
PO15_16 Skull. Adult female. French Congo.
PO15_17 Skull. Subadult female.

La Chaux-De-Fonds

Musée d'Histoire Naturelle de la Chaux-de-Fonds, Avenue Léopold-Robert 63, 2300 La Chaux-de-Fonds

Marcel S. Jacquat; Arnaud Maeder

MHNC_006109 Mounted skin. Adult male. Purchased in the early 1900s.

MHNC_00610 Skull. Juvenile male. Mabiogo, Cameroun. Collected by Dr Albert Monard (1886–1952) in January 1947 (collector's number 120). See Monard (1951).

MHNC_00611 Skeleton. Haute Sanga, Moyen-Congo, French Equatorial Africa. Collected in 1943.

Lausanne

Musée de Zoologie, Palais de Rumine, Place de la Rippone 6, 1014 Lausanne

Olivier Glaizot; Michel Sartori; Claudine Siegenthaler

MZL 21419 Mounted skin. Adult male. Cameroun. Received in exchange for an egg of *Alca impennis*, c.1890.

MZL 21420 Mounted skin. Adult female. Gabon. Collected by Louis Pelot in 1914.

MZL 23603 Skull. Juvenile male. Gabon. Collected and donated by P. Narbel. Received in 1920.

MZL 23678 Skull. Adult male. Libamba, Cameroun. Donated by Bernex in 1951.

MZL 23679 Skull. Adult male. Libamba, Cameroun. Donated by Bernex in 1951.

MZL 23680 Skull. Adult male. Gabon. Collected in 1926 by W. Morton, who donated it in 1951.

MZL 23684 Skull (cast). Adult male. *Gorilla gina.* Donated by Narbel. Received in 1920.

MZL 24048 Skull. Adult male. Gabon. Donated by Morton in 1951.

MZL 24061 Skeleton (cast). *Gorilla gina.*

Neuchâtel

Muséum d'Histoire Naturelle, Rue des Terreaux 14, 2000 Neuchâtel

Celia Bueno; Jean-Paul Haenni

MHNN-94.1337 Mounted skin. Adult male. Origin either 'Congo' or 'Guinée'. It was purchased for £79 1s. in 1881 from the merchant G.A. Frank, London.

MHNN-94.1337A Articulated skeleton. Adult male. Gabon. Acquired in 1863.

MHNN-94.3178 Skull. Adult male. Congo/Haute-Sangha. Collected in 1925. Donated by Walter Kohler-Chevalier, who had obtained it from a professor at the University of Neuchâtel.

St Gallen

Naturmuseum, Museumstrasse 32, 9000 Sankt Gallen

H. Heierli; Lorenzo Vinciguerra

V1687 Mounted skeleton. Adult male. Gaboon. Acquired in 1884 from Hugo von Koppenfels. The skin was mounted by the German taxidermist Kerz; it is no longer in the collection, and its fate is unknown.

V5953 Skull. Male. Both upper canines and 1 lower canine are missing.

V5955 Skull. Adult. It is missing the lower left central incisor.

Winterthur

Naturwissenschaftliche Sammlungen der Stadt Winterthur, Museumsstrasse 52, 8402 Winterthur*

Kurt Madliger

533 Mounted skin. Juvenile male. *Gorilla castaneiceps.* French Congo. Mounted by Ter Meer in 1914. Donated by F. Jmhoof-Blumer on 2 January 1916. Height c. 90 cm.

534 Mounted skeleton. From the same individual as the mounted skin no. 533 above. Mounted by Ter Meer in 1914.

535 Mounted skin. Adult male. *Gorilla gina.* Cameroun. Collected in 1923. Purchased from Umlauff, Hamburg, in 1929. Height c. 150 cm.

536 Skull. *Gorilla gina.*

Zürich

Universität Zürich-Irchel, Anthropologisches Institut und Museum, Bau 10, Winterthurerstrasse 190, 8057 Zürich*

Peter Schmid

AIZ 1224 Skull. Adult female. Received from Linnaea, Berlin, on 31 December 1907.

AIZ 4802 Skull. Adult male. Received on 26 April 1931.

AIZ 4902 Skull. Adult male. Received on 27 June 1931.

AIZ 4903 Skeletal parts (4 cervical and 2 thoracic vertebrae). Adult male. Received on 27 June 1931. They may belong to the skull AIZ 4902 above.

AIZ 5209 Cadaver. Juvenile male. This is 'Bubu', who arrived at Zoologischer Garten Zürich on 3 April 1931, weighing 9 kg. Originally from Cameroun, he was purchased from Herr Schmidt of Basel, with money donated by Herr H.C. Bodmer of Zürich. 'Bubu' died of tuberculosis of the lungs and intestines on 20 January 1935. Received from Zoo Zürich on 15 October 1935. [Some additional information added by GH from correspondence with Zürich Zoo personnel.]

AIZ 5222 Skull. Adult female. Received from Zimmerman on 9 December 1936.

AIZ 5223 Skull. Adult male.

AIZ 5490 Mounted skeleton. Adult male. Iguéla, Ogooué-Maritime, Gabon. Purchased for Fr. 330 from H. Meyer, Schlieren; received on 12 October 1938. It was mounted by A.H. Schultz in March 1952.

AIZ 5491 Skeletal parts (1 hand). Adult male.

AIZ 5562 Skull (sectioned sagittally). Adult male. Port Gentil, Gabon. Received from H. Meyer, Schlieren, on 20 December 1940.

AIZ 5563 Skull. Adult female. Port Gentil, Gabon. Received from H. Meyer, Schlieren, on 20 December 1940.

AIZ 5684 Skull. Adult male. Douala-Sangureima road, Cameroun. Collected in 1948. Prepared by F. Fischer, Zürich. Acquired on 2 April 1949.

AIZ 5685 Skull. Adult female. Douala-Sangureima road, Cameroun. Prepared by F. Fischer, Zürich. Acquired on 2 April 1949.

AIZ 6158 Skull. Infant male. *Gorilla beringei graueri.* Walikale, c. 120 km NW of Lake Kivu, Utu, Belgian Congo. Received from A. Heim on 24 January 1955.

AIZ 6194 Skull, femora, humerus and tibia. Adult female. *Gorilla beringei beringei.* SW of Kisoro, northern foot of the eastern Virunga volcanoes, Uganda. Received from A. Heim on 29 June 1955.

AIZ 6195 Cranium. Infant. *Gorilla beringei beringei.* SW of Kisoro, northern foot of the eastern Virunga volcanoes, Uganda. Received from A. Heim on 29 June 1955.

AIZ 6409 Cadaver (in 10% formalin). Juvenile female. This is 'Epini', who arrived at Zoologischer Garten der Stadt Frankfurt am Main, Germany, on 27 May 1956, aged c. 3–4 years. She came originally from Cameroon. She died on 8 July 1956 (weight 20 kg). Received on 10 July 1956. The facial tissues have been dissected away. [Some additional information added by GH from correspondence with Frankfurt Zoo personnel.]

AIZ 6503 Skull. Adult male. Received on 4 December 1956. It is missing the upper right canine.

AIZ 6504 Skull. Adult male. Cameroun. Received from Phillip J. Carroll, Yaoundé, on 4 December 1956.

AIZ 6592 Skull. Adult female. Cameroun.

AIZ 6593 Skull. Adult female. Cameroun.

AIZ 6594 Skull. Adult female. Cameroun.

AIZ 6595 Skull. Adult female. Cameroun.

AIZ 6596 Skull. Adult female. Cameroun.

AIZ 6597 Skull. Adult female. Cameroun.

AIZ 6598 Skull. Adult male. Cameroun.

AIZ 6599 Skull. Adult male. Cameroun.

AIZ 6600 Skull. Adult female. Cameroun.

AIZ 6601 Skull. Adult female. Cameroun.

AIZ 6602 Skull. Adult male. Cameroun.

AIZ 6603 Skull. Adult male. Cameroun.

AIZ 6604 Skull. Adult male. Cameroun.

The above 13 skulls, AIZ 6592−AIZ 6604 inclusive, were received from Phillip J. Carroll, Yaoundé, on 14 November 1957.

AIZ 6609 Skeleton (incomplete). Juvenile. Cameroun. Received from Phillip J. Carroll, Yaoundé.

AIZ 6610 Skeleton (incomplete). Juvenile. Cameroun. Received from Phillip J. Carroll, Yaoundé.

AIZ 6611 Skeleton (incomplete). Infant. Cameroun. Received from Phillip J. Carroll, Yaoundé.

AIZ 6612 Skeleton (incomplete). Juvenile. Cameroun. Received from Phillip J. Carroll, Yaoundé.

AIZ 6672 Skeleton. Adult male. Equatorial Guinea. Received from J. Sabater Pí on 14 November 1958.

AIZ 6674 Cadaver (internal organs missing). Infant female. Cameroun.

AIZ 6676 Skull. Adult female. Cameroun.

AIZ 6677 Skull. Adult female. Cameroun. It is damaged.

AIZ 6678 Skull. Adult male. Cameroun. It is damaged.

AIZ 6679 Cranium. Adult male. Cameroun.

AIZ 6680 Skeleton. Adult male. Cameroun. Missing terminal phalanges of feet, 4 ribs, and C1.

AIZ 6681 Skull. Infant. Cameroun.

AIZ 6682 Skull. Adult female. Cameroun.

AIZ 6683 Skull. Adult female. Cameroun. It is missing the os occipital.

AIZ 6684 Skull. Adult male. Cameroun.

The above 10 specimens, AIZ 6674−AIZ 6684 inclusive, were received from Phillip J. Carroll, Yaoundé, on 21 November 1958.

AIZ 6699 Skull. Adult female. Equatorial Guinea. Received from J. Sabater Pí on 19 February 1959.

AIZ 6720 Skull. Adult male. Equatorial Guinea. Received from J. Sabater Pí on 19 May 1959. It is damaged.

AIZ 6737 Skull. Adult male. Received from J. Sabater Pí on 1 July 1959.

AIZ 6782 Skeleton. Juvenile male. Equatorial Guinea. Received from J. Sabater Pí on 6 January 1960.

AIZ 6787 Skeleton. Juvenile female. Equatorial Guinea. Received from J. Sabater Pí on 2 February 1960.

AIZ 6788 Skeleton. Juvenile female. Equatorial Guinea. Received from J. Sabater Pí on 2 February 1960.

AIZ 6840 Skeleton. Adult female. Equatorial Guinea. Received from J. Sabater Pí on 26 September 1960.

AIZ 6841 Mandible and postcranial skeleton. Adult male. Equatorial Guinea, on the border with Cameroun. Received from J. Sabater Pí on 17 October 1960. The cranium has been missing since the autumn of 1963.

AIZ 6842 Skeleton. Juvenile female. Equatorial Guinea. Received from J. Sabater Pí on 7 December 1960.

AIZ 6843 Skull. Adult male. Equatorial Guinea. Received from J. Sabater Pí on 7 December 1960.

AIZ 6844 Skull. Infant male. Equatorial Guinea. Received from J. Sabater Pí on 7 December 1960.

AIZ 6845 Skeleton. Infant male. Equatorial Guinea. Received from J. Sabater Pí on 7 December 1960.

AIZ 6846 Skeleton. Juvenile female. Equatorial Guinea. Received from J. Sabater Pí on 7 December 1960.

AIZ 6858 Skull. Juvenile female. Equatorial Guinea. Received from J. Sabater Pí on 17 March 1961.

AIZ 6859 Skull. Adult. Equatorial Guinea. Received from J. Sabater Pí on 17 March 1961.

AIZ 6869 Skull. Adult male. Equatorial Guinea. Received from J. Sabater Pí on 6 July 1961.

AIZ 6870 Skull. Adult female. Equatorial Guinea. Received from J. Sabater Pí on 6 July 1961.

AIZ 6877 Skull. Adult male. Equatorial Guinea. Received from J. Sabater Pí on 27 July 1961.

AIZ 6878 Skeleton. Adult male. Equatorial Guinea. Received from J. Sabater Pí on 27 July 1961. The hands are articulated.

AIZ 6884 Skeleton. Adult male. Equatorial Guinea. Received from J. Sabater Pí on 8 August 1961.

AIZ 6896 Skeleton. Adult female. Equatorial Guinea. Received from J. Sabater Pí on 21 January 1962. The lower left canine was sectioned by G. Schwartz in June 1998.

AIZ 6900 Skeleton. Adult female. Equatorial Guinea. Received from J. Sabater Pí on 21 March 1962. The hands are articulated.

AIZ 6939 Skeleton. Juvenile. Equatorial Guinea. Received from J. Sabater Pí on 21 May 1962.

AIZ 6992 Skull. Adult male. Received from J. Sabater Pí on 27 September 1962.

AIZ 6993 Skull. Subadult female. Received from J. Sabater Pí on 27 September 1962.

AIZ 6994 Skeleton. Juvenile female. Equatorial Guinea. Received from J. Sabater Pí on 27 September 1962.

AIZ 7010 Skeleton. Adult female. Equatorial Guinea. Received from J. Sabater Pí on 3 January 1963. The hands are articulated.

AIZ 7011 Skeleton. Adult male. Equatorial Guinea. Received from J. Sabater Pí on 3 January 1963. The hands are articulated.

AIZ 7035 Skeleton. Subadult male. Received on 22 April 1963.

AIZ 7036 Skeleton. Juvenile female. Equatorial Guinea. Received from J. Sabater Pí on 22 April 1963.

AIZ 7037 Cranium. Adult male. Equatorial Guinea. Received from J. Sabater Pí on 22 April 1963.

AIZ 7038 Skull. Subadult female. Equatorial Guinea. Received from J. Sabater Pí on 22 April 1963.

AIZ 7057 Skeleton. Juvenile female. Equatorial Guinea. Received from J. Sabater Pí on 2 August 1963.

AIZ 7062 Skull. Adult male. Equatorial Guinea. Received from J. Sabater Pí on 24 September 1963.

AIZ 7074 Skull. Juvenile female. Motumoni region, Equatorial Guinea. Received from J. Sabater Pí on 6 November 1963.

AIZ 7075 Skeleton. Adult female. Gabon. Received from J. Sabater Pí on 6 November 1963. The hands are articulated.

AIZ 7079 Skull. Adult male. Ngong region, Equatorial Guinea. Received from J. Sabater Pí on 20 November 1963.

AIZ 7081 Skull. Adult male. Ndjiakom, Nko, Equatorial Guinea. Received from J. Sabater Pí on 20 November 1963.

AIZ 7117 Skull. Adult male. Mbia region, Campo, Equatorial Guinea. Received from J. Sabater Pí on 16 May 1964.

AIZ 7118 Skull. Adult female. Nkin region, Equatorial Guinea. Received from J. Sabater Pí on 16 May 1964.

AIZ 7119 Skull. Subadult male. Campo region, Cameroon. Received from J. Sabater Pí on 16 May 1964.

AIZ 7120 Skull. Subadult male. Mobumuon, Mte Mitra. Received from J. Sabater Pí on 16 May 1964.

AIZ 7121 Skull. Adult male. Received from J. Sabater Pí on 1 June 1964.

AIZ 7122 Skull. Subadult male. Received from J. Sabater Pí on 1 June 1964.

AIZ 7123 Skeleton. Juvenile male. Received from J. Sabater Pí on 1 June 1964. The lower right canine and left M1 were sectioned by G. Schwartz in June 1998.

AIZ 7126 Skull. Juvenile male. Equatorial Guinea. Collected on 27 July 1961. Received from J. Sabater Pí on 15 July 1964.

AIZ 7128 Skeleton. Juvenile male. Niabessan, W of Atonbam, Ntem, Cameroon. Received on 16 July 1964. The mandible is damaged.

AIZ 7129 Skeleton. Infant female. Bitam region, Woleu-Ntem, Gabon. Received on 16 July 1964.

AIZ 7137 Mandible. Adult male. Received on 15 August 1964. The animal died on 21 November 1958.

AIZ 7143 Skull. Adult female. Equatorial Guinea. Received from J. Sabater Pí on 12 October 1964.

AIZ 7144 Skull. Subadult male. Nkin region, Equatorial Guinea. Received from J. Sabater Pí on 26 October 1964.

AIZ 7145 Skull. Subadult female. Nkin region, Equatorial Guinea. Received from J. Sabater Pí on 26 October 1964.

AIZ 7146 Skull. Adult female. Mobumuom region. Received from J. Sabater Pí on 26 October 1964.

AIZ 7148 Pelvis, femur head dex., thorax and C1 – C7. Adult male. Received on 30 October 1964.

AIZ 7200 Skull and 3 caudal vertebrae? Infant male. Received from J. Sabater Pí on 24 May 1965.

AIZ 7203 Skull, left hand, radius and ulna. Adult male. Equatorial Guinea. Received from J. Sabater Pí on 15 June 1965.

AIZ 7204 Skull. Adult male. Equatorial Guinea. Received from J. Sabater Pí on 15 June 1965.

AIZ 7205 Skull. Subadult. Equatorial Guinea. Received from J. Sabater Pí on 15 June 1965. It is pathological (porous).

AIZ 7225 Skull. Adult female. Equatorial Guinea. Received from J. Sabater Pí on 24 June 1965.

AIZ 7279 Cadaver. Juvenile female. Received from the Anatomy Department of Zürich University on 17 January 1966. The animal was about 2–3 years old.

AIZ 7328 Skull. Juvenile. Equatorial Guinea. Received from J. Sabater Pí on 21 February 1966.

AIZ 7329 Four teeth (2 upper molars and 2 premolars). Infant female. Received from Zoologischer Garten Zürich on 15 March 1966.

AIZ 7331 Skull. Adult male. Congo. Received from the Anatomy Department of Zürich University (Bluntschli collection) on 28 March 1966. It is damaged (decalotted).

AIZ 7406 Skull. Adult male. Equatorial Guinea. Received from J. Sabater Pí on 14 April 1966.

AIZ 7407 Skull. Adult male. Equatorial Guinea. Received from J. Sabater Pí on 14 April 1966.

AIZ 7408 Skull. Juvenile. Equatorial Guinea. Received from J. Sabater Pí on 14 April 1966.

AIZ 7420 Skull. Adult female. Equatorial Guinea. Received from J. Sabater Pí on 29 April 1966.

AIZ 7487 Skeleton. Adult male. Equatorial Guinea. Received on 22 November 1966. It is missing C1, distal foot elements sin., and digital phalanges.

AIZ 7488 Skull. Adult female. Equatorial Guinea. Received from J. Sabater Pí on 22 November 1966. It is missing the upper left incisor.

AIZ 7510 Skull. Adult female. Equatorial Guinea. Received from J. Sabater Pí on 29 December 1966. The left mandibular ramus is damaged.

AIZ 7604 Skull. Adult female. Equatorial Guinea. Received from J. Sabater Pí on 29 May 1967. The left canine is missing.

AIZ 7611 Skeleton. Juvenile male. Equatorial Guinea. Received from J. Sabater Pí on 27 July 1967. The pelvis is a wet specimen.

AIZ 7649 Skull. Adult male. Equatorial Guinea. Received from J. Sabater Pí on 4 January 1968.

AIZ 7650 Skull. Adult male. Equatorial Guinea. Received from J. Sabater Pí on 4 January 1968. The lower right 2nd incisor is missing.

AIZ 7658 Skeleton. Juvenile male. Equatorial Guinea. Received from J. Sabater Pí on 13 March 1968.

AIZ 7660 Skeleton. Infant male. Equatorial Guinea. Received from J. Sabater Pí on 13 March 1968.

AIZ 7666 Skull. Adult male. Equatorial Guinea. Received on 9 July 1968.

AIZ 7667 Skull. Adult female. Equatorial Guinea. Received on 9 July 1968. It is damaged (decalotted).

AIZ 7689 Skull. Subadult female. Equatorial Guinea. Received from J. Sabater Pí on 10 September 1968. It is missing the 1st and 2nd right incisors, left canine, and 3rd molar.

AIZ 7690 Skull. Adult female. Equatorial Guinea. Received from J. Sabater Pí on 10 September 1968. It is damaged (decalotted).

AIZ 7692 Skull. Adult male. Equatorial Guinea. Received from J. Sabater Pí on 10 September 1968. The upper 1st incisor and lower right incisor are missing.

AIZ 7693 Skull. Adult female. Equatorial Guinea. Received from J. Sabater Pí on 10 September 1968.

AIZ 7694 Skull. Adult female. Equatorial Guinea. Received from J. Sabater Pí on 10 September 1968. The left incisor is missing; the 2nd right incisor is severely abraded.

AIZ 7908 Skull. Adult male. Received from O. Schlagenhaufen on 9 October 1970.

AIZ 7910 Skull. Adult male. Received on 15 October 1970.

AIZ 8836 Skeleton (mummified). Juvenile female. Received on 8 November 1978.

AIZ 9290 Cadaver. Neonate. Received from Prof. Schmitt, Tübingen, on 14 January 1981. It had been born in Zoologischer Garten der Stadt Frankfurt am Main. The left incisors, premaxilla and nasal cavity (vomer) have been removed.

AIZ 9348 Cadaver (in 4% formalin). Adult male. This is 'Stefi', who arrived at Zoologischer Garten Basel on 16 September 1954, weighing 36 kg. He came originally from Matamba, Cameroun. He was received in exchange from Columbus Zoological Gardens, Powell, Ohio, USA, through the offices of L. Ruhe, Gelsenkirchen. He had arrived at the Columbus Zoo (where he was known as 'Christopher') on 8 January 1951 (weight 18 lb.), along with another male ('Macombo') and a female ('Christina'). All three were captured by William Said in Cameroun, and purchased for $10,000 from Ruhe, Hannover. 'Stefi' died on 6 September 1981 (weight 118 kg); the cause was thought to be meningitis. Received on 7 September 1981. The neck was dissected, eyes removed, and facial tissues dissected away by J. Gurchie in June 1994. [Some additional information added by GH from correspondence with Basel Zoo and Columbus Zoo personnel.]

AIZ 9637 Cadaver (frozen). Infant female. Received from Zoologischer Garten Zürich on 25 October 1982. This is 'Donga', who was born in Zologischer Garten Basel on 16 April 1981. She was sold to Zürich Zoo and arrived there on 18 May 1981. She died of strongyloidiasis on 19 October 1982 (weight 8.5 kg).

AIZ 9787 Skeleton. Subadult female. Fernan-Vaz, Gabon. Received on 13 December 1983 from Museum Bally-Prior, Oltenerstrasse 80, 5012 Schönenwerd. It was originally prepared by Schneider in 1926.

AIZ 9788 Skeleton. Juvenile. Fernan-Vaz, Gabon. Received on 13 December 1983 from Museum Bally-Prior, Oltenerstrasse 80, 5012 Schönenwerd. It was originally prepared by Schneider in 1926.

AIZ 9964 Skeletal components. Infant. Received on 17 April 1986?

AIZ 9965 Skeletal components. Infant. Cameroun. Received from Phillip J. Carroll, Yaoundé. The specimen was packed in a plastic bag with AIZ 6681, and probably belongs to it.

AIZ 10045 Postcranial skeleton. Fetus. Received on 5 September 1986.

AIZ 10074 Ears. Juvenile male. Received on 10 September 1986.

AIZ 10144 Cadaver. Neonate male. Received from Zoo Zürich on 10 July 1987. It had been stillborn on 22 June 1987. The dam was 'Nache' and the sire was 'Ngola'. The cause of death was interuterinary aspiration of placental fluid and entanglement of the umbilical cord. Weight 2400 g. The head was removed from the cadaver in August 1993 to be displayed in the Anthropology Museum.

AIZ 10145 Placenta. Received from Zoo Zürich on 6 July 1987. It is from the above specimen, AIZ 10144. Weight 324 g.

AIZ 10148 Skeleton. Infant male. Received from Zoo Zürich on 24 September 1987. This is 'Moja', who was born on 5 May 1986 after a gestation period of 252 days. The dam was 'Mamitu' and the sire was 'Ngola'. The infant died on 17 September 1987 of megacolon, fibrosis of the dura mater and circulatory collapse. [Some additional information added by GH from correspondence with Zürich Zoo personnel.]

AIZ 10203 Placenta. Received from Zoo Zürich on 10 June 1988.

AIZ 10217 Placenta (frozen). Received from Zoo Zürich on 28 November 1988.

AIZ 10218 Skeleton. Neonate female. Received from Zoo Zürich on 5 December 1988. This is 'Onji', who was born on 26 November 1988. The dam was 'Inge' and the sire was 'Ngola'. Birth weight 1900 g. The infant was not accepted by the mother, and underwent surgery for blocked intestines, but did not survive, dying of peritonitis and pleuritis on 2 December 1988. Weight at death 1750 g. [Some additional information added by GH from correspondence with Zürich Zoo personnel.]

AIZ 10288 Cadaver (frozen). Infant male. Received from Zoo Zürich on 7 August 1989. This is 'Pendo', who was born on 17 May 1989 after a gestation period of 242 days. The dam was 'Mamitu' and the sire was 'Ngola'. The infant was mother-reared, but died on 6 August 1989. [Some additional information added by GH from correspondence with Zürich Zoo personnel.]

AIZ 10544 Skeleton and dermoplastic of hand, right foot and face. Infant male. Received from Zoo Zürich on 14 December 1992. This is 'Sitawi', who was born on 15 August 1992 and was reared by his mother, 'Inge'. The sire was 'Ngola'. The infant died of multiple traumata (rupture of liver, massive haemorrhage in liver and lung, abdominal cavity) on 12 December 1992. [Some additional information added by GH from correspondence with Zürich Zoo personnel.]

AIZ 10616 Skeleton. Subadult female. *Gorilla beringei beringei*. Received from Schwarz on 30 May 1989.

AIZ 10769 Skull. Adult male. Received from Schwarz on 30 May 1989.

AIZ 11307 Fetus (in alcohol). Male. Received from Zoo Zürich on 3 September 1993. It was spontaneously aborted at 3 months by 'Mamitu' on 1 September 1993.

AIZ 11428 Cadaver (frozen). Infant female. Received from Zoo Zürich on 4 July 1994. This is 'Uzima', who was born on 15 February 1994 and died of salmonellosis on 2 July 1994, weighing 2280 g. The dam was 'Sandra' and the sire was 'Ngola'.

AIZ 11440 Cadaver (frozen). Infant female. Received from Zoo Zürich on 19 June 1995. This is 'Vijana', born on 5 June 1995 to 'Sandra'. The infant died on 16 June 1995 of pneumonia after the aspiration of amniotic fluid (weight 1800 g).

AIZ 11455 Skeleton. Fetal female. Received from Zoo Zürich on 9 April 1996. It was stillborn to 'Sandra' on 27 March 1996.

AIZ 12101 Fetus (frozen). Male. Received from Zoo Zürich on 4 January 1999. It was aborted by 'Sandra' on 22 December 1998 (weight 390 g). The sire was 'Ngola' (copulations observed on 5 April 1998 and 27 May 1998).

AIZ 12242 Skeleton. Infant male. Received from Zoo Zürich on 14 January 1998. This is 'Tambo', born on 31 October 1993 to 'Inge' and 'Ngola'. He died of a fracture

of the skull on 9 February 1994 (weight 4.5 kg). The skull has been opened and the sternum dissected.

AIZ 12498 Skull. Adult male. Received from the Institute of Zoology, Fribourg, on 24 February 1998. The cranium was opened at the Musée d'Histoire Naturelle, Genève.

AIZ 12499 Skull. Adult female. Río Muni. Received from the Institute of Zoology, Fribourg, on 24 February 1998. Collected by J. Sabater Pí, received from Bata on 3 January 1963.

AIZ 12500 Skull. Adult female. Río Muni. Received from the Institute of Zoology, Fribourg, on 24 February 1998. Collected by J. Sabater Pí, received from Bata on 3 January 1963.

AIZ 12501 Skull. Adult female. Received from the Institute of Zoology, Fribourg, on 24 February 1998. The cranium was opened at the Anthropological Laboratory, Amsterdam.

AIZ 12502 Skull. Adult female. Received from the Institute of Zoology, Fribourg, on 24 February 1998. The cranium was opened at the Anthropological Laboratory, Amsterdam.

AIZ 12503 Skull. Adult female. Received from the Institute of Zoology, Fribourg, on 24 February 1998. The cranium was opened at the Anthropological Laboratory, Amsterdam.

AIZ 12504 Skull. Subadult male. Received from the Institute of Zoology, Fribourg, on 24 February 1998. The cranium was opened at the Anthropological Laboratory, Amsterdam.

AIZ 12505 Skull. Adult female. Received from the Institute of Zoology, Fribourg, on 24 February 1998. The cranium was opened at the Anthropological Laboratory, Amsterdam. Lower left canine sectioned by G. Schwartz in June 1998.

AIZ 12506 Skull. Infant male. Obala, Batouri district, Cameroun. Received from the Institute of Zoology, Fribourg, on 24 February 1998. The cranium was opened at the Musée d'Histoire Naturelle, Genève.

AIZ 12507 Skull. Infant. Río Muni. Received from the Institute of Zoology, Fribourg, on 24 February 1998. Collected by J. Sabater Pí, received from Bata.

AIZ 12518 Skull. Subadult male. Received from the Institute of Zoology, Fribourg, on 15 October 1998. Cranium opened at the Anthropological Laboratory, Amsterdam.

AIZ 13651 Endocast. Adult. Received from the Institute of Zoology, Fribourg, on 15 October 1998.

AIZ 13652 Endocast. Adult. Received from the Institute of Zoology, Fribourg, on 15 October 1998.

AIZ 13653 Endocast. Subadult. Received from the Institute of Zoology, Fribourg, on 15 October 1998.

AIZ 13654 Endocast. Adult. Received from the Institute of Zoology, Fribourg, on 15 October 1998.

AIZ 13655 Endocast. Adult. Received from the Institute of Zoology, Fribourg, on 15 October 1998.

AIZ 13656 Endocast. Adult. Received from the Institute of Zoology, Fribourg, on 15 October 1998.

NB: The specimens listed below form the A.H. Schultz collections. The collections numbers have the prefixes AS and PAL.

AS 26 Skeleton. Juvenile female. *Gorilla beringei beringei*. Captive animal, weight 72.576 kg. Foot missing.

AS 40 Cadaver. Fetus.

AS 48 Skull. Infant male. Acquired by A.H. Schultz from a taxidermist in München in 1922. 'Claimed to have weighed 11 kg.' Deciduous dentition (DC max. almost erupted). Missing CLI, I2 RE mand.

AS 300 Cadaver parts (hand, foot sin., ears, thorax and death mask). Infant female.

AS 760 Skull. Adult male. Cameroon. Foramen magnum damaged. Acquired by A.H. Schultz from the Kny-Scheerer Co., New York, NY, USA.

AS 764 Skull. Juvenile female. First molars erupting. Missing 2nd lower right incisor.

AS 917 Skull. Adult male. Cameroun. Missing I1 mand. dex. Dentition slightly abraded.

AS 1444 Skull. Adult male. Campo River, extreme SW coast of Cameroun. Acquired by A.H. Schultz from J.W. Underhill in October 1936. The incisors are missing.

AS 1604 Skull. Adult male. Mbimow, Haute Sangha district, French Congo. It was received from Ward's (Rochester, NY, USA) in exchange for 29 other mammalian skulls in 1939. Os zygomaticum damaged; sagittal crest with right asymmetry.

AS 1605 Skeletal parts (humerus, radius, ulna and scapula dex.). Adult male. Acquisition same as for AS 1604 above.

AS 1648 Skull. Adult male. Cameroun. Purchased by A.H. Schultz in 1936. Missing I, C max./L1, C mand. Pathology: additional maxillary molar.

AS 1689 Skull. Juvenile male. Cameroun. Purchased by A.H. Schultz in 1943. Occipital and foramen magnum damaged. Porous bones. M3 max. breaking through.

AS 1690 Skeleton. Adult male. Cameroon. Acquired from W.E. Buck in 1943. Missing 2 distal phalanges (hands), 6 phalanges (feet), os trapezoideum sin. Severely abraded dentition. Pathology: I1 max.

AS 1691 Skull. Adult female. Cameroon. Acquired by A.H. Schultz in 1943. Mandible damaged. Missing I max.; I1; C, M2 dex./mand.; M3 sin./mand.

AS 1692 Cranium. Adult female. Cameroon.

AS 1693 Cranium. Juvenile male. Cameroon. Acquired by A.H. Schultz in 1943. The occipital and foramen magnum are damaged. Dentition: without C, M3. Missing I1, 2 sin./max.

AS 1696 Skeleton (incomplete); hand, foot, and ear sin. (wet specimen). Subadult female.

AS 1698 Skeleton (missing left hand and foot); hand and foot (wet specimen in formalin). Infant female. Deciduous dentition. Captive. Weight 4660 g; length 234 mm.

AS 1765 Skull. Adult female. Cameroon. Exchanged against AS 162, 676, 1178, 1450, 1656. On permanent loan C.E. Snow, Anthropology/Kentucky 1948. Missing: I max.; C max. sin., I2 mand. dex. Pathology: sutura metopica u.a.

AS 1880 Skull. Adult male. S Cameroon. Received in exchange from Prof. Haberer, US National Museum (no. 176212). Dentition: I1, M2 mand.

AS 2145 Cadaver part (left foot).

AS 2146 Cadaver parts. Male. Dja Posten, Cameroun. African Expedition of Columbia University and American Museum of Natural History, 1930, H.C. Raven III.

AS 2148 Thorax. Female.

AS 2149 Death mask. Subadult male. Died in New York, NY, USA, in 1921.

AS 2152 Death mask. Juvenile male. Died in 1915, received in that year from the American Museum of Natural History, New York.

AS 2229 Death mask. Adult female. Specimen was aged 13 years at death.

AS 2231 Death mask. Adult male. Received from the American Museum of Natural History, New York.

AS 2234 Facial cast. Adult.

PAL 1 Skeleton. Adult male. Equatorial Guinea. Postcranial incomplete: missing atlas, sternum, patella, humerus, and scapula sin., radius dex. Some of the hand and foot bones must belong to PAL 2. Orbital and mastoid damaged. Pathology: fractures of ulna and phalanges. Dentition: I1, 2 mand. dex.

PAL 2 Skeleton. Adult female. Equatorial Guinea. From hands and feet only carpus sin., tali, calcaneus sin., 4 metacarpals, 5 phalanges. Damage to calotte, occiput and temporal region. Pathology: unhealed fracture of ulna and thumb. Dentition: I1 sin.

PAL 3 Skeleton. Juvenile. Equatorial Guinea. Sacrum damaged. Dentition: M1.

PAL 5 Skeleton. Infant female. Cameroon. Hand: dist. phal., med. phal. V sin., all dex. Foot: med. phal. I, II, V; distal phal. sin.; med. dist. phal. dex.; several tarsals. Pathology: 5 mand. incisors. Dentition: complete deciduous (I2, C sin. mand. damaged).

PAL 6 Skeleton. Juvenile female. Cameroon. Damaged angulus mandibula. Missing several phalanges. Right tarsus glued together. Dentition: M3 not fully in occlusion; damaged C dex. max.

PAL 8 Skeleton. Adult male. South Cameroon. Hands: distal phalanges I, III, os triquetrum dex. Feet: only phalanges missing. Damaged sphenoid, incisors, canines, P3 sin. max., M1 dex. mand. Massive nuchal-sagittal fusion.

PAL 10 Skeleton. Juvenile female. Cameroon. Missing 3 phalanges of foot. Tarsus glued together. Dentition: M3 maxillary breaking through.

PAL 11 Skeleton. Adult male. Cameroon. Missing tarsus. Robust sagittal/nuchal crest fusion.

PAL 12 Skeleton. Adult male. Cameroon. Damaged right eye socket, cracks in calotte. Pathology: C/P3 mandibular. Dentition: I1 dex. max.

PAL 13 Skeleton. Adult male. Cameroon. Missing: dist. phalanges II, III foot sin., dist. phal. I foot dex. Tarsus mounted. Nuchal/mastoid damaged; radius dex. glued together. Pathology: asymmetrical maxillary arch. Dentition: I dex. max.

PAL 14 Skeleton. Adult female. Cameroon. Cavum nasi damaged.

PAL 233 Cranium. Adult female. Cameroun. Damaged maxilla, teeth, os palatinum. Dentition: I, C max.

PAL 234 Cranium. Adult female. Cameroun. Damaged maxilla and teeth (C, P3 sin., M1 dex., alveoli molaris dex.).

PAL 235 Cranium. Juvenile male. Cameroun. Teeth missing except M1.

PAL 236 Skull. Juvenile male. Cameroun.

PAL 237 Skull. Juvenile male. Cameroun. Pathology: wormian. Dentition: secondary, including M2, eruption P4 mand.

PAL 238 Skull. Adult male. Cameroun. Slight right asymmetry of sagittal crest. Incisors missing.

PAL 239 Skull. Adult male. Cameroun. Damage: occipital and foramen magnum. Pathology: M3, P4 mand.

PAL 240 Skull. Adult male. Cameroun. Incisors missing. Pathology: P3.

PAL 241 Skeleton. Adult male. Cameroun. Incisors missing. Only left hand is complete.

PAL 242 Cranium. Adult male. Cameroun. Incisors missing.

PAL 243 Skull. Adult male. Cameroun. Damage to base of skull and os zygomaticum sin. Some teeth missing. Dentition: I1 max., mand. sin.; I2 mand. dex.

PAL 244 Cranium. Adult male. Cameroun. Damage: C, M1, M2 max. dex. Pathology: M3 > sphenoid/maxilla dex. Dentition: I, M3 max.

PAL 245 Cranium. Adult male. Cameroun. Damage: maxilla, os zygomaticum and mastoid. Dentition: I, C, M3 max. sin.

PAL 246 Skull. Adult male. Cameroun. Damage: mastoid, occipital, os zygomaticum, maxilla, and mandible (last without ascending ramus). Dentition: I1, C max., M3 max. dex.

PAL 247 Cranium. Adult male. Cameroun. Damage: mastoid, occipital and maxilla. Pathology: exostosis orbital torus. Dentition: excl. M2/M3 max. dex.

PAL 248 Skull. Adult male. Cameroun. Damage: occipital rami. Dentition: I, C.

PAL 249 Skull. Juvenile male. Cameroun. Damage: margo orbitalis lateralis sin. Pathology: porous. Dentition: second, including P3.

PAL 250 Skull. Adult male. Cameroun. Occipital damaged. Conspicuous sagittal/nuchal crests. Dentition hardly abraded: I1 max.

PAL 252 Skull. Adult male. Cameroun.

Universität Zürich, Institut für Veterinärpathologie, Winterthurerstrasse 268, 8057 Zürich

Nicole Borel; Peter Ossent

S87-1121 Blocks and sections of heart, lung, liver, kidney, spleen, gut and placenta. Stillborn male. It was stillborn in Zoo Zürich on 22 June 1987, weighing 2.73 kg. The dam was 'Nache' and the sire was 'Ngola'. The cause of death was intrauterine asphyxia (aspiration of placental fluid; entanglement of umbilical cord). NB: blocks a (liver, spleen, kidney), b (lung, heart) and c (intestine, placenta) are available in the archive. [Some additional information added by GH from correspondence with Zürich Zoo personnel.]

S87-1729 Blocks and sections of brain and large intestine; all inner organs in formalin. Infant male. This is 'Moja', who was born in Zoo Zürich on 5 May 1986, after a gestation period of 252 days, and was reared by his mother 'Mamitu' (sire 'Ngola'). He died of chronic megacolon on 18 September 1987. Weight 5.7 kg; length crown to rump 28 cm. NB: blocks a−u (brain, large intestine) are available in the archive, but the formalin material appears to have been lost. [Some additional information added by GH from correspondence with Zürich Zoo personnel.]

D 4728 Blocked up organs and sections of lung, liver, gut × 4, mesenteric lymph, brain × 7, heart, spleen, kidney, pancreas and bone marrow. Juvenile female. This is 'Donga', who was born in Zoologischer Garten Basel on 16 April 1981 and was sold to Zoo Zürich, where she arrived on 18 May 1981. She died of strongyloidiasis on 19 October 1982, weighing 7 kg. See Fehlman et al. (1983). NB: No blocks remain in the archive.

Universität Zürich, Zoologisches Museum, Karl Schmid-Strasse 4, 8006 Zürich

Caesar Claude; Barbara Oberholzer

ZMZ 11781 Mounted skeleton. Subadult male. Lambaréné, Gabon. Collected by Dr Albert Schweitzer. Donated by Prof. Strohl in February 1932.

ZMZ 11870 Skull. Adult male. Purchased from G. Schneider, Basel, in 1882.

ZMZ 11871 Skull. Juvenile female? Lambaréné, Gabon. Purchased from G. Schneider, Basel, in 1905.

ZMZ 11872 Skull. Adult male. Lambaréné, Gabon.

ZMZ 11873 Mounted skin. Adult male. Acquired from Frank, London, in 1884. Owing to its poor condition, it was discarded in c.1950.

ZMZ 11874 Mounted skeleton. Juvenile female. Gomo, Gabon. Collected by Ernest Leu. Donated by E. Bürli, Zürich, in 1922. The animal was approximately 5 years old.

ZMZ 11875 Mounted skeleton. Adult female. Gomo, Gabon. Purchased from Walter Leu, Emmishofen, on 30 October 1923. It is presumed to be the mother of the above specimen, no. ZMZ 11874.

ZMZ 11876 Mounted skeleton. Adult male. French Congo. Purchased from G. Schneider, Basel, in 1909.

ZMZ 11877 Postcranial skeleton (incomplete). Adult male. Cap Lopez, Gabon. Collected on 28 June 1911. Acquired from G. Schneider, Basel, in August 1930. It lacks the extremities.

ZMZ 11878 Skull. Adult male. Received in 1947.

ZMZ 11879 Skull. Adult male. Lambaréné, Gabon. Donated by Prof. Strohl, Zürich, in 1922.

ZMZ 11880 Mounted skeleton. Adult male. Received in 1892.

ZMZ 11881 Skull. Juvenile female? Lambaréné, Gabon. Received in 1922.

ZMZ 13270 Mounted skeleton and skin. Subadult male. *Gorilla beringei graueri*. Kahuzi region, Democratic Republic of Congo. Collected by Uwe Goepel on 13 June 1964. Received from the Jardin Zoologique, Tshibati. Head and body 140 cm; foot length 29 cm; ear 5 cm; weight 112 kg.

ZMZ 13840 Mounted skeleton and dermoplastic. Adult male. Spanish Guinea. Received from Circus Knie, Rapperswil. This is 'Tarzan', who arrived at Knie's Kinderzoo on 7 March 1961, aged c. 8 years. He was purchased from L. Ruhe, who had acquired him from Parque Zoologico de Barcelona, Spain, on 10 July 1959. He died on 4 November 1967. Head and body 112 cm; foot length 27 cm; ear 5.5 cm; weight 189 kg. [Some additional information added by GH from correspondence with R. Knie.]

ZMZ 14262 Skeleton and skin. Infant male. Received, in a dissected condition, from the Institut für Veterinärpathologie, Zürich, on 17 October 1967. It had belonged to Karl Künzler, an animal trainer and dealer in Romanshorn. It died as a result of alimentary tract salmonellosis on 17 September 1967, when 8 months old. Head and body 61 cm; foot 15.3 cm; ear 3.9 cm; weight (without viscera and brain) 5.78 kg.

ZMZ 14770 Mounted skin. Adult male. Gabon. Donated by the Society of Natural Sciences, St Gallen, in March 1973. It dates back to c.1860—90.

ZMZ 125336 Brain (in alcohol). Adult female. Lambaréné, Gabon.

Zoo Zürich AG, Zürichbergstrasse 221, 8044 Zürich

Gabriela Hürlimann

Some frozen blood samples are held (serum and heparin plasma):

Male 'N'Gola' (date of birth 21 June 1977). Collected 1 June 2006: 3 × serum. Coll. 6 March 2012: 2 × EDTA-blood; heparin-plasma; heparin. Coll. 15 August 2013: 4 × heparin-plasma; 2 × serum.

Female 'Ybana' (d.o.b. 20 July 1998). Coll. 24 May 2005: serum; plasma.

Female 'Yangu' (d.o.b. 2 July 1998). Coll. 28 January 2005: 3 × serum. Coll. 21 February 2005: serum/whole blood.

Female 'Inge' (d.o.b. 2 March 1980). Coll. 13 May 1997: serum.

Male 'Bonsega' (d.o.b. 9 September 2001). Coll. 4 January 2010: serum. Coll. 19 January 2010: 2 × serum.

Male 'Binga' (d.o.b. 11 December 2001). Coll. 19 January 2010: 2 × EDTA-blood; 2 × serum.

Male 'Azizi' (d.o.b. 17 May 2000). Coll. 27 July 2008: 2 × serum.

Female 'N'Yokumi' (d.o.b. 27 February 2001). Coll. 27 June 2011: 2 × serum.

Female 'Enea' (d.o.b. 29 November 2004). Coll. 31 January 2012: 3 × plasma; serum.

Female 'Haiba' (d.o.b. 15 July 2007). Coll. 6 December 2013: 3 × serum; EDTA-blood. Coll. 12 January 2016: 2 × serum.

Female 'Eyenga' (d.o.b. 18 July 2004). Coll. 3 August 2012: EDTA-blood.

Female 'Habibu' (d.o.b. 29 July 2007). Coll. 5 August 2014: 3 × serum; EDTA-blood. Coll. 12 September 2014: 2 × serum.

Female 'Mahiri' (d.o.b. 13 August 2012). Coll. 3 June 2014: serum; EDTA-blood. Coll. 11 June 2014: serum.

UGANDA

Entebbe

Conservation Through Public Health, Plot 3 Mapeera Close, Uringi Crescent, Entebbe

Stephen Rubanga

Samples available at GRC (Gorilla Research Centre) Freezer Bank, Bwindi; Veterinary Store, Entebbe and NADDEC (National Animal Disease Diagnostics and Epidemiology Centre) Laboratory, MAAIF (Ministry of Agriculture, Animal Industry and Fisheries), Entebbe:

Approximately 6000 dung samples, analysed with different laboratory methods, are preserved for future research and study. Samples are preserved fresh or frozen, fixed in 10% formalin, 96% ethanol or RNA*later* solution. Both sexes and all age classes are represented. The dung has been collected from different habituated groups since 2005 to date. CTPH and rangers from BINP (Bwindi Impenetrable National Park) started collecting dung from a few habituated gorilla groups, and the numbers have increased as more groups have been habituated over the years. Through collecting gorilla dung, CTPH initiated an early warning system for disease outbreaks among wildlife, livestock and humans. The results from laboratory examination of the dung are used to make recommendations to Park management regarding their decision-making and veterinary interventions. Fifty-one representative samples, collected during the 2011 gorilla census, are kept separately at room temperature, specifically to guide the outcome of the census results and rule out any possible cross-infection between wild and habituated gorillas.

Organs/tissues (liver, spleen, heart, skin, muscle, kidney, tongue, nodules and lymph nodes) from an infant gorilla are fixed in 10% formalin and kept at room temperature.

DNA-sequenced materials from gorilla faeces are stored in a deep freezer for future research and study.

In addition, sera from cattle, goats, sheep and dogs are kept for future study and use in various local community programmes.

UNITED KINGDOM

Aberdeen

University of Aberdeen, Zoology Museum, Tillydrone Avenue, Aberdeen AB24 2TZ

Jennifer Downes

ABDUZ:20131 Partial skull.
ABDUZ:20132 Skull.
ABDUZ:20133 Skull.
ABDUZ:20134 Cranium (in pieces).
ABDUZ:41875 Mounted skeleton. Adult male. Transferred from the Anatomy Department in 2002.
ABDUZ:41877 Mounted skeleton. Adult female. Transferred from the Anatomy Department in 2002.
ABDUZ:72337 Lower jaw. Some teeth missing.
ABDUZ:72341 Skull. Full set of teeth; lower jaw wired to upper.
ABDUZ:72342 Skull. Adult female. Duala, Kamerun. Prepared and presented by Dr R. Semple, of the West Africa Medical Staff, in 1916. A few teeth missing; lower jaw wired to upper. Plastic rod glued in place of top vertebra.

Aberystwyth

University of Wales, Institute of Biological Sciences, Edward Llwyd Building, Aberystwyth SY23 3DA

John Barrett; Gareth Owen

Mounted skeleton. Adult male. Purchased from T. Gerrard & Co. Ltd, 46–48 Pentonville Road, London N1, in 1959.

Bangor

Bangor University, School of Biological Sciences, Museum of Natural History, Brambell Building, Deiniol Road, Bangor LL57 2UW

R.W. Arnold; Rosanna L. Robinson; Wolfgang Wüster

433 Skull. Juvenile male.
435 Skull. Subadult male.
452 Skull. Juvenile.
453 Skull. Male.
1149 Mounted skeleton. Subadult male. One uncut wisdom tooth in the left upper jaw gives 6 cheek teeth on that side; there are the normal 5 on the right side.

Bekesbourne

Howletts Wild Animal Park, Bekesbourne, Kent CT4 5EL; & Port Lympne Reserve, Nr Ashford, Kent CT21 4LR

Jane Hopper

Since 2007, Howletts and Port Lympne Wild Animal Parks have done all their postmortem examinations in-house. The protocol includes sections of all tissue types taken and placed into formalin for histopathology using an external laboratory. A duplicate set of tissues is also retained in formalin at the zoo. The stored tissue samples are usually 1- to 2-cm cubes. These stored samples are recorded on a database for further clinical testing or research purposes. EDTA blood samples are preserved on FTA cards stored at room temperature. These are also recorded on a database.

Belfast

Ulster Museum, Botanic Gardens, Belfast BT9 5AB

Terry Bruton; Angela Ross

BELUM.Lh100063 Skull. Adult male. Purchased from Gerrard & Haig Ltd, East Preston, West Sussex, in February 1972. It has what appears to be a bullet hole above the left ear canal, exiting from the right side of the occipital bone.

BELUM.Lh3001 Mounted skeleton. Adult male. Received from Belfast Zoological Gardens, Antrim Road, Belfast, in 1997. This is 'Kéké', who arrived at the Belfast Zoo on 1 August 1991. He was received from Arnhem Zoo, on loan from the Foundation for the Protection of Gorillas and other Threatened Species. He died on 27 March 1997.

Birchington

Powell-Cotton Museum, Quex Park, Birchington, Kent CT7 0BH

L.R. Barton; Hazel Basford; Inbal Livne; K.W. Nicklin

NB: The gorilla specimens housed here were collected in West Africa by 3 hunters: Major Percy Horace Gordon Powell-Cotton (1866−1940), the museum's founder; K. Zenker, and Frederick George Merfield (1889−1960). Many of the specimens were killed by local people and purchased from them afterwards. The specimens are listed in chronological order within 3 separate groups, according to their collector. Several complete series of field numbers were used by each collector to record not only gorillas but also a wide variety of other creatures. Recent digitisation of the museum records has resulted in the collectors' original field numbers being incorporated into newly formulated accession numbers, which are used here. As mentioned in the lists, some skulls were sold by Merfield to Sir Frank Colyer (1866−1954), curator of the Odontological Museum at the Royal College of Surgeons of England, London. All the postcranial material was transferred to the Medical School, Birmingham University, in the early 1950s, where it was gradually cleaned, numbered, and boxed over the next 10−12 years. All the bones were subsequently returned to the Powell-Cotton Museum for permanent storage. Profs Eric Ashton and Solly Zuckerman, of Birmingham University, studied the museum's material and published several papers based upon it.

To save space, the following abbreviations have been used in the measurement lists:

a. ankle; ak. above knee; b. belly; bc. biceps; bk. below knee; bt. big toe; c. chest; cf. calf; e. ear; f. finger; ft. foot; fm. forearm; fta. finger-tip to armpit; g. girth; h. hand; hb. head and body; hf. head to fork; hrc. head round chin; ht. height; mf. middle finger; mt. middle toe; n. neck; ns. nose; sp. span; tb. thumb; th. thigh; w. wrist; wt. weight. All measurements are in imperial inches (in.), and while nearly all the weights are in pounds avoirdupois (lb.), a few are in kilograms (kg).

P.H.G. Powell-Cotton Collection

NH.FC.111 Skull. Adult male. Molémé, approx. 0°20′N, 15°E, French Congo. Purchased for 2 francs on 22 April 1927. Total length 12½ in.

NH.FC.112 Skull. Juvenile male. Near Molémé, approx. 0°20′N, 15°E, French Congo. Purchased for 2 francs on 23 April 1927.

NH.FC.113 Cranium. Adult male. Near Molémé, approx. 0°20′N, 15°E, French Congo. Purchased on 23 April 1927. The incisors are missing.

NH.FC.114 Skeleton and skin. Juvenile female. Mambili, 0°40′N, 15°30′E, French Congo. Collected on 26 April 1927. Ht. 43 in.; sp. 63 in.; c. 36 in.; h. 7¼ in. × 3½ in.; ft. 9 in. × 3¾ in.; wt. clean but with heart, lungs, and skin 60 lb. Stomach much larger than chest. One of two or more individuals. Atlas present, other neck vertebrae missing; sternum and 1 patella missing. Dental age approx. 8 years.

NH.FC.115 Skeleton and skin. Juvenile male. Mambili, 0°40′N, 15°30′E, French Congo. Collected on 26 April 1927. Ht. 42 in.; sp. 82 in.; c. 43 in.; h. 9¼ in. × 4¼ in.; ft. 11¼ in. × 4¾ in.; wt. clean but with heart, lungs and skin 94 lb. Not from the same family as no. 114. The left forearm was withered and stiff.

NH.FC.120 Skull. Female. On the way to Keba, approx. 0°40′N, 15°E, French Congo. Purchased for 8 francs on 6 May 1927.

NH.FC.122 Skeleton (in bits and incomplete) and skin. Juvenile male. Keba, 0°40′N, 15°E, French Congo. Collected on 8 May 1927. Ht. 29½ in.; sp. 48½ in.; c. 24¾ in.; wt. clean, with skin 27 lb. One of a troop. All bones in bad condition. Missing left ulna, all bones of hands and feet, 5 ribs (others broken); only small part of left scapula present. Dental age approx. 3½ years.

NH.FC.123 Skeleton and mounted skin. Adult male. Keba, 0°40′N, 15°E, French Congo. Collected on 8 May 1927. Ht. 57½ in.; sp. 92½ in.; c. 52½ in.; hb. 39½ in.; b. 60 in.; n. 28½ in.; bc. 16⅜ in.; h. 11⅛ in. × 4½ in.; ft. 12 in. × 4½ in.; wt. clean, with head 158 lb.; wt., no heart or lungs 205 lb.; wt. of skin and 1 foot 47 lb. Master of troop. Missing 2 right finger bones, right patella and coccyx. On exhibit in the monkey case.

NH.FC.124 Skeleton and skin. Juvenile female. Keba, 0°40′N, 15°E, French Congo. Collected on 8 May 1927. Ht. 33 in.; sp. 49 in.; c. 26 in. Missing 3 upper incisors, bones of hands and feet, 1st right rib; left tibia and fibula incomplete, also several ribs; bones in rather bad condition.

NH.FC.129 Skeleton (in bits and incomplete) and mounted skin. Infant male. Keba, 0°40′N, 15°E, French Congo. Collected on 14 May 1927. It was about 2 months old; had alive from 6 May 1927. It is mounted with the female no. 146.

NH.FC.130 Skeleton and skin. Adult male. Mambili, 0°40′N, 15°E, French Congo. Collected on 15 May 1927. Ht. 57½ in.; sp. 87½ in.; hb. 38 in.; b. 49 in.; c. 56 in.; n. 29 in.; bc. 17½ in.; fm. 16 in.; h. 9¾ in. × 5 in.; ft. 11⅛ in. × 4½ in.; wt. of skin, hands and feet 49 lb.; body clean with head 187 lb.; heart, liver and lungs 12 lb.; total wt. 248 lb. Master of troop. Missing atlas and axis vertebrae, 2 left toe bones and patellae.

NH.FC.133 Skeleton and mounted skin. Subadult male. Mambili, 0°40′N, 15°E, French Congo. Collected on 21 May 1927. Ht. 51 in.; sp. 72 in.; hb. 34 in.; b. 52 in.; c. 49 in.; n. 20 in.; h. 9⅜ in. × 3⅞ in.; ft. 10¾ in. × 3¾ in. With troop. Missing atlas and axis vertebrae, right distal radial epiphysis; 11 ribs broken, but with pieces present. Dental age approx. 9–11 years. The skin is exhibited in the monkey case.

NH.FC.146 Skeleton and mounted skin. Adult female. Keba, 0°40′N, 15°E, French Congo. Collected on 31 May 1927. Ht. 50¾ in.; sp. 80¼ in.; hb. 31 in.; c. 43 in.; b. 42½ in.; n. 19½ in.; h. 8½ in. × 4¾ in.; ft. 10¾ in. × 4 in.; wt. with head and skin, no intestines, heart, lungs, etc. 88 lb. With troop. Missing 2 ribs (6 broken), 1 clavicle (other broken), 2 metacarpals and 9 finger bones of right hand, and 2 toe bones of right foot. The skin is mounted with the baby, no. 129.

NH.FC.147 Skeleton and skin. Juvenile female. Keba, 0°40′N, 15°E, French Congo. Collected on 2 June 1927. Ht. 38¼ in.; sp. 61 in.; hb. 24¼ in.; c. 31½ in.; n. 15½ in.; h. 7¾ in. × 3¾ in.; ft. 8½ in. × 3¼ in. Solitary, found sick, shot. Blind in both eyes. Missing atlas and 1 other cervical vertebra (all cervical vertebrae broken), 1 right rib, end of sacrum, 1 right finger bone, 3 left toe bones, 1 right metatarsal and 2 right toe bones. Dental age approx. 5 years.

NH.FC.151 Skeleton and mounted skin. Juvenile female. Keba, 0°40′N, 15°E, French Congo. Collected on 3 June 1927. Ht. 38 in.; sp. 60¾ in.; hb. 25¼ in.; c. 35½ in.; n. 15 in.; h. 7¼ in. × 3½ in.; ft. 8½ in. × 3½ in. With troop. Missing all right hand bones, left clavicle, 2 vertebrae (1 cervical, 1 lumbar), 1 rib, 2 left carpals, 2 right toe bones and patellae. Dental age 5–5½ years. Skin mounted climbing tree in the monkey case.

NH.FC.154 Skull. Adult female. Makumba, 1°20′N, 14°40′E, French Congo. Purchased for 2 francs on 9 June 1927.

NH.FC.159 Skull. Adult male. Arfamaso, 1°20′N, 14°40′E, French Congo. Purchased for 2 francs on 14 June 1927. Six front teeth are missing.

NH.FC.162 Skull. Adult male. Zalangoye, 1°20′N, 14°40′E, French Congo. Purchased for 2 francs on 15 June 1927. The base is damaged.

NH.FC.163 Skull. Adult male. Bade, 1°20′N, 14°40′E, French Congo. Purchased for 2 francs on 18 June 1927. The base is damaged.

NH.FC.195 Skull. Adult male. Mubulu, approx. 3°N, 16°E, French Congo. Purchased for 5 francs on 23 August 1927. Ten teeth are missing.

NH.FC.196 Skull. Adult male. Mubulu, approx. 3°N, 16°E, French Congo. Purchased for 5 francs on 23 August 1927. The base is damaged.

NH.FC.197 Cranium. Female. Mubulu, approx. 3°N, 16°E, French Congo. Purchased for 3.50 francs on 23 August 1927. The base is damaged.

NH.FC.206 Skull. Subadult male. Nola, 3°30′N, 16°E, French Congo. Purchased, with no. 207, for 5 francs on 1 September 1927. Twelve teeth are missing.

NH.FC.207 Cranium. Subadult male. Nola, 3°30′N, 16°E, French Congo. Purchased, with no. 206, for 5 francs on 1 September 1927. Eight teeth are missing.

NH.FC.208 Cranium. Female. Jendi, 3°30′N, 15°20′E, French Congo. Purchased for 1 franc on 3 September 1927. All but 2 teeth are missing, and the base is broken.

NH.FC.215 Cranium. Female. Jendi, 3°30′N, 15°20′E, French Congo. Purchased for 1 franc on 3 September 1927. Eight teeth are missing, and the base is broken.

NH.FC.216 Cranium. Subadult male. Moey, 3°30′N, 15°20′E, French Congo. Purchased for 1 franc on 4 September 1927. Ten teeth are missing, the base is broken, and the zygomatic arches are missing.

NH.FC.217 Cranium. Adult female. Moey, 3°30′N, 15°20′E, French Congo. Collected on 4 September 1927. Five teeth are missing and the base is broken.

NH.FC.218 Cranium. Female. Mavonja, 3°30′N, 15°20′E, French Congo. Purchased for 1 franc on 4 September 1927. Six teeth are missing and the base is broken.

NH.FC.219 Cranium. Female. Mavonja, 3°30′N, 15°20′E, French Congo. Purchased for 1 franc on 4 September 1927. Seven teeth are missing and the base is broken.

NH.FC.224 Skull. Adult female. Dumbela, 3°30′N, 15°20′E, French Congo. Purchased for 1.50 francs on 5 September 1927. One upper tooth is missing.

NH.FC.225 Cranium. Adult male. Dumbela, 3°30′N, 15°20′E, French Congo. Purchased for 1 franc on 5 September 1927. All the teeth are missing and the base is broken.

NH.FC.226 Cranium. Female. Dumbela, 3°30′N, 15°20′E, French Congo. Purchased for 1 franc on 5 September 1927. Eleven teeth are missing and the base is broken.

NH.CAM.1.14 Skeleton and skin. Juvenile female. Belar, approx. 3°5′N, 10°5′E, Cameroun. Collected on 8 February 1929. Ht. 41½ in.; sp. 65½ in.; c. 38 in.; bc. 10 in.; fm. 10 in.; n. 17 in.; h. 7⅞ in. × 3¼ in.; ft. 9¼ in. × 2⅝ in.; e. 2⅛ in. × 1⅜ in.; 3rd f. 4⅞ in.; sp. tip of bt. to tip of 2nd 9½ in. Face black, lips yellow, ears black, brown top of head, cheeks and under chin greyish, arms black, legs black brown, back grey, chest and belly brown black, 2 white patches inside shoulders, eyes clear brown with ring of darker. Good condition, no fat, no fetus. Wt. whole 93 lb. One of two. Missing end of sacrum and coccyx. Dental age approx. 7–7½ years.

NH.CAM.1.41 Skeleton and skin. Adult male. SE of Kribi, approx. 2°50′N, 10°30′E, Cameroun. Purchased from Herr Klett on 29 March 1929. Missing 2 carpals of left hand, 3 carpals of right hand; 3 left toe bones not correct; sternum incomplete. Dental age approx. 9–11 years.

NH.CAM.1.42 Skeleton and skin. Adult female. SE of Kribi, approx. 2°50′N, 10°30′E, Cameroun. Purchased from Herr Klett on 29 March 1929. Missing 5 bones of right foot, coccyx and atlas vertebra; left femur broken.

NH.CAM.1.43 Skeleton and skin. Infant female. SE of Kribi, approx. 2°50′N, 10°30′E, Cameroun. Purchased from Herr Klett on 29 March 1929. Dental age approx. 6–9 months.

NH.CAM.1.44 Skeleton and skin. Female. SE of Kribi, approx. 2°50′N, 10°30′E, Cameroun. Purchased from Herr Klett on 29 March 1929. The front of the skull is shot away, and the skin is bad. Missing 1 clavicle, 1 vertebra, coccyx, patellae, left hand; right hand and both feet incomplete. Dental age approx. 9–11 years.

NH.CAM.1.45 Skin. Juvenile male. SE of Kribi, approx. 2°50′N, 10°30′E, Cameroun. Purchased from Herr Klett on 29 March 1929. It is in poor condition.

NH.CAM.1.46 Skin. Juvenile male. SE of Kribi, approx. 2°50′N, 10°30′E, Cameroun. Purchased from Herr Klett on 29 March 1929. It is in poor condition.

NH.CAM.1.47 Skeleton and skin. Adult male. SE of Kribi, approx. 2°50′N, 10°30′E, Cameroun. Purchased from Herr Klett on 29 March 1929. Tarsals at feet fused together, carpals absent, most phalanges missing. Skin is in poor condition.

NH.CAM.1.48 Skeleton and skin. Adult male. SE of Kribi, approx. 2°50′N, 10°30′E, Cameroun. Purchased from Herr Klett on 29 March 1929. Missing clavicles, patellae, coccyx, 2 bones of left hand, 6 toe bones of left foot, 4 toe bones of right foot; only 1 piece of sternum present. The skin of the right hand is missing. Dental age 9–11 years.

NH.CAM.1.49 Skin. Juvenile male. SE of Kribi, approx. 2°50′N, 10°30′E, Cameroun. Purchased from Herr Klett on 29 March 1929. It is in poor condition.

NH.CAM.1.50 Skin. Juvenile female. SE of Kribi, approx. 2°50′N, 10°30′E, Cameroun. Purchased from Herr Klett on 29 March 1929. It is in poor condition.

NH.CAM.1.73 Skin. Juvenile female. SE of Kribi, approx. 2°50′N, 10°30′E, Cameroun. Purchased from Herr Klett on 29 March 1929. It is in poor condition; the skin of the left hand is missing.

NH.CAM.1.83 Cranium. Male. Olangina, approx. 3°45′N, 12°15′E, Cameroun. Given by chief Ăngulă Ănsennă on 10 May 1929. Six teeth are missing.

NH.CAM.1.95 Skeleton and skin. Adult female. Olangina, approx. 3°45′N, 12°15′E, Cameroun. Collected on 17 May 1929. Ht. 49 in.; sp. 78 in.; hb. 32 in.; n. 21 in.; c. 40 in.; bc. 14 in.; fm. 13½ in.; h. 8⅞ in. × 4½ in.; ft. 11 in. × 4½ in.

NH.CAM.1.96 Skeleton and skin. Adult female. Olangina, approx. 3°45′N, 12°15′E, Cameroun. Collected on 17 May 1929. Ht. 55½ in.; sp. 79 in.; c. 45 in.; hb. 35½ in.; n. 23 in. Atlas vertebra missing.

NH.CAM.1.97 Skull and skin. Female. Olangina, approx. 3°45′N, 12°15′E, Cameroun. Collected on 17 May 1929. Shot by a native, men chopped off the hands and feet.

NH.CAM.1.98 Skeleton and skin. Adult female. Olangina, approx. 3°45′N, 12°15′E, Cameroun. Collected on 17 May 1929. Ht. 40½ in.; sp. 75½ in.; hb. 33¾ in.; c. 47 in.; n. 28½ in.; bc. 18 in.; fm. 15 in.; h. 9 in. × 4 in.; ft. 10¾ in. × 4½ in.; sp. between toes 6½ in.

NH.CAM.1.99 Skeleton and skin. Adult male. Olangina, approx. 3°45′N, 12°15′E, Cameroun. Collected on 17 May 1929. Ht. 60 in.; sp. 92 in.; hb 50 in.; c. 55 in.; h. 10 in. × 5½ in.; ft. 12¾ in.; sp. between toes 9 in. Bright brown top of head, very grey back and thighs. The old male of the party; 2 escaped and 2 captured of a family of 9.

NH.CAM.1.105 Skeleton and skin. Adult male. Olangina, approx. 3°45′N, 12°15′E, Cameroun. Collected on 7 June 1929. Ht. 60½ in.; sp. 95¼ in.; hb. 40¾ in.; c. 56 in.; hrc. 37 in.; n. 29 in.; bc. 18½ in.; fm. 17¼ in.; th. 29½ in.; cf. 15 in.; h. 10¾ in. × 5⅜ in.; ft. 12⅜ in. × 5 in.; sp. between toes 9½ in.; sp. between tb. and forefinger 8½ in.; wt. about 321 lb. clean (cut up and bits weighed separately).

NH.CAM.1.106 Skeleton and skin. Adult male. Akonolinga, approx. 3°40′N, 12°15′E, Cameroun. Collected on 10 June 1929. Ht. 64 in.; sp. 102½ in.; hb. 45¼ in.; c. 62 in.; hrc 40½ in.; n. 33 in.; bc. 19½ in.; fm 18 in.; th. 31 in.; cf. 18 in.; h. 12 in. × 6⅛ in.; ft. 14 in. × 5¾ in.; sp. between toes 11 in.; sp. between tb. and forefinger 9 in.; wt. clean about 401½ lb. (cut up and bits weighed separately). Redder on top of head and much older than no. 105. Solitary beast. Had fasted for 40 hours.

NH.CAM.1.107 Skeleton and skin. Adult male. Akonolinga, approx. 3°40′N, 12°15′E, Cameroun. Collected on 10 June 1929. Ht. 58 in.; sp. 93 in.; hb. 40 in.; c. 49½ in.; n. 24¾ in.; h. 11¼ in. × 6 in.; ft. 13 in.; sp. bt. to toe 10¼ in. Coccyx missing. Dental age approx. 7½–8 years.

NH.CAM.1.108 Skeleton and skin. Juvenile male. Akonolinga, approx. 3°40′N, 12°15′E, Cameroun. Collected on 10 June 1929. Ht. 51 in.; sp. 81 in.; hb. 32½ in.; n. 19¼ in.; c. 41 in. Missing patellae and sternum. Dental age approx. 7 years.

NH.CAM.1.109 Skeleton and skin. Juvenile female. Akonolinga, approx. 3°40′N, 12°15′E, Cameroun. Collected on 10 June 1929. Missing atlas and 1 left carpal; bones of feet not correct; all bones in rather bad condition.

NH.CAM.1.110 Skeleton and skin. Juvenile female. Akonolinga, approx. 3°40′N, 12°15′E, Cameroun. Collected on 10 June 1929. Missing all bones of left leg, astragalus of left foot, 1 right rib, 1 left rib, 3 left carpals, 3 right carpals, 2 cervical vertebrae (3 others very broken), sternum and patellae. Dental age approx. 5½ years.

NH.CAM.1.111 Skeleton and skin. Juvenile female. Akonolinga, approx. 3°40′N, 12°15′E, Cameroun. Collected on 10 June 1929. Missing 2 cervical vertebrae (atlas and axis broken), 1 right tarsal, 1 right and 1 left carpal and 2 finger or toe bones. Dental age approx. 5½ years.

NH.CAM.1.128 Cranium. Male. Near Olangina, approx. 3°45′N, 12°15′E, Cameroun. Collected on 20 July 1929. Occiput broken; 6 teeth missing.

NH.CAM.1.134 Skeleton and skin. Adult male. Olangina, approx. 3°45′N, 12°15′E, Cameroun. Collected on 29 July 1929. Ht. 61 in.; sp. 92 in.; hb. 39¾ in.; c. 56 in.; n. 30 in.; hrc. 38 in.; h. 10¾ in. × 5½ in.; sp. from tb. 7⅜ in.; ft. 12¾ in. × 5¼ in.; sp. from bt. 8¼ in.; bc. 18 in.; fm. 16 in.; th. 29 in.; e. 2¼ in. × 1⅞ in.; wt. clean 389 lb. (cut up and bits weighed separately). Arms black, top of head bright brown, legs and back grey, fine thick coat. Sternum incomplete. Dental age approx. 9−11 years.

NH.CAM.1.139 Skeleton and skin. Adult female. Beycar, approx. 3°40′N, 12°E, Cameroun. Collected on 6 August 1929. Wt. whole 150 lb.

NH.CAM.1.149 Skeleton and skin. Female. South of Yaoundé, approx. 3°30′N, 11°30′E, Nyong district, Cameroun. Purchased for 75 francs on 23 August 1929. Missing 1 finger bone of left hand, 1 patella and sternum.

NH.CAM.1.150 Skeleton and skin. Female. South of Yaoundé, approx. 3°30′N, 11°30′E, Nyong district, Cameroun. Purchased for 75 francs on 23 August 1929. Missing astragalus of right foot and part of sternum.

NH.CAM.1.198 Skull. Juvenile female. South of Yaoundé, approx. 3°30′N, 11°30′E, Nyong district, Cameroun. Given by F.G. Merfield on 23 August 1929.

NH.CAM.1.208 Skeleton and skin. Adult male. Near E-Bey, approx. 3°40′N, 12°15′E, Cameroun. Shot by hunters on 6 September 1929. Ht. 56½ in.; sp. 89½ in.; hb. 38¾ in.; hrc. 33 in.; c. 50 in.; n. 25½ in.; h. 11 in. × 5¾ in.; e. 2⅜ in. × 1¾ in.; sp. from tb. 7½ in.; ft.

12 in. × 5¼ in.; ft. sp. 8 in.; wt. clean 246 lb. Dental age approx. 9−11 years.

NH.CAM.1.224 Skeleton and skin. Infant male. Olangina, approx. 3°45′N, 12°15′E, Cameroun. Captured on 17 May 1929 and purchased alive for 800 francs. Died of dysentery on 27 September 1929. Dental age approx. 1½ years.

NH.CAM.1.229 Skeleton. Male. South of Yaoundé, approx. 3°30′N, 11°30′E, Cameroun. Given by F.G. Merfield on 6 October 1929. Missing 10 right and 7 left ribs, left femur, both fibulae, right scapula, both clavicles, both hands and feet, sacrum; only axis vertebra, 5 other cervical vertebrae and 3 dorsal vertebrae are present. Cranium badly damaged, half mandible missing.

NH.CAM.1.230 Skeleton. Male. South of Yaoundé, approx. 3°30′N, 11°30′E, Nyong district, Cameroun. Purchased for 50 francs on 15 October 1929. One patella is missing.

NH.CAM.1.231 Skeleton. Male. Cameroun. Given by F.G. Merfield on 21 October 1929. Missing right femur and fibula, 1 metacarpal and 3 finger bones of left hand, 1 right finger bone, 1 right carpal, 1 left carpal, 2 left metatarsals, 4 left toe bones (left tarsals deformed), 3 right tarsals and 11 right toe bones.

NH.CAM.2.278 Cranium. Adult male. Lelo, 4¼°N, 14¼°E, Batouri district, Cameroun. Purchased for 2 francs on 14 March 1932. The incisors and 1 canine are missing.

NH.CAM.2.282 Cranium. Female. Lelo, 4¼°N, 14¼°E, Batouri district, Cameroun. Given by chief Mboor on 16 February 1932. Most of the teeth are missing.

NH.CAM.2.315 Skull. Adult female. Beri, 4°30′N, 14°15′E, Batouri district, Cameroun. Purchased for 2.50 francs on 24 February 1932.

NH.CAM.2.323 Skull and skin. Adult male. Beri, 4°30′N, 14°15′E, Batouri district, Cameroun. Purchased for 15 francs on 4 March 1932. One of six killed in round-up. Skin is in very bad condition.

NH.CAM.2.324 Skin. Female. Beri, 4°30′N, 14°15′E, Batouri district, Cameroun. Purchased on 4 March 1932. One of six killed in round-up.

NH.CAM.2.325 Skin. Female. Beri, 4°30′N, 14°15′E, Batouri district, Cameroun. Purchased on 4 March 1932. One of six killed in round-up. The hands and feet are missing.

NH.CAM.2.331 Skull and skin. Juvenile male. Beri, 4°30′N, 14°15′E, Batouri district, Cameroun. Given by chief Mboor on 7 March 1932.

K. Zenker Collection

NH.ZEN.1.17 Skeleton and skin. Adult male. Bakoko, approx. 3½°N, 10°E, Njong, Cameroun. Collected in April 1929. Ht. 55½ in.; hb. 51½ in.; g. 52¼ in.

NH.ZEN.1.30 Postcranial skeleton. Adult male. Miabogo by Bipindi, 3°10′N, 10°20′E, Cameroun.

Collected on 2 May 1929. Hb. 37 in.; g. 42 in. Skull sold, skin sold. Missing astragalus of right foot.

NH.ZEN.2.63 Skeleton and skin. Juvenile female. River Nlonking, approx. 3°10′N, 10°20′E, Bulu Bush, Cameroun. Collected on 14 May 1930. Ht. 43 in.; hb. 38 in.; g. 39 in.; sp. 72½ in.

NH.ZEN.2.64 Skeleton and skin. Adult male. River Bikiango, approx. 3°10′N, 10°20′E, Bulu Bush, Cameroun. Collected on 24 May 1930. Ht. 63 in.; hb. 59 in.; g. 56¾ in.; sp. 98 in.

NH.ZEN.2.65 Skeleton and skin. Adult male. Azija, Bakoko, approx. 3½°N, 10°E, Cameroun. Collected on 2 March 1930. Ht. 57½ in.; hb. 50 in.; g. 49⅝ in.; sp. 79½ in.

NH.ZEN.3.31 Skeleton and skin. Adult male. Bipindi, 3°10′N, 10°20′E, Cameroun. Collected on 17 September 1930. Ht. 58¼ in.; hb. 55½ in.; g. 51⅛ in.; sp. 95¼ in. Dental age approx. 9−11 years.

NH.ZEN.5.91 Skull. Female. Bipindi district, approx. 3°10′N, 10°20′E, Cameroun. Collected in October 1931.

NH.ZEN.6.30 Skeleton. Adult male. Tyanga, approx. 3°10′N, 10°20′E, Bipindi district, Cameroun. Collected on 5 July 1932. Ht. 58¼ in.; hb. 52 in.; g. 57½ in.; sp. 86⅝ in.

NH.ZEN.6.32 Skeleton and skin. Adult male. Ebodonku road, approx. 2°50′N, 11°10′E, Ebolowa, Cameroun. Collected on 30 January 1933. Ht. 57½ in.; hb. 50 in.; g. 57 in.; sp. 85 in.

NH.ZEN.6.33 Skeleton and skin. Female. Bikiango road, approx. 3°10′N, 10°20′E, Bipindi, Cameroun. Collected on 13 March 1933. Ht. 40½ in.; hb. 41 in.; g. 32⅝ in.; sp. 74¼ in.

Merfield Collection

Skeleton (incomplete). Probably a specimen preserved in spirit.

NH.MER.1.28 Skeleton and skin. Adult male. Yaoundé-Kribi road, approx. 3°30′N, 11°30′E, Cameroun. Purchased for 200 francs in 1930. The skin is in very poor condition.

NH.MER.1.29 Skeleton and skin. Adult female. Yaoundé-Kribi road, approx. 3°30′N, 11°30′E, Cameroun. Purchased for 75 francs in 1930. Missing left clavicle and 1 patella. The skin is in poor condition.

NH.MER.1.30 Skin. Juvenile female. Yaoundé district, approx. 3°30′N, 11°30′E, Cameroun. Given by F.G. Merfield in 1930. It is in poor condition.

NH.MER.2.1 Skeleton. Adult female. Yaoundé district, approx. 3°30′N, 11°15′E, Cameroun. Purchased for 100 francs in 1930. Missing 3 ribs, end of sacrum and 1 left carpal.

NH.MER.2.2 Skeleton. Adult female. Southeast of Yaoundé, approx. 3°30′N, 11°30′E, Cameroun. Purchased for 100 francs in 1930. It is missing 13 finger bones and all the tips.

NH.MER.2.3 Skull. Adult male. Yaoundé-Akonolinga road, approx. 3°45′N, 11°45′E, Cameroun. Purchased for 30 francs in 1930.

NH.MER.2.6 Skin. Juvenile male. Akonolinga district, approx. 3°45′N, 12°10′E, Cameroun. Purchased for 10 francs in 1930. It is beetle-eaten.

NH.MER.2.23 Odd bones. Cameroun. Given by F.G. Merfield in 1930. Bones present: all of left leg, right femur, left humerus and ulna, right humerus and part of ulna, both clavicles, scapulae, pelvics, 6 cervical vertebrae (no atlas), 13 thoracic and 4 lumbar vertebrae, sacrum, 13 ribs (right and left, 3 broken), sternum and patellae. Left hand: 5 carpals and 4 other bones missing; right hand: 7 carpals and 5 other bones missing. Right and left feet: tarsals present, 3 toe bones missing.

NH.MER.2.25 Skin. Female. Cameroun. Collected in 1930. Poor condition, no feet.

NH.MER.1.32 Skull. Juvenile female. Lelo, 4¼°N, 14¼°E, Batouri district, Cameroun. Picked up on 11 May 1932. Eleven teeth are missing.

NH.MER.1.119 Skeleton and skin. Adult male. Lumbindu, 4¼°N, 14¼°E, Batouri district, Cameroun. Collected on 4 June 1932. H. 10½ in. × 7¼ in.; ft. 12½ in. × 7½ in. Note penis bone. Dental age approx. 9−11 years.

NH.MER.1.180 Skeleton and skin. Juvenile female. Kanyol, 4°N, 14°E, Batouri district, Cameroun. Collected on 13 June 1932. Ht. 47¼ in.; hf. 31¼ in.; sp. 67¼ in.; g. 37¾ in.; hrc. 25¼ in.; n. 17¾ in.; bc. 10¾ in.; fm. 10½ in.; th. 17 in.; cf. 10 in.; h. 8¼ in. × 5 in.; ft. 9¾ in. × 5½ in. Killed by a poisoned dart from a bow. Missing upper left 2nd molar.

NH.MER.1.183 Odd bones, leg, arms and pelvis. Adult male. Kanyol, 4°N, 14°E, Batouri district, Cameroun. Collected on 13 June 1932. Present: bones of both arms and legs, scapulae, pelvic bones, clavicles, sacrum, 2 lumbar vertebrae and hyoid. Skull sold to Colyer.

NH.MER.1.256 Skull. Female. Kanyol, 4°N, 14°E, Batouri district, Cameroun. Collected on 22 June 1932.

NH.MER.1.281 Skull. Female. Kanyol, 4°N, 14°E, Batouri district, Cameroun. Collected on 29 June 1932. All teeth missing; zygomatic arches chopped off.

NH.MER.1.288 Skull. Adult male. Meyoss, 4°N, 14°E, Batouri district, Cameroun. Collected on 30 June 1932. Seven teeth are missing.

NH.MER.1.300 Skeleton and skin. Adult female. Meyoss, 4°N, 14°E, Batouri district, Cameroun. Collected on 2 July 1932. Ht. 57 in.; hf. 36½ in.; sp. 85 in.; h. 8 in. × 5 in.; ft. 10½ in. × 6½ in. Very old beast, right eye scarred. Had fetus. Missing coccyx and left ulna; sternum incomplete.

NH.MER.1.319 Skull. Juvenile female. Meyoss, 4°N, 14°E, Batouri district, Cameroun. Collected on 5 July 1932. Three teeth are missing.

NH.MER.1.320 Mandible. Male. Meyoss, 4°N, 14°E, Batouri district, Cameroun. Collected on 5 July 1932. The incisors are missing.

NH.MER.1.329 Skeleton and skin. Adult female. Meyoss, 4°N, 14°E, Batouri district, Cameroun. Collected on 6 July 1932. Ht. $55\frac{1}{2}$ in.; hf. 37 in.; sp. $80\frac{1}{2}$ in.; g. $50\frac{1}{2}$ in. Missing 2 toe bones of left foot and 2 toe bones of right foot; sternum incomplete.

NH.MER.1.333 Skeleton and skin. Juvenile female. Meyoss, 4°N, 14°E, Batouri district, Cameroun. Collected on 6 July 1932. Ht. 34 in.; hf 22 in.; g. $27\frac{1}{2}$ in.; sp. $52\frac{1}{2}$ in. Missing atlas and 1 thoracic vertebrae, 1 carpal of right hand, and 1 carpal of left hand. Dental age approx. $2\frac{1}{2}$–3 years.

NH.MER.1.349 Cranium. Female. Meyoss, 4°N, 14°E, Batouri district, Cameroun. Collected on 10 July 1932. One canine is missing.

NH.MER.1.372 Skeleton and skin. Adult male. Meyoss, 4°N, 14°E, Batouri district, Cameroun. Collected on 12 July 1932. Ht. 64 in.; hf. 44 in.; g. $55\frac{1}{2}$ in.; sp. $96\frac{3}{4}$ in.; hrc. $35\frac{1}{4}$ in.; h. $10\frac{1}{2}$ in. × $7\frac{1}{2}$ in.; ft. 12 in. × $7\frac{3}{4}$ in. Back quite bare for about 16 in. × 26 in. Old wound and a cut at back of neck. Not a very old beast.

NH.MER.1.387 Skeleton and skin. Juvenile female. Meyoss, 4°N, 14°E, Batouri district, Cameroun. Collected on 16 July 1932. Ht. 48 in.; hf. 32 in.; g. $37\frac{1}{2}$ in.; sp. $73\frac{1}{4}$ in. Missing all bones of left hand and left foot, and 1 patella. Dental age approx. 7 years.

NH.MER.1.450 Skeleton. Juvenile male. Kanyol, 4°N, 14°E, Batouri district, Cameroun. Collected on 25 July 1932. Missing 3 carpals from each hand, 1 tarsal from each foot and 1 right toe bone. Dental age approx. $3\frac{1}{2}$–4 years.

NH.MER.1.474 Skull. Adult male. Kanyol, 4°N, 14°E, Batouri district, Cameroun. Collected on 25 July 1932. Back part broken away. One lower canine and 1 incisor missing.

NH.MER.1.476 Skeleton and skin. Infant male. Kanyol, 4°N, 14°E, Batouri district, Cameroun. Collected on 25 July 1932. Had alive, died. Dental age approx. 6 months.

NH.MER.1.487 Skull and skin. Juvenile male. Lelo, $4\frac{1}{4}$°N, $14\frac{1}{4}$°E, Batouri district, Cameroun. Collected on 2 August 1932.

NH.MER.1.631 Skull and skin. Juvenile male. Obala, Batouri district, $3\frac{3}{4}$°N, $14\frac{1}{4}$°E, Cameroun. Collected on 29 September 1932. The skin is in poor condition.

NH.MER.1.667 Skeleton and skin. Juvenile female. Obala, $4\frac{1}{4}$°N, $14\frac{1}{4}$°E, Batouri district, Cameroun. Collected on 7 October 1932. Ht. $45\frac{1}{2}$ in.; hf. $29\frac{1}{2}$ in.; g. $33\frac{1}{4}$ in.; sp. $63\frac{1}{2}$ in.; hrc. 24 in.; b. $36\frac{1}{2}$ in.; bc. $10\frac{1}{2}$ in.; fm. $10\frac{1}{2}$ in.; h. $7\frac{1}{4}$ in. × $4\frac{5}{8}$ in.; ft. 9 in. × $5\frac{1}{4}$ in.; mt. $1\frac{1}{2}$ in. × 2 in.; tb. $1\frac{1}{4}$ in. × 2 in.; f. $2\frac{3}{4}$ in. × $3\frac{1}{4}$ in.; e. 2 in. × $1\frac{3}{8}$ in. Missing atlas and 1 other vertebra. Dental age approx. 6 years.

NH.MER.1.690 Skeleton and skin. Juvenile male. Obala, $4\frac{1}{4}$°N, $14\frac{1}{4}$°E, Batouri district, Cameroun. Collected on 5 October 1932. Had alive in captivity since 15 June 1932; called 'Lumbindou'. Dental age approx. 2 years.

NH.MER.1.696 Skeleton and skin. Adult female. Obala, $4\frac{1}{4}$°N, $14\frac{1}{4}$°E, Batouri district, Cameroun. Collected on 5 October 1932. Ht. $54\frac{3}{4}$ in.; hf. $34\frac{3}{4}$ in.; g. $44\frac{1}{2}$ in.; b. $45\frac{1}{2}$ in.; n. 22 in.; hrc. $29\frac{3}{4}$ in.; bc. $14\frac{1}{2}$ in.; fm. $13\frac{3}{4}$ in.; h. 9 in. × $6\frac{1}{4}$ in.; th. 23 in.; cf. $12\frac{3}{4}$ in.; mt. $2\frac{1}{4}$ in. × $2\frac{1}{4}$ in.; tb. $1\frac{3}{4}$ in. × 2 in.; mf. $3\frac{1}{8}$ in. × $3\frac{5}{8}$ in.; e. 2 in. × $1\frac{1}{4}$ in. Milk at breasts. Very hairy beast. Nasty wound on lower part of right shoulder said to have been made by male. Missing 1 right carpal; right trapezoid labelled as left. Dental age approx. 9–11 years.

NH.MER.1.720 Skeleton and skin. Adult male. Obala, $4\frac{1}{4}$°N, $14\frac{1}{4}$°E, Batouri district, Cameroun. Collected on 8 October 1932. Ht. $69\frac{1}{4}$ in.; hf. $45\frac{1}{2}$ in.; sp.106 in.; g. $65\frac{1}{2}$ in.; b. 66 in.; hrc. $38\frac{3}{4}$ in.; n. 32 in.; bc. 19 in.; fm 19 in.; h. 11 in. × 8 in.; tb. $2\frac{1}{2}$ in. × $3\frac{3}{4}$ in.; mf. $4\frac{1}{4}$ in. × 4 in. Measurements broken off owing to accident, see letter 14. This beast was located in a plantation and followed into the forest, where after a time it turned on chief Key-ar-Bar of the village Bey-cum-a-Dee, who threw a spear at it and killed it.

NH.MER.1.729 Skeleton and skin. Adult male. Obala, $4\frac{1}{4}$°N, $14\frac{1}{4}$°E, Batouri district, Cameroun. Collected on 9 October 1932. Ht. $66\frac{1}{2}$ in.; hf. $42\frac{1}{2}$ in.; g. 57 in.; b. $58\frac{1}{2}$ in.; hrc. $39\frac{1}{4}$ in.; n. $29\frac{3}{4}$ in.; bc. 17 in.; fm. 17 in.; h. $10\frac{1}{2}$ in. × $7\frac{3}{4}$ in.; tb. $3\frac{1}{2}$ in. × 5 in.; mf. $3\frac{3}{4}$ in. × $5\frac{1}{2}$ in.; th.29 in.; cf. 17 in.; ft. $13\frac{3}{4}$ in. × $8\frac{1}{2}$ in. Small toe of right foot missing. Mate of no. 696. For an account of the hunt for this beast see letter 14.

NH.MER.1.780 Postcranial skeleton. Infant male. Obala, $3\frac{3}{4}$°N, $14\frac{1}{4}$°E, Batouri district, Cameroun. Collected on 14 October 1932. Skull sold. Atlas and axis missing. Found alone in same bush as the female no. 696 and the male no 729. As 696 had milk at her breasts I take it this was her young. The natives, when they killed 696, said there was a young one.

NH.MER.1.840 Skeleton and skin. Adult female. Obala, $4\frac{1}{4}$°N, $14\frac{1}{4}$°E, Batouri district, Cameroun. Collected on 18 October 1932. Ht. $55\frac{1}{4}$ in.; hf. $36\frac{1}{2}$ in.; sp. $80\frac{1}{4}$ in.; g. 46 in.; b. $50\frac{1}{2}$ in.; hrc. 31 in.; n. $22\frac{1}{2}$ in.; bc. $12\frac{1}{2}$ in.; fm. 13 in.; h. $9\frac{1}{4}$ in. × $6\frac{1}{4}$ in.; th. 22 in.; cf. $12\frac{1}{2}$ in.; ft. $11\frac{1}{4}$ in. × 7 in.; e. $2\frac{3}{4}$ in. × $1\frac{1}{4}$ in. Had fetus. Very grey on back. Skull smashed through falling from high tree and hitting stump.

NH.MER.1.841 Skeleton and skin. Juvenile female. Obala, 4¼°N, 14¼°E, Batouri district, Cameroun. Collected on 18 October 1932. Ht. 46¼ in.; hf. 29¾ in.; sp. 67½ in.; hrc. 25½ in.; n. 17½ in.; g. 37 in.; b. 37½ in.; bc. 7¾ in.; fm. 11 in.; h. 8½ in. × 5¾ in.; th. 16¾ in.; cf. 9½ in.; ft. 10 in. × 6½ in.; e. 2 in. × 1½ in. Dental age approx. 6 years.

NH.MER.1.847 Skeleton and skin. Adult female. Obala, 4¼°N, 14¼°E, Batouri district, Cameroun. Collected on 21 October 1932. Ht. 55¾ in.; hf. 34½ in.; sp. 78 in.; g. 41¼ in.; hrc. 28½ in.; n. 20¼ in.; bc. 10¾ in.; fm. 12 in.; h. 8¾ in. × 7¾ in.; th. 19½ in.; cf. 9¾ in.; ft. 11 in. × 8¼ in.; e. 2⅜ in. × 1½ in. Lower lip deformed and turned down as is the case of a dead gorilla, but the lip is grafted to the chin. Missing 4 left carpals, 4 right carpals and patellae. Dental age approx. 9−11 years.

NH.MER.1.875 Skeleton and skin. Adult male. Obala, 4¼°N, 14¼°E, Batouri district, Cameroun. Collected on 24 October 1932. Ht. 67 in.; hf. 43¾ in.; sp. 97½ in.; g. 63 in.; b. 64½ in.; hrc. 40¾ in.; n. 33¼ in.; bc. 21½ in.; fm. 19½ in.; h. 11¼ in. × 8 in.; th. 33½ in.; cf. 18 in.; ft. 13½ in. × 9 in.; e. 2½ in. × 1½ in.; mf. 5 in.; bt. 5¼ in. Very grey back and legs. Note very high eyebrows. Bullet hole right eye. Skull sold to Colyer on 17 November 1933, now in the Hunterian Museum, Royal College of Surgeons of England (their no. RCSOM/A 64.16).

NH.MER.1.902 Skeleton and skin. Adult female. Obala, 4¼°N, 14¼°E, Batouri district, Cameroun. Collected on 27 October 1932. Ht. 60 in.; hf. 36¾ in.; sp. 78 in.; g. 44½ in.; b. 49½ in.; hrc. 29½ in.; n. 19¼ in.; bc. left arm 12¼ in.; bc. right arm 9½ in.; fm. left 12½ in.; fm. right 10½ in.; th. 20¾ in.; cf. 10¾ in.; h. 8½ in. × 6¼ in.; ft. 11 in. × 7 in.; e. 2 in. × 1¾ in. Left eye looks blind and right one going? Right hand deformed and very bad sores between fingers, big sore on top of hand, old scars on wrist, hand or arm not been used for some time. Forearm skin and bone, first finger looks as if occasionally put on ground. Sexual organs diseased, also surrounding parts. The running sores known to natives as 'marjal': they suffer considerably themselves with this disease, they say sores on hands is the same thing. Missing parts of sternum, coccyx, 1 toe bone of left foot and 2 toe bones of right foot. Right pelvis, ribs and sacrum all broken.

NH.MER.1.932 Skeleton and skin. Adult female. Obala, 3¾°N, 14¼°E, Batouri district, Cameroun. Collected on 31 October 1932. Ht. 56 in.; hf. 37 in.; sp. 78¼ in. No other measurements taken owing to beast being diseased. Upper lip, chin, right cheek, underside of right elbow and right ankle white. Chest and arms have big white blotches. Right eye bunged up. All the right side of neck badly diseased, and large open sores eaten right through skin, in other places the epidermis is eaten away. The disease is known as 'marjal' (in Yaoundé, 'mabada'): see no. 902. The diseases of various gorilla

should be very interesting to medical men if on the spot. It is missing the atlas vertebra and the right clavicle.

NH.MER.1.962 Skeleton and skin. Adult male. Obala, 4¼°N, 14¼°E, Batouri district, Cameroun. Collected on 6 November 1932. No measurements, cut up in the bush.

NH.MER.1.985 Skeleton and skin. Adult female. Meyoss, 4°N, 14°E, Batouri district, Cameroun. Collected on 14 November 1932. Ht. 45 in.; hf. 30½ in.; sp. 64 in.; g. 37½ in.; b. 45 in.; hrc. 24½ in.; n. 15½ in.; bc. 10½ in.; fm. 11 in.; h. 8 in. × 5½ in.; th. 16 in.; cf. 10¼ in.; ft. 9½ in. × 5¼ in.; e. 1¼ in. × 2 in. General colour grey. It is missing the atlas vertebra, all left carpals, and 1 right toe bone.

NH.MER.01 Skull. Adult male. Bimba, 4°10′N, 14°10′E, Batouri district, Cameroun. Collected on 8 December 1932. The back part is broken.

NH.MER.02 Skull. Adult male. Bimba, 4°10′N, 14°10′E, Batouri district, Cameroun. Collected on 11 December 1932. The lower incisors are worn away.

NH.MER.03 Skull. Female. Bimba, 4°10′N, 14°10′E, Batouri district, Cameroun. Collected on 11 December 1932. Damaged, teeth missing.

NH.MER.04 Skull. Female. Bimba, 4°10′N, 14°10′E, Batouri district, Cameroun. Collected on 11 December 1932. Damaged, teeth missing. It has a supernumerary left canine.

NH.MER.2.109 Skeleton and skin. Subadult male. Obala, 4¼°N, 14¼°E, Batouri district, Cameroun. Collected on 4 December 1932. Missing part of left pelvis; atlas, axis, and 1 lumbar vertebrae.

NH.MER.2.204 Skeleton and skin. Adult male. Lelo, 4¼°N, 14¼°E, Batouri district, Cameroun. Collected on 16 February 1933. Sternum incomplete; coccyx missing. Dental age approx. 9−11 years.

NH.MER.2.206 Skeleton and skin. Subadult male. Lelo, 4¼°N, 14¼°E, Batouri district, Cameroun. Collected on 16 February 1933. Dental age approx. 9−11 years.

The above 2 males were found up one tree with a 3rd male (no. 205 in Merfield's list) but there was no female with them. No. 205, an adult male, was sold by Merfield.

NH.MER.2.241 Skeleton. Subadult male. Obala, 3¾°N, 14¼°E, Batouri district, Cameroun. Collected on 21 February 1933. Missing the atlas and axis and 1 other cervical vertebrae, 4 toe bones of right foot, and 4 toe bones of left foot. Dental age approx. 9−11 years.

NH.MER.2.340 Skeleton and skin. Adult male. Obala, 4¼°N, 14¼°E, Batouri district, Cameroun. Collected on 18 March 1933. Very old beast, 2 upper incisors and 1 lower canine missing, all teeth very much worn, canine teeth in upper jaw worn level with other teeth. Missing 1 left toe bone; sacrum, ribs, and vertebrae all very broken.

All gorillas coming from the forest northwest and west from here (Arteck, 3¾°N, 14¼°E) seem to have smaller crests in comparison with those coming from the east and south districts.

NH.MER.2.409 Skeleton and skin. Juvenile female. Obala, 3¾°N, 14¼°E, Batouri district, Cameroun. Collected on 30 March 1933. Ht. 30 in.; hf. 21 in.; sp. 40½ in.; hrc. 19¾ in.; e. 1⅞ in. × 1⅝ in. Dental age approx. 2½ years.

NH.MER.2.460 Skeleton and skin. Adult female. Obala, 3¾°N, 14¼°E, Batouri district, Cameroun. Collected on 8 April 1933. Missing 1 lumbar vertebra.

NH.MER.2.461 Skeleton and skin. Adult male. Njabada, 4¼°N, 14¼°E, Batouri district, Cameroun. Collected on 8 April 1933. Left arm just below elbow missing. Only deformed ulna present; no radius or bones of left hand. Some bones in bad condition; vertebrae and sacrum chipped.

NH.MER.2.462 Skeleton and skin. Juvenile male. Obala, 4¼°N, 14¼°E, Batouri district, Cameroun. Collected on 8 April 1933. Dental age approx. 9–11 years.

NH.MER.2.463 Skeleton and skin. Juvenile male. Obala, 4¼°N, 14¼°E, Batouri district, Cameroun. Collected on 8 April 1933. Missing 1 rib, 3 left toe bones, patellae; sacrum broken. Dental age approx. 4 years.

NH.MER.2.464 Skeleton and skin. Adult male. Obala, 4¼°N, 14¼°E, Batouri district, Cameroun. Collected on 8 April 1933. Missing 2 right finger bones and 2 right tarsals; pelvic bones broken. Dental age approx. 9–11 years.

All the above 5 gorillas were killed in one round-up and were not skinned until the next day.

NH.MER.2.470 Skeleton and skin. Adult female. Obala, 3¾°N, 14¼°E, Batouri district, Cameroun. Collected on 11 April 1933. Ht. 50 in.; hf. 33 in.; sp. 72 in.; g. 44 in.; b. 48¼ in.; hrc. 26½ in.; e. 2⅛ in. × 1⅝ in.; n. 20 in.; bc. 12½ in.; fm. 12½ in.; th. 20½ in.; cf. 11½ in.; h. 8 in. × 5¾ in.; ft. 10⅛ in. × 6½ in. Missing atlas vertebra and 1 left carpal.

NH.MER.2.471 Skeleton and skin. Juvenile male. Obala, 4¼°N, 14¼°E, Batouri district, Cameroun. Collected on 11 April 1933. Ht. 34½ in.; hf. 23 in.; sp. 50 in.; g. 27 in.; b. 30¼ in.; hrc. 21 in. The young of no. 470. Missing 1 patella and 1 finger bone of left hand.

NH.MER.2.490 Skeleton and skin. Adult female. Obala, 3¾°N, 14¼°E, Batouri district, Cameroun. Collected on 18 April 1933. Ht. 53 in.; hf. 34¼ in.; g. 42 in.; b. 46½ in.; hrc. 27⅞ in.; e. 1⅞ in. × 1⅝ in.; n. 19½ in.; bc. 12 in.; fm. 12 in.; th. 19½ in.; cf. 11¼ in.; h. 7⅝ in. × 5¼ in.; ft. 9⅝ in. × 6 in.

NH.MER.2.497 Skeleton and skin. Juvenile male. Obala, 4¼°N, 14¼°E, Batouri district, Cameroun. Collected on 26 April 1933. Ht. 31 in.; hf. 20½ in.; g. 23¼ in.; sp. 43½ in.; hrc. 20 in.; e. 1⅞ in. × 1⅛ in. Most bones crumbling and in bad condition, especially pelvics, ribs, and vertebrae. Left scapula missing, also half of right femur. Dental age approx. 2–2½ years.

NH.MER.2.505 Skeleton and skin. Subadult male. Obala, 4¼°N, 14¼°E, Batouri district, Cameroun. Collected on 29 April 1933. Ht. 54 in.; hf. 37 in.; sp. 79 in. The coccyx is missing. Dental age approx. 9–11 years.

NH.MER.2.532 Skull and skin. Juvenile female. Lelo, 4¼°N, 14¼°E, Batouri district, Cameroun. Collected on 10 September 1933. Ht. 33 in.; hf. 22½ in.; g. 25½ in.; sp. 49 in.

NH.MER.2.625 Skeleton and skin. Juvenile female. Obala, 3¾°N, 14¼°E, Batouri district, Cameroun. Collected on 12 October 1933. Ht. 36 in.; hf. 23 in.; sp. 53¾ in.; g. 30½ in.; hrc. 21½ in.; n. 14¼ in. Head very reddish, grey patches on cheek. Missing 2 left toe bones and 1 right toe bone. Sacrum broken, also some of the vertebrae. Dental age approx. 4–5 years.

[Another specimen, the skeleton of an adult female, is no. 660 in Merfield's list. It is not in this museum, having been sold to Sir Frank Colyer on 16 May 1934.]

NH.MER.2.674 Skeleton and skin. Subadult male. Obala, 4¼°N, 14¼°E, Batouri district, Cameroun. Collected on 22 October 1933. Missing 1 left toe bone and patellae.

NH.MER.2.687 Skeleton and skin. Adult male. Obala, 4¼°N, 14¼°E, Batouri district, Cameroun. Collected on 7 November 1933.

NH.MER.2.688 Skeleton and skin. Adult female. Obala, 3¾°N, 14¼°E, Batouri district, Cameroun. Collected on 7 November 1933. Missing sternum and 1 toe bone of right foot.

NH.MER.2.689 Skeleton and skin. Subadult male. Obala, 4¼°N, 14¼°E, Batouri district, Cameroun. Collected on 7 November 1933. Ht. 51 in.; hf. 32½ in.; sp. 62 in.; g. 43 in.; b. 51½ in.; hrc. 28½ in.; e. 2½ in. × 1⅓ in.; n. 19½ in.; bc. 11 in.; fm. 12 in.; w. 8½ in.; th. 18½ in.; cf. 11 in.; a. 10½ in.; h. 8½ in. × 5½ in.; ft. 10¼ in. × 6¼ in.; fta. 31 in. Missing 1 left carpal.

NH.MER.2.690 Skeleton and skin. Adult male. Obala, 4¼°N, 14¼°E, Batouri district, Cameroun. Collected on 9 November 1933. The 1st right rib is missing.

NH.MER.2.691 Skeleton and skin. Subadult female. Obala, 3¾°N, 14¼°E, Batouri district, Cameroun. Collected on 9 November 1933. Diseased lower lip, said to be caused by 'nguar-zock'. The sternum is missing. The right scapula, right clavicle and right hand bones are all marked 261; the left hand bones are not marked.

NH.MER.2.698 Skeleton and skin. Juvenile male. Obala, 4¼°N, 14¼°E, Batouri district, Cameroun. Collected on 10 November 1933. Dental age approx. 5 years.

Skull and skin. Juvenile male. Ndima, 4¼°N, 14¼°E, Batouri district, Cameroun. Collected on 14 November 1933. Skin dried by natives.

Skull and skin. Juvenile female. Ndima, 4¼°N, 14¼°E, Batouri district, Cameroun. Collected on 14 November 1933. Skin dried by natives.

Skull and skin. Juvenile female. Ndima, 4¼°N, 14¼°E, Batouri district, Cameroun. Collected on 14 November 1933. Skin dried by natives.

NH.MER.2.716 Skeleton and skin. Adult female. Batouri district, 4¼°N, 14¼°E, Cameroun. Collected on 16 November 1933. Ht. 56½ in.; hf. 36 in.; g. 42 in.; b. 52 in.; hrc. 29¼ in.; e. 2½ in. × 1½ in.; n. 19¼ in.; bc. 13 in.; fm. 12 in.; w. 8¼ in.; th. 22¼ in.; cf. 12¼ in.; a. 11¼ in.; h. 8¼ in. × 5¾ in.; ft. 11 in. × 6½ in.; fta. 31½ in. Left forearm very much diseased and withered, the hand being bent inwards and the arm at the elbow could not be straightened. The hunters say the beast did not use the arm when up a tree. The disease is called 'ma-chully-chang' and is said to be caused by the liana 'ngua-zock'. A sore inside the left thigh, the right foot is crippled, and a big sore on the outside ankle. Right pelvis missing; sacrum completely broken.

NH.MER.2.717 Skeleton and skin. Subadult male. Batouri district, 4¼°N, 14¼°E, Cameroun. Collected on 16 November 1933. Ht. 56½ in.; hf. 37 in.; sp. 84¾ in.; g. 41 in.; b. 51 in.; hrc. 29 in.; e. 2¼ in. × 1½ in.; n. 18½ in.; bc. 12 in.; fm. 12¾ in.; w. 9½ in.; th. 19¼ in.; cf. 11¾ in.; a. 10½ in.; h. 9¾ in. × 5½ in.; ft. 11¼ in. × 6½ in.; fta. 35¼ in. Bad wound on upper side of right wrist said to be caused by a fight with an old male, with the object of driving him out of the troop. Note the extraordinary holes in both shoulder blades. Missing 1 right carpal. Dental age approx. 9–11 years.

NH.MER.2.729 Skeleton and skin. Adult female. Batouri district, 4¼°N, 14¼°E, Cameroun. Collected on 22 November 1933. Ht. 57¼ in.; hf. 37¼ in.; sp. 69¼ in.; g. 42 in.; b. 48 in.; hrc. 28½ in.; e. 2¼ in. × 1⅛ in.; n. 20 in.; bc. 12¼ in.; fm. 12 in.; w. 8½ in.; th. 21½ in.; cf. 11 in.; a. 10¼ in.; h. 8¾ in. × 6 in.; ft. 10¾ in. × 6¼ in.; fta. 32 in. All hair grey, a fine red 'cap'. The first gorilla whose fingers I could easily stretch out with the wrist straight. A fine specimen. Dental age approx. 9–11 years.

NH.MER.2.755 Skeleton and skin. Adult female. Batouri district, 4¼°N, 14¼°E, Cameroun. Collected on 30 November 1933. Ht. 57 in.; hf. 37 in.; sp. 81 in. A very fine old beast. Note the prominent eyebrows and crest of skull. Missing 1 left carpal. Bones of left arm, left clavicle, left scapula, and left hand not marked.

NH.MER.2.756 Skeleton and skin. Infant male. Batouri district, 4¼°N, 14¼°E, Cameroun. Collected on 30 November 1933. The baby of no. 755. The hyoid bone is missing.

NH.MER.2.760 Skeleton and skin. Subadult female. Batouri district, 4¼°N, 14¼°E, Cameroun. Collected on 2 December 1933. Missing left toe bone, 2 right toe bones, 1 lumbar vertebra, and coccyx. Dental age approx. 5 years.

NH.MER.2.766 Skeleton and skin. Adult male. Batouri district, 4¼°N, 14¼°E, Cameroun. Collected on 4 December 1933. It was killed by a spear thrust by 'Ell-Lay', a young man about 24 years of age. The man says he was walking along a stream when the beast followed him. He hid behind a tree and speared the beast as it went by, which shows how little fear the Memjims have for the gorilla. The man was quite alone. A fine old specimen. It has irregular teeth after the 3rd molars.

NH.MER.2.786 Skeleton and skin. Adult female. Batouri district, 4¼°N, 14¼°E, Cameroun. Collected on 13 December 1933. A very old beast. No signs of a crest on the skull as in no. 755. The skin is badly cut. Missing 1 right toe bone.

NH.MER.2.791 Skull. Female. Batouri district, 4¼°N, 14¼°E, Cameroun. Collected on 14 December 1933. It is missing 3 incisors and 3 molars.

NH.MER.2.798 Skeleton and skin. Adult female. Batouri district, 4¼°N, 14¼°E, Cameroun. Collected on 25 February 1934. Missing 1 right carpal, 1 left tarsal, and 1 right toe bone.

NH.MER.2.799 Skeleton and skin. Adult female. Batouri district, 4¼°N, 14¼°E, Cameroun. Collected on 25 February 1934. Atlas vertebra missing, 2 other cervical vertebrae broken, no coccyx.

NH.MER.2.801 Skull. Female. Batouri district, 4¼°N, 14¼°E, Cameroun. Collected on 25 February 1934.

NH.MER.2.808 Skull. Female. Batouri district, 4¼°N, 14¼°E, Cameroun. Collected on 27 February 1934. The lower jaw and a zygomatic arch are broken.

NH.MER.2.828 Skeleton and skin. Juvenile female. Batouri district, 4¼°N, 14¼°E, Cameroun. Collected on 13 March 1934. Ht. 35½ in.; hf. 24 in.; sp. 50 in.; g. 30 in.; b. 34 in.; hrc. 21 in.; e. 2 in. × 1⅜ in.; n. 14½ in.; bc. 8¼ in.; fm. 8½ in.; w. 6 in.; th. 13 in.; cf. 7¾ in.; a. 6¾ in.; h. 6 in. × 3⅞ in.; ft. 6½ in. × 4½ in.; fta. 20½ in. Diseased around the nose. Missing 1 vertebra (3 broken), patellae, and several bones from both hands and both feet. Dental age approx. 3½ – 4 years.

NH.MER.2.835 Skeleton and skin. Adult male. Batouri district, 4¼°N, 14¼°E, Cameroun. Collected on 18 March 1934. Ht. 69 in.; hf. 43½ in.; sp. 101½ in.; g. 60 in.; b. 49½ in.; hrc. 37½ in.; e. 1 in. × 2 in.; n. 37 in.; bc. 19 in.; fm. 19½ in.; w. 12 in.; th. 33 in.; cf. 18¼ in.; a. 15 in.; h. 11½ in. × 8¼ in.; ft. 13½ in. × 8⅞ in.; fta. 39 in.; wt. 168 kg; skin wt. 19 kg. Very hairy arms, hairs 5½ to 6 in. Sundry spear holes in skin.

NH.MER.2.847 Skeleton and skin. Subadult male. Batouri district, 4¼°N, 14¼°E, Cameroun. Collected on 25 March 1934. Dental age approx. 5½ years.

NH.MER.2.854 Skeleton and skin. Adult male. Batouri district, 4¼°N, 14¼°E, Cameroun. Collected on 30 March 1934. Ht. 61¾ in.; hf. 40 in.; sp. 91 in.; g. 55 in.; b. 57 in.; hrc. 23 in.; e. 2½ in. × 1¾ in.; n. 22½ in.; bc. 15½ in.; fm. 15¾ in.; w. 10½ in.; th. 25½ in.; cf. 14½ in.; a. 12⅝ in.; h. 9 in. × 6¾ in.; ft. 12 in. × 6¾ in.; tb. 3⅛ in.; mf. 4¾ in.; bt. 3⅞ in.; mt. 2¾ in.; fta. 39 in.; wt. 108 kg; skin wt. 13 kg. Herd bull. Whole back hairy and grey. Arms nothing like so hairy as no. 835. Very little brown on top of head. It would appear that the left cheek bone was broken some time before death. It has supernumerary 3rd molars. The scapulae and pelvic bones are broken and parts of both are missing. Most of the sacrum is missing. The bones are in very bad condition.

NH.MER.2.855 Skeleton and skin. Subadult female. Batouri district, 4¼°N, 14¼°E, Cameroun. Collected on 30 March 1934. Ht. 42 in.; hf. 26 in.; sp. 60¼ in.; g. 35½ in.; b. 40 in.; hrc. 23½ in.; e. 2 in. × 1¼ in.; n. 15¼ in.; bc. 9¼ in.; fm. 10 in.; w. 6⅝ in.; th. 16 in.; cf. 9¼ in.; a. 8⅕ in.; h. 7¾ in. × 5 in.; ft. 8½ in. × 2¼ in. [*sic*]; fta. 25¼ in.; wt. 33 kg. The sternum and coccyx are missing.

NH.MER.2.856 Skeleton and skin. Adult female. Batouri district, 4¼°N, 14¼°E, Cameroun. Collected on 2 April 1934. Ht. 54 in.; hf. 34½ in.; sp. 65¼ in.; g. 42½ in.; b. 49 in.; hrc. 29¼ in.; e. 2 in. × 1½ in.; n. 21 in.; bc. 12¼ in.; fm. 13½ in.; w. 8⅜ in.; th. 21 in.; cf. 12¼ in.; a. 10⅛ in.; h. 8¾ in. × 5⅜ in.; ft. 10⅝ in. × 6¼ in.; fta. 32 in.; wt. 65 kg; skin wt. 9 kg. Very grey, fine red top to head.

NH.MER.2.857 Skeleton and skin. Adult female. Batouri district, 4¼°N, 14¼°E, Cameroun. Collected on 3 April 1934. Ht. 50 in.; hf. 31½ in.; sp. 66⅓ in.; g. 39½ in.; b. 49 in.; hrc. 28 in.; e. 2¼ in. × 1½ in.; n. 16¼ in.; bc. 10¾ in.; fm 12 in.; w. 7¾ in.; th. 20½ in.; cf. 10 in.; a. 8¾ in.; h. 8⅝ in. × 5¾ in.; ft. 10 in. × 6½ in.; fta. 31½ in.; wt. 52 kg; skin wt. 7 kg. Beast all black, red crest almost unnoticeable. The 13th left rib is missing. Dental age approx. 7½ years.

NH.MER.2.862 Skeleton and skin. Adult male. Batouri district, 4¼°N, 14¼°E, Cameroun. Collected on 9 April 1934. Ht. 61 in.; hf. 38 in.; sp. 86½ in.; g. 51½ in.; b. 54¼ in.; hrc. 32¾ in.; e. 2⅜ in. × 1⅝ in.; n. 21 in.; bc. 14½ in.; fm. 14 in.; w. 10 in.; th. 24 in.; cf. 12¾ in.; a. 12 in.; h. 9¾ in. × 7⅛ in.; ft. 11½ in. × 7¾ in.; fta. 35 in. Said to be solitary. Almost no very noticeable red crest. Wisdom teeth not up.

Native report says that if a female gorilla is killed when its male young is old enough to shift for itself, he becomes a 'solitary' for the time being, until he is big and strong enough either to oust a reigning bull or hold his own in a troop; no. 862 is said to be one of the young solitary ones. I should imagine that the deserted young male would be about a quarter grown before he would shift for himself and not allowed in a troop. The above information I believe to be quite correct considering the age of no. 862. It is interesting to note that although no. 862 is a young beast, wisdom teeth not yet up, his measurements are considerably more than the average adult female.

NH.MER.2.865 Skeleton and skin. Juvenile female. Batouri district, 4¼°N, 14¼°E, Cameroun. Collected on 11 April 1934. Ht. 44¼ in.; hf. 28½ in.; sp. 61 in.; g. 35½ in.; hrc. 25⅛ in.; e. 2⅛ in. × 1⅛ in.; n. 17¼ in.; bc. 10¼ in.; fm. 9¼ in.; w. 6⅞ in.; th. 16¾ in.; cf. 9¾ in.; a. 8¾ in.; h. 7¼ in. × 5 in.; ft. 8¼ in. × 5 in.; fta. 35 in.; wt. 37 kg. Dental age approx. 6 years.

NH.MER.2.868 Skeleton and skin. Juvenile female. Batouri district, 4¼°N, 14¼°E, Cameroun. Collected on 13 April 1934. Ht. 40¼ in.; hf. 25¾ in.; sp. 58¼ in.; g. 33¼ in.; b. 36 in.; hrc. 22⅜ in.; e. 2 in. × 1½ in.; n. 12¾ in.; bc. 8½ in.; fm. 9 in.; w. 6¼ in.; th. 14¼ in.; cf. 8 in.; a. 7¼ in.; h. 7⅛ in. × 5 in.; ft. 8⅝ in. × 5 in.; fta. 23¼ in.; wt. 26 kg. Red crest very noticeable and large, going well down into nape of neck and whole breadth of head between ears; fringe of white hairs all round face particularly across eyebrow ridge. Whole beast grey with slightly brownish tinge in places, rather darker on arms and legs; under hair very grey and skin also greyish. Hair on buttocks and bottom of stomach reddish brown. The first specimen of this colouration procured.

NH.MER.2.875 Skeleton and skin. Subadult female. Kanyol, 4°N, 14°E, Batouri district, Cameroun. Collected on 26 April 1934.

NH.MER.2.877 Skeleton and skin. Adult female. Batouri district, 4¼°N, 14¼°E, Cameroun. Collected on 26 April 1934. Note small crest.

NH.MER.2.878 Skeleton and skin. Adult female. Batouri district, 4¼°N, 14¼°E, Cameroun. Collected on 26 April 1934. Missing 1 left toe bone.

NH.MER.2.879 Skeleton and skin. Adult male. Batouri district, 4¼°N, 14¼°E, Cameroun. Collected on 26 April 1934. Big, no measurements or notes. The left patella is missing.

NH.MER.2.880 Skeleton and skin. Juvenile male. Batouri district, 4¼°N, 14¼°E, Cameroun. Collected on 26 April 1934. Missing 4 left carpals, 4 right carpals, 2 right tarsals and 2 left tarsals.

NH.MER.2.887 Skeleton and skin. Infant female. Batouri district, 4¼°N, 14¼°E, Cameroun. Collected on 26 April 1934. The hyoid bone is missing.

NH.MER.2.889 Skin. Infant female. Batouri district, 4¼°N, 14¼°E, Cameroun. Collected on 26 April 1934.

NH.MER.2.890 Skin. Infant female. Batouri district, 4¼°N, 14¼°E, Cameroun. Collected on 26 April 1934.

NH.MER.3.2 Skull. Female. Between Batouri and Lomié, 3¾°N, 13¾°E, Cameroun. Collected on 28 February 1935. Slightly damaged at back.

NH.MER.3.10 Skull. Adult male. Between Batouri and Lomié, 3¾°N, 13¾°E, Cameroun. Collected on 5 March 1935. Teeth missing.

NH.MER.3.11 Skull. Female. Between Batouri and Lomié, 3¾°N, 13¾°E, Cameroun. Collected on 5 March 1935.

NH.MER.3.20 Skeleton (incomplete). Adult male. Between Batouri and Lomié, 3¾°N, 13¾°E, Cameroun. Collected on 20 March 1935.

NH.MER.3.21 Postcranial skeleton. Adult male. Between Batouri and Lomié, 3¾°N, 13¾°E, Cameroun. Collected on 20 March 1935. It is missing 2 right toe bones and the right 13th rib. The skull was sold, and is in the Hunterian Museum of the Royal College of Surgeons of England (no. RCSOM/A 64.21).

NH.MER.3.22 Skeleton. Juvenile male. Between Batouri and Lomié, 3¾°N, 13¾°E, Cameroun. Collected on 20 March 1935. The clavicles are missing. Dental age approx. 7 years.

NH.MER.3.28 Postcranial skeleton. Adult male. Between Batouri and Lomié, 3¾°N, 13¾°E, Cameroun. Collected on 24 March 1935. The skull was sold, and is in the Hunterian Museum of the Royal College of Surgeons of England (no. RCSOM/G 67.13).

NH.MER.3.29 Skeleton and skin. Juvenile female. Between Batouri and Lomié, 3¾°N, 13¾°E, Cameroun. Collected on 26 March 1935. Missing 7 carpals and some other bones of left hand, 2 carpals of right hand and patellae. Dental age approx. 3–3½ years.

NH.MER.3.31 Skull. Male. Between Batouri and Lomié, 3¾°N, 13¾°E, Cameroun. Collected on 28 March 1935.

NH.MER.3.32 Skull. Juvenile male. Between Batouri and Lomié, 3¾°N, 13¾°E, Cameroun. Collected on 28 March 1935.

NH.MER.3.33 Skull. Juvenile male. Between Batouri and Lomié, 3¾°N, 13¾°E, Cameroun. Collected on 28 March 1935.

NH.MER.3.34 Skeleton and skin. Juvenile male. Between Batouri and Lomié, 3¾°N, 13¾°E, Cameroun. Collected on 28 March 1935. The atlas is missing. Dental age approx. 4 years.

NH.MER.3.35 Skeleton and skin. Adult female. Between Batouri and Lomié, 3¾°N, 13¾°E, Cameroun. Collected on 28 March 1935. Missing all bones of right leg and right foot, atlas and clavicles; hands incomplete. Dental age approx. 8 years.

NH.MER.3.36 Skin. Subadult female. Between Batouri and Lomié, 3¾°N, 13¾°E, Cameroun. Collected on 28 March 1935. The skull was sold, and is in the Hunterian Museum of the Royal College of Surgeons of England (no. RCSOM/G 67.14).

NH.MER.3.59 Skin. Juvenile male. Between Batouri and Lomié, 3¾°N, 13¾°E, Cameroun. Collected on 13 April 1935.

NH.MER.3.80 Skin. Subadult male. Between Batouri and Lomié, 3¾°N, 13¾°E, Cameroun. Collected on 25 April 1935. Diseased lower jaw. The skeleton was sold to the Royal College of Surgeons of England; some elements of the dentition are preserved in the odontological collection (nos. RCSOM/A 64.14 and RCSOM/A 64.141).

NH.MER.3.89 Skeleton and skin. Adult female. Between Batouri and Lomié, 3¾°N, 13¾°E, Cameroun. Collected on 10 May 1935. Ht. 52¾ in.; hf. 34¾ in.; hrc. 28¼ in.; n. 17¾ in.; e. 2¼ in. × 1½ in.; c. 42 in.; b. 45½ in.; bc. 12¼ in.; fm. 12½ in.; w. 8⅝ in.; h. 9 in. × 5¾ in.; th. 20 in.; ak. 14¾ in.; bk. 10¼ in.; cf. 10½ in.; a. 10⅛ in.; ft. 10 in. × 6¼ in.; bt. 3⅜ in.; fta. 31¾ in.; sp. 76 in. The coccyx is missing.

NH.MER.3.95 Skeleton and skin. Adult female. Between Batouri and Lomié, 3¾°N, 13¾°E, Cameroun. Collected on 14 May 1935. It had a female baby, no. 97 in Merfield's list.

NH.MER.3.96 Skeleton and skin. Adult female. Between Batouri and Lomié, 3¾°N, 13¾°E, Cameroun. Collected on 14 May 1935. Ht. 53 in.; hf. 35 in.; hrc. 27½ in.; e. 2 in. × 1¼ in.; c. 44¼ in.; b. 51½ in.; bc. 11½ in.; fm. 12⅛ in.; w. 8 in.; h. 8¾ in. × 5⅜ in.; tb. 2⅝ in.; mf. 3¼ in.; n. 19½ in.; th. 22½ in.; ak. 14⅜ in.; bk. 11¼ in.; cf. 11¼ in.; a. 9¾ in.; ft. 10⅝ in. × 5⅞ in.; bt. 3⅜ in.; fta. 31½ in.; sp. 79¼ in. Left shoulder badly grazed. The atlas and coccyx are missing.

NH.MER.3.98 Skeleton and skin. Juvenile female. Between Batouri and Lomié, 3¾°N, 13¾°E, Cameroun. Collected on 14 May 1935. Ht. 37½ in.; hf. 24¼ in.; sp. 55 in. Dental age approx. 3½ years.

NH.MER.3.99 Skeleton and skin. Juvenile male. Between Batouri and Lomié, 3¾°N, 13¾°E, Cameroun. Collected on 14 May 1935. Ht. 31¼ in.; hf. 21¼ in.; sp. 49 in. It is missing the carpals of both hands, 1 left tarsal, and the atlas vertebra. Dental age approx. 2½ years.

NH.MER.3.117 Skeleton and skin. Juvenile female. Between Batouri and Lomié, 3¾°N, 13¾°E, Cameroun. Collected on 27 May 1935. Ht. 29¼ in.; hf. 19¼ in.; sp. 42 in. Hair on back curly. Hairs on and round head much lighter than rest of body. Fourth and fifth fingers of right hand from the 1st joint including nails light yellow. Thumb and 1st finger of left hand including nails, from 1st joint, light yellow. Big toe of right foot all light yellow, 3rd and smaller toe of left foot light yellow. Quite a white patch including the hair and skin of the lower side of the left forearm, the patch 3¼ in. × 3½ in. The sternum is missing.

NH.MER.3.119 Skeleton and skin. Subadult male. Between Batouri and Lomié, 3¾°N, 13¾°E, Cameroun. Collected on 29 May 1935. It is missing the atlas and axis vertebrae, 1 left carpal and the sternum. Dental age approx. 3½ years.

NH.MER.3.120 Skin. Subadult female. Between Batouri and Lomié, 3¾°N, 13¾°E, Cameroun. Collected on 29 May 1935. Skin poor. The skull was sold to the Royal College of Surgeons of England, and is no. RCSOM/G 67.12 in their collection.

NH.MER.3.121 Skin. Adult female. Between Batouri and Lomié, 3¾°N, 13¾°E, Cameroun. Collected on 29 May 1935. The skull was sold to the Royal College of Surgeons of England, and is no. RCSOM/G 67.11 in their collection.

NH.MER.3.123 Skin. Juvenile female. Between Batouri and Lomié, 3¾°N, 13¾°E, Cameroun. Collected on 29 May 1935. The skull was sold to the Royal College of Surgeons of England, and is no. RCSOM/G 67.1 in their collection.

NH.MER.3.124 Skin. Infant male. Between Batouri and Lomié, 3¾°N, 13¾°E, Cameroun. Collected on 29 May 1935. The skull was sold to the Royal College of Surgeons of England, and is no. RCSOM/A 64.13 in their collection.

NH.MER.3.135 Skeleton and skin. Adult male. Between Batouri and Lomié, 3¾°N, 13¾°E, Cameroun. Collected on 7 June 1935.

NH.MER.3.136 Skeleton and skin. Adult female. Between Batouri and Lomié, 3¾°N, 13¾°E, Cameroun. Collected on 7 June 1935.

NH.MER.3.137 Skeleton and skin. Infant male. Between Batouri and Lomié, 3¾°N, 13¾°E, Cameroun. Collected on 7 June 1935. Bones in bad condition and most broken. Ribs, pelvic bones and vertebrae completely broken; scapulae missing. Dental age approx. 1−1½ years.

The above 3 specimens formed a family.

NH.MER.3.138 Skeleton and skin. Adult female. Between Batouri and Lomié, 3¾°N, 13¾°E, Cameroun. Collected on 7 June 1935. Skinned in the bush. Skin badly speared. Long- and narrow-faced. No sign of red crest, beast all black. Missing 1 lumbar vertebra and coccyx.

NH.MER.3.139 Skeleton and skin. Adult female. Between Batouri and Lomié, 3¾°N, 13¾°E, Cameroun. Collected on 7 June 1935. Skinned in the bush. Plump- and round-faced, red crest, and hair tipped with grey. These 2 beasts show the difference in the shape of the face, as shown so well in Carl Akeley's 'In Brightest Africa', p. 222. Missing 1 left toe bone.

NH.MER.3.141 Skeleton and skin. Juvenile male. Between Batouri and Lomié, 3¾°N, 13¾°E, Cameroun. Collected on 8 June 1935. Missing 1 right rib. Dental age approx. 2½ years.

NH.MER.3.150 Skeleton and skin. Adult female. Between Batouri and Lomié, 3¾°N, 13¾°E, Cameroun.

Collected on 18 June 1935. Missing 2 left finger bones and 2 right finger bones.

NH.MER.3.160 Skeleton and skin. Subadult female. Between Batouri and Lomié, 3¾°N, 13¾°E, Cameroun. Collected on 1 July 1935. Dental age approx. 5½−6 years.

NH.MER.3.169 Skeleton and skin. Subadult male. Between Batouri and Lomié, 3¾°N, 13¾°E, Cameroun. Collected on 11 July 1935. Missing 1 right tarsal and 1 left tarsal. Dental age approx. 5 years.

NH.MER.3.170 Skin. Juvenile female. Between Batouri and Lomié, 3¾°N, 13¾°E, Cameroun. Collected on 11 July 1935.

NH.MER.3.174 Skeleton. Adult female. Between Batouri and Lomié, 3¾°N, 13¾°E, Cameroun. Collected on 16 July 1935.

NH.MER.3.177 Skull. Female. Between Batouri and Lomié, 3¾°N, 13¾°E, Cameroun. Collected on 31 July 1935.

NH.MER.3.180 Skin. Infant male. Gadji, southwest of Batouri, approx. 4°N, 14°E, Cameroun. Collected on 2 September 1935. The skull was sold to the Royal College of Surgeons of England, and is no. RCSOM/A 64.11 in their collection.

NH.MER.3.184 Skin. Infant male. Gadji, southwest of Batouri, approx. 4°N, 14°E, Cameroun. Collected on 11 September 1935. The skull was sold to the Royal College of Surgeons of England, and is no. RCSOM/A 64.12 in their collection.

NH.MER.3.57 Skeleton and skin. Adult female. Between Batouri and Lomié, 3¾°N, 13¾°E, Cameroun. Collected on 10 December 1935. Missing 1 lumbar vertebra. Dental age approx. 8 years.

NH.MER.3.58 Skeleton and skin. Adult female. Between Batouri and Lomié, 3¾°N, 13¾°E, Cameroun. Collected on 10 December 1935. The end of the sacrum is missing.

NH.MER.3.211 Skull. Adult male. Bertoua district, 4°35′N, 13¾°E, Cameroun. Collected on 2 March 1936.

NH.MER.3.212 Skull. Adult male. Bertoua district, 4°35′N, 13¾°E, Cameroun. Collected on 2 March 1936.

NH.MER.3.264 Skeleton and skin. Adult male. Lomié district, 3¼°N, 13½°E, Cameroun. Collected on 25 May 1936. Ht. 64¾ in.; hf. 44¾ in.; hrc. 14½ in. [sic]; n. 30½ in.; c. 62 in.; b. 65¼ in.; bc. 23½ in.; fm. 20¼ in.; w. 13 in.; h. 10 in. × 7¼ in.; th. 31½ in.; ak. 22½ in.; bk. 17⅞ in.; cf. 18½ in.; a. 15⅝ in.; ft. 12 in. × 8 in.; fta. 35 in.; sp. 89¾ in.; sitting ht. 40¼ in. The sternum is missing.

NH.MER.3.283 Skull. Adult male. Lomié district, 3¼°N, 13½°E, Cameroun. Collected on 13 June 1936. Remarkably light.

NH.MER.3.342 Skeleton and skin. Adult male. Lomié district, 3¼°N, 13½°E, Cameroun. Collected on 12 July 1936. Dental age approx. 9–11 years.

NH.MER.3.502 Skull. Male. Lomié district, 3¼°N, 13½°E, Cameroun. Collected on 14 September 1936. The top right canine and lower right incisor are missing.

NH.MER.3.503 Skull. Adult male. Lomié district, 3¼°N, 13½°E, Cameroun. Collected on 14 September 1936. Teeth worn. Very big crest.

The following specimens listed by Merfield are not in the Powell-Cotton Museum:

II 199 Adult female. Collected in 1933? It was sold to Rowland Ward Ltd, 167 Piccadilly, London W1, on 20 April 1934 (along with young male no. 780).

II 201 Adult female. Collected in 1933? Ht. 53½ in.; hf. 36 in.; sp. 80½ in. Sold by Merfield.

II 202 in.fant male. The baby of no. 201. Sold by Merfield.

III 40 Adult male. Collected in. 1935? Ht. 65¾ in.; hf. 43 in.; hrc. 36 in.; n. 25¾ in.; e. 2⅝ in. × 1½ in.; ns. 3½ in.; c. 58¼ in.; b. 56½ in.; bc. 16 in.; fm. 15½ in.; w. 10¾ in.; h. 10½ in. × 7 in.; tb. 3⅜ in.; mf. 4⅞ in.; th. 27⅞ in.; ak. 21 in.; bk. 14¼ in.; cf. 15 in.; a. 13½ in.; ft. 13 in × 9 in.; bt. 4¾ in.; fta. 37½ in.; sp. 95 in. No signs of red crest. Skin and skeleton sold to Rowland Ward Ltd on 18 May 1936 (along with female no. 122 and young female no. 90).

III 97 Infant female. The baby of no. 95. Collected on 14 May 1935. Ht. 26⅝ in.; hf. 18½ in.; sp. 39 in. Sold to Rowland Ward Ltd on 30 June 1936 (skeleton incomplete).

III 132 Juvenile female. Collected in 1935? Ht. 34¼ in.; hf. 22¾ in.; sp. 52¾ in. The 3rd, 4th, and 5th finger nails of the right hand are white. Sold to Sir James Frank Colyer on 2 October 1936 (along with 2 other young specimens, nos. 59 and 170).

Other sales by Merfield are known only in brief detail, as shown below.

J.E. Cave, Royal College of Surgeons of England, London: three female postcranial skeletons (nos. 36, 120 and 123) on 21 November 1941.

J.F. Colyer, Odontological Museum, Royal College of Surgeons of England, London: 6 skulls (nos. 183–875) and 2 skulls (no. 80) on 17 November 1933; skeleton (no. 660) on 16 May 1934; nos. 7, 21, 28, 36, 80, 120, 121, 123, 124, 180 and 184 on 29 September 1936; 1 male (no. 210) and 3 females (nos. 209, 213 and 214) on 22 October 1936; 2 males (nos. 269 and 284), 1 young male (no. 500), and 1 young female (no. 501) on 28 July 1937; 3 adult males (nos. 224, 249 and 251), 3 females (nos. 281, 282 and 458a), and 1 young female (no. 283a) on 28 January 1937.

Edward Gerrard & Sons, 61 College Place, London: adult female (no. 140) and young male (no. 151) on 27 June 1936.

J.P. Hill, Department of Anatomy and Embryology, University College London: no. 300, and eye in spirit on 15 October 1932; no. 840, in dried state, on 15 February 1933; nos. 409, 460, 470, 490, and 1 marked I, on 19 September 1933; nos. 625, 688, 716,729 and 755 (embryological material preserved in formalin and in good condition) on 4 April 1934; nos. 786, 855, 856 and 857 on 3 September 1934.

Birmingham

University of Birmingham, School of Earth Sciences, Lapworth Museum of Geology, Edgbaston, Birmingham B15 2TT*

Jon Clatworthy

BIRUG 18620 Half skull (left side). Adult female. It was originally part of the Humphreys Odontological Museum at the University, which is now defunct. Its specimens went to the Chamberlain Museum in the University's Medical School. The Chamberlain Museum still exists, but contains only human-related material. All the nonhuman comparative anatomy material was transferred to the Lapworth Museum in 1996. The half skull is mounted on a board, with bone cut away to show details of the dentition.

MG 1 Postcranial skeleton. No hands or feet.

MG 2 Skeleton. Juvenile male. Minus 1 femur.

MG 3 Postcranial skeleton. Adult male. Minus 1 femur.

The above 3 disarticulated skeletons were obtained by the Chamberlain Museum during the inter-war period (not later than 1938) and all have been articulated and mounted at some time in the past.

Bolton

Bolton Museum, Art Gallery and Aquarium, Le Mans Crescent, Bolton BL1 1SE

Kathryn Berry; Don Stenhouse

INV:12859 Mounted skeleton. Male. On display.

INV:6812 Cranium. Juvenile. It lacks the incisors, the last 2 molars on the left side, the 1st premolar on the right side, and the last 2 molars on the right side.

INV:6813 Cranium. Adult male. Near Benito, 1°50′N, Spanish Guinea. No date, but probably prior to 1883.

Brighton

Booth Museum of Natural History, 194 Dyke Road, Brighton BN1 5AA

Jeremy M. Adams; John A. Cooper

BC100108 Mounted skeleton. Juvenile male.

BC100317 Skull. Juvenile. There are two longitudinal cuts in the cranium.

BC100318 Skull. Infant. Deciduous dentition (20 teeth).

BC100319 Skull. Juvenile.

BC100322 Skull. Adult female. The upper right I1 is missing.

BC100323 Skull. Adult male.

BC100338 Skull. Adult male. Dentition incomplete; abscesses on lower right M1 and M2.

BC100339 Skull. Adult female. All incisors missing.

BC100340 Skull. Subadult female. Dentition incomplete.

BC100341 Skull. Subadult female. Dentition incomplete.

BC100342 Skull. Subadult female. Cameroun. Dentition incomplete.

BC100343 Skull. Juvenile female. Dentition incomplete.

BC100344 Skull. Adult male. Dentition incomplete, foramen magnum artificially enlarged, mandibular condyles missing.

BC100345 Skull. Juvenile female. Dentition incomplete.

BC100346 Skull. Subadult male. Dentition incomplete.

BC100347 Skull. Adult male. Gabon. Collected in 1887. Dentition incomplete, right mandibular ramus broken (by a bullet?). It belongs to the postcranial skeleton no. 101787.

BC100348 Skull. Adult male. Dentition incomplete, abscesses present, foramen magnum artificially enlarged.

BC100349 Skull. Adult male. Upper right canine missing, bone growth over socket.

BC100350 Skull. Subadult male. Upper left canine missing.

BC100351 Skull. Adult male. Dentition incomplete.

The above 20 specimens, BC100108–BC100351 inclusive, were donated by Frederick William Lucas (1842–1932) in November 1925.

BC100352 Skull. Adult female. Donated. Lower right canine missing, bone growth over socket; upper right I1 and I2 missing, bone broken away in this region; left I1 broken off; lower right I2 set at angle and has caused concavity in I1.

BC100395 Mounted skeleton. Juvenile. Prepared by Brazenor Brothers. Donated by F.W. Lucas in November 1925.

BC101693 Right humerus (sectioned). Adult. Donated by F.W. Lucas in November 1925.

BC101694 Right femur (sectioned). Adult. Donated by F.W. Lucas in November 1925.

BC101787 Postcranial skeleton (incomplete). Adult male. Gabon. Donated by A.F. Brazenor, Brighton. It is from the same individual as the skull no. BC100347. Mr Brazenor was a commercial taxidermist and osteological preparator, and had been left with the postcranial elements by Mr Lucas, who it would appear was interested only in the cranial material from this specimen.

BC101788 Postcranial skeleton (incomplete). Adult male. Donated by F.W. Lucas in November 1925. The original label states that there are 'one or two odd bones from another gorilla' in the box.

Bristol

Bristol Museum and Art Gallery, Queen's Road, Bristol BS8 1RL

Bonnie Griffin; Samantha Hallett

BRSMG A4837 Skull, skeleton of thorax and 1 arm (partially disarticulated and smoke damaged), and assorted other limb and foot bones (very burnt). Adult male. North Gabon. Donated by Mr Arthur Gordon on 30 December 1865. They appear to be the burnt remains of a mounted skeleton, salvaged when the museum was bombed during a wartime raid. The skull is in good condition, with articulated lower jaw.

BRSMG A4890 Mounted skin. Adult male. Donated by Mr Arthur Gordon on 30 December 1865. It is probably from the same individual as no. A 4837 above, and is presumed destroyed during the war. The skin was sent to England from the Gaboon River, west coast of Africa, in a cask of rum, whilst the skeleton was in a puncheon filled with salt. Air raids on 24 November 1940 and 2 December 1940 are presumed to have destroyed the mounted specimen and others in the main hall.

BRSMG Ab1993 Cranium. Adult male. Gaboon River. Collected by Captain George Wagstaff, of the ship 'John Cabot', in December 1847. Donated by Samuel Stutchbury on 3 August 1948. It is described and figured in Owen (1849). 'Full adult dentition with merely the summits of the cusps of the molars and the margins of the incisors slightly worn'.

BRSMG Ab1994 Cranium. Adult male. Gaboon River. Collected by Captain George Wagstaff, of the ship 'John Cabot', in December 1847. Donated by Samuel Stutchbury on 3 August 1948. See Owen (1849). 'The molars are worn nearly to stumps and the crown of the canine is reduced, partly by fracture, partly by attrition, to its basal portion: its pulp had been inflamed, and had produced ulceration of the alveolus'.

BRSMG Ab1995 Cranium. Adult female. Gaboon River. Collected by Captain George Wagstaff, of the ship 'John Cabot', in December 1847. Donated by Samuel Stutchbury on 3 August 1948. The dentition is mature and shows slight wear. The top of the cranium has been divided from the rest of the skull.

The above 3 crania had been tribal fetishes, painted red and white and probably mounted on poles, hence the lack of lower jaws. Wagstaff died soon after his arrival in Bristol, having been in poor health. The only information obtained from him was that the local people, when killing one, made a fetish of the cranium. These specimens bore indications of the sacred marks in broad red stripes crossed by a white stripe, of some pigment which could be washed off.

BRSMG Ab1996 Cranium (bisected by Prof. Owen). Adult male. River Danger (=Muni River). Collected by Captain Harris, of the ship 'Englishman'. Donated on 3 August 1848. It is described and figured in Owen (1862). This cranium was bisected vertically, and longitudinally later.

BRSMG Ab1997 Cranium. Male. River Danger (=Muni River). Collected by Captain Harris, of the ship 'Englishman'. Donated on 3 August 1848.

The above 5 gorilla crania are the first ever brought to England. Samuel Stutchbury FLS (1798−1859) was the curator of the Bristol Institution for the Advancement of Science, Literature, and the Arts, and Government Surveyor, Australia. Most probably he was the donor of Ab 1996 and Ab 1997. It is not known whether he paid the collector(s) for the crania.

BRSMG Ab4335 Mounted skin. Adult male. Received from the Bristol, Clifton and West of England Zoological Society in 1948. This is 'Alfred', who arrived at Bristol Zoo Gardens on 5 September 1930 (weight 25 lb.). His parents were shot as they 'raided' a farmer's field for food, and he was captured and raised by hand (suckled by a local woman) at Yaoundé, Cameroun. He was purchased by a Greek merchant in 1929 and taken to Mbalmayo, where he was encountered and filmed by an expedition from Columbia University and the American Museum of Natural History. In 1930 he was sold to an Italian who brought him to Europe and sold him to the animal dealer C. Blazer, for whom he was held in trust at Rotterdam Zoo. He was finally purchased for £350 by Bristol Zoo. At the time of arrival in Bristol, his upper central incisors were missing: these permanent teeth were cut at the age of 6. 'Alfred' died of miliary tuberculosis on 9 March 1948 (weight 224 lb.). The height was given as 5 ft. 11 in., and the arm span 9 ft. 6 in. [Foregoing information added by GH from correspondence with Reginald E. Greed and June Sherborne, Bristol Zoo.] The carcase was sent to the Department of Anatomy, University of Bristol. The pelt was mounted by Rowland Ward Ltd, London. The flayed and eviscerated carcase

was kept in a formalin solution for many years, but as its condition deteriorated over time, and it was unlikely that it could be successfully macerated, it was eventually incinerated. [Information added by GH from correspondence with Jonathan H. Musgrave, Bristol Zoo, and Graham Frankcom, Bristol University School of Medicine.] The death mask is also in the museum.

Bristol Zoo Gardens, Clifton, Bristol BS8 3HA

Michelle Barrows

Fixed blood smears: 'Djengi', male (taken on 26 October 2000 and 10 April 2001); 'Salome', female (taken on 12 July 2005, 7 December 2006, and 8 July 2015); identity unrecorded (taken on 31 March 2009); 'Kera', female (taken on 12 February 2016).

Formalin-fixed tissue samples: 'Claus', male, multiple organs (taken on 1 May 2001); male stillbirth ('Salome's' infant), multiple organs (taken on 12 July 2005 and 1August 2005).

Frozen tissues (−80°C): 'Claus', lung (taken on 31 July 2001); 'Salome', serum (taken on 18 July 2003, 19 November 2003, 19 January 2007, 12 June 2008, and 15 August 2008); 'Romina', female, serum (taken on 12 May 2004); 'Salome', milk (taken on 12 June 2008), 'Namoki', female, plasma and whole blood (taken on 12 September 2014); 'Kera', serum (taken on 3 March 2016); 'Komale', male, serum (taken on 4 March 2016); 'Kera', urine (taken 2014−16); 'Romina', urine (taken 2014−16).

Radiographs: 'Bobby', male, thorax and forelimb (taken on 5 September 2002); 'Romina', thorax (taken on 13 August 2002), pelvic region (taken on 7 July 2004), thorax and pelvis (taken on 2 October 2006), thorax and abdomen (taken on 1 May 2008); 'Komale', forelimb (taken on 19 January 2007), whole body (taken on 12 June 2008); 'Salome', hand (date unrecorded).

Education biofact: Silverback pelt ('Daniel' or 'Samson').

University of Bristol, Faculty of Medical and Veterinary Sciences, Centre for Comparative and Clinical Anatomy, Southwell Street, Bristol BS2 8EJ

Steve Gaze; Kate A. Sparey

Mounted skeleton. Adult male. Received from Bristol Zoo c.1994. This is 'Daniel', who was born in Bristol Zoo on 10 April 1971 and was reared by his mother, 'Delilah'. The sire was 'Samson'. 'Daniel' died in Bristol Zoo on 22 August 1994. It is missing the upper right incisors, premolars, and 1st molar; lower left central incisor and 2nd molar; and the lower right incisors, premolars, and 1st and 2nd molars.

Skull. Adult male.

Skull. Adult female.

University of Bristol, School of Veterinary Sciences, Langford, BS40 5DU

Michael J. Day

According to Prof. Day, there should be retained full written pathology reports, wax blocks of tissues and stained slides for the following cases:

79-0003 Stillborn. No abnormalities detected.

79-0003a Necrotising enteritis, fibrinous peritonitis, focal hepatic necrosis.

83-0378 Multiple abscess. Mucopurulent rhinitis. Bronchiopneumonia.

84-0009 Arterial degeneration (ovary and uterus). Normal tissues.

84-0009a Focal interstitial nephritis.

84-0009b Pneumonia (*Klebsiella pneumonia* and *Staphylococcus aureus*).

84-0009c Normal tissues.

84-0009d Pneumonia (*Klebsiella pneumonia* and *Staphylococcus aureus*).

84-0145 Cystic thymus. Pancreatic fibrosis.

85-0281 Bronchopneumonia. Normal tissues. Fibrosis and 'mucosis' pancreas.

85-0281a Cystic involution thymus.

85-0281b Normal tissues.

85-0281c Fibrosis and 'mucosis' pancreas.

86-0280 Necrotic and inflamed tissue − oesophageal biopsy.

86-0297 Inhalation bronchopneumonia. Normal tissues − various (unsectioned). Oesophageal leiomyoma and reflux oesophagitis.

86-0297a Normal tissues various (unsectioned).

86-0297b Oesophageal leiomyoma and reflux oesophagitis.

V1019-94 Peritoneal fibrosis/adhesions. Gastric mucosal hyperplasia. Vascular amyloidosis − testis. Telangiectasia − liver.

V219-89 Severe ulcerative colitis.

V564-94 Bronchopneumonia/pleurisy. Neck/mediastinal/thymic abscesses/Osteomyelitis. *Salmonella typhimurium.*

V564-94a Bronchopneumonia/pleurisy. Neck/mediastinal/thymic abscesses/Osteomyelitis. *Salmonella typhimurium. Pseudomonas aeruginosa.*

V564-94b Bronchopneumonia/pleurisy. Neck/mediastinal/thymic abscesses/Osteomyelitis. *Salmonella typhimurium. Pseudomonas aeruginosa.*

V698-94 Atheromatous plaque formation within coronary arteries. Myocardial fibrosis.

V886-00 Female (10 years old). Strongyloides/emphysematous colitis/granulomatous inflammation (TB?) spleen.

Cambridge

University of Cambridge, Faculty of Archaeology and Anthropology, Department of Biological Anthropology, Downing Street, Cambridge CB2 3DZ

Maggie Bellatti; J.W. Cash; C. Bernard Denston

NB: Prof. Wynfrid Laurence Henry Duckworth (1870−1956) was a demonstrator in human anatomy, then lecturer in physical anthropology, at Cambridge University. He combined the separate collections of human and nonhuman primate materials from three institutions within the University (the Museum of Zoology, the Anatomy Department and the Museum of Archaeology and Anthropology) into a single collection, under the aegis of the Laboratory of Physical Anthropology, which was established by Dr Jack C. Trevor and was named in honour of Duckworth following his retirement in 1940 at the age of 70. The combined collections were initially housed in the basement of the Oriental Studies building at the Sedgwick site, and in 1988−89 were moved to the basement of Keynes House, at the Old Addenbrooke's site in Trumpington Street. From 2006, the collections were gradually transferred to a new, purpose-built research facility in Fitzwilliam Street, Cambridge CB2 1QH, where the Duckworth Laboratory now forms part of the Leverhulme Centre for Human Evolutionary Studies.

G 1 Mounted skeleton. Adult male. Studied and reported on by Prof. W.L.H. Duckworth (his no. 14).

G 2 Postcranial skeleton. Adult male. Obtained from Rouilly in July 1931. Some bones are missing.

G 3 Skeleton. Adult male. Obtained from Rouilly in November 1931.

G 4 Skeleton. Adult male. Obtained from Rouilly in November 1931. Gunshot wound in cranium.

G 5 Postcranial skeleton (almost complete). Juvenile female. Obtained from Rouilly in November 1931.

G 6 Postcranial skeleton (almost complete) and mandible. Adult female. Obtained in exchange from Gerrard in December 1905.

G 7 Skeleton. Subadult male. From the Duckworth private collection (his no. 1). Epiphyses not yet fully united.

G 8 Skeleton. Juvenile female. From the Duckworth private collection (his no. 2). It was probably acquired by him in 1916.

G 9 Postcranial skeleton (almost complete). Adult. From the Duckworth private collection (his no. 3).

G 11 Skull. Juvenile. Setté Cama, Gabon. From the Duckworth private collection (his no. 12). It was acquired by him in August 1898. Milk dentition: 1st permanent molars not erupted. Lower central incisors and lower left canine missing. There is a bilateral subdivision of the greater wing of the sphenoid.

G 12 Two juvenile postcranial skeletons. Setté Cama, Gabon. From the Duckworth private collection (his no. 11). Acquired by him in August 1898. The skull G 11 belongs to one of these skeletons. The skull of the other skeleton was also in Duckworth's collection, but its identity is uncertain now. A sectioned skull present with these bones may be the one in question.

G 13 Skull. Adult male. From the Duckworth private collection. The maxilla lacks both canines, both central incisors and the left 2nd molar; the mandible lacks the incisors.

G 14 Skull. Adult male. Setté Cama, French Congo. Obtained by Duckworth for his collection from Perks on 4 August 1908. The base is damaged and several teeth are missing.

G 15 Skull. Adult female. The teeth are worn but complete except for the upper left central incisor. The cranium is now missing, leaving the mandible only.

G 16 Skull. Adult male. Ouesso region, Moyen Congo. Collected in 1928. The upper incisors and two lower incisors are missing. The mandible is articulated.

G 17 Skull. Juvenile male. From the Duckworth private collection; it was obtained by him from Gerrard in 1899 or 1900. The permanent canines and 3rd molars are not yet erupted. The upper right central incisor is missing. It has a narrow, cramped premaxilla with rotated lateral incisors.

G 18 Skull. Juvenile. From the Duckworth private collection (his no. 9). Acquired by Duckworth from Gerrard. It has the milk dentition with the 1st permanent molars, although the milk teeth are missing, except for the milk molars. The occipital bone is missing.

G 19 Skull. Adult female. Obtained from Gerrard. All the lower incisors, 3 upper incisors and the upper left canine are missing. There are 4th molars in both jaws. It is damaged near the foramen magnum.

G 20 Skull. Adult female. Obtained from Gerrard. The upper jaw lacks the right canine and both lateral incisors; the lower lacks both canines, both lateral incisors and the left central incisor. It is slightly damaged near the foramen magnum.

G 21 Skull. Adult female. From the Duckworth private collection (his no. 4). Obtained by Duckworth from Gerrard. It is damaged round the foramen magnum.

G 22 Skull. Adult female. Obtained from Gerrard. The lower left lateral incisor is missing. The lower left canine and 1st premolar are damaged. The mandible is repaired across the symphysis with wire. The mandile is missing now.

G 23 Skull. Adult female. Obtained from Gerrard. The teeth are but little worn; 2 lower incisors and the upper left 3rd molar are missing.

G 24 Skull. Juvenile female. From the Duckworth private collection (his no. 7). Obtained by Duckworth from Gerrard. The permanent canines and 3rd molars are unerupted. The floors of the orbits are damaged.

G 25 Skull. Juvenile. From the Duckworth private collection (his no. 10). Originally obtained from Gerrard. The 1st permanent molars are just erupting. It is missing the lower right 1st molar (permanent), the upper right milk canine, the upper left central milk incisor and the basioccipital bone.

G 26 Skull. Adult male. Obtained from Rouilly in July 1931.

G 27 Skull. Adult male. Obtained from Rouilly in July 1931.

G 28 Skull. Adult female. Obtained from Rouilly in July 1931. The upper right canine is missing.

G 29 Skull. Adult male. Obtained from Rouilly in July 1931. The upper dentition is damaged on the left side.

G 30 Skull. Adult female. Obtained from Rouilly in July 1931. It has a large 4th upper right molar. It is damaged round the foramen magnum.

G 31 Skull. Adult female. Obtained from Rouilly in July 1931. The upper incisors, canines and 3rd molars are missing. It is damaged round the foramen magnum.

G 32 Skull. Juvenile. Obtained from Rouilly in February 1931. The 1st permanent molars are erupted; the central incisors are partly erupted.

G 33 Cranium. Male. Obtained from Rouppert in August 1928. The base is damaged, and the incisors are missing.

G 34 Cranium. Adult male. Obtained from Rouppert in August 1928. All the teeth are missing. The right side of the face and palate is badly damaged, as is also the region of the foramen magnum.

G 35 Cranium. Adult male. Obtained from Rouppert in August 1928. The incisors are missing, except for the right lateral. It is damaged round the foramen magnum.

G 36 Cranium. Adult male. Obtained from Rouppert in August 1928. The incisors are missing, except for the left central. The base and right cheek region are damaged.

G 37 Cranium. Adult male. Obtained from Rouppert in August 1928. The incisors and canines are missing, and it is damaged near the foramen magnum.

G 38 Cranium. Adult female. Obtained from Rouppert in August 1928. The incisors are missing.

G 39 Facial skeleton only. Male. Obtained from Rouppert in August 1928. All the postcanine teeth are present.

G 40 Cranium. Adult male. Obtained from Rouppert in August 1928. This specimen is now missing. Prof.

Duckworth noted that it is a large specimen, with an overall length well over 310 mm, thus exceeding the largest skull measured by Coolidge, which is in the Leeds Museum Discovery Centre (with a cast at Tring), and has an overall length of 305 mm only.

G 41 Cranium. The vault is sawn off, and bone has been removed to display air sinuses. The teeth are complete but loose.

G 43 Mandible. Adult. Obtained from Gerrard. From W.L.H. Duckworth's student series; it was acquired by him in 1913. The body is fractured on the left side and wired. All the postcanine teeth and 1 incisor are present.

G 44 Mandible. Adult. The right canine and right central incisor were lost during life; the alveoli were not absorbed but filled up. The left canine is broken.

G 45 Mandible. Obtained from Rouppert in August 1928. Duckworth private collection no. 8. The incisors are missing.

G 46 Mandible. Adult male. Obtained from Rouppert in August 1928. All the incisors, the right 3rd molar, the left 1st molar and the left coronoid process are missing.

G 47 Mandible. Adult male. Obtained from Rouppert in August 1928. The dentition is complete and little worn.

G 48 Mandible. Adult male. Obtained from Rouppert in August 1928. The incisors are lost, except for the right lateral.

G 49 Mandibular fragment, including symphyseal region and several postcanine teeth. Adult. Obtained from Rouppert in August 1928.

G 50 Mandible. Obtained from Rouppert in August 1928.

G 51 Postcranial skeleton (almost complete). Adult male. Obtained from Rouppert in August 1928.

G 52 Postcranial skeleton. Adult male. Obtained from Rouppert in August 1928. The epiphyses are not united.

G 53 Postcranial skeleton (almost complete). Adult male. Obtained from Rouppert in August 1928.

G 54 Postcranial skeleton. Obtained from Rouppert in October 1928. It is imperfect, and from a small specimen.

G 55 Vertebral column: 6 cervical (no atlas), 13 thoracic, 3 lumbar, 6 sacral (labelled 1205); 2 innominate bones (labelled 1219) and lower limb long bones (labelled 1221). Adult. These belong to a single young adult (the epiphyses are not united).

G 56 Vertebral column: 7 cervical, 13 thoracic, 4 lumbar, 4 sacral (labelled 1204); 2 ilia, 2 pubes and 2 ischia (labelled 1218). These bones are probably from the same immature animal.

G 57 Vertebral column: 6 cervical (no axis), 12 thoracic, 4 lumbar, 5 sacral; right hip bone. These are from the same adult individual (labelled 1206).

G 58 Thirteen ribs of right side (labelled 1209); ribs of left side, missing nos. 1 and 13 (labelled 1208).

Obtained from Rouilly on 10 November 1927. These may belong to a single animal.

G 59 Thirteen ribs of right side (labelled 1210); 13 ribs of left side (labelled 1207). Obtained from Rouilly in November 1927. These may belong to a single animal.

G 60 Eight ribs of right side and nine ribs of left side (labelled 1211). Obtained from Rouilly in November 1927. These are from a single, large animal.

G 61 Pelvis (articulated) and various limb bones (3 femora, 2 humeri, 1 radius, 1 ulna, 1 fibula). Purchased from Gerrard in August 1928. Right scapula, obtained in 1931. An additional articulated pelvis, acquired in 1928, is probably of a chimpanzee.

G 62 Limb bones (humerus, radius, ulna and scapula of right side; femur, tibia and fibula of left side). Obtained from Hose in 1912, now missing. Left innominate bone and part of skull cap. Purchased from Frank, now missing. All are from the Duckworth private collection.

G 63 Upper limb bones: 2 humeri, 2 radii, 2 ulnae (labelled 1217); lower limb bones: 2 femora, 2 tibiae, 2 fibulae, 2 calcanei, 1 left talus (labelled 1222). These are from a single animal. The epiphyses are almost united.

G 64 Upper limb bones: 2 scapulae, 2 humeri, 2 radii, 2 ulnae (labelled 1212, 1216). Juvenile. Some of the epiphyses are missing.

G 65 Lower limb bones (2 femora, 2 tibiae, 2 fibulae, 2 calcanei, 2 tali). Juvenile. Most of the epiphyses are still attached. All these bones are now missing, apart from the ankle bones.

G 66 Limb bones (2 femora, 2 tibiae, left fibula, left radius). Adult. The epiphyses are largely distinct.

G 67 Right pubis and ischium and both ilia. Juvenile.

G 68 Left pubis and ischium and parts of both ilia. The epiphyses are not united.

G 69 Pair of hip bones. Adult. Although labelled as 'gorilla hip bones', they appear more likely to belong to a chimpanzee.

G 70 Pair of scapulae (labelled 1213). The right one is damaged. Scale-like epiphyses are evident.

G 71 Pair of scapulae (labelled 1214, 1215). Adult. They appear to be from the same individual. The left scapula is perforated and injured.

G 72 Radius and ulna of right side. Juvenile. They appear to belong to one individual. The epiphyses are not quite fused.

G 73 Bones of manus and pes. A mixed 'bag' seemingly from two individuals at least.

G 74 Skull. Female.

G 75 Skull. Adult male. Obtained from Gerrard. Calvarium sliced.

G 77 Cranium (left half). Subadult male. Pathology: some periosteal reaction combined with erosion to produce appearance of porosity.

G 78 Calvarium (sectioned). Juvenile.

Cranium. Male. Obtained from Gerrard in 1906. Calvarium broken and parts missing.

The following specimens are part of the 'Pearson Collection' of human and nonhuman material, which was transferred from University College, London in late 1945 or early 1946.

Pr. 52.0.1 Mounted skeleton. Juvenile male. Kivu, Belgian Congo.

Pr. 52.0.2 Skull (cast). Adult male.

Pr. 52.0.3 Skull. Adult male. Western Congo.

Pr. 52.0.4 Cranium. Adult female. Western Congo.

Pr. 52.0.5 Skull. Infant.

Pr. 52.0.6 Skeleton (almost complete). Infant.

Pr. 52.0.7 Femur. Adult male.

Pr. 52.0.8 Left femur. Adult male.

Pr. 52.0.9 Femur. Adult male. It is missing now.

Pr. 52.0.10 Right femur. Adult female.

Pr. 52.0.11 Right femur. Adult female.

Pr. 52.0.12 Right femur. Adult female.

Pr. 52.0.13 Right and left ulnae, right and left scapulae. Adult female?

Pr. 52.0.14 Femur. Adult female. It is missing now.

University of Cambridge, Department of Zoology, Downing Street, Cambridge CB2 3EJ

Adrian E. Friday; Mathew Lowe

E.7126.A Skull (cast). Male. Presented by J.W. Clark. The original skull was collected by P.B. Du Chaillu, and is in the Natural History Museum in London.

E.7126.B Mounted skeleton. Adult male. Presented by J.W. Clark. It had been purchased at Du Chaillu's sale. [This may refer to the auction by John Crace Stevens of a large number of bird and mammal skins and skeletons, all collected by Du Chaillu in Equatorial West Africa, which took place at 38 King Street, Covent Garden, London W.C., on 12 June 1863.]

E.7126.C Skull (cast). Female. Presented by Edwin R. Lankester.

E.7126.D Skull (cast). Juvenile female. Presented by Edwin R. Lankester.

E.7126.E Left pes (articulated). Received in exchange from the Royal College of Surgeons of England.

E.7126.F Right manus (disarticulated). Received in exchange from the Royal College of Surgeons of England.

E.7126.G Skull. Juvenile. Presented by Prof. Lewis in 1881.

E.7126.H Mounted skeleton. Adult female. Ogooué River, about 150 mi. from Cap Lopez. Collected by R.W. Stone on 4 July 1875. Stone gave the skeleton, skin and fetus of this female, plus the skeleton of a small male gorilla which had died whilst in his possession, to Captain Hopkins, the Consul at Loanda, who in turn presented the female skeleton to Cambridge University Museum of Zoology in 1876. Length of body from crown of head to sole of foot 58 in.; neck 19½ in.; width across the shoulders 23 in.; chest 42½ in.; abdomen 54 in.; waist 30 in.; biceps 13 in.; arm length from shoulder to elbow 14 in.; arm length from shoulder to wrist 26 in.; hand length from the wrist to the forefinger 11 in.; length of palm 6¾ in.; width of palm 4 in.; middle finger length 4 in.; middle finger circumference 3½ in.; thumb circumference 3 in.; thigh 24 in.; above knee 11½ in.; calf 10½ in.; length of leg to sole of foot 22½ in.; width of foot below toes 5 in.; width of foot near heel 3½ in.; length of middle toe 2 in. One tooth is missing from the lower jaw.

E.7126.I Skull. Female. Mtene, Cameroun. Presented by Ivan Terence Sanderson (1911—73), who collected it on 1 June 1933, during his 1932/33 expedition to Cameroun.

Canterbury

Comparative Pathology, Forensic and Wildlife Health Services (WHS), c/o The Durrell Institute of Conservation and Ecology (DICE), School of Anthropology and Conservation, The University of Kent, Canterbury CT2 7NR

John E. Cooper

Skull. Adult female. Reference JEC/MC/RCS/17 December 1998 (see radiograph Figure 2.1 in this work). It came from a box of bones (mostly human) given to Michael Meredith Brown c.1935 by an old doctor, just retired from general practice in Peppard, near Reading. It was donated by Milford Hospital to the Royal College of Surgeons of England in 1997, and was passed into the care of Prof. John E. Cooper in 1999.

Stained and unstained sections of skin from gorilla and chimpanzee specimens in the Powell-Cotton Museum. Reference JEC/GL 131238 (see description and photomicrographs Figure 9.4 in Chapter 9).

Stained sections from gorilla (London Zoo), postmortem inflation of lungs M32. Slides 164, 1363, 1365 (see Figure 10.6 in this work and 1978 reference to the technique employed).

Ten mounted slides of hairs, from named parts of the body, of lowland gorillas at Bristol Zoo Gardens. Acquired in 2015. Reference JEC/BZG/2015. See Figure 9.1 and Figure 9.2 in Chapter 9.

Temporarily (in due course will be passed to the Natural History Museum, London): Slides of osseous samples from 'Guy' (London Zoo), plus relevant paraffin blocks (all prepared by Department of Veterinary Medicine, University of Cambridge) See Figure 14.5 and Figure 14.6 in this work.

Many records — clinical, postmortem and laboratory — from the period 1993 to 1995, when the Coopers were based in Rwanda working with *Gorilla beringei*. Some of these papers were rescued after the Rwanda genocide and are therefore damaged or soiled.

In addition to the specific items above, the WHS collection includes a large amount of gorilla-related literature, including: (1) reprints, articles and photocopies about the diseases and pathology of gorillas and other species (some signed or inscribed by the author(s)); (2) records and correspondence relating to the period 1993−95 when John and Margaret Cooper were based in Rwanda and administering the Centre Vétérinaire des Volcans (VVC) and (3) records and correspondence from 1995 to the present. All these items can be viewed by bona fide scientists and investigators, including colleagues from the range states in Central and West Africa.

Cardiff

Cardiff Metropolitan University, Cyncoed Campus, Cyncoed Road, Cardiff CF23 6XD

Robert Shave

The International Primate Heart Project (initially established as the European Great Ape Heart Project) is based here. The Project has so far assessed 29 gorillas over the last 3 years: 17 males (aged 4−23 years) and 12 females (aged 4−50 years). For each of these animals, there is a full echocardiogram and electrocardiogram, and also some blood pressure data. Over time, it is hoped that the database will grow in number and that DNA will be collected, so that the cardiac phenotype being characterised can be compared with the genotype. See Shave et al. (2014). See also Chapter 10: Respiratory and Cardiovascular Systems in Part I: Gorilla Pathology and Health.

National Museum of Wales, Cathays Park, Cardiff CF1 3NP

Jennifer Gallichan; Cynthia M. Merrett

NMW. Z.1908.090 Mounted skin. Kamerun. Purchased from Rowland Ward, 167 Piccadilly, London W1, in 1908.

NMW. Z.1909.014 Mounted skeleton. Purchased from Gerrard & Sons, 61 College Place, Camden Town, London N.W., in 1909.

NMW. Z.1927.409.020 Mounted skeleton. Juvenile. Received from the Science Museum, South Kensington, London, in 1927. It was part of a larger donation of various vertebrate material, dissections and skeletal.

NMW. Z.2015.015 Skeleton (partly articulated). Adult male. This may be the specimen presented by the Imperial College of Science and Technology, South Kensington, London SW7, in 1976−77. During its time at Imperial College, it was believed to be one of the artefacts used for teaching by Prof. Thomas Henry Huxley (1825−95). Imperial College had a large collection of osteological material amassed by Huxley when he was Professor of Biology (1854−85) and later Dean of the Royal College of Science (1881−95). This collection was added to through the years until the middle 1970s, when the emphasis of teaching moved away from comparative anatomy to other aspects of biology. The museum was turned into research laboratories and some specimens (mainly the larger and duplicate ones) were disposed of, whilst others are stored within the Department of Pure and Applied Biology. It was then that the National Museum of Wales was given quite a number of specimens, including the gorilla skeleton. Unfortunately, a lot of the Department's records were lost both during and after the last war, when it moved from the Huxley Building (now part of the Victoria and Albert Museum in Exhibition Road) to its present site in Prince Consort Road. [Some additional information added by GH from correspondence with Doris M. Kermack.]

Cardiff University, Cardiff School of Biosciences, Sir Martin Evans Building, Museum Avenue, Cardiff CF10 3AX

E.J. Evans; R. Presley; Hannah Shaw; Swaran G.D.A. Yarnell

Os.A.6 Mounted skeleton.

Os.A.192 Skull (sagittal section). Adult male. Ngouli, Akele, Gabon. Collected in 1928. Obtained from Tramond, Paris.

Os.A.214 Skull. Female. Purchased from Adam, Rouilly in 1930.

Os.A.215 Skull. Female. Purchased from Adam, Rouilly in 1930. The constituent bones are coloured.

Os.A.229 Skull. Juvenile male.

Os.A.257 Skull (sagittal section). Juvenile female. Purchased in 1932.

Os.A.258 Skull. Adult male. Purchased in 1932. The constituent bones are coloured.

Os.A.259 Skull (sagittal section). Adult male. Purchased in 1932.

The following specimens cannot be located:

Os.A.225 Skull. Adult male. Loaned (with Os.A. 282) to Prof. John R. Napier, Birkbeck College, University College London, in May 1962. Upon Prof. Napier's retirement in 1983 his entire collection was transferred to the Department of Anthropology, University College London, where the skulls are now kept, still on loan.

Os.A.226 Skull. Adult male. It has been missing since November 1936.

Os.A.227 Skull. Adult male.

Os.A.228 Skull. Juvenile male.

Os.A.256 Skull. Juvenile. Purchased in 1932. It is dissected to show the temporary and permanent dentition.

Os.A.260 Skull. Adult male. Purchased in 1932.

Os.A.261 Skull. Adult male. Purchased in 1932.

Os.A.282 Skull. Adult male. Out on loan (see no. Os. A.225 above).

Os.A.399 Skull (sagittal section, right half only). Juvenile.

Derby

Museum and Art Gallery, The Strand, Derby DE1 1BS

Spencer Bailey; Mrs S.J. Patrick

DBYMU 1890-234/2 Skull. Adult male. Donated by Dr Alfred Lingard. Accessioned in 1890.

DBYMU 1928-124 Mounted skin. Female. Purchased for £15 from the taxidermist Albert Adsetts, 67 London Road, Derby. Height (upright) approx. 43 in. Accessioned in 1928.

Dundee

University of Dundee, College of Life Sciences, D'Arcy Thompson Zoology Museum, Carnelley Building, Park Place, Dundee DD1 4HN

Matthew H. Jarron

DUNUC 2149 Mounted skeleton. Adult male. It is probably from a gorilla acquired by D'Arcy Wentworth Thompson (1860–1948) from William Cross, Liverpool, in 1893. There are brief references in D'Arcy Thompson's letters: W. Cross to D'Arcy T. 11 July 1893 offers '1 Gorilla in Barrel good for skeleton but I think not for stuffing 40/-'; and D'Arcy T. to W.T. Calman 22 July 1893 'The gorilla has come from Cross.' This skeleton is shown in a photograph of D'Arcy Thompson's museum taken in 1902.

DUNUC 2150 Skull. Female. It has been missing for many years now.

DUNUC 2152 Partial skeleton (thoracic region, articulated). D'Arcy Thompson received gorilla bones from Edward Gerrard in 1898 and 1909, so this may be from one of those purchases. Letter from Edward Gerrard to D'Arcy T on 26 July 1898 says he has today sent the Gorilla bones, birds & co, bought at Stevens sale; letter from Edward Gerrard to D'Arcy T. 11 March 1909: 'I will reduce the Gorillas to £4.00 and hope that will be satisfactory.' NB: This specimen is catalogued as 'possibly gorilla', but further identification is needed.

University of Dundee, Centre for Anatomy and Human Identification, MSI/WTB/JBC Complex, Dow Street, Dundee DD1 5EH

Lucina Hackman

CO20572 Skull. Juvenile.

University of Dundee, Dental Hospital and School, Comparative Anatomy Collection, Park Place, Dundee DD1 4HR

Derrick M. Chisholm

DUNUC 3239 (cranium) and 3239/1 (mandible) Skull. Adult male. Donated in the 1970s, original source unknown. The upper right lateral incisor and and the upper left 3rd molar are missing from their sockets. The lower left 2nd and 3rd molars have never erupted into the mouth. It is in this region that the mandible shows signs of a healed fracture, with a reparative ossifying fibroma on the lingual and occlusal surfaces.

Durham

Durham University, Faculty of Science, Department of Anthropology, Bilsborough Laboratory, Dawson Building, South Road, Durham DH1 3LE

Alan Bilsborough; Kris Kovarovic; Judith Manghan

Skull. Adult male.

Skull. Adult female.

The above specimens were transferred from the Department of Anthropology, Queen's Campus, University Boulevard, Thornaby, Stockton-on-Tees TS17 6BH, in October 2014.

Postcranial skeleton. Adult.

Edinburgh

National Museum of Scotland, Chambers Street, Edinburgh EH1 1JF

Andrew Kitchener

NMS.Z.1862.4.1 Brain (cast). Purchased for 10s 6d from Alex Stewart on 16 January 1862. It is from a specimen lent by Du Chaillu.

NMS.Z.1865.3.21 Model of skull and brain. Purchased for 80 francs (= £3 4s 0d) from Dr Louis Thomas Jérôme Auzoux (1797–1880), Paris, in 1865. It appears to be no longer in the collection.

NMS.Z.1868.33.1 Mounted skin. Adult male. Purchased from E. Gerrard Jnr for £100 in 1868. The skull is inside the skin.

NMS.Z.1868.33.2 Postcranial skeleton (articulated). Adult male. Purchased from E. Gerrard Jnr for £100 in 1868. It is from the same individual as the mounted skin, no. NMSZ 1868.33.1.

NMS.Z.1881.23.1 Skull (bisected). Adult male. Purchased for £5 from Henry Gibson, Leith, in June 1881. The alveoli of the molars are dissected out.

NMS.Z.1881.28 Mounted skin. Juvenile. Purchased for £10 from E. Gerrard Jnr in August 1881. Lent to St

Andrews University on 29 January 1915; donated on 6 April 1944 (therefore it is no longer in the NMS collection).

NMS.Z.1882.68 Skull. Adult male. Purchased for £5 from Henry Gibson, 3 Gibson Street, Broughton Road, in December 1882.

NMS.Z.1885.33 Mounted postcranial skeleton. Adult male. Gabon. Donated by J. Sanderson, Gresham Buildings, Basinghall Street, London, in October 1885. It is on loan to Edinburgh Zoo.

NMS.Z.1889.130 Skeleton and mounted skin. Adult female. Purchased for £20 from J. Hecht, Hamburg, in 1889. It is suckling an unregistered young.

NMS.Z.1891.55.8 Skull. Female. Purchased for £37 15s 0d from E. Gerrard Jnr in 1891. It has been missing from the collection since before 1980.

NMS.Z.1893.34.2 Skull. Adult male. Central Africa. Collected in 1884. Purchased for £3 in Dowell's Rooms in 1893. It has been missing from the collection since before 1980.

NMS.Z.1914.29 Mounted skin. Juvenile female. Bipindi, Kamerun. Purchased for £8 from Rowland Ward Ltd in 1914.

NMS.Z.1914.87.12 Mounted skin. Juvenile. Kamerun. Purchased for £8 from Rowland Ward Ltd in 1914. It is not present in the collection, and may be a duplicate registration number for NMS.Z.1914.29.

NMS.Z.1923.69 Mounted skin. Juvenile. Bapindi, Cameroon. Purchased for £25 from Rowland Ward Ltd, 167 Piccadilly, London. It is not present in the collection, and may be a duplicate registration number for NMS. Z.1914.29.

NMS.Z.1950.34 Mounted skin. Juvenile male. Purchased for £35 from Watkins & Doncaster, 36 Strand, London WC2.

Mounted skin. Adult male. It may be from the same animal as the skeleton no. NMS.Z. 1885.33.

Mounted skin. Infant.

Skull (cast). Adult male. No. 310.

Skull (cast). Adult male. No. 311.

Skull (cast). Adult female. No. 312.

Skull (cast). Adult female. It is from NMS. Z.1889.130.

The following 6 specimens were included in a large collection of miscellaneous osteological material received in 1963 from the Anatomy Department of the University of Edinburgh, having previously been in their museum.

Skull (cranium bisected). Adult male. No. 2C.64.

Skull (cranium bisected). Adult male. M. Toupperch, Paris. No. C.25.4 No. 11.

Skull (cranium bisected). Adult female. M. Toupperch, Paris. No. C.25.4 No. 11.

Skull and part skeleton. Infant. Manders' Collection 1860. NB: This may be a chimpanzee, but was tentatively reidentified as a gorilla on the basis of the position of the foramen magnum.

NMS.Z.1993.108 Skeleton. Adult female. Donated by the Royal Zoological Society of Scotland in 1993. This is 'Killa Killa', who was born at Howletts Zoo Park, Bekesbourne, Kent, on 5 December 1978, after a gestation period of 253 days, and was reared by her mother, 'Mouila'. The sire was 'Kisoro', who was on loan from the Lincoln Park Zoo, Chicago, Illinois, USA. 'Killa Killa' was also the property of Lincoln Park Zoo, and was sent on loan to Edinburgh Zoo on 10 November 1989. She was euthanased due to a long-standing perforation of the dorsal wall of the rectum on 2 June 1993. [Some additional information added by GH from correspondence with Howletts Zoo personnel.]

NMS.Z.1995.238 Skin and skeleton. Adult male. Donated by the Jersey Wildlife Preservation Trust on 16 September 1992. This is 'Jambo', who was born in Zoologischer Garten Basel, Switzerland, on 17 April 1961 after a gestation period of 252 days, and was raised by his mother, 'Achilla'. The sire was 'Stefi'. 'Jambo' was sent to Jersey Zoo on 28 April 1972, and died there of a dissecting aneurysm of the aorta on 16 September 1992. [Some additional information added by GH from correspondence with Basel Zoo personnel.]

NMS.Z.2000.106 Cadaver (preserved in spirit). Infant female. Donated by the East Midlands Zoological Society Ltd in 1995. It died 15 hours after birth at Twycross Zoo on 4 February 1995 (weight 685 g). The sire was 'Daniel' (Bristol Zoo). Head and body length 210 mm; foot length 61 mm; ear length 17 mm.

NMS.Z.2000.370.1 Skeleton. Adult male. This is 'Jeremiah', who was born in Bristol Zoo on 6 March 1984 (sire 'Daniel', dam 'Delilah'), and was removed for hand rearing on 25 August 1984. He was sent to London Zoo on 12 September 1995, then to Belfast Zoo on 29 September 1998, where he died on 27 August 2000. [Some additional information added by GH from the International Studbook for the Western Lowland Gorilla, 2010.]

NMS.Z.2001.42.001 Skeleton. Adult male. This is 'Jomie', who was born in Howletts Zoo Park, Bekesbourne, on 13 November 1980 (sire 'Djoum', dam 'Lomie') and was mother-reared. He was owned by London Zoo, and was transferred there on 6 October 1998, where he died on 26 January 2001. [Some additional information added by GH from International Studbook for the Western Lowland Gorilla, 2010.]

NMS.Z.2001.78 Skeleton. Adult male. This is 'Claus', who was born in Frankfurt Zoo on 6 July 1982 (sire 'Matze', dam 'Dorette'), and was hand-reared. He was sent on loan to Cologne Zoo on 25 February 1984, and was owned by Cologne since 2 August 1984. He was sent on loan to Paignton Zoo on 22 April 1997, and to Bristol

Zoo on 23 January 1998. He was euthanased in Bristol Zoo on 27 April 2001 (weight 160 kg), after suffering an acute bout of pneumonia. [Some additional information added by GH from International Studbook for the Western Lowland Gorilla, 2010.]

NMS.Z.2001.156 Skeleton and skin. Adult male. *Gorilla beringei graueri*. This is 'Mukisi', who arrived at Chester Zoo (North of England Zoological Society, Upton by Chester) on 17 November 1960 (weight c. 55 lb.), with a female named 'Noelle' (see also entry for University of Liverpool, Institute of Ageing and Chronic Disease). The pair had been purchased from Charles Cordier, Utu, Belgian Congo. 'Mukisi' was sent to Antwerp Zoo on 13 May 1985, and died there of purulent necrotising pleuropneumonia of the right lung (bacteriology: polybacterial, Ziehl-Neelsen negative), on 17 December 2000. Head and body length 1560 mm; foot length 323 mm/330 mm; ear length 65 mm. [Some additional information added by GH from correspondence with George S. Mottershead, Chester Zoo.]

NMS.Z.2002.185.002 Skeleton. Adult female. This is 'Diana', who arrived at Bristol Zoo on 2 August 1976, having been purchased from G. van den Brink. She was quarantined at Twycross Zoo with 'Biddy' and 'Eva'. She was sent on loan from Bristol to London Zoo on 12 September 1995, and died there on 11 May 2002 (weight 90 kg). Weight 127.9 kg on 7 January 1999. [Some additional information added by GH from correspondence with Bristol Zoo personnel.]

NMS.Z.2002.230 Skeleton. Adult male. Ngovie, Fernan Vas, West Africa (=Omboué, Gabon, 1°34′S, 9°15′E). Donated by L. Mennie in June 1881.

NMS.Z.2003.70 Skeleton and skin. Adult female. This is 'Kukee', who came originally from Cameroon, and arrived at the Municipal Zoological Gardens, Blackpool, on 4 November 1972, when aged about 10 months (weight 5 kg). She died on 7 January 2003 (weight 65 kg). [Some additional information added by GH from correspondence with Blackpool Zoo personnel.]

NMS.Z.2006.045 Skin. Adult male. This is 'Jason', who arrived at Chester Zoo on 30 July 1965, when aged about 2¼ years, having been purchased from the Jerez de la Frontera Zoo. He was sent on loan to Bristol Zoo on 10 October 1986, and died there on 9 December 1994. [Some additional information added by GH from correspondence with George S. Mottershead, Chester Zoo.]

NMS.Z.2006.60.1 Cadaver (in spirit) and muscle tissue (frozen). Neonate male. Born in Port Lympne Zoo Park on 1 May 2003; died on 4 May 2003. Weight 2400 g on 4 May 2003.

NMS.Z.2009.1.1 Skeleton and muscle tissue (frozen). Adult male. This is 'Joe Lumumba', who arrived at Twycross Zoo on 15 August 1965, aged about 1½ years (22 lb.), having been purchased from Robert Jackson,

Welsh Mountain Zoo, Colwyn Bay; he had arrived there on 27 July 1965. He died on 21 September 2008 (192 kg): elderly; peritonitis (colon rupture) and subclinical atherosclerosis. Weight 50 lb. on 24 November 1966; 144 lb. on 15 April 1970. Head and body length 1452 mm; ear length 51 mm. [Some additional information added by GH from correspondence with Natalie Evans, Twycross Zoo.]

NMS.Z.2009.1.2 Skeleton, muscle tissue (frozen) and skin of head. Adult male. This is 'Sam-Sam', who arrived at Rotterdam Zoo on 26 June 1974 (weight c. 10 kg), having been purchased from G. van den Brink, who had imported him from Sangmélima, Cameroon. He was sent on loan to Edinburgh Zoo on 3 July 1985, transferred to Blackpool Zoo on 30 March 1993, back to Edinburgh on 27 March 1999 and then to Twycross Zoo on 10 August 2006, where he died on 9 October 2008: elderly; hip problems and heart failure. Weight 213 kg on 29 April 2005; 180 kg on 10 August 2006; 127.5 kg on 16 February 2007. Head and body length 1473 mm; foot length 315 mm/318 mm; ear length 37 mm. [Some additional information added by GH from correspondence with Rotterdam Zoo personnel.]

NMS.Z.2009.3 Skeleton, skin, and muscle tissue (frozen). Adult female. This is 'Undi', who was born in Cologne Zoo on 20 September 1990 (sire 'Kim', dam 'Ulca'), and was mother-reared. She was sent on loan to Bristol Zoo on 3 March 1998, and died there on 25 September 2000. Head and body length 880 mm; foot length 255 mm; ear 42 mm.

NMS.Z.2009.22 Skeleton, mounted skin, and muscle tissue (frozen). Adult male. This is 'Bongo Jr', who came originally from Equatorial Guinea, and had been sequestrated by CITES authority from a circus, and placed in the judicial custody of Rome Zoo, where he arrived on 10 December 1992, aged about 9 years. He was sent on loan to Bristol Zoo on 28 November 2001, and to London Zoo on 25 June 2003, where he died of heart failure on 5 December 2008. Weight 205 kg on 23 January 2006; 188.5 kg on 21 July 2006; 175.5 kg on 16 February 2007; 134.75 kg on 7 March 2008; 144 kg on 28 July 2008; 152.8 kg on 17 September 2008. [Some additional information added by GH from correspondence with Rome Zoo personnel.]

NMS.Z.2009.134 Skeleton and skin. Adult female. This is 'Eva', who arrived at Twycross Zoo on 19 December 1975, aged about 2 years (weight 27 lb.). She had been purchased, with another female named 'Biddy', from G. van den Brink. She died on 18 October 2007: acute bloat and aspiration pneumonia. Weights 28 lb. on 28 December 1975; 54 lb. on 2 November 1976; 66 lb. on 13 December 1977; 43.5 kg on 4 May 1978; 48 kg on 17 August 1978; 135 lb. on 19 January 1979; 92 kg on 16 November 2005; 86 kg on 3 January 2006. Head and body length 980 mm; foot length 255 mm/258 mm; ear length 43 mm. Received on 28 November 2007. [Some additional information added by GH from correspondence with Natalie Evans, Twycross Zoo.]

NMS.Z.2010.082 Skeleton, skin, and muscle tissue (frozen). Adult male. This is 'Yeboah', who was born in Hanover Zoo on 10 January 1997 (sire 'Artis', dam 'Evouma', also known as 'Kathrin'), and was mother-reared. He was sent on loan to London Zoo on 27 November 2009, and died there on 25 March 2010 (weight 101 kg). Weights: 119 kg on 27 November 2009; 109 kg on 23 March 2010; 108.5 kg on 23 March 2010. Head and body length 1110 mm; foot length 282 mm; ear length 44 mm.

NMS.Z.2012.61.1 Skeleton and muscle tissue (frozen). Adult male. This is 'Mamfe', who was born in Jersey Zoo on 11 September 1973 (sire 'Jambo', dam 'Npongo'), after a gestation period of 265–267 days. He was removed for hand-rearing at 1 day (weight 2550 g). He was sold to Twycross Zoo, and arrived there on 23 November 1976 (weight 27.3 kg). He died on 20 September 2006: heart disease, acute death (weight 175 kg). Weights 29.1 kg on 13 February 1977; 41 kg on 4 May 1978; 44 kg on 17 August 1976; 47.7 kg on 19 January 1979. Head and body length 1200 mm; foot length 290 mm; ear length 55 mm. [Some additional information added by GH from correspondence with Durrell Wildlife Park personnel.]

NMS.Z.2012.61.2 Skeleton. Adult male. This is 'Ti', who was born in Edinburgh Zoo on 23 March 1989 (sire 'Sam-Sam', dam 'Yinka') and was hand-reared. He was sent on loan to Twycross Zoo on 26 May 1989, and died there on 26 February 2007: acute cerebral ischaemia (weight 152 kg). Weights 22.7 kg on 14 April 1991; 27.3 kg on 5 November 1991. Sent to Birmingham University, returned in August 2010.

NMS.Z.2013.100 Cadaver (in spirit). Neonate female. It was stillborn in Howletts Zoo Park on 16 June 1977 (sire 'Kisoro', dam 'Shamba') after a gestation period of 262 days. Weight 4 lb. 9 oz.

NMS.Z.2014.148 Skeleton and skin. Neonate female. It was born in Belfast Zoo on 10 December 1996 (sire 'Keke', dam 'Kamili') and died the same day. Head and body length 375 mm; foot length 160 mm/160 mm; ear length 31 mm. Received on 1 July 1998.

NMS.Z.2016.25 Tanned skin (for mounting). Adult male. This is 'Kéké', who was one of seven gorillas received in Burgers' Zoo and Safari, Arnhem, on 6 June 1984, from M and Mme Robert Roy, Sangmélima, Cameroon. The International Primate Protection League thwarted efforts by 3 United States zoos to import the gorillas. Although housed at Burgers' Zoo, ownership and responsibilities for exchanges with other zoos and participation in breeding programmes was bound to the Foundation for the Protection of Gorillas and other

Threatened Species of Animals, realised under IUCN Netherlands. The Roys acquired 'Kéké' on 15 March 1980; upon arrival in Arnhem he weighed 33 kg. He was sent on loan to Belfast Zoo on 1 August 1991, and died there on 27 March 1997. Received from the Ulster Museum. [Some additional information added by GH from correspondence with Arnhem Zoo personnel.]

Z.2016.26.1 Skull (cranium bisected). Adult male. Collected by P.B. Du Chaillu. Donated by Dr Burt. Mandible incomplete.

Z.2016.26.2 Skull. Adult female. Collected by P.B. Du Chaillu. Donated by Dr Burt.

Z.2016.126 Skeleton and skin. Infant female. This is "Ndoki", who was born in Twycross Zoo on 3 May 2007 (sire "Ti", dam "Ozala"). She died on 29 August 2007 (weight at post-mortem 1.8 kg). Head and body length 295 mm; foot length 94 mm; ear length 26 mm.

Z.2016.127 Skeleton. Infant male. This is "Tiny" (also known as "Yobie"), who was born in London Zoo on 28 October 2010 (sire "Yeboah", dam "Mjukuu"). He died on 12 May 2011 (weight 4.4 kg), after being injured by a new silverback male in the group, "Kesho". Head and body length 380 mm; foot length 117 mm; ear length 34 mm.

The following specimens are not yet registered:

Adult male. This is 'Kukuma', who was born in Stichting Apenheul, Apeldoorn, on 29 September 1989 (sire 'Kibabu', dam 'Frala'), and was hand-reared. He was sent on loan to Belfast Zoo on 4 October 1993, and died there of anaemia on 25 February 2011. He had been castrated. Cadaver on loan to Liverpool University since September 2011; it will be prepared as a skeleton when returned.

Cadaver (in freezer). Neonate male. Stillborn in Bristol Zoo (sire 'Jock', dam 'Salome') on 10 July 2005 (weight 1550 g).

Cadaver (in freezer). Neonate male. Stillborn in Jersey Zoo (sire 'Jambo', dam 'Kishka') on 7 March 1992, after a gestation period of 263 days (weight 2750 g). Assisted delivery, under sedation, owing to malpresentation of infant. Anoxia owing to umbilical cord presentation prior to assisted delivery.

Cadaver (in freezer). Neonate female. Stillborn in Jersey Zoo (sire 'Ya Kwanza', dam 'Hlala Kahili') on 19 September 2000.

University of Edinburgh, Institute of Cell Biology, School of Biological Sciences, Ashworth Laboratories, The King's Buildings, West Mains Road, Edinburgh EH9 3JJ*

Julie Tansey

Skull. Male.

Skull. Female.

These are thought to have been acquired in the 1930s.

University of Edinburgh, Anatomical Museum, Old Medical School, Teviot Place, Edinburgh EH8 9AG

Malcolm MacCallum; Rudolph Sprinz

2082 Endocranial cast (plaster). Juvenile? It is in two parts, joined with a metal flange.

2121 Endocranial cast (plaster). Juvenile? It is in two parts, joined with a metal flange.

2122 Endocranial cast (plaster). It is in two parts, joined with a metal flange.

2188 Endocranial cast (plaster). It is in two parts, joined with a metal flange.

2193 Endocranial cast (plaster).

2764 Skull. Adult male. The cranium has been bisected sagittally. The right mandible has glued repairs, there is a rounded chip of bone missing from the lower margin of the left mandible, and the maxilla has glued repairs. These are taken to be signs that the animal was shot. The dentition is complete.

2764.1 Mandible. Adult male. There are glued repairs on the right side.

2766 Mandible. Juvenile. It has a complete deciduous dentition.

2767 Skull. Adult female. Received from N. Rouppert, Paris, France. The cranium has been bisected sagittally.

2768 Skull. Juvenile. The cranium has been bisected sagittally. Eleven teeth are missing, mostly at the front.

2769 Skull. Juvenile. The cranium has been bisected sagittally. The dentition is complete. There are also three plaster casts of the cranium, numbered 1722, 2163 and 4801.

2769.1 Mandible. Juvenile. Complete dentition.

3314 Endocranial cast (plaster). Adult male?

3321 Endocranial cast (plaster).

3363 Mounted skeleton. Adult male. *Troglodytes gorilla*. The upper 3rd of the left humerus shows perforation and splintering (from a bullet?); the splinters have been stapled into place postmortem. The spine and acromion of the left scapula have been repaired by a suitably shaped piece of wood.

3912 Cranium (plaster cast).

4173 Section of skin with hair (in a glass cylinder). Juvenile male. *Troglodytes gorilla*.

4174 Section of skin with hair (suspended with clear thread in a glass cylinder). Date of collection or accession: July 1904.

4800 Endocranial cast (plaster). Adult male?

Exeter

Royal Albert Memorial Museum and Art Gallery, Queen Street, Exeter EX4 3RX

David Bolton; Kelvin Boot; Holly Morgenroth

EXEMS M236 Skull (cast). Adult male. Donated by W.R. Bayley on 25 October 1872. It belonged to the Revd R. Kirwan.

EXEMS M237 Skull. Female. *Troglodytes gorilla*. Donated by Robert F. Cooke on 3 May 1869. It came from the Du Chaillu collection. The foramen magnum has been artificially enlarged; both mandibular rami are broken off and missing; and the lower right lateral incisor is missing postmortem.

EXEMS M238 Endocranial cast. Purchased for 3s 0d from the Royal College of Surgeons of England in June 1875. It is marked '11. 5178 Royal College of Surgeons'.

EXEMS M239 Brain (cast). Donated by W.R. Bayley on 25 October 1872. It belonged to the Revd R. Kirwan. It is badly executed.

EXEMS M240 Left hand (cast). Donated by W.R. Bayley on 25 October 1872. It belonged to the Revd R. Kirwan.

EXEMS M241 Right foot (cast). Donated by W.R. Bayley on 25 October 1872. It belonged to the Revd R. Kirwan.

EXEMS M242 Left hand (cast). Donated by John Kendall on 24 March 1875. It is the same as EXEMS M240, and is a copy of the right hand, taken from a specimen imported in spirits, made by Edwin Ward, Wigmore Street.

University of Exeter, College of Life and Environmental Sciences, Department of Biosciences, The Geoffrey Pope Building, Stocker Road, Exeter EX4 4QD

Lindsey Beasley; Anna Davey; David Nichols

C 20008 Skull. Adult male.

Glasgow

Glasgow Museum and Art Gallery, Argyle Street, Kelvingrove, Glasgow G3 8AG

C.E. Palmar; Richard Sutcliffe

NB: The X-rays (radiographs) were taken at Glasgow Dental Hospital on 15 May 1981 by Mr D. Caplan and Miss Bubb, Department of Oral Biology, University of Glasgow.

1872.54 Skull. Acquired from Mr Baillie, Gaboon. Accession date 21 August 1872. It has been missing since before 1980.

1885.5 Skeleton (semiarticulated). Adult female. *Gorilla savagei*. Purchased for £25 on 5 February 1885. X-ray of skull.

1893.134 Skeleton (semiarticulated). Adult male. Donated by Thomas Welsh (Lagos, West Africa) per N. Cameron (Gowanbrae, Pollokshaws, Glasgow) on 2 October 1893. X-ray of skull.

1906.97 Mounted skin and skeleton (semiarticulated). Juvenile. *Gorilla savagei*. Purchased for £5 from dealer William Cross, Liverpool, on 20 September 1906.

Z.1939.30.v Skull. Adult male.

Z.1959.124.a Cranium. Adult male. X-ray of cranium.

Z.1959.124.b Skeleton. Juvenile male. X-ray of skull.

Z.1959.124.d Skull. Adult female. The back of the skull and one end of the mandible are missing. X-ray of skull.

Z.1959.124.e Cranium. Juvenile. The front teeth are missing.

Z.1959.124.f Skull. Adult male. X-ray of skull. The canines are missing.

University of Glasgow, Hunterian Museum and Art Gallery, Zoology Museum, Graham Kerr Building, University Avenue, Glasgow G12 8QQ

Maggie Reilly

NB: This list covers the Hunterian as a whole, that is, Zoological, Anatomical, and History of Science collections, where any gorilla material is likely to reside.

GLAHM:Z1143 Mounted skeleton. Adult male. *Gorilla beringei beringei*. Purchased in April 1915 from Rouppert, Paris.

GLAHM:1300-5 Skull. Adult male. Purchased from N. Rouppert, Paris. The mandible has been sectioned to show the dentition. Teeth perfect.

GLAHM:Z4467 Skull. Adult male. Teeth missing: upper right C and I1 and 2; lower right M3 and I1 and 2; upper left C and I1; lower left M3.

GLAHM:Z4468 Cranium (bisected). Donated by B.B. Morton in 1890. The incisors are missing.

GLAHM:Z4469 Mandible. It does not match any of the crania.

GLAHM:Z4470 Cranium. Male. The only teeth present are the left PM1 and PM2 and the right PM1, PM2 and M1. Parts of the base are missing. It is sun-bleached.

GLAHM:Z4471 Skull. Adult female. It lacks the upper left PM2 and M1, the lower canines, and both lower central incisors.

GLAHM:Z4472 Cranium. Male.

GLAHM:Z5786 Mandible. Adult male. Ex Glasgow Dental School comparative anatomy collection. Teeth perfect.

GLAHM:105570 Microscope slide of four unsectioned fibres of gorilla hair. Juvenile female.

GLAHM:105571 Microscope slide of cross sections of four bundles of fibres of gorilla hair. Juvenile female.

GLAHM:105572 Microscope slide of cross sections of four bundles of fibres of gorilla hair.

The above standard 3 × 1 glass microscope slides are each mounted in Canada balsam, and were manufactured by MANU [Glasgow Police] during the first half of the 20th century. They are associated with Prof. John Glaister Jr. (1892−1971), who provided lectures on forensic medicine to the Glasgow Police. His publications include *A Study of Hairs and Wools Belonging to the Mammalian Group of Animals, Including a Special Study of Human Hair, Considered From the Medico-Legal Aspect* (Misr Press, Cairo, 1931).

GLAHM:125080 Skull. Adult male. Presented by Dr Th. Forrest in 1889. It is in the Cleland Collection, Museum of Anatomy, Thomson Building.

GLAHM:Z140728 Endocranial cast. It is of poor quality.

Harrow-on-the-Hill

Harrow School, Biology Schools, High Street, Harrow-on-the-Hill, HA1 3HP*

Michael R. Etheridge

Cranium. Adult male. The foramen magnum is enlarged/damaged.

Mounted skeleton. Adult male.

Hastings

Hastings Museum and Art Gallery, Johns Place, Bohemia Road, Hastings, TN34 1ET

Catherine Walling

HASMG: SK.3 Brain (cast). Adult male. It was probably acquired in 1900−10.

Ipswich

The Museum, High Street, Ipswich IP1 3QH

David J. Frost; David J. Lampard; Sophie Stevens

IPSMG 1904-24 Mounted skin. Adult male.

IPSMG 1904-24 Mounted skin. Adult female.

IPSMG 1904-24 Mounted skin. Juvenile.

The above 3 specimens were collected in Gabon by P.B. Du Chaillu . They were donated in 1884 by the estate of Dr Edwards Crisp (1806−82) of London (formerly of The Red House, Rendlesham, Suffolk). The exhibit was renovated by Rowland Ward Ltd in 1906, and the case to contain it was presented by E. Herbert Fison. The specimens were restored again in 1989 by a team of specialist natural history conservators.

IPSMG: R.1922-26 Mounted skeleton. Adult female. Fernan Vaz. Donated by E.B. Ellis in 1885. The bones were found in storage in 1922 and articulated the same year by Gerrard, London. It is possibly the same specimen as IPSMG 1949-95.2 (some of the very early specimens that came into the collections have been reaccessioned).

IPSMG: R.1922-26a Skull. Adult male. Purchased from Gerrard. It was found in storage in 1922. Some teeth are missing. It is possibly the same specimen as IPSMG 1949-95.1.

IPSMG: R.1922-26b Skull (cast). Adult female. Donated.

IPSMG: R.1936-121a Skull. Presented by Mrs G. Day. This specimen cannot now be located.

IPSMG 1949-95.1 Skull. Adult male. Donation. On public display. It is possibly the same (reaccessioned) specimen as IPSMG: R. 1922-26a.

IPSMG 1949-95.2 Mounted skeleton. Adult female. Fernan Vaz. On public display. It is possibly the same (reaccessioned) specimen as IPSMG: R.1922-26.

Keighley

International Zoo Veterinary Group Pathology, Station House, Parkwood Street, Keighley, West Yorkshire BD21 4NQ

Mark Stidworthy

Archived samples may be released for valid scientific research on request, provided that the written consent of the submitting collection has been obtained and can be demonstrated, including a formal research proposal where appropriate. IZVG Pathology should be contacted in the first instance to enable contact with the submitting collection to be made. Appropriate offers of authorship/ acknowledgement in scientific publications are expected where the prior work of the IZVG pathologist makes a significant intellectual or practical contribution to the study. For histology samples, referring collections may hold additional archived fixed, frozen or otherwise preserved specimens from these cases, of which IZVG Pathology is unaware.

08-1959 Female (age 29 years). Single tissue biopsy — skin (1 paraffin block, 1 slide). No diagnosis is made.

08-1980 Male (age 32 years). Multiple tissue histology — spleen, liver, intestines, heart (2 × blocks, 2 × slides). Histological diagnoses: (1) Mild myocardial fibrosis with fibre anisocytosis and anisokaryosis, heart. (2) Haemosiderosis, liver and spleen.

04-0402 Female (age 23 years). Cytology slides only (× 4), vagina and milk. No diagnosis is made.

04-0796 Adult female. Cytology slide only, skin mass. Histological diagnosis: Acute to subacute suppurative inflammation with intralesional bacteria.

05-0987 Adult male. Single tissue biopsy — tendon (1 × block, 1 × slide). Histological diagnosis: Severe subacute fibrinosuppurative tenosynovitis, with multifocal collagen degeneration, perivascular neutrophil cuffing and fibrinoid change, and numerous intralesional mixed bacteria, tendon.

05-1010 Adult male. Single tissue histology — digital resection (4 × blocks, 12 × slides). Histological diagnoses: (1) Severe focally extensive chronic-active ulcerative and suppurative dermatitis with extensively inflamed dermal granulation tissue and fibrosis, and superficial bacterial colonisation, digital skin. (2) Severe focally extensive chronic-active suppurative osteomyelitis, phalangeal bone.

06-1015 Male (age 38 years). Multiple tissue histology — heart, spleen, liver, kidney, stomach, intestines, lung. Slides only received (× 8). Histological diagnoses: (1) Severe fibrosing cardiomyopathy, left ventricle, heart. (2) Moderate chronic multifocal (pyo)granulomatous hepatitis, liver. (3) Severe haemosiderosis, liver, spleen. (4) Chronic aneurysm and haemorrhage, 'lung nodule'. (5) Minimal chronic interstitial nephritis, kidney.

06-1140 Male (age 33 years). Multiple tissue histology — brain, liver, lung, kidney, thyroid, adrenal, tongue, salivary gland, testis, spleen, stomach, intestines, pancreas, tonsil, lymph node, oesophagus, urinary bladder, heart (16 × blocks, 16 × slides). Frozen tissues — lung, liver, kidney, brain. Full postmortem, remainder of carcase submitted to National Museums of Scotland. Histological diagnoses: (1) Severe chronic fibrosing cardiomyopathy, heart. (2) Focal nodular regeneration with fibrosis and biliary hyperplasia, liver. (3) Diffuse alveolar oedema and congestion with siderophages, lung. (4) Mild to moderate multifocal chronic fibrosing nonsuppurative interstitial nephritis, kidney. (5) Marked haemosiderosis, liver and spleen. (6) Mild bilateral modular adrenocortical hyperplasia, adrenal. (7) Mild chronic lymphocytic gastritis, stomach.

07-0048 Female (age 34 years 11 months). Multiple tissue histology — brain, bone marrow, heart, kidney, lung, liver, spleen, lymph nodes, thyroid, pituitary, pancreas, adrenal, labial skin, axillary skin, tarsal joint (16 × blocks, 16 × slides). Full postmortem, remainder of carcase submitted to National Museums of Scotland. Histological diagnoses: (1) Severe subacute to chronic necrotising and suppurative synovitis and fasciitis, with vascular thrombosis and intralesional bacterial colonisation, tarsal joint. (2) Chronic hyperplastic lymphoplasmacytic perivascular dermatitis, with multifocal superficial pustular epidermitis and epidermal ulceration, skin. (3) Haemosiderosis, liver and spleen. (4) Mild multifocal candidiasis, labial mucosa. (5) Lipofuscinosis/ceroid accumulation, choroid plexus corpora amylacea-like bodies and vascular mineralisation of basal ganglia, brain. Case published as: Masters et al. (2010).

07-0325 Male (age 17 years 11 months). Multiple tissue histology — brain, trachea, tongue, tonsil, heart, lung, mesenteric lymph node, spleen, kidney, liver, pancreas, adrenal, thyroid, stomach, intestines (16 × blocks, 16 × slides). Frozen tissues — spleen, pancreas, mesenteric lymph node, stomach, kidney, liver, tonsil, trachea and cerebrum. Full postmortem, remainder of carcase submitted to National Museums of Scotland. Histological diagnoses: (1) Severe, focally extensive, acute ischaemic necrosis and haemorrhage, right cerebral cortex, brain. (2) Mild multifocal acute haemorrhage, cerebellum. (3) Severe multifocal acute ulcerative typhlocolitis, with intralesional amoebal trophozoites and Gram-positive bacteria consistent with *Clostridium* spp., colon and caecum. (4) Severe multifocal acute haemorrhagic gastritis, with intralesional yeasts and Gram-positive bacteria consistent with *Clostridium* spp., stomach. (5) (Agonal) congestion, oedema and haemorrhage, with terminal aspiration, lungs. (6) Ventricular hypertrophy, heart. (7) Mild haemosiderosis, liver.

07-1334 Female (age 4 months). Multiple tissue histology − tongue, digital skin, heart, kidney, liver, spleen, lung, brain, trachea, thyroid, adrenal, oesophagus, urinary bladder, pancreas, stomach, intestines, lymph node (11 × blocks, 11 × slides). Frozen tissues − liver, kidney, spleen, brain, tongue and lung. Full postmortem, remainder of carcase submitted to National Museums of Scotland. Histological diagnoses: (1) Severe hyperplastic and erosive glossitis, with intralesional *Candida* and bacteria, tongue. (2) Acute aspiration and diffuse acute congestion and oedema, lungs. (3) Subacute epidermal ulceration/laceration with florid coccoid bacterial colonisation, skin of digit 3.

07-1624 Female (age 4 years). Multiple histology − spleen, pancreas, large intestine, stomach, lung. Slides only submitted (× 3). Histological diagnosis: Mild multifocal necrotising bronchiolitis and alveolitis, lung.

08-0938 Female (age 31 years). Cytology slide − vagina. Histological diagnosis: Mild neutrophilic vaginitis, with intercellular coccoid bacteria and occasional yeasts.

08-1021 Male (age 24 years 10 months). Single tissue histology − testis (3 × blocks, 3 × slides). Histological diagnoses: (1) Leydig (interstitial) cell tumour(s), testis. (2) Tubular atrophy and interstitial cell hyperplasia, testis.

08-1516 Male (approximate age 44 years). Multiple tissue histology − brain, extramural coronary arteries, heart, testes, liver, spleen, kidney, lung, thyroid, pituitary, submandibular lymph node, pancreas, adrenal, abdominal aorta, intestines (27 × blocks, 27 × slides). Frozen tissues − liver, kidney, lung, spleen and brain. Full postmortem, remainder of carcase submitted to National Museums of Scotland. Histological diagnoses: (1) Severe focally extensive subacute fibrinosuppurative peritonitis, with intralesional bacteria, plant debris and presumptive bile crystals, descending/sigmoid colon, ileum and distal jejunum. (2) Marked atherosclerosis, aorta, coronary arteries. (3) Atrophy with interstitial cell hyperplasia, bilateral, testes. (4) Steatosis with mild to moderate multifocal

myocardial fibrosis, heart. (5) Haemosiderosis, liver and spleen. (6) Mild lipofuscinosis, brain and heart.

08-1595 Male (age 35 years 8 months). Multiple tissue histology − heart, kidney, testes, pancreas, spleen, lung, liver, urinary bladder, thyroid, adrenal, intestines, skin, abdominal aorta (19 × blocks, 19 × slides). Frozen tissues − liver, kidney, lung and spleen. Full postmortem, remainder of carcase submitted to National Museums of Scotland. Histological diagnoses: (1) Endocardiosis, left atrioventricular valve. (2) Moderate myocardial fibrosis, fibre anisokaryosis, focal endocardial erosion with intraluminal thrombus, left ventricle, heart. (3) Marked subacute to chronic centrilobular congestion, liver. (4) Acute to subacute oedema and congestion, lung. (5) Mild multifocal chronic interstitial fibrosis with occasional cystic tubules, kidney. (6) Seminoma, right testis. (7) Atrophy, fibrosis and interstitial cell hyperplasia, bilateral, testes. (8) Chronic multifocal ulcerative dermatitis, with secondary coccoid bacterial colonisation, skin. (9) Follicular cyst, eyelid.

09-0847 Female (age 22 years). Single tissue histology − skin (1 × block, 1 × slide). No diagnosis is made.

09-0865 Female (age 23 years). Single tissue histology − salivary gland (1 × block, 1 × slide). Histological diagnosis: Mild multifocal neutrophilic and lymphoplasmacytic sialoadenitis, salivary gland.

09-1339 Male (age 4 years 11 months). Multiple tissue histology − femoral bone marrow, heart, adrenal, mesenteric lymph node, testis, liver, kidney, lung, aorta, trachea, striated muscle, oesophagus, spleen, stomach, intestines, urinary bladder (8 × blocks, 8 × slides). Histological diagnoses: (1) Acute undifferentiated leukaemia, with minimal peripheral infiltration, bone marrow. (2) Subacute ulcerative to transmural typhlocolitis with peritonitis, with *Balantidium coli* and secondary bacterial colonisation, large intestine/mesenteric connective tissues. (3) Accessory spleen.

09-1457 Female (approximate age 25 years). Multiple tissue histology − spleen, aorta, heart, pancreas, liver, lung, kidney, stomach, ovary/uterus, intestines (4 × blocks, 4 × slides). Histological diagnoses: (1) Severe chronic-active fibrinosuppurative and fibrosing peritonitis/retroperitonitis, with embedded bacteria and gut content, and chronic fibrosis/adhesion formation, ovary/uterus/large intestine. (2) Moderate chronic multifocal to coalescent tubulointerstitial nephritis and cortical fibrosis, with tubular proteinuria, kidney. (3) Arteriosclerosis, multiple arteries. (4) Haemosiderosis, liver, spleen, intestinal villi, lung. (5) Serous atrophy, adipose tissues. (6) Lipofuscinosis and fibre anisokaryosis, myocardium.

10-0132 Female (age 4 years 2 months). Single tissue histology − brain (5 × blocks, 5 × slides). Same animal as IZVG ref. 07-1624. Histological diagnosis: Multifocal nonsuppurative meningoencephalitis, immunohistochemistry positive for *Listeria*, brain.

10-0180 Female (age 7 years). Multiple tissue histology − tarsal bone and joint capsule, lung, heart, kidney, liver, small intestine (4 × blocks, 4 × slides). Histological diagnosis: Degenerative joint disease, tarsus.

10-0352 Female (age 38 years). Multiple tissue histology − spleen, heart, intestines, lung, liver, abdominal fat, kidney, uterus (3 × blocks, 3 × slides). Histological diagnosis: (1) Fibrinous peritonitis, intestine/mesenteric fat. (2) Haemosiderosis, liver, spleen, intestine.

10-1648 Adult male. Cytology slides only − urine (× 2). No diagnosis is made.

11-0187 Male (age 7 years). Cytology slide (× 1) only − bone marrow aspirate. No diagnosis is made.

11-0220 Male (age 1 year). Multiple tissue histology − mesenteric lymph node, stomach, heart, intestines, lung, spleen, kidney, adrenal, liver (10 × blocks, 10 × slides). Histological diagnoses: (1) Severe acute segmental neutrophilic enteritis, intestine. (2) Acute multifocal fibrinonecrotic splenitis, spleen. (3) Presumptive ductal plate malformation with mild-to-moderate fibrosis, liver. (4) Minimal cortical dysplasia, kidney.

11-1226 Female (age 37 years). Single tissue histology − skin (2 × blocks, 2 × slides). Histological diagnosis: Chronic mixed cellular dermatitis and perifolliculitis with intrafollicular fungal hyphae, skin.

11-1336 Adult male. Multiple tissue histology − lymph node, lung, kidney, heart, liver, intestines (3 × blocks, 3 × slides). Histopathological diagnoses: (1) Severe diffuse chronic lymphoplasmacytic and neutrophilic fibrosing colitis/typhlocolitis, with intralesional *Balantidium coli*, large intestine. (2) Moderate diffuse oedema and sinus histiocytosis, lymph node. (3) Moderate diffuse chronic lymphoplasmacytic and fibrosing enteritis, small intestine. (4) Moderate diffuse chronic fibrosing cardiomyopathy, heart.

11-1502 Female (approximate age 46 years). Multiple tissue histology − heart, lung, liver, pancreas, spleen, urinary bladder, aorta, thyroid, adrenal, parathyroid, uterus, oesophagus, stomach, intestines, appendix, tongue, tonsil, gallbladder, mesenteric lymph node, ureter, ovaries, kidney, pituitary (16 × blocks, 16 × slides). Frozen tissues − liver, kidney, lung, spleen, brain, tongue and oesophagus. Full postmortem, remainder of carcase submitted to National Museums of Scotland. Histopathological diagnoses: (1) Left ventricular hypertrophy and right ventricular dilatation, with mild multifocal interstitial myocardial fibrosis, heart. (2) Atherosclerosis, multifocal, coronary and systemic arteries. (3) Membranoproliferative glomerulonephritis, kidney. (4) Early organising infarct, spleen. (5) Lymphoid hyperplasia with mild neutrophilic tonsillitis, tonsil. (6) Adenoma, parathyroid. (7) Follicular cysts, thyroid. (8) Cyst, with minimal adenomatous hyperplasia

of pars distalis, pituitary. (9) Multifocal subacute mucosal erosion and haemorrhage, with mild fibrosis, glandular stomach. (10) Paraovarian cyst, right ovary. (11) Leiomyoma, uterus. (12) Nodular cortical hyperplasia, adrenal. (13) Haemosiderosis, liver, spleen, small intestine. (14) Focal nodular hyperplasia, liver.

11-1655 Female (age 2.5 years). Multiple tissue histology − kidney, heart, intestines, spleen, liver, lung (2 × blocks, 2 × slides). Histopathological diagnosis: Lymphoid hyperplasia, spleen, intestine.

12-0162 Male (age 23 years). Multiple tissue histology − kidney, urinary bladder, testis, spleen, liver, gallbladder, air sac, pancreas, lymph nodes, stomach, intestines, adrenal, lung, aorta, trachea, heart (20 × blocks, 20 × slides). Histopathological diagnoses: (1) Fibrosing (interstitial-type) cardiomyopathy, heart. (2) Anatomically variable congestion and oedema with intraalveolar siderophages and mild type II pneumocyte hyperplasia, lung. (3) Minimal glomerulosclerosis, kidney. (4) Atrophy with interstitial cell hyperplasia, testis. (5) Haemosiderosis, (mild) spleen, (moderate) liver. (6) Acute submucosal haemorrhage, trachea.

12-0191 Adult female. Single tissue histology − placenta (3 × blocks, 3 × slides). Histopathological diagnosis: Moderate acute suppurative placentitis with thrombosis and neutrophilic vasculitis, placenta.

12-0430 Male (age 17 years). Single tissue histology − testis (1 × block, 1 × slide). Histopathological diagnosis: Seminiferous tubular atrophy and interstitial cell hyperplasia, testis.

12-1164 Female (age 40 years). Single tissue histology − skin (2 × blocks, 2 × slides). Histopathological diagnosis: Mild (shoulder) to moderate (thigh) chronic fibrosing mononuclear dermatitis with adnexal atrophy, epidermal hyperplasia, orthokeratotic hyperkeratosis and mild superficial pustule formation, skin.

13-0109 Male (age 11 months). Multiple tissue histology − heart, pancreas, mesenteric lymph node, lung, liver, spleen, kidney, urinary bladder, stomach, intestines (4 × blocks, 4 × slides). Histopathological diagnoses: (1) Moderate to severe multifocal subacute ulcerative colitis, large intestine. (2) Moderate to severe multifocal subacute haemorrhagic and necrotising interstitial pneumonia, with thrombosing vasculitis, lung. (3) Focal severe subacute haemorrhagic and necrotising hepatitis, with thrombosing vasculitis, liver. (4) Acute to subacute lymphoid depletion, spleen.

13-0512 Male (age 45 years). Multiple tissue histology − heart, liver, kidney, stomach, intestines, spleen, lung, urinary bladder (14 × blocks, 14 × slides). Histopathological diagnoses: (1) Severe multifocal to confluent chronic fibrosing cardiomyopathy, heart. (2) Mild to moderate subacute alveolar oedema with numerous 'heart-failure cells' (presumed) and multifocal mineral deposition, lung. (3) Mild to moderate subacute centrilobular congestion with hepatocellular depletion and haemosiderosis, liver. (4) Multifocal subacute to chronic splenic haematomas with mineralisation and focal osseus metaplasia, spleen. (5) Focally extensive severe protozoal colonisation with suspected mild to moderate chronic lymphoplasmacytic colitis/typhlitis, large intestine. (6) Focal chronic renal infarct and focal renal cyst, kidney. (7) Multifocal vascular medial hypertrophy, kidney. (8) Haemosiderosis, intestine and spleen.

13-1444 Adult female. Single tissue histology — eyelid (1 × block, 1 × slide). Histopathological diagnosis: Severe diffuse chronic proliferative eosinophilic and lymphoplasmacytic ulcerative conjunctivitis with 'collagen flame figure' formation and conjunctival epithelial hyperplasia, eyelid/conjunctiva (per sender).

14-0762 Female (age 54 years). Multiple tissue histology — heart, intestines, stomach, lung, liver (3 × blocks, 3 × slides). Histopathological diagnoses: (1) Moderate fibrosing (interstitial-type) cardiomyopathy, heart. (2) Moderate irregular congestion and oedema with intraalveolar siderophages (suspected 'heart-failure cells'), lung. (3) Mild to moderate chronic-active mixed cellular enterocolitis, small and large intestine.

15-2206 Male neonate (stillborn). Multiple tissue histology — placenta, fetal lung, kidney, urinary bladder, heart, brain, intestines, umbilical cord, liver, spleen, stomach (3 × blocks, 3 × slides). Histopathological diagnoses: (1) Extramedullary haemopoiesis, liver. (2) Irregular atelectasis, lung. (3) Oedema, multiple tissues.

Kendal

Kendal Museum, Station Road, Kendal, LA9 6BT

Carol Davies; W.M. Grange

K.M.B. 1983.234 Mounted skin. Subadult male. Donated by Dr William Rushton Parker in 1942. It was originally in a separate glass case, with a modelled base, foliage, etc., by H. Murray and Sons, Carnforth, Lancashire. The face had been inaccurately modelled, and was completely remodelled by Mr I. Hughes in 1984. The mount is now included in the African Diorama display.

Leeds

Leeds Museum Discovery Centre, Carlisle Road, Leeds LS10 1LB

Rebecca Machin; Adrian Norris; J.H. Nunney

Hand (cast). Purchased in 1862 or 1863. It no longer exists.

Foot (cast). Purchased in 1862 or 1863. It no longer exists.

LEEDM.C.1926.9.4164 Skull. Adult male. Purchased in 1926 or 1927. A cast of this skull in Lord Rothschild's collection at Tring is figured in Coolidge (1929), plate 4. By measurement, it was the largest skull he found in any collection. He stated, in error, that the original skull is in the Leicester Museum.

LEEDM.C.1938.40.1.4079 Mounted skeleton and mounted skin. Male. Purchased from Gerrard in 1938. This is 'Mok' (a shortened version of his full name, 'Mo Koundje'), who arrived at London Zoo on 22 August 1932, when aged about 2½ years. He arrived with a female named 'Moina' (a shortened version of 'Moina Moassi'). The pair came originally from French Congo, and was purchased for £1200 from Monsieur A. Capagorry, Hôtel des Grands Hommes, Bordeaux. They were housed first in the Lemur House, and later in the specially built Tecton Gorilla House, which was designed by the architect Berthold Lubetkin and completed in 1934. 'Mok's' weight was recorded as 70 lb. on 28 April 1933; 210 lb. on 19 September 1935; 240 lb. in August 1936; 266 lb. in January 1937; 274 lb. on 10 February 1937; and 320 lb. on 18 September 1937. He died of nephrosis on 14 January 1938. His dimensions at death were: circumference of head 32 in.; length from chin to base of skull 25 in.; length of arm 36 in.; length of leg 31 in.; length of body 36 in.; skeleton circumference of head 22 in.; weight 340 lb. [Some additional data added by GH from Zoological Society of London records.]

LEEDM.C.1938.40.2.4080 Mounted skeleton. From the same individual as the above mounted skin, LEEDM.C.40.1.4079.

LEEDM.C.1982.96.4162 Skull. Female. Ogooué, Gabon.

LEEDM.C.1982.318.4163 Skull. Female.

The above 2 skulls were transferred, with a lot of other material, from the Salford Natural History Museum (now the Buile Hill Mining Museum) in 1982. They probably came to Salford from H. Brazenor.

LEEDM.C.2009.459 Mounted skeleton. Juvenile female. *Anthropopithecus gorilla.*

LEEDM.C.2010.345 Cranium. Juvenile. The corresponding lower jaw is numbered as LEEDM.C.2010.712. It may be a chimpanzee.

LEEDM.C.2010.346 Cranium (cast). Juvenile female. The corresponding lower jaw cast has been numbered separately as LEEDM.C.2010.916.

LEEDM.C.2010.351 Skull (cast). Adult female.

LEEDM.C.2010.800 Brain (cast). Adult female/juvenile male? It may be a chimpanzee.

LEEDM.C.2010.801 Brain (cast). Juvenile female. Received from Alexander Falkner in 1862 or 1863.

LEEDM.C.2010.828 Skull (cast). The lower jaw cast is broken.

Leicester

New Walk Museum and Art Gallery, 53 New Walk, Leicester LE1 7EA*

Miss J.E. Dawson

1372 Mounted skeleton. Male. Purchased from Edward Gerrard Jr on 9 December 1884.

1373 Mounted skeleton. Female. Purchased from Edward Gerrard Jr.

12 1930-1 Mounted skin and partially articulated skeleton. Male. Spanish Guinea. Purchased for £250 from Messrs Rowland Ward.

12 1930-2 Mounted skin and partially articulated skeleton. Female. Spanish Guinea. Purchased for £200 from Messrs Rowland Ward.

Liverpool

World Museum, William Brown Street, Liverpool L3 8EN

Clemency Fisher; Anthony Parker

LIV.20.5.62.1 Skeleton. Gabon. Received from R.B. Walker CMZS on 20 May 1862. See Burton (1876). According to Burton, Mr R.B.N. Walker was an agent of Messrs Hatton and Cookson (palm oil traders) at the Gaboon River, and was 'the first to send home a young specimen bodily, stowed away in spirits; two boiled skeletons of large grey animals, whose skins I saw at the factory, and rum-preserved brains, intestines and other interesting parts, which had vainly been desired by naturalists.' See also McMillan (1996).

LIV.2.7.79.1 Skull. Ngoré, West Gabon. Received from I.G.C. Harrison on 2 July 1879.

LIV.4.12.31 Skull. Female. Gambia. Received from Rowland Ward on 4 December 1931.

LIV.1963.173 Skull. Juvenile. Gabon. From the G.W. Otter collection. Mr Otter, of Horsham, West Sussex, inherited a large natural history collection from his grandfather, Sir Edmund Giles Loder (1849–1920), of Leonardslee Gardens, Horsham, upon his death. The osteological part was given to the museum in 1963.

LIV.1963.173.15 Skull. Adult male. From the G.W. Otter collection.

LIV.1979.408 Skull (cast).
LIV.1979.409 Mandible.
LIV.1979.410 Mandible.
LIV.1979.411 Mandible.
LIV.1979.420 Skull.
LIV.1979.421 Cranium.
LIV.1979.422 Skull.
LIV.1979.445 Postcranial skeleton.
LIV.1979.534 Clavicle.

LIV.1980.778 Mounted skin. Juvenile.
LIV.1981.1272 Mounted skeleton. Male. Gabon. From the G.W. Otter collection.
LIV.1981.1273 Mounted skin. Male. Obtained from Rowland Ward Ltd.
LIV.1981.1274 Mounted skin. Female.
LIV.1984.54.2 Skull.
LIV.1984.54.3 Skull.
LIV.1984.54.4 Postcranial skeleton.
LIV.1984.54.5 Skull (cast).
LIV.1984.54.6 Skull (cast).
LIV.1984.54.7 Skull (cast).
LIV.1984.54.8 Skull (cast).
LIV.1984.54.9 Skull (cast).

University of Liverpool, Institute of Ageing and Chronic Disease, Department of Musculoskeletal Biology, The Apex Building, 6 West Derby Street, Liverpool L7 8TX

Andrew T. Chamberlain; Robert Connolly; Robin Huw Crompton; Kristiaan D'Août; Colleen Goh

Carcase (in freezer). Adult female. This is 'Bongo Bongo', who arrived at Twycross Zoo on 29 October 1965, aged about 10 months (weight 18 lb. on 8 December 1965). She was purchased from the Far East Animal Co., Lindeplein 11, Tilburg, Netherlands, where she had been hand-reared from the age of 4 months by Mrs de Souza, wife of the proprietor. She was euthanased on 5 October 2011 after a period of age-related illness (dental issues, stifle and elbow arthritis). Both legs and feet, right thigh (including right pelvis), and left arm (including left shoulder) have been dissected down to the bone. [Some additional information added by GH from correspondence with Natalie Evans and Sarah Chapman, Twycross Zoo.]

The specimens below were transferred from the Department of Human Anatomy and Cell Biology when it closed down.

LA.20.86 Mounted skeleton. Adult female. *Gorilla beringei graueri*. This is 'Noelle', who arrived at Chester Zoo, with a male named 'Mukisi', on 17 November 1960, at which time she weighed about 36 lb. The pair had been purchased from Charles Cordier, Utu, Belgian Congo. She died of bronchial pneumonia and pleurisy on 18 March 1975 (weight c. 120 kg). The skull has supernumerary (4th) molars in the upper and lower left jaws. The postcranial skeleton is currently being used for teaching purposes. [Some additional information added by GH from correspondence with George Saul Mottershead, Chester Zoo.]

LA.22.86 Skull. Adult male.

London

Horniman Museum and Gardens, 100 London Road, London SE23 2PQ

Joanne Hatton; Helen Steers

NB: Many of the primates were originally registered as physical anthropology comparative specimens, forming part of the ethnography collections.

NH.H.2 Skull. Adult male. Frederick J. Horniman Collection. Purchased for £5.

NH.H.3 Skull (bisected sagittally). Adult male. *Gorilla savagei*. Frederick J. Horniman Collection. Purchased for £2. Other no. NH.A967.

NH.3.101 Left hand (cast). Adult male. Gaboon.

NH.3.102 Foot (cast). Adult male. Gaboon. From the same animal as NH.3.101.

Both the above casts were made and donated by F.W. Wilson, Sydenham. They were taken from a specimen shot and sent home in the flesh by Du Chaillu.

NH.70.239 Small carved wooden model of a gorilla. Made by F.W. Wilson, from a specimen sent home in spirits from Gaboon by Du Chaillu. The specimen went to the British Museum. Received from C. Kranz.

9.36 Skull. Juvenile. Purchased for £3 from Edward Gerrard. Ethnography accession no. 318.

9.54 Brain (cast). Ethnography accession no. 322.

9.366 Mounted skeleton. Adult male. Purchased from Edward Gerrard & Sons before 1909. Ethnography accession no. 353.

9.367 Mounted skeleton. Adult female. Purchased for £15 from Edward Gerrard & Sons before 1909. Ethnography accession no. 354.

12.79 Skull. Adult female. Ethnography accession no. 465.

Hunterian Museum, Royal College of Surgeons of England, 35–43 Lincoln's Inn Fields, London WC2A 3PX

Elizabeth Allen; Caroline Grigson; Kristin Hussey

NB: The museums at the Royal College of Surgeons will close to the public in spring 2017, as part of a major refurbishment project due to be completed in the summer of 2020. During the closure it is hoped that managed access to collections for study and research will be available through partner institutions and organisations at other sites in London. For further information please contact museums@rcseng.ac.uk, or visit https://www.rcseng.ac.uk/.

The Hunterian collection was formed by John Hunter (1728–93) between c.1760 and his death. In 1799 his whole collection of plant and animal specimens was purchased by the Government and offered to the Company of Surgeons, which became the College of Surgeons of London the following year, and took its present name in 1843. The museum was officially opened in 1813, later demolished, and rebuilt, officially reopening in 1837. It was largely destroyed in an air raid on 10 May 1941, and over two-thirds of the collection was lost. The surviving collection reopened as the Hunterian Museum in 1963 (the pre-war museum was normally referred to as the Museum of the Royal College of Surgeons). A list of the current material is given first, followed by a list of the pre-1941 material, in so far as can be ascertained by referral to the museum's annotated copy of Flower and Garson (1884). Most of these specimens were completely destroyed by enemy bombing, and it is possible that some others were transferred to the British Museum (Natural History), but there is no list of transferred material.

Current material:

RCSHM/D 656 Left cerebral hemisphere (wet specimen). Juvenile male. *Anthropopithecus gorilla*. The animal from which the brain was taken weighed 33 lb. and was 34 in. tall.

RCSHM/D 657 Left half of brain, minus the cerebral hemisphere (wet specimen). Juvenile male. *Anthropopithecus gorilla*. This is the remainder of the above specimen, RCSHM/D 656 .

RCSHM/D 658 Right half of brain (wet specimen). Juvenile male. *Anthropopithecus gorilla*. From the same animal as RCSHM/D 656 and RCSHM/D 657 above.

RCSHM/D 659 Brain, cut in mesial sagittal section (wet specimen). *Anthropopithecus gorilla*.

RCSHM/D 660 Brain, cut in mesial sagittal section (wet specimen). *Anthropopithecus gorilla*.

RCSHM/D 661 Brain (wet specimen). *Anthropopithecus gorilla*.

RCSHM/D 661.2 Left hemisphere of brain (cast). Fetus (probably 6th or 7th month of gestation). French Congo. Presented by Dr Raoul Anthony (1874–1941) in 1916.

RCSHM/D 661A Brain (wet specimen).

RCSHM/D 662 Endocranial casts (2). Juvenile.

RCSHM/D 663 Endocranial casts (2). Adult ?

RCSHM/D 779 Spinal cord with dura mater removed and nerve roots of cauda equina separated to show filum terminale (wet specimen). Juvenile. *Anthropopithecus gorilla*.

RCSHM/E 670 Section of cranium, dissected to show cranial sinuses. Received from Sir Victor Ewings Negus (1887–1974) in 1961. For deciduous and permanent dentition see RCSOM/A 64.14 and RCSOM/A 64.141.

RCSHM/F 227.1 Epidermis from sole of foot and toes (wet specimen). Adult male. *Troglodytes gorilla*.

RCSHM/H 72.1 Right foot (wet specimen). Juvenile. *Troglodytes savagii* [*sic*]. Acquired in 1888.

RCSHM/H 73.1 Right hand (wet specimen). Juvenile. *Troglodytes savagii*. Acquired in 1888.

RCSHM/J 1321.2 Caecum with parts of ileum and colon (wet specimen). Adult female. Received from the

Prosectorium, Zoological Society of London, in 1939. This is from the captive specimen 'Moina'.

RCSHM/J 1357.2 Part of colon (wet specimen). Received from the Prosectorium, Zoological Society of London, in 1939. This is presumably from 'Moina'.

RCSHM/J 1684.1 Gall bladder (wet specimen). Received from the Prosectorium, Zoological Society of London, in 1939. This is presumably from 'Moina'.

RCSHM/L 200.1 Spleen (wet specimen). *Troglodytes*. Acquired in 1888.

RCSHM/L 439.1 Kidneys and adrenal bodies with blood vessels (wet specimen). *Gorilla savagii*. Purchased in 1888.

RCSHM/N 36 Sagittal section of head to demonstrate organs of vocalisation (wet specimen). Presented by Sir Victor Ewings Negus in 1961.

RCSHM/CO 29.1 Part 1 Hand bones (mounted on a wooden base).

RCSHM/CO 29.1 Part 2 Pelvis, femur, tibia and fibula (mounted separately).

RCSHM/CO 29.1 Part 3 Scapula, radius and ulna (mounted separately).

RCSHM/CO 29.1 Part 4 Humerus (mounted on a base).

RCSOM/OH/118 Postcranial skeleton. Adult male. Bequeathed by William Charles Osman Hill (1901−75).

RCSOM/OH/119 Skull and partial skeleton. Juvenile male. Bequeathed by W.C.O. Hill. Collector's tag no. 90, possibly originally acquired from F.G. Merfield (series III).

RCSOM/OH/123 Skull (cast). *Gorilla beringei graueri*. South-west Kivu. Bequeathed by W.C.O. Hill. Maker: F.T. Dawes, London. The original was presented to the British Museum (Natural History) by Lord Rothschild.

RCSOM/OH/398 Skeleton. Female. Bequeathed by W.C.O. Hill.

RCSOM/OH/W128/01 Half of brain (wet specimen). Female. Bequeathed by W.C.O. Hill.

RCSOM/OH/W071/02 Portion of pectoral skin (wet specimen). Juvenile. Bequeathed by W.C.O. Hill.

RCSOM/OH/W071/08 Axillary skin (wet specimen). Bequeathed by W.C.O. Hill.

RCSOM/OH/W129/03 Brain (wet specimen). Bequeathed by W.C.O. Hill.

RCSMS/OsmanHill 3 Microscope slide of hair. Received from Yvonne Hill (wife of W.C.O. Hill).

RCSMS/OsmanHill 603 Microscope slide of section through penis. Received from Yvonne Hill.

RCSMS/OsmanHill 604 Microscope slide of section through penis. Received from Yvonne Hill.

RCSMS/OsmanHill 605 Microscope slide of section through skin of penis. Juvenile. Received from Yvonne Hill.

RCSMS/OsmanHill 606 Microscope slide of section through penis. Received from Yvonne Hill.

RCSMS/OsmanHill 1897 Microscope slide of hair. Received from Yvonne Hill.

RCSMS/OsmanHill 1898 Microscope slide of hair (cross sections). Received from Yvonne Hill.

RCSMS/OsmanHill 1899 Microscope slide of hair. Received from Yvonne Hill.

RCSMS/OsmanHill 1900 Microscope slide of hair. Received from Yvonne Hill.

RCSMS/OsmanHill 1901 Microscope slide of hair. Received from Yvonne Hill.

RCSMS/OsmanHill 1902 Microscope slide of hair. Received from Yvonne Hill.

RCSMS/OsmanHill 1903 Microscope slide of hair. Received from Yvonne Hill.

RCSMS/OsmanHill 1904 Microscope slide of hair. Juvenile male. *Gorilla beringei beringei*. Received from Yvonne Hill. This was extracted from the right thigh of the captive specimen 'Reuben' (London Zoo) on 13 June 1960.

NB: In 1907 the Odontological Society of Great Britain became the Odontological Section of the Royal Society of Medicine (founded in 1805; received its Royal Charter in 1834), located in Hanover Square. Two years later, its collection was moved to the Royal College of Surgeons under Deed of Trust and arranged with other odontological material belonging to the RCS. Ownership of the RSM collection was transferred to the RCS in 1943. After the war, the collection was displayed in the Odontological Museum, situated next to the Hunterian Museum. The material from this small museum was subsequently integrated into the new Hunterian Museum, which reopened in 2005 after extensive refurbishment. The specimens listed below were originally part of the Odontological Museum.

RCSOM/A 64.1 Skull. Infant. The deciduous dentition is in position and the eruption of the permanent 1st molars has commenced.

RCSOM/A 64.2 Skull. Adult male. Purchased by the Odontological Society of Great Britain for its museum in 1884. Almost the entire skeleton was purchased, and the postcranial portion was later transferred to the Osteological Series of the Royal College of Surgeons. The animal had evidently met with some injury to the mouth and neck, since the hyoid bone had been fractured and reunited by a mass of callus. According to a note by Mr Storer Bennett in the Transactions of the Odontological Society 17 (1884), pp. 35−36, the jaws also showed evidence of the effects of violence, for one of the right upper incisor teeth had been knocked out and another broken off; the alveolus of the first had been absorbed. However, both teeth are in fact visible in the jaw. The teeth, especially the 3rd molars, exhibit a fair degree of wear.

RCSOM/A 64.3 Skull. Adult female. Purchased in 1932. The upper 3rd molars are abnormal in shape. The anteroexternal and anterointernal cusps are not united by a ridge as is normal, but between these cusps there is an additional small cusp. On the internal edge there are three cusps, one of which is prominent and conical, and internal to these there is another cusp. The posterior cusps are normal. There are three roots not fully formed.

RCSOM/A 64.4 Skull. Adult male. Obala village, 3¾°N, 14¼°E, Batouri district, Cameroun. Collected by F. G. Merfield on 22 October 1933 (collector's no. II 660): locally hunted (Mendjim Mey tribe) in old banana plantations. Purchased from the Powell-Cotton Museum in 1934. The upper 1st incisors are a little in advance of the 2nd incisors.

RCSOM/A 64.5 Cranium. Juvenile. *Gorilla beringei beringei*. Kayonza region, 1°S, 29°E, western Kigezi, Uganda. Donated by Charles Robert Senhouse Pitman, 1934—35. The basioccipital is missing. The right maxilla is broken off, but present, and the canine is missing. There are Carabelli cusps on the lower 3rd molars.

RCSOM/A 64.11 Skull. Infant male. Gadji, 4¼°N, 14¼°E, Batouri district, Cameroun. Collected by F.G. Merfield on 2 September 1935 (collector's no. III 180): locally hunted (Kaka tribe) in old banana plantations. Purchased from the Powell-Cotton Museum in 1936. Colyer (1936, p. 674), mentions that the animals from Batouri district were collected by F.G. Merfield in an area covered with old banana plantations and that the animals might venture down at times to maul over refuse heaps and so some acquired gum infections.

RCSOM/A 64.12 Skull. Infant male. Gadji, 4¼°N, 14¼°E, Batouri district, Cameroun. Collected by F.G. Merfield on 11 September 1935 (collector's no. III 184): locally hunted (Kaka tribe) in old banana plantations. Purchased from the Powell-Cotton Museum in 1936. The deciduous dentition is in position.

RCSOM/A 64.13 Skull. Infant male. Between Batouri and Lomié, 3¾°N, 13¾°E, Cameroun. Collected by F.G. Merfield on 29 May 1935 (collector's no. III 124): locally hunted in old banana plantations. Purchased from the Powell-Cotton Museum in 1936. The deciduous dentition is in position.

RCSOM/A 64.14 Developing permanent dentition right, dissected out of the jaw. Juvenile male. Between Batouri and Lomié, 3¾°N, 13¾°E, Cameroun. Collected by F.G. Merfield on 25 April 1935 (collector's no. III 80): locally hunted in old banana plantations. Purchased from the Powell-Cotton Museum in 1936. The skeleton was sold by the Powell-Cotton Museum to the Royal College of Surgeons Museum in 1939. The skull was given to Mr V.E. Negus on 26 April 1952, and is now RCSHM/E 670 in the Hunterian Museum. The deciduous dentition of the same animal was preserved separately as RCSOM/A 64.141. The anterior part of the mandible had an infective osteitis.

RCSOM/A 64.15 Skull. Juvenile female. Dja Posten, 3¼°N, 13½°E, Lomié district, Cameroun. Collected by F. G. Merfield on 18 August 1936 (collector's no. III 458a): locally hunted. Purchased from the Powell-Cotton Museum in 1937. It has mixed deciduous and permanent dentitions.

RCSOM/A 64.16 Skull. Adult male. Obala village, 3¾°N, 14¼°E, Batouri district, Cameroun. Collected by F. G. Merfield on 24 October 1932 (collector's no. I 875): locally hunted in old banana plantations. Purchased by Colyer from the Powell-Cotton Museum on 17 November 1933; donated by Colyer in 1947—48. The left orbit and temporal regions are damaged: according to Merfield there was a bullet hole in the right eye.

RCSOM/A 64.21 Skull. Adult male. Between Batouri and Lomié, 3¾°N, 13¾°E, Cameroun. Collected by F.G. Merfield on 20 March 1935 (collector's no. III 21): locally hunted in old banana plantations. Purchased from the Powell-Cotton Museum in 1936. There was an odontome (G61.2) in the position of the lower right 3rd molar, but this seems to have been lost. The upper incisors are slightly abnormal in position, and in occlusion are a little in front of the lower incisors.

RCSOM/A 64.22 Skull. Adult male. Touki village, 4½°N, 13¾°E, Bertoua district, Cameroun. Collected by F. G. Merfield on 2 March 1936 (collector's no. III 210): locally hunted (Banya tribe). Purchased from the Powell-Cotton Museum in 1936. There has been a loss of bone around the external roots of the upper premolars and 1st molar; in the 1st molars the external roots are exposed. The occipital region and mandibular ramus are damaged postmortem.

RCSOM/A 64.31 Skull. Adult female. Lumbindou village, 4¼°N, 14¼°E, Batouri district, Cameroun. Collected by F.G. Merfield on 28 May 1932 (collector's no. I 80): locally hunted (Kaka tribe) in old banana plantations. Donated by J.F. Colyer in 1947—48: it was no. 1.11 in his private collection, and almost certainly one of the two skulls from Batouri district acquired by him in 1933 (see RCSOM/A 64.16). The bone of the upper and lower jaw on the left side has been removed to expose the roots of the teeth.

RCSOM/A 64.32 Skull. Adult male. Nganda Mossollo: locally hunted, collected from a site 35 km east of Etoumbi, 0°00′N, 15°00′E, Democratic Republic of Congo. Presented by Theo S. Jones in 1991. There is an apical abscess with a drainage channel over the lower right canine, and a large ulcer associated with the loss of the lower left canine. The upper right canine is fractured. The teeth are heavily pigmented.

RCSOM/A 64.41 Skull. Adult female. Touki village, 4½°N, 13¾°E, Bertoua district, Cameroun. Collected by F. G. Merfield on 2 March 1936 (collector's no. III 209): locally hunted (Banya tribe). Purchased from the Powell-

Cotton Museum in 1937. The lower 1st incisors have been lost and the 2nd incisors have drifted towards the midline. There is a space about 5 mm in occlusion between the upper and lower incisors.

RCSOM/A 64.42 Skull. Adult female. Touki village, 4½°N, 13¾°E, Bertoua district, Cameroun. Collected by F.G. Merfield on 2 March 1936 (collector's no. III 213 or 214): locally hunted (Banya tribe). Purchased from the Powell-Cotton Museum in 1937. There has been some destruction of bone around the roots of all the molars and premolars. This is particularly marked on the left side around the 1st and 2nd molars, and in the upper jaw the roots of the first two molars are exposed externally. The incisors are very worn. The mandibular rami are cut and the occipital region is damaged.

RCSOM/A 64.43 Skull. Adult male. Moyer, Congo. Donated by J.F. Colyer in 1947–48. Height 5 ft. 6 in.; chest 60 in.; span 100 in. The lower left 2nd premolar is rotated mesially, so that the external surface faces forwards and is in contact with the posterior surface of the 1st premolar. The occipital region is damaged postmortem.

RCSOM/A 64.44 Skull. Adult male. Meymajong village, 3¼°N, 13½°E, Upper Dja River, Lomié district, Cameroun. Collected by F.G. Merfield on 14 September 1936 (collector's no. III 500): locally hunted. Purchased by Colyer from the Powell-Cotton Museum on 28 July 1937; donated by Colyer in 1947–48. Each lower 3rd molar has an extra (6th) cusp between the anterointernal and posterointernal cusps.

RCSOM/A 64.45 Skull. Juvenile female. Meymajong village, 3¼°N, 13½°E, Upper Dja River, Lomié district, Cameroun. Collected by F.G. Merfield on 14 September 1936 (collector's no. III 501): locally hunted. Donated by Colyer in 1947–48. The canines and the 3rd molars are not in position. The left maxillary 2nd incisor is somewhat spear-shaped on the internal aspect and there is a slight development of the cingulum. The right maxillary 2nd incisor is cone-shaped.

RCSOM/A 64.46 Skull. Adult male. Purchased in 1949. The cranium is sectioned to display the braincase. The sagittal crest is very weak and only partially united.

RCSOM/A 64.47 Skull. Adult male. Purchased in 1949. The lower right 1st premolar is rotated, the internal surface facing partly forwards.

RCSOM/A 64.48 Skull. Adult male. Purchased in 1949.

RCSOM/A 64.49 Upper jaw (cast). Adult male. Transferred to the Odontological Museum from the Museum of the Royal College of Surgeons, to which it had been presented by the Bristol Philosophical Institution in 1848. The date of transfer is uncertain, however the first record is in the Folio of 1947–54. The 1st premolars are displaced bucally.

RCSOM/A 64.51 Skull. Adult male. *Gorilla beringei beringei*. SW Kigezi, approx. 1°S, 29¾°E, Uganda. Donated by C.R.S. Pitman in 1950. The postcranial skeleton was given to the Anatomy Department on 11 January 1951. The supraorbital ridges have been chopped off. The cranium is very asymmetrical, the face bending to the left. The teeth are very worn. There is a large food pocket between the right 2nd and 3rd molars and smaller pockets between the 2nd and 3rd lower molars on both sides. There is very little bone over the roots of the upper 2nd and 3rd molars.

RCSOM/A 64.141 Teeth (deciduous molars and canines). Juvenile male. These are the deciduous teeth removed from the same animal as the permanent dentition RCSOM/A 64.14.

RCSOM/A 64.491 Mandible (bisected). Adult female. A section of bone along the midline has been removed.

RCSOM/G 52.3 Skull. Adult male. Lomié district, 3¼°N, 13½°E, Cameroun. Collected by F.G. Merfield on 13 June 1936 (collector's no. III 284): locally hunted (Njem tribe). Purchased by Colyer from the Powell-Cotton Museum on 28 July 1937; donated by Colyer in 1947–48. The lower right 1st incisor is absent and the lower left 1st incisor is rotated, the mesial surface being directed slightly backwards. The external roots of the upper 1st molars are exposed. The outer halves of the heads of the mandibular condyles are abnormal in shape, the abnormality being possibly the result of injury. The roots of the lower canines are not fully formed. The enamel covering the lower thirds of the crowns of the maxillary and mandibular canines is abnormal in appearance, in the case of the mandibular right canine the form of the tooth is suggestive of injury during development. Dental caries are noticeable on the upper and lower 1st molars and upper central incisors.

RCSOM/G 67.1 Skull. Juvenile female. Between Batouri and Lomié, 3¾°N, 13¾°E, Cameroun. Collected by F.G. Merfield on 29 May 1935 (collector's no. III 123): locally hunted in old banana plantations. Purchased from the Powell-Cotton Museum in 1936. The upper incisors are abnormal in position, the 1st being slightly rotated and the 2nd posterior to the line of the 1st.

RCSOM/G 67.5 Cranium. Adult male.

RCSOM/G 67.11 Skull. Adult female. Between Batouri and Lomié, 3¾°N, 13¾°E, Cameroun. Collected by F.G. Merfield on 29 May 1935 (collector's no. III 121): locally hunted in old banana plantations. Purchased from the Powell-Cotton Museum in 1936. The upper 1st incisors are slightly rotated and are overlapped by the 2nd incisors. The occluding surfaces of the upper 3rd molars are directed slightly backwards. The posterointernal cusp of each lower 3rd molar is weak.

RCSOM/G 67.12 Skull. Adult female. Between Batouri and Lomié, 3¾°N, 13¾°E, Cameroun. Collected by F.G. Merfield on 29 May 1935 (collector's no. III 120): locally hunted in old banana plantations. Purchased from the Powell-Cotton Museum in 1936. Both the upper 1st incisors and the right 2nd incisor are rotated, the 1st incisors overlap one another. The upper 3rd molars are tilted backwards and the lower 3rd molars are tilted forwards and are slightly impacted.

RCSOM/G 67.13 Skull. Adult male. Between Batouri and Lomié, 3¾°N, 13¾°E, Cameroun. Collected by F.G. Merfield on 24 March 1935 (collector's no. III 28): locally hunted in old banana plantations. Purchased from the Powell-Cotton Museum in 1936. The roots of the lower canines are directed towards the midline and they are erupting a little labially. A slide of a longitudinal section through a canine tooth made by UCL during a loan is also retained.

RCSOM/G 67.14 Skull. Adult female. Between Batouri and Lomié, 3¾°N, 13¾°E, Cameroun. Collected by F.G. Merfield on 28 March 1935 (collector's no. III 36): locally hunted in old banana plantations. Purchased from the Powell-Cotton Museum in 1936. The teeth are very worn, and the external roots of the upper 2nd and 3rd premolars are partly exposed. The right side of the face is damaged.

RCSOM/G 67.15 Skull. Adult male. Between Batouri and Lomié, 3¾°N, 13¾°E, Cameroun. Collected by F.G. Merfield on 2 March 1935 (collector's no. III 7): locally hunted in old banana plantations. Purchased from the Powell-Cotton Museum in 1936. The left zygomatic arch and the occipital region are damaged. The upper 2nd incisors are rotated, and the lower right 3rd molar is tilted forwards and is impacted. The upper 3rd molars have Carabelli cusps. There is periodontal disease of the upper right premolars and molars, and caries of the upper and lower central incisors.

RCSOM/G 67.16 Skull. Adult female. Touki, 4½°N, 13¾°E, Bertoua district, Cameroun. Collected by F.G. Merfield on 2 March 1936 (collector's no. III 213 or 214): locally hunted (Banya tribe). Purchased from the Powell-Cotton Museum in 1937. The upper 1st incisors are rotated and so is the lower left 1st premolar. The lower 3rd molar is erupting a little externally to the rest of the dentition on the left side.

RCSOM/G 67.17 Skull. Adult female. Lomié district, 3¾°N, 13½°E, Cameroun. Collected by F.G. Merfield on 13 June 1936 (collector's no. III 282): locally hunted (Njem tribe). Purchased from the Powell-Cotton Museum in 1937. The upper 1st incisors are slightly rotated, and the upper 2nd premolars are slightly inside the line of the cheek teeth. The upper 3rd molars are very high in the tuberosity and the occlusal surfaces are directed backwards and outwards (this was true of both but the left molar has been lost). The lower 2nd premolars are slightly rotated. There is postmortem damage to the supraorbital ridge.

RCSOM/G 67.121 Skull. Adult male. Lomié district, 3¾°N, 13½°E, Cameroun. Collected by F.G. Merfield on 5 May 1936 (collector's no. III 224): locally hunted (Njem tribe). Purchased from the Powell-Cotton Museum in 1937. The lower right 1st incisor, the upper right 2nd premolar and left 3rd molar are rotated and the lower right 2nd incisor is well posterior to the 1st incisor. The lower right 1st molar's anterior root is wedged between the 1st molar and the 2nd premolar and its anterior root is represented by two tapering remnants. The enamel of part of the crown of this tooth is imperfect. The lower 2nd and 3rd molars have additional cusps. There is slight asymmetry of the face.

RCSOM/G 67.161 Skull. Adult female. Cameroons Province, Nigeria (now NW and SW Province of Cameroun). Collected by Frederic Wood Jones (1879–1954). Transferred from stores in April 1949; probably originally acquired by the Royal College of Surgeons in 1943. The upper 1st incisors overlap one another slightly, and the left 2nd incisor is slightly rotated. The upper left canine shows some antemortem damage. There is some periodontal disease. The occipital region is damaged.

RCSOM/G 82.1 Skull. Adult male. Donated by Dr J. Leon Williams c.1897. The mandible was fractured in about the region of the right canine, and the fracture has not healed. The lower 1st premolars, canines, and incisors and the upper 1st incisors were all lost as a result. On the right side the remaining teeth of the lower jaw are far in advance of the upper teeth in occlusion. It is thought that the injuries were sustained during sexual fighting.

RCSOM/G 82.11 Cranium. Adult male. Purchased in 1885. The mandible was formerly catalogued as G 10.1. It presents evidence of considerable injury to the facial bones received during life. The incisors have been lost, and their sockets absorbed; the nares have also been damaged, especially on the right side, and the right zygomatic arch has been fractured. Probable osteomyelitis has affected the damaged zygomatic and nasals. This infection has possibly spread through the sinuses and up to the frontal bone where purulent discharge has created small perforations. The palate is misaligned towards the right, which may be as a result of the trauma suffered. There are two conical, supernumerary teeth in the right ramus of the mandible: the bone has been removed to display them. The injuries were probably caused by sexual fighting, although the necrosis of the nasal bones may be the result of disease, rather than of injury.

RCSOM/G 97.4 Skull (cast). Adult male. Mamfé, British Mandated Cameroons. Donated by Ivan T. Sanderson in 1942. The upper left canine was broken off in life at the level of the neck, the lower half of the root is exposed on the outer surface and there is a sinus in the position of the apex of the root. The outer surfaces of the right 1st and 2nd incisors are bare of bone.

RCSOM/G 119.3 Skull. Adult male. Lomié district, 3¼°N, 13½°E, Cameroun. Collected by F.G. Merfield on 1 June 1936 (collector's no. III 269): locally hunted (Njem tribe). Donated by J.F. Colyer in 1947–48. There is an additional small, molariform tooth behind the upper left 3rd molar, the extra tooth is about two-thirds the size of the 3rd molar. Part of the anterior portion of the upper right 1st premolar has been lost by injury, the exposed surface is polished by wear. The incisors and canines are very worn, the loss of tissue has led to the exposure of the pulp cavities of the upper and lower left canines and infection of the parodontal tissues. In the case of the upper left canine, the pus has found an exit through the outer alveolar process about 6 mm above the margin of the bone covering the mesial aspect of the root, the opening is about 7 mm × 4.5 mm. The suppuration around the root of the upper right canine has been severe; there is an opening in the anterior nares, where the bone shows signs of considerable osteitis, and one through the outer alveolar process over the distal aspect of the root, the opening is circular in shape and is about 3.5 mm in diameter. The suppuration in connection with the mandibular left canine has found an exit through the outer alveolar bone around the apex of the root. The upper right 2nd incisor was lost antemortem, in the position of the healed socket there is an opening which leads to the distal aspect of the root of the right 1st incisor. A portion of the root of this tooth appears to have been flaked off, the covering bone seems to have been lost at the same time, the margin of the bone has a smooth regular appearance. The alveolar bone has receded slightly, the divisions between the external roots are exposed, the recession is most marked in the regions of the upper left 2nd premolar and the 1st molar and the corresponding teeth in the right side of the mandible. In the latter position the interdental bone has been destroyed almost to the

apices of the roots. The position of the roots of the majority of the maxillary premolars and molars are visible and the bone has disappeared over the posteroexternal roots of the left 1st molar and the right 2nd molar.

RCSOM/G 139.21 Skull (sagittally sectioned and repaired; braincase sectioned). Adult female. Hinterland, Cameroun. Transferred from the R.C.S. Museum to the Odontological Museum, 1947–54. This is 'Moina', who arrived at London Zoo (with a male named 'Mok') on 22 August 1932, and died there on 7 July 1939, aged 11 years. There is a slight, and deformed, sagittal crest. There is an osteophyte at the location of the right parietal. The lower 3rd molars are slightly impacted. The lower left 2nd premolar is unerupted. The lower right 2nd premolar is rotated and impacted against the 1st molar. Trauma to the lower left 1st molar has led to suppuration and there are holes for the exit of pus on the inner and outer surfaces of the bone both near the upper part of the root and over the root apex. Apart from the incisors the teeth are scarcely worn. The combined weekly rations for 'Mok' and 'Moina' were: 28 lb. grapes; 84 bananas; 28 lb. apples; 42 oranges; 4 lb. prunes; 7 bundles of rhubarb; 3 lb. tomatoes; 12 lettuces; ¾ bushel of greens; 5 lb. onions; 3 lb. carrots; 1 lb. Quaker Oats; ¼ lb. tea; 5 lb. sugar; 7 half-quartens of bread; ¾ lb. butter; 21 eggs; 6 tins of condensed milk; 14 quarts of TB-free, irradiated milk; and 1½ bottles of cod-liver oil cream.

RCSOM/G 165.1 Skull. Adult female. Gabon. Donated by Morton Alfred Smale (1847–1916), 1888–1909. The margins of the alveolar process show rarefying osteitis and much of the bone here has been resorbed. This has also occurred between the upper 2nd and 3rd molars, especially on the right. There are deep food pockets between the lower left 2nd and 3rd molars. The teeth are very worn; several caries are visible and 9 teeth have been lost, probably antemortem. There is some antemortem trauma noticeable on the left parietal.

RCSOM/G 165.2 Skull. Adult female. Gabon? Received from Morton A. Smale, 1888–1909. The bone around the lower 1st and 2nd molars and the lower right 1st premolar has been completely destroyed and the teeth have been lost in life. Consequently there is very little wear on the upper molars, although there is strong wear on the incisors in both jaws. The alveolar bone has been removed postmortem across the surface of all 4 canines. This skull was used by Charles Dawson to provide evidence to support his claim for the discovery of a transitional fossil hominid at Piltdown in Sussex in 1911–12.

RCSOM/G 165.3 Cranium. Adult female. Ossidinge division, British Cameroon. It was transferred from the RCS. Museum to the Odontological Museum. This is 1 of 42 skulls collected by Dr Neville Alexander Dyce Sharp (1885–1942) from remains of feasts at which gorillas and chimpanzees were eaten by local people. All 42 were presented to the RCS Museum. The teeth are extremely worn, and the outer alveolar margin is very much thickened in the region of the premolars and molars. This marked hyperplasia must have been associated with gross gingival enlargement. The occipital region is damaged.

RCSOM/G 165.11 Mandible. Adult male. Lomié district, 3¼°N, 13½°E, Cameroun. Collected by F.G. Merfield on 17 May 1936 (collector's no. III 249): locally hunted (Njem tribe). Purchased from the Powell-Cotton Museum in 1937. Tooth wear is noticeable across the dentition. The bone between the right 1st and 2nd premolars has been destroyed, and the roots are exposed to the apex. The right condyle appears worn.

Pre-1941 material:

C 121.5 Dissection of right groin. Juvenile male.

C 121.51 Dissection of left groin. Juvenile male.

C 167.1 Superficial muscles of right arm. Acquired before 1900.

C 168.1 Deep muscles of left arm. Acquired before 1900.

C 238.1 Superficial muscles of right leg. Acquired before 1900.

C 239.1 Deep muscles of left leg. Acquired before 1900.

C 240.1 Deep plantar tendons of left foot.

E 168.5 Right half of head. Juvenile male.

E 168.51 Left half of head. Juvenile male. Same individual as above.

E 210.1 Skin of head.

E 400.5 External ears. Juvenile male.

E 1040.1 Head of full-term fetus.

E 1041.1 Tympanic bones of above specimen.

E 1042.1 Head, to show malleus and incus. Juvenile.

E 1043.1 Left malleus.

J 358.1 Tongue. Juvenile.

J 898.1 Stomach. Juvenile. Acquired in 1888.

J 1088.1 Portion of duodenum. Acquired in 1888.

J 1089.1 Coil of small intestine. Acquired in 1888.

J 1321.1 Caecum. Acquired in 1888.

J 1357.1 Colon. Acquired in 1888.

J 1474.1 Pancreas. Acquired in 1888.

J 1642.1 Liver. Acquired in 1888.

J 1642.11 Liver. Adult male. Received from the British Museum in 1864.

K 183.1 Heart. Acquired in 1888.

L 37.1 Right wall of pharynx. Juvenile.

Q 509.1 Bladder, urethra and penis. Juvenile male. Acquired in 1888.

O.C. 5178 Mounted skeleton. Adult male. Presented by Captain Harris in 1851. This is the first skeleton of a gorilla brought to Europe. It is described in great detail in Owen (1866). Paper read on 9 September 1851. The animal had just attained maturity: all the permanent teeth are in place; but some of the larger epiphyses are still ununited, and the sagittal and occipital crests have not reached their full development. The greater part of the sternum, the seventh cervical vertebra, the left leg and foot and the right manus are wanting and have been replaced by plaster models. Vertebrae: C 7, D 13, L 4, S 5, C wanting. The right transverse process of the last lumbar vertebra is expanded and articulates with the ilium. Cranial capacity: 530 cc.

O.C. 5178A Skeleton. Female. Purchased from P.B. Du Chaillu in 1864.

O.C. 5178 B Skeleton. Juvenile. Purchased from P.B. Du Chaillu in 1864.

O.C. 5179 Cranium. Male. Presented by Captain Harris in 1851. The right anterior premolar, with the incisors and canines of both sides, has been lost. The base is broken away, exposing the interior cavity.

O.C. 5180 Cranium (cast). Adult male. Gaboon River. Presented by the Bristol Philosophical Institution in 1848. Though the amount of wear of the teeth indicates an animal of advanced age, the sagittal crest is small and divided by a longitudinal median groove.

O.C.5181 Cranium (cast). Adult male. Gaboon River. Presented by the Bristol Philosophical Institution in 1848. Though the teeth indicate a younger animal than O.C. 5180, the sagittal crest is more fully developed.

O.C. 5182 Cranium (cast). Adult female. Gaboon River. Presented by the Bristol Philosophical Institution in 1848.

O.C. 5183 Cranium (cast). Adult male. River Danger, West Coast of Africa. Presented by the Bristol Philosophical Institution in 1848.

At the time of presentation, the originals of the above four crania were in the Museum of the Philosophical Institution of Bristol. Shortly afterwards, they were given to the Bristol City Museum, where they are still kept. The above casts were presented to Prof. Richard Owen of the Royal College of Surgeons by Samuel Stutchbury, Curator of the Bristol Philosophical Institution. Specimens O.C. 5178, O.C. 5179, O.C. 5180, O.C.5181, O.C. 5182 and O.C. 5183 are described in great detail in Owen (1853).

Skeleton. Male. Purchased in 1860. The alveolar wall has been removed on the right side to show the roots of the teeth. Though by no means an aged individual, the supraorbital ridges and cranial crests are largely developed. Many of the bones of this skeleton are mounted in the separate series.

Skeleton. Female. Purchased from P.B. Du Chaillu in 1863. The skull is perfect, with complete unworn dentition. Many of the bones are mounted in the separate series.

Mounted skeleton. Juvenile. Purchased from P.B. Du Chaillu in 1863. All the milk teeth are in place, with the first permanent molars. Vertebrae: C 7, D 13, L 4, S 5, C 4 (complete). The vertebra here reckoned as the 1st lumbar, in accordance with its characters in the other skeletons, has a pair of short movable ribs, and might therefore be considered as belonging to the dorsal region. The 4th lumbar vertebra has no union with the ilia or sacrum.

Mounted skeleton. Adult female. Gabon. Purchased from P.B. Du Chaillu in 1863. This specimen shows well the inferiority in size and in development of the canine teeth and of the cranial crests in the female sex of the gorilla. Vertebrae: C 7, D 13, L 4, S 5, C 2 (incomplete). The vertebra corresponding to the last lumbar of man and the younger skeletons of the gorilla has both its transverse processes expanded to articulate with the ilium, and is by its body united with the sacrum; so that functionally it is converted into a sacral vertebra, as is partially the case in the male skeleton O.C. 5178.

Basi-hyal. Male. Purchased in 1872.

Skull. Subadult female. Gabon. Collected by R.B. Walker. Purchased in 1876. The basal suture is still open. The upper posterior molars (M3) are not fully in place.

Skeleton. Juvenile. Purchased in 1879. The milk dentition is complete, and the crowns of the first permanent molars are just appearing. Vertebrae: C 7, D 13, L 4, S & C 10.

Skeleton. Male. Gabon. Collected by R.B. Walker. Purchased from Barnard Davis in 1880. The left radius and ulna, 2 vertebrae and some of the bones of the manus and pes are wanting. Traces of immaturity remain in the absence of union of the basal suture of the cranium and the freedom of the epiphyses of the crest of the ilium, suprascapular border and upper end of the humerus. All the other epiphyses of the long bones are united. There are but 3 lumbar vertebrae; the one corresponding to the 4th in the younger specimens, and the 5th in man, is united with the sacrum.

Cranium. Female. Gabon. Collected by R.B. Walker. Purchased from Barnard Davis in 1880.

Skull. Juvenile male. Eloby, West Africa. Purchased in 1881. The canines alone of the milk teeth are retained. The lower 3rd molars are in place, but not those of the upper jaw. The temporal ridges do not quite meet at the vertex; and there is consequently no sagittal crest.

Skeleton. Adult male. River Congo, 300 or 400 mi. from the sea. Collected by Dr Ralph Leslie, a member of General Sir F. Goldsmith's expedition, in 1883. Presented by Sir Joseph Fayrer, MD, KCSI (1824 – 1907), in 1884. It is wanting the clavicles and most of the bones of the hands and feet.

Skull. Juvenile. Acquired in 1909.

Skull. Adult female. It is laid open to show the sinuses of the nose and especially the dilatation of the nasal duct. Presented by Charles Hose (1863–1929) in 1912.

Mounted skeleton. Juvenile female. Presented (in 1912?) by Sir John Bland-Sutton (1855–1936). The animal was 2 years old.

Skull. Juvenile. Presented by Charles Hose in 1912. The milk dentition is complete. The incisors and canines are missing.

Cranium. Juvenile male. Presented by Sir J. Bland-Sutton in 1917.

Skull. Female. Presented by Sir J. Bland-Sutton in 1917.

Cranium. Female. Okondja, 30′S, 13°40′E, Gabon. Presented by Frederick William Hugh Migeod (1872–1952) in 1921.

Cranium. Male. Makoua, lat. 0°, long. 15°14′E, Gabon. Presented by F.W.H. Migeod in 1921.

Cranium. Juvenile female. Ossidinge, British Cameroon. Presented by F.W.H. Migeod in 1924. Capacity 390 cc?

Skull. Juvenile female. Lake Kivu neighbourhood. Presented by Lt Col Robert Henry Elliot (1864–1936) in 1924. Capacity 480 cc.

Skull. Adult male. Lake Kivu neighbourhood. Presented by Lt Col Robert Henry Elliot in 1925. Capacity 610 cc.

Skull. Adult male. Lake Kivu neighbourhood. Presented by Lt Col Robert Henry Elliot in 1925.

Skull. Male. Ossidinge, British Cameroon. Presented by F.W.H. Migeod in 1925.

Skull. Adult male. Ossidinge division, British Cameroon. Presented by F.W.H. Migeod in 1925.

Skeleton. Infant. Presented by Sir J. Bland-Sutton in 1926.

Cranium (longitudinally trisected). Juvenile. Acquired in 1927.

Cranium. Cameroon. Presented by Dr N.A. Dyce Sharp in 1927.

Cranium. Female. Boji Hills, Nigeria. Presented by Dr N.A. Dyce Sharp in 1929.

Cranium. Female. Boji Hills, Nigeria. Presented by Dr N.A. Dyce Sharp in 1929.

Cranium. Female. Ogoja, S. Provinces, Nigeria. Presented by Dr Percy Amaury Talbot (1877–1945) in 1931.

Skull. Adult male. Ogoja, S. Provinces, Nigeria. Presented by Dr P. Amaury Talbot in 1931.

Cranium. Adult male. Ogoja, S. Provinces, Nigeria. Presented by Dr P. Amaury Talbot in 1931.

Cranium. Juvenile. Ogoja, S. Provinces, Nigeria. Presented by Dr P. Amaury Talbot in 1931.

Skull. Adult male. Batouri district, 3¾°N, 14¼°E, Cameroun. Purchased in 1937. It has a pathological chin.

Skull. Adult male. Batouri district, 3¾°N, 14¼°E, Cameroun. Purchased in 1937. It is imperfect.

Skull. Adult female. Batouri district, 3¾°N, 14¼°E, Cameroun. Purchased in 1937.

Postcranial skeleton. Male. Batouri district, Cameroun. Collected in 1935. Purchased in 1939. The skull is number A 64.14 in the Odontological Series.

Skeleton. Female. Cameroun hinterland. Purchased in 1939. This specimen is 'Moina', who was several years in captivity in Africa and later in London Zoo. She died on 7 July 1939 of an injury leading to septic wounds of the foot, gangrene and septicaemia. Her recorded weights were 112 lb. on 28 April 1933; 238 lb. on 19 September 1935; 245 lb. in August 1936 and January 1937; 250 lb. on 10 February 1937; 252 lb. on 18 September 1937; 245 lb. in October 1938 and 187½ lb. (after death) on 8 July 1939. She was aged about 3½ years on arrival at London Zoo on 22 August 1932. There is a misplaced unerupted left mandibular PM2 and an enlarged costal element (right) on lumbar vertebra 1. There are traces of immaturity in the skull, spine, pelvis and long bones. The brain and other viscera are in the museum or museum store. The skull was transferred to the Odontological Museum. [Some additional data added by GH from ZSL records].

Skeleton. Juvenile female. Purchased from Rowland Ward. Presented by Sir J. Bland-Sutton.

The following five plaster casts were presented by the American Museum of Natural History, through Dr Henry Fairfield Osborn. The originals were made by Carl Ethan Akeley when studying and hunting gorillas in the Kivu region of the Belgian Congo.

Face (cast). Adult male. Short face.

Face (cast). Adult male. Long face.

Hand (cast). Adult male.

Hand (cast). Juvenile male.

Foot (cast). Juvenile male.

The following crania, 40 in number (20 males and 20 females) were collected by Dr N.A. Dyce Sharp in villages situated in the northern upland forest area of the Ossidinge Division, British Cameroon. The crania are those of animals which had been shot and eaten by the local villagers. An extreme degree of individual variation is to be noted.

Cranium. Juvenile male.The 1st permanent molar teeth are in place. Cranial cavity 125 mm (internal length) × 101 mm (internal width).

Cranium. Juvenile female. The 1st permanent molars are in place. Nasal bridge is narrow and prominent. Cranial cavity 115 mm × 90 mm. Cubic capacity 420 cc?

Cranium. Juvenile female. The 1st permanent molars are in place. Epipteric bone on left side. Cranial cavity 113 mm × 89 mm. Cubic capacity 420 cc?

Skull. Juvenile female. The 1st permanent molars are in place. Cranial cavity 110 mm? × 96 mm.

Cranium. Juvenile female. The 1st permanent molars are in place. Cranial cavity 123 mm? × 94 mm.

Cranium. Subadult male. All permanent teeth in process of eruption. Cranial cavity 118 mm × 104 mm. Cubic capacity 550 cc.

Cranium. Subadult male. All permanent teeth in process of eruption. Cranial cavity 111 mm × 90 mm. Cubic capacity 400 cc?

Cranium. Subadult male. All permanent teeth in process of eruption. Cranial cavity 112 mm × 91 mm. Cubic capacity 400 cc?

Cranium. Subadult male. All permanent teeth in process of eruption. Cranial cavity 125 mm? × 101 mm.

Cranium. Subadult male. All permanent teeth in process of eruption. Cranial cavity 123 mm × 88 mm. Cubic capacity 500 cc.

Cranium. Subadult female. All permanent teeth in process of eruption. Cranial cavity 117 mm × 92 mm. Cubic capacity 475 cc.

Cranium. Adult male. Permanent teeth fully erupted, molars only slightly worn, skull sutures open. Cranial cavity 126 mm × 99 mm. Cubic capacity 500 cc?

Cranium. Adult male. Permanent teeth fully erupted, molars only slightly worn, skull sutures open. Cranial cavity 119 mm × 100 mm. Cubic capacity 505 cc.

Cranium. Adult male. Permanent teeth fully erupted, molars only slightly worn, skull sutures open. Cranial cavity 120 mm? × 99 mm. Cubic capacity 480 cc?

Cranium. Adult male. Permanent teeth fully erupted, molars only slightly worn, skull sutures open. Cranial cavity 120 mm × 93 mm. Cubic capacity 445 cc.

Cranium. Adult male. Permanent teeth fully erupted, molars only slightly worn, skull sutures open. Cranial cavity 122 mm × 96 mm. Cubic capacity 510 cc.

Cranium. Adult male. Permanent teeth fully erupted, molars only slightly worn, skull sutures open. Cranial cavity 111 mm × 89 mm. Cubic capacity 355 cc.

Cranium. Adult female. Permanent teeth fully erupted, molars only slightly worn, skull sutures open. Cranial cavity 114 mm × 92 mm. Cubic capacity 460 cc.

Cranium. Adult female. Permanent teeth fully erupted, molars only slightly worn, skull sutures open. Cranial cavity 112 mm × 87 mm. Cubic capacity 400 cc?

Cranium. Adult female. Permanent teeth fully erupted, molars only slightly worn, skull sutures open. Cranial cavity 115 mm × 96 mm. Cubic capacity 455 cc.

Cranium. Adult male. Cusps of 1st molars worn down to base, sutures fused. Cranial cavity 128 mm × 97 mm. Cubic capacity 530 cc.

Cranium. Adult male. Cusps of 1st molars worn down to base, sutures fused. Cranial cavity 128 mm × 96 mm. Cubic capacity 510 cc?

Cranium. Adult male. Cusps of 1st molars worn down to base, sutures fused. Cranial cavity 117 mm × 99 mm. Cubic capacity 510 cc?

Cranium. Adult male. Cusps of 1st molars worn down to base, sutures fused. Cranial cavity 125 mm? × 90 mm.

Cranium. Adult male. Cusps of 1st molars worn down to base, sutures fused. Cranial cavity 120 mm × 91 mm. Cubic capacity 510 cc.

Cranium. Adult female. Cusps of 1st molars worn down to base, sutures fused. Cranial cavity 110 mm? × 90 mm.

Cranium. Adult female. Cusps of 1st molars worn down to base, sutures fused. Cranial cavity 114 mm × 85 mm. Cubic capacity 395 cc?

Cranium. Adult female. Cusps of 1st molars worn down to base, sutures fused. Cranial cavity 115 mm × 95 mm. Cubic capacity 430 cc?

Cranium. Adult male. Teeth much worn, the dentine being exposed widely on the chewing surfaces of the 1st and 2nd molars. Cranial cavity 122 mm × 93 mm. Cubic capacity 495 cc?

Cranium. Adult male. Teeth much worn, the dentine being exposed widely on the chewing surfaces of the 1st and 2nd molars. Cranial cavity 124 mm × 97 mm. Cubic capacity 560 cc.

Cranium. Adult male. Teeth much worn, the dentine being exposed widely on the chewing surfaces of the 1st and 2nd molars. Cranial cavity 119 mm × 98 mm. Cubic capacity 485 cc?

Cranium. Adult female. Teeth much worn, the dentine being exposed widely on the chewing surfaces of the 1st and 2nd molars. Cranial cavity 115 mm × 100 mm. Cubic capacity 530 cc.

Cranium. Adult female. Teeth much worn, the dentine being exposed widely on the chewing surfaces of the 1st and 2nd molars. Cranial cavity 116 mm × 92 mm. Cubic capacity 475 cc.

Cranium. Adult female. Teeth much worn, the dentine being exposed widely on the chewing surfaces of the 1st and 2nd molars. Cranial cavity 111 mm × 90 mm. Cubic capacity 425 cc.

Cranium. Adult female. Teeth much worn, the dentine being exposed widely on the chewing surfaces of the 1st and 2nd molars. Cranial cavity 110 mm × 92 mm. Cubic capacity 425 cc?

Cranium. Adult female. Teeth much worn, the dentine being exposed widely on the chewing surfaces of the 1st and 2nd molars. Cranial cavity 113 mm × 90 mm. Cubic capacity 400 cc?

Cranium. Adult female. Teeth much worn, the dentine being exposed widely on the chewing surfaces of the 1st and 2nd molars. Cranial cavity 113 mm × 90 mm. Cubic capacity 410 cc.

Cranium. Adult female. Teeth much worn, the dentine being exposed widely on the chewing surfaces of the 1st and 2nd molars. Cranial cavity 115 mm × 91 mm. Cubic capacity 420 cc.

Cranium. Adult female. Teeth much worn, the dentine being exposed widely on the chewing surfaces of the 1st and 2nd molars. Cranial cavity 107 mm × 91 mm. Cubic capacity 400 cc? The nasal canal of the left orbit is completely absent.

Cranium. Adult female. Teeth much worn, the dentine being exposed widely on the chewing surfaces of the 1st and 2nd molars. Cranial cavity 105 mm × 88 mm. Cubic capacity 370 cc? Transferred to the Odontological Museum.

Imperial College of Science, Technology and Medicine, Department of Life Sciences, Prince Consort Road, London SW7 2BB

Doris M. Kermack; Denis J. Wright

S 271 Mounted skeleton. Juvenile.

S 411 Skull. Adult male.

NB: The above specimens belonged to the old Department of Pure and Applied Biology (now the Department of Life Sciences), which was located at South Kensington, in the Beit Quad (until 1998) and at Silwood Park. There were museum items, both invertebrate and vertebrate, used for undergraduate teaching, which were kept in Beit, but their use in teaching declined from the 1970s, and when the move was made to the new building in 1998, much remaining material was discarded or given away. The gorilla material is thought to have been disposed of at this time, but there is no record of what happened to it.

The Natural History Museum, Cromwell Road, London SW7 5BD

[Information collated by GH directly from NHM accession registers, and from correspondence with Paulina D. Jenkins]

This museum originated in the former Natural History Department of the British Museum in Great Russell Street, Bloomsbury, London. As the British Museum expanded, it eventually became necessary to house its natural history collections apart from the rest, and they were moved to part of the site of the 1862 Great International Exhibition in South Kensington. They were transferred to the present quarters between 1881 and 1885. The British Museum and the British Museum (Natural History) were separated by the British Museum Act, 1963, which made the latter completely independent, with its own body of Trustees. Its name was formally changed to The Natural History Museum in 1990.

When Lord Rothschild's collection was accessioned in 1939 the study material was removed to the British Museum (Natural History) whilst the mounted specimens remained at the Tring Museum. However, in this catalogue all the Rothschild specimens are listed under the Zoological Museum, Tring, purely for historical reasons.

The prefix ZD is for material accessioned in the Zoological Accessions Register; ZE is for skeletal material accessioned in the Osteology Register. For further explanation of the museum's numbering system, see Jenkins (1990).

SS.28.4.1919 Fourteen incomplete male and female, adult, juvenile and infant skeletons. [The code SS stands for Study Series and consists only of these 14 skeletons.]

ZD.1848.12.22.2 Cranium (cast). *Troglodytes gorilla*. Presented by Mr Samuel Stutchbury, Bristol.

ZD.1852.2.26.2 Skeleton. Adult male. *Troglodytes gorilla*. Gabon. Purchased from Monsieur Ch. Parzudaki, 'Naturaliste Preparateur', 2, rue du Bouloi, Paris.

ZD.1852.2.26.3 Skull. Purchased in 1852.

ZD.1857.11.2.2 Skeleton. Adult male. *Troglodytes gorilla*. Gabon. Purchased (from France?). According to Owen (1859c), this skeleton shows an extensive fracture, badly united, of the left arm bone, which has been shortened, and gives evidence of long suffering from abscess and partial exfoliation of bone. The upper canines have been lost some time before death, for their sockets have become absorbed.

ZD.1857.11.2.3 Skull. Adult female. *Troglodytes gorilla*. Gabon. Purchased from Paris.

ZD.1858.3.16.1 Skull. Female. *Troglodytes gorilla*. Gabon. Presented by R B. Walker.

ZD.1858.12.30.3 Mounted skin. Subadult male. *Troglodytes gorilla*. Purchased from Samuel Stutchbury, Bristol. This specimen reached the British Museum preserved in spirit on 10 September 1858. Prof. Owen placed it in the possession of Abraham Dee Bartlett (1812–97) in November 1858 to preserve and mount for the museum. According to Owen (1865), the gorilla 'was killed by natives in the interior of the Gaboon, and brought down to the port entire: it was at once immersed in a cask of spirits; but no antiseptic having been applied to the skin when fresh, as in the case of the gorillas killed by Mr. Du Chaillu, decomposition had made some advance; and when the cask was opened on its arrival in London, a great part of the cuticle with the hair had become detached from the specimen. It had, however, come off in large patches; and, as the texture had acquired a certain hardness through the action of the alcohol, their replacement was practicable, and the characteristic shades of colour of the different parts of the body and limbs could be determined.' After having preserved the skin, Bartlett, by permission of the Trustees, exhibited it at the Crystal Palace, where he was engaged. A portrait of A.D. Bartlett with the gorilla, still in a barrel of spirits, was taken by Peter Ashton and appears on page 20 of Bartlett (1900). The specimen is fully described in Owen (1865b). The substance of this article, including the plates, appeared shortly afterwards in Owen (1865a).

ZD.1858.12.30.4 Skeleton. Subadult male. *Troglodytes gorilla*. From the same animal as the mounted skin ZD.1858.12.30.3 above.

ZD.1861.5.14.1 Mounted skin. Juvenile male. *Troglodytes gorilla*. Gabon. Presented by Dr Philip Lutley Sclater (1829–1913), Secretary of the Zoological Society of London. It had been shipped alive from Africa by John Buchanan in 1859, destined for London Zoo, but died en route.

ZD.1861.5.14.2 Skeleton. Juvenile male. *Troglodytes gorilla*. Presented by Dr P.L. Sclater, Zoological Society of London. Collected on 9 November 1852. The mandible is missing.

ZD.1861.7.29.1 Mounted skin. Adult male. *Troglodytes gorilla*.

ZD.1861.7.29.2 Skeleton. Adult male. *Troglodytes gorilla*. The atlas, 1 patella and a portion of the collar bone are wanting.

ZD.1861.7.29.3 Skin. Female. *Troglodytes gorilla*. The right arm and leg are injured.

ZD.1861.7.29.4 Skeleton. Juvenile female. *Troglodytes gorilla*.

ZD.1861.7.29.5 Skin. Juvenile male. *Troglodytes gorilla*.

ZD.1861.7.29.6 Skeleton. Juvenile male. *Troglodytes gorilla*. The skull and bones of the feet are in the skin.

ZD.1861.7.29.7 Skin. Infant female. *Troglodytes gorilla*. It is in a bad state.

ZD.1861.7.29.8 Skull. Juvenile female. *Troglodytes gorilla*.

ZD.1861.7.29.27 Skeleton. Juvenile. *Troglodytes gorilla*. The hands, feet, atlas and 1st cervical vertebra are wanting.

The above 9 specimens, ZD.1861.7.29.1 to ZD.1861.7.29.27 inclusive, were purchased from P.B. Du Chaillu.

ZD.1861.12.5.1 Skull (plaster cast). *Gorilla savagei*. Purchased from Maison Verreaux, Place Royale 9, Paris.

ZD.1862.6.25.1 Skeleton. Adult male. *Troglodytes gorilla*. Presented by R.B.N. Walker Esq., Gabon. Skeleton, and viscera (in spirits) received on 10 June 1862.

ZD.1862.6.28.1 Skin. *Troglodytes gorilla*. Gabon. Purchased from W. Burton.

ZD.1864.12.1.1 Mounted skin and skeleton. Adult female. *Troglodytes gorilla*. Gabon.

ZD.1864.12.1.2 Skeleton and skin. Infant. *Troglodytes gorilla*.

ZD.1864.12.1.3 Skeleton. Infant. *Troglodytes gorilla*.

ZD.1864.12.1.4 Skeleton. Infant. *Troglodytes gorilla*.

ZD.1864.12.1.5 Skeleton. Adult female. *Troglodytes gorilla*.

The above 5 specimens, ZD.1864.12.1.1–ZD.1864.12.1.5 inclusive, were purchased from P.B. Du Chaillu.

ZD.1864.12.1.13 Skeleton and skin. Adult male. *Troglodytes gorilla*. Presented by P.B. Du Chaillu.

Letters from Du Chaillu are included in the Owen Collection of correspondence, housed in the Rare Book Room of the Natural History Museum's General Library. A letter from Du Chaillu, Fernan-Vaz river, dated 19 August 1864, informs Prof. Owen that: 'I sent the "Renshaw" back to London two days ago and among the curiosities is a live gorilla. Should you believe that a few days ago I had three of them alive? An adult female and her young! She had been wounded in the arm, then beaten almost to death and then secured from hand to foot, her roars were terrific to hear, unfortunately she lived only one night and died the next morning and her babe died three days afterwards. The one I send is a pretty good sized young male and is as savage as a tiger. In order to avoid the same trouble that I had with the chimpanzee (for I begged the Captain, Berridge, to dispose of it to the Zoological Gardens) I have sent the gorilla to Messrs Baring Bros & Co. and promised £100 to the Captain should the animal reach London alive, and I have requested them to let the Zoological Gardens have the animal saying that I have no doubt that the Gardens would give the full value for the animal. Among the six gorilla skins I sent to the British Museum (the one of adult live female included that came alive and wounded in the arm) there are two skins of large males, one of the two is a much larger specimen than the one in the British Museum. I hope the Museum will have it stuffed. Please present the specimen in my name and take whatever you may like.' Du Chaillu shipped on board the 'Renshaw' a large box containing 2 male skeletons, one barrel containing 5 skins, one barrel containing 3 skeletons, one box containing an incomplete skeleton, and one jar containing the skin and skeleton of a young gorilla. Of these, E. Gerrard Jr purchased 1 male skin and skeleton for £30, 1 female skin and skeleton for £25, and 1 young skin for £5. The British Museum purchased 1 female skin and skeleton for £25, 1 young skin and skeleton for £10, 2 young skeletons for £10, and 1 incomplete skeleton. One male skin and skeleton was presented by Du Chaillu. In a follow-up letter to Owen dated 20 August 1864, Du Chaillu added: 'Should the live gorilla die on board the "Renshaw" I have asked the Captain to take care of the skin and skeleton.'

ZD.1867.10.5.25 Mounted skin. *Troglodytes gorilla*. Listed by Jenkins (1990) as a *Pan troglodytes*.

ZD.1878.12.14.1 Skull. Adult male. *Troglodytes gorilla*. Purchased in December 1878 from Mr Rosenbush, 15 Clement's Inn, Strand, London W.C. (Consul for Italy at Sierra Leone, West Africa), by whose father it was killed on one of his journeys in the interior.

ZD.1883.3.29.3 Mounted skeleton. Adult female. *Troglodytes gorilla*. Gabon. Purchased from E. Gerrard Jr.

ZD.1905.1.2.2 Skull (cast). Adult male. *Gorilla gorilla matschiei* (syntype). Yaunde, South Kamerun. Presented by the Hon. Walter Rothschild. The original was collected by Hans Paschen, and was in the Zoological Museum, Tring.

ZD.1907.1.8.1 Cranium. Subadult male.

ZD.1907.1.8.2 Cranium. Adult male.

ZD.1907.1.8.3 Cranium. Adult female.

ZD.1907.1.8.4 Cranium. Adult female.

ZD.1907.1.8.5 Cranium. Adult female.

ZD.1907.1.8.6 Cranium. Adult female.

ZD.1907.1.8.7 Cranium. Adult female.

The above 7 crania, ZD.1907.1.8.1–ZD.1907.1.8.7 inclusive, are from Munshi Country in the NE corner of the Okuni district, Cross River division, southern Nigeria, 7°N, 9°W, and were presented by Elphinstone Dayrell (District Officer, Ikom, southern Nigeria), Beverley, Wickham, Hampshire.

ZD.1907.8.5.1 Mounted skin. Infant. French Congo. Presented by Rowland Ward. In very bad condition. It was handed over to the Exhibition Section for destruction on 29 November 1956 (the feet were required for a further exhibit).

ZD.1908.6.16.1 Skull and skin. Juvenile male. Fernan Vaz, Gabon. Presented by Dr William John Ansorge, 12 Addison Road, Bedford Park, Chiswick, West London. It had been deposited, with a chimpanzee, in London Zoo by Dr Ansorge on 18 March 1908. It died of acute bronchitis, fatty infiltration of the liver and anaemia on 25 March 1908.

ZD.1910.11.27.1 Skull. Juvenile female. Okwa, 6°20'N, 9°12'E, Ikom, Eastern Province, South Nigeria. Presented by Elphinstone Dayrell.

ZD.1912.4.11.1 Mounted skin. Male. Fernan Vaz, Gabon. Presented by the Hon. Walter Rothschild, Tring.

ZD.1913.2.2.1 Cranium. Adult male. Ikom, Nigeria. Presented by Elphinstone Dayrell, Junior Army and Navy Club, Whitehall, London.

ZD.1913.2.2.2 Cranium. Adult male. Ikom, Nigeria. Presented by Elphinstone Dayrell, Junior Army and Navy Club, Whitehall, London.

ZD.1914.5.19.1 Mounted skin and skull (cast). Adult male. *Gorilla beringei graueri*. Baraka, west of Lake Tanganyika, Belgian Congo. Presented by the Trustees of the Rowland Ward Bequest in 1913. The original skull is in the Belgian Congo Museum; an additional cast is in the mount. This specimen was formerly exhibited in the museum's central hall.

ZD.1916.11.1.1 Skeleton and mounted skin. Adult female. *Gorilla gorilla matschiei*. Bipindi district, Cameroun. Presented by the Trustees of the Rowland Ward Bequest in 1916.

ZD.1917.5.1.1 Mounted skin. Juvenile female. *Gorilla gorilla matschiei*. Bipindi district, Cameroun. Presented by the Rowland Ward Trustees in 1917.

ZD.1920.4.13.3 Skull. Adult female? *Gorilla beringei beringei*. Ruanda-Congo frontier, 29°30'E, 1°S. Presented in April 1920 by Capt. J.E.T. Philipps, District Com., Kigezi, British Ruanda through Uganda.

ZD.1920.4.13.4 Skull. Subadult male. *Gorilla beringei beringei*. Ruanda-Congo frontier, 29°30'E, 1°S. Presented in April 1920 by Capt. J.E.T. Philipps, District Com., Kigezi, British Ruanda through Uganda.

ZD.1920.4.13.5 Skull and skin. Juvenile. *Gorilla beringei beringei*. Ruanda-Congo frontier, 29°30'E, 1°S. Presented in April 1920 by Capt. J. E. T. Philipps, District Com., Kigezi, British Ruanda through Uganda.

ZD.1920.8.9.1 Skull (cast). Male. *Gorilla beringei beringei*. Mount Mikeno, Kivu district. Presented by Lord Rothschild, FRS, Tring.

ZD.1920.10.21.1 Skin. Juvenile. *Gorilla beringei beringei*. Mount Sabinio.

ZD.1922.2.10.1 Skeleton. Adult male. *Gorilla beringei beringei*. Kabale station, Kigezi district, Western Province, Uganda. Collected on Mount Sabinio. Presented by Capt. J.E.T. Philipps MC. Sent by Sir Robert T. Coryndon, Governor of Uganda, on behalf of Capt. Philipps. Received on 30 January 1922.

ZD.1922.2.10.2 Skeleton. Adult female. *Gorilla beringei beringei*. Kabale station, Kigezi district, Western Province, Uganda. Collected on Mount Sabinio. Presented by Capt. J.E.T. Philipps MC. Sent by Sir Robert T. Coryndon, Governor of Uganda, on behalf of Capt. Philipps. Received on 30 January 1922.

ZD.1922.12.19.3 Skeleton and mounted skin. Female. *Gorilla beringei graueri*. Belgian Congo. Presented by the Trustees of the Rowland Ward Bequest. It may be incorrectly numbered, and the same specimen as ZD.1922.12.19.4.

ZD.1922.12.19.4 Skeleton and mounted skin. Juvenile. *Gorilla beringei graueri*. Belgian Congo. Presented by the Trustees of the Rowland Ward Bequest.

ZD.1923.11.29.1 Skull. Adult male. Ololi, Moyen-Congo.

ZD.1923.11.29.2 Skull. Adult male. Ololi, Moyen-Congo.

ZD.1923.11.29.3 Skull. Adult male. Ololi, Moyen-Congo.

ZD.1923.11.29.4 Skull. Adult male. Ololi, Moyen-Congo.

ZD.1923.11.29.5 Skull. Adult female. Ololi, Moyen-Congo.

ZD.1923.11.29.6 Skull. Adult male. Ewo, Moyen-Congo.

ZD.1923.11.29.7 Skull. Adult male. Ewo, Moyen-Congo.

ZD.1923.11.29.8 Skull. Adult female. Ewo, Moyen-Congo.

ZD.1923.11.29.9 Skull. Juvenile. Ololi, Moyen-Congo.

ZD.1923.11.29.10 Skull. Juvenile. Ololi, Moyen-Congo.

The above 10 skulls, ZD.1923.11.29.1 to ZD.1923.11.29.10 inclusive, were purchased from Dr J.J. Vassal, Directeur du Société de Santé de l'Afrique Equatoriale Français, Brazzaville. They were sent from Bordeaux and were received on 10 July 1923.

ZD.1923.12.17.1 Mounted skin. Juvenile. *Gorilla beringei beringei*. Virunga volcanoes, Belgian Congo. Collected by T. Alexander Barns. Presented by the Trustees of the Rowland Ward Bequest.

ZD.1924.12.15.1 Mounted skin. Juvenile. *Gorilla beringei beringei*. Virunga volcanoes, Belgian Congo. Presented by the Trustees of the Rowland Ward Bequest.

ZD.1925.1.4.1 Skull. Adult male. French Congo.

ZD.1925.1.4.2 Skull. Adult male. French Congo.

ZD.1925.1.4.3 Skull. Adult male. French Congo.

The above 3 skulls were purchased from Dr J.J. Vassal, care of Mme Vassal, Lyceum, 17, rue de Bellechasse, Paris.

ZD.1926.11.10.1 Skull. Juvenile male. Gabon. Collected by P.B. Du Chaillu. Presented by A.H. Hallam-Murray, Sandling, Kent.

ZD.1929.1.1.1 Skull. Adult male. *Gorilla beringei beringei*. Kumbi, Kigezi, SW Uganda. Presented by Capt. C.R.S. Pitman.

ZD.1929.1.1.2 Skull. Juvenile male. *Gorilla beringei beringei*. Kumbi, Kigezi, SW Uganda. Presented by Capt. C.R.S. Pitman.

ZD.1929.1.1.3 Skull. Juvenile male. *Gorilla beringei beringei*. Kumbi, Kigezi, SW Uganda. Presented by Capt. C.R.S. Pitman.

ZD.1929.12.29.1 Skeleton and mounted skin. Adult male. *Gorilla beringei graueri*. Walikale, SW Kivu, altitude 9000 ft., Belgian Congo. Collected by Lt Col H.F. Fenn D.S.O. in 1927. Purchased from Lord Rothschild Tring. Approximate measurements: height 5 ft. 8 in.; chest 62 in.; biceps 18 in.; middle finger girth 4 in.; appendix 18 in long.

ZD.1935.3.19.1 Skull. Adult male. Obudu, Ogoja Province, South Nigeria. Presented by W.R. Hatch, 2b Clanricarde Gardens, Tunbridge Wells, Kent.

ZD.1935.3.19.2 Skull. Juvenile female. Obudu, Ogoja Province, South Nigeria. Presented by W.R. Hatch, 2b Clanricarde Gardens, Tunbridge Wells, Kent.

ZD.1935.12.16.1 Skull. Adult male. French Congo. Collected in 1933. Purchased from T.A. Glover.

ZD.1935.12.16.2 Skull. Adult male. French Congo. Collected in 1933. Purchased from T.A. Glover.

ZD.1936.7.7.1 Skull. Adult male. Cameroun. Presented by Mrs G.M. de L. Dayrell, 113 Warwick Road, London SW5.

ZD.1936.7.7.3 Cranium. Juvenile. Cameroun.

ZD.1936.7.14.1 Skull. Adult male. British Cameroon. Presented by M.H.W. Swabey, Nigerian Political Service, Mamfé, British Cameroon, via Mrs Swabey, 21a First Avenue, Hove, Sussex.

ZD.1947.3 Skin. Adult female. From the same individual as the skeleton ZE.1948.3.31.1.

ZD.1947.4 Skin. Adult female. From the same individual as the skeleton ZE.1948.3.31.2.

ZD.1948.435 Skull. Adult male. Mtene, Mamfé division, Cameroon. Collected by Ivan T. Sanderson, Percy Sladen Expedition, on 5 December 1932.

ZD.1948.436 Skeleton and skin. Adult male. Tinta, Mamfé division, Cameroon. Collected by Ivan T. Sanderson, Percy Sladen Expedition, on 4 April 1933.

ZD.1948.437 Skull and skin. Infant female. Mtene, Mamfé division, Cameroon. Collected by Ivan T. Sanderson, Percy Sladen Expedition, on 15 June 1933.

ZE.1948.2.27.1 Skeleton. Adult male.

ZE.1948.3.3.2 Skeleton. Adult male.

ZE.1948.3.11.1 Skeleton. Adult.

ZE.1948.3.20.1 Mounted skeleton. Adult female.

ZE.1948.3.31.1 Skeleton. Adult female. Spanish Guinea, 2°0′N, 11°10′E, forest country 700 m. Collected on 8 September 1946. Presented in 1947 by Laurence Hardy Esq., Old Cottage, Sutton Lane, Chiswick, London W4. It had a young one at the breast. Height 140 cm; chest circumference 100 cm. The skin from this specimen is no. ZD.1947.3.

ZE.1948.3.31.2 Skeleton. Adult female. Spanish Guinea, 1°50′N, 11°30′E, forest country 700 m. Collected on 21 September 1946. Presented in 1947 by Laurence Hardy Esq., Old Cottage, Sutton Lane, Chiswick, London W4. Height 135 cm; chest circumference 96 cm. The skin from this specimen is no. ZD.1947.4.

ZE.1948.4.1.1 Skeleton. Adult male.

ZE.1948.4.1.2 Skeleton. Adult female.

ZE.1948.5.4.1 Skeleton. Adult male. French Congo.

ZE.1948.5.4.2 Skeleton. Adult male.

ZE.1948.12.20.2 Skeleton. Adult female. French Congo.

ZD.1949.603 Skull. Adult male. Presented by Miss E.M. White, 49 Old London Road, Hastings, Sussex.

ZD.1949.663 Skull. Adult male. Ziendi, Oubangui, French Congo. Collected on 7 February 1948. Purchased for £5 from Armand Vast, rue Jussieu, Paris, through Miss Trewavas.

ZD.1949.664 Skull. Adult female. Ziendi, Oubangui, French Congo. Collected on 7 February 1948. Purchased for £5 from Armand Vast, rue Jussieu, Paris, through Miss Trewavas.

ZE.1949.12.30.2 Skeleton. Adult.

ZE.1951.9.27.10 Skeleton. Adult male.

ZE.1951.9.27.11 Skeleton. Adult.

ZE.1951.9.27.12 Skull. Adult female. French Congo.

ZE.1951.9.27.13 Skull. Adult female. French Congo.

ZE.1951.9.27.14 Skull. Adult female. French Congo.

ZE.1951.9.27.15 Mounted skeleton. Adult male.

ZE.1951.9.27.17 Cranium. Adult female.

ZE.1951.9.27.18 Cranium. Adult female.

ZE.1951.9.27.19 Skull. Adult male.

ZE.1951.9.27.20 Skull, hyoid, and atlas. Adult female.

ZE.1961.4.5.1 Skeleton. Adult male. *Gorilla beringei beringei*. Uganda. Purchased for £50 from the Game and Fisheries Department, P.O. Box 4, Entebbe, Uganda. The specimen was found by John Mills, Game Ranger, Mbarara, Uganda. It was despatched in a box on 7 November 1960, shipped through the African Mercantile Co. Ltd, P.O. Box 27, Kampala, and was received on 27 March 1961.

ZE.1963.2.11.1 Skeleton. Juvenile male. *Gorilla beringei graueri*. Central Congo. Received from the Zoological Society of London. This is 'Rundi', who arrived at London Zoo on 11 July 1962, accompanied by a female named 'Tanga'. They had been captured by Charles Cordier in the Tshibinda forest, Bukavu territory, Kivu district. They were put in the care of London Zoo by the Congolese Government, and later purchased. Upon arrival, 'Rundi' weighed 22 lb. and was about 1½ years old. He died of bronchopneumonia with red hepatization on 3 December 1962, weighing 13.75 kg. [Some additional information added by GH from London Zoo records.]

ZE.1963.3.25.1 Skeleton. Juvenile male. *Gorilla beringei beringei*. Mount Mgahinga, Virunga volcanoes, Kigezi district,Uganda. Received from the Zoological Society of London. This is 'Reuben', who was discovered sitting next to the body of his recently dead father by Roveni (alias Reuben) Rwanzagire, a Bakiga tribesman and local guide. The young gorilla was captured on 23 February 1960 and taken to the 'Travellers' Rest' safari lodge at Kisoro (owned and run by Walter Baumgartel), then by land rover to Mbarara, then on to the Animal Orphanage at Entebbe, where he was given a Schweinfurth's chimpanzee (*Pan troglodytes schweinfurthii*) for company. Both were sold by the Game

Warden, Entebbe, to London Zoo for 50,000 Ugandan shillings. They arrived at London Airport at 1:15 a.m. on 19 May 1960, where they were met by a heated Zoo van. They were installed in a cage at the Zoo by 3 a.m. The chimpanzee was named 'Cleo', and the gorilla was named 'Reuben', after the man who found him. 'Reuben' was about 2¼ years old, and was placed with 'Cleo' until 9 October 1962. His recorded weights were: 44 lb. on 24 May 1960; 62 lb. on 3 December 1960; 67 lb. in December 1960; 70½ lb. in February 1962 and 71½ lb. in August 1962. He had a 'severe chill' on 5 December 1962, but was much better by 18 December. He had a relapse on 23 December, became very thin, refused all food and aid, but had the energy to move around his cage. He died of lobar pneumonia and septicaemia on 30 December 1962 (weight 92 lb.). [Some additional information added by GH from London Zoo records.]

Z?.??65.3.3.12 Skull. Adult male. Collected on 11 June 1937. Number not traced.

ZE.1968.8.6.1 Skeleton. Cameroon.

ZD.1972.162 Face (cast). Adult male.

ZD.1972.1343 Mounted skin. Adult female. *Gorilla beringei graueri*. It may be the same specimen as ZD.1922.12.19.3.

ZD.1976.439 Skeleton. Adult female. Between Batouri and Lomié, 4°30′N, 14°30′E, Batouri district, Cameroun. Collected by F.G. Merfield on 29 May 1935; collector's number M 121 (3). Merfield's lists: 3rd series, 6th trip, list 2, page 4. Skinned in bush; local hunter's name N'Gillie. Received from Alexander J.E. Cave (1900−2001) in 1976 (his collection number C1.5). Terminal phalanges missing, hyoid present.

ZD.1976.440 Skeleton. Adult female. Obala, 4°30′N, 14°30′E, Batouri district, Cameroun. Collected by F.G. Merfield on 23 October 1933; collector's number M 660 (2). Merfield's lists: 2nd series, 4th trip, list 2, page 7. Local hunter's name N'Gillie. Received from A.J.E. Cave in 1976 (his collection number C1.6). Left arm missing, hyoid present.

ZD.1978.1226 Skeleton and mounted skin. Adult male. Presented by the Zoological Society of London. This is 'Guy', who arrived at London Zoo on 5 November 1947, aged about 1½ years. His 'official' weight upon arrival is 10.635 kg, but in fact this weight was recorded one week after arrival. A systematic record of 'Guy's' weight increases and general development was maintained by his keeper, Lawrence George Smith (1907−73) over a period of more than 20 years. This showed that 'Guy's' weight upon arrival was 10.450 kg (he was undernourished at the time). Unfortunately, Mr Smith's records are now lost, but the highest weight he recorded for 'Guy' was 488½ lb., on 31 July 1967 (L.G. Smith, personal communication). 'Guy' came originally from Cameroun and spent nearly 4 months in captivity in the Ménagerie du Jardin des

Plantes, Paris, before being sent to London Zoo in exchange for a tiger, born in Calcutta, and a zebra. He was accompanied on the trip to England by Francis Edward Fooks (1892−1967). 'Guy' was at first housed in the chimpanzee nursery, and was then moved to the Monkey House on 5 November 1949. He suffered a bout of chronic diarrhoea on 24 July 1950, and pneumonia was diagnosed on 15 August 1950 (weight 89½ lb.), which was treated with M&B 693 (sulphapyridine). His weight dropped to 80 lb. on 26 September 1950, but recovered to 91 lb. on 9 November 1950. He suffered with a cold from 29 April 1951 to 19 May 1951. Up to the age of 6 or 7 years, he suffered intermittently from lesions on the soles and heels of both feet, attributed to a lack of oil and vitamin A in his diet, which was subsequently corrected. Dental development: lost lower incisor, probably broken, on 15 February 1951; upper central incisors erupted over milk teeth on 18 May 1951, milk teeth discarded 1−2 July 1951 and were never found; upper left milk canine discarded on 2 February 1954, upper right on 6 February 1954; lower right canine discarded on 14 February 1954, under which could be seen the point of the new tooth in the cavity, no sign of uppers; both upper canines and lower right canine all showing level with the gum on 8 March 1954. He had infected teeth, and would take no food or drink on 17 February 1971. On 5 October 1971 he was vomiting food, and his weight had dropped to 367 lb. On 12 October he was sedated and removed to the zoo hospital, where his molars were extracted under general anaesthesia. He was returned to new quarters in the Michael Sobell Pavilions (which were officially opened on 4 May 1972), and was fully recovered on 13 October 1971. He was first introduced to the female 'Lomie' on 21 October. His weight on 5 November 1971 was 408 lb. From 7 September 1973 he and 'Lomie' were left together at night. On 2 January 1975 he had a cough, on 3 January a cough and a cold. There was no change by 5 January but he was eating small amounts, same on 6 January (he was not on exhibition from 3 January to 7 January). On 7 January he had a cough, but now was not eating; this remained the same until 16 January. Further dental problems led to the extraction of 4 incisors and 3 premolars on 20 January 1976. Three premolars and 4 incisors were removed on 8 June 1978, but he died at 1:30 p.m. while coming round from the anaesthetic (weight 238 kg). The cause of death was cardiac arrest consequent on arteriosclerosis of the coronary arteries and pulmonary congestion. The skin was mounted by Natural History Museum staff taxidermists Roy Hale, Arthur G. Hayward and Barry Sutton. Measurements taken for taxidermy: shoulder width 22½ in.; chest 68 in.; biceps 19 in.; elbow 20 in.; forearm 16 in.; wrist 10¼ in.; width of back 25½ in.; width at waist 19½ in.; stump of tail to atlas 33 in.; thigh 32 in.; knee 17 in.; ankle 13½ in.; hand 10½ in. × 5¼ in.; foot 13½ in × 6 in. The skin was 'pickled' in a

1.5% formalin solution, then rinsed in cold water, drained, folded and stored in a deep-freeze unit. In April 1981 it was taken out of deep-freeze storage and allowed to thaw out, before commencing the interrupted process of dressing the skin, which up till then had only been fixed and preserved. Due to long storage, the connective tissue situated under the corium had tightened and had dried up more than usual, and had shrunk by 12%. The process of dressing had been interrupted originally because of work-load pressures, and when an attempt was made after more than 2 years to soften the skin, it did not react to the normal dressing method. Mr Hayward was able to reconstitute the skin after consulting experts at the National Leather Sellers Centre at Nene College, Northampton. Mr Hayward made a full-size clay model, cast a glass-fibre mould from it, filled the mould with polyurethane foam and sewed the skin around it. [Some additional data added by GH directly from London Zoo records, and from correspondence with Michael K. Boorer, Michael Carman, and Joan Crammond, London Zoo.]

ZD.1979.36 Hand (cast) and foot (cast). Adult male. *Gorilla beringei beringei.*

ZD.1979.764 Dentition; portion of right maxilla; posterior portion of cranium. Adult.

ZD.1979.1322 Skeleton. Juvenile female. Collected by J.C. Trevor. Acquired from the World Health Organisation. Right ulna broken, damaged during life.

ZD.1979.2055 Fetus (wet specimen). Male.

ZD.1981.757 Skeleton. Adult. Received from the Zoological Society of London.

ZD.1981.758 Skeleton. Juvenile female. Received from the Zoological Society of London.

ZD.1981.898 Mounted skin (on display). Juvenile. *Gorilla beringei beringei.* Virunga mountains, Belgian Congo. This may be the same specimen as either ZD.1923.12.17.1 or ZD.1924.12.15.1.

ZD.1981.980 Mounted skin. Adult male. *Gorilla beringei beringei.* Southwest corner of Lake Kivu, Belgian Congo.

ZD.1981.2634 Mounted skin. Infant. Acquired from Lord Rothschild, Tring.

ZD.1981.2635 Mounted skin. Juvenile female. Acquired from Lord Rothschild, Tring.

ZD.1981.2636 Mounted skin. Male. Acquired from Lord Rothschild, Tring.

ZD.1981.2637 Mounted skin. Juvenile female.

ZD.1986.533 Cranium. Adult female. Sangha and Likouala, between affluents of the Oubangui River, French Congo.

ZD.1986.534 Skull. Adult male. Sangha and Likouala, between affluents of the Oubangui River, French Congo.

ZD.1986.535 Skull. Adult male. Sangha and Likouala, between affluents of the Oubangui River, French Congo.

ZD.1986.536 Cranium. Adult male. Sangha and Likouala, between affluents of the Oubangui River, French Congo.

ZD.1986.537 Skull. Juvenile male. Sangha and Likouala, between affluents of the Oubangui River, French Congo.

ZD.1986.538 Mandible. Adult female. Sangha and Likouala, between affluents of the Oubangui River, French Congo.

ZD.1986.539 Mandible. Adult female. Sangha and Likouala, between affluents of the Oubangui River, French Congo.

ZD.1986.540 Left mandibular ramus. Adult. Sangha and Likouala, between affluents of the Oubangui River, French Congo.

ZD.1986.757 Mounted skeleton. Adult male. Equatorial Guinea. Bequest of Lord Rothschild, Tring. This may be from the same specimen as the mounted skin ZD.1939.3415 (holotype of *Gorilla gorilla halli*).

ZD.1986.758 Skull. Adult female. Kamerun. Collected in March 1904. Bequest of Lord Rothschild, Tring.

ZD.1986.759 Cranium (cast). Adult female? Ndenoa, North Kamerun. Bequest of Lord Rothschild, Tring.

ZD.1986.760 Skin and toe-nails. Adult male. *Gorilla savagei.* Gabon. Donated by Du Chaillu. This specimen or ZD.1986.761 may be the same specimen as ZD.1861.7.29.1 (mounted skin) and ZD.1861.7.29.2 (skeleton).

ZD.1986.761 Skin. Adult male. *Gorilla savagei.* Gabon. Donated by Du Chaillu. See ZD.1986.760 above.

ZD.1986.762 Mounted skeleton. Adult female. *Gorilla beringei graueri.* Kivu, Belgian Congo. Bequest of Lord Rothschild, Tring.

ZD.1986.763 Skeleton (partially articulated). Juvenile male. *Troglodytes gorilla.*

ZD.1986.764 Mounted skin. Infant. Received from the Haslar Hospital, Gosport, Hampshire.

ZD.1986.765 Skull. Adult male. Collected on 6 July 1905. Bequest of Lord Rothschild, Tring.

ZD.1986.766 Skull. Adult male. *Gorilla castaneiceps.* Bequest of Lord Rothschild, Tring.

ZD.1986.767 Skull. Adult male. Bequest of Lord Rothschild, Tring.

ZD.1986.768 Cranium. Adult male.

ZD.1986.769 Skull (cast, mounted). *Gorilla beringei graueri.* Kivu, Belgian Congo. Bequest of Lord Rothschild, Tring.

ZD.1986.770 Skull (cast, mounted). Adult male. North Kamerun. Bequest of Lord Rothschild, Tring.

ZD.1986.771 Skull (cast). Adult male. The original is in the Anatomical Museum, University of Sri Lanka, Colombo. Donated by W.C.O. Hill.

ZD.1989.328 Skeleton. Adult male.

ZD.1989.329 Cranium. Adult female.

ZD.1989.330 Limb bones. Adult.

ZD.1989.331 Mandible. Juvenile.

ZD.1989.749 Skull. Juvenile. Near Etoumbi, 0°15′E, altitude 350 m, Congo Republic. Donated by T.S. Jones, who collected it in 1986 (collector's number RC 38).

ZD.1990.536 Skull (cast). Ex teaching collection, St Mary's Hospital Medical School, London.

ZD.1992.155 Pelvis and limb bones.

University College London, Department of Cell and Developmental Biology, Anatomy Laboratory, Rockefeller Building, University Street, London WC1E 6JJ

Samuel Cobb; John Pegington; Fred Spoor

CA 01 Skull (sagittally bisected, wired together). Adult male?

CA 01A Mounted skeleton. Adult male?

CA 01B Skull. Juvenile. I 1/1; RI 2/2; DC 1/1; LPM 1/ ; Ldm /1; dm/2; Ldm 2/ ; M 1/1; M 2/2.

CA 01 E Skull. Infant. di 1/1; di 2/2; dc 1/1; dm 1/1; dm 2/2. No basioccipital.

CA 01f Skull. Infant. di 1/1 (erupted but missing); di 2/2 (erupted but missing); dm 1/1. No right exoccipital.

CA 01g Skull. Adult male? Previous no. 1474.

CA 01g* Skull. Infant. di 1/1; di 2/2; dm 1/1. No basioccipital, exoccipitals or left temporal.

CA 04 Skull (sagittally bisected). Subadult. I 1/1; I 2/2; PM 1/1; PM 2/2; M 1/1; M 2/2.

CA 07 Skull. Adult male?

CA 08 Mounted skeleton. Adult female?

CA 09 Skull. Adult male?

CA 09a Skull. Adult female?

CA 09b Cranium. Adult male? Part of occipital and temporals missing. Previous no. RADM 2933-4; 1145.

CA 10 Skull. Juvenile. di 1/1; di 2/2; dc 1/1; dm 2/2; M 1/ . Basioccipital and exoccipital missing.

CA 1146 Skull. Adult.

University College London, Department of Anthropology, 14 Taviton Street, London WC1H 0BW

Leslie C. Aiello; Gemma Price

PA617 Maxillary dentition and palate; mandibular dentition (cast). Adult male. From the Napier collection. (When Prof. John Napier retired from Birkbeck College in 1983, his entire collection of Primate material was transferred to the Department of Anthropology. Those specimens not indicated to be part of the Napier collection are part of the Department's original collection).

PA634 Left foot, plantar aspect (cast). Subadult female. *Gorilla beringei beringei*. East Congo. From the Napier collection.

PA954 Postcranial skeleton. Juvenile female. From the Napier collection. Comprises manubrium; 28 ribs wired; 20 vertebrae wired; 3 sacral; both scapulae, humeri, radii, ulnae, innominates, femora, tibiae, fibulae; left clavicle. The hands and feet are missing.

PA955 Postcranial skeleton. Juvenile. From the Napier collection. Cartilage preserved and some elements wired together: 25 vertebral bodies; 4 sacral bodies; 26 ribs; possible hyoid; both scapulae, clavicles, humeri, radii, ulnae, innominates, femora, tibiae, fibulae. The hands and feet are missing.

PA956 Bones of right foot. Juvenile (estimated age 3½ years). From the Napier collection. It died in London Zoo on 23 March 1927. NB: The Zoological Society of London records do not show any gorilla as having died on this date; therefore, this may be from one of two gorillas which arrived at the London Zoo on 18 November 1927 (locality Central Africa, sexes unrecorded). They had been deposited, along with three chimpanzees from West Africa, by G. Bruce Chapman. Both gorillas died on 23 November 1927. [Additional information added by GH directly from London Zoo records.]

PA957 Postcranial skeleton and right hand and foot (fleshed). Juvenile. Comprises 26 ribs; 23 vertebrae; articulated pelvis; hyoid; both scapulae, clavicles, humeri, ulnae, radii, patellae; sternum; right femur; left tibia, fibula, calcaneus, hand; 3rd tarsal digit. Missing 8 hand phalanges. From the Napier collection.

PA958 Bones of right foot. Juvenile. From the Napier collection.

PA998 Postcranial skeleton. Comprises 18 vertebrae; 25 ribs; sacrum; sternum; coccyx; both humeri, radii, ulnae, tibiae, fibulae, clavicles, scapulae; left patella. From the Napier collection. The hands and feet are missing.

PA999 Appendicular skeleton with ribs. Adult male. Comprises both humeri, radii, scapulae, tibiae, fibulae, femora; ribs; innominates; 1 patella; 1 xiphoid process; 3rd and 2nd cuneiform. From the Napier collection: obtained from the British Museum (Natural History).

PA1000 Left forearm and hand (fleshed). From the Napier collection. Currently classed as 'missing'.

PA1073 Left hand, palmar aspect (cast). Adult male. From the Napier collection.

PA1149 Endocranial cast. Male. From the Napier collection.

PA1151 Endocranial cast. Female. From the Napier collection.

PA1158 Endocranial cast. Female. From the Napier collection.

PA1159 Endocranial cast. Male. From the Napier collection.

PA1182 Bones of right and left feet. Adult. Left: 5 metatarsals, 8 phalanges, talus, lateral and medial cuneiform. Right: 5 metatarsals, 7 phalanges, talus, calcaneus, navicular. From the Napier collection.

PA1185 Postcranial skeleton (partial). Juvenile. Comprises both humeri, radii, ulnae, scapulae, innominates, femora, tibiae, fibulae; feet; left clavicle. Missing: hands; right NT5; 12 phalanges; ribs; vertebrae. From the Napier collection.

PA1190 Postcranial skeleton. Juvenile. Comprises 26 ribs; 24 vertebral bodies; 4 sacral bodies; coccyx; manubrium; both clavicles, humeri, radii, ulnae, innominates, femora, tibiae, fibulae; left lateral and basilar occipital. From the Napier collection.

PA1192 Skull. Adult male. Missing lower right I2. From the Napier collection: it belongs to the Department of Anatomy, University of Wales, Cardiff College.

PA1193 Skull. Adult male. From the Napier collection: it belongs to the Department of Anatomy, University of Wales, Cardiff College.

PA1273 Right innominate bone (cast).

PA1282 Skull. Adult. Missing skullcap; both I1, I2. Left M1 lost antemortem.

PA1283 Skull. Juvenile female. Cranium missing left I2; canines and M3 erupting. Mandibular canines and M3 erupting.

PA1305 Endocranial cast.

PA1308 Endocranial cast.

PA1310 Endocranial cast.

PA1311 Endocranial cast of parieto-occipital region.

PA1312 Endocranial cast (left half of brain only).

PA1313 Endocranial cast (left half of brain only).

PA1319 Endocranial cast.

PA2016 Postcranial skeleton. Adult. Comprises both scapulae, clavicles, ulnae, radii, right humerus, left rib. From the Napier collection.

PA2209 Articulated pelvis. Adult.

PA2215 Left femur (gorilla?). Adult.

PA2231 Articulated torso. Juvenile.

PA2568 Articulated hand (cast, rigid).

PA2569 Articulated foot (cast, rigid).

PA2641 Pelves × 2.

PA2642 Skull.

PA2674 Skull.

Skull (bisected). Adult male. It is mounted against a Perspex sheet, with one-half of the skull staggered on each side of the sheet. Received as a personal gift from Prof. Michael Day to Prof. Leslie C. Aiello.

University College London, Grant Museum of Zoology and Comparative Anatomy, Rockefeller Building, 21 University Street, London WC1E 6JJ

Mark Carnall; Miss Jo Hatton

LDUCZ-Z459a Skull. Adult female. The occipital region is missing.

LDUCZ-Z459b Skull. Adult male. It is missing the upper left lateral incisor and lower right 1st molar.

According to the old card catalogue for the collection, the above two skulls were received when the museum took on the University of London Loan Collection in 1910–11.

LDUCZ-Z463 Mandible (plaster cast). Adult male.

LDUCZ-Z467 Left manus (painted plaster cast). Adult male. *Gorilla beringei beringei*. Lake Kivu region.

LDUCZ-Z468 Right pes (painted plaster cast). Adult male. *Gorilla beringei beringei*. Lake Kivu region.

The above casts were obtained by the Columbia University/American Museum of Natural History African Expedition of 1929. They were purchased (with other casts of orangutan) at a total cost of £2 12s 0d from R.E. Damon & Co.

LDUCZ-Z476 Mounted skeleton. Male. Purchased by E. Ray Lankester from Maison Tramond, Paris, in 1880/81. Left foot: missing 2 terminal phalanges on digit 2. Right foot: all phalanges missing on digits 2, 3, 4 and 5.

LDUCZ-Z487 Mounted skeleton. Adult female.

LDUCZ-Z2742 Postcranial skeleton. It is composed of remains from at least two individuals. It was purchased from Adam, Rouilly, London, possibly in the 1920s.

LDUCZ-Z2914 Postcranial skeleton (partially articulated). Juvenile.

In addition to the above specimens, there are a number of specimens which seem to have left the collection or have been misplaced. The 1898 catalogue of the collection lists a cast of skull (male) and cast of skull (female), as well as a cast of hand and cast of foot.

University College London, Institute of Archaeology, 31–34 Gordon Square, London WC1H 0PY

Sandra Bond; D.R. Brothwell

5.63 4/D/V Skull. Adult male. The earlier accession number (marked in pencil) is 1936.121.

University of London, The Royal Veterinary College, Veterinary Museum, Royal College Street, London NW1 0TU

Andrew R. Crook

Skull. Adult male. It is missing all incisors and canines.

Skull. Adult male. Presented by Prof. C.M. West.

Limb bones (left femur, left tibia, left fibula, left clavicle, left scapula, left humerus, left radius, left ulna). Adult.

Zoological Society of London, Veterinary Department, Regent's Park, London NW1 4RY

Tilly Dallas; Edmund J. Flach; Tom Kearns

Z/316/86 Fixed tissues: oesophagus, stomach, duodenum, small intestine, large intestine, caecum, pancreas, liver, spleen, thymus, thyroid, parathyroid, kidney, adrenal, heart, lung, cerebrum, cerebellum, aorta, testes, muscle and skin. Frozen samples: heart and liver. Male. Stillborn 23 May 1986 (weight 3.775 kg): dystocia resulting from fetal oversize and posterior malpresentation.

ZF/531/85 Fetus (fixed in formalin). Female. Aborted 28 August 1985 (weight 255 g).

I/Z/4723/78 Fixed tissues: pancreas, liver, lung, spleen, bladder, kidney, reproductive tract, adrenal, thyroid, cerebrum, cerebellum, heart and aorta. Male ('Guy'). Died 8 June 1978: cardiac arrest following massive pulmonary congestion.

ZM/075/01 Fixed tissues: stomach (cardiac), duodenum, caudal small intestine, colon and rectum, pancreas, 2 × liver, spleen, kidney, adrenal, cerebrum and cerebellum, heart whole with greater vessels, lung, mesenteric lymph nodes, trachea, bronchial lymph nodes, laryngeal mucosa (with nodules) and gall bladder. Frozen samples: 4 × abdominal fluid, 3 × pericardial fluid, bile ex-gall bladder, 4 × thoracic fluid, 2 × kidney, 2 × trachea, 2 × bronchial lymph nodes, 2 × mediastinal lymph nodes, gall bladder, 3 × spleen, 4 × liver, 2 × lungs and bronchus, 4 × rostral cerebrum, 2 × skeletal muscles and subcutaneous adipose tissue. Male ('Jomie'). Born (Howletts Zoo) 13 November 1980; died 26 January 2001.

ZM/181/10 Fixed tissues: eyes, rib, 2 × stomach, duodenum, ileum, 2 × small intestine, caecum, colon, liver, gall bladder, pancreas, lung, heart left and right ventricle, abdominal aorta, spleen, left and right kidney, urinary bladder, thyroid, adrenal, pituitary, epididymis, testes, brain dura, retropharyngeal lymph node, air sac, salivary gland, mucosa tongue base, firm cyst mesentery, left submandibular lymph node, mesenteric lymph node, 2 × histology and duplicates. Frozen samples: 2 × liver, 2 × kidney, 2 × lung, 2 × spleen, cerebellum and brainstem (rabies), 2 × skin and muscle, laryngeal air sac fluid, mucosa from tongue base, urine (refrigerated), left submandibular lymph node, pancreas, retropharyngeal lymph node, and cerebrum. Male ('Yeboah'). Born (Hanover Zoo) 10 January 1997; died 25 March 2010: aspirated food, pulmonary oedema and congestion, haemoglobinurea, possible pancreatic congestion/haem., chronic peritoneal adhesions, pericardial effusion.

ZM/422/11 Fixed tissues: eyes, stomach, duodenum, jejunum, caecum, colon, 2 × liver, pancreas, 3 × lung, heart, aorta, thymus, spleen, kidney, adrenal, pituitary, testes, brain and left radial nerve. Frozen samples: 2 × liver, kidney, lung, 2 × spleen, skin + muscle, peritoneal fluid and thymus. Male ('Tiny'). Born 28 October 2010; died 12 May 2011: fractured left humerus, ruptured left carpus, extreme bite wounds, marked pallor, dehydration, pulmonary haemorrhage and collapse, pleural and peritoneal effusions, post anaesthetic death.

ZM/481/02 Fixed tissue: duodenum, 2 × jejunum, pancreas, 2 × liver, spleen, 3 × kidney, adrenal, cerebrum, cerebellum, pituitary, 2 × heart, anterior and posterior aorta, 2 × lung, axillary lymph node, shoulder mass, aortic valve, trachea, accessory spleen, gingiva, right sciatic nerve, left sciatic nerve, and mammary gland. Frozen samples: retropharyngeal lymph node, 2 × heart, 2 × lung, 4 × liver, 2 × kidney and 4 × rostral cerebrum. Female ('Diana'). Died 11 May 2002: gingivitis, gastroenteritis, aortic valvular vegetative endocarditis, probable myocarditis, probable embolic pneumonia and nephritis, generalised lymphadenomegaly, chronic fibrous peritonitis, probable hepatic passive venous congestion, aortic atherosclerosis, atrophy of the left cerebral hemisphere, calvarial defect with meningocele and degenerative joint disease of hips and knees.

ZM/1115/08 Fixed tissues: tongue muscle, rib, duodenum, jejunum, ileum, caecum, colon, liver, gall bladder, 2 × lung, heart, spleen, kidney, urinary bladder, thyroid, adrenal, left testis, 2 × iliac lymph node, perirenal lymph node, mesenteric lymph node and soft palate nasopharynx. Frozen samples: soft palate, 2 × lung, heart, 2 × kidney, 2 × spleen, rib, 2 × liver, iliac lymph node, abdominal fluid, pericardial fluid and spleen.

Serum and plasma samples saved frozen:

Male 'Kumba' (date of birth 11 April 1976). Collected 25 May 1980: unspecified sample. 22 October 1993: 4 × plasma; serum; 2 × EDTA. 11 September 1995: unspecified sample. No date: spent cells.

Female 'Zaire' (d.o.b. 23 October 1974). Coll. 31 August 1985: 2 × serum post drip; serum pre drip; 4 × plasma post drip; 3 × plasma pre drip. 1 September 1985: plasma pre drip; plasma post drip; 2 × serum post drip; serum pre drip. 3 September 1985: 2 × plasma. 6 September 1985: 3 × serum; 3 × plasma. 23 September 1985: 3 × plasma. 15 April 1989: 2 × plasma. 29 September 1998: 4 × serum. 6 January 1999: 4 × serum. 25 January 1999: 5 × plasma; 6 × serum. 16 December 1999: 2 × serum. 28 December 1999: 6 × serum. 11 October 2002: 5 × unspecified. 12 December 2005: 4 × serum. 23 January 2006: 15 × serum. 16 February 2007: 3 × serum. 23 March 2011: 3 × heparin; EDTA; coagulation; 15 × serum. 13 September 2011: 2 × heparin; 15 × serum; 11 × EDTA plasma; 5 × EDTA spent cells. 30 May 2012: EDTA plasma; EDTA spent cells. 7 May 2014: 5 × plasma; 8 × serum; 4 × EDTA spent cells; 4 × plain spent cells. 22 August 2014: 8 × serum; 3 × heparin plasma; 6 × EDTA plasma;

7 × spent cells plain; 3 × heparin spent; 6 × EDTA spent cells; 2 × serum gel; 3 × serum gel spent.

Female 'Salome' (d.o.b. 16 July 1976). Coll. 23 May 1988: plasma. 11 September 1995: unspecified sample.

Female 'Kamili' (d.o.b. 1 June 1987). Coll. 15 April 1989: 3 × plasma.

Male 'Jomie' (d.o.b. 13 November 1980). Coll. 6 October 1990: unspecified sample. 6 October 1998: 8 × unspecified sample. 6 April 2000: unspecified sample; 5 × serum.

Male 'Jeremiah' (d.o.b. 6 March 1984). Coll. 12 September 1995: plasma; serum.

Female 'Diana' (estimated d.o.b. 1972). Coll. 12 September 1995: plasma; serum. 23 July 1998: 6 × serum; 2 × plasma. 7 January 1999: 15 × serum. 26 April 2002: 13 × serum.

Female 'Messy' (est. d.o.b. 1972). Coll. 14 July 2000: 13 × serum. 31 July 2002: serum. 15 October 2005: unspecified sample. 16 May 2006: 7 × serum. 16 June 2006: 2 × serum.

Female 'Minouche' (est. d.o.b. 1972). Coll. 26 November 2001: 2 × plasma; 4 × serum. 6 December 2001: 8 × serum. 10 January 2003: 3 × serum; 3 × plasma; 3 × unspecified sample. 13 December 2005: serum. 8 February 2006: 2 × serum. 21 September 2006: 6 × serum.

Male 'Jock' (d.o.b. 31 May 1983). Coll. 5 June 2003: 5 × unspecified sample.

Male 'Bobby' (est. d.o.b. 1983). Coll. 8 December 2005: 10 × serum. 23 January 2006: 2 × serum. 21 July 2006: 3 × heparin plasma. 16 February 2007: 2 × serum. 7 March 2008: 5 × serum. 2 December 2008: serum.

Male 'Yeboah' (d.o.b. 10 January 1997). Coll. 23 March 2010: heparin plasma; EDTA plasma. 24 March 2010: 12 × postmortem serum, very dark; 3 × spent cells plain.

Female 'Effie' (d.o.b. 5 March 1993). Coll. 9 June 2010: 12 × serum. 7 December 2010: 4 × serum; 5 × heparin; 3 × EDTA; spent cells plain. 16 September 2011: 2 × heparin; 14 × serum; 18 × EDTA plasma; 5 × EDTA spent cells; 2 × serum gel spent. 29 June 2012: plasma; 12 × serum; heparin spent; 2 × EDTA spent cells.

Male 'Kesho' (d.o.b. 16 February 1999). Coll. 7 October 2010: 12 × serum; clot; EDTA spent cells; 3 × heparin spent; heparin. 29 November 2010: 5 × serum; 4 × heparin; heparin spent; EDTA spent cells. 8 March 2011: 13 × serum; heparin; 2 × EDTA plasma; 2 × heparin plasma; EDTA spent cells; plasma spent. 2 November 2011: 22 × serum; 5 × EDTA plasma; heparin; 3 × EDTA spent cells. 10 May 2012: EDTA; 2 × heparin; 7 × serum. 13 July 2012: serum; unspecified sample.

Male 'Tiny' (d.o.b. 28 October 2010). Coll. 12 February 2011: spent cells plain. 12 May 2011: EDTA; 2 × serum; heparin.

Female 'Mjukuu' (d.o.b. 2 January 1999). Coll. 9 September 2011: 13 × serum; 14 × EDTA plasma; 3 × EDTA spent cells; heparin. 10 July 2012: 4 × EDTA spent cells; 9 × EDTA plasma; 9 × serum.13 September 2012: serum; 2 × EDTA plasma; 2 × EDTA spent cells. 11 December 2012: serum; 3 × plasma; 3 × EDTA. 16 January 2013: 3 × serum; 6 × EDTA plasma; 2 × EDTA spent cells.

Female 'Flossy' Longleat; (est. d.o.b. 1954). No date: plasma.

Manchester

University of Manchester, Manchester Museum, Oxford Road, Manchester M13 9PL

Henry McGhie

A.210.1 Skull. Male. Congo, SWC Africa. Purchased on 24 December 1895.

A.210.2 Skull. Female. Congo, SWC Africa. Purchased on 24 December 1895.

A.813.1 Mounted skeleton. Adult male. Donated by R. B. Knight. It was accessioned on 27 November 1918.

A.813.2 Mounted skin. Male. Gaboon. Collected in 1864? Donated by R.B. Knight. It was accessioned on 27 November 1918. It may be from the same animal as the skeleton, as it has cast teeth.

A.824 Skull. Adult male. Donated by Dr A.G. Anderson. It was accessioned on 23 June 1919. This specimen is unlocated, and may be the same as A.210.1.

A.2289 Skull. Male.

A.2296.143 Skull.

A.2296.144 Skull.

A.2296.149 Cranium.

A.2325.29 Rib.

Neston

University of Liverpool, Department of Veterinary Pathology, Veterinary Field Station, Leahurst Campus, Neston L64 7TE

J.R. Baker; Julian Chantrey

78L-1334 Histological blocks of lungs, kidney, liver, spleen, heart, colon, and omentum. Adult male. These are from 'Jo-Jo', who arrived at Belle Vue Zoological Gardens, Manchester, on 2 July 1963, aged about 2½ years (weight 45 lb.). He arrived with a female named 'Susan'. They were purchased from Carl Krag (Raymond E. Legge, personal communication). Belle Vue Zoo closed on 11 September 1977, and the pair was sold to Bristol Zoo, although 'Jo-Jo' was sent directly to Chester Zoo on breeding loan. He arrived at Chester on 3 October 1977 and died there on 31 March 1978 of chronic pneumonia of several weeks' duration. The primary aetiology was probably of a nonbacterial nature, with eventual secondary bacterial invasion. The animal was about 70–80 lb. underweight, and evidence of considerable fat resorption was present throughout the body. A small umbilical hernia was found which contained a little abdominal fat. Brain weight 414 g.

Newbury

St Bartholomew's School, Andover Road, Newbury RG14 6JP

B.E.D. Cooper; Claire Marchetti

Skull. Female. It is part of a natural history collection donated by Newbury Museum in the early 1960s, and is on display in a glass cabinet for the students to see. A number of the skulls originated from Edward Gerrard & Sons, 61 College Place, London N.W.

Newcastle Upon Tyne

Newcastle University, School of Dental Sciences, Department of Oral Biology, Framlington Place, Newcastle Upon Tyne NE2 4BW

A.D. Beynon; Pamela Walton

Skull. Adult female. It shows considerable occlusal tooth wear. The maxillary anterior teeth are absent, with the exception of the right lateral incisor, and the mandibular left canine is also missing.

Nottingham

Nottingham Natural History Museum, Wollaton Hall, Nottingham NG8 2AE

B.R.P. Playle; Adam S. Smith; Sheila Wright

NOTNH V3011M Skull. Presented by Consul T.J. Hutchinson in 1857.

NOTNH 5822M Mounted skeleton. Adult male. Acquired in 1881?

Mounted skin. Adult male. It was purchased for 2095 francs (£100) by members of the then Nottingham Museum Committee, at the Paris Exposition of 1878, from Monsieur Vasseur, 9, rue de l'École de Médecine, Paris.

University of Nottingham, School of Life Sciences, Life Sciences Building, University Park, Nottingham NG7 2RD

John Brookfield; Mrs Morag M. Kingshott; Miss K. Lyon

Skull. Adult male. Complete dentition; there is damage to both upper canines. The acquisition date is unknown, but is pre-1969.

Oxford

Oxford University Museum of Natural History, Parks Road, Oxford OX1 3PW

J. Hull; Gillian M. King; Darren J. Mann

OUMNH-Z 982 Skeleton. Adult male. *Gorilla savagei*. Gabon. Collected in 1882. Purchased. The skull, vertebral column and pelvis are articulated.

OUMNH-Z 985 Mounted skeleton. Adult male. *Gorilla savagei*. Gabon. Purchased in 1889.

OUMNH-Z 1602 Portions of intestinal tract (in ethyl alcohol). Juvenile male. *Gorilla savagei*. Purchased from A. Cross, Liverpool. Accession date 11 May 1885. The brain was thrown away in 1952, having rotted. The Reference Catalogue records the following measurements: length, head to heel 34 in.; trunk and head 23 in.; across arms 48½ in.; arm, from shoulder to tip of 3rd finger 21 in.; leg, from hip 15¼ in.; length of hand, palmar surface 5¾ in.; length of foot 6⅝ in.; side of trunk, axilla to head of femur 12 in.; back, from occiput to anus 19 in.; chest circumference 22 in.; round length of head, chin to back 7¼ in.; breadth of head, ears included 4⅝ in.; breadth of pelvis 8¼ in.; length of small intestine 10 ft.

OUMNH-Z 6172 Mounted skin. Juvenile.

OUMNH-Z 8753 Skull (cranium bisected vertically). Adult male. Acquired in 1860. It was transferred with the Christ Church College collections in 1874.

OUMNH-Z 13116 (Ost.Cat. no. 2075A) Left hand (cast). Purchased from the Crystal Palace, Sydenham, in 1873.

OUMNH-Z 17067 Skull (cast). Female.

OUMNH-Z 17462 Mounted skeleton. Male. *Gorilla savagei*. Acquired in 1876.

OUMNH-Z 17535 (Ost.Cat.2052C) Postcranial skeleton (partial). It was transferred with the Christ Church College collections in 1881.

OUMNH-Z 17560 Skull.

OUMNH-Z 19475 Interior of skull (cast). Presented by J.E. Gray, British Museum.

OUMNH-Z 19480 Interior of skull (cast).

OUMNH-Z 19488 Interior of right half of skull (cast). Male. Presented by W.J. Sollas on 4 June 1931.

OUMNH-Z 19490 Interior of right half of skull (cast).

OUMNH-Z 19497 Interior of cranium (cast). *Troglodytes gorilla*. Acquired from the Royal College of Surgeons in exchange for a *Platanista* skeleton in 1874.

OUMNH-Z 19500 Brain (cast). Presented by J.E. Gray Esq. in 1877.

OUMNH-Z 19501 Right ovoid of brain (cast). Acquired in 1877.

OUMNH-Z 21650 (Ost.Cat.2051) Mounted skeleton. Subadult female. *Gorilla savagei*. Purchased from the Du Chaillu collection on 12 June 1863. It is kept in the Department of Zoology teaching collection.

OUMNH-Z 21710 (Ost.Cat.2052A) Skull. Infant. Purchased from the Christ Church College collections in 1881. The postcranial skeleton is believed to be in storage.

Ost.Cat.2051B Mounted skeleton. Juvenile female. Donated by Arthur W. Bateman in 1876. It is kept at the Nuneham Courtenay outstation. Height approx. 24 in.

Ost.Cat.2052 Skull. Adult male. Received from the collection of Sir William Henry Flower in 1883. It is kept in the Department of Zoology teaching collection.

Ost.Cat.2052B Cranium (cast). Adult male.

Ost.Cat.2075B Right foot (cast). Purchased from the Crystal Palace, Sydenham, in 1873. It is from the same animal as the left hand, OUMNH-Z 13116 (Ost.Cat. no. 2075A).

Perth

Perth Museum and Art Gallery, 78 George Street, Perth PH1 5LB

Mark Simmons; Michael A. Taylor

PERGM:1978.1691.9 Skull. Adult male. It has been in the collection since 1913 at least.

Plymouth

City Museum and Art Gallery, Drake Circus, Plymouth PL4 8AJ

David Curry; Helen Fothergill; Jan Freedman

2850 Mounted skeleton. Adult female. Purchased for £10 0s 0d from Edward Gerrard & Sons, London, on 27 February 1903. Height approx. 51 in.; femur approx. 11 in.

Reading

University of Reading, Department of Pure and Applied Zoology, Cole Museum of Zoology, Whiteknights Park, Reading RG6 2AJ

M.G. Hardy; Vincent Morris

1000 Mounted skeleton. Adult male. Purchased for £45 in 1916. Height approx. 63 in. Numbers of vertebrae: C 7, T 13, L 3, S 4, Ca 2.

Sheffield

Sheffield City Museum, Weston Park, Sheffield S10 2TP

Alistair McLean; Paul S. Smith; Derek Whiteley

SHEFM:A.87.4 Skull. Female. Donated by Thomas Hoyland, 5 Surrey Street, Sheffield, on 26 May 1887. It is missing 2 incisors from the upper jaw, and both canines and all 4 incisors from the lower jaw.

SHEFM:A.94.7 Mounted skeleton. Adult male. Acquired in 1894. Height approx. 63 in.

University of Sheffield, Department of Biomedical Science, Western Bank, Sheffield S10 2TN

A.W. Rogers; Alistair Warren

Mounted skeleton. Adult male.

Mounted skeleton. Adult female.

The skeletons are on display, along with an orangutan skeleton, just outside the Medical Teaching Unit.

University of Sheffield, Department of Animal and Plant Sciences, Alfred Denny Museum, Western Bank, Sheffield S10 2TN

Timothy R. Birkhead

ZM159 Model of skull. Central Africa.

ZM160 Model of skull. Adult. Removable top of cranium.

ZM161 Model of skull. Female. Removable top of cranium.

The above 3 models were purchased from Anatomie Clastique du Dr Auzoux, Paris, in 1905.

ZM162 Skull. Female. Cameroon. Purchased from E. Gerrard & Sons, 61 College Place, London N.W., in June 1931. The incisors are missing.

ZM163 Cranium (sectioned).

ZM 164 Foot (plaster cast). Adult male. Lake Kivu region, Belgian Congo.

ZM165 Hand (plaster cast). Adult male. Lake Kivu region, Belgian Congo.

ZM209 Long bones, possibly gorilla. These cannot be located.

ZM1738 Mounted skeleton.

ZM1983 Left scapula. Partially damaged.

St Andrews

University of St Andrews, School of Biology, Bell-Pettigrew Museum, Bute Medical Buildings, Queen's Terrace, St Andrews KY16 9TS

P.J.B. Slater; Carl Smith; P.G. Willmer

5051 Mounted skin. Infant.

2294 Mounted skeleton. Subadult. Acquired before 1910 from Adam, Rouilly, London.

25511 Skull. Adult male. Acquired before 1910. It is used as teaching material.

Swansea

Swansea University, Department of Biosciences, Singleton Park, Swansea SA2 8PP

Sarah Walmsley

Cranium. Adult male.

Cranium. Adult female.

Tring

Zoological Museum, Akeman Street, Tring HP23 6AP

The collections were formed by Baron Lionel Walter Rothschild. Tring Park museum opened to the public for the first time in 1892. In 1931 Lord Rothschild had to sell his bird specimens to the American Museum of Natural History, but all the rest of his material, plus the buildings and grounds, was bequeathed to the British Museum (Natural History) upon his death in 1938. The mounted specimens remained at Tring, while the study material was transferred to the BM(NH), and accessioned in 1939. The study specimens are listed here, rather than under their current location, purely for historical reasons.

ZD.1939.900 Skin. Juvenile female. *Gorilla gorilla diehli*. Efulen, Bulu, Kamerun, altitude 1500 ft. Collected by George Latimer Bates (1863−1940) in May 1903.

ZD.1939.901 Skin. Adult. Evoira. Collected on 11 July 1905.

ZD.1939.902 Skeleton and skin. Infant female. Received from the Zoological Society of London.

ZD.1939.903 Cranium. Adult male. Spanish Guinea. Collected by T. Alexander Barns in 1926.

ZD.1939.904 Cranium. Adult female. Spanish Guinea. Collected by T. Alexander Barns in 1926.

ZD.1939.905 Cranium. Adult female. Spanish Guinea. Collected by T. Alexander Barns in 1926.

ZD.1939.906 Cranium. Adult male. Benito River, Spanish Guinea. Collected by George L. Bates.

ZD.1939.907 Skull. Adult female. Benito River, Spanish Guinea. Collected by George L. Bates.

ZD.1939.908a Skull. Adult male. *Gorilla gorilla diehli*. Efulen, South Kamerun. Collected by George L. Bates.

ZD.1939.909a Skull. Adult male. *Gorilla gorilla diehli*. Efulen, South Kamerun. Collected by George L. Bates.

ZD.1939.910 Skull. Adult male. *Gorilla gorilla diehli*. Efulen, South Kamerun. Collected by George L. Bates.

ZD.1939.911 Skull. Adult male. Batolo, British Cameroon. Collected by N. Sharp in 1919.

ZD.1939.912 Skull. Adult male. North Cameroon. Received from Edward Gerrard & Sons, London.

ZD.1939.913 Skull. Adult male. North Cameroon. Received from Edward Gerrard & Sons, London.

ZD.1939.914 Skull. Adult female. North Cameroon. Received from Edward Gerrard & Sons, London.

ZD.1939.915a Skull. Adult male. *Gorilla gorilla diehli*. Kamerun. Collected by George L. Bates.

ZD.1939.916 Cranium. Adult male. *Gorilla gorilla diehli*. Kamerun.

ZD.1939.917 Cranium. Adult male. Kamerun.

ZD.1939.918 Cranium. Adult female. Kamerun.

ZD.1939.919 Skull. Adult male. Setté Cama, Gabon.

ZD.1939.920 Skull. Adult male. Setté Cama, Gabon.

ZD.1939.921 Skull. Adult male. Setté Cama, Gabon.

ZD.1939.922 Skull. Adult female. Setté Cama, Gabon.

ZD.1939.923 Skull. Adult male. French Congo. Received from Edward Gerrard & Sons, London.

ZD.1939.924 Skull. Adult male. Fernan Vaz, Gabon.

ZD.1939.925 Skull. Adult female. Fernan Vaz, Gabon.

ZD.1939.926 Skull. Adult male. Ngounié River, Ogooué, Gabon.

ZD.1939.927 Skull. Adult female. Ombrokua, Ogooué, Gabon. Collected on 15 August 1907.

ZD.1939.928 Skull. Adult male. Gabon.

ZD.1939.929 Skull. Adult male. Gabon.

ZD.1939.930 Skull. Adult male. Gabon.

ZD.1939.931 Skull. Adult male. Gabon.

ZD.1939.932 Skull. Adult male. Gabon.

ZD.1939.933 Skull. Juvenile female. Gabon.

ZD.1939.934 Skull. Adult female. Gabon.

ZD.1939.935 Skull. Adult female. Gabon.

ZD.1939.936 Skull. Adult female. Gabon.

ZD.1939.937 Skull. Female. Gabon.

ZD.1939.938 Skull. Adult male. Ngounié River, Ogooué, Gabon.

ZD.1939.939 Skull. Adult male.

ZD.1939.940 Skull. Adult male.

ZD.1939.941 Skull. Adult male.

ZD.1939.942 Skull. Adult male.

ZD.1939.943 Skull. Adult male.

ZD.1939.944 Skull. Adult male.

ZD.1939.945 Skull. Adult male. *Gorilla gorilla manyema* Rothschild, 1908: possible holotype.

ZD.1939.946 Skull. Adult male.

ZD.1939.947 Cranium. Adult female.

ZD.1939.948 Skull. Adult female.

ZD.1939.949 Skull. Female.

ZD.1939.950 Skull. Female.

ZD.1939.950 bis Skull, adult male; skull, adult female.

ZD.1939.952 Skull. Juvenile female.

ZD.1939.953 Skull. Adult female.

ZD.1939.954 Skull. Adult female.

ZD.1939.955 Skull. Female.

ZD.1939.956 Skull. Adult female.

ZD.1939.958 Skull. Female.

ZD.1939.959 Cranium. Juvenile female.

ZD.1939.960 Skull. Infant.

ZD.1939.961 Skull. Infant.

ZD.1939.962 Skull. Infant.

ZD.1939.963 Skull. Infant.

ZD.1939.964 Skull (cast). Adult male. Received from Edward Gerrard & Sons, London. The original is in the Leeds Museum Discovery Centre (LEEDM. C.1926.9.4164).

ZD.1939.1322 Skeleton. Juvenile female.

ZD.1939.3401 Mounted skin. Juvenile.

ZD.1939.3402 Mounted skin. Juvenile.

ZD.1939.3403 Mounted skin. Juvenile.

ZD.1939.3404 Mounted skin. Juvenile.

ZD.1939.3405 Mounted skin. Adult male. *Gorilla gorilla matschiei* (syntype). Yaunde, South Kamerun. Collected by Hans Paschen on 15 April 1900.

ZD.1939.3406 Mounted skeleton. Adult male. *Gorilla gorilla matschiei* (syntype). It is from the same animal as the mounted skin 1939.3405 above. The skin and skeleton were prepared by Johannes Umlauff, Hamburg, and sold to Lord Rothschild for 20,000 marks in 1901. The following skeletal measurements were recorded by Prof. Dr Lenz of Lübeck: height 165 cm; span of arms from middle finger to middle finger 280 cm; cranial capacity 562 cc; diagonal length of skull 318 mm; greatest breadth of skull 170 mm; auricular breadth 156 mm; jugal breadth 190 mm; nasion to foramen magnum 145 mm; greatest height of sagittal crest 60 mm; length of spinous process of 4th cervical vertebra 100 mm; length of humerus 450 mm; length of radius 350 mm; total length of hand to end of middle finger 270 mm; length of femur 385 mm; length of tibia 315 mm; height of pelvis 380 mm; greatest breadth of pelvis 400 mm; total length of foot to end of 3rd toe 310 mm. (These measurements were included in a catalogue, *Der Riesen-Gorilla des Museum Umlauff Hamburg. Schilderung seiner Erlegung und wissenschaftliche Beschreibung*, which Umlauff had printed privately to accompany the exhibition of the specimen in Germany, before it was sent to Rothschild). See also Lange (2005, 2006).

ZD.1939.3407 Mounted skin. Adult male. *Gorilla beringei beringei*. Lake Kivu, Belgian Congo. Collected by T. Alexander Barns.

ZD.1939.3408 Mounted skin. Adult male. *Gorilla gorilla diehli*. North Kamerun.

ZD.1939.3413 Mounted skin. Juvenile male.

ZD.1939.3414a Mounted skin. Adult female. *Gorilla beringei beringei*. Lake Kivu, Belgian Congo.

ZD.1939.3414b Mounted skin. Infant. *Gorilla beringei beringei*. Lake Kivu, Belgian Congo.

ZD.1939.3415 Mounted skin. Adult male. *Gorilla gorilla halli* (holotype). Punta Mbonda, Spanish Guinea.

ZD.1939.3416 Mounted skin. Adult male. Gabon.

Twycross

East Midland Zoological Society, Twycross Zoo, Burton Road, Atherstone CV9 3PX

Sarah Chapman

Since 2011, Twycross Zoo have done all their post-mortem examinations in-house. The protocol includes sections of all tissue types taken and placed into formalin for histopathology using an external laboratory. A duplicate set of tissues is also retained in formalin at the zoo. Certain tissue sections are also retained at the zoo at −80°C, for example, muscle, liver, kidney. The stored tissue samples are usually 1- to 2-cm cubes. These stored samples are recorded on a database for further clinical testing or research purposes.

Wakefield

Wakefield One Library and Museum, Burton Street, Wakefield WF1 2DD

Mrs G. Spencer; John Whitaker

Mounted skin. Juvenile female. This is believed to relate to 'Jenny', who arrived at Wombwell's Royal Number One Menagerie in September 1855, whilst it was at Burton on Trent. [George Wombwell was born in 1777 and established his travelling menagerie in 1805. When this grew to 14 wagons, he established a second menagerie, known as Wombwell's Number Two. He eventually had three menageries, but continued to live with Number One. When he died in 1850, Number One stayed in the care of his wife, Numbers Two and Three with other relatives. Mrs Wombwell handed Number One over to her nephew, Alexander Fairgrieve, in 1865. He auctioned off the whole of the livestock and equipment (186 lots) at auctioneers Buist in Edinburgh on 9 April 1872. Wombwell's circus continued until the last years of the 19th century]. 'Jenny' had been imported from the Congo to Liverpool earlier that year by Mr Hulse, an animal dealer, who sold her to Mrs Wombwell. During her short lifetime, 'Jenny' was thought to be a chimpanzee. She died at Warrington, Lancashire, in March 1856. Her body was sent to Charles Waterton (1782−1865), an enthusiastic naturalist and the Squire of Walton Hall, near Wakefield. Out of the skin, Waterton manufactured a crouching figure with two horns on the head, which he called 'Martin Luther After His Fall' and exhibited in his gallery at Walton Hall. He sent the skeleton to a museum in Leeds. In 1861 the stuffed skin was seen at Walton Hall by Abraham Dee Bartlett, Superintendent of the Zoological Society of London's Gardens, who immediately recognised it as being a young gorilla, rather than a chimpanzee. Mrs Wombwell's daughter, Mrs Fairgrieve, subsequently lent him a Daguerrotype photograph of the animal, taken when it was alive. The portrait was in turn lent to Mr Joseph Wolf (1820—99), who used it to assist him in preparing a chalk drawing of the gorilla. Bartlett then returned the photograph to the owner, who was at that time living in Lauriston in Edinburgh. Waterton's entire collection was at Ushaw College, Durham, probably from 1865 (the year of his death) until about 1881. It was then at Alston Hall, near Longridge, the home of a Mr Mercer, who had married a Waterton. On 15 July 1908 it arrived on loan at Stonyhurst College, Blackburn, where Waterton had been educated. The loan was converted into a formal deed of gift, signed, sealed and delivered by Monica Mary Collette Paula Waterton on 4 May 1915. The collection finally came into the guardianship of Wakefield Museums Service on long loan from Stonyhurst College in 1969. In 2009 the pattern of the cuticles of 5 hairs taken from various parts of 'Martin Luther' were examined by Dr Phil Greaves, of Microtex International, Otley, West Yorkshire. Three hairs were too brittle for analysis, but it was established that a hair from the left forearm was chimpanzee, while the horns are made from the hair of a donkey, goat, or horse. This is in accordance with Waterton's common practice of creating composite mounts from different species. [Some additional information added by GH from correspondence with Revd B. Payke, Ushaw College, and Revd C.K. Macadam, Stonyhurst College.]

UNITED STATES OF AMERICA

Albany

New York State Museum, 3140 Cultural Education Center, Empire State Plaza, Albany, NY 12230

Joseph Bopp; David W. Steadman

NYSM 3922 Skeleton and skin. Adult male. This is 'Samson', who arrived at Buffalo Zoological Gardens, Delaware Park, Buffalo, NY 14214 on 14 March 1962 (weight 38¾ lb). He was purchased from F.J. Zeehandelaar, Inc., New Rochelle, NY, USA. He was sent on loan to the Chicago Zoological Park, 3300 Golf Road, Brookfield, IL 60513, USA on 28 January 1980, and died there on 17 March 1988. He was euthanased after a debilitating stroke caused by a large brain tumour. [Some additional information added by GH from correspondence with Buffalo Zoo personnel.]

Albuquerque

University of New Mexico, Department of Biology, Museum of Southwestern Biology, Albuquerque, NM 87131

Jonathan L. Dunnum; William L. Gannon

MSB:Mamm:50566 Skeleton. Juvenile. Received from the Oregon Regional Primate Research Center.

MSB:Mamm:54411 Skull. Adult male. Mbini, Nkin, Equatorial Guinea. On permanent loan from Clyde J. Jones, to whom it was gifted by the regional governor.

Amherst

Amherst College, Beneski Museum of Natural History, 11 Barrett Hill Road, Amherst, MA 01002-5000

Linda L. Thomas; Kate M. Wellspring

ACM-OS 776 Skull. Adult female. Mt Bimow, High Sangha district, French Middle Congo. Jaws are spring-mounted; surface of specimen has been coated with shellac or other consolidant that is now yellowing.

ACM-OS 777 Skull (on display) and mounted (suspended) postcranial skeleton (on display in laboratory). Adult male. Ashangoland, Gaboon. Donated by the Revd William Walker, no later than 1863. This specimen was reputedly received (with a chimpanzee) pickled in a barrel of rum. The skin and skeleton were mounted, but the skin was later stolen or destroyed. Dentary is spring-mounted; cranium has drilled hole through sagittal crest (formerly mounted on the skeleton?); right hindlimb and some phalanges detached; sternum is a leather reconstruction; lots of dental wear.

ACM-OS 1154 Skull. Adult female. Mt Bimow, High Sangha district, French Middle Congo. Surface of specimen has been coated with shellac or other consolidant that is now yellowing. Some incisors damaged; foramen magnum enlarged, shows cut marks; also shallow cut marks on dentary.

ACM-OS 1155 Skull. Adult male. Mt Bimow, High Sangha district, French Middle Congo. (Bullet?) hole through right parietal. Surface of specimen has been coated with shellac or other consolidant that is now yellowing.

ACM-OS 1434 Mounted (suspended) skeleton. Subadult male. Donated by Amherst College Department of Anthropology in September 2006. Transfer paperwork states that it was thought to be a composite of several individuals, but supporting evidence is not cited.

Ann Arbor

University of Michigan, Exhibit Museum of Natural History, 1109 Geddes Avenue, Ann Arbor, MI 48109-1079

William A. Lunk

UMMZ 38952 Mounted skeleton. Adult female. Kribi, Kamerun. Collected by Dr W.S. Lehman in 1908. The flat skin of this specimen is in the University's Museum of Zoology.

University of Michigan, Museum of Paleontology, 1109 Geddes Avenue, Ann Arbor, MI 48109-1079

Philip D. Gingerich; Adam N. Rountrey

Cranium. Male. The dentition is largely missing, and a portion of the basiocciput is also missing.

M8 Skull. Juvenile female. *Gorilla savagei*. The mandible is articulated. The upper and lower 1st molars are just beginning to erupt. Collected in 'W. Cent. Africa' prior to 1936. This specimen is also numbered 413.

193 Upper right cuspid. Collected prior to 1936. It is now missing.

208 Mandible. Collected prior to 1936. It appears that the M2 had erupted, but they are not present.

University of Michigan, College of Literature, Science and the Arts, Museum of Zoology, 1109 Geddes Avenue, Ann Arbor, MI 48109-1079

G. Ken Creighton; Stephen H. Hinshaw; Philip Myers; Cody W. Thompson

UMMZ 38369 Skull. Kribi, Kamerun. Collected by George Schwab in 1908. The dentition is in bad condition.

UMMZ 38952 Skin. Female. Kribi, Kamerun. Collected by Dr W.S. Lehman in 1908. The skeleton of this specimen is on display in the University's Exhibit Museum.

UMMZ 39154 Skin. Efulan, Kribi, Kamerun. Collected by George Schwab in 1908–09.

UMMZ 168000 Cadaver (preserved in fluid). Neonate male. A zoo specimen, received on 24 April 1981 (no other associated data).

UMMZ 176886 Skeleton; tissues stored in liquid nitrogen freezer (at −190°C). Adult male. This is 'Colossus', who arrived at Gulf Breeze Zoo, Florida, on 14 March 1988, having been purchased from Benson Wild Animal Farm. He was sent on loan to Cincinnati Zoo on 18 July 1993 (weight 432 lb. on 12 October 1994), and died there on 11 April 2006 (weight 497 lb.), during general anaesthesia for root canal treatment. Posthumous measurements: length, crown−rump (linear) 118 cm; cranial circumference (at brow ridge) 95 cm; length of head (tip of jaw to top of crest) 52 cm; width of brow ridge 18 cm; chest circumference (at nipples) 175.5 cm; abdominal circumference (at umbilicus) 152.5 cm; left arm: shoulder−elbow 42 cm, elbow−wrist 45 cm, wrist−tip of middle finger 27 cm, pollex 9 cm; right arm: shoulder−elbow 51 cm, elbow−wrist 44 cm, wrist−tip of middle finger 26 cm, pollex 8 cm; left leg: hip−knee 51 cm, knee−ankle 40 cm, ankle−tip of big toe 22 cm, heel−tip of big toe 28.5 cm, hallux 13 cm; right leg: hip−knee 47 cm, knee−ankle 41 cm, ankle−tip of big toe 21 cm, heel−tip of big toe 27 cm, hallux 13 cm; spleen 470 g; liver 4060 g; right kidney 330 g; left kidney 330 g; right adrenal 4 g; heart 1140 g; heart length 20 cm; heart circumference 34 cm. [Additional information added by GH from correspondence with Mark Campbell and Mary Noell, Cincinnati Zoo.]

Atlanta

Zoo Atlanta, 800 Cherokee Avenue SE, Atlanta, GA 30315-1440

Joseph R. Mendelson III; Hayley Murphy

Zoo Atlanta has a long history of gorilla research, and also a long history of providing samples or animal-access to qualified internal and external researchers. All biomaterials are sent out to researchers, pending availability, feasibility of collection and after a review by their internal Scientific Review Committee. They do not maintain significant collections of biomaterials at the zoo, and they have not designated a specific repository institution for future samples. Thus, gorilla biomaterials may be available from Zoo Atlanta for approved projects and researchers.

The Great Ape Heart Project is also based at Zoo Atlanta (see Chapter 10: Respiratory and Cardiovascular Systems in Part I: Gorilla Pathology and Health).

Austin

University of Texas at Austin, Department of Anthropology, Austin, TX 78712*

Claud A. Bramblett

L 106 Mounted skeleton. Adult female. It was acquired before 1930. It is mounted for exhibit in knuckle-walking posture; there are minor errors in articulation.

L 112 Cranium. Adult male. It was acquired before 1930. All the incisors and both 3rd molars are missing postmortem. There is minor damage to the occipital area. It has remarkably robust crests.

L 114 Skull. Juvenile male. It was acquired before 1930. The upper left central incisor, upper left 3rd molar, upper and lower right central and lateral incisors are all missing postmortem. The canines and 3rd molars are unerupted. Very large crests are beginning. There is minor damage to the mandibular condyles and floor of the orbits.

Mounted skeleton. Adult male. It was purchased (with at least one chimpanzee) by Dr Pearce from a sea captain at a port near Corpus Christi, on the Texas coast, some time prior to the autumn of 1927. At the time it was a dried specimen in a bad state of decay, and was not skeletonised until about 10 years later. It may be the 'Great Gorilla of Ambam' mentioned by Merfield and Miller (1956). That specimen was shot on 22 May 1926, but was dead for 2 days before Merfield found the body, by which time it was in an advanced state of decomposition. It weighed 214 kg (171 kg clean). The skeleton is mounted for exhibit in a knuckle-walking posture, but there are minor errors in articulation. The dentition is very worn. [Additional information added by GH from correspondence with Trudie Jarrett, née Merfield.]

Baltimore

The Johns Hopkins University, Department of Cell Biology and Anatomy, 725 North Wolfe Street, Baltimore, MD 21205*

Linda Szabelski

Head (in alcohol). Adult male. It was acquired by Dr Adolph Hans Schultz before 1978.

Baton Rouge

Louisiana State University, Museum of Natural Science, 119 Foster Hall, Baton Rouge, LA 70803

Jacob A. Esselstyn

LSUMZ 32845 Skull. Male. Ngui, Cameroon.

Berkeley

University of California at Berkeley, Museum of Vertebrate Zoology, 3101 Valley Life Sciences Building, Berkeley, CA 94720-3160

Phillip Brylski; Carla Cicero; Christopher J. Conroy

MVZ 4819 Skull. Male. Acquired from Kny-Sheerer Company, New York, on 12 February 1909.

MVZ 4820 Skull. Female. Acquired from Kny-Sheerer Company, New York, on 12 February 1909. It is now missing.

MVZ 38930 Skeleton. Male. Two miles south of Ebolowa, Littoral Province, Cameroun. Collected by E. Cozzens on 16 June 1927.

MVZ 38931 Skeleton. Male. Two miles south of Ebolowa, Littoral Province, Cameroun. Collected by E. Cozzens on 9 July 1927.

MVZ 125979 Skull, partial postcranial skeleton, and skin. Male. Received from San Diego Zoological Garden on 14 October 1959. This is 'Scoop', who arrived at San Diego Zoo on 25 April 1958, aged about 1½ years (weight 22 lb.). He came originally from Cameroun, and was purchased for $3000 from Phillip J. Carroll, Yaoundé. Funds for the purchase were donated by James S. Copley,

publisher of the *San Diego Union* and *Evening Tribune*. Recorded weights: 21¾ lb. on 24 April 1958 (weighed in New York by Dr L. Goss); 26 lb. on 24 June 1958; 27 lb. on 1 July 1958; 28 lb. on 8 July 1958; 27½ lb. on 16 July 1958; 29 lb. on 22 July 1958; 29½ lb. on 29 July 1958; 31 lb. on 13 August 1958; 30 lb. on 19 August 1958; 30½ lb. on 26 August 1958; 32 lb. on 2 September 1958; 32½ lb. on 9 September 1958; 32 lb. on 17 September 1958; 32 lb. on 24 September 1958; 33 lb. on 7 October 1958; 33 lb. on 15 October 1958; 33½ lb. on 4 November 1958; 35 lb. on 4 December 1958; 37 lb. on 2 January 1959; 38 lb. on 15 January 1959; 40½ lb. on 6 February 1959; 40 lb. on 10 February 1959; 41½ lb. on 24 February 1959; 42 lb. on 11 March 1959; 43 lb. on 17 March 1959; 43 lb. on 24 March 1959; 43 lb. on 10 April 1959; 43½ lb. on 14 April 1959; 45 lb. on 7 May 1959; 42½ lb. on 20 May 1959; 42 lb. on 18 June 1959; 42½ lb. on 26 June 1959; 43 lb. on 7 July 1959; 41½ lb. on 15 July 1959; 42 lb. on 22 July 1959; 39 lb. on 9 August 1959. He died of terminal acute pneumonitis on 13 August 1959 (weight 32.6 lb.). Total length 585 mm; foot length 190mm; ear from notch 40 mm. [Additional information added by GH from correspondence with San Diego Zoo personnel.]

MVZ 126801 Skull. Female. Ogooué River, Gabon. Collected by R.S. Shackell from a local inhabitant in 1920.

MVZ 126875 Skull, partial postcranial skeleton and skin. Male. Received from the San Diego Zoological Garden on 12 June 1960. This is 'Copy', who arrived at San Diego Zoo on 6 December 1959, aged about 1½ years (weight 28¼ lb). He was captured by Phillip J. Carroll and purchased for $4500 from Hermann Ruhe, Hannover, through Carl Krag, Copenhagen, with funds donated by James S. Copley. Appeared in good spirits on arrival; slight cold, abdomen large. As at 9 December 1959: 22 teeth in total, canines projecting well above incisor row. Given name by the children of San Diego on 13 February 1960 (Belle Benchley selected the name). Recorded weights: 28¼ lb on 7 December 1959; 27¾ lb on 8 December 1959; 29½ lb. on 11 December 1959; 31 lb. on 18 December 1959; 32 lb. on 24 December 1959; 33 lb. on 5 January 1960; 33 lb. on 18 January 1960; 33 lb. on 28 January 1960; 32 lb. on 5 February 1960; 33½ lb. on 10 February 1960; 35 lb. on 11 February 1960; 35 lb. on 3 March 1960; 35 lb. on 10 March 1960; 36 lb. on 18 March 1960; 37 lb. on 25 March 1960; 38 lb. on 31 March 1960; 38 lb. on 7 April 1960; 39 lb. on 14 April 1960; 40 lb. on 21 April 1960. He died of acute verminous pneumonia, intestinal Strongyloidosis and acute dehydration and electrolyte loss on 2 June 1960. Total length 570 mm; foot length 183 mm; ear from notch 40 mm. [Additional information added by GH from correspondence with San Diego Zoo personnel.]

MVZ 174521 Skeleton. Female. This animal was captive in the San Francisco Zoological Gardens, and died on 26 April 1983. The skeleton was formerly in the California Academy of Sciences (accession no. 3523).

MVZ 183656 Skull, spine and bones of one arm and leg. Adult female. Received on 5 January 1995 from Milton Hildebrand, Department of Zoology, University of California at Davis (his collection no. 1365), who obtained it from Sacramento Zoo, William Land Drive and Sutterville Road, Sacramento, CA 95822, on 14 February 1980. The animal was named 'Suzie', and arrived at the zoo with a male named 'Chris' on 17 June 1965 (weight 34 lb.). They were purchased from G. van den Brink. 'Suzie' died on 13 February 1980 of wounds inflicted by the male when the pair could not be separated (weight 207 lb.). The hand and foot are articulated. The specimen was part of the private collection of Milton Hildebrand, Emeritus Professor of Zoology. [Some additional information added by GH from correspondence with Sacramento Zoo personnel.]

MVZ 184532 Skull. Vicinity of Uvira, Belgian Congo. Collected c.1960.

MVZ 192697 Frozen tissue (kidney, heart, 6 × liver). Male. Received from San Diego Zoo.

University of California at Berkeley, Museum of Paleontology, 1101 Valley Life Sciences Building, Berkeley, CA 94720

Robert G. Dundas; Patricia A. Holroyd

UCMP 23971 Manus (cast). Missing, fate unknown.

UCMP 23985 Skull (cast). Missing, fate unknown.

UCMP 117177 Right and left upper and lower dentition (cast). Adult. Received from Bobbitt Labs in 1975.

Bloomington

Indiana University, Department of Anthropology, Student Building 130, 701 E. Kirkwood Avenue, Bloomington, IN 47405-7100*

Kevin D. Hunt

9010288 Skeleton. Captive animal. Acquired in 1990.

9410343 Skeleton. Captive animal. Acquired in 1994.

9510003 Skeleton. Captive animal. Acquired in 1995.

9510181 Skeleton. Captive animal. Acquired in 1995.

9610158 Skeleton. Captive animal. Acquired in 1996.

9610235 Skeleton. Captive animal. Acquired in 1996.

9710009 Skeleton. Captive animal. Acquired in 1997.

9710012 Skeleton. Captive animal. Acquired in 1997.

9710030 Skeleton. Captive animal. Acquired in 1997.

Boston

Boston University, School of Medicine, Department of Anatomy and Neurobiology, 72 East Concord Street, Boston, MA 02118-2394*

Martin L. Feldman

Brain sections. Adult male. Received from Dr John T. McGrath, Laboratory of Pathology, School of Veterinary Medicine, University of Pennsylvania, Philadelphia. This is a series of about 70 mounted 6-μm paraffin sections, all from the captive specimen 'Massa', of Philadelphia Zoo. About one-fifth of the sections have been stained by Dr Martin L. Feldman of the Department of Anatomy, Boston University School of Medicine. Most of the staining is with thionine or with the Kluver–Barrera method. These sections provide samples, not precisely localised, of neocortex, cerebellum, hippocampus, basal ganglia, pons and lower brain stem. All slides are labelled '8510001' and are in three slide boxes, '32', '33' and '100'. Presumably, these designations are all Dr McGrath's.

Boulder

University of Colorado, University of Colorado Museum of Natural History, Zoology Section, Bruce Curtis Building, Campus Box 265, Boulder, CO 80309-0315

Emily Braker; Jennifer E. Haessig

UCM 3778 Skull. Adult male. Purchased in Equatorial Africa by Dr K.K. Cross, c.1940. The teeth are moderately worn.

Buffalo

Buffalo Museum of Science and Tifft Nature Preserve, 1020 Humboldt Parkway, Buffalo, NY 14211

Arthur R. Clark

BSNS 1682 Mounted skeleton and mounted skin. Female. This is 'Jonesie II', who arrived at Buffalo Zoological Gardens, Delaware Park, Buffalo, NY 14214, USA, on 21 April 1963 (weight 36 lb.). She was originally from Cameroon, and was purchased from Dr Deets Pickett, Trans World Animal, Inc., 8505 Lee Boulevard, Kansas City 15, Missouri [information added by GH from correspondence with Buffalo Zoo personnel]. She was sent on breeding loan to the Columbus, Ohio, Zoo on 22 September 1979 (weight 200 lb.). Whilst still at Columbus, during the last few weeks of her third pregnancy, she developed a purulent nasal discharge that did not respond to antibiotics, increasing difficulty in swallowing and feeding, and progressive weight loss. Subsequent examination of biopsies of extensive nasopharyngeal and cervical tumours indicated a large-cell, histiocytic lymphoma. She was returned to Buffalo Zoo on 29 January 1983 (weight 159 lb.) and received radiotherapy which reduced the tumour size. Eight weeks after the last therapy sessions the masses recurred, and did not respond to chemotherapy. She was euthanased on 7 June 1983 (weight 122 lb.). Postmortem examination confirmed that the tumour could be classified as a T-cell lymphoma. She had antibody titres against several retroviruses. The origin of the antibodies in this gorilla is not clear (see also Chapter 3: Infectious Disease and Host Responses and Chapter 12: Lymphoreticular and Haemopoietic Systems and Allergic Conditions in Part I: Gorilla Pathology and Health). The distal phalanges are still in the mounted skin and cast replacements (from specimens in the Museum of Comparative Zoology, Harvard University) are on the skeleton.

BSNS 1683 Skeleton and mounted skin. Male. This is 'Chuma', who arrived at New York Zoological Park (Bronx Zoo) on 10 May 1966 (weight 26.4 lb.). He was purchased, with a female named 'Sukari', from F.J. Zeehandelaar, Inc. He was sent on loan to Buffalo Zoo on 23 January 1984 (weight 321 lb.), where he was euthanased on 20 December 1984 (weight 264 lb.) owing to a mucous adenocarcinoma of the gall bladder with multiple metastases. The eyes are preserved in fluid. [Some additional information added by GH from correspondence with NY Bronx Zoo and Buffalo Zoo personnel.]

Cambridge

Harvard University, Museum of Comparative Zoology, 26 Oxford Street, Cambridge, MA 02138

Judith M. Chupasko; John A.W. Kirsch; Maria E. Rutzmoser

NB: The prefix BOM indicates that the specimen was originally catalogued in 'Bones of Mammals'.

MCZ BOM 6339 Mounted skeleton. Gabon. Received from Alexander Emmanuel Rodolphe Agassiz (1835–1910) in May 1880. Originally purchased from E. Gerrard Jr.

MCZ 6370 Mounted skin. Male. *Gorilla savagei*. Gabon. Received from Alexander Agassiz in 1882. Originally purchased from Henry Augustus Ward.

MCZ 6371 Mounted skin. Juvenile. *Gorilla savagei*. Gabon. Received from Alexander Agassiz in 1882. Originally purchased from Henry A. Ward.

MCZ 6445 Mounted skin. Female. *Gorilla savagei*. Received from Alexander Agassiz on 29 January 1884. Originally purchased from Henry A. Ward.

MCZ BOM 6912 Mounted skeleton. Male. *Gorilla savagei*. Received from Alexander Agassiz in 1883. Originally purchased from Henry A. Ward. On exhibit in the Systematic Hall.

MCZ BOM 9311 Skull. Adult female. *Troglodytes gorilla* (syntype). Empongwe, near the mouth of the Gaboon River. Collected by Thomas S. Savage in 1847. Received from the Boston Society of Natural History on 30 March 1915.

MCZ BOM 9312 Skeleton (incomplete). Male. Gaboon River. Received from the Boston Society of Natural History on 30 March 1915.

MCZ BOM 9313 Skeleton (incomplete). Male. Received from the Boston Society of Natural History on 30 March 1915.

MCZ BOM 9488 Pelvis with sacrum. Male. *Troglodytes gorilla* (syntype). Gaboon. Collected by Thomas S. Savage in 1847. Received from the Boston Society of Natural History. It belongs to the same individual as MCZ BOM 9587.

MCZ BOM 9489 Pelvis with sacrum. Female. *Troglodytes gorilla* (syntype). Gaboon. Collected by Thomas S. Savage in 1847. Received from the Boston Society of Natural History. It belongs to the same individual as MCZ BOM 9311.

MCZ BOM 9490 Half cranium (sagittally sectioned, left half present). Female. Gaboon. Collected in 1847. Received from the Boston Society of Natural History.

MCZ BOM 9491 Skull. Juvenile. It belonged to the Wyman collection. Received from the Boston Society of Natural History.

MCZ BOM 9492 Skeleton (incomplete). Female. Cama River, Gabon. Collected by Dr J.H. Otis, United States Navy. Received from the Boston Society of Natural History.

MCZ BOM 9587 Skull. Adult male. *Troglodytes gorilla* (syntype). Cape Palmas, Gaboon. Collected by Thomas S. Savage in 1847. Received from the Boston Society of Natural History on 26 September 1916. The upper jaw lacks 3 incisors and the right canine.

MCZ BOM 10686 Atlas and 2 dorsal vertebrae. Female. *Troglodytes gorilla* (syntype). Cape Palmas, Gaboon. Collected by Thomas S. Savage in 1847. Received from the Boston Society of Natural History. They are from the same animal as MCZ BOM 9311 and MCZ BOM 9489.

MCZ BOM 10687 Twelve ribs. Male and female. Cape Palmas, Gaboon. Collected by Thomas S. Savage in 1847. Received from the Boston Society of Natural History.

MCZ BOM 10688 Vertebrae (2 cervical, 3 dorsal, 2 lumbar). Male. *Troglodytes gorilla* (syntype). Cape Palmas, Gaboon. Collected by Thomas S. Savage in 1847. Received from the Boston Society of Natural History. They are from the same animal as MCZ BOM 9488 and MCZ BOM 9587.

MCZ BOM 10689 Scapula. Female. Received from the Boston Society of Natural History. It has been missing since August 1977.

MCZ BOM 10690 Limb bones (both scapulae, humeri, radii, femora; 1 ulna, 1 tibia). Male. *Troglodytes gorilla* (syntype). Gaboon. Collected by Thomas S. Savage in 1847. Received from the Boston Society of Natural History on 22 October 1917. From the same individual as MCZ BOM 9587.

MCZ BOM 10691 Clavicle, radius, femur and tibia. Female. Collected by Thomas S. Savage in 1847. Received from the Boston Society of Natural History.

MCZ 14750 Skull. Female? Nellafup, 18 miles east of Efulan, Kamerun. Collected by Revd George W. Schwab in 1912. Purchased.

MCZ 17684 Skull and skin. Adult female. Metet, Kamerun. Collected by Revd George W. Schwab in 1916.

MCZ 20038 Skeleton. Adult male. East of Akonolinga, Cameroun. Collected by Revd George Schwab in 1921–22; received from him in October 1922. This specimen has various deformities. There is a growth on the right zygomatic bone which is about as long as a hen's egg but more narrow. The left innominate is completely fused to the sacrum. The right tibia and fibula may have been broken at one time, the two ends of each bone sliding past each other before re-fusing. This is one possible scenario to explain their appearance. Near the centre of each bone there is a considerably thickened area, and they are both shorter than their counterparts on the left side. The left tibia shows evidence of fresh disease: there are swollen, porous areas along the length of the bone. The left ulna is bent at a greater than normal curve when compared to the right.

MCZ 20039 Skeleton. Adult male. Nlong River, about 70 miles east of Métet, Cameroun. Collected by Revd George Schwab in 1921–22; received from him in October 1922.

MCZ 20043 Skeleton and skin. Female. Métet, Cameroun. Collected by Revd Finley McCorvey Grissett in 1922. Purchased in November 1922.

MCZ 20089 Skeleton and skin. Juvenile female. Métet, Cameroun. Collected in 1922; purchased in 1922.

MCZ 23160 Skeleton. Adult male. Sakbayémé, Cameroun. Collected by Revd George W. Schwab in 1924; received from him in December 1926. This animal allegedly killed three men. Femur pathological, healed fracture.

MCZ 23161 Skeleton. Juvenile male. Sakbayémé, Cameroun. Collected by Revd George W. Schwab; received from him in December 1926.

MCZ 23162 Skeleton. Adult male. Sakbayémé, Cameroun. Collected by Revd George W. Schwab in April 1925; received from him in December 1926. The cervical vertebrae are missing.

MCZ 23182 Skeleton and mounted skin. Male. *Gorilla beringei graueri*. Nakalongi, Belgian Congo, altitude 9000 ft. Collected by Harold J. Coolidge, Harvard African Expedition, on 16 February 1927. Received in July 1927.

MCZ 23187 Skull. Juvenile male. *Gorilla beringei graueri*. Near Nakalongi, Belgian Congo. Collected by Harold J. Coolidge, Harvard African Expedition, on 11 February 1927. Received in August 1927.

MCZ 23990 Skin. Female. Sakbayémé, Cameroun. Collected by Revd George W. Schwab on 31 January 1921. Received from the Peabody Museum on 29 August 1928.

MCZ 25949 Skeleton. Sakbayémé, Cameroun. Collected by Revd George W. Schwab in 1930; received from him in June 1930.

MCZ 26850 Skeleton. Female. Sakbayémé, Cameroun. Collected by Revd Finley McCorvey Grissett in May 1930.

MCZ 27323 Skin. Male. Near Sakbayémé, Cameroun. Collected by Revd George Schwab in 1929; received from him in 1930. The head, hands and feet were accessioned in the Peabody Museum of Archaeology and Ethnology, another part of Harvard University, but to date they have not been located.

MCZ 29047 Skeleton. Female. Elat, 1 mi. from Ebolowa, Cameroun. Collected by R. Evans. Purchased on 2 December 1932. Upper left canine lost in life.

MCZ 29048 Skeleton. Male. Elat, 1 mi. from Ebolowa, Cameroun. Collected by R. Evans. Purchased on 2 December 1932.

MCZ 29049 Skeleton. Male. Elat, 1 mi. from Ebolowa, Cameroun. Collected by Rowland Evans. Purchased on 2 December 1932.

MCZ 37261 Skull. Male. Métet, Cameroun. Purchased from Silas E. Johnson in May 1937.

MCZ 37262 Skull. Male. Métet, Cameroun. Purchased from Silas E. Johnson in May 1937.

MCZ 37263 Cranium. Male. Métet, Cameroun. Purchased from Silas E. Johnson in May 1937. Missing since August 1977.

MCZ 37264 Skeleton. Female. Métet, Cameroun. Purchased from Silas E. Johnson in May 1937.

MCZ 37265 Skeleton. Subadult female. Métet, Cameroun. Purchased from Silas E. Johnson in May 1937.

MCZ 37266 Cranium. Female. Métet, Cameroun. Purchased from Silas E. Johnson in May 1937. Imperfect condition, bone around foramen magnum missing.

MCZ 38017 Skeleton. Adult male. *Gorilla beringei graueri*. Thirty kilometres south of Lubero, 7000 ft. altitude, Belgian Congo. Collected by Duncan M. Hodgson and William F. Coultas on 27 November 1938; received from Hodgson on 5 May 1939. Height, top of head to heel 169.3 cm; top of head to end of extended toe 178 cm; length of foot 31.4 cm; chest girth 143.9 cm; neck girth 80.2 cm; belly girth 164 cm; upper arm girth 52 cm (left), 51.5 cm (right); lower arm girth 46.1 cm (left), 46.2 cm (right); wrist to end of longest finger 25.3 cm; width of palm 17 cm; girth of palm outside thumb 39.2 cm; wrist girth 31.1 cm (left), 32.8 cm (right); foot length along outer side 27.9 cm, along inner side 30.8 cm; foot width at ball 13.2 cm, at heel 12.3 cm; arm spread (rigor mortis) 253.5 cm; top of head to base of spine 99.6 cm; circumference top of head around lower jaw 96 cm, at the mouth 92.2 cm.

MCZ 38326 Skeleton. Female. Métet, Cameroun. Purchased from Silas E. Johnson in May 1937.

MCZ 46325 Skull. Donated by Robert Barbour on 17 January 1947. Lower M2 chipped; hole in right bulla; 3 lower incisors missing. Mandible articulated.

MCZ 46327 Skeleton. Received from Robert Barbour on 17 January 1947. From the Grandidier collection. Plaster casts of hands and feet on exhibit.

MCZ 46413 Skull. Collected by G.A. Peirson. Received from the Peabody Museum, Salem, in 1942. Lower jaw missing 1996.

MCZ 48618 Mounted skeleton. Male. Gabon. Received from Ward's.

MCZ 49006 Skeleton. Juvenile female. Métet, Cameroun. Purchased from Silas E. Johnson in 1939.

MCZ 57482 Skeleton. Adult female. Campo, 2°20′N, Cameroon. Collected by Julie Calvert in May 1977; received from her on 15 September 1977. The animal had been seen the year before, alive and diseased. The right arm was open, pink and frozen in position. The face was twisted, the lower left open and pink. The pelvis was broken. Pelvis and humerus pathological. The animal had been dead for about 2 weeks when collected.

MCZ 61072 Hand (in ethanol).

MCZ 62393 Skeleton and partial skin. Adult male. Belgian Congo. This is 'Bobbie', who arrived at Dierenpark Wassenaar, Netherlands, on 25 August 1960, aged about 1 year. He was transferred to Milwaukee Zoo on 12 May 1985; sent on loan to Fort Worth Zoo on 12 May 1987 and finally sent on loan to Franklin Park Zoo, Boston, on 19 August 1989. He was euthanased due to deteriorating condition on 6 August 1998 (weight 362 lb.). Total length (head and body) 1210 mm; foot length 345 mm; ear length 62 mm.

MCZ 63102 Humerus.

MCZ 65933 Plaster cast of head. Juvenile. Received from Dr Thomas W. French, Massachusetts Division of Fisheries and Wildlife, on 3 November 2008. He had received it from the Boston Children's Museum, who deaccessioned it on 19 October 2006. The BCM had received it from the Wyman collection on 9 November 1932.

MCZ 67244 Death mask, hands and feet (plaster casts). Male?

Harvard University, Department of Anthropology, Peabody Museum of Archaeology and Ethnology, 11 Divinity Avenue, Cambridge, MA 02138

Gloria Polizzotti Greis; Michèle E. Morgan

N817 Cast of head. Cameroun. From the Museum of Comparative Zoology collection.

N818 Cast of left foot (duplicate of N 3507). From the Museum of Comparative Zoology collection.

N819 Cast of left hand (duplicate of N 3506). From the Museum of Comparative Zoology collection.

N3385 Mounted skeleton. Adult male. Purchased from the Kny-Scheerer Co. in 1930. Right foot detached, teeth broken.

N3444 Skull. Adult male. Collected by Dr Kent Cross, 2341½ E. Evans Avenue, Denver, CO, in 1938. Purchased from him, received on 13 June 1939.

N3472 Skull. Adult male. Gabon. Collected c.1883–85. Received on loan from the Museum of Comparative Zoology in 1934, became a permanent loan on 19 December 1939. Missing 2 canines and 2 molars.

N7341 Cast of face (duplicate). Belgian Congo.

N8855 Skull. Adult female. It has a supernumerary molar. Missing since 1966.

N57479 Cranium. Subadult male. Cameroun. Collected and donated by George Schwab. Received in 1919.

59937 Skeleton (partially articulated). Adult female. Purchased from the Derby Museum, Liverpool, in 1898. It is missing 1 incisor and some bones.

N60351 Postcranial skeleton and lower face fragment. Female. Cameroun. Collected and donated by George Schwab. Received in 1919.

Z83963 Cast of left hand. Donated by Charles H. Ward in 1911.

N98198 Cast of face.

N98199 Cast of face.

N98200 Cast of face. Immature. Mount Mikeno.

N98201 Cast of right hand. Female.

N98202 Cast of left hand. Male.

N98203 Cast of left foot. Female.

The above 6 casts, N98198–N98203 inclusive, were donated by Carl Akeley, of the American Museum of

Natural History, in 1926. He collected the specimens near Lake Kivu, East Congo, in 1921 and made the casts in the field. See *Journal of the American Museum of Natural History* 23: 429–447(1923).

Carbondale

Southern Illinois University, Department of Anthropology, Carbondale, IL 62901-4502

George A. Feldhamer; Susan M. Ford

SIU 036668 Mounted postcranial skeleton. Adult male. It was purchased commercially for $273.00 on 1 August 1956. The skull was stolen sometime in the 1960s. Both hands are missing (also stolen at some point). One of the toes is broken off the articulation, but the bones are stored in a box affixed to the mount.

Carmichael

Zoo/Exotic Pathology Service, 6020 Rutland Drive #14, Carmichael, CA 95608-0515

Drury R. Reavill

All tissue samples are embedded in paraffin blocks.

V701981-9 Punch biopsy skin sections. Adult male (27 years old). Housed in Florida (1997). Multifocal depigmented areas in the axillae and a moist dermatitis. The diagnosis is of a severe necrotising dermatitis; no microbes identified.

V087971-1 Punch biopsy skin sections. Adult male (40 years old). Housed in Florida (2008). There is a chronic condition of skin eruptions on the ventrum that are dry, scaled and pruritic. Previous biopsy is V701981-9. The diagnosis is of a severe acute erosive and ulcerative dermatitis with acanthosis, parakeratosis and hyperkeratosis. It is suspected to represent a drug or food hypersensitivity reaction.

V902442-4 Punch biopsy skin sections. Adult male (age not provided). Housed in a zoo in Memphis, Tennessee (1999). There are 1 cm lesions scattered over most of the body. The diagnosis is of moderate to severe superficial pustular dermatitis and inflammatory crust formation. A hypersensitivity reaction is suspected.

V040609-8 Punch biopsy skin sections. Adult male (17 years old). Housed in a zoo in Memphis, Tennessee (2004). There is a chronic skin condition with some pruritus. Multiple round scabby lesions that are generalised with regional alopecia are present. The diagnosis is: (1) multifocal moderate ulcerative dermatitis with suppurative and septic folliculitis; and (2) mild multifocal eosinophilic and acute perifolliculitis. It is suspected to be a hypersensitivity lesion with a secondary bacterial dermatitis/folliculitis.

Chicago

Field Museum of Natural History, Roosevelt Road at Lake Shore Drive, Chicago, IL 60605-2496

Sarah A.D. Bruner; Lawrence R. Heaney; Robert J. Izor; William Stanley, Harold K. Voris; Christyna Zielinska

The primate collection is available for nondestructive research within the museum.

FMNH 336 Skull and skin. Received from Ward's Natural Science Establishment, Rochester, NY.

FMNH 15514 Mounted skeleton. Gabon. Received from Ward's Natural Science Establishment, Rochester, NY.

FMNH 16344 Skeleton and mounted skin. Male. West of Ebolowa, Kamerun. Collected by F.B. Guthrie on 12 January 1907. Cranium sectioned. The skeleton lacks 1 cervical vertebra, 4 thoracic vertebrae, 9 ribs, sternum and costal cartilages, 2 clavicles, 2 patellae, 2 pisiformes and coccyx. The sacrum and 3 lumbars are cut, and 3 ribs are broken. The right innominate and sacrum were loaned on invoice Z-12198 and not returned.

FMNH 18396 Skeleton (hands, feet and lower arms articulated). Male. Between Lomié and Molunda, Cameroun. Collected by G.F. Porter.

FMNH 18397 Skeleton (hands and feet partially articulated). Female. Doumé River, Cameroun. Collected by G.F. Porter.

FMNH 18398 Skeleton (partially articulated). Juvenile. Doumé River, Cameroun. Collected by G.F. Porter. Formerly mounted, dismantled October 1941.

FMNH 18399 Skull and skin. Male. Cameroun/French Congo frontier. Collected by G.F. Porter. The skin is missing.

FMNH 18400 Mounted skeleton and skin. Male. Yaoundé, Cameroun. Collected by G.F. Porter. The skin was noted as missing in 1960.

FMNH 18401 Skeleton (hands and feet articulated). Male. Ogooué River, Gabon. Collected by G.F. Porter.

FMNH 18402 Skeleton (hands and feet articulated). Male. Yaoundé, Cameroun. Collected by G.F. Porter.

FMNH 26065 Skeleton and skin. Adult male. *Gorilla beringei beringei*. Virunga volcanoes, pass between Mounts Sabinio and Mgahinga, bamboo zone 8500 ft., Kigezi district, Uganda. Collected by Edmund Heller on 27 December 1925 (collector's no. 9854). It was shot from a group of 10 or more, 3 other adult males were seen but their backs were less greyish white. This old male was the last of the group to retreat. Height, soles to top of crown 63½ in.; arm stretch 100 in.; chest girth 64 in.; belly girth 72 in.; thigh girth 25 in.; biceps girth 17½ in.; length of arm from axilla 39 in.; length of leg 24 in.; length of head, apex of sagittal crest to snout tip

16 in.; width of chest 22½ in.; length of longest hair on arms 6 in.; weight 375 lb. Skin on crown of head and back of neck 1 in thick. Penis bone discovered and labelled, length 14 mm. The skeleton lacks 3 carpals, 1 terminal phalanx and entire digit I from the right manus and 1 proximal phalanx from the left pes. Articulated bones of left pes loaned on invoice Z-18207 and not returned; 2 vertebrae (last thoracic and last lumbar) loaned on invoice Z-14218 and not returned; skin missing.

FMNH 27525 Cranium. Adult female. Kitunda village, 30 mi. south of Walikale, Belgian Congo. Collected by Edmund Heller from a village local on 17 March 1924 (collector's no. 8037). The teeth are wanting except for worn canines and 2 molars.

FMNH 27550 Skeleton and skin. Adult male. Kitunda village, 30 mi. south of Walikale, Belgian Congo. Collected by Edmund Heller on 14 March 1924 (collector's no. 8035). Age 7 years; back black without any whitish hairs. Teeth worn and adult. Skin salted, ultimate phalanges in skin. Stomach contained figs, nut-like seeds and a few green leaves. Height, sole to crown 55 in.; stretch of arms 83½ in.; vertical reach 83 in.; chest girth 50 in.; belly girth 56 in.; thigh girth 19 in.; stretch of legs 63½ in.; head and body 990 mm; foot length 300 mm; ear from notch 45 mm; estimated weight 200 lb. The skeleton lacks the epiphyses from the left ulna, 43 epiphyses from the digits, all vertebral epiphyses, the terminal phalanges and the coccyx.

FMNH 27551 Skeleton, skin and stomach contents (in alcohol). Adult male. Kitunda village, 30 miles south of Walikale, Belgian Congo. Collected by Edmund Heller on 15 March 1924 (collector's no. 8036). A solitary old male which lived in second growth bush or old shambas. It was hunted for 3 days and charged the hunters 20 times or more. It was shot in the skull and the occipital part was shattered. Another shot struck the shoulder, breaking it badly. The stomach contents were the soft white heart of young banana plants and green leaves of some tree or shrub. Height, soles to crown 66½ in.; stretch of arms 92 in.; length of arm 34 in.; chest girth 63 in.; belly girth 68 in.; thigh girth 26¾ in.; biceps girth 17¾ in.; forearm girth 17¼ in.; head and body 1135 mm; foot length 312 mm; ear from notch 50 mm; weight of all parts after skinning 335 lb., loss of blood 15 lb., total 350 lb. The skeleton lacks the left pisiform.

FMNH 53701 Cadaver (in alcohol). Female. Received from the Chicago Zoological Society in October 1942. Injected; right leg and forearm loaned on invoice Z-12189 with no record of return, but may be in tank 33.

FMNH 57131 Skeleton, skin and head (embalmed). Female. Received from the Chicago Zoological Society. It died in captivity on 8 November 1948, aged approximately 14 years. This is probably 'Suzette', who arrived at the Chicago Zoological Park, Brookfield, Illinois, on 24 October 1936, aged about 1½ years. Bones of left foot loaned on invoice Z-18207 and not returned.

FMNH 57201 Skeleton and skin. Male. Sixty miles west of Ebolowa, Cameroun. Received from Irvin L. Young in 1945.

FMNH 57202 Skeleton and skin. Male. Sixty miles west of Ebolowa, Cameroun. Received from Irvin L. Young in 1945. Skull on loan to the Chicago Academy of Sciences and not returned; presumed lost.

FMNH 57394 Cadaver (in alcohol). Infant male. Received from Lincoln Park Zoological Gardens, Chicago. It died in captivity in October 1960. The finger bones are in a lightwell case.

FMNH 57408 Skeleton, left arm (embalmed) and left leg (embalmed). Female. Received from Lincoln Park Zoological Gardens on 20 September 1961. This is 'Lotus', who arrived at Lincoln Park Zoo on 1 October 1948 (weight 42 lb.), along with another female ('Rajah') and 2 males ('Irving Young' and 'Sinbad'). All 4 specimens came originally from Cameroun, and were donated by Mr Irvin B. Young of Chicago. 'Lotus' died of a lingering chronic cardiac condition on 15 September 1961 (weight 268 lb.). The right knee is missing from the skeleton. Loaned on invoice Z-12189 but no record of return, however upper arm with skin may be in tank 4. Also loaned on invoice Z-12121 with no record of return: right hand and foot, portion of limbs in dry collection, cranium section. [Some additional information added by GH from correspondence with Lincoln Park Zoo personnel.]

FMNH 57738 Cadaver (in alcohol).

FMNH 59001 Cadaver (in alcohol). Male. Received from the Chicago Zoological Society.

FMNH 60272 Skeleton. Adult male. Received from the Chicago Zoological Society. This is 'Baby', who arrived at the Chicago (Brookfield) Zoo on 27 April 1951 (weight 17½ lb.), with a female named 'Sappho'. He died on 18 December 1973, of apparent heart failure whilst engaged in vigorous courtship (weight 419 lb.). The lower long bones, hands and feet are missing. [Some additional information added by GH from correspondence with Brookfield Zoo personnel.]

FMNH 60418 Skeleton. Male. Received from the Chicago Zoological Society. This is 'Weaver', who was born in captivity on 17 May 1971. The dam was 'Alpha' and the sire was 'Omega'. He died of colitis (shigellosis) on 9 July 1974.

FMNH 60596 Cadaver (in alcohol). Male. Received from Lincoln Park Zoo.

FMNH 60578 Skull and skin. Female. Received from Lincoln Park Zoo. Specimen not found on 15 January 1998. One foot loaned for dissection on invoice Z-18836 and no record of return found.

FMNH 72801 Skull and skin. Female. Received from the Chicago Zoological Society. It died in captivity on 10 August 1950.

FMNH 72805 Mounted skin. Adult male. Received from Lincoln Park Zoo, Chicago. This is 'Bushman', who arrived at Lincoln Park Zoo on 18 August 1930 (weight 38 lb.). He had been acquired from local people by the Revd Dr William C. Johnston, head of the American Presbyterian Mission at Yaoundé, Cameroun, when a few months old (weight 8 lb.). He was looked after for 2 years by a man named Belinge, and also by James Allen and his wife Annie Mary, who were stationed there from the Fourth Presbyterian Church in Chicago. He was eventually purchased for $500 by the animal dealer Julius L. Buck. The money was used to commission a stained glass window for the newly built Presbyterian church in Yaoundé. Buck sold him to Lincoln Park Zoo for $3500. 'Bushman' died on 1 January 1951, of terminal pneumonia, predisposed by a degenerative neurological problem. The skin was mounted by staff taxidermists Leon L. Walters and Frank C. Wonder, with the assistance of staff sculptor Joseph B. Krstolich. The face, hands and feet were made as celluloid models using the process invented by Walters. The skeleton was sent to Father Jurica, for his museum at the Illinois Benedictine College. [Some additional information added by GH from correspondence with Lincoln Park Zoo personnel.]

FMNH 81532 Skeleton (arms and legs articulated). Adult female. Received from Ward's Natural Science Establishment on 30 September 1954.

FMNH 81534 Skull. Female. Received from Northwestern University; originally from N. Rouppert Co., Paris.

FMNH 81535 Skull. Received from Northwestern University; originally from N. Rouppert Co., Paris.

FMNH 81536 Skull. Received from Northwestern University; originally from N. Rouppert Co., Paris.

FMNH 81537 Skull. Received from Ward's Natural Science Establishment.

FMNH 81538 Skull. Received from Ward's Natural Science Establishment.

FMNH 81539 Skull. Received from Northwestern University; originally from N. Rouppert Co., Paris.

FMNH 99092 Skeleton. Female? Received from the Chicago Zoological Society. It died in captivity on 13 September 1965. The skin was discarded.

FMNH 123073 Skeleton. Female. Prep. Lab. Catalogue no. 39. Received on 25 November 1975.

FMNH 124558 Skeleton. Female. Received from Lincoln Park Zoo. Prep. Lab. Catalogue no. 225.

FMNH 126045 Skeleton. Adult male. Received from Lincoln Park Zoo. This is 'Sinbad', who arrived at the zoo on 1 October 1948 (weight 11 lb.), and died on 19 March 1985.

FMNH 134482 Skeleton and skin. Adult male. Received from Lincoln Park Zoo on 23 June 1988. This is 'Otto', who arrived at the zoo on 19 July 1968 aged about 3–4 years (estimated weight 70 lb.). He was purchased (with a female named 'Mary') from the animal dealer Pouillox, Cameroon, with funds donated by Mr Franklin Schmick of Chicago. He died on 23 June 1988. [Some additional information added by GH from correspondence with Lincoln Park Zoo personnel.]

FMNH 135290 Skeleton. Adult male. Donated by the Chicago (Brookfield) Zoo in December 1988. This is 'Sheldon', who arrived at the Brookfield Zoo, on loan from the Philadelphia Zoo, on 30 October 1988. He had arrived at Philadelphia Zoo on 11 July 1969, aged about 1 year (weight 21 lb.), having been purchased (with a female named 'Snickers') from G. van den Brink. He died of acute haemmorhagic necrotising typhlitis on 19 December 1988. Length, crown to rump 1170 mm; ear length 63 mm; foot length 300 mm; weight 326 lb. [Some additional information added by GH from correspondence with Philadelphia Zoo personnel.]

FMNH 140890 Mounted skeleton. Received from Ward's Natural Science Establishment. It is a zoo specimen. Originally catalogued by the Department of Anthropology as FMNH 43900; transferred to the Department of Mammals on 7 October 1991.

FMNH 147992 Skeleton. Male. Prep. Lab. Catalogue no. 1005. Received in 1990. It is a zoo specimen.

FMNH 150704 Cadaver (in alcohol). Received on 26 December 1989. It is a zoo specimen (catalogue no. 643).

FMNH 153779 Skeleton. Female. Received on 2 September 1993. It is a zoo specimen. The cranium is sectioned.

FMNH 156679 Sectioned skull (cast). Prep. Lab. Catalogue no. 4414. It is a zoo specimen.

FMNH 156680 Endocranial cast. Prep. Lab. Catalogue no. 4415. It is a zoo specimen.

FMNH 156681 Skull (cast) and endocranial cast. Male. Prep. Lab. Catalogue no. 4416. It is a zoo specimen.

FMNH 156682 Skull (cast). Male. Prep. Lab. Catalogue no. 4417. It is a zoo specimen.

FMNH 156683 Skull (cast). Male. Prep. Lab. Catalogue no. 4418. It is a zoo specimen.

FMNH 156684 Skull (cast). Female. Prep. Lab. Catalogue no. 4419. It is a zoo specimen.

FMNH 156685 Femur (cast). Male. Prep. Lab. Catalogue no. 444420. It is a zoo specimen.

FMNH 156686 Tibia (cast). Male. Prep. Lab. Catalogue no. 4421. It is a zoo specimen.

FMNH 156687 Radius (cast). Male. Prep. Lab. Catalogue no. 4422. It is a zoo specimen.

FMNH 156688 Humerus (cast). Male. Prep. Lab. Catalogue no. 4423. It is a zoo specimen.

FMNH 156689 Femur (cast). Male. Prep. Lab. Catalogue no. 4424. It is a zoo specimen.

FMNH 156690 Femur (cast). Male. Prep. Lab. Catalogue no. 4425. It is a zoo specimen.

FMNH 156691 Ulna (cast). Female. Prep. Lab. Catalogue no. 4426. It is a zoo specimen.

FMNH 156692 Skull. Male. Prep. Lab. Catalogue no. 4427. It is a zoo specimen.

FMNH 163212 Skeleton. Male. Prep. Lab. Catalogue no. 5085. Received on 22 August 1998. It is a zoo specimen.

FMNH 180665 Skeleton. Female. Prep. Lab. Catalogue no. 6585. Received on 12 April 2004. This is 'Baraka', who was born in the Chicago Brookfield Zoo on 19 August 1990 to sire 'Ndume' and dam 'Babs', and was mother-reared. She died of an abdominal infection on 12 April 2004.

FMNH 180677 Skeleton. Male. Prep. Lab. Catalogue no. 3550. Received on 14 May 1995. It was prepared without data, but based on information received from the Chicago Zoological Society in 2005, it is assumed to be 'Abe'.

FMNH 186434 Skeleton. Female. Prep. Lab. Catalogue no. 6851. Received on 7 December 2004. It is a zoo specimen.

University of Illinois, College of Dentistry, Department of Oral Anatomy, 801S. Paulina Street, Chicago, IL 60612

Robert P. Scapino; Clark M. Stanford

Skull. It is kept in the osteological collection of the DuBrul Archives.

Head (partly dissected, embalmed). Adult male. This is 'Irvin Young', who arrived at the Lincoln Park Zoological Gardens on 1 October 1948 (weight 29 lb.) and died of pneumonia on 20 December 1958 (weight 367 lb.). Using portable embalming apparatus, Robert Scapino and Glen Boas from the College of Dentistry perfused the head through the carotid arteries with normal saline and then 10% formalin. It was then removed and taken to the University and immersed in a large crock of 10% formalin. In due course Prof. E. Lloyd DuBrul (1909–96) removed the cranial calotte to expose the brain. He planned to carry out some studies on the brainstem, but apparently never did so. NB: The Department of Oral Anatomy used to have a large anatomical collection of primate and other mammalian specimens stored in formalin containers. The specimens were mostly Lincoln Park and Brookfield zoo animals that had expired or had been euthanased. The specimens were largely of the head and neck only and were variously used by numerous students in comparative studies of the temporomandibular joint and head and neck musculature. Much of this material was stored in the College basement. Complaints arose about this storage site and as a consequence most of the material was destroyed by incineration. It is believed that the head of 'Irvin Young' was destroyed along with the other specimens.

Cincinnati

Cincinnati Museum Center, 1301 Western Avenue, Cincinnati, OH 45203

R.A. Davis; Glenn W. Storrs

CMC M3290 Skull. It was acquired in 1938 or 1939.

Plaster prints (2 left manus and 1 left pes). Female. *Gorilla beringei graueri*. These were made from a live gorilla, named 'Susie', in the Cincinnati Zoo in the early 1930s. 'Susie' was captured in the lowlands of the Belgian Congo by pygmies who were part of a French expedition, and was sold to the animal dealer Ruhe in Hannover in 1927. She was purchased by H.W. Reiners, a department store in New York, and was sent to the United States in cabin one of the *Graf Zeppelin*, on its second trans-Atlantic crossing, from Friedrichshafen to Lakehurst, New Jersey, 1–4 August 1929. After being on show in Reiners for 3 days, she was taken on tour of the United States and Canada with the Ringling Bros. Circus and the 101 Ranch Show. After arriving at Cincinnati zoo on 11 June 1931 (weight 85 lb.), she was purchased from Reiners by Mr Robert J. Sullivan, and remained at the zoo on loan. Some measurements of 'Susie' were taken by her trainer, Mr Gillaume Dressman: 9 January 1938 – height 57 in.; neck 23 in.; chest 49 in.; waist 52 in.; biceps 17½ in.; wrist 12 in.; breadth of hand 7 in.; sitting height 36 in.; calf 15 in.; standing reach 84 in.; weight 275 lb. 9 January 1939 – height 59 in.; span 84 in.; neck 25 in.; chest 50 in.; waist 56 in.; biceps 18 in.; wrist 12½ in.; breadth of hand 7½ in.; sitting height 36½ in.; calf 15 in.; standing reach 86 in.; weight 301 lb. 9 January 1940 – height 60 in.; chest 51 in.; waist 60 in.; weight 315 lb. 9 January 1941 – height 60½ in.; chest 52 in.; waist 62 in.; biceps 19 in.; span 86 in.; weight 335 lb. 'Susie' died of leptospirosis on 29 October 1947. At the time of her death, she was owned jointly by Mrs Stanley Crothers, sister of the late Robert Sullivan, and Mrs Robert F. Romell, his niece. 'Susie's' skeleton was mounted and donated to the Department of Biology, University of Cincinnati but was destroyed in a laboratory store-room fire in the Brodie Science Complex on 17 March 1974. [Some additional information added by GH from local newspaper reports and correspondence with Cincinnati Zoo personnel.]

Cleveland

Cleveland Museum of Natural History, Department of Physical Anthropology, 1 Wade Oval Drive, University Circle, Cleveland, OH 44106-1767

Lyman M. Jellema

The Department houses the Hamann-Todd osteological collection, consisting of nearly 3000 human and nonhuman primates, which was transferred from the Department of Zoology, Case Western University, Cleveland. The gorilla specimens were all collected in the wild between 1900 and 1930. Most were purchased from or donated by the Brush Foundation, the Embalmers' Fund, the Weber Fund, Lehman, Hope and Gerrard.

Some of the specimens listed as 'skull' lack the lower jaw, but this information is currently unavailable on an individual basis.

NB: The CMNH Department of Anthropology is preparing for a series of moves that will culminate in the primate collections being housed in a brand new laboratory. At present, it is hoped that the move will be completed by sometime in 2020. For an overview of these developments, please go to https://www.cmnh.org/centennialhome. To ascertain the possibility of scheduling a visit, please contact: ljellema@cmnh.org or Yhaileselassie@cmnh.org.

HTB 0169 Postcranial skeleton. Infant. Obtained from Gerrard, London, in 1915.

HTB 0170 Skull. Juvenile male. Obtained from Kny-Scheerer Co. in 1914.

HTB 0405 Skull. Adult female. Gabon. Obtained from Gerrard, London, in 1919.

HTB 0406 Skull. Adult male. Gabon. Obtained from Gerrard, London, in 1919.

HTB 0407 Skull. Adult male? Gabon. Obtained from Gerrard, London, in 1919.

HTB 0509 Postcranial skeleton. Adult. Obtained from Gerrard, London, in 1919.

HTB 0510 Postcranial skeleton. Adult. Obtained from Gerrard, London, in 1919.

HTB 0514 Postcranial skeleton. Adult. Obtained from Gerrard, London, in 1919.

HTB 0515 Postcranial skeleton. Adult. Obtained from Gerrard, London, in 1919.

HTB 0516 Postcranial skeleton. Adult. Obtained from Gerrard, London, in 1919.

HTB 0517 Postcranial skeleton. Adult. Obtained from Gerrard, London, in 1919.

HTB 0518 Postcranial skeleton. Adult. Obtained from Gerrard, London, in 1919.

HTB 0519 Postcranial skeleton. Adult. Obtained from Gerrard, London, in 1919.

HTB 0520 Postcranial skeleton. Adult. Obtained from Gerrard, London, in 1919.

HTB 0521 Postcranial skeleton. Adult. Obtained from Gerrard, London, in 1919.

HTB 0522 Postcranial skeleton. Adult. Obtained from Gerrard, London, in 1919.

HTB 0523 Postcranial skeleton. Adult. Obtained from Gerrard, London, in 1919.

HTB 0524 Postcranial skeleton. Adult. Obtained from Gerrard, London, in 1919.

HTB 0525 Postcranial skeleton. Adult. Obtained from Gerrard, London, in 1919.

HTB 0526 Postcranial skeleton. Adult. Obtained from Gerrard, London, in 1919.

HTB 0527 Postcranial skeleton. Adult. Obtained from Gerrard, London, in 1919.

HTB 0528 Postcranial skeleton. Adult. Obtained from Gerrard, London, in 1919.

HTB 0542 Postcranial skeleton. Adult. Obtained from Gerrard, London, in July 1920.

HTB 0624 Postcranial skeleton. Adult male. Gabon. Obtained from Gerrard, London, in 1920.

HTB 0626 Postcranial skeleton. Adult female. Gabon. Obtained from Gerrard, London, in 1921.

HTB 0627 Postcranial skeleton. Adult female? Gabon. Obtained from Gerrard, London, in 1921.

HTB 0638 Skull. Juvenile female? Obtained from Kny-Scheerer Co. in 1914.

HTB 0647 Skull. Adult male. Cameroun. Obtained from Gerrard, London, in 1921.

HTB 0650 Skull. Adult male. Cameroun. Obtained from Gerrard, London, in 1921.

HTB 0842 Skeleton. Adult female. Obtained from Gerrard, London, in 1923.

HTB 1020 Skeleton. Adult male. Lolodorf, Cameroun. Obtained from Dr W.S. Lehman in 1924.

HTB 1057 Skeleton. Adult male. Obtained from Gerrard, London, in 1925.

HTB 1075 Skull. Adult male. Cameroun. Obtained from Gerrard, London, in 1925.

HTB 1076 Skull. Adult male. French West Africa. Obtained from Gerrard, London, in 1925.

HTB 1077 Skull. Adult male. Obtained from Gerrard, London, in 1925.

HTB 1078 Skull. Adult male. Obtained from Gerrard, London, in 1925.

HTB 1079 Skull. Adult male. Obtained from Gerrard, London, in 1925.

HTB 1080 Skull. Adult male. Obtained from Gerrard, London, in 1925.

HTB 1178 Skull. Juvenile male? Obtained from Gerrard, London, in 1926.

HTB 1179 Skull. Juvenile female? Obtained from Gerrard, London, in 1926.

HTB 1180 Skull. Juvenile female. French West Africa. Obtained from Gerrard, London, in 1926.

HTB 1181 Skull. Adult male. French West Africa. Obtained from Gerrard, London, in 1926.

HTB 1182 Skull. Adult male. French West Africa. Obtained from Gerrard, London, in 1926.

HTB 1196 Skull. Adult male. French Congo. Obtained from Gerrard, London, in 1926.

HTB 1197 Skull. Adult female. Sette Kama, Gabon. Obtained from Gerrard, London, in 1917.

HTB 1398 Skull. Adult female. French Congo. Obtained from Gerrard, London, in 1928.

HTB 1399 Skull. Adult female. French Congo. Obtained from Gerrard, London, in 1928.

HTB 1400 Skull. Adult female. French Congo. Obtained from Gerrard, London, in 1928.

HTB 1401 Skull. Adult male. French Congo. Obtained from Gerrard, London, in 1928.

HTB 1402 Skull. Adult male. French Congo. Obtained from Gerrard, London, in 1928.

HTB 1403 Skull. Juvenile male. French Congo. Obtained from Gerrard, London, in 1928.

HTB 1404 Skull. Adult male. French Congo. Obtained from Gerrard, London, in 1928.

HTB 1405 Skull. Adult male. French Congo. Obtained from Gerrard, London, in 1928.

HTB 1406 Skull. Adult male. French Congo. Obtained from Gerrard, London, in 1928.

HTB 1407 Skeleton. Adult male. Dja Posten, Cameroun. Obtained from Dr W.S. Lehman in 1928.

HTB 1408 Skeleton. Adult male. Dja Posten, Cameroun. Obtained from Dr W.S. Lehman in 1928.

HTB 1409 Skeleton. Adult male. Dja Posten, Cameroun. Obtained from Dr W.S. Lehman in 1928.

HTB 1410 Skull. Adult male. Dja Posten, Cameroun. Obtained from Dr W.S. Lehman in 1928.

HTB 1411 Skull. Adult male. Dja Posten, Cameroun. Obtained from Dr W.S. Lehman in 1928.

HTB 1412 Skull. Adult female. Dja Posten, Cameroun. Obtained from Dr W.S. Lehman in 1928.

HTB 1413 Postcranial skeleton. Juvenile. Obtained from Dr W.S. Lehman in 1928.

HTB 1416 Skeleton. Adult male? Ebolowa, Cameroun. Obtained from Mr F.H. Hope in 1928.

HTB 1417 Skeleton. Adult male? Ebolowa, Cameroun. Obtained from Mr F.H. Hope in 1928.

HTB 1418 Skeleton. Adult male? Ebolowa, Cameroun. Obtained from Mr F.H. Hope in 1928.

HTB 1419 Skeleton. Adult female? Ebolowa, Cameroun. Obtained from Mr F.H. Hope in 1928.

HTB 1420 Skeleton. Juvenile male? Ebolowa, Cameroun. Obtained from Mr F.H. Hope in 1928.

HTB 1421 Skeleton. Juvenile male? Ebolowa, Cameroun. Obtained from Mr F.H. Hope in 1928.

HTB 1422 Skeleton. Adult female? Ebolowa, Cameroun. Obtained from Mr F.H. Hope in 1928.

HTB 1423 Skeleton. Adult female. Ebolowa, Cameroun. Obtained from Mr F.H. Hope in 1928.

HTB 1424 Skeleton. Juvenile female? Ebolowa, Cameroun. Obtained from Mr F.H. Hope in 1928.

HTB 1425 Skeleton. Adult male? Ebolowa, Cameroun. Obtained from Mr F.H. Hope in 1929.

HTB 1427 Skull. Adult male. Obtained from Gerrard, London, in 1929.

HTB 1428 Skull. Adult male. Obtained from Gerrard, London, in 1929.

HTB 1429 Skull. Adult male. Obtained from Gerrard, London, in 1929.

HTB 1430 Skeleton. Adult male. Obtained from Gerrard, London, in 1929.

HTB 1431 Skeleton. Adult male. Obtained from Gerrard, London, in 1929.

HTB 1432 Skull. Adult male. Obtained from Gerrard, London, in 1929

HTB 1689 Skull. Adult male. Ebolowa, Cameroun. Obtained from Mr F.H. Hope in 1928.

HTB 1690 Skull. Adult female. Ebolowa, Cameroun. Obtained from Mr F.H. Hope in 1928.

HTB 1704 Skeleton. Adult female. Ebolowa, Cameroun. Obtained from Mr F.H. Hope in 1930.

HTB 1709 Skeleton. Adult male. Ebolowa, Cameroun. Obtained from Mr F.H. Hope in March 1930.

HTB 1710 Skeleton. Adult female. Ebolowa, Cameroun. Obtained from Mr F.H. Hope in March 1930.

HTB 1711 Skeleton. Juvenile male. Ebolowa, Cameroun. Obtained from Mr F.H. Hope in March 1930.

HTB 1712 Skeleton. Adult male. Ebolowa, Cameroun. Obtained from Mr F.H. Hope in March 1930.

HTB 1714 Postcranial skeleton. Juvenile male. Ebolowa, Cameroun. Obtained from Mr F.H. Hope.

HTB 1717 Skeleton. Adult male. Ebolowa, Cameroun. Obtained from Mr F.H. Hope in March 1930.

HTB 1725 Skeleton. Adult female. Ebolowa, Cameroun. Obtained from Mr F.H. Hope in March 1930.

HTB 1727 Skeleton. Juvenile female. Ebolowa, Cameroun. Obtained from Mr F.H. Hope in March 1930.

HTB 1728 Skeleton. Adult male. Ebolowa, Cameroun. Obtained from Mr F.H. Hope in 1929.

HTB 1729 Skeleton. Adult male. Ebolowa, Cameroun. Obtained from Mrs E. Cozzens in 1929.

HTB 1730 Skeleton. Adult male. Ebolowa, Cameroun. Obtained from Mr F.H. Hope in 1929.

HTB 1731 Skeleton. Adult male. Ebolowa, Cameroun. Obtained from Mr F.H. Hope in 1929.

HTB 1732 Skeleton. Adult male. Ebolowa, Cameroun. Obtained from Mrs E. Cozzens in 1929.

HTB 1733 Skeleton. Adult male. Ebolowa, Cameroun. Obtained from Mrs E. Cozzens in 1929.

HTB 1734 Skeleton. Juvenile male. Ebolowa, Cameroun. Obtained from Mr F.H. Hope.

HTB 1736 Skeleton. Adult male. Ebolowa, Cameroun. Obtained from Mr F.H. Hope in March 1930.

HTB 1740 Skeleton. Juvenile female. Ebolowa, Cameroun. Obtained from Mr F.H. Hope in 1929.

HTB 1743 Postcranial skeleton. Adult female. Ebolowa, Cameroun. Obtained from Mr F.H. Hope in 1929.

HTB 1746 Skeleton. Adult male. Ebolowa, Cameroun. Obtained from Mr F.H. Hope in 1929.

HTB 1751 Skeleton. Adult male. Ebolowa, Cameroun. Obtained from Mr F.H. Hope in 1929.

HTB 1752 Skeleton. Adult female. Ebolowa, Cameroun. Obtained from Mr F.H. Hope in 1929.

HTB 1753 Skeleton. Juvenile female. Ebolowa, Cameroun. Obtained from Mr F.H. Hope in 1929.

HTB 1754 Skeleton. Adult male. Ebolowa, Cameroun. Obtained from Mrs E. Cozzens in 1929.

HTB 1756 Skeleton. Adult female. Ebolowa, Cameroun. Obtained from Mr F.H. Hope in 1929.

HTB 1760 Skeleton. Juvenile female? Ebolowa, Cameroun. Obtained from Mr F.H. Hope in 1929.

HTB 1764 Skeleton. Adult female. Ebolowa, Cameroun. Obtained from Mr F.H. Hope in 1929.

HTB 1765 Postcranial skeleton. Adult female. Ebolowa, Cameroun. Obtained from Mr F.H. Hope in 1929.

HTB 1772 Skeleton. Infant. Ebolowa, Cameroun. Obtained from Mr F.H. Hope in 1930.

HTB 1776 Skull. Adult male. Ebolowa, Cameroun. Obtained from Mr F.H. Hope in 1929.

HTB 1780 Skeleton. Adult male. Abong Mbang, Cameroun. Obtained from Dr W.S. Lehman in 1930.

HTB 1781 Skeleton. Adult male? Ebolowa, Cameroun. Obtained from Mr F.H. Hope in March 1932.

HTB 1782 Skeleton. Juvenile female. Ebolowa, Cameroun. Obtained from Mr F.H. Hope.

HTB 1783 Skeleton. Juvenile female. Ebolowa, Cameroun. Obtained from Mr F.H. Hope.

HTB 1784 Skull. Adult male. Ebolowa, Cameroun. Obtained from Mr F.H. Hope.

HTB 1787 Skeleton. Adult male. Ebolowa, Cameroun. Obtained from Mr F.H. Hope in July 1930.

HTB 1788 Skull. Adult male. Abong Mbang, Cameroun. Obtained from Dr W.S. Lehman in 1930.

HTB 1789 Skull. Adult male. Abong Mbang, Cameroun. Obtained from Dr W.S. Lehman in 1930.

HTB 1790 Skull. Adult male. Abong Mbang, Cameroun. Obtained from Dr W.S. Lehman in 1930.

HTB 1794 Skeleton. Adult female? Ebolowa, Cameroun. Obtained from Mr F.H. Hope in 1930.

HTB 1795 Skeleton. Adult male? Ebolowa, Cameroun. Obtained from Mr F.H. Hope in 1930.

HTB 1796 Skeleton. Adult male. Ebolowa, Cameroun. Obtained from Mr F.H. Hope in 1930.

HTB 1797 Skeleton. Adult male. Ebolowa, Cameroun. Obtained from Mr F.H. Hope in 1930.

HTB 1798 Skeleton. Adult female. Ebolowa, Cameroun. Obtained from Mr F.H. Hope in 1930.

HTB 1799 Skeleton. Infant. Ebolowa, Cameroun. Obtained from Mr F.H. Hope in 1930.

HTB 1801 Postcranial skeleton. Adult female? Ebolowa, Cameroun. Obtained from Mr F.H. Hope in 1930.

HTB 1806 Skeleton. Adult female. Ebolowa, Cameroun. Obtained from Mr F.H. Hope in 1930.

HTB 1844 Skeleton. Juvenile male. Ebolowa, Cameroun. Obtained from Mr F.H. Hope in 1930.

HTB 1845 Skeleton. Juvenile male? Ebolowa, Cameroun. Obtained from Mr F.H. Hope in 1930.

HTB 1846 Skeleton. Adult female. Ebolowa, Cameroun. Obtained from Mr F.H. Hope in 1930.

HTB 1847 Skeleton. Adult male. Ebolowa, Cameroun. Obtained from Mr F.H. Hope in 1930.

HTB 1849 Skeleton. Adult female. Ebolowa, Cameroun. Obtained from Mr F.H. Hope in 1930.

HTB 1850 Skeleton. Juvenile male. Ebolowa, Cameroun. Obtained from Mr F.H. Hope in 1930.

HTB 1851 Skeleton. Adult female. Ebolowa, Cameroun. Obtained from Mr F.H. Hope in 1930.

HTB 1852 Skeleton. Adult female. Ebolowa, Cameroun. Obtained from Mr F.H. Hope in 1930.

HTB 1854 Skeleton. Adult female. Ebolowa, Cameroun. Obtained from Mr F.H. Hope in 1930.

HTB 1856 Skeleton. Adult female. Ebolowa, Cameroun. Obtained from Mr F.H. Hope in 1930.

HTB 1857 Skeleton. Adult male. Ebolowa, Cameroun. Obtained from Mr F.H. Hope in 1930.

HTB 1858 Skeleton. Infant. Ebolowa, Cameroun. Obtained from Mr F.H. Hope in 1930.

HTB 1859 Skeleton. Adult male. Ebolowa, Cameroun. Obtained from Mr F.H. Hope in 1930.

HTB 1860 Skeleton. Adult male. Ebolowa, Cameroun. Obtained from Mr F.H. Hope in 1930.

HTB 1872 Skull. Adult male. French Congo. Obtained from J. Fleming, Hamburg.

HTB 1873 Skull. Adult male. French Congo. Obtained from J. Fleming, Hamburg.

HTB 1874 Skull. Adult male. French Congo. Obtained from J. Fleming, Hamburg.

HTB 1876 Skull. Adult female. French Congo. Obtained from J. Fleming, Hamburg.

HTB 1877 Skull. Adult female. French Congo. Obtained from J. Fleming, Hamburg.

HTB 1878 Skull. Adult male. French Congo. Obtained from J. Fleming, Hamburg.

HTB 1879 Skull. Adult male. French Congo. Obtained from J. Fleming, Hamburg.

HTB 1891 Skull. Juvenile. Abong Mbang, Cameroun. Obtained from Dr W.S. Lehman in 1932.

HTB 1892 Skull. Adult male. Abong Mbang, Cameroun. Obtained from Dr W.S. Lehman in 1932.

HTB 1893 Skull. Adult male. Abong Mbang, Cameroun. Obtained from Dr W.S. Lehman in 1932.

HTB 1894 Skull. Adult male. Abong Mbang, Cameroun. Obtained from Dr W.S. Lehman in 1932.

HTB 1895 Skull. Adult male. Abong Mbang, Cameroun. Obtained from Dr W.S. Lehman in 1932.

HTB 1896 Postcranial skeleton. Juvenile female. Abong Mbang, Cameroun. Obtained from Dr W.S. Lehman in 1932.

HTB 1897 Skeleton. Adult female. Abong Mbang, Cameroun. Obtained from Dr W.S. Lehman in 1932.

HTB 1898 Skull. Adult male. Abong Mbang, Cameroun. Obtained from Dr W.S. Lehman in 1932.

HTB 1899 Skull. Adult male. Abong Mbang, Cameroun. Obtained from Dr W.S. Lehman in 1932.

HTB 1900 Skull. Juvenile male. Abong Mbang, Cameroun. Obtained from Dr W.S. Lehman in 1932.

HTB 1901 Skull. Adult male. Abong Mbang, Cameroun. Obtained from Dr W.S. Lehman in 1932.

HTB 1902 Skull. Juvenile male. Abong Mbang, Cameroun. Obtained from Dr W.S. Lehman in 1932.

HTB 1904 Skull. Adult male. Abong Mbang, Cameroun. Obtained from Dr W.S. Lehman in 1932.

HTB 1905 Skull. Adult female. Abong Mbang, Cameroun. Obtained from Dr W.S. Lehman in 1932.

HTB 1906 Skull. Juvenile female. Abong Mbang, Cameroun. Obtained from Dr W.S. Lehman in 1932.

HTB 1907 Skull. Adult female. Abong Mbang, Cameroun. Obtained from Dr W.S. Lehman in 1932.

HTB 1908 Skull. Juvenile female. Abong Mbang, Cameroun. Obtained from Dr W.S. Lehman in 1932.

HTB 1909 Skull. Adult male. Abong Mbang, Cameroun. Obtained from Dr W.S. Lehman in 1932.

HTB 1910 Skull. Adult male. Abong Mbang, Cameroun. Obtained from Dr W.S. Lehman in 1932.

HTB 1911 Skull. Adult male. Abong Mbang, Cameroun. Obtained from Dr W.S. Lehman in 1932.

HTB 1912 Skull. Adult female. Abong Mbang, Cameroun. Obtained from Dr W.S. Lehman in 1932.

HTB 1913 Skull. Adult female. Abong Mbang, Cameroun. Obtained from Dr W.S. Lehman in 1932.

HTB 1914 Skull. Adult female. Abong Mbang, Cameroun. Obtained from Dr W.S. Lehman in 1932.

HTB 1915 Skull. Adult female. Abong Mbang, Cameroun. Obtained from Dr W.S. Lehman in 1932.

HTB 1916 Skull. Juvenile male. Abong Mbang, Cameroun. Obtained from Dr W.S. Lehman in 1932.

HTB 1917 Skull. Juvenile male. Abong Mbang, Cameroun. Obtained from Dr W.S. Lehman in 1932.

HTB 1918 Skull. Adult male. Abong Mbang, Cameroun. Obtained from Dr W.S. Lehman in 1932.

HTB 1919 Skull. Adult male. Abong Mbang, Cameroun. Obtained from Dr W.S. Lehman in 1932.

HTB 1920 Skull. Adult female. Abong Mbang, Cameroun. Obtained from Dr W.S. Lehman in 1932.

HTB 1921 Skull. Juvenile female. Abong Mbang, Cameroun. Obtained from Dr W.S. Lehman in 1932.

HTB 1922 Skull. Adult female. Abong Mbang, Cameroun. Obtained from Dr W.S. Lehman.

HTB 1923 Skull. Adult female. Abong Mbang, Cameroun. Obtained from Dr W.S. Lehman in 1932.

HTB 1924 Skull. Adult female. Abong Mbang, Cameroun. Obtained from Dr W.S. Lehman in 1932.

HTB 1928 Skeleton. Juvenile female. Abong Mbang, Cameroun. Obtained from Dr W.S. Lehman in 1932.

HTB 1929 Skeleton. Juvenile. Abong Mbang, Cameroun. Obtained from Dr W.S. Lehman in 1932.

HTB 1930 Skeleton. Adult male. Abong Mbang, Cameroun. Obtained from Dr W.S. Lehman in 1932.

HTB 1931 Skeleton. Juvenile. Abong Mbang, Cameroun. Obtained from Dr W.S. Lehman in 1932.

HTB 1932 Skeleton. Adult female. Abong Mbang, Cameroun. Obtained from Dr W.S. Lehman in 1932.

HTB 1933 Skeleton. Juvenile male? Abong Mbang, Cameroun. Obtained from Dr W.S. Lehman in 1932.

HTB 1934 Postcranial skeleton. Juvenile. Abong Mbang, Cameroun. Obtained from Dr W.S. Lehman in 1932.

HTB 1935 Skeleton. Juvenile female? Abong Mbang, Cameroun. Obtained from Dr W.S. Lehman in 1932.

HTB 1936 Postcranial skeleton. Juvenile male. Abong Mbang, Cameroun. Obtained from Dr W.S. Lehman in 1932.

HTB 1937 Skull. Juvenile. Abong Mbang, Cameroun. Obtained from Dr W.S. Lehman in 1932.

HTB 1938 Skull. Juvenile male. Abong Mbang, Cameroun. Obtained from Dr W.S. Lehman in 1932.

HTB 1940 Skull. Juvenile female? Abong Mbang, Cameroun. Obtained from Dr W.S. Lehman.

HTB 1941 Skull. Juvenile. Abong Mbang, Cameroun. Obtained from Dr W.S. Lehman.

HTB 1942 Skull. Juvenile female? Abong Mbang, Cameroun. Obtained from Dr W.S. Lehman.

HTB 1943 Skull. Juvenile female? Abong Mbang, Cameroun. Obtained from Dr W.S. Lehman.

HTB 1944 Skull. Juvenile male? Abong Mbang, Cameroun. Obtained from Dr W.S. Lehman in 1932.

HTB 1945 Skull. Adult female. Abong Mbang, Cameroun. Obtained from Dr W.S. Lehman in 1932.

HTB 1946 Skull. Adult male. Abong Mbang, Cameroun. Obtained from Dr W.S. Lehman in 1932.

HTB 1947 Skull. Adult male. Abong Mbang, Cameroun. Obtained from Dr W.S. Lehman.

HTB 1948 Skull. Adult male. Abong Mbang, Cameroun. Obtained from Dr W.S. Lehman in 1932.

HTB 1949 Skull. Juvenile female. Abong Mbang, Cameroun. Obtained from Dr W.S. Lehman in 1932.

HTB 1950 Skull. Adult female. Abong Mbang, Cameroun. Obtained from Dr W.S. Lehman in 1932.

HTB 1951 Skull. Juvenile male. Abong Mbang, Cameroun. Obtained from Dr W.S. Lehman in 1932.

HTB 1952 Skull. Juvenile female. Abong Mbang, Cameroun. Obtained from Dr W.S. Lehman in 1932.

HTB 1953 Skeleton. Adult female? Abong Mbang, Cameroun. Obtained from Dr W.S. Lehman in 1932.

HTB 1954 Skeleton. Adult male. Abong Mbang, Cameroun. Obtained from Dr W.S. Lehman in 1932.

HTB 1955 Skull. Adult female. Abong Mbang, Cameroun. Obtained from Dr W.S. Lehman in 1932.

HTB 1960 Skull. Juvenile female. Abong Mbang, Cameroun. Obtained from Dr W.S. Lehman in 1932.

HTB 1961 Skull. Juvenile female? Abong Mbang, Cameroun. Obtained from Dr W.S. Lehman in 1932.

HTB 1962 Skull. Adult male. Abong Mbang, Cameroun. Obtained from Dr W.S. Lehman in 1932.

HTB 1963 Skull. Adult male. Abong Mbang, Cameroun. Obtained from Dr W.S. Lehman in 1932.

HTB 1964 Skull. Adult male. Abong Mbang, Cameroun. Obtained from Dr W.S. Lehman in 1932.

HTB 1965 Skull. Adult female. Abong Mbang, Cameroun. Obtained from Dr W.S. Lehman in 1932.

HTB 1966 Skull. Adult male. Abong Mbang, Cameroun. Obtained from Dr W.S. Lehman in 1932.

HTB 1967 Skull. Adult male. Abong Mbang, Cameroun. Obtained from Dr W.S. Lehman in 1932.

HTB 1968 Skull. Adult female. Abong Mbang, Cameroun. Obtained from Dr W.S. Lehman in 1932.

HTB 1969 Skull. Adult male. Abong Mbang, Cameroun. Obtained from Dr W.S. Lehman in 1932.

HTB 1970 Skull. Adult female. Abong Mbang, Cameroun. Obtained from Dr W.S. Lehman in 1932.

HTB 1971 Skull. Adult male. Abong Mbang, Cameroun. Obtained from Dr W.S. Lehman in 1932.

HTB 1972 Skull. Adult female. Abong Mbang, Cameroun. Obtained from Dr W.S. Lehman in 1932.

HTB 1973 Skull. Adult female. Abong Mbang, Cameroun. Obtained from Dr W.S. Lehman in 1932.

HTB 1974 Skull. Adult female. Abong Mbang, Cameroun. Obtained from Dr W.S. Lehman in 1932.

HTB 1975 Skull. Adult female. Abong Mbang, Cameroun. Obtained from Dr W.S. Lehman in 1932.

HTB 1976 Skull. Adult male. Abong Mbang, Cameroun. Obtained from Dr W.S. Lehman in 1932.

HTB 1977 Skull. Adult male. Abong Mbang, Cameroun. Obtained from Dr W.S. Lehman in 1932.

HTB 1978 Skull. Adult male. Abong Mbang, Cameroun. Obtained from Dr W.S. Lehman in 1932.

HTB 1979 Skull. Adult male. Abong Mbang, Cameroun. Obtained from Dr W.S. Lehman in 1932.

HTB 1980 Skull. Adult male. Abong Mbang, Cameroun. Obtained from Dr W.S. Lehman in 1932.

HTB 1981 Skull. Adult male. Abong Mbang, Cameroun. Obtained from Dr W.S. Lehman in 1932.

HTB 1982 Skull. Adult male. Abong Mbang, Cameroun. Obtained from Dr W.S. Lehman in 1932.

HTB 1983 Skull. Adult male. Abong Mbang, Cameroun. Obtained from Dr W.S. Lehman in 1932.

HTB 1984 Skull. Adult male. Abong Mbang, Cameroun. Obtained from Dr W.S. Lehman in 1932.

HTB 1985 Skull. Adult female. Abong Mbang, Cameroun. Obtained from Dr W.S. Lehman in 1932.

HTB 1986 Skull. Adult male. Abong Mbang, Cameroun. Obtained from Dr W.S. Lehman in 1932.

HTB 1987 Skull. Adult male. Abong Mbang, Cameroun. Obtained from Dr W.S. Lehman in 1932.

HTB 1988 Skull. Adult male. Abong Mbang, Cameroun. Obtained from Dr W.S. Lehman in 1932.

HTB 1989 Skull. Adult female. Abong Mbang, Cameroun. Obtained from Dr W.S. Lehman in 1932.

HTB 1990 Skull. Adult male. Abong Mbang, Cameroun. Obtained from Dr W.S. Lehman in 1932.

HTB 1991 Skeleton. Adult male. Abong Mbang, Cameroun. Obtained from Dr W.S. Lehman in 1932.

HTB 1992 Skeleton. Adult female. Abong Mbang, Cameroun. Obtained from Dr W.S. Lehman in 1932.

HTB 1994 Skeleton. Adult male. Abong Mbang, Cameroun. Obtained from Dr W.S. Lehman in 1932.

HTB 1995 Skeleton. Adult male. Abong Mbang, Cameroun. Obtained from Dr W.S. Lehman in 1932.

HTB 1996 Skeleton. Adult female. Abong Mbang, Cameroun. Obtained from Dr W.S. Lehman in 1932.

HTB 1997 Skeleton. Adult female. Abong Mbang, Cameroun. Obtained from Dr W.S. Lehman in 1932.

HTB 1998 Skeleton. Adult male. Abong Mbang, Cameroun. Obtained from Dr W.S. Lehman in 1932.

HTB 1999 Skeleton. Adult female. Abong Mbang, Cameroun. Obtained from Dr W.S. Lehman in 1932.

HTB 2000 Skeleton. Adult male. Abong Mbang, Cameroun. Obtained from Dr W.S. Lehman in 1932.

HTB 2015 Postcranial skeleton. Adult. Abong Mbang, Cameroun. Obtained from Dr W.S. Lehman in 1932.

HTB 2016 Postcranial skeleton. Adult male? Abong Mbang, Cameroun. Obtained from Dr W.S. Lehman in 1932.

HTB 2017 Postcranial skeleton. Adult male? Abong Mbang, Cameroun. Obtained from Dr W.S. Lehman in 1932.

HTB 2018 Postcranial skeleton. Adult female? Abong Mbang, Cameroun. Obtained from Dr W.S. Lehman in 1932.

HTB 2019 Postcranial skeleton. Juvenile male? Abong Mbang, Cameroun. Obtained from Dr W.S. Lehman in 1932.

HTB 2022 Postcranial skeleton. Juvenile male? Abong Mbang, Cameroun. Obtained from Dr W.S. Lehman in 1932.

HTB 2023 Postcranial skeleton. Juvenile male? Abong Mbang, Cameroun. Obtained from Dr W.S. Lehman in 1932.

HTB 2024 Postcranial skeleton. Adult male? Abong Mbang, Cameroun. Obtained from Dr W.S. Lehman in 1932.

HTB 2025 Skeleton. Adult male. Ebolowa, Cameroun. Obtained from Mr F.H. Hope in 1933.

HTB 2028 Skeleton. Adult male. Ebolowa, Cameroun. Obtained from Mr F.H. Hope in 1933.

HTB 2029 Skeleton. Adult male. Ebolowa, Cameroun. Obtained from Mr F.H. Hope in 1933.

HTB 2031 Skull. Adult male. Ebolowa, Cameroun. Obtained from Mr F.H. Hope in 1933.

HTB 2035 Skull. Juvenile female? Ebolowa, Cameroun. Obtained from Mr F.H. Hope in 1933.

HTB 2036 Skull. Adult female. Ebolowa, Cameroun. Obtained from Mr F.H. Hope in 1933.

HTB 2039 Postcranial skeleton. Juvenile. Ebolowa, Cameroun. Obtained from Mr F.H. Hope in 1933.

HTB 2069 Postcranial skeleton. Adult female. Ebolowa, Cameroun. Obtained from Mr F.H. Hope in 1933.

HTB 2070 Postcranial skeleton. Juvenile male. Ebolowa, Cameroun. Obtained from Mr F.H. Hope in 1933.

HTB 2739 Skeleton. Adult male. Abong Mbang, Cameroun. Obtained from Dr W.S. Lehman in 1934.

HTB 2741 Skeleton. Adult male. Abong Mbang, Cameroun. Obtained from Dr W.S. Lehman in 1934.

HTB 2742 Skeleton. Adult female? Abong Mbang, Cameroun. Obtained from Dr W.S. Lehman in 1934.

HTB 2744 Skull. Juvenile female. Abong Mbang, Cameroun. Obtained from Dr W.S. Lehman in 1934.

HTB 2745 Skeleton. Adult male. Abong Mbang, Cameroun. Obtained from Dr W.S. Lehman in 1934.

HTB 2749 Postcranial skeleton. Female. Abong Mbang, Cameroun. Obtained from Dr W.S. Lehman in 1934.

HTB 2750 Skull. Adult female. Abong Mbang, Cameroun. Obtained from Dr W.S. Lehman in 1934.

HTB 2752 Skull. Juvenile female? Abong Mbang, Cameroun. Obtained from Dr W.S. Lehman in 1934.

HTB 2753 Skull. Adult male. Abong Mbang, Cameroun. Obtained from Dr W.S. Lehman in 1934.

HTB 2754 Skull. Infant. Abong Mbang, Cameroun. Obtained from Dr W.S. Lehman in 1934.

HTB 2755 Skull. Juvenile female. Abong Mbang, Cameroun. Obtained from Dr W.S. Lehman in 1934.

HTB 2757 Skull. Juvenile male? Abong Mbang, Cameroun. Obtained from Dr W.S. Lehman in 1934.

HTB 2758 Skull. Juvenile male. Abong Mbang, Cameroun. Obtained from Dr W.S. Lehman in 1934.

HTB 2766 Skull. Adult male. Abong Mbang, Cameroun. Obtained from Dr W.S. Lehman in 1934.

HTB 2767 Skeleton. Adult male. Abong Mbang, Cameroun. Obtained from Dr W.S. Lehman in 1934.

HTB 2768 Skull. Infant. Abong Mbang, Cameroun. Obtained from Dr W.S. Lehman in 1934.

HTB 2772 Skeleton. Juvenile female? Abong Mbang, Cameroun. Obtained from Dr W.S. Lehman in 1934.

HTB 2774 Skull. Adult female. Abong Mbang, Cameroun. Obtained from Dr W.S. Lehman in 1934.

HTB 2775 Postcranial skeleton. Adult male? Abong Mbang, Cameroun. Obtained from Dr W.S. Lehman in 1934.

HTB 2776 Skull. Adult male. Abong Mbang, Cameroun. Obtained from Dr W.S. Lehman in 1934.

HTB 2778 Skull. Adult female. Abong Mbang, Cameroun. Obtained from Dr W.S. Lehman in 1934.

HTB 2781 Skeleton. Juvenile female. Abong Mbang, Cameroun. Obtained from Dr W.S. Lehman in 1934.

HTB 2782 Skeleton. Adult female. Abong Mbang, Cameroun. Obtained from Dr W.S. Lehman in 1934.

HTB 2784 Skeleton. Juvenile female. Abong Mbang, Cameroun. Obtained from Dr W.S. Lehman in 1934.

HTB 2785 Skeleton. Adult female. Abong Mbang, Cameroun. Obtained from Dr W.S. Lehman in 1934.

HTB 2786 Skeleton. Adult male. Abong Mbang, Cameroun. Obtained from Dr W.S. Lehman in 1934.

HTB 2787 Skeleton. Adult female. Abong Mbang, Cameroun. Obtained from Dr W.S. Lehman in 1934.

HTB 2788 Skull. Adult male? Abong Mbang, Cameroun. Obtained from Dr W.S. Lehman in 1934.

HTB 2790 Skull. Adult male. Abong Mbang, Cameroun. Obtained from Dr W.S. Lehman in 1934.

HTB 2791 Skull. Adult female. Abong Mbang, Cameroun. Obtained from Dr W.S. Lehman in 1934.

HTB 2792 Skull. Adult male. Abong Mbang, Cameroun. Obtained from Dr W.S. Lehman in 1934.

HTB 2793 Skull. Adult male. Abong Mbang, Cameroun. Obtained from Dr W.S. Lehman in 1934.

HTB 2794 Skull. Adult male. Abong Mbang, Cameroun. Obtained from Dr W.S. Lehman in 1934.

HTB 2795 Skull. Adult male. Abong Mbang, Cameroun. Obtained from Dr W.S. Lehman in 1934.

HTB 2796 Skull. Adult female. Abong Mbang, Cameroun. Obtained from Dr W.S. Lehman in 1934.

HTB 2797 Skull. Adult female. Abong Mbang, Cameroun. Obtained from Dr W.S. Lehman in 1934.

HTB 2798 Skull. Adult female. Abong Mbang, Cameroun. Obtained from Dr W.S. Lehman in 1934.

HTB 2799 Skull. Adult female. Abong Mbang, Cameroun. Obtained from Dr W.S. Lehman in 1934.

HTB 2801 Skull. Adult male. Abong Mbang, Cameroun. Obtained from Dr W.S. Lehman in 1934.

HTB 2802 Skull. Adult male. Abong Mbang, Cameroun. Obtained from Dr W.S. Lehman in 1934.

HTB 2803 Skull. Juvenile male. Abong Mbang, Cameroun. Obtained from Dr W.S. Lehman in 1934.

HTB 2812 Skull. Adult female. Abong Mbang, Cameroun. Obtained from Dr W.S. Lehman in 1934.

HTB 2813 Skull. Adult female. Abong Mbang, Cameroun. Obtained from Dr W.S. Lehman in 1934.

HTB 2816 Postcranial skeleton. Adult male. Abong Mbang, Cameroun. Obtained from Dr W.S. Lehman in 1934.

HTB 2817 Skeleton. Juvenile female. Abong Mbang, Cameroun. Obtained from Dr W.S. Lehman in 1934.

HTB 2818 Skeleton. Adult female. Abong Mbang, Cameroun. Obtained from Dr W.S. Lehman in 1934.

HTB 2819 Skeleton. Juvenile male. Abong Mbang, Cameroun. Obtained from Dr W.S. Lehman in 1934.

HTB 2820 Skull. Adult female. Abong Mbang, Cameroun. Obtained from Dr W.S. Lehman in 1934.

HTB 2821 Postcranial skeleton. Adult male. Abong Mbang, Cameroun. Obtained from Dr W.S. Lehman in 1934.

HTB 2822 Skeleton. Juvenile male? Abong Mbang, Cameroun. Obtained from Dr W.S. Lehman in 1934.

HTB 2824 Skull. Infant female? Abong Mbang, Cameroun. Obtained from Dr W.S. Lehman in 1934.

HTB 2825 Skull. Juvenile male? Abong Mbang, Cameroun. Obtained from Dr W.S. Lehman in 1934.

HTB 2826 Skeleton. Adult male. Abong Mbang, Cameroun. Obtained from Dr W.S. Lehman in 1934.

HTB 2829 Postcranial skeleton. Adult male. Abong Mbang, Cameroun. Obtained from Dr W.S. Lehman in 1934.

HTB 2830 Postcranial skeleton. Juvenile male? Abong Mbang, Cameroun. Obtained from Dr W.S. Lehman in 1934.

HTB 3356 [9.886-0183A] Postcranial skeleton. Adult male. Obtained from Dr W.S. Lehman in 1936.

HTB 3356 [9.886-0183B] Postcranial skeleton. Adult. Obtained from Dr W.S. Lehman in 1936.

HTB 3391 Postcranial skeleton. Adult male. Abong Mbang, Djaposten. Obtained from Dr W.S. Lehman.

HTB 3392 Skeleton. Adult male. Abong Mbang, Djaposten. Obtained from Dr W.S. Lehman.

HTB 3393 Skeleton. Adult female. Abong Mbang, Djaposten. Obtained from Dr W.S. Lehman.

HTB 3394 Postcranial skeleton. Juvenile male. Abong Mbang, Djaposten. Obtained from Dr W.S. Lehman.

HTB 3396 Postcranial skeleton. Juvenile male. Abong Mbang, Djaposten. Obtained from Dr W.S. Lehman.

HTB 3397 Skeleton. Adult male? Abong Mbang, Djaposten. Obtained from Dr W.S. Lehman.

HTB 3400 Skeleton. Adult male. Abong Mbang, Djaposten. Obtained from Dr W.S. Lehman.

HTB 3402 Skeleton. Juvenile male. Abong Mbang, Djaposten. Obtained from Dr W.S. Lehman.

HTB 3403 Skull. Juvenile male? Abong Mbang, Djaposten. Obtained from Dr W.S. Lehman.

HTB 3404 Postcranial skeleton. Adult male. Abong Mbang, Djaposten. Obtained from Dr W.S. Lehman.

HTB 3405 Skeleton. Adult female. Abong Mbang, Djaposten. Obtained from Dr W.S. Lehman.

HTB 3410 Skull. Adult male. Abong Mbang, Djaposten. Obtained from Dr W.S. Lehman.

HTB 3414 Skull. Adult female. Abong Mbang, Djaposten. Obtained from Dr W.S. Lehman.

HTB 3415 Skull. Adult male. Abong Mbang, Djaposten. Obtained from Dr W.S. Lehman.

HTB 3416 Skull. Adult female. Abong Mbang, Djaposten. Obtained from Dr W.S. Lehman.

HTB 3417 Skull. Adult male. Abong Mbang, Djaposten. Obtained from Dr W.S. Lehman.

HTB 3419 Skull. Adult male. Abong Mbang, Djaposten. Obtained from Dr W.S. Lehman.

HTB 3420 Skull. Adult male. Abong Mbang, Djaposten. Obtained from Dr W.S. Lehman.

HTB 3421 Skull. Adult female. Abong Mbang, Djaposten. Obtained from Dr W.S. Lehman.

HTB 3422 Skull. Adult female. Abong Mbang, Djaposten. Obtained from Dr W.S. Lehman.

HTB 3424 Skull. Adult female. Abong Mbang, Djaposten. Obtained from Dr W.S. Lehman.

HTB 3425 Skull. Adult male. Abong Mbang, Djaposten. Obtained from Dr W.S. Lehman.

HTB 3426 Skull. Adult female. Abong Mbang, Djaposten. Obtained from Dr W.S. Lehman.

HTB 3428 Skull. Adult male. Abong Mbang, Djaposten. Obtained from Dr W.S. Lehman.

HTB 3429 Skull. Adult male. Abong Mbang, Djaposten. Obtained from Dr W.S. Lehman.

HTB 3430 Skull. Adult female. Abong Mbang, Djaposten. Obtained from Dr W.S. Lehman.

HTB 3431 Skull. Adult male. Abong Mbang, Djaposten. Obtained from Dr W.S. Lehman.

HTB 3432 Skull. Adult male. Abong Mbang, Djaposten. Obtained from Dr W.S. Lehman.

HTB 3435 Skull. Adult male. Abong Mbang, Djaposten. Obtained from Dr W.S. Lehman.

HTB 3439 Skull. Juvenile female. Abong Mbang, Djaposten. Obtained from Dr W.S. Lehman.

HTB 3440 Skull. Juvenile male. Abong Mbang, Djaposten. Obtained from Dr W.S. Lehman.

HTB 3441 Skull. Adult male. Abong Mbang, Djaposten. Obtained from Dr W.S. Lehman.

HTB 3442 Skull. Adult male. Abong Mbang, Djaposten. Obtained from Dr W.S. Lehman.

HTB 3443 Skull. Juvenile female. Abong Mbang, Djaposten. Obtained from Dr W.S. Lehman.

HTB 3446 Postcranial skeleton. Adult. Abong Mbang, Djaposten. Obtained from Dr W.S. Lehman.

HTB 3447 Skull. Adult female. Abong Mbang, Djaposten. Obtained from Dr W.S. Lehman.

HTB 3449 Skull. Adult male. Abong Mbang, Djaposten. Obtained from Dr W.A. Fritz.

HTB 3474 Postcranial skeleton. Adult male. Abong Mbang, Djaposten. Obtained from Dr W.S. Lehman.

HTB 3542 Skull. Adult female. Hope shipment.

HTB 3543 Skeleton. Adult male? Hope shipment.

HTB 3544 Skeleton. Adult male. Hope shipment.

HTB 3545 Skeleton. Adult male? Hope shipment.

HTB 3546 Skeleton. Adult male. Hope shipment.

HTB 3547 Skeleton. Adult male. Hope shipment.

HTB 3548 Skeleton. Adult male. Hope shipment.

HTB 3549 Skeleton. Adult male. Hope shipment.

HTB 3550 Skeleton. Juvenile male. Hope shipment.

HTB 3554 Skeleton. Infant. Hope shipment.

HTB 3556 Skeleton. Adult male. Second Hope shipment.

HTB 3557 Skeleton. Adult male. Second Hope shipment.

HTB 3558 Skeleton. Adult male. Second Hope shipment.

HTB 3559 Skeleton. Adult male. Second Hope shipment.

HTB 3560 Skeleton. Adult male. Received from Mr G.C. Beanland, Oxford, MS, 4 January 1939.

HTB 3561 Skeleton. Adult male. Received from Mr G.C. Beanland, Oxford, MS, 4 January 1939.

HTB 3562 Skeleton. Adult male. Received from Mr G.C. Beanland, Oxford, MS, 4 January 1939.

Columbia

University of Missouri, Department of Anthropology, 100 Swallow Hall, Columbia, MO 65211

Audrey Gayou; Paige Gordon; Sam D. Stout

Mounted skeleton. Adult male. It was obtained from the University of Toronto in 1969.

Columbus

Ohio State University, Department of Anthropology, Smith Laboratory, 174 West 18th Avenue, Columbus, OH 43210-1106

W. Scott McGraw

Mounted skeleton. Adult female. Donated.

Davis

Gorilla Doctors, U.S. Headquarters, Karen C. Drayer Wildlife Health Center, UC Davis, One Shields Avenue, Davis, CA 95616

Michael R. Cranfield

Gorilla Doctors has a 30-year history of working with gorillas, starting with one veterinarian in Rwanda and rising to 15 veterinarians in Uganda, Rwanda and the Democratic Republic of Congo. Their mission is: To conserve wild mountain and eastern lowland (or Grauer's) gorillas through life-saving veterinary medicine and a One Health approach. This has been successfully accomplished through routine health monitoring, clinical interventions, necropsies and orphan care. Gorilla Doctors is also involved with other wildlife, prioritising primates, but when resources allow responding to requests from the governments to work on other species. Other initiatives include capacity building, research and an employee health program for all conservation workers. Gorilla Doctors has developed secure cold chains and has biological resource centres in Goma (DRC), Kampala (Uganda), Kigali (Rwanda), Baltimore (USA) and Davis (USA). The centres contain mountain gorilla, eastern lowland gorilla, golden monkey and domestic animal samples. There are over 125 sets of wet tissues, blocks and histology slides as well as serum samples, blood clots, buffy coats, RBCs (red blood cells), plasma and blood smears from hundreds of individuals of various primate species. Also included are parasites in various preservatives and hair samples. Although Gorilla Doctors manages the resource, all samples are owned by the governments of the countries of origin and thus access to them is through the Uganda Wildlife Authority (UWA), the Rwanda Development Board (RDB) and the Institut Congolais pour la Conservation de la Nature (ICCN).

See also Mountain Gorilla Veterinary Project (MGVP) referred to frequently in Part I: Gorilla Pathology and Health.

University of California, Department of Anthropology, 1 Shields Avenue, Davis, CA 95616

Christyann M. Darwent; Henry M. McHenry; Timothy D. Weaver

Skeleton. Adult male. It was received from the UC Davis Veterinary School, and came originally from Roeding Park Zoo, 894 West Belmont Avenue, Fresno, CA 93728. This is 'Freddie', who arrived at the zoo on 27 January 1967, aged about 2 years (weight 40 lb.). He was purchased, with a female named 'Nina', from G. van den Brink. They came originally from Cameroon. 'Freddie' died of cardiac failure and atherosclerosis on 24 April 1981. [Some additional information added by GH from correspondence with Roeding Park Zoo personnel]

ANT34 Partial skeleton. Juvenile. It is missing the carpals, metacarpals, hand phalanges, tarsals, metatarsals and foot phalanges.

University of California , Department of Wildlife, Fish and Conservation Biology, Museum of Wildlife and Fish Biology, 1 Shields Avenue, Davis, CA 95616

Irene E. Engilis

WFB-6597z Skeleton, skin (tanned, poor condition), right arm and leg (saved as freeze-dried bone-muscle preparations after dissection). Female. Transferred from the University of California, Department of Evolution and Ecology (formerly Department of Zoology), Storer Hall, Davis, CA 95616 (former catalog no. 6597) in December 1993. It was originally received from Roeding Park Zoo, Fresno, on 16 September 1977. This is 'Betsi', who arrived at the zoo on 18 April 1975, having been purchased for $8500 from Frankfurt Zoo. She had arrived in Frankfurt on 11 April 1961, her place of origin being 80 km from Ebolowa, in the direction of Kribi, Cameroon. She died on 8 September 1977 at the School of Veterinary Medicine, University of California at Davis, within 24 hours of arriving for treatment of chronic coccidioidomycosis, using a new experimental derivative of amphotericin B (weight 195 lb.). [Some additional information added by GH from correspondence with Roeding Park Zoo personnel and Dr Murray E. Fowler, UC Davis.]

De Kalb

Northern Illinois University, Department of Anthropology, De Kalb, IL 60115

Daniel L. Gebo

66-13-1 (NIU 82133) Skull. Subadult male. The lower canines and 3rd molars are not fully erupted.

66-13-2 (NIU 82132) Skull. Adult male. Dentition fully erupted, but with a 4th upper left molar.

68-19-1 (AC 799) Skull. Adult female. The upper and lower 3rd molars are unerupted. There are cut marks on the occipital condyles and left orbit.

The above 3 skulls were purchased from J. Sabater Pí, Río Muni, in 1968.

NIU 084396 Mounted skeleton. Adult male. Purchased from Ward's Scientific, Rochester, NY, in 1967. The dentition is fully erupted and all epiphyses are fused. The right femur is replaced by a plastic cast.

NIU 093095 Mounted skeleton. Male. Purchased from Paris, France, in 1967. The dentition is fully erupted but many epiphyses are unfused (proximal humerus, distal radius and ulna, greater trochanter, distal femur, distal fibula and tibia and proximal tibia).

Denver

Denver Museum of Nature and Science, 2001 Colorado Boulevard, Denver, CO 80205-5798

John R. Demboski; Logan D. Ivy; Cheri A. Jones; Carron Meaney; Paisley A. Seyfarth

NB: The prefix 'ZM' refers to the Denver Museum of Nature and Science Mammal Collection. The institutional prefix 'DMNS' is not used for reporting catalog numbers in publications, because it does not differentiate between other collections in the museum with the same catalog number.

ZM.6456 Skeleton. Juvenile female. Acquired on 3 January 1977 (accession no. T-109). It is a zoo specimen.

ZM.6866 Skeleton and skin. Adult male. Donated by the Zoological Society of San Diego on 26 October 1978 (accession no. 1978-167). This is 'Albert', who arrived at the San Diego Zoo on 10 August 1949, aged about 4−5 months (weight 6 lb.). He came originally from Cameroun, and was purchased with 2 females ('Bata' and 'Bouba') from Henry Trefflich [information added by GH from correspondence with San Diego Zoo personnel]. 'Albert' died of chronic heart failure on 18 October 1978. The following observations were made by Sue Ware, research associate at the museum: 'There is much evidence of osteoarthritis and rheumatoid arthritis in the skeleton. There are varying amounts of lipping and fusion and degenerative resorption of bone. The parts affected are as follows: (i) Vertebral column. The cervical and thoracic vertebrae show extreme lipping and evidence of bone rubbing on bone. The spinous processes are disfigured and both anterior and posterior surfaces are affected. The lumbar vertebrae show severe lipping with total fusion of the last 3 segments. There is lipping on the sacrum also, but not as severe. (ii) Pelves. There is lipping in moderation on the iliac ridge, and moderate lipping and bony exostoses on the ischium. There is lipping on the outer surface ridge of the left acetabulum. There is total closure of, and massive deposits on the acetabulum on the right side. Articulation of femur with acetabulum is dysfunctional at this locale. (iii) Skull. There is an over-abundance of lipping on the sagittal crest, and moderate lipping on top of the orbital ridge. (iv) Upper limbs. The proximal ends of the humeri are affected with lipping and moderate amounts in the epiphyseal areas. The distal ends show some lipping, but not a severe amount. The proximal and distal ends of the ulnae and radii exhibit arthritis, but with lipping to a lesser degree than the humeri. (v) Hands and feet. All phalanges exhibit lipping at every joint, both anterior and posterior. All carpals have some lipping. The calcaneus has lipping but it is not severe. The metacarpals have some lipping but not as much as seen in other areas.'

ZM.7812 Skeleton. Adult male. Cameroon. Received from Denver Zoological Gardens, City Park, Denver, CO 80205, on 1 October 1986 (accession no. 1986-79). This is 'Kisoro', who arrived at Lincoln Park Zoo, Chicago, on 2 June 1964 (weight 34 lb.), having been purchased from F.J. Zeehandelaar. He was sent on breeding loan to Howletts Zoo Park, Bekesbourne, UK, where he arrived on 18 October 1973 (weight 314 lb.). He was sent on loan to Denver Zoo and arrived there on 5 October 1983 (weight 298½ lb.). He died on 24 September 1986 (weight 335½ lb.) of a dissecting aneurysm causing a cardiac tamponade leading to heart failure. [Additional information added by GH from correspondence with Lincoln Park Zoo, Howletts Zoo, and Denver Zoo personnel.]

East Lansing

The Michigan State University Museum, 409 West Circle Drive, East Lansing, MI 48824-1045

Laura Abraczinskas; A. Christopher Carmichael

MR.33606 Skeleton. Adult male. Donated by Detroit Zoological Park, 8450W. Ten Mile Road, Royal Oak, MI 48068-0039, on 1 August 1983. This is 'Jim-Jim', who arrived at the Detroit Zoo on 6 May 1955 (weight 18 lb.). He was captured in the French Congo and purchased from Trefflich's Pet Dept Store, Inc., 228 Fulton Street, New York, NY 10007. He died on 3 July 1983 of fibrotic cardiomyopathy of long duration (weight 406 lb.). [Additional information added by GH from correspondence with Detroit Zoo personnel.]

Fayetteville

University of Arkansas, The University Museum, Museum Building Room 2, Fayetteville, AR 72701*

Nancy Glover McCartney

00-2-80 Skeleton. Adult male. It is thought to have been acquired by Prof. Samuel Claudius Dellinger

(1892–1973), a former curator of the museum, from Ward's Scientific Supplies in the 1940s.

Fort Worth

Fort Worth Museum of Science and History, 1600 Gendy Street, Fort Worth, TX 76107

Lacie Ballinger; Ms Wesley Hathaway; William J. Voss

FWMSH 1982.S.0001 Skin. Adult female. Donated in 1982 by the Department of Biology, University of Texas at Arlington. This specimen is 'Shamba', who arrived at Dallas Zoo on 22 May 1965, aged about 8 years (weight 142 lb.). She was purchased for $4200 from Africa USA, Inc., Saugus, CA. She died on 13 September 1982 (weight 285 lb.) of acute heart failure, about 13 hours after giving birth to her 7th offspring, a 5 lb. 10 oz. male named 'Kanda' (sire 'Fubo'). Death occurred during anaesthesia administered for the removal of her infant. It appears that 'Shamba' suffered from Addison's disease, a disorder due to chronic insufficiency of the adrenocortical gland, which leaves an individual incapable of handling stressful situations. Atrophy of the gland is usually due to autoimmunity and occasionally to destructive processes such as tuberculosis. She was originally donated to the University by Dallas Zoo. The skin is rolled and too dry (stiff) to unroll. [Additional information added by GH from correspondence with Dallas Zoo personnel and Allan Clay Clark, John L. Darling, and Joseph R. Mendelson III, Dept of Biology, University of Texas, Arlington.]

FWMSH 2006.S.0001 Skull. Adult female. Received in 2006 from the Department of Biology, University of Texas at Arlington. It is from the same individual as FWMSH 1982.S.0001.

Gainesville

University of Florida, Department of Natural Sciences, Florida Museum of Natural History, Dickinson Hall, Museum Road, Gainesville, FL 32611-7800

Verity L. Mathis; Candace McCaffery; Laurie Wilkins

UF 11926 Skull. Adult male. Received from the University of Vermont teaching collection. Catalogued on 6 February 1980. Previous collection number 28 is written on it.

UF 32692 Postcranial skeleton, skin and frozen tissue sample. Adult male. Donated by the Jacksonville Zoo, Florida; catalogued on 22 April 2014. This is 'Quito', who was born in the Walter D. Stone Memorial Zoo, Stoneham, MA, on 1 June 1981 to dam 'Gigi' and sire 'Sam'. He was hand-reared (weight 6 lb. 12 oz.). He was sent on loan to Jacksonville Zoo on 11 March 1998, and died there on 26 January 2013 (weight 204 kg).

Greenville

Delaware Museum of Natural History, 4840 Kennett Pike, Greenville, DE 19807

Jean L. Woods

MAM0041 Mounted skin. Subadult male? Donated by the estate of R.R.M. Carpenter, Wilmington, Delaware, in 1973. It is one of a great many trophies obtained by the Carpenters virtually worldwide from the 1920s to the early 1950s.

Honolulu

Bernice Pauahi Bishop Museum, Department of Zoology, 1525 Bernice Street, Honolulu, HI 96817-0916

Carla H. Kishinami; Gail Wine

BBM-X 148214 Skeleton. Adult male. Received from Honolulu Zoo, 151 Kapahula Avenue, Queen Kapiolani Park, Honolulu. This is 'Cameroun', who arrived at the Honolulu Zoo on 25 February 1961 (weight 23 lb.). He had been captured in October 1960 near Djaposten, Haut-Nyong Department, Cameroun, and was purchased from Dr Deets Pickett, Trans World Animals Inc. 'Cameroun' died on 26 December 1975 (weight 370 lb.) of acute haemorrhagic enteritis of the entire gastrointestinal tract, but concentrated near the caecum. [Additional information added by GH from correspondence with Honolulu Zoo personnel.]

Houston

Houston Museum of Natural Science, 1 Hermann Circle Drive, Houston, TX 77030-1799

Daniel M. Brooks; Donna Meadows; T.E. Pulley

HMNS.VM.5 Skull. Adult female. Gabon. Donated by Paul Geiger Jr on 5 June 1986. It was given to the donor's father, Paul L. Geiger Sr, while he was working in Gabon for Bethlehem Steel in the 1970s.

HMNS.VM.198 Mounted skin. Adult male. Congo River basin, Gabon. Collected by Charles B. Greer Jr in 1951 and donated by him in 1952. It was shot with a 375 H and H Magnum and a 12-gauge shotgun loaded with buckshot. Height 63 in.; arm span 90 in.; neck 23½ in.; chest 70 in.; wrist 11 in.

HMNS.VM.253 Mounted skin. Adult female. Collected by Charles B. Greer Jr, in 1951 and donated by him in 1952.

The above 2 specimens were mounted by Louis Paul Jonas, Churchtown-Hollowville Road, Hudson, NY. They have been on display in the Frensley-Graham Hall of African Wildlife since May 2003.

Iowa City

University of Iowa, Museum of Natural History, 10 Macbride Hall, Iowa City, IA 52242-1322

Julia Golden; Cindy Opitz; George D. Schrimper

SUI 19836 Mounted skeleton. Juvenile male? Received from the Kny-Scheerer Company, New York, on 5 June 1905. It now appears to be missing.

SUI 27526 Skin and mounted skeleton. Juvenile female. Cameroun. Received from Dr Frank Ratcliff Senska (1875−1970) on 10 October 1929. Dr Senska was a Presbyterian missionary at Sakbayémé from 1922 to 1929. His specimens were collected by local people (the Bassa).

SUI 30564 Mounted skin. Subadult male. Cameroun. Donated by Dr Senska in 1929. It was taken off display and destroyed c.1988. The skull was removed from the mount (see SUI 30565 below).

SUI 30565 Skeleton. Subadult male. From the same individual as the above mount; the skull is on display and the postcranial skeleton is in storage.

SUI 30566 Skeleton. Adult male. Cameroun. Collected by Dr Senska in 1928; donated in 1929 (but apparently not received at the museum until 6 October 1930). The skeleton has two first ribs on the right side. Skull: length prosthion−opisthocranion 297 mm; bizygomatic width 184 mm; vault length glabella−opisthocranion 192 mm; greatest height of sagittal crest 37 mm; mandibular length 195 mm; mandibular height 114 mm; upper facial height nasion−intradentale superius 129 mm; interorbital width 42.1 mm; maximum inside biorbital width 114 mm; mandibular width between most lateral points of condyles 149 mm.

Ithaca

Cornell University, Department of Anthropology, Biological Anthropology Laboratory, B-65 McGraw Hall, Ithaca, NY 14853

Frederic Wright Gleach

9120 Cranium. Adult female.

9123 Cranium. Adult male.

Cornell University, Department of Ecology and Evolutionary Biology, General Biology Laboratory, Stimson Hall, Ithaca, NY 14853

Charles M. Dardia; Frederic Wright Gleach; Kenneth A. R. Kennedy

CUMV:Mamm:149 Mounted skeleton. Adult male. It may have been obtained from Ward's Natural Science Establishment, Rochester, NY, at the end of the 19th century.

Cornell University, Museum of Vertebrates, Johnson Center for Birds and Biodiversity, 159 Sapsucker Woods Road, Ithaca, NY 14850

Charles M. Dardia; Kenneth A.R. Kennedy

Skeleton. Male. Obtained from the Willard State Hospital, a mental institution in Ovid, NY, in 1968. It suffered a lesion of the mandible which may have been due to rifle fire.

Skull. Female. Obtained with the above skeleton.

Cerebellum and right hemisphere of brain (in alcohol). Infant. This is all that remains of the first live gorilla to enter the United States. It was previously in the Wilder Brain Collection, the bulk of which is held by the Department of Psychology at Cornell University. The animal was transported from Liverpool to Boston on the steamer 'ss Pavonia', which reached port on 2 May 1897. It was about 1 year old, weighed 14½ lb. and died within 5 days of arrival. The body was purchased by Cornell University and the brain was removed by Prof. Burt Green Wilder, who found it to have a volume of 322 cc. The skin was saved and prepared by a taxidermist to form a stuffed specimen, but its present whereabouts is unknown.

Knoxville

University of Tennessee, College of Arts and Sciences, Department of Anthropology, Paleoanthropology Laboratory, South Stadium Hall, Knoxville, TN 37996

Andrew Kramer; Paul W. Parmalee

NB: The Anthropology Department is slated to move into a new building, Strong Hall, by early 2017.

Skull. Adult male.

Laramie

University of Wyoming, Department of Anthropology, 12th and Lewis Street, 1000 E. University Avenue, Laramie, WY 82071

James C. Ahern; George W. Gill

4903 Skull. Adult male. There is some destruction by large calibre bullet to the left zygomatic and partial loss of the right occipital condyle from cutting/decapitation.

Lawrence

University of Kansas, Biodiversity Institute and Natural History Museum, 1345 Jayhawk Boulevard, Lawrence, KS 66045

Maria Eifler; Sandra Johnson

KU 8476 Skeleton. Adult male. Received as part of a large accession from the Kansas City Museum. It probably came from the Kansas City Zoo. Foot length 306 mm; ear length 50 mm.

Lexington

University of Kentucky, College of Arts and Sciences, William S. Webb Museum of Anthropology, 211 Lafferty Hall, Lexington, KY 40506-0024

James P. Fenton; George R. Milner; Mary Lucas Powell; Sissel Schroeder

84-56 Skull. Adult female. Cameroun. Killed by local people in February 1945. Purchased for $50 from the Revd Norman A. Horner, American Presbyterian Mission, Lolodorf. It appears to be missing.

84-57 Skull. Adult male. Cameroun. Killed by local people in October 1945. Purchased for $50 from the Revd Norman A. Horner, American Presbyterian Mission, Lolodorf. It appears to be missing.

84-59 (48-6-3) Postcranial skeleton. Adult female. Cameroun. Killed by local people in 1943. Purchased for $50 in 1948 from the Revd Norman A. Horner, American Presbyterian Mission, Lolodorf.

84-68 (48-6-1B) Skull. Adult male. Cameroun. Purchased for $50 in 1948 from the Revd Norman A. Horner, American Presbyterian Mission, Lolodorf.

84-69 Skeleton. Subadult male. Bipindi, Cameroun. Collected in 1946.

48-6-5 Skeleton. Adult female. Cameroun. Killed by local people in June 1945. Purchased from the Revd Norman A. Horner in 1948. There is a bullet hole in the skull. This skeleton appears to be missing.

University of Kentucky, College of Medicine, Department of Anatomy and Neurobiology, Lexington, KY 40536-0098*

Andrew S. Deane

Carcase (embalmed). Adult male. This specimen is being used in the pilot phase of a project aimed at creating a photographic atlas of gorilla anatomy, which will take full advantage of modern digital imaging modalities, including CT and MRI. Although most of the soft tissue will eventually be discarded, there are plans to preserve (either in formalin or as plastinated specimens) a number of structures (i.e., brain, heart, lung, maybe the liver). There are also hair samples from the individual that were harvested prior to the embalming process. The skeletal materials from the animal will remain at the University of Kentucky. This is 'Timmy', who arrived at Memphis Zoo and Aquarium, 2000 Galloway Avenue, Memphis, TN 38112, on 16 May 1960 (weight 23 lb.). He came originally from Cameroon, and was purchased from Alton Freeman, Spruce Pine, NC. He was sold to Cleveland Metroparks Zoo, 3900 Brookside Park Drive, Cleveland, OH 44109, and arrived there on 14 December 1966 (weight 220 lb.). He was sent on breeding loan to the New York Zoological Park (Bronx Zoo) in 1991. He arrived on loan at the Louisville Zoo on 22 April 2004, and was euthanased there on 2 August 2011, owing to a number of chronic medical problems, including heart disease (cardiomyopathy), heart arrhythmia (atrial fibrillation) and osteoarthritis. He was reputedly the oldest male gorilla in captivity (estimated age 53 years) and had sired 13 offspring. His last recorded weight was 147.9 kg on 31 July 2011. [Some additional information added by GH from correspondence with Memphis Zoo and Cleveland Zoo personnel.]

Skeletal elements from the upper limb, vertebral column, pelvic girdle and head; fixed lower limb (mid-thigh down). Adult male. This is 'Frank', who arrived at the Lincoln Park Zoological Gardens, Chicago, on 17 May 1966, having been donated by Mr Franklin Schmick, Chicago. He was sold to the Louisville Zoo and arrived there on 20 March 2002. He was euthanased on 14 August 2008 (estimated age 44 years). His last recorded weight was 131.5 kg on 15 July 2008. He had sired 15 offspring. Once the lower limb has been dissected and documented, it will be skeletonised and the material will remain at the University of Kentucky. [Some additional information added by GH from correspondence with Lincoln Park Zoo and Louisville Zoo personnel.]

Lincoln

University of Nebraska State Museum, Systematics Research Collections, Zoology Division, W504 Nebraska Hall, 900 N. 16th Street, Lincoln, NE 68588-0514

Thomas E. Labedz

UNSM ZM-14549 Skeleton and skin. Adult female. Cameroun. This is 'Baluba', who died in the Henry Doorly Zoo, Omaha, NE, on 2 May 1969, when aged about 9 years. She had been received from International

Animal Exchange, Inc., Ferndale, MI. The cause of death was a Metazoan infection of the digestive system. [Additional information added by GH from correspondence with Omaha Zoo personnel.]

UNSM ZM-30372 Skull. Adult male. Donated by the Henry Doorly Zoo, Omaha. This is 'Gerry', who was born in San Diego Zoo on 20 May 1993 to dam 'Kubwa Kiwa' and sire 'Memba'. He was transferred to Omaha Zoo on 13 October 2004, and died there on 26 September 2011. [Additional information added by GH from the International Studbook for the Western Lowland Gorilla, 2010.]

Lisle

Jurica-Suchy Nature Museum, Benedictine University, Department of Biology, 5700 College Road, Birck Hall of Science, Room 222, Lisle, IL 60532

Fr Theodore D. Suchy; Karly E. Tumminello

Mounted skeleton.

72805 Skeleton. Adult male. Obtained by Fr Hilary Jurica from the Field Museum of Natural History, Chicago. This is the captive specimen 'Bushman', from the Lincoln Park Zoo, Chicago. Fr Hilary Jurica (1892–1970) served on the board of the Field Museum. The College museum represents the combined efforts of Dr Hilary Jurica and his brother Dr Edmund Jurica, who collected specimens for their students to study. There appears to be only one reference pertaining to the skeleton: Dwight Davis (1951).

Los Angeles

Natural History Museum of Los Angeles County, 900 Exposition Boulevard, Los Angeles, CA 90007

Jim Dines; Sarah B. George

LACM 30561 Skull. Adult female. Collected by J.S. Edwards.

LACM 30566 Skeleton. Male. Collected by Dr Silas Johnson in 1929. It is articulated and hanging on a stand.

LACM 54567 Skull. Adult male. Received from the San Diego Zoo on 3 October 1978. This is 'Massa', who arrived at San Diego Zoo on 28 July 1968, aged about 3 years (weight 45 lb.). He was collected at Makokou, approximately 200 mi. northeast of Lambaréné, Gabon (very close to the Congo Brazzaville border), by Craig Kinzelman of Cleveland, Ohio, Peace Corps Volunteer. The gorilla's parents were travelling as a pair, and were killed by shotgun on capture. 'Massa' was donated to the San Diego Zoo by Mrs Rhena Eckert-Schweitzer, Lambaréné. He died of coccidioidomycosis on 9

September 1978. [Additional information added by GH from correspondence with San Diego Zoo personnel.]

University of California, Department of Biology, Los Angeles, CA 90024

UCLA 1788 Skull. Adult male. Purchased from Ward's Biological Supply Company in June 1940.

UCLA 2367 Cranium. Adult female. Acquired from the Frank C. Clark collection.

Louisville

Louisville Zoo, 1100 Trevilian Way, Louisville, KY 40233

Roy B. Burns

The listed materials are in the cryopreservation bank.

Female (estimated date of birth 2 June 1958). Collected 20 March 2002: serum, clear. 29 April 2003: 4 × serum, clear; red blood cells (heparin, sodium); 2 × plasma (heparin, sodium), clear; 2 × urine. 16 May 2003: urine. 3 November 2004: urine. 19 January 2005: urine. 14 April 2005: urine. 27 May 2005: urine. 21 June 2005: urine. 20 July 2005: urine. 16 September 2005: urine. 5 October 2005: urine. 14 December 2005: urine. 22 March 2006: urine. 19 May 2006: 2 × urine; 2 × serum, clear. 10 December 2006: urine. 20 March 2007: urine. 4 May 2007: urine. 7 June 2007: urine. 28 August 2007: urine. 10 March 2008: urine. 9 April 2008: urine. 24 September 2008: 2 × serum, haemolysis; urine; serum, slight haemolysis; serum, haemolysis. 24 November 2008: urine. 18 April 2009: 3 × serum, clear; 2 × urine.

Male (est. d.o.b. 16 May 1964). Coll. 6 November 2002: 5 × serum, icteric; 2 × urine. 28 June 2003: 3 × serum, haemolysis; urine; 2 × serum, icteric; plasma (heparin, lithium), icteric. 5 April 2004: urine. 17 June 2004: 2 × plasma (heparin, sodium), icteric; 3 × serum, icteric; 2 × urine. 2 July 2004: urine. 13 September 2004: 2 × urine. 7 January 2005: urine. 18 January 2005: urine. 8 February 2005: urine. 8 March 2005: urine. 1 April 2005: urine. 10 June 2005: urine. 8 July 2005: 2 × urine. 15 August 2005: urine. 20 September 2005: urine. 5 October 2005: urine. 18 November 2005: urine. 22 March 2006: 2 × serum, icteric; urine. 11 August 2006: 2 × urine. 18 October 2006: 2 × urine. 14 November 2006: urine. 17 January 2007: urine. 12 February 2007: urine. 2 March 2007: urine. 18 April 2007: urine. 19 July 2007: urine. 15 November 2007: 9 × serum, icteric; 3 × serum (defrosted samples), icteric. 15 June 2008: sciatic nerve. 13 August 2008: other samples (tracheal wash); serum, icteric; trace element serum; 2 × plasma (heparin, sodium), icteric; 15 × serum, icteric. 15 August 2008: pancreas tissue; perineal fat tissue; 2 × bone marrow tissue; 3 × liver tissue; stomach tissue; 4 × kidney tissue; 2 × lung tissue; 4 × spleen tissue.

Female (est. d.o.b. 11 March 1966). Coll. 20 March 2002: serum, clear. 23 August 2003: 5 × serum, clear; urine; plasma (heparin, sodium), clear; plasma (EDTA, sodium), clear. 25 August 2003: 3 × serum, lipaemia/ icteric; 2 × plasma (heparin, sodium), lipaemia. 25 August 2003: plasma (EDTA, sodium), lipaemia. 26 August 2003: urine; 3 × serum, clear. 27 August 2003: 3 × serum, clear. 28 August 2003: 2 × serum, clear; 7 × serum, clear; urine. 30 August 2003: 6 × serum, haemolysis; large intestine tissue; thoracic fluid, left; spleen tissue; 2 × skeletal muscle tissue; sciatic nerve tissue; 5 × heart tissue. 19 September 2003: skin.

Female (d.o.b. 20 June 1996). Coll. 20 March 2002: serum, clear. 25 February 2004: urine; serum, slight haemolysis; 5 × serum, clear.

Female (d.o.b. 3 October 1996). Serum, clear. Coll. 4 February 2004: 2 × serum, slight haemolysis; 11 × serum, clear; urine.

Female (d.o.b. 7 December 1997). Coll. 20 March 2002: serum, clear. 4 February 2004: urine; 6 × serum, clear.

Male (d.o.b. 6 January 1997). Coll. 10 May 2002: 2 × serum, icteric. 15 December 2004: 3 × urine; 5 × serum, clear. 19 April 2005: 7 × serum, clear; 2 × urine. 24 April 2005: urine. 1 March 2006: urine. 1 November 2011: 12 × serum, icteric; blood, whole (EDTA); 2 × plasma (heparin), icteric; 2 × plasma (EDTA), icteric; red blood cells (heparin). 9 November 2011: serum, clear; 6 × serum, clear; serum, haemolysis.

Male (d.o.b. 1 June 1998). Coll. 10 May 2002: 2 × serum, icteric. 15 November 2003: 2 × serum, icteric; 2 × urine. 22 March 2006: 2 × urine. 1 September 2010: 9 × serum, clear. 5 September 2010: 7 × serum, clear. 26 January 2012: 13 × serum, icteric; plasma (heparin), icteric; 2 × plasma (EDTA), icteric; 2 × red blood cells (EDTA). 1 May 2013: urine; 2 × plasma (heparin), icteric; 7 × serum, icteric.

Male (d.o.b. 30 April 1980). Coll. 25 September 2002: 3 × serum, clear. 30 August 2003: serum, clear; 89 × plasma, clear. 12 February 2004: 10 × serum, clear; 3 × plasma (heparin, sodium), clear; 2 × urine; tracheal wash, 2 × red blood cells.

Female (d.o.b. 20 September 1990). Coll. 25 September 2002: 3 × serum, clear. 15 November 2003: 4 × serum, icteric; 2 × urine. 2 April 2004: urine.

Female (d.o.b. 12 July 1987). Coll. 25 September 2002: 3 × serum, clear. 22 January 2003: urine. 3 June 2003: urine. 9 July 2003: urine. 8 August 2003: urine. 27 August 2003: urine. 4 September 2003: urine. 17 September 2003: 2 × urine. 30 September 2003: 2 × urine.

Female (d.o.b. 30 April 1992). Coll. 25 September 2002: 3 × serum, clear. 19 October 2003: urine. 25

February 2004: 2 × red blood cells; 2 × plasma (heparin, sodium), clear; serum, slight haemolysis; 3 × serum, clear; urine.

Male (d.o.b. 25 March 1997). Coll. 4 March 2004: serum, icteric; plasma (heparin, lithium), icteric; red blood cells. 15 February 2006: 5 × serum, clear; 2 × plasma (heparin, sodium), clear. 7 January 2010: 7 × serum, icteric; 2 × urine; plasma (EDTA, sodium), icteric. 3 February 2011: plasma (heparin), icteric; plasma (EDTA), icteric; blood, whole (EDTA); 2 × serum, icteric and haemolysis; 8 × serum, icteric.

Male (d.o.b. 3 November 1998). Urine. Coll. 4 March 2004: 2 × serum, icteric; plasma (heparin, lithium), icteric; red blood cells. 25 January 2006: 4 × serum, clear; 2 × urine. 24 February 2010: 8 × serum, icteric; plasma (heparin, sodium), icteric. 12 August 2010: 4 × urine; 8 × serum, clear; 3 × blood, whole (heparin); 2 × plasma (EDTA); plasma (heparin); blood, whole (EDTA). 21 October 2010: 8 × serum, icteric. 2 April 2011: 2 × urine; 5 × serum, clear; 8 × serum, slight haemolysis; plasma (heparin), clear; red blood cells (heparin). 3 April 2012: serum, haemolysis; 10 × serum, icteric; plasma (heparin, sodium), icteric; plasma (heparin, sodium), icteric; plasma (EDTA, sodium), icteric; red blood cells.

Male (est. d.o.b. 15 January 1959). Coll. 10 June 2004: 2 × urine; 5 × serum, icteric; 2 × urine. 31 August 2004: urine. 8 December 2004: urine. 4 February 2005: urine. 16 March 2005: urine. 3 May 2005: urine. 7 June 2005: 2 × urine. 6 July 2005: 2 × urine. 1 August 2005: urine. 9 September 2005: urine. 14 October 2005: urine. 7 December 2005: urine. 12 January 2006: urine. 8 February 2006: urine. 20 April 2006: urine. 26 April 2006: 2 × plasma (heparin, sodium), haemolysis; 5 × serum, icteric; blood, whole; urine. 23 June 2006: urine. 1 August 2006: urine. 16 October 2006: 2 × urine. 3 November 2006: urine. 7 December 2006: 2 × urine. 8 February 2007: urine. 6 March 2007: urine. 4 May 2007: urine. 13 June 2007: urine. 5 July 2007: urine. 2 September 2007: urine. 13 December 2007: 4 × serum, icteric; serum; urine; serum, icteric (defrosted sample). 9 December 2010: blood, whole (EDTA); urine; 4 × serum, icteric and slight haemolysis. 24 July 2011: 13 × serum, clear; blood, whole (EDTA, sodium); plasma (heparin), clear. 31 July 2011: plasma (heparin), icteric; blood, whole (EDTA); red blood cells (heparin); 8 × serum, icteric.

Female (est. d.o.b. 15 June 1963). Coll. 10 June 2004: 2 × urine; 4 × serum, icteric. 25 July 2004: urine. 13 September 2004: urine. 10 November 2004: urine. 7 December 2004: urine. 4 January 2005: urine. 1 February 2005: urine. 10 March 2005: 2 × urine. 1 April 2005: urine. 3 May 2005: urine. 3 August 2005: urine. 16 September 2005: urine. 4 January 2006: urine. 10 March 2006: urine. 13 August 2006: urine. 24 August 2006: serum, clear. 6 September 2006: 2 × urine. 10 October

2006: 2 × urine. 22 November 2006: urine. 5 December 2006: 2 × urine. 10 January 2007: urine. 9 February 2007: urine. 2 March 2007: urine. 27 April 2007: urine. 4 May 2007: urine. 1 June 2007: urine. 12 July 2007: urine. 3 August 2007: 2 × urine. 2 September 2007: urine. 17 January 2008: urine.

Female (d.o.b. 26 May 1989). Coll. 8 June 2004: urine. 23 June 2004: 2 × plasma (heparin, sodium), clear; 4 × serum, clear; 2 × urine. 12 October 2004: urine. 12 May 2005: urine. 17 December 2007: serum, clear; urine. 22 March 2008: 5 × serum, icteric; serum, icteric. 17 December 2008: serum, icteric. 22 February 2011: serum, icteric; light sensitive samples: blood, whole (heparin); blood, whole (EDTA); 6 × serum, icteric; 2 × serum, icteric and slight haemolysis.

Female (d.o.b. 18 March 1989). Coll. 24 October 2005: 2 × urine; serum, clear. 15 March 2006: urine. 19 May 2006: urine. 24 April 2007: urine. 1 May 2007: 2 × urine. 1 June 2007: urine. 20 July 2007: urine. 18 September 2007: urine. 28 November 2007: urine (defrosted sample); 2 × serum, clear. 13 April 2010: urine. 14 April 2010: serum, icteric; milk; 2 × milk. 16 March 2011: plasma (heparin, lithium), clear; 2 × urine; blood, whole (EDTA, sodium); serum, icteric; light sensitive: 3 × serum, icteric; red blood cells (heparin).

Male (d.o.b. 17 October 1987). Coll. 1 December 2005: serum, clear; 2 × urine. 6 November 2008: urine; serum, slight haemolysis; 10 × serum, icteric. 16 November 2011: serum, clear; thawed and refrozen: 8 × serum, clear; serum, clear (defrosted sample); 2 × plasma (EDTA), clear; plasma (EDTA).

Female (d.o.b. 10 July 1970). Coll. 12 January 2007: plasma (heparin, sodium), slight haemolysis; serum, clear; blood, whole (heparin, sodium). 14 May 2009: urine. 15 July 2009: urine. 17 February 2011: 2 × urine; 12 × serum, icteric; blood, whole (EDTA); blood, whole (heparin). 4 December 2013: 11 × serum, icteric; 2 × plasma (heparin), icteric; plasma (EDTA), icteric.

Female (d.o.b. 1 November 1983). Coll. 2 May 2008: 9 × serum, icteric. 28 March 2012: plasma (heparin, lithium), clear; plasma (EDTA), clear; blood, whole (EDTA); 13 × serum, clear.

Female (d.o.b. 6 February 2010). Coll. 1 April 2010: bone, skin tissue.

Lubbock

Texas Tech University, The Museum, 3301 4th Street, Lubbock, TX 79415

Heath J. Garner; Richard Monk; Robert D. Owen

TTU-M-42641 Skeleton. Adult male. Acquired from Fort Worth Zoological Park, 2727 Zoological Park Drive, Forest Park, Fort Worth, TX 76110, on 7 November 1985. This is 'Mike', who arrived at the Fort Worth Zoo on 5 May 1956, aged about 8 months (weight 11 lb.). He was purchased for $4000 from Fred J. Zeehandelaar, with funds donated by the North Atlantic Fertilizer and Chemical Co. Inc. He died on 6 November 1985 (weight 368 lb.) [foregoing additional information added by GH from correspondence with Fort Worth Zoo personnel]. Skull length 332.3 mm (including incisors); 318.7 mm (not including incisors); zygomatic width 180.4 mm; humerus 391.5 mm (minor arthritic lipping on head); femur 355 mm (lipping on head); radius 357.8 mm (some lipping at proximal end); tibia 306.7 mm; ulna 334.4 mm; fibula 267.7 mm; scapula 252.9 mm (not including coracoid process); calcaneum 97.2 mm (some out-growths). There is some lipping in the patellae, podials and proximal ends of the metapodials. There is minor lipping in the vertebrae, and considerable lipping in the acetabulum fossae. There is minor fusion of the tibia and fibula on one side.

TTU-M-42641 Frozen tissues of the above animal: portions of liver, heart, kidney, and skeletal muscle, maintained in an Ultracold freezer at about −80°C, and available in small amounts for research purposes (such as protein or DNA work). There are several 2 g tubes and a large block (of c. 4 oz.) of each tissue (heart, kidney and liver). The samples are available for loan to appropriate, qualified researchers, attached to an approved institution. Tissue sample field number: TK 26847.

Madison

University of Wisconsin, Zoological Museum, Noland Hall, 250 North Mills Street, Madison, WI 53706

Mary K. Jones; Paula J. Iwen Landers; Holly McEntee; Laura A. Halverson Monahan

UWZM S. 21532 Skull. Subadult male. *Gorilla beringei beringei.* Two kilometres north of Lubango, near Lubero, Kivu Province, Belgian Congo. Killed by local people on 9 May 1959.

UWZM S. 21533 Mandible. Adult male. *Gorilla beringei beringei.* Matembe, near Lubero, Kivu Province, Belgian Congo. Killed by local people in May 1959.

UWZM S. 21534 Cranium. Subadult female. *Gorilla beringei beringei.* Matembe, near Lubero, Kivu Province, Belgian Congo. Killed by local people in February 1959.

UWZM S. 21535 Cranium. Juvenile. *Gorilla beringei beringei.* Kasinga, near Lubero, Kivu Province, Belgian Congo. Killed by local people in April 1959.

UWZM S. 21536 Skull. Juvenile. *Gorilla beringei beringei.* Kasinga, near Lubero, Kivu Province, Belgian Congo. Killed by local people in May 1959.

UWZM S. 21537 Cranium. Subadult male. *Gorilla beringei beringei.* Kasinga, near Lubero, Kivu Province, Belgian Congo. Killed by local people.

UWZM S. 21538 Cranium. Juvenile. *Gorilla beringei beringei.* Kasinga, near Lubero, Kivu Province, Belgian Congo. Killed by local people in November 1958.

UWZM S. 21539 Cranium. Subadult male. *Gorilla beringei beringei.* Kasinga, near Lubero, Kivu Province, Belgian Congo. Killed by local people.

UWZM S. 21540 Cranium. Juvenile. *Gorilla beringei beringei.* Lumboku River, near Lubero, Kivu Province, Belgian Congo. Killed by local people.

UWZM S. 21541 Cranium. Subadult female. *Gorilla beringei beringei.* Lumboku River, near Lubero, Kivu Province, Belgian Congo. Killed by local people.

UWZM S. 21542 Cranium. Subadult female. *Gorilla beringei beringei.* Lumboku River, near Lubero, Kivu Province, Belgian Congo. Killed by local people.

UWZM S. 21543 Cranium. Subadult female. *Gorilla beringei beringei.* Lumboku River, near Lubero, Kivu Province, Belgian Congo. Killed by local people.

UWZM S. 21544 Cranium. Juvenile. *Gorilla beringei beringei.* Lumboku River, near Lubero, Kivu Province, Belgian Congo. Killed by local people.

UWZM S. 21545 Cranium. Subadult male. *Gorilla beringei beringei.* Lumboku River, near Lubero, Kivu Province, Belgian Congo. Killed by local people.

The above 14 specimens, UWZM S. 21532–UWZM S. 21545 inclusive, were obtained by Prof. John T. Emlen Jr, 23–30 May 1959.

UWZMS S. 21546 Cranium. Juvenile. *Gorilla beringei beringei.* Miya, near Kasese, Kivu Province, Belgian Congo. Killed by local people. Obtained by George Beals Schaller on 30 August 1960.

UWZM S. 21547 Skull. Juvenile male? *Gorilla beringei beringei.* Miya, near Kasese, Kivu Province, Belgian Congo. Killed by a European. Obtained by George B. Schaller on 30 August 1960.

UWZM S. 21548 Skull (no left mandible). Subadult female. *Gorilla beringei beringei.* Miya, near Kasese, Kivu Province, Belgian Congo. Killed by local people. Obtained by George B. Schaller on 30 August 1960.

UWZM S. 21549 Skeleton. Infant male. *Gorilla beringei beringei.* Kabara, Virunga volcanoes, Kivu Province, Belgian Congo. It was born on 25 April 1960, died 2 days later, and and carried by its mother until 1 May 1960, on which date it was collected by George B. Schaller.

UWZM S. 21550 Skeleton. Subadult male. *Gorilla beringei beringei.* Kabara, Virunga volcanoes, Kivu Province, Belgian Congo. Found in the forest. Obtained by George B. Schaller on 11 August 1960.

UWZM S. 22290 Cranium. Adult. Purchased from Ward's Natural Science Establishment by Edward A. Birge in 1888.

UWZM M. 27466 Skin and skeleton. Adult female. Captive animal, donated by Milwaukee County Zoo in June 1990. Weight 340 lb. This is 'Terra', who arrived at the Milwaukee Zoo, with a male named 'Tanga', on 24

April 1960. She weighed 16 lb. upon arrival. The pair were purchased from Alton V. Freeman, Miami Rare Bird Farm, Inc., Kendall, Florida. 'Terra' died on 14 June 1990. [Additional information added by GH from correspondence with Milwaukee Zoo personnel.]

UWZM M. 27614 Skin and skeleton. Adult male. Captive animal, donated by Milwaukee County Zoo on 18 November 1995. This is 'Obsus', who arrived in Milwaukee from Stuttgart Zoo on 4 April 1984. He was born in Stuttgart on 4 March 1981 to dam 'Dina' (sire uncertain, may be 'Banjo'), and was hand-reared. He died on 18 November 1995. [Additional information added by GH from the International Studbook for the Western Lowland Gorilla, 2010.]

UWZM S. 34286 Skeleton. Adult male. This is 'Joe Willy', who was born in New York Bronx Zoo on 6 September 1973 (dam 'Sukari', sire 'Bendera'). He died while on loan to Milwaukee Zoo, on 4 December 1997 (weight 360 lb.). Donated by Milwaukee County Zoo on 8 December 1997. [Some additional information added by GH from the International Studbook for the Western Lowland Gorilla, 2010.]

UWZM S. 34353 Skeleton. Juvenile female. This is 'Mgbali', who was born in Milwaukee Zoo on 20 February 1992 (dam 'Femelle', sire 'Obsus'), and died on 10 January 1998 (weight 52.0 kg.). Donated by Milwaukee County Zoo on 10 January 1998. [Some additional information added by GH from the International Studbook for the Western Lowland Gorilla, 2010.]

Memphis

LeMoyne-Owen College, Division of Natural and Mathematical Sciences, 807 Walker Avenue, Memphis, TN 38126*

Mounted skeleton. Adult male. Acquired from the Department of Anatomy, College of Medicine, University of Tennessee, Memphis, c.1977. The University had purchased it from Charles Ward, Rochester, NY.

Milwaukee

Milwaukee County Zoo, 10001 W. Bluemound Road, Milwaukee, WI 53226

Roberta Wallace

The following samples are preserved frozen (individual animal numbers are given):

MCZ#2600 Serum; whole blood (EDTA); urine.

MCZ#2783 Serum; whole blood (EDTA); plasma (potassium EDTA).

MCZ#3065 Serum; urine; whole blood (EDTA); plasma (EDTA).

MCZ#3066 Serum; plasma (potassium EDTA); whole blood (potassium EDTA, lithium heparin); urine; red blood cells (potassium EDTA); formalin-fixed heart and various other tissues.

MCZ#3647 Serum; whole blood (potassium EDTA); variety of formalin-fixed tissues collected postmortem.

MCZ#3810 Serum; whole blood (sodium heparin, potassium EDTA); cerebrospinal fluid (CSF).

MCZ#3959 Serum; plasma (sodium heparin, sodium citrate); urine.

MCZ#4353 Serum; urine; plasma (lithium heparin, sodium EDTA).

MCZ#4378 Serum; whole blood (sodium EDTA); urine.

MCZ#4758 Serum; red blood cells (potassium EDTA); urine.

MCZ#5768 Serum; urine.

MCZ#5848 Whole blood (heparin lithium, serum); various formalin-preserved tissues and paraffin blocks.

The following samples were frozen but had been accidentally defrosted for some days because of a freezer mishap that went unnoticed for an undetermined length of time (perhaps 10−14days). The refrozen samples may still be useful if used judiciously:

MCZ#102 Serum.

MCZ#2326 Serum.

MCZ#2498 Serum.

MCZ#2547 Serum.

MCZ#2695 Serum.

MCZ#2834 Serum.

MCZ#3063 Serum.

MCZ#3064 Serum.

MCZ#3067 Serum.

The following are paraffin blocks and/or formalin-fixed tissue samples (hearts are fixed and examined separately for the Great Ape Heart Project, which is based at Zoo Atlanta):

MCZ#102 Blocks.

MCZ#2326 Blocks.

MCZ#2600 Blocks and tissues.

MCZ#2990 1 block.

MCZ#3065 Blocks and tissues.

MCZ#3066 Blocks and tissues.

MCZ#3471 Blocks and tissues.

MCZ#3647 Blocks and tissues.

MCZ#3810 Blocks (heart and lung) and tissues.

MCZ#5848 Blocks and tissues.

MCZ#5886 Blocks and tissues.

Milwaukee Public Museum, 800 W. Wells Street, Milwaukee, WI 53233

Julia Colby

NB: Apart from the specimens listed below, the Museum has other skeletal material and one skin, but because of poor record-keeping in the past, collections staff can only guess that this additional material represents two to three other gorillas, but cannot associate remains with a specific animal, nor can they associate remains together (partial skeletal lots together, skin to skeleton, etc.).

E40127/11070 Articulated skeleton. Purchased from Ward's Natural Science Company in 1902.

M4252/18185 Mounted skin. Adult male. Cameroun. This is 'Sambo', who arrived at Milwaukee County Zoological Park on 15 October 1950 (weight 15 lb.). He was purchased from Henry Trefflich, New York, with funds donated by the Pabst Brewing Company. He died of pulmonary tuberculosis on 2 November 1959 (weight 350 lb.). Received from the Milwaukee County Zoo on 4 November 1959. [Additional information added by GH from correspondence with Milwaukee Zoo personnel.]

Skin and partial carcase. Adult female. Democratic Republic of Congo. This is 'Ngajji', who arrived at Dierenpark Wassenaar, Netherlands, on 24 May 1972, aged about 6 years. She had been purchased from a private individual. She was one of 5 gorillas received by Milwaukee Zoo from Wassenaar Zoo on 12 May 1987, all of which were loaned to Fort Worth Zoo until quarters were built for them. She returned to Milwaukee on 21 September 1988 and was euthanased after a stroke on 20 July 2010. [Some additional information added by GH from correspondence with Wassenaar Zoo personnel.]

Skeleton. Adult male. This is 'Tanga', who arrived at Milwaukee Zoo on 24 April 1960 (weight 18 lb.), with a female named 'Terra'. They were purchased from Alton V. Freeman, Miami Rare Bird Farm, Inc., Kendall, Florida. 'Tanga' died on 20 October 1989. [Additional information added by GH from correspondence with Milwaukee Zoo personnel.]

Skin (portions), articulated skeleton, mount and death mask. Adult male. Cameroun. This is 'Samson', who arrived at Milwaukee Zoo on 15 October 1950 (weight 12½ lb.), aged about 1 year. He was purchased and donated with 'Sambo' (the Pabst Brewing Co. donated $7000 for the 2 young gorillas). They were moved to the new zoo from Washington Park on 2 October 1959. Recorded weights: 17 lb. on 30 November 1950; 19½ lb. on 18 December 1950; 20½ lb. on 1 January 1951; 21 lb. on 14 January 1951; 24 lb. on 12 February 1951; 26½ lb. on 28 March 1951; 34½ lb. on 24 April 1951; 32 lb. on 25 May 1951; 33 lb. on 25 June 1951; 38 lb. on 27 July 1951; 38½ lb. on 27 August 1951; 40 lb. on 29 September 1951; 45 lb. on 3 November 1951; 47 lb. on 7 December 1951; 47 lb. on 22 January 1952; 80 lb. approx. on 20 October 1952; 150 lb. approx. on 8 January 1954; 275 lb. approx. on 10 April 1957; 360 lb. on 5 October 1959 (new zoo scales); 360 lb. on 30 October 1959; 371 lb. on 1 January 1960; 390 lb. on 14 January 1960; 390 lb. on 21 January 1960; 410 lb. on 23 February 1960; 412 lb. on

1 March 1960; 422 lb. on 18 March 1960; 424 lb. on 26 March 1960; 432 lb. on 8 April 1960; 442 lb. on 14 April 1960; 440 lb. on 20 April 1960; 445 lb. on 23 April 1960; 450 lb. on 27 April 1960; 460 lb. on 3 May 1960; 456 lb. on 12 May 1960; 450 lb. on 30 May 1960; 470 lb. on 8 June 1960; 468 lb. on 25 June 1960; 460 lb. on 25 July 1960; 462 lb. on 9 October 1960; 460 lb. on 27 October 1960; 505 lb. on 18 February 1961; 507 lb. on 11 March 1961; 510 lb. on 27 March 1961; 520 lb. on 24 April 1961; 525 lb. on 19 May 1961; 530 lb. on 29 May 1961; 552 lb. on 25 June 1961; 552 lb. on 29 December 1961; 552 lb. on 3 January 1962; 566 lb. on 19 January 1962; 580 lb. on 4 March 1962; 575 lb. on 1 July 1962; 580 lb. on 2 September 1962; 575 lb. in October 1962; 570 lb. in November 1962; 565 lb. in December 1962; 585 lb. in January 1963; 590 lb. in March 1963; 600 lb. in April 1963; 584 lb. on 13 May 1963; 605 lb. in July 1963; 595 lb. on 20 September 1963; 580 lb. in October 1963; 584 lb. in November 1963; 580 lb. in December 1963; 582 lb. in January 1964; 580 lb. in February 1964; 578 lb. in March 1964; 578 lb. in April 1964; 574 lb. in May 1964; 578 lb. in June 1964; 572 lb. in July 1964; 576 lb. in August 1964; 573 lb. in September 1964; 568 lb. in October 1964; 578 lb. in November 1964; 572 lb. in December 1964; 605 lb. in November 1966; 614 lb. on 6 September 1967; 610 lb. in December 1967; 625 lb. in July 1968; 630 lb. on 5 November 1968; 630 lb. in June 1969; 642 lb. in May 1970; 644 lb. on 1 November 1970; 652 lb. in April 1971 (put on diet); 648 lb. on 15 May 1971; 638 lb. on 13 June 1971; 652 lb. on 15 October 1971; 575 lb. on 25 December 1972; 580 lb. on 11 March 1973; 550 lb. on 17 June 1974; 484 lb. on 25 September 1976; 506 lb. on 20 November 1976; 496 lb. on 21 March 1979; 472 lb. on 28 August 1980. He died of an acute heart attack at 3:45 p.m. on 27 November 1981 (weight 532 lb.). Posthumous measurements: arm spread, finger-tip to finger-tip 95½ in.; circumference of forearm (right) 22 in.; circumference of biceps (right) 22 in.; length of arm (right) 36 in.; circumference of chest 62 in.; circumference of abdomen at navel 58 in.; length of foot 13 in.; circumference of thigh (right) 32 in.; circumference of calf (right) 17 in.; length of leg 27 in.; circumference of head at sagittal ridge 33½ in.; sagittal crest to base of abdomen 45 in.; hip socket to edge of foot pad 27 in (gives total body length 72 in.); ear (meatus to outer edge of pinna) 48 mm; heart 760 g; lungs 810 g; spleen 320 g; kidneys approx. 420 g each; testes 19.6 g and 20.8 g. The hide was stored in a freezer for 3 years and suffered freezer burn, so could not be used in a taxidermy mount. For exhibition purposes, a simulacrum was created by museum taxidermist and artist Wendy Christensen-Senk, using fur from National Fiber Technologies, which is a mix of acrylic, yak and goat hair. [Additional information added by GH from correspondence with Milwaukee Zoo personnel Sam LaMalfa and George Speidel.]

Minneapolis

Bell Museum of Natural History, University of Minnesota, 10 Church Street SE, Minneapolis, MN 55455

Sharon Jansa

MMNH 6323 Mounted skeleton. Juvenile.
MMNH 6336 Mounted skeleton. Subadult female.

Morgantown

West Virginia University, College of Arts and Sciences, Department of Biology, Life Sciences Building, Morgantown, WV 26506

Susan Philhower Raylman; Leah A. Williams

Mounted skeleton. Adult male. The sagittal crest is markedly deflected to the right.

Newark

The Newark Museum, 49 Washington Street, Newark, NJ 07102

Carol J. Bossert; Sule Oygur; Heidi Warbasse

Z 38.141 Skull. Adult male. Length, top of sagittal crest to anterior point of alveolar portion 335 mm; glabella−opisthocranion 165 mm; prosthion−opisthocranion 270 mm; mandibular length, posterior point of condyles to anterior point of alveolar portion, between the central incisors 190 mm; mandibular height 115 mm; greatest height of sagittal crest 25 mm; upper facial height, nasion−intradentale superius 113 mm; bizygomatic width 180 mm; maximum inside biorbital width 118 mm; mandibular width 119 mm. There is a hole measuring 18 mm horizontally and 21 mm vertically in the left maxilla. The lower left lateral incisor is broken.

Z 38.142 Skull. Adult female. Glabella−opisthocranion 150 mm; prosthion−opisthocranion 220 mm; mandibular length 145 mm; mandibular height 85 mm; mandibular width 99 mm; upper facial height 89 mm; bizygomatic width 140 mm; maximum inside biorbital width 94 mm.

Both the above skulls were donated, along with other skulls, by Newark dentist Dr A. Wolfson on 10 November 1938.

New Haven

Peabody Museum of Natural History, Yale University, 170 Whitney Avenue, New Haven, CT 06520-8118

Kristof Zyskowski

YPM MAM 005956 Six pieces of skin (stored in ethanol). Apparently received from the P.T. Barnum menagerie. Original catalog number MAM.P.01457.

YPM MAM 006794 Skull and postcranial skeleton (incomplete). Male? Gabon. Purchased from E. Gerrard, Jr, on 28 March 1874. Donated by Othniel Charles Marsh (1831−99). Original catalog number Osteo 295 (MAM.O.00295).

YPM MAM 006859 Cranium. Juvenile male. Deposited by Dr A.O' Leary, Boston. Original catalog number Osteo 966 (MAM.O.00966).

YP MAM 006860 Skull. Female. Deposited by Dr A. O' Leary on 31 December 1874. Original catalog number Osteo 967 (MAM.O.00967). It is now missing.

YPM MAM 006861 Cranium and brain cast. Female? Gabon. Collected by the Revd A. Bushnell. Purchased from Dr H.W. Boyd, Chicago, on 8 April 1875. Donated by O.C. Marsh. Original catalog number Osteo 968 (MAM.O.00968).

YPM MAM 006862 Skeleton (incomplete). Female? Gabon. Collected by the Revd A. Bushnell. Purchased from Dr H.W. Boyd on 8 April 1875. Donated by O.C. Marsh. Original catalog number Osteo 969 (MAM.O.00969).

YPM MAM 006865 Cranium. Male. Gabon. Collected by the Revd A. Bushnell. Purchased from Dr H.W. Boyd on 8 April 1875. Donated by O.C. Marsh. Original catalog number Osteo 972 (MAM.O.00972).

YPM MAM 007545 Skull (cast). Donated by Dr J.H. Mack. Original catalog number Osteo 1712 (MAM.O.01712).

YPM MAM 007705 Skeleton (formerly mounted, many elements still articulated). Male. Gabon. Collected by the Revd A. Bushnell. Purchased from Dr H.W. Boyd on 8 June 1874. Donated by O.C. Marsh. Original catalog number Osteo 2380 (MAM.O.02380).

YPM MAM 007706 Mounted skeleton. Adult male. Gabon. Collected by the Revd A. Bushnell. Purchased from Dr H.W. Boyd in 1875. Donated by O.C. Marsh. Original catalog number Osteo 2381 (MAM.O.02381).

YPM MAM 007707 Skull. Adult. Gabon. Collected by the Revd A. Bushnell. Purchased from Dr H.W. Boyd on 17 December 1875. Donated by O.C. Marsh. Original catalog number MAM.O.02382. It is now missing.

YPM MAM 007708 Skull (sectioned). Male. Gabon. Collected by the Revd A. Bushnell. Purchased from Dr H. W. Boyd on 17 December 1875. Donated by O.C. Marsh. Original catalog number Osteo 2383 (MAM.O.02383).

YPM MAM 007709 Cast of brain cavity. Male. Purchased from the Royal College of Surgeons of England. Donated by O.C. Marsh. Original catalog number Osteo 2384 (MAM.O.02384).

YPM MAM 007882 Cranium. Male. Gabon. Purchased from E. Gerrard Jr, on 21 July 1875. Donated by O.C. Marsh. Original catalog number Osteo 2539 (MAM.O.02539).

YPM MAM 007890 Skull. Female. Gabon. Purchased from E. Gerrard, Jr on 21 July 1875. Donated by O.C. Marsh. Original catalog number Osteo 2546 (MAM. O.02546).

YPM MAM 007896 Femora and tibiae. Juvenile. Gabon. Purchased from Dr H.W. Boyd on 23 May 1878. Donated by O.C. Marsh. Original catalog number Osteo 2552 (MAM.O.02552).

YPM MAM 007904 Head (cast). Donated? by the Museum of Leyden. It was loaned to North Sheffield Hall (a classroom building) from where it was stolen in 1934 or 1935. The building was demolished in 1967. Original catalog number Osteo 2560 (MAM.O.02560).

YPM MAM 008096 Nine long bones. Purchased from Dr H.W. Boyd on 23 May 1878. Donated by O.C. Marsh. Original catalog number Osteo 2775 (MAM.O.02775).

YPM MAM 009752 Mounted skeleton. Adult male. Donated by Mr North, Ringling Bros Barnum & Bailey Combined Shows, Inc., on 21 December 1949. This is the circus-owned gorilla, 'Gargantua'. He was originally given by local inhabitants to a missionary in Cameroun in November 1930, when only a few months old. He was sold for $400 to Arthur Phillips, captain of the 'West Key Bar', an American freighter, when he called at Kribi nearly a year later. When the freighter reached Boston, a sailor who had been sacked by the captain reportedly took revenge by throwing nitric acid over the gorilla's face and chest. This caused permanent disfigurement, with deformation of the upper lip, and temporary damage to the eyes. The infant gorilla was purchased by Mrs Gertrude Ada Lintz (née Davies), who lived at 8365 Shore Road, Brooklyn, New York, on 28 December 1932 (weight 22 lb.). She named him 'Buddha' (later changed to 'Buddy'). The wounds healed slowly by granulation, but he suffered another setback in September 1936, when a boy employed by Mrs Lintz for a few days as a cleaner was sacked, but came back at night and offered him a bottle of strong disinfectant, sweetened with chocolate syrup. It burned the lining of his stomach and intestines, and he lost 80 lb. before he could take food again. This incident took place in the Miami area of Florida, where Mrs Lintz had a second home, to which she had taken her apes for the winter of each year since 1934. See Lintz (1942). By 1937 'Buddy's' growth and sheer strength dictated that his status as a pet be terminated and that he be caged. He was sold to John Ringling North for a figure reported to be at or near $10,000, and arrived at the circus on 4 December 1937. A special all-steel, glass-enclosed cage, air-conditioned by Carrier Air Conditioning, was built at a cost of about $25,000 to house, transport and exhibit the gorilla, whom the circus authorities renamed 'Gargantua'. Richard Kroener, who had worked for Mrs Lintz for 20 years and looked after all her animals, was hired as his keeper. When the strike at Scranton, PA, closed the show prematurely in 1938, 'Gargantua' and the other Ringling contracted features completed the season on the Ringling-owned Al G. Barnes Sells-Floto Circus. In 1939 'Gargantua' was again featured on the Ringling Bros Barnum & Bailey where he remained a continuous major attraction until his death at Miami, FL, on 25 November 1949 (weight 312 lb.). The cause of death was bilateral lobar pneumonia, complicated by a severe kidney disease, 5 carious cavities in the molars and several alveolar abscesses. See also Plowden (1972). Original catalog number Osteo 4470 (MAM.O.04470). [Some additional information added by GH from correspondence with circus historian Richard J. Reynolds III.]

YPM MAM 10383 Cranium. Male. Collected by Warren Buck and J. van L. Bell. Purchased from George S. Rennie III. Original catalog number MAM O.05345. It is now missing.

YP MAM 013478 Humerus, radius and ulna. Adult.

New York

American Museum of Natural History, Central Park West at 79th Street, New York, NY 10024-5192

Department of Biological Anthropology:

Jaymie L. Brauer; Gisselle Garcia-Pack

L1 G1 Skull. Adult male. Ogooué.

L2 G2 Skull. Adult male. Luembe, Belgian Congo.

L3 Skull. Adult female.

L4 Skull. Adult male. Kamerun.

L5 G5 Skull (back half only). Adult male? Cranial capacity 475 cc.

L6 G6 Cranium. Juvenile male. Yaunde, Kamerun. Collected by Zenker.

L7 G7 Skull. Juvenile male. Yaunde, Kamerun. Collected by Zenker.

L9 G9 Skull. Adult female. Kamerun.

L10 G10 Skull. Adult female.

L11 G11 Cranium. Juvenile female. Gabon.

L12 G12 Skull. Juvenile.

L20 G8 Skull. Adult female. Cranial capacity 470 cc.

L201 G25 Cranium (broken). Adult male?

L213 G19 Skull. Subadult male.

L214 G20 Skull. Subadult male.

L215 G21 Skull. Subadult female.

L216 G13 Skull. Juvenile female. Gabon.

L217 G14 Skull. Adult female? Gabon.

L218 G15 Skull. Juvenile male. Gabon.

L219 G16 Skull. Subadult male? Gabon.

L220 G17 Skull. Adult male. Gabon.

L223 G23 Skull (cranium broken). Adult male?

L224 G24 Skull. Adult male?

L264 G27 Cranium. Adult male?

L265 G28 Cranium. Adult male?

L266 G29 Cranium. Adult female.

L267 G30 Cranium. Adult female?

L280 Mandible. Juvenile.

The above 28 specimens, L1 G1−L280 inclusive, were part of a large collection, which included thousands of human remains as well, purchased from Prof. Felix von Luschan (1854−1924) in 1924. Von Luschan died during the process, and the sale was completed after his death by his wife. The funds to enable the purchase were donated by Felix M. Warburg (1871−1937).

99/97 Skeleton. Infant. Purchased from Giffort Brothers in 1896.

99/9425 Skeleton (partially mounted). Adult male. The mandible is missing. Purchased from the Kny-Scheerer Company.

99/9688 Skull. Subadult female?

99/9690 Skull. Subadult female. Gabon.

99/9691 Skull. Adult male? Cameroun.

99/9693 Skull. Adult male?

The above 4 specimens, 99/9688−99/9693 inclusive, were donated by J. Flemming in 1931.

99.1/1577 Skeleton. Adult male. Mt Mitulu, Nsok, Spanish Guinea. Collected on 31 May 1956. Purchased from J. Sabater Pí in 1957. Height 177 cm; arm span 235 cm; weight 169 kg.

99.1/1578 Skeleton. Adult female. At foot of Mt Mitulu, Mokula forest, Nsok, Spanish Guinea. Collected on 31 May 1956. Purchased from J. Sabater Pí in 1957. Height 142 cm; arm span 195 cm; weight 72.5 kg. It was nursing a female baby about 2 months old.

99.1/1579 Skeleton. Subadult male. Mt Mitulu, Nsok, Spanish Guinea. Collected on 31 May 1956. Purchased from J. Sabater Pí in 1957. Height 137 cm; weight 53 kg. Left arm wounded, caused by a wire snare, deformed and 'rickety'. Body fat much reduced.

99.1/1580 Skeleton. Subadult male. Mokula, Nsok district, Spanish Guinea. Collected on 1 June 1956. Purchased from J. Sabater Pí in 1957. Height 142 cm; arm span 203 cm; weight 63 kg. Face full of disease spots symptomatic of framboesia (yaws).

For further details of the above 4 specimens see Grzimek (1956). According to Grzimek, 99.1/1577−99.1/1579 were collected on 29 May 1956 [rather than 31 May 1956, as stated in the museum records].

99.1/1581 Skeleton. Subadult. Donated by Cornell University Medical School.

99.1/2042 Skull. Adult male. Anizok, Spanish Guinea.

99.1/2043 Skull. Adult male. Anizok, Spanish Guinea.

99.1/2044 Skull. Adult male. Anizok, Spanish Guinea.

99.1/2045 Skull. Adult female. Anizok, Spanish Guinea.

99.1/2046 Skull. Adult male. Anizok, Spanish Guinea.

99.1/2055 Skeleton. Adult female. Nsok, Spanish Guinea.

The above 6 specimens, 99.1/2042−99.1/2055 inclusive, were purchased from J. Sabater Pí in 1962.

99.1/2211 Skull. Adult female. Mbia Campo, Ikunde, Spanish Guinea. It is now missing.

99.1/2212 Skull. Subadult male. Mbia Campo, Ikunde, Spanish Guinea.

99.1/2251 Skull. Juvenile male. Mbia Campo, Ikunde, Spanish Guinea.

99.1/2255 Skeleton. Infant female. Ombam, Cameroon.

99.1/2258 Skull. Subadult male?

The above 5 specimens, 99.1/2211−99.1/2258, were purchased from J. Sabater Pí in 1964.

99.1/2773 Skull. Adult male. Anizok, Spanish Guinea. Purchased from J. Sabater Pí in 1960.

99.1/2774 Skull. Adult female. Rio Benito, Spanish Guinea. Purchased from J. Sabater Pí in 1960.

99.1/2775 Skull. Adult female. Spanish Guinea. Purchased from J. Sabater Pí in 1960.

Department of Mammalogy:

Neil Duncan; Christopher A. Norris

AMNH 123 Cranium (cast). Female. From the Verreaux collection.

AMNH 198 Skull (cast). Male. From the Verreaux collection?

AMNH 959 Mounted skin. Male. Gabon. Acquired from Ward's Natural Science Establishment. Sent as a gift to the Saudi Natural History Museum on 28 February 1980.

AMNH 15967 Skeleton. Male. Donated by the Kny-Scheerer Co.

AMNH 15968 Skeleton. Female. Donated by the Kny-Scheerer Co.

AMNH 15969 Skeleton. Juvenile. Purchased from the Kny-Scheerer Co.

AMNH 22832 Skeleton and skin. Female. Received from the New York Zoological Society on 22 September 1905.

AMNH 28273 Skull. Upper Nasabbe River, French Congo. Collected by Gustave Sabille. Purchased on 29 March 1907.

AMNH 35400 Skeleton and skin. Female. Received from the New York Zoological Society on 7 October 1911. This is 'Madam Ningo', who arrived at the New York Bronx Zoo on 23 September 1911, when she was estimated to be 2½ years old. She had been collected in Fernan Vaz, Gabon, by the Zoological Society's expedition, under the direction of Richard L. Garner. She died of long continued malnutrition on 5 October 1911 (weight

25 lb.). [Some additional information added by GH from correspondence with NY Bronx Zoo personnel.]

AMNH 35617 Skin. Received from the New York Zoological Society on 30 March 1916.

AMNH 38428 Femur. Purchased from Ward's Natural Science Establishment in 1915.

AMNH 54084 Mounted skin and skeleton. Male. Donated by the Ringling Bros and Barnum & Bailey circus on 18 April 1921. This is 'John Daniel'. He came originally from Port-Gentil, Gabon, and was brought to Le Havre by the captain of a French ship. He was purchased there by the London animal dealer John Daniel Hamlyn (1858–1922), who shipped him to England with a lot of monkeys. He was sold by Hamlyn to Derry and Toms department store, 99–121 Kensington High Street, London W8, as a Christmas attraction. There he was purchased for £300 in December 1918, by RAF Major Rupert Penny as a present for his niece, Miss Alyse Mary Cunningham (b.1882). At that time the gorilla weighed 32 lb. and was in a rickety condition. He lived with Miss Cunningham in a flat above shop premises at 14 Sloane Street, Chelsea, London SW1. Miss Cunningham owned a millinery shop next door at 15 Sloane Street. She named the gorilla 'John Daniel', and during the summer months each year, took him by taxi cab to London Zoo on three afternoons each week, so that he could be put on show to the public in one of the big outdoor lion cages. By March 1921 he stood 40½ in. tall and weighed 112 lb., and was becoming difficult to manage in a private home. Consequently, he was sold to the American animal dealer John T. Benson (1879–1943), who in turn sold him to the Ringling Bros and Barnum & Bailey circus for the reported sum of $30,000. He became ill on the sea voyage to the United States, and died of pneumonia on 17 April 1921, soon after arrival in Madison Square Garden Tower, New York. For further information, see Cunningham (1921) and Hornaday (1922).

AMNH 54089 Mounted skin and skeleton. Male. *Gorilla beringei beringei*. Kivu forest, eastern Belgian Congo. Collected by Carl Ethan Akeley in November 1921. Received on 13 March 1922.

AMNH 54090 Mounted skin and skeleton. Male. *Gorilla beringei beringei*. Kivu forest, eastern Belgian Congo. Collected by Carl Ethan Akeley in November 1921. Received on 13 March 1922.

AMNH 54091 Mounted skin and skeleton. Female. *Gorilla beringei beringei*. Kivu forest, eastern Belgian Congo. Collected by Carl Ethan Akeley in November 1921. Received on 13 March 1922.

AMNH 54092 Mounted skin and skeleton. Female. *Gorilla beringei beringei*. Kivu forest, eastern Belgian Congo. Collected by Carl Ethan Akeley in November 1921. Received on 13 March 1922.

AMNH 54093 Mounted skin. Juvenile male. *Gorilla beringei beringei*. Kivu forest, eastern Belgian Congo. Collected by Carl Ethan Akeley in November 1921. Received on 13 March 1922. The pickled body was transferred to the Department of Comparative Anatomy. This department no longer exists, but the mammal material from its collection has been incorporated into the research collections of the Department of Mammalogy.

AMNH 54327 Skeleton and skin. Female. Ebole Bengon, Cameroun. Collected by the Revd Finley McCorvey Grissett (1889–1951) on 12 June 1923.

AMNH 54328 Skeleton and skin. Juvenile male. Ebole Bengon, Cameroun. Collected by the Revd Finley McCorvey Grissett on 12 June 1923. The skull was in the Dept of Comparative Anatomy, but only the skin and mandible could be found on 6 July 1979.

AMNH 54329 Skeleton and skin. Infant male. Métet, Cameroun. Collected by the Revd Finley McCorvey Grissett on 22 July 1923. Estimated date of birth February 1923. Only the skin could be found on 6 July 1979.

AMNH 54355 Skeleton and skin. Male. Sakbayémé, Edéa, Cameroun. Collected by the Revd Jacob A. Reis. Purchased in November 1924.

AMNH 54356 Skeleton and skin. Female. Sakbayémé, Edéa, Cameroun. Collected by the Revd Jacob A. Reis. Purchased in November 1924.

AMNH 69398 Skeleton. Gabon. Purchased from the von Luschan collection in April 1925. Both hands and the right foot are articulated.

AMNH 81651 Skeleton and mounted skin. Male. Nola, French Congo. Collected by Edward Kenneth Hoyt, 15 April–18 May 1932. Entered on 15 August 1932.

AMNH 81652 Skeleton and skin. Female. Nola, French Congo. Collected by Edward Kenneth Hoyt, 15 April–18 May 1932. Entered on 15 August 1932.

AMNH 81653 Skin. Female. Nola, French Congo. Collected by Edward Kenneth Hoyt, 15 April–18 May 1932. Entered on 15 August 1932.

AMNH 90169 Cranium. *Gorilla beringei graueri*. Alimbongo Mountains, 30 mi. south of Lubero, west of south end of Lake Edward, Kivu, Belgian Congo. Locally hunted. Collected by Martin Johnson. Entered on 23 September 1931.

AMNH 90170 Cranium. *Gorilla beringei graueri*. Alimbongo Mountains, 30 mi. south of Lubero, west of south end of Lake Edward, Kivu, Belgian Congo. Locally hunted. Collected by Martin Johnson. Entered on 23 September 1931.

AMNH 90171 Cranium. *Gorilla beringei graueri*. Alimbongo Mountains, 30 mi. south of Lubero, west of south end of Lake Edward, Kivu, Belgian Congo. Locally hunted. Collected by Martin Johnson. Entered on 23 September 1931.

AMNH 90172 Cranium. *Gorilla beringei graueri*. Alimbongo Mountains, 30 mi. south of Lubero, west of south end of Lake Edward, Kivu, Belgian Congo. Locally hunted. Collected by Martin Johnson. Entered on 23 September 1931.

AMNH 90173 Mandible. *Gorilla beringei graueri*. Alimbongo Mountains, 30 mi. south of Lubero, west of south end of Lake Edward, Kivu, Belgian Congo. Locally hunted. Collected by Martin Johnson. Entered on 23 September 1931.

AMNH 90194 Skeleton and skin. Adult male. Nguilili, Moloundou sub-division, Cameroun. Collected by Ernest de Léon in July 1927. Received in exchange for *Pan* skeleton 51382 from the Dept of Comparative Anatomy in July 1932.

AMNH 90289 Skeleton. Male. Métet, Cameroun. Collected by the Revd Finley McCorvey Grissett. Received in June 1924.

AMNH 90290 Skeleton. Male. Métet, Cameroun. Collected by the Revd Finley McCorvey Grissett. Received in June 1924.

AMNH 114217 Skeleton. Métet, Cameroun. Collected by S.E. Johnson in 1936; purchased from same.

AMNH 115609 Skeleton. Adult male. *Gorilla beringei graueri*. This is 'Ngagi', who arrived at San Diego Zoological Garden, Balboa Park, San Diego, CA on 5 October 1931, aged about 6 years. He arrived with another male, 'Mbongo'. They were purchased for $11,000 the pair from Martin and Osa Johnson, who had captured them in November 1930 in the Alimbongo mountains, west of Lake Edward, Kivu Province, Belgian Congo. 'Ngagi' died of a coronary artery occlusion on 12 January 1944 (weight 639 lb.). His carcase was kept in the cold room for 7 years.

AMNH 119224 Skull. Found in collection, 1992.

AMNH 119225 Mandible. Found in collection, 1992.

AMNH 119226 Skull. Found in collection, 1992.

AMNH 145600 Skeleton and skin. Male. Donated by the College of Physicians and Surgeons Medical School?

AMNH 147176 Skeleton and skin. Juvenile male. Donated by the Trefflich Bird and Animal Co. on 31 October 1950.

AMNH 149579 Carcase (in alcohol). Infant female. Imported by the Trefflich Bird and Animal Co. Received from the New York Zoological Society on 29 December 1952. The animal was less than 1 year old and weighed 8 lb.

AMNH 150285 Skeleton and skin. Donated by the New York Zoological Society on 21 April 1949.

AMNH 167325 Skull. Male.

AMNH 167326 Skull. Male.

AMNH 167327 Skull. Male. One lower incisor is missing.

AMNH 167328 Skull. Male.

AMNH 167329 Cranium. Male. It lacks 1 incisor and 1 canine.

AMNH 167330 Skull. Female. It is missing an upper incisor.

AMNH 167331 Skull. It is missing 1 lower canine.

AMNH 167332 Skull. Male.

AMNH 167333 Skull. Juvenile male.

AMNH 167334 Skull. Male. It is missing 3 lower incisors.

AMNH 167335 Skeleton. Male.

AMNH 167336 Postcranial skeleton. Male.

AMNH 167337 Skeleton. Female.

AMNH 167338 Skeleton. Male.

AMNH 167339 Skeleton. Female.

AMNH 167340 Skeleton. Female.

AMNH 167347 Postcranial skeleton. Previously misidentified as chimpanzee.

AMNH 167368 Bones of 2 hands and 1 foot.

AMNH 167369 Skeleton.

The above 19 specimens, AMNH 167325−AMNH 167369 inclusive, were collected in Cameroun by Fred Hope and purchased from Elizabeth C. Hope in 1946.

AMNH 167672 Skull. Efulan, 53 mi. E of Kribi, Cameroun. Collected by the Revd Finley McCorvey Grissett in 1934; received from the same.

AMNH 167675 Skull (part). Efulan, 53 mi. E of Kribi, Cameroun. Collected by the Revd Finley McCorvey Grissett in 1934; received from the same.

AMNH 170362 Skeleton. Female. Bafia, Cameroun. Collected by the Revd Finley McCorvey Grissett.

AMNH 170363 Mandible. Female. Bafia, Cameroun. Collected by the Revd Finley McCorvey Grissett.

AMNH 173649 Skeleton. Juvenile. Received from the New York Zoological Society. Entered on 22 August 1956.

AMNH 183131 Skull. Yokadouma, SE Cameroun. Collected by the Revd Finley McCorvey Grissett. Entered on 2 September 1958.

AMNH 200045 Brain (mummified). Juvenile. *Gorilla beringei beringei*. Mounts Mikeno/Karisimbi, Kivu, Belgian Congo. Collected by Carl E. Akeley in 1921.

AMNH 200046 Genitourinary tract, brain and eye (in alcohol). Adult female. *Gorilla beringei beringei*. Mounts Mikeno/Karisimbi, Kivu, Belgian Congo. Collected by Carl E. Akeley in 1921.

AMNH 200047 Genitourinary tract, tongue, hands and feet (in alcohol). Adult female. *Gorilla beringei beringei*. Kivu, Belgian Congo. Collected by Carl E. Akeley in 1921.

AMNH 200048 Genitourinary tract, brain, tongue and eye (in alcohol). Adult female. *Gorilla beringei beringei*. Kivu, Belgian Congo. Collected by Carl E. Akeley in 1921.

AMNH 200056 Skeleton. Female. SE side of Lake Ngové basin, 40 km north of Setté Cama, Gabon.

Received from the New York Zoological Society. This is 'Dinah', who arrived at the New York Bronx Zoo on 24 August 1914, aged about 3 years. She had been captured by Prof. R.L. Garner while he was on an expedition for the New York Zoological Society. Her dimensions on 25 August 1914 were: span of arms 50½ in.; height standing 36½ in.; around chest 26 in.; height sitting 24 in. Her recorded weights are: 8 October 1914 40½ lb.; 14 December 1914 38 lb.; 24 January 1915 39 lb.; 9 February 1915 40¼ lb; 24 March 1915 41½ lb.; 24 April 1915 43 lb.; 19 May 1915 44 lb.; 18 June 1915 42¼ lb; 2 July 1915 42½ lb. She died on 31 July 1915 of ananition (exhaustion from starvation), malnutrition and rickets. General condition of organs healthy. Appendix congested. Thorax shrunken, ribs and sternum depressed. Muscles much shrunken. Sections made of vertebrae shows signs of osteomalachia. For related information, see Rich (2011). [Some additional information added by GH from correspondence with NY Bronx Zoo personnel.]

AMNH 200061 Left femur.

AMNH 200063 Carcase (in alcohol). Male. *Gorilla beringei beringei*. Kivu, Belgian Congo. Collected by Carl E. Akeley. Skin and brain removed.

AMNH 200252 Skull (cast).

AMNH 200315 Tongue (in alcohol). Belgian Congo.

AMNH 200479 Bust of 'Dinah'. Received from Prof. J.H. McGregor on 4 January 1924.

AMNH 200500 Skull.

AMNH 200501 Skull.

AMNH 200502 Skull.

AMNH 200503 Skull.

AMNH 200504 Skull.

AMNH 200505 Skull.

AMNH 200506 Skull.

AMNH 200507 Skull. Not found in 1976, loan no. 34747.

AMNH 200508 Skull.

The above 9 skulls, AMNH 200500−AMNH 200508 inclusive, were all collected in the Ouesso or Mossaka region, Middle Congo, c.1921−22, and were purchased from the Kny-Scheerer Co. in May 1924.

AMNH 200514 Bust of 'John Daniel'. Received from Prof. J.H. McGregor in January 1924.

AMNH 200515 Bust of 'John Daniel'. Received from Prof. J.H. McGregor in January 1924.

AMNH 200516 Cast of left hand, closed. *Gorilla beringei beringei*. Received from Carl Akeley in January 1924.

AMNH 200517 Cast of left hand, closed. *Gorilla beringei beringei*. Received from Carl Akeley in January 1924.

AMNH 200518 Cast of right hand, partly closed. *Gorilla beringei beringei*. Received from Carl Akeley in January 1924.

AMNH 200519 Cast of right hand, partly closed. *Gorilla beringei beringei*. Received from Carl Akeley in January 1924.

AMNH 200520 Cast of sole of left foot, hallux abducted. *Gorilla beringei beringei*. Received from Carl Akeley in January 1924.

AMNH 200521 Cast of sole of left foot, hallux abducted. *Gorilla beringei beringei*. Received from Carl Akeley in January 1924.

AMNH 200599 Hand and foot (casts). Departmental preparations from 'Dinah'.

AMNH 200602 No indication of nature of material.

AMNH 200839 Male. *Gorilla beringei beringei*. No indication of nature of material.

AMNH 201105 Carcase (injected with carbolic; in alcohol). Infant female. Collected on 31 July 1938. Received from C.V. Noback. Entered on 28 April 1981.

AMNH 201216 Female. No indication of nature of material.

AMNH 201335 Male. *Gorilla beringei beringei*. No indication of nature of material.

AMNH 201336 Male. *Gorilla beringei beringei*. No indication of nature of material.

AMNH 201337 Female. No indication of nature of material.

AMNH 201338 Male. No indication of nature of material.

AMNH 201449 Heart (in fluid). Male. *Gorilla beringei beringei*.

AMNH 201453 Carcase (in alcohol). Male. Emvan, 80 km from Yadunda, Cameroun. Received from the Columbia University/American Museum of Natural History joint expedition.

AMNH 201457 Skeleton (partial). Juvenile female. Vimili, Mbalmayo, Cameroun. Collected by the Revd William C. Johnston in January 1930.

AMNH 201458 Carcase (in alcohol). Female. Djaposten, Cameroun. Received from the Columbia University/American Museum of Natural History joint expedition.

AMNH 201459 Skeleton, skin and cast of head. Male. Djaposten, Cameroun. Collected by the Columbia University/American Museum of Natural History joint expedition in March 1930. Not found in 1976, loan no. 34747.

AMNH 201460 Postcranial skeleton. Adult male. Bilulu, near Djaposten, Cameroun. Collected by the Columbia University/American Museum of Natural History joint expedition in April 1930. Brain discarded in April 1984.

AMNH 201470 Carcase (in alcohol). Male. Djaposten, Cameroun. Received from the Columbia University/American Museum of Natural History joint expedition.

AMNH 201471 Postcranial skeleton. Adult male. Sangmélima, Cameroun. Collected by the Columbia University/American Museum of Natural History joint expedition in 1929.

AMNH 201472 Skull. Adult female. Bilulu, near Djaposten, Cameroun. Collected by the Columbia University/American Museum of Natural History joint expedition in 1930.

AMNH 201473 Skull. Juvenile. Djaposten, Cameroun. Collected by the Columbia University/American Museum of Natural History joint expedition in 1930.

AMNH 202395 Humerus. Female. Djaposten, Cameroun. Collected by Henry Cushier Raven (1889–1944) in 1930.

AMNH 202586 Skull. Received from the College of Physicians and Surgeons, Columbia University, New York, NY 10032.

AMNH 202595 Penis. *Gorilla beringei beringei.* Collected by Carl Akeley. Received on 26 March 1941.

AMNH 202845 Juvenile female. No indication of nature of material.

AMNH 202854 *Gorilla beringei beringei.* No indication of nature of material.

AMNH 202889 No indication of nature of material.

AMNH 202932 Postcranial skeleton. Male. Collected in 1929. Received from the College of Physicians and Surgeons, Columbia University, New York, NY 10032, on 16 April 1946.

AMNH 202933 Skull. Received from the College of Physicians and Surgeons, Columbia University, New York, NY 10032, on 16 April 1946.

AMNH 202934 Skull. Received from the College of Physicians and Surgeons, Columbia University, New York, NY 10032, on 16 April 1946.

AMNH 204158 Skin and skeleton. Female. Collected on 30 March 1916. Received from Samuel H. Chubb, New York, on 19 December 1916. Formerly no. 40048.

AMNH 204658 Bones of right foot. Received from R. R. Patterson. Found in the Osteology Department on 4 March 1963.

AMNH 209086 Skull. Taken from the Age of Man Hall on 17 February 1966.

AMNH 213182 Mummy. Male. Cameroun. Received from Dr F. Ryan. Originally a specimen in alcohol.

AMNH 214103 Postcranial skeleton.
AMNH 214104 Skull.
AMNH 214105 Skull.
AMNH 214106 Skull.
AMNH 214107 Skull. Male.
AMNH 214108 Skull.
AMNH 214109 Cranium.
AMNH 214110 Skull.
AMNH 214111 Skull.
AMNH 214112 Skull.

AMNH 214113 Skull.
AMNH 214114 Skull.
AMNH 214115 Skull.
AMNH 214116 Skull.

The above 14 specimens, AMNH 214103–AMNH 214116 inclusive, were collected by Armand Denis (1896–1971) near the village of Oka, west of Okio, French Congo, in March 1944.

AMNH 235603 Skeleton and skin. Male. Received from the New York Zoological Society. Entered on 16 February 1973.

AMNH 238083 Skeleton and skin. Female. Mt Okra, West Africa. Received from Central Park Zoo, 830 Fifth Avenue, New York, NY 10021. This is 'Jo Anne', who arrived at the Central Park Zoo with two other females ('Arty' and 'Carolyn') on 1 July 1943. Her weight on arrival was estimated to be 60 lb. She died on 31 December 1975 after anaesthesia due to the nonfunction of her kidneys (weight 422 lb.).

AMNH 239121 Skeleton and skin. Female infant (stillborn). Received from the New York Zoological Society. Weight 5 lb. 3oz. Entered on 30 April 1973.

AMNH 239597 Skeleton and skin. Male. Received from the New York Zoological Society. Entered on 29 October 1971.

AMNH 244695 Pieces of skin from centre of head and part of arm (in alcohol). Found in the Comparative Anatomy collection in March 1981.

City University of New York, Brooklyn College, Department of Geology, Brooklyn, NY 11210

John A. Chamberlain

Skeleton. Adult male. It is incomplete and in very bad condition.

City University of New York, Hunter College, Department of Anthropology, 695 Park Avenue, New York, NY 10065

Christopher C. Gilbert; Suzanne Walker

Mounted skeleton. Adult male. The mandible and dentition are damaged.

City University of New York, Lehman College, Department of Anthropology, 250 Bedford Park Boulevard West, Bronx, NY 10468

Eric Delson; Richard Eisner

Skull. Adult female. It shows evidence of a fatal gunshot wound in the orbital-temporal area.

Mounted postcranial skeleton (incomplete). Adult female.

City University of New York, Queensborough Community College, Department of Biological Sciences and Geology, 222-05 56th Avenue, Bayside, NY 11364

Eugene E. Harris

Embalmed cadaver (no head), in freezer. Juvenile.

Embalmed cadaver (lower half only), in freezer. Juvenile.

Both the above specimens were originally held captive in the Yerkes Regional Primate Research Center.

Icahn School of Medicine at Mount Sinai, Fishberg Department of Neuroscience, Hess Center for Science and Medicine, 1470 Madison Avenue, New York, NY 10029

Patrick R. Hof

92-0044 Brain samples (formalin-fixed, frozen). Adult male. These are from 'Massa', who died in the Philadelphia Zoological Garden on 31 December 1984, aged about 54 years. The samples have been used by Dr Linda C. Cork (The Johns Hopkins University, School of Medicine, Division of Comparative Medicine, Baltimore, MD 21205) in histopathological studies aimed at finding evidence of some of the ageing changes detected in other nonhuman primates, for example, senile plaques and neurofibrillary tangles. These are not present in the samples: the major brain lesions are those associated with atherosclerosis. When Dr Cork retired, she sent the samples (along with some other neural tissues of comparative interest) to Dr Patrick Hof. [Additional information added by GH from correspondence with Linda C. Cork.]

New York University, Center for the Study of Human Origins, Department of Anthropology, 25 Waverly Place, New York, NY 10003

Hannah Taboada

NYU 110 Carpals (5) and metacarpals (2).

NYU 115 Skull. Female.

NYU 220 Phalanx (1).

NYU 226 Partial skeleton (no skull).

Norman

Sam Noble Oklahoma Museum of Natural History, University of Oklahoma, 2401 Chautauqua Avenue, Norman, OK 73072-7029

Janet K. Braun; Brandi S. Coyner

OMNH 2708 Skeleton. Female. Received from the Oklahoma City Zoological Park.

OMNH 23903 Skeleton. Adult male. Received from the Oklahoma City Zoo. This is 'Moemba', who arrived at Oklahoma City Zoo on 28 November 1962 (weight 24 lb.). He came from Cameroon, and was purchased from F.J. Zeehandelaar, Inc. He was euthanased, due to cardiomyopathy, on 5 May 1997 (weight 148.3 kg). [Some additional information added by GH from correspondence with Oklahoma City Zoo personnel.]

OMNH 23904 Skeleton. Juvenile female. Received from Oklahoma City Zoo. This is 'Asha', who was born in Oklahoma City Zoo on 15 December 1992 to dam 'Emily' and sire 'Tatu', and was mother-reared. She died on 25 February 1996 (weight 26.8 kg) of cardiopulmonary arrest, having been electrocuted by a containment wire.

OMNH 23905 Skeleton. Infant male. Received from Oklahoma City Zoo. This is 'Mojo', who was born in Oklahoma City Zoo on 2 July 1995 to dam 'Fern' and sire 'Moemba'. He was diagnosed at 6 months as having acute lymphocytic leukaemia, which was treated by chemotherapy under the supervision of Dr Ruprecht Nitschke, a paediatric neurosurgeon at the Children's Hospital of Oklahoma, who administered the drugs via the gorilla's spinal cord. 'Mojo' was euthanased on 15 April 1996 (weight 6.58 kg).

OMNH 23906 Skeleton. Infant female. Received from Oklahoma City Zoo. Born in Oklahoma City Zoo on 5 December 1996 to dam 'Kokamo' and sire 'Tatu'; died on 14 March 1997 of septicaemia caused by salmonellosis.

OMNH 37983 Skeleton; 2 feet and skin of face (in alcohol). Adult female. Received from Oklahoma City Zoo. This is 'Kali', who was born in Oklahoma City Zoo on 24 March 1994 to dam 'Kathryn' and sire 'Moemba', and was mother-reared. She died of systemic shock on 24 June 2007 (weight 75 kg). Height 1300 mm; foot length 260 mm; ear 53 mm.

OMNH 50619 Mounted skin; skeleton; 2 feet (in alcohol). Adult male. Received from Oklahoma City Zoo. This is 'Bombom', who was born in the Yerkes Regional Primate Research Center, Atlanta, GA, on 30 March 1976, to dam 'Segou' and sire 'Rann', and was mother-reared. He arrived on loan in Oklahoma on 1 May 2002, and died there on 25 June 2012 of an aortic aneurysm (weight 129 kg). He had been diagnosed with heart failure (heart muscle thickened, blood pressure very high) in January 2010.

OCGR 7710 Kidney, liver and muscle tissues (frozen). Adult female. They are from OMNH 37983.

OCGR 11654 Muscle tissue (frozen). Adult male. It is from OMNH 50619.

NB: Tissues are maintained in a separate collection within the museum, and thus have their own OCGR (Oklahoma Collection of Genomic Resources) catalogue numbers. All of the above tissues were frozen immediately upon collection, were never stored in ethanol, lysis buffer, or anything else, and are stored frozen.

Ogden

Weber State University, College of Science,
Museum of Natural Science, Lind Lecture Hall,
2504 University Circle, Ogden, UT 84408*

H. Keith Harrison

630 Mounted skeleton. Subadult male. Cameroon mountains. Received from Hogle Zoological Gardens, 2600 Sunnyside Avenue, Salt Lake City, UT 84108. This is 'Dan', who was purchased with a female ('Elaine') from G. van den Brink. They arrived together on 17 June 1967, weighing 16 lb. and 12 lb. respectively. 'Dan' died of kidney failure on 3 July 1975. The skeleton is the property of the Department of Zoology. [Some additional information added by GH from correspondence with Hogle Zoo personnel.]

Oklahoma City

Oklahoma City Zoological Park, 2000
Remington Place, Oklahoma City, OK 73111

Jennifer D'Agostino

Frozen tissue sets:

ISIS No. 647730 Spleen, kidney, liver, spinal cord, heart, skin, stomach, lung, muscle (22 May 1994).

ISIS No. 672432 Spleen, liver, muscle (15 April 1996).

ISIS No. 701333 Right kidney, lung (14 March 1997).

ISIS No. 642528 Vaginal biopsy (23 June 2005).

ISIS No. 184304 Heart, liver (29 November 2005).

ISIS No. 642528 Spleen, skin, heart, hair (24 June 2007).

ISIS No. 769742 Testicles, sex glands (25 June 2012).

Gorilla serum cryopreservation bank:

Male (estimated date of birth 28 August 1960). Collected 15 June 1993: 9 × serum, clear; blood, whole (unknown anticoagulant); 3 × blood, whole (EDTA); 4 × plasma (heparin), clear. 19 January 1995: 5 × serum, clear; 2 × plasma (EDTA, potassium), clear; 6 × white blood cells and red blood cells (EDTA, potassium); wbc and rbc (heparin, sodium); 2 × plasma (EDTA, potassium), clear; plasma (EDTA, potassium), slight haemolysis. 19 March 1997: 3 × serum, clear. 5 May 1997: 10 × serum, clear; 2 × plasma (heparin, lithium), clear; plasma (EDTA, sodium), clear; 2 × rbc (EDTA, sodium).

Female (est. d.o.b. 7 March 1959). Coll. 2 March 1993: 2 × serum, clear. 15 June 1993: 14 × serum, clear; 4 × blood, whole (EDTA); 4 × plasma (heparin), clear. 6 June 1994: 5 × serum. 28 September 1994: 5 × serum, icteric; serum, slight haemolysis. 3 November 1994: 5 × serum, clear.

Female (est. d.o.b. 11 May 1960). Coll. 15 June 1993: 6 × serum, clear; 4 × blood, whole (EDTA); 4 × plasma

(heparin), clear. 23 August 1994: 12 × serum, clear. 10 January 1995: 3 × serum, clear; serum, slight haemolysis; 4 × plasma (EDTA, potassium), clear; 4 × wbc and rbc (EDTA, potassium); 2 × wbc and rbc (heparin, sodium). 14 December 1995: 7 × serum, clear; 2 × plasma (heparin), clear. 25 January 1996: 2 × serum, slight haemolysis; 2 × serum, clear. 25 February 1997: 2 × serum, clear. 7 May 1998: blood, whole (EDTA, sodium); 3 × serum, clear. 24 August 2001: 2 × serum, clear. 15 January 2005: serum, slight haemolysis. 28 November 2005: 2 × serum, icteric.

Female (est. d.o.b. 24 December 1962). Coll. 18 August 1992: 2 × rbc (heparin); 2 × plasma (heparin); normal repro check: plasma (heparin), clear; 7 × serum, clear. 7 November 1992: rbc (heparin); IVF: 3 × rbc (heparin); 2 × serum, clear; 5 × plasma (heparin), clear. 2 March 1993: 2 × serum, clear; 2 × plasma (heparin), clear; blood, whole (EDTA). 18 May 1993: 3 × blood, whole (EDTA); 8 × serum, clear; 5 × plasma (heparin), clear; blood, whole (EDTA). 15 June 1993: 7 × serum, clear. 18 August 1993: rbc (heparin); serum, clear. 11 June 1996: 2 × serum, clear. 13 May 1997: serum, clear. 30 May 1997: 2 × serum, clear. 21 September 1998: blood, whole (EDTA, sodium); 3 × serum, clear. 27 August 2001: 4 × serum, clear. 29 October 2012: plasma (heparin, lithium), clear; 3 × plasma (heparin, lithium); blood, whole (heparin, lithium); 7 × serum, clear. 19 February 2013: 2 × serum, clear; serum, icteric.

Male (d.o.b. 29 January 1975). Coll. 18 May 1993: 5 × plasma (heparin), clear; 5 × blood, whole (EDTA). 9 January 1995: 2 × wbc and rbc (heparin, sodium); 5 × wbc and rbc (EDTA, potassium); 5 × plasma (EDTA, potassium), clear; 8 × serum, clear. 17 September 1996: serum, slight haemolysis; serum, haemolysis. 14 July 1998: blood, whole (EDTA, sodium); 3 × serum, clear. 23 October 1998: 2 × serum, clear; rbc (EDTA, sodium). 18 December 1998: blood, whole (EDTA, sodium); 16 × serum, clear; 2 × serum, slight haemolysis; 4 × rbc (EDTA, sodium). 8 November 1999: 3 × urine.

Female (d.o.b. 30 April 1985). Coll. 18 May 1993: 2 × blood, whole (EDTA); 7 × serum, clear. 17 September 1996: serum, slight haemolysis; 2 × serum, clear. 3 June 1998: 2 × blood, whole (EDTA, sodium); 3 × serum, clear. 29 April 2010: 4 × serum, clear; plasma (EDTA, sodium), clear.

Male (d.o.b. 12 August 1995). Coll. 2 March 1993: 2 × plasma (heparin), clear; serum, clear; blood, whole (EDTA). 15 June 1993: 18 × serum, clear; 4 × blood, whole (EDTA); 3 × plasma (heparin), clear. 28 September 1994: 4 × serum, slight haemolysis; serum, haemolysis. 3 November 1994: 6 × serum, clear.

Female (d.o.b. 5 June 1974). Coll. 18 May 1993: 5 × blood, whole (EDTA); 4 × plasma (heparin), clear; 5 × serum, clear. 11 January 1995: 7 × serum, clear;

8 × plasma (EDTA, potassium), clear; 7 × wbc and rbc (EDTA, potassium); 3 × wbc and rbc (heparin, sodium). 25 July 1995: 3 × serum, clear; 2 × serum, slight haemolysis. 24 August 1995: 4 × serum, lipaemia; rbc (EDTA).

Female (d.o.b. 20 December 1988). Coll. 2 March 1993: plasma (heparin), clear; serum, clear. 18 May 1993: 4 × blood, whole (EDTA); 7 × serum, clear; 4 × plasma (heparin), clear. 9 January 1995: 8 × plasma (EDTA, potassium), clear; rbc (unknown anticoagulant); 3 × rbc (heparin, sodium); rbc (EDTA, potassium); 5 × wbc and rbc (EDTA, potassium); 8 × serum, clear. 17 September 1996: 3 × serum, clear; serum, slight haemolysis. 23 June 1998: 3 × serum, clear. 23 June 2007: serum, clear. 20 February 2010: 5 × serum, clear.

Female (d.o.b. 3 March 1994). Coll. 3 June 1998: 2 × blood, whole (EDTA, sodium); 2 × serum, clear. 5 September 2000: blood, whole (EDTA, sodium); 7 × serum, clear.

Female (d.o.b. 24 March 1994). Coll. 11 June 1996: 2 × serum, slight haemolysis; serum, clear. 23 June 2005: serum, icteric. 1 March 2006: serum, clear; serum, haemolysis. 13 June 2006: serum, clear; serum, slight haemolysis. 7 June 2007: serum, clear. 23 June 2007: blood, whole (EDTA).

Male (d.o.b. 11 June 1993). Coll. 5 October 1999: 3 × serum, clear; rbc (EDTA, sodium). 10 December 1999: 2 × serum, clear. 28 January 2000: 4 × serum, clear. 19 October 2000: 2 × serum, clear; blood, whole (EDTA, sodium). 11 May 2004: serum, clear.

Female (d.o.b. 14 January 1995). Coll. 13 August 1996: 3 × serum, clear. 16 February 1998: serum, clear. 1 February 1999: 3 × serum, clear; blood, whole (EDTA, sodium). 30 November 2000: 7 × serum, clear. 23 January 2010: 3 × serum, clear; plasma (heparin), clear.

Male (d.o.b. 2 July 1995). Coll. 13 December 1995: serum, clear.

Male (d.o.b. 9 October 83). Coll. 16 December 1997: 4 × serum, clear; 2 × serum, slight haemolysis. 3 August 1999: 2 × serum, haemolysis; rbc (EDTA, sodium). 16 November 1999: 4 × serum, clear; plasma (heparin, lithium), haemolysis; blood, whole (EDTA, sodium).

Male (d.o.b. 9 January 1998). Coll. 12 May 1998: 2 × serum, clear.

Male (d.o.b. 25 November 1998). Coll. 24 August 2001: 3 × serum, clear. 10 November 2004: serum, slight haemolysis.

Male (d.o.b. 30 November 1998). Coll. 11 May 2004: serum, clear.

Male (d.o.b. 21 March 1999). Coll. 10 November 2004: serum, haemolysis.

Female (d.o.b. 21 March 1999). Coll. 20 February 2010: 3 × serum, slight haemolysis. 7 February 2013: 3 × serum, clear. 23 July 2014: 4 × serum, icteric; plasma (EDTA, sodium), icteric.

Female (d.o.b. 23 January 1994). Coll. 3 September 2002: 2 × serum, icteric; serum, slight haemolysis. 9 January 2010: serum, clear; plasma (heparin, sodium), clear. 28 May 2010: 4 × serum, clear; serum, slight haemolysis; plasma (EDTA), clear. 27 March 2013: 2 × serum, haemolysis; 10 × serum, clear; 3 × serum, slight haemolysis; plasma (EDTA, sodium), haemolysis; 2 × plasma (heparin, sodium), clear; 2 × plasma (heparin, lithium), slight haemolysis; 4 × blood, whole (EDTA, sodium); 4 × blood, whole (heparin, lithium).

Female (d.o.b. 14 December 2003). Coll. 13 March 2010: 3 × serum, clear.

Male (d.o.b. 25 January 2004). Coll. 13 March 2010: 6 × serum, clear. 12 February 2013: blood, whole (EDTA).

Female (d.o.b. 1 March 1996). Coll. 27 March 2012: plasma (heparin, lithium), clear. 29 March 2012: blood, whole (EDTA, sodium); 8 × serum, clear; serum, slight haemolysis. 7 February 2013: serum, clear. 12 February 2013: 2 × serum, clear.

Female (d.o.b. 5 September 1981). Coll. 19 June 2012: 2 × serum, haemolysis.

Museum of Osteology, Skulls Unlimited International, Inc., 10313 South Sunnylane Road, Oklahoma City, OK 73160

Jay Villemarette; Joey Williams

002014 Skeleton. Adult male. Received in 2000. This is 'Kongo', who died in Gulf Breeze Zoo on 19 August 1998.

002015 Mounted skeleton. Adult male. Received in 2001. This is 'Tzambo', who arrived in the small animal collection privately owned by Gordon Mills (1935–86) in the grounds of his home, 'Little Rhondda', St George's Hill, Weybridge, Surrey, UK, on 13 May 1972. He was purchased with another male, named 'Winston', from G. van den Brink, and both were less than 1 year old. They were sent to San Diego Zoo, along with two other males ('Ollie' and 'Memba'), in March 1984. 'Tzambo' was sent on loan to Los Angeles Zoo, where he arrived on 21 March 1985. He died there on 10 April 1999. [Additional information added by GH from correspondence with Jeremy Keeling, 'Little Rhondda', and from the International Studbook for the Western Lowland Gorilla, 2010.]

002016 Skeleton. Adult male. Received in 2006. This is 'Jabari', who was born in Metro Toronto Zoo, West Hill, Ontario, M1E 4R5, on 15 May 1990 to dam 'Josephine' and sire 'Charlie'. He was hand-reared from the second day due to insufficient milk production by the mother. While on loan to Dallas Zoo, he escaped from his enclosure on 18 March 2004 and, after having attacked three people within the zoo grounds, was shot dead by police.

002017 Skeletal elements. Adult female. Received c.2002. This is 'Pojo', who was born in the Yerkes Regional Primate Research Center, Emory University, Atlanta, GA 30322, on 2 October 1978 (sire 'Calabar', dam 'Paki'). She was mother-reared. She was sent on loan to Busch Gardens Zoo, Tampa, FL, on 15 April 1992, and died there on 5 November 1999, after an immobilisation procedure.

002018 Skull. Male.

002019 Skull. Male.

002020 Skull. Female.

The above 3 specimens came from the wild, but have no associated data.

002021 Skeleton. Adult female. Received in 1994. This is 'Jungle Jeannie', who arrived at Kansas City Zoological Gardens, 6700 Zoo Drive, Kansas City, MO 64132, on 21 June 1959, aged about 1 year (weight 18 lb.). She had been captured in a village 150 mi. south of Yaoundé, Cameroun, and was donated by Dr Deets Pickett. She died on 4 September 1978 of cardiac arrest due to an overwhelming sympathetic response to 4-bromo-3-methoxyamphetamine. It is assumed that a member of the general public threw a food substance containing this street drug into the animal's enclosure. [Additional information added by GH from correspondence with Kansas City Zoo personnel.]

002022 Skeleton. Adult female. Received in 2006. This is 'Maguba', who was captured in Gabon and arrived at Denver Zoological Gardens on 24 July 1971, aged about 9 months−1 year (weight 14 lb.), having been purchased from Rare Feline Breeding Compound, Inc., Center Hill, FL. She was sent on breeding loan to Dallas Zoo, where she arrived on 25 June 1991. She died there on 25 June 2003. See Knapp et al. (2007). [Additional information added by GH from correspondence with Denver Zoo personnel.]

002023 Skeleton. Adult female. Received in November 2005. This is 'Fern', who arrived at the Philadelphia Zoological Garden on 11 May 1961 (weight 13 lb.). She came originally from Cameroon. She was sold to Oklahoma City Zoological Park, Oklahoma City, OK 73111, arriving there on 24 January 1972. She died on 29 November 2005. [Additional information added by GH from correspondence with Philadelphia Zoo personnel.]

002024 Skeleton. Adult female. Received in 2006. This is 'Aquilina', who was born in the Chicago Zoological Park, Brookfield, Illinois, on 15 December 1981 to dam 'Alpha' and sire 'Samson'. She was mother-reared. She died in the Dallas Zoo, Texas, on 24 January 2000.

002025 Skeleton. Adult male. Received in November 2006. This is 'Ben', who was born in Oklahoma City Zoological Park on 12 August 1985 to dam 'Frederika' and sire 'Tatu'. He was mother-reared. He was sent on loan to Jacksonville Zoo and Gardens, 370 Jacksonville Zoo Train, Jacksonville, FL 32218, in 1998, and died there on 8 November 2006. Death was by drowning, after he fell into a moat while being chased by another male gorilla, named 'Quito' (who died of natural causes on 26 January 2013).

002026 Skeleton. Adult male. Received in 2000 from Loma Linda University World Museum of Natural History. This is 'Bum' (previously known as 'Cloudy'), who arrived at the Los Angeles Zoo on 7 March 1965 (estimated weight 12 lb.). He was purchased, with a female named 'Betsy Babinga', from Rider Animal Co., R.R.-2, Box 270, Brooksville, FL 33512. He was euthanased on 28 January 1978 because of pyogranulomatous parasitic hepatitis due to echinococcosis. [Additional information added by GH from correspondence with Los Angeles Zoo personnel.]

002049 Skeleton. Adult male. Received in August 2011. This is 'Hercules', who arrived at Baltimore Zoo, Druid Hill Park, Baltimore, MD, on 17 February 1966 (weight 19¾ lb). He came originally from Diamaré, Cameroon, and was purchased from G. van den Brink. He was received on loan at Dallas Zoo, via Pittsburgh Zoo, on 5 October 1993, and died there on 13 August 2008 after undergoing a medical procedure for spinal disease. [Additional information added by GH from correspondence with Baltimore Zoo personnel.]

002050 Skeleton. Adult female. Received in September 2010. This is 'Jenny', who arrived at Dallas Zoo, 621 East Clarendon Drive, Dallas, TX on 28 July 1957, aged about 3 years (weight 47¾ lb). She had been purchased, with a male named 'Jimmie', from F.J. Zeehandelaar, Inc., New Rochelle, at a cost of $5000 each. She was euthanased on 4 September 2008 because of an inoperable tumour in her stomach. She gave birth to a 4 lb. 2 oz. female named 'Vicki' on 18 December 1965 (sire 'Jimmie'), who was hand-reared, but never conceived again after that. [Additional information added by GH from correspondence with Dallas Zoo personnel.]

002793 Mounted skeleton. Adult male. Received in 1992. This is 'Bulu', who arrived at Monkey Jungle, Inc., 14805 SW 216th Street, Miami, FL 33170, on 18 June 1950 (weight 31 lb.). He had been captured in Cameroun by Phillip J. Carroll and was purchased from Henry Trefflich, New York, NY. He had a history of yearly episodes of upper respiratory tract infections since 1953. On 21 August 1979 he was observed with upper respiratory distress and poor appetite, and was dyspnoeic on 7 September 1979. He died on 12 September 1979 (weight 180 kg). Summary of significant necropsy findings: (1) loss of structure of liver; (2) nephritis; (3) pneumonia, numerous old scars, small area of emphysema. Preliminary diagnosis: Pneumonia (microbiology results confirmed presence of *Klebsiella pneumoniae*); compromised liver and kidney function. Final diagnosis: Heart, extensive myocardial fibrosis; liver, chronic congestion and haemosiderosis; gall bladder, fibrosis; lung, chronic pneumonia, emphysema and cartilaginous nodule; kidney, haemorrhage; spleen, congestion. Pathology report comments: The main problem in this animal appears to have been progressive cardiac failure. The myocardial fibrosis resulted in a decreased cardiac output with pulmonary and hepatic congestion. The large cartilaginous nodule probably represents a healed inflammatory lesion. [Additional information added by GH from correspondence with Sharon Dumond, Monkey Jungle, Inc.]

004456 Skeleton. Adult female. Received in 2013. This is 'Kathryn', who arrived at Philadelphia Zoo on 24 June 1964 (weight 15 lb.), having been purchased from P. H. Hastings Ltd, 182 Sultan Road, Portsmouth, UK. She was sold to Oklahoma City Zoo, where she arrived on 12 March 1971. She died on 19 February 2013. [Additional information added by GH from correspondence with Philadelphia Zoo personnel.]

004536 Skeleton. Adult female. Received in July 2011. This is 'Timbo', who arrived at Fort Worth Zoological Park, Park, Fort Worth, TX on 6 January 1964. She had been purchased, with a male named 'Mimbo', from Dr Deets Pickett, Trans World Animal Inc., 8505 Lee Boulevard, Kansas City 15, MO. The country of origin was Cameroon. Both animals were sent to San Diego Zoological Garden in exchange for the adult female 'Bata'. They arrived on 9 January 1964, at which time 'Timbo' was aged about 1¼ years (weight 29 lb.). She arrived at Dallas Zoo on 3 April 1991 on breeding loan from San Diego, via Fresno Zoo, and died there on 6 September 2011. [Additional information added by GH from correspondence with Fort Worth Zoo personnel.]

005444 Skull and neck vertebrae. Adult male. Received in July 2003. This is 'Dan II' ('Kong'), who was born in Kansas City Zoological Gardens on 16 September 1969 to the dam 'Kribi Kate' and the sire 'Big Man'. Known originally as 'The Colonel', he was hand-reared, and was sent to International Animal Exchange, Ferndale, MI, in September 1970. He died in Henry Doorly Zoological Gardens, 3701 South 10th Street, Omaha, NE 68107, on 4 March 2003. [Additional information added by GH from correspondence with Kansas City Zoo personnel.]

005482 Bones of hands and feet. Adult female.

005484 Bones of hands and feet. Adult female.

006576 Partial skeleton. Male. Received in 1991. This is 'Mickey', who was acquired by the Ringling Bros Barnum & Bailey Circus in 1971. He was sold to Arthur Jones of De Land, Florida, and deposited with Noell's Ark Gorilla Show and Chimpanzee Farm,

Tarpon Springs, Florida, in 1976. In November 1983, he was transferred to Mr Jones's Jumbolair Ranch, near Ocala, Florida, and died there of congestive heart failure on 15 September 1988 (weight 225 kg). He was a twin, probably of 'King', who was originally owned by John Berosini Jr, of Las Vegas, who sold him to L.B. Tucker, owner of the Hoxie Brothers Circus (with winter quarters in Miami). 'King' was sold to Monkey Jungle, Inc., Goulds, Florida, and arrived there on 19 December 1979, aged about 10 years. [Additional data added by GH from correspondence with Richard J. Reynolds III, Atlanta, GA.]

006577 Skull. Adult male. Received in 2002. This is 'Calabar', who arrived at the Yerkes Regional Primate Research Center, Emory University, Atlanta, GA 30322, on 16 July 1965 (weight 6.6 kg), having been purchased from Far East Animal Co., Tilburg. He died on 2 January 2002, while on loan to The Zoo, 5801 Gulf Breeze Parkway, Gulf Breeze, FL 32561. [Additional information added by GH from correspondence with Yerkes RPRC personnel.]

006793 Skeleton. Neonate female. Received in 2014. This is 'Enzi', who was born in Jacksonville Zoo, FL, on 28 March 2014 (sire 'Lash', dam 'Madini Thamani'), and died of birth trauma on the same day.

006891 Skeleton. Adult female. Received in 2014 from Loma Linda University World Museum of Natural History. This is 'Penelope', who arrived at the Cincinnati Zoo on 8 August 1957 (weight 40 lb.), having been donated by Dr Albert Schweitzer (1875–1965), Lambaréné, Gabon. She died on 3 May 1989. [Additional information added by GH from correspondence with Cincinnati Zoo personnel.]

Oxford

Miami University, College of Arts and Science, Department of Biology, Robert A. Hefner Museum of Natural History, 110 Upham Hall, Oxford, OH 45056

Paul M. Daniel

Mounted skeleton. Adult male. It was collected in West Africa by William Said of Columbus, Ohio, and was stored in a carton in his basement in Columbus at the time of his death in an automobile accident near Léopoldville in April 1952. It was given to R.A. Hefner, the museum curator, in April 1953 by K.C. Said Sr, of Columbus. The bones had been only roughly fleshed and the intercostal muscles still held the ribs together. The hard, black muscle tissues were rancid and evidence of dermestid action showed in the bottom of the carton. The skeleton was cleaned and mounted by Ward's Natural Science Establishment for $200. One clavicle and 1 patella were missing, and were replaced by a plaster model for the clavicle and a patella from a human skeleton.

Philadelphia

The Academy of Natural Sciences of Drexel University, 1900 Benjamin Franklin Parkway, Philadelphia, PA 19103-1195

Edward B. Daeschler; Ned Gilmore; Charles L. Smart Jr

ANSP 1441 Life-sized, sculpted and painted head of adult in a ferocious pose. *Gorilla castaneiceps* Slack. It was donated by Dr J.H. Slack on 9 December 1862. This 'plastotype' is mentioned in Slack (1862).

ANSP 1442 Skull (cast). Received from the Vienna Museum.

ANSP 2151 Skull (cast). Received from the Bristol Institution.

ANSP 2152 Skull (cast). Received from the Bristol Institution.

ANSP 2153 Skull (cast). Received from the Bristol Institution.

ANSP 2154 Skull. Adult female. *Troglodytes gorilla.* Collected by the Revd Dr Wilson prior to 1850. Donated by Dr Wier Mitchell on 2 December 1851.

ANSP 2157 Cranium. Adult female. *Troglodytes gorilla.* Collected by the Revd Dr Wilson prior to 1850. Donated by Dr Wier Mitchell on 2 December 1851.

The donation information for the above 2 specimens is taken from the *Proc. Acad. Nat. Sci. Philadel.* 5 (1851): 357. ANSP 2154 has 'Rev. Dr Wilson' written on the tag and in the museum catalogue. ANSP 2157 does not have the data, but it is very similar to 2154. ANSP 2155, also from Revd Dr Wilson, could be the third gorilla skull mentioned in the *Proceedings*, but is now identified as a chimpanzee.

ANSP 2566 Skeleton (partially articulated). Adult male. Gaboon country. Collected in 1851. Donated by Dr Henry A. Ford, a medical missionary, Glasstown, Gaboon River, on 3 February 1852. Prepared by F. Schafhirt, University of Pennsylvania.

ANSP 2862 Skeleton (partially articulated). Adult male. Ogooué River, French Equatorial Africa. Donated by Dr Thomas George Morton (1835–1903) before 1900.

ANSP 2863 Cranium. Adult female. Calvariotomised. Some tooth fragments in a box. Note found with the specimen reads: 'Skull of adult female gorilla. Brain case long and very low. Supraorbital ridges very large. Jaws very projecting. From Africa.' [No lower jaw.]

ANSP 3143 Skull and brain (in alcohol). Juvenile male. Kangwe, Ogooué River, French Equatorial Africa.

Collected by the Revd Robert Hamill Nassau (1835–1921). Donated by Dr Thomas G. Morton on 29 March 1892. The auditory ossicles were removed on 30 December 1940, and are in vials with the specimen. Calvarium missing. Missing teeth.

ANSP 3144 Partial cranium and brain (in alcohol). Juvenile male. Kangwe, Ogooué River, French Equatorial Africa. Collected by the Revd R.H. Nassau. Donated by Dr Thomas G. Morton on 29 March 1892. Skull sectioned; centre section of skull and roof remain, no lower jaw.

ANSP 3145 Cranium and brain (in alcohol). Juvenile male. Kangwe, Ogooué River, French Equatorial Africa. Collected by the Revd R.H. Nassau. Donated by Dr Thomas G. Morton on 29 March 1892. Skull cut in several places, some parts repaired but most of the roof missing.

ANSP 3319 Partial skeleton.

ANSP 3320 Partial skeleton. It is missing.

ANSP 5530 Skeleton. Adult male. Donated on 17 March 1900. From the E.D. Cope collection.

ANSP 11805 Mounted skin, skull removed. Male. Lokundje, South Cameroon. Collected by George Zeuker. Donated by Dr Thomas Biddle on 3 November 1903. Sent in exchange to the South African Museum in 1939.

ANSP 12601 Mounted skin. Female. Gabon. Donated by Dr Thomas Biddle on 9 November 1912.

ANSP 12602 Mounted skin. Juvenile. Gabon. Donated by Dr Thomas Biddle on 9 November 1912.

ANSP 16980 Skull. Adult male. Nola region, Moyen-Congo, French Equatorial Africa. Purchased from local inhabitants by James A.G. Rehn (George W. Vanderbilt African Expedition) on 29 October 1934. Received in 1935. Left malleus and incus, right malleus removed and stored in separate vials.

ANSP 16981 Skeleton and mounted skin. Adult male. Barundu, 22 mi. northeast of Nola, 15 mi. east of Sangha River, Moyen-Congo. Killed by local inhabitants; collected by Harold T. Green (George W. Vanderbilt African Expedition) on 6 November 1934. Received in 1935. Height 68½ in.; arm spread 92 in.

ANSP 16982 Skeleton and mounted skin. Adult male. Aboghi, 40 mi. southwest of Nola, near west bank of Sangha River, French Equatorial Africa. Killed by local inhabitants; collected by Harold T. Green (George W. Vanderbilt African Expedition) on 7 November 1934. Received in 1935. Height 62 in.; arm spread 88¼ in. Left malleus and incus removed and stored in separate vial. Holes drilled through back of skull.

ANSP 16983 Skeleton and mounted skin. Subadult male. Aboghi, 40 mi. southwest of Nola, near west bank of Sangha River, French Equatorial Africa. Killed by local inhabitants; collected by Harold T. Green (George

W. Vanderbilt African Expedition) on 8 November 1934. Received in 1935. Height 55 in.; arm spread 79 in. Left incus and stapes removed and stored in separate vial.

ANSP 16984 Skeleton and mounted skin. Adult male. Aboghi, 40 mi. southwest of Nola, near west bank of Sangha River, French Equatorial Africa. Killed by local inhabitants; collected by Harold T. Green (George W. Vanderbilt African Expedition) on 9 November 1934. Received in 1935. Height 72 in.; arm spread 107 in. Left malleus, right malleus and stapes removed and stored in separate vials.

For further information on the above 4 specimens, ANSP 16981−ANSP 16984 inclusive, see Coolidge (1936).

Taxidermy by Louis Paul Jonas, started 15 March 1937, finished 20 October 1937. Display case and contents finished 13 December 1938, group open to public 16 December 1938.

ANSP 21587 Skeleton and skin. Adult female. One hundred miles northwest of Fort Rousset, 1°N, 15°E, Oka, French Equatorial Africa, altitude 1200 ft. Collected by the William K. Carpenter Africa Expedition on 28 January 1948. It is missing.

ANSP 21588 Skeleton and skin. Adult male. One hundred miles northwest of Fort Rousset, 1°N, 15°E, Oka, French Equatorial Africa, altitude 1200 ft. Collected by the William K. Carpenter Africa Expedition on 28 January 1948. The skin is missing.

ANSP 21589 Skeleton and skin. Adult female. One hundred miles northwest of Fort Rousset, 1°N, 15°E, Oka, French Equatorial Africa, altitude 1200 ft. Collected by the William K. Carpenter Africa Expedition on 28 January 1948. The skin is missing.

ANSP 21633 Skull. Adult male. This is 'Bamboo', who arrived at the Philadelphia Zoo on 5 August 1927 (weight 11 lb.). He died on 21 January 1961, due to a coronary occlusion (weight 281 lb.). Donated by Frederick A. Ulmer Jr. Right malleus and incus removed and stored in a separate vial. [Some additional information added by GH from correspondence with Fred A. Ulmer Jr]

ANSP 23653 Incisor tooth. Male. 'Bamboo II' shed this tooth and handed it to his keeper. Donated by Frederick A. Ulmer Jr, on 2 August 1966. 'Bamboo II' arrived at the Philadelphia Zoo on 11 May 1961 (weight 19 lb.) and died of chronic ulcerative colitis on 25 September 1967. [Some additional information added by GH from correspondence with Fred A. Ulmer Jr]

ANSP 23656 Premolar tooth. Female. 'Fern' shed this tooth on 30 January 1967; it was picked up by 'Bamboo II', who handed it to his keeper. Donated by Frederick A. Ulmer Jr.

ANSP 23659 Molar tooth. Female. Shed by 'Fern'. Donated by Frederick A. Ulmer Jr, on 6 March 1967.

The College of Physicians of Philadelphia, Mütter Museum, 19 S. 22nd Street, Philadelphia, PA 19103

Anna N. Dhody; Hanna Polasky

1660.07 Transverse section of skull, including the maxilla. *Gorilla?* Donated by the M.H. Cryer Anatomical Collection. The canines are missing.

1660.107 Skull. Adult male. *Troglodytes gorilla*. The upper central incisors are missing and have been replaced by wooden replicas.

1660.109 Skull. Adult female. *Troglodytes gorilla*. The upper right central incisor and both lower central incisors are missing and have been replaced by wooden replicas.

The above two skulls were purchased for £12 by Dr Joseph Leidy in London on 15 June 1876. Leidy was paid $69.40 by the museum for these skulls.

1990.1660.105A Cranium. Adult male. Donated by the M.H. Cryer Anatomical Collection. Several teeth are missing.

Philadelphia Zoo, 3400 West Girard Avenue, Philadelphia, PA 19104

Keith C. Hinshaw

Paraffin-embedded tissue blocks, histopathology glass slides and written pathology records on 18 individual gorillas dating back to 1961; frozen samples of serum and/or plasma from 24 individual gorillas dating back to 1979; stained blood smears on glass slides from 23 individual gorillas dating back to 1980.

Research enquiries involving these materials should be directed to the chairman of the Philadelphia Zoo's Conservation and Science Committee, and access to any material is dependent upon approval of a completed research proposal by the Committee.

University of Pennsylvania, Museum of Archaeology and Anthropology, 3260 South Street, Philadelphia, PA 19104

Janet M. Monge

Cranium. Adult male.

Skull and partially articulated postcranial skeleton. Adult male. The lower jaw was fractured and pinned. There is extensive dental pathology.

4491 Skull. Adult male. Received on loan from The Wistar Institute Museum, 3601 Spruce Street, Philadelphia, PA 19104, in 1988. It had been donated to The Wistar Institute by the Medical Department, University of Pennsylvania, in 1893. It was apparently part of a complete, mounted skeleton when donated, but the present whereabouts of the postcranial skeleton is unknown. Some teeth are missing. [Additional information from The Wistar Institute: April Miller; Richard J. Walsh]

LI-2012-13 Skeleton. Female. Received on long-term loan from the Philadelphia Zoo.

Wagner Free Institute of Science, Natural History Museum, 1700 W. Montgomery Avenue, Philadelphia, PA 19121*

John Graham

Mounted skeleton. It is believed to have been purchased by Joseph Leidy in England sometime during the late 1860s or 1870s.

Skull. Adult male. Gabon. Collected in 1931. The jaw is articulated.

Pittsburgh

Carnegie Institute, Carnegie Museum of Natural History, 4400 Forbes Avenue, Pittsburgh, PA 15213-4080

Suzanne B. McLaren

NB: All collection data files of the Section of Mammals, Carnegie Museum of Natural History, have been computerised, and selected data for the entire collection of Recent primates have been published. See McLaren et al. (1984).

CM 145 Mounted skin. Adult male. Kangvé, Ogooué River, Gabon. From the same animal as the mounted skeleton, no. 92137.

CM 1786 Skull and skin. Adult female. Kamerun, possibly in the vicinity of Butanga. Collected by W.E. Johnstone on 23 July 1908.

CM 5103 Skull. Adult male. Métet, Cameroun. Collected by the Revd A.I. Good on 25 April 1919.

CM 5359 Fetus (in formalin). Female. Near Efulen, Kamerun. Collected by medical missionary Dr Hymen L. Weber in 1910. It corresponds in development to human fetuses of the beginning of the 15th week. Vertex to coccyx 85 mm; sitting height 88.5 mm; upper arm length 25 mm; leg length 40 mm. See Schultz (1927). The placenta, uterus and ovaries were studied and reported on separately: see Wislocki (1929).

CM 5360 Skull. Adult male. Cameroun. Collected by the Revd A.I. Good on 27 April 1935.

CM 20985 Skeleton. Adult male. Ebolowa, SW Cameroun. Collected by P.H. Combs on 7 February 1921. The facial region is heavily damaged.

CM 20986 Skeleton. Adult female. Mindini, Ebolowa, Cameroun. Collected by P.H. Combs on 10 May 1921.

CM 20987 Skeleton. Juvenile female. Mindini, Ebolowa, Cameroun. Collected by P.H. Combs on 9 October 1921. The left side of the skull is damaged.

CM 40596 Skeleton and skin. Juvenile male. Donated by the Pittsburgh Zoological Garden, where it died on 21 December 1965.

CM 42753 Skull. Adult male? Twenty-five miles southwest of Ebolowa, Cameroun. Collected by the Revd A.I. Good in 1941. The left side is damaged.

CM 42754 Skull. Adult male. Ebolowa, Cameroun. Collected by the Revd A.I. Good on 10 June 1941. Some teeth are missing.

CM 42755 Skull. Adult male. Twenty-five miles southeast of Ebolowa, Mékomo, Cameroun. Collected by the Revd A.I. Good in 1941.

CM 42756 Skull. Adult male. Twenty-five miles southeast of Ebolowa, Mékomo, Cameroun. Collected by the Revd A.I. Good in 1941. The basisphenoid is damaged.

CM 59606 Skin. Juvenile female. Donated by the Pittsburgh Zoological Garden.

CM 60283 Postcranial skeleton and mounted skin. Adult male. Gabon. Donated by the Pittsburgh Zoological Garden. This is 'George', who arrived at the Pittsburgh Zoo on 9 June 1966, aged about 2 years (weight estimated at 47 lb.). He had been purchased for $3750 from F.J. Zeehandelaar, Inc. He died on 23 December 1978 of brain inflammation resulting from a dental abscess (weight 325 lb.). The skin was mounted by the Otto M. Epping Taxidermy Studio, Chicora, PA. The skull was sent to the University of Pittsburgh School of Dental Medicine, but now appears to be missing. [Some additional information added by GH from correspondence with Carol Glick, Pittsburgh Zoo.]

CM 92137 Mounted skeleton. Adult male. Kangvé, Ogové River, Gabon. Purchased by W.J. Holland (with the skin CM 145) from the Revd Dr A.C. Good in 1886, and transferred to the Carnegie Museum in 1898. This specimen has an injured left humerus. See Holland (1924).

University of Pittsburgh, Faculty of Arts and Sciences, Department of Anthropology, Pittsburgh, PA 15260

Mark P. Mooney

Skull. Adult male. The lower jaw is articulated.

Skeleton. Adult male. Acquired from the Pittsburgh Zoo. This is 'Sultan', who arrived at the Pittsburgh Zoo on 12 August 1986, having been obtained from Alberta Game Farm, Polar Park, Edmonton, Alberta, Canada. Alberta Game Farm had bought him from a dealer (van den Brink?) in 1970. The dealer reportedly captured him in Zaïre, but the zoo vet suggested that his markings were similar to gorillas from Cameroon. 'Sultan' died on 8 October 1988, aged 19 years. Following autopsy no specific cause of death could be ascertained, but it was suggested that a pericardial infection may have been the cause. The skeleton is complete, with the exception of the entire sternum, proximal ends of the clavicles and sternocostal articulations and the pubic symphysis (the anterior wall of the thoracic and abdominal cavities was removed at the time of autopsy and was never recovered from the zoo). The specimen was embalmed and used initially as comparative prosection material for primate anatomy classes, after which the skeleton was macerated for classroom use.

Princeton

Princeton University, Museum of Natural History, Guyot Hall, Princeton, NJ 08544-1003*

David C. Parris, Curator of Natural History, New Jersey State Museum, Trenton

NB: The museum was closed in September 2000 to allow for the renovation of Guyot Hall. Most of the museum contents will remain in storage until a new facility can be made available.

PU-Osteo-13 Mounted skeleton. Male. *Gorilla savagei.* Purchased from H.A. Ward. It lacks the lower incisors and has damage to various other teeth. The upper right 2nd molar is missing, apparently due to an abscess.

PU-Osteo-14 Skull. Female. *Gorilla savagei.* Ogove River, 100 mi. from coast, 1°S. Presented by the Revd R. H.Nassau.

PU-Osteo-15 Skull. Female. *Gorilla savagei.* Ogove River, 100 mi. from coast, 1°S. Presented by the Revd R. H. Nassau.

PU-Osteo-16 Skull. Male. *Gorilla savagei.* Ogove River, 100 mi. from coast, 1°S. Presented by the Revd R. H. Nassau.

PU-Osteo-668 Skull. Female. From the M.F. Ashley Montagu collection. The anterior teeth are lacking.

PU-Osteo-12889 Mandible.

Bronze bust, by Carl Akeley, of the 'Old Man of Mikeno', c.1923. It measures 30 in. wide by 25½ in. high, and was presented by Mrs James Barnes of Princeton on 10 May 1968. It has long been used in conjunction with the skeleton for instructional purposes.

In addition to the above, there was a skull from Cameroun, but it was stolen from a display case quite some decades ago.

Riverside

La Sierra University, World Museum of Natural History, E.E. Cossentine Hall, 4500 Riverwalk Parkway, Riverside, CA 92515*

E.A. Hankins III

Carcase (freeze-dried). Infant male. Captive born in the San Diego Zoo, where it died c.1976, aged 1 week. The brain and viscera were removed for the lyophilisation (freeze-dry) process so that the specimen consists of skin, muscle and skeleton.

Carcase (freeze-dried). Adult male. This is 'Captain Jack', who arrived at the Los Angeles Zoo, along with another male and 3 females, on 17 November 1965. They were all purchased from Far East Animal Co., Lindeplein 11, Tilburg, Netherlands. 'Captain Jack' was euthanased on 27 January 1978 (weight 159 kg) because of advanced infestation with hydatid cysts (*Echinococcus granulosus*). The entire body cavity and internal organs were infested with hydatid cysts, some up to 10 cm diameter. The brain and viscera were removed for the freeze-dry taxidermy process. [Additional information added by GH from correspondence with Los Angeles Zoo personnel.]

Rochester

Rochester Museum and Science Center, 657 East Avenue, Rochester, NY 14607

Charles F. Hayes III; George C. McIntosh

AP 936 Skull. Adult female. Purchased from William Kruse, Rochester, NY, in 1952.

AP 937 Skull. Adult male. Purchased from William Kruse, Rochester, NY, in 1952.

Rock Hill

Museum of York County, 4621 Mount Gallant Road, Rock Hill, SC 29732

Steven E. Fields; C.W. Hall

1963.1.1 Mounted skin. Adult male. About 100 mi. east of Nioungou, Setté Cama, Gabon. Collected by Frank E. Delano on 17 January 1963. The skin was donated by Mr Delano to the Los Angeles County Museum of Natural History in June 1963 but was not considered good enough for mounting. The donor did not want it kept in the study collection, so it was returned to him on 10 May 1965 for shipment to the Nature Museum of York County, where it was mounted by resident taxidermist John Schneider in 1967. The animal was killed by 2 shots from a .44 carbine. The skull, pelvis, 1 shoulder, 1 arm and both bones from the left leg were saved for use in the taxidermy. Height, top of head to bottom of feet 64 in.; chest girth 62 in.; length of arm, shoulder to wrist, outside of arm 38 in.; shoulder width 22 in.; length of leg, pelvis to ankle 22 in.; belly girth 53 in.; circumference of thigh, under crotch 26 in.; depth of chest 22 in.; full height, bottom of feet to knuckle with arms overhead 102 in.

Saint Louis

Saint Louis Science Center, 5050 Oakland Avenue, St Louis, MO 63110

Ronald R. Beer; Lynn C. Fendler; Kristina Hampton

1965-20 Mounted skeleton. Adult male. Donated by St Louis Zoological Park, Forest Park, St Louis, MO 63110, in 1965. This is 'Phil', who arrived at the St Louis Zoo on 9 September 1941 (weight 26 lb.), along with another male ('Kuyon') and two females ('Mussie' and 'Mattite'). They had been captured by Phillip J. Carroll near Okenga, French Equatorial Africa, and were purchased for $14,000 from Henry Trefflich. 'Phil' suffered an attack of pneumonia in November 1948, which was treated with penicillin. Early in his captivity, he developed the habit of chewing on the ball of his left heel. His keepers tried to break the habit by smearing the foot with bitter compounds. A round-the-clock watch was set up by keepers armed with an air rifle, and when 'Phil' leaned back to nibble, he would be peppered with shot. All attempts failed, and by the time of his death the foot bore a round hole on the ball from years of gnawing. His final illness began on 18 October 1958, and from 25 October he ate little. He died on 1 December 1958 of strongyloidiasis, ulcerative colitis, and chronic and acute regional peritonitis. His skin was mounted by Louis Paul Jonas, Hudson, NY, at a cost of $3500, plus a further $3500 for the native jungle display setting, and has been on exhibition within the St Louis Zoo since May 1959. His skeleton was set up for display in the Museum of Science and Natural History, Oak Knoll Park (now known as the St Louis Science Center) by Turtox, Chicago. [Additional data added by GH from correspondence with St Louis Zoo personnel, and from copies of local newspaper reports.]

1983-45 Mounted skin. Adult male. Donated by St Louis Zoological Park on 15 September 1983. This is 'Mzuri', who was born on 16 October 1966 to dam 'Trudy' and sire 'Rudy'. He was removed for hand-rearing 1 day after birth (weight 4 lb. 10 oz.). He died of cardiomyopathy on 24 January 1983 (weight 370 lb.). His skin was mounted by Schwarz Studio. His skeleton is preserved in the St Louis Zoo.

Washington University, Department of Anthropology, 1 Brookings Drive, St Louis, MO 63130

Jessica L. Joganic; Stephen Molnar

M-2100 Skull. Adult male.

M-2101 Skeleton. Subadult male. Donated by St Louis Zoo in 2009. This is 'Muchana', who was born in Columbus Zoo on 23 June 2000 (to dam 'Jumoke' and sire 'Anakka'), and died of strangulation (by a rope hammock) in St Louis Zoo on 28 February 2009 (weight 149 kg). He was owned by Philadelphia Zoo. None of the bones are articulated but all (and accompanying epiphyses) are labelled. See Joganic (2016).

Saint Paul

University of Minnesota, College of Biological Sciences, James Ford Bell Museum of Natural History, 10 Church Street SE, St Paul, MN 55455

Sharon A. Jansa; Kristin M. Kramer; Gerda E. Nordquist

MMNH 6323 Mounted skeleton. Juvenile.

MMNH 6326 Skeletal elements: a few phalanges and the last 2 pieces of the coccyx.

MMNH 6327 Skeleton. Male.

MMNH 6336 Mounted skeleton. Subadult female.

San Angelo

Angelo State University, Department of Biology, ASU Station No. 10890, San Angelo, TX 76909-0890

Robert C. Dowler; Marcia A. Revelez

ASNHC 5670 Skull. Adult female. This is 'Sukari', who arrived at the New York Bronx Zoo on 10 May 1966 (weight 22.68 lb.), having been purchased for $4000 from F.J. Zeehandelaar, Inc. She was euthanased on 6 February 1985. Top of cranium removed, but now glued together. [Additional information added by GH from correspondence with NY Bronx Zoo personnel.]

ASNHC 5671 Skull (bisected sagittally) and skeleton of right arm. Adult male. This is 'Bendera', who arrived at the New York Bronx Zoo on 6 April 1965 (weight 23 lb.), having been purchased for $4125 from F.J. Zeehandelaar, Inc. He died on 24 October 1986 during recovery from anaesthetic: pulmonary haemorrhage; myocardial degeneration; pleural fibrosis; pericardial effusion; gastric mucosal petechiation. [Additional information added by GH from correspondence with NY Bronx Zoo personnel.]

San Diego

San Diego Museum of Man, 1350 El Prado, Balboa Park, San Diego, CA 92101

Tori D. Randall; Rose A. Tyson

1221 Skull. Infant. Brought from the Smithsonian Institution in 1915 for inclusion in the Panama-California International Exposition. It bears the Smithsonian number 174721.

1223 Skull. Adult male. Kamerun. Brought from the Smithsonian Institution in 1915 for inclusion in the Panama-California International Exposition.

2000 Skeleton, mounted skin, and knuckle and foot prints (in plaster). Adult male. *Gorilla beringei graueri*. Alimbongo Mountains, Belgian Congo. All materials are from 'Mbongo', who arrived at the San Diego Zoological Garden (with another male, 'Ngagi') on 5 October 1931, aged about 5 years. He attained a weight of 618 lb. on 1 June 1941. He died of coccidioidal granuloma on 16 March 1942, having been ill for a total of 45 days (weight 580½ lb.). Height 77 in. [?]; arm span 97½ in.; chest circumference (below nipples) 69 in.; waist 72 in.; upper arm 18¾ in.; forearm 15⅜ in.; wrist 14⅜ in.; thigh 27¼ in.; calf 15⅜ in.; brain weight 565 g; heart 1134 g; liver 5000 g; spleen 680 g. The teeth were mature, but showed considerable erosion; the incisor teeth were in malocclusion, the upper cuspids occluded distal to the lower cuspids. See McKenney et al. (1944). The skeleton, prepared by Ward & Co., New York, and the brain were donated by the San Diego Zoological Society in October 1942. The skin was mounted by Elton Green at the San Diego Natural History Museum and is on permanent loan to the San Diego Museum of Man, being the property of the San Diego Zoological Society. The following principal dimensions of the skeleton were recorded by David P. Willoughby on 16 October 1942: Skull. Greatest length, prosthion—inion 310 mm; zygomatic width 194 mm; cranial length, glabella—inion 190 mm; cranial width, biparietal 111 mm; basal length, prosthion—inion 230 mm; upper face height, nasion—prosthion 152 mm; total face height, nasion—gnathion 225 mm; length nasion—basion 162 mm; greatest width across jaws posteriorly 173 mm; auricular height, vertex—porion 103 mm; basal height, vertex—basion 133 mm; cranial capacity (using sand) 600 cc. Sternum. Maximum length, without xiphoid process 225 mm; manubrium, midsagittal length 75 mm; manubrium, greatest breadth 103 mm; corpus length 150 mm; corpus breadth 74 mm. Scapula. Morphological length 187 mm; spine length 248 mm; acromion breadth 43 mm; fossa supraspinata 104 mm; breadth, fossa infraspinata 156 mm; axillo-spinal angle 33 degrees. Pelvis (left os coxae). Ilium length 282 mm; ischium length 133.5 mm; pubis length 122 mm; ilium breadth 236 mm; fossa iliaca width 205 mm; sacral surface width 40 mm; sacral surface length 128 mm; fossa iliaca height 26 mm. Limb bones. Humerus length r. 451 mm, l. 448 mm; humerus, bicondylar breadth r. 123 mm, l. 126 mm; radius length r. 345 mm, l. 350mm; femur length, trochanteria r. 394 mm, l. 397 mm; femur, bicondylar breadth r. 119 mm, l. 116 mm; tibia length r. 321 mm, l. 322mm; clavicle length r. 184.5 mm, l. 184 mm; clavicle girth, at middle r. and l. 60 mm. Limb bone indices.

Brachial index, r. 76.5, l. 78.1; crural index r. 81.5, l. 81.1; intermembral index r.111.3, l. 111.0 (thus 'Mbongo' had relatively short forearms and long tibiae). Hand skeleton (left). Carpus length 25 mm; 3rd metacarpal length 90 mm; 3rd digit, basal phalanx 65 mm; 3rd digit, middle phalanx 44 mm; 3rd digit, terminal phalanx c. 21 mm? total c. 130 mm; 1st metacarpal length 49 mm; 1st digit, proximal phalanx 27 mm; 1st digit, distal phalanx 21 mm; total 97 mm; total hand length c. 245 mm. Foot skeleton (left). Tarsus length 123 mm; 3rd metatarsal length 78 mm; 3rd toe, basal phalanx length 47 mm; 3rd toe, middle phalanx length 28 mm; 3rd toe, terminal phalanx length c. 21 mm?; 3rd ray length c. 174 mm; 1st metatarsal length 65 mm; 1st metatarsal, proximal phalanx 31 mm; 1st metatarsal, distal phalanx c. 26 mm; 1st ray length c. 122 mm; total foot length c. 297 mm. Restored living dimensions: stature 1714 mm; sitting height 1096 mm; anterior trunk height 683 mm.

1954-45-1 Pickled brain of the above specimen ('Mbongo'). The brain is about 4½ in. long, 3½ in. wide, and 2½ in. thick.

San Diego State University, College of Sciences, Department of Biology, North Life Sciences Room 102, 5500 Campanile Drive, San Diego, CA 92182

Robert J. Thiltgen

S-504 Mounted skeleton. Stillborn female. Received from the San Diego Zoological Garden on 17 October 1978.

San Francisco

California Academy of Sciences, Natural History Museum, Golden Gate Park, San Francisco, CA 94118-4599

Stephen F. Bailey; John P. Dumbacher; Moe Flannery; Douglas J. Long; Jacqueline Schonewald

CAS:MAM: 4980 Skeleton and mounted skin. Adult male. *Gorilla beringei beringei*. Virunga volcanoes, Belgian Congo, altitude 10,000 ft. Purchased for 1000 francs in Belgian Congo by A.K. Macomber, and received from him on 25 July 1924. The skeleton lacks 1 patella.

CAS:MAM: 6491 Mounted skeleton. Adult female. Gabon? Received from the Memorial Museum on 28 April 1931. The Memorial Museum, now defunct, was in the adjacent building, which is now occupied by the De Young Museum, which is an art museum. The animal was about 11 years old.

CAS:MAM: 20943 Skeleton. Adult female. Received from San Francisco Zoological Gardens, 1 Zoo Road, San Francisco, CA 94132. This is 'Jacqueline', who arrived at the San Francisco Zoo on 7 April 1964. She was purchased from the Brookfield Zoo, Chicago, where she had arrived on 6 July 1957 (weight 24 lb.), having been purchased from Henry Trefflich. She was originally thought to be male, and was known as 'Jacob'. She died on 9 May 1978. [Additional information added by GH from correspondence with San Francisco Zoo personnel.]

CAS:MAM: 23931 Skeleton. Adult male. Cameroun. This is 'Bwana', who arrived at the San Francisco Zoo, along with a female named 'Missus', on 21 October 1959 (weight 45 lb.). The pair had been purchased from Dr Deets Pickett. 'Bwana' died on 3 September 1994. [Additional information added by GH from correspondence with San Francisco Zoo personnel.]

CAS:MAM:30497 Frozen carcase (sent to Adrienne Zihlman, University of California, Santa Cruz, for use in research; once completed, the prepared specimen will be returned to CAS). Infant female. This is 'Kabibe', who was born in San Francisco Zoo on 17 July 2013 (sire 'Oscar Jonesy', dam 'Nneka'). She was killed in an accident on 7 November 2014.

University of the Pacific, Arthur A. Dugoni School of Dentistry, P&S Comparative Anatomy Collection, 155 Fifth Street, San Francisco, CA 94103

Mitchell B. Day; Dorothy Dechant
CV 103 Skull. Adult male.
Skull (cast). Adult female.
Both were part of a private collection transferred to the University in the mid-1960s.

Seattle

Seattle Historical Society, Museum of History and Industry, 860 Terry Avenue North, Seattle, WA 98109

Betsy Bruemmer; Elizabeth Furlow; Mrs S. Gustison
1968.4393 Mounted skin. Adult male. Received from Woodland Park Zoological Gardens, 5500 Phinney Avenue North, Seattle, WA 98103, in April 1968. This is 'Bobo', who was captured when 2 weeks old in July 1951 by William Said at Bangui Basse-Kotto, north of Macombo, Oubangui-Chari. Said took the infant gorilla back to Columbus, Ohio, where he was placed in the care of Said's mother. There he was seen and purchased for $4000 by Bill Lowman, who took him home to Anacortes, WA, in November 1951. He was looked after by Bill's mother and father, Jean and Raymond Lowman, for the next 2 years. He was then purchased for $5500 by Woodland Park Zoo, and arrived there on 5 December 1953 (weight 61 lb.). He died on 21 February 1968

(weight 460 lb.) of a fracture of the larynx with haemorrhage into the neck, resulting in asphyxia. He also had thrombophlebitis involving the left leg, pelvic veins and inferior vena cava, estimated at 1 month's duration. The skin was mounted by Jonas Bros of Seattle, Inc.

University of Washington, The Burke Museum of Natural History and Culture, Box 353010, Seattle, WA 98195-3010

Jeffrey E. Bradley; John Rozdilsky
UWBM 35403 Skeleton. Adult female. Cameroon. This is 'Caboose', who arrived at the Woodland Park Zoo with 2 other females ('Nina' and 'Engine') on 26 November 1968. They were all aged about 5–6 months, and were purchased from Frans M. van den Brink (Gabria van den Brink), Soest, Netherlands. 'Caboose' died on 25 October 1974 (weight 42.7 kg). She was euthanased following the results of an exploratory laparotomy which revealed massive metastases, a nonoperable malignant tumour. Total length 914 mm; foot length 270 mm. [Some additional information added by GH from correspondence with Woodland Park Zoo personnel.]

UWBM 35427 Skeleton. Adult male. This is 'Bobo' from Woodland Park Zoo (see above, under the Seattle Historical Society Museum of History and Industry). It is missing the atlas, hyoid and two distal phalanges.

UWBM 39024 Scalp and skeleton. Adult male. South Cameroon. This is 'Kiki', who arrived at Woodland Park Zoo (with another male named 'Pete') on 27 June 1969. Both were aged about 6 months and were purchased from G. van den Brink. 'Kiki' had testicular surgery in 1986, when a Leydig cell tumour was removed. He died on 7 September 1991 of an apparent heart attack. He had ankylosing hyperostosis. Total length 1115 mm; foot length 315 mm; ear length 57 mm; weight 193.2 kg (on 19 April 1991). [Some additional information added by GH from correspondence with Woodland Park Zoo personnel.]

Skeleton (in preparation). Adult female. This is 'Nina', who arrived at Woodland Park Zoo on 26 November 1968 (see UWBM 35403 above). She was euthanased on 22 May 2015.

Sioux Falls

The Great Plains Zoo and Delbridge Museum of Natural History, 805 South Kiwanis Avenue, Sioux Falls, SD 57104-3714*

Christine I. Anderson
2086460 Mounted skin. Adult female. It was collected by Henry Brockhouse, a local businessman and hunter, probably between 1970 and 1975. Brockhouse died in

1978 and in 1981 Mr C.J. Delbridge purchased his entire collection and donated it to the city. The skin was mounted over a fibreglass model by Jonas Brothers Taxidermy Studios, Bloomfield, CO.

Tacoma

University of Puget Sound, James R. Slater Museum of Natural History, Thompson/Harned Hall, Room 295, Union Avenue at N. 15th Street, Tacoma, WA 98416

Ellen B. Kritzman; Gary Shugart

PSM 9962 Skeleton and skin. Infant female. Republic of the Congo. Obtained from Mr Ron Irwin, B & I discount merchandise store, 8012S. Tacoma Way, Tacoma, WA 98499, which had a large pet section, where it died on 5 January 1965, aged c. 6 months, after suffering from pneumonia for 2 weeks. Total length 450 mm; leg length 350 mm; foot length 145 mm; arm length 410 mm; hand length 130 mm; ear 47 mm × 30 mm; weight 31 lb. A male obtained with the female grew to adulthood at the B & I Store, and was named 'Ivan'. He was transferred to Woodland Park Zoo, Seattle, on 6 October 1994, and was sent on loan to Zoo Atlanta on 12 October 1994. He died there while under anaesthetic on 20 August 2012. NB: 'Ivan' was listed as a *Gorilla gorilla gorilla*, therefore both specimens presumably came from the extreme southwestern region of the Republic of the Congo.

Tempe

Arizona State University, School of Human Evolution and Social Change, Industrial Arts Building, 900 Cady Mall, Tempe, AZ 85287

Roy A. Barnes; Arleyn W. Simon

Skull. Subadult male. Received from the Phoenix Zoo in February 1969. This is 'Mongo', who arrived at the Phoenix Zoo, with a female named 'Hazel', on 9 October 1961 (weight 22 lb.). They were purchased from Dr Deets Pickett, Trans World Animal, Inc., Kansas City, MO. 'Mongo' died of coccidioidal granuloma on 3 February 1969. [Additional information added by GH from correspondence with Wayne G. Homan, Phoenix Zoo.]

Mounted skeleton. Adult male. Received from the Phoenix Zoo in September 1972. This is 'Jackie', who arrived at the Phoenix Zoo on 22 July 1970, having been purchased from Baltimore Zoo. He was obtained on a Baltimore Zoo expedition at Yaoundé, Cameroun, in March 1954, and arrived at the zoo on 9 April 1954 (weight 15 lb.). He died of coccidioidal granuloma on 6 September 1972. [Some additional information added by GH from correspondence with Baltimore Zoo personnel.]

Urbana

University of Illinois at Urbana-Champaign, College of Liberal Arts and Sciences, Department of Anthropology, 109 Davenport Hall, 607 S. Mathews Avenue, Urbana, IL 61801

Donald F. Hoffmeister, Museum of Natural History; Wayne T. Pitard, Spurlock Museum

304 Mounted skeleton. Adult male. Cameroun. Purchased from Turtox, Chicago, in the early 1900s. It was formerly kept in the University's Museum of Natural History, 1301 West Green Street, Urbana, which closed in 1998.

Washington

Yakovlev-Haleem Collection, National Museum of Health and Medicine, US Army Garrison Forest Glen, 2500 Linden Lane, Silver Spring, MD 20910*

Este Armstrong

Brain (sectioned, stained and mounted). It came from a gorilla shot by a hunting party in West Africa in 1939. It was immersed in formalin, but after 2 years it was cut into 5 blocks and then the blocks were sectioned and stained by Paul Ivan Yakovlev (1894−1983). Because the fixation was poor, the staining is poor. The Loyez's lithium haematoxylin stained sections are the only sections that are of use.

Brain (sectioned, stained and mounted). Adult male. This is from 'Makoko', who arrived at the New York Zoological Park (Bronx Zoo) on 7 September 1941, aged about 2 years (weight 28.16 lb.). He was purchased for $4000 from Henry Trefflich, 215 Fulton Street, New York, NY. A female named 'Oka' arrived with him, having been purchased from Trefflich for $3500. 'Makoko' died of submersion on 13 May 1951: he accidentally slipped into the moat of his outside enclosure and drowned. Height (heel to crown) 58 in.; arm span 92 in.; leg length (heel to centre of inguinal) 26 in.; sitting height 39 in.; greatest chest circumference 51 in.; weight 448 lb.; brain weight 540 g. The brain was blocked, and sections were stained with both Nissl's iron-haematoxylin and myelin stains. The staining is excellent and this brain has been used in many different comparative neuroanatomic studies. [Some additional information added by GH from correspondence with NY Bronx Zoo personnel.]

Smithsonian Institution, National Museum of Natural History, 10th Street and Constitution Avenue, NW, Washington, DC 20560-0108

Craig Ludwig

This museum's collection of great ape skeletal material has been examined by Nancy C. Lovell as part of her research on skeletal pathology in anthropoid apes, and the results have been published. See Lovell (1990a,b). See also Chapter 4, Noninfectious Diseases and Host Responses in Part I, Gorilla Pathology and Health.

A 3646 Skull (cast). Gabon. Received from Ford.

A 12288 Skull (cast). Received from J. Palmer.

A 22460 Skeleton. Female. Collected by Dr Beauregard in 1883; field no. 2029.

USNM 154553 Skeleton. Adult male. Collected by V. Fric.

USNM 154554 Skull. Subadult male. Collected by V. Fric.

USNM 174697 Skeleton. Female. Nkomi River, Gabon. Collected by Richard L. Garner in 1904. It is incomplete.

USNM 174698 Skeleton. Adult female. Gabon. Collected by Richard L. Garner in 1904. It is incomplete.

USNM 174705 Cranium. Adult female. Lake Nkomi, Gabon. Collected by Richard L. Garner in 1905.

USNM 174711 Skull (sectioned sagittally) and cast of whole (=not sectioned) skull. Male. Gabon. Collected by Richard L. Garner in 1905.

USNM 174712 Tarsals of 1 foot. Male. Lac Fernan Vaz, Gabon. Collected by Richard L. Garner in 1905.

USNM 174713 Skull. Adult male. Lac Fernan Vaz, Gabon. Collected by Richard L. Garner in 1905.

USNM 174714 Skull. Adult male. Lac Fernan Vaz, Gabon. Collected by Richard L. Garner in 1905.

USNM 174715 Skull. Adult male. Gabon. Collected by Richard L. Garner in 1905.

USNM 174716 Skull. Adult male. Gabon. Collected by Richard L. Garner in 1904.

USNM 174717 Skull. Adult male. Gabon. Collected by Richard L. Garner in 1904.

USNM 174718 Cranium. Adult male. Gabon. Collected by Richard L. Garner in 1904.

USNM 174719 Cranium. Adult male. Gabon. Collected by Richard L. Garner in 1904.

USNM 174720 Skull. Adult male. Gabon. Collected by Richard L. Garner in 1904.

USNM 174722 Skeleton. Adult male. Gabon. Collected by Richard L. Garner in 1904. It is incomplete, and the skull is a cast.

USNM 174723 Postcranial skeleton. Adult male. Gabon. Collected by Richard L. Garner in 1904.

USNM 174724 Sacrum. Gabon. Collected by Richard L. Garner in 1904.

USNM 176205 Skull. Adult male. South Kamerun. Collected by Karl Albert Haberer in 1913.

USNM 176206 Skull. Adult male. Kamerun. Collected by K.A. Haberer in 1913.

USNM 176207 Skull. Adult male. Kamerun. Collected by K.A. Haberer in 1913.

USNM 176208 Skull. Adult male. Kamerun. Collected by K.A. Haberer in 1913.

USNM 176209 Skull. Male. South Kamerun. Collected by K.A. Haberer in 1913.

USNM 176210 Skull. Adult male. South Kamerun. Collected by K.A. Haberer in 1913.

USNM 176211 Skull. Adult male. South Kamerun. Collected by K.A. Haberer in 1913. Missing, August 2008.

USNM 176213 Skull. Adult male. South Kamerun. Collected by K.A. Haberer in 1913.

USNM 176214 Skull. Juvenile male. South Kamerun. Collected by K.A. Haberer in 1913.

USNM 176215 Skull. Adult male. South Kamerun. Collected by K.A. Haberer in 1913.

USNM 176216 Skull. Adult male. South Kamerun. Collected by K.A. Haberer in 1913.

USNM 176217 Cranium. Adult male. South Kamerun. Collected by K.A. Haberer in 1913.

USNM 176218 Cranium. Adult female. South Kamerun. Collected by K.A. Haberer in 1913.

USNM 176219 Cranium. Juvenile male. South Kamerun. Collected by K.A. Haberer in 1913.

USNM 176220 Cranium. Adult male. South Kamerun. Collected by K.A. Haberer in 1913.

USNM 176221 Cranium. Adult female. South Kamerun. Collected by K.A. Haberer in 1913.

USNM 176222 Cranium. Adult male. South Kamerun. Collected by K.A. Haberer in 1913.

USNM 176223 Cranium. Adult male. South Kamerun. Collected by K.A. Haberer in 1913.

USNM 176224 Cranium. Adult male. South Kamerun. Collected by K.A. Haberer in 1913.

USNM 176225 Skeleton. Adult male. Kamerun. Collected by K.A. Haberer in 1913.

USNM 197037 Skull (cast). Male. The original skull is in the Senckenberg Museum, Frankfurt (no. 1135), and is a syntype of *Pseudogorilla ellioti* Frechkop, 1943. The cast was obtained by R.L. Garner on 1 October 1913.

USNM 220060 Skeleton and skin. Adult female. Mperi, Fernan Vaz, Gabon. Collected by Charles R. Aschemeier (1893–1973) on 9 December 1917. The skeleton is incomplete.

USNM 220061 Calvarium. Adult male. Alboona, Gabon. Collected by C.R. Aschemeier on 23 January 1918.

USNM 220324 Skeleton and skin. Adult male. Moamba, Sanga, Ngovi, Gabon. Collected by C.R. Aschemeier on 11 July 1918. The skeleton is incomplete.

USNM 220325 Skeleton. Adult male. Ogouma, Gabon. Collected by C.R. Aschemeier on 5 January 1919. It is incomplete.

USNM 220380 Skull. Adult female. Mpiviè, Gabon. Collected by C.R. Aschemeier on 17 September 1918.

USNM 239883 Skeleton and mounted skin. *Gorilla beringei beringei*. Mountains north of Lake Kivu, Belgian Congo. Collected by Benjamin Burbridge.

USNM 239884 Skeleton and skin. Infant male. *Gorilla beringei beringei*. Mountains north of Lake Kivu, Belgian Congo. Collected by Benjamin Burbridge.

USNM 241232 Skull and mounted skin. Juvenile female. *Gorilla beringei beringei*. Mount Karisimbi, Belgian Congo. Collected by Benjamin Burbridge in July 1925.

USNM 241233 Skin. Male. *Gorilla beringei beringei*. Mount Karisimbi, Belgian Congo. Collected by Benjamin Burbridge in August 1925.

USNM 252575 Skull. Adult female. Goko, Souanké region, Sangha Province, French Congo. Received from North Rouppert Co., Paris, France.

USNM 252576 Skull. Adult female. Goko, Souanké region, Sangha Province, French Congo. Received from North Rouppert Co., Paris, France.

USNM 252577 Skull. Adult female. Goko, Souanké region, Sangha Province, French Congo. Received from North Rouppert Co., Paris, France.

USNM 252578 Skull. Juvenile male. Goko, Souanké region, Sangha Province, French Congo. Received from North Rouppert Co., Paris, France.

USNM 252579 Skull. Adult female. Yoho, Souanké region, Sangha, French Congo.

USNM 252580 Skull. Adult female. Goko, Souanké region, Sangha Province, French Congo. Received from North Rouppert Co., Paris, France.

USNM 252581 Mandible. Adult female. Ngoko, Souanké region, Sangha, French Congo.

USNM 252582 Skull. Adult female. Goko, Souanké region, Sangha Province, French Congo.

USNM 252617 Skin. Female. *Gorilla beringei beringei*. Fifty to seventy-five miles northeast of Lake Kivu, Belgian Congo. Collected in July 1925. Received from A. H. Schultz.

USNM 257504 Genitalia, brain and 1 tarsal bone (all fluid preserved). Juvenile male. *Gorilla beringei graueri*. Alimbongo Mountains, 100 mi. inland, west of southern end of Lake Edward, Kivu province, Belgian Congo. Collected alive by Martin Johnson in November 1930. This is 'Okaro', who arrived at the National Zoological Park (Smithsonian Institution), Washington, DC 20008, on 17 September 1931, when aged about 2½ years. He

died on 7 October 1932 (weight 40 lb.), of acute anaemia and a large tumour on the bladder. His body was preserved in alcohol and his brain was extracted by Dr T.D. Stewart and placed in the brain collection of the US National Museum's Division of Physical Anthropology. Brain weight 496.6 g with the dura, 466.6 g without the dura but with the pia mater; maximum length 11.8 cm (right), 11.9 cm (left); maximum breadth 9.7 cm; frontal breadth 8.1 cm; mean maximum height 4.2 cm; temporal–occipital diameter 8.6 cm (r.), 8.8 cm (l.); mean frontal height 3.7 cm. See Connolly (1933). Head missing April 1992.

USNM 260582 Cranium. Adult female. *Gorilla beringei graueri*. Shabunda, Kivu Province, Belgian Congo. Collected by E.I. Burk on 16 July 1936.

USNM 271347 Mandible. Adult female. The rami and some teeth are missing, but the dentition includes 4th molars.

USNM 297857 Skeleton. Adult male. Received from Herbert Ward.

USNM 395636 Skeleton. Adult male. *Gorilla beringei beringei*. Mount Visoke, Virunga volcanoes, Rwanda. Collected by Dian Fossey on 3 May 1968. This animal was given the field study name 'Mgonjwa', also known as 'Whinny'.

USNM 396934 Skeleton. Adult male. *Gorilla beringei beringei*. Tsundura, Rwanda. Collected by Dian Fossey in 1973. Collector's field ID: Tsundura 2. It is incomplete.

USNM 396935 Skeleton. Adult female. *Gorilla beringei beringei*. Tsundura, Rwanda. Collected by Dian Fossey in 1973. Collector's field ID: Tsundura 3. It is incomplete.

USNM 396936 Skeleton. Adult female. *Gorilla beringei beringei*. Tsundura, Rwanda. Collected by Dian Fossey in 1973. Collector's field ID: Tsundura 5. It is incomplete.

USNM 396937 Skeleton. Adult female. *Gorilla beringei beringei*. Mount Karisimbi, Virunga volcanoes, Rwanda. Collected by Dian Fossey in 1973. Collector's field ID: Chini Tembo. It lacks the mandible and various postcranial bones.

USNM 396938 Cranium and left ilium. Juvenile female. *Gorilla beringei beringei*. Bisati Ndube, potato field northeast of Mount Visoke, Virunga volcanoes, Rwanda. Collected by Dian Fossey.

USNM 396942 Skeleton. Adult male. *Gorilla beringei beringei*. Mount Sabinio (Congo side), Virunga volcanoes, Rwanda. Collected by Dian Fossey in 1973. Collector's field ID: Congo. It is incomplete.

USNM 397351 Skeleton. Adult male. *Gorilla beringei beringei*. Mount Muhavura, above the saddle between Mounts Muhavura and Mgahinga, Virunga volcanoes, Rwanda. Collected by Jay Matternes and Dian Fossey in 1973. Collector's field ID: Matata.

USNM 397352 Postcranial fragments. Adult male. *Gorilla beringei beringei.* Southeastern slope of Mount Visoke, Virunga volcanoes, Rwanda. Collected by Dian Fossey in November 1973. Collector's field ID: Noel.

USNM 397353 Skeleton. Adult female. *Gorilla beringei beringei.* Southern slope of Mount Sabinio, near shambas, Rwanda. Collected by Dian Fossey in 1973. Collector's field ID: Munya I. It lacks the mandible and many postcranial bones.

USNM 397354 Shaft of humerus (broken). *Gorilla beringei beringei.* Southern slope of Mount Sabinio, Rwanda. Collected by Dian Fossey.

USNM 397355 Skull fragments and atlas. Female. *Gorilla beringei beringei.* Eastern slope of Mount Sabinio (Karisimbi saddle), Rwanda. Collected by Dian Fossey. Collector's field ID: Kundisara.

USNM 397356 Skeleton. Adult female. *Gorilla beringei beringei.* Eastern slope of Mount Visoke, Rwanda. Collected by Dian Fossey in 1973: identified by her as Guamboegazi-Query. It is very incomplete, and the cranium is among the missing elements.

USNM 397357 Long bone shaft only. Female. *Gorilla beringei beringei.* Eastern slope of Mount Visoke, Rwanda. Collected by Dian Fossey.

USNM 397358 Skeleton. Adult male. *Gorilla beringei beringei.* Saddle area between Mount Sabinio and Mount Mgahinga, slightly north of a straight line between the two peaks in Zaïre. Collected by Dian Fossey in 1973: identified by her as Gahinga-Uganda. It is very incomplete.

USNM 397359 Skull fragments, part humerus, part tibia and part sacrum. Adult male. *Gorilla beringei beringei.* Northern slope of Mount Visoke, Rwanda. Collected by Dian Fossey in 1973: identified by her as Guamboegazi Shamba.

USNM 397360 Distal fragment of femur. *Gorilla beringei beringei.* Eastern slope of Mount Visoke, Rwanda. Collected by Dian Fossey.

USNM 397361 Shaft of femur. *Gorilla beringei beringei.* Eastern slope of Mount Visoke, Rwanda. Collected by Dian Fossey.

USNM 397362 Shaft of femur. *Gorilla beringei beringei.* Eastern slope of Mount Visoke, Rwanda. Collected by Dian Fossey.

USNM 397363 Four ribs only. *Gorilla beringei beringei.* Saddle area between Mount Sabinio and Mount Mgahinga, Rwanda. Collected by Dian Fossey.

USNM 397364 Cranium. Female. *Gorilla beringei beringei.* Collected by Jay Matternes and Dian Fossey.

USNM 397365 Fragments of skull and skeleton. Juvenile. *Gorilla beringei beringei.* Collected by Jay Matternes and Dian Fossey. Identified as Infant B or Infant I.

USNM 398214 Carcase (fluid preserved). Neonate male (2 minutes old). Received from Philadelphia Zoo: dam 'Snickers'; sire 'Bobby' or 'Westy'. Weight at birth 3 lb.

USNM 398409 Skull fragments and incomplete skeleton. Infant. *Gorilla beringei beringei.* Southern slope of Mount Mikeno, one-quarter the distance above the saddle area with Karisimbi to the summit of Mikeno, Rwanda. Collected by Jay Matternes and Dian Fossey. Identified as Mikeno Infant A or Mikeno Infant I.

USNM 497150 Cadaver (fluid preserved, head and viscera removed). Juvenile. Collected and donated by Richard W. Thorington Jr. Field no. RWT JR 2024. Catalogued on 29 October 1975.

USNM 497151 Cadaver (fluid preserved, viscera removed, brain entered, skull reattached). Juvenile. Donated by R.W. Thorington Jr. Catalogued on 29 October 1975.

USNM 498378 Cadaver (fluid preserved). Adult male. Received from Busch Gardens Zoological Park, Tampa, FL 33674. This is 'Hercules', who was purchased from Henry Trefflich and arrived on 18 September 1964 (weight 30 lb.). He died on 24 July 1975 (weight 410 lb.) of aspiration of fluid during recovery from anaesthesia. Catalogued on 24 February 1976. [Additional information added by GH from correspondence with Busch Gardens Zoo personnel.]

USNM 545026 Skull. Adult female. *Gorilla beringei beringei.* Virunga volcanoes, Rwanda. Collected and donated by Dian Fossey: identified by her as Green Lady (Jay Matternes disagrees, but cannot identify the specimen). Catalogued on 1 June 1983. See also specimen no. USNM 545045.

USNM 545027 Skull. Adult female. *Gorilla beringei beringei.* Virunga volcanoes, Rwanda. Collected and donated by Dian Fossey: identified by her as Chini ya Tembo (by Jay Matternes as Nemeye). Catalogued on 1 June 1983.

USNM 545028 Skull. Adult male. *Gorilla beringei beringei.* Virunga volcanoes, Rwanda. Collected and donated by Dian Fossey: identified by her and Jay Matternes as Rafiki. Catalogued on 1 June 1983. See also specimen no. USNM 545039.

USNM 545029 Skull. Adult female. *Gorilla beringei beringei.* Virunga volcanoes, Rwanda. Collected and donated by Dian Fossey: identified by her and Jay Matternes as Old Goat. Catalogued on 1 June 1983. See also specimen no. USNM 545042.

USNM 545030 Skull. Adult female. *Gorilla beringei beringei.* Virunga volcanoes, Rwanda. Collected and donated by Dian Fossey. Identified by Jay Matternes as Madame Putea. Catalogued on 1 June 1983.

USNM 545031 Skull. Adult female. *Gorilla beringei beringei.* Virunga volcanoes, Rwanda. Collected and donated by Dian Fossey. Identified tentatively by Fossey as Old Group 5 Female; by Jay Matternes as Green Lady. Catalogued on 1 June 1983.

USNM 545032 Skull. Adult male. *Gorilla beringei beringei*. Virunga volcanoes, Rwanda. Collected and donated by Dian Fossey. Identified by Fossey as Maganga; by Jay Matternes as Kekana. Catalogued on 1 June 1983.

USNM 545033 Cranium. Male. *Gorilla beringei beringei*. Virunga volcanoes, Rwanda. Collected and donated by Dian Fossey. Catalogued on 1 June 1983.

USNM 545034 Skull. Adult male. *Gorilla beringei beringei*. Virunga volcanoes, Rwanda. Collected and donated by Dian Fossey. Identified by Jay Matternes as Limbo. Catalogued on 1 June 1983. See also specimen no. USNM 545041.

USNM 545035 Skull. Adult male. *Gorilla beringei beringei*. Virunga volcanoes, Rwanda. Collected and donated by Dian Fossey. Identified by Fossey and Matternes as Karisimbi. Catalogued on 1 June 1983.

USNM 545036 Skull. Adult male. *Gorilla beringei beringei*. Virunga volcanoes, Rwanda. Collected and donated by Dian Fossey. Identified by Fossey as Muhavura. Catalogued on 1 June 1983. See also specimen no. USNM 545048 (matching skeleton).

USNM 545037 Skull. Male. *Gorilla beringei beringei*. Virunga volcanoes, Rwanda. Collected and donated by Dian Fossey. Identified by Jay Matternes as Big Nove. Catalogued on 1 June 1983.

USNM 545038 Postcranial skeleton. Adult female. *Gorilla beringei beringei*. Virunga volcanoes, Rwanda. Collected and donated by Dian Fossey. It is incomplete. Catalogued on 1 June 1983.

USNM 545039 Postcranial skeleton. Adult male. *Gorilla beringei beringei*. Virunga volcanoes, Rwanda. Collected and donated by Dian Fossey. It matches the skull no. USNM 545028. It is incomplete. Catalogued on 1 June 1983.

USNM 545040 Postcranial skeleton. Adult male. *Gorilla beringei beringei*. Virunga volcanoes, Rwanda. Collected and donated by Dian Fossey. It is incomplete. Catalogued on 1 June 1983.

USNM 545041 Postcranial skeleton. Adult male. *Gorilla beringei beringei*. It belongs to the skull no. USNM 545034. It is incomplete. Catalogued on 1 June 1983.

USNM 545042 Postcranial skeleton. Adult female. *Gorilla beringei beringei*. It matches the skull no. USNM 545029. It is incomplete. Catalogued on 1 June 1983.

USNM 545043 Postcranial skeleton. Adult female. *Gorilla beringei beringei*. Virunga volcanoes, Rwanda. Collected and donated by Dian Fossey. It is incomplete. Catalogued on 1 June 1983.

USNM 545044 Skeleton. Infant male. *Gorilla beringei beringei*. Virunga volcanoes, Rwanda. Collected and donated by Dian Fossey. Identified by Fossey as Thor (known age 11 months). It is incomplete. Catalogued on 1 June 1983.

USNM 545045 Postcranial skeleton. Adult female. *Gorilla beringei beringei*. It is incomplete, and appears to belong to the skull no. USNM 545026. The bones were received mixed with those of USNM 545046. Some ribs and phalanges may be mixed with those of USNM 545047. Catalogued on 1 June 1983.

USNM 545046 Postcranial skeleton. Female. *Gorilla beringei beringei*. It is incomplete. According to Jay Matternes, it matches the skull no. USNM 545027. The bones were received mixed with those of USNM 545045. Catalogued on 1 June 1983.

USNM 545047 Postcranial skeleton. Adult female. *Gorilla beringei beringei*. Virunga volcanoes, Rwanda. Collected and donated by Dian Fossey. It is incomplete. Fossey said that some phalanges and ribs from this specimen belong to USNM 545045. Catalogued on 1 June 1983.

USNM 545048 Postcranial skeleton. Adult male. *Gorilla beringei beringei*. It is incomplete. It matches the skull no. USNM 545036. Catalogued on 1 June 1983.

USNM 574138 Skull; right leg and arm (in fluid). Adult male. This is 'Nikumba', who arrived at the National Zoological Park, Washington, DC 20008, on 24 February 1955 (weight 17 lb.). He had been trapped by locals in Ewo district, Likouala-Mossaka region, French Equatorial Africa, in September 1954, and secured for the Zoo by members of the Russel M. Arundel Expedition. He arrived with a female named 'Moka', who had been captured by an expedition of a French mining firm in the Likouala-Mossaka region in December 1953, and purchased by members of the Arundel Expedition in Brazzaville. 'Nikumba' died on 23 July 1990. [Some additional information added by GH from correspondence with National Zoological Park personnel.]

USNM 579247 Carcase (eviscerated, in fluid). Female.

USNM 582726 Skull; remainder in fluid (eviscerated, head skin included, right thigh dissected). Collected in April 2000. Field no. BES 728. Weight 185 lb.

USNM 583458 Brain.

USNM 583459 Brain. Adult female. Bakoko, Cameroun. Collected in March 1915.

USNM 583460 Brain. Adult. Bakoko, Cameroun. Collected in October 1914.

USNM 583461 Brain. Juvenile male. Bakoko, Cameroun. Collected in October 1914.

USNM 583462 Brain. Juvenile. Bakoko, Cameroun.

USNM 583463 Brain. Adult female. Bakoko, Cameroun. Field no. VIII.

USNM 583464 Brain. Adult. Bakoko, Cameroun. Collected in September 1914.

USNM 583465 Brain. Juvenile female. Bakoko, Cameroun. Collected in November 1914.

USNM 583466 Brain. Female. Bakoko, Cameroun.

USNM 583467 Brain. Juvenile male. Bakoko, Cameroun. Field no. XIII.

USNM 585487 Skull. Male. This is 'Kuja', who arrived at the National Zoological Park, Washington, DC 20008, on 22 October 1985, on loan from Chicago Brookfield Zoo, via Milwaukee Zoo. He died on 1 July 2006 of severe cardiac disease (weight 121.6 kg).

USNM 586541 Skeleton and frozen tissue. Female. This is 'Haloko', who arrived at the National Zoological Park, Washington, DC 20008, on 12 December 1989, on loan from Philadelphia Zoo, via New York Bronx Zoo. She had arrived at Philadelphia Zoo on 31 July 1970, aged about 1½ years (weight 24 lb.), having been purchased from G. van den Brink. She was euthanased on 17 March 2011 (weight 96.2 kg) due to right-sided heart dilatation (with severe tricuspid regurgitation, moderate mitral regurgitation and mild pulmonary hypertension), exhaustion of medical therapeutic options and concerns for quality of life. [Some additional information added by GH from correspondence with Philadelphia Zoo personnel.]

USNM 588746 Skeleton and mounted skin. Female. Collected on 22 July 1999; field no. BES 1113. The skin has been on exhibit in the Kenneth E. Behring Family Hall of Mammals since November 2003.

USNM 590942 Cranium. Female. *Gorilla gorilla diehli*. Kekpane Hills, Takamanda Forest Reserve, Cameroon, centroid latitude 6.09736, centroid longitude 9.39881. Collected at Kekpane village by Jacqueline Groves, World Wildlife Foundation Takamanda Forest Surveys Project, on 1 February 1998. Field no. TSG-001. Date reported killed: December 1997−January 1998. It is missing the canines and incisors.

USNM 590943 Cranium. Male. *Gorilla gorilla diehli*. Basho Hills, border of Takamanda Forest Reserve, Cameroon, centroid latitude 6.13944, centroid longitude 9.45153. Collected by the World Wildlife Foundation Takamanda Forest Surveys Project on 9 February 1998. Field no. TSG-002. Killed in the early 1990s. It is missing the canines and incisors.

USNM 590944 Cranium. Male. *Gorilla gorilla diehli*. Basho Hills, Takamanda Forest Reserve, Cameroon, centroid latitude 6.13944, centroid longitude 9.45153. Collected by Jacqueline Groves, World Wildlife Foundation Takamanda Forest Surveys Project, on 19 August 1998. Field no. TSG-003. Killed in the early 1990s. It is missing the canines and incisors.

USNM 590945 Cranium. Male. *Gorilla gorilla diehli*. Kekpane locality, Takamanda Forest Reserve, Cameroon, centroid latitude 6.09736, centroid longitude 9.39881. Collected at Kekpane village by Jacqueline Groves on 13 November 1998. Field no. TSG-008. Killed in 1994, hunter's age 40. It is missing the right canine and incisors.

USNM 590946 Cranium. Female. *Gorilla gorilla diehli*. Makone River, Takamanda Forest Reserve, centroid latitude 6.06158, centroid longitude 9.42533. Collected by the World Wildlife Foundation Takamanda Forest Surveys Project at Mbakwe village, east border of Takamanda Forest Reserve, Cameroon, on 27 November 1998. Field no. TSG-010. Reportedly killed in October 1996 in Makone Forest, hunter's age 35. It is missing the left incisor.

USNM 590947 Skull. Female. *Gorilla gorilla diehli*. Obonyi Hills, Takamanda Forest Reserve, centroid latitude 6.13231, centroid longitude 9.25775. Collected by Jacqueline Groves on 20 August 1998. Field no. TSG-004. Killed in November 1996 in the Obonyi Hills by Mr Jasper Obi. Killed along with female TSG-005 from the same group. Silverback of the group reported to have denied hunter access to female carcases for 2 days.

USNM 590948 Skull. Female. *Gorilla gorilla diehli*. Obonyi Hills, Takamanda Forest Reserve, centroid latitude 6.13231, centroid longitude 9.25775. Collected by Jacqueline Groves on 20 August 1998. Field no. TSG-005. Killed in November 1996 in the Obonyi Hills by Mr Jasper Obi. Killed along with female TSG-004 from the same group. Silverback of the group reported to have denied hunter access to female carcases for 2 days.

USNM 590949 Cranium. Female. *Gorilla gorilla diehli*. Obonyi Hills, Takamanda Forest Reserve, Cameroon, centroid latitude 6.13231, centroid longitude 9.25775. Collected by Jacqueline Groves at Obonyi village on 20 August 2004. Field no. TSG-030. Killed by Mr Jasper Obi in November 1996. Originally reported as a chimpanzee.

USNM 590950 Cranium. Male. *Gorilla gorilla diehli*. Takpe, Takamanda Forest Reserve, Cameroon, centroid latitude 6.02111, centroid longitude 9.34031. Collected by the World Wildlife Foundation Takamanda Forest Surveys Project at Takpe village, southern boundary of Takamanda Forest Reserve, on 15 November 1998. Field no. TSG-020. Killed in 1995. It is missing the canines and incisors.

USNM 590951 Cranium. Female. *Gorilla gorilla diehli*. Okpambe, Takpe, Takamanda Forest Reserve, Cameroon, centroid latitude 5.13458, centroid longitude 9.34031. Collected by the World Wildlife Foundation Takamanda Forest Surveys Project at Okpambe (Ekpambe) village, south of Takamanda Forest Reserve, on 21 November 1998. Field no. TSG-021. Killed in November 1985 in the Aworrie Forest, not obtained from hunter. It is missing the canines and incisors.

USNM 590952 Cranium. Female. *Gorilla gorilla diehli*. Okpambe Forest (Ekpambe), Aworri, outside of

Takamanda Forest Reserve, Cameroon, centroid latitude 5.971431, centroid longitude 9.338069. Collected by Jacqueline Groves, World Wildlife Foundation Takamanda Forest Surveys Project, at Okpambe (north Munaya). Field no. TSG-022. Killed November 1966. GPS: 05 58 075, 009 20 400 (Okpambe village); 05 58 826, 009 20 169 (Aworri village). It is missing the canines and incisors.

USNM 590953 Cranium. Male. *Gorilla gorilla diehli*. Hills between Assam and Takpe at Membene Stream. Reportedly killed on 15 December 1998 by Abie Fidelis Ketto. Group of 5 gorillas seen at time of shooting (2 adult males, 2 adult females, 1 infant). Killed for Christmas meat. Obtained by Jacqueline Groves, World Wildlife Foundation Takamanda Forest Surveys Project, at Takpe village (southern boundary of Takamanda Forest Reserve), centroid latitude 6.02111, centroid longitude 9.34031, on 24 January 1999. Field no. TSG-029.

USNM 590954 Cranium. Female (may be subadult male). *Gorilla gorilla diehli*. Shot on Moto Hill, Cameroon, in 1984 (wet season), hunter's age 38. Collected by Jacqueline Groves, World Wildlife Foundation Takamanda Forest Surveys Project, at Matene village (Moto Hill), Takamanda Forest Reserve, centroid latitude 6.26917, centroid longitude 9.35672, on 23 January 1999. Field no. TSG-028. It is missing the incisors and left canine.

USNM 590955 Cranium. Male. *Gorilla gorilla diehli*. Killed in 1995 (dry season). Reported to have been solitary. Reported to have been shot towards Mabe Forest on hill approx. 5 km from Mbu (between Mbu and Ambeshu villages, Cameroon). Collected by Jacqueline Groves, World Wildlife Foundation Takamanda Forest Surveys Project, at Mbu village (north Mone), Mone River Forest Reserve, centroid latitude 6.01311, centroid longitude 9.45647, on 13 September 1998. Field no. TSG-006. Sold by Mr Takie Columbus, Mbu. It is missing the canines and incisors.

USNM 590956 Cranium. Female. *Gorilla gorilla diehli*. Mbu (North Mone), Mone River Forest Reserve, Cameroon. Killed in 1996, hunter's age 42. Collected by the World Wildlife Foundation Takamanda Forest Surveys Project at Mbu village, perimeter of Mone Forest Reserve, centroid latitude 6.01311, centroid longitude 9.45647, on 22 November 1996. Field no. TSG-013. It is missing the canines and incisors.

USNM 590957 Cranium. Female. *Gorilla gorilla diehli*. Shot in 1996 in Eshobi-Mbu Forest, close to Eshobi-Bajuo village, northern Mone. Collected by the World Wildlife Foundation Takamanda Forest Surveys Project at Mbu, North Mone, Mone River Forest Reserve, centroid latitude 6.09903, centroid longitude 9.55769, on 28 November 1998. Field no. TSG-023. It is missing the canines and incisors.

USNM 590958 Cranium. Male. *Gorilla gorilla diehli*. Killed in 1975 in the Mone Forest Reserve, Cameroon, hunter's age 75. Collected by the World Wildlife Foundation Takamanda Forest Surveys Project at Amebeshu village, on the eastern boundary of the Mone Forest Reserve, centroid latitude 5.89828, centroid longitude 9.55783, on 23 October 1998. Field no. TSG-024. It is missing the canines and incisors.

USNM 590959 Cranium. Male. *Gorilla gorilla diehli*. Killed in 1990; shot in Mabe towards Mbu village, centroid latitude 5.92, centroid longitude 9.60, but inside the Mone Forest Reserve. Hunter's age 55. Collected by the World Wildlife Foundation Takamanda Forest Surveys Project at Menda village, outside the Mone Forest Reserve, on 23 October 1998. Field no. TSG-025. It is missing the canines and incisors.

USNM 590960 Cranium. Male. *Gorilla gorilla diehli*. Killed in 1994 in Meh Forest (east of Mone). Collected by the World Wildlife Foundation Takamanda Forest Surveys Project at Mesubia village, east of Mone Forest Reserve, Cameroon, centroid latitude 5.88, centroid longitude 9.63, on 25 October 1998. Field no. TSG-026. It is missing the canines and incisors.

USNM 590961 Cranium. Female. *Gorilla gorilla diehli*. Killed in 1978 in area known as Kewikili. Collected by the World Wildlife Foundation Takamanda Forest Surveys Project at Tambo/Tambu village, Mbulu Forest, centroid latitude 6.15328, centroid longitude 9.61078, on 3 December 1998. Field no. TSG-007. Not obtained from hunter. It is missing the canines and incisors.

USNM 590962 Cranium. Sex unknown. *Gorilla gorilla diehli*. Killed in 1995 in area locally known as Nwoso Forest. Collected by the World Wildlife Foundation Takamanda Forest Surveys Project at Ashunda village, Mbulu Forest, centroid latitude 6.19144, centroid longitude 9.61078, on 4 December 1998. Field no. TSG-009. It is missing the canines and incisors.

USNM 590963 Cranium. Female. *Gorilla gorilla diehli*. Reportedly killed in forest known locally as Akwama. Collected by the World Wildlife Foundation Takamanda Forest Surveys Project at Tambo/Tambu village, Mbulu Forest, Cameroon, centroid latitude 6.15328, centroid longitude 9.607282, on 2 December 1998. Field no. TSG-011. It is missing the canines and incisors.

USNM 590964 Cranium. Female. *Gorilla gorilla diehli*. Killed in 1993 in Kewikili Forest; when collected hunter originally specified male. Collected by the World Wildlife Foundation Takamanda Forest Surveys Project at Ashunda village, Mbulu Forest, Cameroon, centroid latitude 6.19144, centroid longitude 9.61078, on 4 December 1998. Field no. TSG-014. It is missing the canines and incisors.

USNM 590965 Cranium. Female. *Gorilla gorilla diehli*. Killed in 1992 in Ashunda Forest, Keba. Hunter originally specified male. Collected by the World Wildlife Foundation Takamanda Forest Surveys Project at Ashunda village, Mbulu Forest, Cameroon, centroid latitude 6.19144, centroid longitude 9.61078, on 4 December 1998. Field no. TSG-015. It is missing the canines and incisors.

USNM 590966 Cranium. Female. *Gorilla gorilla diehli*. Killed in 1989 in the area known locally as Mbione (Mbionyi) Forest. Collected by the World Wildlife Foundation Takamanda Forest Surveys Project at Ashunda village, Mbulu Forest, Cameroon, centroid latitude 6.19144, centroid longitude 9.61078, on 4 December 1998. Field no. TSG-016. It is missing the canines and incisors.

USNM 590967 Cranium. Male. *Gorilla gorilla diehli*. Killed (date unknown) in Kewikili Forest. Hunter's age 80. Collected by the World Wildlife Foundation Takamanda Forest Surveys Project at Bandolo village, Mbulu Forest, Cameroon, centroid latitude 6.15586, centroid longitude 9.52736, on 28 November 1998. Field no. TSG-017. It is missing the canines and incisors.

USNM 590968 Cranium. Male. *Gorilla gorilla diehli*. Killed in 1975 in area referred to as lower Mbulu. Hunter's age 60. Collected by the World Wildlife Foundation Takamanda Forest Surveys Project at Mbulu village, Mbulu Forest, Cameroon, centroid latitude 6.12225, centroid longitude 9.60156, on 12 December 1998. Field no. TSG-018. It is missing the canines and incisors.

USNM 590969 Cranium. Male. *Gorilla gorilla diehli*. Killed in 1995. Collected by the World Wildlife Foundation Takamanda Forest Surveys Project at Ashunda village, Mbulu Forest, Cameroon, centroid latitude 6.19144, centroid longitude 9.61078, on 4 December 1998. Field no. TSG-019. Left maxilla missing, teeth damaged.

USNM 590970 Cranium. Male. *Gorilla gorilla diehli*. Kagwene Forest, Mbulu-Nijikwa Forest, SW-NW province boundary, Cameroon. Collected by Wildlife Conservation Society, Cross River Gorilla Project, on 26 February 2003. Skull collected from forest, surrounded by gorilla hair and 2 bones. Skull has pellets embedded, teeth marks of rodents. Field no. 001.

USNM 599165 Skull. Male? Engong, Río Muni, Equatorial Guinea.

USNM 599166 Skull. Male? Engong, Río Muni, Equatorial Guinea.

USNM 599167 Skull. Male. Akan, Río Muni, Equatorial Guinea.

USNM 599168 Skull. Male. Nkin, Río Muni, Equatorial Guinea.

USNM 599169 Skeleton. Male. Mont Alen, Río Muni, Equatorial Guinea.

USNM 599170 Skeleton. Female. Nkin, Río Muni, Equatorial Guinea.

USNM 599171 Skeleton. Male. Miyobo, Río Muni, Equatorial Guinea. Collected in 1967. Specimen weight 10 kg.

URUGUAY

Montevideo

Museo Nacional de Historia Natural, Casilla de Correo 399, Montevideo

Enrique M. González; Héctor S. Osario

MNHN 2906 Skeleton. Adult. Spanish Guinea. Donated at the beginning of the 20th century by a military colonel. Local people had hunted it to eat. The spinal column and ribcage are articulated.

ZIMBABWE

Bulawayo

Natural History Museum of Zimbabwe, Centenary Park, Selborne Avenue, Bulawayo*
Mounted skin.

Appendix CA1

Use of Collections and Handling of Biological Material

John E. Cooper and Gordon Hull

Museums are sensitive plants. They require to be nourished and cherished continually, or they become moribund.
Alexander G. Ruthven, "A Naturalist in a University Museum"
(1931)

INTRODUCTION

As will be apparent from the pages of this book (and even more so from Part II: A Catalogue of Preserved Materials), there is a remarkable array of material from gorillas residing in museums, zoos, research institutes and in private collections around the world (Fig. CA1.1). Some of this material has already been subjected to extensive investigation, but much of it still awaits study. While it is usual for examinations to be carried out by someone working within a single discipline, an interdisciplinary or multidisciplinary approach would probably obtain better results (Fig. CA1.2). For example, an osteologist might carefully measure a skull, but fail to look closely at the teeth, meaning that important information is being missed. In cases where a study was last carried out many years ago, new knowledge, understanding and techniques could subsequently be applied.

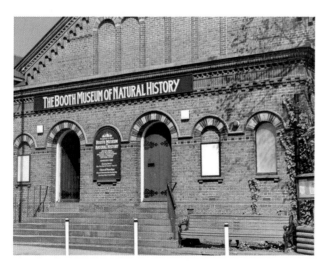

FIGURE CA1.1 Small, locally funded, museums, such as the Booth Museum in southern England, often house interesting material from gorillas.

FIGURE CA1.2 Interdisciplinary studies by an odontologist, osteologist and a veterinary pathologist at the National Museums of Kenya.

THE NEEDS PERTAINING TO MUSEUMS AND COLLECTIONS

Regardless of whether specimens are from *Gorilla* or other taxa, there is clearly a need for archived material to be traced and listed. It should be properly described by experienced specialists, avoiding rash statements and poorly substantiated 'diagnoses'. Ultimately there is the obligation to disseminate the information, so that other researchers can benefit from it.

Cataloguing of specimens and making exhaustive lists available to interested persons has long been good practice in museums. For an early example of this, see the 1862 catalogue of osteological specimens in the British Museum. The list is largely compiled by Edward Gerrard, who also prepared most of the bones. There are four gorilla skeletons and three skulls (including a plaster cast) listed in this very early catalogue.

THE RISE AND FALL OF COLLECTIONS

Paolo Viscardi, in the Introduction to Part II: A Catalogue of Preserved Materials, has succinctly outlined many of the dangers facing natural history collections today. Fortunately, all is not yet lost, and after a period of decline, in which museums fell out of fashion to some extent, there are hopeful signs of a renaissance. One example of this is the Hunterian Museum of the Royal College of Surgeons of England, which saw around two-thirds of its collections destroyed during the Second World War. Acquisitions made since that devastation have broadened the collections (Farrell, 2011). The small Odontological Museum that was part of the Hunterian was dismantled in 2003, after which the majority of the 11,000 plus specimens were kept in storage. But as research demand for the primate collection escalated, it became an increasing priority to provide a stable, secure, and easily accessible environment for the material (Farrell, 2010). A donation by the Royal Society of Medicine, Odontological Section, enabled each of the 3000 skulls in the primate collection to be packed into new lock-lid transparent plastic boxes.

At the time of writing, Wendy Birch, who manages the Anatomy Laboratory at University College London (UCL), has been awarded a Bicentenary Fellowship by the Royal College of Surgeons of England to explore musculoskeletal variations in primates using the Osman Hill primatology collection, a previously under-used resource stored in the Hunterian Museum. This project is a collaboration with Helen Chatterjee (UCL) and Margaret Clegg (NHM, London). Ultimately, they aim to produce a photographic comparative anatomy atlas and an interactive teaching package, illustrating how museum specimens can be reintroduced into modern anatomy teaching. The Osman Hill Collection consists of over 400 specimens, which vary in completeness and level of dissection, each having been fixed and subsequently preserved in 2% formalin (transferred to Kaiserling III prior to dissection).

Museum specimens are now used for many different types of investigations. For instance, Cipriano (2002) looked at incremental lines in dental cementum of gorillas and other great apes and compared them with the structures in five captive specimens of known history. The dental cementum of the eleven free-living animals was regularly structured into alternating light and dark bands, but four out of five captive specimens showed marked irregularities in terms of hypomineralised bands which could be dated to the year 1963. All four captive apes had been kept in a zoo located in the northern hemisphere, where 1963 was an exceptionally cold winter. The author argued that since cold stress is a calcium-consuming process, a lack of available calcium in newly forming cementum could be the cause of the hypomineralisation.

Griffiths and Bates (2002) reported that a molecular study on New World vultures was unable to distinguish between greater (*Cathartes melambrotus*) and lesser (*Cathartes burrovianus*) yellow-headed vultures. Reexamination of the original specimens indicated that some had been misidentified, including one specimen that had been deposited in a collection before the two species were recognised as distinct. Unfortunately, all too many researchers still fail to deposit 'voucher' specimens for studies (Agerer et al., 2000). The need for properly catalogued natural history collections remains as strong now as ever.

THE NEED FOR PATHOLOGY-BASED STUDIES

Most animal specimens in the world's museums have not been subjected to a full pathological investigation; any 'diagnoses' have been made on the basis of gross examination only. This situation applies not only to small collections, but to great ones as well, as evinced by Richard Sabin, Principal Curator of Vertebrates at the NHM (personal communication with the authors, JEC and GH): 'The identification and scoring of pathological lesions from specimens in our collections has largely been neglected'. At the very least, there is a need for skeletal material to be photographed and radiographed (Fig. CA1.3).

When the senior author (JEC) and his wife examined the holotype of *Gorilla beringei* in the Berlin Natural History Museum, an apparently hitherto unrecorded healed fracture of a phalanx was discovered (Fig. CA1.4). Descriptions of holotypes do not always apparently include mention of such lesions, which is regrettable because type specimens are sometimes of interest to pathologists as well as to taxonomists.

An example of the benefits of reappraising old specimens, even without imaging or other modern techniques,

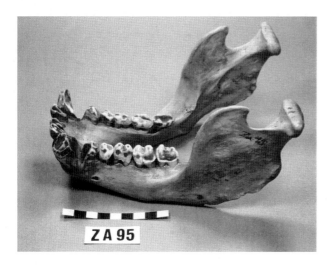

FIGURE CA1.3 Lower jaw of a mountain gorilla, showing dental disease and tartar deposition. This was one of a number of specimens transferred from Uganda to South Africa in the late 1950s.

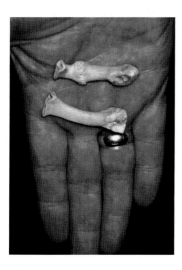

FIGURE CA1.4 Two phalanges from the holotype of the mountain gorilla, shot by Captain von Beringe in 1902. The upper bone shows a healed fracture.

was the reporting by Turk et al. (2000) on John Hunter's animal material at the Royal College of Surgeons of England. That study also illustrated once more the value of an interdisciplinary approach: the authors were a medical pathologist, a biologist (museum curator) and a veterinary pathologist, respectively.

WORKING IN COLLABORATION WITH MUSEUMS

There is a need to standardise the approach to working with specimens so that they are all, as far as possible, dealt with in the same way.

Insofar as work with gorilla material is concerned, the following advice is offered to those interested in examining specimens and in contributing further to the extension of our knowledge:

1. Make contact with the museum or keeper of the collection well in advance. Explain your interest.
2. Complete any necessary forms relating to access and liability on health and safety. Be aware of any in-house rules. Pay laboratory fees if required.
3. Plan your visit carefully. Have a checklist of the equipment that you need and ascertain what is available on site.
4. Follow in-house instructions. If necessary, supplement their health and safety rules and risk assessment with ones of your own. For example, consider wearing gloves and other protective clothing even if the host institution does not insist on this.
5. Handle all material with care and sensitivity. As a general rule, hold bones and other fragile specimens with two hands and over a table or bench rather than where there is a long drop to the floor.
6. Report immediately any damage or leakages (wet specimens). Remember that small pieces of bone may be easily dislodged from skeletons, especially if there is no padding or protection. Take nothing unless authorised to do so.
7. If you want to borrow material, follow the museum's procedures. These may include a formal loan request, in writing, including information such as details of the specimens of interest, the dates needed, the reason for the loan, the potential research output, etc.
8. Seek permission before taking or using photographs of material. Some institutions insist on use of their own photographs or, while allowing pictures to be taken of material, require proper authorisation before they are used in a publication or lecture.
9. Provide the host institution with a copy of your findings. Some (e.g., the Paris Natural History Museum) ask only for a summarised report.
10. Be sure to thank the owners or keepers of the material and offer to acknowledge them in any published work. Be sure to cite them correctly and in full; if in doubt, send a draft of your acknowledgements to them. Provide them with a reprint or pdf of your paper after publication.

MAKING MAXIMUM USE OF MATERIAL IN MUSEUMS

Skeletons and skins of gorillas and other endangered species are so important that maximum use should be made of any samples, large or small. Sometimes material becomes available unexpectedly, during the course of a routine examination. Recent examples of this, experienced by the authors, include hairs that were found during the course of examination of gorilla skeletons in the RCS Osman Hill Collection, and two tiny pieces of bone that fell from the skeleton of 'Guy' while it was being investigated at the Natural History Museum in London (see Appendix 5: Case Studies — Museums and Zoological Collections). In both instances permission was requested, and granted, to remove the material for examination at Cambridge University.

While collaborative research is always encouraged, a dilemma may occur if a specialist technique necessitates, for example, the partial destruction of specimens, as frequently happens in the field of molecular biology.

The ethics of working with museums and respecting their materials should never be disregarded. When the senior author and his wife visited the Mulago Hospital in Uganda to look at gorilla material some years ago, they were welcomed with the words 'We hope you won't steal any of our gorilla skulls, as the last *mzungu* ['European'] visitor did'.

HANDLING PRESERVED MATERIAL

The British Association for Biological Anthropology and Osteoarchaeology (BABAO) is one of a number of bodies that offers information about the use, storage, study, and display of human skeletal remains, which applies equally to nonhuman primate skeletal remains (for further information, see http://www.babao.org.uk/).

Skeletons in museums and universities are sometimes articulated and mounted for display, but from the scientific point of view, unmounted material is far more useful for study purposes, particularly if the bones are bagged and labelled correctly. It is important always to label bags the right way up, so that specimens do not drop out during examination, with the obvious danger of damage being caused. Bones should not be wrapped in newspaper for storage purposes because the paper will quickly become acidic and may damage the bones (Paolo Viscardi, personal communication).

'Wet' tissues in glass or plastic containers ('pots') should always be handled with care, using two hands, over a table or a bench. They should not be shaken or agitated in any way as this may damage the material within or cause sedimented material, such as paraformaldehyde, to obscure the view.

It may be advisable to wear gloves while touching skins, and a face mask will protect a little against chemicals and dust (Fig. CA1.5). Scientists handling postmortem material are at risk of infection by a number of potentially dangerous pathogens. The incidence of such infections is small but that is no reason for complacency (Irvin et al., 1972). If there is concern about pathogens or

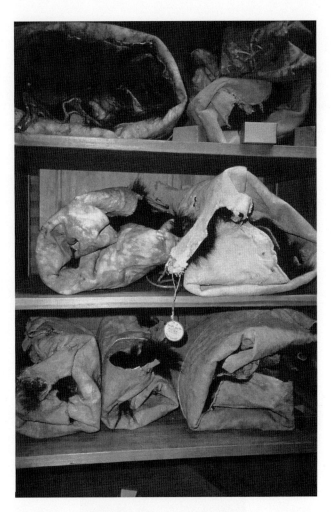

FIGURE CA1.5   Skins of gorillas and other species can be stored rolled but this may be accompanied by growth of mould, invertebrate infestation and the gathering of dust.

chemicals, a safety cabinet should be used. Under less sophisticated conditions, often in Africa, for example, skins are best examined in the open air, not in a dusty, enclosed, inner recess of an office or museum.

Archived sets of microscope slides need to be stored carefully, away from the light. With time, sections fade and sometimes 'bubbles' appear in the mountant. They should be carefully labelled with the species and the tissues, not just the lab reference number, to reduce the risk of transposition. Paper labels may fade or fall off. A useful, but time-consuming precaution is also to mark them with a diamond pen.

Paperwork also demands careful handling. Old paper documents, such as entries in accession books (Fig. CA1.6), become soiled or crack and tear and valuable records in box files (Fig. CA1.7) may deteriorate over time because of mite damage and escape attention.

FIGURE CA1.6   Page of an accession book in a museum in Belgium. The document is of historical value but easily damaged.

FIGURE CA1.7   Paper records of gorillas and other species in museums and zoos are often stored in box files but these may conceal damage by mites or mould.

Further damage to paper records can be minimised if photos are taken of them for analysis later. Ultraviolet light may help to decipher faded script. When transposing information from handwritten or typed documents the original, exact, wording should be copied, even if there are errors; see, for example, the record card depicted in Fig. 18.6, where "beringei" is spelt "berengei".

The Natural Sciences Collections Association (NatSCA) can give guidance on good practice (see http://www.natsca.org/).

Appendix CA2

Retrieval, Preparation and Storage of Skeletal and Other Material

John E. Cooper and Gordon Hull

An animal too long dead could be revivified with hot water or kneading, so that the motions of the tendons might be observed.
For I have always at hand a large number of specially prepared bones of [monkeys].

Galen (*Anat. Admin.* 1.5, 2.243-44K)

INTRODUCTION

This Appendix discusses the retrieval, preparation, preservation and storage of material from gorilla carcases, with particular reference to skeletal material in the field in Africa.

SKELETON

The examination of the skeleton is discussed in detail in Chapter 14: Musculoskeletal System.

Such examination can yield important information but is often neglected in routine diagnostic, health-monitoring and forensic studies on gorillas and other species. It is usually delayed until the bones can be prepared, sometime after the postmortem examination, although radiography and other imaging at the time can provide certain information beforehand (see Chapter 6: Methods of Investigation – Postmortem Examination).

The retrieval of gorilla bones for different purposes is hardly new. Morris and Morris (1966) referred to a Portuguese report concerning the preparation of 'native remedies' from gorilla skeletons and stated that some scientists engaged in collecting gorilla specimens lost bones to local people wanting to concoct remedies!

The preparation of skeletal material was summarised by Martyn Cooke in Cooper and Cooper (2013). He described methods ranging from natural decomposition and chemical maceration to the use of *Dermestes* beetles (Hall and Russell, 1933; Russell, 1947).

SOFT TISSUE PRESERVATION

The recovery, preparation and storage of soft tissues cannot be discussed in detail here. Fixation is referred to elsewhere in the text and embalming is a skilled procedure.

SKINS

The recovery and preparation of these too is a specialised subject. As Akeley (1920) commented: 'The proper care of the skins in the field is itself a subject of infinite ramifications'.

Delay in removing a skin, especially if coupled with poor technique, can result in substantial deterioration. This also often occurred in the past if a gorilla or its skin was inadequately fixed for transportation by sea to Europe or North America – see, for example, the comments in Part II: A Catalogue of Preserved Materials about mounted skin 36.1 in the Muséum National d'Histoire Naturelle, Paris. The catalogue also explains that the term 'removed taxidermy mount' means that the skin was in such poor condition that it had been destroyed and only the bones remain.

THE RECOVERY AND PREPARATION OF SKELETAL REMAINS

There follows an important contribution by Ogeto Mwebi, Head of the Osteology Section of the National Museums of Kenya, who is also a founder member of the Gorilla Pathology Study Group (GPSG) – see Preface and text.

Dr Mwebi's contribution is particularly appropriate at this time as it follows in the wake of an extensive programme to disinter mountain gorilla skeletal specimens in

Rwanda. This project started in March 2008 when a Working Group met at the Kigali Museum of Natural History and formulated a Strategic Plan to prepare, curate and preserve a 'skeletal resource' in the Parc National des Volcans (PNV) in Rwanda. Part of the aim was to establish within Rwanda a collection that could be used for education and research at the Museum of Natural History in Kigali (McFarlin et al., 2009).

During the summer of 2008, skeletal remains from over 70 mountain gorillas, representing both sexes and different ages, were recovered and catalogued from locations within and outside the PNV. The project, supported by the Leakey Foundation and the National Geographic Society, is a partnership between George Washington University, New York University School of Dentistry, the Dian Fossey Gorilla Fund and the Mountain Gorilla Veterinary Project (MGVP).

By December 2015, Antoine Mudakikwa (personal communication) was able to report that the collection, the largest anywhere of known gorillas with a long-term history of each animal, would be open to researchers and scientists for studies but that the team was 'still working on [a] protocol to access the collection'.

Papers that have been published or are in press relating to this resource include McFarlin et al. (2012), Ruff et al. (2013) and Galbany et al. (2016).

This was not the first time that mountain gorilla skeletal specimens had been retrieved in this way in Rwanda. In the late 1970s the skeletons of 31 animals buried by Dian Fossey were disinterred and sent to the Smithsonian Institution in the United States for curation and future study.

In 1994 a study of skeletal remains of *beringei* started at the Volcano Veterinary Center and this included plans for the recovery, following consultation with ORTPN and local people, of known gorilla remains (Fig. CA2.1). Regrettably, because of the outbreak of violence in July 1994, these

FIGURE CA2.1 The genesis of studies on skeletal pathology at the VVC in Rwanda in 1993, a few months before the outbreak of hostilities and subsequent loss or damage of specimens.

plans had to be shelved. Much of the skeletal material collected survived the fighting but some was damaged.

The recent skeletal project in Rwanda is commendable in that the material is remaining in-country, within a range state, and not being sent overseas. As such, it is in keeping with the spirit of the Convention on Biological Diversity (1992) — see Chapter 18: Legal Considerations.

There are two aspects, though, about which more information would have been welcome. One of these is the 'ownership' of the gorilla skeletons. As pointed out in Chapter 18, Legal Considerations, gorillas are listed as 'migratory species' in Appendix I of the Convention on Migratory Species (CMS). In the case of *beringei*, they can and often do roam among three countries. To which of these do the skeletal remains belong?

The other question concerns any ethical issues. In archaeological work in many countries, such as the United Kingdom, the excavation of human remains is subject to codes of good practice relating not only to the remains themselves but also any possible cultural or religious concerns on the part of local people. As pointed out elsewhere in this book, indigenous beliefs about gorillas are often strong in Africa. Some in West Africa relate to death and the fate of these animals, including an injunction amongst some tribes that gorillas are human and should be buried when dead. One should never underestimate the strength of feeling in many African societies, particularly in rural areas, about inhumation, certainly of human cadaver and sometimes of animal carcases too. Cooper and Cooper (2014/15) emphasised the importance of understanding traditional funerary practices in West Africa, especially in the wake of the Ebola outbreak in Guinea, Liberia, Sierra Leone and Nigeria.

One wonders if such cultural mores were taken into account in the project in Rwanda. It requires sensitivity and the ability, both linguistically and practically, to gain the confidence of local people, not just consulting local officials. Comparable pitfalls were described in the context of birds that are sacred to the Maori in New Zealand (Cooper and Cooper, 2007).

INVITED CONTRIBUTION — SKELETAL REMAINS: RECOVERY, PREPARATION AND PRESERVATION

Ogeto Mwebi

INTRODUCTION

This contribution focuses on the location, collection and processing of bones of gorillas that have died in the field. Some may be found in undergrowth, some buried. The recovery procedures outlined below are not only applicable to gorilla remains but to all vertebrate skeletal collections that meet specific needs. It is based on the author's (MO's) experience as an osteologist working in East and Central Africa. His personal experience in the collection, identification and preservation of bone remains has taught him the importance of recovery and preservation of every skeletal element of a given species irrespective of its value to the current users.

A single bone can not only tell a story about the individual taxon it represents but also the general ecological interrelationships (present and past) of the area in which it has been found. As archaeological bone remains can give a glimpse into our ancestors' subsistence economy, fossil bone provides help in palaeoenvironmental reconstruction while modern bone accumulations give information on the current/recent biological diversity of a given area (Behrensmeyer and Dechant, 1980; Kerbis, 1990; Klein and Cruz-Uribe, 1984).

Some animals inhabit specific habitats and have very little tolerance of any changes. Such animals become locally extinct as their habitat is modified. Hence the presence of their bones in a given habitat that is not suitable for them can be an indicator that changes are taking place in the area and this may trigger investigations to ascertain the causes of the changes. The results of such investigations may be used to formulate proper conservation policies.

Skeletal remains are collected and preserved with an overall objective of serving as research material in order to answer various research questions. A single identifiable bone can enable one to reconstruct the life history of the species to which it belongs (Rennick et al., 2005; White and Folkens, 1991, 2005; Mays and Taylor, 2002). However, to reconstruct accurately this life history one has to be thoroughly aware of (1) the normal morphology of the bone and (2) the postdepositional processes (commonly referred to as taphonomic processes) that alter bone morphology (Booth, 2015; de Boer and Maat, 2012; O'Connor, 2004). To answer effectively the various research questions, skeletal researchers seek answers from a bone collection. This has to be carefully prepared with standards that meet and serve diverse research needs.

In addition to the points above, the recovery of skeletal remains of gorillas provides an opportunity to study such material for any evidence of pathology and to take samples for (for example) DNA studies. Gorilla and human skeletal morphology are very similar. Therefore, understanding gorilla skeletal pathology can serve as a model to interpreting modern and ancient human skeletal pathological conditions. Thus, as with any skeletal remains, the recovery of gorilla skeletons has to be done being mindful that, once gathered, the intention is to address both known and unknown scientific and research questions — gathering remains is not done randomly, but with a specific purpose.

PURPOSES OF RECOVERY AND PRESERVATION OF SKELETONS

There are three general approaches (mechanical, chemical and entomological) in skeletal cleaning and preparation. The use of various methods and techniques within these methods is common. Decisions regarding preparation methods include consideration for the types of analyses to be conducted and whether or not long-term curation is the goal. As mentioned above, skeleton collection purposes are various and each will require different processing and preservation precautions. For example:

- In building a comparative anatomy and teaching collection one can use several preparation methods (boiling, maceration and dermestids) that make them attractive/easy to handle and expose their morphological variation.
- Building a research collection will necessitate avoidance of some preparation methods, for example, DNA and other bone tissue studies will require specific techniques that will not cause bone tissue destruction.

Generally, each preparation method is dictated by the source and physical condition of the carcase/specimen to be skeletonised. Some of the sources include:

- Donations from wildlife management authorities when culling problem animals.
- Road kills and opportunistic collection from poached remains or natural deaths.
- Donation of research animals from scientific institutions.
- Dead zoo animals.

Given that gorillas are currently rare and endangered animals, their skeletal sources are mainly zoos and opportunistic collection from poached or natural deaths. Zoo animals' carcases are likely to be reported and collected while fresh, with skin and soft tissue, which will require fleshed specimen skeleton preparation procedures. On the other hand, poached or natural death specimens will be found in various states of decomposition on the surface or excavated from burials carried out and documented by researchers, such as those described by Dian Fossey in Rwanda (Fossey, 1983). Therefore, before deciding which method of skeleton preparation is to be used, the following points have to be taken into consideration:

- Condition of the specimen and its intended future use — freshly dead/euthanased carcases with known health history can be skinned and defleshed, then macerated, boiled or fed to dermestid beetles (see below).
- Whether specimens are on the surface or buried, and if buried whether they were deliberately buried for skeleton recovery or not. Specimens buried for skeleton recovery will have information on the specimens, age, sex, time of burial, cause of death, orientation,

available skeletal parts and precautions taken to ensure complete recovery of all the osseous elements available. Even though recovery methods are the same, specimens buried for disposal demanding more caution because of having no idea of the condition of the buried specimen or its 'grave' size.

- Context, that is, what legal and health and safety conditions are to be observed/met (see Chapter 18: Legal Considerations). Further, having contextual information helps in proper specimen recovery planning and use of appropriate equipment.

RECOVERY OF SKELETONS FROM FLESHED ANIMALS

Skeleton Recovery by Boiling

The procedure demands the following:

- skin the animal
- separate at joints
- drill long bones
- pack different parts in cotton bags
- label as right and left
- boil in water with potassium carbonate (K_2CO_3)
- protect bones from overboiling and cracking or getting charred
- dry bones in well ventilated and shaded areas to avoid unintended bleaching/poisoning or cracking.

While boiling is a relatively fast and 'clean' method of skeleton preparation, it shrinks the bone and denatures bone protein (Antoine et al., 1992; but also see Eriksen et al., 2013), thus limiting the research use of the bone in question.

Skeleton Recovery by Maceration

Using bacterial action in water to clean bone is termed 'maceration'. This is the simplest method for cleaning skeletons and works best for large sturdy bones, for example, mammal skulls and long bones. It is not recommended for very delicate bones such as those found in birds or reptiles, or for skeletons you would like to keep together (articulated). ALL of the bones will separate, as maceration removes the connective tissue that keeps the skeleton together. Thus it is a good method for preparation of comparative study of skeletal material requiring separate bones that expose unique morphological features useful in distinguishing species. Maceration can be used to prepare fleshed specimens as in boiling, following the steps below:

- Remove any remaining tissue or hide as thoroughly as possible. This will speed up the maceration process.
- Fully immerse the bone in a container of plain water.

- Leave the container in a warm location where you won't notice the smell!
- Periodically pour the greasy, smelly, water out and top up with fresh water.
- Clean the specimen in warm water. When the water remains clear, the bacteria have finished breaking down any remaining tissue. At this stage the skeleton is ready for drying and storage unless it is stained or greasy which will necessitate further treatment.

The sternum is never macerated because it is too soft.

Skeleton Recovery Using Invertebrates

- Use insects to eat away the flesh. This means maintaining a colony — species used include beetles (*Dermestes maculatus*) or certain species of ant. The result is an articulated skeleton.

Skeleton Recovery by Exposure to Natural Elements

- Leave the specimens to clean naturally, that is, not by using a particular insect colony, and keep protected from carnivores.
- Allow flesh to rot with or without addition of enzymes.
- Where enzymes are used, separate and label different sides.

Terg-A-Zyme Enzyme Skeleton Cleaning Protocol

All skeletal remains — including those that are already skeletonised and those that have soft tissues present or are still 'greasy' or fatty — are soaked in a weak solution of an enzymatic detergent, called Terg-A-Zyme (produced by Alconox, Inc.), for cleaning. Terg-A-Zyme breaks down soft tissues, without damaging proteins that are bound to mineral in bone. It will also loosen up fats that are located in the marrow cavity and spongy regions of bones. This is a more gentle cleaning strategy than boiling or use of chemical detergents, which can significantly degrade bone surfaces and lead to cracking over time.

Be sure to follow protocols for personal protective equipment (gloves, masks, goggles, tyvec coverings) that have been established by the veterinarians.

- When soft tissues are present, disarticulate and deflesh the skeleton as much as possible, taking care not to damage bone surfaces.
- Soak in a large container filled with a 1% solution of Terg-A-Zyme. (This is equivalent to 10 g of Terg-A-Zyme powder for every 1.0 L of clean water.) Make sure to label the container with the individual name or

identification number of the skeleton, as labels marked directly on the bones often wash away during the cleaning process.

- Take care to keep all bones and teeth belonging to one individual skeleton together, and separated from those belonging to other individuals. Small grouped elements, such as left and right hands and feet, should be cleaned together in small containers (e.g., placed in separate mesh bags, and then soaked in the larger container with the rest of the skeleton), to facilitate sorting and identification later.
- The concentration is not an exact science, and will depend on skeletal age, size and preservation. Use a 1% Terg-A-Zyme solution for specimens that are mostly free of soft tissues, for very small or young infant/juvenile remains, specimens that have been buried (which may exhibit cracking, surface delamination, etc.), or for otherwise fragile specimens. For fleshy large adult specimens that are otherwise good condition, you can start with a 5% Terg-A-Zyme solution until most of the soft tissue has been removed, and then reduce to 1% solution for final cleaning. Cleaning proceeds more slowly in a 1% solution, but this leaves the option to remove infant/juvenile specimens before unfused epiphyses become loosened.
- The enzyme stops working after about 2 days, so more Terg-A-Zyme powder should be added to the water every other day (to restore a 1% solution). (If it goes 3−4 days, there is no harm to the specimen.) The water will also start to look yellow or cloudy if a lot of fat is being released, and changing the water will speed the cleaning process. The water should be replaced with fresh water at least 1−2 times per week, depending on how greasy the specimen is. When emptying the old water, a strainer should be used to ensure that very small elements (e.g., finger bones, loose teeth, unfused epiphyses) are not lost.
- The time required to clean skeletal specimens varies, from a few days (for dry skeletons) to a few weeks (for fleshy adult skeletons), depending on skeletal size and preservation, and temperature. The process will be quicker in warm water, such as if the container is left in a warm place. (The ideal water temperature is 45°C). However, do not heat or boil the water on a stove.
- Remove when the skeleton is cleaned of soft tissues and residual fats, and rinse thoroughly with clean water to remove any enzymatic residue. If there is any doubt, it is best to err on the side of caution:
  - *Infant/juvenile skeletons*: If left in detergent for extended periods of time, unfused long bone epiphyses may become loosened and lost, and there may be some flaking of the outer bone surface. These specimens should be watched carefully, and removed from the Terg-A-Zyme before this occurs.
  - Bones that were previously buried and/or already exhibit some surface damage prior to cleaning should also be checked carefully, and removed from the detergent before additional damage occurs.
- Upon removal, the skeleton must be thoroughly rinsed to remove all Terg-A-Zyme residue.
  - Set up two buckets of clean water for rinsing. As individual skeletal elements are removed from the Terg-A-Zyme solution, rinse them individually and thoroughly in each change of rinse water (i.e., they should be rinsed twice). When cleaning large specimens, or multiple skeletons, make sure to change the rinse water if it becomes soapy.
- Lay the skeleton out on a drying rack in a shaded area, and allow to dry thoroughly. Skeletons that are put away for storage without being allowed to dry completely are more prone to developing mould later.
  - Dry in the shade! Leaving skeletons out in the sun for extended periods will lead to cracking.
  - Drying time depends on specimen size and the weather conditions. Large adults will take longer to dry than infant specimens, and skeletons dry more slowly in humid climates than in dry. Those regions of the skeleton that often take longer to dry include the ends of long bones, bones of the hands and feet, and other areas of the skeleton with a lot of marrow.
  - As a general rule, at the National Museums of Kenya, we allow skeletons to lie out on drying racks for 3 days prior to being put away for storage, but you may modify this time interval depending on the conditions.

An additional note: Terg-A-Zyme achieves a good amount of defatting/degreasing of bones. However, some fats may still remain within the marrow cavity of whole long bones. We can explore final degreasing options later, as necessary.

Skeleton Recovery by Burial and Excavation

One or several fleshed/defleshed carcases are buried for an indefinite period, with the aim of recovering their skeletons at a convenient time, perhaps due to a lack of skeleton preparation facilities or of qualified staff to carry out the preparation.

- Collect and record all the bio-data of the carcase before skinning/defleshing.
- If possible skin and deflesh the carcase without disarticulating it while being careful not to lose any bone, however small, and avoid marking/scratching bones with the defleshing equipment. Lay it in correct

anatomical position and tie each side (right and left) of the extremities (feet and hands = hind and fore limbs) in a cotton bag to avoid these difficult- to -separate (if mixed) skeletal parts. For each extremity, mark its anatomical position with an indelible marker, or by placing it in a metal or plastic container with an inscribed label; keep a separate record indicating which label is for a left/right foot or arm.

- Wrap the entire carcase in a net and place in a cotton bag (old tent material may be used) ensuring opposite extremities are as far apart as possible to avoid mixing during excavation or burial site disturbance.
- If feasible, sketch how you have laid out the carcase and how it is going to be placed in the burial hole. Make sketches and descriptions simple for other people to follow in case you are not present during the reexcavation.
- Identify an appropriate burial site (it should be safe from burrowing animals and excavation by carnivores) and dig the "grave".
- Place a hard board or plastic material on the floor of the burial hole and place the specimen on top. This will help with marking the excavation termination point and with lifting the specimen during excavation without losing bones. Protect the specimen before covering by placing another cover (the same as the one on the bottom) on top. This will also serve as an indicator of the proximity of the specimen during reexcavation.
- If there are several individuals, bury each in its own "grave" and carefully mark each of them with information regarding the specimen it contains bio-data, etc. (species, sex, age, date of burial) or use coded marking for which you have detailed information in your notebook.
- Specimens buried in secure and stable environments can be exhumed from six months to several years later.

PREPARATION, PRESERVATION AND STORAGE OF RECOVERED SKELETONS

Skeletons from fresh carcases that have been carefully prepared by boiling come out ivory white in colour and need no further treatment.

If you wish to whiten or sterilise and bleach the bones, transfer them into a container of hydrogen peroxide (pharmacy strength). Remove them when they have reached the shade of white you desire.

Never use chlorine-based bleach – chlorine-based bleaches will degrade bone. The chlorine remains behind and continues to eat away at the bone, eventually reducing it to a white powder. Hydrogen peroxide is a safer alternative.

- Some teeth may have fallen out during the preparation and recovery process. Once the skull is clean and dry, glue these back into place using a simple white glue (also called PVA glue) such as Weldbond or Elmer's.
- Dip specimen in warm water

- Add potassium carbonate K_2CO_3 and sodium sulphide Na_2S
- Put bones in warm water to soak.

Degreasing

- Chloroform, benzene and petrol have explosive or anaesthetic properties.
- Commercial degreasing plants use trichloroethylene; it is poisonous but not flammable.
- Bones can be soaked in dilute ammonia or carbon tetrachloride.

Preservation

- Dry and store in cartons/boxes
- Repel insects (use naphthalene balls)
- Carry out fumigation, that is, "fogging", on an annual basis.

Labelling

- The specimen (species) should be given its common and scientific names.
- Designate a unique identification number for each specimen.
- A unique identity number should be inscribed on every bone.
- Information should be kept in catalogues (cards, books and electronic forms). See Appendix CA1: Use of Collections and Handling of Biological Material.

Display

- Mounting on boards
- Hanging on walls
- In boxes and on shelves
- Cabinets
- Arranged in taxonomic order.

FACTORS AFFECTING THE PRESERVATION OF MATERIAL

The primary need during skeletal preparation is to contain the body to keep the bones together and to minimise predation by both large and small scavengers. This is to help researchers gain access easily during investigations or studies. There are simple ways of achieving this. For example, when the body of 'Nkuringo', a silverback aged at least 50, was found in Bwindi in 2008, the Chief Warden said 'We wrapped his body in a polythene bag and lowered it into the grave. This is done to help researchers access it easily during investigations or studies'.

COMMENTS ON POSTMORTEM CHANGE

The rate of postmortem decomposition of a gorilla's body was discussed briefly in Chapter 6: Methods of Investigation — Postmortem Examination. During the recent *Ebolavirus* disease outbreaks (see earlier), observers noted that a dead, unburied, gorilla can decompose completely within a month.

The dialogue in the following abstract from William Shakespeare's *Hamlet* is a reminder of the various factors that can influence the decay and preservation of buried carcases:

Hamlet: How long will a man lie i' th' earth ere he rot?
   First Clown: A tanner will last you nine year
   Hamlet: Why he, more than another?
   First Clown: Why sir, his hide is so tann'd with his trade, that he will keep out water a great while. And your water, is a sore decayer of your whoreson dead body...

And, indeed, as Galen pointed out two millennia ago (see quotations at beginning of this Appendix), in order to observe the delicate structures of nerve foramina in bones, it is best to bury the monkey in moist earth for at least four months, so that the bones are clean but not desiccated.

Water is not the only factor that will affect the quality of a specimen. Bioerosion of bones and other tissues by cyanobacteria and algae occurs in marine and lacustrine environments and can also take place in terrestrial environments, especially where it is damp (Davis, 1997; Hackett, 1981).

Dead gorillas that have been partly cremated, perhaps in order to destroy pathogens, will have some bones intact but damaged by the burning. Much has been written on the taphonomy of burned human bones and teeth — and also some on wild animals, especially in the context of poaching (Baker et al., 2009).

It is paradoxical that teeth are so liable to decay in life but persist intact for long periods of time following death. The teeth of gorillas are no exception: dental caries is recognised in captive specimens yet at the same time teeth are probably the most likely tissues to remain intact and available for investigation once the animal has died. Despite their hardness and apparent resilience, however, teeth that have been buried may also deteriorate. They can be damaged by the precipitation of gypsum (calcium sulphate) crystals within the dentine, a condition well recognised by palaeontologists.

Fetal remains rapidly undergo drying and shrinkage and this has to be borne in mind when measuring the bones. Data have not, apparently, been published on the effects of shrinkage on gorilla bones but reference can probably usefully be made to the human literature (see, e.g., Huxley, 1998).

FIGURE CA2.2 Examination and dissection of a gorilla in the field.

Health and safety considerations associated with the handling and preparation of gorilla remains were covered in Appendix 4, Hazards, Including Zoonoses and will not be repeated unnecessarily here. The examination of dead gorillas, especially disinterring animals, in the field presents particular hazards, ranging from security concerns to pathogens (Fig. CA2.2). An example of a consideration in respect of necropsy or retrieval in the field that is rarely encountered in the captive (zoo) environment concerns carcases that contain maggots. These may be of significance insofar as the health of other species, including humans, is considered. Such fly larvae may harbour the toxin of *Clostridium botulinum* (type C). Indeed, the first isolation of this toxin was made from the larvae of *Lucilia caesar* nearly a century ago (Graham, 1978). It is a reason for always wearing and providing protective clothing and using it correctly.

Index

Note: Page numbers followed by '*b*', '*f*', and '*t*' refer to boxes, figures, and tables, respectively.

A

Ab protein. *See* Amyloid beta protein (Ab protein)
Abbreviations, list of, xxxiii
Abortion, 154–156
Abscesses, 20, 107–108, 123, 129, 135, 141
Abu Ja'far Abdullah al Ma'mun, 9. *See also* Islam
Abuse of animals, 60, 178–179. *See also* Welfare of gorillas
Acanthamoeba, 182
Accession number, 361–362
Accidents, 22, 62–63, 163, 172, 225–226, 294*t*
Acknowledgements, xxiii–xxvii
Acquired immunodeficiency syndrome (AIDS), 27–28
Action plans, 213*t*
    regional action plans, 211, 215–218, 221–222
Acute phase proteins (APPs), 24
Acute phase response (APR), 24
Adaptability, 291
Addison's disease, 192–193
Adipose tissue, 175–176
Adrenal, 192–193, 252, 260
Africa, 10, 12–14, 22, 59, 231, 610
    African academic institutions, 232
    African studies, 12–14
    national park in, 211
    role in gorilla conservation, 20, 231–232
African Convention on the Conservation of Nature and Natural Resources (Revised Version 2003), 218–219
Age category and sex, 362–363
Ageing (Age determination), 87
    of animals, 90, 181
    of bruises, 68, 70
    effects, 123–124
    of fractures, 40
    of lesions, 40–41
    of wounds, 40–41
Age-related changes, 87–90, 88*f*
Aggression, 33, 92–93
    behaviour, 92–93
    inter- and intra-specific, 21–22, 92–93
Agreement on the Conservation of Gorillas and their Habitats (Gorilla Agreement), 216*t*, 219

AIDS. *See* Acquired immunodeficiency syndrome (AIDS)
Air sacs, postmortem examination of
    in gorillas, 262–265, 275
    in orangutans, 254–258
Alarm reaction (AR), 195
Alimentary tract, 119
    GIT, 124–134
    oral cavity, 119–121
Allergic conditions, 139
Aluminium, 75
American Association of Zoo Veterinarians (AAZV), 248
American College of Veterinary Pathologists (ACVP), 58
Amputated limbs, 91
Amyloid beta protein (Ab protein), 183
Anaesthesia, 51, 119–120, 141–142, 142*f*, 158–159, 227. *See also* Immobilisation
Anatomical similarity of gorillas to humans, 60
Anatomy, 3–6, 8–9, 12, 22, 37, 119, 149
Angioedema, 141
Angiola, 226, 364, 373
Animal, 19, 150–151, 179, 224
    animal welfare, 43, 93–94, 224–225
Anoplocephala, 126–127
Anoplocephala gorillae, 75, 125, 125*f*
Anorexia, 109–110, 132, 132*f*
Anthracosis, 111*f*
Aorta, 117
Ape cardiac necropsy protocol, 276
APPs. *See* Acute phase proteins (APPs)
APR. *See* Acute phase response (APR)
AR. *See* Alarm reaction (AR)
Archaeogenetics, 60
Archaeologists, 60, 123, 163, 167
Arteriosclerosis, 116
Arthritis, auto-immune, 170–171, 306–307
Arthropathies, 162
Articulated skeleton, 362
Asphyxia, 155
Asymmetry. *See* Fluctuating asymmetry
Atherosclerosis, 113, 116–117
Atopic dermatitis. *See* Neurodermatitis
Attacks by wild animals, 92–93, 174. *See also* Predation
Authorisation, 60–61, 221, 223, 289–290*t*

Authorisation to keeping gorillas, 223
Autolysis, 69*t*, 70*b*, 118, 135–136, 135*b*, 152*f*, 154
Autolysis stages, 70*b*, 156*f*, 179–180
Autopsy, 57. *See also* Postmortem examination
Avicenna, 9
Axillary organ, 97–98

B

BABAO. *See* British Association for Biological Anthropology and Osteoarchaeology (BABAO)
Baboons, 23, 62, 111, 146, 196, 364–365, 374
Bacon, Francis, 6, 121
Bacteria, 25–27, 105–106
Bacteriological tests, 107–108
Bacteriology, 21, 76*t*
Balamuthia mandrillaris, 181
Balantidium, 125
'Basic Cardiac Protocol', 276
Battery-operated equipment, 288, 301–302
BCS. *See* Body condition score (BCS)
Behaviour, 10, 48, 91–92, 132, 177, 187, 376–378
Behavioural considerations, 376–378
Behavioural pathology, 91–92
Biliary system, 136
Biliary tract, 119, 136
Biochemical techniques, 80
Biochemistry, 80
Biodiversity, 30, 220
    convention on, 220
Biofilms, 52, 122
Biological material, handling of, 603
Biological samples, 53, 73, 226, 360
Biology of skin, 97–98
Biomaterials, 73–74
Biopsy, 52, 81
Biosecurity, 57–58, 208
Bite marks and bites, 38, 55*b*
Bite wounds, 39*b*, 45. *See also* Teeth; Wounds
Blast injuries, 44–45
Blood, 16, 21, 28, 43, 57, 74, 79–80, 112, 142, 146–148
    parasites, 147–148
    smears, 80
        methods of producing, 80*b*
    vessels, 112

Blowflies, 238. *See also* Maggots
BMC. *See* Bone mineral content (BMC)
BMD. *See* Bone mineral density (BMD)
Body condition score (BCS), 133
Body parts, 161, 220
Boiling, skeletons recovery, 612
Bone. *See also* Osteology
  density studies, 168*f*
  endoscope permits examination, 166*f*
  examination, 3, 163–169
  fractures and repair, 172–174, 172*f*
  investigation, 165–169
    rules, 165–169
  marrow, 146–147
  pathology, 162–163
  and skeletal lesions, 174
  skeletal material, 165*b*
  transmitted light, 167*f*
Bone mineral content (BMC), 168
Bone mineral density (BMD), 168
Breeding programmes, 223
Bristol Zoo Gardens (BZG), 20, 119
  studies on gorillas at, 155, 171
British Association for Biological
    Anthropology and Osteoarchaeology
    (BABAO), 606
British Veterinary Dermatological Study Group
    (BVDSG), 100
Bronchopneumonia, 143–145, 144*f*
Bruises, 68
Bruising. *See* Contusion
Bullets and bullet wounds, 3, 65, 69, 104.
    See also Wounds
Burial, skeletons recovery, 613–614
Burning, 33, 34*t*
Bushmeat, 22, 27–29, 60
BVDSG. *See* British Veterinary
    Dermatological Study Group
    (BVDSG)
Bwindi Impenetrable National Park (BINP), 88,
    126, 203, 205, 212*f*, 373
BZG. *See* Bristol Zoo Gardens (BZG)

C

Cachexia, 132–133
Cadavers and carcases, use of, 3, 207, 237
Calcitonin, 190
Calcium-consuming process, 604
Calculus, 122
Cameras and film, 49, 64–65. *See also*
    Photography
Cameroon, 221, 371, 374
Campylobacter spp., 30
Capillaria hepatica, 126–127
Captive, 50–51, 83*t*, 225
  captive breeding, 198
  definition of, 5
Captive animals, 11, 49–51, 105, 107, 133,
    185–186, 190, 224, 363
Captive Breeding Specialist Group (CBSG),
    198
Captive gorillas, 50–51, 377–378
Captive living, 211–212
Captive-breeding of gorillas, 154

Captivity
  effect, 93
  gorillas in, 223
    animal welfare, 224–225
    authorisation to keeping gorillas, 223
    database of Gorillas Land, 223
    gorillas in research, 223–224
    law regulating veterinary profession,
      226–227
    movement of gorillas and samples, 226
    occupational health and safety, 225–226
    permission, reference material, data,
      intellectual property and copyright,
      228
    responsibility for damage caused by, 225
    veterinary medicines, 227–228
Carcase(s), 57
  storage and preservation, 69*t*
  storage prior to examination, 609
Cardiac disease, 253–254
Cardiac examination
  for apes and other primates, 261–262
  for Great apes, 253–254
Cardiorenal syndrome (CRS), 87
Cardiovascular disease, 20, 89, 113–115
  pathogenesis of, 116–117
Cardiovascular lesion detection, 113
Cardiovascular pathology, 116–117
Cardiovascular system, 112. *See also*
    Respiratory system
  cardiovascular disease, 113–115
  heart disease in great apes, 115–116, 116*f*,
    117*f*
  investigation, 112–113
  pathology of vascular system, 117–118
  shock, 117–118
Caries, dental, 121, 123, 307
Cartilage, 161, 170
Casts
  brain, 380, 382, 407, 512, 516–517, 521,
    523, 543
  endocranial, 382, 468, 516–517, 523,
    539–540, 556
  face, 531, 537, 553
  femur, 380, 382
  foot, 380, 412, 438, 468, 516, 521, 523, 531,
    538, 540, 544, 553, 581
  hand, 380, 407, 412, 516, 521, 523, 531, 538
  head, 412, 440, 553, 577, 581
  humerus, 382, 556
  radius, 382, 556
  skull, 379–382, 384, 386, 389, 394, 438,
    440–441, 443, 451, 460, 463, 468, 482,
    510, 513, 516, 518, 521–522, 524,
    534–535, 538–539, 543, 546, 550,
    556, 576, 578, 581, 588, 594, 596
  tibia, 382, 556
  ulna, 382, 557
Cause and circumstances of death, 70–71, 73,
    166
Cause of death (COD), 62–63, 71, 90
  specifying, 71
  survey, 63–64
CBD, 216*t*, 220, 226–227

CBSG. *See* Captive Breeding Specialist Group
    (CBSG)
Cellular pathology, 15–16
Central African Republic, 127, 219–221, 371
Central nervous system (CNS) pathology, 33,
    38, 58, 81, 177–181, 185
Centre Vétérinaire des Volcans (CVV).
    See Volcano Veterinary Center (VVC)
Cerebrospinal fluid (CSF), 179
Chain of custody, 68
Chemical analysis, 77*t*
Chilling. *See* Hyperthermia and hypothermia
Chimpanzee, 3, 8, 12–13, 27–28, 89, 93, 111,
    113–114, 122, 177, 183, 207,
    362–365, 367–368, 372, 374,
    377–378
Choriocarcinoma, 155
Chronic idiopathic diarrhoea, 129–130
Chronic kidney disease, 150
Chronic pedal lesions, 101
Chronic pleurisy and collapsed parenchyma,
    108*f*
Chronic renal disease, 89
Chully-Chang disease, 96
Circumstances of death, 73, 166
Circuses, 190
CITES. *See* Convention on International Trade
    in Endangered Species (CITES)
Citrobacter diversus, 159
Civil disturbance, 211
Classification of animals, 7–8
Cleft palate, 90
Climate change, 31, 45
Clinical care, 51
Clinical checklist, 49, 73
Clinical examination, 50–51
Clinical investigation, 48*b*
Clinical signs. *See* Examination of animals
Clinical work, 47
  examination, 66
CMS. *See* Convention on Migratory Species
    (CMS)
CNS. *See* Central nervous system (CNS)
Coccidioidomycosis, 24, 109
COD. *See* Cause of death (COD)
Code
  ethics, 228, 606
Cold, 33, 34*t*, 45
Cold, effects, 33, 36–37, 107, 147, 287, 604.
    See also Chilling; Freezing
Colitis, 129–130, 130*f*
Collaboration, need for in gorilla work, 20, 231
Collection of specimens/samples
  from dead animals, 73
  entomological, 611
  from live animals, 73
  from the surrounding area (environment),
    207
Collections
  date of, 363
  and handling of biological material
    handling preserved material, 606–607,
      607*f*
    interdisciplinary studies, 603*f*

material in museums, 605–606
museums, 603*f*
needs to museums and collections, 604
pathology-based studies, 604–605, 605*f*
rise and fall of collections, 604
working in collaboration with museums, 605
use of, 604
Collector, 363
Colour codes, 294*t*. *See also* Scoring system; Quantification of data
Colour-coded swabs, 75
Communities, 31, 204, 211, 376
Community relations, 288*t*
Comparative anatomy, 5–8
Comparative medicine, inter-disciplinary nature, 234
Comparative pathology, 14–16, 114, 116, 150, 185–186
Competition, 231. *See also* Conflict
Compilation, 359–361
Computed tomography (CT), 8, 63*f*, 90, 119, 164, 164*f*, 175, 191
Condition and condition scoring, 133
Conflict, livestock, 141
Congo, Democratic Republic of; Congo, Republic of the (CR), 28, 105, 141–142, 157, 185, 201*f*, 207, 373
Connective tissue, 175
Conservation, 30, 197, 203–205, 211, 229, 231–232
relevance of studies on pathology and health, 229
Conservation conventions, 215–218
Conservation law, 221
Conservation legislation, 211
Conservation of gorillas, 17, 219. *See also* individual species
holistic approach, 234. *See also* Ecosystem health
Conservation Through Public Health (CTPH), 75, 197, 203–205, 204*f*, 230, 231*f*, 291*f*
Containers for specimens, 267
Contraception, 160
Contusion, 237
Convention Concerning the Protection of the World Cultural and Natural Heritage (World Heritage) (WH), 216*t*
Convention on Biological Diversity (CBD), 19–20, 216*t*, 610
Convention on International Trade in Endangered Species of Wild Fauna and Flora (CITES), 166, 219–220, 222*t*, 237
CITES Appendix I, 219–220
CITES Management Authority, 219
CITES Scientific Authority, 219
derivatives, 219–220
Convention on Migratory Species (CMS), 219, 610
Convention on the Conservation of Migratory Species of Wild Animals (Bonn Convention) (CMS), 216*t*

Convention on Wetlands of International Importance especially as Waterfowl Habitat (Ramsar), 216*t*
Convention relative to the Preservation of Fauna and Flora in their Natural State 1933, 216*t*, 218–219
Conventions, 216*t*, 219–220
Coprophagy, 134
Copulation, 153
Copyright, 228. *See also* Intellectual property and sharing of data
Cortex, 143–145, 145*f*
Corynbacterium pyogenes, 107–108
Cousins, Don, 48, 104
CR. *See* Congo, Democratic Republic of; Congo, Republic of the (CR)
Cranial morphology, 302, 302*f*
table of measurements, 303*t*
Cranium, 362
Crime, investigation, 68
Crime scene, 68, 289–290*t*
Cross River Gorilla, 373
CRS. *See* Cardiorenal syndrome (CRS)
Cruelty to gorillas, 224. *See also* Welfare of gorillas
Cryptosporidia, 197, 204
Cryptosporidium, 126–127, 204
of habituated Nyakagezi mountain gorilla group, 205
identification of, 205
prevalence and means, 206, 206*t*
prevalence of *Cryptosporidium* oocysts, 206
CSF. *See* Cerebrospinal fluid (CSF)
CT scanning. *See* Computed tomography scanning (CT scanning)
Ctenocephalides canis, 104
Ctenocephalides felis, 104
CTPH. *See* Conservation Through Public Health (CTPH)
Cultural considerations, 63, 288*t*, 610. *See also* Fieldwork
Cuvier, Baron Georges, 120, 366
CVV/VVC. *See* Centre Vétérinaire des Volcans; Volcano Veterinary Center (CVV/VVC)
Cystic fibrosis, 134
Cytological techniques, 80
Cytology, 40, 80–81, 81*f*

D

Damage by gorillas
to people, 121
to property, 607*f*
Dangers in gorilla work, 16–18
Dark neurons, 180
Darwin, Charles, 7–8
Data, 16, 156
data collection, 59, 84, 223, 289–290*t*
Dead gorillas, investigation of, 17, 23–24, 66, 74*t*, 100, 111–112, 165, 168, 615
Death
cause of, 21–22, 24, 62–63, 90, 113–114, 143–145, 208, 307

circumstances of, 166
manner of, 71
mechanism of, 71
time of. *See* Ageing; PMI
unexpected. *See* Unexpected death
Death diagnosis causes, 70–71, 72*f*
Decision point, 276–277
Decomposition. *See* Autolysis
Dedication, 94
Definitions, 22, 237. *See also* individual entries
Degreasing, 614
Dehydration, 34*t*, 43, 129, 157
Democratic Republic of the Congo (DRC). *See* Congo, Democratic Republic of
Dental disease and pathology, 121–122
Dental procedures, death during, 49, 122, 184, 301
Dentistry and dental surgeons. *See* Odontology; Teeth
Derivatives of animals, 60, 119, 226, 244
Dermal material
dermoplastic, 368–369, 384–385, 426–428, 430, 432, 438–440, 444–446, 487, 490
mounted skin, 381, 383, 386, 398, 412, 416, 426–428
skin, 384–385, 410, 412, 419, 537–538
skin of face, 386, 583
skin of foot/pes, 415, 431
skin of hand/manus, 415, 431, 453
skin of scalp, 421
skin part, 382
tanned skin, 386–387, 515
Dermatitis, 105
Dermatoglyphics, 97
Dermatoglyphs. *See* Dermatoglyphics
Dermatophytosis, 106
Dermestes beetles, 609
Dermis, 97
of gorillas, 97
Description of lesions, 68
Desiccation of carcases, 615
Destruction of specimens, 359, 606
Determination of age of animal at death, 362. *See also* Ageing
Developmental abnormalities, 90, 158, 169, 172
Developmental pathology, 90
Diabetes mellitus, 193
Diagnosis (morphological/aetiological)
making a diagnosis, 20–21, 71, 71*b*
Diatomaceous pneumoconiosis, 111
Diet. *See* Malnutrition; Obesity; Starvation
Diffuse idiopathic skeletal hyperostosis (DISH), 162
Digital autopsy, 63
Digital images. *See* Photography
Dipetalonema, 125, 185
Diphenylhydantoin, 120
Discomfort, 95. *See also* Welfare of gorillas
Disease, infectious, 19
Disease, non-infectious. *See* Trauma and other headings
Disease and death in gorillas, 19
Disease Risk Analysis (DRA), 198, 200*t*
Disease surveillance, 233

DISH. *See* Diffuse idiopathic skeletal hyperostosis (DISH)

Displaying, 614

Dissection techniques, 5, 58–59, 64, 191, 276, 305, 615*f*

Disseminated idiopathic skeletal hyperostosis (DISH), 170, 307

Distress, 109–110, 184. *See also* Welfare of gorillas

Distributive Shock, 118*b*

'Djombo' (gorilla), 193

DNA and DNA technology, 17, 76*t*, 78–79, 147, 222, 359. *See also* Genetic methodologies

Domesticated, definition of, 237

Domesticated animals, 62, 124, 162, 244. *See also* individual speciesLivestock

DPZ. *See* German Primate Centre (DPZ)

DRA. *See* Disease Risk Analysis (DRA)

Drawings, use in gorilla work, 62, 66, 166

DRC. *See* Congo, Democratic Republic of

Drowning, 19, 34*t*

Drugs, 202, 226
    controlled, 227–228
    dangerous, 227–228

Durrell, Gerald, 11

Durrell Wildlife Conservation Trust (Jersey Zoo, Durrell), 52–53, 101, 113–114, 143–145, 230*f*

E

Ear. *See* Special senses, pathology of

East Africa *See* individual countries

EAZA. *See* European Association of Zoos and Aquaria (EAZA)

Ebola, *Ebolavirus* and Ebolavirus disease (EVD), 24, 28–29, 61, 69, 108, 206–207, 209

Ebola outbreak, Great apes and, 17–18, 20, 206–207
    challenges during Ebolavirus disease, 209
    precautions
      at work, 207–208
      outside sanctuary, 208
      to preventing Ebolavirus disease, 209
    sanctuary during EVDs, 207

Ebolavirus, 28, 206–207

Echinococcus, 127

ECNP. *See* Extrinsic cardiac nerve plexus (ECNP)

Ecohealth approach, 234

Ecosystem Health, 234

Ectoparasites, 55*b*, 74*t*, 76*t*, 99, 102–106, 202, 244

Ectotherms, 36–37

ECVP. *See* European College of Veterinary Pathologists (ECVP)

EID. *See* Emerging infectious disease (EID)

Electrocution, 33, 34*t*

Electronic data, 84

Electron-microscopical, 130, 130*f*

Electron-microscopy, 65, 82, 135, 160
    scanning (SEM), 78
    transmission (TEM), 76*t*, 130*f*

ELISA. *See* Enzyme-linked immunosorbent assay (ELISA)

Embalmed carcase, 569–570

Embryological tissues, 177

Embryologically, 161

Embryos, examination of, 180

Emerging diseases, 25, 31. *See also* Zoonoses

Emerging infectious disease (EID), 31

Enamel hypoplasia, 121, 123, 163

Endangered, 11, 164, 212, 359, 605–606
    Critically Endangered, 203, 212

Endocrine pathology, 189
    effects of, 189*b*

Endocrine system, 189–194
    adrenals, 192–193
    pancreas, 193–194
    parathyroid, 190–192
    pineal, 194
    pituitary, 189–190
    sleep, 194
    stress, 194–196
      stressors in gorillas, 195*b*
    thyroid, 190

Endogenous factors influencing host responses, 24

Endoscopy, clinical, 161, 166
    postmortem, 62

Endotherm, 36–37. *See also* individual taxa

Enforcement, 213*t*, 219
    Enforcement, law, 221–222, 222*t*

English
    British (European/Commonwealth), 237

Enterobius lerouxi, 125

Entomology, collection and packaging of samples, 68

Environment (and environmental)
    assessment, 83
    sampling, 83, 83*t*

Environment monitoring, 83

Environmental enrichment for gorillas, 92*f*

Environmental pathology, 87

Environmental samples, 83, 83*t*

Environmental stressors, 196. *See also* Stress

Enzyme-linked immunosorbent assay (ELISA), 105, 148

Enzyme-specific technology, 83

Epidermis, 97
    of gorillas, 97

Epistaxis, 108

Epithelial cells, 74

Equatorial Guinea, 216*t*, 219, 369, 375–376

Equipment, 65, 167*b*, 289–290*t*, 294*t*
    clinical, 167*b*
    field, 55, 288, 289–290*t*
    laboratory, 84
    postmortem, 243–244

Escherichia coli, 128, 204

Estimating how long a gorilla has been dead, 50. *See also* PMI

Ethical, 94, 157
    ethical codes, 223
    ethical review, 223

Ethical considerations in gorilla, 157, 288*t*

Etymological considerations, 375–376

EU, 61, 219–220, 223, 226

EU CITES Annex A, 219–220

EU. *See* European Union (EU)

European Association of Zoos and Aquaria (EAZA), 115

European College of Veterinary Pathologists (ECVP), 21, 86

European Endangered Species Programme (EEP), 160

European Great Ape Heart Project (IPHP), 115

European Union (EU), 219–220

Euthanasia, 54, 94, 239

EVD. *See* Ebolavirus disease (EVD)

Evidence in legal cases, collecting and preserving, 78

Evidence-based medicine, 52

Ex situ, 211–212, 223, 237

Examination of animals. *See also* Clinical work; Postmortem examination
    clinical signs, 10, 48
    of dead gorillas in the field 172, 288*t*, 615
    of live gorillas in the field, 3, 124, 288–291
    of teeth, 120–121. *See also* Teeth; Oral cavity
    of the gastro-intestinal tract (GIT), 70, 119, 178

Excavation, skeletons recovery, 613–614

Exchange and sharing of information, 20, 231

Exhibits. *See* Museums

Exhumation, 60–61, 239

Exocrine pancreas, 136–137
    histological section, 136*f*

Exostoses, 168–170

Experimental studies, 23–24

Explosions, 44–45. *See also* Blasts

Explosions and unexploded ordnance, 44–45

Explosives. *See* Ordnance

Export permit, 219–220

Exposure to blasts, 44–45

External examination and sampling, 65–66

Extrapolation, advantages and dangers of, 16–18

Extrinsic cardiac nerve plexus (ECNP), 8

Eye. *See* Special senses, pathology of

F

Faeces, 29, 74*t*, 75–78, 220, 237
    assessment, 130–131
      free-living, mountain gorilla defaecates, 131*f*
      in night nest, 131*f*
    examination and assessment, 75–78, 130–131

Farm Animal Welfare Council (FAWC), 94

Fast-track permits, 220

Fat and fat reserves, 133, 175–176. *See also* Condition

Fauna and Flora International report (FFI report), 49

FAWC. *See* Farm Animal Welfare Council (FAWC)

Fetal pathology, 154–155

Fetus, 362
    examination of, 154–155, 254, 262. *See also* Neonates

FFI report. *See* Fauna and Flora International report (FFI report)
Field equipment and kits, 83–84, 287–288. *See also* Fieldwork and field studies
Field pathology
  CTPH permanent field laboratory, 291*f*
  dead gorilla from water-logged valley, 291*f*
  fieldwork, 287
  mountain gorilla, 291*f*
  necropsy of infant gorilla, 288*f*
  Newton microscope in, 288*f*
  versatility and adaptability, 291
  zoonoses awareness, 291*f*
Field studies in pathology and health monitoring
  great apes during Ebola, 206–207
  habituated Nyakagezi mountain gorilla group, 205
  long-term health monitoring systems for mountain gorillas, 203
  prerelease health considerations for gorillas, 197–198
Fieldwork, 197, 203–205, 215, 225, 287
Fieldwork and field studies, 287, 291
  cultural considerations, 63
  equipment, 287–288
  personnel, 58–59, 208
  techniques, 12, 78–83, 287–288
Fighting, 37–38
Filoviruses, 28–29
Firearms, 44–45. *See also* Bullets and bullet wounds; Shooting of gorillas
Firearms and unexploded ordnance, 44–45. *See also* Explosives; Ordnance
Fires and fire damage, 225
Five Freedoms, The, 94, 224–225
Fixation, 65, 81, 207, 609
Fixation of specimens, 187. *See also* Formalin
Fluctuating asymmetry, 169
Food and Agriculture Organization of the United Nations (FAO), 215–218
Food of gorillas, 78, 132, 142
Forensic entomology, 68
Forensic evidence, 220, 222, 237
Forensic investigation of gorillas, 69, 78, 220, 227
Forensic postmortem examinations, 68–69
Formalin and fixation, 79, 187, 252
Fossey, Dian, 12–13, 37, 45, 58, 150, 152
Fossils. *See* Palaeontology
Fractures, 40, 121, 169, 172–174. *See also* Healing
Frass, 294*t*
Free-living, 16–17, 96, 215–222, 302
  definition of, 10, 211–212, 215–222
Free-living gorillas, 21, 211–212, 215–222
  accessing gorillas, 221
  health monitoring, 54*b*
  international legislation, 216*t*, 218–221
  law enforcement, 221–222, 222*t*
  national legislation, 217*t*, 221–222
  Red List status, 215
Free-ranging, 10, 127, 211–212. *See also* free-living gorillas

Freezing
  effects on animals, of samples, 179–180
Fungi, 20*t*, 26*b*, 82, 106

G

Gabon, 3, 127, 213*t*, 364, 375
GAHP. *See* Great Ape Heart Project (GAHP)
Galdikas, Birutė, 12–13
'G-Ann', 141–142, 142*f*
Gastric mucosal hyperplasia, 129, 129*f*
Gastrointestinal tract (GIT), 22, 70, 119, 120*b*, 178, 252
  exocrine pancreas, 136–137
  investigation, 124–125
    assessment of faeces, 130–131
    colitis, 129–130
    condition, 133
    coprophagy, 134
    cystic fibrosis, 134
    diet, feeding and disease, 131–132, 131*f*
    gut fauna, 125
    gut flora, 125
    inanition, starvation and cachexia, 132–133
    malabsorption, 134
    obesity, 133
    parasites, 125–128
    pathology and pathogenesis, 128–129
    R&R, 133–134
  investigation of liver, 134–135
Gastro-intestinal tract (GIT) pathology, 178
Genetic anomalies, 90–91
Genetic methodologies, 79
Genetic resources, 216*t*, 220, 237
Genocide, in Rwanda, xvii, xxv, xxvii, 44, 89–90, 294*t*, 510
Genome amplification, 79
German Primate Centre (DPZ), 105–106
Germinal follicles, 143–145, 145*f*
Giardia, 126–127, 204–205
  of habituated Nyakagezi mountain gorilla group, 205
    identification, 205
    prevalence and means, 206, 206*t*
    prevalence of *Cryptosporidium* oocysts, 206
Gingivitis, 120, 122–123
GIT. *See* Gastrointestinal tract (GIT)
'Global health', 18, 235
Glossary of terms, 237
Gorilla, 60, 119, 179*f*, 180*f*, 211–228, 364–373, 604. *See also* Great apes
  access to, 221
  albino, 102
  brain studies, 178
  characteristic mode of locomotion, 97
  classification/taxonomy of, 215*t*, 364–373
  conservation, 211, 232
  criteria for assessing welfare, 94
  Cross River gorilla, 49–50, 213*t*, 221–222, 232, 372–373, 377, 602
  dermatological investigation, 98
  diagnosis of sarcoptic mange in, 105
  diseases affecting skin, 100

Eastern lowland gorilla, 44*f*, 75, 103, 127, 362, 367–368, 370–371, 373, 565
  examination and dissection, 615*f*
  free-living mountain, 94*f*
  G. beringei, 367–368
  G. beringei graueri, 369
  G. g. diehli, 368
  Gorilla Agreement, 213*t*, 219, 221–222
  Grauer's gorilla, 373
  hair, 98, 99*f*
  hair samples from, 98, 98*b*
  heart disease in, 115–116
  identification of individual free-living, 94*f*
  infant, 98
  keratinous cyst, 100*f*
  life history traits/events, 88
  mountain gorilla, 3–6, 6*f*, 22*t*, 43, 48*f*, 97, 153, 203, 205, 211, 362, 370–371, 373, 565
    history of, 3, 6*f*, 48*f*, 373
  normal arm hair removing, 99*f*
  origin of name, 375
  paper records, 607*f*
  psychological studies, 177
  samples from, 74*t*
  Sarcoptes infection in, 104
  scientific discovery of, 365
  skeletons, 306
  skin, 97, 99*f*
  with skin lesions, 98
  skull, 63*f*, 64*f*
  species/subspecies, 21, 30, 363, 365
  specimens, 359, 363, 368, 370
  storage and preservation of carcases and tissues from, 69*t*
  Western lowland gorilla, 103*f*, 130–132, 149, 201*f*, 214*f*, 215*t*, 223, 362, 370–373, 426, 439, 452–454, 469, 475, 513–514, 570, 574, 585
Gorilla beringei, 3, 178, 213*t*, 367–368
  skull and mandible of holotype, 6*f*
Gorilla beringei beringei, 373
  Gorilla Agreement, 216*t*
Gorilla beringei mikenensis, 369
Gorilla castaneiceps, 367
Gorilla diehli, 368
Gorilla ellioti, 370–373
Gorilla genus, 229
  anatomy, 8–9
  comparative anatomy, 5–8
  habits of, 4*f*
  mountain gorilla, 6*f*
  primate anatomy, 5–8
  problem-based learning, 4
  silverback lowland gorilla, 5*f*
  taxonomic status and anatomical features, 3
  terminology, 9–10
Gorilla gigas, 367
Gorilla gina, 366
Gorilla gorilla, 3, 363
  G. gorilla halli, 369
  G. gorilla manyema, 368
  G. gorilla matschiei, 368
  G. gorilla rex-pygmaeorum, 370

Gorilla gorilla (Continued)
    G. gorilla schwarzi, 368–369
      specimen, 302*f*
Gorilla graueri, 369
Gorilla hansmeyeri, 369
Gorilla jacobi, 368
Gorilla mayêma, 367
Gorilla Pathology Study Group (GPSG), xx,
    xxiv, 18, 20, 24, 232, 243–244, 308,
    609
Gorilla rib. *See* Lumbar rib in humans
GPSG. *See* Gorilla Pathology Study Group
    (GPSG)
Grauer's gorilla, 369, 373
Great Ape Heart Project (GAHP), 113
    cardiac necropsy check sheet, 278
    recommended cardiac
      necropsy, 276
        skeletal examination form, 279–285
        trimming protocol for pathologists,
        277–278
Great Ape Survival Partnership, 213*t*
Great apes. *See* Gorillas; Mountain gorillas
    cardiac examination for, 253–254
    and Ebola outbreak, 206–207
      challenges during Ebolavirus disease, 209
      precautions at work, 207–208
      precautions outside sanctuary, 208
      precautions to preventing Ebolavirus
        disease, 209
      sanctuary during EVDs, 207
    heart disease in, 115–116
Great Hippocampus Question, 7
Gross examination, 21, 57, 174–175
Gross lesions, 68–70, 69*t*, 90, 109
Guanidine thiocyanate (GT), 75
Guerrilla, 57, 221
Guidance, 9, 57, 223, 607
Guidelines, 52, 125, 223–224, 267
Gunshot and gunshot wounds, 44, 71. *See also*
    Wounds
Gut fauna, 30–31, 125
Gut flora, 31, 125
'Guy' gorilla skeleton at NHM, 301–308
    examination of skeleton, 302–307
    history, 301
    materials and methods, 301–302
    notes on cranial morphology, 302, 302*f*
    review of clinical records, 307
    studies on, 59, 163, 165, 301–302

H

H&E. *See* Haematoxylin and eosin (H&E)
Habitat destruction, 31
Habitat(s), 4, 10, 30, 104, 202, 211, 367–368,
    611
Habituated Nyakagezi mountain gorilla group
    Cryptosporidium, *Giardia* and helminths of,
      205
      identification of *Cryptosporidium*, *Giardia*
        and helminths, 205
      prevalence and means, 206, 206*t*
      prevalence of *Cryptosporidium* oocysts,
        206

Habituation, 87–88, 233
Habituation of animals, 87–88
Haematology. *See* Blood
Haematoxylin and eosin (H&E), 22, 193
Haemochorial placenta, 154
Haemopoietic system pathology, 104, 117, 146
Haemopoietic systems, 139
    blood and bone marrow, 146–147
    neoplasia, 148
Haemorrhage. *See* Contusion; Trauma
Haemosiderin, 151
Haemosiderosis, 135, 135*f*
Hair, 78, 97, 100
Hair, examination and assessment, 78, 99*f*
Halitosis, 124
Hand-lens, 54, 67, 124
Handling
    of biological material, 603
    of gorillas, 50, 55*b*, 615
Hand-rearing, gorillas, 157
Hanno's *Gorillai*, 373–375
Hazards, 31, 198, 225, 293, 294*t*. *See also*
    Zoonoses and zoonotic infections
    injured gorilla, 299*f*
    museum specimens, 299*f*
    national instability, 299*f*
    safety cabinets, 300*f*
    safety on training courses, 299*f*
    visibility, 299*f*
    warning notices in zoos, 299*f*
    zoonoses, 31, 206, 293, 294*t*, 297, 300, 300*f*
Healing, 41, 41*b*
Healing of wounds and fractures, 28, 165.
    See also Wounds
Health, 17–18, 52
    African academic institutions, 232
    attention to welfare, 233
    broadening holistic approach, 234
    collaboration and sharing of information, 231
    dissemination of information, 234
    Gorilla genus, 229
    hazards, 31, 61, 206, 293, 294*t*, 297, 300
    human health, 233, 293, 297, 300
    implications and importance of studies on,
      31, 206, 229–230
    monitoring, 49*f*, 53, 53*b*
      free-living gorillas, 54*b*
      of gorillas, 52–53
      rules for, 56*b*
      techniques, 53–54
    pathology, relevance to, 18, 23, 52, 73,
      229–230
    role for Africa, 231–232
    scientific advances, 233
    searching for gorilla material, 233–234
Health hazards, 61, 81. *See also* Zoonoses and
    zoonotic infections
Health monitoring, 52–54, 197, 203, 205, 230
Heart disease in great apes, 115–116, 116*f*, 117*f*
Heart vessels, 112, 117
Helicobacter species, 25
Helminths, 31, 127, 205
Helminths of habituated Nyakagezi mountain
    gorilla group, 205

identification of, 205
prevalence and means, 206, 206*t*
prevalence of *Cryptosporidium* oocysts, 206
Herbal medicines. *See* Medicinal plants
High-throughput sequencing techniques, 79
Histology, 22, 81, 100, 112, 134, 143, 165
Histology and histopathological examination
    scoring of lesions, 63, 81, 141
Histopathological examination, 179–180, 179*f*
History of health monitoring, 47
History, clinical and postmortem, importance,
    11, 18, 59
HIV. *See* Human immunodeficiency virus
    (HIV)
'Hobbit' (gorilla), 193
Holistic approach, broadening, 234
Homo sapiens, 6–8, 24, 30, 140, 229
'Horizon scanning', 226–227, 231
Host responses, 17, 36–37
    comparative approach, 23–24
    endogenous factors influencing, 24
HTLV. *See* Human T cell leukaemia virus
    (HTLV)
Human and Gorilla Conflict Resolution
    (HUGO), 204
Human health, 233
Human immunodeficiency virus (HIV), 27–28
Human medicine and health, 52, 62–63, 227
Human T cell leukaemia virus (HTLV), 148
    HTLV-I, 148
Humans, gorilla attacks on, 38–39
Hunter, John, 12, 13*f*, 604–605
Hunting of gorillas, 28–29, 367, 376–377.
    See also Shooting of gorillas
Hygiene, 183, 204–205, 603, 609
Hyperostosis, 162, 170, 303, 305*f*
Hypersensitivities and allergies of gorillas,
    141–142
Hyperthermia
    in captive gorillas, 101
    and hypothermia, 45, 101
Hypertrophic pulmonary osteoarthropathy, 109
'Hypnic jerks', 194
Hypophysis, 189–190
Hyporexia, 132
Hypostasis, 237
Hypovolaemic Shock causes, 118*b*
Hypoxia. *See* Asphyxia

I

IASP. *See* International Association for Study
    of Pain (IASP)
IATA Regulations, 225
IBD. *See* Inflammatory bowel diseases (IBD)
IBS. *See* Irritable bowel syndrome (IBS)
ICCWC. *See* International Consortium on
    Combating Wildlife Crime (ICCWC)
ICZN. *See* International Committee on
    Zoological Nomenclature (ICZN)
Identification
    of dead gorillas (carcases), 611–612
    of individual gorillas, 193
    of skeletal and non-skeletal remains, 164, 610
    of species, 3, 30, 91

IDRC. *See* International Development Research Centre (IDRC)

IGCP. *See* International Gorilla Conservation Programme (IGCP)

Ill health, indicators of, 50

Illegal trade, 212, 216*t*, 222*t*

Ill-treatment, 109−110, 224. *See also* Welfare of gorillas

Image analysis, 82

Images, 44−45, 49, 112, 228, 306

Imaging, 12*f*, 66, 166. *See also* Radiography and radiology

Imaging techniques, 12, 164

Immobilisation, 21, 34*t*, 50−51, 96

Immobilising equipment, 227−228

Immunisation, 29, 139−140, 202

Immunity of gorillas, 139−140

Immunocytochemistry, 79*f*

Immunodeficiency, 142−146

    ample lymphocytes with germinal follicles, 145*f*

    white pulp contains germinal centres, 145*f*

Immunological disorders of gorillas, 140−141

Immunosuppression, 146, 174

Import permit, 25, 55, 201, 220

Impression smears. *See* Cytology

Improvisation in the field. *See* Fieldwork and field studies

'Impungu' (gorilla), 364−365

In situ, 16, 49, 226, 237

Inanition, 132−133, 133*b*. *See also* Starvation causes, 133*b*

Indigenous knowledge, 220

Infant/juvenile skeletons, 613

Infanticide, 157−158

Infected wounds, 38

Infections from bites. *See* Bites

Infectious disease, 18, 22, 24, 106, 174−175, 198. *See also* Noninfectious disease; Zoonoses and zoonotic infections

    aetiology of disease, 19−23

    bacteria, 25−27, 30, 105−106, 117, 122−123

    and/or causal organisms in gorillas, 26*b*

    and death in gorillas, 19

    emerging diseases, 31

    of gorillas, 20*t*

    infectious conditions, 19, 24−25

    microbiota, microbiome and normal flora, 30−31

    pathology, 23

    predisposing factors of disease, 21*b*

    sampling cow, 23*f*

    sources of disease, 21*b*

    viruses, 27−29

    young gorillas, 22*f*

Infectious lesions, 100−101

Infertility, investigation. *See* Reproductive system

Inflammatory bowel diseases (IBD), 130

Inhalation pneumonia, 108*f*

Injury, 10, 21−22, 33, 42, 181, 223, 237. *See also* Trauma

    inflicted by gorillas, 101

    inflicted by humans, 159

Insects, 141, 612. *See also* Entomology

Institut Congolais pour la Conservation de la Nature (ICCN), 205

Institut für Zoo-und Wildtierforschung (IZW), 142

Institute of Primate Research (IPR), 12−13, 18, 86, 232

Instruments, clinical and postmortem, 61

Integument, 97. *See also* Skin

    areas, 102

    diagnostic procedures to samples from, 100

    system disease of NHPs, 100

Intellectual property, 17, 228

    and sharing of data, 17, 24, 84, 228

Interdisciplinary approach in gorilla work, 17−18, 90, 604−605

Internal examination and sampling, 66−67

International Air Transport Association (IATA), 225

International Association for Study of Pain (IASP), 95

International Commission on Zoological Nomenclature (ICZN), 212, 364

International Consortium on Combating Wildlife Crime (ICCWC), 221−222, 222*t*

International conventions, 219

International Development Research Centre (IDRC), 234

International Gorilla Conservation Programme (IGCP), 44

International legislation, 218−221

    CITES, 219−220

    CMS, 219

    convention on biodiversity, 220

    other international environmental law, 220−221

    relating to gorillas, 216*t*

International Primate Protection League (IPPL), 207

International Species Information System (ISIS), 84, 223

International Union for Conservation of Nature (IUCN), 25, 198, 212, 223−224, 226

    IUCN Guidelines, 223−224, 226

    IUCN Red List, 212

    Red List of threatened species status, 212

Internet, 11, 25, 84, 186, 215−218, 228

Interventions, xix, 39, 50−51, 70, 94, 109−110, 233

Intestinal contents. *See* Faeces

Intraalveolar haemorrhage, 108*f*

Invertebrates, 69, 141, 162, 166, 612

Invertebrates, skeleton recovery, 612

Investigation of wounds, 37. *See also* Wounds

Investigative methods, 73, 98−100

    laboratory tests, 100

    sampling, 98−99

Ionising radiation, 34*t*

IPHP. *See* European Great Ape Heart Project (IPHP)

IPPL. *See* International Primate Protection League (IPPL)

IPR. *See* Institute of Primate Research (IPR)

Irradiation, 33

Irritable bowel syndrome (IBS), 130

ISIS. *See* International Species Information System (ISIS)

Islam, 80

Isotopes. *See* Stable isotopes

IUCN. *See* International Union for Conservation of Nature (IUCN)

IZW. *See* Institutfür Zoo-und Wildtierforschung (IZW); Leibniz-Institut für Zoo-und Wildtierforschung (IZW)

J

Jane Goodall, 12−13, 234, 377

Jersey Zoo. *See* Durrell

Journal of Exotic and Pet Medicine (JEPM), 287

Journals and other publications, 175, 228

K

Kalema-Zikusoka, Gladys, (founder of CTPH), 47, 88, 203, 231*f*

Kenya, 12−13, 17−18, 20, 219*f*, 226, 232, 609, 613

Keratinous structures, 97, 100

Ketamine, 51, 227

Ketamine hydrochloride, 51

Kidneys, 25−26, 149, 381, 408, 419−420, 434, 472, 571

Kigali, 609−610

Killing of gorillas. *See* Euthanasia

Kinshasa Declaration on Great Apes 2005, 216*t*

Knight, Maxwell, xxiv, 56

Knuckle-walking, 97

L

Labelling, 68, 614

Laboratories, use of, 25, 85

Laboratory animals, 13, 15, 52−53, 80, 92

Laboratory investigations

    in the field, 24−25, 287−288

    interpretation and reporting of results, 58

Laboratory techniques, 25, 75, 77*t*, 100

Laboratory tests, 73, 100

Laceration, 40, 237

LAGA. *See* Last Great Ape Organisation (LAGA)

Land, 44, 88, 203−204, 211, 223, 373

Languages, use of, xxvii, 10, 20, 27, 49, 100, 166, 317, 373

Laryngeal sacs, 6, 107, 244

Larynx, 107

Last Great Ape Organisation (LAGA), 219, 221, 222*t*

LATF. *See* Lusaka Agreement Task Force (LATF)

Law, 211, 215−218

    Conservation, 211

    Customary, 215−218

    Law enforcement, 221−222

    Law enforcers, 221

    National, 219

Law enforcement, 221–222, 222*t*
Law regulating veterinary profession, 226–227
Law relevant to gorillas, 213*t*
    animal health, 220, 223, 226
    enforcement, 219, 221–222. *See also*
        Enforcement of legislation
    European, 224*f*
    health and safety, 223, 225–226, 612
    international, 216*t*, 218–221
    keeping gorillas, 223
    liability for gorillas, 223, 225
    local, 211, 215–218
    national, 219, 223, 225
    occupational health and safety, 225–226
    regional, 211, 213*t*, 215–218
    UK, 219, 223
    veterinary, 226–227
    welfare, 223–224
Layout, 361–364
    accession number, 361–362
    age category and sex, 362–363
    diagnoses, 364
    locality, 363
    method and date of acquisition, 364
    name of collector and date of collection, 363
    nature of specimen, 362
    species and subspecies, 363
Leakey, Louis, xv, xviii, xxiii, 7, 7*f*, 15, 18, 23
Legal considerations. *See also* Gorilla(s)
    Bwindi impenetrable National Park, 212*f*
    free-living gorillas, 215–222
    gorilla conservation laws and practices, 211
    gorillas in captivity, 223–225
    information on law and enforcement relating
        to gorillas, 213*t*
    IUCN red list of threatened species status,
        212
    legislation, 212–214
    mountain gorilla, 211, 212*f*
    nomenclature, 212
    Volcanoes National Park, 212*f*
    'wild', 211–212
Legal controls on samples, 220
Legislation, 211–214
    Acts/statutes, 217*t*
    animal health, 223
    Conservation, 215–218, 216*t*
    development of, 12
    International, 216*t*, 218–221
    international, 216*t*
    Multilateral, 218–221
    national, 215–218, 217*t*, 220–222
    Western lowland gorilla, 214*f*
    Wildlife, 221
LEH. *See* Linear enamel hypoplasia (LEH)
Leibniz-Institut für Zoo-und Wildtierforschung
    (IZW), 158
Lens, magnifying. *See* Hand-lens
Lesions, 68, 68*b*, 113
    describing and ageing (age determination),
        68*b*, 98, 113, 181, 196
    dimensions, 63
    scoring of, 82, 82*b*
Leydig cell tumour, 160

Licence, 223
Light, effect on gorillas, 57
Light-for-dates (LFD), 156
Limb bones, 304–305
Linear enamel hypoplasia (LEH), 123
Linnaeus Carolus, 7–9
Linnean Society of London, 45
Lipochromes, 176
Literature searches, 11
Live animals, investigation of, 25, 168
Liver
    biliary system, 136
    examination, 135*b*
    investigation, 134–135
    lowland gorilla, 135*f*
    pathology and pathogenesis of hepatic
        disorders, 135–136
    responses, 135*b*
Livestock, domesticated, 17, 90, 229–230
Locality, 363
Locomotion, 161
Locomotor
    disorders, 161
    pathology, 161
London Zoo, 11, 20, 25, 28, 50, 79–80, 87, 95,
    104, 111, 160, 301, 366
Long-term health monitoring systems for
    mountain gorillas, 203
    assessment of disease risks to gorillas, 203
    future recommendations, 205
    gorilla research clinic for analysis of
        pathogens, 203–204
    impact on conservation, 204–205
    local human health centres
        through community-based health
            promotion, 204
        through comparative disease
            investigations, 204
Loupe, magnifying/dissecting. *See* Hand-lens
Lumbar rib in humans, 165
Lungs, 107
    parasites, 110–111
Lusaka Agreement Task Force (LATF),
    221–222, 222*t*
Lymphocytes, 74
Lymphoid tissue, 139, 140*f*, 144*f*
Lymphoreticular system, 139
    neoplasia, 148
    pathology, 139

M

Maceration, skeletons recovery, 612
MAF. *See* Morris Animal Foundation (MAF)
Maggots, 40, 69, 615
Magnetic resonance imaging (MRI), 62–63,
    178, 181, 190
Malabsorption, 129, 134
Malnutrition, 93, 142–143, 157
Mammary gland, 152
Man and Biosphere, 220–221
Management of gorillas, 178
Mansonella perstans, 125
Marburgvirus, 28
MBD. *See* Metabolic bone disease (MBD)

McMaster technique, 205
Measuring. *See* Morphometrics; Weighing
Meat. *See* Bushmeat
Medical and surgical treatment, 51–52
Medicinal plants, 40, 231. *See also* Traditional
    medicines
Meningoencephalitis, 145–146
Meningothelial meningioma, 183, 184*f*
Metabolic bone disease (MBD), 131, 175, 191
Metabolic disorders, 167–168
Methods of investigation
    additional tests, 82–83
    biochemistry, 80
    biopsies, 81
    blood, 79–80
    cytology, 80–81, 81*f*
    DNA and molecular techniques, 78–79
    environmental testing, 83–84
    faeces, 75–78
    hair, 78
    histology, 81
    microbiology, 82
    personnel and equipment, 85
    quality control (QC), 84
    reading of slides, 81–82
    recording and reporting of findings,
        84–85
    safety in laboratory work, 85
    saliva, 75
    sampling, 73–74
    serology, 82
    storage of samples, 85–86
    swabs, 75
    urine, 78
Mgahinga, Uganda, 197, 205
Mgahinga gorilla National Park
    Cryptosporidium, *Giardia* and helminths of
        habituated Nyakagezi mountain gorilla
        group in, 205
    identification of *Cryptosporidium*, *Giardia*
        and helminths, 205
    prevalence and means, 206, 206*t*
    prevalence of *Cryptosporidium* oocysts, 206
MGVP. *See* Mountain Gorilla Veterinary
    Project (MGVP)
Microbiological safety cabinets, 85
Microbiology, 82
Microbiome, 30–31
Microbiota, 30–31
Microscope, microscopy
    electron, 82, 135, 160
    in the field, 288
    light, 76*t*, 120
Microscopic anatomy, 15–16
Military, 44, 89–90, 221, 367–368
Minimally invasive techniques, 53–54
MNHs. *See* Multinucleated hepatocytes
    (MNHs)
Molecular techniques, 78–79
    diagnostic techniques, 79
Molecular testing use in pathology, 79
Monitoring (screening) of gorillas, 52–53
Monkeypox, 29
Mononuclear cells, 150

Morphological examination, 3–5, 189–190.
    See also Gross examination
Morphology, 3, 8
Morphometrics, 59, 156, 156*b*, 244. *See also*
    Weighing
Morris Animal Foundation (MAF), 13–14,
    18–19, 109, 230, 230*f*
Mountain Gorilla Veterinary Project (MGVP),
    14, 19–20, 22, 27, 47, 186, 610
Mountain gorillas, 6*f*, 48*f*, 50*f*, 97, 211, 212*f*,
    373. *See also* Great apes
    assessment of disease risks to gorillas, 203
    causes of death in, 22*t*
    future recommendations, 205
    gorilla research clinic for analysis of
        pathogens, 203–204
    hair, 98
    immobilised mountain gorilla, 51*f*
    impact on conservation, 204–205
    local human health centres
        through community-based health
            promotion, 204
        through comparative disease
            investigations, 204
    long-term health monitoring systems for, 203
    Mountain gorilla faecal parasites, 206*t*
Mounted skeleton, 362
Movement of gorillas, 226
MRI. *See* Magnetic resonance imaging (MRI)
Mucin-containing cells, 194
Mukono, 44*f*
Multidisciplinary collaboration, 52
Multinucleated hepatocytes (MNHs), 136
Mummies, studies on, 60
Muscle of gorillas, 175
Musculoskeletal pathology, 161
Musculoskeletal system. *See also* Nervous
        system; Reproductive systems; Urinary
        system
    bone
        fractures and repair, 172–174, 172*f*
        investigation, 165–169
        pathology, 162–163
    joints, 165
    movement disorders, 161
    skeletal disease and pathology, 169–172
    skeleton, 163–164, 167*b*, 169*f*
        diagnosis of diseases, 174–175
    skull, 164
    vertebral column, 164–165
Museum, 3, 12, 168–169, 224–225, 228, 359,
    366, 603*f*, 604–606. *See also* Reference
    collections
    collections, 360
    material in, 605–606
    Museum specimens, 123, 167
    needs and collections, 604
    specimens, 301, 604
    working in collaboration with, 605
Museum specimens, 22, 120*f*, 123, 164*f*, 167
    hazards from, 225, 299*f*, 604
    spread of ectoparasites, 99
    studies on gorillas in, 301
Museums and zoological collections

skeletal material in museums, 301
skeleton of 'Guy' at NHM, 301–308
    examination of skeleton, 302–307
    gorilla, 301
    materials and methods, 301–302
    notes on cranial morphology, 302, 302*f*
    review of clinical records, 307
Mycobacteriosis, 26
Mycobacterium leprae, 105–106

N

Nagoya Protocol, 220
Nails, 97, 99
    gorillas, 102
    infectious diseases, 106
Names, scientific, 315
Nasal haemorrhage, 108
National legislation, 221–222
    relating to gorillas, 217*t*
National Museums of Kenya (NMK), 17, 232
Natural history, relevance to gorilla work, 4,
    48, 56
Natural history collections, 359–360
Natural History Museum (NHM), 301, 306
    skeleton of 'Guy' at, 301–308
        examination of skeleton, 302–307
        gorilla, 301
        materials and methods, 301–302
        notes on cranial morphology, 302, 302*f*
        review of clinical records, 307
Natural History Museum, London, 104, 301,
    360, 364, 368
    studies on "Guy", 301–308
Natural history/biology collections
    need for inventorying, 360
    threats to, 359–360
    value of, 73–74, 83*t*, 359–360
Natural Sciences Collections Association
    (NatSCA), 607
Naturalist, 4, 234, 374
Necropsy, 16, 57, 60, 133, 133*f*, 156. *See also*
    Postmortem examination
    of infant gorilla, 288*f*
    nonhuman primate postmortem examination,
        267–268
    precautions, 61*b*
    reports, 118
    stages in, 61*b*
Necrotic lesions, 141–142, 142*f*
Neonatal pathology and health, 155–156
    dead baby free-living gorilla, 156*f*
    morphometrics, 156*b*
Neonates
    and neonatal deaths. *See* Reproductive
        system
    postmortem examination of, 262
Neoplasia, 95–96, 148
Neoplasms in gorillas, 95
Nervous system, 177, 179*b*. *See also*
        Musculoskeletal system; Reproductive
        systems; Urinary system
    gorillas
        brain studies, 178
        psychological studies, 177

investigation, 178–180
pathology and diseases, 180–184
peripheral nerves, 184
Neurofibrillary tangles (NFT), 183
Neurodermatitis of captive gorillas, 102
Newton microscope, 288–291, 288*f*
NFT. *See* Neurofibrillary tangles (NFT)
NHM. *See* Natural History Museum (NHM)
NHPs. *See* Nonhuman primates (NHPs)
Nigeria, 49–50, 215*t*, 217*t*, 219, 221, 226, 373
Night nest, of gorilla, 54, 131
NMK. *See* National Museums of Kenya
    (NMK)
Nomenclature, 9, 212, 364
Non-governmental organisations (NGOs), 20,
    203, 224. *See also* individual bodies
Nonhuman primates (NHPs), 12, 15, 24, 51,
    75, 90, 100, 107, 139, 161, 178, 189,
    206–207
    poisoning, 96
    postmortem examination, 252–253
        fetuses and neonates, 254
        for tissue collection, 267–268
    reproductive tracts, 151
Noninfectious causes, 101–102, 101*f*
Noninfectious disease, 15, 22, 24–25, 34*t*, 175.
    See also Infectious disease
    approach to bite-wounds, 39*b*
    attacks on humans, 38–39
    auscultation, 42*f*
    explosions and unexploded ordnance, 44–45
    exposure
        to blasts, 44–45
        to temperature changes, 45
    fighting, 37–38
    firearms, 44–45
    infected wounds, 38
    Mukono, 44*f*
    noninfectious conditions, 33
    pathogenesis of gorillas, 34*t*
    predation, 45
    sepsis, 40
    snares, 42–44
    stress and stressors, 45
    trauma
        causes and pathogenesis, 33
        specific responses, 33–37
    wounds, 40
        ageing of, 40–41, 41*b*
Noninvasive health monitoring, 55*b*
Nonspecific pathology, 87
    age-related changes, 87–90, 88*f*
    aggression, 92–93
    behavioural pathology, 91–92
    captivity, 93
    developmental abnormalities, 90
    effects of capture, 93
    general considerations in, 87
    genetic anomalies, 90–91
    neoplasia, 95–96
    pain, 95
    poisoning, 96
    transportation, 93
    welfare, 93–95

Nonsteroidal antiinflammatory agents
(NSAIDs), 174
Normal flora, 30—31
Normal values. *See* Reference values
Nose. *See* Special senses, pathology of
Nose prints, of gorilla, 66
NSAIDs. *See* Nonsteroidal antiinflammatory
agents (NSAIDs)
NSH. *See* Nutritional secondary
hyperparathyroidism (NSH)
Nutritional deficiencies and imbalances, 134.
See also Malnutrition
Nutritional disease, 162—163
Nutritional secondary hyperparathyroidism
(NSH), 191

O

Obesity, 31, 133, 307
Observation, importance of, 48—50
Observational studies, 48—50
Occupational health and safety, 225—226
Odontological protocols, 119
Odontology. *See* Teeth
Odzala-Kokoua National Park (PNOK),
105
Office Rwandaise du Tourisme et des Parcs
Nationaux (ORTPN), 610
'One Health', 18, 29, 52, 197—198, 203, 205,
233—234
Ophthalmologic pathology, 187
Oral cavity, 119—121
conditions and pathogenesis, 122—124
dental caries, 123
effects of ageing, 123—124
enamel hypoplasia, 123
halitosis, 124
pathology, 123
periodontal disease, 123
tartar and calculus, 122
dental disease and pathology, 121—122
museum specimen, 120*f*
teeth, 119—121
veterinary pathologist, 120*f*
Orangutan, 363—364, 367
postmortem examination in air sacs of,
254—258
Ordnance, unexploded, 44—45
Organs, 57—58, 70
dimensions, 63
examining body organs in situ, 67
internal, 66
systems, 87, 88*b*
Organs/body parts (in fluid)
appendix, 128, 252, 260, 267, 520, 535,
580—581
axillary skin, 518, 524
brain, 405, 426, 427*f*, 431, 452—453
cadaver, 207, 441, 483—488, 513—515, 548,
555—556, 598
carcase, 98—99, 244, 287—288, 392—394,
396, 400, 402—403, 412, 426, 431, 439,
443, 446, 453, 461, 475, 480, 506,
518—520, 522, 569—570, 575,
580—581, 591, 594, 598—599

ears, 66—67, 98, 124, 187, 453, 473—474,
486, 488, 493—494, 501, 543
eye, 33, 141, 177, 185—187, 504, 580
fetus, 253—254, 261—262, 268, 365, 395,
400, 405, 406*f*, 407, 411, 430—431,
450, 486—488, 493—494, 496—497,
510, 523, 538, 541, 590
foot, 488
gall bladder, 58—59, 119, 125, 136, 252,
268, 400, 524, 541, 551, 586—587
genital, 66, 105, 151, 158, 160, 448
genitourinary tract, 580
hand, 423, 553, 599
head (sagittal section), 524
intestines, 8, 96, 154, 422, 483, 487, 493,
518—522, 577
kidneys and adrenal glands, 472
liver, 489, 491
oesophagus, 453
pectoral skin, 524
penis, 152, 459, 524, 529, 582
placenta, 486, 489
spleen, 421, 453, 489, 491, 584
testes, 20, 159—160, 252, 260, 268, 423,
519, 541
tongue, 474—475, 491, 581
viscera, 6, 59, 67, 154, 175, 192—193, 260,
267, 392—393, 490, 531, 591, 598
Orphaned gorillas. *See* Rehabilitation of
gorillas; Sanctuaries, for gorillas
Orphans, 157
Osman Hill, 604
studies on his collections, 301, 308,
604—606
Osteoarthritis in captive primates, 20, 162,
170—171, 303
Osteological material
arm bones, 59, 396, 440
articulated skeleton, 302, 308, 362, 397,
482—483, 504, 522, 575
bisected skull, 380, 512, 540
calotte, 385, 462, 488—489
calvarium, 164, 166—167, 174*f*, 405—406,
410, 479, 509—510, 588, 596
cranium, 362, 380, 386, 421, 460, 482, 513,
515—516
fetal skeleton, 386
incomplete skeleton, 167, 382, 386,
405—406, 408, 410—412, 421, 425,
427, 435, 440, 463, 469, 484, 488, 490,
496, 502, 505, 551, 576, 588
infant skeleton, 533
juvenile skeleton, 533
leg bones, 440
limb bones, 304—305, 412—413, 415, 456,
509, 539—540, 552, 593
mandible, 164, 397, 426, 436, 484, 548, 569
mounted skeleton, 362, 379, 381, 383, 385,
387—388, 396, 398, 419, 430, 469, 510,
512, 516—517, 529—530, 546,
548—551, 577, 584—588, 592
partially articulated skeleton, 522
pelvis, 8, 303, 304*f*, 404, 415, 425, 435, 441,
448—449, 458, 468, 473, 485—486,

496, 506, 509, 524, 531, 539, 543, 551,
553, 591, 593
postcranial skeleton, 382, 388, 392, 425,
454, 554, 567—568, 576, 589
sectioned cranium, 442*f*, 554
skeletal elements, 120—121, 124, 360, 415,
429, 434, 570, 586, 592
skeleton, 303, 397, 420, 432—433, 442, 448,
454, 458, 471, 536—537, 586—587,
592—595
skull, 388, 391, 412, 425, 454
vertebrae, 161, 164—165, 303, 365—366,
380, 384, 387, 427, 435, 440, 445, 448,
463, 468, 493, 498—499, 503,
529—530, 539—540, 544, 552,
554—555, 572—573, 580—581
Osteologists, 60
Osteology, 3, 162—163, 609
Osteomyelitis, 171
Osteophyte, 162, 170
Owen, Richard, 3, 4*f*, 366

P

Pain, 52, 95. *See also* Welfare of gorillas
Palaeontologists, 60
Palaeontology and palaeopathology, 163
Pan African Sanctuary Alliance (PASA), 26,
160, 213*t*, 231*f*, 318
Pancreas, 136, 193—194
Pancreatic disease, 193—194
Pangorillalges gorilla, 103
Pantothenate kinase 2 gene (PANK2 gene), 181
Pantothenate kinase-associated
neurodegeneration (PKAN), 181
Paralysis. *See* Trauma
Parasites, 29—30, 125—128
Parasiticides, 202
Parasitological examination, 244
Parathyroid, 190—192
Parc National des Volcans (PNV), 609—610
Parturition, 153—154
Pathogenesis, 28—29
of cardiovascular disease, 116—117
of fibrosing cardiomyopathy, 116
noninfectious disease of gorillas, 34*t*
trauma, 33
Pathogens, 19, 24—25, 27, 31, 54, 123,
201—204, 225, 606
Pathologist, 10, 12, 65, 132—133, 220, 222,
604—605
Pathology, 22—23, 231. *See also* Primate
pathology
African academic institutions, 232
attention to welfare, 233
broadening holistic approach, 234
collaboration and sharing of information,
231
dissemination of information, 234
of genus *Gorilla*, 229
human health, 233
implications and studies on, 229—230
pathology-based studies, 604—605, 605*f*
role for Africa, 231—232
scientific advances, 233

searching for gorilla material, 233–234
threats to survival, 229, 230t
of vascular system, 117–118
PCR. *See* Polymerase chain reaction
(PCR)
Pediculus genus, 104
Pelvis, 303, 304f
Penalties, 221
Periodontal disease, 123
Periosteum, 162
Peripheral nerves, 184
Peripheral nervous system (PNS) pathology,
177, 184
Peritonitis, 128
Permission, 60–61, 228
Forms, 168–169, 228
Written, 228
Permit(s), 16, 25, 53, 59, 96, 216t, 219, 223,
227
export, 219–220
import, 219–220
partially-completed, 220
Personnel and training, 58–59
PH. *See* Porotic hyperostosis (PH)
Photography, 49–50, 98
Photomicrograph, 22, 129
lowland gorilla, 133f
PHVA. *See* Population and Habitat Viability
Assessment Workshop for *Gorilla
gorilla beringei* (PHVA)
Physical restraint, 50
Pineal, 194
Pinealocytes, 194
Pithecus gesilla, 366
Pitman, Charles (C.R.S.), 4
Pituitary, 189–190
Pituitary gland of gorilla, 189–190
effects of endocrine pathology, 189b
PKAN. *See* Pantothenate kinase-associated
neurodegeneration (PKAN)
Placenta, 155
Placentae and placental membranes, 68b, 155,
254
Placental pathology, 154–155, 154f
Placentas, postmortem examination, 262
Plants, medicinal, 40, 231
Plasmodium reichenowi, 147
Plasmodium schwetzi, 147
Plastic containers, for specimens, 606
Plastic forceps, 44, 78
Plastinated carcase, 439
PMCT. *See* Postmortem computed tomography
(PMCT)
PMI. *See* Postmortem interval (PMI)
Pneumonyssus duttoni, 111
Pneumonyssus oudemansi, 110–111
Pneumonyssus pangorillae, 111
PNOK. *See* Odzala-Kokoua National Park
(PNOK)
PNS. *See* Peripheral nervous system (PNS)
PNV. *See* Parc National des Volcans (PNV)
Poaching, 21–22, 45, 104, 211, 230t, 615
Poisons and poisoning, 96. *See also* Toxicology
Police, 221, 222t

Pollution (dust, air or water pollution and
dietary contaminants), 87
Polygynous mating system, 151
Polymerase chain reaction (PCR), 30, 73,
78–79, 79b
Pongo, 364
Poorer countries, 203
Population and Habitat Viability Assessment
Workshop for *Gorilla gorilla beringei*
(PHVA), 19–21, 100, 105, 171, 185,
213t, 219–220, 231
Population growth (human), 203
Porotic hyperostosis (PH), 163
Porphyromonas gingivalis, 123
Portable equipment for fieldwork, 16, 288.
See also Fieldwork
Positive reinforcement training (PRT), 92
Postmortem, 62, 179, 237, 243–244
changes, 69–70, 70b
procedures, 61–65
reports, 248–251
samples needing rapid action, 68b
Postmortem change and estimation of time of
death, 69. *See also* Autolysis;
Postmortem interval (PMI)
Postmortem computed tomography (PMCT),
62–63
Postmortem examination, 57
accessing to material for, 60–61, 60f
air sacs
of apes, 275
of gorillas and other apes, 262–265
of orangutans, 254–258
background Information in postmortem case,
58b
causes of death diagnosis, 70–71, 72f
description of lesions, 68
determination of time of death, 69–70
equipment, 65
external examination and sampling, 65–66
forensic, 68–69
gorilla postmortem examination, 260–261
gorillas, 57, 243–251
forensic cases, 245–247
submission forms, 244
interdisciplinary studies, 59–60
internal examination and sampling, 66–67,
67f
nonhuman primate
fetuses and neonates, 254
for tissue collection, 267–268
and other nonhuman primates, 254–258
personnel and training, 58–59
PMI, 69–70
postmortem changes, 69–70
postmortem procedures, 61–65
primate fetuses, neonates and placentas, 262
recording of findings, 65
timing and planning, 65, 66f, 67f
Postmortem examination, necropsy, 57,
243–251
before embarking on, 57–58
of embryos, fetuses and neonates, 254
equipment, choice of, 96

in the field, 288
full versus partial necropsy, 62
imaging, 62–63, 113
purposes, 58b
record-keeping, 68
sampling, 55, 65–67, 74b
toxic substances, 225
Postmortem interval (PMI), 70. *See also*
Autolysis
Powell-Cotton Museum, 98–99, 360, 377
Poxviruses, 29
Predation, of gorillas, *ante-mortem* and
postmortem, 45
Pregnancy in gorillas, 153
Preparation of skeletal and other material,
609–610
Pre-release health considerations for gorillas,
197–198
budgets, 202–203
hazard identification, 198
preventive healthcare
parasiticides, 202
quarantine, 202–203
vaccination, 202
risk assessment, 198–199
disease risk analysis, 200t
risk communication, 203
risk management, 199–201
veterinary protocol as part of, 200t
testing for pathogens and disease, 201–202
Preservation, 614
Prevotella intermedia, 123
Primate anatomy, 5–8
Primate fetuses, postmortem examination of,
262
Primate pathology
access to data, 18
advantages and dangers of extrapolation,
16–18
African studies, 12–14
cellular pathology, 15–16
comparative pathology, 14–15
growth, 11
and health of gorillas, 12f
histological section of brain of gorilla, 15f
microscopic anatomy, 15–16
National Museum of Kenya, 14f
primates in biomedical research, 15
relevance of studies on, 16
role, 17t
studies on, 12
to health, 18
Primate Society of Great Britain (PSGB), 93,
195
Private collections, 223. *See also* Zoos
Problem-based learning, 4
Probstmayria, 126, 134
Prosecuting authorities, 221
Protected areas, 203, 211, 213t, 221
Protected species, 55
Protocols, 115, 243
PRT. *See* Positive reinforcement training
(PRT)
Pseudomelanosis, 151

PSGB. *See* Primate Society of Great Britain (PSGB)
Psychological or behavioural disorders, 91–92
Psychological studies on gorillas, 177
Pthirus gorilla, 103–104, 103*f*
Pthirus pubis, 103–104
Public liability, 223
Publications, 4, 11, 25, 190, 228
Publishing and dissemination of data (findings), 234
Puncture wounds, 101. *See also* Wounds
Putrefaction, 237. *See also* Autolysis
Pyelonephritis, 150

Q
Quality control (QC), 84, 291
Quality control, in laboratories, 16, 84
Quantification of data, 75–78, 94. *See also* Scoring system
Quarantine, 202–203
Quarantine for gorillas, 202–203
Questions that may need to be answered by a necropsy, 62

R
R&R. *See* Regurgitation and reingestion (R&R)
Radiation, ionising, 34*t*
Radiography, 306, 306*f*
Radiography and radiology, 44, 168, 306. *See also* Imaging
digitisation, 171
Radioisotope studies, 76*t*. *See also* Stable isotopes
Ramsar, 216*t*, 220–221
Range states, 9–10, 215, 215*t*, 217*t*, 219, 231
RC. *See* Congo, Democratic Republic of; Congo, Republic of the (CR)
RCS. *See* Royal College of Surgeons of England (RCS)
RCVS. *See* Royal College of Veterinary Surgeons (RCVS)
Recording and collating findings in the field, 65, 288*t*
Record-keeping, 68
Reference collections, 85–86, 86*b*. *See also* Museums
Reference values, 16, 80
References and Further Reading, 11, 25
Regeneration, 41, 287–288
Regurgitation and reingestion (R&R), 133–134
Rehabilitation of gorillas, 30, 83. *See also* Orphaned gorillas; Sanctuaries, for gorillas
Rehydration of carcases, 614
Relative humidity (RH), 69
Releases, Introductions/Re-Introductions and Translocations, 198
Religious considerations, 63
Reports and reporting, 84–85
Reproductive Health Surveillance Program (RSHP), 159
Reproductive system pathology, 151–160. *See also* Infertility

Reproductive systems, 149, 151–160. *See also* Musculoskeletal system; Nervous system; Urinary system
birth control and contraception, 160
copulation, 153
data, 153*b*
developmental abnormalities, 158, 158*f*
female mountain gorilla, 153*f*
female reproductive pathology, 158–159
fetal pathology, 154–155
infanticide, 157–158
male reproductive pathology, 159–160
mammary gland, 152
neonatal pathology and health, 155–156
orphans and hand-rearing, 157
parturition, 153–154
physiology, 152–153
placenta, 155
placental pathology, 154–155, 154*f*
pregnancy, 153
Republic of Congo (CR), 105, 141–142, 207, 213*t*, 373, 376
Research derivatives, 220
Research on gorillas, 231
Respiratory disease, 109, 111
clinical signs, 110
Respiratory pathology, 107–110. *See also* Nonspecific pathology
Respiratory system, 107–110. *See also* Cardiovascular system
investigation of respiratory tract, 111–112, 112*f*
lung parasites, 110–111
respiratory pathology, 107–110
Respiratory tract investigation, 111–112, 112*f*
Respiratory tract of *Homo*, 107
Restraint and handling, 50
Retention of samples from gorillas, 86
Reticuloendothelial system. *See* Lymphoreticular system
Retrieval of skeletal and other material, 609
Retroviruses, 27–28
Reverse zoonoses, 206–207
Reye's syndrome, 181
RH. *See* Relative humidity (RH)
Rhipicephalus appendiculatus, 103
Rifles. *See* Firearms
Rigor mortis, 70, 237
Ringworm. *See* Dermatophytosis
Risk assessment, 165, 198–199, 225–226, 605
Risks, 27, 198, 203, 225, 227, 362
Royal College of Pathologists, 18, 58, 154, 174
Royal College of Surgeons of England (RCS), xxv, 12, 17–18, 25, 301, 308, 318, 523–524, 605–606
historical importance of, 301, 604. *See also* Owen, Richard
studies on gorillas in, 12, 301
Royal College of Veterinary Surgeons (RCVS), 64–65, 93, 307, 318, 523–524
Royal Veterinary College (RVC), 233
RSHP. *See* Reproductive Health Surveillance Program (RSHP)
RVC. *See* Royal Veterinary College (RVC)

Rwanda, 10, 13–14, 17, 41–42, 44, 75, 203, 213*t*, 226, 609–610

S
Safety cabinets, 85, 300*f*
Saliva, 75
Saliva, examination and assessment, 75
Samples, 16, 38, 55*b*, 83*t*, 220, 226, 361
Samples and sampling, 55, 65–67, 73–74, 74*b*, 98–99, 150–151. *See also* Laboratory investigations
chain of custody, 68
dead gorillas, 52, 68
in the field, 288*f*
live gorillas, 80, 124
packing, 226
retention, 85
transportation and storage, 93
Samples/tissues (fixed/frozen)
abdominal fluid, 541
adrenal gland, 192, 195–196, 419–420, 434, 472
aorta, 24, 113–114, 116–117, 181, 186, 253, 261–262, 277–278, 307, 312, 381, 408, 513, 519–520, 541
aortic valve, 114, 541
bile, 78, 136
blood
plasma, 470, 565
red blood cells, 470
serum, 470
white blood cells, 470
whole, 470, 490, 506, 574
bone marrow, 80, 112, 139, 146–148, 252–253, 260–261, 267–268, 489, 518–520
brain, 381, 408, 419–420, 434, 583
caecum, 8, 518–519, 523–524, 541, 568
cerebrospinal fluid, 179, 574
colon, 129–130, 252, 408, 518–519, 523–524, 541–543
diaphragm, 408
DNA, 60, 75, 76*t*, 78–79, 85–86, 103, 105, 147, 164, 222, 308, 309*f*, 310, 312, 359, 370, 372, 402, 430, 433–434, 450, 454, 511
duodenum, 136*f*, 252, 260, 267, 408, 472, 541
epididymis, 160, 541
faecal, 75, 78, 126–127, 147, 170, 203–205, 223, 470
gall bladder, 58–59, 119, 125, 136, 252, 268, 400, 524, 541, 551, 586–587
gingiva, 122, 408, 541
hair, 52, 78, 98, 309, 440, 565, 569–570
heart, 278, 381, 408
ileum, 129, 252, 260, 267, 420, 472, 519, 523–524, 541
intestine, 381, 408, 419–420, 434
jejunum, 252, 260, 267, 472, 519, 541
kidney, 25–26, 381, 408, 419–420, 434, 571
liver, 381, 408, 419–420, 434, 571
lung, 381, 408, 419–420, 434, 571

lymph node, 58, 139, 143–146, 176, 252, 260, 267, 381, 408, 420, 434, 491, 518–520
mammary gland, 97, 133, 152, 252, 541
milk, 312, 409, 497, 518, 585
oesophagus, 27, 108–109, 120, 128–129, 408, 453, 518–520, 541
omentum, 127, 542–543
pancreas, 58, 119, 134, 136, 136f, 189, 193–194, 381, 408, 489, 518–520, 541
parathyroid gland, 190–192
pericardial fluid, 541
perineal fat, 571
pituitary gland, 189–190
placenta, 155, 159, 244, 253–254, 261–262, 268, 387, 400, 486–487, 489, 520–521, 590
radial nerve, 541
rectum, 124, 252, 260, 268, 420, 513, 541
reproductive system, 151–160, 434
salivary gland, 518–519, 541
sciatic nerve, 541, 571
skeletal muscle, 175, 408
skin, 100, 408, 471, 572
spleen, 381, 408, 419–420, 434
stomach, 10, 78, 95, 125–126, 128, 139, 381, 400, 408, 419–420, 434, 492, 518–521, 541, 577, 584, 586
subcutaneous adipose tissue, 541
thoracic fluid, 541, 571
thyroid, 189–191, 191f, 420, 518–520, 541
tongue, 27, 27f, 29, 154, 252, 260, 267, 408, 451, 474–475, 491, 518–520, 580
trachea, 107, 112, 143–145, 252, 260, 262, 267, 275, 307, 518–520, 541
tracheal wash, 571
urinary bladder, 149, 400, 408, 518–521, 541
urine, 150–151, 158–159
Sampling, 55, 74b, 98–99
  from captive gorillas, 73
  cow, 23f
  environmental testing, 83–84
  investigation of samples from gorillas, 76t
  investigative methods, 73
  methods, 75–83
  personnel and equipment, 85
  QC, 84
  recording and reporting of findings, 84–85
  safety in laboratory work, 85
  samples from gorillas, 74t
  storage of samples and reference collections, 85–86, 86b
Sanctuaries, for gorillas, 212. See also Orphaned gorillas; Rehabilitation of gorillas
Sarcoptes
  in free-living gorillas, 104–105
  histology, 105
  infection, 104
Satyrus adrotes, 367
Scales, for weighing, 59
Scanning electron microscopy (SEM), 76t, 78
Scavenging, 70. See also Predation

Scientific (Latin/Greek) names, 315
Scientific advances, 233
Scoring system, 95, 244
Screening, 53. See also Monitoring
SE. See Stage of exhaustion (SE)
'Semiquantitative' methods, 82
Senses in gorillas, 185–187
  ear, 187
  eye, 185–187
  nose, 187
Sepsis, 40
Serology, 82
Severino, Marco Aurelo, 6
SFV. See Simian foamy virus (SFV)
Shigella spp., 25
  S. flexneri, 128, 155
Shock, 41, 87, 117–118
Shock syndrome, 117–118
Shooting of gorillas, 44. See also Gunshot and gunshot wounds; Hunting of gorillas
Shotguns and shotgun wounds, 44–45. See also Wounds
Sierra Leone, 29, 122, 124, 373, 375
Silverback lowland gorilla, 5f
Simian foamy virus (SFV), 28
Simian immunodeficiency virus (SIV), 27–28
Simian T cell leukaemia virus (STLV), 148
SIV. See Simian immunodeficiency virus (SIV)
Skeletal disease and pathology, 17, 169–172
Skeletal material
  factors affecting preservation of material, 614
  genesis of studies on skeletal pathology, 610f
  preparation, preservation and storage of recovered skeletons, 614
    degreasing, 614
    displaying, 614
    labelling, 614
    preservation, 614
  recovery, preparation and preservation
    location, collection and processing of bones of gorillas, 610–611
    preparation method, 611
    processing and preservation precautions, 611
  skeletal remains as research material, 611
skeleton, 609
  recovery and preparation, 609–610
  skins, 609
  soft tissue preservation, 609
skeletons recovery from fleshed animals
  by boiling, 612
  by burial and excavation, 613–614
  by exposure to natural elements, 612–613
  by maceration, 612
  using invertebrates, 612
Skeletal pathology, 610f, 611
Skeleton of gorillas, 163–164, 167b, 169f, 302–307, 362, 605–606
  comment, 306–307
  diagnosis of diseases, 174–175
    adipose tissue, 175–176
    connective tissue, 175
    infectious disease, 174–175

MBD, 175
    muscle, 175
    noninfectious disease, 175
  examination, 302–307
    limb bones, 304–305
    in museums, 606
    pelvis, 303, 304f
    radiography, 306, 306f
    skull, 306
    spine, 303, 304f
    sternum, 303–304
    tarsus, 305
Skeletons and skeletal lesions, 10, 162, 167b, 174–175, 611–612. See also Bones
Skin(s), 97, 609
  biology, 97–98
  diseases and pathology, 100–101
    bacteria, 105–106
    diseases of uncertain aetiology, 106
    ectoparasites, 102–106
    fungi, 106
    infectious causes, 102
    noninfectious causes, 101–102, 101f
    viruses, 106
  of gorillas, 605–606, 606f
  investigative methods, 98–100
    laboratory tests, 100
    sampling, 98–99
  lesions, 98, 98b, 106
Skinning of carcase, 65–66
Skull of gorillas, 164, 174f, 306
Sleep, 194
  disorders, 194
Smuggling, 212, 219, 221–222
Smuggling and smuggled gorillas, 212, 221–222
SNA. See Social network analysis (SNA)
Snares, 42–44
Social network analysis (SNA), 94
Sodium (Na), 192–193
Soft release, 211–212
Soft tissue preservation, 609
Solitary animal, 375–376
SOPs. See Standard operating procedures (SOPs)
Special senses, pathology of, 91, 185–187
Specialists, 134, 227, 301–302, 604
Species, 3, 16, 37, 91, 212, 215t, 363
  gorilla, 30
  Helicobacter species, 25
  Homo species, 6–8, 24
  scientific names of, 315
  and subspecies, 363
Specimen, nature of, 362
Spine, 303, 304f
Spontaneous abortion, 154
SR. See Stage of resistance (SR)
SSI. See Surgical site infection (SSI)
Stab wounds. See Wounds
Stable isotopes, 77t, 83t, 132
Stage of exhaustion (SE), 195
Stage of resistance (SR), 195
Staining techniques, 22, 58
Standard operating procedures (SOPs), 85

Standardisation of laboratory techniques.
    See Quality control
Standards, 65, 223, 225–226, 611
Starvation, 88, 132–133, 133*b*
    causes, 133*b*
Status, 8–9, 24, 52, 212, 215*t*, 221, 375
Stereotypies, 91
Sternum, 303–304
Stinging insects (Hymenoptera), 104
STLV. *See* Simian T cell leukaemia virus
    (STLV)
Stomach contents, examination and assessment,
    74*t*
Storage, 69*t*, 85–86
    of skeletal and other material, 609
Streptococcus faecalis, 128
Stress, 45, 146, 163, 194–196. *See also*
    Environmental stressors
Stressors to gorillas, 45, 195*b*
Stud books, 223
Subcapsular glomerulogenesis, 151, 152*f*
Submission form, for forensic specimens, 159,
    243–244
Subspecies, 98, 215, 215*t*, 216*t*, 237, 363
Sudden death, 58, 113–114
Suffering, 93–94, 222. *See also* Welfare of
    gorilla
Suffocation, 34*t*, 237. *See also* Asphyxia
Supporting investigations, laboratory tests, 63
Surgical site infection (SSI), 52
Surveillance of syndromes, 47
Sustainable use, 216*t*, 220
Swabs, 75
Swahili, use of, 10, 49
Syndactyly, 90
Systemic hypertension, 88

T

TAG. *See* Taxon Advisory Group (TAG)
Tape-recorder and tape-recording, 65, 244
Tarsus, 305
Tartar, 122
Taxa, scientific names of, 315
Taxidermy, 65–66
Taxon Advisory Group (TAG), 115
Taxonomy, 3, 85–86, 215*t*, 220, 364–373.
    See also International Commission on
    Zoological Nomenclature (ICZN)
TB. *See* Tuberculosis (TB)
T-cell lymphoma, 148
Teaching, 4–6, 8, 232
Teeth, 119–122
Temperature changes, exposure to, 45
Terg-A-Zyme enzyme skeleton cleaning
    protocol, 612–613
Terminology, 9–10, 163
Testicular atrophy, 159–160
The International Committee on Zoological
    Nomenclature (ICZN), 212
Theiler, Sir Arnold, 166, 233
Thermal imaging, 50
Thoracic cyst wall, 145*f*
Thrombosis arteriosclerosis, 113
Through-the-lens (TTL), 65

Thymus, 139, 143*f*, 144*f*
    cervical, 144*f*
Thyroid glands, 190
    of lowland gorilla, 191*f*
Time of death, 69–70. *See also* Postmortem
    interval (PMI)
Tissue
    examination of, 71, 82, 142
    for histopathological examination, 81
    responses, 33–37
Tobias, Professor Philip, 7, 7*f*, 45
Tourism, 88, 207, 233
Toxicology, 58, 74*t*, 76*t*, 83*t*, 267. *See also*
    Poisons
Toxicoses, 96. *See also* Toxicology; Poisons
    and poisoning
Trachea, 107
Tracks, trails, and signs of gorillas, 55*b*, 377
Trade in gorilla parts, 222*t*
Traditional medicines, 123
Training of staff, 94
Transillumination, 124–125
Transillumination, in examination and
    detection of lesions, 124, 186
Translation, 5–6, 373. *See also* Language
Translocations. *See* Releases, Introductions/Re-
    Introductions and Translocations
Transport, 75, 207, 225–226
Transportation of gorillas, 93. *See also* IATA
    Regulations; Movement of gorillas
Traps, snares and trapping of gorillas, 42
Trauma, 12, 21–22, 33–37, 123, 155,
    162–163, 179, 184
    causes and pathogenesis, 33
    to the central nervous system, 58
    specific responses, 33–37
Traumatic injury, 33
'Tree of Zoology', 3–4
Treponema pallidum pertenue, 105, 171
'Triage' system, 47
Trichophyton mentagrophytes infection, 106
Trimming, 81
Troglodytella, 125
    T. abrassarti, 126*f*
Troglodytes gorilla, 365–366
Troglodytes niger, 3, 365
Troglodytes savagei, 366
Trypanosoma brucei, 147–148
Trypanosomiasis, 147–148
TTL. *See* Through-the-lens (TTL)
Tuberculosis (TB), 26

U

UCL. *See* University College London (UCL)
UFAW. *See* Universities Federation for Animal
    Welfare (UFAW)
Uganda, 4, 13, 17, 20, 43, 110, 197, 226
Uganda Wildlife Authority (UWA), 110,
    204–205
Ulcerative colitis, 130, 130*f*
Ultrasonographic assessment, 158
Ultrasonography. *See* Imaging
Ultraviolet light, 38
Unexpected death, 113–114

United Kingdom, 20, 39, 49, 64, 86, 93, 223,
    610. *See also* appropriate institutions
United Nations, The (UN), 213*t*, 222, 222*t*
United States of America, 12–14, 184. *See
    also* appropriate institutions
Universities Federation for Animal Welfare
    (UFAW), 93
University College London (UCL), 604
University of Cambridge, 63
University of the Witwatersrand, 168
Unnecessary suffering, 224. *See also* Suffering;
    Welfare of gorillas
Urinary system, 149–151. *See also*
    Musculoskeletal system; Nervous
    system; Reproductive systems
    disease and pathology, 149–150
    methods of investigation, 151
    pathology, 149–151
    sampling techniques, 150–151, 150*f*
    specific considerations, 151
Urine, 78, 150*f*, 307
    examination, 78
Urothelium, 151, 151*f*
Uterine leiomyoma, 158, 158*f*
Utilisation, 220, 301
UWA. *See* Uganda Wildlife Authority (UWA)

V

Vaccination, 202
Vagal inhibition, 237
Vaginitis, 92, 159
Valvular disease, 114
Vascular lesions, 116
Vascular system, pathologies of, 117–118
Vasculogenic Shock. *See* Distributive Shock
Versatility, 291
Vertebral column, 164–165
Vesalius, Andreas, 5–6
Veterinarian, 8, 18, 39, 52, 62, 226–228,
    287–288
Veterinary, 8–9, 11
    care, 223–224
    medicines, 52, 112, 227–228, 287
    pathologist, 15
    registration, 226–227
Veterinary Specialist Group (VSG), 110
Veterinary surgeon (veterinarian), 52, 106, 197
VHCTs. *See* Village Health and Conservation
    Teams (VHCTs)
Village Health and Conservation Teams
    (VHCTs), 204
Viral infections, 110
Virchow, Rudolph, 182
Virunga Mountains/Volcanoes, 390
Virunga National Park, 93, 96, 110, 205, 211
Viruses, 27–29, 106
    filoviruses, 28–29
    parasites, 29–30
    poxviruses, 29
    retroviruses, 27–28
Volcano Veterinary Center (VVC), xxiv, xxvii,
    13–14, 14*f*, 18, 19*f*, 24, 27, 230*f*, 317,
    511, 610
Volcanoes National Park, 212*f*

Von Recklinghausen disease, 191–192
Voucher specimens, 604. *See also* Reference
 collections
VSG. *See* Veterinary Specialist Group (VSG)

W

WAA. *See* Wildlife and Aquatic Animal
 Resources (WAA)
Wallace, Alfred Russel, 7–8
War, 71, 90, 230*t*. *See also* Civil disturbance
WARM. *See* Wildlife and Animal Resources
 Management (WARM)
Waste, 156, 225, 294*t*
Wasting marmoset syndrome (WMS), 134
Water testing, 83
Weapons, 34*t*
Websites. *See also* Internet
Weighing of gorillas and organs, 50, 143
Welfare of gorillas, 5, 17, 93–95, 196, 224, 233
 assessment of, 50, 116
 attention to, 233
 of casualties in sanctuaries, 212
 in zoos, 94
Western lowland gorilla, 214*f*, 370–371, 373
Wild, 10, 15, 54, 121, 211–212, 362
Wild animals, 29, 140
 definition of, 10

free-living (free-ranging), 10, 211–212
Wild plants, 30, 132
Wildlife, 4, 212, 234, 287
 Authority, 205, 223
 Legislation, 221
Wildlife and Animal Resources Management
 (WARM), 232
Wildlife and Aquatic Animal Resources
 (WAA), 197
Wildlife crime, 221
Wildlife forensic investigation, aspects of, 69
Wildlife trade, 213*t*, 222*t*
WMS. *See* Wasting marmoset syndrome
 (WMS)
World Health Organization, 52
World Health Organization of the United
 Nations (WHO), 52, 116, 147, 202,
 288
Wounds, 33, 37–38, 40. *See also* Trauma; Bite
 wounds
 ageing of, 40–41, 41*b*
 classification, 40*b*
 factors influencing wound healing,
  40*b*
 investigation and assessment, 157
 types, 40*b*
 from weapons, 101

X

X-rays. *See* Radiograph; Radiography and
 radiology

Y

Yerkes Regional Primate Research Center
 (YRPRC), 112, 160

Z

Zoo, 4–5, 12, 20, 39, 611–612
Zoo Licensing Act (ZLA), 84, 223
Zooanthroponoses. *See* Reverse zoonoses
Zoological collection, 5, 11, 19–20, 605–606
Zoological Information Management System
 (ZIMS), 84
Zoological nomenclature. *See also* International
 Commission on Zoological
 Nomenclature (ICZN)
Zoological Society of London. *See* London Zoo
Zoonoses, 25, 31, 293, 615
 and zoonotic infections, 31, 54, 63, 107,
  291*f*, 293, 294*t*. *See also* Hazards
Zoos, 19–20, 84, 244, 361. *See also* individual
 collections
 role of, in studies on gorilla, 4–5, 18, 20,
  108, 607*f*, 611–612